Lecture Notes in Computer Science 3884

Commenced Publication in 1973
Founding and Former Series Editors:
Gerhard Goos, Juris Hartmanis, and Jan van Leeuwen

T0180043

Bruno Durand   Wolfgang Thomas (Eds.)

STACS 2006

23rd Annual Symposium
on Theoretical Aspects of Computer Science
Marseille, France, February 23-25, 2006
Proceedings

 Springer

Volume Editors

Bruno Durand
Laboratoire d'Informatique Fondamentale de Marseille CMI
39 rue Joliot-Curie, 13453 Marseille Cedex 13, France
E-mail: bruno.durand@lif.univ-mrs.fr

Wolfgang Thomas
RWTH Aachen
Lehrstuhl Informatik VII
52056 Aachen, Germany
E-mail: thomas@informatik.rwth-aachen.de

Library of Congress Control Number: 2006920575

CR Subject Classification (1998): F, E.1, I.3.5, G.2

LNCS Sublibrary: SL 1 – Theoretical Computer Science and General Issues

ISSN   0302-9743
ISBN-10   3-540-32301-5 Springer Berlin Heidelberg New York
ISBN-13   978-3-540-32301-3 Springer Berlin Heidelberg New York

Springer is a part of Springer Science+Business Media

springer.com

© Springer-Verlag Berlin Heidelberg 2006
Printed in Germany

Typesetting: Camera-ready by author, data conversion by Scientific Publishing Services, Chennai, India
Printed on acid-free paper   SPIN: 11672142   06/3142   5 4 3 2 1 0

Preface

The Symposium on Theoretical Aspects of Computer Science (STACS) is alternately held in France and in Germany. The conference of February 23–25, 2006, held at Marseille was the 23rd in this series. Previous meetings took place in Paris (1984), Saarbrücken (1985), Orsay (1986), Passau (1987), Bordeaux (1988), Paderborn (1989), Rouen (1990), Hamburg (1991), Cachan (1992), Würzburg (1993), Caen (1994), München (1995), Grenoble (1996), Lübeck (1997), Paris (1998), Trier (1999), Lille (2000), Dresden (2001), Antibes (2002), Berlin (2003), Montpellier (2004) and Stuttgart (2005).

The interest in STACS has been increasing continuously during recent years – and even strongly this year. The STACS 2006 call for papers led to 283 submissions from all over the world. We had a two-day physical meeting for the Program Committee in Paris in November 2006 where all members of the Program Committee were present. We would like to thank the Program Committee and all external referees for the valuable work they put into the reviewing process. Each submission was assigned to at least 3 Programme Committee members, hence each member was in charge of about 50 papers. Only 54 papers (i.e., 19 % of the submissions) could be accepted, as we wanted to keep the conference in its standard format with only two parallel sessions.

We would like to thank the three invited speakers, Philippe Flajolet, Leonid Levin and Helmut Seidl, for their contributions to the proceedings.

Special thanks are due to Andrei Voronkov for his EasyChair software (`www.easychair.org`) and for his continuous support in runnning it. This software was used for the process of paper selection and also for preparing the camera-ready copy of this proceedings volume.

December 2005   Bruno Durand and Wolfgang Thomas

Conference Organization

Bruno Durand

Laboratoire d'Informatique Fondamentale de Marseille

Program Chairs

Bruno Durand, Wolfgang Thomas

Program Committee

Mark de Berg
Julian Bradfield
Bruno Durand
Christoph Durr
Fedor V. Fomin
Markus Lohrey
Filippo Mignosi
Madhavan Mukund
Seffi Naor
Rolf Niedermeier
Jean-Eric Pin
Peter Sanders
Luc Segoufin
Alexander Shen
Wolfgang Thomas
Jacobo Toran
Alexander Zvonkin

Local Organization

Grégory Lafitte
Rémi Morin
Nicolas Ollinger, chair
Liva Ralaivola
. . .

External Reviewers

Karen Aardal
Ashkan Aazami
Farid Ablayev
Bharat Adsul
Manindra Agrawal
Dorit Aharonov
Susanne Albers
Gabriele Alessandra
Eric Allender
Jean-Paul Allouche
Noga Alon
Ernst Althaus
Andris Ambainis
Klaus Ambos-Spies
Marcella Anselmo
V. Arvind
Eugene Asarin
David Aspinall
Daniel Augot
Yossi Azar
Franz Baader
Maxim Babenko
Christine Bachoc
Christel Baier
Fabien Baille
Euripide Bampis
Nikhil Bansal
Amotz Bar-Noy
Reuven Bar-Yehuda
Ziv Bar-Yossef
Vince Barany
Jeremy Barbay
David M. Barrington
Anjoy Baruah
Shai Ben-David
Michael A. Bender
Marc Benkert
Lasse Bergroth
Julien Bernet
Jean Berstel
Valrie Berthé
Laurent Bienvenue
Markus Bläser

Stephen Bloch
Luc Boasson
Manuel Bodirsky
Hans L. Bodlaender
Fabrice Bazzaro
Nicolas Bonichon
Paola Bonizzoni
Andreas Brandstädt
Sabine Broda
Andrej Brodnik
Anne Brüggemann-Klein
Niv Buchbinder
Harry Buhrman
Andrei Bulatov
Martin W. Bunder
Marie-Pierre Béal
Gruia Calinescu
Olivier Carton
Julien Cassaigne
Jorge Castro
Bogdan Cautis
Supratik Chakraborty
Timothy Chan
Serge Chaumette
Chandra Chekuri
Joseph Cheriyan
Hubie Chen
Jana Chlebikova
Christian Choffrut
Marek Chrobak
Amin Coja-Oghlan
Richard Cole
Anne Condon
Thierry Coquand
Chris Coulston
Bruno Courcelle
Pierluigi Crescenzi
M. Crochemore
Mary Cryan
Janos Csirik
Felipe Cucker
Artur Czumaj
Edson Norberto Cáceres

Markus Holzer
Falk Hüffner
Paul Hunter
David Ilcincas
Lucian Ilie
Nicole Immorlica
Sandy Irani
Robert Irving
Dmitry Itsykson
Abraham Ittai
Kazuo Iwama
Kamal Jain
Markus Jakobsson
Jesper Jansson
Wojciech Jawor
Emmanuel Jeandel
Mark Jerrum
Jan Johannsen
Peter Jonsson
Antoine Jouglet
Marcin Jurdzinski
Frank Kammer
Iyad Kanj
Bruce Kapron
Juhani Karhumäki
Juha Karkkainen
Claire Kenyon
Alex Kesselman
Sanjeev Khanna
Alexander Khodyrev
Bakhadyr Khoussainov
Daniel Kirsten
Alexey Kitaev
Nils Klarlund
Ralf Klasing
Hartmut Klauck
Christian Knauer
Joachim Kneis
Johannes Köbler
Stavros Kolliopoulos
Petr Kolman
Amos Korman
Darek Kowalski
Evangelos Kranakis
Dieter Kratsch

Matthias Krause
Svetlana Kravchenko
Marc van Kreveld
Andrei Krokhin
Sven O. Krumke
Piotr Krysta
Antonin Kucera
Gregory Kucherov
Manfred Kufleitner
Fabian Kuhn
K. Narayan Kumar
Orna Kupferman
Dietrich Kuske
Gregory Lafitte
Klaus-Jörn Lange
Martin Lange
Sophie Laplante
Troy Lee
Francois Lemieux
Asaf Levin
Liane Lewin-Eytan
Yury Lifshits
Wolfgang Lindner
Andrzej Lingas
Maciej Liskiewicz
Kamal Lodaya
Christof Löding
Sylvain Lombardy
John Longley
Antoni Lozano
Aldo de Luca
Jack Lutz
Alejandro Maass
Alexis Maciel
P. Madhusudan
Frédéric Magniez
Meena Mahajan
Mohammad Mahdian
Jean Mairesse
Yury Makarychev
Konstantin Makarychev
Janos Makowsky
Sebastian Maneth
Yishay Mansour
Sabrina Mantaci

Sciortino Marinella
Eric Martin
Maarten Marx
Daniel Marx
Domagoj Matijevic
Marios Mavronicolas
Elvira Mayordomo
Richard Mayr
Jacques Mazoyer
Klaus Meer
Wolfgang Merkle
Jean-François Mestre
Antoine Meyer
Adam Meyerson
Ilya Mezhirov
Mehdi Mhalla
Loizos Michael
Dimitris Michail
Christian Michaux
Peter Bro Miltersen
Alexandre Miquel
Joseph Mitchell
Petr Mitrichev
Shuichi Miyazaki
Fabien de Montgolfier
F. Morain
Rémi Morin
Mohamed Mosbah
Philippe Moser
Anca Muscholl
Haiko Müller
Jörn Müller-Quade
Pavel Nalivaiko
N.S. Narayanaswamy
Ashwin Nayak
Jaroslav Nesetril
Jesper Buus Nielsen
Bengt Nilsson
Dirk Nowotka
Zeev Nutov
Ryan O'Donnell
Ernst-Rüdiger Olderog
Michel Olivier
Nicolas Ollinger
Ren van Oostrum

Friedrich Otto
Vincent Padovani
Igor Pak
Paritosh Pandya
Christos Papadimitrou
Christophe Papazian
David Peleg
Giuseppe Della Penna
Martin Pergel
Dominique Perrin
Holger Petersen
Gregorio Hernndez Pealver
Jean-Eric Pin
Val Pinciu
Pavithra Prabhakar
Kirk Pruhs
Artem Pyatkin
Arnaud Pêcher
Sandra Quickert
Balaji Raghavachari
Liva Ralaivola
Venkatesh Raman
R. Ramanujam
Dana Randall
C. Pandu Rangan
Ivan Rapaport
Michael Raskin
André Raspaud
Julian Rathke
Dror Rawitz
Alexander Razborov
Igor Razgon
Oded Regev
Klaus Reinhardt
Rüdiger Reischuk
Jean-Pierre Ressayre
Antonio Restivo
Yossi Richter
Mike Robson
Claas Roever
Andrei Romashchenko
Dana Ron
Adi Rosen
Peter Rossmanith
Daniel Molle

Peter Rossmanith
Michel de Rougemont
Sasha Rubin
Andrei Rumyantsev
Jan Rutten
Mohammad R. Salavatipour
Peter Sanders
Miklos Santha
Nicola Santoro
Martin Sauerhoff
Saket Saurabh
Gabriel Scalosub
Nicolas Schabanel
Christian Scheideler
Markus Schmidt
Manfred Schmidt-Schauss
Henning Schnoor
Nicole Schweikardt
Stefan Schwoon
Uwe Schöning
Carlos Seara
Luc Segoufin
Sebastian Seibert
Helmut Seidl
Arunava Sen
Géraud Sénizergues
Anil Seth
Peter Sewell
Jiri Sgall
Hadas Shachnai
Natasha Shakhlevich
Priti Shankar
Abhi Shelat
David Sherma
Detlef Sieling
Sudeshna Sinha
Jozef Siran
Rene Sitters
Martin Skutella
Aleksandrs Slivkins
Shakhar Smorodinsky
Bill Smyth
Sagi Snir
Alexander Sobol
Christian Sohler

Samia Souissi
Francis Sourd
Bettina Speckmann
Daniel Spielman
Andreas Spillner
Aravind Srinivasan
Rob van Stee
Alexey Stepanov
Frank Stephan
Iain Stewart
Colin Stirling
Leen Stougie
Howard Straubing
Georg Struth
K. V. Subrahmanyam
C. R. Subramanian
S. P. Suresh
Maxim Sviridenko
Chaitanya Swamy
Stefan Szeider
Amnon Ta-Shama
Tami Tamir
Till Tantau
Gerard Tel
Kavitha Telikepalli
Alain Terlutte
Pascal Tesson
P. S. Thiagarajan
Aurlie C. Thiele
Thomas Thierauf
Dimitrios M. Thilikos
Torsten Tholey
Rick Thomas
Ashish Tiwari
Laura Toma
Leen Torenvliet
Helene Touzet
Falk Unger
Dominique Unruh
Maxim Ushakov
Patchrawat Uthaisombut
Siavash Vahdati
Kasturi Varadarajan
Yann Vaxs
Santosh Vempala

Nikolai Vereshchagin
Victor Vianu
N. V. Vinodchandran
Mahesh Viswanathan
Berthold Vöcking
Jörg Vogel
Heribert Vollmer
Magnus Wallstrom
Igor Walukiewicz
John Watrous
Karsten Weihe
Pascal Weil

Sebastian Wernicke
Susanne Wetzel
Thomas Wilke
Gerhard Woeginger
Ronald de Wolf
Alexander Wolff
Stefan Wöhrle
Sheng Yu
Mariette Yvinec
Marc Zeitoun
Alex Zelikovsky
Wieslaw Zielonka

Table of Contents

The Ubiquitous Digital Tree

Philippe Flajolet

Algorithms Project, INRIA Rocquencourt, F-78153 Le Chesnay, France
Philippe.Flajolet@inria.fr

Abstract. The *digital tree* also known as *trie* made its first appearance
as a general-purpose data structure in the late 1950's. Its principle is a
recursive partitioning based on successive bits or digits of data items.
Under various guises, it has then surfaced in the management of very
large data bases, in the design of efficient communication protocols, in
quantitative data mining, in the leader election problem of distributed
computing, in data compression, as well as in some corners of computational geometry. The algorithms are invariably very simple, easy to
implement, and in a number of cases surprisingly efficient. The corresponding quantitative analyses pose challenging mathematical problems
and have triggered a flurry of research works. Generating functions and
symbolic methods, singularity analysis, the saddle-point method, transfer operators of dynamical systems theory, and the Mellin transform have
all been found to have a bearing on the probabilistic behaviour of trie
algorithms. We offer here a perspective on the rich algorithmic, analytic,
and probabilistic aspects of tries, culminating with a connection between
a sorting problem and the Riemann hypothesis.

While, in the course of the 1980s and 1990s, a large portion of the theoretical computer science community was massively engaged in worst-case design and
analysis issues, the discovery of efficient algorithms continued to make tangible
progress. Such algorithms are often based on simple and elegant ideas, and, accordingly, their study is likely to reveal structures of great mathematical interest.
Also, efficiency is much better served by probabilistic analyses[1] than by the ter-
atological constructions of worst-case theory. I propose to illustrate this point of
view by discussing a fundamental process shared by algorithmics, combinatorics,
and discrete probability theory—the digital tree process. Because of space-time
limitations, this text, an invited lecture at STACS'06, cannot be but a brief guide
to a rich subject whose proper development would require a book of full length.

1   The Basic Structure

Consider first as domain of our data items the set of all infinitely long binary
strings, $\mathcal{B} = \{0, 1\}^\infty$. The goal is to devise a data structure in which elements
of $\mathcal{B}$ can be stored and easily retrieved. Given a finite set $\omega \subset \mathcal{B}$ like

---

[1] To be mitigated by common sense and a good feel for algorithmic engineering, of
course!

B. Durand and W. Thomas (Eds.): STACS 2006, LNCS 3884, pp. 1–22, 2006.

$$\omega = \{110100\cdots, \quad 01011\cdots, \quad 01101\cdots\},$$

a natural idea is to form a *tree* in which the left subtree will contain all the elements starting with 0, all elements starting with 1 going to the right subtree. (On the example, the last two strings would then go to the left subtree, the first one to the right subtree.) The splitting process is repeated, with the next bit of data becoming discriminant. Formally, given any $\omega \subset \mathcal{B}$, we define

$$\omega \setminus 0 := \{\sigma \mid 0\sigma \in \Omega\}, \qquad \omega \setminus 1 := \{\sigma \mid 1\sigma \in \Omega\}.$$

The motto here is thus simply "filter and shift left". The *digital tree* or *trie* associated to ω is then defined by the recursive rule:

$$\text{trie}(\omega) := \begin{cases} \emptyset & \text{if } \omega = \emptyset \\ \sigma & \text{if } \omega = \{\sigma\} \\ \langle \bullet, \text{ trie}(\omega \setminus 0), \text{ trie}(\omega \setminus 1)\rangle. \end{cases} \quad (1)$$

The tree $\text{trie}(\omega)$ makes it possible to search for elements of ω: in order to access $\sigma \in \mathcal{B}$, simply follow a path in the tree dictated by the successive bits of σ, going left on a 0 and right on a 1. This continues till either an external node containing one element, or being an empty node, is encountered. Insertion proceeds similarly (split an external node if the position is already occupied), while deletion is implemented by a dual process (merging a node with its newly vacant brother). The tree $\text{trie}(\omega)$ can be either constructed from scratch by a sequence of insertions or built by a top down procedure reflecting the inductive definition (1). In summary:

> *Tries serve to implement* dictionaries, *that is, they support the operations of insertion, deletion, and query.*

A trie thus bears some resemblance to the Binary Search Tree (BST), with the basic BST navigation based on relative order being replaced by decisions based on values (bits) of the data items:

$$\text{BST: } \langle x, \text{``}< x\text{''}, \text{``}> x\text{''} \rangle; \qquad \text{Trie: } \langle \bullet, \text{`` } = 0\text{''}, \text{`` } = 1\text{''} \rangle.$$

Equivalently, if infinite binary strings are interpreted as $[0, 1]$ real numbers, the separation at the root is based on the predicates $< \frac{1}{2}, \geq \frac{1}{2}$. Like for the BST, a left to right traversal of the external nodes provides the set ω in sorted order: the resulting sorting algorithm is then essentially isomorphic to *Radix Exchange Sort* [43]. (This parallels the close relationship that BSTs entertain with the Quicksort algorithm.) The books by Knuth [43] and Sedgewick [54] serve as an excellent introduction to these questions.

There are many basic variations on the trie principle (1).

— *Multiway branching.* The alphabet $\{0, 1\}$ has been so far binary. An m-ary alphabet can be accommodated by means of *multiway branching*, with internal nodes being m-ary.

— *Paging.* Recursion may be halted as soon in the set ω has cardinality less than some fixed threshold b. The standard case is $b = 1$. The general case $b \geq 1$ corresponds to "bucketing" or *paging* and is used for retrieval from secondary memory.

— *Finite-length data.* Naturally occurring data tend to be of finite length. The trie can then be implemented by appending a *terminator symbol* to each data item, which causes branching to stop immediately.

— *Digital search trees (DSTs).* These are a hybrid between BSTs and tries. Given a *sequence* of elements of $\mathcal{B}$, place the first element at the root of the tree, partition the rest according to the leading digit and proceed recursively. DSTs are well described and analysed in Knuth's volume [43]. Their interest as a general purpose data structure has faded, but they have been found to play an important rôle in connection with data compression algorithms.

— *Patricia tries.* They are obtained from tries by adding skip fields in order to collapse sequences of one way branches.

Let us point out at this stage that, as a general purpose data structure, tries and their kin are useful for performing not only dictionary operations, but also set intersection and set union. This fact was recognized early by Trabb Pardo [58]. The corresponding algorithms are analysed in [27], which also contains a thorough discussion of the algebra of finite-length models.

Complexity issues. Under the basic model of infinitely long strings, the worst-case complexity of the algorithms can be arbitrarily large. In the more realistic case of finite-length strings, the worst-case search cost may equal the length of the longest item, and this may well be quite a large quantity. Like for many probabilistic algorithms, what is in fact relevant is the "typical" behaviour of the trie, measured either on average or in probability under realistic data models. *Analysis of algorithms* plays here a critical rôle in helping us decide in which contexts a trie can be useful and how parameters should be dimensioned for best effect. This is the topic we address next.

2 Random Tries

The field of analysis of algorithms has evolved over the past two decades. The old-style recurrence approach is nowadays yielding ground to modern "symbolic methods" that replace the study of sequences of numbers (counting sequences, probabilities of events, average-case values or moments of parameters) by the study of *generating functions*. The algebra of series and the analysis of functions, mostly *in the complex domain* $\mathbb{C}$, then provide precise asymptotic information on the original sequence. For an early comparative analysis of tries and digital search trees in this perspective, see [29].

The algebra of generating functions. Let (f_n) be a numeric sequence; its *exponential generating function* (EGF) is by definition the formal sum

$$f(z) = \sum_{n \geq 0} f_n \frac{z^n}{n!}. \tag{2}$$

(We systematically use the same groups of letters for numeric sequences and their EGFs.) Consider a parameter ϕ defined inductively over tries by

$$\phi[\tau] = t[\tau] + \phi[\tau_0] + \phi[\tau_1]. \tag{3}$$

There, the trie τ is of the form $\langle \bullet, \tau_0, \tau_1 \rangle$, and the quantity $t(\tau)$, called the "toll" function, often only depends on the number of items stored in τ (so that $t(\tau) = t_n$ if τ contains n data). Our goal is to determine the expectation ($\mathbb{E}$) of the parameter ϕ, when the set ω on which the trie is built comprises n elements.

The simplest probabilistic model assumes bits in strings to be identically independently distributed,

$$\mathbb{P}(0) = p, \qquad \mathbb{P}(1) = q = 1 - p,$$

the n strings of ω furthermore being drawn independently. This model is known as the *Bernoulli model*. The *unbiased model* also known as *uniform model* corresponds to the further condition $\mathbb{P}(0) = \mathbb{P}(1) = \frac{1}{2}$. Under the general Bernoulli model, the inductive definition (3) admits a direct translation

$$\phi(z) = t(z) + e^{qz}\phi(pz) + e^{pz}\phi(qz). \tag{4}$$

There $\phi(z)$ is the EGF of the sequence (ϕ_n) and $\phi_n = \mathbb{E}_n(\phi)$ is the expectation of the parameter $\phi[\cdot]$ taken over all trees comprising n data items. The verification from simple rules of series manipulation is easy: it suffices to see that, given n elements, the probability that k of them go into the left subtree (i.e, start with a 0) is the binomial probability $p^k q^{n-k} \binom{n}{k}$, so that, as regards expectations,

$$\phi_n = t_n + \sum_k p^k q^{n-k} \binom{n}{k} (\phi_k + \phi_{n-k}).$$

For the number of binary nodes in the tree, a determinant of storage complexity, the toll is $t_n = 1 - \delta_{n0} - \delta_{n1}$. For path length, which represents the total access cost of all elements, it becomes $t_n = n - \delta_{n0}$. The functional equation (4) can then be solved by iteration. Under the unbiased Bernoulli model, we have for instance

$$\phi(z) = t(z) + 2e^{z/2}t(\frac{z}{2}) + 4e^{3z/4}t(\frac{z}{4}) + \cdots .$$

Then, expansion around $z = 0$ yields coefficients, that is expectations. We quote under the unbiased Bernoulli model

The expected size (number of binary nodes) and the expected path length of a trie built out of n uniform independent random keys admit the explicit expressions

$$S_n = \sum_{k \geq 0} 2^k \left(1 - (1 - 2^{-k})^n - \frac{n}{2^k}(1 - 2^{-k})^{n-1}\right), \quad P_n = n \sum_{k \geq 0} \left(1 - (1 - 2^{-k})^{n-1}\right).$$

This result has been first discovered by Knuth in the mid 1960's.

Asymptotic analysis and the Mellin transform. A plot of the averages, S_n and P_n, is instructive. It strongly suggests that S_n is asymptotically linear, $S_n \sim cn(?)$, while $P_n \sim n\lg n(?)$, where $\lg x := \log_2 x$. As a matter of fact, the conjecture on size is false, but by an amazingly tiny amount. What we have is the following property:

The expected size (number of binary nodes) and the expected path length of a trie built out of n uniform independent random keys

$$S_n = \frac{n}{\log 2}(1 + \epsilon(\lg n)) + o(n), \qquad P_n = n\lg n + O(n). \tag{5}$$

There $\epsilon(x)$ is a continuous period function with amplitude $< 10^{-5}$:

We can only give here a few indications on the proof techniques and refer the reader to our long survey [24]. The idea, suggested to Knuth by the great analyst De Bruijn, is to appeal to the theory of integral transforms. Precisely, the *Mellin transform* associates to a function $f(x)$ (with $x \in \mathbb{R}_{\geq 0}$) the function $f^\star(s)$ (with $s \in \mathbb{C}$) defined by

$$f^\star(s) = \mathcal{M}[f(x), x \mapsto s] := \int_0^\infty f(x)x^{s-1}\,dx.$$

For instance $\mathcal{M}[e^{-x}] = \Gamma(s)$, the familiar Gamma function [61]. Mellin tranforms have two strikingly powerful properties. First, they establish a correspondence between the asymptotic expansion of a function at $+\infty$ (resp. 0) and *singularities* of the transform in a right (resp. left) half-plane. Second, they factorize *harmonic sums*, which correspond to a linear superposition of models taken at different scales.

Consider the function $s(x) = e^{-x}S(x)$, where $S(z)$ is the EGF of the sequence (S_n) of expectations of size. (This corresponds to adopting a Poisson rather than Bernoulli model; such a choice does not affect our asymptotic conclusions since, as can be proved elementarily [43, p. 131], $S_n - s(n) = o(n)$.) A simple calculation shows that

$$s(x) = \sum_{k\geq 0} 2^k\left[1 - \left(1 + \frac{x}{2^k}\right)e^{-x/2^k}\right], \qquad s^\star(s) - \frac{(1+s)\Gamma(s)}{1 - 2^{1+s}}.$$

The asymptotic estimates (5) result from there, given that a pole at α of the transform corresponds to a term $x^{-\alpha}$ in the asymptotic expansion of the original function. It is the existence of complex poles at

$$s = -1 + \frac{2ik\pi}{\log 2}, k \in \mathbb{Z},$$

that, in a sense, "creates" the periodic fluctuations present in $s(x)$ (and hence in S_n).

Models. Relative to the unbiased model (unbiased 0-1 bits in independent data), the expected size estimate expresses the fact that storage occupation is at most

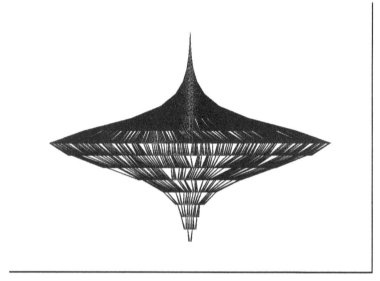

Fig. 1. A random trie of size $n = 500$ built over uniform data

of linear growth, despite the absence of convergence to a constant occupation ratio. The path length estimate means that the trie is nearly optimal in some information theoretic sense, since an element is typically found after $\sim \lg n$ binary questions. The profile of random trie under this model is displayed in Figure 1.

The ϵ-fluctuation, with an amplitude of 10^{-5}, in the asymptotic behaviour of size tends to be quite puzzling to programmers. Undeniably, such fluctuations will never be detected on simulations not to mention executions on real-life data. However, mathematically, their presence implies that most elementary strategies for analysing trie algorithms are doomed to failure. (See however [55, p. 403] for an elementary approach.) It is a fact that *no coherent theory of tries can be developed without taking such fluctuations into account.* For instance, the exact order of the variance of trie size and trie path length *must* involve them [41,42]. As a matter of fact, some analyses, which were developed in the late 1970s and ignored fluctuations, led to wrong conclusions, even regarding the *order of growth* of important characteristics of tries.

Back to modelling issues, the uniform model seems at first sight to be of little value. It is however fully justified in situations where elements are accessed via *hashing* and the indications it provides are precious: see for instance the discussion of dynamic and extendible hashing in Section 4. Also, the Mellin transform technology is equally suitable for extracting asymptotic information from the baised Bernoulli model ($p \neq q$). In that case, it is found that, asymptotically[2]

---

[2] The symbol '$\sim$' is used throughout in the strict sense of asymptotic equivalence; the symbol '$\approx$' is employed here to represent a numerical approximation up to tiny fluctuations.

$$S_n \approx \frac{n}{H}, \qquad P_n \sim \frac{n}{H} \log n, \tag{6}$$

where $H \equiv H(p,q) = -p \log p - q \log q$ is the *entropy* function of the Bernoulli (p,q) model. The formulæ admit natural generalizations to m-ary alphabets.

The estimates (6) indicate that trees become less efficient roughly in proportion to entropy. For instance, for a four symbol alphabet, where each letter has probability larger than 0.10, (this is true of most {A, G, C, T} genomic sequences), the degradation of performance is by less than a factor of 1.5 (i.e., a 50% loss at most). In particular linear storage and logarithmic access costs are preserved. Equally importantly, more realistic and considerably more general data models can be analysed precisely: see Section 8 relative to dynamical sources, which encapsulate the framework of Markov chains as a particular case.

Amongst the many fascinating techniques that have proved especially fruitful for trie analyses, we should also mention: Rice's method from the calculus of finite differences [30,43]; *analytic depoissonization* specifically developed by Jacquet and Szpankowski [40], which has led to marked successes in the analysis of dictionary-based compression algorithms. Complex analysis, that is, the theory of *analytic* (holomorphic) functions is central to most serious works in the area. Books that discuss relevant methods include those of Sedgewick-Flajolet [31], Hofri [35], Mahmoud [45], and Szpankowski [57].

3   Multidimensional Tries

Finkel and Bentley [21] adapted the BST to multidimensional data as early as 1974. Their ideas can be easily transposed to tries. Say you want to maintain sets of points in d-dimensional space. For $d = 2$, this gives rise to the standard quadtrie, which associates to a finite set $\omega \subset [0,1]^2$ a tree defined as follows.

  (*i*) If card$(\omega) = 0$, then quadtrie$(\omega) = \emptyset$;
 (*ii*) if card$(\omega) = 1$, then quadtrie(ω) consists of a single external node containg ω;
(*iii*) else, partition ω into the four subsets determined by their position with respect to the center $(\frac{1}{2}, \frac{1}{2})$ of space, and attach a root to the subtrees recursively associated to the four subsets (NW, NE, SW, SE, where NE stands for North-East, etc).

A moment's reflection shows that the quadtrie is equivalent to the 4-way trie built over an alphabet of cardinality 4: given any point $P = (x, y)$, write its coordinates in binary, $x = x_1 x_2, \ldots$ and $y = y_1 y_2 \cdots$, then encode the pair of coordinates "in parallel" over the alphabet $\{a, b, c, d\}$, where, say, $a = (0, 0)$, $b = (0, 1)$, $c = (1, 0)$, $d = (1, 1)$. The quadtrie is none other than the 4-way trie built over the set of such encodings.

Another idea of Bentley [6] gives rise to *k-d-tries*. For $d = 2$, associate to each point $P = (x, y)$, where $x = x_1 x_2, \ldots$ and $y = y_1 y_2 \cdots$, the binary string $z = x_1 y_1 x_2 y_2 \cdots$ obtained by interleaving bits of both coordinates. The k-d-trie is the binary trie built on the z-codes of points.

Given these equivalences, the analytic methods of Section 2 apply verbatim:

> *Over uniform independent data, the d-dimensional quadtrie requires on aver-*
> *age $\approx cn$ pointers, where $c = 2^d / \log 2^d$; for k-d-tries this number of pointers*
> *is $\approx c'n$, where $c' = 2/\log 2$. The mean number of bit accesses needed by a*
> *search is $\sim \log_2 n$.*

Roughly, multidimensional tries grant us fast access to multidimensional data. The storage requirements of quad-tries may however become prohibitive when the dimension of the underlying data space grows, owing to a large number of null pointers that carry little information but encumber memory.

Quadtries and k-d-tries also serve to implement *partial-match queries* in an elegant way. This corresponds to the situation, in d-dimensional space, where s out of d coordinates are specified and all points matching the s known coordinates are to be retrieved[3]. Put otherwise, one wants to reconstruct data given partial knowledge of their attributes. It is easy to set up recursive procedures reflected by inductive definitions for the cost parameters of such queries. The analytic methods of Section 2 are then fully operational. The net result, due to Flajolet and Puech [26], is

> *The mean number of operations needed to retrieve objects of d-dimensional*
> *space, when s out of d of their coordinates are known, is $O(n^{1-/s/d})$. The*
> *estimate holds both for quadtries and for k-d-tries.*

In contrast, quadtrees and k-d-trees (based on the BST concept) require

$$O(n^{1-s/d+\theta(s/d)})$$

operations for some function $\theta > 0$. For instance, for comparison-based structures, the case $s = 1$ and $d = 2$, entails a complexity $O(n^\alpha)$, where $\alpha = \frac{\sqrt{17}-3}{2} \doteq 0.56155$, which is of higher order than the $O(\sqrt{n})$ attached to bit-based structures. The better balancing on bit based structures pays—at least on uniform enough data.

Devroye [11] has provided an insightful analysis of tries ($d = 1$) under a density model, where data are drawn independently according to a probability density function spread over the unit interval. A study of multidimensional search along similar lines would be desirable.

Ternary search tries (TST). Multiway tries start require a massive amount of storage when the alphabet cardinality is large. For instance, a dictionary that would contain the word APPLE should have null pointers corresponding to the non-existent forms APPLA, APPLB, APPLC, etc. When attempting to address this problem, Bentley and Sedgewick [5] made a startling discovery: it is possible to design a highly efficient hybrid of the trie and the BST. In essence, you build an m-way trie (with m the alphabet cardinality), but implement the local decision structure at each node by a BST. The resulting structure, known as a TST,

---

[3] For an excellent discussion of spatial data structures, we redirect the reader to Samet's books [53,52].

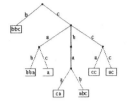 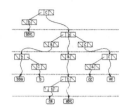 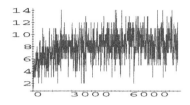

Fig. 2. Left: a trie. Middle: a corresponding TST. Right: cost of TST search on *Moby Dick* (number of letter comparisons against number of words scanned).

is simply a *ternary tree*, where, at each node, a comparison between letters is performed. Upon equality, go down one level, i.e., examine the next letter of the item to be retrieved; else proceed to the left or the right, depending on the outcome of the comparison between letters. It's as simple as that!

The TST was analysed by Clément, Flajolet, and Vallée [7,8]. Quoting from [7]: *Ternary search tries are an efficient data structure from the information theoretic point of view since a search costs typically about* log *n comparisons. For an alphabet of cardinality 26, the storage cost of ternary search tries is about 9 times smaller than standard array-tries.* (Based on extensive natural language data.)

4   Hashing and Height

Paged tries also known as *bucket tries* are digital trees defined like in (1), but with recursion stopped as soon as at most b elements have been isolated. This technique is useful in the context of a two-level memory. The tree itself can then be stored in core-memory as an index. Its end-nodes then point to pages or buckets in secondary memory. The technique can then be applied to hashed values of records, rather than records themselves which may be rather non-uniformly distributed in practice. The resulting algorithm is known as *dynamic hashing* and is due to Larson [44]. It is interesting to note that it was first discovered without an explicit reference to tries, the author viewing it as an evolution of hashing with separate chaining (in a paged environment), and with the splitting of buckets replacing costly chains of linked pages. The analysis methods of Section 2 show that the mean number of pages is

$$\approx \frac{n}{b \log 2}.$$

In other words, the pages are approximately 69% filled, a score that is comparable to the one of B–trees.

For very large data bases, the index of dynamic hashing may become too large to fit in primary memory. Fagin *et al.* [16] discovered an elegant way to remedy the situation, known as *extendible hashing* and based on the following principle:

Perfect tree embedding. Let a tree τ of some height H be given, with only the external nodes of τ containing information. Form the perfect tree P

of height H (i.e., all external nodes are at distance H from the root). The tree τ can be embedded into the perfect tree with any information being pushed to the external nodes of P. (This in general involves duplications.) The perfect tree with decorated external nodes can then be represented as an array of dimension 2^H, thereby granting direct access to its leaves.

In this way, in most practical situations, only two disc accesses suffice to reach any item stored in the structure—one for the index, which is a paged array, the other for the referenced page itself. This algorithm is the definite solution to the problem of maintaining very large hashed tables.

In its time, extendible hashing posed a new problem to analysts. Is the size of the index of linear or of superlinear growth? That question brings the analysis of height into our algorithmic games. General methods of combinatorial enumeration [31,56,33,62] are relevant to derive the basic equations. The starting point is ($[z^n]f(z)$ represents the coefficient of z^n in the expansion of $f(z)$ at 0)

$$\mathbb{P}_n(H \leq h) = n![z^n]e_b\left(\frac{z}{2^h}\right)^{2^h}, \qquad e_b(z) := 1 + \frac{z}{1} + \cdots + \frac{z^b}{b!}.$$

The problem is thus to extract coefficients of large index in the large power of a fixed function (here, the truncated exponential, $e_b(z)$). The saddle point method [10,31] of complex analysis comes to mind. It is based on Cauchy's coefficient formula,

$$[z^n]f(z) = \frac{1}{2i\pi}\int_{O^+} f(z)\,\frac{dz}{z^{n+1}},$$

which relates *values* of an analytic function to its coefficients, combined with the choice of a contour that crosses a saddle point of the integrand (Figure 3). The net result of the analysis [22] is the following:

Height of a paged b-trie is of the form

$$\left(1 + \frac{1}{b}\right)\log n + O(1)$$

both on average and in probability. The limit distributions are in the form of a double exponential function.

Fig. 3. A saddle point of the modulus of an analytic function

The size (2^H) of the extendible-hashing index is on average of the form $C(\log n)n^{1+1/b}$, with $C(\cdot)$ a bounded function. In particular, it grows non-linearly with n.

(See also Yao [63] and Régnier [50] for earlier results under the Poisson model.)

The ideas of extendible hashing are also susceptible of being generalized to higher dimensional data: see Régnier's analysis of *grid-file* algorithms in [51].

Level compressed tries. Nilsson and Karlsson [48] made a sensation when they discovered the *"LC trie"* (in full: Level Compressed trie): they demonstrated that their data structure could handle address lookup in routing tables with a standard PC in a way that can compete favorably with dedicated hardware embedded into routers. One of their beautifully simple ideas consists in compressing the perfect tree contained in a trie (starting from the root) into a single node of high degree—this principle is then used recursively. It is evocative of a partial realization of extendible hashing. The decisive advantage in terms of execution time stems from the fact that chains of pointers are replaced by a single array access, while the search depth decreases to $O(\log \log n)$ for a large class of distributions [12,13,48].

5   Leader Election and Protocols

Tries have found unexpected applications as an abstract structure underlying several algorithms of distributed computing. We discuss here leader election and the tree protocol due to Capetanakis-Tsybakov-Mikhailov (also known as the CTM protocol or the stack protocol). In both cases, what is assumed is a shared channel on which a number of stations are hooked. At any discrete instant, a station can broadcast a message of unit duration. It can also sense the channel and get a ternary feedback: 0 for silence; 1 for a succesful transmission; 2^+ for a collision between an unknown number of individuals.

The leader election protocol in its bare version is as follows:

Basic leader election. At time $t = 0$ the group G of all the n stations[4] on the channel are ready to start a round for electing a reader. Each one transmits its name (an identifier) at time 1. If $n = 0$, the channel has remained silent and nothing happens. If $n = 1$, the channel fedback is 1 and the corresponding individual is elected. Else all contenders in G flip a coin. Let G_H (resp G_T) be the subgroup of those who flipped head (resp. tails). Members of the group G_T withdraw instantly from the competition. Members of G_H repeat the process over the next time slot.

We expect the size of G to decrease by roughly a factor of 2 each time, which suggests that the number of rounds should be close to $\log_2 n$. The basic protocol described above may fail with probability close to 0.27865, see [55, p. 407], but it is easily amended: it suffices to let the last nonempty group of contenders (likely

---

[4] The number n is unknown.

to be of small size) start again the process, and repeat, until a group of size 1 comes out.

This leader election protocol is a perfect case for the analytic methods evoked earlier. The number of rounds for instance coincides with the leftmost branch of tree, a parameter easily amenable to the analytic techniques described in Section 2. The complete protocol has been analysed thoroughly by Prodinger [49] and Fill *et al.* [20]. Fluctuations are once more everywhere to be found.

The tree protocol was invented around 1977 independently in the USA and in the Soviet Union. For background, references, and results, we recommend the special issue of the *IEEE Transactions on Information Theory* edited by Jim Massey [46]. The idea is very simple: instead of developing only the leftmost branch of a trie, develop cooperatively the whole trie.

Basic tree protocol. Let G be the group of stations initially waiting to transmit a message. During the first available slot, all stations of G transmit. If the channel feedback is 0 or 1, transmission is complete. Else, G is split into G_H, G_T. All the members of G_H are given precedence and resolve their collisions between themselves, by a recursive application of the protocol. Once this phase has been completed, the group G_T takes its turn and proceeds similarly.

Our description presupposes a perfect knowledge of the system's state by all protagonists at every instant. It is a notable fact that the protocol can be implemented in a fully decentralized manner, each station only needing to monitor the channel feedback (and maintain a priority stack, in fact, a simple counter). The time it takes to resolve the contention between n initial colliders coincides with the total number of nodes in the corresponding trie (think of stations as having predetermined an infinite sequence of coin flips), that is, $2S_n + 1$ on average. The unbiased Bernoulli model is *exactly* applicable, given a decent random number generator. All in all, the resolution of a collision of multiplicity n takes times asymptotic to (cf Equation (5))

$$\frac{2}{\log 2} n,$$

upon neglecting the usual tiny fluctuations. In other words, the service time per customer is about $2/\log 2$. By standard queuing theory arguments, the protocol is demonstrably stable for all arrival rates λ satisfying $\lambda < \lambda_{\max}$, where

$$\lambda_{\max} = \frac{\log 2}{2} \pm \cdot 10^{-5} \doteq 0.34657.$$

In contrast, the Ethernet protocol has been proved unstable by Aldous in a stunning study [1].

We have described above a simplified version (the one with so-called "blocked arrivals") of the tree protocol. An improved version allows competitors to enter the game as soon as they are ready. This "free arrivals" version leads to nonlocal functional equations of the form

$$\psi(z) = t(z) + \psi(\lambda + pz) + \psi(\lambda + qz),$$

whose treatment involves interesting properties of iterated functions systems (IFS) and associated Dirichlet series: see G. Fayolle *et al* [17,18] and the account in Hofri's book [35]. The best protocol in this class (Mathys-Flajolet [47]) was largely discovered thanks to analytic techniques, which revealed the following: *a throughput of* $\lambda_{\max} = 0.40159$ *is achievable when combining free arrivals and ternary branching.*

6   Probabilistic Counting Algorithms

A problem initially coming from query optimization in data bases led Nigel Martin and me to investigate, at an early stage, the following problem: *Given a multiset M of data of sorts, estimate the number of* distinct *records, also called* cardinality, *that M contains.* The cardinality estimation problem is nowadays of great relevance to data mining and to network management. (We refer to [23] for a general discussion accompanied by references.)

The idea consists in applying a hash function h to each record. Then bits of hashed values are observed. The detection of patterns in observed hashed values can serve as a fair indicator of cardinality. Note that, by construction, such algorithms are totally insensitive to the actual structure of repetitions in the original file (usually, no probabilistic assumption regarding these can be made). Also, once a hash function of good quality has been chosen, the hashed values can legitimately be identified with uniform random strings. This makes it possible to trigger a virtuous cycle, involving probabilistic analysis of observables and suitably tuned cardinality estimators.

The original algorithm, called *probabilistic counting* [25], was based on a simple observable: the length L of the longest initial run of 1-bits in $h(M)$. This quantity can be computed with an auxiliary memory of a single 32 bit word, for reasonable file cardinalities, say $n \leq 10^9$. We expect $L_n \approx \log_2 n$, which suggests 2^{L_n} as a rough estimator of n. The analysis of L_n is attached to that of tries—we are in a way developing the leftmost branch of a pseudo-trie to which the methods of Section 2 apply perfectly. It involves the Thue-Morse sequence, which is familiar to aficionados of combinatorics on words. However, not too surprisingly, the rough estimate just described is likely to be typically off, by a little more than one binary order of magnitude, from the actual (unknown) value of n. Improvements are called for.

The idea encapsulated into the complete Probabilistic Counting algorithm is to emulate at barely any cost the effect of m independent experiments. There is a simple device, called "stochastic averaging" which makes it possible to do so by distribution into buckets and then averaging. The resulting algorithm *estimates cardinalities using m words of memory, with a relative accuracy of about* $\frac{0.78}{\sqrt{m}}$. It is pleasant to note that a multiplicative correction constant, provided by a Mellin transform analysis,

$$\varphi = \frac{e^\gamma}{\sqrt{2}} \frac{2}{3} \prod_{p=1}^{\infty} \left[\frac{(4p+1)(4p+2)}{(4p)(4p+3)} \right]^{\varepsilon(p)}$$

eddcdfdddddfcfdeeeeeefeedfedeffeffdefefeb fedefceffdefd
fefeecfdedeeededffefeffeecddefcfcddccedddcfddedeccdefdd
fcedddfdfedecddfedcfedcdfdeedegddcfededgggfffdggdfgfegdg
ddddegddffededceeeefdedgfgdddeefdceeeefeeddeedefcffffdh
hcgdccgchdfdchdehdgeeegfeedccfdedfddf

Fig. 4. The LogLog algorithm asociates to a text a signature, from which the number of differents words can be inferred. Here, the signature of Hamlet uses $m = 256$ bytes, with which the cardinality of the vocabulary is estimated to an accuracy of 6.6%.

(γ is Euler's constant, $\varepsilon(p) \in \{-1, +1\}$ indicates the parity of the number of 1-bits in the binary representation of p) enters the very design of the algorithm by ensuring that it is free of any systematic bias.

Recently, Marianne Durand and I were led to revisit the question, given the revival of interest in the area of network monitoring and following stimulating exchanges with Estan and Varghese (see, e.g., [15]). We realized that a previously neglected observable, the position $\widetilde{L}$ of the rightmost 1-bit in hashed values, though it has inferior probabilistic properties (e.g., a higher variance), can be maintained as a register in binary, thereby requiring very few bits. Our algorithm [14], called LogLog Counting *estimates cardinalities using m bytes of memory, with a relative accuracy of about* $\frac{1.3}{\sqrt{m}}$. Given that a word is four bytes, the overall memory requirement is divided by a factor of 3, when compared to Probabilistic Counting. This is, to the best of my knowledge, the most efficient algorithm available for cardinality estimation. Once more the analysis can be reduced to trie parameters and Mellin transform as well as the saddle point method play an important part.

For a highly valuable complexity-theoretic perspective on such questions see the study [2] by Alon, Matias, and Szegedy. In recent years, Piotr Indyk and his collaborators have introduced radically novel ideas in quantitative data mining, based on the use of *stable* distributions, but these are unfortunately outside of our scope, since tries do not intervene at all there.

7   Suffix Tries and Compression

Say you want to compress a piece of text, like the statement of Pythagoras' Theorem:

In any right triangle, the area of the square whose side is the hypotenuse (the side of the triangle opposite the right angle) is equal to the sum of the areas of the squares of the other two sides.

It is a good idea to notice that several words appear repeated. They could then be encoded once and for all by numbers. For instance:

1the,2angle,3triangle,4area,5square,6side,7right|In any 7 3, 1 4 of 1 5 whose 6 is 1 hypotenuse (1 6 of 1 3 opposite 1 7 2) is equal to 1 sum of 1 4s of 1 5s of 1 other two 6s.

A dictionary of frequently encountered terms has been formed. That dictionary could even be recursive as in

```
1the,2angle,3tri2,4area,5square,6side,7right|  ...
```

Around 1977–78, Lempel and Ziv developed ideas that were to have a profound impact on the field of data compression. They proposed two algorithms that make it possible to build a dictionary on the fly, and in a way that adapts nicely to the contents of the text. The first algorithm, known as LZ'78, is the following:

LZ'78 algorithm. Scan the text left to right. The text is segmented into phrases. At any given time, the cursor is on a yet unsegmented portion of the text. Find the longest phrase seen so far that matches the continuation of the text, starting from the cursor. A new phrase is created that contains this longest phrase plus one new character. Encode the new phrase with the rank of the previously matching phase and the new character.

For instance, "abracadabra" is segmented as follows

$$^0a|\ ^0b|\ ^0r|\ a^1c|\ a^1d|\ a^1b|\ r^3a|\ ab^6r|\ ac^4a|\ ^0d|\ abr^7a|,$$

resulting in the encoding

```
0a0b0r1c1d1b3a6r4a0d7a,
```

As it is well known, the algorithm can be implemented by means of a trie whose nodes store the ranks of the corresponding phrases.

From a mathematical perspective, the tree built in this way obeys the same laws as a *digital search tree* (DST), so that we'll start with examining them. The DST parameters can be analysed on average by the methods of Section 2, with suitable adjustments [28,29,43,45,57]. For instance, an additive parameter ϕ associated to a toll function t gives rise, at EGF level, to a functional equation of the form (in the unbiased case)

$$\phi(z) = t(z) + 2 \int_0^z e^{t/2} f\left(\frac{t}{2}\right) dt,$$

which is now a difference-differential equation. The treatement is a bit more difficult, but the equation eventually succumbs to the Mellin technology. In particular, path length under a general Bernoulli model is found to be satisfy

$$P_n^\circ = \frac{1}{H} n \log n + O(n), \tag{7}$$

with H the entropy of the model.

Back to the LZ'78 algorithm. Equation (7) means that when n phrases have been produced by the algorithm, the total number of characters scanned is $N \sim H^{-1} n \log n$ on average. Inverting this relation[5] suggest the following relation between number of characters read and number of phrases produced:

$$n \sim H \frac{N}{\log N}.$$

---

[5] This is in fact a renewal type of argument.

Since a phrase requires at most $\log_2 N$ bits to be encoded, the total length of the compressed text should be $\sim hn$ with $h = H/\log 2$ the *binary entropy*. This handwaving argument suggests the true fact: for a memoryless (Bernoulli) source, *the entropic (optimal) rate is achieved by LZ compression*.

The previous argument is quite unrigorous. In addition, information theorists are interested not only in dominant asymptotics but in quantifying *redundancy*, which measures the distance to the entropic optimum. The nature of stochastic fluctuations is also of interest in this context. Jacquet and Szpankowski have solved these difficult questions in an important work [39]. Their treatment starts from the bivariate generating function $P^\circ(z, u)$ of path length in DSTs, which satisfies a nonlinear functional equation,

$$\frac{\partial P^\circ(z, u)}{\partial z} = P^\circ(pz, u) \cdot P^\circ(qz, u).$$

They deduce asymptotic normality via a combination of inductive bounds, bootstrapping, analytic depoissonization, and the Quasi-Powers Theorem of analytic combinatorics [31,36,57]. (Part of the treatment makes use of ideas developed earlier by Jacquet and Régnier in their work establishing asymptotic normality of path length and size in tries [37].) From there, the renewal argument can be put on a sound setting. A full characterization of redundancy results and the fluctuations in the compression rate are determined to be asymptotically normal.

Suffix trees and antidictionaries. Given an infinitely long text $T \in \mathcal{B} \equiv \{0, 1\}^\infty$, the *suffix tree* (or suffix trie) of index n is the trie built on the first n suffixes of T. Such trees are important as an indexing tool for natural language data. They are also closely related to a variant of the LZ algorithm, known as LZ'77, that we have just presented. A new difficulty in the analysis comes from the back that suffix trees are tries built on data that are intrinsically *correlated*, due to the overlapping structure of suffixes of a single text. Jacquet and Szpankowski [38] are responsible for some of the early analyses in this area. Their treatment relies on the autocorrelation polynomial of Guibas and Odlyzko [34] and on complex-analytic techniques. Julien Fayolle [19] has recently extended this methodology. His methods also provide insight on the quantitative behaviour of a new scheme due to Crochemore *et al.* [9], which is based on the surprising idea of using *antidictionaries*, that is, a description of some of the patterns that are *avoided* by the text.

8   Dynamical Sources

So far, tries have been analysed when data are provided by a *source*, but one of a simple type. A new paradigm in this area is Brigitte Vallée's concept of *dynamical sources*. Such sources are most likely constituting the widest class of models, which can be subjected to a complete analytic treatment. To a large extent, Vallée's ideas [59] evolved from the realization that methods, originally developed for the purpose of analysing *continued fraction* algorithms [60], could be of a much wider scope.

Consider a transformation T of the unit interval that is piecewise differentiable and expanding: $T'(x) > 1$. Such a transformation is called a *shift*. It consists of several branches, as does the multivalued inverse function T^{-1}, which is formed of a collection of contractions. Given an initial value x_0, the sequence of iterates $(T^j(x_0))$ can then be encoded by recording at each iteration which branch is selected—this is a fundamental notion of symbolic dynamics. For instance, the function $T(x) = \{2x\}$ (with $\{w\}$ representing the fractional part of w) generates in this way the binary representation of numbers; from a metric (or probabilistic) point of view, it also describes the unbiased Bernoulli model. Via a suitable design, any biased Bernoulli model is associated with a shift, which is a piecewise-linear function. Markov chains (of any order) also appear as a special case. Finally, the continued fraction representation of numbers itself arises from the transformation

$$T(x) := \left\{\frac{1}{x}\right\} = \frac{1}{x} - \left\lfloor\frac{1}{x}\right\rfloor.$$

A dynamical source is specified by a shift (which determines a symbolic encoding) as well as by an initial density on the unit interval. The theory thus unites Devroye's density model, Bernoulli and Markov models, as well as continued fraction representions of real numbers and a good deal more. As opposed to earlier models, such sources take into account correlations between letters at an unbounded distance.

Tries under dynamical sources. Vallée's theory has been applied to tries, in particular in a joint work with Clément [8]. What it brings to the field is the unifying notion of *fundamental intervals* which are the subintervals of $[0, 1]$ associated to places corresponding to potential nodes of tries. Much transparence is gained by this way of viewing a trie process as a succession of refined partitions of the unit interval, and the main parameters of tries can be expressed simply in this framework.

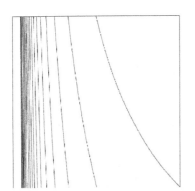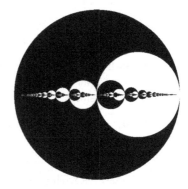

Fig. 5. Dynamical sources: [left] the shift associated with continued fractions; [right] a rendering of fundamental intervals

Technically, a central rôle is played by Ruelle's *transfer operator*. Given a shift T with $\mathcal{H}$ the collection of its inverse branches, the transfer operator is defined over a suitable space of functions by

$$\mathcal{G}_s[f](x) := \sum_{h \in \mathcal{H}} |h'(x)|^s \, f \circ h(x).$$

For instance, in the continued fraction case, one has

$$\mathcal{G}_s[f](x) := \sum_{m \geq 1} \frac{1}{(m+x)^{2s}} f\left(\frac{1}{m+x}\right).$$

The quantity s there is a parameter that is *a priori* allowed to assume complex values. As Vallée showed, by considering iterates of $\mathcal{G}_s$, it then becomes possible to construct generating functions, usually of the Dirichlet type, associated with partitions into fundamental intervals. (In a way, the transfer operator is a supergenerating operator.) Equiped with these, it then becomes possible to express expectations and probability distributions of trie parameters, after a Mellin transform round. Then functional analysis comes into play (the operators have a discrete spectrum), to the effect that *asymptotic properties of tries built on dynamical sources are explicitly related to spectral properties of the transfer operator.*

As an example of unified formulæ, we mention here the mean value estimates of (6) which are seen to hold for an arbitrary dynamical source. The rôle of entropy in these formulæ comes out neatly—entropy is bound to be crucial under any dynamical source model. The analysis of height becomes almost trivial; the characteristic constant turns out to be in all generality none other than $\lambda_1(2)$, the dominant eigenvalue of operator $\mathcal{G}_2$.

We conclude this section with a brief mention of an algorithm due to Gosper, first described in the celebrated "Hacker's memorandum" also known as HAK-MEM [4, Item 101A]. The problem is to compare two fractions $\frac{a}{b}$ and $\frac{c}{d}$. It is mathematically trivial, since

$$\frac{a}{b} - \frac{c}{d} = \frac{ad - bc}{bd},$$

but the algorithms that this last formula suggests either involve going to multiprecision routines or operating with floating point arithmetics at the risk of reaching a wrong conclusion.

Gosper's comparison algorithm. In order to compare $\frac{a}{b}$ and $\frac{c}{d}$, perform a continued fraction expansion of both fractions. Proceed in lazy mode. Stop as soon as a discrepant digit is encountered.

Gosper's solution operates within the set precision of data and is error-free (as opposed to the use of floating point approximations). For this and other reasons[6],

---

[6] The sign question is equivalent to determining the orientation of a triangle in the plane.

it has been found to be of interest by the community of computational geometers engaged in the design of *robust* algorithms [3]. (It also makes an appearance in the source code of Knuth's METAFONT, for similar reasons.)

In our perspective, the algorithm can be viewed as the construction of the digital tree associated to two elements accessible via their continued fraction representations. Vallée and I give a thorough discussions of the fascinating mathematics that surround its analysis in [32]. The algorithm extends to the lazy and robust comparison of a system of n fractions: it suffices to build, by lazy evaluation, the trie associated to continued fraction representations of the entries. What we found in [32] is the following result: *The expected cost of sorting n uniform random real numbers by lazy evaluation of their continued fraction representations satisfies*

$$P_n = K_0 n \log n + K_1 n + Q(n) + K_2 + o(1),$$

where ($\zeta(s)$ is the Riemann zeta function)

$$K_0 = \frac{6 \log 2}{\pi^2}, \qquad K_1 = 18 \frac{\gamma \log 2}{\pi^2} + 9 \frac{(\log 2)^2}{\pi^2} - 72 \frac{\log 2 \, \zeta'(2)}{\pi^4} - \frac{1}{2},$$

and $Q(u)$ is an oscillating function with mean value 0 whose order is

$$Q(u) = O\left(u^{\delta/2}\right), \text{ where } \delta \text{ is any number such that } \delta > \sup\left\{\Re(s) \mid \zeta(s) = 0\right\}.$$

The Riemann hypothesis has just made an entry into the world of tries!

References

1. David J. Aldous, *Ultimate instability of exponential back-off protocol for acknowledgement-based transmission control of random access communication channels*, IEEE Transactions on Information Theory **33** (1987), no. 2, 219–223.

2. Noga Alon, Yossi Matias, and Mario Szegedy, *The space complexity of approximating the frequency moments*, Journal of Computer and System Sciences **58** (1999), no. 1, 137–147.

3. F. Avnaim, J.-D. Boissonnat, O. Devillers, F. P. Preparata, and M. Yvinec, *Evaluating signs of determinants using single-precision arithmetic*, Algorithmica **17** (1997), no. 2, 111–132.

4. M. Beeler, R. W. Gosper, and R. Schroeppel, *HAKMEM*, Memorandum 239, M.I.T., Artificial Intelligence Laboratory, February 1972, Available on the World-Wide Web at http://www.inwap.com/pdp10/hbaker/hakmem/hakmem.html.

5. Jon Bentley and Robert Sedgewick, *Fast algorithms for sorting and searching strings*, Eighth Annual ACM-SIAM Symposium on Discrete Algorithms, SIAM Press, 1997.

6. Jon Louis Bentley, *Multidimensional binary search trees used for associative searching*, Communications of the ACM **18** (1975), no. 9, 509–517.

7. Julien Clément, Philippe Flajolet, and Brigitte Vallée, *The analysis of hybrid trie structures*, Proceedings of the Ninth Annual ACM–SIAM Symposium on Discrete Algorithms (Philadelphia), SIAM Press, 1998, pp. 531–539.

8. Julien Clément, Philippe Flajolet, and Brigitte Vallée, *Dynamical sources in information theory: A general analysis of trie structures*, Algorithmica **29** (2001), no. 1/2, 307–369.

9. M. Crochemore, F. Mignosi, A. Restivo, and S. Salemi, *Text compression using antidictionaries*, Automata, languages and programming (Prague, 1999), Lecture Notes in Computer Science, vol. 1644, Springer, Berlin, 1999, pp. 261–270.

10. N. G. de Bruijn, *Asymptotic methods in analysis*, Dover, 1981, A reprint of the third North Holland edition, 1970 (first edition, 1958).

11. Luc Devroye, *A probabilistic analysis of the height of tries and of the complexity of triesort*, Acta Informatica **21** (1984), 229–237.

12. Luc Devroye, *An analysis of random LC tries*, Random Structures & Algorithms **19** (2001), 359–375.

13. Luc Devroye and Wojtek Szpankowski, *Probabilistic behaviour of level compressed tries*, Random Structures & Algorithms **27** (2005), no. 2, 185–200.

14. Marianne Durand and Philippe Flajolet, LOGLOG *counting of large cardinalities*, Annual European Symposium on Algorithms (ESA03) (G. Di Battista and U. Zwick, eds.), Lecture Notes in Computer Science, vol. 2832, 2003, pp. 605–617.

15. Cristian Estan and George Varghese, *New directions in traffic measurement and accounting: Focusing on the elephants, ignoring the mice*, ACM Transactions on Computer Systems **21** (2003), no. 3, 270–313.

16. R. Fagin, J. Nievergelt, N. Pippenger, and R. Strong, *Extendible hashing: A fast access method for dynamic files*, A.C.M. Transactions on Database Systems **4** (1979), 315–344.

17. Guy Fayolle, Philippe Flajolet, and Micha Hofri, *On a functional equation arising in the analysis of a protocol for a multiaccess broadcast channel*, Advances in Applied Probability **18** (1986), 441–472.

18. Guy Fayolle, Philippe Flajolet, Micha Hofri, and Philippe Jacquet, *Analysis of a stack algorithm for random access communication*, IEEE Transactions on Information Theory **IT-31** (1985), no. 2, 244–254, (Special Issue on Random Access Communication, J. Massey editor).

19. Julien Fayolle, *An average-case analysis of basic parameters of the suffix tree*, Mathematics and Computer Science III: Algorithms, Trees, Combinatorics and Probabilities (M. Drmota *et al.*, ed.), Trends in Mathematics, Birkhäuser Verlag, 2004, pp. 217–227.

20. James Allen Fill, Hosam M. Mahmoud, and Wojciech Szpankowski, *On the distribution for the duration of a randomized leader election algorithm*, The Annals of Applied Probability **6** (1996), no. 4, 1260–1283.

21. R. A. Finkel and J. L. Bentley, *Quad trees, a data structure for retrieval on composite keys*, Acta Informatica **4** (1974), 1–9.

22. Philippe Flajolet, *On the performance evaluation of extendible hashing and trie searching*, Acta Informatica **20** (1983), 345–369.

23. ———, *Counting by coin tossings*, Proceedings of ASIAN'04 (Ninth Asian Computing Science Conference) (M. Maher, ed.), Lecture Notes in Computer Science, vol. 3321, 2004, (Text of Opening Keynote Address.), pp. 1–12.

24. Philippe Flajolet, Xavier Gourdon, and Philippe Dumas, *Mellin transforms and asymptotics: Harmonic sums*, Theoretical Computer Science **144** (1995), no. 1–2, 3–58.

25. Philippe Flajolet and G. Nigel Martin, *Probabilistic counting algorithms for data base applications*, Journal of Computer and System Sciences **31** (1985), no. 2, 182–209.

26. Philippe Flajolet and Claude Puech, *Partial match retrieval of multidimensional data*, Journal of the ACM **33** (1986), no. 2, 371–407.
27. Philippe Flajolet, Mireille Régnier, and Dominique Sotteau, *Algebraic methods for trie statistics*, Annals of Discrete Mathematics **25** (1985), 145–188, In *Analysis and Design of Algorithms for Combinatorial Problems*, G. Ausiello and M. Lucertini Editors.
28. Philippe Flajolet and Bruce Richmond, *Generalized digital trees and their difference-differential equations*, Random Structures & Algorithms **3** (1992), no. 3, 305–320.
29. Philippe Flajolet and Robert Sedgewick, *Digital search trees revisited*, SIAM Journal on Computing **15** (1986), no. 3, 748–767.
30. _____ , *Mellin transforms and asymptotics: finite differences and Rice's integrals*, Theoretical Computer Science **144** (1995), no. 1–2, 101–124.
31. Philippe Flajolet and Robert Sedgewick, *Analytic combinatorics*, October 2005, Chapters I–IX of a book to be published, 688p.+x, available electronically from P. Flajolet's home page.
32. Philippe Flajolet and Brigitte Vallée, *Continued fractions, comparison algorithms, and fine structure constants*, Constructive, Experimental, and Nonlinear Analysis (Providence) (Michel Théra, ed.), Canadian Mathematical Society Conference Proceedings, vol. 27, American Mathematical Society, 2000, pp. 53–82.
33. Ian P. Goulden and David M. Jackson, *Combinatorial enumeration*, John Wiley, New York, 1983.
34. L. J. Guibas and A. M. Odlyzko, *String overlaps, pattern matching, and nontransitive games*, Journal of Combinatorial Theory. Series A **30** (1981), no. 2, 183–208.
35. Micha Hofri, *Analysis of algorithms: Computational methods and mathematical tools*, Oxford University Press, 1995.
36. Hsien-Kuei Hwang, *On convergence rates in the central limit theorems for combinatorial structures*, European Journal of Combinatorics **19** (1998), no. 3, 329–343.
37. Philippe Jacquet and Mireille Régnier, *Trie partitioning process: Limiting distributions*, CAAP'86 (P. Franchi-Zanetacchi, ed.), Lecture Notes in Computer Science, vol. 214, 1986, Proceedings of the 11th Colloquium on Trees in Algebra and Programming, Nice France, March 1986., pp. 196–210.
38. Philippe Jacquet and Wojciech Szpankowski, *Autocorrelation on words and its applications: analysis of suffix trees by string-ruler approach*, Journal of Combinatorial Theory. Series A **66** (1994), no. 2, 237–269.
39. _____ , *Asymptotic behavior of the Lempel-Ziv parsing scheme and digital search trees*, Theoretical Computer Science **144** (1995), no. 1–2, 161–197.
40. _____ , *Analytical de-Poissonization and its applications*, Theoretical Computer Science **201** (1998), no. 1-2, 1–62.
41. Peter Kirschenhofer and Helmut Prodinger, *On some applications of formulæ of Ramanujan in the analysis of algorithms*, Mathematika **38** (1991), 14–33.
42. Peter Kirschenhofer, Helmut Prodinger, and Wojciech Szpankowski, *On the variance of the external path length in a symmetric digital trie*, Discrete Applied Mathematics **25** (1989), 129–143.
43. Donald E. Knuth, *The art of computer programming*, 2nd ed., vol. 3: Sorting and Searching, Addison-Wesley, 1998.
44. P. A. Larson, *Dynamic hashing*, BIT **18** (1978), 184–201.
45. Hosam M. Mahmoud, *Evolution of random search trees*, John Wiley, New York, 1992.
46. James L. Massey (ed.), *Special issue on random-access communications*, vol. IT-31, IEEE Transactions on Information Theory, no. 2, March 1985.

47. Peter Mathys and Philippe Flajolet, *Q–ary collision resolution algorithms in random access systems with free or blocked channel access*, IEEE Transactions on Information Theory **IT-31** (1985), no. 2, 217–243.
48. S. Nilsson and G. Karlsson, *IP–address lookup using LC tries*, IEEE Journal on Selected Areas in Communications **17** (1999), no. 6, 1083–1092.
49. Helmut Prodinger, *How to select a loser*, Discrete Mathematics **120** (1993), 149–159.
50. Mireille Régnier, *On the average height of trees in in digital search and dynamic hashing*, Information Processing Letters **13** (1982), 64–66.
51. _____ , *Analysis of grid file algorithms*, BIT **25** (1985), 335–357.
52. Hanan Samet, *Applications of spatial data structures*, Addison–Wesley, 1990.
53. _____ , *The design and analysis of spatial data structures*, Addison–Wesley, 1990.
54. Robert Sedgewick, *Algorithms in C, Parts 1–4*, third ed., Addison–Wesley, Reading, Mass., 1998.
55. Robert Sedgewick and Philippe Flajolet, *An introduction to the analysis of algorithms*, Addison-Wesley Publishing Company, 1996.
56. Richard P. Stanley, *Enumerative combinatorics*, vol. II, Cambridge University Press, 1998.
57. Wojciech Szpankowski, *Average-case analysis of algorithms on sequences*, John Wiley, New York, 2001.
58. Luis Trabb Pardo, *Set representation and set intersection*, Tech. report, Stanford University, 1978.
59. Brigitte Vallée, *Dynamical sources in information theory: Fundamental intervals and word prefixes*, Algorithmica **29** (2001), no. 1/2, 262–306.
60. Brigitte Vallée, *Euclidean dynamics*, October 2005, 69p. Submitted to *Discrete and Continuous Dynamical Systems*.
61. E. T. Whittaker and G. N. Watson, *A course of modern analysis*, fourth ed., Cambridge University Press, 1927, Reprinted 1973.
62. Herbert S. Wilf, *Generatingfunctionology*, Academic Press, 1990.
63. Andrew Chi Chih Yao, *A note on the analysis of extendible hashing*, Information Processing Letters **11** (1980), 84–86.

Flat Holonomies on Automata Networks[*]

(a more recent version available at:
http://arXiv.org/abs/cs.DC/0512077)

Gene Itkis[1] and Leonid A. Levin[2]

Boston University, Department of Computer Science,
111 Cummington St., Boston, MA 02215

Abstract. We consider asynchronous dynamic networks of identical finite (independent of the network size) automata. A useful data structure
on such networks is a partial *orientation* of its edges. It needs to be *straight*,
i.e. have null holonomy (the difference between the number of up and down
edges in each cycle). It also needs to be *centered*, i.e., have a unique node
with no down edges. Using such orientation, any feasible computational
task can be efficiently implemented with self-stabilization and synchronization. There are (interdependent) self-stabilizing asynchronous finite
automata protocols that straighten and centralize any orientation. Such
protocols may vary in assorted efficiency parameters and it is desirable to
have each replaceable with any alternative responsible for a simple limited
task. We describe an efficient reduction of any computational task to any
set of such protocols compliant with our interface conditions.

1 Introduction

1.1 Dynamic Asynchronous Networks (with Faults)

The computing environment is rapidly evolving into a huge global network. It
is divided into smaller nets, those into computer clusters, individual computers,
their assorted devices, and so on. Even individual silicon chips may evolve into
cheaper structures of reliable circuits build from unreliable elements easy to print
with huge densities. We can expect such networks to penetrate all aspects of our
life and achieve great diversity and complexity. It is interesting to investigate how
such diverse complex unpredictable networks of elementary unreliable nodes can
organize themselves to perform various useful tasks.

Let us view a network as a connected graph of asynchronous finite automata
and try to equip each node with a self-organizing protocol. The automata have
identical transition function independent of size and structure of the network.
So, they have no information about the network, and even no room in their $O(1)$

---

[*] This article supplements a STACS-06 talk "Symmetry Breaking in Computing Media" by Leonid A. Levin, illustrating some of its points.
[1] Supported by NSF grant CCR-0311485.
[2] Supported by NSF grant CCR-0311411.

B. Durand and W. Thomas (Eds.): STACS 2006, LNCS 3884, pp. 23–49, 2006.
© Springer-Verlag Berlin Heidelberg 2006

memories to store, say, its size. The automata run asynchronously with widely varying speeds. Each sees the states of its neighbors but cannot know how many (if any) transitions they made between its own transitions.

The network must recover from any faults. An adversary may initialize the automata in any combination of states without restrictions, after which the protocol must recover some meaningful state and resume normal operation. Such conditions and requirements may seem drastic, but stronger assumptions may be undesirable for the really ubiquitous networks that we came to expect. For instance, the popular assumption that each node grows in complexity with the size of the network and keeps some global information may be too restrictive.

So, which tasks and how efficiently can be solved by such networks? The network's distributed nature, unknown topology, asynchrony, dynamics and faults, etc. complicate this question. The computational power of any network with total memory S is in the obvious class Space(S). In fact, this trivial condition is sufficient as well.

1.1.1 Connections, Slope, and Computing

We consider protocols based on orientation (up, down or horizontal) for each directed edge. It is a somewhat stretched transplantation to graphs of widely used geometric structures, *connections*, that map coordinate features between nearby points of smooth manifolds. Orientation is a simplest analog of such structures, comparing relative heights of adjacent nodes.

An important aspect of a connection is its *holonomy*, i.e. the composition over each circular (often contractible, though in graphs this restriction is mute) path. If this holonomy is null, i.e. an identity mapping, for each cycle, the connection is called *flat*. For our orientations (then called *slope*) this means that every cycle is *balanced* i.e. has equal numbers of up and down edges. Maintaining such property, detecting and repairing its failures is a task useful for many purposes. It may be that studying other examples of connections on graphs can be beneficial for other problems, too.

Networks can deal with asynchrony by keeping in each node a step counter with equal or adjacent values in adjacent nodes. Then a node may advance its counter only if it is a local minimum. Such counters may be reduced mod 3 when no self-stabilization is required. A slope gives such an assignment of mod 3 counters to nodes; the change of its value induces orientation of (directed) edges. Obviously, all cycles are balanced. Faulty configurations, however, can have inconsistent mod 3 counters with *vortices*, i.e., unbalanced, even unidirectional in extreme cases, cycles. Slope is especially useful when *centered*, i.e., having a unique node, *leader*, with no down edges. It then yields a BFS tree, the construction/maintenance of which is known to self-stabilize many basic network management protocols.

1.1.2 Maintaining Centered Slope

The task of initiating an un-centered slope is easier in some aspects, e.g., it can be done deterministically. The other task, centralizing a slope (i.e., modifying it into a centered one), is known to be impossible for deterministic algorithms. A fast randomized algorithm for it, using one byte per node is given in [IL92]. The appendix there gives a collection of deterministic finite automata (i.e., $O(1)$ bits per node) protocols that initiate a slope, running simultaneously in concert with each other and with the slope centralization protocol.

Here, we refer to four separate tasks: build certificates on trees assuring their acyclicity, check such certificates, balance orientation on graphs spanned by forest of such trees, and centralize such a slope merging the forest into a tree. We develop a protocol *Shell* that (using no additional space) coordinates any protocols performing these tasks. Each of this protocols may be arbitrary, as long as it performs the task and complies with simple rules imposed by the interface. The Shell then assures that a centered slope is verified, and repaired if necessary, with the efficiency of the underlying task protocols. Further protocols then make an efficient *Reduction* of any computational task to a self-stabilizing protocol for asynchronous networks.

1.2 Self-stabilizing Protocols

The ability of networks to recover from any faults can be modeled by self-stabilization: A *self-stabilizing* protocol works correctly no matter what state it is initiated in. This approach was pioneered by Dijkstra [Dij74] and has since been a topic of much research in the distributed computation and other areas (see bibliography by T. Herman [Her]).

It was widely believed that self-stabilization is unattainable for most tasks/networks unless the bit-size of network's nodes grows (at least logarithmically) with the size of the network. (See, e.g., [M$^+$92] for discussion of undesirability of such assumptions.) Logarithmic lower-bound for self-stabilization on rings [IJ90] reinforced this believe. However, this lower bound depends on an implicit restriction on the types of protocols: a processor can change state only if a special *token* predicate on the state of the node's neighborhood evaluates to true (this lower bound was later extended to *silent* protocols for which the communication registers do not change values after stabilization [DGS96]).

The first general sub-logarithmic space result, by Awerbuch, Itkis, and Ostrovsky [I$^+$92], gave randomized self-stabilizing protocols using $\log \log n$ space per edge for leader election, spanning tree, network reset and other tasks. This was improved to constant space per node for all linear space tasks by Itkis in [I$^+$92], and by [IL92] using hierarchical constructions similar to those used in other contexts in [Thu12, Ro71, G86]. These results were later modified in [AO94] to extend the scope of tasks solvable deterministically in $O(\log^* n)$ space per edge (beyond forest/slope construction, for which algorithms of [IL92] were already deterministic).

There is a large literature on self-stabilization and similar features in other contexts which we cannot review here. For instance, many difficult and elegant results in related directions were obtained for cellular automata. In most CA contexts, the protocols, similarly to our work, assume finite complexity of each node, independent of the network size. However, the irregular nature of our networks presents major additional complications.

2 Model

2.1 Nodes and Communication

Our *network* is based on an undirected reflexive *communication graph* $G = (V, E)$ of diameter d. Without loss of generality, for the rest of this paper we assume that G is *connected*. Each node v is anonymous and labeled with state consisting of bits and pointers to immediate *neighbors* $w \in \mathbf{E}(v)$. A **link** $[v, w]$ is an isomorphism type of a subnetwork restricted to two nodes v, w (this includes information about equality of the pointers leaving the same node). Nodes **act** as automata changing their states based on the set (without multiplicity) of all incident links. When a node sets a pointer, it can chose a link, but a specific (anonymous) neighbor connected by such link is chosen by the adversary.[1]

In other words, the transition table for the nodes may specify that the active node in state X change its state to Y if it has a neighbor in state Z (or if all the neighbors are in state Z). But it cannot say that this transition should take place if there it has 5 neighbors in state Z. Also, the transition may specify that the node's pointer be set to its neighbor in state Z, but then it will be up to Adversary to select the specific neighbor in the state Z to set the pointer to.

2.2 Asynchrony

Asynchrony is modeled by *Adversary* adaptively determining a sequence of nodes with arbitrary infinite repetitions for each; the nodes act in this order. In other words, at any point in time, Adversary selects the next node to act.

A **step** is the shortest time period since the previous step within which each node acts at least once. We write $s \prec t$ (resp. $s \succ t$) to denote that all of the step s occurs before (resp. after) the time instant t.

For the sake of simplicity, we assume that only one node is active at any time. Assuming atomicity of node steps, we can further relax the asynchrony model to allow adversary activate any set of nodes (including adjacent nodes).

2.2.1 Asynchrony Types

Consider three types of asynchrony: **full**, **independent set** and **singular**, corresponding to the adversary being able to simultaneously activate *any set* of nodes, *any independent set* of nodes, and only *one node* at a time, respectively. Since the transition of a node depends only on its set of incident links, it is obvious that independent set asynchrony is equivalent to singular (a protocol that works in one will work in the other without any changes).

To achieve now full asynchrony, replace each edge uv with a dummy node x and edges ux and xv. This change of the network is used only for our structure fields protocols (see Sec. 3.2, 3.3), for which this change of network is going to be unimportant. The dummy node x is to be simulated by one of the endpoints u, or v —

---

[1] Proposition 4.1 requires deterministic choice of neighbor, e.g., the first qualified, if the edges are ordered.

let it be u. Then u is the *host* of x, x is a *satellite* of u, and v, x are *buddies*. Choosing the host for a dummy node is arbitrary and can be done probabilistically and independently for each edge.

When activated by adversary, a node first performs its own action and then activates all its satellites. Thus, the dummy nodes are never activated simultaneously with their hosts. Now we need to avoid simultaneous activation of any dummy node and its buddy. Let each node (real or dummy) have a binary color — black or white — flipped when the node makes its transaction (even if that transaction does not end up changing anything in the node). A dummy node x makes its transaction only when its buddy v is of the different color, and an actual node v — only when all of its buddies are of the same color as v. This guarantees that the two buddies x and v can never make their transaction simultaneously: when x and v are of different colors — then only x can change, but when their colors are the same, then only v can change (it may still be delayed by its other buddies). If v does not move for one step, then all its buddies are guaranteed to be the same color. Thus each actual node is guaranteed to move at least every second step. If a dummy node x does not move for one step, then its buddy v has all its buddies of the same color and within one more step v has made the move changing its color. And then v cannot change color till x moves. Thus, within two steps x will make its move.

Thus, at the cost of increasing the space of each node to be proportional to its degree we can run any structure protocol designed for singular asynchrony in a fully asynchronous network.

2.3   Faults

The *faults* are modeled by allowing the Adversary to select the initial state of the whole network. This is a standard way of modeling the worst-case but transient, "catastrophic" faults. The same model applies to any changes in the network — since even a non-malicious local change may cause major global change, we treat the changes as faults. After changes or faults introduced by the Adversary, the network takes some time to stabilize (see Sec. 3.3 for the precise definitions) — we assume that Adversary does not affect the transitions during the stabilization period, except by controlling the timing (see asynchrony above).

We discuss the *computational powers* of the Adversary below in subsection 3.3. As a preview, intuitively, *our* protocols require *no computational bounds* on Adversary. However, the protocols given to our protocols for simulation may require that these unreasonable powers of the Adversary be restricted.

Our algorithms (reductions) in this paper are all deterministic. They may interact or even emulate other algorithms which may be either also deterministic or randomized. Therefore we leave out the discussion of the access an adversary may have to the random coins used in the other algorithms — such a discussion is found in [IL92] in the context of the leader election algorithm. Other, emulated, algorithms may have different restrictions on the adversary.

3   Slope, Tasks and Protocols

3.1   Orientation and Slope

Edge **orientation dir**() of G maps each directed edge vw of G to $\mathbf{dir}(vw) \in \{0, \pm 1\}$.

Variance of a path $v_0 \ldots v_k$ is $\sum_{i=0}^{k-1} \mathbf{dir}(v_i v_{i+1})$.

We consider only such orientations for which the variance of any cycle is divisible by 3. These orientations have economical *representations*: Let $v.\mathrm{h3} \in \{-1, 0, 1\}$ be a field in each node. Then, define $\mathbf{dir}(vw) = (w.\mathrm{h3} - v.\mathrm{h3} \bmod 3) \in \{-1, 0, 1\}$, which implies $\mathbf{dir}(vw) = -\mathbf{dir}(wv)$. We say that $w \in \mathbf{E}(v)$ is *under* v (and v is *over* w) if $\mathbf{dir}(vw) = -1$; the directed edge vw points **down** and wv **up**; define $\mathbf{up}(vw) \overset{def}{=} (\mathbf{dir}(vw) = +1)$. A path $v_0 \ldots v_k$ is an **up-path** if for all $0 \le i < k$ v_i is under v_{i+1}.

The *balance* of a directed cycle is $0, 1$ or -1 according to the sign of its variance. We say a cycle with balance 0 is *balanced*, otherwise it is a *vortex*.

An orientation is **flat** if all the cycles are balanced.[2] A flat orientation is **centered** if it has only a unique node with no neighbor under it. Such a local minimum node is referred to as the *leader*.

It is convenient to mark potential leaders — we call them *roots*. Furthermore, we strengthen the flatness condition to require that any path from root must have non-negative variance, and call such orientation with roots a **slope**. In particular, in a slope any root must be a local minimum, and a root-root path must have zero variance. Slope is *centralized* if it is a centered orientation with a single root.

3.2   Tasks

Here, as above, we use a standard probabilistic Turing Machine M with a read-only input tape I, write-only output tape O, and read-write work tape W of size $|W| = n$. We use **reversal number** of times the TM head changes direction as the TM time-complexity measure [Tra64, Bar65] for the definition of the task below.

Let $Q = \{\langle x, y \rangle\}$ be a relation (of question-"correct answer" pairs). A **task** Γ_n for relation Q of input-output pairs is a triplet of probabilistic algorithms $\Gamma_n = \langle S, C, B \rangle$, such that

---

[2] A weaker condition suffices for most applications: absence of long up-paths (especially, cycles). Many protocols work with delays proportional to the maximum up-path length. The zero variance condition assures that no up-path is longer than the graph diameter d. Allowing non-zero variance may increase the maximum up-path length, resulting in the corresponding computational delays.

We make the following observation which is not used in this paper: Many algorithms change orientation gradually, so variance of any path changes at most by 1 at a time. Then the variance of a cycle (being a multiple of 3) never changes. This implies that the variance of any path can change by at most $\pm 2d$, and thus the maximum node-length of up-paths can vary with time within at most a factor of $2d$.

Time-bound T: for input x there is some t_x, such that with probability $> 1/2$, $T(x)$ outputs t_x using $O(\sqrt{t_x})$ time and space $O(n/\lg n)$; [3]

Solver A with probability $> 1/2$ on input x using $O(n)$ space and $O(t_x)$ time outputs y, such that $\langle x, y \rangle \in Q$;

Checker C on input $\langle x, y \rangle \in Q$ never outputs 0 (incorrect), and on input $\langle x, y' \rangle \notin Q$ with probability $> 1/2$ outputs 0 using $O(n)$ space and $O(t_x)$ time.

Our goal is to construct protocols that would take any task (specified for a faultless and synchronous computational model such as TM) and produce a protocol running the task in the tough distributed environment where Adversary controls the timing and initial state of the system. We separate this job into two:

First, we assume that some special protocol generates a centered slope and stabilizes, i.e., the orientation stops changing. The rest of this section discusses how to achieve our goal under this assumption.

Then, the remainder of the paper (starting with Sec. 5) is dedicated to satisfying this assumption: describing the protocol that stabilizes with a centered slope. This standard centered slope generating protocol will run in the fields that we will call *structure*, and will enable the network to solve arbitrary tasks.

3.3 Self-stabilizing Protocols

Let each processor (node) in the network G have the following fields: read-only *input*, write-only *output*, and read/write *work* and *structure*. A *configuration* at time instant t is a quintuple $\langle G, I, O_t, W_t, S_t \rangle$, where functions I, O_t, W_t, S_t on V represent the input, output, work and structure fields respectively. The standard protocol[4] running in S_t and the computation running in the other fields are mutually independent and interact only via reading the slope fields of S_t.

Let Q be a set of correct i/o configurations $\{\langle (G,I), O \rangle\}$, and $\Gamma_n = \langle T, A, C \rangle$ be a corresponding task. A protocol *solves* Γ_n *with self-stabilization in s steps* if starting from any initial configuration, for any time $t \succ s$ the configuration $\langle G, I, O_t \rangle \in Q$.

A protocol, which admits (potentially "incorrect": $\langle (G,I), O' \rangle \notin Q$) halting configurations, cannot be self-stabilizing: adversary can put the network in an incorrect halted configuration and the system cannot correct itself.

A protocol solving Γ_n can repeatedly emulate checker C, invoking A whenever C detects an incorrect configuration (i.e., when C outputs 0). We use here the Las Vegas property of C: it never rejects a good configuration.

---

[3] If desired, the probability (here as well as for Solver and Checker) can be bounded by any constant $\epsilon > 0$ instead of $1/2$. Also, the $O(\sqrt{t_x})$ time and $O(n/\lg n)$ space can be generalized to $O(t_x^{1/c})$ time and $O(n/\log_c n)$ space, for any $c > 1$.

[4] Standard protocols are the same for all tasks, and their main purpose is to maintain the centered slope.

Adversary still might be able to start the network in a bad configuration from which neither A nor C recover within the desired time. For this purpose we use the time-bound algorithm T. In sec. 4.2 we show that the given time and space restrictions on T are sufficient to compute it with self-stabilization. The simplest way to eliminate this issue is to include time-bound t_x as part of the input.

Then the stabilization time will be at most the sum of the three protocols running times (assuming the time-bound T is relatively accurate). This will be formalized in the Proposition 4.1 below.

We use the centered slope to define a bfs spanning tree H of G, and then emulate tree-CA on H. Thus, algorithms T, C, A all run on H-CA (or equivalently on an rTM).

Remark 3.1 (Dynamic Properties). *For the sake of simplicity, we focus on "static" problems. For protocols, however, the dynamic behavior is often of interest as well. We note that many temporal properties can be captured in terms of static (instantaneous) properties. For example, one of the simplest temporal properties is to assure that some field z does not change. This property can be expressed using a flag, raised by transitions changing z, and lowering it by other transitions. Excluding configurations with the raised flag from Γ then guarantees constancy of z. In fact, Proposition 4.1 implies that any temporal property that can be monitored by a linear space TM can be expressed via instantaneous configurations, such as the problem P above.*

Our protocols are Las Vegas. Since they do not halt, this means that after *stabilization* output is independent of the subsequent coin-flips. Stabilization is the repetition of the non-S_t configuration after the orientation stops changing. The Las Vegas *stabilization period* is then the expected stabilization time (from the worst case configuration).

4   Solving Any Task with Centered Slope

One of our results is to show that given a centered slope any task can be solved by a self-stabilizing protocol on an asynchronous network:

Theorem 4.1. *Given an (unchanging) centered slope in S-fields, for any task Γ_n there exists a self-stabilizing randomized protocol solving Γ_n on asynchronous network G. The protocol (stabilizes and) runs in the $O(T(G, I)d\Delta \lg n)$ steps.*

We first define some tools, and then prove the Theorem.

4.1   Tree-CA Time and TM Reversals

We characterize the computational power of the asynchronous dynamic networks in terms of classical complexity notions in two steps: first express it in terms of Cellular Automata on trees (tree-CA). Tree-CA are simpler than asynchronous dynamic networks, but still have significant variability depending on the topology of the trees. To eliminate this variability, we further compare their computational power to TM reversals.

Once our network G stabilizes, it can simulate **tree-CA**: *Cellular Automata on trees*. (We can use the above definition of the network computation simplified to be synchronous and restricted to tree graphs.) Tree-CA in turn can simulate Turing Machines. The subsection 4.1.1 below discusses the simulations between tree-CA and TM, while subsection 4 considers equivalence of these two models and the network.

The efficiency of the tree-CA and TM mutual simulation is expressed more precisely when the **reversal number** $\widetilde{T}$ of times the TM head changes direction is used as the TM time-complexity measure [Tra64, Bar65]. For convenience, we speak about *reversing TM (rTM)* whenever we wish to use the number of reversals as the complexity measure. Without loss of generality, we assume that rTM makes a complete sweep of the tape before changing direction.

Representing G on tree-CA. For a spanning tree H of G, the graph topology of G can be represented on the H-CA by giving each node the list of its neighbors in G as its input. This can be done, for example, as follows. Let each node be identified, say, according to the visitation order of the dfs tour of H. Associate with each node v of the tree-CA a read-only register containing the list of all the neighbors of v. Let each node see only the first bit of its input register. To each node access all of its input bits, we rotate them in the registers. Namely, let each input digit have a binary mark, different from its neighbors. Each node state has a corresponding binary mark. The register makes a cyclic shift (by one) only when the two marks match. This guarantees the input register is not rotated until it is read. In addition, separately delimit each neighbor id as well as the whole list.

Representing G on TM. For a TM (or rTM), we represent the network topology as an adjacency list on the read-only input tape (in addition to the read-write work tape). Each step, both the input and the work tapes are advanced. Assume that the neighbors list for each node is of the same size (pad otherwise). For each node, the bits of the corresponding list of neighbors is stored at intervals n — so that whenever the work-tape head is in cell i, the input-tape head is reading a bit of the i's adjacency list.

The computational power of CA on any tree is the same as that of rTM with the same space and $\widetilde{T}$ close to CA's time (within $O(d\Delta)$ factors, for diameter d and degree Δ).

If we ignore logarithmic factors (and factors of diameter d and degree in the case of tree-CA), both tree-CA and rTM are strictly more powerful than the standard RAM model: Any RAM operation can be simulated by rTM (with logarithmic overhead) and by tree-CA (with overhead $O(d\Delta \lg n)$). But RAM requires linear in n time to, say, flip all bits — an operation that rTM can manage in one step, and tree-CA in $O(d)$ steps.

Neither rTM nor tree-CA capture the full power of the network, however. For example, rTM requires linear time to reverse its tape: change w to w^r (similar task can be defined for tree-CA, also requiring linear time). However, if G is a sorting network then it can achieve it in logarithmic time. In general, the network may be less powerful than a PRAM.

4.1.1   Simulation Between Tree-CA and rTM

In this section, we consider how a tree-CA H can simulate rTM M and how, in turn, tree-CA H can be simulated by an rTM M'.

Let a tree-CA H and a rTM M have the same number n of nodes/cells and an rTM M' have $2n$ cells. Number the tape cells of M and M' from left to right. We will use two ways to number the nodes of H (both based on dfs traversal of H): one more convenient for H simulating M and the other for M' simulating H.

For simulating rTM M on tree-CA H, we number the nodes of H according to the dfs discovery order and in the second — discovery time. Thus the first numbering uses integers from 1 to n not skipping any, while the second — from 1 to $2n$ and some numbers are skipped.

Say that the cell and the node of the same number correspond to each other.

Let H, M and M' have the same inputs: the input register of node number i of H has all the bits in positions $j \equiv i \pmod{n}$ on the input tape of M (and in the same order); and in positions $j' \equiv i \pmod{2n}$ on the input tape of M' (i.e. kth bit of ith register is stored in cell number $i + 2nk$ of M').

Let function g map the automaton states of H to the rTM alphabet characters of M. We say that *the tree-CA H simulates the rTM M with overhead t* if whenever M makes one sweep and H makes t steps, the state of each tape cell of M is determined by the function g applied to the corresponding node of H.

Let function h map the rTM M' alphabet characters to automaton states of H. We say that the *rTM M' simulates the tree-CA H with time overhead t'* if every time M' makes t' sweeps and H makes a single step, the states of H are determined by the function h applied to the corresponding tape cells of M'.

Lemma 1. *Let H be any tree of n nodes, diameter d and degree Δ. Then*
- Any rTM M with n work-tape cells can be simulated by an H-CA with time overhead $O(d\Delta)$.
- Any H-CA can be simulated by an rTM M' using $2n$ work-tape cells with time overhead $O(d)$.

Proof: tree-CA H simulating rTM M. The automata nodes x of each depth in turn (starting from the leaves) compute the transition function f_x. This f_x depends on the current states and inputs of the subtree t_x of x and its offsprings; f_x maps each state in which M may enter t_x (sweeping the tape along the dfs pass of H) to the state in which it would exit t_x. When f_{root} is computed the whole sweep of the TM tape is executed. In our model node x can communicate with only one of its children at a time. Therefore, it takes $O(\Delta)$ steps for x to compute f_x once f_y is computed for each child y of x; once f_x is computed, x may signal (by changing the special mark) the input register to rotate to the next bit (for the next sweep of the M work tape). Since the depth of the tree is d, it takes $O(d\Delta)$ to compute f_{root}, and thus to simulate one sweep of M work tape.

rTM M' Simulating Tree-CA H. M' represents tree H on its work tape tape as a dfs tour of H with parentheses representing direction "away from" or "toward" the root (i.e., parent-child relation). Each tree node x is represented by a pair of matching parentheses enclosing images of all its offspring, the first parenthesis of the pair is located in the cell corresponding to x.

On each sweep M' passes the information between matching parenthesis of certain depth. Once all children of a given node are marked as served, the node itself is served and marked. When the root is marked, all marks are removed and the M' starts simulating the next step (from the leaves).

Now, let us consider how a node gets marked and served in greater detail. Consider the case of the tree-CA node x being served and marked when all its H-children have already been already marked. Let $x_($ denote the tape cell, corresponding to x and containing the corresponding open parenthesis, respectively, cell $x_)$ contains the matching closing parenthesis.

Each tape cell can be either in *waiting*, or *serving*, or *done* states. Initially, all cells are waiting. When $x_($ and $x_)$ are waiting and there are no waiting cells between $x_($ and $x_)$ then we can serve x as follows. All the serving states between $x_($ and $x_)$ correspond to the children of x (grandchildren of x are in done states).

The next sweep carries the state of x to its children allowing them to finish their current transition and enter done (upon the return sweep matching the states of all the serving $y_($ and $y_)$ between $x_($ and $x_)$). The same sweep gathers information from the children of x for the transition of x and carries it to $x_)$ (between $x_($ and $x_)$ at this stage M' is concerned only with serving x). The return sweep brings this information to $x_($; at this point, $x_($ and $x_)$ go into serving state — only the parent of x information is needed to complete the transaction of x.

The input is arranged so that when M' arrives into the position of the next bit of the input register of x, the work tape of M' is at $x_($. It is easiest to space the consecutive bits of each register on the input tape at intervals $O(dn)$ just so that the next bit is read at the beginning of each tree-CA step simulation. If the resulting $O(d)$ factor increase of the input tape size is undesirable, it is possible for M' to maintain a counter which will keep the current register bit number (same for all the registers) and add the corresponding number of sweeps at the beginning (and the end — to keep each cycle of the same length) of each simulation cycle. ∎

4.2   Self-stabilizing Time Bounds

We use time-bounds algorithm T to restart computations that are initiated incorrectly by Adversary. But Adversary might also initiate T incorrectly — we therefore need a self-stabilizing version of T. In this section we show how any algorithm T as defined in sec. 3.2 can be made self-stabilizing.

As above, we refer to a standard probabilistic Turing Machine M with a read-only input tape I, write-only output tape O, and read-write work tape W of size $|W| = n$; Time is measured in tape reversals.

A function f is defined as [time-]*constructible* if there is an algorithm A such that given input x, A computes $A(x) = f(x)$ within $O(f(x))$ steps (tape reversals). We refer to algorithm A as constructible too. We need to extend constructibility to guarantee the time bounds even when the algorithm is initiated in maliciously chosen configurations.

Clearly, if we repeatedly run $A(x)$, properly initializing the TM after each time A halts, the second run of $A(x)$ will result in the output $f(x)$ which would not be changed by any subsequent executions of A. Thus, any constructible (in the extended sense) f has a self-stabilizing implementation, and any constructible algorithm can be converted into a self-stabilizing one.

Lemma 2. *For any real constant $c > 1$ and constructible algorithm A running in space $O(n/\log_c n)$ there is an algorithm B running in space $O(n)$, such that $A(x) = B(x)$, and if T_x is the running time of $A(x)$ starting from the proper initial configuration, then $B(x)$ stabilizes within $O(T_x^c + \lg n)$ reversals starting from any configuration.*

Proof. Our strategy is to run $k = \log_c n$ parallel computations of $A(x)$. The ith computation of $A(x)$ continues for $t_i = 2^{c^i}$ steps and then is reset to start again from the proper initial state. When one of the computations completes (halts, rather than is interrupted) all the computations are stopped as well. Thus, if $A(x)$ takes $t_{i-1} < T_x \leq t_i$ steps, then starting from any configuration, within t_i steps the ith computation of $A(x)$ is restarted from the proper initial configuration and within $T_x \leq t_i$ steps more it completes. Finally, $T_x^c > t_{i-1}^c = t_i$ implies that the ith computation halts (or is interrupted) within $O(T_x^c)$ time.

Using the tape reversal as the time measure allows us in one step to simulate one step of all the parallel computations.

For the sake of simplicity and readability, we focus on the case of $c = 2$, which allows us to skip some of these tedious details. Extending the proof to $c \neq 2$ requires implementing self-stabilizing arithmetic operations, such as multiplication — a tedious and straight-forward task (using techniques similar to those below). Throughout this construction we use repeated recomputing of the same values. Whenever such recomputing discovers an inconsistency we can reset the whole TM and restart computation from the proper initial state (unlike general self-stabilizing networks, such a reset is easy on a TM).

To implement this strategy, we need a few tools. First, we can build a *copy* subroutine, which repeatedly copies bits of a value from the input to output location (overwriting the current output value). Eventually, the output value will be a proper copy of the input.

Next, we build a *timer*: given a value t the timer counts down from t to 0. By repeatedly checking that the input value t is no less than the current timer value (resetting timer to 0 otherwise), we can make sure that the 0 is reached within t decrement steps, no matter what is the initial configuration (implementing the decrement operation requires care in order to avoid potential pitfalls of malicious initial configurations, but is still straight-forward).

Next tool is *counting*. First, (ideally, starting with the tape of all 0s) change every odd 0 to 1 (that is change first encountered 0, and leave the next 0 unchanged). After at most $\lfloor \lg n \rfloor + 1$ sweeps (exactly, if the initial state was correct) there are no more 0s left. If each sweep adds a digit 1 to a counter (starting empty when tape has all 0s), after $2(\lfloor \lg n \rfloor + 1)$ sweeps the counter contains $\lfloor \lg n \rfloor + 1$ digits 1, which makes the counter value u an upper-bound on and an approximation of n: $n \leq u < 2n$. Changing each but the leading 1 of u to 0 changes u into l — a lower-bound approximation of n: $l \leq n \leq u < 2l \leq 2n$.

Similarly, within $O(\lg n)$ we can compute (with self-stabilization) the approximations for $\lfloor \lg \lg n \rfloor$ and $\lfloor \lg n \rfloor - \lfloor \lg \lg n \rfloor$. Starting with all 0s and running the above sweeps replacing every other 0 with 1 for $\lfloor \lg n \rfloor - \lfloor \lg \lg n \rfloor - 1$ times, results with the remaining 0s delimiting intervals of size $2^{\lfloor \lg n \rfloor - \lfloor \lg \lg n \rfloor} \in (n/2 \lg n, n/\lg n]$.

Thus, within $O(\lg n)$ steps we have divided the work tape into at least $\lg n$ intervals of $\Theta(n/\lg n)$ size.

All the intervals now run in parallel (an rTM in a single TM head sweep can emulate a sweep of TM head in all the intervals). In each interval the $A(x)$ is emulated. In addition, the ith interval uses timer initialized to t_i to interrupt and restart from proper initial configuration the interval computation of $A(x)$ if it takes $> t_i$ steps. Initializing the timers is trivial.

Thus, if computing $A(x)$ takes $T_x < t_j$ steps for some j, then within t_j steps after the intervals are properly initialized, the jth interval restarts computing $A(x)$ and so within $t_j + T_x$ steps, $A(x)$ is computed. After that, the TM can halt with the output. Of course, it is possible that the adversary can cause the TM to halt earlier — in which case, the output might be incorrect. But when initialized from the proper initial state (after the restart), the jth interval will complete the computation of $A(x)$ correctly and produce the correct output. ∎

Most functions $f()$ of interest (esp. those which are likely to be used as $T()$ in sec. 3.2) require significantly (often exponentially) less time to compute the value $f(x)$ than $f(x)$ steps. Likewise, the space requirement for these functions is likely to be $O(\lg n)$ or $(\lg n)^{O(1)}$ at most. Thus, the space and time overheads of the lemma are not a problem in most cases.

So, the lemma implies that the task time-bound can be implemented as a self-stabilizing algorithm.

4.3 Proof of Theorem 4.1

A centered slope yields an obvious spanning (bfs) tree H, through the up edges. Consider the tree-CA on H. The synchronization of the tree-CA can be achieved by maintaining a separate slope on the tree, with each node incrementing its slope value whenever making a step, which in turn is allowed only when it has no tree-neighbors under it. H-CA in turn emulates an rTM M.

A read-only input containing the adjacency matrix[5] of G can be simulated as follows. To read the entry (v, w) of the adjacency matrix, find node v and mark it. Then find node w and see if there is an edge coming from a marked node. In the end, clear the mark of the node v. The edge lookup is then reduced to finding a given vertex v.

Assume that vertices are numbered linearly on the tape of M. To find each vertex efficiently, cover the rTM tape with counters, each with the number of the first vertex in the counter. These counters can be initialized in $O(\lg n)$ time by the process similar to the marking of the intervals in Lemma 2 proof. Using this marking, we can find the counter containing v in logarithmic time and then within logarithmically more steps more find v within the counter.

A single look-up of the adjacency matrix can thus be simulated in $O(d\Delta \lg n)$.

The rTM M repeatedly runs constructible (in the extended sense) version of T (see Lemma 2); this gives a self-stabilizing time-bound. Whenever T outputs a value, it is stored on the M tape — denote this constantly recomputed stored value as t. Thus within $O(T(G, I))$ simulated time, $t = T(G, I)$.[6]

M also maintains a counter, c. The counter c is repeatedly compared with t. If $c > t$, then it is set to $c = 0$. Otherwise, it is decremented with every step of M.

Whenever the counter reaches 0, c is set to $c = t$ again and M runs $C(G, I, W_o)$ starting from a proper initial configuration, where W_o are subfields of work fields W which contain a copy of the output.

Whenever $C(G, I, W_o)$ outputs 0 (i.e., W_o contains an incorrect answer), c is set to $c = t$ again and M runs $A(G, I)$, with its output written into W_o fields.

When $C(G, I, W_o)$ outputs 1 (i.e., W_o contains correct answer), then W_o fields are copied to the output fields O, c reset to t and $C(G, I, W_o)$ is run again. ■

5   Protocols for Centered Slope

The task protocols from Theorem 4.1, Sec. 4 use centered slope (given in the structure fields). In this section we consider the protocols initiating a centered slope on a given network.

5.1   Problem Decomposition

As stated in Sec. 1.1.2 we reduce maintaining centered slope to four separate tasks: certificate building **Cb** and checking **Cc** (both working on trees and certifying their acyclicity), balancing orientation **Sb** (of the network spanned by these trees), and centralizing the slope **Sc** (merging the trees into one). We present these tasks below in terms of specifications (specs, for short) and interfaces: an interface defines the fields (with access permissions) and a spec augments the

---

[5] The output must be correct for any numbering of the graph nodes used in the matrix.

[6] The simulation overhead includes maintenance operations such as reseting all fields, counters operations, copying appropriate values, etc. The total *actual* time is within $O(T(G, I)d\Delta \lg n)$.

interface with a "contractual properties". Any protocols satisfying these specs will work for our reduction.

5.1.1 Interfaces, Filters, and Illegal Edges

In general, we structure reduction as a tree of protocols: each subtree corresponds to a protocol; each node corresponds to an interface coordinating interaction of the children sub-protocols, inheriting restrictions of the ancestor interfaces.

An *interface* defines a set of fields shared by the children (and parent) protocols and the corresponding permissions. Consider an interface $\mathcal{I}$ and its child protocol P. An *action* of P at each node v produces the state change of v, depending on the set of v's links. The *environment* of P, denoted $\mathcal{E}_P$, interacts with P via $\mathcal{I}$, performing any actions permitted by $\mathcal{I}$ to programs outside the P subtree. When discussing properties of P, it is often useful to assume that $\mathcal{E}_P$ is controlled by Adversary.

Since Adversary has no restrictions on the combinations of the initial states of the two nodes of an edge, she can create abnormal links, possibly disruptive for P. It might not be easy for P to correct such links (in particular, since P acts at one node at a time, and this action is restricted by $\mathcal{I}$ but affects all the incident links —and not just the abnormal one).

So, P comes with a list of its normal, *P-legal*, links. Any link which is not P-legal is *illegal*. To protect P from illegal links, $\mathcal{I}$ has a special *filter* component $\mathcal{F}_\mathcal{I}$ invoked before any other programs of this interface. $\mathcal{F}_\mathcal{I}$ can read all the fields, and has the lists of P-legal edges from all protocols P of $\mathcal{I}$.

If $\mathcal{F}_\mathcal{I}$, acting at v, finds at least one incident illegal link, then $\mathcal{F}_\mathcal{I}$ preforms *crashing*, also referred to as *closing*: some fields have default values to which they are reset by $\mathcal{F}_\mathcal{I}$, and no other programs are invoked by $\mathcal{I}$ at this node in this activation. The resulting (crashed) node v is called a *crash* or *closed* — we use these terms interchangeably.

The list of P-legal links must include all "closed–closed" links and must be closed under all actions permitted by $\mathcal{I}$ to P and $\mathcal{E}_P$.[7] For all P, any illegal edge is crashed by $\mathcal{F}_\mathcal{I}$ (at both ends, if needed) making it legal; no new illegal edges can be created. Thus, illegal edges disappear in one step and $\mathcal{F}_\mathcal{I}$ stabilizes:

Claim 5.1. *Illegal links disappear within one step.*

Typically, handling and opening the crashes will be responsibility of a special protocol (**Sb** in our case); the other protocols will not act at or near crashes.

5.1.2 Pointers, Roots and Crashes

When slope centralization stabilizes, orientation (defined by the h3 fields, see sec. 3.1) must yield a unique local minimum — the *leader*. Thus, the leader is easy to find by following any down path. We refer to "potential leaders" as *roots*.

---

[7] A slightly less robust version guarantees that P-legality is closed under only the actions that P actually performs rather than all those allowed by $\mathcal{I}$ (closure under $\mathcal{E}_P$ transactions is unchanged).

When the slope is centralized, there remains a single root — the leader. The first task is thus to guarantee existence of roots.

A root r is always a local minimum; we also restrict it to $r.\text{h3} = 0$. However, when many roots exist, the slope centralization protocol needs to "unroot" some of them (eventually, all but one), leading to intermediate configurations where some local minima are not roots, a down-path does not always lead to a root.

To facilitate finding roots (and guarantee their existence) we require the slope centralization protocol **Sc** to maintain a pointer $v.\mathsf{p_c}$ in each node v (supposedly leading to a root). It is to guarantee the acyclicity of such pointer chains that we use the certificate builder and checker (which work on the pointer trees). A **root** r is marked by $r.\mathsf{p_c} = r$.

Let the default value for $v.\mathsf{p_c}$ be *nil*, and v is **closed** whenever $v.\mathsf{p_c} = nil$. A closed v is assumed to have "unreliable" orientation field $v.\text{h3}$, and **Sb** takes over the node until it fixes the problem.

Non-crashes are called **open**. Changing $v.\mathsf{p_c}$ from *nil* to non-*nil* (thus changing v from a closed to open) is referred to as **opening** v. Whenever a closed v is opened, its pointer is set down if v has a down-edge, or to self if v is a local minimum. New roots can only be created from crashes (by opening them).

The shell and **Sb** protocols must keep checking the correctness of the slope, and in particular verify existence of a root, even as **Sc** changes $\mathsf{p_c}$-pointers. To facilitate this we let **Sb** own a pointer, $\mathsf{p_b}$, which copies $\mathsf{p_c}$ before it changes. Furthermore, **Sc** will be invoked only when the shell is ready for the next change. If the shell detects a problem (such as a long edge), all changes are stopped until **Sb** detects and fixes the problem. Thus any open node can have one or two pointers. A closed v has only $v.\mathsf{p_b}$ ($v.\mathsf{p_c} = nil$); and open v has $v.\mathsf{p_c} \neq nil$ and $v.\mathsf{p_b}$ (possibly $v.\mathsf{p_b} = nil$).

In a crash both $\mathsf{p_c} = \mathsf{p_b} = nil$. **Sb** signals that it is done with a crash by setting $\mathsf{p_b}$ (down, or self if at a local minimum). Such node with $\mathsf{p_b} \neq nil = \mathsf{p_c}$ is referred to as *fresh*. A fresh is opened by **Sc** setting $\mathsf{p_c}$ pointer (down, or self if at a local minimum).

We call v a *single, double* or a *split* if $v.\mathsf{p_b} = nil, v.\mathsf{p_b} = v.\mathsf{p_c}, v.\mathsf{p_b} \neq nil \neq v.\mathsf{p_c}$, respectively.

Define the pointer $\mathsf{p}(v)$ to be *nil* if v is closed, $v.\mathsf{p_c}$ if v is a single, and $v.\mathsf{p_b}$ otherwise. This reflects the shell protocol's view which might lag a bit behind that of **Sc**. A maximal acyclic $\mathsf{p}()$-pointer path $v_k, \ldots, v_0 : \mathsf{p}(v_i) = v_{i-1}$ from $v = v_k$ can terminate at a root ($\mathsf{p}(v_0) = v_0$), a crash ($v_0.\mathsf{p_c} = nil$), or a pointer cycle ($\mathsf{p}(v_0) = v_{i>0}$). We call v **rooted** (in v_0), if v_0 is a root and all paths (from the root) $v_0 \ldots v_{i \leq k}$ have non-negative variances. An edge vw is called **balanced** if v and w are rooted in r_v, r_w, respectively, and the path from r_v to r_w —against the pointers, across vw, and down the pointers— has zero variance.

Claim 5.2. *Orientation is a slope with neither crashes nor pointer cycles iff all edges are balanced.*

Indeed, suppose that all the edges are balanced. Then there are neither crashes nor pointer cycles. Consider a path from a root $v_0 \ldots v_k$. Let v_i be rooted in r_i ($r_0 = v_0$), and let p_i be the path from r_i to v_i (against pointers) then across

$v_i v_{i+1}$ and to r_{i+1} (along the pointers). Then the composition of all $p_{i<k}$ equals the path $v_0 \ldots v_k$ followed by the pointer path from v_k to r_k. Since the variance of each p_i is 0 (by the balance of edges), and the variance of the pointer path from v_k to r_k is non-positive (by rootedness of v_k), then the variance of $v_0 \ldots v_k$ is non-negative.

For the converse, consider an edge vw and the pointer paths from v and w to roots r_v, r_w, respectively. If the variance of the path from r_v to r_w via the pointer paths and vw is not 0 then either it or its reverse has a negative variance violating the slope definition. ∎

Unfortunately, the above notion of balanced edge is difficult to use: an edge that is balanced at one instant can become unbalanced (e.g., by crashing its pointer path to root) in the next instant. But what is worse, a process near the edge, which may have just finished verifying its balance, would not be able to find out about the change for some time. This leads us to define a "history-based" notion of height.

Let the network start from some initial (possibly created by Adversary) configuration IC. We use time index i referring to the number of events that had occurred in the network. Its actual value is irrelevant, but it provides a convenient way to unambiguously refer to any instance of the network evolution. We say that v is i-rooted if at the instant i the maximal $\mathsf{p}()$-pointer path from v contains no $\mathsf{p_b}$-pointers that have not change since IC, ends in an open root, and if v is a split, then it was created after IC; then the $\mathsf{p}()$-pointer path is referred to as i-rooting chain of v.

Define height $h_i(v)$ of v at instant i to equal 0: if v is a root (open or crash); else: the variance of the i-rooting chain of v (computed from the root to v), if such exist; and $h_{i-1}(v)$ otherwise. Thus, height might be initially undefined for some nodes. Say an edge vw is long at instant i if both endpoints have defined h_i such that $|h_i(v) - h_i(w)| > 1$.

5.1.3   Slope Balancing (Sb) Spec

Consider environment $\mathcal{E}_b$ interacting with the protocol **Sb**. The **Sb** spec consists of the following

fields:

   $v.\mathsf{p_c}, v.\mathsf{p_b}$:   pointers to neighbors of v
   $v.\mathsf{h3} \in Z_3$:   orientation field

permissions:

   $\mathcal{E}_b$ can

   - crash any open v
   - increment $v.\mathsf{h3}$ in open v with no down edges ($\nexists w \in E(v) : w.\mathsf{h3} = v.\mathsf{h3} - 1$)
   - change pointer $v.\mathsf{p_c} \leftarrow w$ in open v ($w \in E(v)$; $w \neq v$)[8] open a fresh v: if v is over only freshes, set $v.\mathsf{p_c}$ down on a fresh, or self if none

---

[8] Thus, $\mathcal{E}_b$ creates no new roots — they can be created only by **Sb** from crashes.

Sb can

- crash any open or fresh v: all non-**Sb** fields and p_b are set to nil;[9]
- decrement $v.h3$ in any closed v with no open nodes over it
- create a fresh from a closed v not over an open w by setting pointer $v.p_b$ down on a crash, or self if at a local min

commitments:

$\mathcal{E}_b$ guarantees that

($\mathcal{E}_b$.1) There are no pointer cycles (neither old nor new).

($\mathcal{E}_b$.2) Any pointer chain of length L has a segment of variance $\Theta(L)$.

($\mathcal{E}_b$.3) Pointer chains from long edges are unchanged (incl. orientation), except by crashing.[10]

Sb   guarantees that (with the above commitments of $\mathcal{E}_b$)

(**Sb**.1) If $\mathcal{E}_b$ makes no crashes, then after t_{s1} steps there are no crashes and orientation is a slope.

(**Sb**.2) As an addendum, for efficiency: in any unchanged by $\mathcal{E}_b$ (incl. orientation) pointer path with variance greater than $O(d)$ some nodes will be crashed within t_{s0} steps (even as $\mathcal{E}_b$ keeps crashing other nodes).

For simplicity, below we combine t_{s0}, t_{s1} (and a "pointer chain correctness checking time" t_{Cc}, defined in Sec. 5.2 below) into the **Sb** *stabilization time* $t_{Sb} \stackrel{def}{=} 1 + t_{s0} + t_{Cc} + t_{s1}$: this is the time within which **Sb** will stabilize (initiate correct slope, and open all crashes). Intuitively, all external to **Sb** crashes will stop within $1 + t_{s0} + t_{Cc}$ steps: Filter crashes will stop after the first step; then after all long (longer than $(O(d))$ pointer chains disappear within the next t_{s0} steps (commitment **Sb**.2), the remaining short $(O(d))$ pointer chains will be "checked for correctness" by the reduction (protocol **Cc**) and all incorrect ones will be crashed within t_{Cc} steps. If all these times are polynomial in the network diameter d, we say that the protocols stabilize *promptly*. Then, after all external to **Sb** crashes stop, by the commitment **Sb**.1 **Sb** will initiate correct slope, open all crashes and stabilize. The formal details of this process are covered in the Sec. 5.2 below.

5.1.4   Slope Centralization (Sc) Spec

Consider environment $\mathcal{E}_c$ interacting with protocol **Sc**. The $\mathcal{E}_c$-**Sc** interface uses the same **fields** as the $\mathcal{E}_b$-**Sb** interface: $v.h3, v.p_c$. But here $\mathcal{E}_c$ has no permissions to change these fields.

The **Sc** spec is as follows:

fields: (same as in the $\mathcal{E}_b$-**Sb** interface)

$v.p_c$:   pointer (nil in crashes; self in roots)

$v.h3 \in Z_3$:   orientation field

---

[9] $p_b = nil$ denotes **Sb** making p_b unavailable to $\mathcal{E}_b$, while **Sb** may still use it internally.

[10] In particular, nodes near crashes do not change.

permissions: ($\mathcal{E}_c$ has none)
    Sc can
- increment $v.\mathsf{h3}$ if v has no down edges
- can change $v.\mathsf{p}_c$ in open v

commitments:
($\mathcal{E}_c.1$) $\mathcal{E}_c$ guarantees that orientation is a correct slope.
(**Sc**.1) **Sc** guarantees that (provided ($\mathcal{E}_c.1$)) after $t_{\mathbf{Sc}}$ steps the slope is central-
    ized.[11]

5.1.5   Stability of Low Nodes for Sb

Sb can initiate a proper slope and open all nodes only after all other protocols
stop crashing nodes. Filters stop crashing after one step. After that only **Cc**
can crash nodes (other than **Sb** itself). These crashes, however will stop after
all the pre-existing certs are replaced (we assume that **Cb** creates correct certs
which are not crashed by **Cc**). The cert renewal cycle guarantees that each cert
is replaced within time linear in the tree height (ignoring the **Cb** and $\mathbf{S}_{led}$ times,
which are typically nearly linear —or at most polynomial— in the tree-height).
So, if all tree heights are bounded by $O(d)$ then the **Cc** crashes will stop within
poly-d time.

In this section, we show that this indeed is achieved by our algorithms.

Consider all nodes from some initial configuration at time denoted as 0.

We say that node v has (m, h, t)-trajectory if in the 0 to t period (inclusively)
the minimum height of v is m, and $h_t(v) = h$.

Claim 5.3. *If v has (m, h, t)-trajectory and $h > m + 2$ then for any neighbor
$w \in E(v)$ there are $t' < t$, m', h', such that w has (m', h', t')-trajectory and
$|m - m'| \le 2, |h - h'| \le 1$.*

Proof: Let v have (m, h, t)-trajectory and $h > m + 2$. Let t' be the largest such
that $h_{t'+1}(v) = h_{t'}(v) + 1 = h$ (i.e. it is the last float to h of the trajectory of
v). Then v has $(m, h, t' + 1)$-trajectory.

Suppose that the (m', h', t')-trajectory of w violates either $|m - m'| \le 2$
or $|h - h'| \le 1$. Consider the (first) time i when v is at the minimum height
$m = h_i(v)$ and floats at the next step $h_{i+1}(v) = m + 1$ (By R1 below, **Sb** cannot
increase height, so floating is the only way to increment $h_i(v)$). Since $h_i(v)$ is
defined, LEC before this float is not IIA, and therefore w is rooted with variance
m or $m + 1$. However, since w might at this point still be an IC-single, we cannot
conclude that $m' \le m + 1$. However since $h > m + 2$, w must float at least once
before v can increase its height to $m + 3$. At that time its height will be defined
and will have the value $m + 1$ or $m + 2$. Thus, $m' \le m + 2$. Similar argument
considering the first float of w from height w' provides $m \le m' + 2$, showing
$|m - m'| \le 2$.

The above implies that at time t' both $h_{t'}(v)$ and $h_{t'}(w)$ are defined. Furthermore, to permit floating of v (the preceding LEC is not IIA), we must have
$h_{t'}(w)$ be either $h - 1$ or h.   ■

---

[11]   And thus all the changes to the interface fields cease.

Corollary 5.4. *If v raises by $d+1$ while remaining at height $> 2d$ then there are no roots during that period.*

Proof by induction on distance k from v to u (and using Claim for the inductive step).

Internally, **Sb** maintains its own internal height and its own internal version of roots at height 0. $\mathcal{E}_b$ guarantees to **Sb** existence of crashes, freshes or roots. **Sb** should be able to assure that if there are crashes or freshes, then such zeros exist.

However, efficient operation of **Sb** may demand stability of such zeros, or nodes close to them. Then the above corollary guarantees that if a node was ever a root/zero, it will never float above $> 3d+1$.

5.1.6   Certificate Builder and Checker

Certificate builder **Cb** will be invoked by shell on the pointer trees as described in sec. 5.2. The checker **Cc** runs in the pointer fields continuously. The two have the following properties. **Cc** does not crash certificates generated by **Cb** properly invoked on any trees. **Cc** crashes at least one node on any pointer cycle in time poly-logarithmic in the cycle size. **Cc** is suspended at or near crashes.

To illustrate the idea, we briefly sketch a version of certificate and checker used in [IL92]. While there certificate was constructed along the dfs traversal path of a tree, here we define it as a function of tree height: so the node at height i will contain $\alpha(i)$ (defined below) as its certificate.

Thue (or *Thue-Morse*) sequence is defined as $\mu(k) \stackrel{def}{=} \sum_i k_i \bmod 2$, where k_i is the i-th bit of k [Thu12]. We say string $x = x_1 x_2 \dots x_k$ is *asymmetric* if it has a k bits segment of μ embedded in its digits (say, as $x_i \bmod 2$).[12]

Let us cut off the tail of each binary string k according to some rule, say, the shortest one starting with 00 (assume binary representation of k starts with 00). Let us fix a natural representation of all integers $j > 2$ by such tails $\hat{\jmath}$ and call j the *suffix* $\sigma(k)$ of k. For a string χ, define $\rho(\chi, k)$ to be $\chi_{\sigma(k)}$ if $\sigma(k) \le \|\chi\|$, and special symbol $\#$ otherwise. Then $\alpha[k] = \rho(k, k)$, and $\alpha(k) = \langle \alpha[k], \mu(k) \rangle$.[13] Let L_α be the set of all segments of α. L_α can be recognized in polynomial time. Let $L_{T\alpha}$ be a set of asymmetric trees where each root-leaf path contains a string in L_α.

Lemma 3. *Any string of the form ss, $\|s\| > 2$, contains segment $y \notin L_\alpha$, $\|y\| = (\log \|s\|)^2 + o(1)$.*

Other variants of α can be devised to provide greater efficiency or other desirable properties (e.g., one such variant was proposed in [IL92]).

For a language L of strings define a $\Gamma(L)$ be the language of trees, such that any root-leaf path contains a string in L, and any two equal length down-paths ending at the same node are identical.

---

[12] For simplicity, we ignore other ways to break symmetry.

[13] Inclusion of μ in α makes it asymmetric but otherwise is useful only for < 40-bit segments. Also, $\mu(k)$ could be used instead of $\#$ if $i > \|k\|$ in $\alpha[k]$, but this complicates the coding and thus is skipped.

Let $T_A(X_T)$ be a tree T of cellular automata A starting in the initial state with unchanging input X_T. We say that $T_A(X_T)$ *rejects* X_T if some of the automata enter a *reject* state. Language Γ of trees is *t-recognized* by A if for all T of depth $t(k)$, $T_A(X_T)$ (1) rejects within $t(k)$ steps those X_T, which contain a subtree $Y \notin \Gamma$ of depth k; and (2) reject none of the X with all subtrees in Γ. For asynchronous self-stabilizing automata, requirement (1) extends to arbitrary starting configuration and to trees rooted in a cycle; requirement (2) extends to the case when a ancestors or children branches of the tree are cut off during the computation.

Lemma 4. *For any polynomial time language L of asymmetric strings, $\Gamma(L)$ is recognizable in polynomial time by self-stabilizing protocols on asynchronous cellular tree-automata.*

5.2 The $\mathcal{I}_{\mathbf{S}}$-Interface Overview

Our reduction uses shell **Sh** and interface $\mathcal{I}_{\mathbf{S}}$ to run and coordinate the following protocols: **Sb**, **Sc**, **Cb**, **Cc** and **S**$_{led}$. The first four are described in the previous sections. The long-edge check **S**$_{led}$ is invoked on a pointer tree prior to each **Cb** invocation there: if any long edges adjacent to the tree are detected then **Cb** is not invoked. This prevents any changes of the tree pointers and orientation with adjacent long edges, thus satisfying the $\mathcal{E}_{\mathbf{b}}$ commitment of the **Sb** spec. **S**$_{led}$ can be treated as a separate protocol, or be associated with **Cb**, or even **Sh**. Despite its triviality, treating it as a separate protocol is a more flexible option. But we associate it with **Cb** to simplify the paper.

Certs. In order to maintain evidence (certificate) of the slope correctness, **Sb** maintains a collection of smaller certificates —which we refer to as *certs*— each constructed independently along special pointer tree. Thus cert values are associated with specific pointers (one in each node) and we give precise definition of how the next pointer is selected for each cert. Each open node will typically have one or two such certs (and thus cert-pointer chains) passing through it. These certs, will be managed by the **Sh** which will create them using **Cb**. **Cb** can be made self-stabilizing as in sec. 4. However, since **Cb** must halt, rather than stabilize, it might be initialized improperly, and then it might halt with an incorrect certificate (which will be crashed or deleted later). **Cc** runs continuously in the pointers and crashes illegal certs, in particular, on pointer cycles.

Claim 5.5. *If each node has a cert passing through it, and each cert is individually correct and there are no long edges, then the slope is correct.*

Indeed, consider a cycle $v_0 v_1 \ldots v_k$. Define the ***height*** of a rooted pointer to be the variance of its pointer-path to its root. If all certs are correct, then all the pointers are rooted. If a node contains two pointers of a different height then the edge between these pointers is long. Thus, if there are no long edges the height of a pointer uniquely determines the height of its node $h(v)$. Furthermore, the variance of each directed edge must equal the difference of the heights of its endpoints — otherwise the edge is long. But then the variance of the cycle is $(h(v_1)-h(v_0))+(h(v_2)-h(v_1))+\ldots+(h(v_0)-h(v_k))=0$.   ■

Closed nodes are a responsibility of **Sb** and do not have certificates visible to $\mathcal{I_S}$. The transition from a crash to an open node will have an intermediate stage, *fresh*, initially without a cert.

Assuming no incorrect certs are built by **Cb** after some initial period, eventually incorrect tree-certs will disappear, and thus so do pointer cycles. We say that then **Cc** stabilizes.

5.2.1 $\mathcal{I_S}$ Fields

$v.\mathsf{h3}$: Slope.

$v.\mathsf{p_c}, v.\mathsf{p_b}$: pointers belonging to **Sc** and **Sb**, resp.

The following fields are associated with each non-nil pointer (rather than a node) and correspond to an official and a draft version of a cert, where the draft field can also be used for signal:

   cert :   cert digit (possibly *nil*).
   cdft :   draft cert digit (possibly *nil*), or a signal F or W (under specific
            circumstances these signals invoke **Sc** and **Cb**, resp).

5.2.2   Interface Restrictions

The following are formal Interface $\mathcal{I_S}$ restrictions.

$\mathcal{F_{I_S}}$ crashes illegal links.

Sc: can only
     -read h3, $\mathsf{p_c}$;
     -increment $v.\mathsf{h3}$ in open non-root v not over another node;
     -change $v.\mathsf{p_c}$ pointers (to $\neq nil$) for open or fresh v (for open v: to a
     neighbor; for fresh v: down on a fresh, or self if none).

Sb: can only
     -read h3, $\mathsf{p_c}$, $\mathsf{p_b}$, cert fields;
     -**crash** any nodes;
     -decrement to $v.\mathsf{h3}$ of a closed v with no open neighbor over it;
     -change a crash into a *fresh* with following restrictions: (1) a crash cannot
     be opened over a non-crash, (2) after opening, $v.\mathsf{p_b}$ must be down, or to
     self—if there are no nodes under it.

Open, but not fresh, nodes make no $\mathcal{I}_{Main}$ changes (except crashing) near a crash or *fresh*.

Cb: can only
     read h3, $\mathsf{p_c}$, $\mathsf{p_b}$, cdft fields of open and fresh nodes;
     when W, can write to the corresponding cdft field.

Cc : can only

> read h3, p_c, p_b, cert fields of open and fresh nodes;
> crash any nodes.[14]

S$_{led}$: has no $\mathcal{I}_S$-write permissions and can read h3, p_c, p_b[, cert, cdft]. Its only way to communicate with the other protocols is to return —or not return— control after its invocation.

Sh: can only

> invoke Cb, S_{led} and Sc protocols[15];
> move, copy or set to *nil* the p_c, p_b pointers (together with the associated fields);
> can read/write cdft signal values.

Let a pointer from v points on its neighbor w. Then if each w has only one cert, then the cert pointer of v is assumed to point on the cert pointer of w. If w has two certs then v points on the like pointer (p_b on p_b, p_c on p_c). We refer to trees defined by such pointers as *cert-trees*.

An p_b-pointer is said to be **dying** if the node has an p_c cert. A dying p_b-pointer is called **dead** if all its descendants are dying (dead). A pointer is **live** if it is not dead (even if it is dying).

Signals. Intuitively, these signals and draft emulate a variant of the classical fire-water-hay game, where fire burns hay, but is extinguished with water, which in turn is absorbed by hay.

Specifically the "fire" F moves up burning "hay": destroying current cert and moving cdft into its place. The fire is followed up by the "water" or "wait" signal W, which replaces F whenever it has spread to all its possible destination. In turn, W is replaced with the new draft certificate: cdft = *nil* in dead pointers, or obtaining the draft cert value from Cb in live ones. Thus, the dead pointers eventually have no certs. This new draft-hay absorbs the water from the leaves down. When the new fire enters, this new draft cert value is moved to the "official cert" field cert, and so on. This change of cert from one version to the next does not cause Cc to crash.

If the F signal has crossed a $p_b(\neq p_c)$ pointer, then it is considered to be passive (Fp), and it is active (Fa) otherwise. A passive fire Fp does not spread to new pointers (unless its a double and its twin pointer has a certificate there), while active one (Fa) does.

The Cb algorithm is activated from root (after S_{led}) and runs in stabilized water pointer tree: The leaves of the Cb-tree are those W-nodes that have nether W- nor F- children, nor gain such even after each child has a chance to act (at least) once.

Each open node v has a p_c- and, possibly, a p_b- pointers ($v.p_c$, $v.p_b$, resp.).

When a closed v is changed by Sb to a **fresh**, v gains a p_b pointer down (or self). For a *fresh* v, $v.p_c = nil$. Sh calls Sc at fresh v to complete its opening. A

---

[14] Cc is ran constantly on all the cert fields, treating each pointer tree separately.

[15] Cb, S_{led} are invoked from a root of a tree and operate essentially as tree-CA; Sc is invoked at each node separately and makes a single Sc step in that node.

fresh has no cert $v.\mathsf{cert} = nil$, and always points down on another fresh or crash (or is a root).

Sc is always invoked at a single: then $v.\mathsf{p_c}$ is copied to $v.\mathsf{p_b}$ (with all the associated fields), and **Sc** can make its move. If $v.\mathsf{p_c} = v.\mathsf{p_b}$ after this, then **Sh** erases the associated fields of $v.\mathsf{p_b}$. If **Sc** changes $v.\mathsf{p_c}$, then we call such $v.\mathsf{p_b}$ a *split pointer*.

The legal links are defined so a split and single pointers can point only on a single (or closed or fresh). Therefore, any pointer path from leaf to root in any cert-tree crosses at most one split pointer. The rest of pointers on the path coincide with $\mathsf{p_c}$ pointers. So, if **Sb** provides that all pointer chains have variance $O(d)$, then the height of the cert trees is also $O(d)$.

The certs, assumed to be properly constructed by **Cb** (so that **Cc** does not crash them), satisfy the following *requirement* for some sufficiently small $\mathsf{t_{Cc}}$ (added to the overall stabilization time)[16]:

Condition 1. *No cert loops exist for* $> \mathsf{t_{Cc}}$ *steps.*

5.2.3 Cert Renewal Cycle
Certs renewal cycle algorithm works on cert-trees specified by the pointers.

Call a ($\mathsf{p_c}$-)pointer *pre-existing* if it has not changed since the initial configuration. Say a chain of $\mathsf{p_c}$-pointers is *pre-existing* if all its pointers are. Call a $\mathsf{p_c}$-chain *balanced* if all its adjacent edges are.

Claim 5.6. *Whenever* $v.\mathsf{p_c}$ *changes, its* $\mathsf{p_c}$-*chain to root is either balanced or pre-existing.*

Indeed, in order for $v.\mathsf{p_c}$-pointer to change, it must first be a single, and then become a split. If it was pre-existing, this can happen with the $\mathbf{S}_{led}$ skipped or improperly initialized. However, before the next change, $v.\mathsf{p_c}$ must become a single again, and so do all its $\mathsf{p_c}$ ancestors. At least one cert-renewal cycle must pass when v and its $\mathsf{p_c}$-ancestors are turned into singles. This will be accompanied by $\mathbf{S}_{led}$ check,[17] which will guarantee (if it completes) that the $\mathsf{p_c}$-chain from v to the root is balanced. ∎

Claim 5.7. *If any* $\mathsf{p_c}$-*path is bounded by* $O(d)$, *then a root* r *initiates a cert cycle every* $O(d) + t_{\mathbf{S}_{led}} + t_{\mathbf{Cb}}$ *steps, where* $t_{\mathbf{S}_{led}}, t_{\mathbf{Cb}}$ *are the times required by* $\mathbf{S}_{led}, \mathbf{Cb}$ *respectively.*

Indeed, r initiates $\mathbf{S}_{led}$ which completes within $t_{\mathbf{S}_{led}}$ steps. Then, within any two steps any maximal W-path from r either grows, or terminates in a pointer with no eligible F-children. A W-pointer with neither F nor W children cannot gain any such children. Within $t_{\mathbf{Cb}}$ steps more, **Cb** completes its computation

---

[16] In [IL92] we propose certs and **Cc** protocols for which $\mathsf{t_{Cc}} = (\lg n)^{O(1)}$.

[17] In fact, it shows that it is sufficient to execute the long-edge check only once in each cycle when the cert-tree is turned into singles (see below). We chose to execute $\mathbf{S}_{led}$ in every cert cycle to support a stronger notion of a certificate: which includes balance guarantee for adjacent edges.

and simply needs a chance to output them. Then within each step any maximal W-path from r shrinks. Since any W-path can cross at most one split pointer (W follows F, which changes from Fa to Fp on crossing a split pointer, and Fp cannot cross split pointers), the length of the maximal W-path from r is $O(d)$. ∎

5.2.4 Fresh Exiting and Sc-invocation

This part of the **Sh** protocol works on nodes (rather than just pointer trees), and assumes that all nodes have been opened by **Sb** with a correct slope (after filters and **Cc** stop crashing nodes). It deals with fresh nodes changing to normal and then discusses another version of the "fire-water-hay" game, which provides periodic invocation of **Sc** for all nodes. Recall that all the protocols except **Sb** are frozen near crashes, so we ignore closed nodes in this section.

Here we refer to a pointer as *empty* or *full* depending on whether it contains a cert or not.

Node types and stage coding. There can be the following types of nodes: A *single* has one pointer (the other is *nil*). It is a *fresh* if the pointer is p_b. Otherwise, we refer to it as **bot**.

If both pointers are non-*nil*, the node is either a *split* or a *double*. In either case it can be *dry*, *wet* or *twin*: corresponding to having a *nil*-cert in p_c, p_b, or neither, respectively.

Fresh exiting. Consider a fresh node. Only other fresh and closed nodes can be on its down path. On the other hand, any nodes can be rooted in fresh. **Cc** treats such a certificate, where prefix is replaced with fresh, as legal. **Sb** guarantees that when there are no crashes the down path —and in particular the pointer path— from a fresh terminates in a root.

A fresh root initiates the regular cert renewal cycle. The cert pointers of non-fresh nodes on fresh are treated as eligible. When two pointers of a twin are both rooted in the same fresh cert, the p_b-pointer eventually disappears as dead.[18]

Once the illegal for non-fresh configurations disappear, the freshes are changed to **bot** all the way to the root. When a fresh node has a (non-empty) cert and no twins pointing on it, the fresh functions as a regular **bot** single.

Thus the analysis below applies to fresh as well showing that they disappear in polynomial in diameter time after the crashes disappear and the slope is correct.

Sc-invocation cycle. This cycle starts when a root r is dry and the cert cycle initiates F. Then (after S_{led}) r changes to **bot** with Fa.

Any dry v pointing at a **bot** changes to **bot** in one step. A **bot** node v stays unchanged if it is pointed at by a dry, **bot** or full p_b pointer.

When no such pointers on dry v exist, upon arrival of F, v creates a split pointer $v.p_b$ copies $v.p_c$ there (without associated fields), and invokes **Sc**. If **Sc** changes $v.p_c$, then the cert fields of p_c are moved to p_b, and the result is a dry split. If $v.p_c$ is not changed, then v becomes wet.

---

[18] Alternatively a fresh with a twin pointing on it can be viewed as a split with two pointers on two dummy **bot** nodes which both point on another dummy **bot** node.

The signal F is passive in wet or twin nodes: it does not enter new edges (unless they are doubles with a cert on the two pointer). A dry node becomes wet when accepting F from a wet parent. The exception is the case of *dump* nodes: which have cert = *nil* and cdft = W in the twin pointer. Such nodes form when a wet node has bot in the parent but all its children are dry. Dump nodes are treated as wet by the parent (thus preventing it from changing to bot) and as dry by children, allowing F to pass as active and without causing wetness.

The wet node changes to dry if all its children are dry, and the parent is wet.

5.2.5   Sc-invocation Cycle Times

Propagation of bot up the tree is straight-forward: dry changes to bot in one step. A dry node can be pointed at by another dry or by an empty p_c of a split (which will be infected from F in the bot, or ignored if dump).

Consider a split v and a maximal p_c chain P from v to —but not including— bot. Consider the length of P plus the number of non-splits on it. In $O(d)$ steps this sum increases: either the bot under P splits or lowest split merges.

Since P is no longer than $O(d)$, v merges promptly.

Since the tree of single and dry nodes does not increase, and dry promptly changes to single, single leaves promptly loose their split children and can split themselves. Then the tree of single and dry nodes promptly disappears.

Since split promptly merges into a double, it remains to show that any double v promptly changes to single.

Consider P from v to the nearest single. P does not shrink and must grow when the single promptly splits. Eventually but promptly, since P cannot be more than $O(d)$, this path must reach root, become dry, and change to bot.

References

[AKY90]   Yehuda Afek, Shay Kutten, Moti Yung. *Memory-efficient self-stabilization on general networks.* In *Workshop on Distributed Algorithms*, 1990.

[AO94]   B. Awerbuch, R. Ostrovsky. *Memory-efficient and self-stabilizing network reset.* [PODC], 1994.

[Bar65]   J. M. Barzdin. *The complexity of symmetry recognition by Turing machines.* (in Russian) *Problemi Kibernetiki*, **v. 15**, pp.245-248, 1965.

[Dij74]   E. W. Dijkstra. *Self stabilizing systems in spite of distributed control.* *CACM*, 17, 1974.

[DGS96]   Shlomi Dolev, Mohamed G. Gouda, Marco Schneider. *Memory requirements for silent stabilization.* [PODC], 1996.

[FOCS]   Proc. *IEEE Ann. Symp. on the Foundations of Computer Science.*

[G86]   Peter Gács. *Reliable computation with cellular automata.* J. of Comp. System Sci., **32**, 1, 1986.

[GKL78]   P. Gács, G.L. Kurdiumov, L. A. Levin. *One-Dimensional Homogeneous Media Dissolving Finite Islands.* Probl. Inf. Transm., 14/3, 1978.

[Her]   Ted Herman. *Self-stabilization bibliography: Access guide.* Chicago Journal of Theoretical Computer Science, Working Paper WP-1, initiated Nov., 1996. Also available at http://www.cs.uiowa.edu/ftp/selfstab/bibliography/

[IJ90]   Amos Israeli, Marc Jalfon. *Token management schemes and random walks yield self-stabilizing mutual exclusion.* [PODC], 1990.

[I$^+$92] Gene Itkis. *Self-stabilizing distributed computation with constant space per edge.* Colloquia presentations at MIT, IBM, Bellcore, CMU, ICSI Berkeley, Stanford, SRI, UC Davis. 1992. Includes joint results with B. Awerbuch and R. Ostrovsky, and with L. A. Levin (submitted to [FOCS], 1992).

[IL92]   Gene Itkis, Leonid A. Levin. *Self-stabilization with constant space.* Manuscript, Nov. 1992 (submitted to [STOC], 1993). Also in [IL94]. Later versions: *Fast and lean self-stabilizing asynchronous protocols.* TR#829, Technion, Israel, July 1994, and in [FOCS], 1994, pp. 226-239.

[IL94]   Leonid A. Levin. (Joint work with G. Itkis). *Self-Stabilization.* Sunday's Tutorial Lecture. ICALP, July 1994, Jerusalem.

[Joh97]   Colette Johnen. *Memory efficient, self-stabilizing algorithm to construct BFS spanning trees.* [PODC], 1997. Extended version appeared in the Proceedings of *The Third Workshop on Self-Stabilizing System (WSS'97)*, 1997.

[M$^+$92] A. Mayer, Y. Ofek, R. Ostrovsky, M. Yung. *Self-stabilizing symmetry breaking in constant-space.* [STOC], 1992.

[PODC]   Proc. *ACM Ann. Symp. on Principles of Distributed Computing.*

[Ro71]   R. Robinson, *Undecidability and non-periodicity for tiling a plane. Invencione Mathematicae* 12: 177-209, 1971.

[STOC]   Proc. *ACM Ann. Symp. on the Theory of Computation.*

[Thu12]   A. Thue. *Uber die gegenseitige Lage gleicher Teile gewisser Zeichenreichem. Kra.Vidensk.Selsk.I. Mat.-Nat.Kl.,* 10, 1912.   Also in: A.Thue. *Selected Math. Papers.* ed.: T.Nagell, A.Selberg, S.Selberg, K.Thalberg. Universitetsforlaget, 1977.

[Tra64]   B. Trakhtenbrot. *Turing computations with logarithmic delay* (in Russian). *Algebra i Logika,* **3**, pp. 33-48, 1964.

Interprocedurally Analyzing Polynomial Identities

Markus Müller-Olm[1], Michael Petter[2], and Helmut Seidl[2]

[1] Westfälische Wilhelms-Universität Münster, Institut für Informatik,
Einsteinstr. 62, 48149 Münster, Germany
mmo@math.uni-muenster.de
[2] TU München, Institut für Informatik, I2, 80333 München, Germany
seidl@in.tum.de

Abstract. Since programming languages are Turing complete, it is impossible to decide for all programs whether a given non-trivial semantic property is valid or not. The way-out chosen by abstract interpretation is to provide *approximate* methods which may fail to certify a program property on some programs. Precision of the analysis can be measured by providing classes of programs for which the analysis is complete, i.e., decides the property in question. Here, we consider analyses of polynomial identities between integer variables such as $x_1 \cdot x_2 - 2x_3 = 0$. We describe current approaches and clarify their completeness properties. We also present an extension of our approach based on weakest precondition computations to programs with procedures and equality guards.

1 Introduction

Invariants and intermediate assertions are the key to deductive verification of programs. Correspondingly, techniques for automatically checking and finding invariants and intermediate assertions have been studied (cf., e.g., [3,2,22]). In this paper we present analyses that check and find valid polynomial identities in programs. A polynomial identity is a formula $p(x_1, \ldots, x_k) = 0$ where $p(x_1, \ldots, x_k)$ is a multi-variate polynomial in the program variables $x_1, \ldots, x_k$.[1]

Looking for valid polynomial identities is a rather general question with many applications. Many classical data flow analysis problems can be seen as problems about polynomial identities. Some examples are: finding *definite equalities among variables* like $x = y$; *constant propagation*, i.e., detecting variables or expressions with a constant value at run-time; *discovery of symbolic constants* like $x = 5y+2$ or even $x = yz^2+42$; *detection of complex common sub-expressions* where even expressions are sought which are syntactically different but have the same value at run-time such as $xy+42 = y^2+5$; and *discovery of loop induction variables*.

Polynomial identities found by an automatic analysis are also useful for program verification, as they provide non-trivial valid assertions about the program. In particular, loop invariants can be discovered fully automatically. As polynomial identities express quite complex relationships among variables, the discovered assertions may form the backbone of the program proof and thus significantly simplify the verification task.

---

[1] More generally our analyses can handle positive Boolean combinations of polynomial identities.

B. Durand and W. Thomas (Eds.): STACS 2006, LNCS 3884, pp. 50–67, 2006.

In the following, we critically review different approaches for determining valid polynomial identities with an emphasis on their precision. In expressions, only addition and multiplication are treated exactly, and, except for guards of the form $p \neq 0$ for polynomials p, conditional choice is generally approximated by non-deterministic choice. These assumptions are crucial for the design of effective exact analyses [12,13]. Such programs will be called *polynomial* in the sequel.

Much research has been devoted to polynomial programs without procedure calls, i.e., *intraprocedural* analyses. Karr was the first who studied this problem [11]. He considers polynomials of degree at most 1 (*affine* expressions) both in assignments and in assertions and presents an algorithm which, in absence of guards, determines all valid affine identities. This algorithm has been improved by the authors and extended to deal with polynomial identities up to a fixed degree [13]. Gulwani and Necula also re-considered Karr's analysis problem [7] recently. They use randomization in order to improve the complexity of the analysis at the price of a small probability of finding invalid identities.

The first attempt to generalize Karr's method to *polynomial* assignments is [12] where we show that validity of a polynomial identity at a given target program point is decidable for polynomial programs. Later, Rodriguez-Carbonell et al. propose an analysis based on the observation that the set of identities which are valid at a program point can be described by a polynomial *ideal* [20]. Their analysis is based on a constraint system over polynomial ideals whose greatest solution precisely characterizes the set of all valid identities. The problem, however, with this approach is that *descending* chains of polynomial ideals may be infinite implying that no effective algorithm can be derived from this characterization. Therefore, they provide special cases [21] or approximations that allow to infer some valid identities. Opposed to that, our approach is based on effective weakest precondition computations [12,14]. We consider assertions to be checked for validity and compute for every program point weakest preconditions which also are represented by ideals. In this case, fixpoint iteration results in *ascending* chains of ideals which are guaranteed to terminate by Hilbert's basis theorem. Therefore, our method provides a decision procedure for validity of polynomial identities. By using a *generic* identity with unknowns instead of coefficients, this method also provides an algorithm for *inferring* all valid polynomial identities up to a given degree [14].

An interprocedural generalization of Karr's algorithm is given in [15]. Using techniques from linear algebra, we succeed in inferring all interprocedurally valid affine identities in programs with affine assignments and no guards. The method easily generalizes to inferring also all polynomial identities up to a fixed degree in these programs. A generalization of the intraprocedural randomized algorithm to programs with procedures is possible as well [8]. A first attempt to infer polynomial identities in presence of polynomial assignments and procedure calls is provided by Colon [4]. His approach is based on ideals of polynomial *transition invariants*. We illustrate, though, the pitfalls of this approach and instead show how the idea of precondition computations can be extended to an interprocedural analysis. In a natural way, the latter analysis also extends the interprocedural analysis from [15] where only affine assignments are considered.

The rest of the paper is organized as follows. Section 2 introduces basic notions. Section 3 provides a precise characterization of all valid polynomial identities by means of a constraint system. This characterization is based on forward propagation. Section 4 provides a second characterization based on effective weakest precondition computation. This leads to backwards-propagation algorithms. Both Sections 3 and 4 consider only programs without procedures. Section 5 explains an extension to polynomial programs with procedures based on polynomial transition invariants and indicates its limitations. Section 6 presents a possible extension of the weakest-precondition approach to procedures. Section 7 then indicates how equality guards can be added to the analyses. Finally, Section 8 summarizes and gives further directions of research.

2   The General Set-Up

We use similar conventions as in [15,17,16] which we recall here in order to be self-contained. Thus, programs are modeled by systems of non-deterministic flow graphs that can recursively call each other as in Fig. 1. Let $\mathbf{X} = \{\mathbf{x}_1, \ldots, \mathbf{x}_k\}$ be the set of

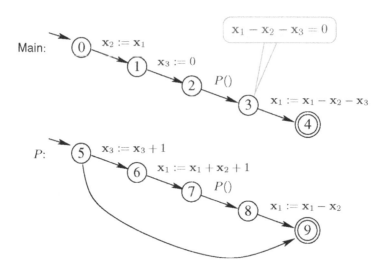

Fig. 1. An interprocedural program

(global) variables the program operates on. For ease of presentation, we assume that variables take values in the field $\mathbb{Q}$. Similar arguments, though, can also be applied in case values are integers from $\mathbb{Z}$ or even when we consider values from a modular ring $\mathbb{Z}_m$ for some $m = 2^w$, $w \geq 1$, as used in programming languages such as Java [17,16].

In the programs we analyze, we assume the assignments to variables to be of the form $\mathbf{x}_j := p$ for some polynomial p from $\mathbb{Q}[\mathbf{X}]$, i.e., the ring of all polynomials with coefficients from $\mathbb{Q}$ and variables from $\mathbf{X}$. Note that this restriction does not come by accident. It is well-known [9,19] that it is undecidable for non-deterministic flow graphs to determine whether a given variable holds a constant value at a given program point in all executions if the full standard signature of arithmetic operators (addition, subtraction, multiplication, and division) is available. Constancy of a variable is obviously a polynomial identity: $\mathbf{x}$ is a constant at program point n if and only if the polynomial identity $\mathbf{x} - c = 0$ is valid at n for some $c \in \mathbb{Q}$. Clearly, we can write all expressions involving addition, subtraction, and multiplication with polynomials. Thus, if we allow also division, validity of polynomial identities becomes undecidable.

Assignments with non-polynomial expressions or input dependent values are therefore assumed to be abstracted with *non-deterministic assignments*. A non-deterministic assignment of the form $\mathbf{x}_j :=?$ (with $\mathbf{x}_j \in \mathbf{X}$) is meant to represent the non-deterministic choice between all assignments $\mathbf{x}_j := c, c \in \mathbb{Q}$. In general, we also assume that conditional branching is abstracted with non-deterministic branching, i.e., either way is possible. Note that in [13] it is pointed out that in presence of equality guards, exact constant propagation again becomes undecidable. The only form of guards at edges which we can handle within our framework precisely are disequality guards of the form $p \neq 0$ for some polynomial p. In order to reduce the number of program points in examples, we sometimes annotate edges with sequences of assignments. Also, we use assignments $\mathbf{x}_j := \mathbf{x}_j$ which have no effect onto the program state as skip-statements and omit these in pictures. For the moment, skip-statements are used to abstract, e.g., equality guards. In Section 7, we will present methods which approximatively deal with equality guards.

A *polynomial program* comprises a finite set Proc of *procedure names* with one distinguished procedure Main. Execution starts with a call to Main. Each procedure $q \in$ Proc is specified by a distinct finite edge-labeled *control flow graph* with a single start point st_q and a single return point ret_q where each edge is labeled with an assignment, a non-deterministic assignment, a disequality guard or a call to some procedure. For simplicity, we only consider procedures without parameters or return values operating on global variables. The framework, though, can straightforwardly be extended to procedures with local variables, call-by-value parameter passing and return values.

The basic approach of [15,13,17] which we take up here is to construct a precise abstract interpretation of a constraint system characterizing the concrete program semantics. For that, we model a *state* attained by program execution when reaching a program point or procedure by a k-dimensional vector $x = [x_1, \ldots, x_k] \in \mathbb{Q}^k$ where x_i is the value assigned to variable $\mathbf{x}_i$. Runs through the program execute sequences of assignments and guards. Each such sequence induces a *partial polynomial transformation* of the program state.

A (total) polynomial transformation τ can be described by a vector of polynomials $\tau = [q_1, \ldots, q_k]$ where τ applied to a vector x equals the vector:

$$\tau(x) = [q_1(x), \ldots, q_k(x)]$$

where we have written $q'(x)$ for the value returned by a polynomial q' for the vector x. A partial polynomial transformation π is a pair $\pi = (q, \tau)$ of a polynomial q and a polynomial transformation τ. If $q(x) \neq 0$ then $\pi(x)$ is defined and returns $\tau(x)$. Otherwise, $\pi(x)$ is undefined. Partial polynomial transformations are closed under composition [14]. The partial polynomial transformations corresponding to single assignments and disequality guards are given by:

$$\llbracket \mathbf{x}_j := p \rrbracket = (0, [\mathbf{x}_1, \ldots, \mathbf{x}_{j-1}, p, \mathbf{x}_{j+1}, \ldots, \mathbf{x}_k])$$
$$\llbracket q \neq 0 \rrbracket = (q, [\mathbf{x}_1, \ldots, \mathbf{x}_k])$$

The definition of a partial polynomial transformation is readily extended to sets of states. Since in general, procedures have multiple runs, we model their semantics by *sets* of partial polynomial transformations.

3   Intraprocedural Analysis: Forward Iteration

Let $\pi = (q, \tau)$ be the partial polynomial transformation induced by some program run. Then, a polynomial identity $p = 0$ is said to be *valid* after this run if, for each initial state $x \in \mathbb{Q}^k$, either $q(x) = 0$ – in this case the run is not executable from x – or $q(x) \neq 0$ and $p(\tau(x)) = 0$ – in this case the run is executable from x and the final state computed by the run is $\tau(x)$. A polynomial identity $p = 0$ is said to be valid at a program point v if it is valid after every run reaching v.

Clearly, if $p = 0$ is valid then also $r \cdot p = 0$ for arbitrary polynomials r. Also, if $p_1 = 0$ and $p_2 = 0$ are valid then also $p_1 + p_2 = 0$ is valid. Thus, the set of polynomials p for which $p = 0$ is valid at v forms a *polynomial ideal*.[2] Recall that, by Hilbert's basis theorem, every polynomial ideal $I \subseteq \mathbb{Q}[\mathbf{X}]$ can be finitely represented by:

$$I = \langle p_1, \ldots, p_m \rangle =_{\mathrm{df}} \{r_1 \cdot p_1 + \ldots + r_m \cdot p_m \mid r_i \in \mathbb{Q}[\mathbf{X}]\}$$

for suitable $p_1, \ldots, p_m \in \mathbb{Q}[\mathbf{X}]$. The set $\{p_1, \ldots, p_m\}$ is also said to *generate* the ideal I. Based on such representations, algorithms have been developed for fundamental operations on ideals [1]. In particular, membership is decidable for ideals as well as containment and equality. Moreover, the set of all ideals $I \subseteq \mathbb{Q}[\mathbf{X}]$ forms a *complete lattice* w.r.t. set inclusion "$\subseteq$" where the least and greatest elements are the zero ideal $\{0\}$ and the complete ring $\mathbb{Q}[\mathbf{X}]$, respectively. The greatest lower bound of a set of ideals is simply given by their intersection while their least upper bound is the ideal *sum*. More precisely, the sum of the ideals I_1 and I_2 is defined by

$$I_1 \oplus I_2 = \{p_1 + p_2 \mid p_1 \in I_1, p_2 \in I_2\}$$

A set of generators for the sum $I_1 \oplus I_2$ is obtained by taking the union of sets of generators for the ideals I_1 and I_2.

For the moment, let us consider *intraprocedural* analysis only, i.e., analysis of programs just consisting of the procedure Main and without procedure calls. Such program consist of a single control-flow graph. As an example, consider the program in Fig. 2.

Given that the set of valid polynomial identities at every program point can be described by polynomial ideals, we can characterize the sets of valid polynomial identities by means of the following constraint system $\mathcal{F}$:

$$
\begin{aligned}
\mathcal{F}(\text{start}) &\subseteq \{0\} & \\
\mathcal{F}(v) &\subseteq [\![\mathbf{x}_i := p]\!]^{\sharp}(\mathcal{F}(u)) & (u, v) \text{ an assignment } \mathbf{x}_i := p \\
\mathcal{F}(v) &\subseteq [\![\mathbf{x}_i := ?]\!]^{\sharp}(\mathcal{F}(u)) & (u, v) \text{ an assignment } \mathbf{x}_i := ? \\
\mathcal{F}(v) &\subseteq [\![p \neq 0]\!]^{\sharp}(\mathcal{F}(u)) & (u, v) \text{ a guard } p \neq 0
\end{aligned}
$$

---

[2] A polynomial ideal I is a set of polynomial which is closed under addition and under multiplication with arbitrary polynomials: $\forall p, q \in I : p + q \in I$ and $\forall p \in I, q \in \mathbb{Q}[\mathbf{X}] : p \cdot q \in I$.

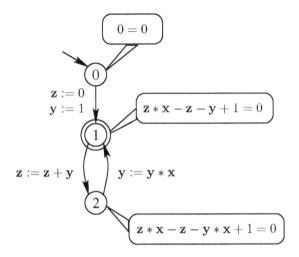

Fig. 2. A program without procedures

where the effects of assignments and disequality guards onto ideals I are given by:

$$[\mathbf{x}_i := p]^\sharp(I) = \{q \mid q[p/\mathbf{x}_i] \in I\}$$
$$[\mathbf{x}_i := ?]^\sharp(I) = \{\textstyle\sum_{j=0}^m q_j \mathbf{x}_i^j \mid q_j \in I \cap \mathbb{Q}[\mathbf{X}\backslash\{\mathbf{x}_i\}]\}$$
$$[p \neq 0]^\sharp(I) = \{q \mid p \cdot q \in I\}$$

Intuitively, these definitions can be read as follows. A polynomial identity q is valid after an execution step iff its weakest precondition was valid before the step. For an assignment $\mathbf{x}_i := p$, this weakest precondition equals $q[p/\mathbf{x}_i] = 0$. For a non-deterministic assignment $\mathbf{x}_i := ?$, the weakest precondition of a polynomial $q = \sum_{j=0}^m q_j \mathbf{x}_i^j$ with $q_j \in \mathbb{Q}[\mathbf{X}\backslash\{\mathbf{x}_i\}]$ is given by:

$$\forall\, \mathbf{x}_i.\, q = 0 \;\equiv\; q_0 = 0 \wedge \ldots \wedge q_m = 0$$

Finally, for a disequality guard $p \neq 0$, the weakest precondition is given by:

$$\neg(p \neq 0) \vee q = 0 \;\equiv\; p = 0 \vee q = 0 \;\equiv\; p \cdot q = 0$$

Obviously, the operations $[\mathbf{x}_i := t]^\sharp$, $[\mathbf{x}_i := ?]^\sharp$, and $[p \neq 0]^\sharp$ are monotonic. Therefore by the fixpoint theorem of Knaster-Tarski, the constraint system $\mathcal{F}$ has a unique greatest solution over the lattice of ideals of $\mathbb{Q}[\mathbf{X}]$. By definition, the operations commute with arbitrary intersections. Therefore, using standard coincidence theorems for completely distributive intraprocedural dataflow frameworks [10], we conclude:

Theorem 1. *Assume p is a program without procedures. The greatest solution of the constraint system $\mathcal{F}$ for p precisely characterizes at every program point v, the set of all valid polynomial identities.* $\qquad\square$

The abstract effect of a disequality guard is readily expressed as an ideal quotient for which effective implementations are well-known. The abstract assignment operations, though, which we have used in the constraint system $\mathcal{F}$ are not very explicit. In order to obtain an effective abstract assignment operation, we intuitively proceed as follows. First, we replace the variable $\mathbf{x}_i$ appearing on the left-hand side of the assignment with a new variable $\mathbf{z}$ both in the ideal I and the right-hand side of the assignment. The variable $\mathbf{z}$ thus represents the value of $\mathbf{x}_i$ *before* the assignment. Then we add the new relationship introduced by the assignment (if there is any) and compute the ideal closure to add all implied polynomial relationships between the variables $\mathbf{X}$ and $\mathbf{z}$. Since the old value of the overwritten variable is no longer accessible, we keep from the implied identities only those between the variables from $\mathbf{X}$. Formally, we verify:

Lemma 1. *For every ideal $I = \langle p_1, \ldots, p_n \rangle \subseteq \mathbb{Q}[\mathbf{X}]$ and polynomial $p \in \mathbb{Q}[\mathbf{X}]$,*

1. $\{q \mid q[p/\mathbf{x}_i] \in I\} = \langle \mathbf{x}_i - s, s_1, \ldots, s_k \rangle \cap \mathbb{Q}[\mathbf{X}]$ *and*
2. $\{\sum_{j=0}^{m} q_j \mathbf{x}_i^j \mid q_j \in I \cap \mathbb{Q}[\mathbf{X} \backslash \{\mathbf{x}_i\}]\} = \langle s_1, \ldots, s_n \rangle \cap \mathbb{Q}[\mathbf{X}]$,

where $s = p[\mathbf{z}/\mathbf{x}_i]$ and $s_j = p_j[\mathbf{z}/\mathbf{x}_i]$ for $i = 1, \ldots, n$.

Note that the only extra operation on ideals we use here is the restriction of an ideal to polynomials with variables from a subset. This operation is also called *elimination* and standard effective algorithms are known [1].

Proof. Assume that the ideal I is generated from the polynomials $p_1, \ldots, p_n$. We only prove statement (1). Assume $q = q_0(\mathbf{x}_i - s) + \sum_{j=1}^{n} q_j s_j$ does not contain variable $\mathbf{z}$ where $q_0, \ldots, q_n \in \mathbb{Q}[\mathbf{X} \cup \{\mathbf{z}\}]$, $s = p[\mathbf{z}/\mathbf{x}_i]$ and for all j, $s_j = p_j[\mathbf{z}/\mathbf{x}_i]$. Since the s_j do not contain $\mathbf{x}_i$,

$$q[p/\mathbf{x}_i] = q_0'(p - s) + \sum_{j=1}^{n} q_j' s_j$$

for suitable polynomials $q_0', \ldots, q_n'$. Substituting again $\mathbf{x}_i$ for $\mathbf{z}$ in this equation, we therefore obtain:

$$q[p/\mathbf{x}_i] = q_0''(p - p) + \sum_{j=1}^{n} q_j'' p_j$$
$$= \sum_{j=1}^{n} q_j'' p_j$$

for suitable polynomials $q_0'', \ldots, q_n''$. Therefore, $q[p/\mathbf{x}_i] \in I$.

For the reverse implication assume $q[p/\mathbf{x}_i] \in I$ which means that $q[p/\mathbf{x}_i] = \sum_{j=1}^{n} q_j p_j$ for suitable polynomials q_j. Substituting $\mathbf{z}$ for $\mathbf{x}_i$ in this equation, therefore gives us $q[s/\mathbf{x}_i] = \sum_{j=1}^{n} q_j' s_j$ for suitable polynomials q_j' where $s = p[\mathbf{z}/\mathbf{x}_i]$ and $s_j = p_j[\mathbf{z}/\mathbf{x}_i]$. Now recall the identity (for $k > 0$):

$$\mathbf{x}_i^k - s^k = g_k \cdot (\mathbf{x}_i - s) \quad \text{for} \quad g_k = \sum_{h=0}^{k-1} \mathbf{x}_i^h s^{k-1-h}$$

and assume that $q = \sum_{k=0}^{d} r_k \mathbf{x}_i^k$ for polynomials $r_k \in \mathbb{Q}[\mathbf{X} \backslash \{\mathbf{x}_i\}]$. Then

$$q = \sum_{k=0}^{d} r_k \cdot (\mathbf{x}_i^k - s^k) + q[s/\mathbf{x}_i]$$
$$= \sum_{k=1}^{d} r_k g_k \cdot (\mathbf{x}_i - s) + \sum_{j=1}^{n} q_j' s_j$$
$$= q_0' \cdot (\mathbf{x}_i - s) + \sum_{j=1}^{n} q_j' s_j \quad \text{for the polynomial} \quad q_0' = \sum_{k=1}^{d} r_k g_k .$$

Therefore, $q \in \langle \mathbf{x}_i - s, s_1, \ldots, s_n \rangle$. Since $q \in \mathbb{Q}[\mathbf{X}]$, the assertion follows.   □

According to Lemma 1, all operations used in the constraint system $\mathcal{F}$ are effective. Nonetheless, this does not in itself provide us with an analysis algorithm. The reason is that the polynomial ring has *infinite decreasing* chains of ideals. And indeed, simple programs can be constructed where fixpoint iteration will not terminate.

Example 1. Consider our simple example from Fig. 2. There, we obtain the ideal for program point 1 as the infinite intersection:

$$
\begin{aligned}
\mathcal{F}(1) \; = \; & \langle \mathbf{z}, \mathbf{y} - 1 \rangle \; \cap \\
& \langle \mathbf{z} - 1, \mathbf{y} - \mathbf{x} \rangle \; \cap \\
& \langle \mathbf{z} - 1 - \mathbf{x}, \mathbf{y} - \mathbf{x}^2 \rangle \; \cap \\
& \langle \mathbf{z} - 1 - \mathbf{x} - \mathbf{x}^2, \mathbf{y} - \mathbf{x}^3 \rangle \; \cap \\
& \cdots
\end{aligned}
$$

□

Despite infinitely descending chains, the greatest solution of $\mathcal{F}$ has been determined precisely by Rodriguez-Carbonell et al. [21] — but only for a sub-class of programs. Rodriguez-Carbonell et al. consider simple loops whose bodies consist of a finite nondeterministic choice between sequences of assignments satisfying additional restrictive technical assumptions. No complete methods are known for significantly more general classes of programs. Based on constraint system $\mathcal{F}$, we nonetheless obtain an effective analysis which infers *some* valid polynomial identities by applying *widening* for fixpoint acceleration [6]. This idea has been proposed, e.g., by Rodriguez-Carbonell and Kapur [20] and Colon [4]. We will not pursue this idea here. Instead, we propose a different approach.

4 Intraprocedural Analysis: Backward Propagation

The key idea of [12,14] is this: instead of propagating ideals of valid identities in a forward direction, we start with a conjectured identity $q_t = 0$ at some program point v and compute weakest preconditions for this assertion by backwards propagation. The conjecture is proven if and only if the weakest precondition at program entry start $=$ st$_{\mathsf{Main}}$ is true. The assertion true, i.e., the empty conjunction is uniquely represented by the ideal $\{0\}$. Note that it is decidable whether or not a polynomial ideal equals $\{0\}$.

Assignments and disequality guards now induce transformations which for every postcondition, return the corresponding weakest precondition:

$$
\begin{aligned}
[\![\mathbf{x}_i := p]\!]^{\mathsf{T}} \, q \; &= \; \langle q[p/\mathbf{x}_i] \rangle \\
[\![\mathbf{x}_i := ?]\!]^{\mathsf{T}} \, q \; &= \; \langle q_1, \ldots, q_m \rangle \quad \text{where } q = \sum_{j=0}^{m} q_j \mathbf{x}_i^j \text{ with } q_j \in \mathbb{Q}[\mathbf{X} \backslash \{x_i\}] \\
[\![p \neq 0]\!]^{\mathsf{T}} \, q \; &= \; \langle p \cdot q \rangle
\end{aligned}
$$

Note that we have represented the disjunction $p = 0 \vee q = 0$ by $p \cdot q = 0$. Also, we have represented conjunctions of equalities by the ideals generated by the respective polynomials. The definitions of our transformers are readily extended to transformers for ideals, i.e., conjunctions of identities. For a given target program point t and conjecture $q_t = 0$, we therefore can construct a constraint system $\mathcal{B}$:

$$
\begin{aligned}
\mathcal{B}(t) &\supseteq \langle q_t \rangle \\
\mathcal{B}(u) &\supseteq [\![\mathbf{x}_i := p]\!]^\mathsf{T}(\mathcal{B}(v)) & (u,v)\ \text{labeled with}\ \mathbf{x}_i := p \\
\mathcal{B}(u) &\supseteq [\![\mathbf{x}_i :=?]\!]^\mathsf{T}(\mathcal{B}(v)) & (u,v)\ \text{labeled with}\ \mathbf{x}_i :=? \\
\mathcal{B}(u) &\supseteq [\![p \neq 0]\!]^\mathsf{T}(\mathcal{B}(v)) & (u,v)\ \text{labeled with}\ p \neq 0
\end{aligned}
$$

Since the basic operations are monotonic, the constraint system $\mathcal{B}$ has a unique least solution in the lattice of ideals of $\mathbb{Q}[\mathbf{X}]$. Consider a single execution path π whose effect is described by the partial polynomial transformation $(q_0, [q_1, \ldots, q_k])$. Then the corresponding weakest precondition is given by:

$$
[\![\pi]\!]^\mathsf{T} p = \langle q_0 \cdot p[q_1/\mathbf{x}_1, \ldots, q_k/\mathbf{x}_k] \rangle
$$

The weakest precondition of p w.r.t. a set of execution paths can be described by the ideal generated by the weakest preconditions for every execution path in the set separately. Since the basic operations in the constraint system $\mathcal{B}$ commute with arbitrary least upper bounds, we once more apply standard coincidence theorems to conclude:

Theorem 2. *Assume p is a polynomial program without procedures and t is a program point of p. Assume the least solution of the constraint system $\mathcal{B}$ for a conjecture $q_t = 0$ at t assigns the ideal I to program point* start. *Then, $q_t = 0$ is valid at t iff $I = \{0\}$.* □

Using a representation of ideals through finite sets of generators, the applications of weakest precondition transformers for edges can be effectively computed. A computation of the least solution of the constraint system $\mathcal{B}$ by standard fixpoint iteration leads to ascending chains of ideals. Therefore, in order to obtain an effective algorithm, we only must assure that *ascending* chains of ideals are ultimately stable. Due to Hilbert's basis theorem, this property indeed holds in polynomial rings over fields (as well as over integral domains like $\mathbb{Z}$). Therefore, the fixpoint characterization of Theorem 2 gives us an effective procedure for deciding whether or not a conjectured polynomial identity is valid at some program point of a polynomial program.

Corollary 1. *In a polynomial program without procedures, it can effectively be checked whether or not a polynomial identity is valid at some target point.* □

Example 2. Consider our example program from Fig. 2. If we want to check the conjecture $\mathbf{z} \cdot \mathbf{x} - \mathbf{z} - \mathbf{y} + 1 = 0$ for program point 1, we obtain:

$$
\begin{aligned}
\mathcal{B}(2) &\supseteq \langle (\mathbf{z} \cdot \mathbf{x} - \mathbf{z} - \mathbf{y} + 1)[\mathbf{y} \cdot \mathbf{x}/\mathbf{y}] \rangle \\
&= \langle \mathbf{z} \cdot \mathbf{x} - \mathbf{z} - \mathbf{y} \cdot \mathbf{x} + 1 \rangle
\end{aligned}
$$

Since,

$$
(\mathbf{z} \cdot \mathbf{x} - \mathbf{z} - \mathbf{y} \cdot \mathbf{x} + 1)[\mathbf{z} + \mathbf{y}/\mathbf{z}] = \mathbf{z} \cdot \mathbf{x} - \mathbf{z} - \mathbf{y} + 1
$$

the fixpoint is already reached for program points 1 and 2. Thus,

$$
\begin{aligned}
\mathcal{B}(1) &= \langle \mathbf{z} \cdot \mathbf{x} - \mathbf{z} - \mathbf{y} + 1 \rangle \\
\mathcal{B}(2) &= \langle \mathbf{z} \cdot \mathbf{x} - \mathbf{z} - \mathbf{y} \cdot \mathbf{x} + 1 \rangle
\end{aligned}
$$

Moreover,

$$
\begin{aligned}
\mathcal{B}(0) &= \langle (\mathbf{z} \cdot \mathbf{x} - \mathbf{z} - \mathbf{y} + 1)[0/\mathbf{z}, 1/\mathbf{y}] \rangle \\
&= \langle 0 \rangle = \{0\}
\end{aligned}
$$

Therefore, the conjecture is proved. □

It seems that the algorithm of testing whether a certain given polynomial identity $p_0 = 0$ is valid at some program point contains no clue on how to infer so far unknown valid polynomial identities. This, however, is not quite true. We show now how to determine all polynomial identities of some arbitrary given form that are valid at a given program point of interest. The form of a polynomial is given by a selection of monomials that may occur in the polynomial.

Let $D \subseteq \mathbb{N}_0^k$ be a finite set of exponent tuples for the variables $x_1, \ldots, x_k$. Then a polynomial q is called a D-polynomial if it contains only monomials $b \cdot \mathbf{x}_1^{i_1} \cdot \ldots \cdot \mathbf{x}_k^{i_k}$, $b \in \mathbb{Q}$, with $(i_1, \ldots, i_k) \in D$, i.e., if it can be written as

$$q = \sum_{\sigma = (i_k, \ldots, i_k) \in D} a_\sigma \cdot \mathbf{x}_1^{i_1} \cdot \ldots \cdot \mathbf{x}_k^{i_k}$$

If, for instance, we choose $D = \{(i_1, \ldots, i_k) \mid i_1 + \ldots + i_k \leq d\}$ for a fixed maximal degree $d \in \mathbb{N}$, then the D-polynomials are all the polynomials up to degree d. Here the *degree* of a polynomial is the maximal degree of a monomial occurring in q where the degree of a monomial $b \cdot \mathbf{x}_1^{i_1} \cdot \ldots \cdot \mathbf{x}_k^{i_k}$, $b \in \mathbb{Q}$, equals $i_1 + \ldots + i_k$.

We introduce a new set of variables $\mathbf{A}_D$ given by:

$$\mathbf{A}_D = \{\mathbf{a}_\sigma \mid \sigma \in D\}.$$

Then we introduce the *generic D-polynomial* as

$$q_D = \sum_{\sigma = (i_k, \ldots, i_k) \in D} \mathbf{a}_\sigma \cdot \mathbf{x}_1^{i_1} \cdot \ldots \cdot \mathbf{x}_k^{i_k}.$$

The polynomial q_D is an element of the polynomial ring $\mathbb{Q}[\mathbf{X} \cup \mathbf{A}_D]$. Note that every concrete D-polynomial $q \in \mathbb{Q}[\mathbf{X}]$ can be obtained from the generic D-polynomial q_D simply by substituting concrete values $a_\sigma \in \mathbb{Q}$, $\sigma \in D$, for the variables $\mathbf{a}_\sigma$. If $a : \sigma \mapsto a_\sigma$ and $\mathbf{a} : \sigma \mapsto \mathbf{a}_\sigma$, we write $q_D[a/\mathbf{a}]$ for this substitution.

Instead of computing the weakest precondition of each D-polynomial q separately, we may compute the weakest precondition of the single generic polynomial q_D once and for all and substitute the concrete coefficients a_σ of the polynomials q into the precondition of q_D later. Indeed, we show in [14]:

Theorem 3. *Assume p is a polynomial program without procedures and let $\mathcal{B}_D(v)$, v program point of p, be the least solution of the constraint system $\mathcal{B}$ for p with conjecture q_D at target t. Then $q = q_D[a/\mathbf{a}]$ is valid at t iff $q'[a/\mathbf{a}] = 0$ for all $q' \in \mathcal{B}_D(\mathsf{start})$.* □

Clearly, it suffices that $q'[a/\mathbf{a}] = 0$ only for a set of generators of $\mathcal{B}_D(\mathsf{start})$. Still, this does not immediately give us an effective method of determining all suitable coefficient vectors, since the precise set of solutions of arbitrary polynomial equation systems are not computable. We observe, however, in [14]:

Lemma 2. *Every ideal $\mathcal{B}_D(u)$, u a program point, of the least solution of the abstract constraint system $\mathcal{B}$ for conjecture q_D at some target node t is generated by a finite set G of polynomials q where each variable $\mathbf{a}_\sigma$ occurs only with degree at most 1. Moreover, such a generator set can be effectively computed.* □

Thus, the set of (coefficient maps) of D-polynomials which are valid at our target program point t can be characterized as the set of solutions of a *linear* equation system. Such equation systems can be algorithmically solved, i.e., finite representations of their sets of solutions can be constructed explicitly, e.g., by Gaussian elimination. We conclude:

Theorem 4. *For a polynomial program p without procedures and a program point t in p, the set of all D-polynomials which are valid at t can be effectively computed.*   □

As a side remark, we should mention that instead of working with the larger polynomial ring $\mathbb{Q}[\mathbf{X} \cup \mathbf{A}_D]$, we could work with *modules* over the polynomial ring $\mathbb{Q}[\mathbf{X}]$ consisting of vectors of polynomials whose entries are indexed with $\sigma \in D$. The operations on modules turn out to be practically much faster than corresponding operations on the larger polynomial ring itself, see [18] for a practical implementation and preliminary experimental results.

5   Interprocedural Analysis: Transition Invariants

The main question of precise interprocedural analysis is this: how can the effects of procedure calls be finitely described? An interesting idea (essentially) due to Colon [4] is to represent effects by polynomial *transition invariants*. This means that we introduce a separate copy $\mathbf{X}' = \{\mathbf{x}'_1, \ldots, \mathbf{x}'_k\}$ of variables denoting the values of variables before the execution. Then we use polynomials to express possible relationships between pre- and post-states of the execution. Obviously, all such valid relationships again form an ideal, now in the polynomial ring $\mathbb{Q}[\mathbf{X} \cup \mathbf{X}']$.

The transformation ideals for assignments, non-deterministic assignments and disequality guards are readily expressed by:

$$
\begin{aligned}
[\![\mathbf{x}_i := p]\!]^{\#\#} &= \langle \{\mathbf{x}_j - \mathbf{x}'_j \mid j \neq i\} \cup \{\mathbf{x}_i - p[\mathbf{x}'/\mathbf{x}]\} \rangle \\
[\![\mathbf{x}_i :=?]\!]^{\#\#} &= \langle \{\mathbf{x}_j - \mathbf{x}'_j \mid j \neq i\} \rangle \\
[\![p \neq 0]\!]^{\#\#} &= \langle \{p[\mathbf{x}'/\mathbf{x}] \cdot (\mathbf{x}_j - \mathbf{x}'_j) \mid j = 1, \ldots, k\} \rangle
\end{aligned}
$$

In particular, the last definition means that either the guard is wrong before the transition or the states before and after the transition are equal. The basic effects can be composed to obtain the effects of larger program fragments by means of a composition operation "∘". Composition on transition invariants can be defined by:

$$
I_1 \circ I_2 \;=\; (I_1[\mathbf{y}/\mathbf{x}'] \oplus I_2[\mathbf{y}/\mathbf{x}]) \cap \mathbb{Q}[\mathbf{X} \cup \mathbf{X}']
$$

where a fresh set $\mathbf{Y} = \{\mathbf{y}_1, \ldots, \mathbf{y}_k\}$ is used to store the intermediate values between the two transitions represented by I_1 and I_2 and the postfix operator $[\mathbf{y}/\mathbf{x}]$ denotes renaming of variables in $\mathbf{X}$ with their corresponding copies in $\mathbf{Y}$. Note that "∘" is defined by means of well-known effective ideal operations. Using this operation, we can put up a constraint system $\mathcal{T}$ for ideals of polynomial transition invariants of procedures:

$$
\begin{aligned}
\mathcal{T}(v) &\subseteq \langle \mathbf{x}_i - \mathbf{x}'_i \mid i = 1, \ldots, k \rangle & &v \text{ is entry point} \\
\mathcal{T}(v) &\subseteq [\![\mathbf{x}_i := p]\!]^{\#\#} \circ \mathcal{T}(u) & &(u, v) \text{ is labeled with } \mathbf{x}_i := p \\
\mathcal{T}(v) &\subseteq [\![\mathbf{x}_i :=?]\!]^{\#\#} \circ \mathcal{T}(u) & &(u, v) \text{ is labeled with } \mathbf{x}_i :=? \\
\mathcal{T}(v) &\subseteq [\![p \neq 0]\!]^{\#\#} \circ \mathcal{T}(u) & &(u, v) \text{ is labeled with } p \neq 0
\end{aligned}
$$

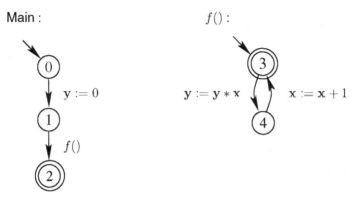

Fig. 3. A simple program with procedures

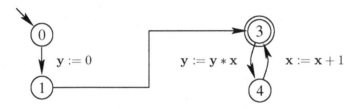

Fig. 4. The inlined version of the example program

$$T(v) \subseteq T(f) \circ T(u) \qquad (u, v) \text{ calls } f$$
$$T(f) \subseteq T(v) \qquad v \text{ exit point of } f$$

Example 3. Consider the program from Fig. 3. We calculate:

$$
\begin{aligned}
T(f) &= \langle x - x', y - y' \rangle \cap \\
&\quad \langle x - x' - 1, y - y' \cdot x' \rangle \cap \\
&\quad \langle x - x' - 2, y - y' \cdot x' \cdot (x' + 1) \rangle \cap \\
&\quad \cdots \\
&= \langle 0 \rangle
\end{aligned}
$$

Using this invariant for analyzing the procedure main, we only find the trivial transition invariant 0. On the other hand, we may instead *inline* the procedure f as in Fig. 4. A corresponding calculation of the transition invariant of main yields:

$$
\begin{aligned}
T(\text{Main}) &= \langle x - x', y \rangle \cap \\
&\quad \langle x - x' - 1, y \rangle \cap \\
&\quad \langle x - x' - 2, y \rangle \cap \\
&\quad \cdots \\
&= \langle y \rangle
\end{aligned}
$$

Thus for this analysis, inlining may gain precision. □

Clearly, using transition invariants incurs the same problem as forward propagation for intraprocedural analysis, namely, that fixpoint iteration may result in infinite decreasing chains of ideals. Our minimal example exhibited two more problems, namely that the composition operation is not *continuous*, i.e., does not commute with greatest lower bounds of descending chains in the second argument, and also that a less compositional analysis through inlining may infer more valid transition invariants.

To be fair here, it should be noted that Colon did not propose to use *ideals* for representing transition invariants. Colon instead considered *pseudo*-ideals, i.e., ideals where polynomials are considered only up to a given degree bound. This kind of further abstraction solves the problems of infinite decreasing chains as well as missing continuity — at the expense, though, of further loss in precision. Colon's approach, for example, fails to find a nontrivial invariant in the example program from Fig. 3 for Main.

6   Interprocedural Analysis: Backward Propagation

Due to the apparent weaknesses of the approach through polynomials as transition invariants, we propose to represent effects of procedures by pre-conditions of generic polynomials. Procedure calls are then dealt with through instantiation of generic coefficients. Thus, effects are still described by ideals — over a larger set of variables (or by modules; see the discussion at the end of Section 4). Suppose we have chosen some finite set $D \subseteq \mathbb{N}_0^k$ of exponent tuples and assume that the polynomial $p = p_D[a/\mathbf{a}]$ is the D-polynomial that is obtained from the generic D-polynomial through instantiation of the generic coefficients with a. Assume further that the effect of some procedure call is given by the ideal $I \subseteq \mathbb{Q}[\mathbf{X} \cup \mathbf{A}_D] = \langle q_1, \ldots, q_m \rangle$. Then we determine a precondition of $p = 0$ w.r.t. to the call by:

$$I\,(p) \;=\; \langle q_1[a/\mathbf{a}], \ldots, q_m[a/\mathbf{a}] \rangle$$

This definition is readily extended to ideals I' generated by D-polynomials. There is no guarantee, though, that all ideals that occur at the target program points v of call edges (u, v) will be generated by D-polynomials. In fact, simple examples can be constructed where no uniform set D of exponent tuples can be given. Therefore, we additionally propose to use an abstraction operator $\mathbf{W}$ that splits polynomials appearing as postcondition of procedure calls which are not D-polynomials.

We choose a maximal degree d_j for each program variable x_j and let

$$D = \{(i_1, \ldots, i_k) \mid i_j \leq d_j \text{ for } i = 1, \ldots, k\}$$

The abstraction operator $\mathbf{W}$ takes generators of an ideal I and maps them to generators of a (possibly) larger ideal $\mathbf{W}(I)$ which is generated by D-polynomials. In order to construct such an ideal, we need a heuristics which decomposes an arbitrary polynomial q into a linear combination of D-polynomials $q_1, \ldots, q_m$:

$$q = r_1 q_1 + \ldots + r_m q_m \tag{1}$$

We could, for example, decompose q according to the first variable:

$$q = q_0' + \mathbf{x}_1^{d_1+1} \cdot q_1' + \ldots + \mathbf{x}_1^{s(d_1+1)} \cdot q_s'$$

where each q_i' contains powers of $\mathbf{x}_1$ only up to degree d_1 and repeat this decomposition with the polynomials q_i' for the remaining variables. Given a decomposition (1), we have $q \in \langle q_1, \ldots, q_m \rangle$. Therefore, we can replace every generator of I by D-polynomials in order to obtain an ideal $\mathbf{W}(I)$ with the desired properties.

We use the new application operator as well as the abstraction operator $\mathbf{W}$ to generalize our constraint system $\mathcal{B}$ to a constraint system $\mathcal{E}$ for the effects of procedures:

$$
\begin{array}{lll}
\mathcal{E}(u) \supseteq \langle q_D \rangle & & u \text{ is exit point} \\
\mathcal{E}(u) \supseteq [\![\mathbf{x}_i := p]\!]^\mathsf{T}(\mathcal{E}(v)) & & (u, v) \text{ labeled with } \mathbf{x}_i := p \\
\mathcal{E}(u) \supseteq [\![\mathbf{x}_i :=?]\!]^\mathsf{T}(\mathcal{E}(v)) & & (u, v) \text{ labeled with } \mathbf{x}_i :=? \\
\mathcal{E}(u) \supseteq [\![p \neq 0]\!]^\mathsf{T}(\mathcal{E}(v)) & & (u, v) \text{ labeled with } p \neq 0 \\
\mathcal{E}(u) \supseteq \mathcal{E}(f)(\mathbf{W}(\mathcal{E}(v))) & & (u, v) \text{ calls } f \\
\mathcal{E}(f) \supseteq \mathcal{E}(u) & & u \text{ entry point of } f
\end{array}
$$

Example 4. Consider again the example program from Fig. 3. Let us choose $d = 1$ where $p_1 = \mathbf{a}\mathbf{y} + \mathbf{b}\mathbf{x} + \mathbf{c}$. Then we calculate for f:

$$
\begin{aligned}
\mathcal{E}(f) = \ & \langle \mathbf{a}\mathbf{y} + \mathbf{b}\mathbf{x} + \mathbf{c} \rangle \ \oplus \\
& \langle \mathbf{a}\mathbf{y}\mathbf{x} + \mathbf{b}(\mathbf{x} + 1) + \mathbf{c} \rangle \ \oplus \\
& \langle \mathbf{a}\mathbf{y}\mathbf{x}(\mathbf{x} + 1) + \mathbf{b}(\mathbf{x} + 2) + \mathbf{c} \rangle \ \oplus \\
& \langle \mathbf{a}\mathbf{y}\mathbf{x}(\mathbf{x} + 1)(\mathbf{x} + 2) + \mathbf{b}(\mathbf{x} + 3) + \mathbf{c} \rangle \ \oplus \\
& \cdots \\
= \ & \langle \mathbf{a}\mathbf{y}, \mathbf{b}, \mathbf{c} \rangle
\end{aligned}
$$

This description tells us that for a linear identity $\mathbf{a}\mathbf{y} + \mathbf{b}\mathbf{x} + \mathbf{c} = 0$ to be valid after a call to f, the coefficients $\mathbf{b}$ and $\mathbf{c}$ necessarily must equal 0. Moreover, either coefficient $\mathbf{a}$ equals 0 (implying that the whole identity is trivial) or $\mathbf{y} = 0$. Indeed, this is the optimal description of the behavior of f with polynomials. $\qquad\square$

The effects of procedures as approximated by constraint system $\mathcal{E}$ can be used to check a polynomial conjecture q_t at a given target node t along the lines of constraint system $\mathcal{B}$. We only have to extend it by extra constraints dealing with function calls. Thus, we put up the following constraint system:

$$
\begin{array}{lll}
\mathcal{R}(t) \supseteq \langle q_t \rangle & & \\
\mathcal{R}(u) \supseteq [\![\mathbf{x}_i := p]\!]^\mathsf{T}(\mathcal{R}(v)) & & (u, v) \text{ labeled with } \mathbf{x}_i := p \\
\mathcal{R}(u) \supseteq [\![\mathbf{x}_i :=?]\!]^\mathsf{T}(\mathcal{R}(v)) & & (u, v) \text{ labeled with } \mathbf{x}_i :=? \\
\mathcal{R}(u) \supseteq [\![p \neq 0]\!]^\mathsf{T}(\mathcal{R}(v)) & & (u, v) \text{ labeled with } p \neq 0 \\
\mathcal{R}(u) \supseteq \mathcal{E}(f)(\mathbf{W}(\mathcal{R}(v))) & & (u, v) \text{ calls } f \\
\mathcal{R}(f) \supseteq \mathcal{R}(u) & & u \text{ entry point of } f \\
\mathcal{R}(u) \supseteq \mathcal{R}(f) & & (u, \_) \text{ calls } f
\end{array}
$$

This constraint system again has a least solution which can be computed by standard fixpoint iteration. Summarizing, we obtain the following theorem:

Theorem 5. *Assume p is a polynomial program with procedures. Assume further that we assert a conjecture q_t at program point t.*

Safety: 1. *For every procedure f, the ideal $\mathcal{E}(f)$ represents a precondition of the identity $p_D = 0$ after the call.*
   2. *If the ideal $\mathcal{R}(\mathsf{Main})$ equals $\{0\}$, then the conjecture q_t is valid at t.*
Completeness: *If during fixpoint computation, all ideals at target program points v of call edges (u, v) are represented by D-polynomials as generators, the conjecture is valid only if the ideal $\mathcal{R}(\mathsf{Main})$ equals $\{0\}$.*

The safety-part of Theorem 5 tells us that our analysis will never assure a wrong conjecture but may fail to certify a conjecture although it is valid. According to the completeness-part, however, the analysis algorithm provides slightly more information: if no approximation steps are necessary at procedure calls, the analysis is precise. For simplicity, we have formulated Theorem 5 in such a way that it only speaks about checking conjectures. In order to infer valid polynomial identities up to a specified degree bound, we again can proceed analogous to the intraprocedural case by considering a generic postcondition in constraint system $\mathcal{R}$.

7   Equality Guards

In this section, we discuss methods for dealing with equality guards $p = 0$. Recall, that in presence of equality guards, the question whether a variable is constantly 0 at a program point or not is undecidable even in absence of procedures and with affine assignments only. Thus, we cannot hope for complete methods here. Still, in practical contexts, equality guards are a major source of information about values of variables. Consider, e.g., the control flow graph from Fig. 5. Then, according to the equality guard, we definitely know that $x = 10$ whenever program point 2 is reached. In order to deal with equality guards, we thus extend forward analysis by the constraints:

$$\mathcal{F}(v) \subseteq [\![p = 0]\!]^{\sharp}(\mathcal{F}(u)) \qquad (u, v) \text{ labeled with } p = 0$$

where the effect of an equality guard is given by:

$$[\![p = 0]\!]^{\sharp}\, I \;=\; I \oplus \langle p \rangle$$

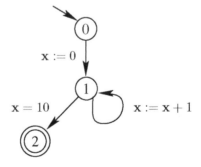

Fig. 5. A simple for-loop

This formalizes our intuition that after the guard, we additionally know that $p = 0$ holds. Such an approximate treatment of equality guards is common in forward program analysis and already proposed by Karr [11]. A similar extension is also possible for inferring transition invariants. The new effect is monotonic. However, it is, no longer distributive, i.e., it does not commute with intersections. Due to monotonicity, the extended constraint systems $\mathcal{F}$ as well as $\mathcal{T}$ still have greatest solutions which provide safe approximations of the sets of all valid invariants and transition invariants in presence of equality guards, respectively.

Example 5. Consider the program from Fig. 5. For program point 1 we have:

$$\mathcal{F}(1) = \langle \mathbf{x} \rangle \cap \langle \mathbf{x} - 1 \rangle \cap \langle \mathbf{x} - 2 \rangle \cap \ldots$$
$$= \{0\}$$

Accordingly, we find for program point 2,

$$\mathcal{F}(2) = \{0\} \oplus \langle \mathbf{x} - 10 \rangle$$
$$= \langle \mathbf{x} - 10 \rangle$$

Thus, given the lower bound $\{0\}$ for the infinite decreasing chain of program point 1, we arrive at the desired result for program point 2.   □

It would be nice if also backward analysis could be extended with some approximate method for equality guards. Our idea for such an extension is based on *Lagrange multipliers*. Recall that the *weakest* precondition for validity of $q = 0$ after a guard $p = 0$ is given by:

$$(p = 0) \Rightarrow (q = 0)$$

which, for every λ, is implied by:

$$q + \lambda \cdot p = 0$$

The value λ is called a *Lagrange*-multiplier and can be arbitrarily chosen. We remark that a related technique has been proposed in [5] for inferring parametric program invariants. Thus, we define:

$$[\![p = 0]\!]^{\mathsf{T}}(q) = \langle q + p \cdot \lambda \rangle \tag{2}$$

where a different formal multiplier λ is chosen for every occurrence of an equality guard. Similar to the treatment of generic postconditions, the parameters λ will occur linearly in a suitably chosen set of generators for the precondition ideal at program start where they can be chosen appropriately.

Example 6. Again consider the program from Fig. 5 and assume that we are interested in identities up to degree 1 at the exit point of the program. Thus we start with the generic polynomial $\mathbf{a}\mathbf{x} + \mathbf{b} = 0$ at node 2. This gives us for program point 1:

$$\mathcal{B}_1(1) = \langle (\mathbf{a} + \lambda) \cdot \mathbf{x} + \mathbf{b} - 10\lambda \rangle \oplus$$
$$\langle (\mathbf{a} + \lambda) \cdot \mathbf{x} + \mathbf{a} + \lambda + \mathbf{b} - 10\lambda \rangle$$
$$= \langle \mathbf{a} + \lambda, \mathbf{b} - 10\lambda \rangle$$

Choosing $\lambda = -\mathbf{a}$, we obtain $\mathbf{b} = -10\mathbf{a}$. Therefore all multiples of the polynomial $\mathbf{x} - 10$ are valid identities for program point 2.   □

Instead of using a single variable λ as a Lagrange multipliers we could also use an entire polynomial. This means that we use in (2) a generic polynomial q_D (for some set D of exponent tuples) instead of the variable λ for each equality guard $p = 0$:

$$[\![p = 0]\!]^{\mathsf{T}}(q) = \langle q + p \cdot q_D\lambda \rangle$$

where we use new variables $A_D = \{\mathbf{a}_\sigma \mid \sigma \in D\}$ in q_D for each equality guard. Now, all the variables in A_D can be adjusted in the computed weakest precondition. This may lead to more precise results – at the price of a more expensive analysis.

8   Discussion

We have summarized forward and backward iteration methods for inferring valid polynomial identities. In absence of procedure calls, we arrived at a rather clear picture: we exhibited a finite constraint system which precisely characterizes the set of all valid polynomial identities in a polynomial program. Due to possibly infinite decreasing chains of ideals, it is currently open whether the greatest solution of this constraint system can effectively be computed. On the other hand, backward analysis based on weakest precondition computations relies on increasing chains of ideals — allowing us to decide whether any given conjecture at a program point is valid. Also, this enables us to effectively find all valid polynomial identities up to a given degree.

In presence of procedure calls, the picture is less clear. The natural extension of the intraprocedural forward propagation suggests to use ideals of polynomial transition invariants to describe effects of procedures. The composition operation for such ideals, though, turned out to be non-continuous. Also, our example shows that using polynomial transition invariants may not be precise, i.e., may miss some valid polynomial identities. Therefore, we considered a generalization of backward analysis which describes effects of procedures by means of preconditions of generic polynomials. Here, we obtained a precise finite characterization of identities of some given form only if in the polynomials occurring during the analysis at procedure exits the degrees of the variables are bounded. Note that this approach can be considered as a smooth generalization of the methods in [15] for affine programs where all occurring polynomials are known to have bounded degree.

It still remains open whether precise techniques can be found for lifting the degree bound in the general intraprocedural case. It is also unclear how to deal with recursive programs precisely if the degrees of weakest preconditions grow arbitrarily.

References

1. T. Becker and V. Weispfenning. *Gröbner Bases*. Springer-Verlag, 1993.
2. S. Bensalem, Y. Lakhnech, and H. Saidi. Powerful Techniques for the Automatic Generation of Invariants. In *8th Int. Conf. on Computer Aided Verification (CAV)*, volume 1102 of *Lecture Notes in Computer Science*. Springer, 1996.
3. N. Bjørner, A. Browne, and Z. Manna. Automatic Generation of Invariants and Intermediate Assertions. *Theoretical Computer Science*, 173(1):49–87–, 1997.

4. M. Colon. Approximating the Algebraic Relational Semantics of Imperative Programs. In *11th Int. Symp. on Static Analysis (SAS)*, pages 296–311. Springer-Verlag, LNCS 3146, 2004.
5. P. Cousot. Proving Program Invariance and Termination by Parametric Abstraction, Lagrangian Relaxation and Semidefinite Programming . In *Verification, Model Checking and Abstract Interpretation (VMCAI)*, pages 1–24. Springer-Verlag, LNCS 3385, 2005.
6. P. Cousot and R. Cousot. Static Determination of Dynamic Properties of Recursive Procedures. In E. Neuhold, editor, *IFIP Conf. on Formal Description of Programming Concepts*, pages 237–277. North-Holland, 1977.
7. S. Gulwani and G. Necula. Discovering Affine Equalities Using Random Interpretation. In *30th ACM Symp. on Principles of Programming Languages (POPL)*, pages 74–84, 2003.
8. S. Gulwani and G. Necula. Precise Interprocedural Analysis Using Random Interpretation. In *32th Ann. ACM Symp. on Principles of Programming Languages (POPL)*, pages 324–337, 2005.
9. M. S. Hecht. *Flow Analysis of Computer Programs*. Elsevier North-Holland, 1977.
10. J. Kam and J. Ullman. Global Data Flow Analysis and Iterative Algorithms. *Journal of the ACM (JACM)*, 23(1):158–171, 1976.
11. M. Karr. Affine Relationships Among Variables of a Program. *Acta Informatica*, 6:133–151, 1976.
12. M. Müller-Olm and H. Seidl. Polynomial Constants are Decidable. In *9th Static Analysis Symposium (SAS)*, pages 4–19. LNCS 2477, Springer-Verlag, 2002.
13. M. Müller-Olm and H. Seidl. A Note on Karr's Algorithm. In *31st Int. Coll. on Automata, Languages and Programming (ICALP)*, pages 1016–1028. Springer Verlag, LNCS 3142, 2004.
14. M. Müller-Olm and H. Seidl. Computing Polynomial Program Invariants. *Information Processing Letters (IPL)*, 91(5):233–244, 2004.
15. M. Müller-Olm and H. Seidl. Precise Interprocedural Analysis through Linear Algebra. In *31st ACM Symp. on Principles of Programming Languages (POPL)*, pages 330–341, 2004.
16. M. Müller-Olm and H. Seidl. A Generic Framework for Interprocedural Analysis of Numerical Properties. In *12th Static Analysis Symposium (SAS)*, pages 235–250. LNCS 3672, Springer-Verlag, 2005.
17. M. Müller-Olm and H. Seidl. Analysis of Modular Arithmetic. In *European Symposium on Programming (ESOP)*, pages 46–60. Springer Verlag, LNCS 3444, 2005.
18. M. Petter. Berechnung von polynomiellen Invarianten, 2004. Diploma Thesis.
19. J. R. Reif and H. R. Lewis. Symbolic Evaluation and the Global Value Graph. In *4th ACM Symp. on Principles of Programming Languages POPL'77*, pages 104–118, 1977.
20. E. Rodriguez-Carbonell and D. Kapur. An Abstract Interpretation Approach for Automatic Generation of Polynomial Invariants. In *11th Int. Symp. on Static Analysis (SAS)*, pages 280–295. Springer-Verlag, LNCS 3146, 2004.
21. E. Rodriguez-Carbonell and D. Kapur. Automatic Generation of Polynomial Loop Invariants: Algebraic Foundations. In *Int. ACM Symposium on Symbolic and Algebraic Computation 2004 (ISSAC04)*, pages 266–273, 2004.
22. S. Sankaranarayanan, H. B. Sipma, and Z. Manna. Non-linear Loop Invariant Generation using Gröbner Bases. In *ACM Symp. on Principles of Programming Languages (POPL)*, pages 318–329, 2004.

External String Sorting:
Faster and Cache-Oblivious

Rolf Fagerberg[1], Anna Pagh[2], and Rasmus Pagh[2]

[1] University of Southern Denmark, Campusvej 55, 5230 Odense M, Denmark
rolf@imada.sdu.dk
[2] IT University of Copenhagen, Rued Langgaards Vej 7, 2300 København S, Denmark
{annao, pagh}@itu.dk

Abstract. We give a randomized algorithm for sorting strings in external memory. For K binary strings comprising N words in total, our algorithm finds the sorted order and the longest common prefix sequence of the strings using $O(\frac{K}{B}\log_{M/B}(\frac{K}{M})\log(\frac{N}{K}) + \frac{N}{B})$ I/Os. This bound is never worse than $O(\frac{K}{B}\log_{M/B}(\frac{K}{M})\log\log_{M/B}(\frac{K}{M}) + \frac{N}{B})$ I/Os, and improves on the (deterministic) algorithm of Arge et al. *(On sorting strings in external memory, STOC '97)*. The error probability of the algorithm can be chosen as $O(N^{-c})$ for any positive constant c. The algorithm even works in the cache-oblivious model under the tall cache assumption, i.e,, assuming $M > B^{1+\epsilon}$ for some $\epsilon > 0$. An implication of our result is improved construction algorithms for external memory string dictionaries.

1 Introduction

Data sets consisting partly or entirely of string data are common: Most database applications have strings as one of the data types used, and in some areas, such as bioinformatics, web retrieval, and word processing, string data is predominant. Additionally, strings form a general and fundamental data model of computer science, containing e.g. integers and multi-dimensional data as special cases.

In internal memory, sorting of strings is well understood: When the alphabet is comparison based, sorting K strings of total length N takes $\Theta(K \log K + N)$ time (see e.g. [8]). If the alphabet is the integers, then on a word-RAM the time is $\Theta(\mathrm{Sort_{Int}}(K) + N)$, where $\mathrm{Sort_{Int}}(K)$ is the time to sort K integers [3].

In external memory the situation is much less clear. Some upper bounds have been given [7], along with matching lower bounds in restricted models of computation. As noted in [7], the natural upper bound to hope for is $O(\frac{K}{B}\log_{M/B}(\frac{K}{M}) + \frac{N}{B})$ I/Os, which is the sorting bound for K single characters plus the complexity of scanning the input. In this paper we show how to compute (using randomization) the sorted order in a number of I/Os that nearly matches this bound.

1.1 Models of Computation

Computers contain a hierarchy of memory levels, with large differences in access time. This makes the time for a memory access depend heavily on what is

B. Durand and W. Thomas (Eds.): STACS 2006, LNCS 3884, pp. 68–79, 2006.

currently the innermost level containing the data accessed. In algorithm analysis, the standard RAM (or von Neumann) model is unable to capture this, and external memory models were introduced to better model these effects. The model most commonly used for analyzing external memory algorithms is the two-level I/O-model [1], also called the External Memory model or the Disk Access model. The I/O-model approximates the memory hierarchy by modeling two levels, with the inner level having size M, the outer level having infinite size, and transfers between the levels taking place in blocks of B consecutive elements. The cost measure of an algorithm is the number of memory transfers, or I/Os, it makes.

The cache-oblivious model, introduced by Frigo et al. [17], elegantly generalizes the I/O-model to a multi-level memory model by a simple measure: the algorithm is not allowed to know the value of B and M. More precisely, a cache-oblivious algorithm is an algorithm formulated in the RAM model, but analyzed in the I/O-model, with an analysis valid for *any* value of B and M. Cache replacement is assumed to take place automatically by an optimal off-line cache replacement strategy. Since the analysis holds for any B and M, it holds for all levels simultaneously. See [17] for the full details of the cache-oblivious model.

Over the last two decades, a large body of results for the I/O-model has been produced, covering most areas of algorithmics. For the newer cache-oblivious model, introduced in 1999, already a sizable number of results exist. One of the fundamental facts in the I/O-model is that comparison-based sorting of N elements takes $\Theta(\text{Sort}(N))$ I/Os [1], where $\text{Sort}(N) = \frac{N}{B} \log_{M/B} \frac{N}{M}$. Also in the cache-oblivious model, sorting can be carried out in $\Theta(\text{Sort}(N))$ I/Os, if one makes the so-called *tall cache* assumption $M \geq B^{1+\varepsilon}$ [10, 17]. This assumption has been shown to be necessary [11]. Another basic fact in the I/O-model is that permutation takes $\Theta(\min\{\text{Sort}(N), N\})$, assuming that elements are indivisible [1].

The subject of this paper is sorting strings in external memory models. Below, we discuss existing results in this area. For a general overview of results for external memory, refer to the recent surveys [4, 21, 23, 24] for the I/O-model, and [6, 9, 13, 21] for the cache-oblivious model.

1.2   Previous Work

Arge et al. [7] were the first to study string sorting in external memory, introducing the size N of the input and the number K of strings as separate parameters. Note that the problem is at least as hard as sorting K words[1], and also requires at least N/B I/Os for reading the input. In internal memory, i.e., for $B = 1$, there are algorithms that meet this lower bound [3, 8]. However, it remains an open problem whether this is possible in external memory for general B.

Arge et al. give several algorithms obeying various indivisibility restrictions. The complexity of these algorithms depends on the number K_1 of strings of less than one block in length, and the number K_2 of strings of at least one block

---

[1] See Section 1.3 below for the model of computation.

in length. The total length of short and long strings is denoted N_1 and N_2, respectively. The fastest algorithm runs in

$$O(\min(K_1 \log_M K_1, \tfrac{N_1}{B} \log_{M/B}(\tfrac{N_1}{M})) + K_2 \log_M K_2 + \tfrac{N}{B}) \text{ I/Os}.$$

Under the tall cache assumption this simplifies to

$$O(\tfrac{N_1}{B} \log_M K_1 + K_2 \log_M K_2 + \tfrac{N_2}{B}) \text{ I/Os}.$$

The first term is the complexity of sorting the short strings using external merge sort. The second term states a logarithmic (base M) I/O cost per long string. The third term is the complexity of reading the long strings. Each of the three terms may be the dominant one. Assume for simplicity that all K strings have the same length. If they are short, the first term obviously dominates. If their length is between B and $B \log_M K$, the second term dominates. For longer strings, the third term dominates, i.e., sorting can be done in scanning complexity. Note that the upper bound is furthest from the lower bound for strings with a length around one block.

Arge et al. also consider "practical algorithms" whose complexity depends on the alphabet size. However, because of the implicit assumption that the alphabet has size $N^{\Theta(1)}$ (it is assumed that a character and a pointer uses the same space, within a constant factor), none of these algorithms are better than the above from a theoretical perspective.

Note that prior to our work there are no direct results on *cache-oblivious* string sorting, except the bound $O(\text{Sort}(N))$, which can be derived from suffix tree/array construction algorithms [14, 19].

1.3   Our Results

Before we state our results, we discuss the model and notation. First of all, we assume the input to be binary strings. This is no restriction in practice, since any finite alphabet can be encoded as binary strings such that replacing each character with its corresponding string preserves the ordering of strings. How- ever, to facilitate a clear comparison with previous results, we will not count the length of strings in bits, but rather in *words* of $\Theta(\log N)$ bits. This is consis- tent with [7] which assumes that a character and a pointer uses the same space, within a constant factor. We will also assume that all strings have length at least one word, which again is consistent with [7]. We will adopt the notation from [7]:

$$
\begin{aligned}
K &= \text{\# of strings to sort,} \\
N &= \text{total \# of words in the K strings,} \\
M &= \text{\# of words fitting in internal memory,} \\
B &= \text{\# of words per disk block,}
\end{aligned}
$$

where $M < N$ and $1 < B \le M/2$. We assume that the input sequence $x_1, \ldots, x_K$ is given in a form such that it can be read in $O(N/B)$ I/Os.

Secondly, we distinguish between standard *value-sorting*, which produces a sequence with the input items in sorted order, and the problem of finding the sorting permutation, i.e., producing a sequence of *references* to the items in sorted order (this is equivalent to what is sometimes referred to as *rank-sorting*, i.e., computing the rank of all input elements). For strings, the latter is often enough, as is e.g. the case for string dictionaries. The *sorting permutation* σ of the input sequence is the permutation such that $\sigma(i) = j$ if x_j has rank i in the sorted order. In this definition, the references to strings are their ranks in input order. If one instead as references wants pointers to the memory locations of the first characters of the strings, conversion between the two representations can be done in $O(\mathrm{Sort}(K) + N/B)$ I/Os by sorting and scanning. The latter representation is commonly called the suffix array in the case where the strings are the suffixes of some base string. Let $\mathrm{lcp}(x_i, x_j)$ be the longest common prefix of x_i and x_j. By the *lcp sequence* we denote the numbers $\mathrm{LCP}(i) = |\mathrm{lcp}(x_{\sigma(i)}, x_{\sigma(i+1)})|$, i.e., the lengths of the longest common prefixes between pairs of strings consecutive in sorted order. For the application of string dictionary construction, we will need this.

Our main result is a Monte Carlo type randomized, cache-oblivious algorithm which computes the sorting permutation and the lcp sequence. The output is the sequences $\sigma(1), \ldots, \sigma(K)$ and $\mathrm{LCP}(1), \ldots, \mathrm{LCP}(K-1)$.

Theorem 1. *Let c and ϵ be arbitrary positive constants, and assume that $M > B^{1+\epsilon}$. Then there is a randomized, cache-oblivious algorithm that, given K strings of N words in total, computes the sorting permutation and the lcp sequence in $O(\frac{K}{B} \log_{M/B}(\frac{K}{M}) \log(\frac{N}{K}) + \frac{N}{B})$ I/Os, such that the result is correct with probability $1 - O(N^{-c})$.*

Note that the I/O bound above coincides with $\mathrm{Sort}(K)$ for strings of length $O(1)$ words, and is never worse than $O(\mathrm{Sort}(K) \log\log_{M/B}(\frac{K}{M}) + \frac{N}{B})$. (If $\frac{N}{K}$ exceeds $(\log_{M/B}(\frac{K}{M}))^{O(1)}$ then the N/B term will be dominant.) Thus, we have optimal dependence on N and are merely a doubly logarithmic factor away from $\mathrm{Sort}(K)$. We prove Theorem 1 in Section 2.

The String B-tree [15] is an external memory string dictionary, which allows (prefix) searches over a set of K strings in $O(\log_B K + P/B)$ I/Os, where P is the length of the search string. Constructing the String B-tree over a set of strings is as hard as finding the sorting permutation. Conversely, from the sorting permutation and the lcp sequence, the String B-tree over the strings can easily be built. Recently, a cache-oblivious string dictionary with the same searching complexity as String B-Trees has been given [12], and the same statement about construction applies to this structure. Hence, one important corollary of Theorem 1 is the following, shown in Section 3:

Corollary 1. *Let c and ϵ be arbitrary positive constants, and assume that $M > B^{1+\epsilon}$. Then there is a randomized, cache-oblivious algorithm that, given K strings of N words in total, constructs a String B-tree or a cache-oblivious string dictionary [12] over the strings in $O(\frac{K}{B} \log_{M/B}(\frac{K}{M}) \log(\frac{N}{K}) + \frac{N}{B})$ I/Os, such that the result is correct with probability $1 - O(N^{-c})$.*

Again, this bound is never worse than $O(\text{Sort}(K)\log\log_{M/B}(\frac{K}{M}) + \frac{N}{B})$. If one wants not only the sorting sequence, but also the strings to appear in memory in sorted order, they must be permuted. In Section 4, we discuss methods for permuting strings based on knowledge of the sorting sequence. The methods are straightforward, but nevertheless show that also for the standard value-sorting problem (i.e. including permuting the strings), our main result leads to asymptotical improvements in complexity, albeit more modest than for finding the permutation sequence.

1.4   Other Related Work

Sorting algorithms for the word RAM model have developed a lot in the last decade. The new RAM sorting algorithms take advantage of the bit represen-tation of the strings or integers to be sorted, in order to beat the $\Omega(K\log K)$ lower bound for sorting K items using comparisons. The currently fastest sorting algorithm for K words runs in time $O(K\sqrt{\log\log K})$, expected [18]. This implies an external memory algorithm running in the same I/O bound, which is better than $O(\text{Sort}(K))$ if K is sufficiently large ($K = M^{\omega(B)}$ is necessary).

Andersson and Nilsson [3] have shown how to reduce sorting of K strings of length N to sorting of $O(K)$ words, in $O(N)$ expected time. This means that the relation between string sorting and word sorting on a word RAM is very well understood. The currently fastest word sorting algorithm gives a bound of $O(K\sqrt{\log\log K} + N)$ expected time. Again, using this directly on external memory gives a bound better than other string sorting bounds (including those in the present paper) for certain extreme instances (very large sets of not too long strings).

If the length w of the machine words to be sorted is sufficiently large in terms of K, there exists an expected linear time word RAM sorting algorithm. Specifically, if $w > (\log K)^{2+\epsilon}$, for some constant $\epsilon > 0$, K words can be sorted in expected $O(K)$ time [2]. To understand the approach of the algorithm it is useful to think of the words as binary strings of length w. The key ingredient of the algorithm is a randomized signature technique that creates a set of "signature" strings having, with high probability, essentially the same trie structure (up to ordering of children of nodes) as the original set of strings. If the word length is large, the signatures can be made considerably shorter than the original strings, and after a constant number of recursive steps they can be sorted in $O(K)$ time. To sort the original strings, one essentially sorts the parts of the original strings that correspond to branching nodes in the trie of signatures.

Our algorithm is inspired by this, and uses the same basic approach as just described. However, the details of applying the technique to external memory strings are quite different. For example, it is easier for us to take advantage of the reduction in string size. This means that the best choice in our case is to use the signatures to decrease the string lengths by a constant factor at each recursive step, as opposed to the logarithmic factor used in [2]. Also, in the cache-oblivious model, it is not clear when to stop the recursion.

Using shorter strings to represent the essential information about longer ones was proposed already in [20]. Indeed, similar to the randomized signature scheme discussed above, the idea of Karp-Miller-Rosenberg labeling [20] is to divide the strings into substrings, and replace each substring with a shorter string. In particular, for each distinct substring one needs a unique shorter string to represent it. This can even be done such that the lexicographical order is preserved. The technique of [20] avoids randomization, but requires in each recursive step the sorting of the substrings that are to be replaced, which takes $O(\text{Sort}(N))$ and hence will not improve on the known external memory upper bounds. The "practical algorithms" in [7] use this technique, but as stated above, these algorithms are asymptotically inferior to the best algorithm in [7].

2   Proof of Main Theorem

In this section we will prove Theorem 1. First we describe in Section 2.1 a recursive algorithm that finds the structure of the *unordered* trie of strings in the set S that is to be sorted. The algorithm is randomized and produces the correct trie with high probability. Each step of the recursion reduces the total length of the strings by a constant factor using a signature method. Section 2.2 then describes how to get from the unordered trie to the ordered trie, from which the sorting permutation and the lcp sequence can easily be found. The crux of the complexity bound is that at each recursive step, as well as at the final ordering step, only the (at most K) branching characters of the current trie are sorted, and the current set of strings is scanned. The use of randomization (hashing) allows the shorter strings of the next recursive step to be computed in scanning complexity (as opposed to the Karp-Miller-Rosenberg method), but also means that there is no relation between the ordering of the strings from different recursive levels. However, for unordered tries, equality of prefixes is all that matters.

Our algorithm considers the input as a sequence of strings $x_1, \ldots, x_K$ over an alphabet of size $\Theta(N^{2+c})$, by dividing the binary strings into chunks of $\lceil (2 + c) \log N \rceil$ bits, where c is the positive constant of Theorem 1. Note that this means that the total length of all strings is $O(N)$ characters. Dividing into chunks may slightly increase the size of the strings, because the length is effectively rounded up to the next multiple of the chunk size. However, the increase is at most a constant factor. To simplify the description of our algorithm we make sure that S is *prefix free*, i.e., that no string in S is a prefix of another string in S. To ensure this we append to $x_1, \ldots, x_K$ special characters $\$_1, \ldots, \$_K$ that do not appear in the original strings. Extending the alphabet with K new characters may increase the representation size of each character by one bit, which is negligible.

2.1   Signature Reduction

We will now describe a recursive, cache-oblivious algorithm for finding the structure of the *blind trie* of the strings in S, i.e., the trie with all unary nodes removed

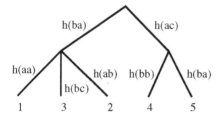 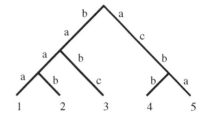

Fig. 1. We consider the set of strings $S = \{$baaa, baab, babc, acbb, acba$\}$. Left is $T(S')$, where each character is a hash function value of two characters from S. Right is $T(S)$, which can be computed from $T(S')$ by considering each branching node in $T(S')$ and its branching characters, each of which corresponds to a unique string of two characters from S.

(blind tries [15] are also known as compressed tries or Patricia tries [22]). Recall that for now, we are only concerned with computing the *unordered* blind trie $T(S)$. We represent each node p as follows (where for brevity, a node is identified with the string represented by the path from the root to the node):

- A unique ID, which is a number in $\{1, \ldots, 2K\}$.
- The ID of its parent node q, which is the longest proper prefix of p in S, if any.
- The number i of a string $x_i \in S$ having p as a prefix (a *representative string*).
- Its *branching character*, i.e., the first character in p after q, and its position in p.

Our algorithm handles strings in S of length 1 separately, in order to be able to *not* recurse on these strings. Because no string is a prefix of another string, the strings of length 1 are leaf children of the root in $T(S)$. Thus we may henceforth assume that S contains only the strings of length $\ell \geq 2$, and add the strings of length 1 to the trie at the end.

The sequence S' is derived from S by hashing pairs of consecutive characters to a single character, using a function chosen at random from a universal family of hash functions [16]. A string of ℓ characters in S will correspond to a string of $\lceil \ell/2 \rceil$ characters in S'. Since all strings have length at least 2, the total length of the strings in S' is at most $\frac{2}{3}N$, as desired. The set S' can be computed in a single scan of S, using $O(N/B)$ I/Os. We denote the strings of S' by $x'_1, \ldots, x'_K$ such that they correspond one by one to the strings $x_1, \ldots, x_K$ of S, in this order. The trie $T(S')$ is computed recursively.

If there are no hash function collisions (i.e., no pair of distinct characters with the same hash function value), the longest common prefix of any two strings $x'_{i_1}, x'_{i_2} \in S'$ is of length $\lfloor |\text{lcp}(x_{i_1}, x_{i_2})|/2 \rfloor$. Intuitively, this means that $T(S')$ has the same structure as $T(S)$, only "coarser". To get $T(S)$ from $T(S')$ we basically need to consider each node and its children, and introduce new branching nodes in this part of the trie by considering the (pairs of) characters of S corresponding to the branching characters. Figure 1 shows an example.

Assuming that the hash function has no collisions, $T(S)$ can be computed from $T(S')$ as follows:

1. Sort the nodes of $T(S')$ according to the numbers of their representative strings in S', and secondly for each representative string according to the position of the branching character. Note that this can be done in a single sorting step (using a cache-oblivious sorting algorithm).

2. By scanning the strings of S in parallel with this sorted list, we can annotate each node p, having representative string x_i, with the two characters $c_{p,1}c_{p,2}$ from x_i that correspond to its branching character (their position can be computed from the position of the branching character in x_i').

3. Sort the annotated nodes of $T(S')$ according to the IDs of their parents, and for each parent ID according to $c_{p,1}$ (using a single sorting step).

4. We can now construct the nodes of $T(S)$ by scanning this sorted list. Consider the children of a node p, occurring together in the list. There are two cases:

   (a) If all children have the same $c_{p,1}$ we can basically copy the structure of $T(S')$. That is, we keep a node for each child p, with the same ID, parent ID, and representative string number as before, and with $c_{p,2}$ as branching character (its position can be computed as above).

   (b) If there are children having different $c_{p,1}$ we introduce a new node for each group of at least two children with the same $c_{p,1}$ (getting new IDs using a global internal memory counter). The branching character for such a node is $c_{p,1}$, the parent ID is that of p, and any representative string of a child can be used. Again, we keep a node for each child p with the same ID as in $T(S')$. If no other child has the same $c_{p,1}$ the node keeps its parent ID, with branching character $c_{p,1}$. If two or more children have the same $c_{p,1}$, their parent is the new node with branching character $c_{p,1}$, and their branching characters are their $c_{p,2}$ characters.

Lemma 1. *The above algorithm uses $O(N/B)$ blocks of external space and $O(\frac{K}{B}\log_{M/B}(\frac{K}{M})\log(\frac{N}{K}) + N/B)$ I/Os. It computes $T(S)$ correctly with probability $1 - O(N^{-c})$,*

Proof sketch. Since the length of the strings is geometrically decreasing during the recursion, the total length of all strings considered in the recursion is $O(N)$. This means that the space usage on external memory is $O(N/B)$ blocks, and that the number of pairs of characters hashed is $O(N)$. In particular, since the collision probability for any pair of inputs to the universal hash function is N^{-2-c}, we have that with probability $1 - O(N^{-c})$ there are no two distinct pairs of characters that map to the same character (i.e., no hash function collisions). In this case, note that all sets of strings in the recursion are prefix free, as assumed by the algorithm. The argument that the trie is correct if there is no hash function collision is based on the fact that the longest common prefixes in S correspond to longest common prefixes of half the length (rounded down) in S'. Details will be given in the full version of this paper.

We now analyze the I/O complexity. At the ith level of the recursion the total length of the strings is bounded by $(\frac{2}{3})^i N$. In particular, for $i > \log_{3/2}(N/K)$

the maximum possible number of strings at each recursive level also starts to decrease geometrically (since the number of strings is bounded by the total length of the strings). Finally note that when the problem size reaches M, the rest of the recursion is completed in $O(M/B)$ I/Os.

Let $j = \lfloor \log_{3/2}(N/K) \rfloor$. We can bound the asymptotic number of I/Os used as follows:

$$\sum_{i=0}^{j} \left(\frac{K}{B} \log_{M/B} \left(\frac{K}{M} \right) + \frac{\left(\frac{2}{3}\right)^i N}{B} \right) + \sum_{i=j+1}^{\infty} \left(\frac{\left(\frac{2}{3}\right)^{i-j} K}{B} \log_{M/B} \left(\frac{\left(\frac{2}{3}\right)^{i-j} K}{M} \right) + \frac{\left(\frac{2}{3}\right)^i N}{B} \right)$$

$$= \left(\frac{K}{B} \log_{M/B} \left(\frac{K}{M} \right) \log \left(\frac{N}{K} \right) + \frac{N}{B} \right) + \left(\frac{K}{B} \log_{M/B} \left(\frac{K}{M} \right) + \frac{N}{B} \right) .$$

The first term is the dominant one, and identical to the bound claimed.   □

2.2   Full Algorithm

We are now ready to describe our string sorting algorithm in its entirety. We start by finding $T(S)$ using the algorithm of Section 2.1. What remains to find the sorting permutation is to order the children of each node according to their branching character and traverse the leaves from left to right. We do this by a reduction to *list ranking*, proceeding as follows:

1. Sort the nodes according to parent ID, and for each parent ID according to branching character, in a single sorting step.
2. We now construct a graph (which is a directed path) having two vertices, v^{in} and v^{out}, for each vertex v of $T(S)$.
   (a) For a node with ID v having d (ordered) children with IDs $v_1, \ldots, v_d$ we construct the edges $(v^{\text{in}}, v_1^{\text{in}}), (v_1^{\text{out}}, v_2^{\text{in}}), \ldots, (v_{d-1}^{\text{out}}, v_d^{\text{in}}), (v_d^{\text{out}}, v^{\text{out}})$. We annotate each "horizontal" edge $(v_i^{\text{out}}, v_{i+1}^{\text{in}})$ by the length of the *lcp* of the representative strings of v_i and v_{i+1} (considered as bit strings). This number can be computed from the branching characters of v_i and v_{i+1} and their positions.
   (b) For a leaf node v we construct the edge $(v^{\text{in}}, v^{\text{out}})$, and annotate it with the number of its representative string.
3. Run the optimal cache-oblivious list ranking algorithm of [5] on the above graph to get the edges in the order they appear on the path.
4. Scan the list, and report the numbers annotated on the edges corresponding to the leaves in $T(S)$ (giving the sorting permutation), and the numbers on the horizontal edges (giving the *lcp* sequence).

The work in this part of the algorithm is dominated by the sorting and list ranking steps, which run in $O(\frac{K}{B} \log_{M/B}(\frac{K}{M}) + N/B)$ I/Os. Again, we postpone the (straightforward) correctness argument to the full version of this paper. This concludes the proof sketch of Theorem 1.

3   Construction of External String Dictionaries

In this section, we prove Corollary 1. A String B-tree [15] is a form of B-tree over pointers to the strings. Each B-tree node contains pointers to $\Theta(B)$ strings, as well as a blind trie over these strings to guide the search through the node. The bottom level of the tree contains pointers (in sorted order) to all strings. These pointers are divided into groups of $\Theta(B)$ consecutive pointers, and a node is built on each group. For each node, the left-most and the right-most pointer are copied to the next level, and these constitute the set of pointers for this next level. Iterating this process defines all levels of the tree.

Building a blind trie on a set of strings is straightforward given the sequence of pointers to the strings in sorted order and the associated lcp sequence: insert the pointers as leaves of the trie in left-to-right order, while maintaining the right-most path of the trie in a stack. The insertion of a new leaf entails popping from the stack until the first node on the right-most path is met which has a string depth at most the length of the longest common prefix of the new leaf and its predecessor. The new leaf can now be inserted, possibly breaking an edge and creating a new internal node. The splitting characters of the two edges of the internal node can be read from the lcp sequence. The new internal node and the leaf is then pushed onto their stack.

For analysis, observe that once a node is popped from the stack, it leaves the right-most path. Hence, each node of the trie is pushed and popped once, for a total of $O(S)$ stack operations, where S is the number of strings represented by the blind trie. Since a stack implemented as an array is I/O-efficient by nature, the number of I/Os for building a blind trie is $O(S/B)$.

Hence, the number of I/Os for building the lowest level of the String B-tree is $O(K/B)$. Finding the pointers in sorted order and their corresponding lcp array for the next level is straightforward, using that the lcp value between any pairs of strings is the minimum of the lcp values for all intervening neighboring (in sorted order) pairs of strings. Hence, the lcp value of the left-most and right-most leaf in the trie in a String B-tree node can be found by scanning the relevant part of the lcp array. As the sizes of each level decreases by a factor $\Theta(B)$, building the entire String B-tree is dominated by the building of its lowest level.

For the cache-oblivious string dictionary [12], it is shown in [12] how to take a blind trie (given as an edge list) for a set of K strings, and build a cache-oblivious search structure for it, using Sort(K) I/Os. The construction algorithm works in the cache-oblivious model. Since an array-based stack is I/O efficient also in the cache-oblivious model, the blind trie can be built cache-obliviously by the method above. The ensuing search structure provides a cache-oblivious dictionary over the strings.

4   Bounds for Permuting Strings

In this section, we discuss methods for permuting the strings into sorted order in memory, based on knowledge of the sorting sequence. Our methods are straightforward, but nevertheless show that also for the standard sorting problem (i.e. including permuting the strings), our main result leads to asymptotical improvements (albeit modest) in complexity.

In the cache-aware case, knowledge of B allows us to follow Arge et al. [7], and divide the strings into short and long strings. We use their terminology K_1, K_2, N_1, N_2 described in Section 1.2. The long strings we permute by direct placement of the strings, based on preprocessing in $O(\text{Sort}(K) + N/B)$ I/Os which calculates the position of each string. The short strings we permute by sorting, using the best algorithm from [7] for short strings. For simplicity of expression, we assume a tall cache. Then the complexity of our randomized sorting algorithm, followed by the permutation procedure above, is

$$O\left(\tfrac{K}{B}\log_M(K)\log(\tfrac{N}{K}) + \tfrac{N}{B} + \tfrac{N_1}{B}\log_M(K_1) + K_2\right).$$

The bound holds for any choice of length threshold between short and long strings. For the threshold equal to B, it is easy to see that the bound improves on the bound of [7] for many parameter sets with long strings. For specific inputs, other thresholds may actually be better.

Turning to the cache-oblivious situation, we can also, on instances with long strings, improve the only existing bound of $O(\text{Sort}(N))$. For simplicity assume a tall cache assumption of $M \geq B^2$. If $N \leq K^2$, we permute by sorting the words of the strings in $O(\text{Sort}(N))$ as usual. Else, we place each string directly (using preprocessing as above). If $M > N$, everything is internal. Otherwise, $N/K \geq \sqrt{N} \geq \sqrt{M} \geq B$, so $N/B \geq K$, and we can afford one random I/O per string placed, leading to permutation in $O(\text{Sort}(K) + N/B)$.

References

1. A. Aggarwal and J. S. Vitter. The Input/Output complexity of sorting and related problems. *Communications of the ACM*, 31(9):1116–1127, 1988.
2. A. Andersson, T. Hagerup, S. Nilsson, and R. Raman. Sorting in linear time? *J. Comput. System Sci.*, 57(1):74–93, 1998.
3. A. Andersson and S. Nilsson. A new efficient radix sort. In *Proceedings of the 35th Annual Symposium on Foundations of Computer Science (FOCS '94)*, pages 714–721. IEEE Comput. Soc. Press, 1994.
4. L. Arge. External memory data structures. In J. Abello, P. M. Pardalos, and M. G. C. Resende, editors, *Handbook of Massive Data Sets*, pages 313–358. Kluwer Academic Publishers, 2002.
5. L. Arge, M. A. Bender, E. D. Demaine, B. Holland-Minkley, and J. I. Munro. Cache-oblivious priority queue and graph algorithm applications. In ACM, editor, *Proceedings of the 34th Annual ACM Symposium on Theory of Computing (STOC '02)*, pages 268–276. ACM Press, 2002.
6. L. Arge, G. S. Brodal, and R. Fagerberg. Cache-oblivious data structures. In D. Mehta and S. Sahni, editors, *Handbook on Data Structures and Applications*. CRC Press, 2005.

7. L. Arge, P. Ferragina, R. Grossi, and J. S. Vitter. On sorting strings in external memory (extended abstract). In ACM, editor, *Proceedings of the 29th Annual ACM Symposium on Theory of Computing (STOC '97)*, pages 540–548. ACM Press, 1997.

8. J. Bentley and R. Sedgewick. Fast algorithms for sorting and searching strings. In *Proc. 8th ACM-SIAM Symposium on Discrete Algorithms (SODA)*, pages 360–369, 1007.

9. G. S. Brodal. Cache-oblivious algorithms and data structures. In *Proc. 9th Scandinavian Workshop on Algorithm Theory*, volume 3111 of *Lecture Notes in Computer Science*, pages 3–13. Springer Verlag, Berlin, 2004.

10. G. S. Brodal and R. Fagerberg. Cache oblivious distribution sweeping. In *Proc. 29th International Colloquium on Automata, Languages, and Programming*, volume 2380 of *Lecture Notes in Computer Science*, pages 426–438. Springer Verlag, Berlin, 2002.

11. G. S. Brodal and R. Fagerberg. On the limits of cache-obliviousness. In *Proc. 35th Annual ACM Symposium on Theory of Computing*, pages 307–315, 2003.

12. G. S. Brodal and R. Fagerberg. Cache-oblivious string dictionaries. In *Proceedings of the 17th Annual ACM-SIAM Symposium on Discrete Algorithms (SODA '06)*, 2006. To appear.

13. E. D. Demaine. Cache-oblivious data structures and algorithms. In *Proc. EFF summer school on massive data sets*, Lecture Notes in Computer Science. Springer, To appear.

14. M. Farach-Colton, P. Ferragina, and S. Muthukrishnan. On the sorting-complexity of suffix tree construction. *J. ACM*, 47(6):987–1011, 2000.

15. P. Ferragina and R. Grossi. The string B-tree: a new data structure for string search in external memory and its applications. *J. ACM*, 46(2):236–280, 1999.

16. M. L. Fredman and D. E. Willard. Trans-dichotomous algorithms for minimum spanning trees and shortest paths. *J. Comput. System Sci.*, 48(3):533–551, 1994.

17. M. Frigo, C. E. Leiserson, H. Prokop, and S. Ramachandran. Cache oblivious algorithms. In *40th Annual IEEE Symposium on Foundations of Computer Science*, pages 285–298. IEEE Computer Society Press, 1999.

18. Y. Han and M. Thorup. Integer sorting in $O(n\sqrt{\log\log n})$ expected time and linear space. In *Proceedings of the 43rd Annual Symposium on Foundations of Computer Science (FOCS '02)*, pages 135–144, 2002.

19. J. Kärkkäinen and P. Sanders. Simple linear work suffix array construction. In *Proc. 30th Int. Colloquium on Automata, Languages and Programming (ICALP)*, volume 2719 of *Lecture Notes in Computer Science*, pages 943–955. Springer Verlag, Berlin, 2003.

20. R. M. Karp, R. E. Miller, and A. L. Rosenberg. Rapid identification of repeated patterns in strings, trees and arrays. In *Proceedings of the 4th Annual ACM Symposium on Theory of Computing (STOC '72)*, pages 125–136, 1972.

21. U. Meyer, P. Sanders, and J. F. Sibeyn, editors. *Algorithms for Memory Hierarchies*, volume 2625 of *Lecture Notes in Computer Science*. Springer Verlag, Berlin, 2003.

22. D. R. Morrison. PATRICIA - practical algorithm to retrieve information coded in alphanumeric. *J. ACM*, 15(4):514–534, Oct. 1968.

23. J. S. Vitter. External memory algorithms and data structures: Dealing with MASSIVE data. *ACM Computing Surveys*, 33(2):209–271, 2001.

24. J. S. Vitter. Geometric and spatial data structures in external memory. In D. Mehta and S. Sahni, editors, *Handbook on Data Structures and Applications*. CRC Press, 2005.

Amortized Rigidness in Dynamic Cartesian Trees

Iwona Bialynicka-Birula and Roberto Grossi

Dipartimento di Informatica, Università di Pisa,
Largo Bruno Pontecorvo 3, 56127 Pisa, Italy
{iwona, grossi}@di.unipi.it

Abstract. Cartesian trees have found numerous applications due to a peculiar rigid structure whose properties can be exploited in various ways. This rigidness, however, is also an obstacle when updating the structure since it can lead to a very unbalanced shape and so up to now most applications either assumed a random distribution of the keys or considered only the static case. In this paper we present a framework for efficiently maintaining a Cartesian tree under insertions and weak deletions in $O(\log n)$ amortized time per operation, using $O(n)$ space. We show that the amortized cost of updating a Cartesian tree is $O(1 + \mathcal{H}(T)/n)$ where $\mathcal{H}(T) = O(n \log n)$ is an entropy-related measure for the partial order encoded by T. We also show how to exploit this property by implementing an algorithm which performs these updates in $O(\log n)$ time per operation. No poly-logarithmic update bounds were previously known.

1 Introduction

Cartesian trees have been introduced 25 years ago by Vuillemin [1]. Due to their unique properties (cf. [2]), they have found numerous applications in priority queue implementations, randomized searching [3], range searching, range maxima queries [2], least common ancestor queries [4], integer sorting, and memory management, to name a few.

A Cartesian tree T is a binary tree storing a set of n points (w.l.o.g. ordered pairs over an unbounded universe with distinct coordinates) according to the following recursive rules: The root stores the point $\langle \bar{x}, \bar{y} \rangle$ with the maximum y-value (priority) in the set. The x-value of the root, $\bar{x}$, induces a partition of the remaining elements into sets $L = \{\langle x, y \rangle : x < \bar{x}\}$ and $R = \{\langle x, y \rangle : x > \bar{x}\}$. The roots of the Cartesian trees obtained from L and R are the left and right children of the root $\langle \bar{x}, \bar{y} \rangle$, respectively (see Figure 1). As follows from this definition, the order on the x-coordinates of the points matches the *inorder* of the nodes in T and the order on the y-coordinates of the points complies with the *heap* order on the nodes in T. The set of points *uniquely* determines the Cartesian tree. This "rigidness" is exploited in applications, but leaves no freedom for balancing operations because the height of the tree can even be $\Theta(n)$ if the distribution of points is skewed.

B. Durand and W. Thomas (Eds.): STACS 2006, LNCS 3884, pp. 80–91, 2006.

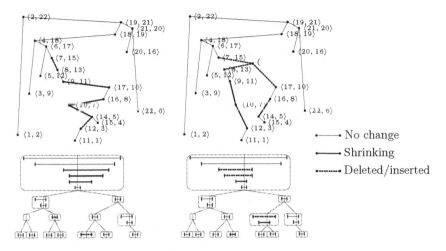

Fig. 1. An example of a Cartesian tree and its companion interval tree before and after the insertion of $\langle 13, 14 \rangle$ with affected edges indicated

Cartesian trees are also known as *treaps* [3] when the priorities (y values) of the nodes are assigned randomly with a uniform distribution and the tree is used as a binary search tree (for the x values). The random distribution of y values guarantees good average dynamic behavior of the tree as the height is expected to stay $O(\log n)$ in this case. It is important to note at this point that the logarithmic expected update time is indeed achieved only under a random uniform distribution (with x values independent from y). In fact, a thorough study of the average behavior of Cartesian trees [5] shows that with most "real life" distributions the expected time is $O(\sqrt{n})$. Figure 2 illustrates the typical worst-case behavior that yields $\Theta(n)$ time for an insertion or a deletion. As far as we know, the amortized complexity of these update operations is currently unknown, except for the fact that a Cartesian tree can be built from scratch in $O(n \log n)$ time, or even in $O(n)$ time when the points are already given in sorted order [2].

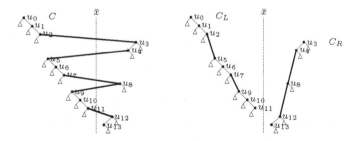

Fig. 2. The outcome of $split(C, \bar{x})$ (or $merge(C_L, C_R)$ if viewed from right to left). Note that the path revealed in the illustration can be proportional in length to the overall tree size in which case an insert or delete operation affects $\Theta(n)$ edges of a tree of size n.

We motivate our study by observing that Cartesian trees are difficult to update and no poly-logarithmic update bounds are currently known. (The alternative solutions proposed, such as priority search trees [6], have weaker topological properties in the Cartesian plane.) Hence, the use of Cartesian trees has been confined to the static case. Along with the least common ancestor (LCA), Cartesian trees are useful for answering the constant-time range maximum queries (RMQ), which are a generalization of priority-queue queries in which the find-min operation is restricted to ranges of values. The implementation of LCA itself in [4] uses Cartesian trees for a reduction from general trees to RMQ, while the dynamic version of LCA [7] does not use Cartesian trees. Hopefully, the efficient dynamization of the Cartesian trees can make a first significant step in finding a suitable array of applications in a dynamic setting (once have it, use it), such as solving the dynamic version of the constant-time RMQ (in logarithmic update time). This can stimulate further research, such as extending the dynamic constant-time LCA to treat cut and link operations among trees (in logarithmic update time).

We study Cartesian trees in a dynamic setting, under insertions and deletions of arbitrary points. Our algorithm provides the actual structure of the Cartesian tree T at all times rather than maintaining some equivalent representation of it. It does not cache the updates so the structure of the tree can be queried between every operation (which makes it significantly more powerful that the $O(n \log n)$ algorithm which first constructs the tree and then allows it to be queried, but not updated). We obtain an amortized update cost of $O(\log n)$ time.

We start by analyzing the behavior of T in a combinatorial setting. We consider the *restructuring cost* of n insertions into an initially empty tree resulting in tree T, that is the total number of structure elements which change due to these insertions. We show that this cost is $O(1 + \mathcal{H}(T))$, where $\mathcal{H}(T) = O(n \log n)$ is an entropy-related measure for the partial order encoded by T. We then show that the cost of locating the elements to update (the *searching cost*) can be reduced to $O(\log n)$ amortized time with the use of a *companion interval tree*, which is based on the interval tree [8] implemented using a weight-balanced B-tree [9]. We take advantage of the special properties of the our problem and the results of our analysis to provide algorithms that match the amortized bounds.

Weak deletions (logically marking nodes as deleted and periodically rebuilding T) can be amortized when coupled with insertions, at the price of having a constant fraction of nodes in T marked as deleted. We can maintain T under insertions and weak deletions in $O(\log n)$ amortized time per operation, using $O(n)$ space. Handling deletions is quite standard and will be explained in detail in the full version of the paper.

1.1 Dynamic Operations on Cartesian Trees

We identify T with the set of points it stores in its nodes and hence write $\langle x, y \rangle \in T$ for a point in T. The main dynamic operations on a Cartesian tree are those of inserting and deleting points. We review these operations as they will be the basis for the analysis in the rest of the paper. It is useful to see them as based on more basic operations, such as splitting and merging Cartesian trees.

Merge and deletion operations can be viewed as the exact opposites of the split and insert operations, so we omit their description.

Split. We define $split(C, \bar{x})$ for a Cartesian tree C and a value $\bar{x}$ as the operation returning $C_L = \{\langle x, y \rangle \in C : x < \bar{x}\}$ and $C_R = \{\langle x, y \rangle \in C : x > \bar{x}\}$. Note that the vertical line at $\bar{x}$ cuts some edges in C (see Figure 2). These edges are each one below the other. They are nested on the x axis and their inclinations are alternating as they become smaller (by inclination we mean rising or falling along the y-axis). The nesting follows directly from the definition of the Cartesian tree. The alternating inclinations are due to the fact all the edges are part of one path and nodes along any path must have a decreasing y value (due to the heap condition).

We classify the edges in C according to their role in the split operation, as this will be important to our analysis. We assume that C_L and C_R are not empty, otherwise the case is trivial. First of all, edges not crossed by the vertical line at $\bar{x}$ are not affected by the split. The lowest edge affected by the split can be considered deleted, since it does not have a counterpart in either C_L or C_R (note that C_L together with C_R have one less edge than C). All the remaining edges in C can be mapped into edges in C_L or C_R so we can view them as altered by the split. The higher endpoint (with greater y value) of each such edge remains the same and only the lower endpoint changes. If the higher endpoint belongs to C_L then the lower endpoint moves from C_R to C_L and vice versa. Note that such a change always shrinks the projection of an edge on the x-axis. We will therefore refer to this type of change as *shrinking* an edge and call such an edge *shrunk*. Note that all the affected edges were nested before the split, but after the split each two overlap by at most one endpoint.

Insertion. Let us consider $T' = T \cup \{\langle \bar{x}, \bar{y} \rangle\}$, where T is a Cartesian tree and $\langle \bar{x}, \bar{y} \rangle$ is a new point to insert into T. If $\bar{y} > y$ for each $\langle x, y \rangle \in T$, then the new point becomes the root of T' and its two children are T_L and T_R, respectively, obtained by invoking $split(T, \bar{x})$. In all other cases $\langle \bar{x}, \bar{y} \rangle$ has a parent in T'. By definition, this parent can be located in the following way. Let $\langle x_L, y_L \rangle \in T$ be the rightmost point such that $x_L < \bar{x}$ and $y_L > \bar{y}$ and let $\langle x_R, y_R \rangle \in T$ be the leftmost point such that $x_R > \bar{x}$ and $y_R > \bar{y}$ (at least one of these two points must exist). The parent of $\langle \bar{x}, \bar{y} \rangle$ in T' is the lower of these two points. Let C be the right subtree of $\langle x_L, y_L \rangle$ in T in the case $\langle x_L, y_L \rangle$ is the parent of $\langle \bar{x}, \bar{y} \rangle$ in T and let it be the left subtree of $\langle x_R, y_R \rangle$ in the other case. Then C_L and C_R obtained by invoking $split(C, \bar{x})$ become the children of $\langle \bar{x}, \bar{y} \rangle$ in T'.

Fact 1. *The insertion (deletion) of a point causes $O(1)$ edges in the tree to be inserted or deleted and causes k edges to shrink (stretch), where $0 \leq k \leq n$.*

2   Bounding the Number of Edge Modifications

We now focus on the number of edge modifications as expressed by Fact 1. In this section, we consider insertions only, so we are interested in bounding the number of inserted, deleted and shrunk edges for a sequence of n insertions of points into an initially empty Cartesian tree. Since the number of shrunk edges for each insertion is $k = O(n)$, it may appear that n insertions cause $\Theta(n^2)$ such edge modifications. However, we make the following observation which will eventually lead to a bound which is lower: Inserting a point into a Cartesian tree does not require performing comparisons on the y-coordinates of the nodes in the Cartesian tree (except pertaining to the node being inserted). On the other hand, reversing this operation by deleting the same node *does* require a number of comparisons proportional to the number of affected edges. This suggests that information is *lost* as a result of an insertion and that entropy can serve as a measure of a tree's potential for costly (affecting many edges) insertions.

We now formalize this intuition. A Cartesian tree T induces a partial order on its elements: $\langle x, y \rangle \prec_T \langle x', y' \rangle$ if and only if $\langle x, y \rangle$ is a descendent of $\langle x', y' \rangle$. The intuition behind this definition is that if $\langle x, y \rangle$ descends from $\langle x', y' \rangle$ then we *know* that $y < y'$ from the heap condition of the tree. In all other cases we can not guess the relative order of y and y' *just* by looking at the position of these two nodes in the tree. Note that even if y values are drawn from a total order, the order $\prec_T$ is only partial[1], so we can use any y-ordering of points satisfying $\prec_T$ without violating the heap condition of the tree.

A *linear extension* of a partially ordered set $\langle P, \prec_P \rangle$ is a permutation of its elements, $p_1, p_2, \ldots, p_n$, such that if $p_i \prec_P p_j$ is defined, then $1 \le i < j \le n$. We will say that an ordering is *valid* for T if it is a linear extension of $\prec_T$. Let $\mathscr{L}(T)$ denote the number of linear extensions of the partial order $\prec_T$ induced by T. We introduce the notion of the *missing entropy* of T as $\mathscr{H}(T) = \log \mathscr{L}(T)$, which is the information needed to sort the set of y values in T starting only from the information inferred from the shape of the tree.[2] Since the number of linear extensions cannot exceed the number of permutations, we trivially have $\mathscr{H}(T) \le \log n! = O(n \log n)$ from order entropy and Stirling's approximation. However, $\mathscr{H}(T)$ can even be zero if the partial order is a total order (this occurs when the Cartesian tree is a single path). We exploit the notion of missing entropy for bounding the number of shrunk edges.

Theorem 1. *Inserting n points into an initially empty Cartesian tree results in $O(n)$ edge insertions and deletions and $O(n + \mathscr{H}(T))$ edges shrinking, where T is the resulting Cartesian tree.*

In order to demonstrate Theorem 1, we use the missing entropy of the Cartesian tree as the potential function in our amortized analysis. We measure the change in the number of linear extensions of T after an insertion $T' = T \cup \{\langle \bar{x}, \bar{y} \rangle\}$. This change can be measured by the ratio $\mathscr{L}(T')/\mathscr{L}(T)$, but it is more convenient to consider its logarithm, which is the change in our potential, $\mathscr{H}(T') - \mathscr{H}(T)$. We consider an insertion $T' = T \cup \{\langle \bar{x}, \bar{y} \rangle\}$ that splits a subtree C of the current Cartesian tree T into C_L and C_R as discussed in Section 1.1. By Fact 1, this operation results in $O(1)$ inserted and deleted edges in T, plus k shrunk edges. We claim that

---

[1] Unlike the y-ordering, the x-ordering induced by the shape of the tree is always total.
[2] All logarithms are to the base of 2.

$$k = O(\mathcal{H}(T') - \mathcal{H}(T)). \tag{1}$$

Equation (1) holds asymptotically and its proof follows from Lemma 2 below: Let $s = \lfloor (k+1)/2 \rfloor$, and take the logarithms obtaining $\log \mathcal{L}(T) + \log \binom{k+1}{s} \leq \log \mathcal{L}(T')$. Since $k = O(\log \binom{k+1}{s}) = O(\log \mathcal{L}(T') - \log \mathcal{L}(T))$, we obtain the claimed bound in (1).

Lemma 1. *Every linear extension of T is also a linear extension of T'.*

Proof. Linear extensions of T do not contain the newly inserted point $\langle \bar{x}, \bar{y} \rangle$, so we only consider relations between points of T. We need to show that for any two nodes $a, d \in T$, we have $d \prec_{T'} a \Rightarrow d \prec_T a$. This can be easily shown by considering all possible locations of a and d with respect to the subtrees C_L and C_R resulting from splitting C.

Lemma 2. *For each linear extension of T, there are at least $\binom{k+1}{s}$ unique linear extensions of T', where $s = \lfloor k + 1 \rfloor / 2 \rfloor$. Hence, $\mathcal{L}(T) \times \binom{k+1}{s} \leq \mathcal{L}(T')$.*

Proof. If $k = 0$ then the proof follows directly from Lemma 1, so we will consider the case where $k > 0$. In this case C_L and C_R are not empty and the split of C causes k edges to shrink. Let us first assume that these k edges form a path whose endpoints alternate between C_L and C_R, where $k = 2s - 1$ for an integer $s > 0$. Let $u_1^L, u_2^L, \ldots, u_s^L$ be the endpoints in C_L, and $u_1^R, u_2^R, \ldots, u_s^R$ be the endpoints in C_R. Without loss of generality, suppose that the path of shrunk edges inside T is $p = \langle u_1^L, u_1^R, u_2^L, u_2^R, \ldots, u_s^L, u_s^R \rangle$. After the split, p becomes two paths $p_L = \langle u_1^L, u_2^L, \ldots, u_s^L \rangle$ and $p_R = \langle u_1^R, u_2^R, \ldots, u_s^R \rangle$ in T'.

Now let us consider any linear extension $\mathcal{L}'$ of $\prec_T$. Notice that $(u' < u'') \in \mathcal{L}'$ for any pair of nodes $u', u'' \in p$ such that u' precedes u'' in p. In particular, $\mathcal{L}'$ reflects the fact that for any pair of nodes $u_i^L \in p_L$ and $u_j^R \in p_R$, either $u_i^L \prec_T u_j^R$ or $u_j^R \prec_T u_i^L$ holds. Now, $\mathcal{L}'$ is also an extension of $\prec_{T'}$ by Lemma 1. However, neither $u_i^L \prec_{T'} u_j^R$ nor $u_j^R \prec_{T'} u_i^L$ holds for $u_i^L \in p_L$ and $u_j^R \in p_R$. We use this property for producing further unique extensions from $\mathcal{L}'$, since $\mathcal{L}'$ enforces just one of the $\binom{k+1}{s}$ distinct ways of merging the nodes in p_L and p_R. Note that all these orderings are valid for T' since they only change the relative order of the nodes in p_L and p_R (for which T' does not define any order). In other words, for each linear extension $\mathcal{L}'$ of $\prec_T$ we can produce $\binom{k+1}{s}$ distinct valid extensions of $\prec_{T'}$ by shuffling the order of elements belonging to p while maintaining the relative order of elements in p_L and the relative order of elements in p_R.

The case in which the path is of length $k = 2s$ for an integer $s > 0$ can be treated analogously by decomposing it into two paths of s and $s + 1$ nodes, respectively. In the general case, the shrunk edges are interspersed along a path of T that is longer than p (e.g. Figure 2). In this case, the number of endpoints is even larger and so is the number of distinct extensions, since there are more distinct ways to merge these endpoints in T'. Hence, $\mathcal{L}(T') \geq \mathcal{L}(T) \times \binom{k+1}{s}$ holds in the general case.

We now complete the proof of Theorem 1. Consider an arbitrary sequence of n insertions into an initially empty Cartesian tree, denoted T_0, where $|T_0| = 0$.

Let $T_1, T_2, \ldots, T_n$ denote the sequence of resulting Cartesian trees, where T_i is formed from T_{i-1} by the ith insert operation in the sequence, which shrinks k_i edges by Fact 1, for $1 \leq i \leq n$. Summing up the numbers of all the shrunk edges, we split the total sum according to the constant k_0 related to equation (1), as

$$\sum_{i=1}^{n} k_i = \sum_{i:k_i \leq k_0} k_i + \sum_{i:k_i > k_0} k_i. \tag{2}$$

We denote the indexes i such that $k_i > k_0$ in (2) by $i_1, i_2, \cdots, i_r$, where $i_1 < i_2 < \cdots < i_r$. Note that $k_{i_j} = O(\mathscr{H}(T_{i_j}) - \mathscr{H}(T_{i_j-1})) = O(\mathscr{H}(T_{i_j}) - \mathscr{H}(T_{i_{j-1}}))$ by equation (1) and Lemma 1, for $1 \leq j \leq r$. Applying these observations to the last term in (2) where $i = i_1, i_2, \ldots, i_r$, we obtain the following bound for (2):

$$\sum_{i:k_i > k_0} k_i = O\left(\sum_{j=1}^{r} \left(\mathscr{H}(T_{i_j}) - \mathscr{H}(T_{i_{j-1}})\right)\right) = O(\mathscr{H}(T_n)). \tag{3}$$

Hence, we obtain a total bound of $O(n + \mathscr{H}(T_n))$ for equation (2). Consequently, for each insertion of a point in the Cartesian tree, there are $O(1)$ inserted and deleted edges and an amortized number of $O(1 + \mathscr{H}(T_n)/n)$ shrunk edges. Since $\mathscr{H}(T_n) = O(n \log n)$, we have $\mathscr{H}(T_n)/n = O(\log n)$, but it can also be significantly lower if the missing entropy of T_n is small.

3   Implementing the Insertions

In this section we show how to exploit the amortized upper bound on the number of edge modifications obtained in Section 2 to achieve the efficient updating of a Cartesian tree. Let T denote the current Cartesian tree and let k denote the number of edges that are shrunk during the insertion $T' = T \cup \{\langle \bar{x}, \bar{y} \rangle\}$ (see Fact 1). We recall that the cost of updating the elements of the Cartesian tree modified by such an insertion is $O(1 + k)$ and that this can be amortized to $O(1 + \mathscr{H}(T)/n)$ (Theorem 1). The searching cost (the cost to locate the elements to update), on the other hand, does not amortize as is, since the tree can have linear height. In order to deal with this, we reduce the maintenance of the edges of T to a *special* instance of the dynamic maintenance of intervals. We do this by mapping each edge $e = (\langle x, y \rangle, \langle x', y' \rangle)$ of T into its *companion* interval, (x, x') (where (x, x') denotes the set of coordinates $\hat{x}$ such that $x < \hat{x} < x'$), and storing the companion intervals in an interval tree [8].

Searching, insertion, deletion and shrinking of T's edges can be rephrased in terms of equivalent operations on their companion intervals (see Figure 1). In the rest of this section, we will exploit the special properties of the companion intervals which result from the fact that they are not arbitrary but are derived from the Cartesian tree. We are able to reduce the searching cost to $O(\log n + k)$ time by a constrained stabbing query on the intervals. We obtain a restructuring cost of $O(\log n + k)$ amortized time by performing the $O(1)$ insertions and the deletions in $O(\log n)$ time and each edge shrink operation in $O(1)$ amortized time on the corresponding intervals. Note that the restructuring cost for the Cartesian tree alone is still $O(1 + k)$. The rest is for the maintenance of the companion intervals, which also need to be updated.

3.1 The Companion Interval Tree

Our *companion interval tree*, W, is implemented using Arge and Vitter's weight-balanced B-tree [9]. We actually need a simpler version, without the leaf parameter. Let the weight $w(u)$ of a node u be the number of its descendent leaves. A weight-balanced B-tree W with branching parameter $a > 4$ satisfies the following constraints: (1) All the leaves have the same depth and are on level 0. (2) An internal node u on level ℓ has weight $(1/2)a^\ell < w(u) < 2a^\ell$. (3) The root has at least two children and weight less than $2a^h$, where h is its level. We fix $a = O(1)$ in our application, so W has height $h = O(\log_a |W|) = O(\log n)$ and each node (except maybe for the root) has between $a/4$ and $4a$ children. We denote the number of children of u by $\deg(u)$. The n leaves of W store elements in sorted order, one element per leaf. Each internal node u contains $d = \deg(u) - 1$ boundaries $b_1, b_2, \ldots, b_d$ chosen from the elements stored in its descendent leaves. In particular, the first child leads to all elements $e \leq b_1$, and the last child to all elements $e > b_d$, while for $2 \leq i \leq d$, the ith child contains all elements $b_{i-1} < e \leq b_i$ in its descendent leaves. Among others, W satisfies an interesting property:

Lemma 3 (Arge and Vitter [9]). *After splitting a node u on level ℓ into two nodes, u' and u'', at least $a^\ell/2$ insertions have to be performed below u' (or u'') before it splits again. After creating a new root in a tree with n elements, at least $3n$ insertions are performed before it splits again.*

A weight-balanced B-tree W with n elements supports leaf insertions and deletions in $O(\log n)$ time per operation. Each operation only involves the nodes on the path from the leaf to the root and their children. We do not need to remove amortization and to split nodes lazily in W, since our bounds are amortized anyway. We refer the reader to [9] for more details on weight-balanced B-trees.

Let $I(T)$ denote the set of companion intervals for the current Cartesian tree T. The leaves of W store the endpoints of the intervals in $I(T)$. We store companion intervals in the nodes of the tree according to standard interval tree rules. Specifically, each node u contains d secondary lists, $L_1(u)$, $L_2(u)$, $\ldots$, $L_d(u)$, where $d = \deg(u) - 1$. For $1 \leq i \leq d$, list $L_i(u)$ is associated with the boundary b_i and stores all intervals $(x, x') \in I$ that contain b_i (i.e., $x < b_i < x'$), but are not stored in an ancestor of u. Since any two intervals in $I(T)$ are either disjoint or one nested within the other (see Fact 2), every internal node $u \in W$ stores a number of intervals that is bounded by $O(w(u))$, which is crucial to amortize the costs by Lemma 3. Note that the same interval can be stored in up to d secondary lists of the same node, but not in different nodes, hence, the space occupancy remains linear. We keep these $O(1)$ copies of each interval in a thread. We can derive the following fact from the properties of the edges in the Cartesian tree.

Fact 2. *For any two intervals in the companion tree, if two intervals overlap, they are nested. Moreover, the edge corresponding to the larger interval is above the edge corresponding to the smaller interval in the Cartesian tree. In particular, for any node $u \in W$, the intervals in $L_i(u) \subseteq I(T)$ are all nested within each other. The order according to which they are nested corresponds to the order of their left endpoints, to the (reverse) order of their right endpoints and also to the vertical y order of their corresponding edges in T.*

We maintain each list $L_i(u)$ of intervals sorted according to the order stated in Fact 2. Each list supports the following operations, where $n_i = |L_i(u)|$ is the number of items in it:

- Insert the smallest item into $L_i(u)$ in $O(1)$, and any item in $O(\log n_i)$ time.
- Delete any item from $L_i(u)$ in $O(1)$ time, provided that we have a pointer to its location in $L_i(u)$.
- Perform a (one-dimensional) range query reporting the f items (in sorted order) between two values in $O(\log n_i + f)$ time. In the case of listing the first items, this takes $O(1 + f)$ time.
- Rebuild $L_i(u)$ from scratch in $O(n_i)$ time, provided that items are given in sorted order.

We implement $L_i(u)$ using a balanced search tree with constant update time [10], in which we maintain a thread of the items linked in sorted order.[3] It is worth giving some detail on how to maintain the secondary lists when a node $u \in W$ splits into u' and u'' (see Lemma 3). Let b_i be the median boundary in u. Node u' gets boundaries $b_1, \ldots, b_{i-1}$ while u'' gets $b_{i+1}, \ldots, b_d$, along with their secondary lists and child pointers. The boundary b_i is inserted into u's parent. Note that no interval in $L_i(u)$ can belong to a secondary list in u's parent by definition. What remains is to use the threads of the copies of the intervals in $L_i(u)$ for removing these copies from secondary lists in u' and u''. But this takes $O(n_i)$ time, which is $O(1)$ amortized by Lemma 3.

We will now see how to use the companion interval tree W to implement the insertion of a new point into a Cartesian tree yielding $T' = T \cup \{\langle \bar{x}, \bar{y} \rangle\}$. As it should be clear at this point, we maintain both T and its auxiliary W. Following the insertion scheme described in Section 1.1, we should perform the following *actions*:

1. Find the node $\langle \hat{x}, \hat{y} \rangle$ in T that will become the parent of $\langle \bar{x}, \bar{y} \rangle$ in T'.[4]
2. Find the edges to shrink in T (and the one to delete) as a result of the split, which is part of the insert operation.
3. For each of the $O(1)$ edges inserted or removed in the Cartesian tree, insert or remove its companion interval from W.

---

[3] Note that we do not need to use the finger search functionality in [10], which requires non-constant update time, as we can easily keep the minimum dynamically. In practice, we can implement $L_i(u)$ as a skip list.

[4] This action only needs to be performed if the new element does not become the root of T' (something that can be easily verified).

4. For each companion interval of the k edges identified in action 2, perform the appropriate shrink of an interval in W.

Action 3 is a standard operation that takes $O(\log n)$ time in W, so in the rest of this section we focus on the remaining.[5] We first show how to find the edges to shrink (action 2).

Fact 3. *The edges affected by the split caused by the insertion of $\langle \bar{x}, \bar{y} \rangle$ into T are the edges in T which cross the vertical line at $\bar{x}$ below $\bar{y}$. Their companion intervals can be identified by a stabbing query $\bar{x}$ with the additional constraint that the y values of the corresponding edges are below $\bar{y}$.*

Suppose that the search for $\bar{x}$ in W from the root visits $u_h, u_{h-1}, \ldots, u_1, u_0$, where $u_\ell \in W$ is the node on level ℓ. Let S denote the set of the k intervals to be identified by the constrained stabbing query. If for some $1 \leq j \leq h$, a list $L_i(u_j)$ in the node u_j contains intervals crossing the line $\bar{x}$, but not contained in S due to the constraint on the y values, then no list $L_{i'}(u_\ell)$ in node u_ℓ above node u_j ($\ell > j$) can contain intervals in S. Moreover, $L_i(u_j) \cap S$ form a contiguous range within $L_i(u_j)$.

As a result of Fact 3, the implementation of action 2 is reduced to the efficient implementation of the constrained stabbing query. Let us recall that a regular, unconstrained, stabbing query traverses a path from the root to the leaf according to the query value $\bar{x}$ and the intervals reported from each considered list always form a contiguous range at the beginning of the list, which allows their efficient extraction. Due to Fact 3 we are able to perform the constrained version of the stabbing query while maintaining the $O(\log n + k)$ time complexity of the operation. The constrained query is performed bottom-up and at each node it checks if the edge corresponding to the first interval on the list satisfies the constraint of being below $\bar{y}$ ($O(1)$ time). This allows us to locate node u_j from Fact 3. The query behaves like a an unconstrained stabbing query before reaching u_j and terminates afterwards. For u_j itself it performs a one-dimensional range query on $L_i(u_j)$ which takes $O(\log |L_i(u_j)| + f)$ time.

We are left with the problem of locating the node $\langle \hat{x}, \hat{y} \rangle$ in T which will become the parent of the new node $\langle \bar{x}, \bar{y} \rangle$ in T' (action 1).

Fact 4. *Let $e = (\langle x_e, y_e \rangle, \langle x'_e, y'_e \rangle)$ be the lowest edge such that $x_e < \bar{x} < x'_e$ and $y_e, y'_e > \bar{y}$. If e exists, then $\langle \hat{x}, \hat{y} \rangle$ is located either on the right downward path which starts from point $\langle x_e, y_e \rangle$ if $y_e < y'_e$, or on the left downward path which starts from point $\langle x'_e, y'_e \rangle$ if $y'_e < y_e$. If e does not exist then $\langle \hat{x}, \hat{y} \rangle$ is located on either the left or right downward path which starts from the root. In all cases, $\hat{y}$ is the minimum satisfying the condition $\hat{y} > \bar{y}$.*

---

[5] Note that a shrinking operation (action 4) cannot be implemented in the trivial way by a deletion and an insertion, since this would give rise to an extra factor of $O(\log n)$, hence $O(\mathscr{H}(T)/n \times \log n) = O(\log^2 n)$ per each point inserted into T and this we want to avoid, so we have to provide an implementation which exploits the special properties of the shrinking phenomenon.

Fact 4 follows from the insertion procedure described in Section 1.1. We exploit it to locate the sought parent node. We defer the details of the procedure to the full version of the paper. The main observation behind it is that due to Fact 4 we can limit the search in the Cartesian tree to a monotonous (along both axis) path of disjoint intervals. Therefore, the search in the companion tree can be performed along a single path and in such a way that only the first intervals in the lists at each node are considered. That way action 1 can be performed in $O(\log n)$ total time.

We are left with the implementation of action 4. We can observe that the shrinking of edges involves some reattachments at their endpoints, while the endpoints themselves stay fixed. Reconnecting the Cartesian tree in $O(1 + k)$ time having located the appropriate edges is just an implementation detail, so we focus on maintaining the companion intervals in W under multiple shrinking operations. We need the crucial properties below to perform this task efficiently.

Fact 5. *Let $(x, x') \in I(T)$ be an interval which shrinks and becomes (x, x''), where $x < x'' < x'$.[6] Let $u \in W$ be the node whose secondary list(s) contain(s) interval (x, x'). The shrinking of (x, x') does not introduce any new endpoints to be stored in W (a leaf of W already stores x''). Moreover, if (x, x'') is relocated to another node, v, then v is a descendent of u and (x, x'') becomes the first interval in the suitable secondary list(s) of v.*

Fact 5 guarantees that no restructuring of the tree shape of W is needed because of a shrinking, so we only need to relocate (x, x') into $O(1)$ secondary lists of W as it shrinks. It also guarantees that this relocation moves the interval downward and requires just $O(1)$ time per node. Consequently, the relocating algorithm is as follows. We consider each node u_ℓ on the path $u_h, u_{h-1}, \ldots, u_1, u_0$ identifying the edges to shrink. For each shrinking interval we execute the following steps:

1. Let (x, x') be the interval shrinking.[6] For each secondary list $L_i(u_\ell)$ that contains (x, x'), let b_i be the boundary associated with the list: if $x'' < b_i$, remove the interval from the list; otherwise leave the intervals as is.
2. If at least one copy of (x, x'') remains in the secondary lists of u_ℓ, stop processing the interval.
3. Otherwise, find the descendent v of u_ℓ, which is the new location of (x, x''). Insert (x, x'') into the secondary lists of v as needed, and create a thread of these copies.

The correctness of the method follows from Fact 5. As for the complexity, each relocation may cost $O(\log n)$ time, but we can show that the amortized cost of the relocations in W is in fact $O(n \log n)$ due to the fact that the intervals can only be relocated downward in the companion tree and so the total journey of each interval is bound by the height of the tree. This amortized analysis is tricky, because we must also consider that nodes of W may split, so the formal

---

[6] Notice that one endpoint always remains unchanged due to shrinking. Here we assume without loss of generality that it is the left endpoint. An analogous result holds when the right endpoint remains unchanged and the shrunk interval is (x'', x').

argument will be detailed in the full version. It is based on two types of credits for each interval e. The interval e is assigned $O(h)$ *downward* credits when it is first inserted into W, where h is the current height of W. Another type, called *floating* credits, are assigned to e by the split operations occurring in W according to Lemma 3. Initially, e has zero floating credits. We maintain the invariant that at any given time the interval e has at least $\ell - \ell_{min}$ floating credits and ℓ_{min} downward credits, where ℓ is the level of the node u_ℓ currently containing e in some of its secondary lists and ℓ_{min} is the deepest (minimum) level reached by e during the lifetime of W ($\ell \geq \ell_{min}$). The intuition behind maintaining this invariant is that if a split pushes e one level up, then it assigns e one more floating credit for relocating e one level down again in the future. Downward credits cannot be recharged but they are sufficient to reach new levels $\ell' < \ell_{min}$ below the ones reached before in W by e. Hence, at any time e has at least ℓ total credits needed to reach the leaves of W.

Theorem 2. *Given a Cartesian tree T, we can maintain it under a sequence of n insertions of points using a modified weight balanced B-tree W as an auxiliary data structure, with an amortized cost of $O(\log n)$ time per insertion. The amortized cost of restructuring T is $O(1 + \mathcal{H}(T)/n)$ per insertion, where $\mathcal{H}(T) = O(n \log n)$ is the missing entropy of T.*

Acknowledgments. The authors wish to thank Marco Pellegrini for some interesting discussions on Cartesian trees and for pointing out reference [5].

References

1. Vuillemin, J.: A unifying look at data structures. Comm. ACM **23** (1980) 229–239
2. Gabow, H.N., Bentley, J.L., Tarjan, R.E.: Scaling and related techniques for geometry problems. In: STOC '84. (1984) 135–143
3. Seidel, R., Aragon, C.R.: Randomized search trees. Algorithmica **16** (1996) 464–497
4. Bender, M.A., Farach-Colton, M.: The LCA problem revisited. In: Latin American Theoretical Informatics '00. (2000) 88–94
5. Devroye, L.: On random Cartesian trees. Random Struct. Algorithms **5** (1994) 305–328
6. McCreight, E.M.: Priority search trees. SIAM J. Comput. **14** (1985) 257–276
7. Cole, R., Hariharan, R.: Dynamic LCA queries on trees. SIAM J. Comput. **34** (2005) 894–923
8. Edelsbrunner, H.: A new approach to rectangle intersections, part I. International Journal Computer Mathematics **13** (1983) 209–219
9. Arge, L., Vitter, J.S.: Optimal external memory interval management. SIAM J. Comput. **32** (2003) 1488–1508
10. Lagogiannis, G., Makris, C., Panagis, Y., Sioutas, S., Tsichlas, K.: New dynamic balanced search trees with worst-case constant update time. J. Autom. Lang. Comb. **8** (2003) 607–632

Distribution-Sensitive Construction of Minimum-Redundancy Prefix Codes

Ahmed Belal and Amr Elmasry

Department of Computer Engineering and Informatics,
Beirut Arab University, Lebanon*
{abelal, elmasry}@alexeng.edu.eg

Abstract. A new method for constructing minimum-redundancy prefix codes is described. This method does not build a Huffman tree; instead it uses a property of optimal codes to find the codeword length of each weight. The running time of the algorithm is shown to be $O(nk)$, where n is the number of weights and k is the number of different codeword lengths. When the given sequence of weights is already sorted, it is shown that the codes can be constructed using $O(\log^{2k-1} n)$ comparisons, which is sub-linear if the value of k is small.

1 Introduction

Minimum-redundancy coding plays an important role in data compression applications [9]. Minimum-redundancy prefix codes give the best possible compression of a finite text when we use one static code for each symbol of the alphabet. This encoding is extensively used in various fields of computer science, such as picture compression, data transmission, etc. Therefore, the methods used for calculating sets of minimum-redundancy prefix codes that correspond to sets of input symbol weights are of great interest [1,4,6].

The minimum-redundancy prefix code problem is to determine, for a given list $W = [w_1, \ldots, w_n]$ of n positive symbol weights, a list $L = [l_1, \ldots, l_n]$ of n corresponding integer codeword lengths such that $\sum_{i=1}^{n} 2^{-l_i} = 1$ (Kraft equality), and $\sum_{i=1}^{n} w_i l_i$ is minimized. Once we have the codeword lengths corresponding to a given list of weights, constructing a corresponding prefix code can be easily done in linear time using standard techniques.

Finding a minimum-redundancy code for $W = [w_1, \ldots, w_n]$ is equivalent to finding a binary tree with minimum-weight external path length $\sum_{i=1}^{n} w(x_i) l(x_i)$ among all binary trees with leaves $x_1, \ldots, x_n$, where $w(x_i) = w_i$ and $l(x_i) = l_i$ is the level of x_i in the corresponding tree. Hence, if we define a leaf as a weighted node, the minimum-redundancy prefix code problem can be defined as the problem of constructing an optimal binary tree for a given list of leaves.

Based on a greedy approach, Huffman algorithm [3] constructs specific optimal trees, which are referred to as Huffman trees. Huffman algorithm starts with a list $\mathcal{H}$ containing n leaves. In the general step, the algorithm selects the two

---

* The two authors are on sabbatical from Alexandria University of Egypt.

B. Durand and W. Thomas (Eds.): STACS 2006, LNCS 3884, pp. 92–103, 2006.

nodes with the smallest weights in the current list of nodes $\mathcal{H}$ and removes them from the list. Next, the removed nodes become children of a new internal node, which is inserted in $\mathcal{H}$. To this internal node is assigned a weight that is equal to the sum of the weights of its children. The general step repeats until there is only one node in $\mathcal{H}$, the root of the Huffman tree. The internal nodes of a Huffman tree are thereby assigned values throughout the algorithm. The value of an internal node is the sum of the weights of the leaves of its subtree. Huffman algorithm requires $O(n \log n)$ time and linear space. Van Leeuwen [8] showed that the time complexity of Huffman algorithm can be reduced to $O(n)$ if the input list is already sorted.

Throughout the paper, we exchange the use of the terms leaves and weights. When mentioning a node of a tree, we mean that it is either a leaf or an internal node. The levels of the tree that are further from the root are considered higher; the root has level 0. We use the symbol k as the number of different codeword lengths, i.e. k is the number of levels that have leaves in the corresponding tree.

A distribution-sensitive algorithm is an algorithm whose running time relies on how the distribution of the input affects the output [5,7]. In this paper, we give a distribution-sensitive algorithm for constructing minimum-redundancy prefix codes. Our algorithm runs in $O(nk)$, achieving a better bound than the $O(n \log n)$ bound of the other known algorithms when $k = o(\log n)$.

The paper is organized as follows. In the next section, we give a property of optimal trees corresponding to prefix codes, on which our construction algorithm relies. In Section 3, we give the basic algorithm and prove its correctness. We show in Section 4 how to implement the basic algorithm to ensure the distribution-sensitive behavior; the bound on the running time we achieve in this section is exponential with respect to k. In Section 5, we improve our algorithm, using a technique that is similar in flavor to dynamic programming, to achieve the $O(nk)$ bound. We conclude the paper in Section 6.

2   The Exclusion Property

Consider a binary tree T^* that corresponds to a list of n weights $[w_1, \ldots, w_n]$ and has the following properties:

1. The n leaves of T^* correspond to the given n weights.
2. The value of a node equals the sum of the weights of the leaves of its subtree.
3. For every level of T^*, let $\tau_1, \tau_2, \ldots$ be the nodes of that level in non-decreasing order with respect to their values, then τ_{2p-1} and τ_{2p} are siblings for all $p \geq 1$.

We define the *exclusion property* for T^* as follows: T^* has the exclusion property if and only if the values of the nodes at level j are not smaller than the values of the nodes at level $j + 1$.

Lemma 1. *Given a prefix code whose corresponding tree T^* has the aforementioned properties, the given prefix code is optimal and T^* is a Huffman tree if and only if T^* has the exclusion property.*

Proof. First, assume that T^* does not have the exclusion property. It follows that there exists two nodes x and y at levels j_1 and j_2 such that $j_1 < j_2$ and $value(x) < value(y)$. Swapping the subtree of x with the subtree of y results in another tree with a smaller external path length and a different list of levels, implying that the given prefix code is not optimal.

Next, assume that T^* has the exclusion property. Let $[x_1, \ldots, x_n]$ be the list of leaves of T^*, with $w(x_i) \leq w(x_{i+1})$. We prove by induction on the number of leaves n that T^* is an optimal binary tree that corresponds to an optimal prefix code. The base case follows trivially when $n = 2$. As a result of the exclusion property, the two leaves x_1, x_2 must be at the highest level of T^*. Also, Property 3 of T^* implies that these two leaves are siblings. Alternatively, there is an optimal binary tree with leaves $[x_1, \ldots, x_n]$, where the two leaves x_1, x_2 are siblings; a fact that is used to prove the correctness of Huffman's algorithm [3]. Remove x_1, x_2 from T^*, replace their parent with a leaf whose weight equals $x_1 + x_2$, and let T' be the resulting tree. Since T' has the exclusion property, it follows using induction that T' is an optimal tree with respect to its leaves $[x_1 + x_2, x_3, \ldots, x_n]$. Hence, T^* is an optimal tree and corresponds to an optimal prefix code.   □

In general, building T^* requires $\Omega(n \log n)$. It is crucial to mention that we do not have to explicitly construct T^*. Instead, we only need to find the values of some of, and not all, the internal nodes at every level.

3   The Main Construction Method

Given a list of weights, we build the tree T^* bottom up. Starting with the highest level, a weight is assigned to a level as long as its value is less than the sum of the two nodes with the smallest values at that level. The Kraft equality is enforced by making sure that the number of nodes at every level is even, and that the number of nodes at the lowest level containing leaves is a power of two.

3.1   Example

For the sake of illustration, consider a list with thirty weights: ten weights have the value 2, ten have the value 3, five the value 5, and five the value 9. To construct the optimal codes, we start by finding the smallest two weights in the list; these will have the values $2, 2$. We now identify all the weights in the list with value less than 4, the sum of these two smallest weights. All these weights will be momentarily placed at the same level. This means that the highest level l will contain ten weights of value 2 and ten of value 3. The number of nodes at this level is even, so we move to the next level $l - 1$. We identify the smallest two nodes at level $l - 1$, amongst the two smallest internal nodes resulting from combining nodes of level l, and the two smallest weights among those remaining in the list; these will be the two internal nodes $4, 4$ whose sum is 8. All the remaining weights with value less than 8 are placed at level $l - 1$. This level now contains an odd number of nodes: ten internal nodes and five weights of value 5. To make this number even, we move the node with the largest weight to the, still empty, next lower level $l - 2$. The node to be moved, in this case, is an internal node with value 6. Moving an internal node one level up implies moving the weights in its subtree one level up. In such case, the subtree consisting of the two weights of value 3 is moved one level up. At the end of this stage, the highest level l contains ten weights of value 2 and eight weights of value 3; level $l - 1$ contains two weights of value 3 and five weights of value 5. For level $l - 2$, the smallest two internal nodes have values $6, 8$ and the smallest weight in the list has value 9. This means that all the five remaining weights in the list will go to level $l - 2$. Since we are done with all the weights, we only need to enforce the condition that the number of nodes at level $l - 3$ is a power of two. Level $l - 2$ now contains eight internal nodes and five weights, for a total of thirteen nodes. All we need to do is to move the three nodes with the largest values, from level $l - 2$, one level up. The largest three nodes at level $l - 2$ are the three internal nodes of values $12, 12$ and 10. So we move eight weights of value 3 and two weights of value 5 one level up. As a result, the number of nodes at level $l - 3$ will be 8. The final distribution of weights will be: the ten weights of value 2 are in the highest level l; level $l - 1$ contains the ten weights of value 3 and three weights of value 5; and level $l - 2$ contains the remaining weights, two of value 5 and five of value 9. The corresponding code lengths are $6, 5$ and 4 respectively.

3.2   The Basic Algorithm

The idea of the algorithm should be clear. We construct the optimal code tree by maintaining the exclusion property for all the levels. Since this property is always satisfied by the internal nodes, the weights are placed at the levels in such a way that the exclusion property is satisfied. Adjusting the number of nodes at each level will not affect this property since we are always moving the largest nodes one level up to a still empty lower level. A formal description follows.

1. The smallest two weights are found, removed from W, and placed at the highest level l. Their sum S is computed. The list W is scanned and all weights less than S are removed and placed in level l. If the number of leaves at level l is odd, the leaf with the largest weight among these leaves is moved to level $l - 1$.

2. In the general iteration, after moving weights from W to level j, determine the weights from W that will go to level $j - 1$ as follows. Find the smallest two internal nodes at level $j - 1$, and the smallest two leaves from W. Find the smallest two nodes amongst these four nodes, and let their sum be S. Scan W for all weights less than S, and move them to level $j - 1$. If the number of nodes at level $j - 1$ is odd, move the subtree of the node with the largest value among these nodes to level $j - 2$.

3. When W is exhausted, let m be the number of nodes at the shallowest level that has leaves. Move the $2^{\lceil \log_2 m \rceil} - m$ subtrees of the nodes with the largest values, from such level, one level up.

3.3 Proof of Correctness

To guarantee its optimality following Lemma 1, we need to show that both the Kraft equality and the exclusion property hold for the constructed tree.

By construction, the number of nodes at every level of the tree is even. At Step 3 of the algorithm, if m is a power of 2, no subtrees are moved to the next level and Kraft equality holds. If m is not a power of two, we move $2^{\lceil \log_2 m \rceil} - m$ nodes to the next level, leaving $2m - 2^{\lceil \log_2 m \rceil}$ nodes at this level other than those of the subtrees that have just been moved one level up. Now, the number of nodes at the next lower level is $m - 2^{\lceil \log_2 m \rceil - 1}$ internal nodes resulting from the higher level, plus the $2^{\lceil \log_2 m \rceil} - m$ nodes that we have just moved. This sums up to $2^{\lceil \log_2 m \rceil - 1}$ nodes, that is a power of 2, and Kraft equality holds.

Throughout the algorithm, we maintain the exclusion property by making sure that the sum of the two nodes with the smallest values is larger than all the values of the nodes at this level. When we move a subtree one level up, the root of this subtree is the node with the largest value at its level. Hence, all the nodes of this subtree at a certain level will have the largest values among the nodes of this level. Moving these nodes one level up will not destroy the exclusion property. We conclude that the resulting tree has the exclusion property.

4 Distribution-Sensitive Construction

Up to this point, we have not shown how to evaluate the internal nodes needed by our basic algorithm, and how to search within the list W to decide which weights are at which levels. The basic intuition behind the novelty of our approach is that it does not require evaluating all the internal nodes of the tree corresponding to the prefix code, and would thus surpass the $\Theta(n \log n)$ bound for several cases, a fact that will be asserted in the analysis. We show next how to implement the basic algorithm in a distribution-sensitive behavior.

4.1 Example

The basic idea is clarified through an example having $1.5n + 2$ weights. Assume that the resulting optimal tree will turn out to have n leaves at the highest level, $n/2$ at the following level, and two leaves at level 2; the $1.5n$ leaves, at the highest two levels, combine to produce two internal nodes at level 2.

In such case, we show how to produce the codeword lengths in linear time. For our basic algorithm, we need to evaluate the smallest node x of the two internal nodes at level 2, which amounts to identifying the smallest $n/2$ nodes amongst the nodes at the second highest level. In order to be able to achieve this in linear time, we need to do it without having to evaluate all $n/2$ internal nodes resulting from the pair-wise combinations of the highest level n weights. We show that this can be done through a simple pruning procedure. The nodes at the second highest level consist of two sets; one set has $n/2$ leaves whose weights are known and thus their median M can be found in linear time [2], and another set containing $n/2$ internal nodes which are not known but whose median M'

can still be computed in linear time, by simply finding the two middle weights of the highest level n leaves and adding them. Assuming without loss of generality that $M > M'$, then the bigger half of the $n/2$ weights at the second highest level can be safely discarded as not contributing to x, and the smaller half of the highest level n weights are guaranteed to contribute to x. The above step is repeated recursively on a problem half the size. This results in a procedure satisfying the recurrence $T(n) = T(n/2) + O(n)$, and hence $T(n) = O(n)$.

If the list of weights is already sorted, the number of comparisons required to find any of the medians M or M' is constant. This results in a procedure satisfying the recurrence $T_s(n) = T_s(n/2) + O(1)$, and hence $T_s(n) = O(\log n)$.

4.2   The Detailed Algorithm

Let $l_1 > l_2 > \ldots l_{k'}$ be the levels that have already been assigned weights at some step of our algorithm (other levels only have internal nodes), n_j be the count of the weights assigned to level l_j, and $\mu_j = \sum_{i=1}^{j} n_i$. At this point, we are looking forward to find the next level $l_{k'+1} < l_{k'}$ that will be assigned weights by our algorithm. Knowing that the weights that have already been assigned to higher levels are the only weights that may contribute to the internal nodes of any level $l \geq l_{k'+1}$, we need to evaluate some internal nodes at these levels.

Finding the splitting node. Consider the internal node x at a level l, $l_{k'} > l \geq l_{k'+1}$, where the count of the weights contributing to the internal nodes of level l, whose values are smaller (larger) than that of x, is at most $\mu_{k'}/2$. We call x the splitting node of l.

The following recursive procedure is used to evaluate x. We find the leaf with the median weight M among the list of the $n_{k'}$ weights already assigned to level $l_{k'}$ (partition the $n_{k'}$ list into two sublists around M), and recursively evaluate the splitting node M' at level $l_{k'}$ using the list of the $\mu_{k'-1}$ weights of the higher levels (partition the $\mu_{k'-1}$ list into two sublists around M'). Comparing M to M', we either conclude that one of the four sublists - the two sublists of the $n_{k'}$ list and the two sublists of the $\mu_{k'-1}$ list - will not contribute to x, or that one of these four sublists contributes to x. If one of the sublists of the $n_{k'}$ list is discarded, find a new median M for the other sublist and compare it with M'. If one of the sublists of the $\mu_{k'-1}$ list is discarded, recursively find the new splitting node M' corresponding to the other sublist and compare it to M. Once one of the two lists becomes empty, we would have identified the weights that contribute to x and hence evaluated x. As a byproduct, we also know which weights contribute to the internal nodes at level l whose values are smaller (larger) than that of x.

Let $T(\mu_{k'}, k')$ be the time required by the above procedure. The total amount of work, in all the recursive calls, required to find the medians among the $n_{k'}$ weights assigned to level k' is $O(n_{k'})$. The time for the i-th recursive call to find a splitting node at level k' is $T(\mu_{k'-1}/2^{i-1}, k'-1)$. The next relations follow:

$$T(\mu_1, 1) = O(n_1),$$
$$T(\mu_{k'}, k') \leq \sum_{i \geq 1} T(\mu_{k'-1}/2^{i-1}, k'-1) + O(n_{k'}).$$

Substitute with $T(a, b) = c \cdot 2^b a$, for $a < \mu_{k'}$, $b < k'$, and some big enough constant c. Then, $T(\mu_{k'}, k') \leq c \cdot 2^{k'-1} \sum_{i \geq 1} \mu_{k'-1}/2^{i-1} + O(n_{k'}) < c \cdot 2^{k'} \mu_{k'-1} + c \cdot n_{k'}$. Since $\mu_{k'} = \mu_{k'-1} + n_{k'}$, it follows that

$$T(\mu_{k'}, k') = O(2^{k'} \mu_{k'}).$$

Consider the case when the list of weights W is already sorted. Let $T_s(\mu_{k'}, k')$ be the number of comparisons required by the above procedure. The number of comparisons, in all recursive calls, required to find the medians among the $n_{k'}$ weights assigned to level k', is at most $\log_2 (n_{k'} + 1)$. The next relations follow:

$$T_s(\mu_2, 2) \leq 2 \log_2 \mu_2,$$
$$T_s(\mu_{k'}, k') \leq \sum_{i \geq 1} T_s(\mu_{k'-1}/2^{i-1}, k' - 1) + \log_2 (n_{k'} + 1).$$

Since the number of internal nodes at level k' is at most $\mu_{k'-1}/2$, the number of recursive calls at this level is at most $\log_2 \mu_{k'-1}$. It follows that $T_s(\mu_{k'}, k') \leq \log_2 \mu_{k'-1} \cdot T_s(\mu_{k'-1}, k' - 1) + \log_2 (n_{k'} + 1) < \log_2 \mu_{k'} \cdot T_s(\mu_{k'}, k' - 1) + \log_2 \mu_{k'}$. Substitute with $T_s(a, b) \leq \log_2^{b-1} a + \sum_{i=1}^{b-1} \log_2^i a$, for $a < \mu_{k'}$, $b < k'$. Then, $T_s(\mu_{k'}, k') < \log_2 \mu_{k'} \cdot \log_2^{k'-2} \mu_{k'} + \log_2 \mu_{k'} \cdot \sum_{i=1}^{k'-2} \log_2^i \mu_{k'} + \log_2 \mu_{k'} = \log_2^{k'-1} \mu_{k'} + \sum_{i=1}^{k'-1} \log_2^i \mu_{k'}$. It follows that

$$T_s(\mu_{k'}, k') = O(\log^{k'-1} \mu_{k'}).$$

Finding the t-th smallest (largest) node. Consider the node x at level $l_{k'}$, which has the t-th smallest (largest) value among the nodes at level $l_{k'}$. The following recursive procedure is used to evaluate x.

As for the case of finding the splitting node, we find the leaf with the median weight M among the list of the $n_{k'}$ weights already assigned to level $l_{k'}$, and evaluate the splitting node M' at level $l_{k'}$ (applying the above recursive procedure) using the list of the $\mu_{k'-1}$ leaves of the higher levels. As with the above procedure, comparing M to M', we conclude that either one of the four sublists - the two sublists of $n_{k'}$ leaves and the two sublists of $\mu_{k'-1}$ leaves - will not contribute to x, or that one of these four sublists contributes to x. Applying the aforementioned pruning procedure, we identify the weights that contribute to x and hence evaluate x. As a byproduct, we also know which weights contribute to the nodes at level $l_{k'}$ whose values are smaller (larger) than that of x.

Let $T'(\mu_{k'}, k')$ be the time required by the above procedure. Then,

$$T'(\mu_{k'}, k') \leq \sum_{i \geq 1} T(\mu_{k'-1}/2^{i-1}, k' - 1) + O(n_{k'}) = O(2^{k'} \mu_{k'}).$$

Let $T'_s(\mu_{k'}, k')$ be the number of comparisons required by the above procedure, when the list of weights W is already sorted. Then,

$$T'_s(\mu_{k'}, k') \leq \sum_{i \geq 1} T_s(\mu_{k'-1}/2^{i-1}, k' - 1) + O(\log n_{k'}) = O(\log^{k'-1} \mu_{k'}).$$

Finding $l_{k'+1}$, the next level that will be assigned weights. Consider level $l_{k'} - 1$, which is the next level lower than level $l_{k'}$. We start by finding the minimum weight w among the weights remaining in W at this point of the algorithm, and use this weight to search within the internal nodes at level $l_{k'} - 1$ in a manner similar to binary search. The basic idea is to find the maximum number of the internal nodes at level $l_{k'} - 1$ with the smallest values, such that the sum of their values is less than w. We find the splitting node x at level $l_{k'} - 1$, and evaluate the sum of the weights contributing to the internal nodes, at that level, whose values are smaller than that of x. Comparing this sum with w, we decide which sublists of the $\mu_{k'}$ leaves to proceed to find its splitting node. At the end of this searching procedure, we would have identified the weights contributing to the r smallest internal nodes at level $l_{k'} - 1$, such that the sum of their values is less than w and r is maximum. We conclude by setting $l_{k'+1}$ to be equal to $l_{k'} - \lceil \log_2 (r+1) \rceil$.

To prove the correctness of this procedure, consider any level l, such that $r > 1$ and $l_{k'} - \lceil \log_2 (r+1) \rceil < l < l_{k'}$. The values of the two smallest internal nodes at level l are contributed to by at most $2^{l_{k'} - l} \leq 2^{\lceil \log_2 (r+1) \rceil - 1} \leq t$ internal nodes from level $l_{k'} - 1$. Hence, the sum of these two values is less than w. For the exclusion property to hold, no weights are assigned to any of these levels. On the contrary, the values of the two smallest internal nodes at level $l_{k'} - \lceil \log_2 (r+1) \rceil$ are contributed to by more than r internal nodes from level $l_{k'} - 1$, and hence the sum of these two values is more than w. For the exclusion property to hold, at least the weight w is assigned to this level.

The time required by this procedure is the $O(n - \mu_{k'})$ time to find the weight w among the weights remaining in W, plus the time for the calls to find the splitting nodes. Let $T''(\mu_{k'}, k')$ be the time required by this procedure. Then,

$$T''(\mu_{k'}, k') \leq \sum_{i \geq 1} T(\mu_{k'}/2^{i-1}, k') + O(n - \mu_{k'}) = O(2^{k'} \mu_{k'} + n).$$

Let $T''_s(\mu_{k'}, k')$ be the number of comparisons required by the above procedure, when the list of weights W is already sorted. Then,

$$T''_s(\mu_{k'}, k') \leq \sum_{i \geq 1} T_s(\mu_{k'}/2^{i-1}, k') + O(1) = O(\log^{k'} \mu_{k'}).$$

Maintaining Kraft equality. After deciding the value of $l_{k'+1}$, we need to maintain Kraft equality in order to produce a binary tree corresponding to the optimal prefix code. This is accomplished by moving the subtrees of the t nodes with the largest values from level $l_{k'}$ one level up. Let m be the number of nodes currently at level $l_{k'}$, then the number of the nodes to be moved up t is $2^{l_{k'} - l_{k'+1}} \lceil m/2^{l_{k'} - l_{k'+1}} \rceil - m$. Note that when $l_{k'+1} = l_{k'} - 1$ (as in the case of our basic algorithm), then t equals one if m is odd and zero otherwise.

To establish the correctness of this procedure, we need to show that both the Kraft equality and the exclusion property hold. For a realizable construction, the number of nodes at level $l_{k'}$ has to be even, and if $l_{k'+1} \neq l_{k'} - 1$, the number of nodes at level $l_{k'} - 1$ has to divide $2^{l_{k'} - l_{k'+1} - 1}$. If m divides $2^{l_{k'} - l_{k'+1}}$, no subtrees are moved to level $l_{k'} - 1$ and Kraft equality holds. If m does not divide $2^{l_{k'} - l_{k'+1}}$, then $2^{l_{k'} - l_{k'+1}} \lceil m/2^{l_{k'} - l_{k'+1}} \rceil - m$ nodes are moved to level $l_{k'} - 1$, leaving $2m - 2^{l_{k'} - l_{k'+1}} \lceil m/2^{l_{k'} - l_{k'+1}} \rceil$ nodes at level $l_{k'}$ other than those of the subtrees that have just been moved one level up. Now, the number of nodes at level $l_{k'} - 1$ is $m - 2^{l_{k'} - l_{k'+1} - 1} \lceil m/2^{l_{k'} - l_{k'+1}} \rceil$ internal nodes resulting from the nodes of level $l_{k'}$, plus the $2^{l_{k'} - l_{k'+1}} \lceil m/2^{l_{k'} - l_{k'+1}} \rceil - m$ nodes that we have just moved. This sums up to $2^{l_{k'} - l_{k'+1} - 1} \lceil m/2^{l_{k'} - l_{k'+1}} \rceil$ nodes, which divides $2^{l_{k'} - l_{k'+1} - 1}$, and Kraft equality holds. The exclusion property holds following the same argument mentioned in the proof of the correctness of the basic algorithm.

The time required by this procedure is basically the time needed to find the weights contributing to the t nodes with the largest values at level $l_{k'}$, which is $O(2^{k'} \mu_{k'})$. If W is sorted, the required number of comparisons is $O(\log^{k'-1} \mu_{k'})$.

Summary of the algorithm

1. The smallest two weights are found, moved from W to the highest level l_1, and their sum S is computed. The rest of W is searched for weights less than S, which are moved to level l_1.
2. In the general iteration of the algorithm, after assigning weights to k' levels, perform the following steps:
   (a) Find $l_{k'+1}$, the next level that will be assigned weights.
   (b) Maintain the Kraft equality at level $l_{k'}$ by moving the t subtrees with the largest values from this level one level up.
   (c) Find the values of the smallest two internal nodes at level $l_{k'+1}$, and the smallest two weights from those remaining in W. Find the two nodes with the smallest values among these four, and let their sum be S.
   (d) Search the rest of W, and move the weights less than S to level $l_{k'+1}$.
3. When W is exhausted, maintain Kraft equality at the last level that has been assigned weights.

4.3   Complexity Analysis

Using the bounds deduced for the described steps of the algorithm, we conclude that the time required by the general iteration is $O(2^{k'} \mu_{k'} + n)$. If W is sorted, the required number of comparisons is $O(\log^{k'} \mu_{k'})$.

To complete the analysis, we need to show the effect of maintaining the Kraft equality on the complexity of the algorithm. Consider the scenario when, as a result of moving subtrees one level up, all the weights at a level move up to the next level that already had other weights. As a result, the number of levels that contain leaves decreases. It is possible that within a single iteration the number of such levels decreases by one half. If this happens for several iterations, the amount of work done by the algorithm would have been significantly large compared to the actual number of distinct codeword lengths k. Fortunately, this scenario will not happen quite often. In the next lemma, we bound the number of iterations performed by the algorithm by $2k$. We also show that, at any iteration, the number of levels that contain leaves is at most twice the number of distinct optimal codeword lengths for the weights that have been assigned so far.

Lemma 2. *Consider the set of weights, all having the j-th largest optimal codeword length. During the execution of the algorithm, such set of weights will be assigned to at most two consecutive levels, among those levels that contain leaves. Hence, these two levels will be the at most $2j - 1$ and $2j$ highest such levels.*

Proof. Consider a set of weights that will turn out to have the same codeword length. During the execution of the algorithm, assume that some of these weights are assigned to three levels. Let $l_i > l_{i+1} > l_{i+2}$ be such levels. It follows that $l_i - 1 > l_{i+2}$. Since we are maintaining the exclusion property throughout the algorithm, there will exist some internal nodes at level $l_i - 1$ whose values are strictly smaller than the values of the weights at level l_{i+2} (some may have the same value as the smallest weight at level l_{i+2}). The only way for all these weights to catch each other at the same level of the tree would be as a result of moving subtrees up (starting from level l_{i+2} upwards) to maintain the Kraft equality. Suppose that, at some point of the algorithm, the weights that are currently at level l_i are moved up to catch the weights of level l_{i+2}. It follows that the internal nodes that are currently at level $l_i - 1$ will accordingly move to the next lower level of the moved weights. As a result, the exclusion property will not hold; a fact that contradicts the behavior of our algorithm. It follows that such set of weights will never catch each other at the same level of the tree; a contradiction.

We prove the second part of the lemma by induction. The base case follows easily for $j = 1$. Assume that the argument is true for $j - 1$. By induction, the levels of the weights that have the $(j - 1)$-th largest optimal codeword length will be the at most $2j - 3$ and $2j - 2$ highest such levels. From the exclusion property, it follows that the weights that have the j-th largest optimal codeword length must be at the next lower levels. Using the first part of the lemma, the number of such levels is at most two. It follows that these weights are assigned to the at most $2j - 1$ and $2j$ highest levels.   □

Using Lemma 2, the time required by our algorithm to assign the set of weights whose optimal codeword length is the j-th largest, among all distinct lengths, is $O(2^{2j}n) = O(4^j n)$. Summing for all such lengths, the total time required by our algorithm is $\sum_{j=1}^k O(4^j n) = O(4^k n)$.

Consider the case when the list of weights W is already sorted. The only step left to mention, for achieving the claimed bounds, is how to find the weights of W smaller than the sum of the values of the smallest two nodes at level l_j. Once we get this sum, we apply an exponential search that is followed by a binary search on the weights of W for an $O(\log n_j)$ comparisons. Using Lemma 2, the number of comparisons performed by our algorithm to assign the weights whose codeword length is the j-th largest, among all distinct lengths, is $O(\log^{2j-1} n)$. Summing for all such lengths, the number of comparisons performed by our algorithm is $\sum_{j=1}^k O(\log^{2j-1} n) = O(\log^{2k-1} n)$. The next theorem follows.

Theorem 1. *If the list of weights is sorted, constructing minimum-redundancy prefix codes can be done using $O(\log^{2k-1} n)$ comparisons.*

Corollary 1. *For $k < c \cdot \log n / \log \log n$, and any constant $c < 0.5$, the above algorithm requires $o(n)$ comparisons.*

5   The Improved Algorithm

The drawback of the algorithm we described in the previous section is that it uses many recursive median-finding calls. The basic idea we use here is to incrementally process the weights throughout the algorithm by partitioning them into unsorted blocks, such that the weights of one block are smaller or equal to the smallest weight of the succeeding block. The time required during the recursive calls becomes smaller when handling these shorter blocks. The details follow.

The invariant we maintain is that during the execution of the general iteration of the algorithm, after assigning weights to k' levels, the weights that have already been assigned to a level $l_j \geq l_{k'}$ are partitioned into blocks each of size at most $n_j/2^{k'-j}$ weights, such that the weights of one block are smaller or equal to the smallest weight of the succeeding block. To accomplish this invariant, once we assign weights to a level, the median of the weights of each block among those already assigned to all the higher levels is found, and each of these blocks is partitioned into two blocks around this median weight. Using Lemma 2, the number of iterations performed by the algorithm is at most $2k$. The amount of work required for this partitioning is $O(n)$ for each of these iterations, for a total of $O(nk)$ time for this partitioning phase.

The basic step for all our procedures is to find the median weight among the weights already assigned to a level l_j. This step can now be done faster. To find such median weight, we can identify the block that has such median in constant time, then we find the required weight in $O(n_j/2^{k'-j})$ time, which is the size of the block at this level. The recursive relations for all our procedures performed at each of the k general iterations of the algorithm can be written as

$$G^{k'}(\mu_1, 1) = O(n_1/2^{k'-1}),$$
$$G^{k'}(\mu_{k'}, k') \leq \sum_{i \geq 1} G^{k'}(\mu_{k'-1}/2^{i-1}, k'-1) + O(n_{k'}).$$

Substitute with $G^{k'}(a, b) = c \cdot a/2^{k'-b}$, for $a < \mu_{k'}$, $b < k'$, and some big enough constant c. Then, $G^{k'}(\mu_{k'}, k') \leq c/2 \cdot \sum_{i \geq 1} \mu_{k'-1}/2^{i-1} + O(n_{k'}) < c \cdot \mu_{k'-1} + c \cdot n_{k'}$. Since $\mu_{k'} = \mu_{k'-1} + n_{k'}$, it follows that

$$G^{k'}(\mu_{k'}, k') = O(\mu_{k'}) = O(n).$$

Since the number of iterations performed by the algorithm is at most $2k$, by Lemma 2. Summing up for these iterations, the running time for performing the recursive calls is $O(nk)$ as well. The next main theorem follows.

Theorem 2. *Constructing minimum-redundancy prefix codes is done in $O(nk)$.*

6 Conclusion

We gave a distribution-sensitive algorithm for constructing minimum-redundancy prefix codes, whose running time is a function of both n and k. For small values of k, this algorithm asymptotically improves over other known algorithms that require $O(n \log n)$; it is quite interesting to know that the construction of optimal codes can be done in linear time when k turns out to be a constant. For small values of k, if the sequence of weights is already sorted, the number of comparisons performed by our algorithm is asymptotically better than other known algorithms that require $O(n)$ comparisons; it is also interesting to know that the number of comparisons required for the construction of optimal codes is poly-logarithmic when k turns out to be a constant.

Two open issues remain; first is the possibility of improving the algorithm to achieve an $O(n \log k)$ bound, and second is to make the algorithm faster in practice by avoiding so many recursive calls to a median-finding algorithm.

References

1. M. Buro. *On the maximum length of Huffman codes*. Information Processing Letters 45 (1993), 219-223.
2. T. Cormen, C. Leiserson, R. Rivest and C. Stein. *Introduction to algorithms, 2nd Edition*. The MIT press (2001).
3. D. Huffman. *A method for the construction of minimum-redundancy codes*. Proc. IRE 40 (1952), 1098-1101.
4. R. Milidiu, A. Pessoa, and E. Laber. *Three space-economical algorithms for calculating minimum-redundancy prefix codes*. IEEE Transactions on Information Theory 47(6) (2001), 2185-2198.
5. J.I. Munro and P. Spira. *Sorting and searching in multisets*. SIAM Journal on Computing 5(1) (1976), 1-8.
6. A. Moffat and A. Turbin. *Efficient construction of minimum-redundancy codes for large alphabets*. IEEE Transactions on Information Theory 44 (1998), 1650-1657.
7. S. Sen and N. Gupta. *Distribution-sensitive algorithms*. Nordic Journal of Computing 6(2) (1999), 194-211.
8. J. Van Leeuwen. *On the construction of Huffman trees*. 3rd International Colloquium for Automata, Languages and Programming (1976), 382-410.
9. J. Zobel and A. Moffat. *Adding compression to a full-text retrieval system*. Software: Practice and Experience 25(8) (1995), 891-903.

On Critical Exponents in Fixed Points of Binary k-Uniform Morphisms

Dalia Krieger

School of Computer Science, University of Waterloo,
Waterloo, ON N2L 3G1, Canada
d2kriege@cs.uwaterloo.ca

Abstract. Let $\mathbf{w}$ be an infinite fixed point of a binary k-uniform morphism f, and let $E(\mathbf{w})$ be the critical exponent of $\mathbf{w}$. We give necessary and sufficient conditions for $E(\mathbf{w})$ to be bounded, and an explicit formula to compute it when it is. In particular, we show that $E(\mathbf{w})$ is always rational. We also sketch an extension of our method to non-uniform morphisms over general alphabets.

MSC: 68R15.
Keywords: Critical exponent; Binary k-uniform morphism.

1 Introduction

Let $\mathbf{w}$ be a right-infinite word over a finite alphabet Σ. The *critical exponent* of $\mathbf{w}$, denoted by $E(\mathbf{w})$, is the supremum of the set of exponents $r \in \mathbb{Q}_{\geq 1}$, such that $\mathbf{w}$ contains an r-power (see Section 2 for the definition of fractional powers). Given an infinite word $\mathbf{w}$, a natural question is to determine its critical exponent.

The first critical exponent to be computed was probably that of the Thue-Morse word, $\mathbf{t}$ [18, 2]. This word, defined as the fixed point beginning with 0 of the Thue-Morse morphism $\mu(0) = 01$, $\mu(1) = 10$, was proved by Thue in the early 20th century to be *overlap-free*, that is, to contain no subword of the form $axaxa$, where $a \in \{0, 1\}$ and $x \in \{0, 1\}^*$. In other words, $\mathbf{t}$ is *r-power-free* for all $r > 2$; and since it contains 2-powers (squares), by our definition $E(\mathbf{t}) = 2$. Another famous word for which the critical exponent has been computed is the Fibonacci word $\mathbf{f}$, defined as the fixed point of the Fibonacci morphism $f(0) = 01$, $f(1) = 0$. In 1992, Mignosi and Pirillo [15] showed that $E(\mathbf{f}) = 2 + \varphi$, where $\varphi = (1 + \sqrt{5})/2$ is the golden mean. This gives an example of an irrational critical exponent.

In a more general setting, critical exponents have been studied mainly with relation to Sturmian words (for the definition, properties and structure of Sturmian words, see e.g. [12, Chapter 2]). In 1989, Mignosi [14] proved that for a Sturmian word $\mathbf{s}$, $E(\mathbf{s}) < \infty$ if and only if the continued fraction expansion of the slope of $\mathbf{s}$ has bounded partial quotients; an alternative proof was given in 1999 by Berstel [3]. In 2000, Vandeth [19] gave an explicit formula for $E(\mathbf{s})$, where $\mathbf{s}$ is a Sturmian word which is a fixed point of a morphism, in terms of the

B. Durand and W. Thomas (Eds.): STACS 2006, LNCS 3884, pp. 104–114, 2006.
© Springer-Verlag Berlin Heidelberg 2006

continued fraction expansion of the slope of **s**. In particular, $E(\mathbf{s})$ is algebraic quadratic. Alternative proofs for the results of Mignosi and Vandeth, with some generalizations, were given in 2000 by Carpi and de Luca [5], and in 2001 by Justin and Pirillo [10]. Carpi and de Luca also showed that $2 + \varphi$ is the minimal critical exponent for any Sturmian word. In 2002, Damanik and Lenz [7] gave a formula for critical exponents of general Sturmian words, again in terms of the continued fraction expansion of the slope. An alternative proof for this result was given in 2003 by Cao and Wen [4].

In this work we consider a different family of words: that of fixed points of binary k-uniform morphisms. Given a binary k-uniform morphism f with a fixed point **w**, we give necessary and sufficient conditions for $E(\mathbf{w})$ to be bounded, and an explicit formula to compute it when it is. In particular, we show that if $E(\mathbf{w}) < \infty$ then it is rational. We also show that, given a rational number $0 < r < 1$, we can construct a binary k-uniform morphism f such $E(f^{\omega}(0)) = n + r$ for some positive integer n. Finally, we sketch a method for extending the results to non-uniform morphisms over an arbitrary finite alphabet, and conjecture that $E(\mathbf{w})$, when bounded, is always algebraic, and lies in the field extension $\mathbb{Q}[s_1, \ldots, s_\ell]$, where $s_1, \ldots, s_\ell$ are the eigenvalues of the incidence matrix of f. For injective morphisms the conjecture has been proved [11].

2   Preliminaries

Let Σ be a finite alphabet. We use the notation Σ^*, Σ^+, Σ^n and Σ^{ω} to denote the sets of finite words, non-empty finite words, words of length n, and right-infinite words over Σ, respectively. We use ϵ to denote the empty word. For words $x \in \Sigma^*$, $y \in \Sigma^* \cup \Sigma^{\omega}$, we use the notation $x \lhd y$ to denote the relation "x is a subword of y". We use $\mathbb{Z}_{\geq r}$ (and similarly $\mathbb{Q}_{\geq r}, \mathbb{R}_{\geq r}$) to denote the integers (similarly rational or real numbers) greater than or equal to r.

Let $z = a_0 \cdots a_{n-1} \in \Sigma^+$, $a_i \in \Sigma$. A positive integer $q \leq |z|$ is a *period* of z if $a_{i+q} = a_i$ for $i = 0, \cdots, n-1-q$. An infinite word $\mathbf{z} = a_0 a_1 \cdots \in \Sigma^{\omega}$ has a period $q \in \mathbb{Z}_{\geq 1}$ if $a_{i+q} = a_i$ for all $i \geq 0$; in this case, $\mathbf{z}$ is *periodic*, and we write $\mathbf{z} = x^{\omega}$, where $x = a_0 \cdots a_{q-1}$. If $\mathbf{z}$ has a periodic suffix, we say it is *ultimately periodic*.

A *fractional power* is a word of the form $z = x^n y$, where $n \in \mathbb{Z}_{\geq 1}$, $x \in \Sigma^+$, and y is a proper prefix of x. Equivalently, z has a $|x|$-period and $|y| = |z| \bmod |x|$. If $|x| = q$ and $|y| = p$, we say that z is an $(n+p/q)$-power, and write $z = x^{n+p/q}$; the rational number $n + p/q$ is the *exponent* of the power. Since q stands for both the exponent's denominator and the period, we use non-reduced fractions to denote exponents: for example, 10101 is a $2\frac{1}{2}$-power, while 1001100110 is a $2\frac{2}{4}$-power.

Let **w** be an infinite word over Σ, and let α be a real number. We say that **w** is α-*power-free* if no subword of it is an r-power for any rational $r \geq \alpha$; we say that **w** *contains an* α-*power* if it has an r-power as a subword for some rational $r \geq \alpha$. The *critical exponent* of **w** is defined by

$$E(\mathbf{w}) = \sup\{r \in \mathbb{Q}_{\geq 1} : \exists x \in \Sigma^+ \text{ such that } x^r \lhd \mathbf{w}\}. \tag{1}$$

By this definition, **w** contains α-powers for all $\alpha \in \mathbb{R}$ such that $1 \leq \alpha < E(\mathbf{w})$, but no α-powers for $\alpha > E(\mathbf{w})$; it may or may not contain $E(\mathbf{w})$-powers.

A morphism $f : \Sigma^* \to \Sigma^*$ is *prolongable* on a letter $a \in \Sigma$ if $f(a) = ax$ for some $x \in \Sigma^+$, and furthermore $f^n(x) \neq \epsilon$ for all $n \geq 0$. If this is the case, then $f^n(a)$ is a proper prefix of $f^{n+1}(a)$ for all $n \geq 0$, and by applying f successively we get an infinite *fixed point* of f, $f^\omega(a) = \lim_{n \to \infty} f^n(a) = axf(x)f^2(x)f^3(x) \cdots$. A morphism $f : \Sigma^* \to \Gamma^*$ is *k-uniform* if $|f(a)| = k$ for all $a \in \Sigma$, where k is a positive integer. In this work we consider powers in fixed points of uniform morphisms defined over a binary alphabet $\Sigma = \{0, 1\}$, therefore we assume that f is prolongable on 0.

Let f be a k-uniform morphism defined over $\Sigma = \{0, 1\}$, and let $\mathbf{w} = f^\omega(0) = w_0 w_1 w_2 \ldots$. Let $z = w_i \cdots w_j \lhd \mathbf{w}$ be an *r-power*, where $r = n + p/q$. We say that z is *reducible* if it contains an r'-power, $r' = n' + p'/q'$, such that $r' > r$, or $r' = r$ and $q' < q$. If $r' > r$ then z is *strictly reducible*. The occurrence of z in $\mathbf{w}$ is *left stretchable (right stretchable)* if $z_L = w_{i-1} \cdots w_j$ ($z_R = w_i \cdots w_{j+1}$) is an $(n + (p+1)/q)$-power.

Since $E(\mathbf{w})$ is an upper bound, it is enough to consider irreducible, unstretchable powers when computing it. Therefore, for a binary alphabet, we can assume $n \geq 2$: since any binary word of length 4 or more contains a square, a $(1 + p/q)$-power over $\{0, 1\}$ is always reducible, save for the $1\frac{1}{2}$-power 101 (010).

The next definition is a key one:

Definition 1. Let f be a binary k-uniform morphism. The *shared prefix* of f, denoted by ρ_f, is the longest word $\rho \in \{0, 1\}^*$ satisfying $f(0) = \rho x$, $f(1) = \rho y$ for some $x, y \in \Sigma^*$. Similarly, the *shared suffix* of f, denoted by σ_f, is the longest word $\sigma \in \{0, 1\}^*$ satisfying $f(0) = x\sigma$, $f(1) = y\sigma$ for some $x, y \in \Sigma^*$. The *shared size* of f is the combined length $\lambda_f = |\rho_f| + |\sigma_f|$.

We can now state our main Theorem:

Theorem 2. *Let f be a binary k-uniform morphism prolongable on 0, and let $\mathbf{w} = f^\omega(0)$. Then:*

1. *$E(\mathbf{w}) = \infty$ if and only if at least one of the following holds:*
   (a) *$f(0) = f(1)$;*
   (b) *$f(0) = 0^k$;*
   (c) *$f(1) = 1^k$;*
   (d) *$k = 2m + 1$, $f(0) = (01)^m 0$, $f(1) = (10)^m 1$.*
2. *Suppose $E(\mathbf{w}) < \infty$. Let $\mathcal{E}$ be the set of exponents $r = n + p/q$, such that $q < k$ and $f^4(0)$ contains an r-power. Then*

$$E(\mathbf{w}) = \max_{n + p/q \in \mathcal{E}} \left\{ n + \frac{p(k-1) + \lambda_f}{q(k-1)} \right\}.$$

In particular, $E(\mathbf{w})$, when bounded, is always rational. The bound $E(\mathbf{w})$ is attained if and only if $\lambda_f = 0$.

Here is an example of an application of Theorem 2:

Example 3. *The Thue-Morse word is overlap-free.*

Proof. The Thue-Morse morphism μ satisfies $\lambda_\mu = 0$; and since the largest power in $\mu^4(0)$ is a square, we get that $E(\mu^\omega(0)) = 2$, and the bound is attained.   □

We end this section by stating a few theorems that will be useful later. Theorem 4 can be found in [12, Thm 8.1.4]; Theorems 5, 6, 7 can be found in [1, Thm 1.5.2, Thm 1.5.3, Thm 10.9.5] In this setting, Σ is any finite alphabet.

Theorem 4 (Fine and Wilf [8]). *Let w be a word having periods p and q, with $p \leq q$, and suppose that $|w| \geq p + q - \gcd(p,q)$. Then w also has period $\gcd(p,q)$.*

Theorem 5 (Lyndon and Schützenberger [13]). *Let $y \in \Sigma^*$ and $x, z \in \Sigma^+$. Then $xy = yz$ if and only if there exist $u, v \in \Sigma^*$ and an integer $e \geq 0$ such that $x = uv$, $z = vu$, and $y = (uv)^e u$.*

Theorem 6 (Lyndon and Schützenberger [13]). *Let $x, y \in \Sigma^+$. Then the following three conditions are equivalent:*

1. *$xy = yx$;*
2. *There exist integers $i, j > 0$ such that $x^i = y^j$;*
3. *There exist $z \in \Sigma^+$ and integers $k, \ell > 0$ such that $x = z^k$ and $y = z^\ell$.*

A word $\mathbf{w} \in \Sigma^\omega$ is *recurrent* if every finite subword of $\mathbf{w}$ occurs infinitely often. It is *uniformly recurrent* if for each finite subword x of $\mathbf{w}$ there exists an integer m, such that every subword of $\mathbf{w}$ of length m contains x. A morphism $h : \Sigma^* \to \Sigma^*$ is *primitive* if there exists an integer n such that for all $a, b \in \Sigma$ we have $b \lhd h^n(a)$.

Theorem 7. *Let $h : \Sigma^* \to \Sigma^*$ be a primitive morphism, prolongable on a. Then $h^\omega(a)$ is uniformly recurrent.*

3   Powers Structure

For the rest of this section, $\Sigma = \{0, 1\}$; $f : \Sigma^* \to \Sigma^*$ is a k-uniform morphism prolongable on 0; and $\mathbf{w} = f^\omega(0)$.

Lemma 8. *Let ρ, σ and λ be the shared prefix, shared suffix, and shared size of f, respectively. Suppose $z = w_i \cdots w_j \lhd \mathbf{w}$ is an $(n + p/q)$-power. Then*

$$E(\mathbf{w}) \geq n + \frac{p(k-1) + \lambda}{q(k-1)} .$$

Proof. If $f(0) = f(1)$ or $f(0) = 0^k$ or $f(1) = 1^k$, it is easy to see that $E(\mathbf{w}) = \infty$. Otherwise, $0 \leq \lambda \leq k - 1$; also, f is primitive, thus $\mathbf{w}$ is recurrent by Theorem 7, and we can assume $i > 0$. Let $z = x^n y$, where $x = a_0 \cdots a_{q-1}$, $y = a_0 \cdots a_{p-1}$. Let $f(w_{i-1}) = t\sigma$ and $f(w_{j+1}) = \rho s$ for some $s, t \in \Sigma^*$. Applying f to $w_{i-1} \cdots w_{j+1}$, we get a subword of $\mathbf{w}$ which is a fractional power with period kq, as illustrated in Fig. 1.

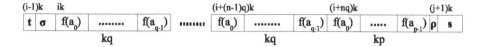

Fig. 1. Applying f to $w_{i-1} \cdots w_{j+1}$

Since σ is a shared suffix, it is a suffix of $f(a_{q-1})$ as well; similarly, ρ is a prefix of $f(a_p)$. Therefore, we can stretch the kq-period of $f(z)$ by σ to the left and ρ to the right. We get that $z' = \sigma f(z)\rho$ is an $(n + (kp + \lambda)/kq)$-power. Note that $kp + \lambda < kq$, since $\lambda < k$ and $p < q$.

The process of applying f and stretching the resulting power can be repeated infinitely. Successive applications of f give a sequence of powers $\{n + p_m/q_m\}_{m \geq 0}$, which satisfy $p_0 = p$, $q_0 = q$, and for $m > 0$, $p_m = kp_{m-1} + \lambda$, and $q_m = kq_{m-1}$. Let $\pi : \Sigma^* \times \mathbb{Q} \to \Sigma^* \times \mathbb{Q}$ be the map defined by

$$\pi\left(z, n + \frac{x}{y}\right) = \left(\sigma f(z)\rho, n + \frac{kx + \lambda}{ky}\right) . \tag{2}$$

We use $\pi(z)$ and $\pi(n + p/q)$ to denote the first and second component, respectively. Iterating π on $(n + p/q)$, we get

$$\pi^m\left(n + \frac{p}{q}\right) = n + \frac{k^m p + \lambda \sum_{i=0}^{m-1} k^i}{k^m q} = n + \frac{k^m p + \lambda \frac{k^m - 1}{k - 1}}{k^m q} \quad \overset{m \to \infty}{}$$

$$n + \frac{p(k-1) + \lambda}{q(k-1)}.$$

The lemma's assertion follows.   □

Our goal is to show that the π mapping defined in (2) is what generates $E(\mathbf{w})$. To do that, we need to rule out arbitrary powers. We start with a few lemmas which describe power behavior in a more general setting, namely, in an infinite word $\mathbf{v} = h(\mathbf{u})$, where h is a k-uniform binary morphism and $\mathbf{u} \in \Sigma^\omega$ is an arbitrary infinite word. We consider 4 cases of block size q: $q \equiv 0 \pmod{k}$; $q \not\equiv 0 \pmod{k}$ and $q > 2k$; $k < q < 2k$; and finally $q < k$.

Definition 9. Let h be a binary k-uniform morphism, and let $\mathbf{v} = h(\mathbf{u})$ for some $\mathbf{u} \in \Sigma^\omega$. We refer to the decomposition $\mathbf{v}$ into images of h as decomposition into *k-blocks*. Let $\alpha = v_i \cdots v_j \lhd \mathbf{v}$. The *outer closure* and *inner closure* of α, denoted by $\hat{\alpha}$, $\check{\alpha}$, respectively, are the following subwords of $\mathbf{v}$:

$$\hat{\alpha} = v_{\hat{i}} \cdots v_{\hat{j}}, \quad \hat{i} = \left\lfloor \frac{i}{k} \right\rfloor k, \quad \hat{j} = \left\lceil \frac{j+1}{k} \right\rceil k - 1 ;$$

$$\check{\alpha} = v_{\check{i}} \cdots v_{\check{j}}, \quad \check{i} = \left\lceil \frac{i}{k} \right\rceil k, \quad \check{j} = \left\lfloor \frac{j+1}{k} \right\rfloor k - 1 .$$

Thus $\hat{\alpha}$ consists of the minimal number of k-blocks that contain $v_i \cdots v_j$; similarly, $\check{\alpha}$ consists of the maximal number of k-blocks that are contained in $v_i \cdots v_j$.

By this definition, both $\hat{\alpha}$ and $\check{\alpha}$ have inverse images under h, denoted by $h^{-1}(\hat{\alpha})$ and $h^{-1}(\check{\alpha})$, respectively. Note that $\check{\alpha}$ may be empty.

Lemma 10. *Let h be an injective binary k-uniform morphism, let $\mathbf{v} = h(\mathbf{u})$ for some $\mathbf{u} \in \Sigma^\omega$, and let $\alpha = v_i \cdots v_j \lhd \mathbf{v}$ be an unstretchable $(n + p/q)$-power, $\alpha = Q^{n+p/q}$ where $Q \in \Sigma^q$. Suppose $q = 0 \pmod{k}$. Then α is an image under the π mapping defined in (2).*

Proof (sketch). Let $\alpha' = h^{-1}(\check{\alpha})$. Since h is injective, we have $h(0) \neq h(1)$, thus α' is an $(n' + p'/q')$-power, where $q' = q/k$ and $n - 1 \leq n' \leq n$. Use the fact that $q \geq k$ to show that $n - n' < 2$; use the maximality of $|\rho|$ and $|\sigma|$ to show that if the q period of α is unstretchable, then $\alpha = \sigma\check{\alpha}\rho$, and thus $n = n'$ and $p = kp' + \lambda$.   □

Lemma 11. *Let h be a binary k-uniform morphism, let $\mathbf{v} = h(\mathbf{u})$ for some $\mathbf{u} \in \Sigma^\omega$, and let $\alpha = v_i \cdots v_j \lhd \mathbf{v}$ be an $(n + p/q)$-power, $\alpha = Q^{n+p/q}$ where $Q \in \Sigma^q$. Suppose $q > 2k$ and $q \not\equiv 0 \pmod{k}$. Then either $h(0) = h(1)$, or α is reducible to a power $n' + r/s$, which satisfies $s \leq k$. If the second case holds, we get one of the following:*

1. $Q = u^c$ *for some* $c \in \mathbb{Z}_{\geq 4}$ *and* $u \in \Sigma^+$ *satisfying* $|u| < k$;
2. $n = 2$ *and* $q < 3k$, *i.e.,* $|\alpha| < 6k + 2$.

Proof (sketch). Consider the decomposition of α into k-blocks. We can assume w.l.o.g. that the decomposition starts from the first character of α (index i). Since $q \not\equiv 0 \pmod{k}$, we get overlaps of k-blocks, as illustrated in Fig. 2.

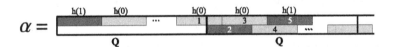

Fig. 2. Overlaps of k-blocks. The bold rectangles denote the power's q-blocks; the light grey and dark grey rectangles stand for $h(0), h(1)$, respectively; the top line of h rectangles stands for the k-decomposition of α; and the bottom line shows the repetition of the q-block.

Since $q > 2k$, there are at least 5 k-blocks involved in the overlap. Each overlap induces a partition on the k-blocks involved, resulting in a system of word equations. The idea of the proof is to do an overlap case analysis, and solve the system for each case. For case notation, we order the k-blocks by their starting index (the numbers 1-5 in Fig. 2), and denote the case by the resulting 5-letter binary word; in the Fig. 2 case, it is 01001. By symmetry arguments, it is enough to consider words that start with 0, therefore we need to consider 16 overlap combinations. We mark the k-block partition induced by the overlaps using the letters x, y, z, t, where x marks the leftmost part; since the k-decomposition starts from the first letter of α, we have $|x| = q \bmod k$, i.e., $|x| > 0$. We give here as an example two simple cases, which are illustrated in Fig. 3.

Fig. 3. Overlaps of k-blocks for cases 00010, 01001

00010: $h(0) = xy = yx = h(1)$.
01001: $h(0) = xy = yz$, $h(1) = yx = zt$. Using Theorems 5, 6, it can be shown that $x = z$ and $y = t$, thus $h(0) = h(1)$. □

Lemma 12. *Let h be a binary k-uniform morphism, let $\mathbf{v} = h(\mathbf{u})$ for some $\mathbf{u} \in \Sigma^\omega$, and let $\alpha = v_i \cdots v_j \lhd \mathbf{v}$ be an $(n + p/q)$-power, $\alpha = Q^{n+p/q}$ where $Q \in \Sigma^q$. Suppose $k < q < 2k$. Then either $h(0) = h(1)$, or α is reducible to a power $n' + r/s$, which satisfies $s \le k$.*

Proof (sketch). If $n \ge 3$, or $n = 2$ and $p \ge 4k - 2q$, then again there are at least 5 k-blocks involved in the overlap. Thus we can assume that $n = 2$, $p < 4k - 2q$. The proof now consists of analyzing 4 overlap cases $(000, 001, 010, 011)$, and showing each of them to be reducible. □

Corollary 13. *In the setting of f and $\mathbf{w}$, assume $E(\mathbf{w}) < \infty$, and let $\mathcal{E}'$ be the set of exponents $r = n + p/q$, such that $q < k$ and $\mathbf{w}$ contains an r-power. Then*

$$E(\mathbf{w}) = \max_{n+p/q \in \mathcal{E}'} \left\{ n + \frac{p(k - 1) + \lambda_f}{q(k - 1)} \right\} . \tag{3}$$

Proof. By Lemmata 10, 11, 12, if $z \lhd \mathbf{w}$ is an irreducible, unstretchable $(n+p/q)$-power for which $q \ge k$, then z is an image under the π mapping; thus the exponent of every such power is an element of a sequence of the form $\{\pi^i(r)\}_{i=0}^\infty$, where $r \in \mathcal{E}'$. The set $\mathcal{E}'$ is finite, since $E(\mathbf{w}) < \infty$, thus there are only finitely many such sequences. By Lemma 8, the limit of each such sequence is given by the expression in (3); and since each of these sequences increases towards its limit, the critical exponent is the maximum of those limits. □

Lemma 14. *Let h be a binary k-uniform morphism, let $\mathbf{v} = h(\mathbf{u})$ for some $\mathbf{u} \in \Sigma^\omega$, and let $\alpha = v_i \cdots v_j \lhd \mathbf{v}$ be an $(n + p/q)$-power, $\alpha = Q^{n+p/q}$ where $Q \in \Sigma^q$. Suppose $q < k$. Then at least one of the following holds:*

1. *$q | k$;*
2. *$q \nmid k$ and $q | 2k$;*
3. *α is reducible;*
4. *$n < 3k/q$.*

Proof (sketch). Assume w.l.o.g. that $i \equiv 0 \pmod{k}$. Suppose $n \ge 3k/q$. Then $|\alpha| \ge 3k$; therefore, $h(a_1 a_2 a_3)$ is a prefix of α for some $a_1, a_2, a_3 \in \Sigma$. We now use the fact that any 3-letter word over a binary alphabet either contains a square, or is a $1\frac{1}{2}$-power; thus $h(a_1 a_2 a_3)$ either contains a subword of length at least $2k$ which has a k-period, or it itself has a $2k$-period. The result now follows from Theorem 4. □

Corollary 15. *If $q < k$ and α is irreducible, then at least one of the following holds:*

1. $h(0) = h(1)$;
2. $h^{-1}(\hat{\alpha}) = ac^\ell b$, *where* $a, b \in \{0, 1, \epsilon\}$, $c \in \{0, 1\}$, *and* $\ell \geq 0$;
3. $h^{-1}(\hat{\alpha}) = ax^\ell b$, *where* $a, b \in \{0, 1, \epsilon\}$, $x \in \{01, 10\}$, *and* $\ell \geq 0$;
4. $|h^{-1}(\hat{\alpha})| \leq 5$.

Corollary 16. *In the setting of f and $\mathbf{w}$, $E(\mathbf{w}) = \infty$ if and only if at least one of the following holds: $f(0) = f(1)$, $f(0) = 0^k$, $f(1) = 1^k$, or $f(0) = (01)^m 0$ and $f(1) = (10)^m 1$, where $k = 2m + 1$.*

Proof. It is easy to see that any of the 4 conditions implies $E(\mathbf{w}) = \infty$. For the converse, suppose $f(0) \neq f(1)$, and $\mathbf{w}$ contains unbounded powers. Then by Lemmata 10, 11, 12, 14 and Corollary 15, $\mathbf{w}$ must contain unbounded powers of the form 0^m, 1^m, or $(01)^m$. If it contains unbounded 0^m powers, then $f(a) = 0^k$ for some $a \in \Sigma$. Suppose $f(1) = 0^k$. Then $\mathbf{w}$ must contain unbounded 1^m powers as well, and so necessarily $f(0) = 1^k$, a contradiction: f is prolongable on 0. Thus $\mathbf{w}$ contains unbounded 0^m powers if and only if $f(0) = 0^k$, and similarly it contains unbounded 1^m powers if and only if $f(1) = 1^k$. Finally, it is easy to see using similar inverse image arguments that $\mathbf{w}$ contains unbounded $(01)^m$ powers if and only if the last condition holds. $\square$

Note. By Frid [9], the four conditions of Corollary 13 characterize *non-circularity* in uniform binary fixed points. More on circularity in Section 4.

To complete the proof of Theorem 2, it remains to show that in order to compute $E(\mathbf{w})$, it is enough to consider $f^4(0)$. We do this in the following lemma.

Lemma 17. *Suppose $E(\mathbf{w}) < \infty$. Let $z \lhd \mathbf{w}$ be an irreducible $(n + p/q)$-power, satisfying $q < k$. Then $z \lhd f^4(0)$ and $n \in O(k^3)$. The bound on the first occurrence of z is tight.*

Proof (sketch). First, we show that any subword of $\mathbf{w}$ of the form ab, a^ℓ, or $(a\bar{a})^\ell$, where ℓ is a positive integer, $a, b \in \Sigma$ and $\bar{a} = 1 - a$, must occur in $f^2(0)$ or $f^3(0)$. We then apply Corollary 15. $\square$

Corollary 18. *Suppose $E(\mathbf{w}) < \infty$. Let $\mathcal{E}$ be the set of exponents $r = n + p/q$, such that $q < k$ and $f^4(0)$ contains an r-power. Then*

$$E(\mathbf{w}) = \max_{n+p/q \in \mathcal{E}} \left\{ n + \frac{p(k-1) + \lambda_f}{q(k-1)} \right\} . \tag{4}$$

Proof. This is an immediate result of Corollary 13 and Lemma 17. $\square$

Corollary 18 completes the proof of Theorem 2. We end this section with a couple of examples.

Example 19. As implied by the tightness assertions of Lemma 17, the prefix $f^4(0)$ is best possible. Consider the morphism $0 \to 010101$, $1 \to 000110$. In this example, $E(\mathbf{w}) = 12\frac{3}{5}$, and the first occurrence of a 12-power is in $f^4(0)$.

Example 20. Let r, s be natural numbers satisfying $0 < r \leq s$. Let f be the following binary $(s+1)$-uniform morphism:

$$f : \begin{array}{l} 0 \to 01^s; \\ 1 \to 01^{r-1}0^{s-r+1}. \end{array}$$

Then f is an $(s+1)$-uniform morphism, satisfying $\rho_f = 01^{r-1}$, $\sigma_f = \epsilon$, and $\lambda_f = r$. Let $\mathbf{w} = f^\omega(0)$. Then 1^s is a subword of $f^1(0)$; also, $0^{s(s+1)+1}$ is a subword of $f^3(0)$ if $r = 1$. Set $z = 1^s$ for $r > 1$ and $z = 0^{s(s+1)+1}$ for $r = 1$. It is easy to check that by applying π to z we get the maximal number in the set $\left\{ n + \frac{p(k-1)+\lambda_f}{q(k-1)} : n + p/q \in \mathcal{E} \right\}$; thus

$$r > 1 \Rightarrow E(\mathbf{w}) = s + \frac{0 \cdot s + r}{1 \cdot s} = s + \frac{r}{s} \;;$$

$$r = 1 \Rightarrow E(\mathbf{w}) = s(s+1) + 1 + \frac{r}{s} \;.$$

Corollary 21. *For any rational number $0 < t < 1$ there exist a binary k-uniform morphism f, such that $E(\mathbf{w}) = n + t$ for some $n \in \mathbb{Z}_{\geq 2}$.*

4   Generalizing the Results

The definitions of ρ, σ, π (Definition 1, Equation (2)) can be generalized to arbitrary morphisms over finite alphabets. Let $\Sigma = \Sigma_t = \{0, \ldots, t-1\}$, let $f : \Sigma^* \to \Sigma^*$ be a morphism prolongable on 0, and let $\mathbf{w} = f^\omega(0)$. For a word $u \in \Sigma^*$, let $[u]$ be the *Parikh vector* of u, i.e., $[u] = (|u|_0, \ldots, |u|_{t-1})^T$, where $|u|_i$ is the number of occurrences of the letter i in u. Let F be the *incidence matrix associated with f*, i.e., $F_{i,j} = |f(j)|_i$, $0 \leq i, j < t$. It is easy to check that for all $u \in \Sigma^*$ we have $[f(u)] = F[u]$.

Let $z = w_i \cdots w_j \lhd \mathbf{w}$ be an $(n + p/q)$-power, $z = x^n y$, and let Q, P be the Parikh vectors of x, y respectively. In order to keep track of the components $|x|_i, |y|_i$, we introduce the notation "z is an $(n+P/Q)$-power", where P/Q stands for $\sum_{i=0}^{t-1} |y|_i / \sum_{i=0}^{t-1} |x|_i$. Under this notation, $f^m(z)$ is an $(n + F^m P/F^m Q)$-power; this power may be stretchable.

Assume $E(\mathbf{w}) < \infty$, and let $z = w_i \cdots w_j \lhd \mathbf{w}$ be an unstretchable $(n+P/Q)$-power. Define

$$\pi\left(z, n + \frac{P}{Q}\right) = \left(\sigma f(z)\rho, n + \frac{FP + \Lambda}{FQ}\right), \tag{5}$$

where $\sigma, \rho \in \Sigma^*$ are the words that stretch the FQ period of $f(z)$ on the left and on the right, respectively, to an unstretchable power, and $\Lambda = [\sigma\rho]$. We call Λ the *stretch vector* of $(f(z), FQ)$. Note that $\pi(z, n+P/Q)$ depends on the context of z, in particular on the letters w_{i-1}, w_{j+1}. Iterating π on the initial power z, we get a sequence of stretch vectors, $\{\Lambda_m\}_{m \geq 0}$, and we have:

$$\pi^m \left(n + \frac{P}{Q} \right) = n + \frac{F^m P + \sum_{i=0}^{m-1} F^{m-1-i} \Lambda_i}{F^m Q}. \tag{6}$$

We call the sequence $\{\pi^m(n + P/Q)\}_{m \geq 0}$ a π-*sequence*. If the sequence $\{\Lambda_m\}_{m \geq 0}$ is ultimately periodic, it can be shown that in the uniform case, the π-sequence converges to a rational number, and in the non-uniform case, its limsup is a rational expression of the eigenvalues of F. In particular, it is algebraic of degree at most t.

To prove that $E(\mathbf{w})$ is algebraic, we need to show that the sequence of stretch vectors is indeed ultimately periodic for every choice of initial power z; we also need to show that $E(\mathbf{w})$ is generated by the π mapping, i.e., we need to rule out arbitrary powers. The overlap analysis method we used in the binary uniform case is too tedious for uniform morphisms over larger alphabets, and will not work at all for non-uniform morphisms. Instead, we use *circularity* arguments.

Basically, a fixed point $f^\omega(0)$ is circular if every sufficiently long subword of it has an unambiguous decomposition into images under f, save maybe for a prefix and a suffix of bounded length. The notion was introduced by Mignosi and Séébold in [16], where they showed that bounded critical exponent implies circularity in fixed points of morphisms over a finite alphabet; see also [17], [6]. In [11], we use circularity arguments to show that when f is an injective morphism and $E(\mathbf{w}) < \infty$, the following holds:

1. the sequence of stretch vectors is ultimately periodic for every choice of initial power z;
2. every unstretchable power with a sufficiently long power block belongs to some π-sequence;
3. there are only finitely many distinct π-sequences occurring in $\mathbf{w}$.

Thus we have proved that if $\mathbf{w}$ is a fixed point of an injective morphism, then either $E(\mathbf{w}) = \infty$, or $E(\mathbf{w})$ is algebraic of degree at most t, where t is the alphabet size. Our method also gives an algorithm for computing $E(\mathbf{w})$, which essentially reduces the problem to computing the Jordan decomposition of the incidence matrix. In particular, computing the critical exponent of the Fibonacci word becomes almost trivial.

It yet remains to extend the results to non-injective morphisms; nevertheless, in light of the observations above, the following conjecture seems reasonable:

Conjecture 22. *Let f be a morphism over a finite alphabet, and let $\mathbf{w}$ be an infinite fixed point of f. Assume $E(\mathbf{w}) < \infty$. Then*

1. *if f is uniform, then $E(\mathbf{w})$ is rational;*
2. *if f is non-uniform, then $E(\mathbf{w}) \in \mathbb{Q}[s_1, \ldots, s_\ell]$, where $s_1, \ldots, s_\ell$ are the eigenvalues of the incidence matrix of f. In particular, $E(\mathbf{w})$ is algebraic.*

Acknowledgements

The author would like to thank Jeffrey Shallit, for his comments and suggestions; Anna Frid, for her comments about circularity; and Kalle Saari and Narad Rampersad, for helpful discussions.

References

1. J.-P. Allouche and J. Shallit. *Automatic Sequences: Theory, Applications, Generalizations*. Cambridge University Press (2003).
2. J. Berstel. Axel Thue's papers on repetitions in words: a translation. Publications du Laboratoire de Combinatoire et d'Informatique Mathématique **20**, Université du Québec à Montréal (1995).
3. J. Berstel. On the index of Sturmian words. In *Jewels are Forever*, Springer, Berlin (1999), 287–294.
4. W.-T. Cao and Z.-Y. Wen. Some properties of the factors of Sturmian sequences. *Theoret. Comput. Sci.* **304** (2003), 365–385.
5. A. Carpi and A. de Luca. Special factors, periodicity, and an application to Sturmian words. *Acta Informatica* **36** (2000), 983–1006.
6. J. Cassaigne. An algorithm to test if a given circular HD0L language avoids a pattern. In *IFIP World Computer Congress'94*, Vol **1**, Elsevier (North-Holland) (1994), 459–464.
7. D. Damanik and D. Lenz. The index of Sturmian sequences. *European J. Combin.* **23** (2002), 23–29
8. N. J. Fine and H. S. Wilf. Uniqueness theorems for periodic functions. *Proc. Amer. Math. Soc.* **16** (1965), 109–114
9. A. E. Frid. On uniform D0L words. *STACS'98, LNCS* **1373** (1998), 544–554.
10. J. Justin and G. Pirillo. Fractional powers in Sturmian words. *Theoret. Comput. Sci.* **255** (2001), 363–376.
11. D. Krieger. On critical exponents in fixed points of injective morphisms. Preprint.
12. M. Lothaire. *Algebraic Combinatorics on Words*, Vol. 90 of *Encyclopedia of Mathematics and Its Applications*. Cambridge University Press (2002).
13. R. C. Lyndon and M. P. Schützenberger. The equation $a^M = b^N c^P$ in a free group. *Michigan Math. J.* **9** (1962), 289–298.
14. F. Mignosi. Infinite words with linear subword complexity. *Theoret. Comput. Sci.* **65** (1989), 221–242.
15. F. Mignosi and G. Pirillo. Repetitions in the Fibonacci infinite word. *RAIRO Inform. Théor.* **26** (1992), 199–204.
16. F. Mignosi and P. Séébold. If a D0L language is k-power free then it is circular. *ICALP'93, LNCS* **700** (1993), 507–518.
17. B. Mossé. Puissances de mots et reconnaissabilité des points fixes d'une substitution. *Theoret. Comput. Sci.* **99** (1992), 327–334.
18. A. Thue. Über die gegenseitige Lage gleicher Teile gewisser Zeichenreihen. *Norske vid. Selsk. Skr. Mat. Nat. Kl.* **1** (1912), 1–67.
19. D. Vandeth. Sturmian words and words with critical exponent. *Theoret. Comput. Sci.* **242** (2000), 283–300.

Equivalence of $\mathbb{F}$-Algebras and Cubic Forms*

Manindra Agrawal and Nitin Saxena

IIT Kanpur, India
{manindra, nitinsa}@cse.iitk.ac.in

Abstract. We study the isomorphism problem of two "natural" algebraic structures – $\mathbb{F}$-algebras and cubic forms. We prove that the $\mathbb{F}$-algebra isomorphism problem reduces in polynomial time to the cubic forms equivalence problem. This answers a question asked in [AS05]. For finite fields of the form $3 \nmid (\#\mathbb{F} - 1)$, this result implies that the two problems are infact equivalent. This result also has the following interesting consequence:

Graph Isomorphism $\leq_m^P$ $\mathbb{F}$-algebra Isomorphism $\leq_m^P$ Cubic Form Equivalence.

1 Introduction

For a field $\mathbb{F}$, $\mathbb{F}$-*algebras* are commutative rings of finite dimension over $\mathbb{F}$. One of the fundamental computational problems about $\mathbb{F}$-algebras is to decide, given two such algebras, if they are isomorphic. When $\mathbb{F}$ is an algebraically closed field, it follows from Hilbert's Nullstellensatz [Bro87] that the problem can be decided in **PSPACE**. When $\mathbb{F} = \mathbb{R}$, the problem is in **EEXP** due to the result of Tarski on the decidability of first-order equations over reals [DH88]. When $\mathbb{F} = \mathbb{Q}$, it is not yet known if the problem is decidable. When $\mathbb{F}$ is a finite field, the problem is in **NP** $\cap$ **coAM** [KS05]. In all of the above results, we assume that an $\mathbb{F}$-algebra is presented by specifying the product of its basis elements over $\mathbb{F}$.

$\mathbb{F}$-*Cubic Forms* are homogeneous degree 3 polynomials over field $\mathbb{F}$. We call two such forms *equivalent* if an invertible linear transformation on the variables makes one equal to the other. The problem of equivalence of $\mathbb{F}$-cubic forms has a very similar complexity to that of $\mathbb{F}$-algebra isomorphism for different $\mathbb{F}$. This follows from the result of [AS05] showing that $\mathbb{F}$-cubic form equivalence reduces, in polynomial time, to $\mathbb{F}$-algebra isomorphism (in case $\mathbb{F}$ is a finite field, the result holds for $3 \nmid (\#\mathbb{F} - 1)$ due to technical reasons).

Both the problems have been well studied in mathematics (for instance see [Har75, MH74], [Rup03]). Over the last ten years, these problems have been found to be useful in computer science as well: [Pat96, CGP98] proposes a cryptosystem based on the hardness of the cubic form equivalence over finite fields, [AS05] show

---

* This work was done while the authors were visiting National University of Singapore. The second author was also supported by research funding from Infosys Technologies Limited, Bangalore.

that the Graph Isomorphism problem reduces to both $\mathbb{F}$-algebra isomorphism and $\mathbb{F}$-cubic form equivalence for any $\mathbb{F}$. Therefore, the two problems are of an intermediate complexity but seemingly harder than Graph Isomorphism.

Of the two problems, cubic form equivalence might appear to be an easier problem because, for example, the reduction from Graph Isomorphism to $\mathbb{F}$-algebra isomorphism is simple while the reduction to $\mathbb{F}$-cubic form equivalence is very involved. In this paper, we show that this is not the case by exhibiting a reduction from $\mathbb{F}$-algebra isomorphism to $\mathbb{F}$-cubic form equivalence. Apart from showing that the two problems are essentially equivalent, this has other interesting implications. For example, this suggests that $\mathbb{Q}$-algebra isomorphism is decidable because $\mathbb{Q}$-cubic form equivalence appears to be decidable due to the rich structure they possess.

Our reduction is a two step process. We first reduce $\mathbb{F}$-algebras to local $\mathbb{F}$-algebras of a special form. Then we use the properties of these local algebras to show that a "natural" construction of $\mathbb{F}$-cubic forms works.

In section 2 we give an overview of the reduction. In section 3 we reduce general $\mathbb{F}$-algebra isomorphism to the isomorphism problem for local $\mathbb{F}$-algebras and in section 4 we reduce $\mathbb{F}$-algebra isomorphism problem to $\mathbb{F}$-cubic form equivalence.

2   The Basics

An $\mathbb{F}$-algebra R is a commutative ring containing field $\mathbb{F}$. We assume that R is specified in terms of its additive generators over $\mathbb{F}$, say $b_1, \ldots, b_n$. Thus, $R = \mathbb{F}b_1 \oplus \ldots \oplus \mathbb{F}b_n$. To completely specify R, the product of pairs of basis elements is given in terms of a linear combination of b's. Thus, a's $\in \mathbb{F}$ are given in the input such that:

$$\forall i, j, \quad b_i b_j = \sum_{1 \le k \le n} a_{ij,k} b_k \tag{1}$$

Let S be another $\mathbb{F}$-algebra with basis elements $b_1, \ldots, b_n$ satisfying:

$$\forall i, j, \quad b_i b_j = \sum_{1 \le k \le n} a'_{ij,k} b_k$$

To specify an isomorphism ψ from R to S it is sufficient to describe $\psi(b_i)$, for each i, as a linear combination of $b_1, \ldots, b_n$ in S.

The isomorphism problem for these $\mathbb{F}$-algebras is related to polynomial equivalence problem over $\mathbb{F}$ because we can combine equations (1), by using new variables $\bar{z}$ for various i, j, to construct:

$$f_R(\bar{z}, \bar{b}) := \sum_{1 \le i \le j \le n} z_{ij} \left(b_i b_j - \sum_{1 \le k \le n} a_{ij,k} b_k \right) \tag{2}$$

In the above expression we consider z_{ij} and b_i as formal variables and thus f_R is a degree-3 or cubic polynomial. Similarly, construct $f_S(\bar{z}, \bar{b})$ from S. It was shown in [AS05] that equivalence of the polynomials f_R and f_S is sufficient to decide whether R and S are isomorphic. If ϕ is an isomorphism from R to S then it is easy to see that there is a linear invertible map τ on $\overline{z}$ such that: $f_R(\tau\overline{z}, \phi\overline{b}) = f_S(\overline{z}, \overline{b})$. More work is needed to show that if f_R is equivalent to f_S then infact $R \cong S$. The main idea being that any equivalence ψ from f_R to f_S will map b_i's to a linear combination of $\overline{b}$'s and hence ψ becomes our natural candidate for an isomorphism from R to S (for details see [AS05]).

The question we resolve in this paper is whether there is a way to construct *homogeneous* cubic polynomials, *i.e.* cubic forms over $\mathbb{F}$ (henceforth referred to as $\mathbb{F}$-*cubic forms*), such that their equivalence implies the isomorphism of R and S. The cubic form we construct looks like:

$$g_R(\overline{z}, \overline{b}, v) := \sum_{1 \leq i \leq j \leq n} z_{ij} \left(b_i b_j - v \cdot \sum_{1 \leq k \leq n} a_{ij,k} b_k \right) \tag{3}$$

Here v is a new formal variable. We reduce $\mathbb{F}$-algebra isomorphism to $\mathbb{F}$-cubic form equivalence by first constructing special $\mathbb{F}$-algebras R', S' from R, S (in section 3) and then showing that equivalence of $g_{R'}, g_{S'}$ implies the isomorphism of R, S (in section 4). The idea again is to show that any equivalence ψ from $g_{R'}(\overline{z}, \overline{b}, v)$ to $g_{S'}(\overline{z}, \overline{b}, v)$ sends b_i's to a linear combination of $\overline{b}$'s and thus ψ leads us to an isomorphism from R to S.

3   Local $\mathbb{F}$-Algebra Isomorphism Problem

An $\mathbb{F}$-algebra is *local* if it cannot be broken into simpler $\mathbb{F}$-algebras *i.e.* if it cannot be written as a direct product of algebras. Given an $\mathbb{F}$-algebra this direct product decomposition can be done by factoring polynomials over the field $\mathbb{F}$. Any non-unit r in a local $\mathbb{F}$-algebra is *nilpotent* i.e., there is an m such that $r^m = 0$ (see [McD74]).

In this section we give a many-to-one reduction from $\mathbb{F}$-algebra isomorphism to local $\mathbb{F}$-algebra isomorphism. Moreover, the local $\mathbb{F}$-algebras that we construct have basis elements most of whose products vanish. We exploit the properties of this local $\mathbb{F}$-algebra to give a reduction from $\mathbb{F}$-algebra to cubic forms in the next section.

Theorem 3.1. $\mathbb{F}$-*algebra isomorphism* $\leq_m^P$ *Local* $\mathbb{F}$-*algebra isomorphism.*

Proof. Given two $\mathbb{F}$-algebras R and S, [AS05] constructs two cubic polynomials p and q respectively such that p, q are equivalent iff R, S are isomorphic. These polynomials look like (as in equation (2))):

$$p(\overline{z}, \overline{b}) = \sum_{1 \leq i \leq j \leq n} z_{ij} \left(b_i b_j - \sum_k a_{ij,k} b_k \right)$$

$$q(\overline{z}, \overline{b}) = \sum_{1 \leq i \leq j \leq n} z_{ij} \left(b_i b_j - \sum_k a'_{ij,k} b_k \right)$$

Let

$$p_3(\bar{z}, \bar{b}) = \sum_{1 \le i \le j \le n} z_{ij} b_i b_j \ \text{and} \ p_2(\bar{z}, \bar{b}) = - \sum_{1 \le i \le j \le n} \left(z_{ij} \sum_k a_{ij,k} b_k \right). \quad (4)$$

Similarly define $q_3(\bar{z}, \bar{b})$ and $q_2(\bar{z}, \bar{b})$ from q. Thus, $p = p_3 + p_2$ and $q = q_3 + q_2$ where p_3, q_3 are homogeneous of degree 3 and p_2, q_2 are homogeneous of degree 2.

Using p, q we construct the following $\mathbb{F}$-algebras:

$$R' := \mathbb{F}[\bar{z}, \bar{b}, u] / \langle p_3, up_2, u^2, \mathcal{I} \rangle$$
$$S' := \mathbb{F}[\bar{z}, \bar{b}, u] / \langle q_3, uq_2, u^2, \mathcal{I} \rangle \quad (5)$$

where, $\mathcal{I}$ is the ideal generated by all possible products of 4 variables.

Note that all the variables in R', S' are nilpotent and hence the two rings are *local* $\mathbb{F}$-algebras (see [McD74]). The following claim tells us that it is enough to consider the isomorphism problem for these local structures. Recall that $R \cong S$ iff p, q are equivalent polynomials.

Claim 3.1.1. $p(\bar{z}, \bar{b}), q(\bar{z}, \bar{b})$ *are equivalent polynomials iff* $R' \cong S'$.

Proof of Claim 3.1.1. If p, q are equivalent then the same equivalence, extended by sending $u \mapsto u$, gives an isomorphism from R' to S'.

Conversely, say ϕ is an isomorphism from R' to S'. Our intention is to show that the *linear part* of ϕ induces an equivalence from p to q. Note that since $\bar{z}, \bar{b}, u$ are nilpotents in R', therefore $\forall i \le j \in [n], k \in [n], \phi(z_{ij}), \phi(b_i), \phi(u)$ can have no constant term.

Let us see where ϕ sends u. Since $\phi(u)^2 = 0$ in S' while for all i, j: $z_{ij}^2, b_i^2 \ne 0$, the linear part of $\phi(u)$ can have no $\bar{z}, \bar{b}$'s. Thus,

$$\phi(u) = c \cdot u + (\text{terms of degree 2 or more}), \ \text{where } c \in \mathbb{F}. \quad (6)$$

Now by the definition of ϕ:

$\phi(p_3) = c_1 \cdot q_3 + c_2 \cdot uq_2 + (\text{linear terms in } \bar{z}, \bar{b}, u) \cdot u^2 + (\text{terms of degree 4 or more}),$ where $c_1, c_2 \in \mathbb{F}$.

By substituting $u = 0$ we get,

$$\phi(p_3) |_{u=0} = c_1 q_3 + (\text{terms of degree 4 or more}) \quad (7)$$

Also,

$\phi(up_2) = d_1 \cdot q_3 + d_2 \cdot uq_2 + (\text{linear terms in } \bar{z}, \bar{b}, u) \cdot u^2 + (\text{terms of degree 4 or more}),$ where $d_1, d_2 \in \mathbb{F}$.

Using eqn (6) we deduce that $d_1 = 0$. Thus,

$$\phi(up_2) = d_2 \cdot uq_2 + (\text{linear terms in } \bar{z}, \bar{b}, u) \cdot u^2 + (\text{terms of degree 4 or more})$$

Again using eqn (6) we deduce:

$$u\phi(p_2) = d_2' \cdot uq_2 + (\text{linear terms in } \bar{z}, \bar{b}, u) \cdot u^2 + (\text{terms of degree 4 or more}),$$
where $d_2' \in \mathbb{F}$.

Factoring out u and substituting $u = 0$ gives us:

$$\phi(p_2)\mid_{u=0} = d_2' \cdot q_2 + (\text{terms of degree 3 or more}) \tag{8}$$

Let ψ be the linear part of ϕ after substituting $u = 0$, that is:

for all $i \leq j$, $\psi(z_{ij}) :=$ linear terms of $\phi(z_{ij})$ other than u and

for all i, $\psi(b_i) :=$ linear terms of $\phi(b_i)$ other than u

By comparing degree 3 and degree 2 terms on both sides of equations (7) and (8) respectively, we get:

$$\psi(p_3) = c_1 q_3 \tag{9}$$
$$\psi(p_2) = d_2' q_2 \tag{10}$$

Note that since ϕ is an isomorphism, ψ has to be an invertible map and thus, $\psi(p_3), \psi(p_2) \neq 0$. As a result c_1 and d_2' are both non-zero. Consider the map $\psi' := (\frac{d_2'}{c_1}) \circ \psi$. The above two equations give us: $\psi'(p_3 + p_2) = \frac{d_2'^3}{c_1^3} \cdot (q_3 + q_2)$. Denote $\frac{d_2'^3}{c_1^3}$ by c. Thus,

$$\psi'(p(\bar{z}, \bar{b})) = c \cdot q(\bar{z}, \bar{b})$$

Now we can get rid of the extra factor of c by defining a map ψ'':

$$\forall i, j, \ \psi''(z_{ij}) := \frac{1}{c} \psi'(z_{ij})$$
$$\forall i, \ \psi''(b_i) := \psi'(b_i)$$

It follows that $\psi''(p) = q$ and thus $p(\bar{z}, \bar{b}), q(\bar{z}, \bar{b})$ are equivalent. $\qquad \square$

Thus, $R \cong S$ iff $R' \cong S'$ and hence it is sufficient to study $\mathbb{F}$-algebra isomorphism over local $\mathbb{F}$-algebras of the form (5).

4   Cubic Form Equivalence

Given two cubic forms $f(\bar{x}), g(\bar{x})$ (homogeneous degree 3 polynomials over a field $\mathbb{F}$) the equivalence problem is to determine whether there is an invertible linear transformation (over the field $\mathbb{F}$) on the variables that makes the two forms equal. When field $\mathbb{F}$ is finite, cubic form equivalence is in $\mathbf{NP} \cap \mathrm{co}\mathbf{AM}$. For an infinite field $\mathbb{F}$ we expect the problem to be *decidable* but it is still open for $\mathbb{F} = \mathbb{Q}$.

Here we show that $\mathbb{F}$-algebra isomorphism reduces to cubic form equivalence. This improves the result of [AS05] that graph isomorphism reduces to cubic form equivalence. The proof involves the use of similar cubic forms as constructed in [AS05] but here we heavily use the properties of the intermediate local $\mathbb{F}$-algebras to study the equivalences of these cubic forms.

Theorem 4.1. $\mathbb{F}$-*algebra isomorphism* $\leq_m^P$ $\mathbb{F}$-*cubic form equivalence.*

Proof. Given $\mathbb{F}$-algebras R, S we will construct cubic forms ϕ_R, ϕ_S such that the cubic forms are equivalent iff the algebras are isomorphic. The construction involves first getting the local $\mathbb{F}$- algebras R', S' (as in thm 3.1) and then the cubic forms out of these local algebras (similar to [AS05]).

Let $b_1, \ldots, b_n$ be the additive basis of R over $\mathbb{F}$. Let the multiplication in the algebra be defined as:

$$\text{for all } i, j \in [n] : \; b_i \cdot b_j = \sum_{k=1}^{n} a_{ij,k} b_k, \text{ where } a_{ij,k} \in \mathbb{F}$$

Consider the following local ring R' constructed from R:

$$R' \; := \; \mathbb{F}[\bar{z}, \bar{b}, u] / \langle p_3, up_2, u^2, \mathcal{I} \rangle \tag{11}$$

where $p_3(\bar{z}, \bar{b}) := \sum_{1 \leq i \leq j \leq n} z_{ij} b_i b_j$ and $p_2(\bar{z}, \bar{b}) := \sum_{1 \leq i \leq j \leq n} z_{ij} (\sum_{k=1}^{n} a_{ij,k} b_k)$. $\mathcal{I}$ is the set of all possible products of 4 variables.

Similarly, construct S' from S and we know from thm 3.1 that $R \cong S$ iff $R' \cong S'$. Now we move on to constructing cubic forms from these local algebras R' and S'.

A natural set of generators of the ring R' is: $\{1\} \cup \{z_{ij}\}_{1 \leq i \leq j \leq n} \cup \{b_i\}_{1 \leq i \leq n} \cup \{u\}$. For simplicity let us call them $1, x_1, \ldots, x_g, u$ respectively, where $g := \binom{n+1}{2} + n$. A natural additive basis of R' over $\mathbb{F}$ is:

$$\{1\} \cup \{x_i\}_{1 \leq i \leq g} \cup \{u\} \cup \{x_i x_j\}_{1 \leq i \leq j \leq g} \cup \{u x_i\}_{1 \leq i \leq g} \cup \{x_i x_j x_k\}_{1 \leq i \leq j \leq k \leq g}$$
$$\cup \{u x_i x_j\}_{1 \leq i \leq j \leq g} \text{ minus oneterm each from } p_3 \text{ and } up_2. \tag{12}$$

For simplicity denote this additive basis by $1, c_1, \ldots, c_d$ respectively, where

$$d := g + 1 + \binom{g+1}{2} + g + \binom{g+2}{3} + \binom{g+1}{2} - 2 = 2g + 2\binom{g+1}{2} + \binom{g+2}{3} - 1$$

Finally, we construct a cubic form ϕ_R using R' as follows:

$$\phi_R(\bar{y}, \bar{c}, v) := \sum_{1 \leq i \leq j \leq d} y_{ij} c_i c_j - v \sum_{1 \leq i \leq j \leq d} y_{ij} \left(\sum_{k=1}^{d} \tilde{a}_{ij,k} c_k \right) \tag{13}$$

where $\forall i, j, \; c_i \cdot c_j = \sum_{k=1}^{d} \tilde{a}_{ij,k} c_k$ in R', for some $\tilde{a}_{ij,k} \in \mathbb{F}$.

Observe that the v terms in this cubic form are "few" because most of the $\tilde{a}$ are zero. This property is useful in analysing the equivalence of such forms. Let us bound the number of v terms in ϕ_R.

Claim 4.1.1. *The number of surviving v terms in the rhs of eqn (13) is $<$ $(3d - 6)$.*

Proof of Claim 4.1.1. The number of surviving v terms in the rhs of eqn (13) is:

$$\leq \#\{(k,l) \mid 1 \leq k \leq l \leq d, \; c_k c_l \neq 0 \text{ in } R'\} + 3\left[\#(\text{terms in } p_3) + \#(\text{terms in } p_2)\right]$$

The first expression above accounts for all the relations in R' of the form $c_k c_l = c_m$. The second expression takes care of the relations that arise from $p_3 = 0$ and $up_2 = 0$. The factor of 3 above occurs because a term $x_i x_j x_k$ in p_3, up_2 can create v terms in atmost 3 ways: from $(x_i) \cdot (x_j x_k)$ or $(x_j) \cdot (x_i x_k)$ or $(x_k) \cdot (x_i x_j)$.

$$\leq \#\left\{(k,l) \mid k \leq l, \; c_k, c_l \in \{x_i\}_{1 \leq i \leq g}\right\} + \#\left\{(k,l) \mid c_k \in \{x_i\}_{1 \leq i \leq g}, c_l = u\right\}$$

$$+ \#\left\{(k,l) \mid c_k \in \{x_i\}_{1 \leq i \leq g}, c_l \in \{x_i x_j\}_{1 \leq i \leq j \leq g}\right\}$$

$$+ \#\left\{(k,l) \mid c_k \in \{x_i\}_{1 \leq i \leq g}, c_l \in \{u x_i\}_{1 \leq i \leq g}\right\}$$

$$+ \#\left\{(k,l) \mid c_k = u, c_l \in \{x_i x_j\}_{1 \leq i \leq j \leq g}\right\} + 3\left[\#(\text{terms in } p_3) + \#(\text{terms in } p_2)\right]$$

$$\leq \left[\binom{g+1}{2} + g + g \cdot \binom{g+1}{2} + g^2 + \binom{g+1}{2}\right] + 3\left[\binom{n+1}{2} + \binom{n+1}{2} \cdot n\right]$$

Note that the dominant term in the above expression is $\frac{g^3}{2}$ while in that of d it is $\frac{g^3}{6}$. Computation gives the following bound:

$$< (3d - 6) \qquad \qquad \square$$

Construct a cubic form ϕ_S from ring S in a way similar to that of eqn (13).

$$\phi_S(\overline{y}, \overline{c}, v) := \sum_{1 \leq i \leq j \leq d} y_{ij} c_i c_j - v \sum_{1 \leq i \leq j \leq d} y_{ij} \left(\sum_{k=1}^{d} \tilde{e}_{ij,k} c_k\right) \tag{14}$$

where $\forall i, j, \; c_i \cdot c_j = \sum_{k=1}^{d} \tilde{e}_{ij,k} c_k$ in S' for some $\tilde{e}_{ij,k} \in \mathbb{F}$.

The following claim is what we intend to prove now.

Claim 4.1.2. $\phi_R(\overline{y}, \overline{c}, v)$ *is equivalent to* $\phi_S(\overline{y}, \overline{c}, v)$ *iff* $R' \cong S'$ *iff* $R \cong S$.

Proof of Claim 4.1.2. The part of this claim that needs to be proved is $\phi_R \sim \phi_S \Rightarrow R' \cong S'$. Suppose ψ is an equivalence from $\phi_R(\overline{y}, \overline{c}, v)$ to $\phi_S(\overline{y}, \overline{c}, v)$. We will show how to extract from ψ an isomorphism from R' to S'.

We have the following starting equation to analyze:

$$\sum_{1 \leq i \leq j \leq d} \psi(y_{ij}) \psi(c_i) \psi(c_j) - \psi(v) \sum_{1 \leq i \leq j \leq d} \psi(y_{ij}) \left(\sum_{k=1}^{d} \tilde{a}_{ij,k} \psi(c_k)\right)$$

$$= \sum_{1 \leq i \leq j \leq d} y_{ij} c_i c_j - v \sum_{1 \leq i \leq j \leq d} y_{ij} \left(\sum_{k=1}^{d} \tilde{e}_{ij,k} c_k\right) \tag{15}$$

The main property of this huge equation that we would like to show is: $\psi(c_i)$ *consists of only $\overline{c}$ terms.* Thus, $\psi(c_i)$ has enough information to extract a ring isomorphism from R' to S'. In the rest of the proof we will rule out the unpleasant cases of $\psi(c_i)$ having $\bar{y}, v$ terms and $\psi(v)$ having $\bar{y}$ terms.

Let for every i, $\psi(c_i) = \sum_j \alpha_{i,j} c_j + \sum_{j,k} \beta_{i,jk} y_{jk} + \gamma_i v$ where α, β, γ's $\in \mathbb{F}$. For obvious reasons we will call the expression $\sum_{j,k} \beta_{i,jk} y_{jk}$ as the $\bar{y}$ *part of* $\psi(c_i)$. $\bar{y}$ parts of $\psi(v)$ and $\psi(y_{ij})$ are defined similarly. We will show that the rank of the $\bar{y}$ part of $\psi(c_1), \ldots, \psi(c_d), \psi(v)$ is less than 3.

Assume that for some i, j, k the $\bar{y}$ parts of $\psi(c_i), \psi(c_j), \psi(c_k)$ are linearly independent over $\mathbb{F}$. By a *term* on the lhs of eqn (15) we mean expressions of the form $\psi(y_{ls})\psi(c_l)\psi(c_s)$ or $\psi(v)\psi(y_{ls})\psi(c_t)$ where $l, s, t \in [d]$. Let T_0 be the set of all terms. There are at least $d + (d-1) + (d-2) = (3d-3)$ terms on the lhs of eqn (15) that have an occurrence of $\psi(c_i), \psi(c_j)$ or $\psi(c_k)$, denote this set of terms by T_1. Let the set of the remaining terms be T_2. Let us build a maximal set Y of linearly independent $\bar{y}$ parts and a set T of terms as follows: Start with keeping $\bar{y}$ parts of $\psi(c_i), \psi(c_j), \psi(c_k)$ in Y and setting $T = T_1$. Successively add a new $\bar{y}$ part to Y that is linearly independent from the elements already in Y and that occurs in a term t in $T_0 \setminus T$, also add t to T. It is easy to see (by claim 4.1.1) that:

$$\#Y \leq 3 + \#T_2$$
$$< 3 + \left[\binom{d+1}{2} + (3d-6) - (3d-3)\right] = \binom{d+1}{2} = \# \{y_{ij}\}_{1 \leq i \leq j \leq d} \quad (16)$$

Now apply an invertible linear transformation τ on the $\bar{y}$ variables in equation (15) such that all the $\bar{y}$ parts in Y are mapped to *single* $\bar{y}$ variables, let $\tau(Y)$ denote the set of these variables. By substituting suitable linear forms, having only $\bar{c}, v$'s, to variables in $\tau(Y)$ we can make all the terms in $\tau(T)$ zero and the rest of the terms, *i.e.* $\tau(T_0 \setminus T)$, will then have no occurrence of $\bar{y}$ variables (as Y is the *maximal* set of linearly independent $\bar{y}$ parts). Thus, the lhs of eqn (15), after applying τ and the substitutions, is completely in terms of $\bar{c}, v$ while the rhs still has at least one free $\bar{y}$ variable (as we fixed only $\#\tau(Y) < \# \{y_{ij}\}_{1 \leq i \leq j \leq d}$ $\bar{y}$ variables and as τ is an invertible linear transformation). This contradiction shows that the $\bar{y}$ part of $\psi(c_i), \psi(c_j), \psi(c_k)$ cannot be linearly independent, for any i, j, k. Using a similar argument it can be shown that the $\bar{y}$ part of $\psi(c_i), \psi(c_j), \psi(v)$ cannot be linearly independent, for any i, j. Thus, the rank of the $\bar{y}$ part of $\psi(c_1), \ldots, \psi(c_d), \psi(v)$ is ≤ 2. For concreteness let us assume that the rank is *exactly* 2, the proof we give below will easily go through even when the rank is 1.

Again let Y be a maximal set of linearly independent $\bar{y}$ parts occurring in $\{\psi(y_{ij})\}_{1 \leq i \leq j \leq d}$ with the extra condition that $\bar{y}$ parts in Y are also linearly independent from that occurring in $\psi(c_1), \ldots, \psi(c_d), \psi(v)$. As we have assumed the rank of the $\bar{y}$ part of $\psi(c_1), \ldots, \psi(c_d), \psi(v)$ to be 2 we get $\#Y = \binom{d+1}{2} - 2$. Let $(i_1, j_1), (i_2, j_2)$ be the two tuples such that the $\bar{y}$ parts of $\psi(y_{i_1 j_1}), \psi(y_{i_2 j_2})$ do not appear in Y. To make things easier to handle let us apply an invertible linear transformation τ_1 on the $\bar{y}$ variables in equation (15) such that:

- the $\bar{y}$ parts of $\tau_1 \circ \psi(c_1), \ldots, \tau_1 \circ \psi(c_d), \tau_1 \circ \psi(v)$ have only $y_{i_1 j_1}$ and $y_{i_2 j_2}$.
- for all (i, j) other than (i_1, j_1) and (i_2, j_2), the $\bar{y}$ part of $\tau_1 \circ \psi(y_{ij})$ has *only* y_{ij}.
- τ_1 is identity on $\bar{c}, v$.

For clarity let $\psi' := \tau_1 \circ \psi$. Rest of our arguments will be based on comparing the coefficients of y_{ij}, for $(i,j) \neq (i_1, j_1), (i_2, j_2)$, on both sides of the equation:

$$\sum_{1 < i < j < d} \psi'(y_{ij}) \left(\psi'(c_i c_j) - \psi'(v) \sum_{k=1}^{d} \tilde{a}_{ij,k} \psi'(c_k) \right)$$

$$= \sum_{1 \leq i \leq j \leq d} y_{ij} (\text{quadratic terms in } \overline{c}, v) \tag{17}$$

For any c_i, choose distinct basis elements c_j, c_k and c_l satisfying $c_i c_j = c_i c_k = c_i c_l = 0$ in R' (note that there is an ample supply of such j, k, l), such that by comparing coefficients of y_{ij}, y_{ik}, y_{il} (assumed to be other than $y_{i_1 j_1}, y_{i_2 j_2}$) on both sides of equation (17) we get:

$$\psi'(c_i c_j) + (e_{ij,1} E_1 + e_{ij,2} E_2) = (\text{quadratic terms in } \overline{c}, v)$$
$$\psi'(c_i c_k) + (e_{ik,1} E_1 + e_{ik,2} E_2) = (\text{quadratic terms in } \overline{c}, v)$$
$$\psi'(c_i c_l) + (e_{il,1} E_1 + e_{il,2} E_2) = (\text{quadratic terms in } \overline{c}, v) \tag{18}$$

where, $e_{ij,1}, e_{ij,2}, e_{ik,1}, e_{ik,2}, e_{il,1}, e_{il,2} \in \mathbb{F}$ and

$$E_1 = \psi'(c_{i_1} c_{j_1}) - \psi'(v) \sum_{k=1}^{d} \tilde{a}_{i_1 j_1, k} \psi'(c_k)$$

$$E_2 = \psi'(c_{i_2} c_{j_2}) - \psi'(v) \sum_{k=1}^{d} \tilde{a}_{i_2 j_2, k} \psi'(c_k)$$

Now there exist $\lambda_1, \lambda_2, \lambda_3 \in \mathbb{F}$ (not all zero) such that equations (18) can be combined to get rid of E_1, E_2 and get:

$$\psi'(c_i) (\lambda_1 \psi'(c_j) + \lambda_2 \psi'(c_k) + \lambda_3 \psi'(c_l)) = (\text{quadratic terms in } \overline{c}, v)$$

This equation combined with the observation that both $\psi'(c_i)$ and $(\lambda_1 \psi'(c_j) + \lambda_2 \psi'(c_k) + \lambda_3 \psi'(c_l))$ are non-zero (as ψ' is invertible) implies that:

$$\forall i, \quad \psi'(c_i) = (\text{linear terms in } \overline{c}, v) \tag{19}$$

This means that the $\overline{y}$-variables are only in $\psi'(y_{ij})$s and possibly $\psi'(v)$. Again apply an invertible linear transformation τ_2 on the $\overline{y}$-variables in equation (17) such that $\tau_2 \circ \psi'(v)$ has only $y_{i_0 j_0}$ in the $\overline{y}$ part and except for one tuple (i_0, j_0), the $\overline{y}$ part of $\tau_2 \circ \psi'(y_{ij})$ has *only* y_{ij} for all other (i,j). For clarity let $\psi'' := \tau_2 \circ \psi'$. Our equation now is:

$$\sum_{1 \leq i \leq j \leq d} \psi''(y_{ij}) \left(\psi''(c_i c_j) - \psi''(v) \sum_{k=1}^{d} \tilde{a}_{ij,k} \psi''(c_k) \right)$$

$$= \sum_{1 \leq i \leq j \leq d} y_{ij} (\text{quadratic terms in } \overline{c}, v) \tag{20}$$

By comparing coefficients of y_{ij} (other that $y_{i_0 j_0}$) on both sides of the above equation we get:

$$\left(\psi''(c_i c_j) - \psi''(v) \sum_{k=1}^{d} \tilde{a}_{ij,k} \psi''(c_k) \right) + e \left(\psi''(c_{i_0} c_{j_0}) - \psi''(v) \sum_{k=1}^{d} \tilde{a}_{i_0 j_0,k} \psi''(c_k) \right)$$

$$= (\text{quadratic terms in } \bar{c}, v), \quad \text{for some } e \in \mathbb{F}.$$

Pick i, j such that $\sum_{k=1}^{d} \tilde{a}_{ij,k} c_k \neq 0$ in R'. Now if $\psi''(v)$ has a nonzero $y_{i_0 j_0}$ term then by comparing coefficients of $y_{i_0 j_0}$ on both sides of the above equation we deduce:

$$\sum_{k=1}^{d} \tilde{a}_{ij,k} \psi''(c_k) + e \sum_{k=1}^{d} \tilde{a}_{i_0 j_0,k} \psi''(c_k) = 0 \tag{21}$$

But again we can pick i, j suitably so that $\left(\sum_{k=1}^{d} \tilde{a}_{ij,k} c_k \right) \notin \left\{ 0, -e \sum_{k=1}^{d} \tilde{a}_{i_0 j_0,k} c_k \right\}$ and hence avoiding equation (21) to hold. Thus, proving that $\psi''(v)$ has no $y_{i_0 j_0}$ term. So we now have:

$$\psi''(v) = (\text{linear terms in } \bar{c}, v)$$

$$\text{and}$$

$$\forall i, \quad \psi''(c_i) = (\text{linear terms in } \bar{c}, v) \tag{22}$$

Since $\bar{y}$-variables are present only in $\psi''(y_{ij})$'s, comparing coefficients of y_{ij}'s on both sides of equation (20) gives:

$$\forall i, j, \quad \psi''(c_i c_j) - \psi''(v) \sum_{k=1}^{d} \tilde{a}_{ij,k} \psi''(c_k) = (\text{quadratic terms in } \bar{c}) - v(\text{linear terms in } \bar{c}) \tag{23}$$

Using this equation we will prove now that $\psi''(c_i)$ has only $\bar{c}$-variables.

Consider a c_i such that $c_i^2 = 0$ in R', then from equation (23):

$$\psi''(c_i)^2 = (\text{quadratic terms in } \bar{c}) - v(\text{linear terms in } \bar{c}) \tag{24}$$

Now if $\psi''(c_i)$ has a nonzero v term then there will be a v^2 term above on the lhs which is absurd. Thus, $\psi''(c_i)$ has only $\bar{c}$-variables when $c_i^2 = 0$ in R'. When $c_i^2 \neq 0$ then $c_i^2 = \sum_{k=1}^{d} \tilde{a}_{ii,k} c_k$ in R' where the c_k's with nonzero $\tilde{a}_{ii,k}$ satisfy $c_k^2 = 0$. This happens because the way $\bar{c}$'s are defined in eqn (12) the expression of c_i^2 will have only quadratic or cubic terms in $\bar{x}$ and the square of these terms would clearly be zero in R'. Thus, again if $\psi''(c_i)$ has a v term then there will be an uncancelled v^2 term on the lhs of the equation:

$$\psi''(c_i)^2 - \psi''(v) \sum_{k=1}^{d} \tilde{a}_{ii,k} \psi''(c_k) = (\text{quadratic terms in } \bar{c}) - v(\text{linear terms in } \bar{c})$$

Thus, we know at this point that $\psi''(v)$ has only $\bar{c}, v$ terms and $\psi''(c_i)$ has only $\bar{c}$ terms. Since τ_1, τ_2 act only on the $\bar{y}$'s we have what we intended to prove from the beginning (recall eqn (15)):

$$\psi(v) = (\text{linear terms in } \bar{c}, v)$$

$$\text{and}$$

$$\forall i, \quad \psi(c_i) = (\text{linear terms in } \bar{c}) \tag{25}$$

We have now almost extracted a ring isomorphism from the cubic form equivalence ψ, just few technicalities are left which we resolve next.

Apply an invertible linear transformation τ_3 on the $\bar{y}$-variables in equation (15) such that the $\bar{y}$ part of $\tau_3 \circ \psi(y_{ij})$ has *only* y_{ij} for all $i \leq j \in [d]$. Of course we assume that τ_3 is identity on the $\bar{c}, v$ variables. So on comparing coefficients of y_{ij} on both sides of the eqn (15) after applying τ_3 we get:

$$\forall i, j, \quad \tau_3 \circ \psi(c_i c_j) - \tau_3 \circ \psi(v) \sum_{k=1}^{d} \tilde{a}_{ij,k} \tau_3 \circ \psi(c_k) = \sum_{i \leq j} \lambda_{ij} \left(c_i c_j - v \sum_{k=1}^{d} \tilde{e}_{ij,k} c_k \right) \tag{26}$$

for some $\lambda_{ij} \in \mathbb{F}$.

Substitute $v = 1$ in the expression for $\tau_3 \circ \psi(v) = \gamma_{vv} v + \sum_i \alpha_{vi} c_i$ and denote the result by m. Observe that $\gamma_{vv} \neq 0$ and $\forall i$, c_i is a nilpotent element in S' and hence m is a *unit* in the ring S'. On substituting $v = 1$ in eqn (26) we get:

$$\forall i, j, \quad \tau_3 \circ \psi(c_i) \tau_3 \circ \psi(c_j) - m \cdot \sum_{k=1}^{d} \tilde{a}_{ij,k} \tau_3 \circ \psi(c_k) = 0 \quad \text{in } S'$$

If we define $\Psi := \frac{\tau_3 \circ \psi}{m}$ then we get:

$$\forall i, j, \quad \Psi(c_i) \Psi(c_j) - \sum_{k=1}^{d} \tilde{a}_{ij,k} \Psi(c_k) = 0 \quad \text{in } S'$$

Now observe that if for some λ_i's $\in \mathbb{F}$, $\Psi(\sum_{i=1}^{d} \lambda_i c_i) = 0$ in S' then $\tau_3 \circ \psi(\sum_{i=1}^{d} \lambda_i c_i) = 0$ in S'. Since $\tau_3 \circ \psi$ is an invertible linear map this means that $\sum_{i=1}^{d} \lambda_i c_i = 0$ in R'. Thus, showing that Ψ is an *injective* map from R' to S'. Since R' and S' are of the same dimension over $\mathbb{F}$, Ψ becomes *surjective* too. Thus, Ψ is an isomorphism from R' to S'. $\qquad \square$

This completes the reduction from $\mathbb{F}$-algebra isomorphism to cubic form equivalence.

5   Conclusion

In this paper we gave a reduction from $\mathbb{F}$-algebra isomorphism to $\mathbb{F}$-cubic form equivalence for any field $\mathbb{F}$. Thus, cubic form equivalence, in addition to being a

natural algebraic problem, is also directly related to isomorphism problems for
$\mathbb{F}$-algebras and graphs. We would like to pose the following questions related to
cubic forms:

- Is there a subexponential algorithm for $\mathbb{F}$-cubic forms for any field $\mathbb{F}$? Such an
  algorithm will result in a subexponential time algorithm for Graph
  Isomorphism.
- Is $\mathbb{Q}$-cubic form equivalence decidable? This will make $\mathbb{Q}$-algebra isomorphism decidable too.

References

[AS05]   M. Agrawal, N. Saxena. *Automorphisms of Finite Rings and Applications
          to Complexity of Problems.* STACS'05, Springer LNCS 3404, 2005, 1-17.
[Bro87]   W. D. Brownawell. *Bounds for the degrees in the Nullstellensatz.* Annals
          of Maths, 126, 1987, 577-591.
[CGP98]   N. Courtois, L. Goubin, J. Patarin. *Improved Algorithms for Isomorphism
          of Polynomials.* Eurocrypt'98, Springer LNCS 1403, 1998, 184-200.
[DH88]   J. Davenport, J. Heintz. *Real Quantifier Elimination Is Doubly Exponen-
          tial.* Journal of Symbolic Computation, 5, 1988, 29-35.
[Har75]   D. Harrison. *A Grothendieck ring of higher degree forms.* Journal of Alge-
          bra, 35, 1975, 123-128.
[KS05]   N. Kayal, N. Saxena. *On the Ring Isomorphism and Automorphism Prob-
          lems.* IEEE Conference on Computational Complexity, 2005, 2-12.
[Lang]   S. Lang. *Algebra.* 3rd edition, Addison Wesley.
[McD74]   Bernard R. McDonald. *Finite Rings with Identity.* Marcel Dekker, Inc.,
          1974.
[MH74]   Y. I. Manin, M. Hazewinkel. *Cubic forms: algebra, geometry, arithmetic.*
          North-Holland Publishing Co., Amsterdam, 1974.
[Pat96]   J. Patarin. *Hidden Field Equations (HFE) and Isomorphisms of Polyno-
          mials (IP): Two new families of asymmetric algorithms.* Eurocrypt'96,
          Springer LNCS 1070, 1996, 33-48.
[Rup03]   C. Rupprecht. *Cohomological invariants for higher degree forms.* PhD
          Thesis, Universität Regensburg, 2003.
[Sch88]   U. Schoning. *Graph isomorphism is in the low hierarchy.* Journal of Com-
          puter and System Science, **37**, 1988, 312-323.

Complete Codes in a Sofic Shift

Marie-Pierre Béal and Dominique Perrin

Institut Gaspard-Monge, University of Marne-la-Vallée, France
{beal, perrin}@univ-mlv.fr

Abstract. We define a code in a sofic shift as a set of blocks of symbols of the shift such that any block of the shift has at most one decomposition in code words. It is maximal if it is not strictly included in another one. Such a code is complete in the sofic shift if any block of the shift occurs within some concatenation of code words. We prove that a maximal code in an irreducible sofic shift is complete in this shift. We give an explicit construction of a regular completion of a regular code in a sofic shift. This extends the well known result of Ehrenfeucht and Rozenberg to the case of codes in sofic systems. We also give a combinatorial proof of a result concerning the polynomial of a code in a sofic shift.

1 Introduction

In this paper, we continue the study of codes in sofic shifts initiated in [1]. This generalization of the theory of (variable length) codes extends previous works of Reutenauer [2], Restivo [3] and Ashley *et al.* [4]. The main result of this paper is an extension of a classical result of Schützenberger [5] relating the notions of completeness and maximality of codes.

Let S be a sofic shift, *i.e.* the set of bi-infinite sequences of symbols labelling paths in a finite automaton. The set of factors of S, denoted by $\mathrm{Fact}(S)$, is the set of blocks appearing in the elements of S. We call S-code a set of elements of $\mathrm{Fact}(S)$ such that any element of $\mathrm{Fact}(S)$ has at most one decomposition in code words. A set of words X is S-complete if any element of $\mathrm{Fact}(S)$ occurs within some concatenation of elements of X. An S-code is maximal if it is maximal for inclusion.

We prove that, for any irreducible sofic shift S, any maximal S-code is S-complete. Moreover, we give an effective embedding of a regular S-code into an S-complete one. This extends the well known theorem of Ehrenfeucht and Rozenberg [6] to codes in a sofic shift.

Our definition of S-codes generalizes the notion introduced by Restivo [3] and Ashley *et al.* [4]. In the first place, they consider subshifts of finite type instead of the more general notion of sofic shifts. Although shifts of finite type can also be described by a finite automaton, there is a real gap between the two classes. Indeed, representations of shifts of finite type have nice strong properties of synchronization that do not have general sofic shifts. These properties are used to complete the codes. In the second place, they consider codes such that all concatenations of code words are in $\mathrm{Fact}(S)$, a condition that we do not

B. Durand and W. Thomas (Eds.): STACS 2006, LNCS 3884, pp. 127–136, 2006.
© Springer-Verlag Berlin Heidelberg 2006

impose. Our definition here is also slightly more general than the one used in our previous paper [1]. Indeed, we only require the unique factorization for the words of Fact(S) and not for all products of code words. We think that this definition is more natural. The results of [1] all extend straightforwardly to this new class.

In the last section, we give a combinatorial proof of the main result of our previous paper [1] concerning the polynomial of a finite code. This proof is interesting because it is simpler and also because it relates our result to ones due to S. Williams [7] and M. Nasu [8].

The paper is organized as follows. We first recall some basic definitions from the area of symbolic dynamics and from the theory of codes. We introduce the notions of S-code, maximal S-code, and S-complete code when S denotes a sofic shift. In Section 3, we prove that any maximal S-code is S-complete. A combinatorial proof of the result of [1] is given in the last section.

2   Codes and Sofic Shifts

2.1   Sofic Shifts

Let A be a finite alphabet. We denote by A^* the set of finite words, by A^+ the set of nonempty finite words, and by $A^{\mathbb{Z}}$ the set of bi-infinite words on A. A *subshift* is a closed subset S of $A^{\mathbb{Z}}$ which is invariant by the shift transformation σ (*i.e.* $\sigma(S) = S$) defined by $\sigma((a_i)_{i \in \mathbb{Z}}) = (a_{i+1})_{i \in \mathbb{Z}}$.

A finite *automaton* is a finite multigraph labeled on a finite alphabet A. It is denoted $\mathcal{A} = (Q, E)$, where Q is a finite set of states, and E a finite set of edges labeled by A. All states of these automata can be considered as both initial and final states.

A *sofic shift* is the set of labels of all bi-infinite paths in a finite automaton. A sofic shift is *irreducible* if there is such a finite automaton with a strongly connected graph. In this case the automaton also is said to be irreducible. An automaton $\mathcal{A} = (Q, E)$ is deterministic if, for any state $p \in Q$ and any word u, there is at most one path labeled u and going out of p. When it exists, the target state of this path is denoted by $p \cdot u$. An automaton is *unambiguous* if there is at most one path labeled by u going from a state p to a state q for any given triple p, u, q. Irreducible sofic shifts have a unique (up to isomorphisms of automata) *minimal deterministic automaton*, that is a deterministic automaton having the fewest states among all deterministic automata representing the shift. This automaton is called the *Fischer cover* of the shift. A *subshift of finite type* is defined as the bi-infinite words on a finite alphabet avoiding a finite set of finite words. It is a sofic shift. The *full shift* on the finite alphabet A is the set of all bi-infinites sequences on A, *i.e.* the set $A^{\mathbb{Z}}$.

The *entropy* of a sofic shift S is defined as

$$h(S) = \lim_{n \to \infty} \frac{1}{n} \log_2 s_n,$$

where s_n is the number of words of length n of Fact(S). The Fischer cover of a transitive sofic shift of null entropy is made of one cycle.

Example 1. Let S be the irreducible sofic subshift on $A = \{a, b\}$ defined by the automaton on the left of Figure 1. This automaton is the Fischer cover of S. This shift is the so-called *even system* since its bi-infinite sequences are those having an even number of b's between two a's. It is not a shift of finite type.

Let T be the irreducible shift on $A = \{a, b\}$ defined by the forbidden block bb. It is a shift of finite type. Its Fischer cover is given on the right of Figure 1. This shift is the so-called *golden mean system*.

Fig. 1. The Fischer covers of the even system S on the left, and of the golden mean system T on the right

Let S be a subshift on the alphabet A. We denote by $\mathrm{Fact}(S)$ the set of finite factors (or blocks) of elements of S. Each element of $\mathrm{Fact}(S)$ is the label of a finite path of the Fischer cover of S.

Let $\mathcal{A}$ be a finite automaton. A word w is said to be a *synchronizing* word of $\mathcal{A}$ if and only if any path in $\mathcal{A}$ labeled by w ends in a same state depending only on w. If p denotes this states, one says that w *synchronizes* to p. For instance the words a, bab are synchronizing words of the Fischer cover of the even system. In the golden mean shift, which is a shift of finite type, each word of length 1, *i.e.* a or b, is a synchronizing word. For any Fischer cover of a shift of finite type, there is a positive integer k such that any word of length k is synchronizing.

Let L be a language of finite words. A word w is a *synchronizing* word of L if and only if whenever u, v are words such that uw and wv belong to L, one has uwv belongs to L. Note that if w is a synchronizing word of an automaton $\mathcal{A}$ recognizing a sofic shift S, it is a synchronizing word of the language $\mathrm{Fact}(S)$.

It is known that the Fischer cover of an irreducible sofic shift S has a synchronizing word [9, Proposition 3.3.16]. If w is one of them, for any words u, v such that $uwv \in \mathrm{Fact}(S)$, uwv is a synchronizing word also.

2.2   Codes

Let S be a sofic shift. A set of finite words $X \subset \mathrm{Fact}(S)$ on an alphabet A is an *S-code* if and only if whenever $w = x_1 x_2 \ldots x_n = y_1 y_2 \ldots y_m$, where $x_i, y_j \in X$, n, m are positive integers, and $w \in \mathrm{Fact}(S)$, one has $n = m$ and $x_i = y_i$ for $1 \leq i \leq n$. Thus the classical definition of a code corresponds to the case where S is the full shift. Any code is an S-code but the converse is false as shown with the following example.

Example 2. The set $\{a, ab, ba\}$ is not a code but it is not difficult to see that it is an S-code in the even system. Indeed any word with two factorizations contains the block aba.

Let S be a sofic shift. A set X on the alphabet A is said to be *complete* in S, or *S-complete*, if X is an S-code and any word in $\mathrm{Fact}(S)$ is a factor of a word in X^*. For instance the code $X = \{a, bb\}$ is complete in the even system.

An S-code X is *maximal* if it is not strictly included in another S-code.

In [2] is an example of an S-complete code which is not maximal. Indeed, let us consider the shift of finite type S defined on the alphabet $A = \{a, b\}$ and avoiding the blocks aa and bb. The S-code $X = \{ab\}$ is S-complete but not maximal since X is strictly included in the S-code $Y = \{ab, ba\}$.

There is a connection between complete S-codes and a concept which has been studied in symbolic dynamics. This explains why the results proved in Section 4 are related with the results of Williams [7] and Nasu [8]. Let X be a complete S-code. Let $\mathcal{A} = (Q, E)$ be the Fischer cover of S. We build an automaton $\mathcal{B}$ computed from X and $\mathcal{A}$ as follows. The set of states of $\mathcal{B}$ contains the set of states Q of $\mathcal{A}$. For each path in $\mathcal{A}$ labeled by a word in X going from a state p to a state q, we build a path in $\mathcal{B}$ from p to q with dummy states inbetween. Let T be the subshift of finite type made of the bi-infinite paths of the graph of $\mathcal{B}$. The labelling of the paths in the automaton $\mathcal{B}$ defines a block map ϕ from T to S. The set X is an S-code if and only if ϕ is finite-to-one. It is S-complete if and only if ϕ is onto. Thus statements on complete S-codes can be reformulated as statements on finite-to-one factor maps between irreducible sofic shifts.

3   Completion of an S-Code

The following result generalizes the theorem of Ehrenfeucht and Rozenberg [6]. As in the case of the extension to subshifts of finite type obtained in [4], the proof uses the same type of construction as the one of [6]. It requires however, as we shall see, a careful adaptation to extend to sofic shifts.

Theorem 1. *Let S be an irreducible sofic shift. If X is an S-code, there is an S-code Y such that $X \subseteq Y$ and Y is S-complete. If X is moreover regular, Y can be chosen regular and is computable in an effective way.*

A nonempty word w of A^* is called *unbordered* if no proper nonempty left factor of w is a right factor of w. In other words, w is unbordered if and only if $w \in uA^+ \cap A^+u$ implies $u = \varepsilon$, where ε denotes the empty word.

The following lemma provides the construction of an unbordered word in the set of factors of an irreducible sofic shift. It replaces the construction used in [5, Proposition 3.6] for the case of the full shift.

Lemma 1. *Let S be an irreducible sofic shift which has a positive entropy. Let z be a word in $\mathrm{Fact}(S)$. Then there is a word y in $\mathrm{Fact}(S)$ such that z is a factor of y and y is unbordered.*

Proof. Let $\mathcal{A}$ be the Fischer cover of S. Let m be the number of states of $\mathcal{A}$ and let k be the length of z. Since S has a positive entropy, there are two distinct nonempty words u, v labels of first return paths in $\mathcal{A}$ to state p. The words

u and v are not two powers a same word since $\mathcal{A}$ is deterministic. Moreover $\{u, v\}^*$ is a submonoid of $\text{Fact}(S)$. Let $w = u^{k+m}v^{k+m}$. Since $k + m \geq 2$, by [10, Theorem 9.2.4, pp. 166], w is a primitive word. It follows that the Lyndon word w' conjugate to w is unbordered (see for instance [10, Proposition 5.1.2, p. 65]). Since $\mathcal{A}$ is irreducible, there are two words b_1, b_2 of length at most m such that the word $y = w'b_1zb_2w' \in \text{Fact}(S)$.

We claim that y is unbordered. This fact is trivial by considering the length, greater than $2k + 2m$, of w', the length $k + 2m$ of b_1zb_2 and the fact that w' is unbordered.

Proof (Sketch of proof of Theorem 1). Let S be an irreducible sofic shift. We denote by $\mathcal{A}$ the Fischer cover of S. Let X be an S-code.

Let us suppose that X is not S-complete. Consequently there is a word z in $\text{Fact}(S)$ which is not in $\text{Fact}(X^*)$.

We first assume that S has a null entropy. This means that the Fischer cover $\mathcal{A}$ is made of a unique cycle. One can assume that there is a state p such that p has no outgoing path in $\mathcal{A}$ labeled in X. Otherwise X is already S-complete. Since $\mathcal{A}$ is irreducible, one can assume without loss of generality that z is the label of a path in $\mathcal{A}$ going from a state p to itself, and that z is moreover a synchronizing word of $\mathcal{A}$. We set $Y = X \cup \{z\}$. Now we show that Y is an S-code. Assume the contrary and consider a relation

$$x_1x_2 \ldots x_n = y_1y_2 \ldots y_m,$$

with $x_1x_2 \ldots x_n \in \text{Fact}(S)$, $x_i, y_j \in Y$, and $x_n \neq y_m$. The set X being an S-code, at least one of the words x_i, y_j must be z. Hence, for instance $x_1x_2 \ldots x_n = x_1x_2 \ldots x_r z x_{r+1} \ldots x_n$. The word $z x_{r+1} \ldots x_n$ is the label of a path in $\mathcal{A}$ going through the state p after reading the label z. Since p has no outgoing path in $\mathcal{A}$ labeled in X, it follows that $x_{r+1} \ldots x_n = z^{n-r}$. Hence there is a positive integer k such that $x_1x_2 \ldots x_n = x_1x_2 \ldots x_r z^k$ with $x_1, x_2, \ldots, x_r \neq z$. Since z is not a factor of X^*, there is also a positive integer l such that $y_1y_2 \ldots y_m = y_1y_2 \ldots y_t z^l$ with $y_1, y_2, \ldots, y_t \neq z$. The above relation becomes

$$x_1x_2 \ldots x_r z^k = y_1y_2 \ldots y_t z^l,$$

which contradicts the hypothesis that $x_n \neq y_m$ since $z \notin \text{Fact}(X^*)$. It is trivial that Y is S-complete.

We may now assume that S has a positive entropy. Without loss of generality, by extending z on the right, one can moreover assume that z is a synchronizing word. By Lemma 1, we construct a word $y \in \text{Fact}(S)$ which is unbordered and has w as factor. Moreover y is a synchronizing word of $\mathcal{A}$.

If L is a language of finite words, we denote by $u^{-1}L$ (resp. Lu^{-1}) the set of words z such that $uz \in L$ (resp. $zu \in L$).

We define the sets U and Y by

$$U = y^{-1}\text{Fact}(S)y^{-1} - X^* - A^*yA^*, \tag{1}$$

$$Y = X \cup y(Uy)^*. \tag{2}$$

The rest of the proof consists in verifying the following three properties.

- The set Y is a subset of $\mathrm{Fact}(S)$.
- The set Y is an S-code.
- The set Y is S-complete.

It is clear from Equations (1) and (2) that Y is regular when X is regular. It can be computed in an effective way from these equations.   □

Remark 1. Note that our proof shows that, if S is an irreducible sofic shift with a positive entropy, and X is a code, then X can be completed into a code Y (*i.e* a code for the full shift) which is S-complete. We do not know whether this property also holds for irreducible shifs of entropy zero.

In [11,3] (see also [4]), it is proved that if S is an irreducible shift of finite type and X a code with $X^* \subseteq \mathrm{Fact}(S)$ which is not S-complete, X can be embedded into an S-complete set which is moreover a code (*i.e* a code for the full shift). The proof of our theorem allows us to recover this result. Indeed, when $X^* \subseteq \mathrm{Fact}(S)$, our construction build an S-code Y which is a code. Moreover, the S-complete code Y that we have built satisfies also $Y^* \subseteq \mathrm{Fact}(S)$, when $X^* \subseteq \mathrm{Fact}(S)$. This is due to the strong synchronization properties of the Fischer cover of an irreducible shift of finite type.

Example 3. We consider the even system S of Example 1 on the alphabet $A = \{a, b\}$. Let $X = \{a, ba\}$. The set X is an S-code but it is not S-complete since for instance $z = bb$ does not belong to $\mathrm{Fact}(X^*)$. The regular completion of X is obtained following the proof of Theorem 1. We replace z by bba in order to get a synchronizing word. The proof of Lemma 1 says that the word $a^2b^4bbaa^2b^4$ is an unbordered word of $\mathrm{Fact}(S)$. But a smaller y can be chosen. For instance $y = bba$ also is an unbordered word of $\mathrm{Fact}(S)$. We then define U and Y as in Equations (1) and (2). The set Y is a regular S-complete code.

We derive the following corollary which generalizes to codes in irreducible sofic shifts the fact that any maximal code is complete [5, Theorem 5.1].

Corollary 1. *Let S be an irreducible sofic shift. Any maximal S-code is S-complete.*

4   Polynomial of a Code

In the sequel, S is an irreducible sofic shift recognized by its Fischer cover $\mathcal{A} = (Q, E)$. Let μ_A (or μ) be the morphism from A^* into $\mathbb{N}^{Q \times Q}$ defined as follows. For each word u, the matrix $\mu(u)$ is defined by

$$\mu(u)_{pq} = \begin{cases} 1 & \text{if } p \cdot u = q \\ 0 & \text{otherwise.} \end{cases}$$

The matrix $\alpha_A(u)$ (or $\alpha(u)$) is defined by $\alpha(u) = \mu(u)u$. Thus the matrix $\alpha(u)$ is obtained from $\mu(u)$ by replacing its coefficients 1 by the word u. The coefficients of $\alpha(u)$ are either 0 or u. In this way α is a morphism from A^* into the monoid of matrices with elements in the set of subsets of A^*.

The morphism α is extended to subsets of A^* by linearity.

For a finite set X, we denote by p_X the polynomial in commuting variables:

$$p_X = \det(I \quad \alpha(X)).$$

The following result is proved in [1]. It is a generalization of a result of C. Reutenauer [2] who has proved it under more restrictive assumptions.

Theorem 2. *Let S be an irreducible sofic shift and let X be a finite complete S-code. The polynomial p_A divides p_X.*

Example 4. For the even shift and the set $X = \{aa, ab, ba, bb\}$, we have

$$\alpha(A) = \begin{bmatrix} a & b \\ b & 0 \end{bmatrix} \quad \text{and} \quad \alpha(X) = \begin{bmatrix} aa + bb & ab \\ ba & bb \end{bmatrix},$$

and $p_A = 1 - a - bb$, $p_X = 1 - aa - 2bb + b^4 = (1 + a - bb)(1 - a - bb)$.

We present here two combinatorial proofs of this result, which come as an alternative to the analytic proof presented in [1]. Both proofs rely on the reduction of automata with multiplicities.

The first proof goes along the same line as the proof of a result of S. Williams presented in Kitchen's book [12, p. 156], giving a necessary condition to the existence of a finite-to-one factor map between irreducible sofic shifts.

We first build as in Section 2 an automaton $\mathcal{B}$ computed from X and $\mathcal{A}$ as follows. The set of states of $\mathcal{B}$ contains the set of states Q of $\mathcal{A}$. For each path in $\mathcal{A}$ labeled by a word in X going from state p to state q, we build a path in $\mathcal{B}$ from p to q with dummy states inbetween as shown in Example 5. The automaton $\mathcal{B}$ is unambiguous if and only if the set X is an S-code. It represents S if and only if the set X is S-complete.

Example 5. Consider the code $X = \{aa, ab, ba, bb\}$ in the even system S. The automaton $\mathcal{B}$ is represented in the right part of Figure 2.

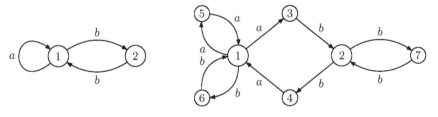

Fig. 2. The automaton $\mathcal{A}$ (on the left), and the automaton $\mathcal{B}$ computed from $\mathcal{A}$ and $X = \{aa, ab, ba, bb\}$ (on the right)

Since X is a complete S-code, $\mathcal{B}$ is unambiguous and represents S. Without loss of generality, one can assume that $\mathcal{B}$ is irreducible. Otherwise, one keeps only a strongly connected component of $\mathcal{B}$ representing S. By construction,

$$p_A = \det(I - \alpha_{\mathcal{A}}(A)) \quad \text{and} \quad p_X = \det(I - \alpha_{\mathcal{B}}(A)).$$

Hence, Theorem 2 is a consequence of the following result.

Proposition 1. *Let S be an irreducible sofic shift and let $\mathcal{A}$ be its Fischer cover. If $\mathcal{B}$ is an unambiguous and irreducible automaton representing S, $\det(I - \alpha_{\mathcal{A}}(A))$ divides $\det(I - \alpha_{\mathcal{B}}(A))$.*

Proof (Sketch of proof). The *degree* of a word u in an automaton is defined as the number of paths labeled by u. The degree of an automaton is the minimal non-null value of the degree of words. Any unambiguous irreducible automaton of degree k has the following property: for any word u of degree k and any word w such that uwu has a non-null degree, uwu has degree k.

We first assume that the Fischer cover $\mathcal{A}$ of S is codeterministic (or left resolving): for any state $p \in Q$ and any word u, there is at most one path labeled by u and ending at p. In this case the degree of $\mathcal{A}$ is $d = 1$. Indeed, since $\mathcal{A}$ is a Fischer cover, it has a synchronizing word. Since $\mathcal{A}$ is codeterministic, each synchronizing word has degree 1.

Let v (resp. w) be a word which has a non-null and minimal degree k (resp. $d = 1$) in $\mathcal{B}$ (resp. in $\mathcal{A}$). Since $\mathcal{B}$ is irreducible, there are words z, z' such that $vzwz'v$ has a non-null degree. Hence $vzwz'v$ has degree k in $\mathcal{B}$ and degree $d = 1$ in $\mathcal{A}$. We set $u = vzwz'v$.

An $\mathbb{N}$-automaton with a set of states Q is a triple $\langle I, \mu, T \rangle$, where I and T are two vectors — respectively initial row vector and final column vector — with entries in $\mathbb{N}$, and where μ is a morphism from A^* into $\mathbb{N}^{Q \times Q}$. It is equivalently defined by the triple $\langle I, \alpha(A), T \rangle$. Two $\mathbb{N}$-automata $\langle I, \mu, T \rangle$ and $\langle J, \mu', F \rangle$ are *equivalent* if and only if, for any word $w \in A^*$, $I\mu(w)T = J\mu'(w)F$.

Let $\mathbf{1}_{\mathcal{A}}$ be the row-vector with all coefficients equal to 1 of size the number of states of $\mathcal{A}$, and $\mathbf{1}_{\mathcal{A}}^t$ its transpose. It follows from the definition of the word u that the two $\mathbb{N}$-automata $\mathcal{C} = \langle k\mathbf{1}_{\mathcal{A}}\mu_{\mathcal{A}}(u), \ \mu_{\mathcal{A}}, \ \mu_{\mathcal{A}}(u)\mathbf{1}_{\mathcal{A}}^t \rangle$, and $\mathcal{D} = \langle d\mathbf{1}_{\mathcal{B}}\mu_{\mathcal{B}}(u), \ \mu_{\mathcal{B}}, \ \mu_{\mathcal{B}}(u)\mathbf{1}_{\mathcal{B}}^t \rangle$, are equivalent.

The standard Schützenberger reductions of the $\mathbb{N}$-automata $\mathcal{C}$ and $\mathcal{D}$ over the field $\mathbb{R}$ are similar. The reduction of each $\mathbb{N}$-automaton is obtained through a left reduction followed by a right reduction (see for instance [13] or [14]).

Since u has degree 1, the initial row (resp. final column) vector of $\mathcal{C}$ has a unique non-null coefficient. Consequently, since $\mathcal{A}$ is deterministic (resp. codeterministic) and irreducible, the automaton $\mathcal{C}$ is left (resp. right) reduced. Hence $\mathcal{C}$ is already reduced.

Finally, one can prove that the transition matrix of $\mathcal{D}$ is similar to a matrix having a principal subblock equal to the transition matrix of its left (or right) reduced form. It follows that $\det(I - \alpha_{\mathcal{A}}(A))$ divides $\det(I - \alpha_{\mathcal{B}}(A))$. The extension to sofic shifts that may not have a codeterministic Fischer cover can be obtained with a specialization argument (see [1]).   □

Example 6. We continue with Example 5. The word bab has degree 2 in $\mathcal{B}$ and 1 in $\mathcal{A}$. Hence the $\mathbb{N}$-automata

$$\mathcal{C} = \langle [0\ 2]\,,\ \mu_{\mathcal{A}}(A) = \begin{bmatrix} a & b \\ b & 0 \end{bmatrix},\ \begin{bmatrix} 0 \\ 1 \end{bmatrix}\rangle,\ \text{and } \mathcal{D} = \langle [0\ 1\ 0\ 0\ 0\ 1\ 0]\,,\ \mu_{\mathcal{B}}(A), [0\ 1\ 0\ 0\ 0\ 1\ 0]^t\rangle,$$

are equivalent. We obtain a right-reduction of the automaton $\mathcal{D} - \langle 1, E -$
$\alpha_{\mathcal{B}}(A), T\rangle$ by computing a basis of the vector space generated by the vectors in
$\mu(A^*)T$. We can choose the basis $(T, \mu(b)T, \mu(ab)T)$ since $\mu(a)T = 0$, $\mu(bb)T = T$, $\mu(bab)T = T$ and $\mu(aab)T = \mu(ab)T$. This basis is extended to a basis of $\mathbb{R}^7$, for instance with the first 4 column vectors $e_1, \ldots e_4$ of the canonical basis of $\mathbb{R}^7$.

Let F and H be the matrices

$$F = \begin{bmatrix} \begin{bmatrix} 0 & b & b \\ b & 0 & 0 \\ 0 & a & a \end{bmatrix} & \begin{bmatrix} b & 0 & 0 & 0 \\ 0 & b & 0 & 0 \\ a & 0 & 0 & 0 \end{bmatrix} \\ \begin{bmatrix} 0 & 0 & 0 \\ 0 & 0 & 0 \\ 0 & 0 & 0 \\ 0 & 0 & 0 \end{bmatrix} & \begin{bmatrix} -a & -b & a & 0 \\ -b & 0 & 0 & b \\ a & 0 & 0 & 0 \\ -a & 0 & 0 & 0 \end{bmatrix} \end{bmatrix}, \qquad H = \begin{bmatrix} \begin{bmatrix} 0 & b \\ b & a \end{bmatrix} & \begin{bmatrix} 0 \\ 0 \end{bmatrix} \\ [1\ 0] & [0] \end{bmatrix}.$$

We get that E is similar to F. Let us denote by G the upper left block matrix of size 3 of F. The right-reduced automaton $\langle (2\ 0\ 0), G = \left(\begin{smallmatrix} 0 & b & b \\ b & 0 & 0 \\ 0 & a & a \end{smallmatrix}\right), \left(\begin{smallmatrix} 1 \\ 0 \\ 0 \end{smallmatrix}\right)\rangle$ can be now reduced on the left side. We get that G is similar to H. The upper left block matrix of size 2 of G is similar to $\alpha_{\mathcal{A}}(A)$. As a consequence, $\det(I - \alpha_{\mathcal{A}}(A)) = 1 - a - bb$ divides $\det(I - H)$ which divides $\det(I - F) = \det(I - \alpha_{\mathcal{B}}(A)) = (1 - a - bb)(1 + a - bb)$.

A variant of the combinatorial proof uses an argument due to Nasu [8].

We denote by M (resp. M') the matrix $M = \sum_{a \in A} \mu_{\mathcal{A}}(a)$ and (resp. $M' = \sum_{a \in A} \mu_{\mathcal{B}}(a)$). It is known from the Perron-Frobenius theory that M and M' have the same positive spectral radius λ, the logarithm of λ being called the topogical entropy of the sofic shift S [9]. Let U, V (resp. U', V') be two real positive left and right eigenvectors of M (resp. of M') for the eigenvalue λ. One can choose these vectors such that $UV = U'V' = 1$. With these settings, the two $\mathbb{R}$-automata $\mathcal{C} = \langle U, \mu_{\mathcal{A}}, V\rangle$ and $\mathcal{D} = \langle U', \mu_{\mathcal{B}}, V'\rangle$ are equivalent.

The proof of this equivalence relies on the following arguments. One first divides $\mu_{\mathcal{A}}$ and $\mu_{\mathcal{B}}$ by λ to assume that $\lambda = 1$.

For any word $x \in A^*$ and any $\mathbb{R}$-automaton $S = \langle I, \mu, T\rangle$, we denote by $\pi_S(x)$ the real coefficient $I\mu(x)T$. Hence $\mathcal{C}$ and $\mathcal{D}$ are equivalent if and only if $\pi_{\mathcal{C}}(x) = \pi_{\mathcal{D}}(x)$ for any $x \in A^*$. The functions $\pi_{\mathcal{C}}$ and $\pi_{\mathcal{D}}$ define two rational probability measures on A^* [15]. These measures satisfy the following properties.

- A right (and left) invariance property: for any $x \in A^*$, with S equal to $\mathcal{C}$ or $\mathcal{D}$.

$$\sum_{w \in A^k} \pi_S(xw) = \pi_S(x).$$

- An ergodic property: for any $x \in A^*$, with $\mathcal{S}$ equal to $\mathcal{C}$ or $\mathcal{D}$.

$$\lim_{n \to \infty} \frac{1}{n} \sum_{i=0}^{n-1} \sum_{w \in A^i} \pi_{\mathcal{S}}(xwy) = \pi_{\mathcal{S}}(x)\pi_{\mathcal{S}}(y).$$

Moreover, since the automata $\mathcal{A}$ and $\mathcal{B}$ are unambiguous, one can show that there are positive real numbers ρ, ρ' such that for any $x \in A^*$, $\pi_{\mathcal{C}}(x) \leq \rho\, \pi_{\mathcal{D}}(x)$ and $\pi_{\mathcal{D}}(x) \leq \rho'\, \pi_{\mathcal{C}}(x)$. The equivalence of $\mathcal{C}$ and $\mathcal{D}$ follows from these inequalities. The reduction of the automata is used to finish the proof as before.

Acknowledgments. The authors would like to thank an anonymous reviewer for detecting an error in Lemma 1 in a preliminary version of this paper.

References

1. Béal, M.P., Perrin, D.: Codes and sofic constraints. Theoret. Comput. Sci. **340**(2) (2005) 381–393
2. Reutenauer, Ch.: Ensembles libres de chemins dans un graphe. Bull. Soc. Math. France **114**(2) (1986) 135–152
3. Restivo, A.: Codes and local constraints. Theoret. Comput. Sci. **72**(1) (1990) 55–64
4. Ashley, J., Marcus, B., Perrin, D., Tuncel, S.: Surjective extensions of sliding-block codes. SIAM J. Discrete Math. **6**(4) (1993) 582–611
5. Berstel, J., Perrin, D.: Theory of codes. Volume 117 of Pure and Applied Mathematics. Academic Press Inc., Orlando, FL (1985) http://www-igm.univ-mlv.fr/~berstel/LivreCodes/Codes.html.
6. Ehrenfeucht, A., Rozenberg, G.: Each regular code is included in a maximal regular code. RAIRO Inform. Théor. Appl. **20**(1) (1986) 89–96
7. Williams, S.: Lattice invariants for sofic shifts. Ergodic Theory and Dynamical Systems **11** (1991) 787–801
8. Nasu, M.: An invariant for bounded-to-one factor maps between transitive sofic subshifts. Ergodic Theory Dynam. Systems **5**(1) (1985) 89–105
9. Lind, D., Marcus, B.: An Introduction to Symbolic Dynamics and Coding. Cambridge University Press, Cambridge (1995)
10. Lothaire, M.: Combinatorics on words. Volume 17 of Encyclopedia of Mathematics and its Applications. Addison-Wesley Publishing Co., Reading, Mass. (1983)
11. Restivo, A.: Codes with constraints. In: Mots. Lang. Raison. Calc. Hermès, Paris (1990) 358–366
12. Kitchens, B.P.: Symbolic dynamics. Universitext. Springer-Verlag, Berlin (1998) One-sided, two-sided and countable state Markov shifts.
13. Berstel, J., Reutenauer, C.: Rational series and their languages. Volume 12 of EATCS Monographs on Theoretical Computer Science. Springer-Verlag, Berlin (1988)
14. Sakarovitch, J.: Éléments de Théorie des Automates. Vuibert, Paris (2003) english translation to appear, Cambridge University Pres.
15. Hansel, G., Perrin, D.: Mesures de probabilités rationnelles. In: Mots. Lang. Raison. Calc. Hermès, Paris (1990) 335–357

Kolmogorov Complexity with Error

Lance Fortnow[1], Troy Lee[2], and Nikolai Vereshchagin[3,*]

[1] University of Chicago, 1100 E. 58th Street, Chicago, IL, USA 60637
fortnow@cs.uchicago.edu
http://people.cs.uchicago.edu/~fortnow
[2] CWI and University of Amsterdam, 413 Kruislaan,
1098 SJ Amsterdam, The Netherlands
tlee@cwi.nl
http://www.cwi.nl/~tlee
[3] Moscow State University, Leninskie Gory, Moscow, Russia 119992
ver@mccme.ru
http://lpcs.math.msu.su/~ver

Abstract. We introduce the study of Kolmogorov complexity with error. For a metric d, we define $C_a(x)$ to be the length of a shortest program p which prints a string y such that $d(x,y) \leq a$. We also study a conditional version of this measure $C_{a,b}(x|y)$ where the task is, given a string y' such that $d(y,y') \leq b$, print a string x' such that $d(x,x') \leq a$. This definition admits both a uniform measure, where the *same* program should work given any y' such that $d(y,y') \leq b$, and a nonuniform measure, where we take the length of a program for the worst case y'. We study the relation of these measures in the case where d is Hamming distance, and show an example where the uniform measure is exponentially larger than the nonuniform one. We also show an example where symmetry of information does not hold for complexity with error under either notion of conditional complexity.

1   Introduction

Kolmogorov complexity measures the information content of a string typically by looking at the size of a smallest program generating that string. Suppose we received that string over a noisy or corrupted channel. Such a channel could change random bits of a string, possibly increasing its Kolmogorov complexity without adding any real information.

Alternatively, suppose that we do not have much memory and are willing to sacrifice fidelity to the original data in order to save on compressed size. What is the cheapest approximation to a string within our level of tolerance to distortion? Such compression where some, less important we hope, information about the original data is lost is known as lossy compression.

* Supported in part by the RFBR grants 03-01-00475, 358.20003.1. Work done while visiting CWI.

B. Durand and W. Thomas (Eds.): STACS 2006, LNCS 3884, pp. 137–148, 2006.

Intuitively, these scenarios are in some sense complementary to one another: we expect that if we lossy compress a string received over a corrupted channel with our level of tolerance equal to the number of expected errors, then the cheapest string within the level of tolerance will be the one with the high complexity noise removed. Ideally we would get back our original string. For certain compression schemes and models of noise this intuition can be made precise [8].

In this paper we explore a variation of Kolmogorov complexity designed to help us measure information in these settings. We define the Kolmogorov complexity of a string x with error a as the length of a smallest program generating a string x' that differs from x in at most a bits. We give tight bounds (up to logarithmic factors) on the maximum complexity of such strings and also look at time-bounded variations.

We also look at conditional Kolmogorov complexity with errors. Traditional conditional Kolmogorov complexity looks at the smallest program that converts a string y to a string x. In our context both x and y could be corrupted. We want the smallest program that converts a string close to y to a string close to x. We consider two variations of this definition, a uniform version where we have a single program that that converts any y' close to y to a string x' close to x and a nonuniform version where the program can depend on y'. We show examples giving a large separation between the uniform and nonuniform definitions.

Finally we consider symmetry of information for Kolmogorov complexity with error. Traditionally the complexity of the concatenation of strings x, y is roughly equal to the sum of the complexity of x and the complexity of y given x. We show that for any values of d and a the complexity of xy with error d is at most the sum of the complexity of x with error a and the complexity of converting a string y with $d - a$ error given x with a bits of error. We show the other direction fails in a strong sense—we do not get equality for any a.

2   Preliminaries

We use $|x|$ to denote the length of a string x, and $\|A\|$ to denote the cardinality of a set A. All logarithms are base 2.

We use $d_H(x, y)$ to denote the Hamming distance between two binary strings x, y, that is the number of bits on which they differ. For $x \in \{0, 1\}^n$ we let $B_n(x, R)$ denote the set of n-bit strings within Hamming distance R from x, and $V(n, R) = \sum_{i=0}^{R} \binom{n}{i}$ denote the volume of a Hamming ball of radius R over n-bit strings. For $0 < \lambda \leq 1/2$ the binary entropy of λ is $H(\lambda) = -\lambda \log \lambda - (1 - \lambda) \log(1 - \lambda)$. The binary entropy is useful in the following approximation of $V(n, R)$ which we will use on several occasions (a proof can be found in [1]).

Lemma 1. *Suppose that $0 < \lambda \leq 1/2$ and λn is an integer. Then*

$$\frac{2^{nH(\lambda)}}{\sqrt{8n\lambda(1 - \lambda)}} \leq V(n, \lambda n) \leq 2^{nH(\lambda)}.$$

3   Defining Kolmogorov Complexity with Error

We consider several possible ways of defining Kolmogorov complexity with error. In this section we present these alternatives in order to evaluate their relative merits in the coming sections. First, we review the standard definition of Kolmogorov complexity. More details can be found in [6].

For a Turing machine T, the Kolmogorov complexity $C_T(x|y)$ of x given y is the length of a shortest program p such that $T(p, y) = x$. The theory of Kolmogorov complexity begins from the following invariance theorem: there is a universal machine U such that for any other Turing machine T, there exists a constant c_T such that $C_U(x|y) \leq C_T(x|y) + c_T$, for all x, y. We now fix such a U and drop the subscript. Now we define also the unconditional Kolmogorov complexity $C(x) = C(x|\text{empty string})$.

Definition 1. *Let $d : (\{0,1\}^n)^2 \to R$ be a metric, and $a \in R$. The complexity of x with error a, denoted $C_a(x)$ is $C_a(x) = \min_{x'}\{C(x') : d(x', x) \leq a\}$.*

We will also consider a time bounded version of this definition, $C_a^t(x) = \min_{x'}\{C^t(x'|\text{empty string}) : d(x, x') \leq a\}$, where $C^t(x|y)$ is the length of a shortest program p such that $U(p, y)$ prints x in less than $t(|x| + |y|)$ time steps. Here we assume that the machine U is universal in the following sense: for any other Turing machine T, there exists a constant c_T and a polynomial q such that $C_U^{q(|x|,|y|,t)}(x|y) \leq C_T^t(x|y) + c_T$, for all x, y, t.

A relative version of Kolmogorov complexity with error is defined by Impagliazzo, Shaltiel and Wigderson [4]. That is, they use the definition $C_\delta(x) = \min\{C(y) : d_H(x, y) \leq \delta|x|\}$. We prefer using absolute distance here as it behaves better with respect to concatenations of strings—using relative distance has the disadvantage of severe nonmonotonicity over prefixes. Take, for example, $x \in \{0, 1\}^n$ satisfying $C(x) \geq n$. Let $y = 0^{2n}$. Then $C_{1/3}(x) \geq n - \log V(n, n/3)$ while $C_{1/3}(xy) \leq \log n + O(1)$. Using absolute error we have that $C_a(xy) \geq C_a(x) - O(\log n)$, that is it only suffers from logarithmic dips as with standard definition.

Defining conditional complexity with error is somewhat more subtle. We introduce both uniform and nonuniform versions of conditional complexity with error.

Definition 2. *For a Turing machine T, the uniform conditional complexity, denoted $(C_{a,b}^u)_T(x|y)$, is the length of a shortest program p such that, for any y' satisfying $d(y, y') \leq b$ it holds that $T(p, y')$ outputs a string whose distance from x is less than a.*

The invariance theorem remains true: there is a universal machine U such that for any other Turing machine T, there exists a constant c_T such that $(C_{a,b}^u)_U(x|y) \leq (C_{a,b}^u)_T(x|y) + c_T$, for all x, y, a, b. We fix such a U and drop the subscript.

Definition 3. *Nonuniform conditional complexity, which we denote $C_{a,b}(x|y)$ is defined as $C_{a,b}(x|y) = \max_{y'} \min_{x'}\{C(x'|y') : d(x', x) \leq a \text{ and } d(y', y) \leq b\}$.*

In section 6 we study the difference between these two measures.

4   Strings of Maximal Complexity

One of the most famous applications of Kolmogorov complexity is the incompressibility method (see [6], Chapter 6). To prove there exists an object with a certain property, we consider an object with maximal Kolmogorov complexity and show that it could be compressed if it did not possess this property.

This method relies on a simple fact about strings of maximal complexity: for every length n, there is a string x of complexity at least n. This follows from simple counting. It is also easy to see that, up to an additive constant, every string has complexity at most its length. What is the behavior of maximal complexity strings in the error case? In this paper we restrict ourselves to the Hamming distance case.

Again by a counting argument, we see that for every n there is an x of length n with $C_a(x) \geq \log 2^n/V(n,a) = n - \log V(n,a)$. Upper bounding the complexity of strings in the error case requires a bit more work, and has a close connection with the construction of covering codes. A covering code $\mathcal{C}$ of radius a is a set of strings such that for every $x \in \{0,1\}^n$ there is an element $y \in \mathcal{C}$ such that $d_H(x,y) \leq a$. Thus an upper bound on the maximum complexity strings will be given by the existence of covering codes of small size. The following Lemma is well known in the covering code literature, (see [1] or [5]).

Lemma 2. *For any n and integer $R \leq n$, there exists a set $\mathcal{C} \subseteq \{0,1\}^n$ with the following properties:*

1. *$\|\mathcal{C}\| \leq n2^n/V(n,R)$*
2. *for every $x \in \{0,1\}^n$, there exists $c \in \mathcal{C}$ with $d_H(x,c) \leq R$*
3. *The set $\mathcal{C}$ can be computed in time $\text{poly}(2^n)$*

Proof. For the first two items we argue by the probabilistic method. Fix a point $x \in \{0,1\}^n$. We uniformly at random choose k elements $x_1, \ldots, x_k$ of $\{0,1\}^n$. The probability P_x that x is not contained in $\cup_{i=1}^k B(x_i, R)$ is precisely

$$P_x = (1 - V(n,R)/2^n)^k \leq e^{-kV(n,R)/2^n}.$$

For the inequality we have used the fact that $e^z \geq 1+z$ for any z. Taking k to be $n2^n/V(n,R)$ makes this probability strictly less than 2^{-n}. Thus the probability of the union of the events P_x over $x \in \{0,1\}^n$ is, by the union bound, less than 1 and there exists a set of $n2^n/V(n,R)$ centers which cover $\{0,1\}^n$. This gives items 1 and 2.

For item 3 we now derandomize this argument using the method of conditional probabilities. The argument is standard as found in [7], and omitted here.   □

To achieve part 3 of Lemma 2 one could alternatively apply a general theorem that the greedy algorithm always finds a covering of a set X of size at most a $\ln \|X\|$ multiplicative factor larger than the optimal covering (see Corollary 37.5 in [2]). This would give the slightly worse bound of $O(n^2 2^n/V(n,R))$.

Theorem 1. *For every n, a and $x \in \{0,1\}^n$, $C_a(x) \leq n - \log V(n, a) + O(\log n)$.*

Proof. Use the lexicographically first covering code of radius a whose existence is given by Lemma 2. □

One nice property of covering codes is that they behave very well under concatenation. Let $\mathcal{C}_1$ be a covering code of $\{0,1\}^{n_1}$ of radius R_1 and $\mathcal{C}_2$ be a covering code of $\{0,1\}^{n_2}$ of radius R_2. Now let $\mathcal{C} = \{cc' : c \in \mathcal{C}_1, c' \in \mathcal{C}_2\}$ be the set of all ordered concatenations of codewords from $\mathcal{C}_1$ with codewords from $\mathcal{C}_2$. Then $\mathcal{C}$ is a covering code over $\{0,1\}^{n_1+n_2}$ of radius $R_1 + R_2$.

We can use this idea in combination with item 3 of Lemma 2 to efficiently construct near-optimal covering codes. This construction has already been used for a complexity-theoretic application in [3].

Theorem 2. *There is a polynomial time bound $p(n)$ such that $C_a^{p(n)}(x) \leq n - \log V(n, a) + O(n \log \log n / \log n)$ for every $x \in \{0,1\}^n$ and every a.*

Proof. We construct a covering code over $\{0,1\}^n$ with radius a such that the ith element of the covering can be generated in time polynomial in n. Let $\ell = \log n$ and divide n into n/ℓ blocks of length ℓ. Let $r = (a/n)\ell$. Now by item 3 of Lemma 2 we can in time polynomial in n construct a covering code over $\{0,1\}^\ell$ of radius r and of cardinality $\ell 2^\ell / V(\ell, r)$. Call this covering $\mathcal{C}_\ell$. Our covering code $\mathcal{C}$ over $\{0,1\}^n$ will be the set of codewords $\{c_1 c_2 \cdots c_{n/\ell} : c_i \in \mathcal{C}_\ell\}$. The size of this code will be:

$$\|\mathcal{C}\| \leq (2^{\ell - \log V(\ell,r) + \log \ell})^{n/\ell} = (2^{\ell - \ell H(a/n) + O(\log \ell)})^{n/\ell}$$
$$= 2^{n - nH(a/n) + O(n \log \ell / \ell)} = 2^{n - \log V(n,a) + O(n \log \ell / \ell)}. \tag{1}$$

The second and last inequalities hold by Lemma 1.

In this proof we assumed that $\log n$, $n/\log n$, and $a \log n / n$ are all integer. The general case follows with simple modifications. □

5   Dependence of Complexity on the Number of Allowed Errors

Both the uniform and the non-uniform conditional complexities $C_{a,b}^u$ and $C_{a,b}$ are decreasing functions in a and increasing in b. Indeed, if b decreases and a increases then the number of y''s decreases and the number of x''s increases, thus the problem to transform every y' to some x' becomes easier. What is the maximal possible rate of this decrease/increase? For the uniform complexity, we have no non-trivial bounds. For the non-uniform complexity, we have the following

Theorem 3. *For all x, y of length n and all $a \leq a'$, $b' \leq b$ it holds*

$$C_{a,b}(x|y) \leq C_{a',b'}(x|y) + \log(V(n, a)/V(n, a')) + \log(V(n, b')/V(n, b)) + O(\log n).$$

Proof. Let y' be a string at distance b from y. We need to find a short program mapping it to a string at distance a from x. To this end we need the following lemma from [9].

Lemma 3. *For all $d \leq d' \leq n$ having the form i/n, every Hamming ball of radius d' in the set of binary strings of length n can be covered by at most $O(n^4 V(n, d')/V(n, d))$ Hamming balls of radius d.*

Apply the lemma to $d' = b$, $d = b'$ and to the ball of radius b centered at y'. Let $B_1, \ldots, B_N$, where $N = O(n^4 V(n, b)/V(n, b'))$, be the covering balls. Let B_i be a ball containing the string y and let y'' be its center. There is a program, call it p, of length at most $C_{a',b'}(x|y)$ mapping y'' to a string at distance a' from x. Again apply the lemma to $d = a$, $d' = a'$ and to the ball of radius d' centered at x'. Let $C_1, \ldots, C_M$, where $M = O(n^4 V(n, a')/V(n, a))$, be the covering balls. Let C_j be a ball containing the string x and let x'' be its center. Thus x'' is at distance a from x and can be found from y', p, i, j. This implies that $K(x''|y') \leq |p| + \log N + \log M + O(\log n)$ (extra $O(\log n)$ bits are needed to separate p, i and j).   □

In the above proof, it is essential that we allow the program mapping y' to a string close to x depend on y'. Indeed, the program is basically the triple (p, i, j) where both i and j depend on y'. Thus the proof is not valid for the uniform conditional complexity. And we do not know whether the statement itself is true for the uniform complexity.

By using Theorem 2 one can prove a similar inequality for time bounded complexity with the $O(\log n)$ error term replaced by $O(n \log \log n / \log n)$.

6   Uniform vs. Nonuniform Conditional Complexity

In this section we show an example where the uniform version of conditional complexity can be exponentially larger than the nonuniform one. Our example will be for $C_{0,b}(x|x)$. This example is the standard setting of error correction: given some x' such that $d_H(x, x') \leq b$, we want to recover x exactly. An obvious upper bound on the nonuniform complexity $C_{0,b}(x|x)$ is $\log V(n, b) + O(1)$—as we can tailor our program for each x' we can simply say the index of x in the ball of radius b around x'.

In the uniform case the same program must work for every x' in the ball of radius b around x and the problem is not so easy. The following upper bound was pointed out to us by a referee.

Proposition 1. $C_{0,b}^u(x|x) \leq \log V(n, 2b) + O(1)$.

Proof. Let $\mathcal{C} \subseteq \{0, 1\}^n$ be a set with the properties:

1. For every $x, y \in \mathcal{C}$: $B_n(x, b) \cap B_n(y, b) = \varnothing$.
2. For every $y \in \{0, 1\}^n$ $\exists x \in \mathcal{C}$: $d_H(x, y) \leq 2b$.

We can greedily construct such a set as if there is some string y with no string $x \in \mathcal{C}$ of distance less than $2b$, then $B_n(y, b)$ is disjoint from all balls of radius b around elements of $\mathcal{C}$ and so we can add y to $\mathcal{C}$.

Now for a given x, let x^* be the closest element of $\mathcal{C}$ to x, with ties broken by lexicographical order. Let $z = x \oplus x^*$. By the properties of $\mathcal{C}$ this string has Hamming weight at most $2b$ and so can be described with $\log V(n, 2b)$ bits. Given input x' with $d_H(x, x') \leq b$, our program does the following: computes the closest element of $\mathcal{C}$ to $x' \oplus z$, call it w, and then outputs $w \oplus z = w \oplus x^* \oplus x$. Thus for correctness we need to show that $w = x^*$ or in other words that $d_H(x' \oplus z, x^*) \leq b$. Notice that $d_H(\alpha \oplus \beta, \beta) = d_H(\alpha, 0)$, thus

$$d_H(x' \oplus z, x^*) = d_H(x' \oplus x \oplus x^*, x^*) = d_H(x' \oplus x, 0) = d_H(x, x') \leq b. \quad \square$$

We now turn to the separation between the uniform and nonuniform measures. The intuition behind the proof is the following: say we have some computable family S of Hamming balls of radius b, and let x be the center of one of these balls. Given any x' such that $d_H(x, x') \leq b$, there may be other centers of the family S which are also less than distance b from x'. Say there are k of them. Then x has a nonuniform description of size about $\log k$ by giving the index of x in the k balls which are of distance less than b from x'.

In the uniform case, on the other hand, our program can no longer be tailored for a particular x', it must work for any x' such that $d_H(x, x') \leq b$. That is, intuitively, the program must be able to distinguish the ball of x from any other ball intersecting the ball of x. To create a large difference between the nonuniform and uniform conditional complexity measures, therefore, we wish to construct a large family of Hamming balls, every two of which intersect, yet that no single point is contained in the intersection of too many balls. Moreover, we can show the stronger statement that $C_{0,b}(x|x)$ is even much smaller than $C_{a,b}^u(x|x)$, for a non-negligible a. For this, we further want that the contractions of any two balls to radius a are disjoint. The next lemma shows the existence of such a family.

Lemma 4. *For every length m of strings and a, b, and N satisfying the inequalities*

$$N^2 V(m, 2a) \leq 2^{m-1}, \quad N^2 V(m, m - 2b) \leq 2^{m-1}, \quad NV(m, b) \geq m2^{m+1} \quad (2)$$

there are strings $x_1, \ldots, x_N$ such that the balls of radius a centered at $x_1, \ldots, x_N$ are pairwise disjoint, and the balls of radius b centered at $x_1, \ldots, x_N$ are pairwise intersecting but no string belongs to more than $NV(m, b)2^{1-m}$ of them.

Proof. The proof is by probabilistic arguments. Take N independent random strings $x_1, \ldots, x_N$. We will prove that with high probability they satisfy the statement.

First we estimate the probability that there are two intersecting balls of radius a. The probability that two fixed balls intersect is equal to $V(m, 2a)/2^m$. The number of pairs of balls is less than $N^2/2$, and by union bound, there are two intersecting balls of radius a with probability at most $N^2 V(m, 2a)/2^{m+1} \leq 1/4$ (use the first inequality in (2)).

Let us estimate now the probability that there are two disjoint balls of radius b. If the balls of radius b centered at x_j and x_i are disjoint then x_j is at distance at most $m - 2b$ from the string $\bar{x}_i$, that is obtained from x_i by flipping all bits. Therefore the probability that for a fixed pair (i, j) the balls are disjoint is at most $V(m, m - 2b)/2^m$. By the second inequality in (2), there are two disjoint balls with probability at most $1/4$.

It remains to estimate the probability that there is a string that belongs to more than $NV(m, b)2^{1-m}$ balls of radius b. Fix x. For every i the probability that x lands in B_i, the ball of radius b centered at x_i, is equal to $p = |B_i|/2^m = V(m, b)/2^m$. So the average number of i with $x \in B_i$ is $pN = NV(m, b)/2^m$. By Chernoff inequality the probability that the number of i such that x lands in B_i exceeds twice the average is at most

$$\exp(-pN/2) = \exp(-NV(m, b)/2^{m+1}) \leq \exp(-m) \ll 2^{-m}$$

(use the third inequality in (2)). Thus even after multiplying it by 2^m the number of different x's we get a number close to 0.   □

Using this lemma we find x with exponential gap between $C_{0,b}(x|x)$ and $C^u_{0,b}(x|x)$ and even between $C_{0,b}(x|x)$ and $C^u_{a,b}(x|x)$ for a, b linear in the length n of x.

Theorem 4. *Fix rational constants α, β, γ satisfying $\gamma \geq 1$ and*

$$0 < \alpha < 1/4 < \beta < 1/2, \quad 2H(\beta) > 1 + H(2\alpha), \quad 2H(\beta) > 1 + H(1 - 2\beta) \quad (3)$$

Notice that if β is close to $1/2$ and α is close to 0 then these inequalities are satisfied. Then for all sufficiently large m there is a string x of length $n = \gamma m$ with $C_{0,\beta m}(x|x) = O(\log m)$ while $C^u_{\alpha m, \beta m}(x|x) \geq m(1 - H(\beta)) - O(\log m)$.

Proof. Given m let $a = \alpha m$, $b = \beta m$ and $N = m2^{m+1}/V(m, b)$. Let us verify that for large enough m the inequalities (2) in the condition of Lemma 4 are fulfilled. Taking the logarithm of the first inequality (2) and ignoring all terms of order $O(\log m)$ we obtain

$$2(m - mH(\beta)) + mH(2\alpha) < m$$

This is true by the second inequality in (3). Here we used that, ignoring logarithmic terms, $\log V(m, b) = mH(\beta)$ and $\log V(m, 2a) = mH(2\alpha)$ as both $\beta, 2\alpha$ are less than $1/2$. Taking the logarithm of the second inequality (2) we obtain

$$2(m - mH(\beta)) + mH(1 - 2\beta) < m.$$

This is implied by the third inequality in (3). Finally, the last inequality (2) holds by the choice of N.

Find the first sequence $x_1, \ldots, x_N$ satisfying the lemma. This sequence has complexity at most $C(m) = O(\log m)$. Append 0^{n-m} to all strings $x_1, \ldots, x_N$. Obviously the resulting sequence also satisfies the lemma. For each string x_i we have $C_{0,b}(x_i|x_i) = O(\log m)$, as given any x' at distance at most b from x_i we can specify x_i by specifying its index among centers of the balls in the family containing x' in $\log(NV(m, b)2^{1-m}) = \log 4m$ bits and specifying the family itself in $O(\log m)$ bits.

It remains to show that there is x_i with $C^u_{a,b}(x_i|x_i) \geq \log N$. Assume the contrary and choose for every x_i a program p_i of length less than $\log N$ such that $U(p, x')$ is at distance a from x_i for every x' at distance at most b from x_i. As N is strictly greater than the number of strings of length less than $\log N$, by the Pigeon Hole Principle there are different x_i, x_j with $p_i = p_j$. However the balls of radius b with centers x_i, x_j intersect and there is x' at distance at most b both from x_i, x_j. Hence $U(p, x')$ is at distance at most a both from x_i, x_j, a contradiction. $\qquad\square$

Again, at the expense of replacing $O(\log m)$ by $O(m \log \log m / \log m)$ we can prove an analog of Theorem 4 for time bounded complexity. We defer the proof to the final version.

Theorem 5. *There is a polynomial p such that for all sufficiently large m there is a string x of length $n = \gamma m$ with $C^{p(n)}_{0,\beta m}(x|x) = O(m \log \log m / \log m)$ while $C^u_{\alpha m, \beta m}(x|x) \geq m(1 - H(\beta)) - O(m \log \log m / \log m)$. (Note that C^u has no time bound; this makes the statement stronger.)*

7 Symmetry of Information

The principle of symmetry of information, independently proven by Kolmogorov and Levin [10], is one of the most beautiful and useful theorems in Kolmogorov complexity. It states $C(xy) = C(x) + C(y|x) + O(\log n)$ for any $x, y \in \{0, 1\}^n$. The direction $C(xy) \leq C(x) + C(y|x) + O(\log n)$ is easy to see—given a program for x, and a program for y given x, and a way to tell these programs apart, we can print xy. The other direction of the inequality requires a clever proof.

Looking at symmetry of information in the error case, the easy direction is again easy: The inequality $C_d(xy) \leq C_a(x) + C_{d-a,a}(y|x) + O(\log n)$ holds for any a — let p be a program of length $C_a(x)$ which prints a string x^* within Hamming distance a of x. Let q be a shortest program which, given x^*, prints a string y^* within Hamming distance $d - a$ of y. By definition, $C_{d-a,a}(y|x) = \max_{x'} \min_{y'} C(y'|x') \geq \min_{y'} C(y'|x^*) = |q|$. Now given p and q and a way to tell them apart, we can print the string xy within d errors.

For the converse direction we would like to have the statement

$$\text{For every } d, x, y \text{ there exists } a \leq d \text{ such that} \qquad (*)$$
$$C_d(xy) \geq C_a(x) + C_{d-a,a}(y|x) - O(\log n).$$

We do not expect this statement to hold for every a, as the shortest program for xy will have a particular pattern of errors which might have to be respected in the programs for x and y given x. We now show, however, that even the formulation $(*)$ is too much to ask.

Theorem 6. *For every n and all $d \leq n/4$ there exist $x, y \in \{0,1\}^n$ such that for all $a \leq d$ the difference*

$$\Delta(a) = (C_a(y) + C_{d-a,a}(x|y)) - C_d(xy)$$

is more than both

$$\log V(n,d) - \log V(n,a), \qquad \log V(n,d+a) - \log V(n,d-a) - \log V(n,a),$$

up to an additive error term of the order $O(\log n)$.

Since $C^u_{d-a,a}(x|y) \geq C_{d-a,a}(x|y)$, Theorem 6 holds for uniform conditional complexity as well.

Before proving the theorem let us show that in the case, say, $d = n/4$ it implies that for some positive ε we have $\Delta(a) \geq \varepsilon n$ for all a. Let $\alpha < 1/4$ be the solution to the equation

$$H(1/4) = H(1/4 + \alpha) - H(1/4 - \alpha).$$

Note that the function in the right hand side increases from 0 to 1 as α increases from 0 to 1/4. Thus this equation has a unique solution.

Corollary 1. *Let $d = n/4$ and let x, y be the strings existing by Theorem 6. Then we have $\Delta(a) \geq n(H(1/4) - H(\alpha)) - O(\log n)$ for all a.*

The proof is simply a calculation and is omitted. Now the proof of Theorem 6.

Proof. Coverings will again play an important role in the proof. Let $\mathcal{C}$ be the lexicographically first minimal size covering of radius d. Choose y of length n with $C(y) \geq n$, and let x be the lexicographically least element of the covering within distance d of y. Notice that $C_d(xy) \leq n - \log V(n,d)$, as the string xx is within distance d of xy, and can be described by giving a shortest program for x and a constant many more bits saying "repeat". (In the whole proof we neglect additive terms of order $O(\log n)$). Let us prove first that $C(x) = n - \log V(n,d)$ and $C(y|x) = \log V(n,d_1) = \log V(n,d)$, where d_1 stands for the Hamming distance between x and y. Indeed,

$$n \leq C(y) \leq C(x) + C(y|x) \leq n - \log V(n,d) + C(y|x)$$
$$\leq n - \log V(n,d) + \log V(n,d_1) \leq n.$$

Thus all inequalities here are equalities, hence $C(x) = n - \log V(n,d)$ and $C(y|x) = \log V(n,d_1) = \log V(n,d)$.

Let us prove now the first lower bound for $\Delta(a)$. As y has maximal complexity, for any $0 \leq a \leq d$ we have $C_a(y) \geq n - \log V(n,a)$. Summing the inequalities

$$-C_d(xy) \geq -n + \log V(n,d),$$
$$C_a(y) \geq n - \log V(n,a),$$
$$C_{d-a,a}(x|y) \geq 0,$$

we obtain the lower bound $\Delta(a) \geq \log V(n,d) - \log V(n,a)$. To prove the second lower bound of the theorem, we need to show that

$$C_{d-a,a}(x|y) \geq \log V(n,d+a) - \log V(n,d-a) - \log V(n,d). \qquad (4)$$

To prove that $C_{d-a,a}(x|y)$ exceeds a certain value v we need to find a y' at distance at most a from y such that $C(x'|y') \geq v$ for all x' at distance at most $d - a$ from x. Let y' be obtained from y by changing a random set of a bits on which x and y agree. This means that $C(y'|y,x) \geq \log V(n - d_1, a)$. It suffices to show that

$$C(x|y') \geq \log V(n,d+a) - \log V(n,d).$$

Indeed, then for all x' at distance at most $d - a$ from x we will have

$$C(x'|y') + \log V(n,d-a) \geq C(x|y')$$

(knowing x' we can specify x by its index in the ball of radius $d - a$ centered at x'). Summing these inequalities will yield (4).

We use symmetry of information in the nonerror case to turn the task of lower bounding $C(x|y')$ into the task of lower bounding $C(y'|x)$ and $C(x)$. This works as follows: by symmetry of information,

$$C(xy') = C(x) + C(y'|x) = C(y') + C(x|y').$$

As $C(y')$ is at most n, using the second part of the equality we have $C(x|y') \geq C(x) + C(y'|x) - n$. Recall that $C(x) = n - \log V(n,d)$. Thus to complete the proof we need to show the inequality $C(y'|x) \geq \log V(n,d+a)$, that is, y' is a random point in the Hamming ball of radius $d + a$ with the center at x. To this end we first note that $\log V(n,d+a) = \log V(n,d_1+a)$ (up to a $O(\log n)$ error term). Indeed, as $a + d \leq n/2$ we have $\log V(n,d+a) = \log \binom{n}{d+a}$ and $\log V(n,d) = \log \binom{n}{d}$. The same holds with d_1 in place of d. Now we will show that $\log V(n,d) - \log V(n,d_1) = O(\log n)$ implies that $\log V(n,d+a) - \log V(n,d_1 + a) = O(\log n)$. It is easy to see that $\binom{n}{d+1}/\binom{n}{d_1+1} \leq \binom{n}{d}/\binom{n}{d_1}$ provided $d_1 \leq d$. Using the induction we obtain $\binom{n}{d+a}/\binom{n}{d_1+a} \leq \binom{n}{d}/\binom{n}{d_1}$.

Thus we have

$$\log V(n,d+a) - \log V(n,d_1+a) = \log\left(\binom{n}{d+a}\Big/\binom{n}{d_1+a}\right)$$

$$\leq \log\left(\binom{n}{d}\Big/\binom{n}{d_1}\right) = \log V(n,d) - \log V(n,d_1) = O(\log n).$$

Again we use (the conditional form of) symmetry of information:

$$C(y'y|x) = C(y|x) + C(y'|y,x) = C(y'|x) + C(y|y',x).$$

The string y differs from y' on a bits out of the $d_1 + a$ bits on which y' and x differ. Thus $C(y|y',x) \leq \log\binom{d_1+a}{a}$. Now using the second part of the equality we have

$$C(y'|x) = C(y|x) + C(y'|y, x) - C(y|y', x)$$
$$\geq \log V(n, d_1) + \log V(n - d_1, a) - \binom{d_1 + a}{a}.$$

We have used that $\log V(n - d_1, a) = \log \binom{n-d_1}{a}$, as $a \leq (n - d_1)/2$. Hence,

$$C(y'|x) \geq \log \binom{n}{d_1} + \log \binom{n - d_1}{a} - \log \binom{d_1 + a}{a} = \log V(n, d + a).$$

$\square$

Again, at the expense of replacing $O(\log n)$ by $O(n \log \log n / \log n)$ we can prove an analog of Theorem 6 for time bounded complexity.

Acknowledgment

We thank Harry Buhrman for several useful discussions and the anonymous referees for valuable remarks and suggestions.

References

1. G. Cohen, I. Honkala, S. Litsyn, and A. Lobstein. *Covering Codes*. North-Holland, Amsterdam, 1997.
2. T. Cormen, C. Leiserson, and R. Rivest. *Introduction to Algorithms*. MIT Press, 1990.
3. E. Dantsin, A. Goerdt, E. Hirsch, and U. Schöning. Deterministic algorithms for k-SAT based on covering codes and local search. In *Proceedings of the 27th International Colloquium On Automata, Languages and Programming*, Lecture Notes in Computer Science, pages 236–247. Springer-Verlag, 2000.
4. R. Impagliazzo, R. Shaltiel, and A. Wigderson. Extractors and pseudo-random generators with optimal seed length. In *Proceedings of the 32nd ACM Symposium on the Theory of Computing*, pages 1–10. ACM, 2000.
5. M. Krivelevich, B. Sudakov, and V. Vu. Covering codes with improved density. *IEEE Transactions on Information Theory*, 49:1812–1815, 2003.
6. M. Li and P. Vitányi. *An Introduction to Kolmogorov Complexity and its Applications*. Springer-Verlag, New York, second edition, 1997.
7. R. Motwani and P. Raghavan. *Randomized Algorithms*. Cambridge University Press, 1997.
8. B. Natarajan. Filtering random noise from deterministic signals via data compression. *IEEE transactions on signal processing*, 43(11):2595–2605, 1995.
9. N. Vereschagin and P. Vitányi. Algorithmic rate-distortion theory. http://arxiv.org/abs/cs.IT/0411014, 2004.
10. A. Zvonkin and L. Levin. The complexity of finite objects and the algorithmic concepts of information and randomness. *Russian Mathematical Surveys*, 25:83–124, 1970.

Kolmogorov Complexity
and the Recursion Theorem

Bjørn Kjos-Hanssen[1], Wolfgang Merkle[2], and Frank Stephan[3,*]

[1] University of Connecticut, Storrs
bjorn@math.uconn.edu
[2] Ruprecht-Karls-Universität Heidelberg
merkle@math.uni-heidelberg.de
[3] National University of Singapore
fstephan@comp.nus.edu.sg

Abstract. We introduce the concepts of complex and autocomplex sets, where a set A is complex if there is a recursive, nondecreasing and unbounded lower bound on the Kolmogorov complexity of the prefixes (of the characteristic sequence) of A, and autocomplex is defined likewise with recursive replaced by A-recursive. We observe that exactly the autocomplex sets allow to compute words of given Kolmogorov complexity and demonstrate that a set computes a diagonally nonrecursive (DNR) function if and only if the set is autocomplex. The class of sets that compute DNR functions is intensively studied in recursion theory and is known to coincide with the class of sets that compute fixed-point free functions. Consequently, the Recursion Theorem fails relative to a set if and only if the set is autocomplex, that is, we have a characterization of a fundamental concept of theoretical computer science in terms of Kolmogorov complexity. Moreover, we obtain that recursively enumerable sets are autocomplex if and only if they are complete, which yields an alternate proof of the well-known completeness criterion for recursively enumerable sets in terms of computing DNR functions.

All results on autocomplex sets mentioned in the last paragraph extend to complex sets if the oracle computations are restricted to truth-table or weak truth-table computations, for example, a set is complex if and only if it wtt-computes a DNR function. Moreover, we obtain a set that is complex but does not compute a Martin-Löf random set, which gives a partial answer to the open problem whether all sets of positive constructive Hausdorff dimension compute Martin-Löf random sets.

Furthermore, the following questions are addressed: Given n, how difficult is it to find a word of length n that (a) has at least prefix-free Kolmogorov complexity n, (b) has at least plain Kolmogorov complexity n or (c) has the maximum possible prefix-free Kolmogorov complexity among all words of length n. All these questions are investigated with respect to the oracles needed to carry out this task and it is shown that (a) is easier than (b) and (b) is easier than (c). In particular, we argue that for plain Kolmogorov complexity exactly the PA-complete sets compute incompressible words, while the class of sets

---

* F. Stephan is supported in part by NUS grant number R252–000–212–112.

B. Durand and W. Thomas (Eds.): STACS 2006, LNCS 3884, pp. 149–161, 2006.

that compute words of maximum complexity depends on the choice of the universal Turing machine, whereas for prefix-free Kolmogorov complexity exactly the complete sets allow to compute words of maximum complexity.

1   Introduction and Overview

The Recursion Theorem, one of the most fundamental results of theoretical computer science, asserts that — with a standard effective enumeration $\varphi_0, \varphi_1, \ldots$ of all partial recursive functions understood — every recursive function g has a fixed point in the sense that for some index e, the partial recursive functions φ_e and $\varphi_{g(e)}$ are the same. The question of which type of additional information is required in order to compute a fixed-point free function g is well understood; in particular, it is known that a set can compute such a function iff it can compute a diagonally nonrecursive (DNR) function g, that is, a function g such that for all e the value $\varphi_e(e)$, if defined, differs from $g(e)$ [6].

By a celebrated result of Schnorr, a set is Martin-Löf random if and only if the length n prefixes of its characteristic sequence have prefix-free Kolmogorov complexity H of at least $n - c$ for some constant c. From this equivalence, we obtain easily a proof for Kučera's result [4, Corollary 1 to Theorem 6] that any Martin-Löf random set R computes a DNR function, and hence also computes a fixed-point free function; simply let $f(e) = R(0)R(1)\ldots R(e-1)$ (where the prefixes of R are interpreted as binary expansions of natural numbers), then for appropriate constants c and c' and for all e, the prefix-free Kolmogorov complexity of the function values $\varphi_e(e)$ and $f(e)$ is at most $2\log e + c'$ and at least $e - c$, respectively, hence for almost all e these two values differ, that is, changing f at at most finitely many places yields a DNR function. Similarly, if one lets $g(e)$ be an index for the constant function with value $f(e)$, then φ_e and $\varphi_{g(e)}$ differ at place e for almost all e, and thus a finite variant of g is fixed-point free.

We will call a set A complex if there is a nondecreasing and unbounded recursive function h such that the length $h(n)$ prefix of A has plain Kolmogorov complexity of at least n and autocomplex is defined likewise with recursive replaced by A-recursive. Obviously, any autocomplex sets computes a function f such the plain Kolmogorov complexity of $f(n)$ is at least n and as observed in Proposition 3, this implication is in fact an equivalence; a similar equivalence is stated in Proposition 4 for complex sets and for computing such a function f in truth-table or weak truth-table style. By an argument similar to the one given in the last paragraph, any autocomplex set computes a DNR function. In fact, Theorem 5 asserts that the reverse implication holds too, that is, the class of autocomplex sets coincides with the class of sets that compute a DNR function, or, equivalently, with the class of sets that compute fixed-point free functions. This means that the sets relative to which the Recursion Theorem does not hold can be characterized as the autocomplex sets, that is, like for Schnorr's celebrated characterization of Martin-Löf random sets as the sets with incompressible prefixes, we obtain a characteriziation of a fundamental concept of theoretical computer science in terms of Kolmogorov complexity. By similar arguments, the class of complex sets coincides with the class of sets that compute a DNR function via a truth-table or weak truth-table reduction, where the DNR function then automatically is recursively bounded.

From the mentioned results on complex sets and by work of Ambos-Spies, Kjos-Hanssen, Lempp and Slaman [2], we obtain in Proposition 7 that there is a complex set that does not compute a Martin-Löf random set, thus partially answering an open problem by Reimann [13] about extracting randomness.

Theorem 8 states that recursively enumerable (r.e.) sets are complete if and only if it they are autocomplex, and are wtt-complete if and only if they are complex. Arslanov's completeness criteria in terms of DNR functions are then immediate from Theorems 5 and 8, which in summary yields simplified proofs for these criteria.

Theorem 10 asserts that the complex sets can be characterized as the sets that are not wtt-reducible to a hyperimmune set. Miller [8] demonstrated that the latter property characterizes the hyperavoidable sets, thus we obtain as corollary that a set is complex if and only if it is hyperavoidable.

In the characterization of the autocomplex sets as the sets that compute a function f such that the complexity of $f(n)$ is at least n, the values $f(n)$ of such a function f at place n might be very large and hence the complexity of the function value $f(n)$ might be arbitrarily small compared to its length. Theorem 14 states that the sets that allow to compute a function f such that the length and the plain Kolmogorov complexity of $f(n)$ are both equal to n are just the PA-complete sets and that, furthermore, these two identical classes of sets coincide with the class of sets that compute a lower bound b on plain Kolmogorov complexity such that b attains values strictly smaller than n not more often than $2^n - 1$ times. Recall in this connection that by definition a set is PA-complete if and only if it computes a complete extension of Peano arithmetic and that the concept of PA-completeness is well understood and allows several interesting characterizations, for example, a set is PA-complete if and only if it computes a $\{0, 1\}$-valued DNR function [6].

For a word $f(n)$ of length n, in general plain Kolmogorov complexity n is not maximum, but just maximum up to an additive constant. In Theorem 15 it is demonstrated that the class of sets that allow to compute a function f such that $f(n)$ has indeed maximum plain Kolmogorov complexity depends on the choice of the universal machine used for defining plain Kolmogorov complexity; more precisely, for any recursively enumerable set B there is a universal machine such that this class coincides with the class of sets that are PA-complete and compute B. In contrast to this, Theorem 17 asserts that in the case of prefix-free Kolmogorov complexity exactly the sets that compute the halting problem allow to compute a function f such that $f(n)$ has length n and has maximum complexity among all words of the same length.

In the remainder of this section we describe some notation and review some standard concepts. If not explicitly stated differently, a set is always a subset of the natural numbers $\mathbb{N}$. Natural numbers are identified with binary words in the usual way, hence we can for example talk of the length $|n|$ of a number n. A set A will be identified with ist characteristic sequence $A(0)A(1)\ldots$, where $A(i)$ is 1 iff i is in A and $A(i)$ is 0, otherwise; this way for example we can speak of the length n prefix $A\upharpoonright n$ of a set A, which consists just of the first n bits of the characteristic sequence of A.

We write φ_e for the partial recursive functional computed by the $(e+1)$st Turing machine in some standard effective enumeration of all Turing machines. Similarly, φ_e^X denotes the partial function computed by the $(e+1)$st oracle Turing machine on oracle X.

Recall that a set A is weak truth-table reducible (wtt-reducible) to a set B if A is computed with oracle B by some Turing machine M such that for some recursive function g, machine M will access on input n at most the first $g(n)$ bits of its oracle and that A is truth-table reducible (tt-reducible) to B if A is computed with oracle B by some Turing machine which is total for all oracles.

A function f is called fixed-point free iff $\varphi_x \neq \varphi_{f(x)}$ for all x. The partial recursive function $x \mapsto \varphi_x(x)$ is called the diagonal function and a function g is called diagonally nonrecursive (DNR) iff g is total and differs from the diagonal function at all places where the latter is defined.

We write $C(x)$ and $H(x)$ for the plain and the prefix-free Kolmogorov complexity of x, see Li and Vitányi [7] (who write "K" instead of "H").

2   Autocomplex and Complex Sets

Definition 1 (Schnorr). A function $g\colon \mathbb{N} \to \mathbb{N}$ is an ORDER if g is nondecreasing and unbounded.

Definition 2. A set A is COMPLEX if there is a recursive order g such that for all n, we have $C(A\upharpoonright n) \geq g(n)$.

A set A is AUTOCOMPLEX if there is an A-recursive order g such that for all n, we have $C(A\upharpoonright n) \geq g(n)$.

The concepts complex and autocomplex remained the same if one would replace in their definitions plain Kolmogorov complexity C by its prefix-free variant H, and similarly the following Propositions 3 and 4 remain valid with C replaced by H. The reason is that the two variants of Kolmogorov complexity differ by less than a multiplicative constant.

Proposition 3. *For any set A, the following conditions are equivalent.*

(1) *The set A is autocomplex.*
(2) *There is an A-recursive function h such that for all n, $C(A\upharpoonright h(n)) \geq n$.*
(3) *There is an A-recursive function f such that for all n, $C(f(n)) \geq n$.*

Proof. We show $(1) \Rightarrow (2) \Rightarrow (3) \Rightarrow (1)$. Given an autocomplex set A, choose an A-recursive order g where $C(A\upharpoonright n) \geq g(n)$ and in order to obtain a function h as required by (2), let
$$h(n) = \min\{l\colon g(l) \geq n\}.$$

Given a function h as in (2), in order to obtain a function f as required by (3), simply let $f(n)$ be equal to (an appropriate encoding of) the prefix of A of length $h(n)$. Finally, given an A-recursive function f as in (3), let $u(n)$ be an A-recursive order such that some fixed oracle Turing machine M computes f with oracle A such that M queries on input n only bits $A(m)$ of A where $m \leq u(n)$. Then for any $l \geq u(m)$, the value of $f(n)$ can be computed from n and $A \restriction l$, hence

$$n \leq C(f(n)) \leq^+ C(A \restriction l) + 2\log n,$$

and thus for almost all n and all $l \geq u(n)$, we have $n/2 \leq C(A \restriction l)$. As a consequence, a finite variation of the A-recursive order

$$g: n \mapsto \max\{l: u(l) \leq n\}/2$$

witnesses that A is autocomplex.   □

Proposition 4. *For any set A, the following conditions are equivalent.*

(1) *The set A is complex.*
(2) *There is a recursive function h such that for all n, $C(A \restriction h(n)) \geq n$.*
(3) *The set A tt-computes a function f such that for all n, $C(f(n)) \geq n$.*
(4) *The set A wtt-computes a function f such that for all n, $C(f(n)) \geq n$.*

We omit the proof of Proposition 4, which is very similar to the proof of Proposition 3, where now when proving the implication $(4) \Rightarrow (1)$ one exploits that the use of f is bounded by a recursive function.

Theorem 5. *A set is autocomplex if and only if it computes a DNR function.*

Proof. First assume that A is autocomplex and choose an A-recursive function f as in assertion (3) of Proposition 3. Then we have for some constant c and almost all n,

$$C(\varphi_n(n)) \leq C(n) + c \leq \log n + 2c < n \leq C(f(n)),$$

and consequently some finite variation of f is a DNR-function.

Next suppose that A is not autocomplex. Assume for a contradiction that A computes a DNR function, that is, for some r, φ_r^A is DNR. For any z there is an index $e(z)$ such that on every input x, $\varphi_{e(z)}(x)$ is computed as follows: first, assuming that z is a code for a prefix w of an oracle (meant to be equal to A), try to decode this prefix by simulating the universal Turing machine used to define C on input z; on success, simulate $\varphi_r(x)$ with the prefix w as oracle; if the latter computation converges with the given prefix, output the computed value.

Now consider the A-recursive function h, where $h(n)$ is the maximum of the uses of all computations of values $\varphi_r^A(e(z))$ with $|z| < n$ on oracle A. Because A is not autocomplex and by Proposition 3, there are infinitely many n such that the complexity of the length $h(n)$ prefix of A is less than n, say, witnessed by a code z_n. Then for all such n and z_n, we have

$$\varphi_r^A(e(z_n)) = \varphi_{e(z_n)}(e(z_n)),$$

hence the function computed from A by φ_r is not DNR, which contradicts our assumption. □

The complex sets are just the sets that wtt-compute a DNR function.

Theorem 6. *For a set A the following statements are equivalent.*

(1) *The set A is complex.*
(2) *The set A tt-computes a DNR function.*
(3) *The set A wtt-computes a DNR function.*

In particular, the sets that permit to wtt-compute DNR-functions and to wtt-compute recursively bounded DNR-functions coincide.

Proof. (1) implies (2): Similar to the proof of Theorem 5, the set A tt-computes a DNR function where $f(n)$ is equal to the prefix of A of length $l(n)$, where l is defined as above but can now be chosen to be recursive.

(2) implies (3): This is immediate from the definition.

(3) implies (1): Follows as in the proof of Theorem 5, where now the function h can be chosen to be recursive.

The concluding remark follows because a DNR function as in (2) is bounded by the maximum of the function values over all possible oracles. □

The constructive Hausdorff dimension of a set A can be defined as the limit inferior of the relative Kolmogorov complexities $C(A{\restriction}n)/n$ of the prefixes of A. Reimann [13] has asked whether one can "extract randomness" in the sense that any set with constructive Hausdorff dimension $\alpha > 0$ computes a set with constructive Hausdorff dimension 1 or even a Martin-Löf random set. By Theorem 6 and work of Ambos-Spies, Kjos-Hanssen, Lempp and Slaman [2] (which is in turn based on an unpublished construction by Kumabe), we obtain as a partial answer to this question that there is a complex set that does not compute a Martin-Löf random set, that is, in general it is not possible to compute a set of roughly maximum complexity from a set that has a certain minimum and effectively specified amount of randomness. Reimann and Slaman [14] have independently announced a proof of Theorem 7 by means of a direct construction.

Theorem 7. *There is a complex set that does not compute a Martin-Löf random set.*

Proof. Ambos-Spies, Kjos-Hanssen, Lempp and Slaman [2] demonstrate that for any recursive function h there is a DNR function d that cannot compute a DNR function that is bounded by $h(n)$. In their construction it is implicit that the constructed function d can be made recursively bounded, hence the function d is tt-reducible to its graph D and, by Theorem 6, the set D is complex. Apply this construction to the function $h(n) = 2^n$ and assume that the graph D of the resulting function d computes a Martin-Löf random set R. Then D also computes the function $f: n \mapsto R(0)\dots R(n)$, which is $h(n)$-bounded. Now $f(n)$ has H-complexity of at least n for almost all n, hence $f(n)$ differs from $\varphi_n(n)$ for almost all n, that is, by changing f at finitely many places, we obtain an $h(n)$-bounded DNR function recursive in D, contradicting the choice of D. □

Theorem 8. *An r.e. set is Turing complete if and only if it is autocomplex. An r.e. set is wtt-complete if and only if it is complex.*

Proof. Fix any r.e. set A and let A_s be the finite set of all elements that have been enumerated after s steps of some fixed effective enumeration of A; similarly, fix finite approximations K_s to the halting problem K. We demonstrate the slightly less involved second equivalence before the first one.

First assume that A is wtt-complete. That is, A wtt-computes the halting problem, which in turn wtt-computes the function f that maps n to the least word that has C-complexity of at least n. Since wtt-reductions compose, A wtt-computes f, hence A is complex by Proposition 4. Conversely, if the set A is complex, by Proposition 4 fix a recursive function h such that for all n we have $C(A\restriction h(n)) \geq n$. For all n, let

$$k(n) = \min\{s \in \mathbb{N}\colon K(n) = K_s(n)\}, \quad a(n) = \min\{s \in \mathbb{N}\colon A\restriction h(n) = A_s \restriction h(n)\}.$$

Obviously A wtt-computes a, hence in order to show that A wtt-computes K it suffices to show that for almost all n we have $k(n) \leq a(n)$. For a proof by contradiction, assume that there are infinitely many n such that $k(n) > a(n)$. Each such n must be in K because otherwise $k(n)$ were equal to 0; consequently, given any such n, the word

$$w_n = A\restriction h(n) = A_{a(n)} \restriction h(n) = A_{k(n)} \restriction h(n)$$

can be computed as follows: first compute $k(n)$ by simulating the given approximation to K, then compute $h(n)$ and let w_n be equal to $A_{k(n)} \restriction h(n)$. Consequently, up to an additive constant we have $C(w_n) \leq C(n)$, that is, $C(w_n)$ is in $O(\log n)$, which contradicts the choice of w_n for all sufficiently large n.

The proof that Turing complete sets are autocomplex is almost literally the same as for the corresponding assertion for wtt-complete sets and is omitted. In order to prove the reverse implication, assume that A is autocomplex and according to Proposition 3, fix an A-recursive function h such that for all n we have $C(A\restriction h(n)) \geq n$. Let $a(n)$ be equal to the least s such that A_s agrees with A on all elements that are queried while computing $h(n)$ and on all numbers up to $h(n)$. Like in the case of wtt-complete sets, we can argue that A computes a and that for all n where $k(n) > a(n)$, the word $A\restriction h(n)$ can be computed from n, hence for almost all n we have $k(n) \leq a(n)$ and A computes K.   □

As immediate corollaries to Theorems 5, 6, and 8, we obtain the following well-known characterizations of T- and wtt-completeness [11, Theorem III.1.5 and Proposition III.8.17], where the latter characterization is known as Arslanov's completeness criterion [1, 6].

Corollary 9. *An r.e. set is Turing complete if and only if it computes a DNR-function. An r.e. set is wtt-complete if and only if it wtt-computes a DNR-function.*

A set $A = \{a_0, a_1, \ldots\}$ with $a_0 < a_1 < \ldots$ is called hyperimmune if A is infinite and there is no recursive function h such that $a_n \leq h(n)$ for all n [11]. A further characterization is that there is no recursive function f such that A intersects almost all sets of the form $\{n+1, n+2, \ldots, f(n)\}$. Intuitively speaking, hyperimmune sets have large gaps that exceed all recursive bounds, hence if a set A is wtt-reducible to a hyperimmune set, that is, is reducible by a reduction with recursively bounded use, then A must have prefixes of very low complexity. This intuition is made precise by the next proposition which gives another characterization of complex sets.

Theorem 10. *A set is complex if and only if it is not wtt-reducible to a hyperimmune set.*

Proof. First assume that f, g are recursive functions, A is wtt-reducible to a hyperimmune set $B = \{b_0, b_1, \ldots\}$ with use f and $C(A(0)A(1)\ldots A(g(n))) \geq n$ for all n. There are infinitely many n such that $f(g(4^{b_n})) < b_{n+1}$. Thus there is a constant c with $C(A(0)A(1)\ldots A(g(4^{b_n}))) \leq 2^{b_n+c}$ for infinitely many n; this happens at those n where at the computation of $A(0)A(1)\ldots A(g(4^{b_n}))$ relative to B only the characteristic function of B up to b_n has to be taken into account since it is 0 afterwards up to the query-bound $f(g(4^{b_n}))$. On the other hand, $C(A(0)A(1)\ldots A(g(4^{b_n}))) \geq 4^{b_n}$ for all n. This gives $4^{b_n} \leq 2^{b_n+c}$ for infinitely many n and contradicts to the fact that the sequence $b_0, b_1, \ldots$ is strictly increasing as it is the ascending sequence of elements of an infinite set. Thus A is not complex.

Next assume that A is not wtt-reducible to any hyperimmune set. Let $p(m)$ be that word σ such that 1σ has the binary value $m+1$, so $p(0) = \lambda$, $p(1) = 0$, $p(2) = 1$, $p(3) = 00$ and so on. Let U be the universal machine on which C is based and assume that U is such that $C(A(0)A(1)\ldots A(n)) \leq n+1$ for all n. Now let $f(n)$ be the first m such that $U(p(m))$ is a word extending $A(0)A(1)\ldots A(n)$ and let

$$B = \{(n, m) : f(n) = m \wedge \forall k < n\, (f(k) \neq m)\}.$$

Now $A \leq_{wtt} B$ since $A(n) = U(p(m))(n)$ for the maximal m such that $(k, m) \in B \wedge k \leq n$; one can find this m by querying B at (i, j) for all $(i, j) \in \{0, 1, \ldots, n\} \times \{0, 1, \ldots, 2^{n+1}\}$. Therefore B is not hyperimmune and there is a recursive function g such that B has more than 2^{n+1} elements falling into the rectangle $\{0, 1, \ldots, h(n)\} \times \{0, 1, \ldots, g(n)\}$. Now one knows that $f(A(0)A(1)\ldots A(g(n))) \geq 2^{n+1}$ and thus $C(A(0)A(1)\ldots A(g(n))) \geq n$. So A is complex. $\qquad\square$

Remark 11. Post [12] introduced the notion of hyperimmune sets and demonstrated that every r.e. set with a hyperimmune complement is wtt-incomplete and that there are such sets, that is, hyperimmunity was used as a tool for constructing an r.e. wtt-incomplete set. Moreover, Post showed that if an r.e. set A is wtt-complete then A is not wtt-reducible to any hyperimmune set. This implication is in fact an equivalence, that is, an r.e. set is wtt-complete if and only if it is not wtt-reducible to a hyperimmune set, as is immediate from Theorems 6 and 8.

Miller [8] introduced the notion of hyperavoidable set. A set is hyperavoidable iff it differs from any characteristic function of a recursive set on a prefix of length computable from an index for that recursive set. Formally, a set A is hyperavoidable if there is a nondecreasing and unbounded recursive function h such that for all e, whenever $\varphi_e(x)$ is defined for all $y < h(e)$, then we have

$$A(0) \ldots A(h(e) - 1) \neq \varphi_e(0) \ldots \varphi_e(h(e) - 1) .$$

Among other results on hyperavoidable sets, Miller showed that hyperavoidability can be characterized by not being wtt-reducible to any hyperimmune set.

Theorem 12 ([8, Theorem 4.6.4]). *A set is hyperavoidable if and only if it is not wtt-reducible to any hyperimmune set.*

The following proposition is then immediate from Theorems 10 and 12.

Proposition 13. *A set is hyperavoidable if and only if it is complex.*

3 Plain Kolmogorov Complexity and Completions of Peano Arithmetic

By Propositions 3 and Theorem 5, computing a DNR function is equivalent to the ability to compute on input n a word $f(n)$ of C-complexity of at least n. The next theorem shows that if one enforces the additional constraint that the word $f(n)$ has length n, that is, if one requires $f(n)$ to be an incompressible word of length n, then one obtains a characterization of the strictly smaller class of PA-complete sets. Recall that a set A is PA-complete if and only if A computes a DNR function with finite range, which by a result of Jockusch [6] in turn is equivalent to computing a $\{0, 1\}$-valued DNR function.

Theorem 14. *The following is equivalent for every set A.*

(1) *A computes a $\{0, 1\}$-valued DNR function.*
(2) *A computes a function f such that for all n, $f(n)$ has length n and satisfies $C(f(n)) \geq n$.*
(3) *A computes a lower bound b on plain complexity C such that for all n there are at most $2^n - 1$ many x with $b(x) < n$.*

Proof. (3) implies (2): Just let $f(n)$ be the lexicographically first word y of length n such that $b(y) \geq n$. This word exists by the condition that there are at most $2^n - 1$ words x with $b(x) < n$. Since b is a lower bound for C, one has that $C(f(n)) \geq n$ for all n. Furthermore, f is computed from b.

(2) implies (1): Let the partial recursive function ψ be defined by $\psi(x) = x\varphi_n(n)$ where n is the length of x and ψ is defined if and only if $\varphi_n(n)$ is defined. Then there is a constant c such that $C(\psi(x)) < n + c$ for all x, n with $x \in \{0, 1\}^n$. In order to obtain a DNR function d with finite domain that is computed by f, let $d(n)$ consists of the last c bits of $f(n + c)$. Now consider any n such that $\varphi_n(n)$ is defined. If we let x be the first n bits of $f(n + c)$, then we have

$$xd(n) = f(n + c) \neq \psi(x) = x\varphi_n(n)$$

where the inequality holds by assumption on f and because of $C(\psi(x)) < n + c$. Thus d is a DNR function and its range is the finite set $\{0, 1\}^c$. By the already mentioned result of Jockusch [6], this implies that f computes a $\{0, 1\}$-valued DNR function.

(1) implies (3): Assume that A computes a $\{0, 1\}$-valued DNR function and that hence A is PA-complete. Consider the Π_1^0 class of all sets G that satisfy the following two conditions:

- $\forall p, x, s\, (U_s(p)\downarrow = x \Rightarrow (p, x) \in G)$;
- $\forall p, x, y\, ((p, x) \in G \land (p, y) \in G \Rightarrow x = y)$.

Since A is PA-complete, by the Scott Basis Theorem (see Odifreddi [11, Theorem V.5.35]) this Π_1^0 class contains a set that is computable in A; fix such a set G. Now one defines $b(x) = \min\{|p| : |p| < n \ \& \ (p, x) \in G\}$. By the first condition, the function b is a lower bound for C. By the second condition, any word p can occur in at most one pair (p, x) in G, hence there are at most $2^n - 1$ many x where there is such a pair with $|p| < n$, or equivalently, where $b(x) < n$.   □

One might ask whether one can strengthen Theorem 14 such that any PA-complete set A, that is, any set that satisfies the first condition in the theorem, computes a function f such $f(n)$ is a word of length n that, instead of just being incompressible as required by the second condition, has maximum plain Kolmogorov complexity among all words of the same length. The answer to this question depends on the universal machine that is used to define plain complexity; more precisely, for every r.e. oracle B one can compute a corresponding universal machine which makes this problem hard not only for PA but also for B. The proof of Theorem 15 is omitted due to space considerations.

Theorem 15. *For every recursively enumerable oracle B there is a universal machine U_B such that the following two conditions are equivalent for every oracle A:*

(1) *A has PA-complete degree and $A \geq_T B$.*
(2) *There is a function $f \leq_T A$ such that for all n and for all $x \in \{0, 1\}^n$, $f(n) \in \{0, 1\}^n$ and $C_B(f(n)) \geq C(x)$, where C_B is the plain Kolmogorov complexity based on the universal machine U_B.*

4   Computing Words with Maximum Prefix-Free Kolmogorov Complexity

While PA-completeness can be characterized in terms of C, an analogous result for H fails. First, one cannot replace C-incompressibility by H-incompressibility since relative to any Martin-Löf random set A, which includes certain non-PA-complete sets, one can compute the function mapping n to the H-incompressible word $A(0)...A(n)$. So one might consider the sets A which permit to compute words of maximal complexity in order to obtain a characterization of PA-complete sets. However, instead the corresponding notion gives a characterization of the sets that compute the halting problem K. The proof of this result makes use of the following proposition, which we state without proof because lack of space. Note that the following proposition does not hold with C in place of H.

Proposition 16. *Let f be a partial recursive surjective function with $|f(x)| < |x|$ for all x in the domain of f. Then there is a constant c and an enumeration U_s of the universal prefix-free machine U such that*

$$\forall n \, \forall x \, \forall s \, (f(x) = n \wedge H_s(x) = H(x) \Rightarrow H_s(n) \leq H(n) + c)$$

where H_s is the approximation to H based on U_s.

In contrast to Theorem 15, the following result does not depend on the choice of the universal machine that is used when defining Kolmogorov complexity.

Theorem 17. *A set A computes the halting problem K if and only if there is a function $f \leq_T A$ such that for all n, the word $f(n)$ has length n and has maximum H-complexity among all words of the same length, that is, $H(x) \leq H(f(n))$ for all words x of length n.*

Proof. Since H is K-recursive, a function f as required can obviously be computed if $A \geq_T K$. For the reverse direction, assume that $f \leq_T A$ is a function as stated in the theorem. Let U be the universal prefix-free machine on which H is based. Given any n, m and any $o < 2^m$, let h map every word of length $2^{n+m+1} + 2^m + o$ to n; h is undefined on words of length $0, 1, 2$. Note that $|h(x)| < |x|$ for all x. By Proposition 16 there is an enumeration U_s of U and a constant, here called c_1, such that

$$\forall n \, \forall x \, \forall s \, (h(x) = n \wedge H_s(x) = H(x) \Rightarrow H_s(n) \leq H(n) + c_1).$$

Let $P_0, P_1, \ldots$ be an enumeration of all primitive-recursive partitions of the natural numbers and let $P_{m,0}, P_{m,1}, \ldots$ be the members of partition P_m enumerated such that the member $P_{m,o}$ exists whenever $P_{m,o+1}$ exists. We can assume that every partition in the enumeration has infinitely many indices. Now define a family of partial recursive function $T_0, T_1, \ldots$ such that each T_k on input p does the following steps:

- Compute the first s such that $U_s(p)$ is defined;
- If $U_s(p)$ is defined then check whether there are values n, m, ℓ, x, y, z such that $U(p) = xz$, $|x| = 2^{n+m+1} + 2^m$, $\ell = 2^{2^{|z|}}$ and $\ell < 2^m$.
- If this also goes through and $H_s(n) \geq k$ then search for o such that $H_s(n) - k \in P_{m,o}$.
- If all previous steps have gone through and $o < 2^\ell$ then let $T_k(p) = xy$ for the unique $y \in \{0, 1\}^\ell$ with $bv(y) = o$.

Here $bv(y)$ is the binary value of y, for example, $bv(00101) = 5$. The machines T_k are prefix-free and there is a constant c_2 such that for all $k \leq c_1$, $H(T_k(p)) \leq |p| + c_2$. Furthermore, there is a constant c_3 such that $H(F(o + c_3)) + c_2 < H(F(o + 2^{2^{c_3}}))$ for all o. In particular one has

$$\forall n \, \forall m > c_3 + 2 \, \forall p \, (|U(p)| = 2^{n+m+1} + 2^m + c_3 \Rightarrow$$
$$\forall k \leq c_1 \, (H(T_k(p)) < H(F(2^{n+m+1} + 2^m + c_4)))).$$

where $c_4 = 2^{2^{c_3}}$. Given any n and $m > 2^{c_3}$, let y be the last c_4 bits of $F(2^{n+m+1} + 2^m + c_4)$. Then $P_{m,bv(y)}$ does either not exist or not contain $H(n)$.

Thus one can run the following A-recursive algorithm to determine for any given n a set of up to $2^{c_4} - 1$ elements which contains $H(n)$ by the following algorithm.

- Let $E = \{0, 1, \ldots, 2n + 2\}$ and $m = c_4 + 1$.
- While $|E| \geq 2^{c_4}$ Do Begin $m = m + 1$,
  Determine the word y consisting of the last c_4 bits of $f(2^{n+m+1} + 2^m + c_4)$,
  If $P_{m,bv(y)}$ exists and intersects E then let $E = E - P_{m,bv(y)}$ End.
- Output E.

This algorithm terminates since whenever $|E| \geq 2^{c_4}$ at some stage m then there is $o > m$ such that P_o has 2^{c_4} members all intersecting E and one of them will be removed so that E loses an element in one of the stages $m+1, \ldots, o$. Thus the above algorithm computes relative to A for input n a set of up to $2^{c_4} - 1$ elements containing $H(n)$. By a result of Beigel, Buhrman, Fejer, Fortnow, Grabowski, Longpré, Muchnik, Stephan and Torenvliet [5], such an A-recursive algorithm can only exist if $K \leq_T A$.   □

Remark 18. Theorem 17 and its proof answer a question of Calude, who had asked whether the statement of the theorem is true with the condition that $H(f(n))$ is maximum (among all words of the same length) replaced by the condition that $H(f(n))$ is maximum up to an additive constant. This variant of Theorem 17 can be obtained by a minor adaptation of the proof of the theorem given above.

Nies [10] pointed out that for any n the word x of maximum prefix-free Kolmogorov complexity among all words of length n satisfies, up to an additive constant, the equality $H(x) = n + H(n)$, hence the variant of Theorem 17 can be rephrased as follows. For any oracle A, $A \geq_T K$ if and only if there is a function $f \leq_T A$ and a constant c such that for all n, $f(n) \in \{0,1\}^n$ and $H(f(n)) \geq f(n) + H(f(n)) - c$.

Acknowledgements. We would like to thank Cristian Calude and André Nies for helpful discussion.

References

1. Marat M. Arslanov, On some generalizations of the Fixed-Point Theorem, *Soviet Mathematics (Iz. VUZ)*, Russian, 25(5):9–16, 1981, English translation, 25(5):1–10, 1981.
2. Klaus Ambos-Spies, Bjørn Kjos-Hanssen, Steffen Lempp and Theodore A. Slaman. Comparing DNR and WWKL, *Journal of Symbolic Logic*, 69:1089-1104, 2004.
3. Cristian S. Calude. Private Communication, 2005.
4. Antonín Kučera. Measure, Π_1^0-classes and complete extensions of PA. In *Recursion theory week 1984*, *Lecture Notes in Mathematics* 1141:245–259, 1985.

5. Richard Beigel, Harry Buhrman, Peter Fejer, Lance Fortnow, Piotr Grabowski, Luc Longpré, Andrej Muchnik, Frank Stephan, and Leen Torenvliet. Enumerations of the Kolmogorov function. *Electronic Colloquium on Computational Complexity*, TR04-015, 2004.
6. Carl G. Jockusch, Jr. Degrees of functions with no fixed points. In *Logic, methodology and philosophy of science, VIII (Moscow, 1987)*, volume 126 of *Stud. Logic Found. Math.*, pages 191–201. North-Holland, Amsterdam, 1989.
7. Ming Li and Paul Vitányi. *An Introduction to Kolmogorov Complexity and its Applications*. Graduate texts in Computer Science, Springer, Heidelberg, 1997.
8. Joseph Stephen Miller. Π_1^0 *classes in Computable Analysis and Topology*. PhD thesis, Cornell University, 2002.
9. Bjørn Kjos-Hanssen, André Nies and Frank Stephan. Lowness for the class of Schnorr random reals. SIAM Journal on Computing, to appear.
10. André Nies. Private Communication, 2004.
11. Piergiorgio Odifreddi. *Classical Recursion Theory*. North-Holland, Amsterdam, 1989.
12. Emil Post. Recursively enumerable sets of positive integers and their decision problems, *Bulletin of the American Mathematical Society*, 50:284–316, 1944.
13. Jan Reimann. *Computability and Fractal Dimension*. Doctoral dissertation, Fakultät für Mathematik und Informatik, Universität Heidelberg, INF 288, D-69120 Heidelberg, Germany, 2004.
14. Jan Reimann. Extracting randomness from sequences of positive dimension. Post of an open problem in the recursion theory section of the Mathematical Logic Forum at math.berkeley.edu/Logic/problems/, 2004.

Entanglement in Interactive Proof Systems with Binary Answers

Stephanie Wehner[*]

CWI, Kruislaan 413, 1098 SJ Amsterdam, The Netherlands
wehner@cwi.nl

Abstract. If two classical provers share an entangled state, the resulting interactive proof system is significantly weakened [6]. We show that for the case where the verifier computes the XOR of two binary answers, the resulting proof system is in fact no more powerful than a system based on a single quantum prover: $\oplus\text{MIP}^*[2] \subseteq \text{QIP}(2)$. This also implies that $\oplus\text{MIP}^*[2] \subseteq \text{EXP}$ which was previously shown using a different method [7]. This contrasts with an interactive proof system where the two provers do not share entanglement. In that case, $\oplus\text{MIP}[2] = \text{NEXP}$ for certain soundness and completeness parameters [6].

1 Introduction

Interactive proof systems have received considerable attention [2,3,4,8,14,10] since their introduction by Babai [1] and Goldwasser, Micali and Rackoff [11] in 1985. An interactive proof system takes the form of a protocol of one or more rounds between two parties, a verifier and a prover. Whereas the prover is computationally unbounded, the verifier is limited to probabilistic polynomial time. Both the prover and the verifier have access to a common input string x. The goal of the prover is to convince the verifier that x belongs to a pre-specified language L. The verifier's aim, on the other hand, is to determine whether the prover's claim is indeed valid. In each round, the verifier sends a polynomial (in x) size query to the prover, who returns a polynomial size answer. At the end of the protocol, the verifier decides to accept, and conclude $x \in L$, or reject based on the messages exchanged and his own private randomness. A language has an interactive proof if there exists a verifier V and a prover P such that: If $x \in L$, the prover can always convince V to accept. If $x \notin L$, no strategy of the prover can convince V to accept with non-negligible probability. IP denotes the class of languages having an interactive proof system. Watrous [30] first considered the notion of *quantum* interactive proof systems. Here, the prover has unbounded quantum computational power whereas the verifier is restricted to quantum polynomial time. In addition, the two parties can now exchange quantum messages. QIP is the class of languages having a quantum interactive proof system. Classically, it is known that IP $=$ PSPACE [22,23]. For the quantum case, it has

[*] Supported by EU project RESQ IST-2001-37559 and NWO Vici grant 2004-2009.

B. Durand and W. Thomas (Eds.): STACS 2006, LNCS 3884, pp. 162–171, 2006.

been shown that PSPACE $\subseteq$ QIP $\subseteq$ EXP [30,12]. If, in addition, the verifier is given polynomial size quantum advice, the resulting class QIP/qpoly contains all languages [21]. Let QIP(k) denote the class where the prover and verifier are restricted to exchanging k messages. It is known that QIP = QIP(3) [12] and QIP(1) $\subseteq$ PP [29,15]. We refer to [15] for an overview of the extensive work done on QIP(1), also known as QMA. Very little is known about QIP(2) and its relation to either PP or PSPACE.

In multiple-prover interactive proof systems the verifier can interact with multiple, computationally unbounded provers. Before the protocol starts, the provers are allowed to agree on a joint strategy, however they can no longer communicate during the execution of the protocol. Let MIP denote the class of languages having a *multiple*-prover interactive proof system. In this paper, we are especially interested in two-prover interactive proof systems as introduced by Ben-Or, Goldwasser, Kilian and Widgerson [3]. Feige and Lovász [10] have shown that a language is in NEXP if and only if it has a two-prover one-round proof system, i.e. MIP[2] = MIP = NEXP. Let $\oplus$MIP[2] denote the restricted class where the verifier's output is a function of the XOR of two binary answers. Even for such a system $\oplus$MIP[2] = NEXP, for certain soundness and completeness parameters [6]. Classical multiple-prover interactive proof systems are thus more powerful than classical proof systems based on a single prover, assuming PSPACE $\neq$ NEXP. Kobayashi and Matsumoto have considered *quantum* multiple-prover interactive proof systems which form an extension of quantum single prover interactive proof systems as described above. Let QMIP denote the resulting class. In particular, they showed that QMIP = NEXP if the provers do *not* share quantum entanglement. If the provers share at most polynomially many entangled qubits the resulting class is contained in NEXP [13].

Cleve, Høyer, Toner and Watrous [6] have raised the question whether a *classical* two-prover system is weakened when the provers are allowed to share arbitrary entangled states as part of their strategy, but all communication remains classical. We write MIP* if the provers share entanglement. The authors provide a number of examples which demonstrate that the soundness condition of a classical proof system can be compromised, i.e. the interactive proof system is weakened, when entanglement is used. In their paper, it is proved that $\oplus$MIP*[2] $\subseteq$ NEXP. Later, the same authors also showed that $\oplus$MIP*[2] $\subseteq$ EXP using semidefinite programming [7]. Entanglement thus clearly weakens an interactive proof system, assuming EXP $\neq$ NEXP.

Intuitively, entanglement allows the provers to coordinate their answers, even though they cannot use it to communicate. By measuring the shared entangled state the provers can generate correlations which they can use to deceive the verifier. Tsirelson [26,24] has shown that even quantum mechanics limits the strength of such correlations. Consequently, Popescu and Roehrlich [17,18,19] have raised the question why nature imposes such limits. To this end, they constructed a toy-theory based on non-local boxes [17,27], which are hypothetical "machines" generating correlations stronger than possible in nature. In their full generalization, non-local boxes can give rise to any type of correlation as long as they cannot be used to signal. van Dam has shown that sharing certain non-local boxes allows two remote parties to perform any distributed computation using only a single bit of communication [27,28]. Preda [20] showed that sharing non-local boxes can then allow two provers to coordinate their answers perfectly and obtained $\oplus \text{MIP}_{\text{NL}} = \text{PSPACE}$, where we write $\oplus \text{MIP}_{\text{NL}}$ to indicate that the two provers share non-local boxes.

Kitaev and Watrous [12] mention that it is unlikely that a single-prover *quantum* interactive proof system can simulate multiple classical provers, because then from $\text{QIP} \subseteq \text{EXP}$ and $\text{MIP} = \text{NEXP}$ it follows that $\text{EXP} = \text{NEXP}$.

1.1 Our Contribution

Surprisingly, it turns out that when the provers are allowed to share entanglement it can be possible to simulate two such classical provers by one quantum prover. This indicates that entanglement among provers truly leads to a weaker proof system. In particular, we show that a two-prover one-round interactive proof system where the verifier computes the XOR of two binary answers and the provers are allowed to share an arbitrary entangled state can be simulated by a single quantum interactive proof system with two messages: $\oplus \text{MIP}^*[2] \subseteq \text{QIP}(2)$. Since very little is known about $\text{QIP}(2)$ so far [12], we hope that our result may help to shed some light about its relation to PP or PSPACE in the future. Our result also leads to a proof that $\oplus \text{MIP}^*[2] \subseteq \text{EXP}$.

2 Preliminaries

2.1 Quantum Computing

We assume general familiarity with the quantum model [16]. In the following, we will use $\mathcal{V}, \mathcal{P}$ and $\mathcal{M}$ to denote the Hilbert spaces of the verifier, the quantum prover and the message space respectively. $\Re(z)$ denotes the real part of a complex number z.

2.2 Non-local Games

For our proof it is necessary to introduce the notion of (non-local) games: Let S, T, A and B be finite sets, and π a probability distribution on $S \times T$. Let V be a predicate on $S \times T \times A \times B$. Then $G = G(V, \pi)$ is the following two-person cooperative game: A pair of questions $(s, t) \in S \times T$ is chosen at random according to the probability distribution π. Then s is sent to player 1, henceforth called Alice, and t to player 2, which we will call Bob. Upon receiving s, Alice has to reply with an answer $a \in A$. Likewise, Bob has to reply to question t with an answer $b \in B$. They win if $V(s, t, a, b) = 1$ and lose otherwise. Alice and Bob may agree on any kind of strategy beforehand, but they are no longer allowed to communicate once they have received questions s and t. The value $\omega(G)$ of a game G is the maximum probability that Alice and Bob win the game. We will follow the approach of Cleve et al. [6] and write $V(a, b|s, t)$ instead of $V(s, t, a, b)$ to emphasize the fact that a and b are answers given questions s and t.

Here, we will be particularly interested in non-local games. Alice and Bob are allowed to share an arbitrary entangled state $|\Psi\rangle$ to help them win the game. Let $\mathcal{A}$ and $\mathcal{B}$ denote the Hilbert spaces of Alice and Bob respectively. The state $|\Psi\rangle \in \mathcal{A} \otimes \mathcal{B}$ is part of the quantum strategy that Alice and Bob can agree on beforehand. This means that for each game, Alice and Bob can choose a specific $|\Psi\rangle$ to maximize their chance of success. In addition, Alice and Bob can agree on quantum measurements. For each $s \in S$, Alice has a projective measurement described by $\{X_s^a : a \in A\}$ on $\mathcal{A}$. For each $t \in T$, Bob has a projective measurement described by $\{Y_t^b : b \in B\}$ on $\mathcal{B}$. For questions $(s, t) \in S \times T$, Alice performs the measurement corresponding to s on her part of $|\Psi\rangle$ which gives her outcome a. Likewise, Bob performs the measurement corresponding to t on his part of $|\Psi\rangle$ with outcome b. Both send their outcome, a and b, back to the verifier. The probability that Alice and Bob answer $(a, b) \in A \times B$ is then given by

$$\langle\Psi|X_s^a \otimes Y_t^b|\Psi\rangle.$$

The probability that Alice and Bob win the game is given by

$$\Pr[\text{Alice and Bob win}] = \sum_{s,t} \pi(s, t) \sum_{a,b} V(a, b|s, t)\langle\Psi|X_s^a \otimes Y_t^b|\Psi\rangle.$$

The *quantum value* $\omega_q(G)$ of a game G is the maximum probability over all possible quantum strategies that Alice and Bob win. An *XOR game* is a game where the value of V only depends on $c = a \oplus b$ and not on a and b independently. For XOR games we write $V(c|s, t)$ instead of $V(a, b|s, t)$. We will use $\tau(G)$ to denote the value of the trivial strategy where Alice and Bob ignore their inputs and return random answers $a \in_R \{0, 1\}$, $b \in_R \{0, 1\}$ instead. For an XOR game,

$$\tau(G) = \frac{1}{2} \sum_{s,t} \pi(s, t) \sum_{c \in \{0,1\}} V(c|s, t). \tag{1}$$

In this paper, we will only be interested in the case that $a \in \{0, 1\}$ and $b \in \{0, 1\}$. Alice and Bob's measurements are then described by $\{X_s^0, X_s^1\}$ for $s \in S$ and $\{Y_t^0, Y_t^1\}$ for $t \in T$ respectively. Note that $X_s^0 + X_s^1 = I$ and $Y_t^0 + Y_t^1 = I$ and thus these measurements can be expressed in the form of observables X_s and Y_t with eigenvalues ± 1: $X_s = X_s^0 - X_s^1$ and $Y_t = Y_t^0 - Y_t^1$. Tsirelson [26,24] has shown that for any $|\Psi\rangle \in \mathcal{A} \otimes \mathcal{B}$ there exists real vectors $x_s, y_t \in \mathbb{R}^N$ with $N = \min(|S|, |T|)$ such that $\langle\Psi|X_s \otimes Y_t|\Psi\rangle = \langle x_s|y_t\rangle$. Conversely, if $dim(\mathcal{A}) = dim(\mathcal{B}) = 2^{\lceil N/2 \rceil}$ and $|\Psi\rangle \in \mathcal{A} \otimes \mathcal{B}$ is a maximally entangled state, there exist observables X_s on $\mathcal{A}$, Y_t on $\mathcal{B}$ such that $\langle x_s|y_t\rangle = \langle\Psi|X_s \otimes Y_t|\Psi\rangle$. See [25, Theorem 3.5] for a detailed construction.

2.3 Interactive Proof Systems

Multiple Provers. It is well known [6,10], that two-prover one-round interactive proof systems with classical communication can be modeled as (non-local) games. Here, Alice and Bob take the role of the two provers. The verifier now poses questions s and t, and evaluates the resulting answers. A proof system associates with each string x a game G_x, where $\omega_q(G_x)$ determines the probability that the verifier accepts (and thus concludes $x \in L$). The string x, and thus the nature of the game G_x is known to both the verifier and the provers. Ideally, for all $x \in L$ the value of $\omega_q(G_x)$ is close to one, and for $x \notin L$ the value of $\omega_q(G_x)$ is close to zero. It is possible to extend the game model for MIP[2] to use a randomized predicate for the acceptance predicate V. This corresponds to V taking an extra input string chosen at random by the verifier. However, known applications of MIP[2] proof systems do not require this extension [9]. Our argument in Section 3 can easily be extended to deal with randomized predicates. Since V is not a randomized predicate in [6], we here follow this approach.

In this paper, we concentrate on proof systems involving two provers, one round of communication, and single bit answers. The provers are computationally unbounded, but limited by the laws of quantum physics. However, the verifier is probabilistic polynomial time bounded. As defined by Cleve et al. [6],

Definition 1. *For $0 \leq s < c \leq 1$, let $\oplus \text{MIP}_{c,s}[2]$ denote the class of all languages L recognized by a classical two-prover interactive proof system of the following form:*

- *They operate in one round, each prover sends a single bit in response to the verifier's question, and the verifier's decision is a function of the parity of those two bits.*
- *If $x \notin L$ then, whatever strategy the two provers follow, the probability that the verifier accepts is at most s (the* soundness *probability).*
- *If $x \in L$ then there exists a strategy for the provers for which the probability that the verifier accepts is at least c (the* completeness *probability).*

Definition 2. *For $0 \leq s < c \leq 1$, let $\oplus \text{MIP}^*_{c,s}[2]$ denote the class corresponding to a modified version of the previous definition: all communication remains classical, but the provers may share prior quantum entanglement, which may depend on x, and perform quantum measurements.*

A Single Quantum Prover. Instead of two classical provers, we now consider a system consisting of a single quantum prover P_q and a quantum polynomial time verifier V_q as defined by Watrous [30]. Again, the quantum prover P_q is computationally unbounded, however, he is limited by the laws of quantum physics. The verifier and the prover can communicate over a quantum channel. In this paper, we are only interested in one round quantum interactive proof systems: the verifier sends a single quantum message to the prover, who responds with a quantum answer. We here express the definition of QIP(2) [30] in a form similar to the definition of $\oplus \text{MIP}^*$:

Definition 3. *Let $\text{QIP}(2, c, s)$ denote the class of all languages L recognized by a quantum one-prover one-round interactive proof system of the following form:*

- *If $x \notin L$ then, whatever strategy the quantum prover follows, the probability that the quantum verifier accepts is at most s.*

– *If $x \in L$ then there exists a strategy for the quantum prover for which the probability that the verifier accepts is at least c.*

3   Main Result

We now show that an interactive proof system where the verifier is restricted to computing the XOR of two binary answers is in fact no more powerful than a system based on a single quantum prover. The main idea behind our proof is to combine two classical queries into one quantum query, and thereby simulate the classical proof system with a single quantum prover. Recall that the two provers can use an arbitrary entangled state as part of their strategy. For our proof we will make use of the following proposition shown in [6, Proposition 5.7]:

Proposition 1 (CHTW). *Let $G(V, \pi)$ be an XOR game and let $N = min$ $(|S|, |T|)$. Then*

$$w_q(G) - \tau(G) = \frac{1}{2} \max_{x_s, y_t} \sum_{s,t} \pi(s,t) \left(V(0|s,t) - V(1|s,t) \right) \langle x_s | y_t \rangle,$$

where the maximization is taken over unit vectors

$$\{x_s \in \mathbb{R}^N : s \in S\} \cup \{y_t \in \mathbb{R}^N : t \in T\}.$$

Theorem 1. *For all s and c such that $0 \leq s < c \leq 1$, $\oplus\text{MIP}^*_{c,s}[2] \subseteq \text{QIP}(2,c,s)$*

Proof. Let $L \in \oplus\text{MIP}^*_{c,s}[2]$ and let V_e be a verifier witnessing this fact. Let P_e^1 (Alice) and P_e^2 (Bob) denote the two provers sharing entanglement. Fix an input string x. As mentioned above, interactive proof systems can be modeled as games indexed by the string x. It is therefore sufficient to show that there exists a verifier V_q and a quantum prover P_q, such that $w_{sim}(G_x) = w_q(G_x)$, where $w_{sim}(G_x)$ is the value of the simulated game.

Let s,t be the questions that V_e sends to the two provers P_e^1 and P_e^2 in the original game. The new verifier V_q now constructs the following state in $\mathcal{V} \otimes \mathcal{M}$

$$|\Phi_{init}\rangle = \frac{1}{\sqrt{2}} (\underbrace{|0\rangle}_{\mathcal{V}} \underbrace{|0\rangle|s\rangle}_{\mathcal{M}} + \underbrace{|1\rangle}_{\mathcal{V}} \underbrace{|1\rangle|t\rangle}_{\mathcal{M}}),$$

and sends register $\mathcal{M}$ to the single quantum prover P_q[1]

We first consider the honest strategy of the prover. Let a and b denote the answers of the two classical provers to questions s and t respectively. The quantum prover now transforms the state to

$$|\Phi_{honest}\rangle = \frac{1}{\sqrt{2}}((-1)^a \underbrace{|0\rangle}_{\mathcal{V}} \underbrace{|0\rangle|s\rangle}_{\mathcal{M}} + (-1)^b \underbrace{|1\rangle}_{\mathcal{V}} \underbrace{|1\rangle|t\rangle}_{\mathcal{M}}),$$

---

[1] If questions s and t are always orthogonal, it suffices to use $\frac{1}{\sqrt{2}}(|0\rangle|s\rangle + |1\rangle|t\rangle)$.

and returns register $\mathcal{M}$ back to the verifier. The verifier V_q now performs a measurement on $\mathcal{V} \otimes \mathcal{M}$ described by the following projectors

$$P_0 = |\Psi_{st}^+\rangle\langle\Psi_{st}^+| \otimes I$$
$$P_1 = |\Psi_{st}^-\rangle\langle\Psi_{st}^-| \otimes I$$
$$P_{reject} = I - P_0 - P_1,$$

where $|\Psi_{st}^\pm\rangle = (|0\rangle|0\rangle|s\rangle \pm |1\rangle|1\rangle|t\rangle)/\sqrt{2}$. If he obtains outcome "reject", he immediately aborts and concludes that the quantum prover is cheating. If he obtains outcome $m \in \{0, 1\}$, the verifier concludes that $c = a \oplus b = m$. Note that $\Pr[m = a \oplus b|s, t] = \langle\Phi_{honest}|P_{a\oplus b}|\Phi_{honest}\rangle = 1$, so the verifier can reconstruct the answer perfectly.

We now consider the case of a dishonest prover. In order to convince the verifier, the prover applies a transformation on $\mathcal{M} \otimes \mathcal{P}$ and send register $\mathcal{M}$ back to the verifier. We show that for any such transformation the value of the resulting game is at most $w_q(G_x)$: Note that the state of the total system in $\mathcal{V} \otimes \mathcal{M} \otimes \mathcal{P}$ can now be described as

$$|\Phi_{dish}\rangle = \frac{1}{\sqrt{2}}(|0\rangle|\phi_s\rangle + |1\rangle|\phi_t\rangle)$$

where $|\phi_s\rangle = \sum_{u \in S' \cup T'} |u\rangle|\alpha_u^s\rangle$ and $|\phi_t\rangle = \sum_{v \in S' \cup T'} |v\rangle|\alpha_v^t\rangle$ with $S' = \{0s|s \in S\}$ and $T' = \{1t|t \in T\}$. Any transformation employed by the prover can be described this way. We now have that

$$\Pr[m = 0|s, t] = \langle\Phi_{dish}|P_0|\Phi_{dish}\rangle = \frac{1}{4}(\langle\alpha_s^s|\alpha_s^s\rangle + \langle\alpha_t^t|\alpha_t^t\rangle) + \frac{1}{2}\Re(\langle\alpha_s^s|\alpha_t^t\rangle) \quad (2)$$

$$\Pr[m = 1|s, t] = \langle\Phi_{dish}|P_1|\Phi_{dish}\rangle = \frac{1}{4}(\langle\alpha_s^s|\alpha_s^s\rangle + \langle\alpha_t^t|\alpha_t^t\rangle) - \frac{1}{2}\Re(\langle\alpha_s^s|\alpha_t^t\rangle) \quad (3)$$

The probability that the prover wins is given by

$$\Pr[\text{Prover wins}] = \sum_{s,t} \pi(s, t) \sum_{c \in \{0,1\}} V(c|s, t) \Pr[m = c|s, t].$$

The prover will try to maximize his chance of success by maximizing $\Pr[m = 0|s, t]$ or $\Pr[m = 1|s, t]$. We can therefore restrict ourselves to considering real unit vectors for which $\langle\alpha_s^s|\alpha_s^s\rangle = 1$ and $\langle\alpha_t^t|\alpha_t^t\rangle = 1$. This also means that $|\alpha_s^{s'}\rangle = 0$ iff $s \neq s'$ and $|\alpha_t^{t'}\rangle = 0$ iff $t \neq t'$. Any other strategy can lead to rejection and thus to a lower probability of success. By substituting into Equations 2 and 3, it follows that the probability that the quantum prover wins the game when he avoids rejection is then

$$\frac{1}{2}\sum_{s,t,c} \pi(s, t)V(c|s, t)(1 + (-1)^c\langle\alpha_s^s|\alpha_t^t\rangle). \quad (4)$$

In order to convince the verifier, the prover's goal is to choose real vectors $|\alpha_s^s\rangle$ and $|\alpha_t^t\rangle$ which maximize Equation 4. Since in $|\phi_s\rangle$ and $|\phi_t\rangle$ we sum over $|S'| +$

$|T'| = |S|+|T|$ elements respectively, the dimension of $\mathcal{P}$ need not exceed $|S|+|T|$. Thus, it is sufficient to restrict the maximization to vectors in $\mathbb{R}^{|S|+|T|}$. In fact, since we are interested in maximizing the inner product of two vectors from the sets $\{\alpha_s^s : s \in S\}$ and $\{\alpha_t^t : t \in T\}$, it is sufficient to limit the maximization of vectors to $\mathbb{R}^N$ with $N = \min(|S|, |T|)$ [6]: Consider the projection of the vectors $\{\alpha_s^s : s \in S\}$ onto the span of the vectors $\{\alpha_t^t : t \in T\}$ (or vice versa). Given Equation 4, we thus have

$$w_{sim}(G_x) = \max_{\alpha_s^s, \alpha_t^t} \frac{1}{2} \sum_{s,t,c} \pi(s,t) V(c|s,t)(1 + (-1)^c \langle \alpha_s^s | \alpha_t^t \rangle),$$

where the maximization is taken over vectors $\{\alpha_s^s \in \mathbb{R}^N : s \in S\}$, and $\{\alpha_t^t \in \mathbb{R}^N : t \in T\}$. However, Proposition 1 and Equation 1 imply that

$$w_q(G_x) = \max_{x_s, y_t} \frac{1}{2} \sum_{s,t,c} \pi(s,t) V(c|s,t)(1 + (-1)^c \langle x_s | y_t \rangle)$$

where the maximization is taken over unit vectors $\{x_s \in \mathbb{R}^N : s \in S\}$ and $\{y_t \in \mathbb{R}^N : t \in T\}$. We thus have

$$w_{sim}(G_x) = w_q(G_x)$$

which completes our proof.

Corollary 1. *For all s and c such that $0 \leq s < c \leq 1$, $\oplus\mathrm{MIP}_{c,s}^*[2] \subseteq \mathrm{EXP}$.*

Proof. This follows directly from Theorem 1 and the result that $\mathrm{QIP}(2) \subseteq \mathrm{EXP}$ [12].

4   Discussion

It would be interesting to show that this result also holds for a proof system where the verifier is not restricted to computing the XOR of both answers, but some other boolean function. However, it remains unclear what the exact value of a binary game would be. The approach based on vectors from Tsirelson's results does not work for binary games. Whereas it is easy to construct a single quantum query which allows the verifier to compute an arbitrary function of the two binary answers with some advantage, it thus remains unclear how the value of the resulting game is related to the value of a binary game. Furthermore, mere classical tricks trying to obtain the value of a binary function from XOR itself seem to confer extra cheating power to the provers.

Examples of non-local games with longer answers [6], such as the Kochen-Specker or the Magic Square game, seem to make it even easier for the provers to cheat by taking advantage of their entangled state. Furthermore, existing proofs that MIP = NEXP break down if the provers share entanglement. It is therefore an open question whether MIP* = NEXP or, what may be a more likely outcome, MIP* ⊆ EXP.

Non-locality experiments between two spacelike separated observers, Alice and Bob, can be cast in the form of non-local games. For example, the experiment based on the well known CHSH inequality [5], is a non-local game with binary answers of which the verifier computes the XOR [6]. Our result implies that this non-local game can be simulated in superposition by a single prover/observer: Any strategy that Alice and Bob might employ in the non-local game can be mirrored by the single prover in the constructed "superposition game", and also vice versa, due to Tsirelson's constructions [26,24] mentioned earlier. This means that the "superposition game" corresponding to the non-local CHSH game is in fact limited by Tsirelson's inequality [26], even though it itself has no non-local character. Whereas this may be purely coincidental, it would be interesting to know its physical interpretation, if any. Perhaps it may be interesting to ask whether Tsirelson type inequalities have any consequences on local computations in general, beyond the scope of these very limited games.

Acknowledgments

Many thanks go to Julia Kempe, Oded Regev and Ronald de Wolf for useful discussions. I would also like to thank Richard Cleve for very useful comments on an earlier draft. Thanks to Daniel Preda for his talk at the CWI seminar [20] about interactive provers using generalized non-local correlations which rekindled my interest in provers sharing entanglement. Many thanks also to Boris Tsirelson for sending me a copy of [24] and [26], and to Falk Unger and Ronald de Wolf for proofreading. Finally, thanks to the anonymous referee whose suggestions helped to improve the presentation of this paper.

References

1. L. Babai. Trading group theory for randomness. In *Proceedings of 17th ACM STOC*, pages 421–429, 1985.
2. L. Babai, L. Fortnow, and C. Lund. Non-deterministic exponential time has two-prover interactive protocols. *Computational Complexity*, 1(1):3–40, 1991.
3. M. Ben-Or, S. Goldwasser, J. Kilian, and A. Wigderson. Multi prover interactive proofs: How to remove intractability. In *Proceedings of 20th ACM STOC*, pages 113–131, 1988.
4. J. Cai, A. Condon, and R. Lipton. On bounded round multi-prover interactive proof systems. In *Proceedings of the Fifth Annual Conference on Structure in Complexity Theory*, pages 45–54, 1990.
5. J. Clauser, M. Horne, A. Shimony, and R. Holt. Proposed experiment to test local hidden-variable theories. *Physical Review Letters*, 23:880–884, 1969.
6. R. Cleve, P. Høyer, B. Toner, and J. Watrous. Consequences and limits of nonlocal strategies. In *Proceedings of 19th IEEE Conference on Computational Complexity*, pages 236–249, 2004. quant-ph/0404076.
7. R. Cleve, P. Høyer, B. Toner, and J. Watrous. Consequences and limits of nonlocal strategies. Presentation at 19th IEEE Conference on Computational Complexity, 2004.

8. U. Feige. On the success probability of two provers in one-round proof systems. In *Proceedings of the Sixth Annual Conference on Structure in Complexity Theory*, pages 116–123, 1991.

9. U. Feige. Error reduction by parallel repetition - the state of the art. Technical Report CS95-32, Weizmann Institute, 1, 1995.

10. U. Feige and L. Lovász. Two-prover one-round proof systems; their power and their problems. In *Proceedings of 24th ACM STOC*, pages 733–744, 1992.

11. S. Goldwasser, S. Micali, and C. Rackoff. The knowledge complexity of interactive proof systems. *SIAM Journal on Computing*, 1(18):186–208, 1989.

12. A. Kitaev and J. Watrous. Parallelization, amplification, and exponential time simulation of quantum interactive proof systems. In *Proceedings of 32nd ACM STOC*, pages 608–617, 2000.

13. H. Kobayashi and K. Matsumoto. Quantum Multi-Prover Interactive Proof Systems with Limited Prior Entanglement *J. of Computer and System Sciences*, 66(3):pages 429–450, 2003.

14. D. Lapidot and A. Shamir. Fully parallelized multi prover protocols for NEXPtime. In *Proceedings of 32nd FOCS*, pages 13–18, 1991.

15. C. Marriott and J. Watrous. Quantum Arthur-Merlin games. cs.CC/0506068.

16. M. A. Nielsen and I. L. Chuang. *Quantum Computation and Quantum Information*. Cambridge University Press, 2000.

17. S. Popescu and D. Rohrlich. Quantum nonlocality as an axiom. *Foundations of Physics*, 24(3):379–385, 1994.

18. S. Popescu and D. Rohrlich. Nonlocality as an axiom for quantum theory. In *The dilemma of Einstein, Podolsky and Rosen, 60 years later: International symposium in honour of Nathan Rosen*, 1996. quant-ph/9508009.

19. S. Popescu and D. Rohrlich. Causality and nonlocality as axioms for quantum mechanics. In *Proceedings of the Symposium of Causality and Locality in Modern Physics and Astronomy: Open Questions and Possible Solutions*, 1997. quant-ph/9709026.

20. D. Preda. Non-local multi-prover interactive proofs. CWI Seminar, 21 June, 2005.

21. R. Raz. Quantum information and the PCP theorem. quant-ph/0504075. To appear in FOCS 2005.

22. A. Shamir. IP = PSPACE. *Journal of the ACM*, 39(4):869–877, 1992.

23. A. Shen. IP = PSPACE: simplified proof. *Journal of the ACM*, 39(4):878–880, 1992.

24. B. Tsirelson. Quantum analogues of Bell inequalities: The case of two spatially separated domains. *Journal of Soviet Mathematics*, 36:557–570, 1987.

25. B. Tsirelson. Some results and problems on quantum Bell-type inequalities. *Hadronic Journal Supplement*, 8(4):329–345, 1993.

26. B. Cirel'son (Tsirelson). Quantum generalizations of Bell's inequality. *Letters in Mathematical Physics*, 4:93–100, 1980.

27. W. van Dam. *Nonlocality & Communication Complexity*. PhD thesis, University of Oxford, Department of Physics, 2000.

28. W. van Dam. Impossible consequences of superstrong nonlocality. quant-ph/0501159, 2005.

29. M. Vyalyi. QMA=PP implies that PP contains PH. *Electronic Colloquium on Computational Complexity*, TR03-021, 2003.

30. J. Watrous. PSPACE has constant-round quantum interactive proof systems. In *Proceedings of 40th IEEE FOCS*, pages 112–119, 1999. cs.CC/9901015.

Quantum Algorithms for Matching and Network Flows

Andris Ambainis[1,*] and Robert Špalek[2,**]

[1] Institute for Quantum Computing and University of Waterloo
ambainis@math.uwaterloo.ca
[2] CWI and University of Amsterdam
sr@cwi.nl

Abstract. We present quantum algorithms for some graph problems: finding a maximal bipartite matching in time $O(n\sqrt{m}\log n)$, finding a maximal non-bipartite matching in time $O(n^2(\sqrt{m/n}+\log n)\log n)$, and finding a maximal flow in an integer network in time $O(\min(n^{7/6}\sqrt{m}\cdot U^{1/3}, \sqrt{nU}m)\log n)$, where n is the number of vertices, m is the number of edges, and $U \leq n^{1/4}$ is an upper bound on the capacity of an edge.

1 Introduction

Network flows is one of the most studied problems in computer science. We are given a directed graph with two designated vertices: a source and a sink. Each edge has assigned a capacity. A network flow is an assignment of flows to the edges such that the capacity of an edge is never exceeded and the total incoming and outgoing flow are equal for each vertex except for the source and the sink. A size of the flow is the total flow going from the source. The task is to find a flow of maximal size.

After the pioneering work of Ford and Fulkerson [1], many algorithms have been proposed. Let n denote the number of vertices and let m denote the number of edges. For networks with real capacities, the fastest algorithms run in time $O(n^3)$ [2,3]. If the network is sparse, one can achieve a faster time $O(nm(\log n)^2)$ [4]. If all capacities are integers bounded by U, the maximal flow can be found in time $O(\min(n^{2/3}m, m^{3/2})\log(n^2/m)\log U)$ [5]. For unit networks, the log-factor is not necessary and the fastest algorithm runs in time $O(\min(n^{2/3}m, m^{3/2}))$ [6]. For undirected unit networks, the fastest known deterministic algorithm runs in time $O(n^{7/6}m^{2/3})$ and the fastest known probabilistic algorithm runs in time $O(n^{20/9})$ [7].

Another well studied problem is finding a matching in a graph. We are given an undirected graph. A matching is a set of edges such that every vertex is connected to at most one other vertex. The task is to find a matching of maximal size. The simplest classical algorithm based on augmenting paths runs in time

* Supported by NSERC, ARDA, CIAR and IQC University Professorship.
** Supported in part by the EU fifth framework project RESQ, IST-2001-37559. Work conducted while visiting University of Waterloo and University of Calgary.

B. Durand and W. Thomas (Eds.): STACS 2006, LNCS 3884, pp. 172–183, 2006.
© Springer-Verlag Berlin Heidelberg 2006

$O(n^3)$ [8,9]. If the graph is bipartite, then the simple algorithm finds a maximal matching in faster time $O(n^{5/2})$ [10]. Finding a bipartite matching can be reduced to finding a maximal flow in a directed unit network, hence one can apply the same algorithms and achieve a running time $O(\min(n^{2/3}m, m^{3/2}))$ [6]. The fastest known algorithm for general sparse graphs runs in time $O(\sqrt{n}m)$ [11]. Recently, Mucha and Sankowski published a new algorithm [12] based on matrix multiplication that finds a maximal matching in general graphs in time $O(n^\omega)$, where $2 \leq \omega \leq 2.38$ is the exponent of the best matrix multiplication algorithm.

In our paper, we analyze the quantum time complexity of these problems. We use Grover's search [13,14] to speed up searching for an edge. A similar approach has been successfully applied by Dürr et al. [15] to the following graph problems: connectivity, strong connectivity, minimum spanning tree, and single source shortest paths. Our bipartite matching algorithm is polynomially faster than the best classical algorithm when $m = \Omega(n^{1+\varepsilon})$ for some $\varepsilon > 0$, and the network flows algorithm is polynomially faster when $m = \Omega(n^{1+\varepsilon})$ and U is small. Out non-bipartite matching algorithm is worse than the best known classical algorithm [11].

There is an $\Omega(n^{3/2})$ quantum adversary lower bound for the bipartite matching problem [16,17]. Since the bipartite matching problem is a special case of the other problems studied in this paper, this implies an $\Omega(n^{3/2})$ quantum lower bound for all problems in this paper.

2   Preliminaries

An excellent book about quantum computing is the textbook by Nielsen and Chuang [18]. In this paper, we only use two quantum sub-routines and otherwise our algorithm are completely classical. The first one is a generalization of Grover's search that finds k items in a search space of size ℓ in total time $O(\sqrt{k\ell})$ [13,14]. An additional time $O(\sqrt{\ell})$ is needed to detect that there are no more solutions; this term is only important when $k = 0$. The second one is quantum counting that estimates the number of ones in a string of length n within additive constant $\sqrt{n}$ with high probability in time $O(\sqrt{n})$ [19, Theorem 13].

Each of those algorithms may output an incorrect answer with a constant probability. Our algorithms may use a polynomial number n^c of quantum sub-routines. Because of that, we have to repeat each quantum subroutine $O(\log n)$ times, to make sure that the probability of an incorrect answer is less than $1/n^{c+1}$. Then, the probability that all quantum subroutines in our algorithm output the correct answer is at least $1 - 1/n$. This increases the running time of all our algorithms by a $\log n$ factor. We omit the log-factors in the proofs, but we state them in the statements of our theorems.

A very good book about network flows is the classical book by Ahuja, Magnanti, and Orlin [20]. It, however, does not contain most of the newest algorithms that we compare our algorithms to. We use the following concepts: A *layered network* is a network whose vertices are ordered into a number of layers, and whose edges only go from the i-th layer to the $(i+1)$-th layer. A *residual network* is a network whose capacities denote the residual capacity of the edges in the original network. When an edge has a capacity c and carries a flow f, then its residual capacity is either $c - f$ or $c + f$ depending on the direction. An *augmenting path* in a network is a path from the source to the sink whose residual capacity is bigger than 0. An augmenting path for the matching problem is a path that consists of alternated non-edges and edges of the current matching, and starts and ends in a free vertex. A *blocking flow* in a layered residual network is a maximal flow with respect to inclusion. A blocking flow cannot be increased by one augmenting path. A *cut* in a network is a subset of edges such that there is no path from the source to the sink if we remove these edges. The size of a cut is the sum of the capacities of its edges. Any flow has size smaller or equal to the size of any cut.

Let us define our computational model. Let V be a fixed vertex set of size $n \geq 1$ and let $E \subseteq \binom{V}{2}$ be a set of edges. E is a part of the input. Let m denote the number of edges. We assume that $m \geq n$, since one can eliminate zero-degree vertices in classical time $O(n)$. We consider the following two black-box models for accessing directed graphs:

- *Adjacency* model: the input is specified by an $n \times n$ Boolean matrix A, where $A[v, w] = 1$ iff $(v, w) \in E$.
- *List* model: the input is specified by n arrays $\{N_v : v \in V\}$ of length $1 \leq d_v \leq n$. Each entry of an array is either a number of a neighbor or a hole, and $\{N_v[i] : i = 1, \ldots, d_v\} - \{\text{hole}\} = \{w : (v, w) \in E\}$.

The structure of the paper is as follows: In Section 3, we present a quantum algorithm for computing a layered network from a given network. It is used as a tool in almost all our algorithms. In Section 4, we present a simple quantum algorithm for bipartite matching. In Section 5, we show how to quantize the classical algorithm for non-bipartite matching. In Section 6, we present a quantum algorithm for network flows.

3   Finding a Layered Subgraph

We are given a connected directed black-box graph $G = (V, E)$ and a starting vertex $a \in V$, and we want to assign layers $\ell : V \to \mathbb{N}$ to its vertices such that $\ell(a) = 0$ and $\ell(y) = 1 + \min_{x:(x,y) \in E} \ell(x)$ otherwise. The following quantum algorithm computes layer numbers for all vertices:

1. Set $\ell(a) = 0$ and $\ell(x) = \infty$ for $x \neq a$.
   Create a one-entry queue $W = \{a\}$.
2. While $W \neq \emptyset$,
   - take the first vertex x from W,
   - find by Grover's search all its neighbors y with $\ell(y) = \infty$,
     set $\ell(y) := \ell(x) + 1$, and append y into W,
   - and remove x from W.

Theorem 1. *The algorithm assigns layers in time $O(n^{3/2} \log n)$ in the adjacency model and in time $O(\sqrt{nm} \log n)$ in the list model.*

Proof. The algorithm is a quantum implementation of breadth-first search. The initialization costs time $O(n)$. Every vertex is processed at most once. In the adjacency model, every vertex contributes by time at most $O(\sqrt{n})$, because finding a vertex from its ancestor costs time at most $O(\sqrt{n})$ and discovering that a vertex has no descendant costs the same.

In the list model, processing a vertex v costs time $O(\sqrt{n_v d_v} + \sqrt{d_v + 1})$, where n_v is the number of vertices inserted into W when processing v. Let $f \le \min(n, m)$ be the number of found vertices. Since $\sum_v n_v \le f \le n$ and $\sum_v (d_v + 1) \le m + f = O(m)$, the total running time is upper-bounded by the Cauchy-Schwarz inequality as follows:

$$\sum_v \sqrt{n_v d_v} \le \sqrt{\sum_v n_v} \sqrt{\sum_v d_v} = O(\sqrt{nm}),$$

and $\sum_v \sqrt{d_v + 1} \le \sqrt{f}\sqrt{m + f}$ is upper-bounded in the same way.

4   Bipartite Matching

We are given an undirected bipartite black-box graph $G = (V_1, V_2, E)$ and we want to find a maximum matching among its vertices. This can be done classically in time $O(n^{5/2})$ [10] as follows:

1. Set M to an empty matching.
2. Let $H = (V', E')$ denote the following graph:

$$V' = V_1 \cup V_2 \cup \{a, b\}$$
$$E' = \{(a, x) : x \in V_1, x \notin M\}$$
$$\cup \{(x, y) : x \in V_1, y \in V_2, (x, y) \in E, (x, y) \notin M\}$$
$$\cup \{(y, x) : x \in V_1, y \in V_2, (x, y) \in E, (x, y) \in M\}$$
$$\cup \{(y, b) : y \in V_2, y \notin M\},$$

where the shortcut $x \notin M$ means that x is not matched.

Find a maximal (with respect to inclusion) set S of vertex-disjoint augmenting paths of minimal length. This is done as follows: First, construct a layered subgraph H' of H. Second, perform a depth-first search for a maximal set of vertex-disjoint paths from a to b in H'. Every such a path is an augmenting path in M, and they all have the same minimal length.
3. Augment the matching M by S.
4. If $S \ne \emptyset$, go back to step 2, otherwise output the matching M.

The algorithm is correct because (1) a matching is maximal iff there is no augmenting path, and (2) the minimal length of an augmenting path is increased by at least one after every iteration. The construction of H' classically and the depth-first search both cost $O(n^2)$. The maximal number of iterations is $O(\sqrt{n})$ due to the following statement:

Lemma 1. [10] *If M_1 and M_2 are two matchings of size s_1 and s_2 with $s_1 < s_2$, then there exist $s_2 - s_1$ vertex-disjoint augmenting paths in M_1.*

Let s be the size of the maximal matching M in G, and let s_i be the size of the found matching M_i after the i-th iteration. Let j be the number of the first iteration with $s_j \geq s - \sqrt{n}$. The total number of iterations is at most $j + \sqrt{n}$, because the algorithm finds at least one augmenting path in every iteration. On the other hand, by Lemma 1, there are $s - s_j \geq \sqrt{n}$ vertex-disjoint augmenting paths in M_j. Since all augmenting paths in the j-th iteration are of length at least $j + 2$, it must be that $j < \sqrt{n}$, otherwise the paths would not be disjoint. We conclude that the total number of iterations is at most $2\sqrt{n}$.

Theorem 2. *Quantumly, a maximal bipartite matching can be found in time $O(n^2 \log n)$ in the adjacency model and $O(n\sqrt{m} \log n)$ in the list model.*

Proof. We present a quantum algorithm that finds all augmenting paths in one iteration in time $O(n^{3/2})$, resp. $O(\sqrt{nm})$, times a log-factor for Grover's search. Since the number of iterations is $O(\sqrt{n})$, the upper bound on the running time follows. Our algorithm works similarly to the classical one; it also computes the layered graph H' and then searches in it.

The intermediate graph H is generated on-line from the input black-box graph G and the current matching M, using a constant number of queries as follows: the sub-graph of H on $V_1 \times V_2$ is the same as G except that some edges have been removed; here we exploit the fact that the lists of neighbors can contain holes. We also add two new vertices a and b, add one list of neighbors of a with holes of total length n, and at most one neighbor b to every vertex from V_2. Theorem 1 states how long it takes to compute H' from H. It remains to show how to find the augmenting paths in the same time.

This is simple once we have computed the layer numbers of all vertices. We find a maximal set of vertex-disjoint paths from a to b by a depth-first search. A descendant of a vertex is found by Grover's search over all unmarked vertices with layer number by one bigger. All vertices are unmarked in the beginning. When we find a descendant, we mark it and continue backtracking. Either the vertex will become a part of an augmenting path, or it does not belong to any and hence it needs not be tried again. Each vertex is thus visited at most once.

In the adjacency model, every vertex costs time $O(\sqrt{n})$ to be found and time $O(\sqrt{n})$ to discover that it does not have any descendant. In the list model, a vertex v costs time $O(\sqrt{n_v d_v} + \sqrt{d_v})$, where n_v is the number of unmarked vertices found from v. The sum over all vertices is upper-bounded like in the proof of Theorem 1. Note that $\sum_v d_v$ has been increased by at most $2n$.

5   Non-bipartite Matching

We are given an undirected graph $G = (V, E)$ and we want to find a maximal matching among its vertices. There is a classical algorithm [8,9] running in total time $O(n^3)$ in n iterations of time $O(n^2)$.

Each iteration consists of searching for an augmenting path. The algorithm performs a breadth-first search from some free vertex. It browses paths that consist of alternated non-edges and edges of the current matching. The matching is specified by pointers *mate*. Let us call a vertex v *even* if we have found such an alternated path of even length from the start to v; otherwise we call it *odd*. Newly discovered vertices are considered to be odd. For each even vertex, we store two pointers *link* and *bridge* used for tracing the path back, and a pointer *first* to the last odd vertex on this path. The algorithm works as follows and its progress on an example graph is outlined in Figure 1:

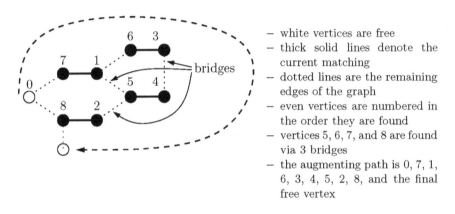

- white vertices are free
- thick solid lines denote the current matching
- dotted lines are the remaining edges of the graph
- even vertices are numbered in the order they are found
- vertices 5, 6, 7, and 8 are found via 3 bridges
- the augmenting path is 0, 7, 1, 6, 3, 4, 5, 2, 8, and the final free vertex

Fig. 1. The classical non-bipartite matching algorithm [8,9]

1. Initialize a queue of even vertices $W = \{a\}$ with some free vertex a.
2. Take the first vertex v from W and delete it from W.
3. If there exists an free vertex w connected to v, then augment the current matching by the path $a \rightarrow v$ plus the edge (v, w), and quit. A general subpath $\rho : b \rightarrow v$ is traced recursively using v's pointers as follows:
   - If *bridge* is nil, then *link* points to the previous even vertex on ρ. Output 2 edges from v to *mate* and *link*, and trace ρ from *link* to b.
   - Otherwise v was discovered via a bridge, *link* points to v's side of the bridge, and *bridge* to the other side. Trace ρ from *link* to v in the opposite direction, and then from *bridge* to b in the normal direction.
4. For every odd vertex w connected to v, do the following:
   - Let w be connected to a mate w'. If w' is even, do nothing.
   - Otherwise mark w' as even, append it to W, and set its pointers as follows: *link* to v, *bridge* to nil, and *first* to w.
5. For every even vertex w connected to v, do the following:
   - Compare the pointers *first* of v and w. If they are equal, do nothing.
   - Now, v and w lie on a circle of odd length, and the edge (v, w) is a *bridge* between the two subpaths. Find the nearest common odd ancestor p of v and w by tracing the pointers *first*. Collapse the circle as follows:

- Mark all odd vertices between v and p as even, append them to W, and set their pointers as follows: *link* to v, *bridge* to w, and *first* to p.
- Do the same for odd vertices between w and p.
- Finally, rewrite all links *first* pointing to odd vertices that have just become even to p.

6. If W is empty, then there is no augmenting path from a and we quit, otherwise go back to step 2.

It holds that if an augmenting path from some vertex has not been found, then it would not be found even later after more iterations of the algorithm. Hence it suffices to search for an augmenting path from each vertex once.

Theorem 3. *Quantumly, a maximal non-bipartite matching can be found in time $O(n^{5/2} \log n)$ in the adjacency model and $O(n^2(\sqrt{m/n} + \log n) \log n)$ in the list model.*

Proof. The algorithm iteratively augments the current matching by single augmenting paths, like the classical algorithm. An augmenting path is found using Grover's search in faster time $O(n^{3/2})$, resp. $O(n(\sqrt{m/n} + \log n))$, times the usual log-factor. This implies the bound on the total running time, since there are n vertices and each of them is used as the starting vertex a at most once. Let us prove the time bound for the list model.

Let $f \leq \min(n, m)$ denote the number of even vertices. For every even vertex v, we perform the following 3 Grover's searches: First, we look for a free neighbor of v in time $O(\sqrt{d_v})$. Second, we process all odd neighbors of v whose mate is still odd in total time $O(\sqrt{e_v d_v})$, where e_v is the number of odd vertices that are found during processing v. Third, we process all even neighbors of v whose pointer *first* is different from v's pointer *first*, in time $O(\sqrt{b_v d_v})$, where b_v is the number of bridges that are found during processing v. Clearly $\sum_v e_v \leq f$ and $\sum_v b_v \leq f$, and, since $\sum_v d_v \leq m$, by the Cauchy-Schwarz inequality, the total time spent in all Grover's searches is $O(\sqrt{nm})$.

Let us estimate the running time of collapsing one circle. Let p_1 be the length of the link-list of pointers *first* from one side of the bridge into the nearest common parent, let p_2 be the other one, and let $p = \max(p_1, p_2)$. The nearest common parent is found in time $O(p \log p)$ as follows: we maintain two balanced binary trees for each link-list, add vertices synchronously one-by-one, and search for every newly inserted vertex in the opposite tree, until we find a collision. Let r_v be the number of odd vertices collapsed during processing a vertex v. It holds that $r_v = p_1 + p_2 = \Theta(p)$ and $\sum_v r_v \leq f$. Hence the total time spent in collapsing circles is $O(f \log f)$.

Rewriting the pointers *first* of all even vertices inside a collapsed circle would be too slow. We instead maintain aside a Union-tree of all these pointers, and for every odd vertex converted to even, we append its subtree to the node of the nearest common ancestor. The total time spent in doing this is $O(f \log f)$.

The augmenting path has length at most n and it is traced back in linear time. We conclude that the total running time of finding an augmented time is

$O(\sqrt{nm} + n\log n) = O(n(\sqrt{m/n} + \log n))$, which is $O(\sqrt{nm})$ for $m \geq n(\log n)^2$. The running time in the adjacency model is equal to the running time in the list model with $m = n^2$, that is $O(n^{5/2})$.

It would be interesting to quantize the fastest known classical algorithm by Micali and Vazirani [11] running in total time $O(\sqrt{nm})$

6   Integer Network Flows

We are given a directed network with real capacities, and we want to find a maximal flow from the source to the sink. There are classical algorithms running in time $O(n^3)$ [2,3]. They iteratively augment the current flow by adding blocking flows in layered residual networks [21] of increasing depth. Since the depth is increased by at least one after each iteration, there are at most n iterations. Each of them can be processed in time $O(n^2)$. For sparse real networks, the fastest known algorithm runs in time $O(nm(\log n)^2)$ [4].

Let us restrict the setting to integer capacities bounded by U. There is a simple capacity scaling algorithm running in time $O(nm\log U)$ [22,21]. The fastest known algorithm runs in time $O(\min(n^{2/3}m, m^{3/2})\log(n^2/m)\log U)$ [5]. For unit networks, i.e. for $U = 1$, a simple combination of the capacity scaling algorithm and the blocking-flow algorithm runs in time $O(\min(n^{2/3}m, m^{3/2}))$ [6]. For undirected unit networks, there is an algorithm running in time $O(n^{3/2}\sqrt{m})$ [23], and the fastest known algorithm runs in worst-case time $O(n^{7/6}m^{2/3})$ and expected time $O(n^{20/9})$ [7].

Lemma 2. [6] *Let us have an integer network with capacities bounded by U, whose layered residual network has depth k. Then the size of the residual flow is at most $\min((2n/k)^2, m/k) \cdot U$.*

Proof. (1) There exist layers V_ℓ and $V_{\ell+1}$ that both have less than $2n/k$ vertices. This is because if for every $i = 0, 1, \ldots, k/2$, at least one of the layers V_{2i}, V_{2i+1} had size at least $2n/k$, then the total number of vertices would exceed n. Since V_ℓ and $V_{\ell+1}$ form a cut, the residual flow has size at most $|V_\ell| \cdot |V_{\ell+1}| \cdot U \leq (2n/k)^2 U$.
(2) For every $i = 0, 1, \ldots, k - 1$, the layers V_i and V_{i+1} form a cut. These cuts are disjoint and they together have at most m edges. Hence at least one of them has at most m/k edges, and the residual flow has thus size at most $O(mU/k)$.

Theorem 4. *Let $U \leq n^{1/4}$. Quantumly, a maximal network flow with integer capacities at most U can be found in time $O(n^{13/6} \cdot U^{1/3}\log n)$ in the adjacency model and in time $O(\min(n^{7/6}\sqrt{m} \cdot U^{1/3}, \sqrt{nU}m)\log n)$ in the list model.*

Proof. The algorithm iteratively augments the current flow by blocking flows in layered residual networks [21], until the depth of the network exceeds $k = \min(n^{2/3}U^{1/3}, \sqrt{mU})$. Then it switches to searching augmenting paths [22], while there are some. The idea of switching the two algorithms comes from [6]. Our algorithm uses classical memory of size $O(n^2)$ to store the current flow and its direction for every edge of the network, and a 1-bit status of each vertex. A blocking flow is found as follows:

1. Compute a layered subgraph H' of the *residual* network H. The capacity of each edge in H is equal to the original capacity plus or minus the current flow depending on the direction. Edges with zero capacities are omitted.
2. Mark all vertices as enabled.
3. Find by a depth-first search a path ρ in H' from the source to the sink that only goes through enabled vertices. If there is no such a path, quit. During back-tracking, disable all vertices from which there is no path to the sink.
4. Compute the minimal capacity μ of an edge on ρ.
   Augment the flow by μ along ρ.
5. Go back to step 3.

The layered subgraph H' is computed from H using Theorem 1, and the capacities of H are computed on-line in constant time. When the flow is augmented by μ along the path ρ, the saturated edges will have been automatically deleted. This is because the algorithm only stores layer numbers for the vertices, and the edges of H' are searched on-line by Grover's search.

Let us compute how much time the algorithm spends in a vertex v during searching the augmenting paths. Let a_v denote the number of augmenting paths going through v and let $e_{v,i}$ denote the number of outgoing edges from v at the moment when there are still i remaining augmenting paths. The capacity of every edge is at most U, hence $e_{v,i} \geq \lceil i/U \rceil$. The time spent in Grover's searches leading to an augmenting path in v is thus at most

$$\sum_{i=1}^{a_v} \sqrt{\frac{d_v}{e_{v,i}}} \leq \sqrt{U} \cdot \sum_{i=1}^{a_v} \sqrt{\frac{d_v}{i}} = O(\sqrt{U a_v d_v}).$$

Let c_v denote the number of enabled vertices found from v that do not lie on an augmenting path and are thus disabled. The time spent in Grover's searches for these vertices is at most $O(\sqrt{c_v d_v})$. Furthermore, it takes additional time $O(\sqrt{d_v + 1})$ to discover that there is no augmenting path from v, and in this case v is disabled and never visited again.

Let j denote the depth of the network and let A_j be the size of its blocking flow. The total number of augmenting paths going through vertices in any given layer is at most A_j. We conclude that $\sum_v a_v \leq j A_j$. We also know that $\sum_v c_v \leq n$. Since $\sum_v d_v \leq m$, by the Cauchy-Schwarz inequality, the total time spent by finding one blocking flow is

$$\sum_v (\sqrt{U a_v d_v} + \sqrt{c_v d_v} + \sqrt{d_v + 1}) \leq \sqrt{U} \sqrt{\sum_v a_v} \sqrt{\sum_v d_v} + 2\sqrt{nm}$$

$$= O(\sqrt{jm A_j U} + \sqrt{nm}).$$

Our algorithm performs at most $k = \min(n^{2/3} U^{1/3}, \sqrt{mU})$ iterations of finding the blocking flow in total time at most $\sqrt{mU} \cdot \sum_{j=1}^{k} \sqrt{j A_j} + k\sqrt{nm}$. Let us assume that the algorithm has not finished, and estimate the size of the residual flow and thus upper-bound the number of augmenting paths that need to be found. The algorithm has constructed in this iteration a layered network of depth bigger than k. By Lemma 2, the residual flow has size $O(\min((n/k)^2, m/k) \cdot U) = O(k)$, hence the algorithm terminates in $O(k)$ more iterations. From this point on, the algorithm only looks for one augmenting path in each layered network, hence its complexity drops to $O(\sqrt{j'm}) = O(\sqrt{nm})$ per iteration, omitting the factor $\sqrt{A_{j'}U}$. The total running time is thus at most

$$O\left(\sqrt{mU} \cdot \sum_{j=1}^{k} \sqrt{jA_j} + k\sqrt{nm}\right) + O(k\sqrt{nm}).$$

Let us prove that $\sum_j \sqrt{jA_j} = O(k^{3/2})$. We split the sequence into $\log k$ intervals $S_i = \{2^i, 2^i + 1, \ldots, 2^{i+1} - 1\}$ of length 2^i. By Lemma 2, the residual flow after $\ell = k/2^i$ iterations is at most $O(\min((n/k)^2 \cdot 2^{2i}, m/k \cdot 2^i) \cdot U) \le O(2^{2i}k) = O((k/\ell)^2\ell) = O(k^2/\ell)$. Since the total size of all blocking flows cannot exceed the residual flow, $\sum_{j=\ell}^{2\ell-1} A_j = O(k^2/\ell)$. By applying the Cauchy-Schwarz inequality independently on each block, we get

$$\sum_{j=1}^{k} \sqrt{jA_j} = \sum_{i=0}^{\log k} \sum_{j=2^i}^{2^{i+1}-1} \sqrt{jA_j} \le \sum_{i=0}^{\log k} \sqrt{2^i \cdot 2^{i+1}} \sqrt{\sum_{j=2^i}^{2^{i+1}-1} A_j}$$

$$\le \sqrt{2} \sum_{i=0}^{\log k} 2^i \sqrt{k^2/2^i} = \sqrt{2} \cdot k \sum_{k=0}^{\log k} 2^{i/2} = O(k^{3/2}).$$

The total running time is thus $O(k\sqrt{m}(\sqrt{kU} + \sqrt{n}))$. Now, $kU \le n$, because $U \le n^{1/4}$ and $kU = \min(n^{2/3}U^{4/3}, \sqrt{m} \cdot U^{3/2}) \le n^{2/3}n^{1/3} = n$. The running time is therefore $O(k\sqrt{nm}) = O(\min(n^{7/6}\sqrt{m} \cdot U^{1/3}, \sqrt{nU}m))$, times a log-factor for Grover's search. The time for the adjacency model follows from setting $m = n^2$ and it is $O(n^{13/6} \cdot U^{1/3} \log n)$.

It is not hard to compute an upper bound on the running time of the network flows algorithm for $U > n^{1/4}$ by the same techniques. One obtains $O(\min(n^{7/6}\sqrt{m}, \sqrt{nm}) \cdot U \log n)$ for arbitrary U by setting $k = \min(n^{2/3}, \sqrt{m})$. It would be interesting to apply techniques of [5] to improve the multiplicative constant in Theorem 4 from poly(U) to $\log U$. If $m = \Omega(n^{1+\varepsilon})$ for some $\varepsilon > 0$ and U is small, then our algorithm is polynomially faster than the best classical algorithm. For constant U and $m = O(n)$, it is slower by at most a log-factor. The speedup is biggest for dense networks with $m = \Omega(n^2)$.

Theorem 5. *Any bounded-error quantum algorithm for network flows with integer capacities bounded by $U = n$ has quantum query complexity $\Omega(n^2)$.*

Proof. Consider the following layered graph with $m = \Theta(n^2)$ edges. The vertices are ordered into 4 layers: the first layer contains the source, the second and third layer contain $p = \frac{n}{2} - 1$ vertices each, and the last layer contains the sink. The source and the sink are both connected to all vertices in the neighboring layer by p edges of full capacity n. The vertices in the second and third layer are connected by either $\frac{p^2}{2}$ or $\frac{p^2}{2} + 1$ edges of capacity 1 chosen at random. The edges between these two layers form a minimal cut. Now, deciding whether the maximal flow is $\frac{p^2}{2}$ or $\frac{p^2}{2} + 1$ allows us to compute the majority on p^2 bits. There is an $\Omega(p^2) = \Omega(n^2)$ lower bound for majority, hence the same lower bound also holds for the computation of the maximal flow.

Acknowledgments

We thank Marek Karpinski for discussions that lead to the quantum bipartite matching algorithm and help with literature on classical algorithms for bipartite matchings.

References

1. Ford, L.R., Fulkerson, D.R.: Maximal flow through a network. Canadian Journal of Mathematics **8** (1956) 399–404
2. Karzanov, A.V.: Determining the maximal flow in a network by the method of preflows. Soviet Mathematics Doklady **15** (1974) 434–437
3. Malhotra, V.M., Kumar, P., Maheshwari, S.N.: An $O(V^3)$ algorithm for finding the maximum flows in networks. Information Processing Letters **7** (1978) 277–278
4. Galil, Z., Naamad, A.: Network flow and generalized path compression. In: Proc. of 11th ACM STOC. (1979) 13–26
5. Goldberg, A.V., Rao, S.: Beyond the flow decomposition barrier. Journal of the ACM **45** (1998) 783–797
6. Even, S., Tarjan, R.E.: Network flow and testing graph connectivity. SIAM Journal on Computing **4** (1975) 507–518
7. Karger, D.R., Levine, M.S.: Finding maximum flows in undirected graphs seems easier than bipartite matching. In: Proc. of 30th ACM STOC. (1998) 69–78
8. Edmonds, J.: Paths, trees, and flowers. Canadian Journal of Mathematics **17** (1965) 449–467
9. Gabow, H.N.: An efficient implementation of Edmonds' algorithm for maximum matching on graphs. Journal of the ACM **23** (1976) 221–234
10. Hopcroft, J.E., Karp, R.M.: An $n^{5/2}$ algorithm for maximum matchings in bipartite graphs. SIAM Journal on Computing **2** (1973) 225–231
11. Micali, S., Vazirani, V.V.: An $O(\sqrt{|V|} \cdot |E|)$ algorithm for finding maximum matching in general graphs. In: Proc. of 21st IEEE FOCS. (1980) 17–27
12. Mucha, M., Sankowski, P.: Maximum matchings via Gaussian elimination. In: Proc. of 45th IEEE FOCS. (2004) 248–255
13. Grover, L.K.: A fast quantum mechanical algorithm for database search. In: Proc. of 28th ACM STOC. (1996) 212–219
14. Boyer, M., Brassard, G., Høyer, P., Tapp, A.: Tight bounds on quantum searching. Fortschritte der Physik **46** (1998) 493–505 Earlier version in Physcomp'96.
15. Dürr, C., Heiligman, M., Høyer, P., Mhalla, M.: Quatum query complexity of some graph problems. In: Proc. of 31st ICALP. (2004) 481–493 LNCS 3142.

16. Berzina, A., Dubrovsky, A., Freivalds, R., Lace, L., Scegulnaja, O.: Quantum query complexity for some graph problems. In: Proc. of 30th SOFSEM. (2004) 140–150

17. Zhang, S.: On the power of Ambainis's lower bounds. Theoretical Computer Science **339** (2005) 241–256 Earlier version in ICALP'04.

18. Nielsen, M.A., Chuang, I.L.: Quantum Computation and Quantum Information. Cambridge University Press (2000)

19. Brassard, G., Høyer, P., Mosca, M., Tapp, A.: Quantum amplitude amplification and estimation. In: Quantum Computation and Quantum Information: A Millennium Volume. Volume 305 of AMS Contemporary Mathematics Series. (2002) 53–74

20. Ahuja, R.K., Magnanti, T.L., Orlin, J.B.: Network Flows. Prentice-Hall (1993)

21. Dinic, E.A.: Algorithm for solution of a problem of maximum flow in networks with power estimation. Soviet Mathematics Doklady **11** (1970) 1277–1280

22. Edmonds, J., Karp, R.M.: Theoretical improvement in algorithmic efficiency for network flow problems. Journal of the ACM **19** (1972) 248–264

23. Goldberg, A.V., Rao, S.: Flows in undirected unit capacity networks. SIAM Journal on Discrete Mathematics **12** (1999) 1–5

The Number of Runs in a String: Improved Analysis of the Linear Upper Bound[*]

Wojciech Rytter

Instytut Informatyki, Uniwersytet Warszawski,
Banacha 2, 02–097, Warszawa, Poland
Department of Computer Science, New Jersey Institute of Technology
rytter@mimuw.edu.pl

Abstract. A *run* (or a *maximal repetition*) in a string is an inclusion-maximal periodic segment in a string. Let $\rho(n)$ be the maximal number of runs in a string of length n. It has been shown in [8] that $\rho(n) = O(n)$, the proof was very complicated and the constant coefficient in $O(n)$ has not been given explicitly. We propose a new approach to the analysis of runs based on the properties of subperiods: the periods of periodic parts of the runs. We show that $\rho(n) \leq 5\,n$. Our proof is inspired by the results of [4], where the role of new periodicity lemmas has been emphasized.

1 Introduction

Periodicities in strings were extensively studied and are important both in theory and practice (combinatorics of words, pattern-matching, computational biology). The set of all runs in a string corresponds to the structure of its repetitions. Initial interest was mostly in repetitions of the type xx (so called *squares*), [1, 10]. The number of squares, with *primitive* x, is $\Omega(n \log n)$, hence the number of periodicities of this type is not linear. Then, it has been discovered that the number of runs (also called maximal repetitions or repeats) is linear and consequently linear time algorithms for runs were investigated [8, 7]. However the most intriguing question remained the asymptotically tight bound for the number of runs. The first bound was quite complicated and has not given any *concrete* constant coefficient in $O(n)$ notation. This subject has been studied in [12, 13, 2]. The lower bound of approximately $0.927\,n$ has been given in [2]. The exact number of runs has been considered for special strings: *Fibonacci words* and (more generally) *Sturmian words*, [6, 5, 11]. In this paper we make a step towards better understanding of the structure of runs. The proof of the linear upper bound is simplified and small *explicit* constant coefficient is given in $O(n)$ notation.

Let $period(w)$ denote the size of the smallest period of w. We say that a word w is *periodic* iff $period(w) \leq \frac{|w|}{2}$.

---

[*] Research supported by the grants 4T11C04425 and CCR-0313219.

B. Durand and W. Thomas (Eds.): STACS 2006, LNCS 3884, pp. 184–195, 2006.

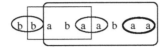

Fig. 1. $RUNS(\,b\,b\,a\,b\,a\,a\,b\,a\,a\,) = \{[1,2],[2,5],[3,9],[5,6],[8,9]\}$

A *run* in a string w is an inclusion-maximal interval $\alpha = [i...j]$ such that the substring $w[i...j] = w[i]w[i+1]...w[j]$ is periodic. Denote by $RUNS(w)$ the set of runs of w. For example we have 5 runs in an example string in Figure 1.

Denote: $\rho(n) = \max\{|RUNS(w)| : |w| = n\}$.

The most interesting conjecture about $\rho(n)$ is: $\rho(n) < n$.

We make a small step towards proving validity of this conjecture and show that $\rho(n) \leq 5\,n$. The proof of linear upper bound in [8] does not give any explicit constant coefficient at all.

The value of the run $\alpha = [i...j]$ is $val(\alpha) = w[i...j]$. When it creates no ambiguity we identify sometimes runs with their values although two different runs could correspond to the identical subwords, if we disregard positions of these runs. Hence runs are also called maximal *positioned* repetitions.

Each value of the run α is a string $x^k y = w[i...j]$, where $|x| = period(\alpha) \geq 1$, $k \geq 2$ is an integer and y is a proper prefix of x (possibly empty). The subword x is called the periodic part of the run and denoted by $PerPart(\alpha) = x$. Denote $SquarePart(\alpha) = [i...i + 2\,period(\alpha) - 1]$.

We also introduce terminology for the starting position of the second occurrence of periodic part: $center(\alpha) = i + |x|$.

The position i is said to be the *occurrence* of this run and is denoted by $first(\alpha)$. We write $\alpha \prec \beta$ iff $first(\alpha) < first(\beta)$.

Example. In Figure 2 we have: $first(\alpha) = 2$, $first(\beta) = 4$ and $center(\alpha) = 22$, $center(\beta) = center(\gamma) = 21$, $PerPart(\gamma) = (aba)^4 ab$.

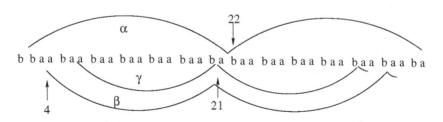

Fig. 2. Example of three highly periodic runs $\alpha \prec \beta \prec \gamma$ with subperiod 3. The runs β, γ are left-periodic (the subperiod 3 continues to the left), α is not. The runs α, β (as well as β, γ) are "neighbors" in sense of Lemma 1. The occurrences (starting positions) of very large runs can be very close. The periodic parts are indicated by the arcs.

In the paper the crucial role is played by the runs α with highly periodic $PerPart(\alpha)$. Denote

$$\mathbf{subperiod}(\alpha) = period(PerPart(\alpha)).$$

In Figure 2 we have: $subperiod(\alpha) = subperiod(\beta) = subperiod(\gamma) = 3$.

We say that a word w is **highly periodic** (*h-periodic*) if $period(w) \le \frac{|w|}{4}$. A run is said to be a **highly periodic run** (an *hp-run*, in short) iff $PerPart(\alpha)$ is h-periodic. The run which is not h-periodic is called a **weakly-periodic** run (*wp-run*). In Figure 2 α, β, γ are a highly periodic runs.

Denote $\mathbf{\Delta} = \frac{5}{4}$. We say that two different runs α, β are **neighbors** iff there is a positive number η such that:

$$|first(\alpha) - first(\beta)| \le \frac{1}{4}\eta \ \text{ and } \ \eta \le period(\alpha), period(\beta) \le \Delta\,\eta$$

Informally, two runs are neighbors iff they have similar periods and are positioned close to each other relatively to their sizes, in particular this means that

$$period(\alpha), period(\beta) \ge 4\,|first(\alpha) - first(\beta)|.$$

It is *"intuitively obvious"* that if we have many neighbors gathered together then such situation forces one of them to be highly periodic. The tedious proof of the following key-lemma is given in Section 3.

Lemma 1 [The Three-Neighbors]. *Lemma] If we have three distinct runs which are pairwise neighbors with the same number η then at least one of them is h-periodic.*

We cannot replace Three-Neighbors Lemma with *Two-Neighbors Lemma*, see Figure 3.

We show that *hp-runs* are also *sparse* in a certain sense. Another tedious proof of the following lemma is given in Section 4. Figure 2 shows that "two" cannot be replaced by "single", the runs α, β have subperiod 3 and start in the interval $[2 \ldots 4]$ of size 3.

b b a a a a a a a a a a a b b a a a a a a a a a a a b b a

Fig. 3. Two weakly-periodic runs which are neighbors

Lemma 2 [HP-Runs Lemma]. *For a given $p > 1$ there are at most two occurrences of hp-runs with subperiod p in any interval of length p.*

2   Estimating the Number $\rho(n)$

The analysis is based on the *sparsity* properties of *hp-runs* and *wp-runs* expressed by Lemmas 1 and 2.

Denote by $\mathbf{WP}(n, k)$ the maximal number of wp-runs α in a string of length n with $period(\alpha) \geq k$.

Let $\mathbf{HP}(n)$ be the maximal number of all hp-runs in a string of length n. It can be shown that $HP(n) \geq \frac{1}{3}n - c_0$, where c_0 is a constant (take $w = (ab)^m b(ab)^m b(ab)^m$). However we are interested in the upper bound.

Let $\rho(n, k)$ be the maximal number of all runs α with $period(\alpha) \leq k$, in a string of length n. We separately estimate the numbers $WP(n, k)$, $HP(n)$, $\rho(n, k)$.

2.1   Estimating the Number of Weakly Periodic Runs

We group wp-runs into groups of potential neighbors. Denote

$$\mathcal{G}(k) = \{\alpha : \alpha \text{ is a weakly periodic run of } w, \; \Delta^k \leq period(\alpha) < \Delta^{k+1}\};$$

Lemma 3. $WP(n, \lceil \Delta^r \rceil) \leq 40\Delta^{-r} \times n.$

Proof. Let w be a string of length n. If $\alpha, \beta \in \mathcal{G}(k)$ for the same k, and $|first(\alpha) - first(\beta)| \leq \Delta^k/4$ then α, β are neighbors with $\eta = \Delta^k$.

Now Lemma 1 can be reformulated as follows: $|\mathcal{G}(k)| \leq 2 \cdot (1/(\Delta^k \cdot \frac{1}{4}) \cdot n = 8\Delta^{-k} \cdot n.$

The last inequality follows directly from Lemma 1, which implies that there are at most two elements of $\mathcal{G}(k)$ in any interval of size $\frac{1}{4}\Delta^k$.
Consequently we have

$$WP(n, \lceil \Delta^r \rceil) \leq \sum_{k=r}^{\infty} |\mathcal{G}(k)| \leq \sum_{k=r}^{\infty} 8 \cdot \Delta^{-k} \cdot n = 8\Delta^{-r} \times \frac{1}{1 - \Delta^{-1}} = 40 \cdot \Delta^{-r}$$

2.2   Estimating the Number of Highly Periodic Runs

Denote by $\mathbf{hp}(n, p)$ the maximal number hp-runs α with $p \leq subperiod(\alpha) \leq 2p$, maximized over strings of length n.

Lemma 4. *If $p \geq 2$ then $hp(n, p) \leq \frac{2}{p} n$.*

Proof. It is easy to see the following claim (using the periodicity lemma).

Claim. If α, β are two hp-runs which satisfy
$$|first(\alpha) - first(\beta)| < p \text{ and } p \leq subperiod(\alpha), subperiod(\beta) \leq 2p,$$
then $subperiod(\alpha) = subperiod(\beta)$.

It follows from the claim and Lemma 2 that for any interval of length p there are at most two hp-runs occurring in this interval and having subperiods in $[p\ldots 2p]$, since such hp-runs should have the same subperiod $p' \geq p$. Therefore there are at most $\frac{2}{p'} n \leq \frac{2}{p} n$ hp-runs with subperiods in $[p\ldots 2p]$. This completes the proof.

Lemma 5. $HP(n) \leq 1.75\, n$.

Proof. Observe that there are no hp-runs with subperiod 1.
According to Lemma 4 we have:

$$HP(n) \leq hp(n,2) + hp(n,5) + hp(n,11) + hp(n,23) + hp(n,47) + hp(n,95) + \ldots$$

$$= 2\,n \times \left(\frac{1}{2} + \frac{1}{5} + \frac{1}{11} + \frac{1}{23} + \frac{1}{47} + \ldots\right) \times n = 2\,n \times \sum_{k=1}^{\infty} \frac{1}{p_k},$$

where $p_k = 2^k + 2^{k-1} - 1$. A rough estimation gives:

$$2 \times \sum_{k=1}^{\infty} \frac{1}{p_k} < 1.75$$

Hence $HP(n) \leq 1.75\, n$.

2.3 The Runs with Periods Bounded by a Constant

We estimate the number of runs with small periods in a rather naive way.

Lemma 6. *For any given $k \geq 1$ there are at most $\frac{1}{k+1}\, n$ runs with $period(\alpha) = k$ or $period(\alpha) = 2k$.*

Proof. We omit the proof of the following simple fact.

Claim. If u, v are primitive words and $|u| = 2|v|$, then vv is not contained in uu as a subword.

Assume that $\alpha \prec \beta$ are two different runs with periods k or $2k$.
  If $period(\alpha) = period(\beta) = k$ then α, β can have an overlap of size at most $k - 1$, otherwise α, β could be merged into a single run. Hence $first(\beta) - first(\alpha) \geq k + 1$.
  If $period(\alpha) = k$ and $period(\beta) = 2k$ then it is possible that $first(\beta) - first(\alpha) = 1$. Due to the claim the distance from $first(\beta)$ to the occurrence of the next hp-run γ with period k or $2k$ is at least $2k + 1$. Then two consecutive distances give together $(first(\beta) - dirst(\alpha) + (first(\gamma) - first(\beta)) \geq 2k + 2$, and "on average" the distance is $k + 1$. Therefore there are at most $\frac{n}{k+1}$ runs with a period k or $2k$.

The last lemma motivates the introduction of the infinite set Φ, generated by the following algorithm (which never stops).

$\Phi := \emptyset;\ \Psi := \{1,2,3,\ldots\};$
repeat forever
    $k :- \min \Psi;$
    remove k and $2k$ from $\Psi;$
    insert k into $\Phi;$

Define the set $\Phi(p) = \{k \in \Phi : k \leq p\}$. For example:

$$\Phi(34) = \{1,3,4,5,7,9,11,12,13,15,16,17,19,20,21,23,25,27,28,29,31,33\}$$

For $p \geq 1$ define the numbers:

$$\mathcal{H}(p) = \sum_{k \in \Phi(p)} \tfrac{1}{k+1}.$$

The next lemma follows directely from Lemma 6 and from the structure of the set Φ.

Lemma 7. $\rho(n,p) \leq \mathcal{H}(p) \times n.$

2.4   Estimating the Number of all Runs

Our main result is a *concrete* constant coefficient in $O(n)$ notation for $\rho(n)$.

Theorem 1. $\rho(n) \leq 5\,n.$

Proof. Obviously, for each $r \geq 1$ we have:

$$\rho(n) \leq HP(n) + WP(n, \lceil \Delta^r \rceil) + \rho(n, \lfloor \Delta^r \rfloor)$$

$$\leq (1.75 + 40\,\Delta^{-r} + \mathcal{H}(\lceil \Delta^r \rceil)) \times n.$$

If we choose $r = 20$, then

$$\lfloor \Delta^{20} \rfloor = 86,\quad \mathcal{H}(86) \leq 2.77,\quad 40\Delta^{-20} \leq 0.4612.$$

Due to Lemmas 5,6,7 we have:

$$\rho(n) \leq (1.75 + \mathcal{H}(86) + 40\Delta^{-20}) \times n \leq$$

$$(1.75 + 2.77 + 0.4612) \times n < 5\,n.$$

This completes the proof of the main result.

3   The Proof of Lemma 1

If $\alpha \prec \beta$ and the *square part* of β is not contained in the *square part* of α then we write $\alpha \prec\prec \beta$ (see Figure 5). More formally:

$\alpha \sqsupset \beta$ iff   *SquarePart*(β) is contained in *SquarePart*(α) as an interval

$$\alpha \prec\prec \beta \quad \text{iff} \quad [\, \alpha \prec \beta \text{ and } not \ (\alpha \sqsupset \beta)\,]$$

Lemma 8. (a) *If $\alpha \sqsupset \beta$ are distinct neighbors then β is highly periodic.*
(b) *If $\alpha \prec\prec \beta$ are distinct neighbors then the prefix of β of size $period(\alpha) - \delta$ has a period $|q-p|$, where $\delta = first(\beta) - first(\alpha)$ and $p = period(\alpha)$, $q = period(\beta)$.*

Proof. **Point (a).** We refer the reader to Figure 4, where the case $center(\beta) > center(\alpha)$ is illustrated. Obviously $p > q$. It is easy to see that the whole $PerPart(\beta)$ has a period $period(\alpha) - period(\beta)$.

Let η be the constant from the definition of neighbors, then

$$period(\alpha) - period(\beta) \leq \frac{1}{4}\eta \ \text{ and } \ |PerPart(\beta)| \geq \eta \ ,$$

hence $PerPart(\beta)$ is h-periodic. The case $center(\beta) \leq center(\alpha)$ can be considered similarly.

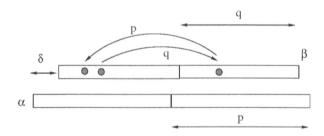

Fig. 4. Two neighbors with $\alpha \sqsupset \beta$, a case $center(\beta) > center(\alpha)$. The *square part* of β is contained in the *square part* of α. The periodic part of β is h-periodic, so it should have a period $p - q$, where $p = period(\alpha)$, $q = period(\beta)$.

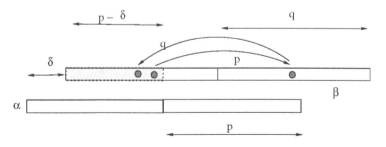

Fig. 5. Two neighbors with $\alpha \prec\prec \beta$, the case $p < q$. The shaded part has the period $|q - p|$, where $p = period(\alpha)$, $q = period(\beta)$.

Point (b). We refer to Figure 5, when only the case $p < q$ is shown. For each position i in the shaded area we have $w[i] = w[i+p] = w[i+p-q]$. The opposite case $p > q$ can be considered similarly. This completes the proof.

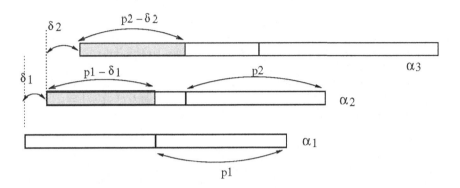

Fig. 6. The Three-Neighbors Lemma, a situation when $\alpha_1 \prec\prec \alpha_2 \prec\prec \alpha_3$. α_2 should be h-periodic, since both its large suffix and large prefix have small periods.

The Proof of the Three-Neighbors Lemma
Assume we have 3 runs $\alpha_1 \prec \alpha_2 \prec \alpha_3$ which are pairwise neighbors, with periods $p1, p2, p3$, respectively. Let $\delta_1 = first(\alpha_2) - first(\alpha_1)$, and $\delta_2 = first(\alpha_3) - first(\alpha_2)$. Then, due to Lemma 8 the "middle" run α_2 has a suffix $\gamma 2$ of size $p2 - \delta_2$ with a period $|p3 - p2|$ and a prefix $\gamma 1$ of size $p1 - \delta 1$ with a period $|p2 - p1|$, see Figure 6.

Let η be the number from the definition of neighbors. We have

$$\delta_1 + \delta_2 \leq \tfrac{1}{4}\eta, \ \ p1 \geq \eta, \text{ and } |\gamma 1 \cup \gamma 2| = p_2.$$

Hence:

$$|\gamma_1 \cap \gamma 2| \geq (p_2 - \delta_2) + (p1 - \delta 1) - p2 = p1 - \delta 1 - \delta 2 \geq \frac{3}{4}\eta$$

We have $|p3 - p2|, |p2 - p1| \leq \tfrac{1}{4}\eta$, hence $period(\gamma 1), period(\gamma 2) \leq \tfrac{1}{4}\eta$. Due to the periodicity lemma $\gamma_1 \cap \gamma 2$ has a period which divides periods of $\gamma 1$ and $\gamma 2$, and the whole $\alpha_2 = \gamma_1 \cup \gamma 2$ has a period of size not larger than $\tfrac{1}{4}\eta$. Consequently, the run α_2 is h-periodic. This completes the proof of our key lemma.

4   The Proof of Lemma 2

The proof is based on the following simple lemma.

Lemma 9. *Assume we have two distinct hp-runs α, β with the same subperiod p and such that periodic part of one of them is a prefix of the periodic part of another. Then $|first(\alpha) - first(\beta)| \geq p$.*

Proof. If $|first(\alpha) - first(\beta)| < p$ then, due to *periodicity lemma* [9, 3, 12], the periodic part of one of the runs would have subperiod smaller than p, which contradicts the assumption that p is the smallest subperiod.

We say that a hp-run $\alpha = [i \dots j]$ of a string w is **left-periodic** iff $w[i-1] = w[i-1+subperiod(\alpha)]$. The runs β, γ in Figure 2 are left-periodic. We also say that a position i in a word w *breaks* period p iff $w[i] \neq w[i+p]$. Hence a *hp-run* α of a word w is *left-periodic* iff $first(\alpha) - 1$ does not break $subperiod(\alpha)$. In other words the subperiod of $PerPart(\alpha)$ continues to the left.

Example. In Figure 2 the runs α, β, γ are shown, the first one is not left periodic and the other two are. The position $center(\beta) - 1 = center(\gamma) - 1 = 21$ breaks subperiod 3. The periodic part of β is a prefix of a periodic part of γ.

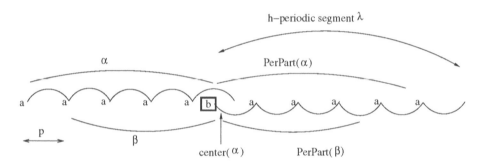

Fig. 7. Two left-periodic runs. The position $center(\alpha) - 1 = center(\beta) - 1$ breaking subperiod p is placed in a small square. $subperiod(\alpha) = subperiod(\beta) = p$, $center(\alpha) = center(\beta)$. The second occurrences of periodic parts of α and β start at the same position $center(\alpha)$, consequently $PerPart(\beta)$ is a prefix of $PerPart(\alpha)$.

Lemma 10. *Assume two neighbors α, β are left-periodic and h-periodic. Then $center(\alpha) = center(\beta)$.*

Proof. We first prove that positions $center(\alpha) - 1, center(\beta) - 1$ break $subperiod(\alpha)$, see Figure 7. The proof is by contradiction. If it is not true then one of these runs can be extended one position to the left. This contradicts the definition of the run as a left non-extendible segment. The positions $center(\alpha)$ and $center(\beta)$ are positions in the same h-periodic segment λ, see Figure 7. They should be equal to the first position of this segment, because the next position to the left breaks the period. Hence they should be the same position, consequently $center(\alpha) = center(\beta)$.

The Proof of the HP-Runs Lemma

For a given $p > 1$ there are at most two occurrences of hp-runs with subperiod p in any interval of length p.

Proof. The proof is by contradiction. Assume we have three distinct highly periodic runs $\alpha_1 \prec \alpha_2 \prec \alpha_3$ with the same subperiod p such that $|first(\alpha_i) - first(\alpha_j)| \leq p$ for $1 \leq i, j \leq 3$. Then all of them are neighbors. We show that $\alpha_2 = \alpha_3$. Both α_2, α_3 should be left-periodic since their subperiods extend to the left at least to $first(\alpha_1)$.

Therefore the runs α_2, α_3 are h-periodic and they are neighbors. Due to Lemma 10 $center(\alpha_2) = center(\alpha_3)$. Consequently periodic parts of α_2 and α_3 have occurrences starting at the same position $center(\alpha_2)$. If two words start at a same position then one should be a prefix of another. Consequently $PerPart(\alpha_3)$ is a prefix of $PerPart(\alpha_2)$. Now, due to Lemma 9, if $\alpha_2 \neq \alpha_3$ then $first(\alpha_3) - first(\alpha_2) \geq p$. However $first(\alpha_3) - first(\alpha_2) < p$. This implies that all of $\alpha_1, \alpha_2, \alpha_3$ cannot be pairwise distinct. This contradicts the assumption and completes the proof.

5   The Sum of Exponents of Periodicities

We define the *exponent of periodicity* of a run α as $exp(\alpha) = |\alpha|/period(\alpha)$.

The linear bound on $\rho(n)$ gives, almost automatically, a linear upper bound on the sum of exponents of periodicities. The run α is called a *long* run iff $exp(\alpha) \geq 4$. Denote by $Exp(w)$ the sum of exponents of periodicity of all runs of w, and by $L\text{-}Exp(w)$ the sum of exponents of all long runs of w.

Let $\mu(n)$ be maximum $Exp(w)$ and $\mu(n, 4)$ be maximum $L\text{-}Exp(w)$ of a string w of length n. Denote by $\gamma(n)$ the maximum number of long runs in a string of size n.

Lemma 11. (a) $\mu(n, 4) \leq 5\,n;$ (b) $\gamma(n) \leq 1.25\,n;$ (c) $\mu(n) \leq \mu(n, 4) + 4\,\rho(n)$.

Proof. Denote

$$\mathcal{G}'(k) = \{\alpha : 2^k \leq period(\alpha) < 2^{k+1}, \ exp(\alpha) \geq 4\}$$

If $\alpha = [i...j]$ then denote $\Gamma(\alpha) = [i + 3\,period(\alpha) - 1 \ ... \ j]$.

Claim. If $\alpha \neq \beta$ are in a same $\mathcal{G}'(k)$, for some integer k, then $\Gamma(\alpha) \cap \Gamma(\beta) = \emptyset$.

Proof (of the claim). The following inequality follows from the *periodicity lemma*:

$$|\alpha \cap \beta| \leq \min \{3\,period(\alpha), \ 3\,period(\beta)\}$$

The claim follows easily from this inequality.

Observe now that $|\Gamma(\alpha)| = (exp(\alpha) - 3)\,period(\alpha)$.

Denote by $\mathcal{L}$ the set of long runs with $period(\alpha) > 1$. In other words $\mathcal{L} = \sum_{k>0} \mathcal{G}'(k)$. Due to the claim and the inequality $period(\alpha) \geq 2^k$ we have:

$$\sum_{\alpha \in \ \mathcal{G}'(k)} (exp(\alpha) - 3)\,period(\alpha) \leq n, \quad \text{hence} \quad \sum_{\alpha \in \mathcal{G}'(k)} (exp(\alpha) - 3) \leq \frac{n}{2^k} \quad \text{and}$$

$$\sum_{\alpha \ \in \ \mathcal{L}} (exp(\alpha) - 3) \ \leq \ n \sum_{k=1}^{\infty} \frac{1}{2^k} \ \leq \ n. \tag{1}$$

We have that $exp(\alpha) - 3 \geq 1$, hence $|\mathcal{L}| \leq n$, and we have at most n long runs with $period(\alpha) > 1$. There are at most $\frac{1}{4}\,n$ long runs with period 1. Altogether we have $\gamma(n) \leq 1.25\,n$. This proves point (b).

We now prove point (a). Due to Equation 1 we have:

$$\sum_{\alpha \in \mathcal{L}} exp(\alpha) \leq n + \sum_{\alpha \in \mathcal{L}} 3 \;\leq\; n + 3\,|\mathcal{L}| \;\leq\; 4n$$

On the other hand all runs with period 1 are pairwise disjoint, so the sum of exponents of these runs is at most n. Hence the total sum of exponents of all long α's is at most $n + 4\,n = 5\,n$. This completes the proof of point (a). Point (c) follows directly from definitions.

6   Final Remarks

We gave an estimation $\rho(n) \leq 5\,n$. The important part of our contribution is also a new approach based on subperiods. The proof is completely different from the one in [8], where the proof was by induction on n. The only complicated parts of our proof are the proofs of Lemma 1 and Lemma 2, which can be viewed as *new periodicity lemmas* of independent interest. The proofs of these lemmas are tedious but the lemmas are intuitively almost obvious. In a certain sense we demystified the whole proof of the linear upper bound for $\rho(n)$. The point (c) of Lemma 11 gives directly linear bound on $\mu(n)$ (the sum of exponents of periodicities of all runs), though the constant coefficient is still not satisfactory. Experimental evidence suggests $\mu(n) \leq 2n$. One should possibly rewrite the whole proof of Theorem 1, proving the linear bound on $\rho(n)$ in terms of $\mu(n)$, to improve the coefficient in the linear bound for $\mu(n)$. However this would hideously obscure the proof of Theorem 1.

References

1. *M. Crochemore*, An optimal algorithm for computing the repetitions in a word, Inf. Proc. Letters 42:5(1981) 244-250
2. *F. Franek, R.J.Simpson, W.F.Smyth*, The maximum number of runs in a string, Proc. 14-th Australian Workshop on Combinatorial Algorithms, M.Miller, K. Park (editors) (2003) 26-35
3. *M. Crochemore, W.Rytter*, Jewels of stringology: text algorithms, World Scientific 2003
4. *Kangmin Fan, William F. Smyth, R. J. Simpson*: A New Periodicity Lemma. CPM 2005: 257-265
5. *F. Franek, A. Karaman, W.F.Smyth*, Repetitions in Sturmian strings, TCS 249-2 (2000) 289-303
6. *C. Iliopoulos, D. Moore, W.F.Smyth*, A characterization of the squares in a Fibonacci string, TCS 172 (1997) 281-291
7. *R.Kolpakov, G.Kucherov*, On maximal repetitions in words, Journal of Discr. Algorithms 1 (2000) 159-186

8. *R.Kolpakov, G.Kucherov*, Finding maximal repetitions in a word in linear time, FOCS (1999) 596-604

9. *Lothaire*, Algebraic combinatorics on words, Cambridge University Press

10. *M.G.Main, R.J.Lorentz*, An $O(n \log n)$ algorithm for finding all repetitions in a string, Journal of Algorithms 5 (1984) 422-432

11. *W.Rytter*, The structure of subword graphs and suffix trees of Fibonacci words, in Colloquium on Implementation and Application of Automata, CIAA (2005)

12. *W.F.Smyth*, Computing patterns in strings, Addison-Wesley (2003)

13. *W.F.Smyth*, Repetitive perhaps, but certainly not boring, TCS 249-2 (2000) 343-355.

Estimating Entropy and Entropy Norm
on Data Streams

Amit Chakrabarti[1,*], Khanh Do Ba[1,**], and S. Muthukrishnan[2,***]

[1] Department of Computer Science, Dartmouth College, Hanover, NH 03755, USA
[2] Department of Computer Science, Rutgers University, Piscataway, NJ 08854, USA

Abstract. We consider the problem of computing information theoretic functions such as entropy on a data stream, using sublinear space.

Our first result deals with a measure we call the "entropy norm" of an input stream: it is closely related to entropy but is structurally similar to the well-studied notion of frequency moments. We give a polylogarithmic space one-pass algorithm for estimating this norm under certain conditions on the input stream. We also prove a lower bound that rules out such an algorithm if these conditions do not hold.

Our second group of results are for estimating the empirical entropy of an input stream. We first present a sublinear space one-pass algorithm for this problem. For a stream of m items and a given real parameter α, our algorithm uses space $\widetilde{O}(m^{2\alpha})$ and provides an approximation of $1/\alpha$ in the worst case and $(1 + \varepsilon)$ in "most" cases. We then present a two-pass polylogarithmic space $(1+\varepsilon)$-approximation algorithm. All our algorithms are quite simple.

1   Introduction

Algorithms for computational problems on data streams have been the focus of plenty of recent research in several communities, such as theory, databases and networks [1, 5, 2, 12]. In this model of computation, the input is a stream of "items" that is too long to be stored completely in memory, and a typical problem involves computing some statistics on this stream. The main challenge is to design algorithms that are efficient not only in terms of running time, but also in terms of space (i.e., memory usage): sublinear space is mandatory and polylogarithmic space is often the goal.

The seminal paper of Alon, Matias and Szegedy [1] considered the problem of estimating the *frequency moments* of the input stream: if a stream contains m_i occurrences of item i (for $1 \leq i \leq n$), its kth frequency moment is denoted F_k and is defined by $F_k := \sum_{i=1}^{n} m_i^k$. Alon et al. showed that F_k could be estimated arbitrarily well in sublinear space for all nonnegative integers k and

---

   * Supported by an NSF CAREER award and Dartmouth College startup funds.
  ** Work partly done while visiting DIMACS in the REU program, supported by NSF ITR 0220280, DMS 0354600, and a Dean of Faculty Fellowship from Dartmouth College.
*** Supported by NSF ITR 0220280 and DMS 0354600.

B. Durand and W. Thomas (Eds.): STACS 2006, LNCS 3884, pp. 196–205, 2006.

in polylogarithmic (in m and n) space for $k \in \{0, 1, 2\}$. Their algorithmic results were subsequently improved by Coppersmith and Kumar [3] and Indyk and Woodruff [9].

In this work, we first consider a somewhat related statistic of the input stream, inspired by the classic information theoretic notion of entropy. We consider the entropy norm of the input stream, denoted F_H and defined by $F_H :=$ $\sum_{i=1}^{n} m_i \lg m_i$.[1] We prove (see Theorem 2.2) that F_H can be estimated arbitrarily well in polylogarithmic space provided its value is not "too small," a condition that is satisfied if, e.g., the input stream is at least twice as long as the number of distinct items in it. We also prove (see Theorem 2.5) that F_H cannot be estimated well in polylogarithmic space if its value *is* "too small."

Second, we consider the estimation of entropy itself, as opposed to the entropy norm. Any input stream implicitly defines an *empirical* probability distribution on the set of items it contains; the probability of item i being m_i/m, where m is the length of the stream. The *empirical entropy* of the stream, denoted H, is defined to be the entropy of this probability distribution:

$$H := \sum_{i=1}^{n}(m_i/m)\lg(m/m_i) = \lg m - F_H/m. \tag{1}$$

An algorithm that computes F_H exactly clearly suffices to compute H as well. However, since we are only able to approximate F_H in the data stream model, we need new techniques to estimate H. We prove (see Theorem 3.1) that H can be approximated using sublinear space. Although the space usage is not polylogarithmic in general, our algorithm provides a tradeoff between space and approximation factor and can be tuned to use space arbitrarily close to polylogarithmic.

The standard data stream model allows us only one pass over the input. If, however, we are allowed *two passes* over the input but still restricted to small space, we have an algorithm that approximates H to within a $(1+\varepsilon)$ factor and uses polylogarithmic space.

Both entropy and entropy norm are natural statistics to approximate on data streams. Arguably, entropy related measures are even more natural than L_p norms or frequency moments F_k. In addition, they have direct applications. The quintessential need arises in analyzing IP network traffic at packet level on high speed routers. In monitoring IP traffic, one cares about anomalies. In general, anomalies are hard to define and detect since there are subtle intrusions, sophisticated dependence amongst network events and agents gaming the attacks. A number of recent results in the networking community use entropy as an approach [6, 13, 14] to detect sudden changes in the network behavior and as an indicator of anomalous events. The rationale is well explained elsewhere, chiefly in Section 2 of [13]. The current research in this area [13, 6, 14] relies on full space algorithms for entropy calculation; this is a serious bottleneck in high speed routers where high speed memory is at premium. Indeed, this is

---

[1] Throughout this paper "lg" denotes logarithm to the base 2.

the bottleneck that motivated data stream algorithms and their applications to
IP network analysis [5, 12]. Our small-space algorithms can immediately make
entropy estimation at line speed practical on high speed routers.

To the best of our knowledge, our upper and lower bound results for the entropy norm are the first of their kind. Recently Guha, McGregor and Venkatasubramanian [7] considered approximation algorithms for the entropy of a given
distribution under various models, including the data stream model. They ob-
tain a $\left(\frac{e}{e-1} + \varepsilon\right)$-approximation for the entropy H of an input stream provided
H is at least a sufficiently large constant, using space $\widetilde{O}(1/(\varepsilon^2 H))$, where the $\widetilde{O}$-
notation hides factors polylogarithmic in m and n. Our work shows that H can
be $(1 + \varepsilon)$-approximated in $\widetilde{O}(1/\varepsilon^2)$ space for $H \geq 1$ (see the remark after Theorem 3.1); more importantly, we obtain efficient sublinear space approximations
when $H < 1$ (the most challenging case). Our space bounds are independent of
H. Guha et al. also give a *two-pass* $(1+\varepsilon)$-approximation algorithm for entropy,
using $\widetilde{O}(1/(\varepsilon^2 H))$ space. We do the same using only $\widetilde{O}(1/\varepsilon^2)$ space. Finally, Guha
et al. consider the entropy estimation problem in the *random streams model*,
where it is assumed that the items in the input stream are presented in a uniform random order. Under this assumption, they obtain a $(1 + \varepsilon)$-approximation
using $\widetilde{O}(1/\varepsilon^2)$ space. We study adversarial data stream inputs only.

2   Estimating the Entropy Norm

In this section we present a polylogarithmic space $(1 + \varepsilon)$-approximation algorithm for entropy norm that assumes the norm is sufficiently large, and prove a
matching lower bound if the norm is in fact not as large.

2.1   Upper Bound

Our algorithm is inspired by the work of Alon et al. [1]. Their first algorithm,
for the frequency moments F_k, has the following nice structure to it (some of the
terminology is ours). A subroutine computes a *basic estimator*, which is a random
variable X whose mean is exactly the quantity we seek and whose variance is
small. The algorithm itself uses this subroutine to maintain $s_1 s_2$ independent
basic estimators $\{X_{ij} : 1 \leq i \leq s_1, 1 \leq j \leq s_2\}$, where each X_{ij} is distributed
identically to X. It then outputs a *final estimator* Y defined by

$$Y := \operatorname*{median}_{1 \leq j \leq s_2} \left(\frac{1}{s_1} \sum_{i=1}^{s_1} X_{ij} \right)$$

The following lemma, implicit in [1], gives a guarantee on the quality of this final
estimator.

Lemma 2.1. *Let $\mu := E[X]$. For any $\varepsilon, \delta \in (0, 1]$, if $s_1 \geq 8 \operatorname{Var}[X]/(\varepsilon^2 \mu^2)$ and
$s_2 = 4 \lg(1/\delta)$, then the above final estimator deviates from μ by no more than
$\varepsilon \mu$ with probability at least $1 - \delta$. The above algorithm can be implemented to use*

space $O(S \log(1/\delta) \operatorname{Var}[X]/(\varepsilon^2 \mu^2))$, *provided the basic estimator can be computed using space at most* S.

Proof. The claim about the space usage is immediate from the structure of the algorithm. Let $Y_j = \frac{1}{s_1} \sum_{i=1}^{s_1} X_{ij}$. Then $\operatorname{E}[Y_j] = \mu$ and $\operatorname{Var}[Y_j] = \operatorname{Var}[X]/s_1 \leq \varepsilon^2 \mu^2/8$. Applying Chebyshev's Inequality gives us

$$\Pr[|Y_j - \mu| \geq \varepsilon\mu] \leq 1/8.$$

Now, if fewer than $(s_2/2)$ of the Y_j's deviate by as much as $\varepsilon\mu$ from μ, then Y must be within $\varepsilon\mu$ of μ. So we upper bound the probability that this does not happen. Define s_2 indicator random variables I_j, where $I_j = 1$ iff $|Y_j - \mu| \geq \varepsilon\mu$, and let $W = \sum_{j=1}^{s_2} I_j$. Then $\operatorname{E}[W] \leq s_2/8$. A standard Chernoff bound (see, e.g. [11, Theorem 4.1]) gives

$$\Pr\left[|Y - \mu| \geq \varepsilon\mu\right] \leq \Pr\left[W \geq \frac{s_2}{2}\right] \leq \left(\frac{e^3}{4^4}\right)^{s_2/8} = \left(\frac{e^3}{4^4}\right)^{\frac{1}{2}\lg(1/\delta)} \leq \delta,$$

which completes the proof. □

We use the following subroutine to compute a basic estimator X for the entropy norm F_H.

Input stream: $A = \langle a_1, a_2, \ldots, a_m \rangle$, where each
$a_i \in \{1, \ldots, n\}$.

1 Choose p uniformly at random from $\{1, \ldots, m\}$.
2 Let $r = |\{q : a_q = a_p, p \leq q \leq m\}|$. Note that $r \geq 1$.
3 Let $X = m(r\lg r - (r-1)\lg(r-1))$, with the convention that
$0\lg 0 = 0$.

Our algorithm for estimating the entropy norm outputs a final estimator based on this basic estimator, as described above. This gives us the following theorem.

Theorem 2.2. *For any* $\Delta > 0$, *if* $F_H \geq m/\Delta$, *the above one-pass algorithm can be implemented so that its output deviates from* F_H *by no more than* εF_H *with probability at least* $1 - \delta$, *and so that it uses space*

$$O\left(\frac{\log(1/\delta)}{\varepsilon^2} \log m(\log m + \log n)\Delta\right).$$

In particular, taking Δ *to be a constant, we have a polylogarithmic space algorithm that works on streams whose* F_H *is not "too small."*

Proof. We first check that the expected value of X is indeed the desired quantity:

$$\operatorname{E}[X] = \frac{m}{m} \sum_{v=1}^{n} \sum_{r=1}^{m_v} (r\lg r - (r-1)\lg(r-1))$$

$$= \sum_{v=1}^{n} (m_v \lg m_v - 0\lg 0) = F_H.$$

The approximation guarantee of the algorithm now follows from Lemma 2.1. To bound the space usage, we must bound the variance $\mathrm{Var}[X]$ and for this we bound $E[X^2]$. Let $f(r) := r \lg r$, with $f(0) := 0$, so that X can be expressed as $X = m(f(r) - f(r-1))$. Then

$$E[X^2] = m \sum_{v=1}^{n} \sum_{r=1}^{m_v} \left(f(r) - f(r-1) \right)^2$$

$$\leq m \cdot \max_{1 \leq r \leq m} \left(f(r) - f(r-1) \right) \cdot \sum_{v=1}^{n} \sum_{r=1}^{m_v} \left(f(r) - f(r-1) \right)$$

$$\leq m \cdot \sup \left\{ f'(x) : x \in (0, m] \right\} \cdot F_H \tag{2}$$

$$= (\lg e + \lg m) \, m F_H \tag{3}$$

$$\leq (\lg e + \lg m) \, \Delta F_H^2 \,,$$

where (2) follows from the Mean Value Theorem.

Thus, $\mathrm{Var}[X]/E[X]^2 = O(\Delta \lg m)$. Moreover, the basic estimator can be implemented using space $O(\log m + \log n)$: $O(\log m)$ to count m and r, and $O(\log n)$ to store the value of a_p. Plugging these bounds into Lemma 2.1 yields the claimed upper bound on the space of our algorithm.

Let F_0 denote the number of distinct items in the input stream (this notation deliberately coincides with that for frequency moments). Let $f(x) := x \lg x$ as used in the proof above. Observe that f is convex on $(0, \infty)$ whence, via Jensen's inequality, we obtain

$$F_H = \frac{F_0}{F_0} \sum_{v=1}^{n} f(m_v) \geq F_0 f\left(\frac{1}{F_0} \sum_{v=1}^{n} m_v \right) = m \lg \frac{m}{F_0} \,. \tag{4}$$

Thus, if the input stream satisfies $m \geq 2F_0$ (or the simpler, but stronger, condition $m \geq 2n$), then we have $F_H \geq m$. As a direct corollary of Theorem 2.2 (for $\Delta = 1$) we obtain a $(1 + \varepsilon)$-approximation algorithm for the entropy norm in space $O((\log(1/\delta)/\varepsilon^2) \log m (\log m + \log n))$. However, we can do slightly better.

Theorem 2.3. *If $m \geq 2F_0$ then the above one-pass, $(1 + \varepsilon)$-approximation algorithm can be implemented in space*

$$O\left(\frac{\log(1/\delta)}{\varepsilon^2} \log m \log n \right)$$

without a priori knowledge of the stream length m.

Proof. We follow the proof of Theorem 2.2 up to the bound (3) to obtain $\mathrm{Var}[X] \leq (2 \lg m) m F_H$, for m large enough. We now make the following claim

$$\frac{\lg m}{\lg(m/F_0)} \leq 2 \max\{\lg F_0, 1\} \,. \tag{5}$$

Assuming the truth of this claim and using (4), we obtain

$$\mathrm{Var}[X] \leq (2\lg m)mF_H \leq \frac{2\lg m}{\lg(m/F_0)}F_H^2 \leq 4\max\{\lg F_0, 1\}F_H^2 \leq (4\lg n)F_H^2\,.$$

Plugging this into Lemma 2.1 and proceeding as before, we obtain the desired space upper bound. Note that we no longer need to know m before starting the algorithm, because the number of basic estimators used by the algorithm is now independent of m. Although maintaining each basic estimator seems, at first, to require prior knowledge of m, a careful implementation can avoid this, as shown by Alon et al [1].

We turn to proving our claim (5). We will need the assumption $m \geq 2F_0$. If $m \leq F_0^2$, then $\lg m \leq 2\lg F_0 = 2\lg F_0 \lg(2F_0/F_0) \leq 2\lg F_0 \lg(m/F_0)$ and we are done. On the other hand, if $m \geq F_0^2$, then $F_0 \leq m^{1/2}$ so that $\lg(m/F_0) \geq \lg m - (1/2)\lg m = (1/2)\lg m$ and we are done as well.

Remark 2.4. Theorem 2.2 generalizes to estimating quantities of the form $\hat{\mu} = \sum_{v=1}^{n} \hat{f}(m_v)$, for any monotone increasing (on integer values), differentiable function $\hat{f}$ that satisfies $\hat{f}(0) = 0$. Assuming $\hat{\mu} \geq m/\Delta$, it gives us a one-pass $(1+\varepsilon)$-approximation algorithm that uses $\widetilde{O}(\hat{f}'(m)\Delta)$ space. For instance, this space usage is polylogarithmic in m if $\hat{f}(x) = x\,\mathrm{polylog}(x)$.

2.2 Lower Bound

The following lower bound shows that the algorithm of Theorem 2.2 is optimal, up to factors polylogarithmic in m and n.

Theorem 2.5. *Suppose Δ and c are integers with $4 \leq \Delta \leq o(m)$ and $0 \leq c \leq m/\Delta$. On input streams of size at most m, a randomized algorithm able to distinguish between $F_H \leq 2c$ and $F_H \geq c+2m/\Delta$ must use space at least $\Omega(\Delta)$. In particular, the upper bound in Theorem 2.2 is tight in its dependence on Δ.*

Proof. We present a reduction from the classic problem of (two-party) Set Disjointness in communication complexity [10].

Suppose Alice has a subset X and Bob a subset Y of $\{1, 2, \ldots, \Delta - 1\}$, such that X and Y either are disjoint or intersect at exactly one point. Let us define the mapping

$$\phi : x \longmapsto \left\{\frac{(m-2c)x}{\Delta} + i : i \in \mathbb{Z},\, 0 \leq i < \frac{m-2c}{\Delta}\right\}\,.$$

Alice creates a stream A by listing all elements in $\bigcup_{x \in X} \phi(x)$ and concatenating the c special elements $\Delta + 1, \ldots, \Delta + c$. Similarly, Bob creates a stream B by listing all elements in $\bigcup_{y \in Y} \phi(y)$ and concatenating the same c special elements $\Delta + 1, \ldots, \Delta + c$. Now, Alice can process her stream (with the hypothetical entropy norm estimation algorithm) and send over her memory contents to Bob, who can then finish the processing. Note that the length of the combined stream $A \circ B$ is at most $2c + |X \cup Y| \cdot ((m-2c)/\Delta) \leq m$.

We now show that, based on the output of the algorithm, Alice and Bob can tell whether or not X and Y intersect. Since the set disjointness problem has communication complexity $\Omega(\Delta)$, we get the desired space lower bound.

Suppose X and Y are disjoint. Then the items in $A \circ B$ are all distinct except for the c special elements, which appear twice each. So $F_H(A \circ B) = c \cdot (2 \lg 2) = 2c$. Now suppose $X \cap Y = \{z\}$. Then the items in $A \circ B$ are all distinct except for the $(m - 2c)/\Delta$ elements in $\phi(z)$ and the c special elements, each of which appears twice. So $F_H(A \circ B) = 2(c + (m - 2c)/\Delta) \geq c + 2m/\Delta$, since $\Delta \geq 4$.

Remark 2.6. Notice that the above theorem rules out even a polylogarithmic space *constant factor* approximation to F_H that can work on streams with "small" F_H. This can be seen by setting $\Delta = m^\gamma$ for some constant $\gamma > 0$.

3   Estimating the Empirical Entropy

We now turn to the estimation of the empirical entropy H of a data stream, defined as in equation (1): $H = \sum_{i=1}^{n}(m_i/m) \lg(m/m_i)$. Although H can be computed exactly from F_H, as shown in (1), a $(1 + \varepsilon)$-approximation of F_H can yield a poor estimate of H when H is small (sublinear in its maximum value, $\lg m$). We therefore present a different sublinear space, one-pass algorithm that directly computes entropy.

Our data structure takes a user parameter $\alpha > 0$, and consists of three components. The first (A1) is a sketch in the manner of Section 2, with basic estimator

$$X = m \left(\frac{r}{m} \lg \frac{m}{r} - \frac{r-1}{m} \lg \frac{m}{r-1} \right),$$ (6)

and a final estimator derived from this basic estimator using $s_1 = (8/\varepsilon^2)m^{2\alpha}$ $\lg^2 m$ and $s_2 = 4 \lg(1/\delta)$. The second component (A2) is an array of $m^{2\alpha}$ counters (each counting from 1 to m) used to keep exact counts of the first $m^{2\alpha}$ distinct items seen in the input stream. The third component (A3) is a Count-Min Sketch, as described by Cormode and Muthukrishnan [4], which we use to estimate k, defined to be the number of items in the stream that are *different* from the most frequent item; i.e., $k = m - \max\{m_i : 1 \leq i \leq n\}$. The algorithm itself works as follows. Recall that F_0 denotes the number of distinct items in the stream.

---

1 Maintain A1, A2, A3 as described above. When queried (or at end of input):

2 **if** $F_0 \leq m^{2\alpha}$ **then return** exact H from A2.

3 **else**

4   let $\hat{k}$ = estimate of k from A3.

5   **if** $\hat{k} \geq (1 - \varepsilon)m^{1-\alpha}$ **then return** final estimator, Y, of A1.

6   **else return** $(\hat{k} \lg m)/m$.

7 **end**

Theorem 3.1. *The above algorithm uses*

$$O\left(\frac{\log(1/\delta)}{\varepsilon^2}m^{2\alpha}\log^2 m\left(\log m + \log n\right)\right)$$

space and outputs a random variable Z that satisfies the following properties.

1. *If $k \leq m^{2\alpha} - 1$, then $Z = H$.*
2. *If $k \geq m^{1-\alpha}$, then $\Pr\left[|Z - H| \geq \varepsilon H\right] \leq \delta$.*
3. *Otherwise (i.e., if $m^{2\alpha} \leq k < m^{1-\alpha}$), Z is a $(1/\alpha)$-approximation of H.*

Remark 3.2. Under the assumption $H \geq 1$, an algorithm that uses only the basic estimator in A1 and sets $s_1 = (8/\varepsilon^2)\lg^2 m$ suffices to give a $(1+\varepsilon)$-approximation in $O(\varepsilon^{-2}\log^2 m)$ space.

Proof. The space bound is clear from the specifications of A1, A2 and A3, and Lemma 2.1. We now prove the three claimed properties of Z in sequence.

PROPERTY 1: This follows directly from the fact that $F_0 \leq k + 1$.

PROPERTY 2: The Count-Min sketch guarantees that $\hat{k} \leq k$ and, with probability at least $1 - \delta$, $\hat{k} \geq (1 - \varepsilon)k$. The condition in Property 2 therefore implies that $\hat{k} \geq (1 - \varepsilon)m^{1-\alpha}$, that is, $Z = Y$, with probability at least $1 - \delta$. Here we need the following lemma.

Lemma 3.3. *Given that the most frequent item in the input stream A has count $m - k$, the minimum entropy $H_{\min}$ is achieved when all the remaining k items are identical, and the maximum $H_{\max}$ is achieved when they are all distinct. Therefore,*

$$H_{\min} = \frac{m-k}{m}\lg\frac{m}{m-k} + \frac{k}{m}\lg\frac{m}{k}, \quad and$$

$$H_{\max} = \frac{m-k}{m}\lg\frac{m}{m-k} + \frac{k}{m}\lg m.$$

Proof. Consider a minimum-entropy stream $A_{\min}$ and suppose that, apart from its most frequent item, it has at least two other items with positive count. Without loss of generality, let $m_1 = m - k$ and $m_2, m_3 \geq 1$. Modify $A_{\min}$ to A' by letting $m_2' = m_2 + m_3$ and $m_3' = 0$, and keeping all other counts the same. Then

$$\begin{aligned}
H(A') - H(A_{\min}) &= (\lg m - F_H(A')/m) - (\lg m - F_H(A_{\min})/m) \\
&= (F_H(A_{\min}) - F_H(A'))/m \\
&= m_2\lg m_2 + m_3\lg m_3 - (m_2 + m_3)\lg(m_2 + m_3) \\
&< 0,
\end{aligned}$$

since $x\lg x$ is convex and monotone increasing (on integer values), giving us a contradiction. The proof of the maximum-entropy distribution is similar.

Now, consider equation (6) and note that for any r, $|X| \leq \lg m$. Thus, if $E[X] = H \geq 1$, then $\mathrm{Var}[X]/E[X]^2 \leq E[X^2] \leq \lg^2 m$ and our choice of s_1 is sufficiently large to give us the desired $(1+\varepsilon)$-approximation, by Lemma 2.1.[2] On the other hand, if $H < 1$, then $k < m/2$, by a simple argument similar to the proof of Lemma 3.3. Using the expression for $H_{\min}$ from Lemma 3.3, we then have

$$H_{\min} = \lg \frac{m}{m-k} + \frac{k}{m} \lg \frac{m-k}{k} \geq -\lg\left(1 - \frac{k}{m}\right) \geq \frac{k}{m} \geq m^{-\alpha},$$

which gives us $\mathrm{Var}[X]/E[X]^2 \leq E[X^2]/m^{-2\alpha} \leq (\lg^2 m)m^{2\alpha}$. Again, plugging this and our choice of s_1 into Lemma 2.1 gives us the desired $(1+\varepsilon)$-approximation.

PROPERTY 3: By assumption, $k < m^{1-\alpha}$. If $\hat{k} \geq (1-\varepsilon)m^{1-\alpha}$, then $Z = Y$ and the analysis proceeds as for Property 2. Otherwise, $Z = (\hat{k} \lg m)/m \leq (k \lg m)/m$. This time, again by Lemma 3.3, we have

$$H_{\min} \geq \frac{k}{m} \lg \frac{m}{k} \geq \frac{k}{m} \lg(m^{\alpha}) = \frac{\alpha k}{m} \lg m,$$

and

$$\begin{aligned} H_{\max} &= \frac{m-k}{m} \lg \frac{m}{m-k} + \frac{k}{m} \lg m \\ &= \lg \frac{m}{m-k} + \frac{k}{m} \lg(m-k) \\ &\leq \frac{k}{m} \lg m + O\left(\frac{k}{m}\right), \end{aligned}$$

which, for large m, implies $H - o(H) \leq Z \leq H/\alpha$ and gives us Property 3. Note that we did not use the inequality $m^{2\alpha} \leq k$ in the proof of this property.

The ideas involved in the proof of Theorem 3.1 can be used to yield a very efficient *two-pass* algorithm for estimating H, the details of which will be provided in the full version of the paper.

4   Conclusions

Entropy and entropy norms are natural measures with direct applications in IP network traffic analysis for which one-pass streaming algorithms are needed. We have presented one-pass sublinear space algorithms for approximating the entropy norms as well as the empirical entropy. We have also presented a two-pass algorithm for empirical entropy that has a stronger approximation guarantee and space bound. We believe our algorithms will be of interest in practice of data stream systems. It will be of interest to study these problems on streams in the presence of inserts and deletes. **Note:** Very recently, we have learned of a work in progress [8] that may lead to a one-pass polylogarithmic space algorithm for approximating H to within a $(1+\varepsilon)$-factor.

---

[2] This observation, that $H \geq 1 \implies \mathrm{Var}[X] \leq \lg^2 m$, proves the statement in the remark following Theorem 3.1.

References

1. N. Alon, Y. Matias and M. Szegedy. The space complexity of approximating the frequency moments. *Proc. ACM STOC*, 20–29, 1996.
2. B. Babcock, S. Babu, M. Datar, R. Motwani and J. Widom. Models and issues in data stream systems. *ACM PODS, 2002*, 1–16
3. D. Coppersmith and R. Kumar. An improved data stream algorithm for frequency moments. *ACM-SIAM SODA*, 151–156, 2004.
4. G. Cormode and S. Muthukrishnan. An improved data stream summary: the count-min sketch and its applications. *J. Algorithms*, 55(1): 58–75, April 2005.
5. C. Estan and G. Varghese. New directions in traffic measurement and accounting: Focusing on the elephants, ignoring the mice. *ACM Trans. Comput. Syst.*, 21(3): 270–313, 2003.
6. Y. Gu, A. McCallum and D. Towsley. Detecting Anomalies in Network Traffic Using Maximum Entropy Estimation. *Proc. Internet Measurement Conference*, 2005.
7. S. Guha, A. McGregor, and S. Venkatasubramanian. Streaming and Sublinear Approximation of Entropy and Information Distances. *ACM-SIAM SODA*, to appear, 2006.
8. P. Indyk. Personal e-mail communication. September 2005.
9. P. Indyk and D. Woodruff. Optimal approximations of the frequency moments of data streams. *ACM STOC*, 202–208, 2005.
10. E. Kushilevitz and N. Nisan. *Communication Complexity*. Cambridge University Press, Cambridge, 1997.
11. R. Motwani and P. Raghavan. *Randomized Algorithms*. Cambridge University Press, New York, 1995.
12. S. Muthukrishnan. Data Streams: Algorithms and Applications. *Manuscript*, Available online at http://www.cs.rutgers.edu/~muthu/stream-1-1.ps
13. A. Wagner and B. Plattner Entropy Based Worm and Anomaly Detection in Fast IP Networks. *14th IEEE International Workshops on Enabling Technologies: Infrastructures for Collaborative Enterprises* (WET ICE), STCA security workshop, Linkping, Sweden, June, 2005
14. K. Xu, Z. Zhang, and S. Bhattacharya. Profiling Internet Backbone Traffic: Behavior Models and Applications. *Proc. ACM SIGCOMM* 2005.

Pay Today for a Rainy Day: Improved Approximation Algorithms for Demand-Robust Min-Cut and Shortest Path Problems

Daniel Golovin[1,*], Vineet Goyal[2,**], and R. Ravi[2,**]

[1] Computer Science Department, Carnegie Mellon University, Pittsburgh PA 15213, USA
dgolovin@cs.cmu.edu
[2] Tepper School of Business, Carnegie Mellon University, Pittsburgh PA 15213, USA
{vgoyal, ravi}@andrew.cmu.edu

Abstract. Demand-robust versions of common optimization problems were recently introduced by Dhamdhere et al. [4] motivated by the worst-case considerations of two-stage stochastic optimization models. We study the demand robust min-cut and shortest path problems, and exploit the nature of the robust objective to give improved approximation factors. Specifically, we give a $(1 + \sqrt{2})$ approximation for robust min-cut and a 7.1 approximation for robust shortest path. Previously, the best approximation factors were $O(\log n)$ for robust min-cut and 16 for robust shortest paths, both due to Dhamdhere et al. [4].

Our main technique can be summarized as follows: We investigate each of the second stage scenarios individually, checking if it can be independently serviced in the second stage within an acceptable cost (namely, a guess of the optimal second stage costs). For the costly scenarios that cannot be serviced in this way ("rainy days"), we show that they can be fully taken care of in a near-optimal first stage solution (i.e., by "paying today").

We also consider "hitting-set" extensions of the robust min-cut and shortest path problems and show that our techniques can be combined with algorithms for Steiner multicut and group Steiner tree problems to give similar approximation guarantees for the hitting-set versions of robust min-cut and shortest path problems respectively.

1 Introduction

Robust optimization has been widely studied to deal with the data uncertainty in optimization problems. In a classical optimization problem, all parameters such as costs and demands are assumed to be precisely known. A small change in these parameters can change the optimal solution considerably. As a result, classical optimization is ineffective in those real life applications where robustness to uncertainty is desirable.

Traditional approaches toward robustness have focused on uncertainty in data [3,12,13]. In a typical data-robust model, uncertainty is modeled as a finite set of scenarios, where a scenario is a plausible set of values for the data in the model. The objective is to find a feasible solution to the problem which is "good" in all or most scenarios, where various notions of "goodness" have been studied. Some of them include

* Supported in part by NSF ITR grants CCR-0122581 and IIS-0121678.
** Supported in part by NSF grant CCF-0430751 and ITR grant CCR-0122581.

B. Durand and W. Thomas (Eds.): STACS 2006, LNCS 3884, pp. 206–217, 2006.

1. Absolute Robustness (min-max): The objective is to find a solution such that the maximum cost over all scenarios is minimized.
2. Robust Deviation (min-max regret): For a given solution, regret in a particular scenario is the difference between cost of this solution in that scenario and the optimal cost in that scenario. In the robust deviation criteria, the objective is to minimize the maximum regret over all scenarios.

More recent attempts at capturing the concept of robust solutions in optimization problems include the work of Rosenblatt and Lee [19] in the facility design problem, and Mulvey *et al.* [14] in mathematical programming. Even more recently, an approach along similar lines has been advocated by Bertsimas *et al.* [1,2]. Other related works in the data-robust models include heuristics such as branch and bound and surrogate relaxation for efficiently solving the data-robust instances. A research monograph by Kouvelis and Yu [11] summarizes this line of work. An annotated bibliography available online is a good source of references for work in data-robustness [16].

Most of the prior work addresses the problem of robustness under data uncertainty. In this paper, we consider a model which also allows uncertainty in the problem constraints along with the uncertainty in data. We call this model of robustness as *demand-robust* model since it attempts to be robust with respect to problem demands (constraints). Our model is motivated by the recent work in two-stage stochastic programming problems with recourse [7,5,9,17,20]. In a two-stage stochastic approach, the goal is to find a solution that minimizes the expected cost over all possible scenarios. While the expected value minimization is reasonable in a repeated decision-making framework, one shortcoming of this approach is that it does not sufficiently guard against the worst case over all the possible scenarios. Our demand-robust model for such problems is a natural way to overcome this shortcoming by postulating a model that minimizes this worst-case cost.

Let us introduce the new model with the demand-robust min-cut problem: Given an undirected graph $G = (V, E)$, a root vertex r and costs c on the edges. The uncertainty in demand and costs is modeled as a finite set of scenarios, one of which materializes in the second stage. The ith scenario is a singleton set containing only the node t_i. We call the nodes specified by the scenarios *terminals*. An edge costs $c(e)$ in the first stage and $\sigma_i \cdot c(e)$ in the recourse (second) stage if the ith scenario is realized. The problem is to find a set $E_0 \subseteq E$ (edges to be bought in the first stage) and for each scenario i, a set $E_i \subseteq E$ (edges to be bought in the recourse stage if scenario i is realized), such that removing $E_0 \cup E_i$ from the graph G disconnects r from the terminal t_i. The objective is to minimize the cost function $\max_i \{c(E_0) + \sigma_i \cdot c(E_i)\}$.

Note that in the above model, each scenario has a different requirement (in scenario i, t_i is required to be separated from r). Such a scenario model allows to handle uncertainty in problem constraints. Another point of difference with the previous data-robust models is that the demand-robust model is two-stage i.e. solution is bought partially in first stage and is then augmented to a feasible solution in the second stage after the uncertainty is realized. However, cost uncertainty in the our demand-robust model is restrictive, as each element becomes costlier by the same factor in a particular scenario in the second stage unlike the data-robust models which handle general cost uncertainties.

1.1 Our Contributions

In this paper we consider the shortest path and min-cut problems in the two-stage demand-robust model. In a recent paper Dhamdhere *et al.* [4] introduced the model of demand robustness and gave approximation algorithms for various problems such as min-cut, multicut, shortest path, Steiner tree and facility location in the framework of two-stage demand-robustness. They use rounding techniques recently developed for stochastic optimization problems [7,8,17] for many of their results and obtain similar guarantees for the demand-robust versions of the problem. In this paper we crucially exploit and benefit from the structure of the demand-robust problem: *in the second stage, every scenario can pay up to the maximum second stage cost without worsening the solution cost*. This is not true for the stochastic versions where the objective is to minimize the expected cost over all scenarios. At a very high level, the algorithms for the problems considered in this paper are as follows: Guess the maximum second stage cost C in some optimal solution. Using this guess identify scenarios which do not need any first stage "help" i.e. scenarios for which the best solution costs at most a constant times C in the second stage. Such scenarios can be ignored while building the first stage solution. For the remaining scenarios or a subset of them, we build a low-cost first stage solution and prove the constant bounds by a charging argument.

We give the first constant factor approximation for the demand-robust min-cut problem. The charging argument leading to a constant factor argument crucially uses the laminarity of minimum cuts separating a given root node from other terminals. The previous best approximation factor was $O(\log n)$ due to Dhamdhere *et al.* [4].

Theorem 1.1. *There is a polynomial time algorithm which gives a* $(1 + \sqrt{2})$ *approximation for the robust min-cut problem.*

For the demand-robust shortest path problem, we give an algorithm with an improved approximation factor of 7.1 as compared to the previous 16-approximation [4].

Theorem 1.2. *There is a polynomial time algorithm which gives a 7.1 approximation for the robust shortest path problem.*

Demand-robust shortest path generalizes the Steiner tree problem and is thus **NP**-hard. The complexity of demand-robust min-cut is still open. However, in section 4 we present **NP**-hard generalizations of both problems, together with approximation algorithms for them. In particular, we consider "hitting set" versions of demand-robust min-cut and shortest path problems where each scenario is a set of terminals instead of a single terminal and the requirement is to satisfy at least one terminal (separate from the root for the min-cut problem and connect to the root for the shortest path problem) in each scenario. We obtain approximation algorithms for these "hitting set" variants by relating them to two classical problems, namely Steiner multicut and group Steiner tree.

2 Robust Min-Cut

In this section, we present a constant factor approximation for this problem. To motivate our approach, let us consider the robust min-cut problem on trees. Suppose we know the maximum cost that some optimal solution pays in the second stage (say C). Any terminal t_i whose min-cut from r costs more than $\frac{C}{\sigma_i}$ should be cut away from r in the first stage. Thus, if we know C, we can identify exactly which terminals U should be cut in the first stage. The remaining terminals pay at most C to buy a cut in the second stage. If there are k scenarios, then there are only $k + 1$ choices for C that matter, as there are only $k + 1$ possible sets that U could be. Though we may not be able to guess C, we can try all possible values of U and find the best solution. This algorithm solves the problem exactly on trees.

The algorithm for general graphs has a similar flavor. In a general graph if for any terminal the minimum r-t_i cut costs more than $\frac{C}{\sigma_i}$, then we can only infer that the first stage should "help" this terminal i.e. buy some edges from a r-t_i cut. In the case of trees, every minimal r-t_i cut is a single edge, so the first stage cuts t_i from the root. However, this is not true for general graphs. But we prove that a similar algorithm gives a constant factor approximation using a charging argument. As in the algorithm for trees, we reduce the needed non-determinism by guessing a set of terminals rather than C itself.

Algorithm for Robust Min-Cut

$T = \{t_1, t_2, \ldots, t_k\}$ are the terminals, $r \leftarrow$ root .
$\alpha \leftarrow (1 + \sqrt{2})$.

1. For each terminal t_i, compute the cost (with respect to c) of a minimum r-t_i cut, denoted $\mathrm{mcut}(t_i)$.
2. Let C be the maximum second stage cost of some optimal solution.
   Guess $U := \{t_i : \sigma_i \cdot \mathrm{mcut}(t_i) > \alpha \cdot C\}$.
3. First stage solution: $E_0 \leftarrow$ minimum r-U cut.
4. Second stage solution for scenario i: $E_i \leftarrow$ any minimum r-t_i cut in $G \setminus E_0$.

If we relabel the scenarios in decreasing order of $\sigma_i \cdot \mathrm{mcut}(t_i)$, then for every choice of C, $U = \emptyset$ or $U = \{t_1, t_2, \ldots, t_j\}$ for some $j \in \{1, 2, \ldots, k\}$. Thus, we need to try only $k + 1$ values for C. This algorithm runs in $\tilde{O}(k^2 mn)$ time on undirected graphs using the max flow algorithm of Goldberg and Tarjan [6] to find min cuts. The above algorithm $(1 + \sqrt{2})$-approximates the robust min-cut problem.

Proof of Theorem 1.1. Let OPT be an optimal solution, let E_0^* be the edge set it buys in stage one, and let C_0^* and C be the amount it pays in the first and second stage, respectively. Let α be a constant to be specified later, and let $U := \{t_i : \sigma_i \cdot \mathrm{mcut}(t_i) > \alpha \cdot C\}$, where $\mathrm{mcut}(t_i)$ is the cost of minimum r-t_i cut in G with respect to the cost function c. Note that we can handle every terminal $t_i \notin U$ by paying at most αC in the second stage. We will prove that the first stage solution E_0, given by the algorithm has cost $c(E_0) \leq (1 + \frac{2}{\alpha - 1})C_0^*$. The output solution is thus a $\max\{\alpha, (1 + \frac{2}{\alpha - 1})\}$-approximation. Setting $\alpha := (1 + \sqrt{2})$ then yields the claimed approximation ratio.

To show $c(E_0) \leq (1 + \frac{2}{\alpha - 1})C_0^*$, we exhibit an r-U cut of cost at most $(1 + \frac{2}{\alpha - 1})C_0^*$. Recall that OPT buys E_0^* in the first stage. Since $\sigma_i \cdot \mathrm{mcut}(t_i) > C$ for all $t_i \in U$, E_0 must "help" each such t_i reduce its second stage cost by a large fraction. The high level idea is as follows: we show how to group terminals of U into equivalence classes such that each edge of E_0^* helps at most two such classes and then cut away each equivalence class from the root using a cut that can be charged to its portion of E_0^*.

Formally, let $G = (V, E)$ be our input. Define $G' := (V, E \setminus E_0^*)$. The goal is to construct a low-cost r-U cut, C. We include E_0^* in C. This allows us to ignore terminals that E_0^* separates from the root. U is the set of remaining terminals with $\sigma_i \cdot \mathrm{mcut}(t_i) > \alpha \cdot C$. For a terminal $t \in U$, let $Q_t \subset V$ be the t side of some minimum r-t cut in G'. Lemma 2.1 proves that there exist min cuts such that $\mathcal{F} := \{Q_t : t \in U\}$ is a laminar family (see figure 1). Let F be all the node-maximal elements of $\mathcal{F}$, that is, $F = \{Q \in \mathcal{F} : \forall Q' \in \mathcal{F}, \text{ either } Q' \subseteq Q, \text{ or } Q' \cap Q = \phi\}$. For $Q \in F$, we say Q *uses* edges $\{(u, v) \in E_0^* \mid Q \cap \{u, v\} \neq \emptyset\}$. Since $\mathcal{F}$ is laminar, all the sets in F are disjoint. It follows that each edge $e \in E_0^*$ can be used by at most two sets of F. For each $Q \in F$, we include the edges of G' incident to Q in the cut C, and charge it to the edges of E_0^* it uses as follows:

For a graph $G = (V, E)$ and $Q \subset V$, let $\delta_G(Q) := \{(q, w) \mid q \in Q, w \in V \setminus Q\} \cap E$ be the boundary of Q in graph G. Fix $Q_{t_i} \in F$, let $X = \delta_G(Q_{t_i}) \cap E_0^*$ (edges that Q_{t_i} uses) and let $Y = \delta_G(Q_{t_i}) \setminus E_0^*$ (edges of G' incident to Q_{t_i}). Since $\delta_G(Q_{t_i})$ is a r-t_i cut in G,

$$c(\delta_G(Q_{t_i})) = c(X) + c(Y) \geq \mathrm{mcut}(t_i) \tag{2.1}$$

Since $t_i \in U$, $\sigma_i \cdot \mathrm{mcut}(t_i) > \alpha \cdot C$ so with (2.1) we have $(c(X) + c(Y)) > \frac{\alpha \cdot C}{\sigma_i}$. Also, we know that OPT pays at most C in second stage costs for any scenario which implies $\sigma_i \cdot c(Y) \leq C$. Thus, we have $c(Y) < \frac{1}{\alpha - 1} c(X)$ Thus we can pay for $c(Y)$ by charging it to the cost of $X \subseteq E_0^*$ and incurring an overhead of $1/(\alpha - 1)$ on the charged edges. Since each edge in E_0^* is charged at most twice, the total charge to buy all edges in $\bigcup_{Q \in F}(\delta_G(Q) \setminus E_0^*)$ is at most $\frac{2c(E_0^*)}{\alpha - 1} = \frac{2C_0^*}{\alpha - 1}$. Thus, a minimum r-U cut costs at most $(1 + \frac{2}{\alpha - 1})C_0^*$. $\square$

Lemma 2.1. *Let U, Q_t be defined as in the proof of Theorem 1.1 Then there exists a minimum r-t cut in G' for each terminal $t \in U$ such that $\mathcal{F} := \{Q_t : t \in U\}$ is a laminar family.*

Proof. We start with minimally sized sets Q_t. That is, for each $t \in U$, Q_t is the t side of a minimum r-t cut in G', and every vertex set Q' containing t but not the root such that $|Q'| < |Q_t|$ satisfies $c(\delta_{G'}(Q')) > c(\delta_{G'}(Q_t))$. We claim this family is laminar. Suppose not, then there exists $A := Q_a$, $B := Q_b$, $a, b \in U$ that violate the laminar property. Thus, $A \cap B \neq \emptyset$, $A \not\subseteq B$, and $B \not\subseteq A$.

Case 1: $a \in A \setminus B$, $b \in B \setminus A$. Let $X := A \cap B$, $A' := A \setminus X$, and $B' := B \setminus X$. Note the cut capacity function of G', defined $f(Q) := c(\delta_{G'}(Q))$, is submodular. We claim that $f(A') \leq f(A)$ or $f(B') \leq f(B)$, contradicting the minimality of A and B. Let $c(V_1, V_2)$ denote the sum of costs of edges from V_1 to V_2 in G', where $V_1, V_2 \subseteq V$. Then

$$f(A) < f(A') \implies c(X, B') + c(X, (V \setminus (A \cup B))) < c(A', X) \tag{2.2}$$
$$f(B) < f(B') \implies c(X, A') + c(X, (V \setminus (A \cup B))) < c(B', X) \tag{2.3}$$

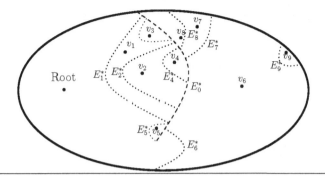

Fig. 1. Once the edges bought in the first stage, E_0^*, are fixed, there exists an optimal (w.r.t. E_0^*) second stage solution $\{E_i^* \mid i = 1, 2, \ldots, k\}$ such that the t_i sides of the cuts $\{E_0^* \cup E_i^*\}$ are a laminar family. Here, the labeled vertices are all terminals, the dashed contour corresponds to E_0^*, and the dotted contours correspond to second stage edge sets for various terminals. The node-maximal elements of this family are the terminal side cuts for $v_1, v_5,$ and v_6.

Adding inequalities (2.2) and (2.3), we get $c(X, (V \setminus (A \cup B))) < 0$, which is clearly impossible.

Case 2: $a \in B$ (equivalently, $b \in A$). Since A and B are terminal sides of min-cuts,

$$\max\{f(A), f(B)\} \leq f(A \cup B) \tag{2.4}$$

$$f(A \cap B) + f(A \cup B) \leq f(A) + f(B) \tag{2.5}$$

where (2.5) follows from submodularity. Inequalities (2.4) and (2.5) together imply $f(A \cap B) \leq \min\{f(A), f(B)\}$. But $f(A \cap B) \leq f(A)$ contradicts the minimality of A. $\qquad\qquad\square$

3   Demand-Robust Shortest Path Problem

The problem is defined on a undirected graph $G = (V, E)$ with a root vertex r and cost c on the edges. The ith scenario S_i is a singleton set $\{t_i\}$. An edge e costs $c(e)$ in the first stage and $c_i(e) = \sigma_i \cdot c(e)$ in the ith scenario of the second stage. A solution to the problem is a set of edges E_0 to be bought in the first stage and a set E_i in the recourse stage for each scenario i. The solution is feasible if $E_0 \cup E_i$ contains a path between r and t_i. The cost paid in the ith scenario is $c(E_0) + \sigma_i \cdot c(E_i)$. The objective is to minimize the maximum cost over all scenarios.

The following structural result for the demand-robust shortest path problem can be obtained from a lemma proved in Dhamdhere et al. [4].

Lemma 3.1. [4] Given a demand-robust shortest path problem instance on an undirected graph, there exists a solution that costs at most twice the optimum such that the first stage solution is a tree containing the root.

The above lemma implies that we can restrict our search in the space of solutions where first stage is a tree containing the root and lose only a factor of two. This property is exploited crucially in our algorithm.

Algorithm for Robust Shortest Path

Let C be the maximum second stage cost of some fixed connected optimal solution.
$T = \{t_1, t_2, \ldots, t_k\}$ are the terminals, $r \leftarrow$ root, $\alpha \leftarrow 1.775$, $V' \leftarrow \phi$.

1. $V' := \{t_i | \operatorname{dist}_c(t_i, r) > \frac{2\alpha \cdot C}{\sigma_i}\}$
2. $\mathcal{B} := \{B_i = B(t_i, \frac{\alpha \cdot C}{\sigma_i}) | \ t_i \in V'\}$, where $B(v, d)$ is a ball of radius d around v with respect to cost c. Choose a maximal set $\mathcal{B}_\mathcal{I}$ of non-intersecting balls from $\mathcal{B}$ in order of non-decreasing radii.
3. Guess $R^0 := \{t_i | B_i \in \mathcal{B}_\mathcal{I}\}$.
4. First stage solution: $E_0 \leftarrow$ The Steiner tree on terminals $R^0 \cup \{r\}$ output by the best approximation algorithm available.
5. Second stage solution for scenario i: $E_i \leftarrow$ Shortest path from t_i to the closest node in the tree E_0.

3.1 Algorithm

Lemma 3.1 implies that there is a first stage solution which is a tree containing the root r and it can be extended to a final solution within twice the cost of an optimum solution. We call such a solution as a *connected solution*. Fix an optimal connected solution, say $E_0^*, E_1^*, \ldots, E_k^*$. Let C be the maximum second stage cost paid by this solution over all scenarios, i.e. $C = \max_{i=1}^{k}\{\sigma_i \cdot c(E_i^*)\}$. Therefore, for any scenario i, either there is path from t_i to root r in E_0^*, or there is a vertex within a distance $\frac{C}{\sigma_i}$ of t_i which is connected to r in E_0^*, where distance is with respect to the cost function c, denoted $\operatorname{dist}_c(\cdot, \cdot)$. We use this fact to obtain a constant factor approximation for our problem.

The algorithm is as follows: Let C be the maximum second stage cost paid by the connected optimal solution (fixed above) in any scenario. We need to try only $k \cdot n$ possible values of C [1], so we can assume that we have correctly guessed C. For each scenario t_i, consider a shortest path (say P_i) to r with respect to cost c. If $c(P_i) \leq \frac{2\alpha \cdot C}{\sigma_i}$, then we can handle scenario i in the second stage with cost only a factor 2α more than the optimum. Thus, t_i can be ignored in building the first stage solution. Here $\alpha > 1$ is a constant to be specified later. Let $V' = \{t_i \mid \operatorname{dist}_c(r, t_i) > \frac{2\alpha \cdot C}{\sigma_i}\}$.

For each $t_i \in V'$, let B_i be a ball of radius $\frac{\alpha \cdot C}{\sigma_i}$ around t_i. Here, we include internal points of the edges in the ball. We collectively refer to vertices in V and internal points on edges as *points*, V_P. Thus, $B_i = \{v \in V_P \mid \operatorname{dist}_c(t_i, v) \leq \frac{\alpha \cdot C}{\sigma_i}\}$.

The algorithm identifies a set of terminals $R^0 \subseteq V'$ to connect to the root in the first stage such that the remaining terminals in V' are close to some terminal in R^0 and thus, can be connected to the root in the second stage paying a low-cost.

Proposition 3.1. *There exist a set of terminals $R^0 \subseteq V'$ such that:*

1. *For every $t_i, t_j \in R^0$, we have $B_i \cap B_j = \phi$; and*
2. *For every $t_i \in V' \setminus R^0$, there is a representative $\operatorname{rep}(t_i) = t_j \in R^0$ such that $B_i \cap B_j \neq \phi$ and $\frac{\alpha \cdot C}{\sigma_j} \leq \frac{\alpha \cdot C}{\sigma_i}$.*

---

[1] For each scenario i, the second stage solution is a shortest path from t_i to one of the n vertices (possibly t_i), so there are at most $k \cdot n$ choices of C.

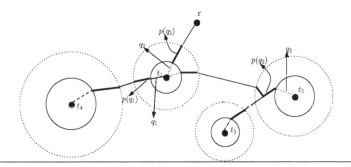

Fig. 2. Illustration of first-stage tree computation described in Lemma 3.2. The balls with solid lines denote $B(t_i, \frac{C}{\sigma_i})$, while the balls with dotted lines denote $B(t_i, \frac{\alpha \cdot C}{\sigma_i})$.

Proof. Consider terminals in V' in non-decreasing order of the radii $\frac{\alpha \cdot C}{\sigma_t}$ of the corresponding balls B_t. If terminal t_i is being examined and $B_i \cap B_j = \phi$, $\forall t_j \in R^0$, then include t_i in R^0. If not, then there exists $t_j \in R^0$ such that $B_i \cap B_j \neq \phi$; define $rep(t_i) = t_j$. Note that $\frac{\alpha \cdot C}{\sigma_j} \leq \frac{\alpha \cdot C}{\sigma_i}$ as the terminals are considered in order of non-decreasing radii of the corresponding balls.   □

The First Stage Tree. The first stage tree is a Steiner tree on the terminal set $R^0 \cup \{r\}$. However, in order to bound the cost of first stage tree we build the tree in a slightly modified way. For an illustration, refer to Figure 2.

Let G' be a new graph obtained when the balls $B(t_i, \frac{C}{\sigma_i})$ corresponding to every terminal $t_i \in R^0$ are contracted to singleton vertices. We then build a Steiner tree E_{01} in G' with the terminal set as the shrunk nodes corresponding to terminals in R^0 and the root vertex r. In Figure 2, E_{01} is the union of solid edges and the thick edges. Now, for every shrunk node corresponding to $B(t_i, \frac{C}{\sigma_i})$, we connect each tree edge incident to $B(t_i, \frac{C}{\sigma_i})$ to terminal t_i using a shortest path; these edges are shown as dotted lines in Figure 2 and are denoted by E_{02}. Our first stage solution is the Steiner tree $E_0 = E_{01} \cup E_{02}$.

Lemma 3.2. *The cost of E_0 is at most $\frac{1.55\alpha}{\alpha-1}$ times $c(E_0^*)$, the first stage cost of the optimal connected solution.*

Proof. We know that the optimal first stage tree, E_0^* connects some vertex in the ball $B(t_i, \frac{C}{\sigma_i})$ to the root r for every $t_i \in R^0$, for otherwise the maximum second stage cost of OPT would be more than C. Thus, E_0^* induces a Steiner tree on the shrunk nodes in G'. We build a Steiner tree on the shrunk nodes as terminals using the algorithm due to Robins and Zelikovsky [18]. Thus,

$$c(E_{01}) \leq 1.55 \, c(E_0^*) \tag{3.6}$$

Now, consider edges in E_{02}. Consider a path $q \in E_{02}$ connecting some edge incident to $B(t_i, \frac{C}{\sigma_i})$ to t_i. Since q is the shortest path between its end points, we have $c(q) \leq \frac{C}{\sigma_i}$. Now, consider a path from terminal t_i along q until it reaches $B(t_i, \frac{\alpha \cdot C}{\sigma_i})$ and label the portion between $B(t_i, \frac{C}{\sigma_i})$ and $B(t_i, \frac{\alpha \cdot C}{\sigma_i})$ as $p(q)$. By construction, we have $c(p(q)) \geq \frac{(\alpha-1) \cdot C}{\sigma_i}$, so $c(q) \leq \frac{1}{\alpha-1} \cdot c(p(q))$.

For any two paths $q_1, q_2 \in E_{02}$, the paths $p(q_1)$ and $p(q_2)$ are edge-disjoint. Clearly, if q_1 and q_2 are incident to distinct terminals of R^0, then $p(q_1)$ and $p(q_2)$ are contained in disjoint balls and thus are edge-disjoint. If q_1 and q_2 are incident to the same terminal, then it is impossible that $p(q_1) \cap p(q_2) \neq \phi$ as E_{01} is a tree on the shrunk graph. Hence, we have

$$\sum_{e \in E_{02}} c(e) = \sum_{q \in E_{02}} c(q) \leq \sum_{q \in E_{02}} \frac{1}{\alpha-1} \cdot c(p(q)) \leq \sum_{e \in E_{01}} \frac{1}{\alpha-1} \cdot c(e) \qquad (3.7)$$

where the last inequality is due to edge-disjointness of $p(q_1)$ and $p(q_2)$ for any two paths $q_1, q_2 \in E_{02}$. Thus, $c(E_0) = c(E_{01}) + c(E_{02}) \leq c(E_{01}) + \frac{1}{\alpha-1} \cdot c(E_{01}) \leq \frac{1.55\alpha}{\alpha-1} \cdot c(E_0^*)$, where the last inequality follows from (3.6). □

Second Stage. The second stage solution for each scenario is quite straightforward. For any terminal t_i, E_i is the shortest path from t_i to the closest node in E_0.

Lemma 3.3. *The maximum second stage cost for any scenario is at most $2\alpha \cdot C$.*

Proof. We need to consider the following cases:

1. $t_i \in R^0$: Since the first stage tree E_0 connects t_i to r, $E_i = \phi$. Thus, $c(E_i) = 0$.
2. $t_i \in V' \setminus R^0$: By Proposition 3.1, there exists a representative terminal $t_j \in R^0$ such that $B_i \cap B_j \neq \phi$ and $\sigma_j \geq \sigma_i$. Therefore, $\text{dist}_c(t_i, t_j) \leq \frac{\alpha \cdot C}{\sigma_i} + \frac{\alpha \cdot C}{\sigma_j} \leq \frac{2\alpha \cdot C}{\sigma_i}$. We know that t_j is connected to r in E_0. Thus, the closest node to t_i in the first stage tree is at a distance at most $\frac{2\alpha \cdot C}{\sigma_i}$. Hence, $\sigma_i \cdot c(E_i) \leq 2\alpha \cdot C$.
3. $t_i \notin V'$: Then the shortest path from t_i to r with respect to cost c is at most $\frac{2\alpha \cdot C}{\sigma_i}$. Hence, the closest node to t_i in the first stage tree is at a distance at most $\frac{2\alpha \cdot C}{\sigma_i}$ and $\sigma_i \cdot c(E_i) \leq 2\alpha \cdot C$. □

Proof of Theorem 1.2. From Lemma 3.2, we get that $c(E_0) \leq \frac{1.55\alpha}{\alpha-1} c(E_0^*)$. From Lemma 3.3, we get that the second stage cost is at most $2\alpha \cdot C$. Choose $\alpha = \frac{3.55}{2} = 1.775$. Thus, we get $c(E_0) \leq (3.55) \cdot c(E_0^*)$ and $\max_{i=1}^k \{\sigma_i \cdot c(E_i)\} \leq (3.55) \cdot C$. From Lemma 3.1 we know that $c(E_0^*) + C \leq 2 \cdot \text{OPT}$, where OPT is the cost of optimal solution to the robust shortest path instance. Together the previous three inequalities imply $c(E_0) + \max_{i=1}^k \{\sigma_i \cdot c(E_i)\} \leq (7.1) \cdot \text{OPT}$. □

4 Extensions to Hitting Versions

In this problem, we introduce generalizations of demand-robust min-cut and shortest path problems that are closely related to Steiner multicut and group Steiner tree, respectively. In a Steiner multicut instance, we are given a graph $G = (V, E)$ and k sets of vertices $X_1, X_2, \ldots, X_k$ and our goal is to find the cheapest set of edges S whose removal *separates* each X_i, i.e. no X_i lies entirely within one connected component of

$(V, E \setminus S)$. If $\bigcap_{i=1}^{k} X_i \neq \emptyset$, we call the instance *restricted*. In a group Steiner tree instance, we are given a graph $G = (V, E)$, a root r, and k sets of vertices $X_1, X_2, \ldots, X_k$ and our goal is to find a minimum cost set of edges S that connects at least one vertex in each $X_i, i = 1, \ldots, k$ to the root r. We show how approximation algorithms for these problems can be combined with our techniques to yield approximation algorithms for "hitting versions" of demand-robust min-cut and shortest path problems.

In the hitting version of robust min-cut (resp. shortest path), each scenario i is specified by an inflation factor σ_i and a set of nodes $T_i \subset V$ (rather than a single node). A feasible solution is a collection of edge sets $\{E_0, E_1, \ldots, E_k\}$ such that for each scenario i, $E_0 \cup E_i$ contains an root-t cut (resp. path) for some $t \in T_i$. The goal is to minimize $c(E_0) + \max_i \{\sigma_i \cdot c(E_i)\}$.

4.1 Robust Hitting Cuts

Robust hitting cut is $\Omega(\log k)$-hard, where k is the number of scenarios, even when the graph is a star. In fact, if we restrict ourselves to inputs in which the graph is a star, the root is the center of the star, and $\sigma = \infty$ for all scenarios, then robust hitting cut on these instances is exactly the hitting set problem. In contrast, we can obtain an $O(\log k)$ approximation for robust hitting cut on trees, and $O(\log n \cdot \log k)$ in general using results of Nagarajan and Ravi [15] in conjunction with the following theorem.

Theorem 4.1. *If for some class of graphs there is a ρ-approximation for Steiner multicut on restricted instances, then for that class of graphs there is a $(\rho+2)$-approximation for robust hitting cut. Conversely, if there is a ρ-approximation for robust hitting cut then there is a ρ-approximation for Steiner multicut on restricted instances.*

Algorithm: Let $\alpha = \frac{1}{2}(\rho + 1 + \sqrt{\rho^2 + 6\rho + 1})$ and let C be the cost that some optimal solution pays in the second stage. For each terminal t in some group, compute the cost of a minimum root-t cut, denoted $\mathrm{mcut}(t)$. Let $\mathcal{T}' := \{T_i : \forall t \in T_i, \sigma_i \cdot \mathrm{mcut}(t) > \alpha \cdot C\}$. Note that there are only $k + 1$ possibilities, as in the robust min-cut algorithm. For each terminal set $T_i \in \mathcal{T}'$, separate at least one terminal in T_i from the root in the first stage using an ρ-approximation algorithm for Steiner Multicut [10,15].

Proof of Theorem 4.1. We first show that a ρ-approximation for robust hitting cut implies a ρ-approximation for Steiner multicut on restricted instances. Given a restricted instance of Steiner multicut $(G, X_1, X_2, \ldots, X_k)$ build a robust hitting cut instance as follows: use the same graph and costs, set the root r to be any element of $\bigcap_i X_i$, and create scenarios $T_i = X_i \setminus r$ with $\sigma_i = \infty$ for each i. Note that solutions to this instance correspond exactly to Steiner multicuts of the same cost. Thus robust hitting cut generalizes Steiner multicut on restricted instances.

We now show the approximate converse, that a ρ-approximation for Steiner multicut on restricted instances implies a $(\rho + 2)$-approximation for robust hitting cut. Let OPT be an optimal solution, and let E_0^* be the edge set it buys in stage one, and let C_1 and C_2 be the amount it pays in the first and second stage, respectively. Note we can handle every $T_i \notin \mathcal{T}'$ while paying at most $\alpha \cdot C_2$.

We prove that the first stage edges $E_0 \subset E[G]$ given by our algorithm satisfy all scenarios in $\mathcal{T}'$, and have cost $c(E_0) \leq \rho(1 + \frac{2}{\alpha-1})C_1$. Thus, the total solution cost is at most $\rho(1 + \frac{2}{\alpha-1})C_1 + \alpha \cdot C_2$. Compared to the optimal cost, $C_1 + C_2$, we obtain a $\max\{\alpha, \rho(1 + \frac{2}{\alpha-1})\}$-approximation. Setting $\alpha = \frac{1}{2}(\rho + 1 + \sqrt{\rho^2 + 6\rho + 1})$ then yields the claimed $(\rho + 2)$ approximation ratio.

A cut is called a T'-*cut* if it separates at least one terminal in each $T \in T'$ from the root. There exists a T'-cut of cost at most $(1 + \frac{2}{\alpha-1})C_1$, by the same argument as in the proof of Theorem 1.1. Suppose OPT cuts away t_i^* when scenario T_i occurs. Then OPT is also an optimal solution to the robust min-cut instance on the same graph with terminals $\{t_i^* \mid i = 1, 2, \ldots, k\}$ as k scenarios. Since, for all $t \in T$ such that $T \in T'$, we have $\sigma_t \cdot \text{mcut}(t) > \alpha \cdot C$, we can construct a root-$\{t_i^* \mid i = 1, 2, \ldots, k\}$ cut of cost at most $(1 + \frac{2}{\alpha-1})C_1$. Thus, the cost of an optimal T'-cut is at most $(1 + \frac{2}{\alpha-1})C_1$. Now apply the ρ-approximation for Steiner multicut on restricted instances. To build the Steiner multicut instance, we use the same graph and edge costs, and create a groups $X_i = T_i \cup \{\text{root}\}$ for each $T_i \in T'$. Clearly, the instance is restricted. Note that every solution to this instance is a T'-cut of the same cost, and vice-versa. Thus a ρ-approximation for for Steiner multicut on restricted instances yields a T'-cut of cost at most $\rho(1 + \frac{2}{\alpha-1})C_1$. □

Corollary 4.1. *There is a polynomial time $O(\log n \cdot \log k)$-approximation algorithm for robust hitting cut on instances with k scenarios and n nodes, and an $O(\log k)$-approximation algorithm for robust hitting cut on trees.*

4.2   Robust Hitting Paths

Theorem 4.2. *If there is a ρ-approximation for group Steiner tree then there is a 2ρ-approximation for robust hitting path. If there is a ρ-approximation for robust hitting path, then there is a ρ-approximation for group Steiner tree.*

Proof. Note that robust hitting path generalizes group Steiner tree (given a GST instance with graph G, root r and groups $X_1, X_2, \ldots, X_k$, use the same graph and root, make each group a scenario, and set $\sigma_i = \infty$ for all scenarios i). Thus a ρ-approximation for robust hitting path immediately yields a ρ-approximation for group Steiner tree.

Now suppose we have an ρ-approximation for group Steiner tree. Lemma 3.1 guarantees that there exists a solution $\{E_0, E_1, \ldots, E_k\}$ of cost at most 2OPT whose first stage edges, E_0, are a tree containing root r.

The algorithm is as follows. Guess $C := \max_i \{\sigma_i c(E_i)\}$. Note that for each scenario i the tree E_0 must touch one of the balls in $\{B(t, C/\sigma_i) | t \in T_i\}$, where $B(v, x) := \{u | \text{dist}_c(v, u) \leq x\}$. Thus we can construct groups $X_i := \bigcup_{t \in T_i} B(t, C/\sigma_i)$ for each scenario i and use the ρ-approximation for group Steiner tree on these groups to obtain a set of edges E_0' to buy in the first stage.

Note that $c(E_0') \leq \rho c(E_0)$ and any scenario i has a terminal $t \in T_i$ that is within distance C/σ_i of some vertex incident on an edge of tree E_0'. We conclude that the total cost is at most $\rho c(E_0) + C \leq 2\rho \cdot \text{OPT}$. □

5   Conclusion

In this paper we give improved approximation algorithms for robust min-cut and shortest path problems and extend our results to an interesting "hitting-set" variant. It would be interesting to use the techniques introduced in this paper to obtain better approximations for robust minimum multicut and Steiner tree problems. The technique of guessing and pruning crucially uses the fact that each scenario can pay up to the maximum second stage cost without worsening the optimal cost. However, this is not true for the stochastic optimization problems and hence our technique doesn't extend to stochastic versions in a straightforward way. It would be interesting to adapt this idea for stochastic optimization.

References

1. D. Bertsimas and M. Sim. Robust discrete optimization and network flows. *Mathematical Programming Series B*, 98:49–71, 2003.
2. D. Bertsimas and M. Sim. The price of robustness. *Operation Research*, 52(2):35–53, 2004.
3. G. B. Dantzig. Linear programming under uncertainty. *Management Sci.*, 1:197–206, 1955.
4. K. Dhamdhere, V. Goyal, R. Ravi, and M. Singh. How to pay, come what may: Approximation algorithms for demand-robust covering problems. In *FOCS*, pages 367–378, 2005.
5. K. Dhamdhere, R. Ravi, and M. Singh. On two-stage stochastic minimum spanning trees. In *IPCO*, pages 321–334, 2005.
6. A. V. Goldberg and R. E. Tarjan. A new approach to the maximum-flow problem. *J. ACM*, 35(4):921–940, 1988.
7. A. Gupta, M. Pál, R. Ravi, and A. Sinha. Boosted sampling: approximation algorithms for stochastic optimization. In *STOC*, pages 417–426, 2004.
8. A. Gupta, R. Ravi, and A. Sinha. An edge in time saves nine: LP rounding approximation algorithms for stochastic network design. In *FOCS*, pages 218–227, 2004.
9. N. Immorlica, D. Karger, M. Minkoff, and V. Mirrokni. On the costs and benefits of procrastination: Approximation algorithms for stochastic combinatorial optimization problems. In *SODA*, pages 684–693, 2004.
10. Philip N. Klein, Serge A. Plotkin, Satish Rao, and Eva Tardos. Approximation algorithms for steiner and directed multicuts. *J. Algorithms*, 22(2):241–269, 1997.
11. P. Kouvelis and G. Yu. *Robust Discrete Optimisation and Its Applications*. Kluwer Academic Publishers, Netherlands, 1997.
12. R. D. Luce and H. Raiffa. *Games and Decision*. Wiley, New York, 1957.
13. J. W. Milnor. Games against nature. In R. M. Thrall, C. H. Coombs, and R. L. Davis, editors, *Decision Processes*. Wiley, New York, 1954.
14. J. M. Mulvey, R. J. Vanderbei, and S. J. Zenios. Robust optimization of large-scale systems. *Operations Research*, 43:264–281, 1995.
15. V. Nagarajan and R. Ravi. Approximation algorithms for requirement cut on graphs. In *APPROX + RANDOM*, pages 209–220, 2005.
16. Y. Nikulin. Robustness in combinatorial optimization and scheduling theory: An annotated bibliography, 2004. http://www.optimization-online.org/DB_FILE/2004/11/995.pdf.
17. R. Ravi and A. Sinha. Hedging uncertainty: Approximation algorithms for stochastic optimization problems. In *IPCO*, pages 101–115, 2004.
18. G. Robins and A. Zelikovsky. Improved steiner tree approximation in graphs. In *SODA*, pages 770–779, 2000.
19. M. J. Rosenblatt and H. L. Lee. A robustness approach to facilities design. *International Journal of Production Research*, 25:479–486, 1987.
20. D. Shmoys and C. Swamy. Stochastic optimization is (almost) easy as deterministic optimization. In *FOCS*, pages 228–237, 2004.

Exact Price of Anarchy for Polynomial Congestion Games[*]

Sebastian Aland, Dominic Dumrauf, Martin Gairing,
Burkhard Monien, and Florian Schoppmann

Department of Computer Science, Electrical Engineering and Mathematics,
University of Paderborn, Fürstenallee 11, 33102 Paderborn, Germany
{sebaland, masa, gairing, bm, fschopp}@uni-paderborn.de

Abstract. We show exact values for the price of anarchy of weighted and unweighted congestion games with polynomial latency functions. The given values
also hold for weighted and unweighted *network* congestion games.

1 Introduction

Motivation and Framework. Large scale communication networks, like e.g. the internet, often lack a central regulation for several reasons: The size of the network may
be too large, or the users may be free to act according to their private interests. Even
cooperation among the users may be impossible due to the fact that users may not
even know each other. Such an environment—where users neither obey some central
control instance nor cooperate with each other—can be modeled as a *non-cooperative
game* [18].

One of the most widely used solution concepts for non-cooperative games is the concept of *Nash equilibrium*. A Nash equilibrium is a state in which no player can improve
his objective by unilaterally changing his strategy. A Nash equilibrium is called *pure* if
all players choose a pure strategy, and *mixed* if players choose probability distributions
over strategies.

Rosenthal [25] introduced a special class of non-cooperative games, now widely
known as *congestion games*. Here, the strategy set of each player is a subset of the power
set of given resources. The players share a private cost function, defined as the sum (over
their chosen resources) of functions in the number of players sharing this resource. Later
Milchtaich [20] considered *weighted congestion games* as an extension to congestion
games in which the players have weights and thus different influence on the congestion
of the resources. Weighted congestion games provide us with a general framework for
modeling any kind of non-cooperative resource sharing problem. A typical resource
sharing problem is that of routing. In a routing game the strategy sets of the players
correspond to paths in a network. Routing games where the demand of the players
cannot be split among multiple paths are also called *(weighted) network congestion
games*. Another model for selfish routing—the so called *Wardrop* model—was already

---

[*] This work has been partially supported by the DFG-SFB 376 and by the European Union
within the 6th Framework Programme under contract 001907 (DELIS).

B. Durand and W. Thomas (Eds.): STACS 2006, LNCS 3884, pp. 218–229, 2006.

studied in the 1950's (see e.g. [3,29]) in the context of road traffic systems, where traffic flows can be split arbitrarily. The Wardrop model can be seen as a special network congestion game with infinitely many players each carrying a negligible demand.

In order to measure the degradation of social welfare due to the selfish behavior of the players, Koutsoupias and Papadimitriou [16] introduced a global objective function, usually coined as *social cost*. They defined the *price of anarchy*, also called coordination ratio, as the worst-case ratio between the value of social cost in a Nash equilibrium and that of some social optimum. Thus, the price of anarchy measures the extent to which non-cooperation approximates cooperation. The price of anarchy directly depends on the definition of social cost. Koutsoupias and Papadimitriou [16] considered a very simple weighted network congestion game on parallel links, now known as KP-model. For this model they defined the social cost as the expected maximum latency. For the Wardrop model, Roughgarden and Tardos [28] considered social cost as the *total latency*, which is a measure for the (weighted) total travel time. Awerbuch et al. [1] and Christodoulou and Koutsoupias [5] considered the total latency for congestion games with a finite number of players with non-negligible demands. In this setting, they show asymptotic bounds on the price of anarchy for weighted (and unweighted) congestion games with polynomial latency (cost) functions. Here, all polynomials are of maximum degree d and have non-negative coefficients. For the case of linear latency functions they give exact bounds on the price of anarchy.

Contribution and Comparison. In this work we prove *exact* bounds on the price of anarchy for unweighted and weighted congestion games with polynomial latency functions. We use the total latency as social cost measure. This improves on results by Awerbuch et al. [1] and Christodoulou and Koutsoupias [5], where non-matching upper and lower bounds are given.
We now describe our findings in more detail.

- For *unweighted congestion games* we show that the price of anarchy (PoA) is exactly
$$\text{PoA} = \frac{(k+1)^{2d+1} - k^{d+1}(k+2)^d}{(k+1)^{d+1} - (k+2)^d + (k+1)^d - k^{d+1}},$$
  where $k = \lfloor \Phi_d \rfloor$ and Φ_d is a natural generalization of the golden ratio to larger dimensions such that Φ_d is the solution to $(\Phi_d + 1)^d = \Phi_d^{d+1}$. Prior to this paper the best known upper and lower bounds were shown to be of the form $d^{d(1-o(1))}$ [5]. However, the term $o(1)$ still hides a gap between the upper and the lower bound.
- For *weighted congestion games* we show that the price of anarchy (PoA) is exactly
$$\text{PoA} = \Phi_d^{d+1}.$$

This result closes the gap between the so far best upper and lower bounds of $O(2^d d^{d+1})$ and $\Omega(d^{d/2})$ from [1].

We show that the above values on the price of anarchy also hold for the subclasses of unweighted and weighted network congestion games.

For our upper bounds we use a similar analysis as in [5]. The core of our analysis is to determine parameters c_1 and c_2 such that

$$y \cdot f(x+1) \le c_1 \cdot x \cdot f(x) + c_2 \cdot y \cdot f(y) \tag{1}$$

for all polynomial latency functions of maximum degree d and for all reals $x, y \geq 0$. For the case of unweighted demands it suffices to show (1) for all integers x, y. In order to prove their upper bound Christodoulou and Koutsoupias [5] looked at (1) with $c_1 = \frac{1}{2}$ and gave an asymptotic estimate for c_2. In our analysis we optimize both parameters c_1, c_2. This optimization process requires new ideas and is non-trivial.

Table 1 shows a numerical comparison of our bounds with the previous results of Awerbuch et al. [1] and Christodoulou and Koutsoupias [5].

For $d \geq 2$, the table only gives the respective lower bounds that are given in the cited works (before any estimates are applied). Values in parentheses denote cases in which the bound for linear functions is better than the general case.

In [1, Theorem 4.3], a construction scheme for networks is described with price of anarchy approximating $\frac{1}{e} \sum_{k=1}^{\infty} \frac{k^d}{k!}$ which yields the d-th Bell number. In [5, Theorem 10], a network with price of anarchy $\frac{(N-1)^{d+2}}{N}$ is given, with N being the largest integer for which $(N-1)^{d+2} \leq N^d$ holds.

The column with the upper bound from [5] is computed by using (1) with $c_1 = \frac{1}{2}$ and optimizing c_2 with help of our analysis. Thus, the column shows the best possible bounds that can be shown with $c_1 = \frac{1}{2}$.

Table 1. Comparison of our results to [5] and [1]

| | | | unweighted PoA | | weighted PoA | |
d	Φ_d	Our exact result	Upper Bound [5]	Lower bound [5]	Our exact result	Lower bound [1]
1	1.618	**2.5**	2.5	2.5	**2.618**	2.618
2	2.148	**9.583**	10	(2.5)	**9.909**	(2.618)
3	2.630	**41.54**	47	(2.5)	**47.82**	5
4	3.080	**267.6**	269	21.33	**277.0**	15
5	3.506	**1,514**	2,154	42.67	**1,858**	52
6	3.915	**12,345**	15,187	85.33	**14,099**	203
7	4.309	**98,734**	169,247	170.7	**118,926**	877
8	4.692	**802,603**	1,451,906	14,762	**1,101,126**	4,140
9	5.064	**10,540,286**	20,241,038	44,287	**11,079,429**	21,147
10	5.427	**88,562,706**	202,153,442	132,860	**120,180,803**	115,975

Related Work. The papers most closely related to our work are those of Awerbuch et al. [1] and Christodoulou and Koutsoupias [5,4]. For (unweighted) congestion games and social cost defined as average private cost (which in this case is the same as total latency) it was shown that the price of anarchy of pure Nash equilibria is $\frac{5}{2}$ for linear latency functions and $d^{\Theta(d)}$ for polynomial latency functions of maximum degree d [1,5]. The bound of $\frac{5}{2}$ for linear latency function also holds for the correlated and thus also for the mixed price of anarchy [4]. For *weighted* congestion games the mixed price of anarchy for total latency is $\frac{3+\sqrt{5}}{2}$ for linear latency functions and $d^{\Theta(d)}$ for polynomial latency functions [1].

The *price of anarchy* [24], also known as *coordination ratio*, was first introduced and studied by Koutsoupias and Papadimitriou [16]. As a starting point of their investigation they considered a simple weighted congestion game on parallel links, now known as KP-model. In the KP-model latency functions are linear and social cost is defined as the maximum expected congestion on a link. In this setting, there exist *tight* bounds on the price of anarchy of $\Theta(\frac{\log m}{\log \log m})$ for identical links [7,15] and $\Theta(\frac{\log m}{\log \log \log m})$ [7] for related links. The price of anarchy has also been studied for variations of the KP-model, namely for non-linear latency functions [6,12], for the case of restricted strategy sets [2,10], for the case of incomplete information [14] and for different social cost measures [11,17]. In particular Lücking et al. [17] study the total latency (they call it quadratic social cost) for routing games on parallel links with linear latency functions. For this model they show that the price of anarchy is exactly $\frac{4}{3}$ for case of identical player weights and $\frac{9}{8}$ for the case of identical links and arbitrary player weights.

The class of *congestion games* was introduced by Rosenthal [25] and extensively studied afterwards (see e.g. [8,20,21]). In Rosenthal's model the strategy of each player is a subset of resources. Resource utility functions can be arbitrary but they only depend on the number of players sharing the same resource. Rosenthal showed that such games always admit a pure Nash equilibrium using a potential function. Monderer and Shapley [21] characterize games that possess a potential function as potential games and show their relation to congestion games. Milchtaich [20] considers weighted congestion games with player specific payoff functions and shows that these games do not admit a pure Nash equilibrium in general. Fotakis et al. [8,9] consider the price of anarchy for symmetric weighted network congestion games in layered networks [8] and for symmetric (unweighted) network congestion games in general networks [9]. In both cases they define social cost as expected maximum latency. For a survey on weighted congestion games we refer to [13].

Inspired by the arisen interest in the price of anarchy Roughgarden and Tardos [28] re-investigated the Wardrop model and used the *total latency* as a social cost measure. In this context the price of anarchy was shown to be $\frac{4}{3}$ for linear latency functions [28] and $\Theta(\frac{d}{\log d})$ [26] for polynomial latency functions of maximum degree d. An overview on results for this model can be found in the recent book of Roughgarden [27].

Roadmap. The rest of this paper is organized as follows. In Section 2 we give an exact definition of weighted congestion games. We present exact bounds on the price of anarchy for unweighted congestion games in Section 3 and for weighted congestion games in Section 4. Due to lack of space we omit some of the proofs.

2 Notations

General. For all integers $k \geq 0$, we denote $[k] = \{1, \ldots, k\}$, $[k]_0 = \{0, \ldots, k\}$. For all integers $d > 0$, let $\Phi_d \in \mathbb{R}^+$ denote the number for which $(\Phi_d + 1)^d = \Phi_d^{d+1}$. Clearly, Φ_1 coincides with the *golden ratio*. Thus, Φ_d is a natural generalization of the golden ratio to larger dimensions.

Weighted Congestion Games. A *weighted congestion game* Γ is a tuple $\Gamma = \left(n, E, (w_i)_{i \in [n]}, (S_i)_{i \in [n]}, (f_e)_{e \in E}\right)$. Here, n is the number of *players* (or *users*) and E is the finite set of *resources*. For every player $i \in [n]$, $w_i \in \mathbb{R}^+$ is the *weight* and $S_i \subseteq 2^E$ is the *strategy set* of player i. Denote $S = S_1 \times \ldots \times S_n$ and $S_{-i} = S_1 \times \ldots \times S_{i-1} \times S_{i+1} \ldots \times S_n$. For every resource $e \in E$, the *latency function* $f_e : \mathbb{R}^+ \to \mathbb{R}^+$ describes the *latency* on resource e. We consider only *polynomial latency functions* with maximum degree d and non-negative coefficients, that is for all $e \in E$ the latency function is of the form $f_e(x) = \sum_{j=0}^{d} a_{e,j} \cdot x^j$ with $a_{e,j} \geq 0$ for all $j \in [d]_0$.

In a (unweighted) *congestion game*, the weights of all players are equal. Thus, the private cost of a player only depends on the *number* of players choosing the same resources.

Strategies and Strategy Profiles. A *pure strategy* for player $i \in [n]$ is some specific $s_i \in S_i$ whereas a *mixed strategy* $P_i = (p(i, s_i))_{s_i \in S_i}$ is a probability distribution over S_i, where $p(i, s_i)$ denotes the probability that player i chooses the pure strategy s_i.

A *pure strategy profile* is an n-tuple $\mathbf{s} = (s_1, \ldots, s_n) \in S$ whereas a *mixed strategy profile* $\mathbf{P} = (P_1, \ldots, P_n)$ is represented by an n-tuple of mixed strategies. For a mixed strategy profile $\mathbf{P}$ denote by $p(\mathbf{s}) = \prod_{i \in [n]} p(i, s_i)$ the probability that the players choose the pure strategy profile $\mathbf{s} = (s_1, \ldots, s_n)$. Following standard game theory notation, we denote $\mathbf{P}_{-i} = (P_1, \ldots, P_{i-1}, P_{i+1}, \ldots, P_n)$ as the (mixed) strategy profile of all players except player i and $(\mathbf{P}_{-i}, Q_i)$ as the strategy profile that results from $\mathbf{P}$ if player i deviates to strategy Q_i.

Private Cost. Fix any pure strategy profile $\mathbf{s}$, and denote by $l_e(\mathbf{s}) = \sum_{i \in [n], s_i \ni e} w_i$ the *load* on resource $e \in E$. The *private cost* of player $i \in [n]$ in a pure strategy profile $\mathbf{s}$ is defined by $PC_i(\mathbf{s}) = \sum_{e \in s_i} f_e(l_e(\mathbf{s}))$. For a mixed strategy profile $\mathbf{P}$, the *private cost* of player $i \in [n]$ is

$$PC_i(\mathbf{P}) = \sum_{\mathbf{s} \in S} p(\mathbf{s}) \cdot PC_i(\mathbf{s}).$$

Social Cost. Associated with a weighted congestion game Γ and a mixed strategy profile $\mathbf{P}$ is the *social cost* $SC(\mathbf{P})$ as a measure of social welfare. In particular we use the expected total latency, that is,

$$SC(\mathbf{P}) = \sum_{\mathbf{s} \in S} p(\mathbf{s}) \sum_{e \in E} l_e(\mathbf{s}) \cdot f_e(l_e(\mathbf{s}))$$

$$= \sum_{\mathbf{s} \in S} p(\mathbf{s}) \sum_{i \in [n]} \sum_{e \in s_i} w_i \cdot f_e(l_e(\mathbf{s}))$$

$$= \sum_{i \in [n]} w_i \cdot PC_i(\mathbf{P}).$$

The *optimum* associated with a weighted congestion game is defined by $OPT = \min_{\mathbf{P}} SC(\mathbf{P})$.

Nash Equilibria and Price of Anarchy. We are interested in a special class of (mixed) strategy profiles called Nash equilibria [22,23] that we describe here. Given a weighted congestion game and an associated mixed strategy profile $\mathbf{P}$, a player $i \in [n]$ is *satisfied* if he can not improve his private cost by unilaterally changing his strategy. Otherwise, player i is *unsatisfied*. The mixed strategy profile $\mathbf{P}$ is a *Nash equilibrium* if and only if all players $i \in [n]$ are satisfied, that is, $PC_i(\mathbf{P}) \leq PC_i(\mathbf{P}_{-i}, s_i)$ for all $i \in [n]$ and $s_i \in S_i$.

Note, that if this inequality holds for all pure strategies $s_i \in S_i$ of player i, then it also holds for all mixed strategies over S_i. Depending on the type of strategy profile, we differ between *pure* and *mixed* Nash equilibria.

The *price of anarchy*, also called *coordination ratio* and denoted PoA, is the maximum value, over all instances Γ and Nash equilibria $\mathbf{P}$, of the ratio $\frac{SC(\mathbf{P})}{OPT}$.

3   Price of Anarchy for Unweighted Congestion Games

In this section, we prove the exact value for the price of anarchy of unweighted congestion games with polynomial latency functions. We start with two technical lemmas which are crucial for determining c_1 and c_2 in (1) and thus for proving the upper bound Theorem 1. In Theorem 2 we give a matching lower bound which also holds for unweighted network congestion games (Corollary 1).

Lemma 1. *Let $0 \leq c < 1$ and $d \in \mathbb{N}_0$ then*

$$\max_{x \in \mathbb{N}_0, y \in \mathbb{N}} \left\{ \left(\frac{x+1}{y} \right)^d - c \cdot \left(\frac{x}{y} \right)^{d+1} \right\} = \max_{x \in \mathbb{N}_0} \left\{ (x+1)^d - c \cdot x^{d+1} \right\}.$$

Lemma 2. *Let $d \in \mathbb{N}$ and*

$$\mathcal{F}_d = \{ g_r^{(d)} : \mathbb{R} \to \mathbb{R} \mid g_r^{(d)}(x) = (r+1)^d - x \cdot r^{d+1}, r \in \mathbb{R}_{\geq 0} \}$$

be an infinite set of linear functions. Furthermore, let $\gamma(s,t)$ for $s,t \in \mathbb{R}_{\geq 0}$ and $s \neq t$ denote the intersection abscissa of $g_s^{(d)}$ and $g_t^{(d)}$. Then it holds for any $s,t,u \in \mathbb{R}_{\geq 0}$ with $s < t < u$ that $\gamma(s,t) > \gamma(s,u)$ and $\gamma(u,s) > \gamma(u,t)$.

Theorem 1. *For unweighted congestion games with polynomial latency functions of maximum degree d and non-negative coefficients, we have*

$$\mathsf{PoA} \leq \frac{(k+1)^{2d+1} - k^{d+1}(k+2)^d}{(k+1)^{d+1} - (k+2)^d + (k+1)^d - k^{d+1}}, \quad \text{where } k = \lfloor \Phi_d \rfloor.$$

Proof. Let $\mathbf{P} = (P_1, ..., P_n)$ be a (mixed) Nash equilibrium and let $\mathbf{Q} = (Q_1, ..., Q_n)$ be a pure strategy profile with optimum social cost. Since $\mathbf{P}$ is a Nash equilibrium, player $i \in [n]$ cannot improve by switching from strategy P_i to strategy Q_i. Thus,

$$\mathsf{PC}_i(\mathbf{P}) = \sum_{\mathbf{s} \in S} p(\mathbf{s}) \sum_{e \in s_i} f_e(l_e(\mathbf{s})) \leq \mathsf{PC}_i(\mathbf{P}_{-i}, Q_i)$$

$$= \sum_{\mathbf{s} \in S} p(\mathbf{s}) \left[\sum_{e \in Q_i \cap s_i} f_e(l_e(\mathbf{s})) + \sum_{e \in Q_i \setminus s_i} f_e(l_e(\mathbf{s}) + 1) \right]$$

$$\leq \sum_{\mathbf{s} \in S} p(\mathbf{s}) \sum_{e \in Q_i} f_e(l_e(\mathbf{s}) + 1).$$

Summing up over all players $i \in [n]$ yields

$$\mathsf{SC}(\mathbf{P}) = \sum_{i \in [n]} \sum_{\mathbf{s} \in S} p(\mathbf{s}) \sum_{e \in s_i} f_e(l_e(\mathbf{s})) \leq \sum_{i \in [n]} \sum_{\mathbf{s} \in S} p(\mathbf{s}) \sum_{e \in Q_i} f_e(l_e(\mathbf{s}) + 1)$$

$$= \sum_{\mathbf{s} \in S} p(\mathbf{s}) \sum_{e \in E} l_e(\mathbf{Q}) \cdot f_e(l_e(\mathbf{s}) + 1).$$

Now, $l_e(\mathbf{Q})$ and $l_e(\mathbf{s})$ are both integer, since $\mathbf{Q}$ and $\mathbf{s}$ are both pure strategy profiles. Thus, by choosing c_1, c_2 such that

$$y \cdot f(x+1) \leq c_1 \cdot x \cdot f(x) + c_2 \cdot y \cdot f(y) \tag{2}$$

for all polynomials f with maximum degree d and non-negative coefficients and for all $x, y \in \mathbb{N}_0$, we get

$$\mathsf{SC}(\mathbf{P}) \leq \sum_{s \in S} p(s) \sum_{e \in E} [c_1 l_e(s) f_e(l_e(s)) + c_2 l_e(\mathbf{Q}) f_e(l_e(\mathbf{Q}))] = c_1 \cdot \mathsf{SC}(\mathbf{P}) + c_2 \cdot \mathsf{SC}(\mathbf{Q}).$$

With $c_1 < 1$ it follows that $\frac{\mathsf{SC}(\mathbf{P})}{\mathsf{SC}(\mathbf{Q})} \leq \frac{c_2}{1-c_1}$. Since $\mathbf{P}$ is an arbitrary (mixed) Nash equilibrium we get

$$\mathsf{PoA} \leq \frac{c_2}{1 - c_1}. \tag{3}$$

In fact, c_1 and c_2 depend on the maximum degree d, however, for the sake of readability we omit this dependence in our notation.

We will now show how to determine constants c_1 and c_2 such that Inequality (2) holds and such that the resulting upper bound of $\frac{c_2}{1-c_1}$ is minimal. To do so, we will first show that it suffices to consider Inequality (2) with $y = 1$ and $f(x) = x^d$.

Since f is a polynomial of maximum degree d with non-negative coefficients, it is sufficient to determine c_1 and c_2 that fulfill (2) for $f(x) = x^r$ for all integers $0 \leq r \leq d$.

So let $f(x) = x^r$ for some $0 \leq r \leq d$. In this case (2) reduces to

$$y \cdot (x + 1)^r \leq c_1 \cdot x^{r+1} + c_2 \cdot y^{r+1}. \tag{4}$$

For any given constant $0 \leq c_1 < 1$ let $c_2(r, c_1)$ be the minimum value for c_2 such that (4) holds, that is

$$c_2(r, c_1) = \max_{x \in \mathbb{N}_0, y \in \mathbb{N}} \left\{ \frac{y(x+1)^r - c_1 \cdot x^{r+1}}{y^{r+1}} \right\} = \max_{x \in \mathbb{N}_0, y \in \mathbb{N}} \left\{ \left(\frac{x+1}{y} \right)^r - c_1 \cdot \left(\frac{x}{y} \right)^{r+1} \right\}.$$

Note that (4) holds for any c_2 when $y = 0$. By Lemma 1 we have

$$c_2(r, c_1) = \max_{x \in \mathbb{N}_0} \left\{ (x+1)^r - c_1 \cdot x^{r+1} \right\}. \tag{5}$$

Now, $c_2(r, c_1)$ is the maximum of infinitely many linear functions in c_1; one for each $x \in \mathbb{N}_0$. Denote $\mathcal{F}_r$ as the (infinite) set of linear functions defining $c_2(r, c_1)$:

$$\mathcal{F}_r := \{ g_x^{(r)} : (0,1) \to \mathbb{R} \mid g_x^{(r)}(c_1) = (x+1)^r - c_1 \cdot x^{r+1}, x \in \mathbb{N}_0 \}$$

For the partial derivative of any function $(x, r, c_1) \mapsto g_x^{(r)}(c_1)$ we get

$$\frac{\partial((x+1)^r - c_1 \cdot x^{r+1})}{\partial r} = (x+1)^r \cdot \ln(x+1) - c_1 \cdot x^{r+1} \cdot \ln(x)$$

$$> \ln(x+1) \left[(x+1)^r - c_1 \cdot x^{r+1} \right] \geq 0,$$

for $(x+1)^r - c_1 \cdot x^{r+1} \geq 0$, that is, for the positive range of the chosen function from $\mathcal{F}_r$. Thus, the positive range of $(x+1)^d - c_1 \cdot x^{d+1}$ dominates the positive range of $(x+1)^r - c_1 \cdot x^{r+1}$ for all $0 \leq r \leq d$. Since $c_2(r, c_1) > 0$ for all $0 \leq r \leq d$, it follows that $c_2(d, c_1) \geq c_2(r, c_1)$, for all $0 \leq r \leq d$. Thus, without loss of generality, we may assume that $f(x) = x^d$.

For $s, t \in \mathbb{R}_{\geq 0}$ and $s \neq t$ define $\gamma(s, t)$ as the intersection abscissa of $g_s^{(d)}$ and $g_t^{(d)}$ (as in Lemma 2). Now consider the intersection of the two functions $g_v^{(d)}$ and $g_{v+1}^{(d)}$ from $\mathcal{F}_d$ for some $v \in \mathbb{N}$. We show that this intersection lies above all other functions from $\mathcal{F}_d$.

- First consider any function $g_z^{(d)}$ with $z > v + 1$. We have $g_z^{(d)}(0) > g_{v+1}^{(d)}(0) > g_v^{(d)}(0)$. Furthermore, by Lemma 2 we get $\gamma(v, z) < \gamma(v, v + 1)$. It follows that $g_v^{(d)}(\gamma(v, v + 1)) > g_z^{(d)}(\gamma(v, v + 1))$.
- Now consider any function $g_z^{(d)}$ with $z < v$. We have $g_{v+1}^{(d)}(0) > g_v^{(d)}(0) > g_z^{(d)}(0)$. Furthermore, by Lemma 2 we get $\gamma(v, z) > \gamma(v, v + 1)$. Again, it follows that $g_v^{(d)}(\gamma(v, v + 1)) > g_z^{(d)}(\gamma(v, v + 1))$.

Thus, all intersections of two consecutive linear functions from $\mathcal{F}_d$ lie on $c_2(d, c_1)$.

By (3), any point that lies on $c_2(d, c_1)$ gives an upper bound on PoA. Let k be the largest integer such that $(k + 1)^d \geq k^{d+1}$, that is $k = \lfloor \Phi_d \rfloor$. Then $(k + 2)^d < (k + 1)^{d+1}$. Choose c_1 and c_2 at the intersection of the two lines from $\mathcal{F}_d$ with $x = k$ and $x = k + 1$, that is $c_2 = (k + 1)^d - c_1 \cdot k^{d+1}$ and $c_2 = (k + 2)^d - c_1 \cdot (k + 1)^{d+1}$. Thus,

$$c_1 = \frac{(k + 2)^d - (k + 1)^d}{(k + 1)^{d+1} - k^{d+1}} \quad \text{and} \quad c_2 = \frac{(k + 1)^{2d+1} - (k + 2)^d \cdot k^{d+1}}{(k + 1)^{d+1} - k^{d+1}}.$$

Note that by the choice of k we have $0 < c_1 < 1$.

It follows that

$$\mathsf{PoA} \leq \frac{(k + 1)^{2d+1} - k^{d+1}(k + 2)^d}{(k + 1)^{d+1} - (k + 2)^d + (k + 1)^d - k^{d+1}}.$$

This completes the proof of the theorem. □

Theorem 2. *For unweighted congestion games with polynomial latency functions of maximum degree d and non-negative coefficients, we have*

$$\mathsf{PoA} \geq \frac{(k + 1)^{2d+1} - k^{d+1}(k + 2)^d}{(k + 1)^{d+1} - (k + 2)^d + (k + 1)^d - k^{d+1}}, \quad \text{where } k = \lfloor \Phi_d \rfloor.$$

Proof. Given the maximum degree $d \in \mathbb{N}$ for the polynomial latency functions, we construct a congestion game for $n \geq k + 2$ players and $|E| = 2n$ facilities.

We divide the set E into two subsets $E_1 := \{g_1, \ldots, g_n\}$ and $E_2 := \{h_1, \ldots, h_n\}$. Each player i has two pure strategies, $P_i := \{g_{i+1}, \ldots, g_{i+k}, h_{i+1}, \ldots, h_{i+k+1}\}$ and $Q_i := \{g_i, h_i\}$ where $g_j := g_{j-n}$ and $h_j := h_{j-n}$ for $j > n$. I.e. $S_i = \{Q_i, P_i\}$.

Each of the facilities in E_1 share the latency function $x \mapsto ax^d$ for an $a \in \mathbb{R}_{>0}$ (yet to be determined) whereas the facilities in E_2 have latency $x \mapsto x^d$.

Obviously, the optimal allocation $\mathbf{Q}$ is for every player i to choose Q_i. Now we determine a value for a such that the allocation $\mathbf{P} := (P_1, \ldots, P_n)$ becomes a Nash Equilibrium, i.e., each player i is satisfied with $\mathbf{P}$, that is $\mathsf{PC}_i(\mathbf{P}) \leq \mathsf{PC}_i(\mathbf{P}_{-i}, Q_i)$ for all $i \in [n]$, or equivalently $k \cdot a \cdot k^d + (k + 1) \cdot (k + 1)^d \leq a \cdot (k + 1)^d + (k + 2)^d$. Resolving to the coefficient a gives

$$a \geq \frac{(k + 1)^{d+1} - (k + 2)^d}{(k + 1)^d - k^{d+1}} > 0. \tag{6}$$

Because $(k + 1)^d \neq k^{d+1}$, due to either $k + 1$ or k being odd and the other being even, a is well defined and positive. Now since for any player i the private costs are $\mathsf{PC}_i(\mathbf{Q}) = a + 1$ and $\mathsf{PC}_i(\mathbf{P}) = a \cdot k^{d+1} + (k + 1)^{d+1}$, it follows that

$$\frac{\mathsf{SC}(\mathbf{P})}{\mathsf{SC}(\mathbf{Q})} = \frac{\sum_{i \in [n]} \mathsf{PC}_i(\mathbf{P})}{\sum_{i \in [n]} \mathsf{PC}_i(\mathbf{Q})} = \frac{a \cdot k^{d+1} + (k + 1)^{d+1}}{a + 1}. \tag{7}$$

Provided that $(k+1)^d \geq k^{d+1}$, it is not hard to see that (7) is monotonically decreasing in a. Thus, we assume equality in (6), which then gives

$$\mathsf{PoA} \geq \frac{\mathsf{SC}(\mathbf{P})}{\mathsf{SC}(\mathbf{Q})} = \frac{(k+1)^{2d+1} - k^{d+1}(k+2)^d}{(k+1)^{d+1} - (k+2)^d + (k+1)^d - k^{d+1}}.$$

This completes the proof of the theorem. □

Corollary 1. *The lower bound in Theorem 2 on* PoA *also holds for unweighted network congestion games.*

4   Price of Anarchy for Weighted Congestion Games

In this section, we prove the exact value for the price of anarchy of weighted congestion games with polynomial latency functions. The proof of the upper bound in Theorem 3 has a similar structure as the one for the unweighted case (cf. Theorem 1). In Theorem 4 we give a matching lower bound which also holds for weighted network congestion games (Corollary 2). Corollary 3 shows the impact of player weights to the price of anarchy.

Theorem 3. *For weighted congestion games with polynomial latency functions of maximum degree d and non-negative coefficients we have* $\mathsf{PoA} \leq \Phi_d^{d+1}$.

Theorem 4. *For weighted congestion games with polynomial latency functions of maximum degree d and non-negative coefficients, we have* $\mathsf{PoA} \geq \Phi_d^{d+1}$.

Proof. Given the maximum degree $d \in \mathbb{N}$ for the polynomial latency functions, set $k \geq \max\{\binom{d}{\lfloor d/2 \rfloor}, 2\}$. Note, that $\binom{d}{\lfloor d/2 \rfloor} = \max_{j \in [d]_0}\binom{d}{j}$. We construct a congestion game for $n = (d+1) \cdot k$ players and $|E| = n$ facilities.

We divide the set E into $d+1$ partitions: For $i \in [d]_0$, let $E_i := \{g_{i,1}, \ldots, g_{i,k}\}$, with each $g_{i,j}$ sharing the latency function $x \mapsto a_i \cdot x^d$. The values of the coefficients a_i will be determined later. For simplicity of notation, set $g_{i,j} := g_{i,j-k}$ for $j > k$ in the following.

Similarly, we partition the set of players $[n]$: For $i \in [d]_0$, let $N_i := \{u_{i,1}, \ldots, u_{i,k}\}$. The weight of each player in set N_i is Φ_d^i, so $w_{u_{i,j}} = \Phi_d^i$ for all $i \in [d]_0, j \in [k]$.

Now, for every set N_i, each player $u_{i,j} \in N_i$ has exactly two strategies:

$$Q_{u_{i,j}} := \{g_{i,j}\} \quad \text{and} \quad P_{u_{i,j}} := \begin{cases} \{g_{d,j+1}, \ldots, g_{d,j+\binom{d}{i}}, g_{i-1,j}\} & \text{for } i = 1 \text{ to } d \\ \{g_{d,j+1}\} & \text{for } i = 0 \end{cases}$$

Now let $\mathbf{Q} := (Q_1, \ldots, Q_n)$ and $\mathbf{P} := (P_1, \ldots, P_n)$ be strategy profiles. The facilities in each set E_i then have the following loads for $\mathbf{Q}$ and $\mathbf{P}$:

		load on every facility $e \in E_i$	
i	$l_e(\mathbf{Q})$	$l_e(\mathbf{P})$	
d	Φ_d^d	$\sum_{l=0}^{d}\binom{d}{l}\Phi_d^l = (\Phi_d + 1)^d = \Phi_d^{d+1}$	
0 to $d-1$	Φ_d^i	Φ_d^{i+1}	

For $\mathbf{P}$ to become a Nash Equilibrium, we need to fulfill the following Nash inequalities for each set N_i of players:

i	Nash inequality to fulfill
1 to d	$PC_{u_{i,j}}(\mathbf{P}) = \binom{d}{i} a_d (\Phi_d^{d+1})^d \mid u_{i-1} \cdot (\Phi_d^i)^d$ $\leq a_i \cdot (\Phi_d^{i+1} + \Phi_d^i)^d = PC_{u_{i,j}}(\mathbf{P}_{-u_{i,j}}, Q_{u_{i,j}})$
0	$PC_{u_{0,j}}(\mathbf{P}) = a_d \cdot (\Phi_d^{d+1})^d \leq a_0 \cdot (\Phi_d + 1)^d = PC_{u_{0,j}}(\mathbf{P}_{-u_{0,j}}, Q_{u_{0,j}})$

Replacing "$\leq$" by "$=$" yields a homogeneous system of linear equations, i.e., the system $B_d \cdot a = 0$ where B_d is the following $(d+1) \times (d+1)$ matrix:

$$B_d = \begin{pmatrix} -\Phi_d^{d^2+d+1} + \Phi_d^{d^2+d} & \Phi_d^{d^2} & 0 & \cdots & & & \cdots & 0 \\ \binom{d}{d-1}\Phi_d^{d^2+d} & -\Phi_d^{d^2+1} & \ddots & & & & & \vdots \\ \vdots & 0 & \ddots & & & & & \\ \vdots & \vdots & & \ddots & \ddots & & & \vdots \\ \binom{d}{i}\Phi_d^{d^2+d} & 0 & \cdots & 0 & -\Phi_d^{id+d+1} & \Phi_d^{id} & 0 & \cdots & 0 \\ \vdots & \vdots & & & 0 & \ddots & \ddots & & \vdots \\ \vdots & \vdots & & & \vdots & & \ddots & 0 \\ \vdots & \vdots & & & \vdots & & & \ddots & \Phi_d^d \\ \Phi_d^{d^2+d} & 0 & \cdots & & 0 & \cdots & & 0 & -\Phi_d^{d+1} \end{pmatrix} \qquad (8)$$

and $a := (a_d \ldots a_0)^t$. Obviously, a solution to this system fulfills the initial Nash inequalities. Note that

$$(\Phi_d^{i+1} + \Phi_d^i)^d = (\Phi_d^i)^d \cdot (\Phi_d + 1)^d = \Phi_d^{id+d+1}.$$

Claim. The $(d+1) \times (d+1)$ matrix B_d from (8) has rank d.

Proof. We use the well-known fact from linear algebra that if a matrix C results from another matrix D by adding a multiple of one row (or column) to another row (or column, respectively) then $\mathrm{rank}(C) = \mathrm{rank}(D)$.

Now consider the matrix C_d that results from adding row j multiplied by the factor Φ_d^{-1} to row $j-1$, sequentially done for $j = d+1, d, \ldots, 2$. Obviously, C_d is a lower triangular matrix with nonzero elements only in the first column and on the principal diagonal.

For the top left element of C_d we get

$$-\Phi_d^{d^2+d+1} + \sum_{j=0}^{d} \binom{d}{j} \Phi_d^{d^2+j} = \Phi_d^{d^2} \cdot \left(-\Phi_d^{d+1} + \underbrace{\sum_{j=0}^{d} \binom{d}{j} \Phi_d^j}_{(\Phi_d+1)^d} \right) = 0.$$

Since all elements on the principal diagonal of C_d—with the just shown exception of the first one—are nonzero, it is easy to see that C_d (and thus also B_d) has rank d. $\quad\square$

By the above claim it follows that the column vectors of B_d are linearly dependent and thus there are—with degree of freedom 1—infinitely many linear combinations of them yielding 0. In other words, $B_d \cdot a = 0$ has a one-dimensional solution space.

We now show (by induction over i) that all coefficients a_i, $i \in [d]_0$ must have the same sign and thus we can always find a valid solution. From the last equality, for $i = 0$, we have that a_d and a_0 must have the same sign. Now for $i = 1, \ldots, d - 1$, it follows that a_i must have the same sign as a_{i-1} and a_d, for $(\Phi_d^{d+1})^d$, $(\Phi_d^i)^d$, and $(\Phi_d^{i+1} + \Phi_d^i)^d$ are all positive.

Choosing $a \neq 0$ with all components being positive, all coefficients of the latency functions are positive. We get,

$$\mathsf{PoA} \geq \frac{\mathsf{SC}(\mathbf{P})}{\mathsf{SC}(\mathbf{Q})} = \frac{k \cdot \sum_{i=0}^{d} a_i (\Phi_d^{i+1})^{d+1}}{k \cdot \sum_{i=0}^{d} a_i (\Phi_d^i)^{d+1}} = \Phi_d^{d+1}.$$

$\square$

Corollary 2. *The lower bound in Theorem 4 on* PoA *also holds for weighted network congestion games.*

Corollary 3. *The exact price of anarchy for* unweighted congestion games

$$\mathsf{PoA} = \frac{(k + 1)^{2d+1} - k^{d+1}(k + 2)^d}{(k + 1)^{d+1} - (k + 2)^d + (k + 1)^d - k^{d+1}},$$

where $k = \lfloor \Phi_d \rfloor$, *is bounded by* $\lfloor \Phi_d \rfloor^{d+1} \leq \mathsf{PoA} \leq \Phi_d^{d+1}$.

References

1. B. Awerbuch, Y. Azar, and A. Epstein. The Price of Routing Unsplittable Flow. In *Proc. of the 37th Annual ACM Symposium on Theory of Computing (STOC'05)*, pages 57–66, 2005.
2. B. Awerbuch, Y. Azar, Y. Richter, and D. Tsur. Tradeoffs in Worst-Case Equilibria. *Proc. of the 1st Int. Workshop on Approximation and Online Algorithms (WAOA'03)*, LNCS 2909, pages 41–52, 2003.
3. M. Beckmann, C. B. McGuire, and C. B. Winsten. *Studies in the Economics of Transportation*. Yale University Press, 1956.
4. G. Christodoulou and E. Koutsoupias. On The Price of Anarchy and Stability of Correlated Equilibria of Linear Congestion Games. In *Proc. of the 13th Annual European Symposium on Algorithms (ESA'05)*, LNCS 3669, pages 59–70, 2005.
5. G. Christodoulou and E. Koutsoupias. The Price of Anarchy of Finite Congestion Games. In *Proc. of the 37th Annual ACM Symposium on Theory of Computing (STOC'05)*, pages 67–73, 2005.
6. A. Czumaj, P. Krysta, and B. Vöcking. Selfish Traffic Allocation for Server Farms. In *Proc. of the 34th Annual ACM Symposium on Theory of Computing (STOC'02)*, pages 287–296, 2002.
7. A. Czumaj and B. Vöcking. Tight Bounds for Worst-Case Equilibria. In *Proc. of the 13th Annual ACM-SIAM Symposium on Discrete Algorithms (SODA'02)*, pages 413–420, 2002. Also accepted to *Journal of Algorithms* as Special Issue of SODA'02.
8. D. Fotakis, S. Kontogiannis, and P. Spirakis. Selfish Unsplittable Flows. *Proc. of the 31st Int. Colloquium on Automata, Languages, and Programming (ICALP'04)*, LNCS 3142, pages 593–605, 2004.

9. D. Fotakis, S. Kontogiannis, and P. Spirakis. Symmetry in Network Congestion Games: Pure Equilibria and Anarchy Cost. In *Proc. of the 3rd Int. Workshop on Approximation and Online Algorithms (WAOA'05)*, 2005.

10. M. Gairing, T. Lücking, M. Mavronicolas, and B. Monien. Computing Nash Equilibria for Scheduling on Restricted Parallel Links. In *Proc. of the 36th Annual ACM Symposium on Theory of Computing (STOC'04)*, pages 613–622, 2004.

11. M. Gairing, T. Lücking, M. Mavronicolas, and B. Monien. The Price of Anarchy for Polynomial Social Cost. *Proc. of the 29th Int. Symposium on Mathematical Foundations of Computer Science (MFCS'04)*, LNCS 3153, pages 574–585, 2004.

12. M. Gairing, T. Lücking, M. Mavronicolas, B. Monien, and M. Rode. Nash Equilibria in Discrete Routing Games with Convex Latency Functions. *Proc. of the 31st Int. Colloquium on Automata, Languages, and Programming (ICALP'04)*, LNCS 3142, pages 645–657, 2004.

13. M. Gairing, T. Lücking, B. Monien, and K. Tiemann. Nash Equilibria, the Price of Anarchy and the Fully Mixed Nash Equilibrium Conjecture. In *Proc. of the 32nd Int. Colloquium on Automata, Languages, and Programming (ICALP'05)*, LNCS 3850, pages 51–65, 2005.

14. M. Gairing, B. Monien, and K. Tiemann. Selfish Routing with Incomplete Information. In *Proc. of the 17th Annual ACM Symposium on Parallel Algorithms and Architectures (SPAA'05)*, pages 203–212, 2005.

15. E. Koutsoupias, M. Mavronicolas, and P. Spirakis. Approximate Equilibria and Ball Fusion. *Theory of Computing Systems*, 36(6):683–693, 2003.

16. E. Koutsoupias and C. H. Papadimitriou. Worst-Case Equilibria. *Proc. of the 16th Int. Symposium on Theoretical Aspects of Computer Science (STACS'99)*, LNCS 1563, pages 404–413, 1999.

17. T. Lücking, M. Mavronicolas, B. Monien, and M. Rode. A New Model for Selfish Routing. *Proc. of the 21st Int. Symposium on Theoretical Aspects of Computer Science (STACS'04)*, LNCS 2996, pages 547–558, 2004.

18. A. Mas-Colell, M. D. Whinston, and J. R. Green. *Microeconomic Theory*. Oxford University Press, 1995.

19. R. D. McKelvey and A. McLennan. Computation of Equilibria in Finite Games. *Handbook of Computational Economics*, 1996.

20. I. Milchtaich. Congestion Games with Player-Specific Payoff Functions. *Games and Economic Behavior*, 13(1):111–124, 1996.

21. D. Monderer and L. S. Shapley. Potential Games. *Games and Economic Behavior*, 14(1):124–143, 1996.

22. J. F. Nash. Equilibrium Points in n-Person Games. *Proc. of the National Academy of Sciences of the United States of America*, 36:48–49, 1950.

23. J. F. Nash. Non-Cooperative Games. *Annals of Mathematics*, 54(2):286–295, 1951.

24. C. H. Papadimitriou. Algorithms, Games, and the Internet. In *Proc. of the 33rd Annual ACM Symposium on Theory of Computing (STOC'01)*, pages 749–753, 2001.

25. R. W. Rosenthal. A Class of Games Possessing Pure-Strategy Nash Equilibria. *Int. Journal of Game Theory*, 2:65–67, 1973.

26. T. Roughgarden. How Unfair is Optimal Routing. In *Proc. of the 13th Annual ACM-SIAM Symposium on Discrete Algorithms (SODA'02)*, pages 203–204, 2002.

27. T. Roughgarden. *Selfish Routing and the Price of Anarchy*. MIT Press, 2005.

28. T. Roughgarden and É. Tardos. How Bad Is Selfish Routing? *Journal of the ACM*, 49(2):236–259, 2002.

29. J. G. Wardrop. Some Theoretical Aspects of Road Traffic Research. In *Proc. of the Institute of Civil Engineers, Pt. II, Vol. 1*, pages 325–378, 1952.

Oblivious Symmetric Alternation

Venkatesan T. Chakaravarthy[1] and Sambuddha Roy[2]

[1] IBM India Research Lab, New Delhi
vechakra@in.ibm.com
[2] Department of Computer Science, Rutgers University
samroy@paul.rutgers.edu

Abstract. We introduce a new class O_2^p as a subclass of the symmetric alternation class S_2^p. An O_2^p proof system has the flavor of an S_2^p proof system, but it is more restrictive in nature. In an S_2^p proof system, we have two competing provers and a verifier such that for any input, the honest prover has an irrefutable certificate. In an O_2^p proof system, we require that the irrefutable certificates depend only on the length of the input, not on the input itself. In other words, the irrefutable proofs are oblivious of the input. For this reason, we call the new class *oblivious symmetric alternation*. While this might seem slightly contrived, it turns out that this class helps us improve some existing results. For instance, we show that if $NP \subset P/poly$ then $PH = O_2^p$, whereas the best known collapse under the same hypothesis was $PH = S_2^p$.

We also define classes YO_2^p and NO_2^p, bearing relations to O_2^p as NP and coNP are to P, and show that these along with O_2^p form a hierarchy, similar to the polynomial hierarchy. We investigate other inclusions involving these classes and strengthen some known results. For example, we show that $MA \subseteq NO_2^p$ which sharpens the known result $MA \subseteq S_2^p$ [16]. Another example is our result that $AM \subseteq O_2 \cdot NP \subseteq \Pi_2^p$, which is an improved upper bound on AM. Finally, we also prove better collapses for the 2-queries problem as discussed by [12,1,7]. We prove that $P^{NP[1]} = P^{NP[2]} \Rightarrow PH = NO_2^p \cap YO_2^p$.

1 Introduction

The symmetric alternation class (S_2^p) was introduced independently by Russell and Sundaram [16] and by Canetti [5]. The class S_2^p contains languages having an interactive proof system of the following type. The proof system consists of two mutually adversarial and computationally all-powerful provers called the YES-PROVER and the NO-PROVER, and a polynomial time verifier. The verifier interacts with the two provers to ascertain whether or not an input string x belongs to a language L. The YES-PROVER and the NO-PROVER make contradictory claims: $x \in L$ and $x \notin L$, respectively. Of course, only one of them is honest. To substantiate their claims, the provers give strings y and z as certificates. The verifier analyzes the input x and the two certificates and votes in favor of one of the provers. The requirement is that, if $x \in L$, the YES-PROVER has a certificate y using which he can win the vote, for any certificate z of the NO-PROVER.

B. Durand and W. Thomas (Eds.): STACS 2006, LNCS 3884, pp. 230–241, 2006.
© Springer-Verlag Berlin Heidelberg 2006

Similarly, if $x \notin L$, the NO-PROVER has a certificate z using which he can win the vote, for any certificate y of the YES-PROVER. We call certificates satisfying the above requirements as *irrefutable certificates*. We can rephrase the requirements as follows. If $x \in L$, the YES-PROVER has an irrefutable certificate and if $x \notin L$, then the NO-PROVER has an irrefutable certificate. The class S_2^p consists of languages having a proof system of the above type. We will provide a formal definition of S_2^p later. Symmetric alternation has been gaining attention recently and several nice results involving the class are known (see [3,4,5,6,7,8,16,18]).

In this paper, we define a class called *oblivious symmetric alternation*, denoted O_2^p, as a subclass of S_2^p by incorporating a few additional requirements to the S_2^p proof system. We show that some of the earlier S_2^p-related results can be strengthened by using O_2^p and related classes. We study these classes and show that they enjoy some interesting properties. Our results seem to indicate that these classes are worthy of further investigation. We start with an informal description of O_2^p.

Similar to S_2^p, an O_2^p proof system for a language L consists of two competing all powerful provers, the YES-PROVER and the NO-PROVER and a polynomial time verifier. For an input x the two provers make contradictory claims: the YES-PROVER claims $x \in L$ and NO-PROVER claims $x \notin L$. To substantiate their claims the provers present polynomially long certificates and the verifier analyzes the input and the two certificates and votes in favor of one of the provers. The requirement is that, for any n, there exists a pair of certificates (y^*, z^*) such that y^* serves as an irrefutable certificate for the YES-PROVER, for all strings in $L \cap \{0,1\}^n$ and similarly, z^* serves as an irrefutable certificate for the NO-PROVER, for all strings in $\overline{L} \cap \{0,1\}^n$. The difference between S_2^p and O_2^p is as follows. Fix an input length n. In an S_2^p proof system, we require that for any input x of length n, if $x \in L$, then the YES-PROVER should have an irrefutable certificate, and if $x \notin L$, then the NO-PROVER should have an irrefutable certificate. Whereas, in an O_2^p proof system, the YES-PROVER should have a single string y^* which is an irrefutable certificate for all strings in L of length n, and similarly, the NO-PROVER should have a string z^* which is an irrefutable certificate for all strings not in L of length n. In a nutshell, the irrefutable certificates of in an S_2^p proof system may depend on the input, whereas in an O_2^p proof system, they depend only on the length of the input – the certificates are *oblivious* of the input. Borrowing terminology from the theory of non-uniform computation, we call y^* and z^* as *irrefutable advice* at length n for the YES-PROVER and the NO-PROVER, respectively.

The class O_2^p can be used to strengthen some of the earlier results involving S_2^p. The first such result we consider is the classical Karp–Lipton theorem. Karp and Lipton [14] showed that if NP $\subset$ P/poly then the polynomial time hierarchy (PH) collapses to $\Sigma_2^p \cap \Pi_2^p$. Köbler and Watanabe [15] improved the collapse as PH $=$ ZPP^{NP}. Sengupta observed that under the same hypothesis, we have PH $= S_2^p$ (see [3]), which is an improvement over the Köbler–Watanabe result, since by Cai's result [3], $S_2^p \subseteq$ ZPP^{NP}. We strengthen the collapse further. We show that if NP $\subset$ P/poly then PH $= O_2^p$. By definition, $O_2^p \subseteq S_2^p$ and thus, the collapse of PH to O_2^p improves the earlier collapse to S_2^p. But, how much is this an improvement? To answer this question, we compare S_2^p and O_2^p.

O_2^p seems to be much weaker than S_2^p as evidenced by our observation that $O_2^p \subset$ P/poly. Hence, if S_2^p, or even NP, is contained in O_2^p, then PH collapses to O_2^p. Thus, while S_2^p is a stronger class containing P^{NP} [16], O_2^p is unlikely to contain even NP.

We next consider the related issue of whether NP is contained in coNP/poly (or equivalently, coNP $\subset$ NP/poly). Yap [19] showed that if coNP $\subset$ NP/poly, then PH $= \Sigma_3^p \cap \Pi_3^p$. Köbler and Watanabe improved the collapse as PH $=$ $\text{ZPP}^{\text{NP}^{\text{NP}}}$. This was further strengthened as PH $= S_2^{\text{NP}}$ [4,17]. We improve the result further by showing that PH $= O_2^{\text{NP}}$. In fact, we show that if coNP $\subseteq$ NP/poly then PH $= O_2 \cdot \text{P}^{\text{NP}[2]}$. (In $O_2 \cdot \text{P}^{\text{NP}[2]}$, we allow the verifier to make only two queries to an NP oracle).

We also investigate the lowness properties of S_2^p and O_2^p and show that O_2^p nicely connects some earlier results regarding low sets of S_2^p. It is known that BPP is low for S_2^p [16,5]. We observe that this result can be generalized as follows: the class IP[P/poly] [10], which contains BPP, is low for S_2^p. On a different note, Cai et al. [4] showed that the class of Turing self-reducible languages having polynomial size circuits is low for S_2^p. Using O_2^p, we connect these two seemingly unrelated results. We prove that O_2^p contains both IP[P/poly] and the class of Turing self-reducible languages having polynomial size circuits. Then, we argue that O_2^p is low for S_2^p. Moreover, we observe that O_2^p is low for O_2^p.

We then proceed to study one-sided versions of O_2^p and define two classes YO_2^p and NO_2^p. In the YO_2^p proof system, only the YES-PROVER is required to present an irrefutable advice, whereas it is sufficient for the NO-PROVER to present irrefutable certificates. In other words, the irrefutable certificates of the NO-PROVER may depend on the input, whereas those of the YES-PROVER must depend only the length of the input. The class NO_2^p is defined analogously by interchanging the above requirements of the YES-PROVER and NO-PROVER. Notice that YO_2^p and NO_2^p are complementary classes: $L \in \text{YO}_2^p$ if and only if $\overline{L} \in \text{NO}_2^p$.

We study the properties of YO_2^p and NO_2^p and use these classes to sharpen some of the earlier results involving S_2^p. It is known that MA and coMA are contained in S_2^p [16]. We strengthen the above result as MA $\subseteq \text{NO}_2^p$ and coMA $\subseteq$ YO_2^p. Building on the work of Buhrman and Fortnow [1], Fortnow, Pavan and Sengupta [7] showed that if $\text{P}^{\text{NP}[1]} = \text{P}^{\text{NP}[2]}$ then PH $= S_2^p$. Under the same hypothesis, we improve the collapse to PH $= \text{YO}_2^p \cap \text{NO}_2^p$.

We illustrate the relationships among O_2^p, YO_2^p, NO_2^p and other classes in Figure 1.

Finally, we use O_2^p, NO_2^p and YO_2^p to build an hierarchy of classes similar to the polynomial time hierarchy. In this new hierarchy, which we call the *oblivious symmetric alternation hierarchy* (OSH), O_2^p, NO_2^p and YO_2^p play the role of P, NP and coNP, respectively. As it turns out, the *oblivious symmetric alternation hierarchy* behaves similarly to PH and we discuss other features of this hierarchy in Sec. 6. For instance, it is true that OSH is finite if and only if PH is finite.

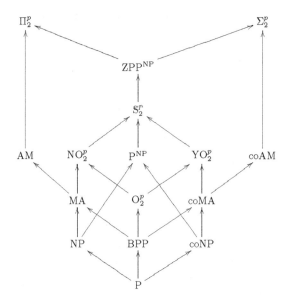

Fig. 1. O_2^p and other classes

Note: Due to the lack of space, we are unable to present proofs for most of the results. These will be included in the full version of the paper, which is under preparation.

2   Definitions

We start with the formal definition of S_2^p. For the sake of brevity, we straightaway define the relativized version.

Definition 1 ([5,16]). *Let C be a complexity class. A language L is said to be in the class $S_2 \cdot C$, if there exists a 3-argument predicate $V \in C$ such that for all n, for all $x \in \{0,1\}^n$,*

$$x \in L \implies \exists^m y \forall^m z \ [V(x,y,z) = 1] \quad and$$
$$x \notin L \implies \exists^m z \forall^m y \ [V(x,y,z) = 0],$$

where m is polynomial in n. For an $x \in L$ (resp. $x \notin L$), any y (resp. z) satisfying the first (resp. second) requirement above is called an irrefutable certificate *of the* Yes-prover *(resp. the* No-prover*).*

We next define O_2^p, YO_2^p and NO_2^p. Again, we straightaway present the relativized versions.

Definition 2. *Let C be a complexity class. A language L is said to be in the class $O_2 \cdot C$, if there exists a 3-argument predicate $V \in C$ such that, for all n, there exist* *strings* y^* *and* z^*, *with* $|y^*| = |z^*| = m$, *satisfying the following requirements, where m is polynomial in n. For all $x \in \{0,1\}^n$,*

$$x \in L \Longrightarrow \forall^m z \ [V(x, y^*, z) = 1] \quad and$$
$$x \notin L \Longrightarrow \forall^m y \ [V(x, y, z^*) = 0].$$

We call y^* *and* z^* irrefutable advice *at length n, for the* YES-PROVER *and* NO-PROVER, *respectively.*

Definition 3. *Let $\mathcal{C}$ be a complexity class. A language L is said to be in the class* $\mathrm{NO}_2 \cdot \mathcal{C}$, *if there exists a 3-argument predicate $V \in \mathcal{C}$ such that, for all n, there exist a string z^*, with $|z^*| = m$, satisfying the following requirements, where m is polynomial in n. For all $x \in \{0,1\}^n$,*

$$x \in L \Longrightarrow \exists^m y \forall^m z \ [V(x, y, z) = 1] \quad and$$
$$x \notin L \Longrightarrow \forall^m y \ [V(x, y, z^*) = 0].$$

We call z^* irrefutable advice *at length n for the* NO-PROVER *and for an $x \in L$, any string y satisfying the first requirement is called an* irrefutable certificate of *the* YES-PROVER *for x.*

We define $\mathrm{YO}_2 \cdot \mathcal{C}$ similarly : in this case the YES-PROVER has an *irrefutable advice* string, and the NO-PROVER has an *irrefutable certificate*.

We denote by S_2^p, O_2^p, YO_2^p, and NO_2^p, the classes $\mathrm{S}_2 \cdot \mathrm{P}$, $\mathrm{O}_2 \cdot \mathrm{P}$, $\mathrm{YO}_2 \cdot \mathrm{P}$ and $\mathrm{NO}_2 \cdot \mathrm{P}$, respectively. For any oracle A, S_2^A, O_2^A, YO_2^A, NO_2^A, denote the classes $\mathrm{S}_2 \cdot \mathrm{P}^A$, $\mathrm{O}_2 \cdot \mathrm{P}^A$, $\mathrm{YO}_2 \cdot \mathrm{P}^A$ and $\mathrm{NO}_2 \cdot \mathrm{P}^A$, respectively. Similarly, for a class $\mathcal{C}$, $\mathrm{S}_2^{\mathcal{C}}$, $\mathrm{O}_2^{\mathcal{C}}$, $\mathrm{YO}_2^{\mathcal{C}}$ and $\mathrm{NO}_2^{\mathcal{C}}$ denote the classes $\mathrm{S}_2 \cdot \mathrm{P}^{\mathcal{C}}$, $\mathrm{O}_2 \cdot \mathrm{P}^{\mathcal{C}}$, $\mathrm{YO}_2 \cdot \mathrm{P}^{\mathcal{C}}$ and $\mathrm{NO}_2 \cdot \mathrm{P}^{\mathcal{C}}$, respectively.

3   Oblivious Symmetric Alternation and Other Classes

In this section, we compare O_2^p, NO_2^p and YO_2^p with other standard complexity classes. We start with the following easy (syntactic) observations:

Proposition 1. *The following (relativizing) inclusions hold for the classes* O_2^p, NO_2^p, YO_2^p :

(i) $\mathrm{O}_2^p \subseteq \mathrm{NO}_2^p \cap \mathrm{YO}_2^p$
(ii) $\mathrm{coNO}_2^p = \mathrm{YO}_2^p$
(iii) $\mathrm{NP} \subseteq \mathrm{NO}_2^p \subseteq \mathrm{S}_2^p$
(iv) $\mathrm{coNP} \subseteq \mathrm{YO}_2^p \subseteq \mathrm{S}_2^p$

So we know thereby that $\mathrm{O}_2^p \subseteq \mathrm{S}_2^p$ - how much weaker is O_2^p as compared to S_2^p? We observe below that $\mathrm{O}_2^p \subset \mathrm{P}/\mathrm{poly}$. On the other hand, S_2^p contains NP, and NP is unlikely to have polynomial size circuits (for that would collapse PH). This gives credence to the belief that O_2^p might indeed be quite weaker than S_2^p.

Theorem 1. $\mathrm{O}_2^p \subset \mathrm{P}/\mathrm{poly}$. *In general, for any oracle A, $\mathrm{O}_2^A \subset \mathrm{P}^A/\mathrm{poly}$. Similarly,* $\mathrm{NO}_2^p \subset \mathrm{NP}/\mathrm{poly}$, $\mathrm{YO}_2^p \subset \mathrm{coNP}/\mathrm{poly}$, *and these inclusions relativize as* $\mathrm{NO}_2^A \subset \mathrm{NP}^A/\mathrm{poly}$, $\mathrm{YO}_2^A \subset \mathrm{coNP}^A/\mathrm{poly}$.

We can ask other warm-up questions about these new classes, for instance, is coNP $\subseteq$ NO$_2^p$? This is unlikely, because from the above we know that NO$_2^p$ $\subset$ NP/poly, so such a containment would imply that coNP $\subset$ NP/poly, which is known to imply a collapse of the polynomial time hierarchy: PH $=$ S$_2^{NP}$ [4,17]. We obtain a better collapse under the hypothesis coNP is contained in NO$_2^p$.

Theorem 2. *If* coNP $\subseteq$ NO$_2^p$, *then* PH $=$ NO$_2^p$ $\cap$ YO$_2^p$.

We now compare Arthur-Merlin classes with the O$_2$ classes, thereby improving some of the syntactic inclusions in Proposition 1.

Theorem 3. MA $\subseteq$ NO$_2^p$, coMA $\subseteq$ YO$_2^p$.

While the intuition gained so far about the class O$_2^p$ is that it is a relatively small class, we are not able to show containment of O$_2^p$ inside the class AM. On the other hand, we are able to prove an upper bound on AM better than the previously known upper bound of Π_2^p.

Theorem 4. AM $\subseteq$ O$_2 \cdot$ NP $\subseteq \Pi_2^p$.

4   Low Sets of S$_2^p$ and O$_2^p$

In this section, we connect some earlier lowness results of S$_2^p$ using O$_2^p$. We also give a precise characterization of low sets of O$_2^p$.

Russell and Sundaram [16] and Canetti [5] showed that BPP is low for S$_2^p$. We can prove that IP[P/poly] is low for S$_2^p$, where IP[P/poly] consists of languages having an interactive proof system in which the power of the honest prover lies in P/poly, see [10]. Since BPP $\subseteq$ IP[P/poly], this latter lowness result implies the former. On a different note, Cai et al. [4] showed that any Turing self-reducible language in P/poly is low for S$_2^p$. We connect these two seemingly unrelated lowness results using O$_2^p$. We show that IP[P/poly] $\subseteq$ O$_2^p$ and that any Turing self-reducible language in P/poly belongs to O$_2^p$. Then, we prove that O$_2^p$ is low for S$_2^p$.

Theorem 5. IP[P/poly] $\subseteq$ O$_2^p$.

Theorem 6. *If a language A is Turing self-reducible and $A \in$ P/poly then $A \in$ O$_2^p$.*

Cai et al. [4] prove a theorem similar to Theorem 6: the set A is low for S$_2^p$. Our lowness result is proved via an argument similar to theirs. We next show:

Theorem 7. O$_2^p$ *is low for* S$_2^p$, *i.e.,* S$_2^{O_2^p}$ $=$ S$_2^p$.

Theorem 7 gives a partial characterization of low sets of S$_2^p$. We can do a better job in the case of O$_2^p$. Below, we show that O$_2^p$ is low for O$_2^p$. As a consequence, we can precisely specify the low sets of O$_2^p$: **Low**(O$_2^p$) $=$ O$_2^p$.

Theorem 8. O_2^p is low for O_2^p, i.e., $O_2^{O_2^p} = O_2^p$.

Also, similar lowness results hold for NO_2^p, YO_2^p :

Theorem 9. O_2^p is low for NO_2^p and YO_2^p. The claim relativizes with respect to any oracle A: $NO_2^{O_2^A} = NO_2^A$ and $YO_2^{O_2^A} = YO_2^A$.

5   Strengthening Collapses Using O_2^p

In this section, we show that some earlier collapse results involving S_2^p and related classes can be improved by using O_2^p and its relativized versions.

The first result we consider is the classical Karp–Lipton theorem [14], which deals with the issue of NP having polynomial size circuits. The theorem states that if $NP \subset P/poly$ then $PH = \Sigma_2^p \cap \Pi_2^p$. The collapse has been strengthened subsequently. Under the same hypothesis, the collapse was improved as $PH = ZPP^{NP}$ [15] and further as $PH = S_2^p$ (see Cai [3]). (Note $S_2^p \subseteq ZPP^{NP}$ [3]). We further strengthen the collapse to $PH = O_2^p$.

For this section and Section 7, we define following:

Definition 4. A circuit (purportedly computing SAT for a specific input length) is called nice if it does not accept unsatisfiable formulas. We note that, via self-reducibility, we can design a polynomial time algorithm that converts a given circuit C into a nice circuit C', such that if C is a correct circuit for SAT at a certain length m, then so is C' (at length m).

Theorem 10. If $NP \subset P/poly$ then $PH = O_2^p$.

Proof. Assuming $NP \subset P/poly$, we show that $\Sigma_2^p \subseteq O_2^p$, which implies that $PH = O_2^p$. Let $L \in \Sigma_2^p$. We have polynomials $p(\cdot)$ and $q(\cdot)$, and a polynomial time algorithm which given an x and a $y \in \{0,1\}^{p(|x|)}$, produces a formula $\varphi_{x,y}$ of length at most $q(|x|)$ such that the following is true. For any x,

$$x \in L \iff \exists^{p(|x|)} y [\varphi_{x,y} \notin SAT].$$

Our O_2^p proof system works as follows. Let $x \in \{0,1\}^n$ be the input. The YES-PROVER and the No-PROVER will provide circuits C_Y and C_N respectively, that are purported to compute SAT at lengths $r(n)$, $q(n)$ respectively ($r(n)$ to be defined shortly). The verifier converts the circuits C_Y, C_N into nice (cf. Definition 4) circuits C_Y', C_N' respectively. The verifier's algorithm is as follows. Convert the following NP question:

$$\exists^{p(n)} y [C_N'(\varphi_{x,y}) = 0].$$

into a boolean formula $\tilde{\varphi}_x$. The length of $\tilde{\varphi}_x$ is at most $r(n)$, where $r(\cdot)$ is a polynomial. We then use C_Y' to check the satisfiability of $\tilde{\varphi}_x$. If C_Y' accepts the formula, we vote in favor of the YES-PROVER and otherwise vote in favor of the No-PROVER.

Consider an input length n. Let C_Y^* and C_N^* be correct circuits for SAT at lengths $r(n)$ and $q(n)$, respectively, whose existence is guaranteed by our assumption that $\text{NP} \subset \text{P/poly}$. We can then argue that at length n, C_Y^* and C_N^* are irrefutable advice strings for the YES-PROVER and NO-PROVER, respectively. The argument is based on the niceness transformation described above.   □

It is curious to note that the above implication is in fact an equivalence. As a corollary, we have the following improvement of [13]. Here, $\text{Size}(n^k)$ refers to languages which have deterministic circuits of size n^k for some fixed constant k.

Corollary 1. $\text{NP}^{\text{O}_2^p} \not\subset \text{Size}(n^k)$ *and* $\text{NEXP}^{\text{O}_2^p} \not\subset \text{P/poly}$.

The previous best known result was that $\text{S}_2^p \not\subset \text{Size}(n^k)$ (attributed to Sengupta in Cai [3]). Also, Impagliazzo and Kabanets [11] prove conditionally that $\text{NEXP}^{\text{RP}} \not\subset \text{P/poly}$. On the other hand, [2] prove that $\text{MA}_{exp} \not\subset \text{P/poly}$. which already implies our result for $\text{NEXP}^{\text{O}_2^p}$ not being in P/poly (since $\text{MA}_{exp} \subseteq \text{NEXP}^{\text{O}_2^p}$).

Theorem 10 studies the consequence of NP being contained in P/poly. We now consider a weaker hypothesis, namely $\text{NP} \subset \text{coNP/poly}$. Under the above hypothesis, Yap [19] showed that PH collapses to $\Sigma_3^p \cap \Pi_3^p$, which was improved as $\text{PH} = \text{ZPP}^{\text{NP}^{\text{NP}}}$ [15], and further improved to $\text{PH} = \text{S}_2^{\text{NP}}$ [4,17]. We bring down the collapse further to O_2^{NP}. In fact, we show that the verifier needs to make only two queries to the NP oracle.

Theorem 11. *If* $\text{coNP} \subset \text{NP/poly}$ *then* $\text{PH} = \text{O}_2 \cdot \text{P}^{\text{NP}[2]}$.

6   Oblivious Symmetric Alternation Hierarchy

In this section, we use O_2^p, NO_2^p and YO_2^p to construct a hierarchy of classes akin to the polynomial time hierarchy. We call this hierarchy, the *oblivious symmetric alternation hierarchy* and denote it OSH. Here, O_2^p, NO_2^p and YO_2^p play a role analogous to that of P, NP and coNP, respectively. We start with the definition of various levels of OSH.

Definition 5. *Define* $\text{NO}_2^{(0)} = \text{YO}_2^{(0)} = \text{O}_2^p$. *For* $k \geq 1$, *we define*

$$\text{NO}_2^{(k)} = \text{NO}_2^{\text{NO}_2^{(k-1)}} \quad and \quad \text{YO}_2^{(k)} = \text{YO}_2^{\text{NO}_2^{(k-1)}}$$

The classes in OSH enjoy a simpler characterization, as shown by the following theorem. We can prove it using induction, along with our observation that $\text{NO}_2^A \subseteq \text{O}_2^{\text{NP}^A}$, for any oracle A.

Theorem 12. *For any* $k \geq 1$, $\text{NO}_2^{(k)} = \text{NO}_2^{\Sigma_{k-1}^p}$ *and* $\text{YO}_2^{(k)} = \text{YO}_2^{\Sigma_{k-1}^p}$.

The hierarchies OSH and PH are intertwined, as shown by the following result, which is a generalization of Proposition 1.

Theorem 13. *For any $k \geq 1$,*

$$\Sigma_k^p \subseteq \mathrm{NO}_2^{(k)} \subseteq \Sigma_{k+1}^p \cap \Pi_{k+1}^p \ and \ \Pi_k^p \subseteq \mathrm{YO}_2^{(k)} \subseteq \Sigma_{k+1}^p \cap \Pi_{k+1}^p.$$

As a consequence of Theorem 13, we have that if PH is infinite then OSH is infinite.

Theorem 14. *(i) If $\mathrm{OSH} = \mathrm{NO}_2^{(k)}$ then $\mathrm{PH} = \Sigma_{k+1}^p$.*
(ii) If $\mathrm{PH} = \Sigma_k^p$ then $\mathrm{OSH} = \mathrm{NO}_2^{(k)}$.

It is well known that if $\Pi_k^p = \Sigma_k^p$, then PH collapses to the kth level: $\mathrm{PH} = \Sigma_k^p \cap \Pi_k^p$. The following theorem provides a similar result for OSH. We can prove it using a generalization of Theorem 2.

Theorem 15. *For any $k \geq 1$, if $\mathrm{NO}_2^{(k)} = \mathrm{YO}_2^{(k)}$ then $\mathrm{OSH} = \mathrm{NO}_2^{(k)} \cap \mathrm{YO}_2^{(k)}$.*

7   Application to the Two Queries Problem

In this section, we consider the two queries problem: Is $\mathrm{P}^{\mathrm{NP}[1]} = \mathrm{P}^{\mathrm{NP}[2]}$? We refer the reader to an excellent article by Hemaspaandra et al. [9] for a survey on the above and related issues. It is known that if $\mathrm{P}^{\mathrm{NP}[1]} = \mathrm{P}^{\mathrm{NP}[2]}$, then the polynomial time hierarchy collapses and the best known collapse result is $\mathrm{PH} = \mathrm{S}_2^p$ [7] (building on [1]). We strengthen the collapse consequence as follows.

Theorem 16. *If $\mathrm{P}^{\mathrm{NP}[1]} = \mathrm{P}^{\mathrm{NP}[2]}$ then $\mathrm{PH} = \mathrm{NO}_2^p \cap \mathrm{YO}_2^p$.*

For this, as in [7], we use the following theorem from [1]:

Theorem 17. *If $\mathrm{P}^{\mathrm{NP}[1]} = \mathrm{P}^{\mathrm{NP}[2]}$, then there exists a polynomial-time predicate R and a constant $k > 0$ such that for every n, one of the following holds :*

1. *Locally $\mathrm{NP} = \mathrm{coNP}$: For every unsatisfiable formula ϕ of length n, there is a short proof of unsatisfiability w, i.e. $\phi \notin \mathrm{SAT} \Leftrightarrow \exists w \ R(\phi, w) = 1$, where $|w|$ is polynomial in n.*
2. *There exists a circuit of size n^k that decides SAT at length n.*

We also use the key lemma in [7]. The following notation is needed in the lemma and in the rest of the proof.

    Given a circuit C claiming to compute SAT, we convert it into a *nice* (cf. Definition 4) circuit C'. We assume henceforth that after receiving circuits from the YES-PROVER and the NO-PROVER, the verifier makes the above "niceness" transformation.

Lemma 1. *([7]) Fix $n > 0$. For every $k > 0$, if SAT does not have n^{k+2} size circuits at length n, then there exists a set S of satisfiable formulas of length n, called counterexamples, such that every nice circuit of size n^k is wrong on at least one formula from S. The cardinality of S is polynomial in n.*

Proof. (of Theorem 16) The intuition behind the proof is clear. Buhrman and Fortnow [1] prove that under the premise, it holds that either NP is (locally) like coNP or SAT has small circuits. Now under the first condition, clearly coNP $\subseteq$ NO$_2^p$, while under the second condition, the NO-PROVER can give the existing small circuit for SAT to prove coNP $\subseteq$ NO$_2^p$. Altogether, we would have a NO$_2^p$ collapse of the hierarchy. We formalize this intuition in the following.

We show that $\Sigma_2^p \subseteq$ NO$_2^p$. Let language L be a language in Σ_2^p. So, for any x, we have that

$$x \in L : \quad \exists y \ \varphi_{x,y} \notin \text{SAT}$$
$$x \notin L : \quad \forall y \ \varphi_{x,y} \in \text{SAT},$$

where $\varphi_{x,y}$ is a formula computable in polynomial time, given x and y. In the following we denote the length of $\varphi_{x,y}$ by m, noting that we can make the lengths of $\varphi_{x,y}$ corresponding to different y's uniformly m.

YES-PROVER gives the following strings : bit b_1, y, X. If $b_1 = 0$ then it means that YES-PROVER claims there is a circuit of size m^k computing SAT on inputs of length m, and then $X = C_Y$, a circuit of size m^k (purportedly solving SAT). If $b_1 = 1$, that means YES-PROVER claims that there is no such circuit, so it gives $X = w$ corresponding to Case 1 of Theorem 17.

NO-PROVER gives the following strings : bit b_2, X. If $b_2 = 0$ then it means that NO-PROVER claims there is a circuit of size m^{k+2} solving SAT on inputs of length m, and then $X = C_N$, a circuit of size m^{k+2} (purportedly solving SAT). If $b_2 = 1$, then NO-PROVER claims that there is no such circuit, and it gives $X = $ a list $\mathcal{L}$ of counterexamples ψ as guaranteed by Lemma 1. Formally, $\mathcal{L}$ is a list of satisfiable formulas along with proofs of satisfiability (satisfying truth assignments).

For the verifier's computation, we have the following cases:

1. $b_2 = 0$. In this case, the verifier uses the circuit C_N to check if $\varphi_{x,y}$ is satisfiable (by self-reducibility of SAT). If $C_N(\varphi_{x,y}) = 1$, then the verifier rejects x, else accepts x.

2. $b_2 = 1$. This breaks down into two subcases
   (a) $b_1 = 0$. In this case, the verifier converts the circuit C_Y into a nice circuit (while preserving correctness). The verifier then checks if $\mathcal{L}$ is valid, meaning all the formulas in the list are satisfied by the corresponding truth assignments. If this check fails, the verifier accepts the input and halts. In the next step, the verifier checks if C_Y is correct against the counterexamples in $\mathcal{L}$, i.e. it checks if for every $\psi \in \mathcal{L}$, $C_Y(\psi) = 1$. If C_Y fails this test, then the verifier rejects x, else accepts x.
   (b) $b_1 = 1$. In this case, neither YES-PROVER nor NO-PROVER is claiming that there are small circuits for SAT. So the verifier just checks if $R(\varphi_{x,y}, w) = 1$. If so, then it accepts x, else rejects x.

Let us prove that this proof system indeed accepts the language L. Again we have cases:

1. $x \in L$. This case breaks down into two subcases.
   (a) There exists circuits of size m^k solving SAT. In this case, YES-PROVER gives the appropriate y, $b_1 = 0$, and a circuit C_Y of size m^k solving SAT. Since $x \in L$, NO-PROVER cannot make the verifier reject x by showing a circuit C_N such that $C_N(\varphi_{x,y}) = 1$, because then C_N has to exhibit a satisfying assignment for $\varphi_{x,y}$ and in this case of the case analysis, $\varphi_{x,y} \notin$ SAT. So then the verifier checks C_Y against the list of counterexamples ψ in $\mathcal{L}$, and C_Y will pass all these tests. So the verifier accepts x.
   (b) There do not exist circuits of size m^k solving SAT, and YES-PROVER gives w satisfying Case 1 of Theorem 17. If $b_2 = 0$, and NO-PROVER gives a circuit C_N of size m^{k+2} solving SAT, then clearly $C_N(\varphi_{x,y}) = 0$. On the other hand if both $b_2 = b_1 = 0$, then $R(\varphi_{x,y}, w) = 1$. In either case, the verifier accepts x.
2. $x \notin L$. Again, the two subcases are as follows:
   (a) There exist circuits of size m^k solving SAT, and so also circuits of size m^{k+2}. NO-PROVER gives $b_2 = 0$ and one such circuit C_N. Since $x \notin L$, for any choice of y, it holds that $\varphi_{x,y} \in$ SAT. So $C_N(\varphi_{x,y}) = 1$ and the verifier rejects x.
   (b) There does not exist circuits of size m^{k+2} solving SAT. NO-PROVER gives $b_2 = 1$ and a list of counterexamples $\mathcal{L}$ which catches any "spurious" circuit C_Y of size m^k given by YES-PROVER (in case YES-PROVER gives $b_1 = 0$). If YES-PROVER gives $b_1 = 1$, then $R(\varphi_{x,y}, w) = 0$ and the verifier rejects x.

Clearly we observe that in the above, NO-PROVER needs to give only a single "proof" for any $x \notin L$. Thus, $\Sigma_2^p \subseteq \text{NO}_2^p$. By Proposition 1, we therefore have that $\text{PH} = \text{NO}_2^p \cap \text{YO}_2^p$. $\qquad \square$

8 Conclusions and Open Problems

We introduced a new class O_2^p and showed some of the nice properties it has (improved lowness results, improved collapses). A natural question is whether $\text{O}_2^p \subseteq \text{AM}$. Note that such an inclusion would prove $\text{NP} \subset \text{P/poly} \Rightarrow \text{PH} = \text{MA}$, a longstanding open problem. Whether O_2^p is contained in P^{NP} is also an interesting question. Since $\text{BPP} \subseteq \text{O}_2^p$, a positive answer to the above question would imply $\text{BPP} \subseteq \text{P}^{\text{NP}}$, which is a well-known open problem. So, it is worthwhile to show that $\text{O}_2^p \subseteq \text{P}^{\text{NP}}$, under some hardness assumption. Shaltiel and Umans [18] studied a similar question for S_2^p. They obtained $\text{S}_2^p = \text{P}^{\text{NP}}$, under the hypothesis that E^{NP} requires SV-nondeterministic circuits of size $2^{\Omega(n)}$. Since $\text{O}_2^p \subseteq \text{S}_2^p$, we have $\text{O}_2^p \subseteq \text{P}^{\text{NP}}$, under the same hypothesis. Can one show $\text{O}_2^p \subseteq \text{P}^{\text{NP}}$, under some weaker hypothesis? Another challenge is to compare O_2^p with the counting classes. For instance, is $\text{O}_2^p \subseteq \text{PP}$?

We observed that O_2^p is low for NO_2^p. Is $\text{NO}_2^p \cap \text{YO}_2^p$ low for NO_2^p? A positive answer would imply that $\text{O}_2^{\text{NP} \cap \text{coNP}} \subseteq \text{S}_2^p$, which in turn would settle a open question raised in [4]: if NP is contained in $(\text{NP} \cap \text{coNP})/\text{poly}$, does PH collapse to S_2^p? (The best known collapse is to $S_2^{NP \cap coNP}$). We conclude with an open problem that we find interesting and challenging: can we put the graph isomorphism problem in $NO_2^p \cap YO_2^p$?

Acknowledgments. We thank Eric Allender for insightful discussions and suggestions. We thank the anonymous referees for their useful comments. The second author gratefully acknowledges support of NSF Grant CCF-0514155.

References

1. H. Buhrman and L. Fortnow. Two queries. *Journal of Computer and System Sciences*, 59(2), 1999.
2. H. Buhrman, L. Fortnow, and T. Thierauf. Nonrelativizing separations. CCC, 1998.
3. J. Cai. $S_2^p \subseteq ZPP^{NP}$. FOCS, 2001.
4. J. Cai, V. Chakaravarthy, L. Hemaspaandra, and M. Ogihara. Competing provers yield improved Karp–Lipton collapse results. STACS, 2003.
5. R. Canetti. More on BPP and the polynomial-time hierarchy. *Information Processing Letters*, 57(5):237–241, 1996.
6. L. Fortnow, R. Impagliazzo, V. Kabanets, and C. Umans. On the complexity of succinct zero-sum games. CCC, 2005.
7. L. Fortnow, A. Pavan, and S. Sengupta. Proving SAT does not have small circuits with an application to the two queries problem. CCC, 2003.
8. O. Goldreich and D. Zuckerman. Another proof that BPP $\subseteq$ PH (and more). Technical Report TR97–045, ECCC, 1997.
9. E. Hemaspaandra, L. Hemaspaandra, and H. Hempel. What's up with downward collapse: Using the easy-hard technique to link boolean and polynomial hierarchy collapses. Technical Report TR-682, University of Rochester, 1998.
10. V. Arvind, J. Köbler, and R. Schuler. On Helping and Interactive Proof Systems. *International Journal of Foundations of Computer Science*, 6(2):137–153, 1995.
11. Valentine Kabanets and Russell Impagliazzo. Derandomizing polynomial identity tests means proving circuit lower bounds. STOC, 2003.
12. Jim Kadin. The polynomial time hierarchy collapses if the boolean hierarchy collapses. *SIAM Journal on Computing*, 17(6):1263–1282, 1988.
13. Ravi Kannan. Circuit-size lower bounds and non-reducibility to sparse sets. *Information and Control*, 55:40–56, 1982.
14. R. Karp and R. Lipton. Some connections between nonuniform and uniform complexity classes. STOC, 1980.
15. J. Köbler and O. Watanabe. New collapse consequences of NP having small circuits. *SIAM Journal on Computing*, 28(1):311–324, 1998.
16. A. Russell and R. Sundaram. Symmetric alternation captures BPP. *Computational Complexity*, 7(2):152–162, 1998.
17. A. Selman and S. Sengupta. Polylogarithmic-round interactive protocols for coNP collapse the exponential hierarchy. CCC, 2004.
18. R. Shaltiel and C. Umans. Pseudorandomness for approximate counting and sampling. CCC, 2005.
19. C. Yap. Some consequences of non-uniform conditions on uniform classes. *Theoretical Computer Science*, 26(3):287–300, 1983.

Combining Multiple Heuristics[*]

Tzur Sayag[1], Shai Fine[2], and Yishay Mansour[1]

[1] School of Computer Science, Tel Aviv University, Tel Aviv, Israel
tzurs@post.tau.ac.il, mansour@post.tau.ac.il
[2] IBM Research Laboratory in Haifa, Israel
shai@il.ibm.com

Abstract. In this work we introduce and study the question of combining multiple heuristics. Given a problem instance, each of the multiple heuristics is capable of computing the correct solution, but has a different cost. In our models the user executes multiple heuristics until one of them terminates with a solution. Given a set of problem instances, we show how to efficiently compute an optimal fixed schedule for a constant number of heuristics, and show that in general, the problem is computationally hard even to approximate (to within a constant factor). We also discuss a probabilistic configuration, in which the problem instances are drawn from some unknown fixed distribution, and show how to compute a near optimal schedule for this setup.

1 Introduction

Many important optimization and decision problems are computationally intractable and are sometimes even hard to approximate. The computational aspect of these problems is often depressing from the theoretical perspective. In reality however, in many cases, some of these hard problems are central and require a solution. Given that an efficient algorithm for these problems is (probably) out of the question, practical research is devoted to the discovery of heuristic methods which would optimistically be efficient over a large variety of problem instances.

One example of the above is the constraint satisfaction problem (SAT), which is highly important in the field of hardware and software verification. The fact that in general it admits no efficient algorithm has not stopped people from routinely solving impressively large instances of SAT problems. Quite to the contrary, the research community has come up with a large variety of heuristic methods for solving SAT problems (see [10, 9, 6] for example). A live proof of the efforts and interest in these problems is the fact that there are even organized competitions between the different academic and industrial implementations (see http://www.satcompetition.org/). One major difficulty concerting SAT solvers as well as heuristic methods for other problems is that the running time over a particular problem instance cannot be guaranteed. For this reason, once a heuristic consumes too much resources (e.g., time or space) one might suspend

[*] This research was supported on part by an IBM faculty award.

it and start running a different heuristic instead. The main goal of our model is to abstract this switching between different heuristics for solving the same instance of the problem.

Our model includes k distinct heuristics. Given a problem instance each heuristic has a different cost for solving it (for simplicity, the cost can be thought as the time needed by the heuristic to compute a solution). The cost of solving an instance of a problem is the total cost invested until one of the heuristics terminates with a solution. If we had only one problem instance then it would clearly be optimal to simply run the best heuristic (assuming it was known). We however, are interested in a case where we have n different problem instances, and the same schedule is used for all the problem instances. The total cost of a set of problem instances is the sum of the costs of the various problem instances. As a solution, we study two types of fixed schedulers. The first is a *resource sharing* scheduler that devotes a fixed fraction of its resources to each heuristic until some heuristic terminates with a solution. The second is a *task switching* scheduler in which only one heuristic is executed in each time unit. In both models, the schedulers are kept fixed for all problem instances, and the optimization problem is to find an optimal fixed scheduler for a given set of problem instances (and their related costs).

Our main result is an algorithm that computes an optimal resource sharing schedule in time $O(n \log n)$ for two heuristics (e.g., $k = 2$) and $O(n^{k-1})$ for a constant $k \geq 3$. For computing an optimal task switching schedule we briefly report an $O(n^{k+1})$ time algorithm. Note that both algorithms run in polynomial time algorithm for a constant number of heuristics. We also show that in general finding the optimal resource sharing schedule is NP-hard even to approximate. To make our setup more realistic, we study a probabilistic model in which the problem instances are drawn from some fixed unknown distribution. In this setup, one can search for a near optimal scheduler by sampling a number of problem instances and finding an optimal scheduler for them (in such case, the assumption that the cost of every heuristics over each problem instance in the sample is known in advance, becomes reasonable). Conceptually, in the initial learning phase we invest in learning the cost of each heuristic and the goal is to select a near optimal fixed scheduler. For the probabilistic model we show that if the number of problem instances (i.e., n) is a large enough sample, then, with high probability, the performance of any scheduler on the sample is close to its expected cost on the distribution (simultaneously for all schedulers). We note that since the number of schedulers might be infinite, we employ tools from computational learning theory to ensure the uniform convergence of the best scheduler on the sample to the optimal scheduler for the (unknown) distribution.

Our scheduling models have a similar flavor to the classical job scheduling models [11, 8]. The problem instances can be viewed as different jobs while the heuristics can be viewed as different machines. Since each pair of heuristic and problem instance have some arbitrary cost, the setup resembles the parallel unrelated machines model of job scheduling. The main conceptual difference lies in the optimization function, which as we will see, is not linear, and hence, introduces new computational challenges.

Another related research topic is competing with the best expert in an online setting (e.g., [3]). In that setting one competes with the best single expert, which is in our case the best single heuristic. One should note that the performance of the best single heuristic can be significantly inferior even with respect to a round robin task switching (or equal weight resource sharing) schedule. This is since each heuristic might have a problem instance on which it performs very poorly. We note that both of our scheduling models do compare, implicitly, to single best heuristic (by simply assigning all the weight to one heuristic). However, we would like to stress that our scheduling models encompass a much wider class of schedulers, which is especially beneficial for cases in which problem instances have significantly different processing costs by various heuristics.

The rest of this paper is organized as follows. Section 2 introduces the model and the two scheduling schemes. The algorithm for the resource sharing scheduling problem appears in Section 3 in which we also state the hardness of approximation result. Section 4 introduces and discusses the probabilistic setup. We conclude with some empirical results in Section 5. Due to space limitations the proofs are deferred to the full version of the paper.

2   Model

We consider computing solutions to a set J of n problem instances using k different heuristics h_j, for $j \in [1, k]$. The cost (e.g., time) of solving the problem instance x_i using heuristic h_j is $\tau_{i,j}$, therefore, the input to the problem is modelled by a *processing cost matrix* $T = [\tau_{i,j}] \in (\mathbb{R}^+)^{n \times k}$.

The *resource sharing* model encapsulates a scenario in which we execute all the heuristics concurrently, while devoting to each heuristic a fixed fraction of the resources. A *resource sharing* scheduler has a vector of shares $s \in \triangle_k$, where $\triangle_k = \{(s_1, \ldots, s_k) : \sum_{i=1}^{k} s_i = 1; s_i \in [0, 1]\}$. Conceptually, s_i is the fraction of resources we devote to execute heuristic h_i. Assume that heuristic h_i has cost τ_i, this implies that if we invest a total cost of $c = \tau_i / s_i$ then heuristic h_i receives τ_i resources and terminates with a solution. For the global cost we consider the first heuristic that terminates, which for a given problem instance x and schedule s we call the *completing heuristic*. Formally, the cost of a resource sharing schedule s for a problem instance x with costs $\tau = (\tau_1, \ldots, \tau_k)$ is given by

$$cost_{rs}(x, s) := \min_{1 \leq j \leq k} \frac{\tau_j}{s_j}$$

and the completing heuristic is $\arg\min_{1 \leq j \leq k} \frac{\tau_j}{s_j}$.

The cost of a resource sharing schedule s for a set of problem instances J is $\sum_{x \in J} cost_{rs}(x, s)$. The optimization task is, given a processing cost matrix $T \in (\mathbb{R}^+)^{n \times k}$, compute a schedule s^* of minimum cost, i.e., $cost_{rs}(T, s^*) = \min_{s \in \triangle_k} cost_{rs}(T, s)$.

The *task switching* model divides the resources (e.g., time) to units and at each unit assigns all the resources to a single heuristic. (We assume here that all the costs are also integral number of units.) A *task switching* scheduler for k heuristics is a function $W(t) : \mathbb{N} \mapsto [1, k]$. The cost of a task switching scheduler W for a problem instance x with processing costs $\mathcal{T} = (\tau_1, \ldots \tau_k)$ is defined recursively,

$$cost_{ts}^{W}(x, t) = \begin{cases} 1 + cost_{ts}^{W}(\mathcal{T} - e_{W(t)}, t + 1), & \forall i, \tau_i > 0 \\ 0, & \text{otherwise.} \end{cases}$$

where $\vec{e}_1, \ldots \vec{e}_k$ be the standard unit vectors in $\mathbb{R}^k$. In other words, we count the total number of resource units until one of the heuristics, say h_i, was executed at least τ_i times and thus terminates with a solution. Again, the cost for a set of problem instances J with processing cost matrix $\mathcal{T}$ is the sum of costs, i.e., $cost_{ts}(\mathcal{T}, W) = \sum_{x \in J} cost_{ts}^{W}(x, 0)$.

To see the difference between the performance of the different schedulers, consider the following processing cost matrix:

$$\mathcal{T} = \begin{pmatrix} 2 & 10 \\ 10 & 1 \end{pmatrix}.$$

The best single heuristic has cost 11. The optimal resource sharing model is not the trivial $(0.5, 0.5)$ but rather $(2 - \sqrt{2}, \sqrt{2} - 1)$ which has cost of about 5.8284. The optimal task switching schedule has cost 4.

The following lemma relates the optimal cost of the two models,

Lemma 1. *Let s be a resource sharing schedule, for any processing cost matrix $\mathcal{T}$ there is a task switching schedule W such that $cost_{ts}(\mathcal{T}, W) \le cost_{rs}(\mathcal{T}, s)$.*

We can also compare the performance of the above fixed scheduling schemes with the optimal dynamic scheme that selects the best heuristic for each problem instance, i.e., $best(\mathcal{T}) = \sum_{i=1}^{n} \min_{k \ge j \ge 1} \{\tau_{i,j}\}$.

Theorem 2. *Let $s^* = s^*(\mathcal{T})$ ($W^* = W^*(\mathcal{T})$, respectively), be an optimal resource sharing (task switching) scheduler for the processing cost matrix $\mathcal{T}$. The following bounds hold: (1) For any $\mathcal{T}$ we have $\frac{cost_{ts}(\mathcal{T}, W^*)}{best(\mathcal{T})} \le \frac{cost_{rs}(\mathcal{T}, s^*)}{best(\mathcal{T})} \le k$. (2) There is a matrix $\mathcal{T}$ such that $\frac{cost_{rs}(\mathcal{T}, s^*)}{best(\mathcal{T})} \ge \frac{cost_{ts}(\mathcal{T}, W^*)}{best(\mathcal{T})} \ge \frac{k+1}{2}$.*

3   Resource Sharing Scheduler

In this section we present an algorithm for the resource sharing scheduling problem, whose performance is stated in the following theorem,

Theorem 3. *Given a set of problem instances J and $\mathcal{T}$, the related processing cost matrix, an optimal resource sharing schedule can be computed in time[1] $O(n \log n)$ for $k = 2$, and time $O(n^{k-1})$ for a constant $k \ge 3$.*

---

[1] We remark that our "O" notation has a constant that depends exponentially on the number of heuristics k.

Algorithm overview: Consider a set of problem instances J. Fix a resource sharing schedule s, and observe that for each $x \in J$, the minimum cost for x is attained on one of the k heuristics, i.e., the completing heuristic. The key idea of the algorithm is to divide the space of possible resource sharing schedules into disjoint cells with the following special property. In each such cell c, for each problem instance $x \in J$ and for any schedule $s \in c$, the same heuristic achieves the minimum cost, i.e., for every $x \in J$ the completing heuristic is the same for every $s \in c$. This implies that we can rewrite the total cost function for any schedule $s \in c$ as $cost_{rs}(\mathcal{T}, s) = \frac{A_1}{s_1} + \frac{A_2}{s_2} + \ldots + \frac{A_k}{s_k}$, where $A_1, \ldots, A_k$ are constants which depend only on the cell c. The main benefit is that we can efficiently minimize such a function over $s \in c$. This suggests that an influential complexity term would be the number of cells that we need in order to cover the entire space of resource sharing schedules.

For the resource sharing schedules, we show that the cells are obtained from hyperplane arrangements, for which there is a well established theory that allows us to bound the number of cells and their complexity. In addition, it allows us an efficient traversal of the cells.

Our arguments are organized as follows: we first introduce some basic facts about hyperplane arrangements, which are the main theoretical tool that we use. We then introduce a specific hyperplane arrangement we use to analyze our cost function. Finally, we describe the algorithm that finds an optimal schedule s^* and state its time complexity.

Hyperplane Arrangements: Hyperplane arrangements are well studied objects in computational geometry (see [4, 7, 5]). Informally, given a finite collection H of hyperplanes in $\mathbb{R}^d$, the hyperplanes in H induce a decomposition of $\mathbb{R}^d$ into connected open cells. We denote the arrangement of H by $\mathcal{A}(H)$. We now briefly review some definitions and basic results regarding hyperplanes arrangements. A *hyperplane* $H \subset \mathbb{R}^d$ is defined by an equation of the form $L(x) = b$ where $L : \mathbb{R}^d \to \mathbb{R}$ is a linear functional and b is a constant. A *hyperplane arrangement* $\mathcal{A}$ is a collection of hyperplanes in $\mathbb{R}^d$. Such a collection defines *cells* or *regions* (sometimes also called *faces*), namely the connected components of the set $\mathbb{R}^d - \cup_{H \in \mathcal{A}} H$. More formally, let $H_1, H_2, \ldots H_n$ be the set of hyperplane in $\mathcal{A}$. For a point p define

$$u_i(p) =: \begin{cases} +1, & \text{if } p \in H_i^+; \\ 0, & \text{if } p \in H_i; \\ -1, & \text{if } p \in H_i^-, \end{cases}$$

for $1 \leq i \leq n$. The vector $u(p) = (u_1(p), u_2(p), \ldots u_n(p))$ is called the position vector of p. If $u(p) = u(q)$, then we say that the points p and q are *equivalent*, and the equivalence classes thus defined are called the *cells* of the arrangement $\mathcal{A}$. A *d-dimensional cell* in the arrangement is a maximal connected region of $\mathbb{R}^d$ that is not intersected by any hyperplane in H. An *i-dimensional cell* in $\mathcal{A}(\mathcal{H})$, for $1 \leq i \leq d-1$ is a maximal connected region in the intersection of a subset of the hyperplanes in H, that is not intersected by any other hyperplane in H. Special names are used to denote i-cells for special i values: a 0-cell is called a *vertex*, a 1-cell is an *edge*, $(d-1)$-cell is a *facet* and a d-cell is called a *cell*. For each (d-dimensional) cell c let $f_i(c)$ denote the number of i-dimensional cells on the boundary of c. We refer to $f(c) = \sum_{i=0}^{d-1} f_i(c)$ as the *complexity of the cell* c. The *complexity of an arrangement* $\mathcal{A}$ is defined as $\sum_{c \in \mathcal{A}} f(c)$.

The following results from [5] and [2] bound the hyperplane arrangement complexity.

Theorem 4 ([5]). *Let $\mathcal{A}$ be an arrangement of n hyperplanes in $\mathbb{R}^d$, for a constant $d \geq 1$. Then, the number of cells is $O(n^d)$, $\sum_{c \in \mathcal{A}} f(c) = O(n^d)$ and for every cell $c \in \mathcal{A}$ we have $f(c) = O(n^{\lfloor \frac{d}{2} \rfloor})$.*

Representing Hyperplane Arrangements: The common data structure that represents an arrangement is the incidence graph $I(\mathcal{A})$. (We briefly describe it and refer the interested reader to [5] for a more detailed exposition.) The incidence graph $I(\mathcal{A})$ contains a node $v(c)$ for each i-cell of $\mathcal{A}$; if two i-cells c and c' are incident upon each other then $v(c)$ and $v(c')$ are connected by an edge. When implemented, each node $v(c)$ of an incidence graph has a record that contains some auxiliary information (see [5]) among which is the coordinates of some point $p(v(c))$ inside the cell c and two lists containing pointers $L(v(c))$ to the sub-cells[2] and to the super-cells of c. (When convenient, we will use the term cell to represent the data structure $v(c)$.) Note, that the size of the representation of a cell $c \in \mathcal{A}$, is $O(f(c))$. The following theorem from [5] builds the incidence graph for a hyperplane arrangement.

Theorem 5 ([5], Theorem 7.6). *Let $\mathcal{A}$ be an arrangement of n hyperplanes in $\mathbb{R}^d$, for a constant $d \geq 2$. There is an algorithm that constructs the incidence graph $I(\mathcal{A})$ in time $O(n^d)$.*

Two Heuristics: Before we describe our algorithm for the general case of $k \geq 3$ heuristics, it is instructive to review the above definitions and the main steps of our algorithm for the simple case of two heuristics. Further, this case is of interest since it does not admit the same time complexity result we obtain later for $k \geq 3$, instead, we have an $O(n \log n)$ algorithm for this case.

Recall that for $k = 2$ we have a processing cost matrix $\mathcal{T} \in (\mathbb{R}^+)^{n \times 2}$, and we seek to find a resource sharing schedule s^* such that $cost_{rs}(s^*) = min_{s \in [0,1]} \sum_{i=1}^{n} \min(\frac{\tau_{i,1}}{s}, \frac{\tau_{i,2}}{1-s})$. (Note that for $k = 2$ we have effectively only one parameter for the scheduler.) We describe our algorithm in terms that would be useful for the general case of $k \geq 3$. Our algorithm iterates the high dimensional cells of the arrangement one by one, minimizing the cost function for the cell and its sub-cells (we slightly abuse the notation here and use sub-cells to denote all cells of lower dimension that are on the boundary of c).

We now describe the above procedure in more detail (for the special case of $k = 2$). Given a set of problem instances J and $\mathcal{T}$ the related processing cost matrix, we construct a hyperplane arrangement of dimension $k - 1 (= 1)$

---

[2] A cell c' is said to be a *sub-cell* of another cell c if the dimension of c' is one less then the dimension of c and c' is contained in the boundary of c. Similarly, a cell c' is a *super-cell* of c if c is a *sub-cell* of c'.

consisting of $n + 2$ hyperplanes of dimension 0 (i.e., points). Our hyperplanes (i.e., the points) define where the cost of two heuristics is identical. Namely, the hyperplanes:

$$H_i : \frac{T_{i,1}}{s} = \frac{T_{i,2}}{1-s},$$

for $i = 1, \ldots n$. Additionally, we add the hyperplanes $s = 0$ and $s = 1$. The resulting hyperplane arrangement $\mathcal{A}(\mathcal{T})$ has two types of cells: at most $n + 2$ points (which are the 0-cells) that lie on the segment $[0, 1]$ and at most $n + 1$ line segments (which are the 1-cells). It is easy to see that the line segments are exactly the cells we are after. Specifically, for each schedule s inside each such line segment or cell c, the minimal cost for each problem instance $x \in J$ is attained by the same heuristic. Our algorithm now needs to traverse the cells and write the cost function for each cell. Computing A_1 and A_2 for a specific cell directly requires to iterate through the problem instances and decide which of the two heuristics has the minimal cost (for all schedules inside c), we can then accumulate the relevant quantities to obtain A_1 and A_2 for the cell. This procedure takes $O(n)$ steps for each cell, and since there are $O(n)$ cells, this gives an $O(n^2)$ algorithm.

To improve the time complexity, observe that for schedules that are in the left most line segment $[0, a]$, for all problem instances the second heuristic is the completing heuristic. In the segment $[a, b]$, which is just to the right of $[0, a]$, only the problem instance corresponding to the point a changes its completing heuristic. Thus, given the cost function for $[0, a]$ we can use a simple update rule and in time $O(1)$ write the cost function for $[a, b]$. This method takes $O(n)$ steps for the first segment and only $O(1)$ for each of the remaining segments. To support this update scheme, we need a simple data structure that supports a sorted traversal of the cells. This adds a factor of $O(n \log n)$ time which turns out to be dominant in our algorithm.

Having shown how to write the cost function for each line segment, it is left to verify that the resulting functions can be easily minimized. Direct calculation shows that a local minimum for $\frac{A}{s} + \frac{B}{1-s}$ is attained at $\bar{s} = \sqrt{A}/(\sqrt{A} + \sqrt{B})$. This means that for each of the $O(n)$ segments, we consider at most three points as possible global minimum ($\bar{s}$ and the two end points – the subcells of the interval). The following lemma summarizes the above,

Lemma 6. *Given a processing cost matrix $\mathcal{T}$, an optimal resource sharing schedule can be found in time $O(n \log n)$ for $k = 2$.*

Hyperplane Arrangement for Resource Sharing: We now return to the general case of a constant $k \geq 3$ and show that the ideas used for $k = 2$ can be applied here as well. One difference is that instead of sorting the cells we use Theorem 5 and a special data structure that allows the cell traversal that is required for the efficient update of the cost function.

Given the processing cost matrix $\mathcal{T}$ we construct a hyperplane arrangement $\mathcal{A}(\mathcal{T})$ consisting of $\binom{k}{2} \cdot n = O(k^2 \cdot n)$ hyperplanes. We first consider the set of $O(k^2 \cdot n)$ functionals indexed by $1 \leq i < j \leq k$ and $1 \leq r \leq n$. They are define by

$$L_{i,j,r}(s_1, \ldots, s_{k-1}) = \frac{s_i}{\tau_{r,i}} - \frac{s_j}{\tau_{r,j}}$$

Note that when $j = k$ we interpret s_k as $1 - s_1 - \ldots - s_{k-1}$. We consider the hyperplane $H_{i,j,r}$ in $\mathbb{R}^{k-1}$ defined by $L_{i,j,r}(x) = 0$. We also consider the hyperplanes $s_i = 0$ for $1 \le i \le k$ and the hyperplane $s_1 + \ldots + s_{k-1} = 1$. Thus the arrangement $\mathcal{A}$ consists of $N = \binom{k}{2}n + k + 1 - O(k^2 n)$ hyperplanes in $\mathbb{R}^{k-1}$.

The Cost Function in Each Cell: To motivate the choice of this particular hyperplane arrangement we study the cost function on each cell $c \in \mathcal{A}(\mathcal{T})$. The key observation is that for each schedule $s \in c$, exactly one unique heuristic has minimal cost for $x \in J$, which is true by definition of c. (If for some problem instance $x_r \in J$ there are two schedules $s_i, s_j \in c$ with different completing heuristics, then the hyperplane $L_{i,j,r}$ must intersect the cell.) The following lemma formalizes this argument.

Lemma 7. *In each cell c of the arrangement $\mathcal{A}(\mathcal{T})$ the function $cost_{rs}(\mathcal{T}, s)$ has the form $\frac{A_1}{s_1} + \ldots + \frac{A_k}{s_k}$. Moreover, given a cell c, we can obtain this representation (i.e., the constants $A_1, \ldots A_k$) in time $O(kn)$.*

To support a more efficient update of the cost functions for the cells, we define a *partition graph* $\mathcal{P}(\mathcal{A})$, which we use to efficiently compute the cost functions in the simplified form for the highest dimensional cells,

Definition 8. *For $I(\mathcal{A}(\mathcal{T}))$, the incidence graph of $\mathcal{A}(\mathcal{T})$, we define the partition graph $P(\mathcal{A}(\mathcal{T})) = (V, E)$ where $V = \{v(c) \in I(\mathcal{A}(\mathcal{T})), \text{ s.t. } dim(c) = k-1\}$ (i.e., V has a vertex representing each highest dimensional cell in $\mathcal{A}(\mathcal{T})$), and an edge $(u, v) \in E$ if there is a path (u, w, v) in the incidence graph and $dim(w) = k-2$ (i.e., two vertices $v(c), v(c')$ are connected in $\mathcal{P}(\mathcal{A}(\mathcal{T}))$ if there is a cell of dimension $k-2$ that is on the boundary of c and c' in the arrangement). As an auxiliary information, each vertex v in the partition graph $\mathcal{P}(\mathcal{A}(\mathcal{T}))$ points to the corresponding vertex in $I(\mathcal{A}(\mathcal{T}))$.*

Lemma 9. *Given an incidence graph $I(\mathcal{A}(\mathcal{T}))$, (1) there is an algorithm that constructs its partition graph $\mathcal{P}(\mathcal{A}(\mathcal{T}))$ in time $O(n^{k-1})$ and, (2) $\mathcal{P}(\mathcal{A}(\mathcal{T}))$ is connected.*

Recall that for any cell Lemma 7 computes the values of $A_1, \ldots A_k$ in time $O(kn)$. To reduce the time complexity, we note that given these values for an initial $(k-1)$-cell, along with the partition graph, it is straight forward to compute these constants for a neighbor $(k-1)$-cell in $O(k)$ time. This implies that this computation, for all cells, can be done in time $O(k \cdot n^{k-1}) = O(n^{k-1})$. The idea is that we use the partition graph to update the cost function for a neighboring $(k-1)$-cell (i.e., a highest dimensional cell). Once the constants for a $(k-1)$-cell are known, we can compute the cost function in any subcell of $(k-1)$-cell using an algebraic manipulation taking advantage of the additional hyperplane equation that is satisfied on the boundary (the details are deferred to the full version of this paper).

Next, we show how to find the minimal value of the specific cost function in a given cell,

Observation 10. *Given an i-cell c and the constants $A_1, \ldots A_d$ of the cost function $cost_{rs}(\mathcal{T}, s) = \sum_j A_j / s_j$, there is an algorithm that rejects all schedules s inside c as possible minimal points of $cost_{rs}(\mathcal{T}, s)$ and works in time $O(f_{i-1}(c))$ where $f_{i-1}(c)$ denotes the number of faces on the boundary of c.*

Our algorithm visits each $(k - 1)$ dimensional cell c (a vertex in the partition graph) and finds the minimal value of the specific cost function inside c and on its boundaries (which are sub-cells that are reachable from c in the incidence graph). The correctness follows from the fact that we visit all the $(k - 1)$-cells, and test each of them along with its sub-cells.

Theorem 11. *There is an algorithm that finds an optimal resource sharing schedule in time $O(n^{k-1})$, for a constant $k \geq 3$.*

We also derive an algorithm to compute the optimal task switching schedule. Our algorithm uses a dynamic programming like approach, and the following theorem summarizes our result,

Theorem 12. *Given a set of problem instances J and $\mathcal{T}$, the related processing cost matrix, an optimal task switching schedule can be found in time $O(n^{k+1})$.*

Both our algorithms are polynomial only for a constant k. For an arbitrary k we show a hardness result. Namely, we show that finding a near optimal resource sharing schedule is NP-hard. First we formalize the related decision problem.

Definition 13 (RS). *The (decision version) of a resource sharing scheduling problem is a pair $(\mathcal{T}, C)$ where $\mathcal{T} = [\tau_{i,j}] \in \mathbb{R}^{n \times k}$ is a processing time matrix and $C \in \mathbb{R}$ is a goal cost. The question is whether there is a resource sharing scheduler $s \in \Delta_k$ with $cost_{rs}(\mathcal{T}, s) = \sum_{i=1}^{n} \min_{j=1}^{k} (\frac{\tau_{i,j}}{s_j}) \leq C$.*

Using a reduction from 3-dimensional matching (3-DM) we show that the resource sharing scheduling problem is NP-hard to approximate.

Theorem 14. *RS is NP-complete, and there is a constant $\alpha > 1$ such that it is hard to approximate RS to within α, i.e., to find a schedule s such that $cost_{rs}(\mathcal{T}, s) \leq \alpha \cdot cost_{rs}(\mathcal{T}, s^*)$, where s^* is the optimal resource sharing schedule.*

4   Probabilistic Model

The main reason for studying a resource sharing or a task switching schedules is the fact that we hope to be able to fix such a schedule, based on a small number of problem instances, and expect it to behave similarly on future unseen problem instances. A crucial point is that once a schedule was fixed, the various processing costs of the problem instance are not needed in order to run the schedule. This methodology can be viewed as having a *learning phase* (in which we determine the schedule) and an *execution phase* (where we use the schedule). For the above methodology to work, we need that the performance in the learning phase and execution phase is similar, something which is not always guaranteed.

In this section we extend our scheduling models and describe a probabilistic model of generation of problem instances, this helps demonstrate the feasibility of the above methodology. Namely, first observing a small sample of problem instances (for each problem instance in the sample we have full information about all its processing costs), and then, computing an optimal schedule (resource sharing or task switching). The main contribution of this section is that we show that if the sample size is large enough then the observed cost on the sample is close to the expected cost on the distribution (simultaneously for all schedulers). The results and techniques that we use in this section are in the spirit of the classical uniform convergence results in computational learning theory (see [1]).

We first define the probabilistic model. Let $\mathcal{X}$ be the set of all possible problem instance and $\mathcal{H}$ be the set of all heuristics. For each pair of heuristic $h \in \mathcal{H}$ and problem instance $x \in \mathcal{X}$ let $\tau_{x,h} \in [1, B]$ be the cost of heuristic h on problem instance x, where B is an upper bound on the processing cost. (Note that the processing costs are arbitrary!)

Let Q be some fixed unknown probability measure on $\mathcal{X}$. When we draw a problem instance $x \in \mathcal{X}$, we are given an access to an oracle that given a heuristic $h \in \mathcal{H}$ returns $\tau_{x,h}$. Let $z = (x_1, \ldots x_m) \in \mathcal{X}^m$ be a *sample* of m problem instances identically distributed drawn independently from Q.

Let $\mathcal{W}$ be the set of all possible task switching schedules for k heuristics, i.e., $\mathcal{W} = \{W : W \in [1, kB] \mapsto [1, k]\}$ for problem instances in $\mathcal{X}$. (Note that any problem instance in $\mathcal{X}$ is completed after at most kB steps.)

Slightly abusing our previous notation, let $cost_{ts}(Q, W) = E_{x \sim Q}[cost_{ts}(W, x)]$ be the expected processing cost of W under the distribution Q. Similarly, let $cost_{ts}(z, W) = \frac{1}{m} \sum_{i=1}^{m} cost_{ts}(W, x_i)$ be the average cost of schedule W on the set of problem instances z. Finally, denote by $W^* = \arg\min_W cost_{ts}(Q, W)$ the optimal schedule with respect to the probability distribution Q and by $\hat{W}^*(z) = \arg\min_W cost_{ts}(z, W)$ the optimal task switching schedule on the set of problem instances z. Our main result is that with high probability, simultaneously for all schedules, the error by estimating the cost of a schedule on the sample is at most ε.

Theorem 15. *Let Q be a fixed distribution over $\mathcal{X}$ and $\epsilon, \delta > 0$ be fixed parameters. There exists a constant $m_0(\epsilon, \delta) = O(\frac{k^3 B^3}{\epsilon^2} \log(k/\delta))$, such that for any sample z of size $m \geq m_0(\epsilon, \delta)$, with probability at least $1 - \delta$ we have, $|cost_{ts}(z, W) - cost_{ts}(Q, W)| \leq \varepsilon$, for all $W \in \mathcal{W}$.*

An immediate implication of Theorem 15 is that computing an optimal task switching for a sample results in a near optimal task switching schedule. Note that our result has the accuracy (denoted by ε) converging to zero at rate of $O(\frac{1}{\sqrt{m}})$, as in the classical convergence bounds in computational learning theory [1]. We also show a similar result for the case of resource sharing scheduling,

Theorem 16. *For any fixed probability distribution Q on $\mathcal{X}$, any $\varepsilon, \delta \in (0, 1)$, there is a constant $m_0 = \tilde{\Omega}(\frac{k^2 B^2}{\varepsilon^2} \ln \frac{1}{\delta})$ [3] such that with probability at least $1 - \delta$, for any sample z of size $m > m_0(\varepsilon, \delta)$ we have, $|cost_{rs}(z, s) - cost_{rs}(Q, s)| \leq \varepsilon$, for every resource sharing schedule s.*

5   Empirical Results

The original motivation for this line of research was related to scheduling multiple SAT heuristics and improving over the performance of the best single heuristic. In the experiment design here we considered only two heuristics and tested on real verification data the performance of the optimal resource sharing schedule. Even in this simple case one can see the potential for improvements using our methodology. We note that these results are not obtained using the sampling probabilistic algorithm but instead we calculate the exact optimal resource sharing scheduler to emphasize the performance boost.

As input to our experiment we used real data that was generated to test the performance of several SAT solvers. Each data set contains several hundred problem instances that were generated for property testing of hardware circuits composed of various adders and multipliers (The data sets contain real SAT verification problems provided with the curtsey of IBM® research lab). In addition to the different type of circuits, each data sets is also different in the size of the problems (i.e. the number of variables) they contain. Since all these SAT solvers are DPLL based, we used the number of branch actions as the performance measure of a SAT solver on a problem instance (this measure has a very high correlation with real computation time and has the benefit of being machine independent).

Table 1. Resource Sharing vs Best Heuristics on SAT instances

Data Set	instances	#Variables	Best Heu.	2nd Best		RS	Ratio
ADD	441	40,000	287,579,911	740,846,088	286,561,199		0.99
MADD	162	40,000	4,414,139,866	8,401,158,224	2,981,405,433		0.67
ADDI	281	1,500	1,178,621	1,178,621	1,178,621		1.00
MULI	265	< 5,000	541,765	541,765	541,765		1.00

To obtain these empirical results we picked the two best heuristics for each data set and computed the optimal resource sharing scheduler, we then compare its performance to these two best heuristics. Table 1 summarizes the performance of the optimal resource sharing compared to the top two heuristics. Clearly, the resource sharing algorithm is never worse than the best heuristic. In some cases it is significantly better (for the MADD set, the optimal resource sharing schedule has a cost of only 67% of the best heuristic, a very significant improvement). For the two small sets, the optimal resource sharing schedule is simply the best

---

[3] We use the notation $\tilde{\Omega}(f) = \Omega(f \log^c f)$, for some constant $c > 0$.

heuristic (which explains why the ratio is precisely 1), for the two larger data sets the improvements are significant.

References

1. M. Anthony and Peter L. Bartlett, *Neural network learning: Theoretical foundations*, Cambridge University Presss, 1999.
2. Boris Aronov, Jiri Matousek, and Micha Sharir, *On the sum of squares of cell complexities in hyperplane arrangements*, SCG '91: Proceedings of the seventh annual symposium on Computational geometry (New York, NY, USA), ACM Press, 1991, pp. 307–313.
3. Nicolò Cesa-Bianchi, Yoav Freund, David P. Helmbold, David Haussler, Robert E. Schapire, and Manfred K. Warmuth, *How to use expert advice*, Journal of the ACM **44** (1997), no. 3, 427–485.
4. Mark de Berg, Marc van Kreveld, Mark Overmars, and Ptfried Schwarzkopf, *Computational geometry algorithms and applications*, second, revised ed., ch. 8, pp. 172–180, Springer, 2000.
5. Herbert Edelsbrunner, *Algorithms in combinatorial geometry*, Springer-Verlag New York, Inc., New York, NY, USA, 1987.
6. E. Giunchiglia, M. Maratea, A. Tacchella, and D. Zambonin, *Evaluating search heuristics and optimization techniques in propositional satisfiability*, Proceedings of International Joint Conference on Automated Reasoning (IJCAR'01) (Siena), June 2001, to appear.
7. D. Halperin, *Arrangements*, Handbook of Discrete and Computational Geometry (Jacob E. Goodman and Joseph O'Rourke, eds.), CRC Press LLC, Boca Raton, FL, 2004, pp. 529–562.
8. Dorit S. Hochbaum (ed.), *Approximation algorithms for NP-hard problems*, PWS Publishing Co., Boston, MA, USA, 1997.
9. O. Kullmann, *Heuristics for SAT algorithms: Searching for some foundations*, September 1998, 23 pages, updated September 1999 w.r.t. running times, url = http://cs-svr1.swan.ac.uk/ csoliver/heur2letter.ps.gz.
10. Chu-Min Li and Anbulagan, *Heuristics based on unit propagation for satisfiability problems*, Proceedings of the Fifteenth International Joint Conference on Artificial Intelligence (IJCAI'97) (Nagoya, Japan), August 23–29 1997, pp. 366–371.
11. M. Pinedo, *Scheduling theory, algorithms and systems*, Prentice-Hall International Series in Industrial and Systems EngineeringPrentice-Hall: Englewood Cliffs, NJ, 1995.

Conflict-Free Colorings of Rectangles Ranges

Khaled Elbassioni and Nabil H. Mustafa

Max-Planck-Institut für Informatik, Saarbrücken, Germany
{elbassio, nmustafa}@mpi-sb.mpg.de

Abstract. Given the range space $(P, \mathcal{R})$, where P is a set of n points in $\mathbb{R}^2$ and $\mathcal{R}$ is the family of subsets of P induced by all axis-parallel rectangles, the conflict-free coloring problem asks for a coloring of P with the minimum number of colors such that $(P, \mathcal{R})$ is *conflict-free*. We study the following question: Given P, is it possible to add a small set of points Q such that $(P \cup Q, \mathcal{R})$ can be colored with fewer colors than $(P, \mathcal{R})$? Our main result is the following: given P, and any $\epsilon \geq 0$, one can always add a set Q of $O(n^{1-\epsilon})$ points such that $P \cup Q$ can be conflict-free colored using $\tilde{O}(n^{\frac{3}{8}(1+\epsilon)})$[1] colors. Moreover, the set Q and the conflict-free coloring can be computed in polynomial time, with high probability. Our result is obtained by introducing a general probabilistic re-coloring technique, which we call *quasi-conflict-free* coloring, and which may be of independent interest. A further application of this technique is also given.

1 Introduction

A set of points P in $\mathbb{R}^2$, together with a set $\mathcal{R}$ of ranges (say, the set of all discs or rectangles in the plane) is called a range space $(P, \mathcal{R})$. For a given range space $(P, \mathcal{R})$, the goal is to assign a color to each point $p \in P$ such that for any range $R \in \mathcal{R}$ with $R \cap P \neq 0$, the set $R \cap P$ contains a point of unique color. Call such a coloring of the points P a *conflict-free* coloring of $(P, \mathcal{R})$. The problem then is to assign a conflict-free coloring to the range space $(P, \mathcal{R})$ with the *minimum* number of colors.

The study of the above problem was initiated in [ELRS03, Smo03], motivated by the problem of frequency-assignment in cellular networks. The work in [ELRS03] presented a general framework for computing a conflict-free coloring for *shrinkable* range spaces. In particular, for the case where the ranges are discs in the plane, they present a polynomial-time coloring algorithm that uses $O(\log n)$ colors for a conflict-free coloring. They also present an algorithm that conflict-free colors the set of points P when the ranges are scaled translates of a compact convex region in the plane. This result was then extended in [HS05] by considering the case where the ranges are rectangles in the plane. This seems harder than the disc case, and the work in [HS05] presented a simple algorithm that uses $O(\sqrt{n})$[2] colors for a conflict-free coloring. They also show that for the case of random points in a unit square, $O(\log^4 n)$ colors suffice for rectangle ranges. Finally, they show that if the points lie on an *exact* uniform $\sqrt{n} \times \sqrt{n}$ grid, then also $O(\log n)$ colors are sufficient. This result very strongly uses the degeneracy of the

---

[1] Ignoring poly-logarithmic factors.
[2] Ignoring poly-logarithmic improvements [PT03].

B. Durand and W. Thomas (Eds.): STACS 2006, LNCS 3884, pp. 254–263, 2006.
© Springer-Verlag Berlin Heidelberg 2006

grid (i.e., $\sqrt{n}$ points lie on each grid line), and it is not known what is the minimum number of colors needed if each point is perturbed within a small distance of its original location.

In this paper, we study the following question: Given P, is it possible to add a small set of points Q such that $(P \cup Q, \mathcal{R})$ can be colored with fewer colors than $(P, \mathcal{R})$? It is instructive to look at the one-dimensional case as an example. Here we are given n points in $\mathbb{R}$, and would like to color them such that any *interval* contains a point of unique color. It is not hard to see that the geometric location of the points is irrelevant – *any* set of n such points would require the same number of colors. Hence, adding points will only increase the total number of colors required.

Does a similar situation hold in higher dimensions? For example, given n points in $\mathbb{R}^2$, and where the ranges are discs, is it possible to add $f(n)$ points such that the total number of colors for a conflict-free coloring is at most $o(\log(n + f(n)))$ (as mentioned earlier, for disc ranges, it is always possible to conflict-free color using $O(\log n)$ colors). The result of Pach and Toth [PT03] answers the question negatively: *any* set of n points need $\Omega(\log n)$ colors for a conflict-free coloring.

A more interesting case is that of axis-parallel rectangle ranges, where the current best bounds are much worse than for the disc case: $\tilde{O}(n^{1/2})$ colors for n points. We prove here that by adding a small (*sub-linear*, in fact) number of points, the total number of colors can be reduced substantially. We now state our results more precisely.

Our results. Let P be a set of n points in the plane, and $(P, \mathcal{R})$ the axis-parallel rectangle range space over P. Our main theorem is the following, stating that one can always decrease the number of colors needed by adding a small number of new points.

Theorem 1. *Given a set P of n points in $\mathbb{R}^2$, and any $\epsilon \geq 0$, it is always possible to add a set Q of $O(n^{1-\epsilon})$ points such that $P \cup Q$ can be conflict-free colored using $\tilde{O}(n^{\frac{3}{8}(1+\epsilon)})$ colors. Furthermore, the set Q and the conflict-free coloring can be computed in polynomial time with high probability.*

We prove the above theorem by using a probabilistic re-coloring technique that can be used to get a coloring with weaker properties, which we call a *quasi-conflict-free* coloring. As another application of quasi-conflict-free colorings, we show that points that are "regularly placed" can be colored using fewer colors than for the general case. More precisely, if one can partition P by a set of vertical and horizontal lines such that each cell contains at least one point (i.e., a non-uniform grid), then one can color P using few colors:

Theorem 2. *Given any $\sqrt{n} \times \sqrt{n}$ grid G such that each cell contains one point of P, there exists a conflict-free coloring of P using $O(n^{3/8+\epsilon} + 2^{8/3\epsilon})$ colors, for any constant $\epsilon > 0$. Furthermore, this coloring can be computed in expected time $O(n^{5/4})$.*

Outline. We start by definitions and briefly describing the framework for conflict-free coloring proposed in [ELRS03] in Section 2. Section 3 describes our quasi-conflict-free coloring technique. We then prove, using quasi-conflict-free colorings, our main Theorem 1 in Section 4, and Theorem 2 in Section 5.

2   Preliminaries

Definitions. Let $P = \{p_1, \ldots, p_n\}$ be a set of n points in the plane, and $(P, \mathcal{R})$ any shrinkable range space over P. Define the *conflict-free* graph (with vertex set V and edge set E) of the range space $(P, \mathcal{R})$ as follows: each vertex $v_i \in V$ corresponds to the point $p_i \in P$, and $(v_i, v_j) \in E$ iff there exists a range $T \in \mathcal{R}$ such that T contains both p_i and p_j and no other point of P. Call a range $T \in \mathcal{R}$ conflict-free if it contains a point of unique color.

Let G be a $\sqrt{r} \times \sqrt{r}$ (non-uniform) grid consisting of r cells. Let $P(G)$ denote the set of r points when each grid cell contains exactly one such point, and where the point $p_{ij} \in P$ lies in the i-th row and j-th column of G. Let $C_i \subseteq P$ (respectively, $R_i \subseteq P$) denote the *sequence* of $\sqrt{r}$ points lying in the i-th column (respectively, i-th row) of the grid. Namely, $C_i = \langle p_{1i}, p_{2i}, \ldots, p_{\sqrt{r}i} \rangle$, and $R_i = \langle p_{i1}, p_{i1}, \ldots, p_{i\sqrt{r}} \rangle$. Note that C_i and R_i are ordered sequences.

Observe that for axis-parallel rectangle ranges, if we perturb the points such that the sorted order of the points in both the x and y-coordinates is unchanged, then the conflict-free graph would also be unchanged. Hence, for the general case, one can assume that the points lie at the vertices of an $n \times n$ integer grid and no two points have the same x or y coordinates.

A general framework. In [ELRS03], a general framework for computing conflict-free colorings of $(P, \mathcal{R})$ is presented, as follows. Iteratively, compute a large independent set, say I_1, in the conflict-free graph of $(P, \mathcal{R})$ and color all the points in I_1 with the same color, say c_1. Now repeat the above procedure for the range $(P \setminus I_1, \mathcal{R})$, coloring the next independent set with color c_2, and so forth. It is shown in [ELRS03] that firstly, for certain range spaces, discs for example, one can guarantee an independent set of size linear in the number of points. Therefore, for disc ranges, the above procedure finishes after at most a logarithmic number of steps, and hence the number of colors used are logarithmic. Second, the above coloring procedure yields a conflict-free coloring.

The above framework therefore reduces the conflict-free coloring problem to one of showing large independent sets in certain graphs. This can then be applied to computing conflict-free coloring for other, more general range spaces. In [HS05], a conflict-free coloring of rectangle ranges using $\tilde{O}(n^{1/2})$ colors is presented. The underlying lemma used to derive this result is the following.

Lemma 1 (Erdős-Szkeres Theorem). *Given any sequence S of n numbers, there exists a monotone subsequence of S of size at least $\sqrt{n}$. Furthermore, one can partition S into $O(\sqrt{n})$ monotone subsequences in time $O(n^{3/2} \log n)$.*

Now take the set of points P, and sort them by their increasing x-coordinates. Take the corresponding y-coordinates of points in this sequence as the sequence S, and compute the largest monotone subsequence. It is easy to see that the points corresponding to this subsequence form a monotone sequence in both x (because of the initial sorting) and y (monotone subsequence). Now, picking every other point in this sequence forms a set of $\sqrt{n}/2$ points which are independent in the conflict-free graph. Iteratively repeating this gives the above-mentioned result.

3   Quasi-conflict Free Colorings

Given a set P of n points in $\mathbb{R}^2$ and a parameter k, denote by G_k a $\frac{n}{k} \times \frac{n}{k}$ grid such that every row and column of G_k has exactly k points of P. Note that each cell of G_k need not contain any point; see Figure 1(a).

Given P and k, we call a coloring of P *quasi-conflict-free* with respect to k, if every rectangle which contains points from the same row or the same column of G_k is conflict-free. Note that every conflict-free coloring of P is quasi-conflict-free, though the converse might not be true. We now prove that P can be quasi-conflict-free colored with fewer colors. The coloring procedure is probabilistic [MR95]: we show that with high probability, a certain coloring is quasi-conflict-free. The existence of such a coloring then follows.

Theorem 3. *Given any point-set P and a parameter $k \geq n^c$ for $c > 0$, there exists a quasi-conflict-free coloring of G_k using $\tilde{O}(k^{3/4})$ colors. This coloring can be computed, with high probability, in polynomial time.*

PROOF. Set $r = n/k$ to be the number of rows and columns. We use the following coloring procedure:

Step 1. By Lemma 1, partition the points in column j, for $j = 1, \ldots, r$, into $h = \theta(\sqrt{k})$ independent sets, each of size at most $\sqrt{k}$. Note that by simply iteratively extracting the largest monotone subsequence, one can get $O(\sqrt{k})$ independent sets, where, however, some sets can be of size much larger than $\sqrt{k}$. This considerably complicates the analysis later on, by forcing the consideration of large and small independent sets separately. To avoid that (without affecting the worst-case analysis), at each iteration we only extract an independent set of size $\theta(\sqrt{k})$, forcing a decomposition into $\theta(\sqrt{k})$ independent sets, each of size $\theta(\sqrt{k})$.

Step 2. Pick, independently for each column j, $j = 1, \ldots, r$, $\pi_j \in \mathbb{S}_h$ to be a random permutation. Equivalently, we can think of π_j as an assignment of h distinct colors to the independent sets of column j (all columns share the same set of h colors). From here on, it is assumed that assigning a color to a given set of points means assigning the same color to all points in the set. Thus $\pi_j(S)$ is the color assigned to set S and $\pi_j(p)$ is the color assigned to the independent set containing point p. For $l = 1 \ldots h$, let S_j^l be the set of points in column j with the color l (i.e., belonging to the same independent set).

Let $\mathcal{X}_1, \ldots, \mathcal{X}_h$ be a family of h pairwise-distinct sets of colors. We shall recolor the points of G_k using these sets of colors, such that the final coloring is quasi-conflict free: we assign colors from $\mathcal{X}_l$ to points which were colored l above.

Step 3. Fix a row $i \in [r]$, and a color $\ell \in [h]$. Let A_i^l be the set of points with color ℓ, taken from all the cells of row i, i.e.,

$$A_i^l = \bigcup_{j=1}^{r} S_j^l \cap R_i$$

We recolor the set of points A_i^l using colors from $\mathcal{X}_\ell$, such that any rectangle containing points from this set is conflict-free. The number of colors needed for this step is, from Lemma 1,

$$\Delta \overset{\text{def}}{=} \sum_{\ell=1}^{h} \max\{\sqrt{|A_i^\ell|} \ : \ i = 1, \dots, r\}. \tag{1}$$

Now, the theorem follows from the following two lemmas.

Lemma 2. *The coloring procedure described above is quasi-conflict-free.*

PROOF. Take any rectangle T that lies completely inside a row or a column of G_k:

- **The column case.** If T contains only points belonging to a single column C_j of G_k, then the fact that the coloring of Step 1 was conflict-free within each column implies that T contains a point $p \in C_j$ such that $\pi_j(p) \neq \pi_j(p')$ for all $p' \neq p$ inside T. But then p maintains a unique color inside T also after recoloring in *Step 2*, since points with different π_j values are colored using different sets of colors.
- **The row case.** Now assume that T contains only points belonging to a single row i of G_k. If there is an $\ell \in [h]$ such that $T \cap A_i^\ell \neq \emptyset$, then by the conflict-free coloring of A_i^ℓ in *Step 3* above, and the fact that $\mathcal{X}_\ell$ is distinct from all other colors used for row i, we know that there exists a point of A_i^l inside T having a unique color. ◻

Lemma 3. *With probability $1 - o(1)$, the procedure uses $\tilde{O}(k^{3/4})$ colors.*

PROOF. Fix $i \in [r]$ and $\ell \in [h]$. Define $t = k/h$ to be the size of the largest independent set in any column (as defined in *Step 1.*). We now upper-bound Δ by estimating the maximum size of the union of sets of a fixed color ℓ in row i. The sizes of these sets vary from $1, \dots, t$, so we first partition A_i^ℓ into approximately same-sized sets, and estimate each separately as follows.

For $m = 1, 2, \dots, \log t$, let

$$\mathcal{A}_{i,j}^m = \cup_{\ell=1}^{h}\{S \ : \ S = C_j \cap A_i^\ell, \ 2^{m-1} \leq |S| \leq 2^m\}$$

be the family of sets in cell ij with size in the interval $[2^{m-1}, 2^m]$. Note that

$$\sum_{j=1}^{r} |\mathcal{A}_{i,j}^m| \leq \frac{k}{2^{m-1}}, \tag{2}$$

since the total number of points in row i is at most k, and each set in $\mathcal{A}_{i,j}^m$ has at least 2^{m-1} points.

Let $Y_{i,j}^{m,\ell}$ be the indicator random variable that takes value 1 if and only if there exists a set $S \in \mathcal{A}_{i,j}^m$ with $\pi_j(S) = \ell$. Let $Y_i^{m,\ell} = \sum_{j=1}^{r} Y_{i,j}^{m,\ell}$. Then,

$$\mathbb{E}[Y_{i,j}^{m,\ell}] = \Pr[Y_{i,j}^{m,\ell} = 1] = \frac{|\mathcal{A}_{i,j}^m|\binom{h-1}{|\mathcal{A}_{i,j}^m|-1}(|\mathcal{A}_{i,j}^m| - 1)!}{\binom{h}{|\mathcal{A}_{i,j}^m|}|\mathcal{A}_{i,j}^m|!} = \frac{|\mathcal{A}_{i,j}^m|}{h}$$

$$\mathbb{E}[Y_i^{m,\ell}] = \sum_{j=1}^{r} \frac{|\mathcal{A}_{i,j}^m|}{h} \leq \frac{k}{h2^{m-1}} = \frac{t}{2^{m-1}},$$

where the last inequality follows from (2).

Note that the variable $Y_i^{m,\ell}$ is the sum of independent Bernoulli trials, and thus applying the Chernoff bound[3], we get

$$\Pr[Y_i^{m,\ell} > \frac{t \log k}{2^{m-1}}] \le e^{-\frac{t \log k}{4 \cdot 2^{m-1}} \ln\left(\frac{t \log k}{\mathbb{E}[Y_i^{m,\ell}] \cdot 2^{m-1}}\right)}. \tag{3}$$

Using $\mathbb{E}[Y_m^{i,\ell}] \le t/2^{m-1}$ and $2^m \le t$, we deduce from (3) that

$$\Pr[Y_i^{m,\ell} > \frac{t \log k}{2^{m-1}}] \le (\log k)^{-(\log k)/2}.$$

Thus, the probability that there exist i, m, and ℓ such that $Y_i^{m,\ell} > t \log k / 2^{m-1}$ is at most $rh(\log t)(\log k)^{-(\log k)/2} = o(1)$. Therefore with probability $1 - o(1)$, $Y_i^{m,\ell} \le t \log k / 2^{m-1}$ for all i, ℓ, and m. In particular, with high probability,

$$|A_i^\ell| \le \sum_{m=1}^{\log t} Y_i^{m,\ell} \cdot 2^m \le 2t \log k \log t.$$

Combining this with (1), we get $\Delta \le h\sqrt{2t \log k \log t} = \sqrt{2hk \log k \log t} = \tilde{O}(k^{3/4})$, as desired. □

4   Conflict-Free Coloring of General Point Sets

In this section, we prove Theorem 1 which states that by adding a small number of points, the total number of colors for a conflict-free coloring can be significantly reduced.

First note that it is quite easy to add a *super-linear* number of points to get better colorings. Indeed, by adding n^2 points, one can get a conflict-free coloring using $O(\log n)$ colors: simply partition the point set into a $n \times n$ grid, where each row and column contains exactly one point. Add the n^2 points on an exact uniform grid such that each cell contains one point, and conflict-free color these points using $O(\log n)$ colors [4].

We now prove our main theorem.

Proof of Theorem 1. Let k be an integer, and set $r = \frac{n}{k}$. We start by putting the set of points P into an $r \times r$ grid G, in which each row and each column contains exactly k points. See Figure 1(a) for an illustration. We let Q be a set of r^2 points organized in a uniform grid G', such that every cell of G contains a point of G' in its interior. Next we partition G into 4 grids $G_{0,0}$, $G_{0,1}$, $G_{1,0}$ and $G_{1,1}$, where $G_{i,j}$ consists of cells lying in even rows (resp. columns) if i (resp. j) is 0, and odd rows (resp. columns) if i (resp. j) is 1. Finally, we color these grids by 5 pairwise-disjoint sets of colors, such that $Q(G')$ is conflict-free and $G_{0,0}, \ldots, G_{1,1}$ are quasi-conflict-free.

---

[3] In particular, the following version [MR95]: $\Pr[X \ge (1 + \delta)\mu] \le e^{-(1+\delta)\ln(1+\delta)\mu/4}$, for $\delta > 1$ and $\mu = \mathbb{E}[X]$.

[4] In fact, as pointed out by an anonymous referee, it follows from [HS05] that one can add a set Q of $O(n \log n)$ points such that $P \cup Q$ can be colored using $O(\log^2 n)$ colors: project all the points onto the vertical bisector in the x-ordering, color the projected points using $O(\log n)$ colors, and recurse on the two sides.

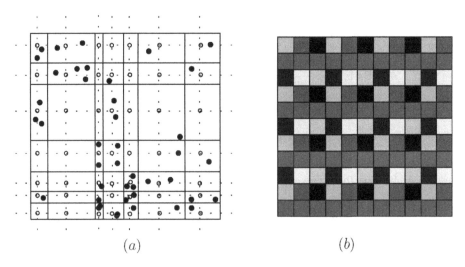

(a) (b)

Fig. 1. (a) The grid G_k, $k = 6$. The solid points are part of the input P, while the empty ones are the newly added ones, Q (that lie on a uniform grid). (b) Partitioning the grid into k^2 grids; here $k = 3$. The points in these grids will be colored with distinct sets of colors (indicated by the cell colors).

Clearly the resulting coloring of $P \cup Q$ is conflict-free: any rectangle spanning more than two rows and two columns of G must contain a point of Q. Otherwise it contains points from a single row or single column of one of $G_{i,j}$, which are quasi-conflict-free colored.

Since Q lies on the uniform grid G', the number of colors needed for G' is $O(\log r)$ [HS05]. Furthermore, by Theorem 3, the number of colors needed for the other four grids is $\tilde{O}(k^{3/4})$. Finally, setting $k = n^{(1+\epsilon)/2}$, it follows that one can add the set Q of $n^2/k^2 = n^{1-\epsilon}$ points to P to get a conflict-free coloring of $P \cup Q$ using $\tilde{O}(k^{3/4}) = \tilde{O}(n^{\frac{3}{8}(1+\epsilon)})$ colors. □

Remark: The set Q of points we add lies on an exact uniform grid. This can be relaxed by instead placing points anywhere inside the cells of G_k, and using Theorem 2 to compute the conflict-free coloring of Q. This allows one greater freedom to place the new points, although the bounds become correspondingly worse.

5   Conflict-Free Colorings of Points in a Grid

We are given a set P of n points and a $\sqrt{n} \times \sqrt{n}$ grid G such that each cell of G contains exactly one point of P. We now first show that Theorem 3 can be used, together with *shifted dissection*, to obtain a conflict-free coloring of $P(G)$ which uses fewer colors than the general case. We also observe that this case is not much easier than the general case, i.e., coloring a grid with fewer than $O(n^{1/4})$ colors would imply that the general case can be colored with fewer than $O(n^{1/2})$ colors.

Given G and an integer $k \leq \sqrt{n}$, we can partition G into a set of k^2 grids in a natural way (see Figure 1(b)): For $x = 0, 1, \ldots, k-1$ and $y = 0, 1, \ldots, k-1$, grid $G_{x,y}$ consists of all the points p_{ij} of G such that $i \bmod k = x$ and $j \bmod k = y$, namely,

$$P(G_{x,y}) = \{p_{x+ik,y+jk} \in P(G) \mid i, j \in \mathbb{Z} : 0 \leq x + ik, y + jk < \sqrt{n}\}.$$

Now we color the points of G as follows. Let $r = \frac{n}{k^2}$. Each of the k^2 different grids (of size $\sqrt{n}/k \times \sqrt{n}/k$) will be colored with a different set of colors. One grid, say $G_{0,0}$ is *recursively* conflict-free colored. For the other $k^2 - 1$ grids, we use Theorem 3 to get a quasi-conflict-free coloring of the points of each grid using $O(r^{3/8})$ colors.

Lemma 4. *The coloring of $P(G)$ described above is conflict-free.*

PROOF. Let T be a rectangle in the plane containing at least one point of $P(G)$. Consider the following two cases:

First, if T contains only points from at most $k-1$ rows of G, then let R be any row of G such that $R \cap T$ is non-empty. Take any grid $G_{x,y}$ such that $P(G_{x,y}) \cap R \cap T \neq \emptyset$. Then T is a rectangle containing points belonging to only one row of $G_{x,y}$ (since T spans less than $k-1$ rows), and consequently a point with a unique color exists inside T, since (i) the grid $G_{x,y}$ is quasi-conflict-free colored, and (ii) the points belonging to all other grids are colored with a different set of colors than the one used for $G_{x,y}$. The case when T contains points from at most $k-1$ columns of G is similar.

Second, if T is a large rectangle, i.e., contains points from at least k rows and at least k columns of G, then T must contain a point from $G_{0,0}$. By the conflict-free coloring of $G_{0,0}$ we know that $T \cap P(G_{0,0})$ contains a point with a unique color. The same point has a unique color among all other points in $T \cap P(G)$ since, again, the different grids are colored with different sets of colors.   □

Proof of Theorem 2. Using the above coloring scheme, the number of colors needed is at most

$$f(n) = \min_{1 \leq k \leq \sqrt{n}} \left\{ k^2 \left(\frac{n}{k^2}\right)^{3/8} + f\left(\frac{n}{k^2}\right) \right\}.$$

We prove by induction that

$$f(n) \leq n^{3/8+\epsilon} + 2^{8/3\epsilon}$$

The base case is when the number of points is small, more precisely, when $n \leq 2^{16/3\epsilon}$. Using Lemma 1, one can color them with at most $\sqrt{n} \leq 2^{8/3\epsilon}$ colors. When $n \geq 2^{16/3\epsilon}$, then by setting $k^2 = n^{\epsilon/(1+\epsilon)}$, and applying the inductive hypothesis, we have

$$f(n) \le k^2 \left(\frac{n}{k^2}\right)^{3/8} + (\frac{n}{k^2})^{3/8+\epsilon} + 2^{8/3\epsilon}$$
$$= (n^{\epsilon/(1+\epsilon)})(n^{1/(1+\epsilon)})^{3/8} + (n^{1/(1+\epsilon)})^{3/8+\epsilon} + 2^{8/3\epsilon}$$
$$= 2(n^{1/(1+\epsilon)})^{3/8+\epsilon} + 2^{8/3\epsilon}$$
$$= (2^{8/(3+8\epsilon)}n^{1/(1+\epsilon)})^{3/8+\epsilon} + 2^{8/3\epsilon} \le (2^{8/3}n^{1/(1+\epsilon)})^{3/8+\epsilon} + 2^{8/3\epsilon}$$
$$\le (n^{\epsilon/(1+\epsilon)}n^{1/(1+\epsilon)})^{3/8+\epsilon} + 2^{8/3\epsilon} = n^{3/8+\epsilon} + 2^{8/3\epsilon}$$

where the last inequality follows from the assumption that $n \ge 2^{16/3\epsilon}$.   ⊡

Remark 1: One can also consider computing an independent set of the points in a grid. In fact, using similar ideas, one can show the existence of an independent set of size $\Omega(n^{5/8})$. However, this does not immediately yield a conflict-free coloring of $P(G)$ using $O(n^{3/8})$ colors. This is because, once we extract an independent set of points P' from $P(G)$ of size $\Omega(n^{5/8})$, the remaining set $P(G) \setminus P'$ is no longer a grid. However, we have seen above, within an arbitrarily small additive constant in the exponent, such a coloring can indeed be achieved.

Remark 2: Having partitioned the grid G into k^2 grids, we need not color all these grids with distinct sets of colors. In fact, we can assign only $O(k)$ sets of colors to the different grids such that the whole grid G can be conflict-free colored.

Finally, observe that this grid case is not much easier than the general case.

Observation 1. *If there always exists an independent set of size at least n^c for a point set $P(G)$ lying on a $\sqrt{n} \times \sqrt{n}$ grid, then, for any general set of points, there exists an independent set of size at least n^{2c-1}.*

PROOF. Assume otherwise. Then there exists a general point set Q of size $\sqrt{n}$ whose maximum independent set has size at most $n^{c-1/2}$. We construct $P(G)$ of size n by putting $\sqrt{n}$ copies of Q in each column, i.e., set each C_i to be a (translated copy) of Q. By our earlier observation, this can be done since we can always move the y-coordinates of each point in Q to lie in a different row. The maximum independent set of this new point set can contain at most $n^{c-1/2}$ points from each column, and therefore has maximum independent set of size less than n^c, a contradiction.   ⊡

In particular, if one could show the existence of a linear sized independent set for the grid case, that would imply a linear sized independent set for the general case, for which the current best bound is roughly $\tilde{O}(n^{1/2})$.

Finally, we have used the algorithm for conflict-free coloring general pointsets as a black-box to get the decomposition into a small number of independent sets (in the conflict-free graph). Therefore, any improved bounds on the general problem imply an improvement in our bounds.

Acknowledgements. We would like to thank Sathish Govindarajan for many helpful discussions, and two anonymous reviewers for useful suggestions that improved the content and presentation of this paper.

References

[ELRS03]   G. Even, Z. Lotker, D. Ron, and S. Smorodinsky. Conflict-free colorings of simple geometric regions with applications to frequency assignment in cellular networks. *SIAM J. Comput.*, 33:94–136, 2003.

[HS05]   S. Har-Peled and S. Smorodinsky. Conflict-free coloring of points and simple regions in the plane. *Discrete & Comput. Geom.*, 34:47–70, 2005.

[MR95]   R. Motwani and P. Raghavan. *Randomized Algorithms*. Cambridge University Press, New York, NY, 1995.

[PT03]   J. Pach and G. Toth. Conflict free colorings. In *Discrete & Comput. Geom., The Goodman-Pollack Festschrift*. Springer Verlag, Heidelberg, 2003.

[Smo03]   S. Smorodinsky. *Combinatorial problems in computational geometry*. PhD thesis, Tel-Aviv University, 2003.

Grid Vertex-Unfolding Orthogonal Polyhedra

Mirela Damian[1], Robin Flatland[2], and Joseph O'Rourke[3]

[1] Comput. Sci., Villanova University, Villanova, PA 19085, USA
mirela.damian@villanova.edu
[2] Comput. Sci., Siena College, Loudonville, NY 12211, USA
flatland@siena.edu
[3] Comput. Sci., Smith College, Northampton, MA 01063, USA
orourke@cs.smith.edu

Abstract. An *edge-unfolding* of a polyhedron is produced by cutting along edges and flattening the faces to a *net*, a connected planar piece with no overlaps. A *grid unfolding* allows additional cuts along grid edges induced by coordinate planes passing through every vertex. A *vertex-unfolding* permits faces in the net to be connected at single vertices, not necessarily along edges. We show that any orthogonal polyhedra of genus zero has a grid vertex-unfolding. (There are orthogonal polyhedra that cannot be vertex-unfolded, so some type of "gridding" of the faces is necessary.) For any orthogonal polyhedron P with n vertices, we describe an algorithm that vertex-unfolds P in $O(n^2)$ time. Enroute to explaining this algorithm, we present a simpler vertex-unfolding algorithm that requires a 3×1 refinement of the vertex grid.

1 Introduction

Two unfolding problems have remained unsolved for many years [DO05]: (1) Can every convex polyhedron be edge-unfolded? (2) Can every polyhedron be unfolded? An *unfolding* of a 3D object is an isometric mapping of its surface to a single, connected planar piece, the "net" for the object, that avoids overlap. An *edge-unfolding* achieves the unfolding by cutting edges of a polyhedron, whereas a *general unfolding* places no restriction on the cuts.

It is known that some nonconvex polyhedra cannot be unfolded without overlap with cuts along edges. However, no example is known of a nonconvex polyhedron that cannot be unfolded with unrestricted cuts. Advances on these difficult problems have been made by specializing the class of polyhedra, or easing the stringency of the unfolding criteria. On one hand, it was established in [BDD+98] that certain subclasses of *orthogonal polyhedra*—those whose faces meet at angles that are multiples of 90°—have an unfolding. In particular, the class of *orthostacks*, stacks of extruded orthogonal polygons, was proven to have an unfolding (but not an edge-unfolding). On the other hand, loosening the criteria of what constitutes a net to permit connection through points/vertices, the so-called *vertex-unfoldings*, led to an algorithm to vertex-unfold any triangulated manifold [DEE+03].

A second loosening of the criteria is the notion of grid unfoldings, which are especially natural for orthogonal polyhedra. A *grid unfolding* adds edges

B. Durand and W. Thomas (Eds.): STACS 2006, LNCS 3884, pp. 264–276, 2006.
© Springer-Verlag Berlin Heidelberg 2006

to the surface by intersecting the polyhedron with planes parallel to Cartesian coordinate planes through every vertex. It was recently established that any orthostack may be grid vertex-unfolded [DIL04]. For orthogonal polyhedra, a grid unfolding is a natural median between edge-unfoldings and unrestricted unfoldings.

Our main result is that any orthogonal polyhedron, without shape restriction except that its surface be homeomorphic to a sphere, has a grid vertex-unfolding. We present an algorithm that grid vertex-unfolds any orthogonal polyhedron with n vertices in $O(n^2)$ time. We also present, along the way, a simpler algorithm for 3×1 *refinement* unfolding, a weakening of grid unfolding that we define below. We believe that the techniques in our algorithms can be further exploited to show that all orthogonal polyhedra can be grid edge-unfolded.

2   Definitions

A $k_1 \times k_2$ *refinement* of a surface [DO04] partitions each face into a $k_1 \times k_2$ grid of faces. We will consider refinements of grid unfoldings, with the convention that a 1×1 refinement is an unrefined grid unfolding.

Following the physical model of cutting out the net from a sheet of paper, we permit cuts representing *edge overlap*, where the boundary touches but no interior points overlap. We also insist as part of the definition of a vertex-unfolding, again keeping in spirit with the physical model, that the unfolding "path" never self-crosses on the surface in the following sense. If (A, B, C, D) are four faces incident in that cyclic order to a common vertex v, then the net does not include both the connections AvC and BvD.[1]

We use the following notation to describe the six type of faces of an orthogonal polyhedron, depending on the direction in which the outward normal points: *front*: $-y$; *back*: $+y$; *left*: $-x$; *right*: $+x$; *bottom*: $-z$; *top*: $+z$. We take the z-axis to define the vertical direction; *vertical* faces are parallel to the xz-plane. We distinguish between an original vertex of the polyhedron, which we call a *corner vertex* or just a *vertex*, and a *gridpoint*, a vertex of the grid (might be an original vertex). A *gridedge* is an edge segment with both endpoint gridpoints.

Let O be a solid orthogonal polyhedron with the surface homeomorphic to a sphere (i.e, genus zero). Let Y_i be the plane $y = y_i$ orthogonal to the y-axis. Let $Y_0, Y_1, \ldots, Y_i, \ldots$ be a finite sequence of parallel planes passing through every vertex of O, with $y_0 < y_1 < \cdots < y_i < \cdots$. We call the portion of O between planes Y_i and Y_{i+1} *layer i*; it includes a collection of disjoint connected components of O. We call each such component a *slab*. Referring to Figure 1a, layer 0, 1 and 2 each contain one slab (C, B and A, respectively). The surface piece that surrounds a slab is called a *band* (shaded in Figure 1a). Each slab is bounded by an outer band, but it may also contain inner bands, bounding holes. Outer bands are called *protrusions* and inner bands are called *dents* (D in Figure 1a). For a band A, $r_i(A)$ is the closed region of Y_i enclosed by A.

---

[1] This was not part of the original definition in [DEE$^+$03] but was achieved by those unfoldings.

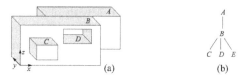

Fig. 1. Definitions: A, B and C are protrusions; D is a dent

3   Dents vs. Protrusions

We observe that dents may be treated exactly the same as protrusions with respect to unfolding, because unfolding of a 2-manifold to another surface (in our case, a plane) depends only on the intrinsic geometry of the surface, and not on how it is embedded in $\mathbb{R}^3$. All that matters for unfolding is the faces sharing an edge, and the cyclic ordering of the faces incident to a vertex, i.e., our unfolding algorithms will make local decisions and will be oblivious to the embedding in $\mathbb{R}^3$. These local relationships are identical if we conceptually "pop-out" dents to become protrusions (this popping-out is conceptual only, for it could produce self-intersecting objects.) Henceforth, we will describe only protrusions in our algorithms, with the understanding that nothing changes for dents.

4   Overview

The two algorithms we present share a common central structure, with the second achieving a stronger result; both are vertex-unfoldings that use orthogonal cuts only. We note that it is the restriction to orthogonal cuts that makes the vertex-unfolding problem difficult: if arbitrary cuts are allowed, then a general vertex unfolding can be obtained by simply triangulating each face and applying the algorithm from [DEE+03].

The (3×1)-algorithm unfolds any genus-0 orthogonal polyhedron that has been refined in one direction 3-fold. The bands themselves are never split (unlike in [BDD+98]). The (1×1)-algorithm also unfolds any genus-0 orthogonal polyhedron, but this time achieving a grid vertex-unfolding, i.e., without refinement. This algorithm is more delicate, with several cases not present in the (3×1)-algorithm that need careful detailing. Clearly this latter algorithm is stronger, and we vary the detail of presentation to favor it. The overall structure of the two algorithms is the same:

1. A band "unfolding tree" T_U is constructed by shooting rays vertically from the top of bands. The root of T_U is a *backmost* band (of largest y-coordinate), with ties arbitrarily broken.
2. A forward and return *connecting* path of vertical faces is identified, each of which connects a parent band to a child band in T_U.
3. Each band is unfolded horizontally as a unit, but interrupted when a connecting path to a child is encountered. The parent band unfolding is suspended at that point, and the child band is unfolded recursively.

4. The vertical front and back faces of each slab are partitioned according to an illumination model, with variations for the more complex (1×1)-algorithm. These vertical faces are attached below and above appropriate sections of the band unfolding.

The final unfolding lays out all bands horizontally, with the vertical faces hanging below and above the bands. Non-overlap is guaranteed by this strict two-direction structure.

Although our result is a broadening of that in [DIL04] from orthostacks to all orthogonal polyhedra, we found it necessary to employ techniques different from those used in that work. The main reason is that, in an orthostack, the adjacency structure of bands yields a path, which allows the unfolding to proceed from one band to the next along this path, never having to return. In an orthogonal polyhedron, the adjacency structure of bands yields a tree (cf. Figure 1b). Thus unfolding band by band leads to a tree traversal, which requires traversing each arc of the tree in both directions. It is this aspect which we consider our main novelty, and which leads us to hope for application to edge-unfoldings as well.

5 (3×1)-Algorithm

5.1   Computing the Unfolding Tree T_U

We first describe finding an unfolding tree T_U that spans all bands in O. For each i, consider in turn each pair of bands A and B separated by Y_i such that regions $r_i(A)$ and $r_i(B)$ are not disjoint. Let e be the highest horizontal gridedge of $r_i(A) \cap r_i(B)$, breaking ties for highest arbitrarily. Since A and B are non-disjoint, e always exists. W.l.o.g., assume e is on B (as in Figure 2a). We partition e into three equal segments, and define the *pivot point* x_b to be the $\frac{1}{3}$-point of e (or, in circumstances to be explained below, the $\frac{2}{3}$-point.) These pivot points are base points of the connecting rays, which determine the unfolding tree T_U (refer to Figure 2b). Note that if e belongs to both A and B, then the ray connecting A and B degenerates to a point. To either side of a connecting ray we have two *connecting paths* of vertical faces, the *forward* and *return* path. In Figure 2a, these connecting paths are the shaded strips on the front face of A.

5.2   Unfolding Bands

Starting from a backmost *root band*, each band is unfolded as a conceptual unit, but interrupted by the connecting rays incident to it from its front and back faces. In Fig. 2, band A is unfolded as a rectangle, but interrupted at the rays connecting to B, B' and C. At each such ray the parent band unfolding is suspended, the unfolding follows the forward connecting path to the child, the child band is recursively unfolded, then the unfolding returns along the return connecting path back to the parent, resuming the parent band unfolding from the point it left off.

Fig. 2 illustrates this unfolding algorithm. The cw unfolding of A, laid out horizontal to the right, is interrupted to traverse the forward path down to B,

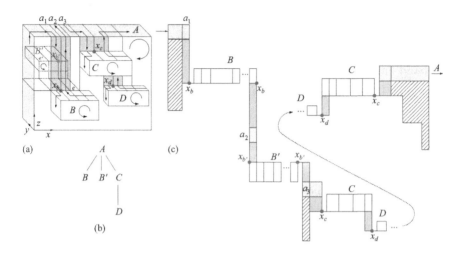

Fig. 2. (a) Orthogonal polyhedron. (b) Unfolding tree T_U. (c) Unfolding of bands and front face pieces. Vertex connection through the pivots points x_b, $x_{b'}$, x_c, x_d is shown exaggerated for clarity.

and B is then unfolded as a rectangle (composed of its contiguous faces). The base x_b of the connecting ray is called a *pivot point* because the unfolding of B is rotated 180° ccw about x_b so that the unfolding of B is also horizontal to the right. The unfolding of B proceeds ccw back to x_b, crosses over A to unfold B', then a cw rotation by 180° around the second image of pivot x'_b orients the return path to A so that the unfolding of A continues horizontal to the right. Note that the unfolding of C is itself interrupted to unfold child D. Also note that there is edge overlap in the unfolding at each of the pivot points.

The reason for the 3×1 refinement is that the upper edge e' of the back child band B' has the same (x, z)-coordinates as the upper edge e of B on the front face. In this case, the faces of band A induced by the connecting paths to B would be "overutilized" if there were only two. Let a_1, a_2, a_3 be the three faces of A induced by the 3×1 refinement of the connecting path to B, as in Fig. 2. Then the unfolding path winds around A to a_1, follows the forward connecting path to B, returns along the return connecting path to a_2, crosses over A and unfolds B' on the back face, with the return path now joining to a_3, at which point the unfolding of A resumes. In this case, the pivot point $x_{b'}$ for B' is the $\frac{2}{3}$-point of e'. Other such conflicts are resolved similarly. It is now easy to see that the unfolding of the bands and the connecting-path faces has the general form illustrated in Fig. 2c. The unfolding is an orthogonal polygon monotone with respect to the horizontal.

5.3   Attaching Front and Back Faces

Finally, we "hang" front and back faces from the bands as follows. The front face of each band A is partitioned by imagining A to illuminate downward lightrays in the front face. The pieces that are illuminated are then hung vertically

downwards from the horizontal unfolding of the A band. The portions unillumi-
nated will be attached to the obscuring bands.

In the example in Figure 2, this illumination model partitions the front face
of A into three pieces (the striped pieces from Figure 2c). These three pieces are
attached under A; the portions of the front face obscured by B but illuminated
downwards by B are hung beneath the unfolding of B (not shown in the figure),
and so on. Because the vertical illumination model produces vertical strips, and
because the strips above and below the band unfoldings are empty, there is
always room to hang the partitioned front face. Thus, any orthogonal polygon
may be vertex-unfolded with a 3×1 refinement of the vertex grid.

Although we believe this algorithm can be improved to 2×1 refinement,
the complications needed to achieve this are similar to what is needed to avoid
refinement entirely, so we instead turn directly to 1×1 refinement.

6 (1×1)-Algorithm

This algorithm follows the same sequence of steps as the 3×1 algorithm, with
each step involving more complex techniques discussed in detail in subsequent
sections:

---

1. Compute T_U (Section 6.1)
2. Select Pivot Points (Section 6.2)
3. Select Connecting Paths (Section 6.3)
4. Recurse:
    a. Unfold Bands (Section 6.4)
    b. Attach Front and Back Faces (Section 6.5)

6.1   Computing the Unfolding Tree T_U

Let A be a band separated by planes Y_i and Y_{i+1} (initially the root band). We
select zero or more connecting vertical rays for A in each of Y_i and Y_{i+1} according
to the following rules (which we describe only for Y_i, as they are identical for
Y_{i+1}). Refer to Fig. 3(a,b):

(1) For each band B that shares a gridpoint u with A, select u to be the (de-
    generate) ray connecting A to B (if several candidates for u, pick any one).
    Add arc (A, B) to T_U.
(2) If there is a band C such that $r_i(A)$ lies interior to $r_i(C)$, do the following.
    Let u be the topmost corner among the vertical leftmost (rightmost) edges
    of A, provided that A unfolds cw(ccw). Shoot a vertical ray r from u upward
    in Y_i, terminating at the first band B it hits (it may be that $B = C$.) Select
    r to be the ray connecting A to B, and add arc (A, B) to T_U.

In either case, if A unfolds cw, B unfolds ccw and vice-versa (the root band
always unfolds cw). We then recurse on the children of A in T_U.

Lemma 1. T_U *is a tree that spans all bands.*

6.2   Selecting Pivot Points

The pivot point x_a for a band A is the gridpoint of A where the unfolding of A starts and ends. The y-edge of A incident to x_a is the first edge of A we cut to unfold A.

The pivot point for the root band is always the front, topmost point among the leftmost faces. For each child B of A that does not intersect A, select x_b to be the base point of the ray connecting B to A. Any other child B shares one or more grid points with A. Our goal is to select pivot points for the children of A such that no two pivots are incident to the same y-edge of A.

For any child B such that $r_i(A)$ intersects $r_i(B)$, define $g(B)$ to be the set of gridpoints shared by noncoplanar faces of A and B; these are candidate pivot points for B. Note that $g(B)$ always contains at least two gridpoints. At each stage of the algorithm, define u to be the pivot point selected in the previous stage; initially, $u = x_a$. Let $s = (u, w)$ be the y-edge extending from u to a point w on the other side of A. Say that a band is *pivoted* if a pivot point for it has been selected, and otherwise *unpivoted*. The algorithm repeats as long as there is some unpivoted band. Each step of the algorithm assigns one child band of A (to either side) a pivot. The goal is to avoid assigning w from the previous iteration as a pivot, for that represents a conflict.

(1) If there is no child X with $w \in g(X)$ (i.e., there is no conflict with w) choose any unpivoted band B, and select u to be any gridpoint of $g(B)$.

(2a) If there is child X with $w \in g(X)$ (potential conflict) but X is already pivoted (necessarily at a gridpoint other than w, so there is no actual conflict), choose any unpivoted band B, and select u to be any gridpoint of $g(B)$.

(2b) If there is an unpivoted child B with $w \in g(B)$ (potential conflict), select u to be any vertex in $g(B)$ other than w. Because $g(B)$ contains two or more points, u always exists.

Band B is pivoted at gridpoint $x_b = u$ and the next iteration of the algorithm begins. Fig. 3(c,d) illustrates Cases (2a) and (2b).

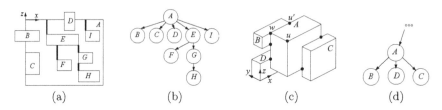

                    (a)   (b)   (c)   (d)

Fig. 3. (a) Bands B, C, D share gridpoints with A; $r_i(E) \dots r_i(I)$ lie interior to $r_i(A)$.
(b) T_U for the example in (a). (c) Selecting pivot points. (d) T_U for the example in (c).

Lemma 2. *The Pivot Selection Algorithm terminates with each child B assigned a unique conflict-free pivot $u = x_b$.*

6.3   Selecting Connecting Paths

Having established a pivot point for each band, we are now ready to define the *forward* and *return* connecting paths for a child band in T_U. Let B be an arbitrary child of a band A. If B intersects A, both forward and return connection paths for B reduce to the pivot point x_b (e.g., u in Fig. 4). If B does not intersect A, then a ray r connects x_b to A (Figs. 5a and 6a). The connecting paths are the two vertical paths separated by r comprised of the grid faces sharing an edge with r (paths a_1 and a_2 in Figs. 5a and 6a). The path first encountered in the unfolding of A is used as a forward connecting path; the other path is used as a return connecting path.

6.4   Unfolding Bands

Let A be a band to unfold, initially the root band. The unfolding of A starts at x_a and proceeds cw or ccw around A (cw for the root band); henceforth we assume cw w.l.o.g. In the following we describe our method to unfold every child B of A recursively, which falls naturally into several cases.

Case 1: B intersects A. Then, whenever the unfolding of A reaches x_b, we unfold B as in Fig. 4a or 4b, depending on whether B is encountered along A after (4a) or before (4b) x_b. The unfolding uses the two band faces of A incident to x_b (a_0 and a_1 in Fig. 4). Because the pivots of any two children that intersect A are conflict-free, there is no competition over the use of these two faces. Note also that the unfolding path does not self-cross. For example, the cyclic order of the faces incident to u in Fig. 4a is $(a_0, A_{front}, b_0, b_1, B_{back}, a_1)$, and the unfolding path follows $(a_0, b_0, \ldots, b_1, a_1)$.

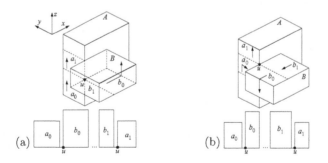

Fig. 4. Unfolding B; $u = x_b$; B extends (a) cw, and (b) ccw from u along A

Case 2: B does not intersect A. This case is more complex, because it involves conflicts over the use of the connecting paths for B; i.e, it may happen that a connecting path for B overlaps a connecting path for another descendant of A. We discuss these conflicts after settling some notation (cf. Figs 5a and 6a): r is the ray connecting B to A; a_1 and a_2 are forward and return connecting paths for B (one to either side of r); u_1 is the endpoint of r that lies on A; and u_2 is the other endpoint of the y-edge of A incident to u_1.

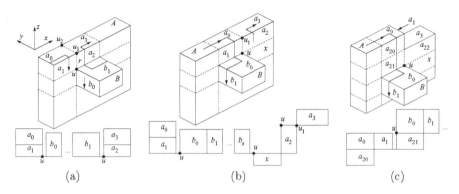

Fig. 5. Unfolding B; $u = x_b$; a_1 and a_2 are forward and return connecting paths

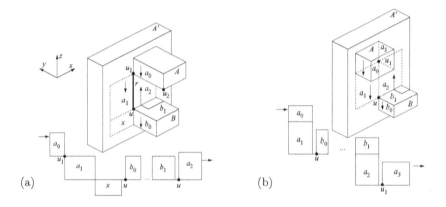

Fig. 6. Unfolding B; (a) leftmost, and (b) rightmost face of A vertically aligned with leftmost face of B; $u = x_b$; a_0 and a_1 are forward and return connecting paths for B

Case 2a: u_2 is neither incident to a connecting ray, nor is it a pivot for a child that intersects A. This means that no other child of A competes over the use of the two band faces of A incident to u_1 (see faces a_0 and a_3 in Fig. 5a); so these faces can be used in the unfolding of B.

Assume first that the forward path a_1 for B does not overlap the return path for another descendant of A. In this case, whenever the unfolding of A reaches a_1, unfold B recursively as follows. If u_1 is not a corner vertex of A, unfold B as in Fig. 5a. Otherwise, if u_1 is a left corner vertex of A, unfold B as in Fig. 6a. Otherwise, u_1 must be a right corner vertex of A: unfold B as in Fig. 6b.

Assume now that the forward path a_1 for B overlaps the return path for another descendant C of A. Then C must be positioned as in Fig. 7a, since x_c must lie on a left face of C. In this case, we unfold B as soon as (or as late as, if the unfolding of A starts at u_2) the unfolding of C along the return path to A meets the left face of B incident to x_b. At that point we recursively unfold B as in Fig. 7b, then continue the unfolding along the return path for C back to A. Fig. 7b shows face a_1 in two positions: we let a_1 hang down only if the next

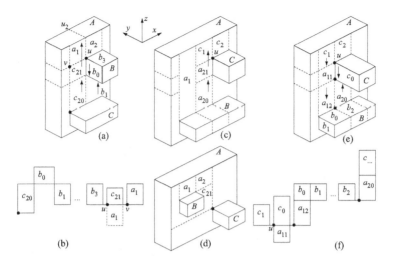

Fig. 7. (a) Return path for C includes c_{20}, c_{21}, a_1; Forward path for B is a_1. (b) Unfolding for (a) (c) Return path for B includes a_{20}, a_{21}, c_1; Forward path for C is c_1. (d) Return path for B is a_2; Forward path for C includes a_2, c_{21}. (e) Forward path for B includes c_1, a_{11}, a_{12}; Forward path for C is c_1. (f) Unfolding for (c).

face to unfold is a right face of a child of A (see the transition from k_7 to c_5 in Fig. 8); for all other cases, use a_1 in the upward position.

The return path a_2 for B always goes back to A. However, if the unfolding of a_2 encounters a child C of A, recursively unfold C as described in above, then continue along a_2 back to A.

Case 2b: u_2 is incident to the connecting ray for a child D that does not intersect A. This situation creates a conflict over the use of the two band faces of A incident to u_1 and u_2 (see a_0 and a_3 in Fig. 5a). We resolve this conflict by arbitrarily picking either B or D to unfold first. Suppose for instance that the method picks D to unfold first. Then, once the unfolding returns from D back to A, we proceed to unfolding B as in Fig. 5b or 5c.

We must again consider the situation in which the return path for B overlaps the forward path for another child C of A. This situation is illustrated in Figs. 7c and 7d. The case depicted in Fig. 7c is similar to the one in Fig. 7a and is resolved in the same manner. For the case depicted in Fig. 7d, notice that a_2 is on both the forward path for C and the return path for B. However, no conflict occurs here: from a_2 the unfolding continues downward along the forward path to C and unfolds C next.

A close look at the situation illustrated in Fig. 5b suggests that, in this case, it is also possible that the forward path for B overlaps the forward path for another descendant C of A, as shown in Fig. 7e. We handle this situation slightly differently: on the way to B, unfold the entire subband of C that extends ccw between the highest and the lowest left faces of C that align with the forward path to B – with the understanding that, if any children of C attach to this subband,

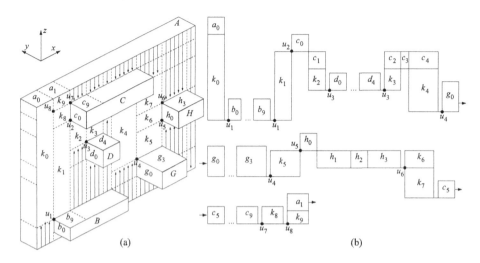

Fig. 8. (a) An example. (b) The vertex-unfolding.

they are recursively unfolded when encountered. After recursively unfolding B, resume the unfolding of C on the return path back for B. This approach is illustrated in Fig. 7f.

Case 2c: u_2 is a pivot point x_d for a child D that intersects A. We resolve this conflict by always unfolding D first as in Case 1; then unfold B as in Case 2b.

Fig. 8 shows a more complex example that emphasizes these subtle unfolding issues. Note that the return path k_1, k_8, k_9 for B overlaps the forward path k_9 for C; and the return path k_5, k_6 and k_7 for G overlaps the forward path for H, which includes k_7. The unfolding produced by the method described in this section is depicted in Fig. 8(b).

6.5 Attaching Front and Back Faces

The front and back faces of a slab are "hung" from bands following the basic idea of the illumination model discussed in Section 5.3. There are three differences, however, caused by the employment of some front and back gridfaces for the connecting paths, which can block illumination from the bands: (1) We illuminate both upward and downward from each band: each x-edge illuminates the vertical face it attaches to. (2) Some gridfaces may be obscured both above and below by paths in connecting faces. Therefore we incorporate connecting faces into the band for the purposes of illumination. The reason this works is that, with two exceptions, each vertical connecting strip remains vertical in the unfolding, and so illuminated strips can be hung safely without overlap. (3) The two exceptions are the return connecting path a_2 in Fig. 5b, and the forward connecting path a_1 in Fig. 6a. These two paths unfold "on their side". Note, however, that the face x below each of these paths (a face always present), is oriented vertically.

We thus consider x to be part of the connecting path for illumination purposes, permitting the strip below to be hung under x.

6.6   Algorithm Complexity

Because there are so few unfolding algorithms, that there is *some* algorithm for a class of objects is more important than the speed of the algorithm. Nevertheless, we offer an analysis of the complexity of our algorithm. Let n be the number of corner vertices of the polyhedron, and $N = O(n^2)$ the number of gridpoints. The vertex grid can be easily constructed in $O(N)$ time, leaving a planar surface map consisting of $O(N)$ gridpoints, gridedges, and gridfaces. The computation of connecting rays (Sec. 6.1) requires determining a common vertex between bands A and B, or, if $A \cap B = \emptyset$, determining whether $r_i(A) \subset r_i(B)$ or $r_i(A) \supset r_i(B)$. The former can be easily read off in $O(n^2)$ from the planar map by running through the n vertices and noting which of the $O(n)$ bands share a vertex. The latter can be accomplished with $O(n)$ ray shooting queries. Each connecting ray determines an arc of T_U, with the band nesting implicit. Although an implementation would employ an efficient data structure, perhaps BSP trees [PY92], for complexity purposes the naive $O(n)$ query cost suffices to lead to $O(n^2)$ time to construct T_U. Selecting pivots (Sec. 6.2) takes $O(n)$ constant-time iterations. Unfolding bands (Sec. 6.4) involves a depth-first traversal of T_U, which takes $O(n)$ time, and laying out the $O(N)$ gridfaces in $O(N)$ time. Thus, the algorithm can be implemented to run in $O(N) = O(n^2)$ time.

7   Further Work

Extending these algorithms to arbitrary genus orthogonal polyhedra remains an interesting open problem. Holes that extend only in the x and z directions within a slab seem unproblematic, as they simply disconnect the slab into several components. Holes that penetrate several slabs (i.e, extend in the y direction) present new challenges. One idea to handle such holes is to place a virtual xz-face midway through the hole, and treat each half-hole as a dent (protrusion).

Acknowledgements. We thank the anonymous referees for their careful reading and insightful comments, which greatly improved this paper. The third author was supported by NSF DTS award DUE-0123154.

References

[BDD+98]   T. Biedl, E. Demaine, M. Demaine, A. Lubiw, J. O'Rourke, M. Overmars, S. Robbins, and S. Whitesides. Unfolding some classes of orthogonal polyhedra. In *Proc. 10th Canad. Conf. Comput. Geom.*, pages 70–71, 1998.

[DEE+03]   E. D. Demaine, D. Eppstein, J. Erickson, G. W. Hart, and J. O'Rourke. Vertex-unfoldings of simplicial manifolds. In Andras Bezdek, editor, *Discrete Geometry*, pages 215–228. Marcel Dekker, 2003. Preliminary version appeared in *18th ACM Symposium on Computational Geometry*, Barcelona, June 2002, pp. 237-243.

[DIL04]   E. D. Demaine, J. Iacono, and S. Langerman. Grid vertex-unfolding of orthostacks. In *Proc. Japan Conf. Discrete Comp. Geom.*, November 2004. To appear in LNCS, 2005.

[DO04]   E. D. Demaine and J. O'Rourke. Open problems from CCCG 2004. In *Proc. 16th Canad. Conf. Comput. Geom.*, 2004.

[DO05]   E. D. Demaine and J. O'Rourke. A survey of folding and unfolding in computational geometry. In J. E. Goodman, J. Pach, and E. Welzl, editors, *Combinatorial and Computational Geometry*. Cambridge Univ. Press, 2005.

[PY92]   M. S. Paterson and F. F. Yao. Optimal binary space partitions for orthogonal objects. *J. Algorithms*, 13:99–113, 1992.

Theory and Application of Width Bounded
Geometric Separator*

Bin Fu

Dept. of Computer Science, University of New Orleans, LA 70148, USA
Research Institute for Children, 200 Henry Clay Avenue, LA 70118
fu@cs.uno.edu

Abstract. We introduce the notion of the width bounded geometric separator and develop the techniques for the existence of the width bounded separator in any d-dimensional Euclidean space. The separator is applied in obtaining $2^{O(\sqrt{n})}$ time exact algorithms for a class of NP-complete geometric problems, whose previous algorithms take $n^{O(\sqrt{n})}$ time [2,5,1]. One of those problems is the well known disk covering problem, which seeks to determine the minimal number of fixed size disks to cover n points on a plane [10]. They also include some NP-hard problems on disk graphs such as the maximum independent set problem, the vertex cover problem, and the minimum dominating set problem.

1 Introduction

The geometric separator has applications in many problems . It plays an important role in the divide and conquer algorithms for geometric problems. Lipton and Tarjan [18] showed the well known geometric separator for planar graphs. They proved that every n vertices planar graph has at most $\sqrt{8n}$ vertices whose removal separates the graph into two disconnected parts of size at most $\frac{2}{3}n$. Their $\frac{2}{3}$-separator was improved to $\sqrt{6n}$ by Djidjev [8], $\sqrt{5n}$ by Gazit [13], $\sqrt{4.5n}$ by Alon, Seymour and Thomas [6] and $1.97\sqrt{n}$ by Djidjev and Venkatesan [9]. Spielman and Teng [25] showed a $\frac{3}{4}$-separator with size $1.82\sqrt{n}$ for planar graphs. Separators for more general graphs were derived in (e.g.[14]). Some other forms of the geometric separators were studied by Miller, Teng, Thurston, and Vavasis [22,21] and Smith and Wormald [24]. For a set of points on the plane, assume each point is covered by a regular geometric object such as circle, rectangle, etc. If every point on the plane is covered by at most k objects, it is called k-thick. Some $O(\sqrt{k \cdot n})$ size separators and their algorithms were derived in [22,21,24].

The planar graph separators were applied in deriving some $2^{O(\sqrt{n})}$-time algorithms for certain NP-hard problems on planar graphs by Lipton, Tarjan [19], Ravi and Hunt [23]. Those problems include computing the maximum independence set, minimum vertex covers and three-colorings of a planar graph, and the number of satisfying truth assignments to a planar 3CNF formula [17].

* This research is supported by Louisiana Board of Regents fund under contract number LEQSF(2004-07)-RD-A-35.

In [24], their separators were applied in deriving $n^{O(\sqrt{n})}$-time algorithms for some geometric problems such as the planar Traveling Salesman and Steiner Tree problems. The separators were applied to the parameterized independent set problem on planar graphs by Alber, Fernau and Niedermeier [3,4] and disk graphs by Alber and Fiala [5].

We introduce the concept of width bounded separator. For a set of points Q on the plane, an a-wide separator is the region between two parallel lines of distance a. It partitions the set Q into two balanced subsets and its size is measured by the number of points of Q in the strip region. Our width bounded separator concept is geometrically natural and can achieve a much smaller constant c for its size upper bound $c\sqrt{n}$ than the previously approaches. Fu and Wang [11] developed a method for deriving sharper upper bound separator for grid points by controlling the distance to the separator line. They proved that for a set of n grid points on the plane, there is a separator that has $\leq 1.129\sqrt{n}$ points and has $\leq \frac{2}{3}n$ points on each side. That method was used to obtain the first sub-exponential time algorithm for the protein folding problem in the HP model. This paper not only generalizes the results of [11], but also substantially improves the techniques in [11]. We would like to mention our new technical developments in this paper. 1) In order to apply the separator to more general geometric problems with arbitrary input points other than grid points, we use weighted points in Euclid space and the sum of weights to measure the separator size instead of counting the number of points close to it. We introduce the local binding method to merge some points into a nearby grid point. This method is combined with our separator in deriving $2^{O(\sqrt{n})}$ time exact algorithms for a class of NP-complete geometric problems, whose previous algorithms take $n^{O(\sqrt{n})}$ time [2,5,1]. One of those problems is the well known disk covering problem, which seeks to determine the minimal number of fixed size disks to cover n points on a plane [10]. They also include some NP-hard problems on disk graphs such as the maximum independent set problem, the vertex cover problem, and minimum dominating set problem. 2) We will handle the case of higher dimension. We develop an area ratio method to replace the previous angle ratio method [11] for obtaining higher dimensional separators. 3) We develop a similar separator theorem for a set of points with distance of at least 1 between any two of them, called 1-separated set, we establish the connection between this problem and the famous fixed size discs packing problem. The discs packing problem in 2D was solved in the combinatorial geometry (see [26]). The 3D case, which is the Kepler conjecture, has a very long proof (see [16]). It is still a very elusive problem at higher dimensions. Our Theorem 2 shows how the separator size depends on packing density. 4) We develop a simple polynomial time algorithm to find the width-bounded separator for a fixed dimensional space. This is a starting point for the algorithms finding the width bounded geometric separator, and it is enough to obtain the $2^{O(\sqrt{n})}$ time exact algorithms for a class of NP-hard geometric problems.

2 Width-Bounded Separators on the d-Dimension

Throughout section 2 we assume the dimensional number d to be fixed. We will use the following well known fact that can be easily derived from Helly theorem (see [15,26]) to obtain our width bounded separator.

Lemma 1. *For an n element set P in d dimensional space, there is a point q with the property that any half-space that does not contain q, covers at most $\frac{d}{d+1}n$ elements of P. (Such a point q is called a centerpoint of P).*

Definition 1. *For two points p_1, p_2 in the d-dimensional Euclidean space R^d, $\mathrm{dist}(p_1, p_2)$ is the Euclidean distance between p_1 and p_2. For a set $A \subseteq R^d$, $\mathrm{dist}(p_1, A) = \min_{q \in A} \mathrm{dist}(p_1, q)$. In particular, if L is a hyper-plane in R^d, $\mathrm{dist}(p, L)$ is the regular distance from the point p to L. For $a > 0$ and a set A of points on d-dimensional space, if the distance between every two points is at least a, the set A is called a-separated. For $\epsilon > 0$ and a set of points $Q \subseteq R^d$, an ϵ-sketch of Q is a set of points $P \subseteq R^d$ such that each point in Q has distance $\leq \epsilon$ to some point in P. We say that P is a sketch of Q if P is an ϵ-sketch of Q for some positive constant ϵ (ϵ does not depend on the size of Q). A sketch set is usually an 1-separated set such as a set of grid points. A weight function $w : P \to [0, \infty)$ is often used to measure the point density of Q near each point of P. A hyper-plane in R^d through a fixed point $p_0 \in R^d$ is defined by the equation $(p - p_0) \cdot v = 0$, where v is normal vector of the plane and " $\cdot$ " is the regular inner product ($u \cdot v = \sum_{i=1}^d u_i v_i$ for $u = (u_1, \cdots, u_d)$ and $v = (v_1, \cdots, v_d)$). For $Q \subseteq R^d$ with sketch $P \subseteq R^d$, the width parameter $a > 0$, and the weight function $w : P \to [0, \infty)$, a $(2a)$-wide-separator is determined by a hyper-plane L. The separator has two measurements for its quality of separation: (1) $\mathrm{balance}(L, Q) = \frac{\max(|Q_1|,|Q_2|)}{|Q|}$, where Q_1 is the set of all points of Q on one side of L and Q_2 is the set of all points of Q on the other side of L (Note: Q_1 or Q_2 does not contain any point on L); and (2) $\mathrm{measure}(L, P, a, w) = \sum_{p \in P, \mathrm{dist}(p,L) \leq a} w(p)$.*

2.1 Volume, Area, Integrations and Probability

We need some integrations in order to compute volume and surface area size at arbitrary dimensional space. Some of the materials can be found in standard calculus books. We will treat the case of any fixed dimension. We recommend the reader understands the cases $d = 2$ and 3 first. We use the standard polar transformation: $x_d = r \cos\theta_{d-1}$; $x_{d-1} = r \sin\theta_{d-1} \cos\theta_{d-2}$; $\cdots$; $x_2 = r \sin\theta_{d-1} \sin\theta_{d-2} \cdots \sin\theta_2 \cos\theta_1$; and $x_1 = r \sin\theta_{d-1} \sin\theta_{d-2} \cdots \sin\theta_2 \sin\theta_1$.

It is a smooth map from $[0, R] \times [0, \pi] \times \cdots \times [0, \pi] \times [0, 2\pi]$ to the d-dimensional ball of radius R with center at the origin. The Jacobian form is

$$J_d(r, \theta_{d-1}, \cdots, \theta_1) = \frac{\partial(x_d, x_{d-1}, \cdots, x_1)}{\partial(r, \theta_{d-1}, \cdots, \theta_1)} = \begin{vmatrix} \frac{\partial x_d}{\partial r} & \frac{\partial x_{d-1}}{\partial r} & \cdots & \frac{\partial x_1}{\partial r} \\ \frac{\partial x_d}{\partial \theta_{d-1}} & \frac{\partial x_{d-1}}{\partial \theta_{d-1}} & \cdots & \frac{\partial x_1}{\partial \theta_{d-1}} \\ \cdots & & & \\ \frac{\partial x_d}{\partial \theta_1} & \frac{\partial x_{d-1}}{\partial \theta_1} & \cdots & \frac{\partial x_1}{\partial \theta_1} \end{vmatrix}.$$

It has the recursive equation: $J_d(r, \theta_{d-1}, \cdots, \theta_1) = r \cdot (\sin\theta_{d-1})^{d-2} \cdot J_{d-1}(r,$ $\theta_{d-2}, \cdots, \theta_1)$ for $d > 2$, which in turn gives the explicit expression: $J_d(r, \theta_{d-1}, \cdots,$ $\theta_1) = r^{d-1} \cdot (\sin\theta_{d-1})^{d-2} \cdot (\sin\theta_{d-2})^{d-3} \cdots (\sin\theta_2)$. Let $B_d(R, o)$ be the d- dimensional ball of radius R and center o. The volume of the d-dimensional ball of radius R is $V_d(R) = \int_{B_d(R,o)} 1 d_z = \int_0^R \int_0^\pi \cdots \int_0^\pi \int_0^{2\pi} |J_d(r, \theta_{d-1}, \cdots, \theta_2, \theta_1)| d_r d_{\theta_{d-1}}$ $\cdots d_{\theta_2} d_{\theta_1} = \frac{2^{(d+1)/2}\pi^{(d-1)/2}}{1 \cdot 3 \cdots (d-2) \cdot d} R^d$ if d is odd, and $\frac{2^{d/2}\pi^{d/2}}{2 \cdot 4 \cdots (d-2) \cdot d} R^d$, otherwise.

Let the d-dimensional ball have the center at o. We also need the integration $\int_{B_d(R,o)} \frac{1}{\text{dist}(z,o)} d_z$, which is equal to $\int_0^R \int_0^\pi \cdots \int_0^\pi \int_0^{2\pi} \frac{|J_d(r, \theta_{d-1}, \cdots, \theta_2, \theta_1)|}{r} d_r d_{\theta_{d-1}} \cdots$ $d_{\theta_2} d_{\theta_1} = \frac{d}{(d-1)R} V_d(R)$. Thus,

$$\int_{B_d(R,o)} \frac{1}{\text{dist}(z, o)} d_z = \frac{d}{(d-1)R} V_d(R) \qquad (1)$$

Let $V_d(r) = v_d \cdot r^d$, where v_d is constant for fixed dimensional number d. In particular, $v_1 = 2$, $v_2 = \pi$ and $v_3 = \frac{4\pi}{3}$. Define $A_d(h, R) = \{(x_1, \cdots, x_d) | \sum_{i=1}^d x_i^2 \le R^2 \text{ and } 0 \le x_1 \le h\}$, which is a horizontal cross section of d-dimensional half ball of the radius R. The volume of $A_d(h, R)$ in the d-dimensional space is calculated by

$$U_d(h, R) = \int_0^h V_{d-1}\left(\sqrt{R^2 - x_1^2}\right) d_{x_1} = v_{d-1} \int_0^h \left(\sqrt{R^2 - x_1^2}\right)^{d-1} d_{x_1} \qquad (2)$$

The surface area size of 3D ball $(4\pi R^2)$ is the derivative of its volume $(\frac{4}{3}\pi R^3)$. The boundary length of circle $(2\pi R)$ is the derivative of its area size (πR^2). This fact can be extended to both a higher dimensional ball and a cross section of a ball. The surface area size of $B_d(R, o)$ is $W_d(R) = \frac{\partial V_d(R)}{\partial R} = d \cdot v_d \cdot R^{d-1}$. The side surface of $A_d(h, R)$ is $\{(x_1, \cdots, x_d) | \sum_{i=1}^d x_i^2 = R^2 \text{ and } 0 \le x_1 \le h\}$. Its area size is

$$S_d(h, R) = \frac{\partial U_d(h, R)}{\partial R} = (d-1)v_{d-1}R \int_0^h \left(\sqrt{R^2 - x_1^2}\right)^{d-3} d_{x_1}$$

When R is fixed and h is small, we have $S_d(h, R) = v_{d-1} \cdot (d-1) \cdot R^{d-2} \cdot h + O(h^2)$. For a $a > 0$, the probability that a d-dimensional point p has the distance $\le a$ to a random plane through origin will be determined. This probability at dimension 3 was not well treated in [11].

Lemma 2. *Let a be a real number ≥ 0. Let p and o be the two points on a d-dimensional space. Then the probability that p has distance $\le a$ to a random plane through o is in $[\frac{h_d \cdot a}{\text{dist}(p,o)} - \frac{c_o \cdot a^2}{\text{dist}^2(p,o)}, \frac{h_d \cdot a}{\text{dist}(p,o)} + \frac{c_o \cdot a^2}{\text{dist}^2(p,o)}]$, where $h_d = \frac{2(d-1)v_{d-1}}{d \cdot v_d}$ and c_o are constants for a fixed d. In particular, $h_2 = \frac{2}{\pi}$ and $h_3 = 1$.*

Proof. Without loss of generality, let o be the origin $(0, \cdots, 0)$ (notice that the probability is invariant under translation). The point p can be moved to an axis via rotation that does not change the probability. Let's assume the point

$p = (x_1, 0, \cdots, 0)$, where $x_1 = \text{dist}(p, o)$. For an unit vector $v = (v_1, \cdots, v_d)$ with $v_1 \geq 0$ in the d-dimensional space, the plane through the origin with normal vector v is defined as $u \cdot v = 0$, where $\cdot$ represents the regular inner product between two vectors. The distance between p to the plane is $|p \cdot v| = x_1 v_1$. If $x_1 v_1 \leq a$, then $v_1 \leq \frac{a}{x_1}$. The area size of $\{(v_1, \cdots, v_d) | \sum_{i=1}^{d} v_i^2 = 1 \text{ and } 0 \leq v_1 \leq \frac{a}{x_1}\}$ is $S_d(\frac{a}{x_1}, 1)$. The probability that p has a distance $\leq a$ to a random hyper-plane through the origin is $\frac{S_d(\frac{a}{x_1}, 1)}{\frac{1}{2} \cdot W_d(1)} = h_d \cdot \frac{a}{\text{dist}(p, o)} + O(\frac{a^2}{\text{dist}^2(p, o)})$.

2.2   Width Bounded Separator

Definition 2. *The diameter of a region R is $\sup_{p_1, p_2 \in R} \text{dist}(p_1, p_2)$. A (b, c)-partition of a d-dimensional space makes the space as the disjoint unions of regions $P_1, P_2, \cdots$ such that each P_i, called a regular region, has the volume to be equal to b and the diameter of each P_i is $\leq c$. A (b, c)-regular point set A is a set of points on a d-dimensional space with (b, c)-partition $P_1, P_2, \cdots$ such that each P_i contains at most one point from A. For two regions A and B, if $A \subseteq B$ ($A \cap B \neq \emptyset$), we say B contains (intersects resp.) A.*

Lemma 3. *Assume that $P_1, P_2, \cdots$ form a (b, c)-partition on a d-dimensional space. Then (i) every d-dimensional ball of radius r intersects at most $\frac{v_d \cdot (r+c)^d}{b}$ regular regions; (ii) every d-dimensional ball of radius r contains at least $\frac{v_d \cdot (r-c)^d}{b}$ regular regions; (iii) every d-dimensional ball of radius $(nb/v_d)^{\frac{1}{d}} + c$ contains at least n (b, c)-regular regions in it; and (iv) every d-dimensional ball of radius $(nb/v_d)^{\frac{1}{d}} - c$ intersects at most n (b, c)-regular regions.*

Proof. (i) If a (b, c)-regular region P_i intersects a ball C of radius r at center o, the regular region P_i is contained by the ball C' of radius $r + c$ at the same center o. As the volume of each regular region is b, the number of regular regions contained by C' is no more than the volume size of the ball C' divided by b. (ii) If a regular region P_i intersects a ball C'' of radius $r - c$ at center o, P_i is contained in the ball C of radius r at the same center o. The number of those regular regions intersecting C'' is at least the volume size of the ball C'' divided by b. (iii) Apply $r = (\frac{bn}{v_d})^{\frac{1}{d}} + c$ to (ii). (iv) Apply $r = (\frac{bn}{v_d})^{\frac{1}{d}} - c$ to (i).

Definition 3. *Let a be a non-negative real number. Let p and o be two points in a d-dimensional space. Define $Pr_d(a, p_0, p)$ as the probability that the point p has $\leq a$ perpendicular distance to a random hyper-plane L through the point p_0. Let L be a hyper-plane. Then define the function $f_{a,p,o}(L) = 1$ if p has distance $\leq a$ to the hyper-plane L and L is through o; and 0 otherwise.*

The expectation of function $f_{a,p,o}$ is $E(f_{a,p,o}) = Pr_d(a, o, p)$. Assume that $P = \{p_1, p_2, \cdots, p_n\}$ is a set of n points in R^d and each p_i has weight $w(p_i) \geq 0$. Define function $F_{a,P,o}(L) = \sum_{p \in P} w(p) f_{a,p,o}(L)$. We give an upper bound for the expectation $E(F_{a,P,o})$ for $F_{a,P,o}$ in the lemma below.

Lemma 4. *Let a be a non-negative real number, b and c be positive constants, and $\delta > 0$ be a small constant. Assume that $P_1, P_2, \cdots$ form a (b, c) partition in R^d. Let $w_1 > w_2 > \cdots > w_k > 0$ be positive weights, and $P = \{p_1, \cdots, p_n\}$ be a set of (b, c)-regular points in R^d. Let w be a mapping from P to $\{w_1, \cdots, w_k\}$ and n_j be the number of points $p \in P$ with $w(p) = w_j$. Let o (the center point) be a fixed point in R^d. Then for a random hyper-plane passing through o, $E(F_{a,P,o}) \leq (\frac{d \cdot h_d \cdot v_d}{(d-1) \cdot b} + \delta) \cdot a \cdot \sum_{j=1}^{k} w_j (r_j^{d-1} - r_{j-1}^{d-1}) + c_2 \sum_{j=1}^{k-1} w_{j+1} \cdot r_j^{d-2} + O((a + c_1)^d) \cdot w_1$, where (1) $r_0 = 0$ and $r_i(i > 0)$ is the least radius such that $B_d(r_i, o)$ intersects at least $\sum_{j=1}^{i} n_j$ regular regions, (2) c_1 and c_2 are constants for a fixed d, and (3) h_d and v_d are constants as defined in section 2.1.*

Proof. Assume p is a point of P and L is a random plane passing through the center o. Let C be the ball of radius r and center o such that C covers all the points in P. Let C' be the ball of radius $r' = r + c$ and the same center o. It is easy to see that every regular region with a point in P is inside C'. The probability that the point p has a distance $\leq a$ to L is $\leq h_d \cdot \frac{a}{\text{dist}(o,p)} + \frac{c_0 \cdot a^2}{\text{dist}(o,p)^2}$ (by Lemma 2).

Let $\epsilon > 0$ be a small constant which will be determined later. Select a large constant $r_0 > 0$ and $R_0 = \frac{c_0 \cdot a}{h_d \cdot \epsilon} + r_0$ such that for every point p with $\text{dist}(o,p) \geq R_0$, $\frac{c_0 \cdot a}{h_d \cdot \text{dist}(o,p)^2} < \frac{\epsilon}{\text{dist}(o,p)}$ and for every point p' with $\text{dist}(p',p) \leq c$, $\frac{1}{\text{dist}(o,p')} \leq \frac{1+\epsilon}{\text{dist}(o,p)}$. Let P_1 be the set of all points p in P such that $\text{dist}(o,p) < R_0$. For each point $p \in P_1$, $Pr_d(a,o,p) \leq 1$. For every point $p \in P - P_1$, $Pr_d(a,o,p) \leq h_d \cdot \frac{a}{\text{dist}(o,p)} + \frac{c_0 \cdot a^2}{\text{dist}(o,p)^2} < \frac{h_d \cdot a(1+\epsilon)}{\text{dist}(o,p)}$. From the transformation $E(F_{a,P,o}) = E(\sum_{i=1}^{n} w(p_i) \cdot f_{a,p_i,o}) = \sum_{i=1}^{n} w(p_i) \cdot E(f_{a,p_i,o}) = \sum_{j=1}^{k} w_j \sum_{w(p_i)=w_j} E(f_{a,p_i,o}) = \sum_{j=1}^{k} w_j \sum_{w(p_i)=w_j} Pr_d(a,o,p_i)$, we have

$$E(F_{a,P,o}) \leq w_1 |P_1| + \sum_{j=1}^{k} w_j \sum_{w(p_i)=w_j} \frac{h_d \cdot a \cdot (1+\epsilon)}{b} \cdot \frac{1}{\text{dist}(o,p_i)} \cdot b \qquad (3)$$

The contribution to $E(F_{a,P,o})$ from the points in P_1 is $\leq w_1 |P_1| \leq w_1 \cdot \frac{v_d (R_0 + c)^d}{b} = w_1 \cdot O((a + c_1)^d)$ for some constant $c_1 > 0$ (by Lemma 3). Next we only consider those points from $P - P_1$. The sum (3) is at a maximum when $\text{dist}(p,o) \leq \text{dist}(p',o)$ implies $w(p) \geq w(p')$. The ball C' is partitioned into k ring regions such that the j-th area is between $B_d(r_j, o)$ and $B_d(r_{j-1}, o)$ and it is mainly used to hold those points with weight w_j. Notice that each regular region has diameter $\leq c$ and holds at most one point in P. It is easy to see that all points of $\{p_i \in P | w(p_i) = w_j\}$ are located between $B_d(r_j, o)$ and $B_d(r_{j-1} - c, o)$ when (3) is maximal.

$$\sum_{w(p_i)=w_j} \frac{h_d \cdot a \cdot (1+\epsilon)}{b} \cdot \frac{1}{\text{dist}(o,p_i)} \cdot b \qquad (4)$$

$$\leq \frac{h_d \cdot a \cdot (1+\epsilon)^2}{b} \int_{B_d(r_j,o) - B_d(r_{j-1}-c,o)} \frac{1}{\text{dist}(o,z)} dz \qquad (5)$$

$$= \frac{h_d a (1+\epsilon)^2}{b} \int_{r_{j-1}-c}^{r_j} \int_0^\pi \cdots \int_0^\pi \int_0^{2\pi} \frac{J_d(r, \theta_{n-1}, \cdots, \theta_1)}{r} dr\, d\theta_{n-1} \cdots d\theta_1 \quad (6)$$

$$= \frac{h_d \cdot a \cdot (1+\epsilon)^2}{b} \cdot \frac{d}{(d-1)} \cdot \left(\frac{V_d(r_j)}{r_j} - \frac{V_d(r_{j-1}-c)}{r_{j-1}-c} \right) \quad (7)$$

$$< \left(\frac{d \cdot h_d \cdot v_d}{(d-1) \cdot b} + \delta \right) a \; (r_j^{d-1} \quad r_{j-1}^{d-1}) \; | \; O(r_{j-1}^{d-2}). \quad (8)$$

Note: $(6) \rightarrow (7) \rightarrow (8)$ follows from (1), and selecting ϵ small enough.

Lemma 5. *Assume* $a = O(n^{\frac{d-2}{d^2}})$. *Let* o *be a point on the plane,* b *and* c *be positive constants, and* $\epsilon, \delta > 0$ *be small constants. Assume that* $P_1, P_2, \cdots$ *form a* (b, c) *partition in* R^d. *The weights* $w_1 > \cdots > w_k > 0$ *satisfy* $k \cdot \max_{i=1}^k \{w_i\} = O(n^\epsilon)$. *Let* P *be a set of* n *weighted* (b, c)-*regular points in a* d-*dimensional space with* $w(p) \in \{w_1, \cdots, w_k\}$ *for each* $p \in P$. *Let* n_j *be the number of points* $p \in P$ *with* $w(p) = w_j$ *for* $j = 1, \cdots, k$. *Then* $E(F_{a,P,o}) \leq (k_d \cdot (\frac{1}{b})^{\frac{1}{d}} + \delta) \cdot a \cdot \sum_{j=1}^k w_j \cdot n_j^{\frac{d-1}{d}} + O(n^{\frac{d-2}{d}+\epsilon})$, *where* $k_d = \frac{d \cdot h_d}{d-1} \cdot v_d^{\frac{1}{d}}$. *In particular,* $k_2 = \frac{4}{\sqrt{\pi}}$ *and* $k_3 = \frac{3}{2} \left(\frac{4\pi}{3} \right)^{\frac{1}{3}}$.

Proof. Let r_j be the least radius such that the ball of radius r_j intersects at least $\sum_{i=1}^j n_i$ regular regions $(j = 1, \cdots, k)$. By Lemma 3, $\left(\frac{(\sum_{i=1}^j n_i) b}{v_d} \right)^{\frac{1}{d}} - c \leq r_j \leq \left(\frac{(\sum_{i=1}^j n_i) b}{v_d} \right)^{\frac{1}{d}} + c$ for $j = 1, \cdots, k$.

$$r_j^{d-1} - r_{j-1}^{d-1} \leq \left(\left(\frac{(\sum_{i=1}^j n_i) b}{v_d} \right)^{\frac{1}{d}} + c \right)^{d-1} - \left(\left(\frac{(\sum_{i=1}^{j-1} n_i) b}{v_d} \right)^{\frac{1}{d}} - c \right)^{d-1} \quad (9)$$

$$= \left(\frac{b}{v_d} \right)^{\frac{d-1}{d}} \left((\sum_{i=1}^j n_i)^{\frac{d-1}{d}} - (\sum_{i=1}^{j-1} n_i)^{\frac{d-1}{d}} \right) + O((\sum_{i=1}^j n_i)^{\frac{d-2}{d}}) \quad (10)$$

$$= \left(\frac{b}{v_d} \right)^{\frac{d-1}{d}} n_j^{\frac{d-1}{d}} + O((\sum_{i=1}^j n_i)^{\frac{d-2}{d}}) \quad (11)$$

By Lemma 4, Lemma 5 is proven.

Definition 4. *Let* $a_1, \cdots, a_d > 0$ *be positive constants. A* $(a_1, \cdots, a_d)$-*grid regular partition divides the* d-*dimensional space into disjoint union of* $a_1 \times \cdots \times a_d$ *rectangular regions. A* $(a_1, \cdots, a_d)$-*grid regular point is a corner point of a rectangular region. Under certain translation and rotation, each* $(a_1, \cdots, a_d)$-*grid regular point has coordinates* $(a_1 t_1, \cdots, a_d t_d)$ *for some integers* $t_1, \cdots, t_d$.

Theorem 1. *Let* $a = O(n^{\frac{d-2}{d^2}})$. *Let* $a_1, \cdots, a_d$ *be positive constants and* $\epsilon, \delta > 0$ *be small constants. Let* P *be a set of* n $(a_1, \cdots, a_d)$-*grid points in* R^d, *and* Q *be another set of* m *points in* R^d *with sketch* P. *Let* $w_1 > w_2 \cdots > w_k > 0$ *be*

positive weights with $k \cdot \max_{i=1}^{k} \{w_i\} = O(n^\epsilon)$, and w be a mapping from P to $\{w_1, \cdots, w_k\}$. Then there is a hyper-plane L such that (1) each half space has $\leq \frac{d}{d+1} m$ points from Q, and (2) for the subset $A \subseteq P$ containing all points in P with $\leq a$ distance to L has the property $\sum_{p \in A} w(p) \leq \left(k_d \cdot \left(\prod_{i=1}^{d} a_i \right)^{\frac{-1}{d}} + \delta \right) \cdot$
$a \cdot \sum_{j=1}^{k} w_j \cdot n_j^{\frac{d-1}{d}} + O(n^{\frac{d-2}{d} + \epsilon})$ for all large n, where $n_j = |\{p \in P | w(p) = w_j\}|$.

Proof. Let $b = \prod_{i=1}^{d} a_i$, $c = \sqrt{\sum_{i=1}^{d} a_i^2}$, and the point o be the center point of Q via Lemma 1. Apply Lemma 5.

Corollary 1. *[11] Let Q be a set of n $(1,1)$-grid points on the plane. Then there is a line L such that each half plane has $\leq \frac{2n}{3}$ points in Q and the number of points in Q with $\leq \frac{1}{2}$ distance to L is $\leq 1.129\sqrt{n}$.*

Proof. Let each weight be 1, $k = 1$, $a = \frac{1}{2}$ and $P = Q$. Apply Theorem 1.

Corollary 2. *Let Q be a set of n $(1,1,1)$-grid points on the 3D Euclidean space. Then there is a plane L such that each half space has $\leq \frac{3n}{4}$ points in Q and the number of points in Q with a $\leq \frac{1}{2}$ distance to L is $\leq 1.209 n^{\frac{2}{3}}$.*

Corollaries 1 and 2 are the separators for the 2D grid graphs and, respectively, 3D grid graphs. An edge connecting two neighbor grid points has a distance of 1. If two neighbor grid points are at different sides of the separator, one of them has distance $\leq \frac{1}{2}$ to the separator. We have a separator for the 1-separated set.

Theorem 2. *Assume that the packing density (see [26]) for the d-dimensional congruent balls is at most D_d. Then for every 1-separated set Q on the d-dimensional Euclidean space, there is a hyper-plane L with balance$(L, Q) \leq \frac{d}{d+1}$ and the number of points with distance $\leq a$ to L is $\leq (2k_d \cdot (D_d/v_d)^{\frac{1}{d}} + o(1))a \cdot |Q|^{\frac{d-1}{d}}$.*

We develop a brute force method to find the width-bounded separator. In order to determine the position of the hyper-plane in d-dimensional space. For every integer pair $s_1, s_2 \geq 0$ with $s_1 + s_2 = d$, select all possible s_1 points $p_1, \cdots, p_{s_1}$ from P and all possible s_2 points $q_1, \cdots, q_{s_2}$ from Q. Try all the hyper-planes L that are through $q_1, \cdots, q_{s_2}$ and tangent to $B_d(a + \delta, p_i)$ $(i = 1, \cdots, s_1)$. Then select the one that satisfies the balance condition and has small sum of weights for the points of P close to L. A more involved sub-linear time algorithm for finding width-bounded separator was recently developed by Fu and Chen [12].

Theorem 3. *Assume $a = O(n^{\frac{d-2}{d^2}})$. Let $a_1, \cdots, a_d$ be positive constants and $\delta, \epsilon > 0$ be small constants. Let P be a set of n $(a_1, \cdots, a_d)$-grid points and Q be another set of points on d-dimensional space. The weights $w_1 > \cdots > w_k > 0$ have $k \cdot \max_{i=1}^{k} \{w_i\} = o(n^\epsilon)$. There is an $O(n^{d+1})$ time algorithm that finds a separator such that balance$(L, Q) \leq \frac{d-1}{d}$, and measure$(L, P, a, w) \leq \left(\frac{k_d}{(a_1 \cdots a_d)^{1/d}} + \delta \right) a \sum_{i=1}^{k} w_i n_i^{\frac{d-1}{d}} + O(n^{\frac{d-2}{d} + \epsilon})$ for all large n, where $n_i = |\{p \in P | w(p) = w_i\}|$.*

3   Application of Width Bounded Separator

In this section, we apply our geometric separator to the well-known disk covering problem: Given a set of points on the plane, find the minimal number of discs with fixed radius to cover all of those points. The d-dimensional ball covering problem is to cover n points on the d-dimensional Euclid space with minimal number of d-dimensional ball of fixed radius.

Before proving Theorem 4, we briefly explain our method. To cover a set of points Q on the plane, select a set P of grid points such that each point in Q is close to at least one point in P. A grid point p is assigned the weight i if there are 2^i to 2^{i+1} points of Q on the 1×1 grid square with p as the center. A balanced separator line for Q also has a small sum of weights ($O(\sqrt{n})$) for the points of P near the line. This gives at most $2^{O(\sqrt{n})}$ ways to cover all points of Q close to the separator line and decompose the problem into two problems Q_1 and Q_2 that can be covered independently. This method takes the total time of $2^{O(\sqrt{n})}$.

Theorem 4. *There is a $2^{O(\sqrt{n})}$-time exact algorithm for the disk covering problem on the $2D$ plane.*

Proof. Assume the diameter of any disk is 1. Assume that Q is the set of n input points on the plane. Let's set up an $(1,1)$-grid regular partition. For a grid point $p = (i,j)$ (i and j are integers) on the plane, define $\text{grid}(p) = \{(x,y) | i - \frac{1}{2} \le x < i + \frac{1}{2}, j - \frac{1}{2} < y \le j + \frac{1}{2}\}$, which is a half close and half open 1×1 square. There is no intersection between $\text{grid}(p)$ and $\text{grid}(q)$ for two different grid points p and q. Our "local binding" method is to merge the points of $Q \cap \text{grid}(p)$ to the grid point p and assign certain weight to p to measure the Q points density in $\text{grid}(p)$. The function Partition(Q) divides the set Q into $Q = Q(p_1) \cup Q(p_2) \cup \cdots \cup Q(p_m)$, where p_i is a grid point for $i = 1, 2, \cdots, m$ and $Q(p_i) = Q \cap \text{grid}(p_i) \neq \emptyset$.

Let n_i be the number of grid points $p_j \in P$ with $g^{i-1} \le |Q(p_j)| < g^i$, where g is a constant > 1 (for example, $g = 2$). From this definition, we have

$$\sum_{i=1}^{\lceil \log_g n \rceil} g^i \cdot n_i \le g \cdot n, \tag{12}$$

where $\lceil x \rceil$ is the least integer $\ge x$. Let $P = \{p_1, \cdots, p_m\}$ be the set grid points derived from partitioning set Q in Partition(Q). Define function $w : P \to \{1, 2, \cdots, \lceil \log_g n \rceil\}$ such that $w(p) = i$ if $g^{i-1} \le |Q(p)| < g^i$.

Select small $\delta > 0$ and $a = \frac{3}{2} + \frac{\sqrt{2}}{2}$. By Theorem 3, we can get a line L on the plane such that $balance(L, Q) \le \frac{2}{3}$ and $measure(L, P, a, w) \le (k_2 + \delta) \cdot a \cdot (\sum_{i=1}^{\lceil \log_g n \rceil} i \cdot \sqrt{n_i})$. Let $J(L) = \{p | p \in P \text{ and } dist(q, L) \le \frac{1}{2} \text{ for some } q \in Q(p)\}$. After those points of Q with distance $\le \frac{1}{2}$ to the separator line L are covered, the rest of points of Q on the different sides of L can be covered independently. Therefore, the covering problem is solved by divide and conquer method as described by the algorithm below.

Algorithm
　　Input a set of points Q on the plane.
　　run Partition(Q) to get $P = \{p_1, \cdots, p_m\}$ and $Q(p_1), \cdots, Q(p_m)$
　　find a separator line L (by Theorem 3) for P, Q with
　　　　balance$(L, Q) \leq \frac{2}{3}$ and measure$(L, P, a, w) \leq (k_2 + \delta)a \sum_{i=1}^{\lceil \log n \rceil} i\sqrt{n_i}$
　　for each covering to the points in Q with $\leq 1/2$ distance to L
　　　　let $Q_1 \subseteq Q$ be the those points on the left of L and not covered
　　　　let $Q_2 \subseteq Q$ be the those points on the right of L and not covered
　　　　recursively cover Q_1 and Q_2
　　　　merge the solutions from Q_1 and Q_2
　　Output the solution with the minimal number of discs covering Q
End of Algorithm

For each grid area grid(p_i), the number of discs containing the points in $Q(p_i)$ is no more than the number of discs covering the 3×3 area, which needs no more than $c_3 = (\lceil \frac{3}{\frac{\sqrt{2}}{2}} \rceil)^2 = 25$ discs. Two grid points $p = (i, j)$ and $p' = (i', j')$ are *neighbors* if $\max(|i - i'|, |j - j'|) \leq 1$. For each grid point p, define $m(p)$ to be the neighbor grid point q of p (q may be equal to p) with largest weight $w(q)$. For a grid point $p = (i, j)$, the 3×3 region $\{(x, y) | i - \frac{3}{2} \leq x < i + \frac{3}{2} \text{ and } j - \frac{3}{2} < y \leq j + \frac{3}{2}\}$ has $\leq 9 \times g^{w(m(p))}$ points in Q. The number of ways to put one disc covering at least one point in $Q(p)$ is $\leq (9 \times g^{w(m(p))})^2$ (let each disc have two points from Q on its boundary whenever it covers at least two points). The number of ways to arrange $\leq c_3$ discs to cover points in $Q(p)$ is $\leq (9 \times g^{w(m(p))})^{2c_3}$. The total number of cases to cover all points with distance $\leq \frac{1}{2}$ to L in $\cup_{p \in J(L)} Q(p)$ is

$$\leq \prod_{p \in J(L)} (9 \cdot g^{w(m(p))})^{2c_3} = \prod_{p \in J(L)} 2^{(\log_2 9 + w(m(p)) \cdot \log_2 g) 2c_3} \tag{13}$$

$$\leq \prod_{p \in J(L)} 2^{2c_3(\log_2 9 + \log_2 g) w(m(p))} \tag{14}$$

$$= 2^{2c_3(\log_2 9 + \log_2 g) \sum_{p \in J(L)} w(m(p))} \leq 2^{2c_3(\log_2 9 + \log_2 g) 9 \cdot \text{measure}(L, P, a, w)} \tag{15}$$

$$\leq 2^{2c_3(\log_2 9 + \log_2 g) 9(k_2 \cdot a + \delta)(\sum_{i=1}^{\lceil \log n \rceil} i \cdot \sqrt{n_i})} \tag{16}$$

This is because that for each grid point q, there are at most 9 grid points p with $m(p) = q$. Furthermore, for each $p \in J(L)$, p has a distance $\leq \frac{1}{2} + \frac{\sqrt{2}}{2}$ to L and $m(p)$ has a distance $\leq \frac{3}{2} + \frac{\sqrt{2}}{2} = a$ to L. Let the exponent of (16) be represented by $u = 2c_3(\log_2 9 + \log_2 g) 9(k_2 + \delta)a(\sum_{i=1}^{\lceil \log n \rceil} i \cdot \sqrt{n_i})$. By Cauchy-Schwarz inequality $(\sum_{i=1}^{m} a_i \cdot b_i)^2 \leq (\sum_{i=1}^{m} a_i^2) \cdot (\sum_{i=1}^{m} b_i^2)$,

$$(\sum_{i=1}^{\lceil \log_g n \rceil} i\sqrt{n_i})^2 = (\sum_{i=1}^{\lceil \log_g n \rceil} \frac{i}{g^{i/2}} \cdot g^{i/2} \sqrt{n_i})^2 \leq (\sum_{i=1}^{\lceil \log_g n \rceil} \frac{i^2}{g^i}) \cdot (\sum_{i=1}^{\lceil \log_g n \rceil} g^i n_i) \tag{17}$$

Using standard calculus, $\sum_{i=1}^{\infty} \frac{i^2}{g^i} = \frac{g(g+1)}{(g-1)^3}$. By (17) and (12), $u \leq e(g)\sqrt{n}$, where $e(g) = 2c_3(\log_2 9 + \log_2 g)(k_2 + \delta)a\sqrt{\frac{g(g+1)}{(g-1)^3}} \cdot \sqrt{g}$. Let $T(n)$ be the maximal computational time of the algorithm for covering n points. The problem $T(n)$ is reduced to two problems $T(\frac{2}{3}n)$. We have $T(n) \leq 2 \cdot 2^{e(g)\sqrt{n}}T(\frac{2n}{3}) \leq 2^{\log_{3/2} n}2^{e(g)(1+\alpha+\alpha^2+\cdots)\sqrt{n}} = 2^{e(g)(\frac{1}{1-\alpha})\sqrt{n}+\log_{3/2} n} = 2^{O(\sqrt{n})}$, where $\alpha = \sqrt{\frac{2}{3}}$.

Definition 5. *We consider undirected graphs $G = (V, E)$, where V denotes the vertex set and E denotes the edge set. An independent set I of a graph $G = (V, E)$ is a set of pairwise nonadjacent vertices of a graph. A vertex cover C of a graph $G = (V, E)$ is a subset of vertices such that each edge in E has at least one end point in C. A dominating set D is a set of vertices such that the rest of the vertices in G has at least one neighbor in D. For a point p on the plane and $r > 0$, $C_r(p)$ is the disk with center at p and radius r. For a set of disks $D = \{C_{r_1}(p_1), C_{r_2}(p_2), \cdots, C_{r_n}(p_n)\}$, the disk graph is $G_D = (V_D, E_D)$, where vertices set $V_D = \{p_1, p_2, \cdots, p_n\}$ and $E_D = \{(p_i, p_j)|C_{r_i}(p_i) \cap C_{r_j}(p_j) \neq \emptyset\}$. DG is the class of all disk graphs. DG_σ is the class of all disk graphs G_D such that D is the set of disks $\{C_{r_1}(p_1), C_{r_2}(p_2), \cdots, C_{r_n}(p_n)\}$ with $\frac{\max_{i=1}^n r_i}{\min_{i=1}^n r_i} \leq \sigma$.*

Several standard graph theoretic problems for GD_1 are NP-hard [7,10,20,27]. The $n^{O(\sqrt{n})}$-time exact algorithm for the maximum independent set problem for DG_σ with constant σ was derived by Alber and Fiala [5] via parameterized approach, which was further simplified by Agarwal, Overmars and Sharir [1] for DG_1. We obtain $2^{O(\sqrt{n})}$-time algorithms for maximum independent set, minimum vertex cover, and minimum dominating set problems for DG_σ with constant σ. Their algorithms are similar each other. The d-dimensional versions of those problems, including the ball covering problem, have algorithms with computational time $2^{O(n^{1-1/d})}$.

Acknowledgments. The author is very grateful to Sorinel A Oprisan for his many helpful comments on an earlier version of this paper, and also the reviewers of STACS'06 whose comments improve the presentation of this paper.

References

1. P. Agarwal, M. Overmars, and M. Sharir, Computing maximally separated sets in the plane and independent sets in the intersection graph of unit graphs, In proceedings of 15th ACM-SIAM symposium on discrete mathematics algorithms, ACM-SIAM 2004, pp. 509-518.
2. P. Agarwal and C. M. Procopiuc, Exact and approximation algorithms for clustering, Algorithmica, 33, 2002, 201-226.
3. J. Alber, H. Fernau, and R. Niedermeier, Graph separators: a parameterized view, In proceedings of 7th Internal computing and combinatorics conference 2001, pp. 318-327.
4. J. Alber, H. Fernau, and R. Niedermeier, Parameterized complexity: exponential speed-up for planar graph problems, In Proceedings of 28st international colloquium on automata, languages and programming, 2001, pp. 261-272.
5. J. Alber and J. Fiala, Geometric separation and exact solution for parameterized independent set problem on disk graphs, Journal of Algorithms, 52, 2, 2004, pp. 134-151.

6. N. Alon, P. Seymour, and R. Thomas, Planar Separator, SIAM J. Discr. Math. 7,2(1990) 184-193.
7. B. N. Clark, C. J. Colbourn, and D. S. Johnson, Unit disk graphs, Discrete mathematics, 86(1990), pp. 165-177.
8. H. N. Djidjev, On the problem of partitioning planar graphs. SIAM journal on discrete mathematics, 3(2) June, 1982, pp. 229-240.
9. H. N. Djidjev and S. M. Venkatesan, Reduced constants for simple cycle graph separation, Acta informatica, 34(1997), pp. 231-234.
10. R. J. Fowler, M. S. Paterson, and S. L. Tanimoto, Optimal packing and covering in the plane are NP-complete, Information processing letters, 3(12), 1981, pp. 133-137.
11. B. Fu and W. Wang, A $2^{O(n^{1-1/d}\log n)}$-time algorithm for d-dimensional protein folding in the HP-model, Proceedings of 31st international colloquium on automata, languages and programming, 2004, pp. 630-644.
12. B. Fu and Z. Chen, Sublinear-time algorithms for width-bounded geometric separator and their application to protein side-chain packing problem, submitted.
13. H. Gazit, An improved algorithm for separating a planar graph, manuscript, USC, 1986.
14. J. R. Gilbert, J. P. Hutchinson, and R. E. Tarjan, A separation theorem for graphs of bounded genus, Journal of algorithm, (5)1984, pp. 391-407.
15. R. Graham, M. Grötschel, and L. Lovász, Handbook of combinatorics (volume I), MIT Press, 1996
16. T. C. Hales, A computer verification of the Kepler conjecture, Proceedings of the ICM, Beijing 2002, vol. 3, 795–804
17. D. Lichtenstein, Planar formula and their uses, SIAM journal on computing, 11,2(1982), pp. 329-343.
18. R. J. Lipton and R. Tarjan, A separator theorem for planar graph, SIAM Journal on Applied Mathematics, 36(1979) 177-189.
19. R. J. Lipton and R. Tarjan, Applications of a planar separator theorem, SIAM journal on computing, 9,3(1980), pp. 615-627.
20. N. Meggido and K. Supowit, On the complexity of some common geometric location problems, SIAM journal on computing, v. 13, 1984, pp. 1-29.
21. G. L. Miller, S.-H. Teng, and S. A. Vavasis, An unified geometric approach to graph separators. In 32nd annual symposium on foundation of computer science, IEEE 1991, pp. 538-547.
22. G. L. Miller and W. Thurston, Separators in two and three dimensions, In 22nd Annual ACM symposium on theory of computing, ACM 1990, pp. 300-309.
23. S. S. Ravi, H. B. Hunt III, Application of the planar separator theorem to computing problems, Information processing letter, 25,5(1987), pp. 317-322.
24. W. D. Smith and N. C. Wormald, Application of geometric separator theorems, The 39th annual symposium on foundations of computer science,1998, 232-243.
25. D. A. Spielman and S. H. Teng, Disk packings and planar separators, The 12th annual ACM symposium on computational geometry, 1996, pp.349-358.
26. J. Pach and P. K. Agarwal, Combinatorial geometry, Wiley-Interscience Publication, 1995.
27. D. W. Wong and Y. S. Kuo, A study of two geometric location problems, Information processing letters, v. 28, 1988, pp. 281-286.

Invariants of Automatic Presentations and Semi-synchronous Transductions

Vince Bárány[*]

Mathematische Grundlagen der Informatik, RWTH Aachen
vbarany@informatik.rwth-aachen.de

Abstract. Automatic structures are countable structures finitely presentable by a collection of automata. We study questions related to properties invariant with respect to the choice of an automatic presentation. We give a negative answer to a question of Rubin concerning definability of intrinsically regular relations by showing that order-invariant first-order logic can be stronger than first-order logic with counting on automatic structures. We introduce a notion of equivalence of automatic presentations, define semi-synchronous transductions, and show how these concepts correspond. Our main result is that a one-to-one function on words preserves regularity as well as non-regularity of all relations iff it is a semi-synchronous transduction. We also characterize automatic presentations of the complete structures of Blumensath and Grädel.

1 Introduction

Automatic structures are countable structures presentable by a tuple of finite automata. Informally, a structure is automatic if it is isomorphic to one consisting of a regular set of words as universe and having only regular relations, i.e., which are recognizable by a finite synchronous multi-tape automaton. Every such isomorphic copy, and any collection of automata representing it, as well as the isomorphism itself may, and will, ambiguously, be called an *automatic presentation* (a.p.) of the structure. It follows from basic results of automata theory [9] that the first-order theory of every automatic structure is decidable. Using automata on ω-words or finite- or infinite trees in the presentations, one obtains yet other classes of structures with a decidable first-order theory [4,6]. This paper is solely concerned with presentations via automata on finite words.

The notion of (ω-)automatic structures first appeared in [12]. Khoussainov and Nerode [13] have reintroduced and elaborated the concept, and [5] has given momentum to its systematic investigation. Prior to that, automatic groups [8], automatic sequences [1], and expansions of Presburger arithmetic [3,7] by relations regular in various numeration systems have been studied thoroughly. These investigations concern only certain naturally restricted automatic presentations of the structures involved.

[*] Part of this work was conducted during the author's visit to LIAFA, Université Paris 7 - Denis-Diderot, supported by the GAMES Network.

The first logical characterization of regularity is due to Büchi. Along the same lines Blumensath and Grädel characterized automatic structures in terms of interpretations [4,6]. They have shown that for each finite non-unary alphabet Σ, $\mathfrak{S}_\Sigma = (\Sigma^*, (S_a)_{a \in \Sigma}, \preceq, \mathsf{el})$ is a complete automatic structure, i.e. such that all automatic structures, and only those, are first-order interpretable in it. (See Example 1 and Prop. 2 below.) In this setting each interpretation, as a tuple of formulas, corresponds to an a.p. given by the tuple of corresponding automata.

An aspect of automatic structures, which has remained largely unexplored concerns the richness of the various automatic presentations a structure may possess. The few exceptions being numeration systems for $(\mathbb{N}, +)$ [3,7], automatic presentations of $(\mathbb{Q}, <)$ [17], as well as of $(\mathbb{N}, <)$ and $(\mathbb{N}, S)$ [16]. The paper at hand presents contributions to this area.

Recently, Khoussainov et al. have introduced the notion of *intrinsic regularity* [16]. A relation is intrinsically regular wrt. a given structure, if it is mapped to a regular relation in *every* automatic presentation of that structure. Khoussainov et al. have shown that with respect to each automatic structure every relation definable in it using first-order logic with counting quantifiers is intrinsically regular. In [20, Problem 3] Rubin asked whether the converse of this is also true. In Section 3 we show that this is not the case. First, we observe that relations *order-invariantly* definable in first-order logic (with counting) are intrinsically regular with respect to each structure. Next, by adapting a technique of Otto [19] we exhibit an automatic structure together with an order-invariantly definable relation, which is not definable in any extension of first-order logic with *unary generalized quantifiers*. In [16, Question 1.4] Khoussainov et al. have called for a logical characterization of intrinsic regularity. Our example shows that it is not sufficient to add only unary generalized quantifiers to the language.

We propose to call two automatic presentations of the same structure equivalent, whenever they map exactly the same relations to regular ones. In Section 4 we investigate automatic presentations of $\mathfrak{S}_\Sigma$, where Σ is non-unary. Due to completeness, every a.p. of $\mathfrak{S}_\Sigma$ maps regular relations to regular ones. Our first result, Theorem 1, establishes, that, conversely, every a.p. of $\mathfrak{S}_\Sigma$ maps non-regular relations to non-regular ones. As a consequence we observe that $\mathfrak{S}_\Sigma$ has, up to equivalence, but one automatic presentation and that non-definable relations over $\mathfrak{S}_\Sigma$ are therefore intrinsically non-regular with respect to it.

Turning our attention to regularity-preserving mappings we introduce *semi-synchronous rational transducers*. These are essentially synchronized transducers in the sense of [10] with the relaxation that they may read each tape at a different, but constant pace. Our main result, Theorem 2 of Section 5, is that a bijection between regular languages preserves regularity of all *relations* in both directions if and only if it is a semi-synchronous transduction. It follows that two automatic presentations of an automatic structure are equivalent precisely when the bijection translating names of elements from one automatic presentation to the other is a semi-synchronous transduction.

I thank Luc Segoufin for our numerous fruitful discussions on the topic, and the anonymous referees for valuable remarks as to the presentation of the paper.

2   Preliminaries

Multi-tape automata. Let Σ be a finite alphabet. The length of a word $w \in \Sigma^*$ is denoted by $|w|$, the empty word by ε, and for each $0 < i \leq |w|$ the ith symbol of w is written as $w[i]$. We consider relations on words, i.e. subsets R of $(\Sigma^*)^n$ for some $n > 0$. Asynchronous n-tape automata accept precisely the *rational relations*, i.e., rational subsets of the product monoid $(\Sigma^*)^n$. Finite *transducers*, recognizing *rational transductions* [2], are asynchronous 2-tape automata. A relation $R \subseteq (\Sigma^*)^n$ is *synchronized rational* [10] or *regular* [15] if it is accepted by a *synchronous n-tape automaton*. We introduce the following generalization.

Definition 1 (Semi-synchronous rational relations). *Let $\square$ be a special end-marker symbol, $\square \notin \Sigma$, and $\Sigma_\square = \Sigma \cup \{\square\}$. Let $\alpha = (a_1, \ldots, a_n)$ be a vector of positive integers and consider a relation $R \subseteq (\Sigma^*)^n$. Its α-convolution is $\boxtimes_\alpha R = \{(w_1 \square^{m_1}, \ldots, w_n \square^{m_n}) \mid (w_1, \ldots, w_n) \in R$ and the m_i are minimal, such that there is a k, with $ka_i = |w_i| + m_i$ for every $i\}$. This allows us to identify $\boxtimes_\alpha R$ with a subset of the monoid $((\Sigma_\square)^{a_1} \times \cdots \times (\Sigma_\square)^{a_n})^*$. If $\boxtimes_\alpha R$ thus corresponds to a regular set, then we say that R is α-synchronous (rational). R is semi-synchronous if it is α-synchronous for some α.*

Intuitively, our definition expresses that although R requires an asynchronous automaton to accept it, synchronicity can be regained when processing words in blocks, the size of which are component-wise fixed by α. As a special case, for $\alpha = \mathbf{1}$, we obtain the regular relations. Recall that a relation $R \subseteq (\Sigma^*)^n$ is *recognizable* if it is saturated by a congruence (of the product monoid $(\Sigma^*)^n$) of finite index, equivalently, if it is a finite union of direct products of regular languages [10]. We denote by Rat, SRat, S_αRat, Reg, Rec the classes of rational, semi-synchronous, α-synchronous, regular, and recognizable relations respectively.

Automatic structures. We take all structures to be relational with functions represented by their graphs.

Definition 2 (Automatic structures). *A structure $\mathfrak{A} = (A, \{R_i\}_i)$ consisting of relations R_i over the universe $\mathrm{dom}(\mathfrak{A}) = A$ is automatic if there is a finite alphabet Σ and an injective naming function $f : A \to \Sigma^*$ such that $f(A)$ is a regular subset of Σ^*, and the images of all relations of $\mathfrak{A}$ under f are in turn regular in the above sense. In this case we say that f is an (injective) automatic presentation of $\mathfrak{A}$. The class of all injective automatic presentations of $\mathfrak{A}$ is denoted $\mathrm{AP}(\mathfrak{A})$. AUTSTR designates the class of automatic structures.*

Example 1. Let Σ be a finite alphabet. Let S_a, $\preceq$ and el denote the a-successor relation, the prefix ordering, and the relation consisting of pairs of words of equal length. These relations are clearly regular, thus $\mathfrak{S}_\Sigma = (\Sigma^*, (S_a)_{a \in \Sigma}, \preceq, \mathrm{el})$ is automatic, having $\mathrm{id} \in \mathrm{AP}(\mathfrak{S}_\Sigma)$. Note that $\mathfrak{S}_{\{1\}}$ is essentially $(\mathbb{N}, \leq)$.

We use the abbreviation FO for first-order logic, and $\mathrm{FO}^{\infty, \mathrm{mod}}$ for its extension by infinity ($\exists^\infty$) and modulo counting quantifiers ($\exists^{(r,m)}$). The meaning of the formulae $\exists^\infty x\,\theta$ and $\exists^{(r,m)}x\,\theta$ is that there are infinitely many elements x, respectively r many elements x modulo m, such that θ holds. We shall make extensive use of the following facts, often without any reference.

Proposition 1. *(Consult [4,6] and [16,20].)*
  i) *Let* $\mathfrak{A} \in \text{AutStr}$, $f \in \text{AP}(\mathfrak{A})$. *Then one can effectively construct for each* $\text{FO}^{\infty,\text{mod}}$-*formula* $\varphi(\boldsymbol{a}, \boldsymbol{x})$ *with parameters* $\boldsymbol{a}$ *from* $\mathfrak{A}$, *defining a k-ary relation R over* $\mathfrak{A}$, *a k-tape synchronous automaton recognizing $f(R)$.*
  ii) *The* $\text{FO}^{\infty,\text{mod}}$-*theory of any automatic structure is decidable.*
  iii) AutStr *is effectively closed under* $\text{FO}^{\infty,\text{mod}}$-*interpretations.*

Moreover, for each non-unary Σ, $\mathfrak{S}_\Sigma$ is *complete*, in the sense of item ii) below, for AutStr with respect to first-order interpretations.

Proposition 2. *[6] Let* Σ *be a finite, non-unary alphabet.*
  i) *A relation R over* Σ^* *is regular if and only if it is definable in* $\mathfrak{S}_\Sigma$.
  ii) *A structure* $\mathfrak{A}$ *is automatic if and only if it is first-order interpretable in* $\mathfrak{S}_\Sigma$.

Intrinsic regularity. Let $\mathfrak{A} = (A, \{R_i\}_i) \in \text{AutStr}$ and $f \in \text{AP}(\mathfrak{A})$. By definition f maps every relation R_i of $\mathfrak{A}$ to a regular one. The previous observations yield that f also maps all relations $\text{FO}^{\infty,\text{mod}}$-definable in $\mathfrak{A}$ to regular ones. Intrinsically regular relations of structures were introduced in [16]. We shall also be concerned with the dual notion of intrinsic non-regularity.

Definition 3 (Intrinsic regularity). *Let* $\mathfrak{A}$ *be automatic. The* intrinsically (non-)regular *relations of* $\mathfrak{A}$ *are those, which are (non-)regular in every a.p. of* $\mathfrak{A}$. *Formally,* $\text{IR}(\mathfrak{A}) = \{R \subseteq A^r \mid r \in \mathbb{N}, \forall f \in \text{AP}(\mathfrak{A})\, f(R) \in \text{Reg}\}$ *and* $\text{INR}(\mathfrak{A}) = \{R \subseteq A^r \mid r \in \mathbb{N}, \forall f \in \text{AP}(\mathfrak{A})\, f(R) \notin \text{Reg}\}$.

For any given logic $\mathcal{L}$ extending FO let $\mathcal{L}(\mathfrak{A})$ denote the set of relations over $\text{dom}(\mathfrak{A})$ definable by an $\mathcal{L}$-formula using a finite number of parameters.

Remark 1. [16] $\text{FO}^{\infty,\text{mod}}(\mathfrak{A}) \subseteq \text{IR}(\mathfrak{A})$ holds, by Prop. 1 i), for every $\mathfrak{A} \in \text{AutStr}$.

Khoussainov et al. asked whether there is a logic $\mathcal{L}$ capturing intrinsic regularity, i.e., such that $\mathcal{L}(\mathfrak{A}) = \text{IR}(\mathfrak{A})$ for all $\mathfrak{A} \in \text{AutStr}$. We address this question in Section 3.

Example 2. Consider the structure $\mathcal{N} = (\mathbb{N}, +)$. For any integer $p \geq 2$ the base p (least-significant digit first) encoding provides an automatic presentation of $\mathcal{N}$. None of these presentations can be considered "canonical". On the contrary, by a deep result of Cobham and Semenov $\text{base}_p^{-1}[\text{Reg}] \cap \text{base}_q^{-1}[\text{Reg}] = \text{FO}(\mathcal{N})$ for any p and q having no common power (cf. [3,7]), hence $\text{FO}(\mathcal{N}) = \text{IR}(\mathcal{N})$.

When studying intrinsic regularity, it is natural to distinguish automatic presentations based on which relations they map to regular ones. To this end we introduce the following notion.

Definition 4 (Equivalence of automatic presentations). *For any $f, g \in$* $\text{AP}(\mathfrak{A})$ *let* $f \sim g \overset{\text{def}}{\Longleftrightarrow} f^{-1}[\text{Reg}] = g^{-1}[\text{Reg}]$.

Translations. Let $\mathfrak{A} \in \text{AUTSTR}$ and $f, g \in \text{AP}(\mathfrak{A})$. The mapping $t = g \circ f^{-1}$ is a bijection between regular languages, which *translates* names of elements of $\mathfrak{A}$ from one presentation to the other. Additionally, t preserves regularity of (the presentation of) all intrinsically regular relations of $\mathfrak{A}$.

Definition 5 (Translations). *A* translation *is a bijection* $t : D \to C$ *between regular sets* $D \subseteq \Sigma^*$, $C \subseteq \Gamma^*$. *If* $D = \Sigma^*$ *then* t *is a total- otherwise a partial translation. A translation* t *preserves regularity (non-regularity) if the image of every regular relation under* t *(respectively under* t^{-1}*) is again regular. Finally,* t *is* weakly regular *if it preserves both regularity and non-regularity.*

Note that by Proposition 2 all automatic presentations of $\mathfrak{G}_\Sigma$ are regularity preserving total translations. In general, one can fix a presentation $f \in \text{AP}(\mathfrak{A})$ of every $\mathfrak{A} \in \text{AUTSTR}$ and consider instead of each presentation $g \in \text{AP}(\mathfrak{A})$ the translation $t = g \circ f^{-1} \in \text{AP}(f(\mathfrak{A}))$ according to the correspondence $\text{AP}(\mathfrak{A}) = \text{AP}(f(\mathfrak{A})) \circ f$. Also observe that $t = g \circ f^{-1}$ is weakly regular if and only if $f \sim g$. This holds, in particular, when t is a bijective synchronized rational transduction, referred to as an "automatic isomorphism" in [17] and [15]. Clearly, every bijective rational transduction qualifies as a translation, however, not necessarily weakly regular. In Section 5 we will show that weakly regular translations coincide with bijective semi-synchronous transductions.

We associate to each translation f its *growth function* G_f defined as $G_f(n) = max\{|f(u)| : u \in \Sigma^*, |u| \leq n\}$ for each n and say that f is *length-preserving* if $|f(x)| = |x|$ for every word x, further, f is *monotonic* if $|x| \leq |y|$ implies $|f(x)| \leq |f(y)|$ for every x and y, finally, f *has bounded delay* if there exists a constant δ such that $|x| + \delta < |y|$ implies $|f(x)| < |f(y)|$ for every x and y.

3   Order-Invariant Logic

In [16, Question 1.4] Khoussainov et al. have called for a logical characterization of intrinsic regularity over all automatic structures. The same question, and in particular, whether $\text{FO}^{\infty,\text{mod}}$ is capable of defining all intrinsically regular relations over any automatic structure is raised in [20, Problem 3]. In this section we answer the latter question negatively. We do this by exhibiting an automatic structure $\mathfrak{B}$ together with a relation, which is order-invariantly definable, but not $\text{FO}^{\infty,\text{mod}}$-definable in $\mathfrak{B}$.

Let $\mathfrak{A}$ be a structure of signature τ. Assume that $<$ is a binary relation symbol not occurring in τ. A formula $\phi(\boldsymbol{x}) \in \text{FO}[\tau, <]$ is *order-invariant over* $\mathfrak{A}$ if for any linear ordering $<_A$ of the elements of $\mathfrak{A}$, when $<$ is interpreted as $<_A$, $\phi(\boldsymbol{x})$ defines the same relation R over $\mathfrak{A}$. The relation R is in this case *order-invariantly definable*. We denote the set of order-invariantly definable relations over $\mathfrak{A}$ by $\text{FO}_{<-\text{inv}}(\mathfrak{A})$, and by $\text{FO}^{\infty,\text{mod}}_{<-\text{inv}}(\mathfrak{A})$ when counting quantifiers are also allowed. Although it is only appropriate to speak of order-invariantly definable relations, rather than of relations definable in "order-invariant logic", we will tacitly use the latter term as well.

The fact that over any $\mathfrak{A} \in$ AUTSTR order-invariantly definable relations are intrinsically regular is obvious. Indeed, given a particular automatic presentation of $\mathfrak{A}$ one just has to "plug in" any regular ordering (e.g. the lexicographic ordering, which does of course depend on the automatic presentation chosen) into the order-invariant formula defining a particular relation, thereby yielding a regular relation, which, by order-invariance, will always represent the same relation.

Proposition 3. $\mathsf{FO}^{\infty,\mathrm{mod}}_{<-\mathrm{inv}}(\mathfrak{A}) \subseteq \mathsf{IR}(\mathfrak{A})$.

Order-invariant first-order logic has played an important role in finite model theory. It is well known that $\mathsf{FO}_{<-\mathrm{inv}}$ is strictly more expressive than FO on finite structures. Gurevich was the first to exhibit an order-invariantly definable class of finite structures, which is not first-order definable [18, Sect. 5.2]. However, his class is $\mathsf{FO}^{\infty,\mathrm{mod}}$-definable. In [19] Otto showed how to use order-invariance to express connectivity, which is not definable even in infinitary counting logic. Both constructions use order-invariance and some auxiliary structure to exploit the power of monadic second order logic (MSO). We adopt Otto's technique to show that $\mathsf{FO}_{<-\mathrm{inv}}$ can be strictly more expressive than $\mathsf{FO}^{\infty,\mathrm{mod}}$ on automatic structures. The proof involves a version of the bijective Ehrenfeucht-Fraïssé games, introduced by Hella [11], which capture equivalence modulo $\mathsf{FO}(\mathbf{Q}_1)$, the extension of FO with *unary generalized quantifiers* [18, Chapter 8]. The simples ones of these are $\exists^\infty$ and $\exists^{(r,m)}$. Therefore, as also observed by Otto, the separation result applies not only to $\mathsf{FO}^{\infty,\mathrm{mod}}$ but to the much more powerful logic $\mathsf{FO}(\mathbf{Q}_1)$.

Consider the structure $\mathfrak{B} = (\mathbb{N} \uplus \mathcal{P}_{\mathrm{fin}}(4\mathbb{N} + \{2,3\}), S, \varepsilon, \iota, \subseteq)$, illustrated in Figure 1, where $\mathcal{P}_{\mathrm{fin}}(H)$ consists of the finite subsets of H, S is the relation $\{(4n, 4n+4), (4n+1, 4n+5) \mid n \in \mathbb{N}\}$, ε is the equivalence relation consisting of classes $\{4n, 4n+1, 4n+2, 4n+3\}$ for each $n \in \mathbb{N}$, ι is the set of pairs $(n, \{n\})$ with $n \in 4\mathbb{N} + \{2,3\}$, and $\subseteq$ is the usual subset inclusion.

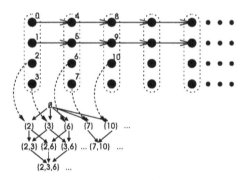

Fig. 1. $\mathfrak{B}$, a separating example

To give an automatic presentation of $\mathfrak{B}$ over the alphabet $\{b, 0, 1\}$ we represent $(\mathbb{N}, S, \varepsilon)$ in the unary encoding using the symbol b, and the finite sets by their (shortest) characteristic words over $\{0, 1\}$. Regularity of ι and $\subseteq$ is obvious.

Proposition 4. *The transitive closure S^* of S is order-invariantly definable, hence intrinsically regular, but not $\mathsf{FO}(\mathbf{Q}_1)$-definable in $\mathfrak{B}$.*

Proof. The proof is a straightforward adaptation of the one presented in [19]. $S^* \in \mathsf{FO}_{<-\mathrm{inv}}(\mathfrak{B})$: Given any ordering $\prec$ of the universe of $\mathfrak{B}$ we can first-order define a bijection $\nu = \nu_{\prec} : 4\mathbb{N} \cup 4\mathbb{N} + 1 \rightarrow 4\mathbb{N} + 2 \cup 4\mathbb{N} + 3$ as follows. Each ε-class contains two points, $4n + 2$ and $4n + 3$, having an outgoing ι edge and two points, $4n$ and $4n + 1$, having an S-successor. Using $\prec$ we can map e.g. the smaller (larger) of the latter to the smaller (larger) of the former. This bijection, regardless of the actual mapping, provides access to the subset structure, thus unleashing full power of weak-MSO. Transform, using ν and the built-in subset structure, any weak-MSO definition of transitive closure into one expressing S^*. $S^* \notin \mathsf{FO}(\mathbf{Q}_1)(\mathfrak{B})$: The proof of this statement involves a fairly standard application of bijective Ehrenfeucht-Fraïssé games as hinted in [19]. $\qquad\square$

Corollary 1. *No extension of FO with unary generalized quantifiers is capable of capturing intrinsic regularity over all automatic structures.*

4   Automatic Presentations of $\mathfrak{S}_{\Sigma}$

Recall $\mathfrak{S}_{\Sigma}$ of Example 1. Let Σ be non-unary. The main result of this section is that automatic presentations of $\mathfrak{S}_{\Sigma}$ are weakly regular translations, hence are all equivalent. We treat the case of length-preserving presentations first.

Proposition 5. *Let $f : \Sigma^* \rightarrow \Gamma^*$ be a length-preserving automatic presentation of $\mathfrak{S}_{\Sigma}$. Then (the graph of) f is regular.*

Proof. Consider $\{(u, v) \in (\Sigma^*)^2 : |u| \leq |v| \wedge v[|u|] = a\}$, which is clearly regular for each $a \in \Sigma$. Their images under f are regular relations over Γ^*. Since only the length of the first component plays a role in these relations, and it is preserved by f, the following "variants" over $\Sigma^* \times \Gamma^*$ are also regular.

$$R_a = \{(u, x) \in \Sigma^* \times f(\Sigma^*) : |u| \leq |x| \wedge f^{-1}(x)[|u|] = a\} \quad (a \in \Sigma)$$

Thus, we may define the graph of f as follows showing that it is indeed regular.

$$\mathrm{graph}(f) = \{(u, x) \in \Sigma^* \times \Gamma^* : |u| = |x| \wedge \forall v \preceq u \bigwedge_{a \in \Sigma} v[|v|] = a \rightarrow R_a(v, x)\} \qquad\square$$

Theorem 1. $f \in \mathsf{AP}(\mathfrak{S}_{\Sigma}) \iff f$ *is a total and weakly regular translation.*

Proof. We only need to prove "$\Rightarrow$". Let $f \in \mathsf{AP}(\mathfrak{S}_{\Sigma})$. In two steps of transformations we will show that f is equivalent to a length-preserving presentation h of $\mathfrak{S}_{\Sigma}$. The claim then follows by Proposition 5.

The relations $|y| \leq |x|$, and S_a ($a \in \Sigma$) are locally finite and regular. A standard pumping argument (e.g. [14, Lemma 3.1]) then shows, that there are constants K and C such that $|y| \leq |x| \rightarrow |f(y)| \leq |f(x)| + K$ and $|f(xa)| \leq |f(x)| + C$ for every $a \in \Sigma$ and $x, y \in \Sigma^*$. It is easily seen, that by suffixing each codeword $f(x)$ by an appropriate ($\leq K$) number of some new padding symbols, we can obtain an equivalent monotonic presentation g.

Lemma 1. $\forall f \in \mathsf{AP}(\mathfrak{S}_\Sigma) \; \exists g \in \mathsf{AP}(\mathfrak{S}_\Sigma) : g \sim f$, g is monotonic and $G_g = G_f$.

Proof. By the choice of K above, we have $G_f(|x|) \leq |f(x)| + K$, and for each $s = 0..K$ the set $D_s = \{x : G_f(|x|) - |f(x)| = s\}$ is regular, being definable. This observation allows us to pad each codeword accordingly. Let us therefore define g by letting $g(x) = f(x)@^{G_f(|x|)-|f(x)|}$ for every $(x \in \Sigma^*)$. where @ is a new padding symbol. The domain of the new presentation, that is $g(\Sigma^*) = \bigcup_{s=1}^{k} f(D_s) \cdot @^s$ is by the above argument regular. Moreover, since this padding is definable f and g map the same relations to regular ones. Finally, it is clear that g is monotonic, because $|g(x)| = G_f(|x|) = G_g(|x|)$ holds for every word x, and the growth function G_f is by definition always monotonic. $\qquad\square$

The decisive step of the construction requires two key lemmas.

Lemma 2. $\forall f \in \mathsf{AP}(\mathfrak{S}_\Sigma)$: f has bounded delay.

Proof. Consider the equivalent presentation g obtained from f by padding each codeword with at most K new symbols as in Lemma 1. If g has bounded delay with bound δ then f has bounded delay with bound $\leq K\delta$. Assume therefore that f is monotonic. Let $D = f(\Sigma^*)$, $s = |\Sigma| \geq 2$, and $D_{\leq n} = \{x \in D \mid |x| \leq n\}$ for each $n \in \mathbb{N}$. Assume, that for some n and t we find the following situation.

$$G_f(n-1) < G_f(n) = G_f(n+1) = \ldots = G_f(n+t-1) < G_f(n+t)$$

Then $|D_{\leq G_f(n-1)}| = (s^n - 1)/(s-1)$ and $|D_{\leq G_f(n)}| = |D_{\leq G_f(n+t-1)}| = (s^{n+t} - 1)/(s-1)$ since they contain (due to monotonicity) precisely the images of words of length at most $n - 1$ and $n + t - 1$ respectively. On the other hand, by the choice of C, we have $G_f(n) \leq G_f(n-1) + C$, hence $D_{\leq G_f(n)} \subseteq D_{\leq G_f(n-1)+C}$ for every $n \in \mathbb{N}$. In [14, Lemma 3.12] it is shown that $|D_{\leq n+C}| \in \Theta(|D_{\leq n}|)$ for each C. Thus, there is a constant β (certainly, $\beta \geq 1$) such that $|D_{\leq G_f(n)}| \leq |D_{\leq G_f(n-1)+C}| \leq \beta \cdot |D_{\leq G_f(n-1)}|$. By simple arithmetic, $t \leq \log_s(\beta)$, which proves that f has bounded delay, namely, bounded by $\delta = \log_s(\beta) + 1$. $\qquad\square$

Lemma 3. For all $f \in \mathsf{AP}(\mathfrak{S}_\Sigma)$ the infinite sequence of increments of the growth function of f, $\partial G_f = \langle G_f(1) - G_f(0), \ldots, G_f(n+1) - G_f(n), \ldots \rangle \in \{0, \ldots, C\}^\omega$, is ultimately periodic.

Proof. Consider the monotonic mapping g obtained from f by padding as in Lemma 1, and the language $L = \{x = g(u) \mid \forall y = g(v)(|u| = |v| \to x \leq_{llex} y)\}$ consisting of the length-lexicographically least g-representants of some word of length n for each non-negative n. L is regular and since g has bounded delay, say with bound δ, it is δ-thin, meaning that there are at most δ many words in L of each length. We can thus write L as disjoint union of the regular languages $L_k = \{x \in L \mid \exists^{=k} y \in L \; |x| = |y|\}$ for $k = 1, \ldots, \delta$. Let us project L as well as L_k's onto 1^* in length-preserving manner. $G_g = G_f$ is a non-decreasing sequence of naturals in which each number can occur at most δ times. The projection of L corresponds, in the unary encoding, to the pruned sequence obtained from G_f by omitting the repetitions, whereas L_k is mapped onto those 1^n for which

n is repeated exactly k times in G_f. All these projections are regular unary languages, which is the same as saying that the corresponding sets of naturals are ultimately periodic. The claim follows.   □

This last result allows us to construct an equivalent length-preserving a.p. h by factoring each word $g(u)$ of length $G_g(|u|)$ into "blocks" according to ∂G_g.

Lemma 4. $\forall g \in \mathsf{AP}(\mathfrak{S}_\Sigma) : g$ *is monotonic* $\rightarrow \exists h \sim g : h$ *is length-preserving.*

Proof. Let $g : \Sigma^* \rightarrow \Gamma^*$ be a monotonic a.p. of $\mathsf{AP}(\mathfrak{S}_\Sigma)$ with $D = g(\Sigma^*)$. The fact that ∂G_g is ultimately periodic allows us to construct an equivalent length-preserving presentation h by subdividing codewords produced by g into blocks according to ∂G_g. (For this we need to assume that $G_g(0) = 0$, i.e. the empty word is represented by itself. Clearly, this is no serious restriction as changing an a.p. on a finite number of words always yields an equivalent a.p.)

Consider some word $u \in \Sigma^*$ of size n and its image $v = g(u) \in \Gamma^*$. Since g is monotonic $|v| = G_g(|u|) = G_g(n)$. Thus we can factorize v as $v_1 v_2 \cdots v_n$ where $|v_i| = \partial G_g[i]$ for each $i \leq n$. Since $\partial G_g[i] \leq C$ for every i, we can consider each v_i as a single symbol of the alphabet $\Theta = \Gamma^{\leq C} = \{w \in \Gamma^* : |w| \leq C\}$. Let β be the natural projection mapping elements of Θ to the corresponding words over Γ, and let $\lambda(w) = |\beta(w)|$ for each $w \in \Theta$.

We define the mapping $h : \Sigma^* \rightarrow \Theta^*$ by setting for each $u \in \Sigma^*$, with factorization as above, $h(u) = v_1 \cdot v_2 \cdot \ldots \cdot v_n$ when considered as a word of length n over Θ. Thus, h is by definition length-preserving and maps Σ^* injectively onto the set $D' = \{x \in \Theta^* \mid \beta(x) \in D \wedge (\forall i = 1..|x|) \, \lambda(x[i]) = \partial G_g(i)\}$. Because β is a homomorphism, D regular and ∂G_g ultimately periodic, D' can clearly be accepted by a finite automaton. Moreover, the fact that any two words w, w' belonging to D' are synchronously blocked (in the sense that $x[i]$ and $x'[i]$ have the same length for all $i \leq |x|, |x'|$) enables us to easily simulate any n-tape automaton $\mathcal{A}$ accepting a relation over D by an automaton $\mathcal{A}'$ accepting the "same" relation over D' and vice versa.   □

This concludes the proof of Theorem 1.   □

Corollary 2. *Non-regular relations are intrinsically non-regular wrt.* $\mathfrak{S}_\Sigma$.

Corollary 3. *Every total translation that preserves regularity also preserves non-regularity, hence is weakly regular.*

Theorem 1 fails for unary alphabets, because, as can easily be checked, the mapping from unary to binary presentation of the naturals does preserve regularity, but also maps some non-regular relations to regular ones. The same argument shows that Corollary 3 does not hold for partial translations: simply take a "variant" of the unary presentation over the partial domain $(ab)^* \subsetneq \{a, b\}^*$.

Corollary 4. *The complete structures* $\mathfrak{S}_\Sigma$ *have, up to equivalence, only a single automatic presentation:* $\mathsf{AP}(\mathfrak{S}_\Sigma)/\sim = \{[\mathrm{id}]\}$.

Not all complete structures have this property. Let $\mathfrak{C} = \mathfrak{A} \uplus \mathfrak{B}$ be the disjoint union of $\mathfrak{A}$ and $\mathfrak{B}$ having an additional unary predicate A identifying elements belonging to $\mathfrak{A}$. Thus, $\mathfrak{A}$ and $\mathfrak{B}$ are trivially FO-interpretable in $\mathfrak{C}$, and $\mathfrak{C} \in$ AUTSTR iff $\mathfrak{A}, \mathfrak{B} \in$ AUTSTR. It follows from Proposition 7 below, that $\mathfrak{A} \uplus \mathfrak{B}$ has infinitely many inequivalent automatic presentations, provided both $\mathfrak{A}$ and $\mathfrak{B}$ are infinite. In particular, this holds for the complete structure $\mathfrak{S}_\Sigma \uplus \mathfrak{S}_\Sigma$. Let us therefore say that a structure is *rigidly automatic* if it has but one automatic presentation up to equivalence. Finite structures are trivially rigidly automatic.

Question 1. Is every infinite, rigidly automatic structure complete?

5   Equivalence Via Semi-synchronous Transductions

Observe that we have proved more than what is claimed in Theorem 1. The above proof shows indeed, that every $f \in \mathsf{AP}(\mathfrak{S}_\Sigma)$ can be decomposed as

$$f = \pi^{-1} \circ \beta^{-1} \circ h$$

where π applies the padding, β the cutting of words into blocks, and where h is length-preserving and regular. Since both π^{-1} and β^{-1} are projections the composition is a rational transduction.[1] Moreover, we know that ∂G_f is ultimately periodic, say from threshold N with period p. Let $q = G_f(N + p) - G_f(N)$ be the total length of *any* p consecutive blocks with sufficiently high indices. This means that after reading the first N input symbols and the first $G_f(N)$ output symbols a transducer accepting f can proceed by reading blocks of p input- and q output symbols in each step, which shows that f is in fact a (p, q)-synchronous transduction. This decomposition is the idea underlying one direction of the main result of this section, the next lemma constitutes the other.

Lemma 5. *For every vector α of nonnegative integers $\mathsf{S}_\alpha \mathsf{Rat}$ is closed under taking images (hence also inverse images) of semi-synchronous transductions.*

Proof. Let T be a (p, q)-synchronous transduction, R an α-synchronous n-ary relation with $\alpha = (a_1, \ldots, a_n)$. $T(R) = \{v \mid \exists u \in R \; \forall i \leq n \; (u_i, v_i) \in T\}$ is the projection of the $(pa_1, \ldots, pa_n, qa_1, \ldots, qa_n)$-synchronous relation $\{(u, v) \mid u \in R \; \forall i \leq n \; (u_i, v_i) \in T\}$. Hence, by Propositions 6 and 7, $T(R)$ is α-synchronous. Closure under taking inverse images follows from the fact, that the inverse of a (p, q)-synchronous transduction is (q, p)-synchronous. □

Theorem 2. *A translation f is weakly regular if and only if it is a semi-synchronous transduction.*

Proof. The "if" part is a special case of Lemma 5. To prove the "only if" part we show that, under the assumption of weak-regularity, all lemmas used to prove

---

[1] Knowing this, Proposition 5 follows from [9, Corollary 6.6] stating that length-preserving transductions are synchronized rational. (See also [10].)

Theorem 1 hold even for partial translations. We only note that lemmas 1, 3 and 4 carry over without modification. To prove that a monotonic, possibly partial, weakly-regular translation g has bounded delay it suffices to consider the inverse image of the predicate $|x| \leq |y|$ under g. A pumping argument shows that there is a constant D such that $|g^{-1}(x)| \leq |g^{-1}(y)| + D$ whenever $|x| \leq |y|$, in other words $|u| > |v| + D$ implies that $|g(u)| > |g(v)|$, i. e. g has bounded delay. This proves the analog of Lemma 2.[2] Finally, note that our proof of Proposition 5 works only assuming that the domain (hence the range) of the length-preserving weakly regular mapping considered contains at least one word of every length. This requirement is not essential. Instead of R_a's ($a \in \Sigma$) one can just as well use the relations R_w defined for each $w \in \Sigma^k$ in the obvious way for a sufficiently large k. Thus, we obtain the same decomposition $f = \pi^{-1} \circ \beta^{-1} \circ h$, which shows, as above, that f is a semi-synchronous transduction.   □

Corollary 5. *Let $\mathfrak{A} \in \text{AutStr}$ and $f, g \in AP(\mathfrak{A})$. Then $f \sim g$ if and only if the translation $g \circ f^{-1}$ (and $f \circ g^{-1}$) is a semi-synchronous transduction.*

Basic properties of semi-synchronous relations. Note that the composition of a (p, q)-synchronous and an (r, s)-synchronous transduction is (pr, qs)-synchronous, thus, the class of semi-synchronous transductions is closed under composition. Alternative to our definition of $\mathsf{S}_\alpha\mathsf{Rat}$ based on α-convolution one can introduce α-*synchronous automata*, defined in the obvious way, accepting α-synchronous relations. These automata, being essentially synchronous, can be determinized, taken product of, etc. Hence the following.

Proposition 6. $\mathsf{S}_\alpha\mathsf{Rat}$ *is an effective boolean algebra for each α. The projection of every $\alpha\beta$-synchronous relation onto the first $|\alpha|$ many components, is α-synchronous.*

Evidently, $\mathsf{Reg} \subset \mathsf{SRat} \subset \mathsf{Rat}$ and both containments are strict as illustrated by the examples $\{(a^n, a^{2n}) \mid n \in \mathbb{N}\}$ and $\{(a^n, a^{2n}), (b^n, b^{3n}) \mid n \in \mathbb{N}\}$. SRat is closed under complement but not under union, as also shown by the latter example. Comparing classes $\mathsf{S}_\alpha\mathsf{Rat}$ and $\mathsf{S}_\beta\mathsf{Rat}$ we observe the following "Cobham-Semenov-like" relationship. Let us say that α and β are dependent if $k \cdot \alpha = l \cdot \beta$ for some $k, l \in \mathbb{N} \setminus \{0\}$, where multiplication is meant component-wise.

Proposition 7. *Let $n, p, q \in \mathbb{N}$ and $\alpha, \beta \in \mathbb{N}^n$.*

i) If α and β are dependent, then $\mathsf{S}_\alpha\mathsf{Rat} = \mathsf{S}_\beta\mathsf{Rat}$.

ii) If (p, q) and (r, s) are independent, then $\mathsf{S}_{(p,q)}\mathsf{Rat} \cap \mathsf{S}_{(r,s)}\mathsf{Rat} = \mathsf{Rec}$.

Adapting techniques from [2,10], used to prove undecidability of whether a given rational relation is synchronized rational, we obtain the following results.

Proposition 8. *For any given $p, q \in \mathbb{N}$ the following problems are undecidable.*

i) Given a rational transduction $R \in \mathsf{Rat}$ is $R \in \mathsf{S}_{(p,q)}\mathsf{Rat}$?

ii) Given a rational transduction $R \in \mathsf{Rat}$ is $R \in \mathsf{SRat}$?

---

[2] Note that this argument was not applicable in the proof of Lemma 2, since at that point we could not rely on the mapping preserving non-regularity but only on it being total.

6   Future Work

Themes of follow-up work include generalizing results to ω-automatic structures, questions concerning the number of automatic presentations modulo equivalence, and a finer analysis of definability of intrinsically regular relations on restricted classes of structures.

References

1. J.-P. Allouche and J. Shallit. *Automatic Sequences, Theory, Applications, Generalizations*. Cambridge University Press, 2003.
2. J. Berstel. *Transductions and Context-Free Languages*. Teubner, Stuttgart, 1979.
3. A. Bès. Undecidable extensions of Büchi arithmetic and Cobham-Seménov theorem. *J. Symb. Log.*, 62(4):1280–1296, 1997.
4. A. Blumensath. Automatic structures. Diploma thesis, RWTH-Aachen, 1999.
5. A. Blumensath and E. Grädel. Automatic Structures. In *Proceedings of 15th IEEE Symposium on Logic in Computer Science LICS 2000*, pages 51–62, 2000.
6. A. Blumensath and E. Grädel. Finite presentations of infinite structures: Automata and interpretations. *Theory of Computing Systems*, 37:641 – 674, 2004.
7. V. Bruyère, G. Hansel, Ch. Michaux, and R. Villemaire. Logic and p-recognizable sets of integers. *Bull. Belg. Math. Soc.*, 1:191 – 238, 1994.
8. J.W. Cannon, D.B.A. Epstein, D.F. Holt, S.V.F. Levy, M.S. Paterson, and W.P. Thurston. *Word processing in groups*. Jones and Barlett Publ., Boston, MA, 1992.
9. C. C. Elgot and J. E. Mezei. On relations defined by generalized finite automata. *IBM J. Research and Development*, 9:47 – 68, 1965.
10. Ch. Frougny and J. Sakarovitch. Synchronized rational relations of finite and infinite words. *Theor. Comput. Sci.*, 108:45–82, 1993.
11. L. Hella. Definability hierarchies of generalized quantifiers. *Ann. Pure Appl. Logic*, 43:235 – 271, 1989.
12. B.R. Hodgson. On direct products of automaton decidable theories. *TCS*, 19:331–335, 1982.
13. B. Khoussainov and A. Nerode. Automatic presentations of structures. In *LCC '94*, volume 960 of *LNCS*, pages 367–392. Springer-Verlag, 1995.
14. B. Khoussainov, A. Nies, S. Rubin, and F. Stephan. Automatic structures: Richness and limitations. In *LICS*, pages 44–53. IEEE Computer Society, 2004.
15. B. Khoussainov and S. Rubin. Automatic structures: Overview and future directions. *Journal of Automata, Languages and Combinatorics*, 8(2):287–301, 2003.
16. B. Khoussainov, S. Rubin, and F. Stephan. Definability and regularity in automatic structures. In *STACS '04*, volume 2996 of *LNCS*, pages 440–451, 2004.
17. D. Kuske. Is cantor's theorem automatic? In M. Y. Vardi and A. Voronkov, editors, *LPAR*, volume 2850 of *LNCS*, pages 332–345. Springer, 2003.
18. L. Libkin. *Elements of Finite Model Theory*. Texts in Theoretical Computer Science. Springer, 2004.
19. M. Otto. Epsilon-logic is more expressive than first-order logic over finite structures. *J. Symb. Logic*, 65:1749–1757, 2000.
20. S. Rubin. Automatic structures. Ph.D. thesis, University of Auckland, NZ, 2004.

On the Accepting Power of
2-Tape Büchi Automata

Olivier Finkel

Equipe de Logique Mathématique, U.F.R. de Mathématiques,
Université Paris 7, 2 Place Jussieu 75251 Paris cedex 05, France
finkel@logique.jussieu.fr

Abstract. We show that, from a topological point of view, 2-tape Büchi automata have the same accepting power than Turing machines equipped with a Büchi acceptance condition. In particular, for every non null recursive ordinal α, there exist some $\mathbf{\Sigma}^0_\alpha$-complete and some $\mathbf{\Pi}^0_\alpha$-complete infinitary rational relations accepted by 2-tape Büchi automata. This surprising result gives answers to questions of Simonnet [Sim92] and of Lescow and Thomas [Tho90, LT94].

Keywords: 2-tape Büchi automata; infinitary rational relations; Cantor topology; topological complexity; Borel hierarchy; complete sets.

1   Introduction

In the sixties, automata accepting infinite words were firstly considered by Büchi in order to study decidability of the monadic second order theory S1S of one successor over the integers [Büc62]. Then the so called ω-regular languages have been intensively studied and have found many applications for specification and verification of non terminating systems, see [Tho90, Sta97, PP04] for many results and references. On the other hand, rational relations on finite words were also studied in the sixties, and played a fundamental role in the study of families of context free languages [Ber79]. Investigations on their extension to rational relations on infinite words were carried out or mentioned in the books [BT70, LS77]. Gire and Nivat studied infinitary rational relations in [Gir81, GN84]. These relations are sets of pairs of infinite words which are accepted by 2-tape finite Büchi automata with asynchronous reading heads. The class of infinitary rational relations, which extends both the class of finitary rational relations and the class of ω-regular languages, and the rational functions they may define, have been much studied, see for example [CG99, BCPS00, Sim92, Sta97, Pri00].

Notice that a rational relation $R \subseteq \Sigma_1^\omega \times \Sigma_2^\omega$ may be seen as an ω-language over the alphabet $\Sigma_1 \times \Sigma_2$.

A way to study the complexity of languages of infinite words accepted by finite machines is to study their topological complexity and firstly to locate them with regard to the Borel and the projective hierarchies. This work is analysed for example in [Sta86, Tho90, EH93, LT94, Sta97]. It is well known that every ω-language accepted by a Turing machine with a Büchi or Muller acceptance

B. Durand and W. Thomas (Eds.): STACS 2006, LNCS 3884, pp. 301–312, 2006.
© Springer-Verlag Berlin Heidelberg 2006

condition is an analytic set and that ω-regular languages are boolean combinations of $\mathbf{\Pi}_2^0$-sets hence $\mathbf{\Delta}_3^0$-sets, [Sta97, PP04].

The question of the topological complexity of relations on infinite words also naturally arises and is asked by Simonnet in [Sim92]. It is also posed in a more general form by Lescow and Thomas in [LT94] (for infinite labelled partial orders) and in [Tho89] where Thomas suggested to study reducibility notions and associated completeness results.

Every infinitary rational relation is an analytic set. We showed in [Fin03a] that there exist some infinitary rational relations which are analytic but non Borel, and in [Fin03c] that there are some $\mathbf{\Sigma}_3^0$-complete and some $\mathbf{\Pi}_3^0$-complete infinitary rational relations, using a coding of ω^2-words by pairs of infinite words. Using a different coding we proved in [Fin03d] that there exist such infinitary rational relations which have a very simple structure and can be easily described by their sections. Using this very simple structure, we constructed also some infinitary rational relations, accepted by 3-tape Büchi automata, which are $\mathbf{\Sigma}_4^0$-complete.

On the other hand we recently proved in [Fin05a, Fin05b] that the Borel hierarchy of ω-languages accepted by Büchi real time 1-counter automata is equal to the Borel hierarchy of ω-languages accepted by Büchi Turing machines. In particular, for each non null recursive ordinal α, there exist some $\mathbf{\Sigma}_\alpha^0$-complete and some $\mathbf{\Pi}_\alpha^0$-complete ω-languages accepted by Büchi real time 1-counter automata.

Using a simulation of real time 1-counter automata we prove in this paper a similar result: the Borel hierarchy of the class of infinitary rational relations is equal to the Borel hierarchy of ω-languages accepted by Büchi real time 1-counter automata which is also equal to the Borel hierarchy of ω-languages accepted by Büchi Turing machines. In particular, for each non null recursive ordinal α, there exist some $\mathbf{\Sigma}_\alpha^0$-complete and some $\mathbf{\Pi}_\alpha^0$-complete infinitary rational relations. This gives answers to questions of Simonnet [Sim92] and of Lescow and Thomas [Tho90, LT94].

The paper is organized as follows. In section 2 we recall the notion of 2-tape automata and of real time 1-counter automata with Büchi acceptance condition. In section 3 we recall definitions of Borel and analytic sets, and we prove our main result in section 4.

2   2-Tape Automata and 1-Counter Automata

We assume the reader to be familiar with the theory of formal (ω)-languages [Tho90, Sta97]. We shall use usual notations of formal language theory.

When Σ is a finite alphabet, a *non-empty finite word* over Σ is any sequence $x = a_1 \ldots a_k$, where $a_i \in \Sigma$ for $i = 1, \ldots, k$, and k is an integer ≥ 1. The *length* of x is k, denoted by $|x|$. The *empty word* has no letter and is denoted by λ; its length is 0. For $x = a_1 \ldots a_k$, we write $x(i) = a_i$ and $x[i] = x(1) \ldots x(i)$ for $i \leq k$ and $x[0] = \lambda$. $\Sigma^\star$ is the *set of finite words* (including the empty word) over Σ.

The *first infinite ordinal* is ω. An *ω-word* over Σ is an ω-sequence $a_1 \ldots a_n \ldots$, where for all integers $i \geq 1$, $a_i \in \Sigma$. When σ is an ω-word over Σ, we write

$\sigma = \sigma(1)\sigma(2)\ldots\sigma(n)\ldots$, where for all i, $\sigma(i) \in \Sigma$, and $\sigma[n] = \sigma(1)\sigma(2)\ldots\sigma(n)$ for all $n \geq 1$ and $\sigma[0] = \lambda$.

The *prefix relation* is denoted $\sqsubseteq$: a finite word u is a *prefix* of a finite word v (respectively, an infinite word v), denoted $u \sqsubseteq v$, if and only if there exists a finite word w (respectively, an infinite word w), such that $v = u.w$. The *set of ω-words* over the alphabet Σ is denoted by Σ^ω. An *ω-language* over an alphabet Σ is a subset of Σ^ω. The complement (in Σ^ω) of an ω-language $V \subseteq \Sigma^\omega$ is $\Sigma^\omega - V$, denoted V^-.

Infinitary rational relations are subsets of $\Sigma^\omega \times \Gamma^\omega$, where Σ and Γ are finite alphabets, which are accepted by 2-tape Büchi automata (2-BA).

Definition 1. *A 2-tape Büchi automaton is a sextuple $T = (K, \Sigma, \Gamma, \Delta, q_0, F)$, where K is a finite set of states, Σ and Γ are finite alphabets, Δ is a finite subset of $K \times \Sigma^\star \times \Gamma^\star \times K$ called the set of transitions, q_0 is the initial state, and $F \subseteq K$ is the set of accepting states.*

A computation C of the 2-tape Büchi automaton T is an infinite sequence of transitions

$$(q_0, u_1, v_1, q_1), (q_1, u_2, v_2, q_2), \ldots (q_{i-1}, u_i, v_i, q_i), (q_i, u_{i+1}, v_{i+1}, q_{i+1}), \ldots$$

The computation is said to be successful iff there exists a final state $q_f \in F$ and infinitely many integers $i \geq 0$ such that $q_i = q_f$.
The input word of the computation is $u = u_1.u_2.u_3 \ldots$
The output word of the computation is $v = v_1.v_2.v_3 \ldots$
Then the input and the output words may be finite or infinite.

The infinitary rational relation $R(T) \subseteq \Sigma^\omega \times \Gamma^\omega$ accepted by the 2-tape Büchi automaton T is the set of couples $(u, v) \in \Sigma^\omega \times \Gamma^\omega$ such that u and v are the input and the output words of some successful computation C of T.
The set of infinitary rational relations will be denoted RAT_ω.

Definition 2. *A (real time) 1-counter machine is a 4-tuple $\mathcal{M} = (K, \Sigma, \Delta, q_0)$, where K is a finite set of states, Σ is a finite input alphabet, $q_0 \in K$ is the initial state, and the transition relation Δ is a subset of $K \times \Sigma \times \{0, 1\} \times K \times \{0, 1, -1\}$. If the machine $\mathcal{M}$ is in state q and $c \in \mathbf{N}$ is the content of the counter then the configuration (or global state) of $\mathcal{M}$ is (q, c).*

For $a \in \Sigma$, $q, q' \in K$ and $c \in \mathbf{N}$, if $(q, a, i, q', j) \in \Delta$ where $i = 0$ if $c = 0$ and $i = 1$ if $c \neq 0$ then we write:

$$a : (q, c) \mapsto_\mathcal{M} (q', c + j)$$

$\mapsto^\star_\mathcal{M}$ *is the transitive and reflexive closure of $\mapsto_\mathcal{M}$.*
Thus we see that the transition relation must satisfy:
if $(q, a, i, q', j) \in \Delta$ and $i = 0$ then $j = 0$ or $j = 1$ (but j may not be equal to -1).

Let $\sigma = a_1 a_2 \ldots a_n$ be a finite word over Σ. A sequence of configurations $r = (q_i, c_i)_{1 \leq i \leq n+1}$ is called a run of $\mathcal{M}$ on σ, starting in configuration (p, c), iff:

(1) $(q_1, c_1) = (p, c)$
(2) for each $i \in [1, n]$, $a_i : (q_i, c_i) \mapsto_{\mathcal{M}} (q_{i+1}, c_{i+1})$

Let $\sigma = a_1 a_2 \ldots a_n \ldots$ be an ω-word over Σ. An ω-sequence of configurations $r = (q_i, c_i)_{i \geq 1}$ is called a run of $\mathcal{M}$ on σ, starting in configuration (p, c), iff:

(1) $(q_1, c_1) = (p, c)$
(2) for each $i \geq 1$, $a_i : (q_i, c_i) \mapsto_{\mathcal{M}} (q_{i+1}, c_{i+1})$

For every such run, $\mathrm{In}(r)$ is the set of all states entered infinitely often during run r.

A run r of M on σ, starting in configuration $(q_0, 0)$, will be simply called "a run of M on σ".

Definition 3. *A (real time) Büchi 1-counter automaton is a 5-tuple*

$$\mathcal{M} = (K, \Sigma, \Delta, q_0, F),$$

where $\mathcal{M}' = (K, \Sigma, \Delta, q_0)$ is a (real time) 1-counter machine and $F \subseteq K$ is the set of accepting states. The ω-language accepted by $\mathcal{M}$ is

$$L(\mathcal{M}) = \{\sigma \in \Sigma^\omega \mid \text{there exists a run } r \text{ of } \mathcal{M} \text{ on } \sigma \text{ such that } \mathrm{In}(r) \cap F \neq \emptyset\}$$

The class of (real time) Büchi 1-counter automata will be denoted **r-BC(1)**.

The class of ω-languages accepted by real time Büchi 1-counter automata will be denoted **r-BCL(1)$_\omega$**.

3   Borel Hierarchy

We assume the reader to be familiar with basic notions of topology which may be found in [Mos80, LT94, Kec95, Sta97, PP04]. There is a natural metric on the set Σ^ω of infinite words over a finite alphabet Σ which is called the *prefix metric* and defined as follows. For $u, v \in \Sigma^\omega$ and $u \neq v$ let $\delta(u, v) = 2^{-l_{\mathrm{pref}(u,v)}}$ where $l_{\mathrm{pref}(u,v)}$ is the first integer n such that the $(n+1)^{st}$ letter of u is different from the $(n+1)^{st}$ letter of v. This metric induces on Σ^ω the usual Cantor topology for which *open subsets* of Σ^ω are in the form $W.\Sigma^\omega$, where $W \subseteq \Sigma^\star$. A set $L \subseteq \Sigma^\omega$ is a *closed set* iff its complement $\Sigma^\omega - L$ is an open set. Define now the *Borel Hierarchy* of subsets of Σ^ω:

Definition 4. *For a non-null countable ordinal α, the classes $\mathbf{\Sigma}^0_\alpha$ and $\mathbf{\Pi}^0_\alpha$ of the Borel Hierarchy on the topological space Σ^ω are defined as follows: $\mathbf{\Sigma}^0_1$ is the class of open subsets of Σ^ω, $\mathbf{\Pi}^0_1$ is the class of closed subsets of Σ^ω, and for any countable ordinal $\alpha \geq 2$:*

$\mathbf{\Sigma}^0_\alpha$ *is the class of countable unions of subsets of Σ^ω in $\bigcup_{\gamma < \alpha} \mathbf{\Pi}^0_\gamma$.*
$\mathbf{\Pi}^0_\alpha$ *is the class of countable intersections of subsets of Σ^ω in $\bigcup_{\gamma < \alpha} \mathbf{\Sigma}^0_\gamma$.*

For a countable ordinal α, a subset of Σ^ω is a Borel set of *rank* α iff it is in $\mathbf{\Sigma}^0_\alpha \cup \mathbf{\Pi}^0_\alpha$ but not in $\bigcup_{\gamma < \alpha}(\mathbf{\Sigma}^0_\gamma \cup \mathbf{\Pi}^0_\gamma)$.

There are also some subsets of Σ^ω which are not Borel. In particular the class of Borel subsets of Σ^ω is strictly included into the class $\mathbf{\Sigma}_1^1$ of *analytic sets* which are obtained by projection of Borel sets, see for example [Sta97, LT94, PP04, Kec95] for more details. The (lightface) class Σ_1^1 of *effective analytic sets* is the class of sets which are obtained by projection of arithmetical sets. It is well known that a set $L \subseteq \Sigma^\omega$, where Σ is a finite alphabet, is in the class Σ_1^1 iff it is accepted by a Turing machine with a Büchi or Muller acceptance condition [Sta97].

We now define completeness with regard to reduction by continuous functions. For a countable ordinal $\alpha \geq 1$, a set $F \subseteq \Sigma^\omega$ is said to be a $\mathbf{\Sigma}_\alpha^0$ (respectively, $\mathbf{\Pi}_\alpha^0$, $\mathbf{\Sigma}_1^1$)-*complete set* iff for any set $E \subseteq Y^\omega$ (with Y a finite alphabet): $E \in \mathbf{\Sigma}_\alpha^0$ (respectively, $E \in \mathbf{\Pi}_\alpha^0$, $E \in \mathbf{\Sigma}_1^1$) iff there exists a continuous function $f : Y^\omega \to \Sigma^\omega$ such that $E = f^{-1}(F)$. $\mathbf{\Sigma}_n^0$ (respectively $\mathbf{\Pi}_n^0$)-complete sets, with n an integer ≥ 1, are thoroughly characterized in [Sta86].

4   Topology and Infinitary Rational Relations

The first non-recursive ordinal, usually called the Church-Kleene ordinal, will be denoted below by ω_1^{CK}.

We have proved in [Fin05a, Fin05b] the following result.

Theorem 5. *For every non null countable ordinal $\alpha < \omega_1^{\mathrm{CK}}$, there exist some $\mathbf{\Sigma}_\alpha^0$-complete and some $\mathbf{\Pi}_\alpha^0$-complete ω-languages in the class* **r-BCL**$(1)_\omega$.

We are going to prove a similar result for the class RAT_ω, using a simulation of 1-counter automata.

We now first define a coding of an ω-word over a finite alphabet Σ by an ω-word over the alphabet $\Gamma = \Sigma \cup \{A\}$, where A is an additionnal letter not in Σ.

For $x \in \Sigma^\omega$ the ω-word $h(x)$ is defined by:

$$h(x) = A.0.x(1).A.0^2.x(2).A.0^3.x(3).A.0^4.x(4).A \ldots A.0^n.x(n).A.0^{n+1}.x(n+1).A \ldots$$

Then it is easy to see that the mapping h from Σ^ω into $(\Sigma \cup \{A\})^\omega$ is continuous and injective.

Lemma 6. *Let Σ be a finite alphabet and $\alpha \geq 2$ be a countable ordinal. If $L \subseteq \Sigma^\omega$ is $\mathbf{\Pi}_\alpha^0$-complete (respectively, $\mathbf{\Sigma}_\alpha^0$-complete) then*

$$h(L) \cup h(\Sigma^\omega)^-$$

is a $\mathbf{\Pi}_\alpha^0$-complete (respectively, $\mathbf{\Sigma}_\alpha^0$-complete) subset of $(\Sigma \cup \{A\})^\omega$.

Proof. Let L be a $\mathbf{\Pi}_\alpha^0$-complete (respectively, $\mathbf{\Sigma}_\alpha^0$-complete) subset of Σ^ω, for some countable ordinal $\alpha \geq 2$.

The topological space Σ^ω is compact thus its image by the continuous function h is also a compact subset of the topological space $(\Sigma \cup \{A\})^\omega$. The set $h(\Sigma^\omega)$ is compact hence it is a closed subset of $(\Sigma \cup \{A\})^\omega$ and its complement

$$(h(\Sigma^\omega))^- = (\Sigma \cup \{A\})^\omega - h(\Sigma^\omega)$$

is an open (i.e. a $\mathbf{\Sigma}_1^0$) subset of $(\Sigma \cup \{A\})^\omega$.

On the other side the function h is also injective thus it is a bijection from Σ^ω onto $h(\Sigma^\omega)$. But a continuous bijection between two compact sets is an homeomorphism therefore h induces an homeomorphism between Σ^ω and $h(\Sigma^\omega)$. By hypothesis L is a $\mathbf{\Pi}_\alpha^0$ (respectively, $\mathbf{\Sigma}_\alpha^0$)-subset of Σ^ω thus $h(L)$ is a $\mathbf{\Pi}_\alpha^0$ (respectively, $\mathbf{\Sigma}_\alpha^0$)-subset of $h(\Sigma^\omega)$ (where Borel sets of the topological space $h(\Sigma^\omega)$ are defined from open sets as in the case of the topological space Σ^ω).

The topological space $h(\Sigma^\omega)$ is a topological subspace of $(\Sigma \cup \{A\})^\omega$ and its topology is induced by the topology on $(\Sigma \cup \{A\})^\omega$: open sets of $h(\Sigma^\omega)$ are traces on $h(\Sigma^\omega)$ of open sets of $(\Sigma \cup \{A\})^\omega$ and the same result holds for closed sets. Then one can easily show by induction that for every ordinal $\beta \geq 1$, $\mathbf{\Pi}_\beta^0$-subsets (resp. $\mathbf{\Sigma}_\beta^0$-subsets) of $h(\Sigma^\omega)$ are traces on $h(\Sigma^\omega)$ of $\mathbf{\Pi}_\beta^0$-subsets (resp. $\mathbf{\Sigma}_\beta^0$-subsets) of $(\Sigma \cup \{A\})^\omega$, i.e. are intersections with $h(\Sigma^\omega)$ of $\mathbf{\Pi}_\beta^0$-subsets (resp. $\mathbf{\Sigma}_\beta^0$-subsets) of $(\Sigma \cup \{A\})^\omega$.

But $h(L)$ is a $\mathbf{\Pi}_\alpha^0$ (respectively, $\mathbf{\Sigma}_\alpha^0$)-subset of $h(\Sigma^\omega)$ hence there exists a $\mathbf{\Pi}_\alpha^0$ (respectively, $\mathbf{\Sigma}_\alpha^0$)-subset T of $(\Sigma \cup \{A\})^\omega$ such that $h(L) = T \cap h(\Sigma^\omega)$. But $h(\Sigma^\omega)$ is a closed i.e. $\mathbf{\Pi}_1^0$-subset (hence also a $\mathbf{\Pi}_\alpha^0$ (respectively, $\mathbf{\Sigma}_\alpha^0$)-subset) of $(\Sigma \cup \{A\})^\omega$ and the class of $\mathbf{\Pi}_\alpha^0$ (respectively, $\mathbf{\Sigma}_\alpha^0$)-subsets of $(\Sigma \cup \{A\})^\omega$ is closed under finite intersection thus $h(L)$ is a $\mathbf{\Pi}_\alpha^0$ (respectively, $\mathbf{\Sigma}_\alpha^0$)-subset of $(\Sigma \cup \{A\})^\omega$.

Now $h(L) \cup (h(\Sigma^\omega))^-$ is the union of a $\mathbf{\Pi}_\alpha^0$ (respectively, $\mathbf{\Sigma}_\alpha^0$)-subset and of a $\mathbf{\Sigma}_1^0$-subset of $(\Sigma \cup \{A\})^\omega$ therefore it is a $\mathbf{\Pi}_\alpha^0$ (respectively, $\mathbf{\Sigma}_\alpha^0$)-subset of $(\Sigma \cup \{A\})^\omega$ because the class of $\mathbf{\Pi}_\alpha^0$ (respectively, $\mathbf{\Sigma}_\alpha^0$)-subsets of $(\Sigma \cup \{A\})^\omega$ is closed under finite union.

In order to prove that $h(L) \cup (h(\Sigma^\omega))^-$ is $\mathbf{\Pi}_\alpha^0$ (respectively, $\mathbf{\Sigma}_\alpha^0$)-**complete** it suffices to remark that

$$L = h^{-1}[h(L) \cup (h(\Sigma^\omega))^-]$$

This implies that $h(L) \cup (h(\Sigma^\omega))^-$ is $\mathbf{\Pi}_\alpha^0$ (respectively, $\mathbf{\Sigma}_\alpha^0$)-complete because L is assumed to be $\mathbf{\Pi}_\alpha^0$ (respectively, $\mathbf{\Sigma}_\alpha^0$)-complete. □

Let now Σ be a finite alphabet such that $0 \in \Sigma$ and let α be the ω-word over the alphabet $\Sigma \cup \{A\}$ which is defined by:

$$\alpha = A.0.A.0^2.A.0^3.A.0^4.A.0^5.A \ldots A.0^n.A.0^{n+1}.A \ldots$$

We can now state the following Lemma.

Lemma 7. *Let Σ be a finite alphabet such that $0 \in \Sigma$, α be the ω-word over $\Sigma \cup \{A\}$ defined as above, and $L \subseteq \Sigma^\omega$ be in $\mathbf{r}$-$\mathbf{BCL}(1)_\omega$. Then there exists an infinitary rational relation $R_1 \subseteq (\Sigma \cup \{A\})^\omega \times (\Sigma \cup \{A\})^\omega$ such that:*

$$\forall x \in \Sigma^\omega \quad (x \in L) \text{ iff } ((h(x), \alpha) \in R_1)$$

Proof. Let Σ be a finite alphabet such that $0 \in \Sigma$, α be the ω-word over $\Sigma \cup \{A\}$ defined as above, and $L = L(\mathcal{A}) \subseteq \Sigma^\omega$, where $\mathcal{A} = (K, \Sigma, \Delta, q_0, F)$ is a 1-counter Büchi automaton.

We define now the relation R_1. A pair $y = (y_1, y_2)$ of ω-words over the alphabet $\Sigma \cup \{A\}$ is in R_1 if and only if it is in the form

$$y_1 = A.u_1.v_1.x(1).A.u_2.v_2.x(2).A.u_3.v_3.x(3).A \ldots A.u_n.v_n.x(n).A \ldots$$
$$y_2 = A.w_1.z_1.A.w_2.z_2.A.w_3.z_3.A \ldots A.w_n.z_n.A \ldots$$

where $|v_1| = 0$ and for all integers $i \geq 1$,

$$u_i, v_i, w_i, z_i \in 0^\star \text{ and } x(i) \in \Sigma \text{ and}$$

$$|u_{i+1}| = |z_i| + 1$$

and there is a sequence $(q_i)_{i \geq 0}$ of states of K such that for all integers $i \geq 1$:

$$x(i) : (q_{i-1}, |v_i|) \mapsto_{\mathcal{A}} (q_i, |w_i|)$$

Moreover some state $q_f \in F$ occurs infinitely often in the sequence $(q_i)_{i \geq 0}$.

Notice that the state q_0 of the sequence $(q_i)_{i \geq 0}$ is also the initial state of $\mathcal{A}$.

Let now $x \in \Sigma^\omega$ such that $(h(x), \alpha) \in R_1$. We are going to prove that $x \in L$.

By hypothesis $(h(x), \alpha) \in R_1$ thus there are finite words $u_i, v_i, w_i, z_i \in 0^\star$ such that $|v_1| = 0$ and for all integers $i \geq 1$, $|u_{i+1}| = |z_i| + 1$, and

$$h(x) = A.u_1.v_1.x(1).A.u_2.v_2.x(2).A.u_3.v_3.x(3).A \ldots A.u_n.v_n.x(n).A \ldots$$

$$\alpha = A.w_1.z_1.A.w_2.z_2.A.w_3.z_3.A \ldots A.w_n.z_n.A \ldots$$

Moreover there is a sequence $(q_i)_{i \geq 0}$ of states of K such that for all integers $i \geq 1$:

$$x(i) : (q_{i-1}, |v_i|) \mapsto_{\mathcal{A}} (q_i, |w_i|)$$

and some state $q_f \in F$ occurs infinitely often in the sequence $(q_i)_{i \geq 0}$.

On the other side we have:

$h(x) = A.0.x(1).A.0^2.x(2).A.0^3.x(3).A \ldots A.0^n.x(n).A.0^{n+1}.x(n+1).A \ldots$
$\alpha = A.0.A.0^2.A.0^3.A.0^4.A \ldots A.0^n.A \ldots$

So we have $|u_1.v_1| = 1$ and $|v_1| = 0$ and $x(1) : (q_0, |v_1|) \mapsto_{\mathcal{A}} (q_1, |w_1|)$. But $|w_1.z_1| = 1$, $|u_2.v_2| = 2$, and $|u_2| = |z_1| + 1$ thus $|v_2| = |w_1|$.

We are going to prove in a similar way that for all integers $i \geq 1$ it holds that $|v_{i+1}| = |w_i|$.

We know that $|w_i.z_i| = i$, $|u_{i+1}.v_{i+1}| = i + 1$, and $|u_{i+1}| = |z_i| + 1$ thus $|w_i| = |v_{i+1}|$.

Then for all $i \geq 1$, $x(i) : (q_{i-1}, |v_i|) \mapsto_{\mathcal{A}} (q_i, |v_{i+1}|)$.

So if we set $c_i = |v_i|$, $(q_{i-1}, c_i)_{i \geq 1}$ is an accepting run of $\mathcal{A}$ on x and this implies that $x \in L$.

Conversely it is easy to prove that if $x \in L$ then $(h(x), \alpha)$ may be written in the form of $(y_1, y_2) \in R_1$.

It remains to prove that the above defined relation R_1 is an infinitary rational relation. It is easy to find a 2-tape Büchi automaton $\mathcal{T}$ accepting the infinitary rational relation R_1.

Lemma 8. *The set*

$$R_2 = (\Sigma \cup \{A\})^\omega \times (\Sigma \cup \{A\})^\omega - (h(\Sigma^\omega) \times \{\alpha\})$$

is an infinitary rational relation.

Proof. By definition of the mapping h, we know that a pair of ω-words over the alphabet $(\Sigma \cup \{A\})$ is in $h(\Sigma^\omega) \times \{\alpha\}$ iff it is in the form (σ_1, σ_2), where
$\sigma_1 = A.0.x(1).A.0^2.x(2).A.0^3.x(3).A\ldots.A.0^n.x(n).A.0^{n+1}.x(n+1).A\ldots$
$\sigma_2 = \alpha = A.0.A.0^2.A.0^3.A\ldots A.0^n.A.0^{n+1}.A\ldots$

where for all integers $i \geq 1$, $x(i) \in \Sigma$.

So it is easy to see that $(\Sigma \cup \{A\})^\omega \times (\Sigma \cup \{A\})^\omega - (h(\Sigma^\omega) \times \{\alpha\})$ is the union of the sets $\mathcal{C}_j$ where:

- $\mathcal{C}_1$ is formed by pairs (σ_1, σ_2) where
  σ_1 has not any initial segment in $A.\Sigma^2.A.\Sigma^3.A$, or σ_2 has not any initial segment in $A.\Sigma.A.\Sigma^2.A$.
- $\mathcal{C}_2$ is formed by pairs (σ_1, σ_2) where
  $\sigma_2 \notin (A.0^+)^\omega$, or $\sigma_1 \notin (A.0^+.\Sigma)^\omega$.
- $\mathcal{C}_3$ is formed by pairs (σ_1, σ_2) where
  $\sigma_1 = A.w_1.A.w_2.A.w_3.A\ldots A.w_n.A.u.A.z_1$
  $\sigma_2 = A.w_1'.A.w_2'.A.w_3'.A\ldots A.w_n'.A.v.A.z_2$

  where n is an integer ≥ 1, for all $i \leq n$ $w_i, w_i' \in \Sigma^\star$, $z_1, z_2 \in (\Sigma \cup \{A\})^\omega$ and

  $$u, v \in \Sigma^\star \text{ and } |u| \neq |v| + 1$$

- $\mathcal{C}_4$ is formed by pairs (σ_1, σ_2) where
  $\sigma_1 = A.w_1.A.w_2.A.w_3.A.w_4\ldots A.w_n.A.w_{n+1}.A.v.A.z_1$
  $\sigma_2 = A.w_1'.A.w_2'.A.w_3'.A.w_4'\ldots A.w_n'.A.u.A.z_2$

  where n is an integer ≥ 1, for all $i \leq n$ $w_i, w_i' \in \Sigma^\star$, $w_{n+1} \in \Sigma^\star$, $z_1, z_2 \in (\Sigma \cup \{A\})^\omega$ and
  $$u, v \in \Sigma^\star \text{ and } |v| \neq |u| + 2$$

Each set $\mathcal{C}_j$, $1 \leq j \leq 4$, is easily seen to be an infinitary rational relation $\subseteq (\Sigma \cup \{A\})^\omega \times (\Sigma \cup \{A\})^\omega$ (the detailed proof is left to the reader). The class RAT_ω is closed under finite union thus

$$R_2 = (\Sigma \cup \{A\})^\omega \times (\Sigma \cup \{A\})^\omega - (h(\Sigma^\omega) \times \{\alpha\}) = \bigcup_{1 \leq j \leq 4} \mathcal{C}_j$$

is an infinitary rational relation. □

We can now state the following result:

Theorem 9. *For every non null countable ordinal $\gamma < \omega_1^{CK}$, there exists some $\mathbf{\Sigma}_\gamma^0$-complete and some $\mathbf{\Pi}_\gamma^0$-complete infinitary rational relations in the class RAT_ω.*

Proof. For $\gamma = 1$ (and even $\gamma = 2$) the result is already true for regular ω-languages.

Let then $\gamma \geq 2$ be a countable non null recursive ordinal and $L = L(\mathcal{A}) \subseteq \Sigma^\omega$ be a $\mathbf{\Pi}_\gamma^0$-complete (respectively, $\mathbf{\Sigma}_\gamma^0$-complete) ω-language accepted by a (real time) Büchi 1-counter automaton $\mathcal{A}$.

Let $\Gamma = \Sigma \cup \{A\}$ and $R_1 \subseteq \Gamma^\omega \times \Gamma^\omega$ be the infinitary rational relation constructed from $L(\mathcal{A})$ as in the proof of Lemma 7 and let

$$R = R_1 \cup R_2 \subseteq \Gamma^\omega \times \Gamma^\omega$$

The class RAT_ω is closed under finite union therefore R is an infinitary rational relation.

Lemma 7 and the definition of R_2 imply that $R_\alpha = \{\sigma \in \Gamma^\omega \mid (\sigma, \alpha) \in R\}$ is equal to the set $\mathcal{L} = h(L) \cup (h(\Sigma^\omega))^-$ which is a $\mathbf{\Pi}_\gamma^0$-complete (respectively, $\mathbf{\Sigma}_\gamma^0$-complete) subset of $(\Sigma \cup \{A\})^\omega$ by Lemma 6.

Moreover, for all $u \in \Gamma^\omega - \{\alpha\}$, $R_u = \{\sigma \in \Gamma^\omega \mid (\sigma, u) \in R\} = \Gamma^\omega$ holds by definition of R_2.

In order to prove that R is a $\mathbf{\Pi}_\gamma^0$ (respectively, $\mathbf{\Sigma}_\gamma^0$)-complete set remark first that R may be written as the union:

$$R = \mathcal{L} \times \{\alpha\} \;\bigcup\; \Gamma^\omega \times (\Gamma^\omega - \{\alpha\})$$

We already know that $\mathcal{L}$ is a $\mathbf{\Pi}_\gamma^0$ (respectively, $\mathbf{\Sigma}_\gamma^0$)-complete subset of $(\Sigma \cup \{A\})^\omega$. Then it is easy to show that $\mathcal{L} \times \{\alpha\}$ is also a $\mathbf{\Pi}_\gamma^0$ (respectively, $\mathbf{\Sigma}_\gamma^0$)-subset of $(\Sigma \cup \{A\})^\omega \times (\Sigma \cup \{A\})^\omega$. On the other side it is easy to see that $\Gamma^\omega \times (\Gamma^\omega - \{\alpha\})$ is an open subset of $\Gamma^\omega \times \Gamma^\omega$. Thus R is a $\mathbf{\Pi}_\gamma^0$ (respectively, $\mathbf{\Sigma}_\gamma^0$)-set because the Borel class $\mathbf{\Pi}_\gamma^0$ (respectively, $\mathbf{\Sigma}_\gamma^0$) is closed under finite union.

Moreover let $g : \Sigma^\omega \to (\Sigma \cup \{A\})^\omega \times (\Sigma \cup \{A\})^\omega$ be the function defined by:

$$\forall x \in \Sigma^\omega \qquad g(x) = (h(x), \alpha)$$

It is easy to see that g is continuous because h is continuous. By construction it turns out that for all ω-words $x \in \Sigma^\omega$, $(x \in L)$ iff $((h(x), \alpha) \in R)$ iff $(g(x) \in R)$. This means that $g^{-1}(R) = L$. This implies that R is $\mathbf{\Pi}_\gamma^0$ (respectively, $\mathbf{\Sigma}_\gamma^0$)-complete because L is $\mathbf{\Pi}_\gamma^0$ (respectively, $\mathbf{\Sigma}_\gamma^0$)-complete. $\qquad\square$

Remark 10. *The structure of the infinitary rational relation R can be described very simply by the sections R_u, $u \in \Gamma^\omega$. All sections but one are equal to Γ^ω, so they have the lowest topological complexity and exactly one section (R_α) is a $\mathbf{\Pi}_\gamma^0$ (respectively, $\mathbf{\Sigma}_\gamma^0$)-complete subset of Γ^ω.*

5 Concluding Remarks

The Wadge hierarchy is a great refinement of the Borel hierarchy and we have proved in [Fin05a, Fin05b] that the Wadge hierarchy of the class $\mathbf{r}\text{-}\mathbf{BCL}(1)_\omega$ is equal to the Wadge hierarchy of the class of ω-languages accepted by Büchi Turing machines. Using the above coding and similar reasoning as in [Fin05b], we can easily infer that the Wadge hierarchy of the class RAT_ω and the Wadge hierarchy of the class $\mathbf{r}\text{-}\mathbf{BCL}(1)_\omega$ are equal. Thus the Wadge hierarchy of the class RAT_ω is also the Wadge hierarchy of the (lightface) class Σ_1^1 of ω-languages accepted by Turing machines with a Büchi acceptance condition. In particular their Borel hierarchies are also equal.

We have to indicate here a mistake in [Fin05a]. We wrote in that paper that it is well known that if $L \subseteq \Sigma^\omega$ is a Σ_1^1 set (i.e. accepted by a Turing machine with a Büchi acceptance condition), and is a Borel set of rank α, then α is smaller than ω_1^{CK}. This fact, which is true if we replace Σ_1^1 by Δ_1^1, seemed to us an obvious fact, and was accepted by many people as true, but it is actually not true. Kechris, Marker and Sami proved in [KMS89] that the supremum of the set of Borel ranks of (lightface) Π_1^1, so also of (lightface) Σ_1^1, sets is the ordinal γ_2^1. This ordinal is defined in [KMS89] and it is proved to be strictly greater than the ordinal δ_2^1 which is the first non Δ_2^1 ordinal. Thus it holds that $\omega_1^{CK} < \gamma_2^1$.

The ordinal γ_2^1 is also the supremum of the set of Borel ranks of ω-languages in the class $\mathbf{r}\text{-}\mathbf{BCL}(1)_\omega$ or in the class RAT_ω. Notice however that it is not proved in [KMS89] that every non null ordinal $\gamma < \gamma_2^1$ is the Borel rank of a (lightface) Π_1^1 (or Σ_1^1) set, while it is known that every ordinal $\gamma < \omega_1^{CK}$ is the Borel rank of a (lightface) Δ_1^1 set. The situation is then much more complicated than it could be expected. More details will be given in the full versions of [Fin05a] and of this paper.

Acknowledgements. Thanks to the anonymous referees for useful comments on a preliminary version of this paper.

References

[BT70]   Ya M. Barzdin and B.A. Trakhtenbrot, Finite Automata, Behaviour and Synthesis, Nauka, Moscow, 1970 (English translation, North Holland, Amsterdam, 1973).

[BC00]   M.-P. Bal and O. Carton, Determinization of Transducers over Infinite Words, in ICALP'2000 (U. Montanari et al., eds.), vol. 1853 of Lect. Notes in Comput. Sci., pp. 561-570, 2000.

[BCPS00] M.-P. Béal , O. Carton, C. Prieur and J. Sakarovitch, Squaring Transducers: An Efficient Procedure for Deciding Functionality and Sequentiality, Theoretical Computer Science, vol. 292, no. 1, pp. 45-63, 2003.

[Ber79]   J. Berstel, Transductions and Context Free Languages, Teubner Verlag, 1979.

[Büc62]   J.R. Büchi, On a Decision Method in Restricted Second Order Arithmetic, Logic Methodology and Philosophy of Science, (Proc. 1960 Int. Congr.), Stanford University Press, 1962, 1-11.

[CDT02]   T. Cachat, J. Duparc and W. Thomas, Solving Pushdown Games with a Σ_3 Winning Condition, proceedings of CSL 2002, Lecture Notes in Computer Science, Springer, Volume 2471, pp. 322-336.

[Cho77]   C. Choffrut, Une Caractérisation des Fonctions Séquentielles et des Fonctions Sous-Séquentielles en tant que Relations Rationnelles, Theoretical Computer Science, Volume 5, 1977, p.325-338.

[CG99]   C. Choffrut and S. Grigorieff, Uniformization of Rational Relations, Jewels are Forever 1999, J. Karhumki, H. Maurer, G. Paun and G. Rozenberg editors, Springer, p.59-71.

[EH93]   J. Engelfriet and H. J. Hoogeboom, X-automata on ω-Words, Theoretical Computer Science, Volume 110, (1993) 1, 1-51.

[Fin03a]   O. Finkel, On the Topological Complexity of Infinitary Rational Relations, RAIRO-Theoretical Informatics and Applications, Volume 37 (2), 2003, p. 105-113.

[Fin03b]   O. Finkel, Undecidability of Topological and Arithmetical Properties of Infinitary Rational Relations, RAIRO-Theoretical Informatics and Applications, Volume 37 (2), 2003, p. 115-126.

[Fin03c]   O. Finkel, On Infinitary Rational Relations and Borel Sets, in the Proceedings of the Fourth International Conference on Discrete Mathematics and Theoretical Computer Science DMTCS'03, 7 - 12 July 2003, Dijon, France, Lecture Notes in Computer Science, Springer, Volume 2731, p. 155-167.

[Fin03d]   O. Finkel, On Infinitary Rational Relations and the Borel Hierarchy, submitted to Logical Methods in Computer Science, 27 pages.

[Fin05a]   O. Finkel, Borel Ranks and Wadge Degrees of Context Free ω-Languages, in the Proceedings of the First Conference on Computability in Europe: New Computational Paradigms, CiE 2005, Amsterdam, The Netherlands, Lecture Notes in Computer Science, Volume 3526, Springer, 2005, p. 129-138.

[Fin05b]   O. Finkel, Borel Ranks and Wadge Degrees of Context Free ω-Languages, long version of [Fin05a], Mathematical Structures in Computer Science, Special Issue on Mathematics of Computation at CIE 2005, to appear.

[Gir81]   F. Gire, Relations Rationnelles Infinitaires, Thèse de troisième cycle, Université Paris 7, Septembre 1981.

[Gir83]   F. Gire, Une Extension aux Mots Infinis de la Notion de Transduction Rationnelle, 6th GI Conf., Lect. Notes in Comp. Sci., Volume 145, 1983, p. 123-139.

[GN84]   F. Gire and M. Nivat, Relations Rationnelles Infinitaires, Calcolo, Volume XXI, 1984, p. 91-125.

[Kec95]   A. S. Kechris, Classical Descriptive Set Theory, Springer-Verlag, 1995.

[KMS89]   A. S. Kechris, D. Marker, and R. L. Sami, Π_1^1 Borel Sets, The Journal of Symbolic Logic, Volume 54 (3), 1989, p. 915-920.

[Kur66]   K. Kuratowski, Topology, Academic Press, New York 1966.

[Lan69]   L. H. Landweber, Decision Problems for ω-Automata, Math. Syst. Theory 3 (1969) 4,376-384.

[LT94]   H. Lescow and W. Thomas, Logical Specifications of Infinite Computations, In: "A Decade of Concurrency" (J. W. de Bakker et al., eds), Lecture Notes in Computer Science, Springer, Volume 803 (1994), 583-621.

[LS77]   R. Lindner and L. Staiger, Algebraische Codierungstheorie - Theorie der Sequentiellen Codierungen, Akademie-Verlag, Berlin, 1977.

[Mos80]   Y. N. Moschovakis, Descriptive Set Theory, North-Holland, Amsterdam 1980.

[PP04]   D. Perrin and J.-E. Pin, Infinite Words, Automata, Semigroups, Logic and Games, Volume 141 of Pure and Applied Mathematics, Elsevier, 2004.

[Pin96]   J-E. Pin, Logic, Semigroups and Automata on Words, Annals of Mathematics and Artificial Intelligence 16 (1996), p. 343-384.

[Pri00]   C. Prieur, Fonctions Rationnelles de Mots Infinis et Continuité, Thèse de Doctorat, Université Paris 7, Octobre 2000.

[Sim92]   P. Simonnet, Automates et Théorie Descriptive, Thèse de Doctorat, Université Paris 7, March 1992.

[Sta86]   L. Staiger, Hierarchies of Recursive ω-Languages, Jour. Inform. Process. Cybernetics EIK 22 (1986) 5/6, 219-241.

[Sta97]   L. Staiger, ω-Languages, Chapter of the Handbook of Formal languages, Vol 3, edited by G. Rozenberg and A. Salomaa, Springer-Verlag, Berlin.

[SW78]   L. Staiger and K. Wagner, Rekursive Folgenmengen I, Z. Math Logik Grundlag. Math. 24, 1978, 523-538.

[Tho89]   W. Thomas, Automata and Quantifier Hierarchies, in: Formal Properties of Finite automata and Applications, Ramatuelle, 1988, Lecture Notes in Computer Science 386, Springer, Berlin, 1989, p.104-119.

[Tho90]   W. Thomas, Automata on Infinite Objects, in: J. Van Leeuwen, ed., Handbook of Theoretical Computer Science, Vol. B (Elsevier, Amsterdam, 1990), p. 133-191.

Weighted Picture Automata and Weighted Logics[*]

Ina Mäurer

Institut für Informatik, Universität Leipzig,
Augustusplatz 10-11, D-04109 Leipzig, Germany
maeurer@informatik.uni-leipzig.de

Abstract. The theory of two-dimensional languages, generalizing formal string languages, was motivated by problems arising from image processing and models of parallel computing. Weighted automata and series over pictures map pictures to some semiring and provide an extension to a quantitative setting. We establish a notion of a weighted MSO logics over pictures. The semantics of a weighted formula will be a picture series. We introduce weighted 2-dimensional online tessellation automata (W2OTA) extending the common automata-theoretic model for picture languages. We prove that the class of picture series defined by sentences of the weighted logics coincides with the family of picture series that are computable by W2OTA. Moreover, behaviours of W2OTA coincide precisely with the recognizable picture series characterized in [18].

1 Introduction

In the literature, a variety of formal models to recognize or generate two-dimensional arrays of symbols, called pictures, have been proposed [2,11,13,15]. This research was motivated by problems arising from the area of image processing and pattern recognition [8,19], and also plays a role in frameworks concerning cellular automata and other models of parallel computing. Different authors obtained an equivalence theorem for picture languages describing languages in terms of types of automata, sets of tiles, rational operations or existential monadic second-order (MSO) logic [10,12,13,15]. New notions of weighted recognizability for picture languages defined by weighted picture automata and picture series were introduced in [3]. The weights are taken from some commutative semiring. In [18], we showed that the family of behaviours of such weighted picture automata coincides with the class of projections of certain rational picture series and can be characterized also by using tiling and domino systems. These equivalent weighted picture devices can be used to model several application examples.

Recently, Droste and Gastin [4] introduced the framework of a weighted logic over words and characterized recognizable formal power series, computed by weighted finite automata, as semantics of monadic second-order sentences within their logic. Here, we will establish a weighted MSO logic for pictures. The semantics of a weighted sentence will be a picture series that maps pictures over the underlying alphabet to elements of a commutative semiring. We also introduce weighted 2-dimensional online tessellation automata (W2OTA). This model extends the known notion of 2-dimensional online tessellation automata (2OTA) [13] for picture languages and is equivalent to the concept

---

[*] Supported by the GK 446/3 of the German Research Foundation.

B. Durand and W. Thomas (Eds.): STACS 2006, LNCS 3884, pp. 313–324, 2006.
© Springer-Verlag Berlin Heidelberg 2006

of recognizability in [18,3]. Our main result proves that for an alphabet and any commutative semiring the family of picture series computable by W2OTA coincides with the family of series that are definable by weighted monadic second-order sentences.

For the syntax, we basically follow classical logic. But additionally, similar to [4], we also let elements of the semiring be atomic formulas, hence are able to formulate quantitative properties of picture languages: imagine for instance the number of a's occurring in a picture. The other atomic formulas will have a semantics with values in $\{0, 1\}$. Problems arise when negation is applied, because it is not clear how to define the semantics of a negated formula. Therefore, we apply negation only to atomic formulas. Universal quantification does not preserve recognizability. Hence, as in [4], we disallow universal set quantification, but here we restrict universal first-order (FO) quantification in a new way to particular formulas, since not every recognizable picture language is determinizable and the proof in [4] does not work for two dimensions.

Crucial for proving the main theorem is to show, that the universal FO quantification of a formula, with restricted semantics, defines a recognizable series. Unlike to the word-case, we build a formula instead of constructing a certain (unweighted) automaton. This formula defines a picture language which is computable by a 2OTA that is unambiguous. Also, we use successor relations instead of built-in relations $\leq_v$ and $\leq_h$, since there are $(\leq_v, \leq_h)$-definable picture languages that are not recognizable [16]. Using successor relations, in contrast to words, not every (unweighted) FO-formulas can be made unambiguous.

Considering (unweighted) logic, our restriction of the formulas is not an essential restriction, since every (unweighted) existential MSO-formula is equivalent (in the sense of defining identical languages) to a formula in which negation is only applied to atomic formulas, and since every recognizable picture language is definable by a restricted formula. We obtain the corresponding classical equivalence when restricting to the Boolean semiring. The main result of the paper indicates that the notion of weighted recognizability for picture series is robust, since it can be characterized in terms of a logic and different automata-theoretic devices and generalizes the common frameworks for picture languages.

2   Pictures and EMSO-Logic

We recall notions and results of two-dimensional languages and MSO-logic over pictures. We assume the reader is familiar with principles in MSO logic and the equivalence theorem for picture languages [11,12,13,21].

Let $\mathbb{N} = \{0, 1, \ldots\}$ and Σ and Γ be finite alphabets. A *picture* over Σ is a non-empty rectangular array of elements in Σ.[1] A *picture language* is a set of pictures. The set of all pictures over Σ is denoted by Σ^{++}. Let $p \in \Sigma^{++}$. We write $p(i, j)$ or $p_{i,j}$ for the component of p at position (i, j) and let $l_v(p)$ $(l_h(p))$ be the number of rows (columns) of p (v stands for vertical, h for horizontal). The pair $(l_v(p), l_h(p))$ is the *size* of p. The set $\Sigma^{m \times n}$ comprises all pictures with size (m, n). The domain of p is $\text{Dom}(p) = \{1, \ldots, l_v(p)\} \times \{1, \ldots, l_h(p)\}$. A mapping $\pi : \Gamma \to \Sigma$ is called *projection*. It can be extended pointwise to pictures and languages as usual.

---

[1] We assume a picture to be non-empty for technical simplicity, as in [2,13,15].

We fix an alphabet Σ. In the literature, there are many equivalent models defining or recognizing picture languages [9,10,11,13,15]. These devices define *recognizable* picture languages and form the class $\text{Rec}(\Sigma^{++})$. The set $\text{MSO}(\Sigma^{++})$ of MSO-formulas over Σ is defined recursively by

$$\varphi ::= P_a(x) \mid xS_vy \mid xS_hy \mid x \in X \mid x = y \mid \varphi \vee \psi \mid \varphi \wedge \psi \mid \neg\varphi$$
$$\mid \exists x.\varphi \mid \exists X.\varphi \mid \forall x.\varphi \mid \forall X.\varphi$$

where $a \in \Sigma$, x, y are FO variables and X is a second-order variable. A picture p is represented by the relational structure $(\text{Dom}(p), S_v, S_h, (R_a)_{a \in \Sigma})$ where $R_a = \{(i, j) \in \text{Dom}(p) \mid p(i,j) = a\}$, $(a \in \Sigma)$ and S_v, S_h are the two successor relations of both directions: $(i,j)S_v(i+1,j), (i,j)S_h(i,j+1)$. Formulas containing no set quantification are collected in $\text{FO}(\Sigma^{++})$. We denote by $\text{EMSO}(\Sigma^{++})$ the set of formulas of the form $\exists X_1, \ldots \exists X_n.\psi$ such that $\psi \in \text{FO}(\Sigma^{++})$. Languages definable by formulas in $Z \subseteq \text{MSO}(\Sigma^{++})$ form the set $\mathcal{L}(Z)$.

For a finite set $\mathcal{V}$ of variables, a $(\mathcal{V}, p)$-assignment σ maps FO variables in $\mathcal{V}$ to elements of $\text{Dom}(p)$ and second-order variables in $\mathcal{V}$ to subsets of $\text{Dom}(p)$. If x is a FO variable and $(i,j) \in \text{Dom}(p)$ then $\sigma[x \to (i,j)]$ coincides with σ on $\mathcal{V} \setminus \{x\}$ and assigns (i,j) to x (similarly $\sigma[X \to I]$ for $I \subseteq \text{Dom}(p)$). We encode (p, σ) where σ is a $(\mathcal{V}, p)$-assignment as a picture over $\Sigma_{\mathcal{V}} = \Sigma \times \{0,1\}^{\mathcal{V}}$. Conversely, an element in $\Sigma_{\mathcal{V}}^{++}$ is a pair (p, σ) where p is the projection over Σ and σ is the projection over $\{0,1\}^{\mathcal{V}}$. Then σ represents a *valid* assignment over $\mathcal{V}$ if for each FO variable $x \in \mathcal{V}$, the projection of σ to the x-coordinate contains exactly one 1. In this case, we identify σ with the $(\mathcal{V}, p)$-assignment. Let $N_{\mathcal{V}} \subseteq \Sigma_{\mathcal{V}}^{++}$ comprise $\{(p, \sigma) \mid \sigma \text{ is valid}\}$. Clearly, $N_{\mathcal{V}}$ is a recognizable picture language. We write $\text{Free}(\varphi)$ for the set of all free variables in φ and $N_{\varphi} = N_{\text{Free}(\varphi)}$. If $\mathcal{V}$ contains $\text{Free}(\varphi)$, the definition that (p, σ) *satisfies* φ, i.e. $(p, \sigma) \models \varphi$ is as usual and we let $\mathcal{L}_{\mathcal{V}}(\varphi) = \{(p, \sigma) \in N_{\mathcal{V}} \mid (p, \sigma) \models \varphi\}$. We say that the formula φ *defines* the picture language $\mathcal{L}_{\text{Free}(\varphi)}(\varphi) =: \mathcal{L}(\varphi)$.

Proposition 2.1 ([12]). *A language is* EMSO*-definable iff it is recognizable.*

The aim of this paper is to generalize this result to a quantitative setting. For this, we will define weighted 2-dimensional online tessellation automata (W2OTA). The weights are taken from a commutative semiring.

3   Weighted Automata over Pictures

A *semiring* $(K, +, \cdot, 0, 1)$ is a structure K such that $(K, +, 0)$ is a commutative monoid, $(K, \cdot, 1)$ is a monoid, multiplication distributes over addition, and $x \cdot 0 = 0 = 0 \cdot x$ for all $x \in K$. If multiplication is commutative, K is called *commutative*. Examples of semirings useful to model problems in operations research and carrying quantitative properties for many devices include e.g. the *Boolean* semiring $\mathbb{B} = (\{0,1\}, \vee, \wedge, 0, 1)$, the natural numbers $\mathbb{N} = (\mathbb{N}, +, \cdot, 0, 1)$, the *tropical* semiring $\mathbb{T} = (\mathbb{R} \cup \{\infty\}, \min, +, \infty, 0)$, the *arctical (or max-plus)* semiring $\text{Arc} = (\mathbb{N} \cup \{-\infty\}, \max, +, -\infty, 0)$, the language-semiring $(\mathscr{P}(\Sigma^*), \cup, \cap, \emptyset, \Sigma^*)$ and $([0,1], \max, \cdot, 0, 1)$ (to capture probabilities).

We will now assign weights to pictures. This provides a generalization of the theory of picture languages to formal power series over pictures, cf. [18] and [1,7,14,20]. Examples are given below. Subsequently, K will always denote a commutative semiring and Σ, Δ, Γ are alphabets.

A *picture series* is a mapping $S : \Sigma^{++} \to K$. We let $K\langle\langle\Sigma^{++}\rangle\rangle$ comprise all picture series. We write (S,p) for $S(p)$, then a series S often is written as a formal sum $S = \sum_{p \in \Sigma^{++}} (S,p) \cdot p$. The set $\text{supp}(S) = \{p \in \Sigma^{++} \mid (S,p) \neq 0\}$ is the *support* of S. For a language $L \subseteq \Sigma^{++}$, the *characteristic series* $\mathbb{1}_L : \Sigma^{++} \to K$ is defined by $(\mathbb{1}_L, p) = 1$ if $p \in L$, and $(\mathbb{1}_L, p) = 0$ otherwise. For $K = \mathbb{B}$, the mapping $L \mapsto \mathbb{1}_L$ gives a natural bijection between languages over Σ and series in $\mathbb{B}\langle\langle\Sigma^{++}\rangle\rangle$.

Definition 3.1. *A weighted 2-dimensional online tessellation automaton over Σ is a tuple $\mathfrak{A} = (\Sigma, Q, I, F, E)$, consisting of a finite set Q of states, sets of initial and final states $I, F \subseteq Q$, respectively, and a finite set of transitions $E \subseteq Q \times Q \times \Sigma \times K \times Q$.*

For $r = (q_v, q_h, a, k, q) \in E$, we set $\sigma_v(r) = q_v, \sigma_h(r) = q_h, \sigma(r) = q, \text{label}(r) = a, weight(r) = k$, and, extending this to pictures, get a function $\text{label} : E^{++} \to \Sigma^{++}$. A *run* (or *computation*) in $\mathfrak{A}$ is an element in $E^{m \times n}$ satisfying natural compatibility properties, more precisely, for $c = (c_{i,j}) \in E^{m \times n}$ we have

$$\forall\, 1 \leq i \leq m, 1 \leq j \leq n :\ \sigma_v(c_{i,j}) = \sigma(c_{i-1,j}), \sigma_h(c_{i,j}) = \sigma(c_{i,j-1}).$$

We put $weight(c) = \prod_{i,j} weight(c_{i,j})$. A run c in $\mathfrak{A}$ is *successful* if for all $1 \leq i \leq m$ and $1 \leq j \leq n$, we have $\sigma_v(c_{1,j}), \sigma_h(c_{i,1}) \in I$ and $\sigma(c_{m,n}) \in F$. The set of all successful runs labelled with p is denoted by $I \overset{p}{\leadsto} F$. The automaton $\mathfrak{A}$ *computes* (or *recognizes*) the picture series $\|\mathfrak{A}\| : \Sigma^{++} \to K$, defined for a picture $p \in \Sigma^{++}$, as $(\|\mathfrak{A}\|, p) = \sum_{c \in I \overset{p}{\leadsto} F} weight(c)$. We call $\|\mathfrak{A}\|$ the *behaviour* of $\mathfrak{A}$ and write $K^{rec}\langle\langle\Sigma^{++}\rangle\rangle$ for the family of series that are computable by W2OTA over Σ.

Considering the unweighted case, instead of E, one could also define a *transition function* $\delta : Q \times Q \times \Sigma \to 2^Q$. If $|I| = 1$ and $\delta : Q \times Q \times \Sigma \to Q$, we call $\mathfrak{A}$ *deterministic*. W2OTA generalize in a straightforward way the automata-theoretic recognizability of 2OTA for picture languages.

For motivation, we now give two examples of functions $S : \Sigma^{++} \to \mathbb{R} \cup \{\infty\}$ and $T : \Sigma^{++} \to \mathbb{N}$.

Example 3.2. Let $D \subset [0,1]$ be a finite set of discrete values and let $L \subseteq D^{++}$ be recognizable. Consider $S : D^{++} \to \mathbb{R} \cup \{\infty\}$, mapping p to $S(p) = \sum_{i,j} p_{i,j}$ if $p \in L$ and to ∞ otherwise. One could interpret the values in D as different levels of gray [6]. Then, for each picture $p \in L$, the series S provides the total value $S(p)$ of light of p.

Example 3.3. Let C be a finite set of colors and consider $T : C^{++} \to \mathbb{N}$, defined by $(T, p) = \max\{l_v(q) \cdot l_h(q) \mid q \text{ is a monochrome subpicture of } p\}, (p \in C^{++})$. Then $T(p)$ gives the largest size of a monochrome rectangle, contained in p.

One can prove that the functions S and T are computable by W2OTA, more precisely $S \in \mathbb{T}^{rec}\langle\langle D^{++}\rangle\rangle$ and $T \in \text{Arc}^{rec}\langle\langle C^{++}\rangle\rangle$.

We define *rational* operations $\oplus$ and $\odot$, referred to as *sum* and *Hadamard product*, and also *scalar multiplications*, in the following way. For $S, T \in K\langle\langle\Sigma^{++}\rangle\rangle, k \in K$

and $p \in \Sigma^{++}$, we set $(S \oplus T, p) := (S, p) + (T, p)$, $(S \odot T, p) := (S, p) \cdot (T, p)$ and $(k \cdot S, p) := k \cdot (S, p)$. Extending projections and inverse projections to series, for $\pi : \Gamma \to \Sigma$, $R \in K\langle\langle \Gamma^{++} \rangle\rangle$ and $q \in \Gamma^{++}$, we set $(\pi(R), p) := \sum_{\pi(p')=p}(R, p')$ and $(\pi^{-1}(S), q) := (S, \pi(q))$. Then, $\pi(R) \in K\langle\langle \Sigma^{++} \rangle\rangle$ and $\pi^{-1}(S) \in K\langle\langle \Gamma^{++} \rangle\rangle$.

Now, similar to common constructions and using ideas in [3], we can prove

Proposition 3.4. *Recognizable picture series are closed under $\odot, \oplus, \cdot$, projections and inverse projections. For languages, inverse projections preserve deterministic devices. If L is deterministically recognizable then $\mathbb{1}_L$ is recognizable.*

4 Weighted Logics

In this section we introduce the syntax and semantics of the weighted MSO-logic on pictures. We fix K and Σ. For $a \in \Sigma$, P_a denotes a unary predicate symbol. Formulas of the *weighted MSO-logic* are defined recursively as follows:

$$\varphi ::= k \mid P_a(x) \mid \neg P_a(x) \mid xS_vy \mid \neg(xS_vy) \mid xS_hy \mid \neg(xS_hy) \mid x \in X \mid \neg(x \in X)$$
$$\mid x = y \mid \neg(x = y) \mid \varphi \vee \psi \mid \varphi \wedge \psi \mid \exists x.\varphi \mid \exists X.\varphi \mid \forall x.\varphi \mid \forall X.\varphi$$

where $k \in K, a \in \Sigma$ and x, y (resp. X) are first (resp. second)-order variables. The class $\mathrm{MSO}(K, \Sigma)$ comprises all such weighted MSO-formulas φ. The formulas $k, P_a(x), xS_vy, xS_hy$ and $x = y$ are referred to as *atomic formulas*. Subsequently, we will also consider the class $\mathrm{FO}(K, \Sigma) \subset \mathrm{MSO}(K, \Sigma)$ of all formulas having no set quantification. Clearly, formulas in $\mathrm{MSO}(K, \Sigma)$, containing no fragment of the form k, may also be regarded as unweighted formula defining a language $\mathcal{L}(\varphi)$. Now, similar to [4] we give the semantics of weighted MSO-formulas φ.

Definition 4.1. *Let $\varphi \in \mathrm{MSO}(K, \Sigma)$ and $\mathcal{V}$ be a finite set of variables containing Free(φ). The semantics of φ is a series $\llbracket \varphi \rrbracket_\mathcal{V} : \Sigma_\mathcal{V}^{++} \to K$. Let $(p, \sigma) \in \Sigma_\mathcal{V}^{++}$. If σ is not a valid $\mathcal{V}$-assignmen, then we set $\llbracket \varphi \rrbracket_\mathcal{V}(p, \sigma) = 0$. Otherwise, we define $\llbracket \varphi \rrbracket_\mathcal{V}(p, \sigma) \in K$ inductively as:*

$$\llbracket k \rrbracket_\mathcal{V}(p, \sigma) = k \qquad \llbracket P_a(x) \rrbracket_\mathcal{V}(p, \sigma) = \begin{cases} 1 & \text{if } p(\sigma(x)) = a \\ 0 & \text{otherwise} \end{cases}$$

$$\llbracket xS_vy \rrbracket_\mathcal{V}(p, \sigma) = \begin{cases} 1 & \text{if } \sigma(x)S_v\sigma(y) \\ 0 & \text{otherwise} \end{cases} \qquad \llbracket xS_hy \rrbracket_\mathcal{V}(p, \sigma) = \begin{cases} 1 & \text{if } \sigma(x)S_h\sigma(y) \\ 0 & \text{otherwise} \end{cases}$$

$$\llbracket x \in X \rrbracket_\mathcal{V}(p, \sigma) = \begin{cases} 1 & \text{if } \sigma(x) \in \sigma(X) \\ 0 & \text{otherwise} \end{cases} \qquad \llbracket x = y \rrbracket_\mathcal{V}(p, \sigma) = \begin{cases} 1 & \text{if } \sigma(x) = \sigma(y) \\ 0 & \text{otherwise} \end{cases}$$

$$\llbracket \neg\varphi \rrbracket_\mathcal{V}(p, \sigma) = \begin{cases} 1 & \text{if } \llbracket \varphi \rrbracket_\mathcal{V}(p, \sigma) = 0 \quad \text{if } \varphi \text{ is of the form } P_a(x), x = y, \\ 0 & \text{if } \llbracket \varphi \rrbracket_\mathcal{V}(p, \sigma) = 1 \quad (xS_vy), (xS_hy) \text{ or } (x \in X) \end{cases}$$

$$\llbracket \varphi \vee \psi \rrbracket_\mathcal{V}(p, \sigma) = \llbracket \varphi \rrbracket_\mathcal{V}(p, \sigma) + \llbracket \psi \rrbracket_\mathcal{V}(p, \sigma)$$
$$\llbracket \varphi \wedge \psi \rrbracket_\mathcal{V}(p, \sigma) = \llbracket \varphi \rrbracket_\mathcal{V}(p, \sigma) \cdot \llbracket \psi \rrbracket_\mathcal{V}(p, \sigma)$$
$$\llbracket \exists x.\varphi \rrbracket_\mathcal{V}(p, \sigma) = \sum_{(i,j) \in \mathrm{Dom}(p)} \llbracket \varphi \rrbracket_{\mathcal{V} \cup \{x\}}(p, \sigma[x \to (i,j)])$$

$$[\exists X.\varphi]_{\mathcal{V}}(p,\sigma) = \sum_{I \subseteq \text{Dom}(p)} [\varphi]_{\mathcal{V} \cup \{X\}}(p, \sigma[X \to I])$$

$$[\forall x.\varphi]_{\mathcal{V}}(p,\sigma) = \prod_{(i,j) \in \text{Dom}(p)} [\varphi]_{\mathcal{V} \cup \{x\}}(p, \sigma[x \to (i,j)])$$

$$[\forall X.\varphi]_{\mathcal{V}}(p,\sigma) = \prod_{I \subseteq \text{Dom}(p)} [\varphi]_{\mathcal{V} \cup \{X\}}(p, \sigma[X \to I]).$$

We write $[\varphi]$ for $[\varphi]_{\text{Free}(\varphi)}$. In case φ is a sentence, then $[\varphi] \in K\langle\!\langle \Sigma^{++} \rangle\!\rangle$. For $Z \subseteq$ MSO(K, Σ), we call a series $S : \Sigma^{++} \to K$ Z-definable if there exists a sentence $\varphi \in Z$ satisfying $[\varphi] = S$.

Example 4.2. Consider the formula $\varphi = \exists x.P_a(x) \in \text{MSO}(\mathbb{N}, \{a, b, c\})$. Then $[\varphi]$ is the series that computes for a picture $p \in \{a, b, c\}^{++}$ the number of occurrences of the letter a in p. Also, consider Example 3.2 of Section 3 again. For $L = D^{++}$, the formula $\psi = \forall x.\big(\bigvee_{d \in D}(P_d(x) \wedge d)\big) \in \text{MSO}(\mathbb{T}, D)$ satisfies $[\psi] = S$.

For different sets of variables $\mathcal{V}$, we show that our semantics are consistent:

Proposition 4.3. *Let* $\varphi \in \text{MSO}(K, \Sigma)$, $\mathcal{V}$ *be finite containing* Free(φ) *and* $(p, \sigma) \in N_{\mathcal{V}}$. *Then* $[\varphi]_{\mathcal{V}}(p, \sigma) = [\varphi](p, \sigma_{|\text{Free}(\varphi)})$, *and* $[\varphi]$ *is recognizable iff* $[\varphi]_{\mathcal{V}}$ *is recognizable.*

Proof. The first claim is proved by induction. Let $[\varphi] \in K^{\text{rec}}\langle\!\langle \Sigma_\varphi^{++} \rangle\!\rangle$ and $\pi : \Sigma_{\mathcal{V}} \to \Sigma_\varphi$ the projection. Then, $[\varphi]_{\mathcal{V}} = (\pi^{-1}[\varphi]) \odot \mathbb{1}_{N_{\mathcal{V}}} \in K^{\text{rec}}\langle\!\langle \Sigma_{\mathcal{V}}^{++} \rangle\!\rangle$ by Proposition 3.4. Now, let $[\varphi]_{\mathcal{V}} \in K^{\text{rec}}\langle\!\langle \Sigma_{\mathcal{V}}^{++} \rangle\!\rangle$ and $\mathcal{V}_1$ (resp. $\mathcal{V}_2$) be the set of first (resp. second)-order variables in $\mathcal{V}$. Then, $N^{\text{norm}} = \left\{ (p, \sigma) \in N_{\mathcal{V}} \,\middle|\, \begin{matrix} \forall x \in \mathcal{V}_1 \setminus \text{Free}(\varphi): \sigma(x) = (1,1), \\ \forall X \in \mathcal{V}_2 \setminus \text{Free}(\varphi): \sigma(X) = \{(1,1)\} \end{matrix} \right\}$ is deterministically recognizable. For $(p, \sigma) \in N_\varphi$, π maps exactly one element $(p, \sigma^{\text{norm}}) \in N^{\text{norm}}$ on (p, σ). With the above and Proposition 3.4, we conclude

$$\big(\pi([\varphi]_{\mathcal{V}} \odot \mathbb{1}_{N^{\text{norm}}}), (p, \sigma)\big) = \sum_{\substack{\pi(p,\sigma') = (p,\sigma) \\ (p,\sigma') \in N^{\text{norm}}}} [\varphi]_{\mathcal{V}}(p, \sigma') = [\varphi]_{\mathcal{V}}(p, \sigma^{\text{norm}}) = [\varphi](p, \sigma). \qquad \square$$

For words, examples show that unrestricted application of universal first-order quantification does not preserve recognizability [4, Ex. 3.3, 3.4]. These settings are contained in our context of the weighted MSO logic and series over pictures.

Definition 4.4. *A picture series* $S : \Sigma^{++} \to K$ *is a* first-order step function (*FO step function*), *if* $S = \bigoplus_{i=1}^{n} k_i \cdot \mathbb{1}_{L_i}$ *for some* $n \in \mathbb{N}$, $k_i \in K$ *and languages* $L_i \in \mathcal{L}(\text{FO}(\Sigma^{++}))$ $(i = 1, \ldots, n)$ *that are definable by* FO *formulas.*

We will call $\varphi \in \text{MSO}(K, \Sigma)$ *restricted*, if φ contains no universal set quantification of the form $\forall X.\psi$, and whenever φ contains a universal quantification $\forall x.\psi$, then $[\psi]$ is a FO step function. We let RMSO(K, Σ) comprise all restricted formulas of MSO(K, Σ). Furthermore, let REMSO(K, Σ) contain all restricted *existential* MSO-formulas φ, i.e. φ is of the form $\varphi = \exists X_1, \ldots, X_n.\psi$ such that $\psi \in \text{FO}(K, \Sigma) \cap$ RMSO(K, Σ). The families $K^{\text{rmso}}\langle\!\langle \Sigma^{++} \rangle\!\rangle$ (resp. $K^{\text{remso}}\langle\!\langle \Sigma^{++} \rangle\!\rangle$) contain all picture series $S \in K\langle\!\langle \Sigma^{++} \rangle\!\rangle$ which are definable by some sentence in RMSO(K, Σ) (resp. in

$REMSO(K, \Sigma)$). The following equivalence theorem states that for an alphabet Σ and any commutative semiring K, the family of recognizable picture series coincides with the families of series defined in terms of weighted RMSO resp. REMSO logic.

Theorem 4.5. *We have* $K^{\mathrm{rec}}\langle\!\langle \Sigma^{++}\rangle\!\rangle = K^{\mathrm{rmso}}\langle\!\langle \Sigma^{++}\rangle\!\rangle = K^{\mathrm{remso}}\langle\!\langle \Sigma^{++}\rangle\!\rangle.$

In parts of our proofs, we follow ideas of [4]. The crucial difference concerns the universal FO quantification. For pictures, not every recognizable language is determinizable, but this is one important property within the proofs of [4]. Here we consider a restriction of this quantification to formulas having a semantics which is a FO step function. But still the proof of the word-case does not work due to the two dimensions of a run in an automaton. We therefore rather build a formula instead of constructing a certain (unweighted) automaton. For the disposition of weights, the key property will be that this unweighted formula defines a language which is computable by a 2OTA that is unambiguous. Also, observe that, in the Theorem 4.5, going from RMSO to REMSO is not at all clear, since unlike to the situation of words [4, Lemma 5.2], in the framework of pictures using successor relations, instead of $\leq_v$ and $\leq_h$, not every (unweighted) FO-formula can be made unambiguous. However, we have to handle successor relations, since there are $(\leq_v, \leq_h)$-definable picture languages that are not recognizable.

5   Unambiguous Picture Languages

We call a possibly weighted 2OTA $\mathfrak{A}$ *unambiguous* if for any input picture there exists at most one successful run in $\mathfrak{A}$. Simulating the proof of Proposition 3.4, if L is unambiguously computable, then $\mathbb{1}_L \in K^{\mathrm{rec}}\langle\!\langle \Sigma^{++}\rangle\!\rangle$. For $L \subseteq \Gamma^{++}$ and a projection $\pi : \Gamma \to \Sigma$, we call π *injective* on L if $\pi : L \to \Sigma^{++}$ is an injective mapping. For $p \in \Sigma^{++}$, $\hat{p}$ denotes the picture that results from p by surrounding it with the (new) boundary symbol $\#$. If p has size (m, n) then $\hat{p}$ has size $(m+2, n+2)$. *Tiles* are pictures of size $(2, 2)$. We denote by $T(p)$ the set of all sub-tiles of p. A language $L \subseteq \Gamma^{++}$ is *local* if there exists a set Θ of tiles over $\Gamma \cup \{\#\}$, such that $L = \{p \in \Gamma^{++} \mid T(\hat{p}) \subseteq \Theta\}$. Then (Γ, Θ) *characterizes* L. We write $L = \mathcal{L}(\Theta)$. In [9], the authors briefly mention the notion of ambiguity for picture languages in the context of tiling systems (TS). We define $L \subseteq \Sigma^{++}$ as *unambiguously tiling recognizable* if there exists a local language $L' \subseteq \Gamma^{++}$, characterized by (Γ, Θ), and a projection $\pi : \Gamma \to \Sigma$ such that π is injective on L' and $\pi(L') = L$. In this case, we call $(\Sigma, \Gamma, \Theta, \pi)$ an *unambiguous* TS *computing* L. If the projection is not necessarily injective, we obtain the known definition of a TS. Unambiguously tiling recognizable languages over Σ are collected in $\mathrm{UPLoc}(\Sigma^{++})$.

Lemma 5.1. $\mathrm{UPLoc}(\Sigma^{++})$ *is closed under injective projections and disjoint union. A language L is recognizable by an unambiguous 2OTA if and only if it is computable by an unambiguous tiling system.*

Proof. Let Σ, Γ, Δ be alphabets and $(\Gamma, \Delta, \Theta, \psi)$ unambiguous for $L \subseteq \Gamma^{++}$. If $\pi : \Gamma \to \Sigma$ is injective on L, then $\tau := (\Sigma, \Delta, \Theta, \psi \circ \pi)$ is unambiguous for $\pi(L)$. Let $L_1, L_2 \in \mathrm{UPLoc}\,\Sigma^{++}, L_1 \cap L_2 = \emptyset$. We follow the construction in [11, Theorem 7.4]. The given TS for $L := L_1 \cup L_2$ is unambiguous since the union is disjoint. For the second claim, the TS, constructed in [11, Lemma 8.1] and also the automaton constructed in [11, Lemma 8.2] are unambiguous. $\square$

We call languages in UPLoc(Σ^{++}) *unambiguous*. We obtain further equivalences by injective projections of unambiguous rational operations and unambiguously domino recognizable series, cf. [17]. We now show, that FO definable picture languages are unambiguous. This will be crucial for proving Lemma 6.3 below. The idea for the course of the proof is to follow constructions in [12]. But, we now need unambiguous picture languages, hence we have to construct injective projections and disjoint unions.

Proposition 5.2. *If L be* FO(Σ^{++})*-definable. Then L is unambiguous.*

Proof. For $d, t \geq 1$, (d, t)-locally threshold testable (LTT) picture languages can be characterized by subsquares of dimension $\leq d$, where the occurrences are counted up to a threshold t. A language is FO-definable iff it is LTT and every LTT language L is recognizable [12]. We show that LTT languages are unambiguous. Let $L \subseteq \Sigma^{++}$ be LTT for (d, t). As in Lemma 3.7 [12], we partition L into a union of strictly LTT languages (where strictly means, only squares of dimension d are considered). This union is easily proved as disjoint. Strictly LTT languages are projections d-local languages (for d-locality we use a set $\Theta^{(d)}$ of $(d \times d)$-tiles instead of (2×2)-tiles for local sets)[12, Lemma 3.9]. For a given (d, t)-strictly LTT language L', in the construction, one performs a scanning of $p' \in L'$ using certain d-squares and counts occurrences up to t. For acceptance, one compares these computed values with the tuples characterizing L'. It defines a d-local language L'' and a projection π satisfying $\pi(L'') = L'$. We can modify the d-tiles (and hence L'') by strengthening their border-conditions in such a way that for every $p' \in L'$ there exists one uniquely determined $p'' \in L''$ with $\pi(p'') = p'$. Hence, the modified projection then is injective on L''.

It remains to show that every d-local language M is unambiguous. For this, let Δ be arbitrary, $d \geq 3$ and M characterized by $(\Delta, \Theta^{(d)})$. We can assume $M \subseteq \Delta^{m \times n}$ such that $m, n \geq d - 2$ ([12, Lemma 3.10]). We prove that M is an injective projection of a local set, that is, M is computable by an unambiguous tiling system. We define $T = (\Delta, \Gamma, \Theta, \pi)$ as

- $\overline{\Theta^{(d)}} := \left\{ \begin{array}{|c|c|}\hline A & B \\\hline C & D \\\hline\end{array} \in (\Delta \cup \{\#\} \cup \{+\})^{d \times d} \mid B = C = D \equiv +, \exists\, \begin{array}{|c|c|}\hline A1 & A2 \\\hline A3 & A \\\hline\end{array} \in \Theta^{(d)} \right\}$

- $\Gamma := \overline{\Theta^{(d)}} \setminus \{p \mid p_{1,1} = \#\}$; border symbols: $\{p \in \overline{\Theta^{(d)}} \mid p_{1,1} = \#\}$

- $\Theta := \left\{ \begin{array}{|c|c|}\hline A & B \\\hline C & D \\\hline\end{array} \in \Gamma^{2 \times 2} \mid A = \begin{array}{|c|c|}\hline a & N \\\hline W & Q \\\hline\end{array}, B = \begin{array}{|c|c|}\hline N & b \\\hline Q & E \\\hline\end{array}, C = \begin{array}{|c|c|}\hline W & Q \\\hline c & S \\\hline\end{array}, D = \begin{array}{|c|c|}\hline Q & E \\\hline S & d \\\hline\end{array} \right\}$

  where $Q \in (\Delta \cup \{\#\} \cup \{+\})^{(d-1) \times (d-1)}$ and W, S, E, N, a, b, c, d accordant.

We set $\pi : \Gamma \to \Delta$, $p \mapsto p_{1,1}$ and show $\pi(\mathcal{L}(\Theta)) = M$. Let $p \in M$. We extend p to $\bar{p} \in (\Delta \cup \{\#\} \cup \{+\})^{(m+d-1) \times (n+d-1)}$ and define $p' \in \Gamma^{m \times n}$, as

$$\bar{p}(i,j) = \begin{cases} p(i,j) & i \leq m, j \leq n \\ \# & \begin{array}{l} i = m+1, j \leq n+1 \\ \text{or } i \leq m+1, j = n+1 \end{array} \\ + & \textit{otherwise.} \end{cases} , \quad p'(i,j) = \begin{array}{|ccc|}\hline \bar{p}(i,j) & & \bar{p}(i,j+d-1) \\ & \cdots & \\ \bar{p}(i+d-1,j) & & \bar{p}(i+d-1,j+d-1) \\\hline\end{array}.$$

Then, $p' \in \mathcal{L}(\Theta)$ and $\pi(p') = p$. Now let $p' \in \mathcal{L}(\Theta)$ and q be a $(d \times d)$-subpicture of $\hat{p}'$. It suffices to show $\pi(q) \in \Theta^{(d)}$. With the construction of $\overline{\Theta^{(d)}}$ we have $q_{1,1} \in \Theta^{(d)}$. But,

$q_{1,1} = \pi(q)$. By the structure of the d-tiles in Θ, one can show that T is unambiguous, i.e., π is injective on $\mathcal{L}(\Theta)$. We constructed unambiguous languages and disjoint unions. With Lemma 5.1, L is unambiguous. $\qquad\qquad\qquad\qquad\qquad\qquad\qquad\qquad\qquad\square$

6 Definable Picture Series Are Recognizable

The aim of this section is to show that semantics of sentences in $\mathrm{RMSO}(K, \Sigma)$ are recognizable series. We prove this implication by structural induction on the formulas in $\mathrm{RMSO}(K, \Sigma)$.

Lemma 6.1. *Let $\mathcal{V}$ be a set of variables. Then the set $N_{\mathcal{V}}$ is FO-definable. The class $\mathcal{L}(\mathrm{FO}(\Sigma^{++}))$ is closed under boolean operations.*

Lemma 6.2. *Let $\varphi, \psi \in \mathrm{MSO}(K, \Sigma)$. Then the following holds.*

(a) If φ is atomic or the negation of an atomic formula, then φ is recognizable.
(b) If $[\![\varphi]\!]$ and $[\![\psi]\!]$ are recognizable, then $[\![\varphi \vee \psi]\!]$ and $[\![\varphi \wedge \psi]\!]$ are recognizable.
(c) If $[\![\varphi]\!]$ is recognizable, then $[\![\exists x.\varphi]\!]$ and $[\![\exists X.\varphi]\!]$ are recognizable.

Proof. (a) We construct W2OTA using Proposition 3.4. The other proofs are similar to the word-case ([4, Lemma 4.1]) and use Propositions 4.3 and 3.4. $\qquad\qquad\qquad\square$

The next Lemma shows that for FO step functions the application of the universal first-order quantification provides a recognizable semantics. We use ideas of [4, Lemma 4.2], but these did not completely work in this setting.

Lemma 6.3. *Let $\varphi \in \mathrm{MSO}(K, \Sigma)$ such that $[\![\varphi]\!]$ is a first-order step function. Then $[\![\forall x.\varphi]\!]$ is a recognizable picture series.*

Proof. As prerequisite, let $\mathcal{W} = \mathrm{Free}(\varphi)$, $\mathcal{V} = \mathrm{Free}(\forall x.\varphi) = \mathcal{W} \setminus \{x\}$ and assume $[\![\varphi]\!] = \bigoplus_{l=1,\dots,n} k_l \cdot \mathbb{1}_{L_l}$ with $n \in \mathbb{N}$, $k_l \in K$ and $L_l \in \mathcal{L}(\mathrm{FO}(\Sigma_{\mathcal{W}}^{++}))$ ($l = 1, \dots, n$) such that the languages L_l form a partition (use Lemma 6.1). Assume $x \in \mathcal{W}$.

The definition of the semantics of the universal FO quantification of a formula maps a picture p to the product over all positions in p of certain values in K. In our setting, the factors are the elements k_l corresponding to the supports of $[\![\varphi]\!]$. We mark positions of p by their respective index l of k_l. Let $\tilde{\Sigma} = \Sigma \times \{1, \dots, n\}$. A picture in $(\tilde{\Sigma}_{\mathcal{V}})^{++}$ will be written as (p, ν, σ) where $(p, \sigma) \in \Sigma_{\mathcal{V}}^{++}$ and $\nu \in \{1, \dots, n\}^{++}$ is interpreted as a mapping from $\mathrm{Dom}(p)$ to $\{1, \dots, n\}$. Let $\tilde{L}$ be the set of $(p, \nu, \sigma) \in (\tilde{\Sigma}_{\mathcal{V}})^{++}$ such that $\nu(i,j) = l \iff (p, \sigma[x \rightarrow (i,j)]) \in L_l$ for all $(i,j) \in \mathrm{Dom}(p)$ and $l \in \{1, \dots, n\}$. We prove $\tilde{L} \in \mathrm{FO}(\tilde{\Sigma}_{\mathcal{V}}^{++})$ by presenting a formula. Let $1 \le l \le n$ and φ_l be an FO-sentence over $\Sigma_{\mathcal{W}}^{++}$ for L_l. We define $\widetilde{\varphi_l} \in \mathrm{FO}((\widetilde{\Sigma_{\mathcal{W}}})^{++})$ as φ_l where all occurrences of $P_{(a,r)}(y)$ (here, $a \in \Sigma, r \in \{0,1\}^{\mathcal{W}}$) are replaced by $\bigvee_{1 \le l \le n} P_{(a,r,l)}(y)$. Then, for $(p, \tau, \nu) \in \widetilde{\Sigma_{\mathcal{W}}}^{++}$, we conclude $(p, \tau, \nu) \in \mathcal{L}(\widetilde{\varphi_l})$ iff $(p, \tau) \in \mathcal{L}(\varphi_l)$. Additionally, we define $\tilde{\varphi}_l{}'$ as $\tilde{\varphi}_l$, modified as follows. Occurrences of $P_{(a,r,l)}(y)$ satisfying $r(x) = 1$ become $P_{(a,r',l)}(y) \wedge (x = y)$ and occurrences of $P_{(a,r,l)}(y)$ with $r(x) = 0$ become $P_{(a,r',l)}(y) \wedge \neg(x = y)$, where r' is the restriction of r to $\mathcal{W} \setminus \{x\}$. Then, $\widetilde{\varphi_l}'$ is an FO-formula over the alphabet $\tilde{\Sigma}_{\mathcal{V}}$ with $\mathrm{Free}(\widetilde{\varphi_l}') = \{x\}$

satisfying for all $(p, \tau, \nu) \in N_{\widetilde{\varphi_l}'}$, that $(p, \tau', \nu) \in \mathcal{L}(\widetilde{\varphi_l}')$ if and only if $(p, \tau, \nu) \in \mathcal{L}(\widetilde{\varphi_l})$. Now, set $\widetilde{\varphi} := \forall x. \bigwedge_{1 \leq l \leq n} \left[(\nu(x) = l) \Leftrightarrow \widetilde{\varphi_l}' \right]$ where $\nu(x) = l$ and $\Leftrightarrow$ are standard abbreviations. Now, $\widetilde{\varphi}$ is an FO-sentence over $\widetilde{\Sigma}_{\mathcal{V}}$. We show $\mathcal{L}(\widetilde{\varphi}) = \widetilde{L}$.

Let $(q, \nu) \in (\widetilde{\Sigma}_{\mathcal{V}})^{++}$ (here, $q \in \Sigma_{\mathcal{V}}^{++}$). Then $(q, \nu) \models \widetilde{\varphi}$ iff for all $(i, j) \in \mathrm{Dom}(q)$ and all $1 \leq l \leq n$, $(q, \nu, [x \rightarrow (i, j)]) \in \mathcal{L}\big((\nu(x) = l) \Leftrightarrow \widetilde{\varphi_l}'\big)$, where $[x \rightarrow (i, j)]$ denotes the assignment defined on $\{x\}$ mapping x to (i, j). Now, $(q, \nu, [x \rightarrow (i, j)]) \in \mathcal{L}(\nu(x) = l)$ iff $\nu(i, j) = l$ and $(q, \nu, [x \rightarrow (i, j)]) \in \mathcal{L}(\widetilde{\varphi_l}')$ iff $(q, \nu, \sigma[x \rightarrow (i, j)]) \in \mathcal{L}(\widetilde{\varphi_l})$ iff $(q, [x \rightarrow (i, j)]) \in \mathcal{L}(\varphi_l)$ iff $(q, [x \rightarrow (i, j)]) \in L_l$. Hence the constructed formula $\widetilde{\varphi}$ defines $\widetilde{L}$.

Now, using Proposition 5.2 and Lemma 5.1, there exists an unambiguous 2OTA $\widetilde{A} = (\widetilde{\Sigma}_{\mathcal{V}}, Q, I, F, E)$ computing $\widetilde{L}$. We obtain a W2OTA $\widetilde{\mathfrak{A}} = (\widetilde{\Sigma}_{\mathcal{V}}, Q, I, F, \bar{E})$ disposing weights along $\widetilde{A}$ as: $(p, q, (a, l, s), r) \in E$ iff $(p, q, (a, l, s), k_l, r) \in \bar{E}$, where $[\![\varphi]\!] = \bigoplus_{l=1,\ldots,n} k_l \cdot \mathbb{1}_{L_l}$. Then $\widetilde{\mathfrak{A}}$ is unambiguous. Similar to [4, Lemma 4.2], using Proposition 3.4 and Lemma 6.2, for the projection $\pi : \widetilde{\Sigma}_{\mathcal{V}} \rightarrow \Sigma_{\mathcal{V}}$, one proves for $(p, \sigma) \in \Sigma_{\mathcal{V}}^{++}$: $\big(\pi(\|\widetilde{\mathfrak{A}}\|), (p, \sigma)\big) = [\![\forall x. \varphi]\!](p, \sigma)$, hence $[\![\forall x. \varphi]\!]$ is recognizable. The case $x \notin \mathcal{W}$ is reduced to above. □

Theorem 6.4. *We have* $K^{\mathrm{rmso}}\langle\!\langle \Sigma^{++} \rangle\!\rangle \subseteq K^{\mathrm{rec}}\langle\!\langle \Sigma^{++} \rangle\!\rangle$.

7   Recognizable Picture Series Are Definable

We want to show that recognizable series are REMSO-definable. Similar to [4,5], for a W2OTA $\mathfrak{A}$ we construct a weighted EMSO-sentence γ such that $\|\mathfrak{A}\| = [\![\gamma]\!]$. It then remains to prove that γ is restricted. We also need the notion of unambiguous formulas. We note that, unlike to the word-cases, we use successor relations, here not every (unweighted) FO-formula can be made unambiguous. The class of *unambiguous* formulas in $\mathrm{FO}(K, \Sigma)$ is defined inductively as follows: All atomic formulas and their negations are unambiguous. If φ, ψ are unambiguous, then $\varphi \wedge \psi, \forall x. \varphi$ are unambiguous. If φ, ψ are unambiguous and $\mathrm{supp}([\![\varphi]\!]) \cap \mathrm{supp}([\![\psi]\!]) = \emptyset$, then $\varphi \vee \psi$ is unambiguous. Let $\mathcal{V} = \mathrm{Free}(\varphi)$. If φ is unambiguous and for any $(p, \sigma) \in \Sigma_{\mathcal{V}}^{++}$ there is at most one element $(i, j) \in \mathrm{Dom}(p)$ such that $[\![\varphi]\!]_{\mathcal{V} \cup \{x\}}(p, \sigma[x \rightarrow (i, j)]) \neq 0$, then $\exists x. \varphi$ is unambiguous.

By qf-$\mathrm{MSO}^-(K, \Sigma)$, we denote formulas in $\mathrm{MSO}(K, \Sigma)$ having no quantification and no subformula of the form k. To make such formulas unambiguous we perform a syntactic transformations $(.)^+$ and $(.)^-$ in a simultaneous induction such that, for $\varphi, \psi \in \mathrm{qf\text{-}MSO}^-(K, \Sigma)$, we have $\mathcal{L}(\varphi^+) = \mathcal{L}(\varphi)$ and $\mathcal{L}(\varphi^-) = \Sigma_{\mathrm{Free}(\varphi)}^{++} \setminus \mathcal{L}(\varphi)$. Now, similar to [4, Prop. 5.1], we get:

Lemma 7.1. *Let* $\varphi \in \mathrm{FO}(K, \Sigma)$ *be unambiguous. Then* $[\![\varphi]\!] = \mathbb{1}_{\mathcal{L}(\varphi)}$. *For* $\psi \in$ qf-$\mathrm{MSO}^-(K, \Sigma)$, *the formula* ψ^+ *is unambiguous.*

Notions like $\min_v(x)$, $\max_h(z)$ or $\mathrm{part}(X_1, \ldots, X_l)$ abbreviate common formulas. We set (analog $\mathrm{init}_{\mathrm{W}}$):

$$\mathrm{init}_{\mathrm{N}} := \forall x. \Big(\big[\min_v(x) \wedge \big(\bigvee_{q_h^x, q^x \in Q, q_v^x \in I, a \in \Sigma} x \in X_{(q_h^x, q_h^x, a, q^x)} \big)^+ \big] \vee \exists s. (s S_v x) \Big).$$

For intuition, the formulas init_N (resp. init_W) simulate accepting conditions of the automaton for the first row (resp. first column) of an input picture.

Theorem 7.2. *We have* $K^{\text{rec}}\langle\!\langle \Sigma^{++}\rangle\!\rangle \subseteq K^{\text{remso}}\langle\!\langle \Sigma^{++}\rangle\!\rangle$.

Proof (sketch). Let $\mathfrak{A} = (\Sigma, Q, I, F, E)$ be a W2OTA. For $a \in \Sigma, q_v, q_h, q \in Q$, we set $\mu_{(q_v,q_h,q)}(a) = \sum_{(q_v,q_h,a,k,q)\in E} k$. Let $\mathcal{V} = \{X_{(q_v,q_h,a,q)} \mid (q_v, q_h, a, q) \in Q^2 \times \Sigma \times Q\}$ the set of set variables, $(X_1, \ldots, X_l)$ an enumeration of $\mathcal{V}$. We set

$$\alpha(X_1, \ldots, X_l) := \text{part}(X_1, \ldots, X_l) \wedge \bigwedge_{q_v,q_h,a,q} \forall x.\Big((x \in X_{(q_v,q_h,a,q)}) \to P_a(x)\Big)$$

$$\wedge\ \forall x \forall z.\Big((xS_v z) \to \bigvee_{q_v^x,q_h^x,q^x,q_h^z,q^z \in Q;a,b\in\Sigma} (x \in X_{(q_v^x,q_h^x,a,q^x)}) \wedge (z \in X_{(q^x,q_h^z,b,q^z)})\Big)^+$$

$$\wedge\ \forall y \forall z.\Big((yS_h z) \to \bigvee_{q_v^y,q_h^y,q^y,q_v^z,q^z \in Q;c,b\in\Sigma} (y \in X_{(q_v^y,q_h^y,c,q^y)}) \wedge (z \in X_{(q_v^z,q^y,b,q^z)})\Big)^+.$$

The formula α qualifies unweighted runs in $\mathfrak{A}$. Now, let $\beta(X_1, \ldots, X_l) :=$

$$\alpha \wedge \bigwedge_{q_v,q_h,a,q} \forall x.\Big(x \in X_{(q_v,q_h,a,q)}) \to \mu_{(q_v,q_h,q)}(a)\Big) \wedge \text{init}_N \wedge \text{init}_W$$

$$\wedge \exists z.\Big(\max_v(z) \wedge \max_h(z) \wedge \bigvee_{q_v^z,q_h^z \in Q, q^z \in F, b\in\Sigma} (z \in X_{(q_v^z,q_h^z,b,q^z)})\Big).$$

Here, β simulates the distribution of weights along transitions and successful runs. Let $\gamma := \exists X_1 \cdots \exists X_l.\beta(X_1, \ldots, X_l)$ and $p \in \Sigma^{++}$. Then $[\![\gamma]\!](p) = (\|\mathfrak{A}\|, p)$, hence $\|\mathfrak{A}\| = [\![\gamma]\!]$. Furthermore, using Lemma 7.1 and remarks above, one can show that the specified formula γ lies in $\text{REMSO}(K, \Sigma)$. $\square$

8 Conclusion

In [18] we assigned weights to tiling systems, domino systems or weighted (quadrapolic) picture automata and proved for an alphabet Σ and any commutative semiring K the coincidence of the corresponding series with the projections of series defined by rational operations. In fact, one can prove that this very class coincides with $K^{\text{rec}}\langle\!\langle \Sigma^{++}\rangle\!\rangle$ [17]. With Theorem 4.5, this implies that the notion of weighted recognizability presented here is robust and extends the main result of [11] to the weighted case. Furthermore, in [3] it is shown, that a picture language is recognizable if and only if it is the support of a recognizable series with coefficients in $\mathbb{B}$. Hence, we obtain the classical equivalence in [12] by restricting to $\mathbb{B}$.

Acknowledgements. I would like to thank Manfred Droste and Dietrich Kuske for their helpful discussions and comments, as well as the unknown referees whose remarks resulted in improvements of this paper.

References

1. J. Berstel and C. Reutenauer. *Rational Series and Their Languages*, volume 12 of *EATCS Monographs on Theoretical Computer Science*. Springer-Verlag, 1984.
2. M. Blum and C. Hewitt. Automata on a 2-dimensional tape. In *IEEE Symposium on Switching and Automata Theory*, pages 155–160. 1967.
3. S. Bozapalidis and A. Grammatikopoulou. Recognizable picture series. In *special volume on Weighted Automata, presented at WATA 2004, Dresden*, Journal of Automata, Languages and Combinatorics. 2005, in print.
4. M. Droste and P. Gastin. Weighted automata and weighted logics. In *32th ICALP, Proceedings*, volume 3580 of *Lecture Notes in Computer Science*, pages 513–525. Springer-Verlag, 2005.
5. M. Droste and H. Vogler. Weighted tree automata and weighted logics. Submitted, 2005.
6. D. Dubois, E. Hüllermeier, and H. Prade. A systematic approach to the assessment of fuzzy association rules. Technical report, Philipps-Universität Marburg, 2004.
7. S. Eilenberg. *Automata, Languages, and Machines*, Vol. A. Academic Press, New York, 1974.
8. K. S. Fu. *Syntactic Methods in Pattern Recognition*. Academic Press, New York, 1974.
9. D. Giammarresi and A. Restivo. Recognizable picture languages. *International Journal Pattern Recognition and Artificial Intelligence*, 6(2-3):31–46, 1992. Issue on *Parallel Image Processing*. M. Nivat, A. Saoudi, and P.S.P. Wangs (Eds.).
10. D. Giammarresi and A. Restivo. Two-dimensional finite state recognizability. *Fundam. Inform.*, 25(3):399–422, 1996.
11. D. Giammarresi and A. Restivo. Two-dimensional languages. In G. Rozenberg and A. Salomaa, editors, *Handbook of Formal Languages, Vol. 3, Beyond Words*, pages 215–267. Springer-Verlag, Berlin, 1997.
12. D. Giammarresi, A. Restivo, S. Seibert, and W. Thomas. Monadic second-order logic over rectangular pictures and recognizability by tiling systems. *Information and Computation*, 125:32–45, 1996.
13. K. Inoue and A. Nakamura. Some properties of two-dimensional on-line tessellation acceptors. *Information Sciences*, 13:95–121, 1977.
14. W. Kuich and A. Salomaa. *Semirings, Automata, Languages*, volume 6 of *EATCS Monographs in Theoretical Computer Science*. Springer-Verlag, Berlin, 1986.
15. M. Latteux and D. Simplot. Recognizable picture languages and domino tiling. *Theoretical Computer Science*, 178:275–283, 1997.
16. O. Matz. On piecewise testable, starfree, and recognizable picture languages. In P. van Emde Boas et al., editors, *FoSSaCS*, volume 1378 of *Lecture Notes in Computer Science*, pages 203–210. Springer-Verlag, Berlin, 1998.
17. I. Mäurer. Characterizations of weighted picture series. In preparation, 2005.
18. I. Mäurer. Recognizable and rational picture series. In *Conference on Algebraic Informatics*, pages 141–155. Aristotle University of Thessaloniki Press, 2005.
19. M. Minski and S. Papert. *Perceptron*. M.I.T. Press, Cambridge, Mass., 1969.
20. A. Salomaa and M. Soittola. *Automata-Theoretic Aspects of Formal Power Series*. Texts and Monographs on Computer Science. Springer-Verlag, 1978.
21. W. Thomas. On logics, tilings, and automata. In *18th ICALP, Proceedings*, volume 510 of *Lecture Notes in Computer Science*, pages 441–453. Springer-Verlag, 1991.

Markov Decision Processes
with Multiple Objectives*

Krishnendu Chatterjee[1], Rupak Majumdar[2], and Thomas A. Henzinger[1,3]

[1] UC Berkeley
[2] UC Los Angeles
[3] EPFL
{c_krish, tah}@eecs.berkeley.edu, rupak@cs.ucla.edu

Abstract. We consider Markov decision processes (MDPs) with multiple discounted reward objectives. Such MDPs occur in design problems where one wishes to simultaneously optimize several criteria, for example, latency and power. The possible trade-offs between the different objectives are characterized by the Pareto curve. We show that every Pareto-optimal point can be achieved by a memoryless strategy; however, unlike in the single-objective case, the memoryless strategy may require randomization. Moreover, we show that the Pareto curve can be approximated in polynomial time in the size of the MDP. Additionally, we study the problem if a given value vector is realizable by any strategy, and show that it can be decided in polynomial time; but the question whether it is realizable by a deterministic memoryless strategy is NP-complete. These results provide efficient algorithms for design exploration in MDP models with multiple objectives.

1 Introduction

Markov decision processes (MDPs) are a widely studied model for dynamic and stochastic systems [2,8]. An MDP models a dynamic system that evolves through stages. In each stage, a controller chooses one of several actions, and the system stochastically evolves to a new state based on the current state and the chosen action. In addition, one associates a cost or reward with each state and transition, and the central question is to find a strategy of choosing the actions that optimizes the rewards obtained over the run of the system, where the rewards are combined using a discounted sum. In many modeling domains, however, there is no unique objective to be optimized, but multiple, potentially dependent and conflicting objectives. For example, in designing a computer system, one is interested not only in maximizing performance but also in minimizing power. Similarly, in an inventory management system, one wishes to optimize several potentially dependent costs for maintaining each kind of product, and in AI planning, one wishes to find a plan that optimizes several distinct goals. The usual MDP model is insufficient to express these natural problems.

---

* This research was supported in part by the AFOSR MURI grant F49620-00-1-0327, and the NSF grants CCR-0225610, CCR-0234690, and CCR-0427202.

B. Durand and W. Thomas (Eds.): STACS 2006, LNCS 3884, pp. 325–336, 2006.

We study MDPs with multiple objectives, an extension of the MDP model where there are several reward functions [4,10]. In MDPs with multiple objectives, we are interested not in a single solution that is simultaneously optimal in all objectives (which may not exist), but in a notion of "trade-offs" called the *Pareto curve*. Informally, the Pareto curve consists of the set of realizable value profiles (or dually, the strategies that realize them) which are not dominated (in every dimension) by any other value profile. Pareto optimality has been studied in co-operative game theory [6], and in multi-criterion optimization and decision making in both economics and engineering [5,9,11]. Finding *some* Pareto-optimal point can be reduced to optimizing a single objective: optimize a convex combination of objectives using a set of positive weights; the optimal strategy must be Pareto-optimal as well (the "weighted factor method") [4]. In design space exploration, however, we want to find not one, but *all* Pareto-optimal points in order to better understand the trade-offs in the design. Unfortunately, even with just two reward functions, the Pareto curve may have infinitely many points, and also contain irrational payoff values. Thus, previous work has focused on constructing a sampling of the Pareto curve, either by choosing a variety of weights in the weighted factor method, or by imposing a lexicographic ordering on the objectives and sequentially optimizing each objective according to the order [1,2]. Unfortunately, this does not provide any guarantee about the quality of the solutions.

Instead, we study the *approximate* version of the problem: the ϵ-approximate Pareto curve [7] for MDPs with multiple discounted reward criteria. Informally, the ϵ-approximate Pareto curve for $\epsilon > 0$ contains a set of strategies (or dually, their payoff values) such that there is no other strategy whose value dominates the values in the Pareto curve by a factor of $1 + \epsilon$. Surprisingly, a polynomial-sized ϵ-approximate Pareto curve always exists. Moreover, we show that such an approximate Pareto curve may be computed efficiently (in polynomial time) in the size of the MDP. Our proof is based on the following characterization of Pareto-optimal points: every Pareto-optimal value profile can be realized by a memoryless (but possibly randomized) strategy. This enables the reduction of the problem to multi-objective linear programs, and we can apply the methods of [7].

We also study the *Pareto realizability* decision problem: given a profile of values, is there a Pareto-optimal strategy that dominates it? We show that the Pareto realizability problem can be solved in polynomial time. However, if we restrict the set of strategies to be pure (i.e., no randomization), then the problem becomes NP-hard. Our complexities are comparable to the single discounted reward case, where linear programming provides a polynomial-time solution [8]. However, unlike in the single-reward case, where pure and memoryless optimal strategies always exist, here, checking pure and memoryless realizability is hard.

The results of this paper provide polynomial-time algorithms for both the decision problem and the optimization problem for MDPs with multiple discounted reward objectives. Since the Pareto curve forms a useful "user interface" for desirable solutions, we believe that these results will lead to efficient design space exploration algorithms in multi-criterion design.

The rest of the paper is organized as follows. In Section 2, we give the basic definitions, and show in Section 3 the sufficiency of memoryless strategies. Section 4 gives a polynomial-time algorithm to construct the ϵ-approximate Pareto curve. Section 5 studies the decision version of the problem. Finally, in Section 6, we discuss the extension to MDPs with limit-average (not discounted) reward objectives, and mention some open problems.

2   Discounted Reward Markov Decision Processes

We denote the set of probability distributions on a set U by $\mathcal{D}(U)$.

Markov decision processes (MDPs). A *Markov decision process* (MDP) $G = (S, A, \delta)$ consists of a finite, non-empty set S of states, a finite, non-empty set A of actions, and a probabilistic transition function $\delta : S \times A \rightarrow \mathcal{D}(S)$ that, given a state $s \in S$ and an action $a \in A$, gives the probability $\delta(s, a)(t)$ of the next state t. We denote by $\mathrm{Dest}(s, a) = \mathrm{Support}(\delta(s, a))$ the set of possible successors of s when the action a is chosen. Given an MDP G, we define the set of edges by $E = \{ (s, t) \mid \exists a \in A.\, t \in \mathrm{Dest}(s, a) \}$, and write $E(s) = \{ t \mid (s, t) \in E \}$ for the set of possible successors of s.

Plays and strategies. A *play* of G is an infinite sequence $\langle s_0, s_1, \ldots \rangle$ of states such that for all $i \geq 0$, we have $(s_i, s_{i+1}) \in E$. A strategy σ is a recipe that specifies how to extend a play. Formally, a *strategy* σ is a function $\sigma \colon S^+ \rightarrow \mathcal{D}(A)$ that, given a finite and non-empty sequence of states representing the history of the play so far, chooses a probability distribution over the set A of actions. In general, a strategy depends on the history and uses randomization. A strategy that depends only on the current state is a *memoryless* or *stationary* strategy, and can be represented as a function $\sigma \colon S \rightarrow \mathcal{D}(A)$. A strategy that does not use randomization is a *deterministic* or *pure* strategy, i.e., for all histories $\langle s_0, s_1, \ldots, s_k \rangle$ there exists $a \in A$ such that $\sigma(\langle s_0, s_1, \ldots, s_k \rangle)(a) = 1$. A *pure memoryless* strategy is both pure and memoryless, and can be represented as a function $\sigma \colon S \rightarrow A$. We denote by Σ, Σ^M, Σ^P, and Σ^{PM} the sets of all strategies, all memoryless strategies, all pure strategies, and all pure memoryless strategies, respectively.

Outcomes. For a strategy σ and an initial state s, we denote by $\mathrm{Outcome}(s, \sigma)$ the set of possible plays that start from s given strategy σ, that is, $\mathrm{Outcome}(s, \sigma) = \{ \langle s_0, s_1, \ldots \rangle \mid \forall k \geq 0.\, \exists a_k \in A.\, \sigma(\langle s_0, s_1, \ldots, s_k \rangle)(a_k) > 0 \text{ and } s_{k+1} \in \mathrm{Dest}(s_k, a_k) \}$. Once the initial state and a strategy is chosen, the MDP is reduced to a stochastic process. We denote by X_i and θ_i random variables for the i-th state and the i-th action, respectively, in this stochastic process. An event is a measurable subset of $\mathrm{Outcome}(s, \sigma)$, and the probabilities of events are uniquely defined. Given a strategy σ, an initial state s, and an event Φ, we denote by $\mathrm{Pr}_s^\sigma(\Phi)$ the probability that a play belongs to Φ when the MDP starts in state s and the strategy σ is used. For a measurable function f

that maps plays to reals, we write $\mathbb{E}_s^\sigma[f]$ for the expected value of f when the MDP starts in state s and the strategy σ is used.

Rewards and objectives. Let $r: S \times A \to \mathbb{R}$ be a *reward function* that associates with every state and action a real-valued reward. For a reward function r the discounted reward objective is to maximize the discounted sum of rewards, which is defined as follows. Given a discount factor $0 \le \beta < 1$, the *discounted reward* or *payoff value* for a strategy σ and an initial state s with respect to the reward function r is $\mathrm{Val}_{dis}^\sigma(r, s, \beta) = \sum_{t=0}^{\infty} \beta^t \cdot \mathbb{E}_s^\sigma[r(X_t, \theta_t)]$.

We consider MDPs with k different reward functions $r_1, \ldots, r_k$. Given an initial state s, a strategy σ, and a discount factor $0 \le \beta < 1$, the discounted reward value vector, or *payoff profile*, at s for σ with respect to $\boldsymbol{r} = \langle r_1, \ldots, r_k \rangle$ is defined as $\mathrm{Val}_{dis}^\sigma(\boldsymbol{r}, s, \beta) = \langle \mathrm{Val}_{dis}^\sigma(r_1, s, \beta), \ldots, \mathrm{Val}_{dis}^\sigma(r_k, s, \beta) \rangle$.

Comparison operators on vectors are interpreted in a point-wise fashion, i.e., given two real-valued vectors $\boldsymbol{v}_1 = \langle v_1^1, \ldots, v_1^k \rangle$ and $\boldsymbol{v}_2 = \langle v_2^1, \ldots, v_2^k \rangle$, and $\bowtie \in \{<, \le, = \}$, we write $\boldsymbol{v}_1 \bowtie \boldsymbol{v}_2$ if and only if for all $1 \le i \le k$, we have $v_1^i \bowtie v_2^i$. We write $\boldsymbol{v}_1 \ne \boldsymbol{v}_2$ to denote that vector $\boldsymbol{v}_1$ is not equal to $\boldsymbol{v}_2$, that is, it is not the case that $\boldsymbol{v}_1 = \boldsymbol{v}_2$.

Pareto-optimal strategies. Given an MDP G and reward functions $r_1, \ldots, r_k$, a strategy σ is a *Pareto-optimal* strategy [6] from a state s if there is no strategy $\sigma' \in \Sigma$ such that both $\mathrm{Val}_{dis}^\sigma(\boldsymbol{r}, s, \beta) \le \mathrm{Val}_{dis}^{\sigma'}(\boldsymbol{r}, s, \beta)$ and $\mathrm{Val}_{dis}^\sigma(\boldsymbol{r}, s, \beta) \ne \mathrm{Val}_{dis}^{\sigma'}(\boldsymbol{r}, s, \beta)$; that is, there is no strategy σ' such that for all $1 \le j \le k$, we have $\mathrm{Val}_{dis}^\sigma(r_j, s, \beta) \le \mathrm{Val}_{dis}^{\sigma'}(r_j, s, \beta)$, and there exists $1 \le j \le k$ with $\mathrm{Val}_{dis}^\sigma(r_j, s, \beta) < \mathrm{Val}_{dis}^{\sigma'}(r_j, s, \beta)$. For a Pareto-optimal strategy σ, the corresponding payoff profile $\mathrm{Val}_{dis}^\sigma(\boldsymbol{r}, s, \beta)$ is referred to as a *Pareto-optimal point*. In case $k = 1$, the class of Pareto-optimal strategies are called *optimal* strategies.

Sufficiency of strategies. Given reward functions $r_1, \ldots, r_k$, a family Σ^C of strategies *suffices* for Pareto optimality for discounted reward objectives if for every discount factor β, state s, and Pareto-optimal strategy $\sigma \in \Sigma$, there is a strategy $\sigma' \in \Sigma^C$ such that $\mathrm{Val}_{dis}^\sigma(\boldsymbol{r}, s, \beta) \le \mathrm{Val}_{dis}^{\sigma'}(\boldsymbol{r}, s, \beta)$.

Theorem 1. [2] *In MDPs with a single reward function r, the pure memoryless strategies suffice for optimality for the discounted reward objective, i.e., for all discount factors $0 \le \beta < 1$ and states $s \in S$, there exists a pure memoryless strategy $\sigma^* \in \Sigma^{PM}$ such that for all strategies $\sigma \in \Sigma$, we have $\mathrm{Val}_{dis}^\sigma(r, s, \beta) \le \mathrm{Val}_{dis}^{\sigma^*}(r, s, \beta)$.*

3   Memoryless Strategies Suffice for Pareto Optimality

In the sequel, we fix a discount factor β such that $0 \le \beta < 1$. Proposition 1 shows the existence of pure memoryless Pareto-optimal strategies.

Proposition 1. *There exist pure memoryless Pareto-optimal strategies for MDPs with multiple discounted reward objectives.*

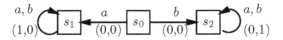

Fig. 1. MDP for Example 1

Proof. Given reward functions $r_1, \ldots, r_k$, consider a reward function $r_+ = r_1 + \cdots + r_k$, that is, for all $s \in S$, we have $r_+(s) = r_1(s) + \cdots + r_k(s)$. Let $\sigma^* \in \Sigma^{PM}$ be a pure memoryless optimal strategy for the reward function r_+ with the discounted reward objective with discount β (such a strategy exists by Theorem 1). We show that σ^* is Pareto-optimal. Assume towards contradiction that σ^* is not a Pareto-optimal strategy, then let $\sigma \in \Sigma$ be such that $\mathrm{Val}_{dis}^{\sigma^*}(\boldsymbol{r}, s, \beta) \leq \mathrm{Val}_{dis}^{\sigma}(\boldsymbol{r}, s, \beta)$, and for some j, $\mathrm{Val}_{dis}^{\sigma^*}(r_j, s, \beta) < \mathrm{Val}_{dis}^{\sigma}(r_j, s, \beta)$. Then we have $\mathrm{Val}_{dis}^{\sigma^*}(r_+, s, \beta) = \sum_{j=1}^{k} \mathrm{Val}_{dis}^{\sigma^*}(r_j, s, \beta) < \sum_{j=1}^{k} \mathrm{Val}_{dis}^{\sigma}(r_j, s, \beta) = \mathrm{Val}_{dis}^{\sigma}(r_+, s, \beta)$. This contradicts that σ^* is optimal for r_+. $\blacksquare$

The above proof can be generalized to any convex combination of the multiple objectives, that is, for positive weights $w_1, \ldots, w_k$, the optimal strategy for the single objective $\sum_i w_i \cdot r_i$ is Pareto-optimal. This technique is called the *weighted factor method* [4,10], and used commonly in engineering practice to find subsets of the Pareto set [5]. However, not all Pareto-optimal points are obtained in this fashion, as the following example shows.

Example 1. Consider the MDP from Fig. 1, with two actions a and b, and two reward functions r_1 and r_2. The transitions and the respective rewards are shown as labeled edges in the figure. Consider the discounted reward objectives for reward functions r_1 and r_2. For the pure memoryless strategies (and also the pure strategies) in this MDP, the possible value vectors are $(\frac{\beta}{1-\beta}, 0)$ and $(0, \frac{\beta}{1-\beta})$. However, consider a memoryless strategy σ_m that at state s_0 chooses action a and b each with probability $1/2$. For $\boldsymbol{r} = (r_1, r_2)$, we have $\mathrm{Val}_{dis}^{\sigma_m}(\boldsymbol{r}, s_0, \beta) = (\frac{\beta}{2 \cdot (1-\beta)}, \frac{\beta}{2 \cdot (1-\beta)})$. The strategy σ_m is Pareto-optimal and no pure memoryless strategy can achieve the corresponding value vector. Hence it follows that the pure strategies (and the pure memoryless strategies) do not suffice for Pareto optimality. Note that for all $0 < x < 1$, the memoryless strategy that chooses a with probability x, is a Pareto-optimal strategy, with value vector $(\frac{x \cdot \beta}{1-\beta}, \frac{(1-x) \cdot \beta}{1-\beta})$. Hence the set of Pareto-optimal value vectors may be uncountable and value vectors may have irrational values. $\blacksquare$

We now show that the family of memoryless strategies suffices for Pareto optimality. We assume the state space S is enumerated as $S = \{1, \ldots, n\}$. For a state $t \in S$, we define the reward function r_t by $r_t(s, a) = 1$ if $s = t$, and 0 otherwise, i.e., a reward of value 1 is gained whenever state t is visited. Similarly, we define the reward function $r_{t,b}$ for a state $t \in S$ and an action $b \in A$ by $r_{t,b}(s, a) = 1$ if $s = t$ and $b = a$, and 0 otherwise, i.e., a reward of value 1 is gained whenever state t is visited and action b is chosen. Given a strategy $\sigma \in \Sigma$

and a state $s \in S$, we define the discounted frequency of the state-action pair (t, a) as

$$\text{Freq}_s^\sigma(t, a, \beta) = \text{Val}_{dis}^\sigma(r_{t,a}, s, \beta) = \sum_{k=0}^{\infty} \beta^k \cdot \mathbb{E}_s^\sigma[\mathbf{1}_{(X_k = t, \theta_k = a)}],$$

and the discounted frequency of the state t as

$$\text{Freq}_s^\sigma(t, \beta) = \text{Val}_{dis}^\sigma(r_t, s, \beta) = \sum_{k=0}^{\infty} \beta^k \cdot \mathbb{E}_s^\sigma[\mathbf{1}_{(X_k = t)}].$$

Observe that $\sum_{a \in A} \text{Freq}_s^\sigma(t, a, \beta) = \text{Freq}_s^\sigma(t, \beta)$ for all $s, t \in S$ and $\sigma \in \Sigma$. For a memoryless strategy $\sigma_m \in \Sigma^M$ and a transition function δ, we denote by $\delta_{\sigma_m}(s, t) = \sum_{a \in A} \sigma_m(s)(a) \cdot \delta(s, a)(t)$ the probability of the transition from s to t given δ and σ_m.

Proposition 2. *Given a memoryless strategy $\sigma_m \in \Sigma^M$, consider a vector $z = \langle z_1, \ldots, z_n \rangle$ of variables, where $S = \{1, \ldots, n\}$. The set of n equations*

$$z_i = r_s(i) + \beta \cdot \sum_{j \in S} \delta_{\sigma_m}(j, i) \cdot z_j, \quad \text{for } i \in S,$$

has the unique solution $z_i = \text{Freq}_s^{\sigma_m}(i, \beta)$ for all $1 \leq i \leq n$.

Proof. To establish the desired claim we show that for all $i \in S$, we have $\text{Freq}_s^{\sigma_m}(i, \beta) = r_s(i) + \beta \cdot \sum_{j \in S} \delta_{\sigma_m}(j, i) \cdot \text{Freq}_s^{\sigma_m}(j, \beta)$. The uniqueness follows from arguments similar to the uniqueness of values under memoryless strategies (see [2]).

$$\begin{aligned}
\text{Freq}_s^{\sigma_m}(i, \beta) &= \sum_{k=0}^{\infty} \beta^k \cdot \mathbb{E}_s^{\sigma_m}[\mathbf{1}_{(X_k = i)}] \\
&= \sum_{k=0}^{\infty} \beta^k \cdot \mathbb{E}_s^{\sigma_m}[\sum_{j \in S} \mathbf{1}_{(X_{k-1} = j)} \, \delta_{\sigma_m}(j, i)] \\
&= r_s(i) + \sum_{k=1}^{\infty} \beta^k \cdot \sum_{j \in S} \mathbb{E}_s^{\sigma_m}[\mathbf{1}_{(X_{k-1} = j)}] \cdot \delta_{\sigma_m}(j, i) \\
&= r_s(i) + \beta \cdot \sum_{j \in S} \left(\sum_{k=1}^{\infty} \beta^{k-1} \mathbb{E}_s^{\sigma_m}[\mathbf{1}_{(X_{k-1} = j)}] \right) \cdot \delta_{\sigma_m}(j, i) \\
&= r_s(i) + \beta \cdot \sum_{j \in S} \left(\sum_{z=0}^{\infty} \beta^z \cdot \mathbb{E}_s^\sigma[\mathbf{1}_{(X_z = j)}] \right) \cdot \delta_{\sigma_m}(j, i) \\
&= r_s(i) + \beta \cdot \sum_{j \in S} \text{Freq}_s^{\sigma_m}(j, \beta) \cdot \delta_{\sigma_m}(j, i). \quad \blacksquare
\end{aligned}$$

Given a strategy $\sigma \in \Sigma$ and an initial state s, we define a memoryless strategy $\sigma_{f,s} \in \Sigma^M$ from the discounted frequency of the strategy σ as follows:

$$\sigma_{f,s}(t)(a) = \frac{\text{Freq}_s^\sigma(t, a, \beta)}{\text{Freq}_s^\sigma(t, \beta)}, \quad \text{for all } t \in S \text{ and } a \in A.$$

Since $\sum_{a \in A} \text{Freq}_s^\sigma(t, a, \beta) = \text{Freq}_s^\sigma(t, \beta)$ and $\text{Freq}_s^\sigma(t, a, \beta) \geq 0$, it follows that $\sigma_{f,s}(t)$ is a probability distribution. Thus $\sigma_{f,s}$ is a memoryless strategy. From Proposition 2, and the identity $\text{Freq}_s^\sigma(i, \beta) = r_s(i) + \beta \cdot \sum_{j \in S} \text{Freq}_s^\sigma(j, \beta) \cdot \delta_{\sigma_{f,s}}(j)$, we obtain the following lemma.

Lemma 1. *For all strategies $\sigma \in \Sigma$ and states $i, s \in S$, we have $\text{Freq}_s^\sigma(i, \beta) = \text{Freq}_s^{\sigma_{f,s}}(i, \beta)$.*

Proof. We show that $\text{Freq}_s^\sigma(i, \beta) = r_s(i) + \beta \cdot \sum_{j \in S} \text{Freq}_s^\sigma(j, \beta) \cdot \delta_{\sigma_{f,s}}(j)$. The result then follows from Proposition 2.

$$
\begin{aligned}
\text{Freq}_s^\sigma(i, \beta) &= \sum_{k=0}^\infty \beta^k \cdot \mathbb{E}_s^\sigma[\mathbf{1}_{(X_k=i)}] \\
&= \sum_{k=0}^\infty \beta^k \cdot \mathbb{E}_s^\sigma[\sum_{j \in S} \sum_{a \in A} \mathbf{1}_{(X_{k-1}=j, \theta_{k-1}=a)} \delta(j, a)(i)] \\
&= r_s(i) + \sum_{k=1}^\infty \beta^k \cdot \sum_{j \in S} \sum_{a \in A} \mathbb{E}_s^\sigma[\mathbf{1}_{(X_{k-1}=j, \theta_{k-1}=a)}] \cdot \delta(j, a)(i) \\
&= r_s(i) + \beta \cdot \sum_{j \in S} \sum_{a \in A}(\sum_{k=1}^\infty \beta^{k-1} \mathbb{E}_s^\sigma[\mathbf{1}_{(X_{k-1}=j, \theta_{k-1}=a)}]) \cdot \delta(j, a)(i) \\
&= r_s(i) + \beta \cdot \sum_{j \in S} \sum_{a \in A} \text{Freq}_s^\sigma(j, a, \beta) \cdot \delta(j, a)(i) \\
&= r_s(i) + \beta \cdot \sum_{j \in S} \sum_{a \in A}\left(\text{Freq}_s^\sigma(j, \beta) \cdot \tfrac{\text{Freq}_s^\sigma(j, a, \beta)}{\text{Freq}_s^\sigma(j, \beta)}\right) \cdot \delta(j, a)(i) \\
&= r_s(i) + \beta \cdot \sum_{j \in S} \text{Freq}_s^\sigma(j, \beta) \cdot \left(\sum_{a \in A} \tfrac{\text{Freq}_s^\sigma(j, a, \beta)}{\text{Freq}_s^\sigma(j, \beta)}\right) \cdot \delta(j, a)(i) \\
&= r_s(i) + \beta \cdot \sum_{j \in S} \text{Freq}_s^\sigma(j, \beta) \cdot \sum_{a \in A} \sigma_{f,s}(j)(a) \cdot \delta(j, a)(i) \\
&= r_s(i) + \beta \cdot \sum_{j \in S} \text{Freq}_s^\sigma(j, \beta) \cdot \delta_{\sigma_{f,s}}(j, i). \qquad\blacksquare
\end{aligned}
$$

Corollary 1. *For all strategies $\sigma \in \Sigma$, all states $i, s \in S$, and all actions $a \in A$, we have $\text{Freq}_s^\sigma(i, a, \beta) = \text{Freq}_s^{\sigma_{f,s}}(i, a, \beta)$.*

Proof. The following equalities follow from the definitions and Lemma 1:

$$
\begin{aligned}
\text{Freq}_s^{\sigma_{f,s}}(i, a, \beta) &= \text{Freq}_s^{\sigma_{f,s}}(i, \beta) \cdot \sigma_{f,s}(i)(a) \\
&= \text{Freq}_s^\sigma(i, \beta) \cdot \tfrac{\text{Freq}_s^\sigma(i, a, \beta)}{\text{Freq}_s^\sigma(i, \beta)} \\
&= \text{Freq}_s^\sigma(i, a, \beta). \qquad\blacksquare
\end{aligned}
$$

Theorem 2. *For all reward functions r, all strategies $\sigma \in \Sigma$, and all states $s \in S$, we have $\text{Val}_{dis}^\sigma(r, s, \beta) = \text{Val}_{dis}^{\sigma_{f,s}}(r, s, \beta)$.*

Proof. The result is proved as follows:

$$
\begin{aligned}
\text{Val}_{dis}^\sigma(r, s, \beta) &= \sum_{k=0}^\infty \beta^k \cdot \mathbb{E}_s^\sigma[r(X_k, \theta_k)] \\
&= \sum_{k=0}^\infty \beta^k \cdot \mathbb{E}_s^\sigma[\sum_{i \in S} \sum_{a \in A} r(i, a) \cdot \mathbf{1}_{(X_k=i, \theta_k=a)}] \\
&= \sum_{i \in S} \sum_{a \in A}(\sum_{k=0}^\infty \beta^k \cdot \mathbb{E}_s^\sigma[\mathbf{1}_{(X_k=i, \theta_k=a)}]) \cdot r(i, a) \\
&= \sum_{i \in S} \sum_{a \in A} \text{Freq}_s^\sigma(i, a, \beta) \cdot r(i, a) \\
&= \sum_{i \in S} \sum_{a \in A} \text{Freq}_s^{\sigma_{f,s}}(i, a, \beta) \cdot r(i, a) \quad \text{(by Corollary 1)}.
\end{aligned}
$$

Similarly, it follows that $\text{Val}_{dis}^{\sigma_{f,s}}(r, s, \beta) = \sum_{i \in S} \sum_{a \in A} \text{Freq}_s^{\sigma_{f,s}}(i, a, \beta) \cdot r(i, a)$. This establishes the result. $\blacksquare$

Theorem 2 yields Theorem 3, and since the set of memoryless strategies is convex, it also shows that the set of Pareto-optimal points is convex.

Theorem 3. *Given an MDP with multiple reward functions r, for all strategies $\sigma \in \Sigma$ and all states $s \in S$, the memoryless strategy $\sigma_{f,s} \in \Sigma^M$ satisfies $\text{Val}_{dis}^\sigma(r, s, \beta) = \text{Val}_{dis}^{\sigma_{f,s}}(r, s, \beta)$. Consequently, the memoryless strategies suffice for Pareto optimality for MDPs with multiple discounted reward objectives.*

4   Approximating the Pareto Curve

Pareto curve. Let M be an MDP with k reward functions $r = \langle r_1, \ldots, r_k \rangle$. The *Pareto curve* $P_{dis}(M, s, \beta, r)$ of the MDP M at state s with respect to discounted reward objectives is the set of all k-vectors of values such that for each $\boldsymbol{v} \in P_{dis}(M, s, \beta, \boldsymbol{r})$, there is a Pareto-optimal strategy σ such that $\mathrm{Val}^{\sigma}_{dis}(\boldsymbol{r}, s, \beta) = \boldsymbol{v}$. We are interested not only in the values, but also in the Pareto-optimal strategies. We often blur the distinction and refer to the Pareto curve $P_{dis}(M, s, \beta, \boldsymbol{r})$ as a set of strategies that achieve the Pareto-optimal values (if there is more than one strategy that achieves the same value vector, then $P_{dis}(M, s, \beta, \boldsymbol{r})$ contains at least one of them). For an MDP M and a real $\epsilon > 0$, an ϵ-*approximate Pareto curve*, denoted $P^{\epsilon}_{dis}(M, s, \beta, \boldsymbol{r})$, is a set of strategies in Σ such that there is no strategy $\sigma' \in \Sigma$ such that for all strategies $\sigma \in P^{\epsilon}_{dis}(M, s, \beta, \boldsymbol{r})$, we have $\mathrm{Val}^{\sigma'}_{dis}(r_i, s, \beta) \geq (1 + \epsilon) \cdot \mathrm{Val}^{\sigma}_{dis}(r_i, s, \beta)$ for all $1 \leq i \leq k$. That is, an ϵ-approximate Pareto curve contains enough strategies such that every Pareto-optimal strategy is "almost" dominated by some strategy in $P^{\epsilon}_{dis}(M, s, \beta, \boldsymbol{r})$.

Multi-objective linear programming. A *multi-objective linear program* L consists of (i) a set of k objective functions $o_1, \ldots, o_k$, where $o_i(\boldsymbol{x}) = \boldsymbol{c}^T_i \cdot \boldsymbol{x}$, for a vector $\boldsymbol{c}_i$ of coefficients and a vector $\boldsymbol{x}$ of variables; and (ii) a set of linear constraints specified by $A \cdot \boldsymbol{x} \geq \boldsymbol{b}$, for a matrix A and a value vector $\boldsymbol{b}$. A valuation of $\boldsymbol{x}$ is a *solution* if it satisfies the set (ii) of linear constraints. A solution $\boldsymbol{x}$ is *Pareto-optimal* if there is no other solution $\boldsymbol{x}'$ such that both $\langle o_1(\boldsymbol{x}), \ldots, o_k(\boldsymbol{x}) \rangle \leq \langle o_1(\boldsymbol{x}'), \ldots, o_k(\boldsymbol{x}') \rangle$ and $\langle o_1(\boldsymbol{x}), \ldots, o_k(\boldsymbol{x}) \rangle \neq \langle o_1(\boldsymbol{x}'), \ldots, o_k(\boldsymbol{x}') \rangle$. Given a multi-objective linear program L, the *Pareto curve* for L, denoted $P(L)$, is the set of k-vectors $\boldsymbol{v}$ of values such that there is a Pareto-optimal solution $\boldsymbol{x}$ of L with $\boldsymbol{v} = \langle o_1(\boldsymbol{x}), \ldots, o_k(\boldsymbol{x}) \rangle$. The definition of ϵ-*approximate Pareto curves* $P^{\epsilon}(L)$ for a multi-objective linear program L and a real $\epsilon > 0$, is analogous to the definition of ϵ-approximate Pareto curves for multi-objective MDPs given above.

Theorem 4. [7] *Given a multi-objective linear program L with k objective functions, the following assertions hold:*

1. *For all $\epsilon > 0$, there exists an ϵ-approximate Pareto curve $P^{\epsilon}(L)$ whose size is is polynomial in $|L|$ and $\frac{1}{\epsilon}$, and exponential in k.*
2. *For all $\epsilon > 0$, there exists an algorithm to construct an ϵ-approximate Pareto curve $P^{\epsilon}(L)$ in time polynomial in $|L|$ and $\frac{1}{\epsilon}$, and exponential in k.*

Proof. Part 1 is a direct consequence of Theorem 1 of [7]. Part 2 follows from Theorem 3 of [7] and the fact that linear programming can be solved in polynomial time.   ∎

Solving MDPs by linear programming. Given an MDP $M = (S, A, \delta)$ with state space $S = \{1, \ldots, n\}$, a reward function r, and a discount factor $0 \leq \beta < 1$, the discounted reward objective can be computed as the optimal solution of a linear program [2]. For multi-objective MDPs, we extend the standard linear programming formulation as follows. Given MDP M an discount factor β as before, an initial state s, and reward functions $r_1, \ldots, r_k$, the multi-objective linear program has the set $\{ x(t, a) \mid t \in S \text{ and } a \in A \}$ of variables. Intuitvely, the variable $x(t, a)$ represents the discounted frequency of the state-action pair

(t, a) when the starting state is s. The constraints of the multi-objective linear program over the variables $x(\cdot, \cdot)$ are given by:

$$\sum_{a \in A} x(t, a) = r_s(t) + \beta \cdot \sum_{u \in S} \sum_{a_1 \in A} \delta(u, a_1)(t) \cdot x(u, a_1), \quad \text{for } t \in S;$$

$$x(t, a) \geq 0, \qquad \qquad \qquad \text{for } t \in S, a \in A.$$
$$(1)$$

Equation (1) provides constraints on the discounted frequencies. The k objective functions are

$$\max \sum_{t \in S} \sum_{a \in A} r_i(t, a) \cdot x(t, a), \quad \text{for } i \in \{1, \ldots, k\}.$$

Consider any solution $x(t, a)$, for $t \in S$ and $a \in A$, of this linear program. Let $x(t) = \sum_{a \in A} x(t, a)$. The solution derives a memoryless strategy that chooses action a at state t with probability $\frac{x(t,a)}{x(t)}$. The linear program with the i-th objective function asks to maximize the discounted reward for the i-th reward function r_i over the set of all memoryless strategies. The optimal solution for the linear program with only the i-th objective also derives an optimal memoryless strategy for the reward function r_i. Furthermore, given a solution of the linear program, or equivalently, the memoryless strategy derived from the solution, we can compute the corresponding payoff profile in polynomial time, because the MDP reduces to a Markov chain when the strategy is fixed.

We denote by $L_{dis}(M, s, \beta, r)$ the multi-objective linear program defined above for the memoryless strategies of an MDP M, state s of M, discount factor β, and reward functions $r = \langle r_1, \ldots, r_k \rangle$. Let $P(L_{dis}(M, s, \beta, r))$ be the Pareto curve for this multi-objective linear program. With abuse of notation, we write $P(L_{dis}(M, s, \beta, r))$ also for the set of memoryless strategies that are derived from the Pareto-optimal solutions of the multi-objective linear program. It follows that the Pareto curve $P(L_{dis}(M, s, \beta, r))$ characterizes the set of memoryless Pareto-optimal points for the MDP with k discounted reward objectives. Since memoryless strategies suffice for Pareto optimality for discounted reward objectives (Theorem 3), the following lemma is immediate. Theorem 5 follows from Theorem 4 and Lemma 2.

Lemma 2. *Given an MDP M with k reward functions r, a state s of M, and a discount factor $0 \leq \beta < 1$, let $L_{dis}(M, s, \beta, r)$ be the corresponding multi-objective linear program. The following assertions hold:*

1. *$P(L_{dis}(M, s, \beta, r)) = P_{dis}(M, s, \beta, r)$, that is, the Pareto curves for the linear program and the discounted reward MDP coincide.*
2. *For all $\epsilon > 0$ and all ϵ-approximate Pareto curves $P^\epsilon(L_{dis}(M, s, \beta, r))$ of $L_{dis}(M, s, \beta, r)$, there is an ϵ-approximate Pareto curve $P^\epsilon_{dis}(M, s, \beta, r)$ such that $P^\epsilon(L_{dis}(M, s, \beta, r)) = P^\epsilon_{dis}(M, s, \beta, r)$.*

Theorem 5. *Given an MDP M with k reward functions r and a discount factor $0 \leq \beta < 1$, the following assertions hold:*

1. *For all $\epsilon > 0$, there exists an ϵ-approximate Pareto curve $P_{dis}^\epsilon(M, s, \beta, r)$ whose size is polynomial in $|M|$, $|\beta|$, $|r|$, and $\frac{1}{\epsilon}$, and exponential in k.*
2. *For all $\epsilon > 0$, there exists an algorithm to construct an ϵ-approximate Pareto curve $P_{dis}^\epsilon(M, s, \beta, r)$ in time polynomial in $|M|$, $|\beta|$, $|r|$, and $\frac{1}{\epsilon}$, and exponential in k.*

Theorem 5 shows that the Pareto curve can be efficiently ϵ-approximated. Recall that it follows from Example 1 that the set of Pareto-optimal points may be uncountable and the values may be irrational. Hence the ϵ-approximation of the Pareto curve is a useful finite approximation. The approximate Pareto curve allows us to answer trade-off queries about multi-objective MDPs. Specifically, given a multi-objective MDP M with k reward functions r and discount factor β, and a value profile $w = \langle w_1, \ldots, w_k \rangle$, we can check whether w is ϵ-close to a Pareto-optimal point at state s by constructing $P_{dis}^\epsilon(M, s, \beta, r)$ in polynomial time, and checking that there is some strategy in $P_{dis}^\epsilon(M, s, \beta, r)$ whose payoff profile is ϵ-close to w.

5   Pareto Realizability

In this section we study two related aspects of multi-objective MDPs: Pareto realizability, and pure memoryless Pareto realizability. The *Pareto realizability problem* asks, given a multi-objective MDP M with reward functions $r = \langle r_1, \ldots, r_k \rangle$ and discount factor $0 \leq \beta < 1$, a state s of M, and a value profile $w = \langle w_1, \ldots, w_k \rangle$ of k rational numbers, whether there exists a strategy σ such that $\mathrm{Val}_{dis}^\sigma(r, s, \beta) \geq w$. Observe that such a strategy exists if and only if there is a Pareto-optimal strategy σ' such that $\mathrm{Val}_{dis}^{\sigma'}(r, s, \beta) \geq w$. Also observe that it follows from Lemma 2 that a value profile w is realizable if and only if it is realizable by a memoryless strategy. The *pure memoryless* Pareto realizability problem further requires this strategy to be pure and memoryless.

The Pareto realizability problem arises when certain target behaviors are required, and one wishes to check if they can be attained on the model. Pure Pareto realizability arises in situations, such as circuit implementations, where the implemented strategy does not have access to randomization.

Theorem 6. *The Pareto realizability problem for MDPs with multiple discounted reward objectives can be solved in polynomial time. The pure memoryless Pareto realizability problem for MDPs with multiple discounted reward objectives is NP-complete.*

Proof. We show that Pareto realizability is in polynomial time by reduction to linear programming. The reduction is obtained as follows: along with the constraints defined by Equation (1) we add the constraints

$$w_i \leq \sum_{t \in S} \sum_{a \in A} x(t, a) \cdot r_i(t, a), \quad \text{for } i \in \{1, \ldots, k\}.$$

The original constraints from Equation (1) provide constraints on the discounted frequencies. The additional new constraints ensure that the payoff value for each reward function r_i is greater than or equal to the corresponding profile value w_i. Thus, if the set is consistent, then the answer to the Pareto realizability problem is "yes," and if inconsistent, the answer is "no." Consistency of this set can be checked in polynomial time using linear programming.

Pure and memoryless Pareto realizability is in NP since we can guess a pure memoryless strategy and compute its payoff values in polynomial time. We can then check that each payoff value is greater than or equal to the given profile value. It is NP-hard by reduction from subset sum. The subset sum problem takes as input natural numbers $\{a_1, \ldots, a_n\}$, and a natural number p, and asks if there exist $v_1, \ldots, v_n$ in $\{0, 1\}$ such that $a_1 \cdot v_1 + \cdots + a_n \cdot v_n = p$. It is NP-complete [3].

For an instance of the subset sum problem, we construct an MDP with two reward functions as follows. We assume for clarity that $\beta = 1$. The construction can be adapted for any fixed discount factor by suitably scaling the rewards. The MDP has $n + 1$ states, numbered from 1 to $n + 1$. We fix the start state to be 1. There are two actions, L and R. The transition relation is deterministic, for state $i \in \{1, \ldots, n\}$, we have $\delta(i, L)(i+1) = \delta(i, R)(i+1) = 1$. For state $n + 1$, we have $\delta(n + 1, L)(n + 1) = \delta(n + 1, R)(n + 1) = 1$. The reward function r_1 is defined as $r_1(i, L) = a_i$, and $r_1(i, R) = 0$ for $i \in \{1, \ldots, n\}$, and $r_1(n + 1, L) = r_1(n+1, R) = 0$. Similarly, the reward function r_2 is defined as $r_2(i, R) = a_i$, and $r_2(i, L) = 0$ for $i \in \{1, \ldots, n\}$, and $r_2(n+1, L) = r_2(n+1, R) = 0$. We now ask if the value profile $(p, \sum_i a_i - p)$ is pure Pareto realizable for this MDP. From the construction, it is clear that this profile is pure memoryless Pareto realizable iff the answer to the subset sum problem is "yes". In fact, the pure strategy that realizes the profile provides the required v_i's: if action L is played at state i, then $v_i = 1$, else $v_i = 0$. Since the MDP is a DAG, the hardness construction holds if we require the realizing strategy to be pure (not necessarily memoryless).   ∎

The *pure* Pareto realizability problem requires the realizing strategy to be pure, but not necessarily memoryless. It follows from the reduction given above that the pure Pareto realizability problem for MDPs with multiple discounted reward objectives is NP-hard; however, we do not have a characterization of the exact complexity of the problem.

6   Limit-Average Reward Objectives

We now briefly discuss the class of limit-average reward objectives, which is widely studied in the context of MDPs. Given a reward function r, the *limit-average reward* for a strategy σ at an initial state s is $\mathrm{Val}^{\sigma}_{avg}(r, s) =$

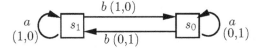

Fig. 2. MDP for Example 2

$\limsup_{k\to\infty} \frac{1}{k} \cdot \sum_{t=0}^{k} \mathbb{E}_s^\sigma [r(X_t, \theta_t)]$. With this definition, Theorem 1 holds for a single limit-average reward objective, and Proposition 1 extends to multiple limit-average reward objectives. Moreover, a simple adaptation of Example 1 shows that the pure strategies do not suffice for Pareto optimality for limit-average reward objectives. Unfortunately, Theorem 3 does not generalize. Example 2 below shows that for limit-average reward objectives, the family of memoryless strategies does not capture all Pareto-optimal strategies. However, it is still possible that the Pareto curve for limit-average reward objectives can be approximated in polynomial time. This remains an open problem.

Example 2. Fig. 2 shows an MDP with two actions a and b, and two reward functions r_1 and r_2. The transitions and the respective rewards are shown as labeled edges in the figure. Consider the limit-average reward objectives for r_1 and r_2. Given any memoryless strategy σ_m, at s_0 we have $\text{Val}_{avg}^{\sigma_m}(s_0, r_1) + \text{Val}_{avg}^{\sigma_m}(s_0, r_2) = 1$. We now consider the following strategy σ, which is played in rounds. In round j, the strategy σ first goes to state s_1, chooses action a (i.e., stays in s_1) unless the average for reward r_1 is at least $1 - \frac{1}{j}$, then goes to state s_0, chooses action a unless the average reward for reward r_2 is at least $1 - \frac{1}{j}$, and then proceeds to round $j + 1$. Given σ, we have $\text{Val}_{avg}^{\sigma}(s_0, r_1) = 1$ and $\text{Val}_{avg}^{\sigma}(s_0, r_2) = 1$. There is no memoryless Pareto-optimal strategy to achieve this value vector. ∎

References

1. O. Etzioni, S. Hanks, T. Jiang, R.M. Karp, O. Madari, and O. Waarts. Efficient information gathering on the internet. In *FOCS 96*, pages 234–243. IEEE, 1996.
2. J. Filar and K. Vrieze. *Competitive Markov Decision Processes*. Springer, 1997.
3. M.R. Garey and D.S. Johnson. *Computers and Intractability*. W.H. Freeman, 1979.
4. R. Hartley. *Finite Discounted Vector Markov Decision Processes*. Technical Report, Department of Decision Theory, Manchester University, 1979.
5. J. Koski. Multicriteria truss optimization. In *Multicriteria Optimization in Engineering and in the Sciences*. 1988.
6. G. Owen. *Game Theory*. Academic Press, 1995.
7. C.H. Papadimitriou and M. Yannakakis. On the approximability of trade-offs and optimal access of web sources. In *FOCS 00*, pages 86–92. IEEE, 2000.
8. M.L. Puterman. *Markov Decision Processes*. Wiley, 1994.
9. R. Szymanek, F. Catthoor, and K. Kuchcinski. Time-energy design space exploration for multi-layer memory architectures. In *DATE 04*. IEEE, 2004.
10. D.J. White. Multi-objective infinite-horizon discounted Markov decision processes. *Journal of Mathematical Analysis and Applications*, 89:639–647, 1982.
11. P. Yang and F. Catthoor. Pareto-optimization based run time task scheduling for embedded systems. In *CODES-ISSS 03*, pages 120–125. ACM, 2003.

The Algorithmic Structure of Group Strategyproof Budget-Balanced Cost-Sharing Mechanisms[*]

Paolo Penna and Carmine Ventre

Dipartimento di Informatica ed Applicazioni "R.M. Capocelli",
Università di Salerno, Italy
{penna, ventre}@dia.unisa.it

Abstract. We study mechanisms for *cooperative* cost-sharing games satisfying: *voluntary participation* (i.e., no user is forced to pay more her valuation of the service), *consumer sovereignty* (i.e, every user can get the service if her valuation is large enough), *no positive transfer* (i.e., no user receives money from the mechanism), *budget balance* (i.e., the total amount of money that users pay is equal to the cost of servicing them), and *group strategyproofness* (i.e., the mechanism is resistant to coalitions).

We show that mechanisms satisfying all these requirements must obey certain *algorithmic properties* (which basically specify how the serviced users are selected). Our results yield a *characterization of upper continuous mechanisms* (this class is interesting as all known general techniques yield mechanisms of this type). Finally, we extend some of our negative results and obtain the first negative results on the existence of mechanisms satisfying all requirements above. We apply these results to an interesting generalization of cost-sharing games in which the mechanism cannot service certain "forbidden" subsets of users. These *generalized cost-sharing games* correspond to natural variants of known cost-sharing games and have interesting practical applications (e.g., sharing the cost of multicast transmissions which cannot be encrypted).

1 Introduction

Consider a set U of n users that wish to buy a certain service from some service providing company P. Each user $i \in U$ valuates the service offered an amount equal to v_i. This value represents how much user i would benefit from being serviced. Alternatively, v_i quantifies the maximum amount of money that user i is willing to pay for getting the service. The service provider must then develop a so called *mechanism*, that is, a policy for deciding (i) which users should be serviced and (ii) the price that each of them should pay for getting the service.

Mechanisms are complex auctions where users are asked to report their willingness to pay which, in the end, determines the mechanism outcome (i.e., the

---

[*] Work supported by the European Project FP6-15964, Algorithmic Principles for Building Efficient Overlay Computers (AEOLUS).

B. Durand and W. Thomas (Eds.): STACS 2006, LNCS 3884, pp. 337–348, 2006.

serviced users and the prices). In particular, the value v_i is known to user i but not to the provider. Hence, users may act *selfishly* and misreport v_i (e.g., trying to get the service for a better price). *Group strategyproof* mechanisms are "resistent" to coalitions of selfish users (see below for a formal definition), and thus particularly appealing for *cost-sharing games* requiring some "reasonable" share of the costs among the (possibly selfish) users.

An instance of a *cost-sharing games* is a pair $\mathcal{I} = (U, C)$, where U is a set of n *users*, and the *cost function* $C : 2^U \rightarrow \mathbf{R}^+ \cup \{0\}$ gives the cost $C(Q) > 0$ of servicing all users in a non-empty set $Q \subseteq U$. Each user is a *selfish agent* reporting some bid value b_i (possibly different from v_i); the true value v_i is *privately known* to agent i. Based on the *reported* values $\mathbf{b} = (b_1, \ldots, b_n)$ a *mechanism* $M = (\mathsf{A}, P)$ uses an algorithm A to select a subset $\mathsf{A}(\mathbf{b}|\mathcal{I}) \in 2^U$ of users to service. Moreover, according to the payment functions $P = (P^1, \ldots, P^n)$, each user $i \in \mathsf{A}(\mathbf{b}|\mathcal{I})$ must pay $P^i(\mathbf{b}|\mathcal{I})$ for getting the service. (Users that do not get serviced do not pay.) Hence, the *utility* of agent i when she reports b_i, and the other agents report $\mathbf{b}_{-i} := (b_1, \ldots, b_{i-1}, b_{i+1}, \ldots, b_n)$, is equal to

$$u_i^M(b_i, \mathbf{b}_{-i}|\mathcal{I}) := \begin{cases} v_i - P^i(b_i, \mathbf{b}_{-i}|\mathcal{I}) & \text{if } i \in \mathsf{A}(b_i, \mathbf{b}_{-i}|\mathcal{I}), \\ 0 & \text{otherwise.} \end{cases}$$

Developing economically viable cost-sharing *mechanisms* is a central problem in (cooperative) game theory. In particular, there is a number of natural constraints/goals that, for every instance $\mathcal{I} = (U, C)$ and for every $\mathbf{v} = (v_1, \ldots, v_n)$, a mechanism $M = (\mathsf{A}, P)$ should satisfy/meet: [1]

1. **α-Approximate Budget Balance (α-BB).** The prices charged to all users should recover the cost of servicing them and, at the same time, should not be more than $\alpha > 1$ times this cost. In particular, we require that

$$C(\mathsf{A}(\mathbf{b})) \leq \sum_{i \in \mathsf{A}(\mathbf{b}|\mathcal{I})} P^i(\mathbf{b}|\mathcal{I}) \leq \alpha \cdot C(\mathsf{A}(\mathbf{b})). \tag{1}$$

   The lower bound guarantees that there is no loss for the provider. The upper bound implies that a competitor could offer a better price to all users only if coming up with payments such that the above condition is satisfied for some $1 \leq \alpha' < \alpha$. Ideally, one wishes the **budget-balance (BB)** condition, that is, the case $\alpha = 1$. In this case, no competitor can offer better prices to all users in $\mathsf{A}(\mathbf{b})$ without running into a loss. (The cost $C(Q)$ is "common" to all providers and represents the "minimum" cost for servicing Q.)
2. **No Positive Transfer (NPT).** No user receives money from the mechanism, i.e., $P^i(\cdot) \geq 0$.
3. **Voluntary Participation (VP).** We never charge a user an amount of money greater than her *reported* valuation, that is, $\forall b_i, \forall \mathbf{b}_{-i}$ it holds that $b_i \geq P^i(b_i, \mathbf{b}_{-i}|\mathcal{I})$. In particular, a user has always the option of not paying for a service for which she is not interested. Moreover, $P^i(\mathbf{b}|\mathcal{I}) = 0$, for all $i \notin \mathsf{A}(\mathbf{b}|\mathcal{I})$, i.e., only the users getting the service will pay.

---

[1] Notice that we need to consider all possible $\mathbf{v} = (v_1, \ldots, v_n)$ since the mechanism does not known these values.

4. **Consumer Sovereignty (CS).** Every user is guaranteed to get the service if she reports a high enough valuation, that is, $\forall\ \mathbf{b}_{-i}$, $\exists\ \bar{b}_i = \bar{b}_i(\mathbf{b}_{-i})$ such that $i \in \mathsf{A}(\bar{b}_i, \mathbf{b}_{-i}|\mathcal{I})$.

5. **Group Strategyproofness (GSP).** We require that a user $i \in U$ that misreport her valuation (i.e., $b_i \neq v_i$) cannot improve her utility nor improve the utility of other users without worsening her own utility (otherwise, a coalition C containing i would secede). Consider a coalition $C \subseteq U$ of users. For any two vectors $\mathbf{x}$ and $\mathbf{y}$ of length n, $(\mathbf{x}_C, \mathbf{y}_{-C})$ denotes the vector $\mathbf{z} = (z_1, \dots, z_n)$ such that $z_i = x_i$ if $i \in C$ and $z_i = y_i$ if $i \notin C$. The group strategyproofness requires that if the inequality

$$u_i^M(\mathbf{b}_C, \mathbf{v}_{-C}) \geq u_i^M(\mathbf{v}_C, \mathbf{v}_{-C}) \tag{2}$$

holds for all $i \in C$ then it must hold with equality for all $i \in C$ as well. Notice that, since we require the condition on Eq. 2 to hold for every $\mathbf{v} = (\mathbf{v}_C, \mathbf{v}_{-C})$, replacing $\mathbf{b}_{-C}$ by $\mathbf{v}_{-C}$ does not change the definition of group strategyproofness. Hence, the special case of $C = \{i\}$ yields the weaker notion of *strategyproofness*: $\forall b_i$ and $\forall \mathbf{b}_{-i}$ it holds that

$$u_i^M(v_i, \mathbf{b}_{-i}) \geq u_i^M(b_i, \mathbf{b}_{-i}), \tag{3}$$

for every user i.

Mechanisms satisfying all requirements above have been deeply investigated (see e.g. [10,9,3]). All known techniques yield mechanisms which select the final set $Q = \mathsf{A}(\mathbf{b})$ among a "sufficiently reach" family $\mathcal{P}^{\mathsf{A}} \subseteq 2^U$ of candidates. More specifically, an invariant of known mechanisms is that one can always find an order $i_1, \dots, i_n$ of the users such that *each* of the following subsets is given in output for some bid vector $\mathbf{b}$:

$$\underbrace{\{i_1, \dots, i_n\}}_{Q_1 = U}, \underbrace{\{i_2, \dots, i_n\}}_{Q_2}, \dots, \underbrace{\{i_j, \dots, i_n\}}_{Q_j}, \dots, \emptyset. \tag{4}$$

In general, an algorithm A may consider all possible subsets of U, that is, $\mathcal{P}^{\mathsf{A}} = 2^U$, meaning that every $Q \subseteq U$ is returned for some bid vector $\mathbf{b}$. In some cases, however, it may be convenient/necessary to *never* output certain subsets. There are (at least) two main reasons for this:

1. *Computational complexity.* Computing $C(Q)$ may be NP-hard for certain $Q \subseteq U$. In this case, it may be good to avoid $Q = \mathsf{A}(\mathbf{b})$ since otherwise $M = (\mathsf{A}, P)$ will not run in polynomial time, or it will only guarantee α-BB condition, for some $\alpha > 1$, unless P=NP.

2. *Generalized cost-sharing games.* In many practical applications, certain subsets $Q \subseteq U$ may be "forbidden" in the sense that it is impossible to service all and only those users in Q. We model these applications by introducing *generalized cost-sharing games* where instances are triples $\mathcal{I} = (U, \mathcal{P}, C)$, with $\mathcal{P} \subseteq 2^U$ and $C : \mathcal{P} \to \mathbf{R}^+ \cup \{0\}$. The set $\mathcal{P}$ contains the non-forbidden sets and thus we require $\mathsf{A}(\mathbf{b}) \in \mathcal{P}$, for all $\mathbf{b}$. (We assume $\emptyset \in \mathcal{P}$ and $C(Q) > 0$ for $Q \neq \emptyset$.)

As mentioned above, all known techniques yield mechanisms which are *sequential*, that is, there exists $\sigma = (i_1, \ldots, i_n)$ such that $\mathcal{P}^\sigma \subseteq \mathcal{P}^A$, where $\mathcal{P}^\sigma$ consists of all the subsets listed in Eq. 4 (see Def. 1). This poses severe limitations on which (generalized) cost-sharing games these techniques can "solve efficiently": (i) Polynomial running time can be achieved only if $C(\cdot)$ can be approximated in polynomial time within a factor α for all sets in $\mathcal{P}^\sigma$; (ii) For generalized cost-sharing games, the instance $\mathcal{I} = (U, \mathcal{P}, C)$ must satisfy $\mathcal{P}^\sigma \subseteq \mathcal{P}$. It is then natural to ask whether there exist mechanisms of a totally different type (i.e., *not* sequential) which are more powerful, that is, they are computationally more efficient and/or solve more (generalized) cost-sharing games.

In this work we prove that, for the natural class of *upper continuous* mechanisms [2] (see also Def. 2), the answer to this question is "no". And it remains "no" even if we allow the α-BB condition for an *arbitrarily large* $\alpha < \infty$ (e.g., $\alpha = n$). More specifically, for every upper continuous mechanism $M = (A, P)$ which is α-BB, VP, CS, NPT and GSP, it must be the case that A is sequential, for *every* $\alpha \geq 1$. Our proofs show an interesting phenomenon: for upper continuous mechanisms satisfying all but the α-BB condition above, the fact that A is not sequential creates a "gap" in the payments which either must be all 0 or cannot be bounded from above (i.e., for every $\beta > 0$ there exists $\mathbf{b}$ such that $P^i(\mathbf{b}) > \beta$).

Our result, combined with a simple upper continuous mechanism given in [14,2], shows that *sequential algorithms characterize upper continuous mechanisms*. This implies that generalized cost-sharing games admits such upper continuous mechanisms if and only if they admit sequential algorithms (see Corollary 5). In particular, relaxing BB to α-BB, for *any* $\alpha > 1$, would not allow for solving a wider class of problems; and the "simple" technique in [14,2] is not less powerful than more complex ones which yield upper continuous mechanisms.

Given our characterization, we can better understand which are the limitations of upper continuous mechanisms satisfying α-BB, NPT, VP, CS and GSP:

1. Polynomial-time mechanisms exist only if $C(\cdot)$ is approximable within polynomial time over $\mathcal{P}^\sigma$, for some σ. If we require BB, then $C(\cdot)$ must be polynomial-time computable over $\mathcal{P}^\sigma$, for some σ.
2. For generalized cost-sharing games, these mechanisms exist only for those instances $\mathcal{I} = (U, \mathcal{P}, C)$ satisfying $\mathcal{P}^\sigma \subseteq \mathcal{P}$, for some σ. Moreover, the factor α in the α-BB condition is totally irrelevant: if α-BB is possible then BB is possible too, for any $\alpha > 1$.

We stress that these are the first lower bounds on (upper continuous) mechanisms satisfying α-BB, NPT, VP, CS and GSP. On one hand, one cannot derive any lower bound on polynomial-time mechanisms from the computational complexity of approximating $C(\cdot)$: indeed, there exists cost-sharing games which admit (upper continuous) polynomial-time BB mechanisms satisfying NPT, VP, CS and GSP [14,15], while the cost function $C(\cdot)$ is NP-hard to approximate within some $\alpha > 1$. On the other hand, generalized cost-sharing games have not been investigated before, though many practical applications require them (see Sect. 4).

We also obtain *necessary* conditions for *general* (i.e., non upper continuous) mechanisms. We use these conditions (Def.s 3 and 4) to prove general lower bounds and that, for two users, upper continuous mechanisms are not less powerful than general ones (basically, every mechanism must be sequential – Corollary 3). We describe several applications of generalized cost-sharing games and of our results in Sect. 4.

Due to lack of space, some of the proofs are omitted; these proofs are available in the full version of the paper [13].

Related Work. Probably the simplest BB, NPT, VP, CS and GSP mechanism is the one independently described in [14,2]: Starting from U, drop users in some fixed order $\sigma = (i_1, \ldots, i_n)$, until some user i_r accepts to pay for the total cost of the current set, that is, $b_{i_r} \geq C(\{i_r, \ldots, i_n\})$.

More sophisticated mechanisms were already known from the seminal works by Moulin and Shenker [10,9]. Their mechanisms employ so called *cross-monotonic cost-sharing methods* which essentially divide the cost $C(Q)$ among all users in Q so that user i would not pay more if the mechanism expands the set Q to some $Q' \supset Q$. Cross-monotonic functions do not exists for several games of interest, thus requiring relaxing BB to α-BB, for some factor $\alpha > 1$ [8,2,1,7,5]. Moreover, cross-monotonicity is difficult to obtain in general (e.g., the works [3,12,6,1,5] derive these cost-sharing methods from the execution of non-trivial primal-dual algorithms).

In [14] the authors prove that Moulin and Shenker mechanisms also work for a wider class of cost-sharing methods termed *self cross-monotonic*. The simple mechanism described above is one of such mechanisms [14]. Also the polynomial-time mechanisms for the Steiner tree game in [14,15] are in this class.

Basically, all known mechanisms are upper continuous, except for the one in [2] which, however, requires $C(\cdot)$ being subadditive. All mechanisms in the literature are either variants of Moulin and Shenker mechanisms [10,9], or have been presented in [2]. In all cases, the mechanisms are sequential (and apart from those in [14,15], they use algorithms such that $\mathcal{P}^A = 2^U$).

Characterizations of BB, NPT, VP, CS and GSP mechanisms are known only for the following two cases: (i) the cost function $C(\cdot)$ is submodular [10,9], or (ii) the mechanism is upper continuous and with no free riders [2] [2]. In both cases, these mechanisms are characterized by cross-monotonic cost-sharing methods.

1.1   Preliminaries and Basic Results

Throughout the paper we let $A_i(\mathbf{b}) = 1$ if $i \in A(\mathbf{b})$, and $A_i(\mathbf{b}) = 0$ otherwise, for all i and all $\mathbf{b}$.

Definition 1. *For any ordering $\sigma = (i_1, \ldots, i_n)$ of the users, we let*

$$\mathcal{P}^\sigma := \{\emptyset\} \cup \{i_j, i_{j+1}, \ldots, i_n\}_{1 \leq j \leq n}.$$

---

[2] Mechanisms without free riders guarantee that all users in $A(\mathbf{b})$ pay something.

An algorithm A *is* sequential *if there exists* σ *such that* $\mathcal{P}^\sigma \subseteq \mathcal{P}^{\mathsf{A}}$. *An instance* $\mathcal{I} = (U, \mathcal{P}, C)$ *of a generalized cost-sharing game admits a sequential algorithm if* $\mathcal{P}^\sigma \subseteq \mathcal{P}$, *for some* σ. *A generalized cost-sharing game admits a sequential algorithm if every instance of the game does.*

Theorem 1 ([14,2]). *For any ordering* σ *of the users, there exists an upper continuous BB, NPT, VP, CS and GSP mechanism* $M = (\mathsf{A}, P)$ *such that* $\mathcal{P}^{\mathsf{A}} = \mathcal{P}^\sigma$. *Hence, every instance of a generalized cost-sharing game which admits a sequential algorithm, admits an upper continuous BB, NPT, VP, CS and GSP mechanism.*

The following lemma is a well-known result in mechanism design. (See also [13] for a proof.)

Lemma 1 ([16,11]). *For any strategyproof mechanism* $M = (\mathsf{A}, P)$ *the following conditions must hold:*

$$\mathsf{A}_i(b_i, \mathbf{b}_{-i}) = 1 \Rightarrow \forall b_i' > b_i, \ \mathsf{A}_i(b_i', \mathbf{b}_{-i}) = 1; \tag{5}$$
$$\mathsf{A}_i(b_i, \mathbf{b}_{-i}) = \mathsf{A}_i(b_i', \mathbf{b}_{-i}) \Rightarrow \ P^i(b_i, \mathbf{b}_{-i}) = P^i(b_i', \mathbf{b}_{-i}). \tag{6}$$

Lemma 1 and the CS condition imply that, for every i and every $\mathbf{b}_{-i}$, there exists a threshold $\theta_i(\mathbf{b}_{-i})$ such that agent i is serviced for all $b_i > \theta(\mathbf{b}_{-i})$, while for $b_i < \theta_i(\mathbf{b}_{-i})$ agent i is not serviced. The following kind of mechanism breaks ties, i.e. the case $b_i = \theta_i(\mathbf{b}_{-i})$, in a fixed manner:

Definition 2 (upper continuous mechanisms [2]). *A mechanism* $M = (\mathsf{A}, P)$ *is upper-continuous if* $\mathsf{A}_i(x, \mathbf{b}_{-i}) = 1$ *for all* $x \geq \theta_i(\mathbf{b}_{-i})$, *where* $\theta_i(\mathbf{b}_{-i}) := \inf\{y| \ \mathsf{A}_i(y, \mathbf{b}_{-i}) = 1\}$ *(This value exists unless the CS condition is violated.)*

We will use the following technical lemma to show that, if payments are bounded from above, then once a user i bids a "very high" b_i, then this user will have to be serviced no matter what the other agents report.

Lemma 2. *Let* $M = (\mathsf{A}, P)$ *be a strategyproof mechanism satisfying NPT, CS and* $\sum_{i \in U} P^i(\mathbf{b}) \leq \beta$, *for all* $\mathbf{b}$. *Then, there exists* $B = B(\beta) \geq 0$ *such that, for all* i *and all* $\mathbf{b}_{-i}$, $\mathsf{A}_i(B, \mathbf{b}_{-i}) = 1$.

2 Cost-Sharing Mechanisms and Strategyproofness

2.1 Two Necessary Conditions

We next show that strategyproof α-BB, NPT, VP, CS mechanisms must be able to service (i) all users and (ii) exactly one out of any pair $i, j \in U$. (Of course, for some bid vector $\mathbf{b}$.)

Definition 3. *An algorithm* A *satisfies the* full coverage *property if* $U \in \mathcal{P}^{\mathsf{A}}$, *that is, the algorithm decides to service all users for some bid vector* $\mathbf{b}$.

We next show that full coverage is a necessary condition for obtaining strate-gyproof mechanisms satisfying NPT and CS and whose prices are bounded from above (a necessary condition for α-BB).

Theorem 2. *If* A *does not satisfy the full coverage property, then any strate-gyproof mechanism* $M = (A, P)$ *satisfying NPT and CS will run in an unbounded surplus, that is, for every* $\beta > 0$, *there exists* b *such that* $\sum_{i \in U} P^i(\mathbf{b}) > \beta$.

Proof. We prove the contraposition. Suppose $\sum_{i \in U} P^i(\mathbf{b}) \leq \beta$, for all b. Then, Lemma 2 implies that $A(\mathbf{B}) = U$, for some constant $B \geq 0$ and for $\mathbf{B} = (B, \ldots, B)$.

Theorem 2 states that, if the mechanism is not able to service all users, then an unbounded surplus must be created. The result we will prove next is a sort of "dual": if the mechanism is not able to selectively service two users, then it will not collect any money.

Definition 4. *An algorithm* A *satisfies the* weak separation *property if, for any* $i, j \in U$, *the algorithm can return a feasible solution to service only one of them, that is, there exists* $Q \in \mathcal{P}^A$ *such that* $|Q \cap \{i, j\}| = 1$.

Condition weak separation is also necessary for strategyproof mechanisms:

Theorem 3. *If* A *does not satisfy the weak separation condition, then any strat-egyproof mechanism* $M = (A, P)$ *satisfying NPT, VP and CS will not collect any money from the users when they report some bid vector* b. *Moreover, mechanism* M *will service a subset* $Q \neq \emptyset$, *thus implying that, mechanism* M *cannot be* α-*BB, for any* $\alpha > 1$.

Proof. Since A does not satisfy the weak separation condition there exist $j, k \in U$ such that

$$\forall \mathbf{b}, \ A_j(\mathbf{b}) = A_k(\mathbf{b}). \tag{7}$$

Let $(x, \mathbf{0}_{-l})$ denote the vector having the l-th component equal to x and all others being equal 0. Consider the following three bid vectors:

$$\mathbf{b}^{(j)} := (\bar{b}_j, \mathbf{0}_{-j}) = (0, \ldots, 0, \bar{b}_j, 0, \ldots, 0, 0, 0, \ldots, 0)$$
$$\mathbf{b}^{(k)} := (\bar{b}_k, \mathbf{0}_{-k}) = (0, \ldots, 0, 0, 0, \ldots, 0, \bar{b}_k, 0, \ldots, 0)$$
$$\mathbf{b}^{(j,k)} := (0, \ldots, 0, \bar{b}_j, 0, \ldots, 0, \bar{b}_k, 0, \ldots, 0)$$

with $\bar{b}_j$ and $\bar{b}_k$ such that $A_j(\mathbf{b}^{(j)}) = 1$ and $A_k(\mathbf{b}^{(k)}) = 1$. (These two values exist by the CS condition.) Then, Eq. 7 implies $A_j(\mathbf{b}^{(k)}) = 1$ and $A_k(\mathbf{b}^{(j)}) = 1$. The CS and NPT conditions imply that $P^j(\mathbf{b}^{(k)}) = 0$ and $P^k(\mathbf{b}^{(j)}) = 0$. We apply Lemma 1 and obtain the following implications:

$$A_j(\mathbf{b}^{(k)}) = 1 \Rightarrow A_j(\mathbf{b}^{(j,k)}) = 1 \tag{8}$$
$$\Rightarrow P^j(\mathbf{b}^{(j,k)}) = P^j(\mathbf{b}^{(k)}) = 0, \tag{9}$$
$$A_k(\mathbf{b}^{(j)}) = 1 \Rightarrow A_k(\mathbf{b}^{(j,k)}) = 1 \tag{10}$$
$$\Rightarrow P^k(\mathbf{b}^{(j,k)}) = P^k(\mathbf{b}^{(j)}) = 0. \tag{11}$$

The VP condition, Eq. 9, and Eq. 11 imply that, $P^i(\mathbf{b}^{(j,k)}) = 0$, for $1 \leq i \leq n$. Taking $\mathbf{b} = \mathbf{b}^{(j,k)}$, we get the first part of the theorem. Moreover, Eq.s 8 and 10 prove the second part.

The following result follows from Theorems 2 and 3:

Corollary 1. *Any α-BB strategyproof mechanism $M = (A, P)$, also satisfying NPT, VP and CS, must use an algorithm A satisfying both the full coverage and weak separation properties.*

The above result implies the first lower bound on *polynomial-time α-BB, NPT, VP, CS, and GSP mechanisms:*

Corollary 2. *If $C(U)$ is NP-hard to approximate within a factor $\alpha \geq 1$, then no α-BB mechanism satisfying NPT, VP, CS and GSP can run in polynomial-time.*

2.2 Characterization of Mechanisms for Two Users

We will prove that, for the case of two users, full coverage and weak separation suffice for the existence of mechanisms. The following fact will be the key property:

Fact 4. *Any algorithm A satisfying the full coverage and weak separation conditions is a sequential algorithm for the case of two users. (Indeed, $U \in \mathcal{P}^A$ and $\{1\} \in \mathcal{P}^A$ or $\{2\} \in \mathcal{P}^A$.)*

The above fact and Corollary 1 imply the following:

Corollary 3. *For generalized cost-sharing games involving two users, the following are equivalent:*

1. *There exists a strategyproof α-BB mechanism $M_\alpha = (A_\alpha, P_\alpha)$ satisfying NPT, VP and CS;*
2. *Every instance $\mathcal{I} = (U, \mathcal{P}, C)$ admits a sequential algorithm A;*
3. *There exists a group strategyproof BB mechanism $M = (A, P)$ satisfying NPT, VP and CS.*

Our next result, whose proof is given in the full version of this work [13], shows that Corollary 3 does not apply to the case of three (or more) users.

Theorem 5. *There exists an instance $\mathcal{I} = (U, \mathcal{P}, C)$, with $|U| = 3$, which does not admit sequential algorithms. However, there exist a strategyproof mechanism $M = (A, P)$ satisfying BB, NPT, VP and CS for this instance. Hence, A is not sequential.*

We stress that the mechanism of the above theorem is not upper continuous nor GSP.

3 Characterization of Upper Continuous Mechanisms

We begin with the following technical lemma.

Lemma 3. *Let $M = (\mathsf{A}, P)$ be a GSP mechanism satisfying NPT and VP. For all $\mathbf{b}' = (b_i', \mathbf{b}_{-i})$ and $\mathbf{b}'' = (b_i'', \mathbf{b}_{-i})$ such that $\mathsf{A}_i(\mathbf{b}') = \mathsf{A}_i(\mathbf{b}'')$, the following holds for all $j \in U$: if $\mathsf{A}_j(\mathbf{b}') - \mathsf{A}_j(\mathbf{b}'')$ then $P^j(\mathbf{b}') = P^j(\mathbf{b}'')$.*

In this section we will consider bid vectors which take values 0 or some "sufficiently large" $B \geq 0$. Recall that a user bidding B is serviced no matter the other agents bids.

Definition 5. *For any mechanism $M = (\mathsf{A}, P)$ such that $\sum_{i \in U} P^i(\mathbf{b}) \leq \beta$, for all $\mathbf{b}$, we let $B = B(\beta)$ be the constant of Lemma 2, and $\mathcal{P}_\beta^\mathsf{A} := \{\mathsf{A}(\mathbf{b}) | \mathbf{b} \in \{0, B\}^n\} \subseteq \mathcal{P}^\mathsf{A}$. Moreover, $\boldsymbol{\beta}_Q$ denotes the vector whose i-th component is equal to B for $i \in Q$, and 0 otherwise.*

Lemma 4. *Let $M = (\mathsf{A}, P)$ be an upper continuous GSP mechanism satisfying NPT, VP and CS. Moreover, let $\sum_{i \in U} P^i(\mathbf{b}) \leq \beta$, for all $\mathbf{b}$. Then, for all $Q \in \mathcal{P}_\beta^\mathsf{A}$, it holds that $Q = \mathsf{A}(\boldsymbol{\beta}_Q)$.*

Theorem 6. *Let $M = (\mathsf{A}, P)$ be an upper continuous α-BB mechanism satisfying GSP, NPT, VP and CS. Then, for every $Q \in \mathcal{P}_\beta^\mathsf{A}$, there exists $i_Q \in Q$ such that $Q \setminus \{i_Q\} \in \mathcal{P}_\beta^\mathsf{A}$.*

Proof. Notice that α-BB implies $\sum_{i \in U} P^i(\mathbf{b}) \leq \beta$ for all bid vectors $\mathbf{b}$ with $\beta = \max_{Q \in \mathcal{P}^\mathsf{A}} \alpha C(Q)$. Thus from Lemma 4, we can assume $Q = \mathsf{A}(\boldsymbol{\beta}_Q)$. First of all, we claim that there is at least one user $i_Q \in Q$ such that $i_Q \notin \mathsf{A}(\boldsymbol{\beta}_{Q \setminus \{i_Q\}})$. Indeed, Lemma 1 implies that, for all $i \in \mathsf{A}(\boldsymbol{\beta}_{Q \setminus \{i\}})$, it must be the case that $P^i(\boldsymbol{\beta}_{Q \setminus \{i\}}) = 0$. Hence, if such an i_Q does not exist, then $\sum_{i \in Q} P^i(\mathbf{b}) = 0$, which contradicts the α-BB condition (i.e. $\sum_{i \in U} P^i(\mathbf{b}) > 0$ for all bid vectors $\mathbf{b}$).

Let us then consider $i_Q \in Q$ such that $i_Q \notin \mathsf{A}(\boldsymbol{\beta}_{Q \setminus \{i_Q\}})$, and let $R := Q \setminus \{i_Q\}$. Lemma 2 implies $R \subseteq \mathsf{A}(\boldsymbol{\beta}_R)$. By contradiction, assume $R \subset \mathsf{A}(\boldsymbol{\beta}_R)$ and let $k \in \mathsf{A}(\boldsymbol{\beta}_R) \setminus R$. We will show that a coalition $C = \{i_Q, k\}$ will violate the GSP condition. To this end, consider the following bid vectors which differ only in the i_Q-th coordinate. We let $* \in \{0, B\}$ denote the coordinates of these two vectors other than i_Q and k:

$$\mathbf{b}^{(1)} = \boldsymbol{\beta}_Q = (*, \ldots, *, B, *, \ldots, *, 0, *, \ldots, *) \tag{12}$$

$$\mathbf{b}^{(2)} = \boldsymbol{\beta}_R = (*, \ldots, *, 0, *, \ldots, *, 0, *, \ldots, *) \tag{13}$$

Since $k \notin R$ and $k \neq i_Q$, it must be the case $k \notin Q = R \cup \{i_Q\} = \mathsf{A}(\mathbf{b}^{(1)})$. From the fact that M is upper continuous, we can choose b_k such that $0 < b_k < \theta_k(\mathbf{b}_{-k}^{(1)})$. Let $b_{i_Q} = P^{i_Q}(\mathbf{b}^{(1)})$ and consider the following bid vector which differs from $\mathbf{b}^{(1)}$ only in the i_Q-th and k-th entries:

$$\mathbf{b}^{(3)} = (*, \ldots, *, b_{i_Q}, *, \ldots, *, b_k, *, \ldots, *).$$

The proof of the following fact is given in the full version [13].

Fact 7. *In the sequel we will use the fact that $u_{i_Q}^M(\mathbf{b}^{(3)}) = u_{i_Q}^M(\mathbf{b}^{(1)})$ and $u_k^M(\mathbf{b}^{(3)})$*
$= u_k^M(\mathbf{b}^{(1)})$, for $v_k = b_k$.

We are now ready to show that, under the hypothesis $k \in \mathsf{A}(\mathbf{b}^{(2)})$, the coalition
$C = \{i_Q, k\}$ violates the GSP condition. Indeed, consider $\mathbf{v}_C = (v_{i_Q}, v_k) =$
(b_{i_Q}, b_k), $\mathbf{b}_C = (0, 0)$ and $\mathbf{v}_{-C} = \mathbf{b}_{-C}^{(1)} = \mathbf{b}_{-C}^{(2)} = \mathbf{b}_{-C}^{(3)}$. Hence, $(\mathbf{v}_C, \mathbf{v}_{-C}) = \mathbf{b}^{(3)}$
and $(\mathbf{b}_C, \mathbf{v}_{-C}) = \mathbf{b}^{(2)}$. Fact 7 implies

$$u_{i_Q}^M(\mathbf{v}_C, \mathbf{v}_{-C}) = u_{i_Q}^M(\mathbf{b}^{(3)}) = u_{i_Q}^M(\mathbf{b}^{(1)}) = v_{i_Q} - P^{i_Q}(\mathbf{b}^{(1)}) = 0 = u_{i_Q}^M(\mathbf{b}^{(2)}),$$

where the last inequality is due to the definition of $\mathbf{b}^{(2)}$ and to the VP condition.
(Observe that it must hold $P^{i_Q}(\mathbf{b}^{(2)}) = 0$.) Similarly,

$$u_k^M(\mathbf{v}_C, \mathbf{v}_{-C}) = u_k^M(\mathbf{b}^{(3)}) = u_k^M(\mathbf{b}^{(1)}) = 0 < v_k = u_{i_Q}^M(\mathbf{b}^{(2)}),$$

where the last equality follows from the definition of $\mathbf{b}^{(2)}$, from the VP condition,
and from $k \in \mathsf{A}(\mathbf{b}^{(2)})$. The above two inequalities thus imply that the coalition
$C = \{i_Q, k\}$ violates the GSP condition. Hence a contradiction derived from the
assumption $R \subset \mathsf{A}(\boldsymbol{\beta}_R)$. It must then hold $Q \setminus \{i_Q\} = R = \mathsf{A}(\boldsymbol{\beta}_R) = \mathsf{A}(\boldsymbol{\beta}_{Q \setminus \{i_Q\}})$.

Corollary 4. *If $M = (\mathsf{A}, P)$ is an upper-continuous mechanism satisfying α-BB, NPT, VP, CS and GSP, then A must be sequential.*

Proof. Let $Q_1 := U$ and observe that, from the proof of Theorem 2, $Q_1 = U \in \mathcal{P}_\beta^{\mathsf{A}}$. We proceed inductively and apply Theorem 6 so to prove that $Q_j \in \mathcal{P}_\beta^{\mathsf{A}}$ and therefore we can define $i_j := i_{Q_j}$ such that $Q_{j+1} := Q_j \setminus \{i_j\} \in \mathcal{P}_\beta^{\mathsf{A}}$.

Corollary 5. *For generalized cost-sharing games involving any number of users, the following are equivalent:*

1. *There exists an upper-continuous mechanism $M_\alpha = (\mathsf{A}_\alpha, P_\alpha)$ satisfying α-BB, NPT, VP, CS and GSP;*
2. *Every instance $\mathcal{I} = (U, \mathcal{P}, C)$ admits a sequential algorithm A;*
3. *There exists an upper-continuous mechanism $M = (\mathsf{A}, P)$ satisfying BB, NPT, VP, CS and GSP.*

4   Applications, Extensions and Open Questions

Cost-sharing games have been studied under the following (underlying) assumption: given *any* subset Q of users, it is possible to provide the service to *exactly* those users in Q.

This hypothesis cannot be taken for granted in several applications. Indeed, consider the following (simple) scenarios:

Fig. 1(a). A network connecting a source node s to another node t, and $n \geq 2$ users all sitting on node t. If the source s transmits to any of them, then all the others will also receive it. (Consider the scenario in which there is no encryption and one users can "sniff" what is sent to the others. This problem is a variant of the games considered in [8,4,3,5].

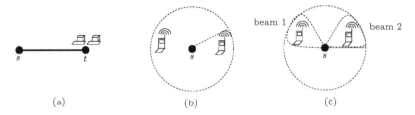

Fig. 1. (a) A variant of the Steiner tree game in [3]; (b) A variant of the wireless multicast game in [1]; (c) Another variant of the wireless multicast game obtained by considering stations with switched-beams antennae and limited battery capacity

Fig. 1(b). The wireless multicast game in which a source station s and $n \geq 2$ other stations/users located all in the transmission range of s. Similarly to the previous example, station s can only choose to transmit to *all* of the them or to none. This game is a variant of the one in [1], where the authors implicitly assume that stations/users receiving a physical signal are not able to get the transmission.

Fig. 1(c). As above, but now the source s uses a switched beam antenna: the coverage area is divided into independent sectors or *beams*. The energy spent by s depends on the number of used sectors. It may be the case that the battery level of s is sufficient to reach one user, but not both.

The first two problems are equivalent to a simple generalized cost-sharing game with $\mathcal{P} = \{U, \emptyset\}$. The latter, instead, corresponds to the case $U \notin \mathcal{P}$. Corollary 1 implies that none of the three instances above admits an α-BB, NPT, VP, CS, and GSP mechanism. The same holds for several natural variants of cost-sharing games studied in the literature [10,9,3,1,12,6,2,5], where connectivity games on graphs allow more than one user per node but no "encryption": either all users in that node are serviced or none.

A similar negative result holds if the service provider is not able to service all of its potential customers (i.e., $U \notin \mathcal{P}^A$), as in the third example. This requirement implies some lower bounds on *polynomial-time* mechanisms which

Table 1. A summary of upper/lower bounds on 'α' for mechanisms satisfying α-BB, NPT, VP, CS and GSP. Quantity $\rho(\mathcal{X})$ is the best approximation guarantee of any polynomial-time algorithm approximating $C(\cdot)$ over $\mathcal{X} \subseteq 2^U$). Results marked '*' holds in general (i.e., for non-upper continuous mechanisms too).

(Generalized) Cost-Sharing Games		Upper Continuous Mechanisms	
		any (non polytime)	poly-time
With Sequential Algorithms	$\mathcal{P} = 2^U$	1 [14,2]	$\alpha \leq \rho(2^U)$ [14] $\alpha \geq \rho(\{U\})$ [Cor. 2]*
	$\mathcal{P}^\sigma \subseteq \mathcal{P}$	1 [14,2]	$\alpha \leq \rho(\mathcal{P}^\sigma)$ [14] $\alpha \geq \rho(\{U\})$ [Cor. 2]*
With No Sequential Algorithm	$\mathcal{P}^\sigma \not\subseteq \mathcal{P}$	unbounded [Cor. 4]	unbounded [Cor. 4]

relate the computational hardness of approximating $C(U)$ to the factor α-BB condition (Corollary 2).

If one ignores computational issues, than Corollary 5 states that, for upper continuous mechanisms, generalized games which are "solvable" are all and only those that admit a sequential algorithm. Here the factor α plays no role. In other words, if we stick to properties NPT, VP, CS and GSP only, then it makes sense to relax BB to α-BB only for computational reasons. This contrasts with prior results in [2] where adding a "fairness" requirement (i.e., no free riders) then paying a factor $\alpha > 1$ is necessary (and sometimes sufficient) for upper continuous mechanisms, regardless of their running time.

References

1. V. Bilò, C. Di Francescomarino, M. Flammini, and G. Melideo. Sharing the cost of muticast transmissions in wireless networks. In *Proc. of SPAA*, pages 180–187. ACM Press, 2004.
2. N. Immorlica, M. Mahdian, and V. Mirrokni. Limitations of cross-monotonic cost-sharing schemes. In *Proc. of the 16th SODA*. ACM Press, January 2005.
3. K. Jain and V.V. Vazirani. Applications of approximation algorithms to cooperative games. In *Proc. of the 33rd STOC*, pages 364–372. ACM Press, 2001.
4. K. Kent and D. Skorin-Kapov. Population monotonic cost allocation on MST's. In *Operational Research Proceedings KOI*, volume 43-48, 1996.
5. J. Könemann, S. Leonardi, and G. Schäfer. A group-strategyproof mechanism for Steiner forests. In *Proc. of the 16th SODA*. ACM Press, January 2005.
6. S. Leonardi and G. Schäfer. Cross-monotonic cost-sharing methods for connected facility location games. In *Proc. of the 5th ACM EC*, pages 242–243, 2004.
7. X. Li, Z. Sun, and W. Wang. Cost sharing and strategyproof mechanisms for set cover games. In *Proc. of the 22nd STACS*, volume 3404 of *LNCS*, pages 218–230, 2005.
8. N. Megiddo. Cost allocation for steiner trees. *Networks*, 8:1–6, 1978.
9. H. Moulin. Incremental cost sharing: characterization by coalition strategy-proofness. *Social Choice and Welfare*, 16:279–320, 1999.
10. H. Moulin and S. Shenker. Strategyproof sharing of submodular costs: budget balance versus efficiency. 1997. http://www.aciri.org/shenker/cost.ps.
11. R. Myerson. Optimal auction design. *Mathematics of Operations Research*, 6:58–73, 1981.
12. M. Pàl and É. Tardos. Strategy proof mechanisms via primal-dual algorithms. In *Proc. of the 44th IEEE FOCS*, 2003.
13. P. Penna and C.Ventre. The algorithmic structure of group strategyproof budget-balanced cost-sharing mechanisms. Available at http://www.dia.unisa.it/~ventre.
14. P. Penna and C. Ventre. More powerful and simpler cost-sharing methods. In *Proc. of the 2nd WAOA*, volume 3351 of *LNCS*, pages 97–110, 2004.
15. P. Penna and C. Ventre. Free-riders in steiner tree cost-sharing games. In *Proc. of 12th SIROCCO*, volume 3499 of *LNCS*, pages 231–245, 2005.
16. W. Vickrey. Counterspeculation, auctions and competitive sealed tenders. *Journal of Finance*, pages 8–37, 1961.

Convergence and Approximation in Potential Games

George Christodoulou[1,*], Vahab S. Mirrokni[2,**],
and Anastasios Sidiropoulos[2]

[1] National and Kapodistrian University of Athens,
Dept. of Informatics and Telecommunications
gchristo@di.uoa.gr
[2] MIT Computer Science and Artificial Intelligence Laboratory
mirrokni@theory.csail.mit.edu
tasos@theory.csail.mit.edu

Abstract. We study the speed of convergence to approximately optimal states in two classes of potential games. We provide bounds in terms of the number of *rounds*, where a round consists of a sequence of movements, with each player appearing at least once in each round. We model the sequential interaction between players by a *best-response walk* in the state graph, where every transition in the walk corresponds to a best response of a player. Our goal is to bound the social value of the states at the end of such walks. In this paper, we focus on two classes of potential games: selfish routing games, and cut games (or party affiliation games [7]).

1   Introduction

The main tool for analyzing the performance of systems where selfish players interact without central coordination, is the notion of the *price of anarchy* in a game [16]; this is the worst case ratio between an optimal social solution and a Nash equilibrium. Intuitively, a high price of anarchy indicates that the system under consideration requires central regulation to achieve good performance. On the other hand, a low price of anarchy does not necessarily imply high performance of the system. One main reason for this phenomenon is that in many games, the repeated selfish behavior of players may not lead to a Nash equilibrium. Moreover, even if the selfish behavior of players converges to a Nash equilibrium, the *rate* of convergence might be very slow. Thus, from a practical and computational viewpoint, it is important to evaluate the rate of convergence to approximate solutions.

By modeling the repeated selfish behavior of the players as a sequence of atomic improvements, the resulting convergence question is related to the running time of local search algorithms. In fact, the theory of PLS-completeness [22]

* Research supported in part by the programme AEOLUS FP6-015964. Research supported in part by the programme pythagoras II, which is funded by the European Social Fund (75%) and the Greek Ministry of Education (25%).
** Supported in part by ONR grant N00014-05-1-0148.

and the existence of exponentially long walks in local optimization problems such as Max-2SAT and Max-Cut, indicate that in many of these settings, we cannot hope for a polynomial-time convergence to a Nash equilibrium. Therefore, for such games, it is not sufficient to just study the value of the social function at Nash equilibria. To deal with this issue, we need to bound the social value of a strategy profile after *polynomially many* best-response improvements by players.

Potential games are games in which any sequence of improvements by players converges to a pure Nash equilibrium. Equivalently, in potential games, there is no cycle of strict improvements of players. This is equivalent to the existence of a potential function that is strictly increasing after any strict improvement. In this paper, we study the speed of convergence to approximate solutions in two classes of potential games: selfish routing (or congestion) games and cut games.

Related Work. This work is motivated by the negative results of the convergence in congestion games [7], and the study of convergence to approximate solutions games [14,11]. Fabrikant, Papadimitriou, and Talwar [7] show that for general congestion and asymmetric selfish routing games, the problem of finding a pure Nash equilibrium is PLS-complete. This implies exponentially long walks to equilibria for these games. Our model is based on the model introduced by Mirrokni and Vetta [14] who addressed the convergence to approximate solutions in basic-utility and valid-utility games. They prove that starting from any state, one round of selfish behavior of players converges to a $1/3$-approximate solution in basic-utility games. Goemans, Mirrokni, and Vetta [11] study a new equilibrium concept (i.e. sink equilibria) inspired from convergence on best-response walks and proved fast convergence to approximate solutions on random best-response walks in (weighted) congestion games. In particular, their result on the price of sinking of the congestion games implies polynomial convergence to constant-factor solutions on random best-response walks in selfish routing games with linear latency functions. Other related papers studied convergence for different classes of games such as load balancing games [6], market sharing games [10], and distributed caching games [8].

A main subclass of potential games is the class of *congestion games* introduced by Rosenthal [18]. Monderer and Shapley [15] proved that congestion games are equivalent to the class of *exact potential games*. In an exact potential game, the increase in the payoff of a player is equal to the increase in the potential function. Both selfish routing games and cut games are a subclass of exact potential games, or equivalently, congestion games. Tight bounds for the price of anarchy is known for both of these games in different settings [19,1,5,4]. Despite all the recent progress in bounding the price of anarchy in these games, many problems about the speed of convergence to approximate solutions for them are still open.

Two main known results for the convergence of selfish routing games are the existence of exponentially long best-response walks to equilibria [7] and fast convergence to constant-factor solutions on random best-response walks [11]. To the best of our knowledge, no results are known for the speed of convergence to approximate solutions on deterministic best-response walks in the general selfish routing game. Preliminary results of this type in some special load balancing games are due to Suri, Tóth and Zhou [20,21]. Our results for general selfish routing games generalize their results.

The Max-Cut problem has been studied extensively [12], even in the local search setting. It is well known that finding a local optimum for Max-Cut is PLS-complete [13,22], and there are some configurations from which walks to a local optimum are exponentially long. In the positive side, Poljak [17] proved that for cubic graphs the convergence to a local optimum requires at most $O(n^2)$ steps. The total happiness social function is considered in the context of correlation clustering [2], and is similar to the total agreement minus disagreement in that context. The best approximation algorithm known for this problem gives a $O(\log n)$-approximation [3], and is based on a semidefinite relaxation.

Our Contribution. Our work deviates from bounding the distance to a Nash equilibrium [22,7], and focuses in studying the rate of convergence to an approximate solution [14,11]. We consider two types of walks of best responses: random walks and deterministic fair walks. On random walks, we choose a random player at each step. On deterministic fair walks, the time complexity of a game is measured in terms of the number of *rounds*, where a round consists of a sequence of movements, with each player appearing at least once in each round.

First, we give tight bounds for the approximation factor of the solution after one round of best responses of players in selfish routing games. In particular, we prove that starting from an arbitrary state, the approximation factor after one round of best responses of players is at most $O(n)$ of the optimum and this is tight up to a constant factor. We extend the lower bound for the case of multiple rounds, where we show that for any constant number of rounds t, the approximation guarantee cannot be better than $n^{\epsilon(t)}$, for some $\epsilon(t) > 0$. On the other hand, we show that starting from an empty state, the state resulting after one round of best responses is a constant-factor approximation.

We also study the convergence in *cut games*, that are motivated by the *party affiliation game* [7], and are closely related to the local search algorithm for the Max-Cut problem [22]. In the party affiliation game, each player's strategy is to choose one of two parties, i.e., $s_i \in \{1, -1\}$ and the payoff of player i for the strategy profile $(s_1, s_2, \ldots, s_n)$ is $\sum_j s_j s_i w_{ij}$. The weight of an edge corresponds to the level of *disagreement* of the endpoints of that edge. This game models the clustering of a society into two parties that minimizes the disagreement within each party, or maximizes the disagreement between different parties. Such problems play a key role in the study of social networks.

We can model the party affiliation game as the following cut game: each vertex of a graph is a player, with payoff its contribution in the cut (i.e. the total weight of its adjacent edges that have endpoints in different parts of the cut). It follows that a player moves if he can improve his contribution in the cut, or equivalently, he can improve the value of the cut. The pure Nash equilibria exist in this game, and selfish behavior of players converges to a Nash equilibrium.

We consider two social functions: the cut and the total happiness, defined as the value of the cut minus the weight of the rest of edges. First, we prove *fast convergence on random walks*. More precisely, the selfish behavior of players in a round in which the ordering of the player is picked uniformly at random, results in a cut that is a $\frac{1}{8}$-approximation in expectation. We complement our positive results by examples that exhibit *poor deterministic convergence*. That is, we show the existence of fair walks with *exponential* length, that result in a poor social value. We also model the selfish behavior of *mildly greedy* players that move if their payoff increases by at least a factor of $1 + \epsilon$. We prove that in contrast to the case of (totally) greedy players, mildly greedy players converge to a constant-factor cut after one round, under any ordering. For unweighted graphs, we give an $\Omega(\sqrt{n})$ lower bound and an $O(n)$ upper bound for the number of rounds required in the worst case to converge to a constant-factor cut.

Finally, for the total happiness social function, we show that for unweighted graphs of large girth, starting from a random configuration, greedy behavior of players in a random order converges to an approximate solution after one round. We remark that this implies a combinatorial algorithm with sub-logarithmic approximation ratio, for graphs of sufficiently large girth, while the best known approximation ratio for the general problem is $O(\log n)$ [3], and is obtained using semidefinite programming.

2   Definitions and Preliminaries

In order to model the selfish behavior of players, we use the notion of a *state graph*. Each vertex in the state graph represents a *strategy state* $S = (s_1, s_2, \ldots, s_n)$, and corresponds to a pure strategy profile (e.g an allocation for a congestion game, or a cut for a cut game). The arcs in the state graph correspond to best response moves by the players.

Definition 1. *A state graph $\mathcal{D} = (\mathcal{V}, \mathcal{E})$ is a directed graph, where each vertex in $\mathcal{V}$ corresponds to a strategy state. There is an arc from state S to state S' with label j iff by letting player j play his best response in state S, the resulting state is S'.*

Observe that the state graph may contain loops. A *best response* walk is a directed walk in the state graph. We say that player i plays in the best response walk $\mathcal{P}$, if at least one of the edges of $\mathcal{P}$ has label i. Note that players play their best responses sequentially, and not in parallel. Given a best response walk starting from an arbitrary state, we are interested in the social value of the last state on the walk. Notice that if we do not allow every player to make a best response on a walk $\mathcal{P}$, then we cannot bound the social value of the final state with respect to the optimal solution. This follows from the fact that the actions of a single player may be very important for producing solutions of high social value[1]. Motivated by this simple observation, we introduce the following models that capture the intuitive notion of a fair sequence of moves.

One-round walk: Consider an arbitrary ordering of all players $i_1, \ldots, i_n$. A walk $\mathcal{P}$ of length n in the state graph is a *one-round walk* if for each $j \in [n]$, the jth edge of P has label i_j.

---

[1] E.g. in the cut social function, most of the weight of the edges of the graph might be concentrated to the edges that are adjacent to a single vertex.

Covering walk: A walk $\mathcal{P}$ in the state graph is a *covering walk* if for each player i, there exists an edge of P with label i.

k-Covering walk: A walk $\mathcal{P}$ in the state graph is a *k-covering walk* if there are k covering walks $\mathcal{P}_1, \mathcal{P}_2, \ldots, \mathcal{P}_k$, such that $\mathcal{P} = (\mathcal{P}_1, \mathcal{P}_2, \ldots, \mathcal{P}_k)$.

Random walk: A walk $\mathcal{P}$ in the state graph is a *random walk*, if at each step the next player is chosen uniformly at random.

Random one-round walk: Let σ be an ordering of players picked uniformly at random from the set of all possible orderings. Then, the one-round walk $\mathcal{P}$ corresponding to the ordering σ, is a *random one-round walk*.

Note that unless otherwise stated, all walks are assumed to start from an arbitrary initial state. This model has been used by Mirrokni and Vetta [14], in the context of extensive games with complete information.

Congestion games. A congestion game is defined by a tuple $(N, E, (\mathcal{S}_i)_{i \in N}, (f_e)_{e \in E})$ where N is a set of players, E is a set of facilities, $\mathcal{S}_i \subseteq 2^E$ is the pure strategy set for player i: a pure strategy $s_i \in \mathcal{S}_i$ for player i is a set of facilities, and f_e is a latency function for the facility e depending on its load. We focus on linear delay functions with nonnegative coefficients; $f_e(x) = a_e x + b_e$.

Let $S = (s_1, \ldots, s_N) \in \times_{i \in N} \mathcal{S}_i$ be a state (strategy profile) for a set of N players. The cost of player i, in a state S is $c_i(S) = \sum_{e \in s_i} f_e(n_e(S))$ where by $n_e(S)$ we denote the number of players that use facility e in S. The objective of a player is to minimize its own cost. We consider as a social cost of a state S, the sum of the players' costs and we denote it by $C(S) = \sum_{i \in N} c_i(S) = \sum_{e \in E} n_e(S) f_e(n_e(S))$.

In weighted congestion games, player i has weighted demand w_i. By $\theta_e(S)$, we denote the total load on a facility e in a state S. The cost of a player in a state S is $c_i(S) = \sum_{e \in s_i} f_e(\theta_e(S))$. We consider as a social cost of a state S, the weighted sum $C(S) = \sum_{i \in N} w_i c_i(S) = \sum_{e \in E} \theta_e(S) f_e(\theta_e(S))$. We will use subscripts to distinguish players and superscripts to distinguish states.

Note that the selfish routing game is a special case of congestion games. Although we state all the results for congestion games with linear latency functions, all of the results (including the lower and upper bounds) hold for selfish routing games.

Cut Games. In a cut game, we are given an undirected graph $G(V, E)$, with n vertices and edge weights $w : E(G) \to \mathbb{Q}^+$. We will always assume that G is connected, simple, and does not contain loops. For each $v \in V(G)$, let $\deg(v)$ be the degree of v, and let $\mathsf{Adj}(v)$ be the set of neighbors of v. Let also $w_v = \sum_{u \in \mathsf{Adj}(v)} w_{uv}$. A cut in G is a partition of $V(G)$ into two sets, T and $\bar{T} = V(G) - T$, and is denoted by $(T, \bar{T})$. The value of a cut is the sum of edges between the two sets T and $\bar{T}$, i.e $\sum_{v \in T, u \in \bar{T}} w_{uv}$.

The *cut game* on a graph $G(V, E)$, is defined as follows: each vertex $v \in V(G)$ is a player, and the strategy of v is to chose one side of the cut, i.e. v can chose $s_v = -1$ or $s_v = 1$. A strategy profile $S = (s_1, s_2, \ldots, s_n)$, corresponds to a cut $(T, \bar{T})$, where $T = \{i | s_i = 1\}$. The payoff of player v in a strategy profile S, denoted by $\alpha_v(S)$, is equal to the contribution of v in the cut, i.e.

$\alpha_v(S) = \sum_{i:s_i \neq s_v} w_{iv}$. It follows that the cut value is equal to $\frac{1}{2} \sum_{v \in V} \alpha_v(S)$. If S is clear from the context, we use α_v instead of $\alpha_v(S)$ to denote the payoff of v. We denote the maximum value of a cut in G, by $c(G)$. The *happiness* of a vertex v is equal to $\sum_{i:s_i \neq s_v} w_{iv} - \sum_{i:s_i = s_v} w_{iv}$.

We consider two social functions: the cut value and the cut value minus the value of the rest of the edges in the graph. It is easy to see that the cut value is half the sum of the payoffs of vertices. The second social function is half the sum of the happiness of vertices. We call the second social function, *total happiness*.

3   Congestion Games

In this section, we focus on the convergence to approximate solutions in congestion games with linear latency functions. It is known [15,18] that any best-response walk on the state graph leads to a pure Nash equilibrium, and a pure equilibrium is a constant-factor approximate solution [1,5,4]. Unless otherwise stated, we assume without loss of generality, that the players' ordering is $1, \ldots, N$.

3.1   Upper Bounds for One-Round Walks

In this section, we bound the total delay after one round of best responses of players. We prove that starting from an arbitrary state, the solution after one round of best responses is a $\Theta(N)$-approximate solution. We will also prove that starting from an *empty* state, the approximation factor after one round of best responses is a constant factor. This shows that the assumption about the initial state is critical for this problem.

Theorem 1. *Starting from an arbitrary initial state S^0, any one-round walk $\mathcal{P}$ leads to a state S^N that has approximation ratio $O(N)$.*

Proof. Let X be the optimal allocation and $S^i = (s_1^N, \ldots, s_i^N, s_{i+1}^0, \ldots, s_N^0)$ an intermediate state. Let $m_e(S^i), k_e(S^i)$ be the number of the players of the final and of the initial state respectively, using facility e in a state S^i, and $M(S^i), K(S^i)$ the corresponding sums. Clearly $n_e(S^i) = m_e(S^i) + k_e(S^i)$ and $K(S^i) = K(S^{i-1}) - \sum_{e \in s_i^0}(a_e - b_e - 2a_e k_e(S^{i-1}))$. By summing over all intermediate states and using the fact $K(S^N) = 0$, it follows that:

$$K(S^0) = C(S^0) = \sum_{e \in E} k_e(S^0) f_e(k_e(S^0)) = \sum_{i \in N} \sum_{e \in s_i^0} (2a_e k_e(S^{i-1}) - a_e + b_e) \quad (1)$$

Since player i in state S^{i-1} prefers strategy s_i^N than x_i, we get

$$\sum_{e \in s_i^N} f_e(n_e(S^{i-1})) + \sum_{e \in s_i^N - s_i^0} a_e \leq \sum_{e \in x_i} f_e(n_e(S^{i-1}) + 1)$$

For every intermediate state S^i, the social cost is

$$C(S^i) = C(S^{i-1}) + \sum_{e \in s_i^N - s_i^0}(2a_e n_e(S^{i-1}) + a_e + b_e) + \sum_{e \in s_i^0 - s_i^N}(a_e - b_e - 2a_e n_e(S^{i-1}))$$

Summing over all intermediate states and using equality (1), we get

$$C(S^N) = \sum_{i\in N}\sum_{e\in s_i^N - s_i^0}(2a_e n_e(S^{i-1}) + a_e + b_e) + \sum_{i\in N}\sum_{e\in s_i^0}(2a_e k_e(S^{i-1}) - a_e + b_e)$$

$$+ \sum_{i\in N}\sum_{o\subset o_i^0\ o_i^N}(a_e - b_e - 2a_e n_e(S^{i-1}))$$

$$= \sum_{i\in N}\sum_{e\in s_i^N - s_i^0}(2a_e n_e(S^{i-1}) + a_e + b_e) + \sum_{i\in N}\sum_{e\in s_i^0\cap s_i^N}(2a_e k_e(S^{i-1}) - a_e + b_e)$$

$$- \sum_{i\in N}\sum_{e\in s_i^0 - s_i^N} 2a_e m_e(S^{i-1})$$

$$\leq 2\sum_{i\in N}\sum_{e\in s_i^N} f_e(n_e(S^{i-1})) + 2\sum_{i\in N}\sum_{e\in s_i^N - s_i^0} a_e$$

$$\leq \sum_{i\in N}\sum_{e\in x_i} 2f_e(n_e(S^{i-1}) + 1)$$

$$\leq \sum_{i\in N}\sum_{e\in x_i} 2f_e(N+1) = \sum_{e\in E} 2n_e(X)f_e(N+1) = O(N)C(X)$$

$\square$

In the next section, we will show that the above bound is tight up to a constant factor. As mentioned earlier, the assumption about the initial state is critical for this problem. We will call a state *empty*, if no player is committed to any of its strategies. Note that the one-round walk starting from an empty state is essentially equivalent to the greedy algorithm for a generalized scheduling problem, where a task may be assigned into many machines. Suri et al. [20,21] address similar questions for the special case of the congestion games where the available strategies are single sets (i.e. each player can choose just one facility). They give a 3.08 lower bound and a 17/3 upper bound. For the special case of identical facilities (equal speed machines) they give an upper bound of $\frac{(\phi+1)^2}{\phi} \approx$ 4.24. We generalize this result for our more general setting.

Theorem 2. *Starting from the empty state S^0, any one-round walk $\mathcal{P}$ leads to a state S^N that has approximation ratio of at most $\frac{(\phi+1)^2}{\phi} \approx 4.24$.*

Now we turn our attention to weighted congestion games with linear latency functions, where player i has weighted demand w_i. Fotakis et al. [9] showed that this game with linear latency functions is a potential game.

Theorem 3. *In weighted congestion games with linear latency functions, starting from the initial empty state S^0, any one-round walk $\mathcal{P}$ leads to a state S^N that has approximation ratio of at most $(1 + \sqrt{3})^2 \approx 7.46$.*

3.2   Lower Bounds

The next theorem shows that the result of Theorem 1 is tight and explains why it is necessary in the upper bounds given above to consider walks starting from an empty allocation.

Theorem 4. *For any $N > 0$, there exists an N-player instance of the un-weighted congestion game, and an initial state S^0 such that for any one-round walk $\mathcal{P}$ starting from S^0, the state at the end of $\mathcal{P}$ is an $\Omega(N)$-approximate solution.*

Proof. Consider $2N$ players and $2N+2$ facilities $\{0, 1, \ldots 2N+1\}$. The available strategies for the first players are $\{\{0\}, \{i\}, \{N+1, \ldots, 2N\}\}$ and for the N last $\{\{2N+1\}, \{i\}, \{1, \ldots, N\}\}$. In the initial allocation, every player plays its third strategy. Consider any order on the players and let them begin to choose their best responses. It is easy to see that in the first steps, the players would prefer their first strategy. If this happens until the end of the round, the resulting cost is $\Omega(N^2)$. Thus, we can assume that at some step, the $(k+1)$-th player from the set $\{1, \ldots, N\}$ prefers his second strategy while all the previous k players of the same set have chosen their first strategies. The status of the game at this step is as follows: k players of the first group play their first strategy, m players of the second group play their first strategy and the remaining players play their initial strategy. Since player $k + 1$ prefers his second strategy, this means $k = N - m$ and so one of the m, N is at least $N/2$. The cost at the end will be at least $m^2 + k^2 + N = \Omega(N^2)$. On the other hand, in the optimal allocation everybody chooses its second strategy which gives cost $2N$. Thus, the approximation ratio is $\Omega(N)$.   □

We extend theorem 4 for the case of t-covering walks, for $t > 1$. We remark that the following result holds only for a fixed ordering of the players.

Theorem 5. *For any $t > 0$, and for any sufficiently large $N > 0$, there exists an N-player instance of the unweighted congestion game, an initial state S^0, and an ordering σ of the players, such that starting from S^0, after t rounds where the players play according to σ, the cost of the resulting allocation is a $(N/t)^\epsilon$-approximation, where $\epsilon = 2^{-O(t)}$.*

4   Cut Games: The Cut Social Function

4.1   Fast Convergence on Random Walks

First we prove positive results for the convergence to constant-factor approximate solutions with random walks. We show that the expected value of the cut after a random one-round walk is within a constant factor of the maximum cut.

Theorem 6. *In weighted graphs, the expected value of the cut at the end of a random one-round walk is at least $\frac{1}{8}$ of the maximum cut.*

Proof. It suffices to show that after a random one-round walk, for every $v \in V(G)$, $\mathbf{E}[\alpha_v] \geq \frac{1}{8} w_v$.

Consider a vertex v. The probability that v occurs after exactly k of its neighbors is $\frac{1}{\deg(v)+1}$. After v moves, the contribution of v in the cut is at least $\frac{w_v}{2}$. Conditioning on the fact that v occurs after exactly k neighbors, for each vertex u in the neighborhood of v, the probability that it occurs after v

is $\frac{\deg(v)-k}{\deg(v)}$, and only in this case u can decrease the contribution of v in the cut by at most w_{uv}. Thus the expected contribution of v in the cut is at least $\max(0, w_v(\frac{1}{2} - \frac{\deg(v)-k}{\deg(v)}))$. Summing over all values of k, we obtain $\mathbf{E}[\alpha_v] \geq \sum_{k=0}^{\deg(v)} \frac{1}{\deg(v)+1} \max(0, w_v(\frac{1}{2} - \frac{\deg(v)-k}{\deg(v)})) = \frac{w_v}{\deg(v)+1} \sum_{k=0}^{\lfloor \frac{\deg(v)}{2} \rfloor + 1} \frac{2k-\deg(v)}{2\deg(v)} \geq \frac{w_v}{8}$. The result follows by the linearity of expectation. □

The next theorem studies a random walk of best responses (not necessarily a one-round walk).

Theorem 7. *There exists a constant $c > 0$ such that the expected value of the cut at the end of a random walk of length $cn \log n$ is a constant-factor of the maximum cut.*

4.2 Poor Deterministic Convergence

We now give lower bounds for the convergence to approximate solutions for the cut social function. First, we give a simple example for which we need at least $\Omega(n)$ rounds of best responses to converge to a constant-factor cut. The construction resembles a result of Poljak [17].

Theorem 8. *There exists a weighted graph $G(V, E)$, with $|V(G)| = n$, and an ordering of vertices such that for any $k > 0$, the value of the cut after k rounds of letting players play in this ordering is at most $O(k/n)$ of the maximum cut.*

We next combine a modified version of the above construction with a result of Schaffer and Yannakakis for the Max-Cut local search problem [22], to obtain an exponentially-long walk with poor cut value.

Theorem 9. *There exists a weighted graph $G(V, E)$, with $|V(G)| = \Theta(n)$, and a k-covering walk $\mathcal{P}$ in the state graph, for some k exponentially large in n, such that the value of the cut at the end of $\mathcal{P}$, is at most $O(1/n)$ of the optimum cut.*

Proof. In [22], it is shown that there exists a weighted graph $G_0(V, E)$, and an initial cut $(T_0, \bar{T}_0)$, such that the length of *any* walk in the state graph, from $(T_0, \bar{T}_0)$ to a pure strategy Nash equilibrium, is exponentially long. Consider such a graph of size $\Theta(n)$, with $V(G_0) = \{v_0, v_1, \ldots, v_N\}$. Let $\mathcal{P}_0$ be an exponentially long walk from $(T_0, \bar{T}_0)$ to a Nash equilibrium in which we let vertices $v_0, v_1, \ldots, v_N$ play in this order for exponential an number of rounds. Let $S_0, S_1, \ldots, S_{|\mathcal{P}_0|}$ be the sequence of states visited by $\mathcal{P}_0$ and let y_i be the vertex that plays his best response from state S_i to state S_{i+1}. The result of [22] guarantees that there exists a vertex, say v_0, which wants to change side (i.e. strategy) an exponential number of times along the walk $\mathcal{P}_0$ (since otherwise we can find a small walk to a pure Nash equilibrium). Let $t_0 = 0$, and for $i \geq 1$, let t_i be the time in which v_0 changes side for the i-th time along the walk $\mathcal{P}_0$. For $i \geq 1$, let $\mathcal{Q}_i$ be the sequence of vertices $y_{t_{i-1}+1}, \ldots, y_{t_i}$. Observe that each $\mathcal{Q}_i$ contains all of the vertices in G_0.

Consider now a graph G, which consists of a path $L = x_1, x_2, \ldots, x_n$, and a copy of G_0. For each $i \in \{1, \ldots, n-1\}$, the weight of the edge $\{x_i, x_{i+1}\}$ is 1.

We scale the weights of G_0, such that the total weight of the edges of G_0 is less than 1. Finally, for each $i \in \{1, \ldots, n\}$, we add the edge $\{x_i, v_0\}$, of weight ϵ, for some sufficiently small ϵ. Intuitively, we can pick the value of ϵ, such that the moves made by the vertices in G_0, are independent of the positions of the vertices of the path L in the current cut.

For each $i \geq 1$, we consider an ordering $\mathcal{R}_i$ of the vertices of L, as follows: If i is odd, then $\mathcal{R}_i = x_1, x_2, \ldots, x_n$, and if i is even, then $\mathcal{R}_i = x_n, x_{n-1}, \ldots, x_1$.

We are now ready to describe the exponentially long path in the state graph. Assume w.l.o.g., that in the initial cut for G_0, we have $v_0 \in T_0$. The initial cut for G is $(T, \bar{T})$, with $T = \{x_1\} \cup T_0$, and $\bar{T} = \{x_2, \ldots, x_n\} \cup \bar{T}_0$. It is now straightforward to verify that there exists an exponentially large k, such that for any i, with $1 \leq i \leq k$, if we let the vertices of G play according to the sequence $\mathcal{Q}_1, \mathcal{R}_1, \mathcal{Q}_2, \mathcal{R}_2, \ldots, \mathcal{Q}_i, \mathcal{R}_i$, then we have (see Figure 1):

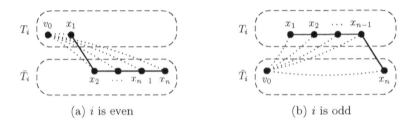

(a) i is even   (b) i is odd

Fig. 1. The cut $(T_i, \bar{T}_i)$ along the walk of the proof of Theorem 9

- If i is even, then $\{v_0, x_1\} \subset T$, and $\{x_2, \ldots, x_n\} \subset \bar{T}$.
- If i is odd, then $\{x_1, \ldots, x_{n-1}\} \subset T$, and $\{v_0, x_n\} \subset \bar{T}$.

It follows that for each i, with $1 \leq i \leq k$, the size of the cut is at most $O(1/n)$ times the value of the optimal cut. The result follows since each walk in the state graph induced by the sequence $\mathcal{Q}_i$ and $\mathcal{R}_i$ is a covering walk.   □

4.3   Mildly Greedy Players

By Theorem 6, it follows that for any graph, and starting from an arbitrary cut, there exists a walk of length at most n to an $\Omega(1)$-approximate cut, while Theorems 8 and 9, show that there are cases where a deterministic ordering of players may result to very long walks that do reach an approximately good cut.

We observe that if we change the game by assuming that a vertex changes side in the cut if his payoff is multiplied by at least a factor $1 + \epsilon$, for a constant $\epsilon > 0$, then the convergence is faster. We call such vertices $(1 + \epsilon)$-*greedy*. In the following, we prove that if all vertices are $(1 + \epsilon)$-greedy for a constant $\epsilon > 0$, then the value of the cut after any one-round walk is within a constant factor of the optimum.

Theorem 10. *If all vertices are $(1 + \epsilon)$-greedy, then the cut value at the end of any one-round walk is within a* $\min\{\frac{1}{4+2\epsilon}, \frac{\epsilon}{4+2\epsilon}\}$ *factor of the optimal cut.*

4.4   Unweighted Graphs

In unweighted simple graphs, it is straight-forward to verify that the value of the cut at the end of an n^2-covering walk is at least $\frac{1}{2}$ of the optimum. The following theorem shows that in unweighted graphs, the value of the cut after any $\Omega(n)$-covering walk is a constant-factor approximation.

Theorem 11. *For unweighted graphs, the value of the cut after an $\Omega(n)$-covering walk is within a constant-factor of the maximum cut.*

Proof. Consider a k-covering walk $\mathcal{P} = (\mathcal{P}_1, \ldots, \mathcal{P}_k)$, where each $\mathcal{P}_i$ is a covering walk. Let $M_0 = 0$, and for any $i \geq 1$, let M_i be the size of the cut at the end of $\mathcal{P}_i$. Note that if $M_i - M_{i-1} \geq \frac{|E(G)|}{10n}$, for all i, with $1 \leq i \leq k$, then clearly $M_k \geq k \frac{|E(G)|}{10n}$, and since the maximum size of a cut is at most $|E(G)|$, the Lemma follows.

It remains to consider the case where there exists i, with $1 \leq i \leq k$, such that $M_i - M_{i-1} < \frac{|E(G)|}{10n}$. Let V_1 be the set of vertices that change their side in the cut on the walk $\mathcal{P}_i$, and $V_2 = V(G) \setminus V_1$. Observe that when a vertex changes its side in the cut, the size of the cut increases by at least 1. Thus, $|V_1| < \frac{|E(G)|}{10n}$, and since the degree of each vertex is at most $n - 1$, it follows that the number of edges that are incident to vertices in V_1, is less than $\frac{|E(G)|}{10}$.

On the other hand, if a vertex of degree d remains in the same part of the cut, then exactly after it plays, at least $\lceil d/2 \rceil$ of its adjacent edges are in the cut. Thus, at least half of the edges that are incident to at least one vertex in V_2, were in the cut, at some point during walk $\mathcal{P}_i$. At most $\frac{|E(G)|}{10}$ of these edges have an end-point in V_1, and thus at most that many of these edges may not appear in the cut at the end of $\mathcal{P}_i$. Thus, the total number of edges that remain in the cut at the end of walk $\mathcal{P}_i$, is at least $\frac{|E(G)| - |E(G)|/10}{2} - \frac{|E(G)|}{10} = \frac{7|E(G)|}{20}$. Since the maximum size of a cut is at most $|E(G)|$, we obtain that at the end of $\mathcal{P}_i$, the value of the cut is within a constant factor of the optimum. □

Theorem 12. *There exists an unweighted graph $G(V, E)$, with $|V(G)| = n$, and an ordering of the vertices such that for any $k > 0$, the value of the cut after k rounds of letting players play in this ordering is at most $O(k/\sqrt{n})$ of the maximum cut.*

5   The Total Happiness Social Function

Due to space limitations, this section has been left to the full version.

Acknowledgments. We wish to thank Michel Goemans and Piotr Indyk for many helpful discussions. We also thank Ioannis Caragiannis for reading an earlier version of this work, and for improving the bound given in Theorem 7.

References

1. B. Awerbuch, Y. Azar, and A. Epstein. The price of routing unsplittable flow. In *STOC*, pages 57–66, 2005.
2. N. Bansal, A. Blum, and S. Chawla. Correlation clustering. *Machine Learning*, 56:89–113, 2004.
3. M. Charikar and A. Wirth. Maximizing quadratic programs: extending Grothendieck's inequality. In *FOCS*, page to appear, 2004.
4. G. Christodoulou and E. Koutsoupias. On the price of anarchy and stability of correlated equilibria of linear congestion games. In *ESA*, 2005.
5. G. Christodoulou and E. Koutsoupias. The price of anarchy of finite congestion games. In *STOC*, pages 67–73, 2005.
6. E. Even-dar, A. Kesselman, and Y. Mansour. Convergence time to Nash equilibria. In *ICALP*, pages 502–513, 2003.
7. A. Fabrikant, C. Papadimitriou, and K. Talwar. On the complexity of pure equilibria. In *STOC*, 2004.
8. L. Fliescher, M. Goemans, V. S. Mirrokni, and M. Sviridenko. (almost) tight approximation algorithms for maximizing general assignment problems. 2004. Submitted.
9. D. Fotakis, S. Kontogiannis, and P. Spirakis. Selfish unsplittable flow. In *ICALP*, 2004.
10. M. Goemans, L. Li, V.S.Mirrokni, and M. Thottan. Market sharing games applied to content distribution in ad-hoc networks. In *MOBIHOC*, 2004.
11. M. Goemans, V. S. Mirrokni, and A. Vetta. Sink equilibria and convergence. In *FOCS*, 2005.
12. M. Goemans and D. Williamson. Improved approximation algorithms for maximum cut and satisfiability problems using semidefinite programming. *J. ACM*, 42:1115–1145, 1995.
13. D. Johnson, C.H. Papadimitriou, and M. Yannakakis. How easy is local search? *J. Computer and System Sciences*, 37:79–100, 1988.
14. V.S. Mirrokni and A. Vetta. Convergence issues in competitive games. In *RANDOM-APPROX*, pages 183–194, 2004.
15. D. Monderer and L. Shapley. Potential games. *Games and Economics Behavior*, 14:124–143, 1996.
16. C. Papadimitriou. Algorithms, games, and the Internet. In *STOC*, 2001.
17. S. Poljak. Integer linear programs and local search for max-cut. *Siam Journal of Computing*, 24(4):822–839, 1995.
18. R. W. Rosenthal. A class of games possessing pure-strategy Nash equilibria. *International Journal of Game Theory*, 2:65–67, 1973.
19. T. Roughgarden and E. Tardos. How bad is selfish routing? *J. ACM*, 49(2):236–259, 2002.
20. C. D. Tóth S. Suri and Y. Zhou. Selfish load balancing and atomic congestion games. In *SPAA*, pages 188–195, 2004.
21. C. D. Tóth S. Suri and Y. Zhou. Uncoordinated load balancing and congestion games in p2p systems. In *IPTPS*, pages 123–130, 2004.
22. A. Schaffer and M. Yannakakis. Simple local search problems that are hard to solve. *SIAM journal on Computing*, 20(1):56–87, 1991.

Fast FPT-Algorithms for Cleaning Grids*

Josep Díaz and Dimitrios M. Thilikos

Departament de Llenguatges i Sistemes Informàtics,
Universitat Politècnica de Catalunya, Campus Nord, Desp. Ω-228,
c/Jordi Girona Salgado, 1-3. E-08034, Barcelona, Spain

Abstract. We consider the problem that, given a graph G and a parameter k, asks whether the edit distance of G and a rectangular grid is at most k. We examine the general case where the edit operations are vertex/edge removals and additions. If the dimensions of the grid are given in advance, we give a parameterized algorithm that runs in $2^{O(\log k \cdot k)} + O(n^3)$ steps. In the case where the dimensions of the grid are not given we give a parameterized algorithm that runs in $2^{O(\log k \cdot k)} + O(k^2 \cdot n^3)$ steps. We insist on parameterized algorithms with running times where the relation between the polynomial and the non-polynomial part is additive. Our algorithm is based on the technique of kernelization. In particular we prove that for each version of the above problem there exists a kernel of size $O(k^4)$.

1 Introduction

We consider the problem of measuring the *degree of similarity* of two graphs G and H, where the degree of similarity is considered to be the minimum number of *edit operations* needed to transform one graph into the other. This measure of similarity is also known as the *edit distance* between graphs. The problem has received a lot of attention due to its multiple applications in patten recognition and computer vision among others topics. In the more usual setting the edit operations are: contract an edge, applying an inverse contraction[1] of a vertex and (in case of labeled graphs) relabel an edge. In general, the problem is *NP*-hard and can be computed in polynomial time on trees (see for ex. [11]). For a recent update on the heuristics developed for the graph edit distance problem see [9].

   In this paper, we examine decision problems associated with the editing distance between G and H when H is a grid of size $p \times q$. We call such a grid (p, q)-*grid* *and we denote it as* $H_{p,q}$. We consider as edit operations the removal or the insertion of either an edge or a vertex. We represent these operations by the members of the set $\mathcal{U} = \{\text{e-out}, \text{e-in}, \text{v-out}, \text{v-in}\}$. Given $\mathcal{E} \subseteq \mathcal{U}$, we denote $\mathcal{E}\text{-}\mathbf{dist}(G, H)$ as

---

* This research was supported by the spanish CICYT project TIN-2004-07925 (GRAMMARS). The first author was partially supported by the *Distinció per a la Promoció de la Recerca de la GC, 2002.*

[1] H' is the result of the inverse contraction of a vertex v of H if H is the result of the contraction of an edge to vertex v.

B. Durand and W. Thomas (Eds.): STACS 2006, LNCS 3884, pp. 361–371, 2006.
© Springer-Verlag Berlin Heidelberg 2006

the *edit distance of G and H with respect to* $\mathcal{E}$, which is the minimum number of operations in the set $\mathcal{E}$ that when applied to G can transform G into H.

Our problem setting has been motivated by the problem of *regularity extraction* in digital systems, where we are looking for regular patterns in complicated circuits (usually VLSI circuits) indicating ways to organize or embed them (see [8] for an example). In particular, we study the decision version of the problem that, given a graph G, a pair $p, q \in \mathbb{N}$ and a set $\mathcal{E} \subseteq \mathcal{U}$, asks for the minimum number of operations needed to transform G to a (p, q)-grid. This problem is NP-complete, as can be seeing considering the particular case where $q = 1$ and $\mathcal{E}$ contains only the *erasing a vertex* operation, is equivalent to the LONGEST INDUCED PATH problem, i.e. the problem of given a graph G and a constant $k \geq 0$ decide if G contains a simple path of length at least k as an induced subgraph (we just set $k \leftarrow |V(G)| - k$).

Our study adopts the point of view of parameterized complexity introduced by Downey and Fellows (see [2]). We consider parameterizations of hard problems, i.e. problems whose input contains some (in general small) parameter k and some main part. A parameterized problem belongs in the class FPT if it can be solved by an algorithm of time complexity $g(k) \cdot n^{O(1)}$ where g is a non-polynomial function of k and n is the size of the problem. We call such an algorithm *FPT-algorithm*. A popular technique on the design of parameterized algorithms is *kernelization*. Briefly, this technique, consists in finding a polynomial-time reduction of a parameterized problem to itself in a way that the sizes of the instances of the new problem, we call it *kernel*, depend *only* on the parameter k. The function that bounds the size of the main part of the reduced problem determines the *size* of the kernel and is usually of polynomial (on k) size. Notice that if a parameterized problem admits a reduction to a problem kernel, then it is in FPT because any brute force algorithm can solve the reduced problem in time that does not depend on the main part of the original problem. Notice also that this technique provides FPT-algorithms of time complexity $g(k) + n^{O(1)}$ where the non-polynomial part $g(k)$ is just additive to the overall complexity.

In Section 3, we prove that the k-ALMOST GRID problem defined below is FPT[2] by giving a kernel of size $O(k^4)$ or $O(k^3)$ depending on the size of the grid we are looking for.

---

k-ALMOST GRID

Input: A graph G, two positive integers p, q, a non-negative integer k, and a set $\mathcal{E} \subseteq \mathcal{U}$.

Parameter: A non-negative integer k.

Question: Can G be transformed to a (p, q)-grid after at most k edit operations from $\mathcal{E}$?

Notice that our parameterization also includes the "dual" parameterization of the LONGEST INDUCED PATH problem, in the sense that now the parameter is not

---

[2] In an abuse of notation, we indistinctly refer to FPT problem or to problem in the FPT class.

the length of the induced path but the number of vertices in G that are *absent* in such a path. The parameterized complexity of the "primal" parameterization of the LONGEST INDUCED PATH problem remains, to our knowledge, an open problem.

In Section 4 we consider the following more general problem:

---

k-ALMOST SOME GRID
Input: A graph G and a set $\mathcal{E} \subseteq \mathcal{U}$.
Parameter: A non-negative integer k.
Question: Decide if there exist some pair p, q such that $\mathcal{E}$-**dist**$(G, H_{p,q}) \leq k$.

The above problem can be solved applying the algorithm for k-ALMOST GRID for all possible values of p and q. This implies an algorithm of time complexity $O(\log n(g(k) + n^{O(1)}))$. In Section 4, we explain how to avoid this $O(\log n)$ overhead and prove that there exists also a time $O(g(k) + n^{O(1)})$ algorithm for the k-ALMOST SOME GRID problem. That way, both of our algorithms can be seen as pre-processing algorithms that reduce the size of the problem input to a function depending *only* on the parameter k and not on the main part of the problem.

A different but somehow related sets of problems, which have received plenty of attention is the following: Given a graph G and a property Π, decide what is the minimum number of edges (nodes) that must be removed in order to obtain a subgraph of G with property Π. In general all these problems are NP-hard [10,4,7]. Some of those problems have been studied from the parameterized point of view [5], however, all these problems have the characteristic that the property Π must be an hereditary property. We stress that the k-ALMOST GRID and k-ALMOST ANY GRID problems completely different nature, as the property of *containing a grid* is not hereditary.

2   Definitions and Basic Results

All graphs we consider are undirected, loop-less and without multiple edges. Define the *neighbourhood* of a vertex $v \in V(G)$ as $N_G(v) = \{u \in V(G) \mid (u, v) \in E(G)\}$, and let $\Delta(G) = \max\{|N_G(v)| \mid v \in V(G)\}$.

Denote by (p, q)-*grid*, any graph $H_{p,q}$ that is isomorphic to a graph with vertex set $\{0, \ldots, p - 1\} \times \{0, \ldots, q - 1\}$ and edge set

$$\{\{(i, j), (k, l)\} \mid (i, j), (k, l) \in V(H) \text{ and } |i - k| + |j - l| = 1\}.$$

In the above definition p is the column number and q is the row number of the grid. The r-*border* of $H_{s,r}$ is defined as $B^{(r)}(H_{s,r}) = ((i, 0), (i, r - 1) \mid i = 1, \ldots, s - 1)$. We call a vertex/edge/column of $H_{s,r}$ *internal* if it is not in/none of its endpoints is in/none of its vertices is in $B^{(r)}(H_{s,r})$.

All the algorithms and results in this paper work for the general case where $\mathcal{E} = \mathcal{U}$. The other cases are straightforward simplifications of our results[3].

---

[3] Our results also hold for more general sets of operations given that they locally change the structure of the input graph (See Section 5).

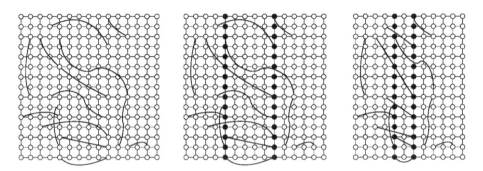

Fig. 1. A graph G whith $\mathbf{dist}(G, H_{15,15}) \leq 13$ with a $(6, 15)$-band J in it and the result of a 3-contraction of J in G

Subsequently, we will also drop the $\mathcal{E}$ part in the $\mathcal{E}\text{-}\mathbf{dist}(G, H)$ notation. If $\mathbf{dist}(G, H) \leq k$ we will denote as $V_{\text{dead}}/E_{\text{dead}}$ the vertices/edge that should be removed and $V_{\text{add}}/E_{\text{add}}$ the vertices/edges that should be added towards transforming G to H. In order to well define $V_{\text{add}}/E_{\text{add}}$ we always fix one of the possible ways that such a transformation can be accomplished. Without loss of generality, we assume that the transformation procedure gives priority first to vertex removals then to edge removals, then to vertex insertions, and finally to edge insertions. We also use the notation $k_1 = |V_{\text{dead}}|$, $k_2 = |E_{\text{dead}}|$, $k_3 = |V_{\text{add}}|$ and $k_4 = |E_{\text{add}}|$. Observe that $k_1 + k_2 + k_3 + k_4 \leq k$.

We say that a graph G contains an r-band J of width s or, alternatively, an (r, s)-band, if the following conditions are satisfied:

a) G contains $H_{s,r}$ as induced subgraph.
b) $\nexists\{x, y\} \in E(G)$ such that $x \in V(G) - V(H_{s,r})$ and $y \in V(H_{s,r}) - B^{(r)}(J)$.

If G is a graph and $H_{p,q}$ is a (p, q)-band in G where $p \geq 3$ and $q \geq 1$, for $0 \leq w \leq p-3$, a w-contraction of $H_{p,q}$ in G is the graph obtained by contracting in G all the edges of $H_{p,q}$ that are in the set $\{\{(i, m), (i + 1, m)\} \mid 2 \leq i \leq w+1, 1 \leq m \leq q\}$. We use the notation contract$(G, H_{p,q}, w)$ to denote the result of the operation just described. Notice that the routine contract$(G, H_{p,q}, w)$ runs in $O(wq)$ in steps. For an example of a c-contraction, see Figure 1.

Lemma 1. *Let G be a graph containing a $(3 + c, q)$-band $H_{3+c,q}$ for some $c \geq \lceil \frac{k}{q} \rceil$, and let $G' = $ contract$(G, H_{3+c,q}, w)$, where $w = c - \lceil \frac{k}{q} \rceil + 1$. Then $\mathbf{dist}(G, H_{p,q}) \leq k$ iff $\mathbf{dist}(G', H_{p-w,q}) \leq k$.*

Given a positive integer q, we call an edge of a graph q-extremal if $q > 1$ and both its endpoints have degree 3; $q = 1$ and both its endpoints have degree 2. For the set of q-extremal edges of a graph G we use the notation $E_{\text{ext}}^{(q)}(G)$.

Lemma 2. *There exists an algorithm* find-edge-max-band$(G, q, e))$ *that, given a graph G, a positive integer q, and a q-extremal edge e of G, returns a maximal length $(3 + c, q)$-band $H_{3+c,q}$ where $c \geq 0$ and e is also an extremal edge of H. If* *such a* $(3 + c, q)$-*band does not exists, then return "NO". If the answer is "NO" the algorithm finishes in* $O(q)$ *steps. If the answer is a vertex set with* $(3 + c)q$ *vertices, then the algorithm finishes in* $O(qc)$ *steps.*

Proof sketch: Algorithm find-edge-max-band(G, q, e) checks first whether e is the "upper" edge of some pair of neighboring columns by "guessing" neighbors of degree that agree to the corresponding sub-grid pattern. If this guess finishes successfully, then the algorithm ties to extend this $(q, 2)$-band to the right or to the left. If this extension is possible to some of the two directions then the output is the corresponding (q, c)-band, where $c \geq 2$. □

The proof of the following lemma is based on a contradiction argument.

Lemma 3. *If* G *is a graph where* $\Delta(G) \leq d$ *and* $\mathbf{dist}(G, H_{p,q}) \leq k$ *where* $p \geq q$, *then* G *contains at most* $2(p - 2) + 2(q - 2) + 2dk$ *q-extremal edges.*

The algorithm of the following lemma makes use of Lemma 3 and Algorithm find-edge-max-band(G, q, e) in Lemma 2.

Lemma 4. *There exists an algorithm* find-max-band(G, q, c)) *that, given a graph* G *and two positive integers* q, c, *returns a maximal length* $(3 + c', q)$-*band* $H_{3+c,q}$ *where* $c' \geq c$. *If such a* $(3 + c', q)$-*band does not exists, then* find-max-band(G, q, c) *returns "NO". In case of negative answer* find-max-band(G, q, c) *runs in* $O(q|E_{\text{ext}}^q(G)|)$ *steps. In case of a positive answer* find-max-band(G, q, c) *runs in* $O(qc|E_{\text{ext}}^q|)$ *steps.*

Suppose that G can be transformed to H after a sequence of edit operations. We call a vertex $v \in G$ *safe* if is not removed by a vertex removal operation in this sequence. If a vertex of G is not safe then we call it *dead*. We also say that an safe vertex v is *dirty* if some of the edit operations alters the set of edges incident to v in G (i.e. either adds or removes some edge incident to v). If a safe vertex is not dirty, then it is *clean*. A vertex of H is *new* if it was introduced during some vertex insertion operation. Notice that if $\mathbf{dist}(G, H) \leq k$, then there are at most k new vertices in H and at most k dead vertices in G, which implies the following lemma.

Lemma 5. *Let* G *and* H *be two graphs where* $\mathbf{dist}(G, H) \leq k$ *and such that the transformation of* G *to* H *involves* k_1 *vertex removals and* k_2 *vertex additions. Then,* $|V(H)| - k_2 \leq |V(G)| \leq |V(H)| + k_1$.

A straightforward counting argument gives the following lemma.

Lemma 6. *Let* G *be a graph such that* $\mathbf{dist}(G, H_{p,q}) \leq k$. *Suppose also that* G *does not contain a* $(3 + c, q)$-*band where* $q \cdot c > k$ *and* $H_{p,q}$ *contains* $\leq d$ *dirty vertices. Then* $p \leq (\lceil \frac{k}{q} \rceil + 3)(d + k + 1)$.

Lemma 7. *Let* G *be a graph with a vertex* v *of degree more than* $k + 4$ *in* G *and let* $H_{p,q}$ *be a grid. Then any transformation of* G *to* $H_{p,q}$ *should involve the operation of the removal of* v. *In particular,* $\mathbf{dist}(G, H_{p,q}) \leq k$ *iff* $\mathbf{dist}(G[V(G) - v], H) \leq k - 1$.

Proof sketch: Notice that more than k of the edges of v should be removed towards transforming G to $H_{p,q}$ while it is possible to apply at most k edit operations on G. Therefore, any sequence on $\mathcal{E}$ able to transform G to $H_{p,q}$, should include the removal of v.   □

Notice that if v is a dirty vertex of H then v is either adjacent to a vertex $v' \in V_{\text{dead}}$ or is incident to an edge in $E_{\text{add}} \cup E_{\text{dead}}$. As $v' \in V_{\text{dead}}$ can be adjacent to at most $\Delta(G)$ dirty vertices of H and an edge in $E_{\text{add}} \cup E_{\text{dead}}$ can be incident to at most two dirty vertices of H, then each edit operation creates at most $\max\{2, \Delta(G)\}$ dirty vertices. Therefore,

Lemma 8. *Let G be a graph and, let H be some grid. If $\mathbf{dist}(G, H) \leq k$, then H contains at most $\max\{\Delta(G), 2\} \cdot k$ dirty vertices.*

3   Looking for a Given Grid $H_{p,q}$

In this section, we show that the k-ALMOST GRID problem is in FPT. For this, we first combine Lemmata 5, 8, and 6 and prove the following bound f to the size of the kernel.

Lemma 9. *Let G be a graph where $\mathbf{dist}(G, H_{p,q}) \leq k$. Suppose also that $\Delta(G) \leq b$ and that G does not contain a $(3+c, q)$-band where $c \geq \lceil k/q \rceil$ or any $(3+c, p)$-band where $c \geq \lceil k/p \rceil$. Then $|V(G)| \leq f(k, b, p, q)$ where*

$$f(k, b, p, q) = \begin{cases} k + k^2 & \text{if } p \leq k \text{ and } q \leq k, \\ k + 5k(b \cdot k + k + 1) & \text{if } \min\{p, q\} \leq k < \max\{p, q\}, \\ k + 16(b \cdot k + k + 1)^2 & \text{if } p > k \text{ and } q > k. \end{cases}$$

Next, we present the algorithm to construct the kernel.

Kernel-for-Check-Grid(G, p, q, k)
Input: A graph G, two positive integers p, q, and a non-negative integer k
Output: Either returns "NO", meaning that that $\mathbf{dist}(G, H_{p,q}) > k$, or returns a graph G' and a triple p', q', k' such that: $\mathbf{dist}(G, H_{p,q}) \leq k$ iff $\mathbf{dist}(G', H_{p',q'}) \leq k'$.
0. Set $k' = k$, $G' = G$, $p' = p$ and $q' = q$.
1. As long as G' has a vertex of degree $\geq k'+4$, remove it from G' and set $k' \leftarrow k'-1$. If at the end of this proccess $k' < 0$,, then **return** "NO".
2. Apply any of the following procedures as long as it is possible:
 • If find-max-band$(G', q', \lceil \frac{k'}{q'} \rceil) = H_{3+c,q'}$,
   then set $w = c - \lceil \frac{k'}{q'} \rceil + 1$, $G' = $ contract$(G', H_{3+c,q'}, w)$ and $p' \leftarrow p' - w$.
 • If find-max-band$(G', p', \lceil \frac{k'}{p'} \rceil) = H_{3+c,p'}$,
   then set $w = c - \lceil \frac{k'}{p'} \rceil + 1$, $G' = $ contract$(G', H_{3+c,p'}, w)$ and $q' \leftarrow q' - w$.
3. If $|V(G')| > f(k', k + 4, p', q')$, then **return** "NO".
4. **return** G', p', q', k'.

Theorem 1. *Algorithm Kernel-for-Check-Grid(G, p, q, k) is correct and if it outputs the graph G' then $|V(G')| \leq f(k, k + 4, p, q)$. Moreover, it runs in $O(pq(p + q + k^2))$ steps.*

Proof: Step **1** is justified by Lemma 7. Notice that before the algorithm enters Step **2**, $\Delta(G) \leq k+4$. Step **2** creates equivalent instances of the problem because of Lemma 1. From Lemmata 3 and 4, each time a contraction of step **2** is applied, the detection and contraction of the corresponding q-band (p-band) requires $O(q'c(p' + q' + k'^2))$ $(O(p'c(p' + q' + k'^2)))$ steps. As the the sum of the lengths of the bands cannot exceed $q(p)$ we obtain that step **2** requires, in total, $O(pq(p + q + k'^2))$ steps. Moreover, before the check of Step **3**, all the conditions of Lemma 9 are satisfied. Therefore, if the algorithm returns G', p', q', k', then $|V(G')| \leq f(k', k + 4, p', q') \leq f(k, k + 4, p, q)$. $\qquad \square$

Theorem 2. *There exists an algorithm* Check-Grid(G, p, q, k) *that given a graph* G, *two positive integers* p, q, *and a non-negative integer* k *either returns "*NO*", meaning that that* $\text{dist}(G, H_{p,q}) > k$, *or it returns a sequence of at most* k *operations that transforms* G *to* $H_{p,q}$. Check-Grid(G, p, q, k) *runs in* $O(16^k \cdot (2n + 3k)^{2k})$ *steps.*

We stress that Algorithm Check-Grid(G, p, q, k) can be replaced by a faster one when we do not consider edge additions in the set of edit operations.

Now comes the main algorithm of this section which implements the kernelization.

Almost-grid(G, p, q, k)
Input: A graph G, two positive integers p, q, and a non-negative integer k
Output: Either returns "NO", meaning that that $\text{dist}(G, H_{p,q}) > k$, or returns a sequence of at most k operations that transforms G to $H_{p,q}$.

1. If Kernel-for-Check-Grid$(G, p, q, k) = (G', p', q', k')$ then return
         Check-Grid(G, p, q, k)
2. return "NO"

We conclude with the following.

Theorem 3. *Algorithm* Almost-grid(G, p, q, k) *solves the* k-ALMOST-GRID *problem in* $2^{O(k \log k)} + O(n^3)$ *steps.*

4 Looking for Any Grid

To solve the k-ALMOST SOME GRID problem it is enough to apply Check-Grid(G, p, q, k) for all $(p, q) \in \mathbf{A}(n, k) = \bigcup_{n-k \leq i \leq n+k}\{(p, q) \mid p \cdot q = i\}$. As $|\mathbf{A}(n, k)| \leq 2k \log(k + n)$, k-ALMOST SOME GRID can be solved after $O(k \log(n + k))$ calls of Almost-grid(G, p, q, k) which gives a running time of $O(\log n)(2^{O(k \log k)}) + O(k^2 n^3 \log n)$ steps. We call this algorithm check-all-cases(G, k). We stress that there are cases where $|\mathbf{A}(n, k)| \geq \frac{1}{2} \log n$ and therefore, that way, we may not avoid the "$\log n$"-overhead. In this section, we will give an alternative approach that gives running times of the same type as the case of k-ALMOST GRID.

We call *r-band collection* $\mathcal{C}$ of a graph G a collection of $|\mathcal{C}|$ r-bands in G, with widths $s_1, \ldots, s_{|\mathcal{C}|}$, where no pair of them have common interior vertices. The *width* of such a collection is $\sum_{i=1,\ldots,|\mathcal{C}|}(s_i - 2)$.

The following algorithm uses the Algorithm find-edge-max-band in order to identify the possible components of an r-band collection. Notice that the algorithm has two levels of "guessing" edges: The first level guesses a starting r-extremal edge, if the edge belongs to some $(3 + c, q)$-band, the algorithm extends the collection by guessing r-extremal edges of new components, excluding r-extremal edges that already belong to bands included in the collection.

The algorithm of the following lemma makes use of Algorithm find-edge-max-band(G, q, e).

Lemma 10. *There exists an algorithm* find-max-band-collection(G, r, w) *that, given a graph G and two positive integers r and w, returns "YES" if G contains an r-band collection of width $\geq w$; otherwise, returns "NO". The algorithm* find-max-band-collection(G, r, w) *runs in $O(rw|E_{\text{ext}}(G)| + n)$ steps.*

The following lemma identifies a bound on the size of a graph without a q-band collection of sufficient big width.

Lemma 11. *Let G be a graph where $\Delta(G) \leq b$ and let $H_{p,q}$ be a grid where* $\text{dist}(G, H) \leq k$. *Then if G does not contain any q-band collection of width $> k$ we have that $|V(H)| \leq (b \cdot k + 2)^2$.*

Check-some-Grid(G, k)

Input: A graph G and a non-negative integer k
Output: The answer to the k-ALMOST-SOME-GRID problem with instance G
    and parameter k.
1. While G has a vertex of degree $\geq k + 4$, remove it from G and
        set $k \leftarrow k - 1$. If at the end of this proccess $k < 0$ **then return** "NO".
2. $R \leftarrow \emptyset$
3. for $i = 1, \ldots, n$, if find-max-band-collection$(G, i, 3k + 1) =$"YES", then
                set $R \leftarrow R \cup \{i\}$
4. **If** $|R| = 0$ **then if** $V(G) > k + ((k + 4)k + 2)^2$, **then return** "NO",
                otherwise, goto Step 7
5. **For any** $r \in R$,
    If $r > 2k$, **then**
        begin
        If $\mathbb{N} \cap [\frac{n-k}{r}, \frac{n+k}{r}] = \emptyset$, **then return** "NO", **otherwise,**
            return Almost-Grid(G, r, s, k) where $\{s\} = \mathbb{N} \cap [\frac{n-k}{r}, \frac{n+k}{r}]$.
        end
        otherwise, for $(i, j) \in \bigcup_{n-k \leq i \leq n+k}\{(p, q) \mid p \leq 2k, p \cdot q \leq i\}$,
        begin
        If Almost-Grid$(G, i, j, k) =$ YES, **then return** Almost-Grid(G, i, j, k).
        end
6. **Return** "NO".
7. check-all-cases(G, k).

Lemma 12. *Let G be a graph where $\Delta(G) \leq b$ and let H be a grid where* **dist**$(G, H) \leq k$. *Then, if G contains an r-band collection of width $> 3k$ then $H = H_{s,r}$ for some $s \in [\frac{n-k}{r}, \frac{n+k}{r}]$. Moreover such an r-band collection can exist for at most two distinct values of r and s will be unique when $r > 2k$.*

We now proceed with the main algorithm of this Section, which states that k-ALMOST ANY GRID is in FPT.

Theorem 4. *Algorithm* Check-some-Grid(G, k) *is correct. Moreover, the k-*ALMOST-ANY-GRID *problem can be solved in time $2^{O(k \log k)} + O(k^2 \cdot n^3)$.*

Proof: Step **1** is justified by Lemma 7 and we may assume that before the algorithm enters Step **2**, $\Delta(G) \leq k + 4$. Steps **2**–**4** are based on Lemma 11. Step **3** involves $O(n)$ calls of find-max-band-collection$(G, i, 3k+1)$ which requires $O(kn^2)$ steps because of Lemma 10. Theorefore, Step **3** requires $O(kn^3)$ steps. Finally the loop in Step **5** is correct because of Lemma 12. Note that the first loop in step **5** is applied at most 2 times and that $|\bigcup_{n-k \leq i \leq n+k} \{(p,q) \mid p \leq 2k, p \cdot q \leq i\}| = O(k^2)$. Therefore, step **5** runs in $2^{O(k \log k)} + O(k^2 \cdot n^3)$ steps. As we noticed in the begining of this section, check-all-cases(G, k) requires $O(\log n')(2^{O(k \log k)} + O(\log n' \cdot n'^3 \cdot k^2))$ steps where n' is the number of vertices of G in step **7**. As $n' \leq O(k^4)$, this means that step **7** requires $2^{O(k \log k)}$ steps.   □

5   Extensions

The ideas of the kernelization algorithm Kernel-for-Check-Grid can be also used for more general "cleanning" problems. The directions in which our results can be extended are the following:

Other Patterns. Algorithm Find-edge-max-band intents to find "regions of regularity" (r-bands) in the input graph that can be safely contracted. This is used by the kernelization Algorithm Kernel-for-Check-Grid in order to output an equivalent instance of the problem (of small size) containing the "essential" part of the "dirtinesss" of the input graph. While Algorithm Find-edge-max-band works for grids, it can be adapted for other "regular" patterns as well. In Figure 2 we depict some examples of such patterns. In each of them one has to define the adequate notion of regularity pattern and design the analogous of Algorithm Find-edge-max-band for its detection. It is enough to adapt Lemma 1 in order

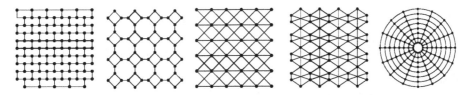

Fig. 2. Patterns of graphs for which the edit distance problem is in FPT

Fig. 3. (i) identification of two vertices, (ii) Δ-Y transformation, (iii) vertex duplication, and (iv) addition of any number of incident edges

to justify the compression of this regular patterns and to modify Lemma 9 to prove that when no further compression is possible. Therefore, the graph has a size that depends only on the parameter k (the proofs are based on the same arguments as in the case of grid).

Other Operations. Observe that the set $\mathcal{U} = \{$e-out, e-in, v-out, v-in$\}$ is not the most general set of edit operations, where the arguments of our proofs can be applied. In fact, all the proofs in this paper exploit the "locality" of these operations. All our results hold also for any set of operations, where each operation is equivalent to the detection of a constant size subgraph H, together with the application of a constant number of edge removals, edge additions, edge contractions, vertex removals, vertex additions (with the addition of an arbitrary number of incident edges), or inverse contractions on H. Some examples of such transformations are depicted in Figure 3.

Acknowledgement

We wish to thank Prof. Jordi Cortadella whose suggestions invited us to study this problem.

References

1. N. Alon, R. Yuster and U. Zwick. Color-Coding. *Journal of the ACM*, 42(4), 844–856, 1995.
2. R. Downey and M. Fellows. *Parameterized complexity*. Springer-Verlag, 1999.
3. H. Kaplan, R. Shamir and R. Tarjan. Tractability of parameterized completion problems on chordal, strongly chordal and proper interval graphs. *SIAM Journal of Computing*, 28, 1906–1922, 1999.
4. J. Lewis and M. Yannakakis. The node-deletion problem for hereditary properties is NP-Complete. *Journal Comput. and Systems Sci.* 20(2), 219–230, 1980.
5. Leizhen Cai. Fixed-parameter tractability of graph modification problems for hereditary properties. *Information Processing Letters*, 58(4)171–176, 1996
6. B. Monien. How to find paths efficiently. *Annals of Discrete Mathematics*, 25, 239–254, 1985.
7. A. Natanzon, R. Shamir and R. Sharan. Complexity classification of some edge modification problems. *Discrete Applied Mathematics*, 113(1), 109–128, 2001.
8. Rao, S. D. and Kurdahi, F. J. On clustering for maximal regularity extraction. *IEEE transactions on Computer-aided Design of Integrated Circuits and Systems*, 12(8), 1198–1208, 1993.

9. A. Robles-Kelly, E. Hankok. Graph edit-distance from spectral seriation. *IEEE Transactions on Pattern analysis and Machine Intelligence*, 27, 365–378, 2005.
10. M. Yannakakis. Node and edge deletion NP-complete problems. *ACM Symposium on Theory of Computing (STOC)*, 253–264, 1978.
11. K. Zhang and D. Sasha. Simple fast algorithms for the editing distance between trees and related problems. *SIAM Journal on Computing* 18, 1245–1262, 1989.

Tradeoffs in Depth-Two Superconcentrators

Chinmoy Dutta[1,*] and Jaikumar Radhakrishnan[1,2]

[1] School of Technology and Computer Science,
Tata Institute of Fundamental Research,
Homi Bhabha Road, Mumbai 400005, India
{chinmoy, jaikumar}@tifr.res.in
[2] Toyota Technological Institute at Chicago,
1427 E 60th Street, Chicago, IL 60637, USA

Abstract. An *N-superconcentrator* is a directed graph with N input vertices and N output vertices and some intermediate vertices, such that for $k = 1, 2, \ldots, N$, between any set of k input vertices and any set of k output vertices, there are k vertex disjoint paths. In a *depth-two* N-superconcentrator each edge either connects an input vertex to an intermediate vertex or an intermediate vertex to an output vertex. We consider tradeoffs between the number of edges incident on the input vertices and the number of edges incident on the output vertices in a depth-two N-superconcentrator. For an N-superconcentrator G, let $a(G)$ be the average degree of the input vertices and $b(G)$ be the average degree of the output vertices. Assume that $b(G) \geq a(G)$. We show that there is a constant $k_1 > 0$ such that

$$a(G) \log \left(\frac{2b(G)}{a(G)} \right) \log b(G) \geq k_1 \cdot \log^2 N \ .$$

We further show a complementary sufficient condition: there is a constant $k_2 > 0$, such that if some a and b ($a \leq b$) satisfy the above inequality with k_1 replaced by k_2, then there is an N-superconcentrator G with $a(G) \leq a$ and $b(G) \leq b$. In particular, these results imply that the minimum size of a depth-two N-superconcentrator is $\theta \left(N \frac{\log^2 N}{\log \log N} \right)$, which was already known [9].

Our results are motivated by the connection between the size of depth-two superconcentrators and the problem of maintaining the Discrete Fourier Transform (DFT) in the straight-line program model [3]. Our necessary condition implies that in this model, for any solution to the problem of maintaining the DFT of a vector of length N over an algebraically closed field of characteristic 0, if each update is processed using at most d atomic operations (for $d \leq \left(\frac{\log N}{\log \log N} \right)^2$), then at least $N^{\Omega(\frac{1}{\sqrt{d}})}$ atomic operations are required to process a query, in the worst case. In particular, if each update is to be processed in constant time, then some query takes $\Omega (N^\epsilon)$ worst-case time (for some constant $\epsilon > 0$). Before this work, it was only known [3] that one of these operations requires $\Omega \left(\frac{\log^2 N}{\log \log N} \right)$ time.

---

* Part of this work was done while the author was visiting IBM India Research Lab, New Delhi.

B. Durand and W. Thomas (Eds.): STACS 2006, LNCS 3884, pp. 372–383, 2006.
© Springer-Verlag Berlin Heidelberg 2006

1   Introduction

Valiant [11] showed that the graphs underlying the algorithms for a number of important arithmetic problems, e.g. polynomial multiplication, matrix multiplication and discrete Fourier transform, have properties closely related to concentration networks. This opened up the intriguing possibility of showing lower bounds on running time, by showing lower bounds on the size of such networks.

Superconcentrators (defined by Valiant [11]) are directed graphs with disjoint sets of input, output and intermediate vertices, such that any subset of inputs is connected to any equinumerous subset of outputs by vertex disjoint paths. The *depth* of a superconcentrator is the maximum length of a path from an input to an output; its *size* is the number of edges.

Superconcentrators have been the subject of intensive research as they are useful in proving lower bounds in complexity theory [11, 3], demonstrating optimality of algorithms [5], and constructing high connectivity networks. Valiant [12, 11] showed the existence of N-superconcentrators of size $O(N)$ and with depth $O(\log^2 N)$. Pippenger [6] gave a simpler recursive construction for an N-supercon-centrator with $O(N)$ edges and depth $O(\log N)$. On the other hand, Pippenger [7] proved a lower bound of $\Omega(N \log N)$ and an upper bound of $O(N \log^2 N)$ for the minimum size of a depth-two N-superconcentrator.

This led researchers to investigate the exact tradeoff between depth and size [2, 1, 8]. These results together determined the size of the smallest depth-d superconcentrator for all depths $d \geq 3$. For depth-two superconcentrators, Alon and Pudlák [1] improved the lower bound of Pippenger to $\Omega(N \log^{\frac{3}{2}} N)$. Radhakrishnan and Ta-Shma [9] showed that the smallest size of a depth-two N-superconcentrator is $\Theta(N \frac{\log^2 N}{\log \log N})$.

The work presented in this paper is motivated by a result of Frandsen et al. [3] who used the tight lower bound for the minimum size of a depth-two superconcentrator to show a lower bound for the time complexity of dynamic evaluation of certain algebraic functions in the straight line program model. They considered the setup of Reif and Tate [10] for dynamic algebraic algorithms where two kinds of on-line operations can be handled: *update* an input variable or *query* an output variable. They proved a lower bound of $\Omega(\frac{\log^2 N}{\log \log N})$ for the worst-case time complexity per operation of any algorithm for maintaining the Discrete Fourier Transform of a vector of length N. Earlier, Reif and Tate [10] had shown an upper bound of $O(\sqrt{N})$ in this model.

1.1   Definitions and Results

Definition 1 (Depth-two superconcentrator). *A depth-two N-superconcentrator is a directed graph $G(A, B, C, E)$, where A is the set of N input vertices, B is the set of N output vertices and C is the set of intermediate vertices. The edge set E is a subset of $A \times C \cup C \times B$. Furthermore, for $k = 1, 2, \ldots, N$, between any two subsets $S \subseteq A$ and $T \subseteq B$ consisting of k vertices each, there are k vertex disjoint paths. Let $a(G)$ be the average degree of the vertices in A and $b(G)$ be the average degree of the vertices in B.*

Remark: Since the roles of the input and output vertices are interchangeable, we will assume without loss of generality that $b(G) \geq a(G)$.

We study the tradeoff between the number of edges in the two levels of a depth-two superconcentrator and obtain matching (but for constants) necessary and sufficient conditions relating them.

Theorem 1 (Necessary condition). *For all large enough N and all depth-two N-superconcentrators G,*

$$a(G) \log \left(\frac{2b(G)}{a(G)} \right) \log b(G) \geq \frac{1}{200} \log^2 N \ .$$

Theorem 2 (Sufficient condition). *For all large enough N and all $b \geq a \geq 1$ such that*

$$a \log \left(\frac{2b}{a} \right) \log b \geq 4000 \log^2 N \ ,$$

there is an N-superconcentrator G with $a(G) \leq a$ and $b(G) \leq b$.

These two theorems have the following corollaries.

Corollary 1. *(a) The minimum number of edges in a depth-two N-superconcentrator is $\theta \left(N \frac{\log^2 N}{\log \log N} \right)$.*
(b) There is a constant $c_1 \geq 1$, such that for all large N and all depth-two N-superconcentrators G with $a(G) \in \left[1, \left(\frac{\log N}{\log \log N} \right)^2 \right]$ we have $b(G) \geq N^{\frac{c_1}{\sqrt{a(G)}}}$. Furthermore, this condition is tight: there is a constant c_2 such that for all large N, if $a \in \left[1, \left(\frac{\log N}{\log \log N} \right)^2 \right]$ and b satisfies $b \geq N^{\frac{c_2}{\sqrt{a}}}$, then there there is depth-two N-superconcentrator G with $a(G) \leq a$ and $b(G) \leq b$.

Theorem 1 has the following consequence for the dynamic DFT_N problem in the straight-line program model (the formal definitions appear in the next section).

Theorem 3 (Tradeoff for DFT). *In any solution to dynamic DFT_N in the straight-line program model, supporting updates and queries over an algebraically closed field of characteristic 0, if one operation is implemented with worst-case time $a = a(N)$, and the other operation with worst-case time $b = b(N) \geq a(N)$, then for all large N,*

$$a \log \left(\frac{2b}{a} \right) \log b \geq c_3 \cdot \log^2 N \ ,$$

where $c_3 > 0$ is a a fixed constant. In particular, as in Corollary 1(b), if $a \in \left[1, \left(\frac{\log N}{\log \log N} \right)^2 \right]$, then $b \geq N^{\frac{c_4}{\sqrt{a}}}$, for some $c_4 > 0$.

The main contributions of this work are the matching necessary and sufficient conditions for the existence of depth-two superconcentrators. Previous studies of depth-two superconcentrators focused on the overall size, and did not consider the tradeoff between the number of edges in the two levels. This work was motivated by the application of this tradeoff result to the Dynamic FFT problem.

Our proof for the necessary condition is based on a probabilistic argument used by Radhakrishnan and Ta-Shma [9] to show an $\Omega(N\frac{\log^2 N}{\log\log N})$ lower bound for the size of depth-two superconcentrators. The proof that the condition is sufficient is similar to the argument in [9] for obtaining the upper bound of $O(N\frac{\log^2 N}{\log\log N})$; in particular, we use Meshulam's [4] sufficient condition.

In the next section, we describe the framework for the dynamic Discrete Fourier Transform problem and derive Thm. 3 assuming that Thm. 1 holds. In Sect. 3, we prove Thm. 1 and in Sect. 4 we prove Thm. 2.

2   The DFT Problem

Reif and Tate [10] and Frandsen et al. [3] considered the following setup for dynamic algebraic algorithms. Let $f_1, f_2, \cdots, f_M$ be a system of N-variate polynomials over a commutative ring or rational functions over a field. Let the variables be $x_1, x_2, \ldots, x_N$. A dynamic algorithm for this system is one that when given an initial input vector, does some pre-processing and then can efficiently handle online requests of the following two kinds: $update_k(v)$ asks for changing the input x_k to the new value v, and $query_k$ asks for the value of the output f_k. There are several natural and important examples for this setup like dynamic polynomial evaluation, dynamic polynomial multiplication, dynamic matrix-vector multiplication, dynamic matrix-matrix multiplication, and dynamic discrete Fourier transform.

Definition 2 (Discrete Fourier Transform). *Fix $N \geq 1$. Let k be an algebraically closed field and $\omega \in k$ be a primitive N-th root of unity. Let F be the $N \times N$ matrix defined by $(F)_{ij} = \omega^{ij}$. The Discrete Fourier Transform, $\mathsf{DFT}_N : k^N \to k^N$, is the map $x \to Fx$, $\forall x \in k^N$.*

The most basic model of computation for dynamic algorithms is the straight line program model. In this model, given the problem of evaluating $f : k^N \to k^M$, we assign a straight line program to each of the operations $update_1$, $update_2$, $\cdots$, $update_N$, and $query_1$, $query_2$, $\cdots$, $query_M$. We allow addition, subtraction, multiplication and division as the basic operations.

Theorem 4 (Frandsen, Hansen and Miltersen [3]). *If there is a solution to the DFT_N problem over an algebraically closed field of characteristic 0 in the straight-line program model, where the update operations need worst-case time u and the query operations need worst-case time q, then there is a depth-two L-superconcentrator G with $L = \Omega(\sqrt[3]{\frac{N}{\log\log N}})$ such that $a(G) \leq min(u, q)$ and $b(G) \leq max(u, q)$.*

Proof (Thm. 3.). Immediate by combining Thm. 4 with Thm. 1 and Cor. 1.   □

3   Necessity: Proof of Theorem 1

Assume N is large, and $G(A, B, C, E)$ is a depth-two N-superconcentrator. From the lower bound of [9], we know that $b(G) \geq \log N$. Also, if $b(G) > N^{\frac{1}{10}}$, it is easy to see that the claim holds. So, in the following, we will assume that $\log N \leq b \leq N^{\frac{1}{10}}$. Also, we will write a instead of $a(G)$ and b instead of $b(G)$. Now, suppose the theorem does not hold, that is,

$$a \log \left(\frac{2b}{a} \right) \log b < \frac{1}{200} \log^2 N \ . \tag{1}$$

Let $L \triangleq \left\lfloor \frac{1}{10} \frac{\log N}{\log b} \right\rfloor$. Using the assumption on b, we conclude that $L \geq 1$ and in general $L \geq \frac{1}{20} \frac{\log N}{\log b}$. We will classify the vertices $v \in C$ according to their number of neighbours in B, which we denote by $\deg_B(v)$. For $i = 1, 2, \ldots, L$, let $K_i \triangleq b^{5i}$, and

$$C_i \triangleq \left\{ v \in C : \frac{N}{b^2 K_i} \leq \deg_B(v) \leq \frac{b^2 N}{K_i} \right\} \ ;$$

$$D_i \triangleq \left\{ v \in C : \deg_B(v) < \frac{N}{b^2 K_i} \right\} \ .$$

The idea of the proof is as follows. When considering subsets of A and B of size K_i, we will examine how these sets get connected. For this task the set D_i will be thought of as *low degree* vertices and the set C_i will be thought of as *medium degree* vertices; those not in D_i or C_i are the *high degree* vertices. We will argue that the paths connecting sets of size K_i cannot all pass though high degree vertices. Also, by averaging, we find an i such that the set C_i has few edges incident on it. This will imply that many sets have common neighbors among *low degree* vertices. This will lead to the final contradiction.

Our definition of K_i ensures that the sets C_i are disjoint. If i is chosen uniformly at random from $[L]$, then the expected number of edges between A and C_i is at most aN/L and the expected number of edges between B and C_i is at most bN/L. Using, Markov's inequality we conclude that there is an $\ell \in [L]$, such that the number of edges between A and C_ℓ is at most $2aN/L$ and the number of edges between B and C_ℓ is at most $2bN/L$. Fix such an ℓ. For $u \in A \cup B$, let d_u be the number of neighbors of u in C_ℓ. Let A' be the set of $\frac{N}{2}$ vertices $u \in A$ with the smallest d_u, and similarly B' be the set of $\frac{N}{2}$ vertices $v \in B$ with the smallest d_v. We thus have that the degree d_u for each vertex $u \in A'$ is at most

$$a_\ell \triangleq \frac{4a}{L} \leq \frac{80a \log b}{\log N} \ .$$

Similarly, the degree d_v for each vertex $v \in B'$ is at most

$$b_\ell \triangleq \frac{4b}{L} \ .$$

Now, consider subsets $S \subseteq A'$ and $T \subseteq B'$, each of size K_ℓ. We claim that S and T have a common neighbor in $C_\ell \cup D_\ell$. For otherwise, there must be K_ℓ (high degree) vertices in $C \setminus (C_\ell \cup D_\ell)$, each with $\deg_B$ more than $\frac{b^2 N}{K_\ell}$. But this implies that the number of edges between B and C is more than $K_\ell \cdot \frac{b^2 N}{K_\ell} > bN$, a contradiction.

We will now analyze the connections between S and T via $C_\ell \cup D_\ell$. As in [9], we identify large sets $A'' \subseteq A'$ and $B'' \subseteq B'$ such that A'' and B'' have no common neighbor in C_ℓ. So, every two subsets $S \subseteq A''$ and $T \subseteq B''$ of size K_ℓ must have a common (low degree) neighbor in D_ℓ. Since, vertices in D_ℓ have few neighbors in B'', we will be able to derive a contradiction from this.

The set A'' and B'' are obtained using the following randomized procedure. For each $v \in C_\ell$, with probability $\frac{a_\ell}{a_\ell + b_\ell}$ delete all its neighbors in B', and with probability $\frac{b_\ell}{a_\ell + b_\ell}$ delete all its neighbors in A'. These operations are executed independently for each $v \in C_\ell$. Thus, for $u \in A'$, the probability that it survives is exactly $\left(\frac{a_\ell}{a_\ell + b_\ell}\right)^{d_u}$. We then delete this vertex u again with probability $1 - \left(\frac{a_\ell}{a_\ell + b_\ell}\right)^{a_\ell - d_u}$, so that the probability that u survives is exactly

$$p_A = \left(\frac{a_\ell}{a_\ell + b_\ell}\right)^{a_\ell} \geq \left(\frac{a_\ell}{2b_\ell}\right)^{a_\ell} \geq N^{-\frac{2}{5}} .$$

To justify the last inequality, first substitute the values for a and b in terms of a_ℓ and b_ℓ in our assumption (1):

$$a_\ell \left(\frac{L}{4}\right) \log \left(\frac{2b_\ell}{a_\ell}\right) \log b < \frac{1}{200} \log^2 N ;$$

next using the bound $L \geq \frac{1}{20} \frac{\log N}{\log b}$, obtain $a_\ell \log \left(\frac{2b_\ell}{a_\ell}\right) < \frac{2}{5} \log N$.

Similarly, we ensure that every vertex in B' survives with probability exactly

$$p_B = \left(\frac{b_\ell}{a_\ell + b_\ell}\right)^{b_\ell} = \left(\frac{a_\ell + b_\ell}{b_\ell}\right)^{-b_\ell} \geq \exp(-a_\ell) \geq N^{-\frac{2}{5}} .$$

Again, for the last inequality, we use our assumption (1) and the definition of a_ℓ.

The (random) set of vertices in A' that survive is our A''. Similarly, the set of vertices in B' that survive is the set B''. Our construction ensures that A'' and B'' don't have any common neighbors in C_ℓ. So, all connections between subsets of size K_ℓ contained in A'' and B'' must go through the (low degree) vertices in D_ℓ. Let

$$F = \{(u,v) \in A'' \times B'' : u \text{ and } v \text{ have a common neighbor in } D_\ell\} .$$

Thus, the pairs in F contains the connections between A'' and B'' that are available via vertices in D_ℓ. These connections must suffice to connect every two subsets $S \subseteq A''$ and $T \subseteq B''$ of size K_ℓ. It can be shown (see Claim 3.8 of [9]) that we then have

$$|F| \geq \frac{|A''||B''|}{K_\ell} - |A''| - |B''| \ .$$

We then have the same inequality between the expected value (note that F, A'' and B'' are random variables) of the two sides:

$$\mathrm{E}[|F|] \geq \mathrm{E}\left[\frac{|A''||B''|}{K_\ell}\right] - \mathrm{E}[|A''|] - \mathrm{E}[|B''|] \ . \tag{2}$$

We will estimate the terms in this inequality separately and derive our final contradiction.

Upper bound for $\mathrm{E}[|F|]$: We first count the number of pairs $(u,v) \in A' \times B'$ that have a common neighbor in D_ℓ: this number is at most the number of edges incident on A', times maximum the number of neighbors in B' for a vertex in D_ℓ, that is, $aN \cdot \left(\frac{N}{b^2 K_\ell}\right) \leq \frac{N^2}{bK_\ell}$. Such a pair is included in F only if both u and v survive, which happens with probability at most $p_A p_B$ (if u and v have a common neighbor in C_ℓ, this probability is zero, otherwise, they survive independently, with probability p_A and p_B respectively). Thus,

$$\mathrm{E}[|F|] \ \leq \ \left(\frac{N^2}{bK_\ell}\right) \cdot p_A p_B \ \leq \ \frac{N^2 p_A p_B}{bK_\ell} \ = \ o\left(\frac{N^2 p_A p_B}{bK_\ell}\right) \ .$$

Lower bound for $\mathrm{E}[\frac{|A''||B''|}{K_\ell}]$: Note that $|A''||B''|$ is precisely the number of pairs $(u,v) \in A' \times B'$ where both u and v survive. Note that if $u \in A'$ and $v \in B'$ have no common neighbor in C_ℓ, then the events $u \in A''$ and $v \in B''$ are independent. Now, a fixed $u \in A'$ has at most a_ℓ neighbors in C_ℓ, and each vertex in C_ℓ has at most $\frac{b^2 N}{K_\ell} \leq \frac{N}{b^3}$ neighbors in B' (by the definition of C_ℓ). Thus, the number of pairs $(u,v) \in A' \times B'$ that do not have a common neighbor in C_ℓ is at least

$$\frac{N}{2}\left(\frac{N}{2} - a_\ell \cdot \frac{N}{b^3}\right) \ \geq \ \frac{N}{2}\left(\frac{N}{2} - \frac{4N}{b^2}\right) \ \geq \ \frac{N^2}{8} \ .$$

The probability that both vertices in such a pair survive is exactly $p_A p_B$. Thus,

$$\mathrm{E}\left[\frac{|A''||B''|}{K_\ell}\right] \geq \left(\frac{N^2}{8K_\ell}\right) p_A p_B \ .$$

Upper bounds for $\mathrm{E}[|A''|]$ and $\mathrm{E}[|B''|]$: We have

$$\mathrm{E}[|A''|] \ = \ Np_A \ = \ o\left(\frac{N^2}{K_\ell}\right) p_A p_B \ ,$$

where, for the last inequality, we used $K_\ell \leq N^{1/2}$ and $p_B \geq N^{-\frac{2}{5}}$. Similarly,

$$\mathrm{E}[|B''|] \ = \ Np_B \ = \ o\left(\frac{N^2}{K_\ell}\right) p_A p_B \ .$$

Now, returning to (2), we see that the first term on the right hand side is $\Omega\left(\left(\frac{N^2}{K_\ell}\right) p_A p_B\right)$, but all other terms are $o\left(\left(\frac{N^2}{K_\ell}\right) p_A p_B\right)$ – a contradiction.

4 Sufficiency

In this section, we give a probabilistic construction of depth-two superconcentrators in order to show that the condition stated in Thm. 2 is sufficient. Instead of working directly with that condition, we will first construct superconcentrators parametrized by a variable ℓ; later we will choose ℓ suitably and establish Thm. 2.

Lemma 1. *For all large N and all $\ell \in [1, \log N]$, there is a depth-two N-superconcentrator $G(A, B, C, E)$ with at most*

$$50 \left(\frac{\log N}{\log(\ell N^{\frac{1}{\ell}})} \right) \ell N$$

edges between A and C and at most

$$50 \left(\frac{(\log N)^2}{\log(\ell N^{\frac{1}{\ell}})} \right) N^{1+\frac{2}{\ell}}$$

edges between C and B.

We now assume this lemma and complete the proof of Thm. 2.

Proof (Thm. 2). Assume that N is large and that a and b satisfy

$$a \log \left(\frac{2b}{a} \right) \log b \geq 4000 \log^2 N \ . \tag{3}$$

The proof will split into three parts depending on whether a is *small*, *large* or *medium*.

Small a: Suppose $a \in [1, 3000]$. Then, (3) implies that $b \geq N$. To justify our theorem consider the superconcentrator with N intermediate vertices, each connected to a different input vertex and to all output vertices.

Large a: Suppose $a \geq 200 \frac{\log^2 N}{\log(2 \log N)}$. In this case, Lemma 1 with $\ell = \log N$, gives us the required superconcentrator.

Medium a: We may now assume that $a \in [3000, 200 \frac{\log^2 N}{\log(2 \log N)}]$. Let $\ell \geq 0$ be such that

$$a = 200 \left(\frac{\log N}{\log \ell N^{\frac{1}{\ell}}} \right) \ell \ . \tag{4}$$

It is easy to see that such an $\ell \in [1, \log N]$ exists. Further, we have $a \leq 200\ell^2$, and since $a \geq 3000$, we have $a \leq \ell^6$. Now, the idea is to invoke Lemma 1 with this value for ℓ. To complete the proof we need to show that (3) implies that

$$b \geq 50 \left(\frac{\log N}{\log \ell N^{\frac{1}{\ell}}} \right) N^{\frac{2}{\ell}} \ . \tag{5}$$

We rewrite (3) as $\log^2 b - \log\left(\frac{a}{2}\right)\log b - 4000\frac{\log^2 N}{a} \geq 0$. Thus, $\log b$ is at least as big as the positive root of the polynomial $X^2 - \log(\frac{a}{2})X - 4000\frac{\log^2 N}{a}$:

$$\log b \geq \frac{1}{2}\left(\log\left(\frac{a}{2}\right) + \sqrt{\log^2\left(\frac{a}{2}\right) + 16000\frac{\log^2 N}{a}}\right)$$

$$= \frac{1}{2}\left(\log\left(\frac{a}{2}\right) + \sqrt{\log^2\left(\frac{a}{2}\right) + 80\log N^{\frac{1}{\ell}}\log(\ell N^{\frac{1}{\ell}})}\right) \quad \text{(using (4))}$$

$$\geq \frac{1}{2}\left(\log\left(\frac{a}{2}\right) + \sqrt{\log^2\left(\frac{a}{2}\right) + 80\log^2 N^{\frac{1}{\ell}} + 80\log(\ell)\log N^{\frac{1}{\ell}}}\right)$$

$$\geq \frac{1}{2}\left(\log\left(\frac{a}{2}\right) + \sqrt{\log^2\left(\frac{a}{2}\right) + 36\log^2 N^{\frac{1}{\ell}} + 12\log\left(\frac{a}{2}\right)\log N^{\frac{1}{\ell}}}\right)$$

$$\geq \frac{1}{2}\left(\log\left(\frac{a}{2}\right) + \sqrt{\left(\log\left(\frac{a}{2}\right) + 6\log N^{\frac{1}{\ell}}\right)^2}\right)$$

$$= \log\left(\frac{a}{2}\right) + 3\log N^{\frac{1}{\ell}} \ .$$

For the third inequality we used $\ell^6 \geq a \geq \frac{a}{2}$. By exponentiating both sides and using (4), we can now derive (5):

$$b \geq \left(\frac{a}{2}\right)N^{\frac{3}{\ell}} \geq \frac{200}{2}\left(\frac{\log N}{\log \ell N^{\frac{1}{\ell}}}\right)\ell \cdot N^{\frac{3}{\ell}} \geq 100\left(\frac{\log^2 N}{\log \ell N^{\frac{1}{\ell}}}\right)N^{\frac{2}{\ell}} \ .$$

[For the last inequality, note that the $\ell N^{\frac{1}{\ell}}$ is minimum when $\ell = \ln N$, and then we have $\ell N^{\frac{1}{\ell}} = e\ln N \geq \log N$.]   □

4.1 Proof of Lemma 1

We will use the following sufficient condition shown by Meshulam [4].

Lemma 2 (Meshulam [4]). *A depth-two network $G(A, B, C, E)$, with A and B of size N, is a superconcentrator if and only if every two subsets $S \subseteq A$ and $T \subseteq B$ of the same size s have at least s common neighbors in C.*

Idea of the construction: Fix a $\delta \in (0, 1]$. Our goal is to produce a network G that satisfies the sufficient condition given in Meshulam's characterization. Now, s can take N values, namely, $1, 2, \ldots, N$. We think of this range as the union of roughly $\frac{1}{\delta}$ intervals, and dedicate one subnetwork for each interval. The first interval will consist of all sizes in $[1, M]$, the second interval will consists of sizes in $[M, M^2]$, and so on, where $M \approx N^\delta$. Our final N-superconcentrator will be obtained by putting these $O(\log N/\log M)$ subnetworks together. This motivates the following definition.

Definition 3 (Connector network). *For $N \geq 1$, $k \in [1, N]$, and $0 < \epsilon \leq 1$, an (N, k, ϵ)-connector is a depth-two network $H(A, B, C, E)$ with $|A|, |B| = N$,* such that every $S \subseteq A$ and $T \subseteq B$ of the same size $s \in [k, kN^\epsilon]$ have at least $\lfloor \epsilon^{-1}s \rfloor + 1$ common neighbors in C.

Lemma 3. If N is large enough, $\frac{1}{\log N} < \epsilon \leq 1$ and $k \in [1, N]$, there is an (N, k, ϵ)-connector with at most $\lceil 8\epsilon^{-1} \rceil N$ edges between A and C and at most $\lceil 20N^{1+2\epsilon} \log N \rceil$ edges between C and B.

Proof. We take C to be a set of size $\lceil 2\epsilon^{-1}kN^{2\epsilon} \rceil$. Our graph will consist of two random bipartite graphs, one between A and C and the other between B and C. In the first graph, each vertex in A is independently assigned $\lceil 8\epsilon^{-1} \rceil$ neighbors chosen uniformly with replacement from $|C|$. In the second graph, each vertex in B is similarly assigned $\lceil 20N^{2\epsilon} \log N \rceil$ neighbors in C. We have two claims.

Claim 1. With probability at least $\frac{3}{4}$, for all $S \subseteq A$ of size $s \in [k, kN^\epsilon]$, we have $|N(S)| \geq 2 \lfloor \epsilon^{-1}s \rfloor + 1$.

Claim 2. With probability at least $\frac{3}{4}$, for all $T \subseteq B$ of size $t \in [k, kN^\epsilon]$, we have $|N(T)| \geq |C| - \lfloor \epsilon^{-1}t \rfloor$.

We will justify these claims below. First, let us see that they immediately imply our lemma. With probability at least $\frac{1}{2}$ the events in both the claims hold simultaneously. But then, every two sets S and T of the same size $s \in [k, kN^\epsilon]$ have at least $\lfloor \epsilon^{-1}s \rfloor + 1$ neighbors in common.   □

We still need to establish the two claims above. These claims are about the existence of expander or disperser-like graphs. In [9], probabilistic constructions of disperser graphs were used for this purpose. But in Claim 1, we require expansion not just for a set of fixed size, but for sets of a whole range of sizes, and this does not immediately fit in the disperser-graph framework. It is simpler, therefore, to provide routine probabilistic arguments to prove these claims, rather than deduce them from similar constructions in the literature.

Proof (Claim 1). We first fix a subset $S \subseteq A$ of size $s \in [k, kN^\epsilon]$ and a subset R of C of size $2 \lfloor \epsilon^{-1}s \rfloor$, and estimate the probability that all neighbors of S lie in R. This probability is at most

$$\left(\frac{2\epsilon^{-1}s}{2\epsilon^{-1}kN^{2\epsilon}} \right)^{\left(\frac{8}{\epsilon}\right)s} \leq \frac{1}{N^{8s}} .$$

There are at most $\binom{N}{s}$ choices for S and at most $\binom{|C|}{2\lfloor \epsilon^{-1}s \rfloor}$ choices for R. Furthermore, s takes integral values in $[k, kN^\epsilon]$. We conclude that the probability that some $S \subseteq A$ of size $s \in [k, kN^\epsilon]$ has $2 \lfloor \epsilon^{-1}s \rfloor$ or fewer neighbors in C is at most

$$\sum_{s \in [k, kN^\epsilon]} \frac{1}{N^{8s}} \cdot \binom{N}{s} \cdot \binom{|C|}{2\lfloor \epsilon^{-1}s \rfloor} \leq \sum_{s=1}^{N} \frac{1}{N^{8s}} \cdot N^s \cdot \left(\frac{e|C|}{2\lfloor \epsilon^{-1}s \rfloor} \right)^{2\lfloor \epsilon^{-1}s \rfloor}$$

$$\leq \sum_{s=1}^{N} \frac{1}{N^{8s}} \cdot N^s \cdot \left(\frac{e \cdot \lceil 2\epsilon^{-1} k N^{2\epsilon} \rceil}{2\epsilon^{-1} s} \right)^{2\epsilon^{-1} s}$$

$$\leq \sum_{s=1}^{N} \frac{1}{N^{0.1s}} \leq \frac{1}{4} .$$

For the second inequality, we used the fact that the function $(eM/x)^x$ is an increasing function of x for $x \leq M$. For the third inequality, we used the assumption that $\epsilon^{-1} \leq \log N$. For the last inequality, we assumed that N is large enough.   □

Proof (Claim 2). This is similar to the proof above. Let t be the smallest integer in $[k, kN^\epsilon]$. Clearly, it is enough to show that the claim holds for sets of size t. Fix a set T of size t and a subset R of C of size $|C| - \lfloor \epsilon^{-1} t \rfloor - 1$, and estimate the probability that all neighbors of T lie in R. This probability is at most

$$\left(1 - \frac{\lfloor \epsilon^{-1} t \rfloor + 1}{|C|} \right)^{(20N^{2\epsilon} \log N)t} \leq \left(1 - \frac{\epsilon^{-1} t}{\lceil 2\epsilon^{-1} k N^{2\epsilon} \rceil} \right)^{(20N^{2\epsilon} \log N)t} \leq \frac{1}{N^{9t}} .$$

There are at most $\binom{N}{t}$ choices for T and at most $\binom{|C|}{\lfloor \epsilon^{-1} t \rfloor + 1}$ choices for R. We conclude that the probability that some set $T \subseteq B$ of size t has at most $|C| - \lfloor \epsilon^{-1} t \rfloor - 1$ neighbors is at most

$$\frac{1}{N^{9t}} \cdot \binom{N}{t} \cdot \binom{|C|}{\lfloor \epsilon^{-1} t \rfloor + 1} \leq \frac{1}{N^{9t}} \cdot N^t \cdot \left(\frac{e \lceil 2\epsilon^{-1} k N^{2\epsilon} \rceil}{\epsilon^{-1} t + 1} \right)^{\epsilon^{-1} t + 1}$$

$$\leq \frac{1}{N^{9t}} \cdot N^t \cdot N^{2t+2} \cdot 8^{\epsilon^{-1} t + 1}$$

$$\leq \frac{8}{N^{3t-2}} \leq \frac{1}{4} .$$

For the third inequality, we used our assumption $\epsilon^{-1} \leq \log N$. For the last inequality, we assumed that N is large.   □

Finally, we put together connectors to obtain the N-superconcentrators as claimed in Lemma 1.

Proof (Lemma 1). Fix $\ell \in [1, \log N]$. Let $M = \ell N^{\frac{1}{\ell}}$, $p = \left\lceil \frac{\log N}{\log M} \right\rceil \leq \frac{2 \log N}{\log M}$, and for $j = 0, 1, \ldots, p-1$, define $k_j = N^{\frac{j}{p}}$. Using Lemma 3 (with $\epsilon = \frac{1}{\ell}$), we conclude that there is an $(N, k_j, \frac{1}{\ell})$-connector H_j with $\lceil 8\ell \rceil N$ edges between A and C and $\left\lceil 20N^{1+\frac{2}{\ell}} \log N \right\rceil$ edges between C and B. Note that in this connector, any two sets, $S \subseteq A$ and $T \subseteq B$, of size $s \in [k_j, k_j M]$ have at least s common neighbors. By putting these p connectors together, we obtain an N-superconcentrator. The total number of edges between A and C is then at most

$$\lceil 8\ell \rceil N p \leq 50 \left(\frac{\log N}{\log(\ell N^{\frac{1}{\ell}})} \right) \ell N ,$$

and between B is C is at most

$$\left\lceil 20N^{1+\frac{2}{\ell}}\log N\right\rceil p \;\leq\; 50\left(\frac{(\log N)^2}{\log(\ell N^{\frac{1}{\ell}})}\right)N^{1+\frac{2}{\ell}}\;.$$

$\square$

Acknowledgments

We are grateful to Peter Bro Miltersen for suggesting this problem. The second author is grateful to Peter Bro Miltersen and Gudmund Frandsen for discussions on the results in [3].

References

[1] N. Alon and P. Pudlák. Superconcentrators of depth 2 and 3; odd levels help (rarely). *J. Comput. System Sci.*, 48:194–202, 1994.

[2] D. Dolev, C. Dwork, N. Pippenger, and A. Wigderson. Superconcentrators, generalizers and generalized connectors with limited depth. In *Proc. 15th Annual ACM Symposium on Theory of Computing, New York*, pages 42–51, 1983.

[3] G. S. Frandsen, J. P. Hansen, and P. B. Miltersen. Lower bounds for dynamic algebraic problems. *Information and Computation*, 171(2):333–349, 2002.

[4] R. Meshulum. A geometric construction of a superconcentrator of depth 2. *Theoret. Comput. Sci.*, 32:215–219, 1984.

[5] W. J. Paul, R. E. Tarjan, and J. R. Celoni. Space bounds for a game on graphs. In *Proc. 8th Annual ACM Symposium on Theory of Computing, Hershey, PA*, pages 149–160, May 1976.

[6] N. Pippenger. Superconcentrators. *SIAM J. Comput.*, 6:298–304, 1977.

[7] N. Pippenger. Superconcentrators of depth 2. *J. Comput. System Sci.*, 24:82–90, 1982.

[8] P. Pudlák. Communication in bounded depth circuits. *Combinatorica*, 14:203–216, 1994.

[9] J. Radhakrishnan and A. Ta-Shma. Bounds for dispersers, extractors and depth-two superconcentrators. *SIAM J. Disc. Math.*, 32:1570–1585, 2000.

[10] J. H. Reif and S. R. Tate. On dynamic algorithms for algebraic problems. *Journal of Algorithms*, 22:347–371, 1997.

[11] L. G. Valiant. On nonlinear lower bounds in computational complexity. In *Proc. 7th Annual ACM Symposium on Theory of Computing, Albuquerque, NM*, pages 45–53, May 1975.

[12] L. G. Valiant. Graph-theoretic properties in computational complexity. *J. Comput. System Sci.*, 13:278–285, 1976.

On Hypergraph and Graph Isomorphism with Bounded Color Classes[*]

V. Arvind[1] and Johannes Köbler[2]

[1] The Institute of Mathematical Sciences, Chennai 600 113, India
arvind@imsc.res.in
[2] Institut für Informatik, Humboldt Universität zu Berlin, Germany
koebler@informatik.hu-berlin.de

Abstract. Using logspace counting classes we study the computational complexity of hypergraph and graph isomorphism where the vertex sets have bounded color classes for certain specific bounds. We also give a polynomial-time algorithm for hypergraph isomorphism for bounded color classes of arbitrary size.

1 Introduction

In this paper we explore the complexity of Graph Isomorphism (GI) and Hypergraph Isomorphism (HGI) in the bounded color class setting. This means that the vertices of the input graphs are colored and we are only interested in isomorphisms that preserve the colors. The restriction of graph isomorphism to graphs with n vertices where the number of vertices with the same color is bounded by $b(n)$ is very well studied (we call this problem GI_b). In fact, for $b(n) = O(1)$ it was the first restricted version of GI to be put in polynomial time using group-theoretic methods [5] (as usual we denote $\text{GI}_{O(1)}$ by BCGI). Later Luks put BCGI in NC using nontrivial group theory [8], whereas Torán showed that BCGI is hard for the logspace counting classes Mod_kL, $k \geq 2$ [11]. Actually, Torán's proof shows that for $k \geq 2$, GI_{k^2} (as well as GA_{2k^2}) is hard for Mod_kL. More recently, in [1] it is shown by carefully examining Luks' algorithm and Torán's hardness result that BCGI is in the Mod_kL hierarchy and is in fact hard for this hierarchy.

For a fixed constant b, there is still a gap between the general upper bound result of [1] for GI_b and Torán's hardness result. More precisely, if GI_b is upper bounded by, say, the t-th level of the Mod_jL hierarchy, and is hard for the s-th level of the Mod_kL hierarchy, the constants t and j are much larger than s and k respectively. In the absence of a general result closing this gap, it is interesting to investigate the complexity of GI_b for specific values of b. In [6] GI_2 and GI_3 are shown to be equivalent to undirected graph reachability implying that they are complete for L [10]. In the present paper, we take a linear-algebraic approach to proving upper and lower bounds for GI_b. This is natural because the Mod_kL classes for prime k have linear-algebraic complete problems. Using linear algebra

---

[*] Work supported by a DST-DAAD project grant for exchange visits.

B. Durand and W. Thomas (Eds.): STACS 2006, LNCS 3884, pp. 384–395, 2006.
© Springer-Verlag Berlin Heidelberg 2006

over $\mathbb{F}_2$ we are able to show that GI_b is $\oplus L$ complete for $b \in \{4,5\}$. Our techniques involve a combination of linear algebra with a partial Weisfeiler-Lehman type of labeling procedure (see e.g. [4]).

Another natural question to investigate is the complexity of Hypergraph Isomorphism when the vertex set is divided into color classes of size at most $b(n)$ (call it HGI_b). Firstly, notice that even for constant b, it is not clear whether HGI_b is reducible to BCGI. At least the usual reduction from HGI to GI does not give a constant bound on the color classes. The reason is that hyperedges can be of unbounded size. Hence a hyperedge's orbit can be exponentially large even under color-preserving vertex permutations. Thus, we need to directly examine the complexity of HGI_b. We show using group theory and linear algebra that HGI_2 is $\oplus L$ complete. Further we show that for any prime p, HGA_p (as well as GA_{p^2}) is $Mod_p L$ hard.

Next we consider HGI_b for arbitrary b. Since HGI is polynomial-time many-one equivalent to GI, complexity-theoretic upper bounds for GI like $NP \cap coAM$ and SPP hold for HGI. However, consider an instance of HGI: a pair of hypergraphs (X_1, X_2), with n vertices and m edges each. The reduction to GI maps it to a pair of graphs (Y_1, Y_2) with vertex sets of size $m+n$. The best known isomorphism testing algorithm (see [2]) which has running time $c^{\sqrt{n \lg n}}$ will take time $c^{\sqrt{(m+n)\lg(m+n)}}$ when combined with the above reduction and applied to HGI. The question whether HGI has a simply exponential time (i.e. c^n time) algorithm was settled positively by Luks by using a dynamic programming approach [9]. In Section 5 we give a polynomial-time upper bound for HGI_b when b is bounded by a constant. This result is based on Luks' polynomial-time algorithm [7] for the set stabilizer problem for a permutation group in the class Γ_d.

2   Preliminaries

We first fix some notation. Let $X = (V,E)$ denote a (finite) hypergraph, i.e., E is a subset of the power set $\mathcal{P}(V)$ of V, and let g be a permutation on V. We can extend g to a mapping on subsets $U = \{u_1, \ldots, u_k\}$ of V by

$$g(U) = \{g(u_1), \ldots, g(u_k)\}.$$

g is an *isomorphism* between hypergraphs $X = (V,E)$ and $X' = (V,E')$, if

$$\forall e \subseteq V : e \in E \Leftrightarrow g(e) \in E'.$$

We also say that g maps X to X' and write $g(X) = X'$. If $g(X) = X$, then g is called an *automorphism* of X. Note that the identity mapping on V is always an automorphism. Any other automorphism is called *nontrivial*.

A *coloring* of a hypergraph (V,E) is given by a partition $\mathcal{C} = (C_1, \ldots, C_m)$ of V into disjoint *color classes* C_i. We call $X = (V, E, \mathcal{C})$ a *colored hypergraph*. In case $\|C_i\| \leq b$ for all $i = 1, \ldots, m$, we refer to X as a *b-bounded hypergraph*. Further, for $1 \leq k \leq b$, we use $\mathcal{C}_k = \{C \in \mathcal{C} \mid \|C\| = k\}$ to denote the set of all color classes having size exactly k.

A permutation g on V is called an *isomorphism* between two colored hypergraphs $X = (V, E, \mathcal{C})$ and $X' = (V, E', \mathcal{C}')$ with colorings $\mathcal{C} = (C_1, \ldots, C_m)$ and $\mathcal{C}' = (C_1', \ldots, C_m')$, if g preserves the hyperedges (i.e., $g(V, E_1) = (V, E_2)$) and g preserves the colorings (i.e., $g(C_i) = g(C_i')$ for all $i = 1, \ldots, m$).

The decision problem HGI_b consists of deciding whether two given b-bounded hypergraphs X_1 and X_2 are *isomorphic*. A related problem is the hypergraph automorphism problem HGA_b of deciding if a given b-bounded hypergraph X has a nontrivial automorphism. For usual b-bounded graphs $X = (V, E)$ (i.e., each edge $e \in E$ contains exactly 2 nodes), we denote the isomorphism and automorphism problems by GI_b and GA_b, respectively.

Let $X = (V, E)$ be a graph. For a subset $U \subseteq V$, we use $X[U]$ to denote the *induced subgraph* $(U, E(U))$ of X, where $E(U) = \{e \in E \mid e \subseteq U\}$. Further, for disjoint subsets $U, U' \subseteq V$, we use $X[U, U']$ to denote the *induced bipartite subgraph* $(U \cup U', E(U, U'))$, where $E(U, U')$ contains all edges $e \in E$ with $e \cap U \neq \emptyset$ and $e \cap U' \neq \emptyset$. For a set U of nodes, we use $\Gamma_X(U)$ to denote the *neighborhood* $\{v \in V \mid \exists u \in U : (u, v) \in E\}$ of U in X.

We denote the *symmetric group* of all permutations on a set A by $\mathrm{Sym}(A)$ and by S_n in case $A = \{1, \ldots, n\}$. Let G be a subgroup of $\mathrm{Sym}(A)$ and let $a \in A$. Then the set $\{b \in A \mid \exists \pi \in G : \pi(a) = b\}$ of all elements $b \in A$ reachable from a via a permutation $\pi \in G$ is called the *orbit* of a in G.

3   Graphs with Color Classes of Size 5

In this section we prove that GA_4 is contained in $\oplus L$. The proof is easily extended to GI_4 as well as to GA_5 and GI_5. Let $X = (V, E, \mathcal{C})$ be a 4-bounded graph (an instance of GA_4) and let $\mathcal{C} = (C_1, \ldots, C_m)$. We use X_i to denote the graph $X[C_i]$ induced by C_i and X_{ij} to denote the bipartite graph $X[C_i, C_j]$ induced by the pair of color classes C_i and C_j. We assume that all vertices in the same color class have the same degree and that the edge set E_i of X_i is either empty or consists of two disjoint edges (only if $\|C_i\| = 4$), since otherwise we can either split C_i into smaller color classes or we can replace X_i by the complement graph without changing the automorphism group $\mathrm{Aut}(X)$ of X. Further, we assume that the edge set E_{ij} of X_{ij} is of size at most $\|C_i\| \cdot \|C_j\|/2$, since otherwise, we can replace X_{ij} by the complement bipartite graph without changing $\mathrm{Aut}(X)$.

Any $\pi \in \mathrm{Aut}(X)$ can be written as $\pi = (\pi_1, \ldots, \pi_m)$ in $\mathrm{Aut}(X_1) \times \cdots \times \mathrm{Aut}(X_m)$, where π_i is an automorphism of X_i. Furthermore, for any pair of color classes C_i and C_j, (π_i, π_j) has to be an automorphism of X_{ij}. These are precisely the constraints that any automorphism in $\mathrm{Aut}(X)$ must satisfy.

Since each $\mathrm{Aut}(X_i)$ is isomorphic to a subgroup of S_4, the only prime factors of $\|\mathrm{Aut}(X)\|$ (if any) are 2 and 3. Thus, $\mathrm{Aut}(X)$ is nontrivial if and only if it has either an automorphism of order 2 or of order 3. By a case analysis, we will show that the problem of testing whether X has a nontrivial automorphism can be reduced to either undirected graph reachability or to solving a system of linear equations over $\mathbb{F}_2$, implying that the problem is in $\oplus L$. In fact, as we will see, it is also possible to compute a generating set for $\mathrm{Aut}(X)$ in $FL^{\oplus L}$.

Let G_i be the intersection of $\mathrm{Aut}(X_i)$ with the projections of $\mathrm{Aut}(X_{ij})$ on C_i for all $j \neq i$. Any subgroup of the symmetric group $\mathrm{Sym}(C_i)$ of all permutations on C_i is called a *constraint* for C_i. We call G_i the *direct constraint* for C_i.

The algorithm proceeds in several preprocessing steps which progressively eliminate different cases and simplify the graph. We describe some base steps of the algorithm in a sequence of claims.

Claim 1. *The direct constraints can be determined in deterministic logspace.*

Proof. Follows easily from the fact that the color classes are of constant size. □

Next we consider a specific way in which the direct constraints get propagated to other color classes in X. To this end, we define a symmetric binary relation $\mathcal{T}$ on the set $\mathcal{C}$. Let $C_i, C_j \in \mathcal{C}$ such that $\|C_i\| = \|C_j\|$. Then $(C_i, C_j) \in \mathcal{T}$ if

- $\|C_i\| \in \{1, 2, 3\}$ and X_{ij} is a perfect matching or
- $\|C_i\| = 4$ and X_{ij} is either a perfect matching or an 8-cycle.

The following easy lemma states a specific useful way in which the constraints get propagated over color classes related via $\mathcal{T}$.

Lemma 2. *For each pair $(C_i, C_j) \in \mathcal{T}$ there is a bijection $f_{ij} : C_i \to C_j$ such that for any automorphism $\pi = (\pi_1, \ldots, \pi_m) \in \mathrm{Aut}(X)$ the permutations $\pi_i \in G_i$ and $\pi_j \in G_j$ are related as follows:*

$$\forall u, v \in C_i : \pi_i(u) = v \iff \pi_j(f_{ij}(u)) = f_{ij}(v).$$

In other words, if $(C_i, C_j) \in \mathcal{T}$ via f_{ij} and $\pi = (\pi_1, \ldots, \pi_m) \in \mathrm{Aut}(X)$ is an automorphism, then $\pi_j(f_{ij}(u)) = f_{ij}(\pi_i(u))$, i.e., π_j is the image of π_i under the bijection $g_{ij} : \mathrm{Sym}(C_i) \to \mathrm{Sym}(C_j)$ defined by $\pi \mapsto f_{ij} \circ \pi \circ f_{ij}^{-1}$ (here we use $g \circ h$ to denote the mapping $x \mapsto g(h(x))$).

We use Lemma 2 to define a symmetric relation on constraints. Let G and H be constraints of two different color classes C_i and C_j, respectively, where $(C_i, C_j) \in \mathcal{T}$. We say that G is *directly induced by* H, if g_{ij} is an isomorphism between G and H. Further, G *is induced by* H, if G is reachable from H via a chain of directly induced constraints. Note that the latter relation is an equivalence on the set of all constraints. We call the intersection of all constraints of C_i that are induced by some direct constraint the *induced constraint* of C_i and denote it by G_i'. Note that $\mathrm{Aut}(X)$ is a subgroup of the m-fold product group $\prod_{i=1}^{m} G_i'$ of all induced constraints.

Claim 3. *The induced constraints can be determined in deterministic logspace.*

Proof. Consider the undirected graph $X' = (V', E')$ where V' consists of all constraints G in X and $E' = \{(G, H) \mid G \text{ is directly induced by } H\}$. In this graph we mark all direct constraints computed by Claim 1 as special nodes. Now, the algorithm outputs for each color class C_i the intersection of all constraints for C_i that are reachable from some special node, and since $\mathrm{SL} = \mathrm{L}$ [10], this can be done in deterministic logspace. □

We define two special types of constraints. We say that C_i is *split*, if G_i' has at least two orbits, and we call the partition of C_i in the orbits of G_i' the *splitting partition* of C_i. Further, a partition $\{H_0, H_1\}$ of C_i with $\|H_0\| = \|H_1\| = 2$ is called a *halving* of C_i, if any $\pi \in G_i'$ either maps H_0 to itself or to H_1. Any class C_i which has a halving, is called *halved*, and all color classes that are neither split nor halved are called *whole*. Now let $\mathcal{C}_s$, $\mathcal{C}_h$ and $\mathcal{C}_w$ denote the subclasses of $\mathcal{C}$ containing all split, halved, and whole color classes, respectively, and consider the following two cases:

Case A: All color classes in $\mathcal{C}_4$ are halved (i.e., $\mathcal{C} \subseteq \mathcal{C}_h \cup \mathcal{C}_3 \cup \mathcal{C}_2 \cup \mathcal{C}_1$).
Case B: All color classes in $\mathcal{C}_4$ are halved and $\mathcal{C}_3$ is empty (i.e., $\mathcal{C} \subseteq \mathcal{C}_h \cup \mathcal{C}_2 \cup \mathcal{C}_1$).

We first show how the general case logspace reduces to Case A. Then we reduce Case A to Case B in logspace, and finally we show how Case B is solved in logspace with a $\oplus L$ oracle. We start by summarizing some properties of whole color classes which are easily proved by a case analysis.

Lemma 4. *Let $C_i, C_j \in \mathcal{C}$ be color classes, where C_i is whole and $E_{ij} \neq \emptyset$.*

- *All vertices in C_i have the same degree in X_{ij}. Likewise, all vertices in the neighborhood $\Gamma_{X_{ij}}(C_i)$ have the same degree in X_{ij}.*
- *If C_j is whole, then $\|C_i\| = \|C_j\|$ and $(C_i, C_j) \in \mathcal{T}$.*
- *If C_j is halved, then $\|C_i\| \leq 3$.*
- *If C_j is halved and $E_j \neq \emptyset$, then $\|C_i\| \leq 2$.*
- *If C_j is split and $\|C_j\| \leq \|C_i\|$, then all vertices in C_i have the same neighborhood in X_{ij}.*

Lemma 4 tells us that the action of an automorphism on a whole color class $C \in \mathcal{C}_4$ is not influenced by its action on color classes that are either smaller or halved or split, i.e., only other whole color classes in $\mathcal{C}_4$ can influence C. This means that we can write $\mathrm{Aut}(X)$ as the product $\mathrm{Aut}(X') \times \mathrm{Aut}(X'')$, where X' is the induced subgraph of X containing the nodes of all color classes in $\mathcal{C}_4 \cap \mathcal{C}_w$ and X'' is induced by the set of all other nodes. Clearly, it suffices to compute generating sets for $\mathrm{Aut}(X')$ and $\mathrm{Aut}(X'')$.

Claim 5. *A generating set for $\mathrm{Aut}(X')$ can be computed in* FL.

Proof. The algorithm will work by reducing the problem to reachability in undirected graphs. For each whole color class $C_i \in \mathcal{C}_4$ we create a set P_i of 4! nodes (one for each permutation of C_i). Consider $C_i, C_j \in \mathcal{C}_4 \cap \mathcal{C}_w$ such that $(C_i, C_j) \in \mathcal{T}$ and let f_{ij} be the bijection from Lemma 2. Recall that for each $\pi \in P_i$ the bijection f_{ij} induces a unique permutation $\psi = g_{ij}(\pi)$ on C_j and hence, we put an undirected edge between π and ψ. We thus get an undirected graph $\hat{X}$ with $4!\|\mathcal{C}_4 \cap \mathcal{C}_w\|$ nodes.

A connected component P in $\hat{X}$ that picks out at most one element π_i from each set P_i defines a valid automorphism π for the graph X', if P contains only elements $\pi_i \in \mathrm{Aut}(X_i)$. On the color classes C_i, for which P contains an element $\pi_i \in P_i$, π acts as π_i, and it fixes all nodes of the other color classes. By collecting these automorphisms we get a generating set for $\mathrm{Aut}(X')$ and since $\mathrm{SL} = \mathrm{L}$ [10], this can be done in deterministic logspace.   □

By using Claim 5 we can already assume that $\mathcal{C}_4$ only contains halved or split color classes. To fulfil the assumption of Case A, we now take all the split color classes and break them up into smaller color classes defined by the split. Then only halved color classes remain in $\mathcal{C}_4$. Further, we can assume that if C_i is halved, then E_i consists of two disjoint edges.

Next we consider the reduction of Case A to Case B. Since $\mathcal{C}_4$ only contains halved color classes with two disjoint edges, Lemma 4 now guarantees that a whole color class $C \in \mathcal{C}_3$ can only influence other whole color classes in $\mathcal{C}_3$. Hence, we can remove all whole color classes in $\mathcal{C}_3$ exactly as we removed the whole color classes in $\mathcal{C}_4$, by applying an analogue of Claim 5. Now we can again break up all split color classes yielding a graph that fulfils Case B. Observe that the splitting partition of a halved color class only contains sets of size 1, 2 or 4.

It remains to compute a generating set for a 4-bounded graph which only has color classes of size 1, 2 or 4, where all the size 4 color classes are halved. Clearly, in this case X can only have nontrivial automorphisms of orders 2 or 4. We continue by defining an encoding of the candidate automorphisms of $\mathrm{Aut}(X)$ as vectors over $\mathbb{F}_2$. Later we show how one can reduce the problem of finding $\mathrm{Aut}(X)$ to the problem of solving linear equations over $\mathbb{F}_2$.

An Encoding Trick

We now describe how to encode the automorphisms of X with vectors over $\mathbb{F}_2$. We introduce some $\mathbb{F}_2$ indeterminates for each color class C in X of size more than 1. More precisely, we encode each automorphism $\pi \in \mathrm{Aut}(X_i)$ by a vector $v_\pi \in \mathbb{F}_2^{n_i-1}$, where $n_i = \|C_i\|$.

1. To each color class $C_i = \{u_0, u_1\}$ of size 2 we introduce a single variable x. Here $x = 1$ denotes the transposition $(u_0\ u_1)$ and $x = 0$ denotes the identity mapping on C_i.

2. If C_i is a halved color class of size 4, let $e_0 = \{u_{00}, u_{01}\}$ and $e_1 = \{u_{10}, u_{11}\}$ be the two disjoint edges in E_i. Notice that each vertex index is encoded with two bits. The first bit encodes the edge and the second bit the vertex in that edge. Then we encode an automorphism $\pi \in \mathrm{Aut}(X_i)$ by a three bit vector $v_\pi = xyz$, where $\pi(u_{00}) = u_{xy}$ and $\pi(u_{10}) = u_{\bar{x}z}$. In other words, the permutation π encoded by v_π maps

$$u_{ab} \mapsto u_{a'b'}, \text{ where } a' = a + x \text{ and } b' = \begin{cases} b + y, & a = 0, \\ b + z, & a = 1. \end{cases}$$

Notice that the addition of the 3-bit representations in $\mathbb{F}_2^3$ *does not* capture the permutation group structure of $\mathrm{Aut}(X_i)$, since the latter is nonabelian.

Let $t = \sum_{i=1}^{m}(n_i - 1)$ denote the sum of the lengths of the $\mathbb{F}_2$ representations for each color class. Thus, every element $\pi = (\pi_1, \ldots, \pi_m)$ of $\mathrm{Aut}(X)$ is encoded as a vector $v_\pi = (v_{\pi_1}, \ldots, v_{\pi_m})$ in $\mathbb{F}_2^t$, and each vector in $\mathbb{F}_2^t$ represents a potential automorphism. Recall that a permutation $\pi = (\pi_1, \ldots, \pi_m) \in \mathrm{Aut}(X_1) \times \cdots \times \mathrm{Aut}(X_m)$ is in $\mathrm{Aut}(X)$ if and only if for each C_i and C_j, (π_i, π_j) is an automorphism of X_{ij}. We now show that each of these constraints yields a set of linear equalities on the indeterminates that encode (π_i, π_j). Hence, we claim that a vector in $\mathbb{F}_2^t$ encodes an automorphism of X if and only if it satisfies the set of all these linear equalities.

Lemma 6. *For any pair of color classes C_i and C_j in X the set*

$$F_{ij} = \{v_\pi v_\varphi \mid \pi \in \mathrm{Aut}(X_i), \varphi \in \mathrm{Aut}(X_j) \ and \ (\pi, \varphi) \in \mathrm{Aut}(X_{ij})\}$$

forms a subspace of $\mathbb{F}_2^{n_i + n_j - 2}$.

Proofsketch. The proof is by a case analysis checking all the possibilities. First note that for any color class C_i of size more than 1, the encodings of all automorphisms in $\mathrm{Aut}(X_i)$ form the space $\mathbb{F}_2^{n_i - 1}$. Hence, $F_{ij} = \mathbb{F}_2^{n_i + n_j - 2}$ if $E_{ij} = \emptyset$.

To simplify the analysis if E_{ij} is not empty, notice that if C_i, C_j are unsplit halved color classes, then either $(C_i, C_j) \in \mathcal{T}$ or E_{ij} is the union of two 4-cycles. In the latter case we can modify X_{ij} by removing the two 4-cycles and introducing a new color class $C = \{c_1, c_2\}$.

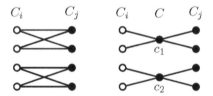

The nodes c_1 and c_2 represent the two deleted 4-cycles in the following sense: the node c_1 is adjacent to the 4 nodes that were part of one of the 4-cycles (two each in C_i and C_j). Similarly, c_2 is adjacent to the 4 nodes on the other 4-cycle (two each in C_i and C_j). Then the modified graph has the same automorphism group as X_{ij}, after we project out the new color class C. Hence, we can assume that if C_i and C_j are both halved, then $(C_i, C_j) \in \mathcal{T}$.

Now consider the case that both C_i and C_j are of size 2. This case is easy, since the addition of the representations $v_\pi v_\varphi$ captures the permutation group structure. In fact, the only interesting case is when E_{ij} is a perfect matching where $F_{ij} = \{x_i x_j \mid x_i = x_j\}$ (see the following picture).

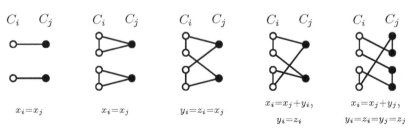

Next we consider the case $n_i = 4$ and $n_j = 2$. It is easy to see by considering the different cases that the elements $x_i y_i z_i x_j$ of F_{ij} form a subspace of $\mathbb{F}_2^4$ (see the picture for three interesting cases).

Finally, if $n_i = n_j = 4$, we can assume that $(C_i, C_j) \in \mathcal{T}$. It is easy to check that exactly one of the following two possibilities can occur: either the projection of $(\mathrm{Aut}(X_i) \times \mathrm{Aut}(X_j)) \cap \mathrm{Aut}(X_{ij})$ on C_i contains all 8 automorphisms (this case is similar to the case $E_{ij} = \emptyset$), or it is a subgroup that contains no automorphisms of order four. An interesting property of the encoding is that all subgroups containing no order four elements are such that permutation composition coincides with vector addition in $\mathbb{F}_2^{n_i-1}$. It follows from this observation that the elements $x_i y_i z_i x_j y_j z_j$ of F_{ij} form a subspace of $\mathbb{F}_2^6$ (see the picture for an example). $\qquad\square$

As a direct consequence of the above lemma and the fact that the solution space for a system of linear equations over $\mathbb{F}_2$ can be computed in $\mathrm{FL}^{\oplus\mathrm{L}}$ (see [3]) we get the following upper bound.

Claim 7. *Let X be a 4-bounded graph fulfilling the assumption of Case B. Then the problem of computing a generating set for its automorphism group is in $\mathrm{FL}^{\oplus\mathrm{L}}$.*

By refining Torán's proof [11] that GA is hard for $\oplus\mathrm{L}$, it can be shown that also GA_4 is hard for $\oplus\mathrm{L}$. By combining this hardness result with the $\oplus\mathrm{L}$ upper bound for GA_4 established in this section we get the following completeness result.

Theorem 8. GA_4 *is complete for* $\oplus\mathrm{L}$.

We end this section by remarking that Theorem 8 easily extends to GA_5 as well as to GI_5. Color classes of size 5 are either whole or split. The halved classes can only be of size 4. Thus, the $\oplus\mathrm{L}$ upper bound goes through for GA_5 with minor changes.

4   Hypergraphs with Color Classes of Size 2

In this section we show that HGA_2 and HGI_2 are complete for $\oplus\mathrm{L}$ under logspace reductions. We only give the proof for HGA_2. The proof for HGI_2 is similar. Let $X = (V, E)$ be a 2-bounded hypergraph. Thus V is partitioned into color classes C_i with $\|C_i\| \le 2$ for $i = 1, \ldots, m$. To each color class C_i of size 2 we associate an indeterminate x_i over $\mathbb{F}_2$ which indicates by its value whether the vertices of C_i flip or not. Thus we can represent any color preserving permutation of X by a vector $x = x_1 \cdots x_m$ in $\mathbb{F}_2^m$.

Our aim is to compute in deterministic logspace a set of linear constraints on $x_1, \ldots, x_m$ over $\mathbb{F}_2$ that determines $\mathrm{Aut}(X)$. This will imply that HGA_2 is in $\oplus\mathrm{L}$. Hyperedges e and e' have the same *type* if $\|e \cap C_i\| = \|e' \cap C_i\|$ for all i. We can partition E into subsets $E_1, \ldots, E_t$ of distinct types. Clearly, automorphisms preserve edge types. Thus, $\mathrm{Aut}(X)$ is expressible as the intersection of $\mathrm{Aut}(V, E_j)$ over all edge types.

Proposition 9. *A vector x in $\mathbb{F}_2^m$ represents an element in $\mathrm{Aut}(X)$ if and only if it represents an element in $\mathrm{Aut}(V, E_j)$ for each E_j.*

If for each E_j we can compute in logspace a set of linear constraints on $x_1, \ldots, x_m$ that determines $\mathrm{Aut}(V, E_j)$, then the union of these constraints will determine $\mathrm{Aut}(X)$. Thus, it suffices to consider the case that all edges in E are of the same type. Further, by ignoring color classes C_i with $\|e \cap C_i\| \in \{0, 2\}$ for all e, we can assume that $\|e \cap C_i\| = 1$ for all $e \in E$ and $i = 1, \ldots, m$.

Let $C_i = \{u_{0i}, u_{1i}\}$ for $i = 1, \ldots, m$. We can represent the hyperedges $e \in E$ by vectors $v_e = v_1 \cdots v_m \in \mathbb{F}_2^m$ with $v_j = 1$ if $u_{1j} \in e$ and $v_j = 0$ if $u_{0j} \in e$. With this representation, a candidate automorphism $x \in \mathbb{F}_2^m$ acts on the hyperedges by vector addition in $\mathbb{F}_2^m$:

$$x : v_e \mapsto v_e + x.$$

Since every automorphism x maps v_e to some hyperedge $v_{e'}$, the candidate automorphisms are in $S = \{v_e + v_{e'} \mid e' \in E\}$, for a fixed $e \in E$. A logspace machine M can easily check whether each vector $x \in S$ represents an automorphism, by testing if $x + v_e \in E$ for each $e \in E$.

Thus, M can compute the set $F \subseteq S$ containing the encodings of all automorphisms. Notice that F is a subspace of $\mathbb{F}_2^m$.

Finally, we can easily see that M can compute the dual space in terms of a matrix A over $\mathbb{F}_2$ such that $x \in F$ if and only if $Ax = 0$. This matrix A provides the desired set of linear constraints. Combining the constraints of all edge types gives the overall system of linear constraints, whose solutions are the automorphisms. In summary, we have proved the following theorem.

Theorem 10. *Let X be a 2-bounded hypergraph. Then the problem of computing a generating set for its automorphism group is in $\mathrm{FL}^{\oplus \mathrm{L}}$. In particular, the problem HGA_2 is in $\oplus \mathrm{L}$.*

By modifying Torán's proof [11] that GA is hard for $\oplus \mathrm{L}$, it can be shown that for any prime p, HGA_p is hard for $\mathrm{Mod}_p \mathrm{L}$. Combined with the above theorem this gives the following completeness result.

Corollary 11. *HGA_2 is complete for $\oplus \mathrm{L}$ under logspace many-one reductions.*

5   Hypergraphs with Constant Size Color Classes

In this section we give a polynomial time upper bound for the problem of computing a generating set of $\mathrm{Aut}(X)$ for a b-bounded hypergraph $X = (V, E)$. Further we show that for any prime $p \leq k$, HGA_k is hard for $\mathrm{Mod}_p \mathrm{L}$ (and hence also hard for $\mathrm{Mod}_j \mathrm{L}$ under logspace conjunctive truth-table reductions, by closure properties of $\mathrm{Mod}_j \mathrm{L}$ classes, where j is the product of all primes $p \leq k$ [3]). Since the orbit size of the hyperedges in the reduced hypergraphs is bounded by p^2, we also get that $\mathrm{GA}_{p^2})$ is $\mathrm{Mod}_p \mathrm{L}$ hard. For prime k, this slightly improves Torán's result that GA_{2k^2} is hard for $\mathrm{Mod}_k \mathrm{L}$.

Theorem 12. *For any prime p, HGA_p is logspace many-one hard for Mod_pL.*

Proofsketch. It is well-known that the evaluation problem $CirVal_p$ for arithmetic circuits with $\oplus_p$-gates is complete for Mod_pL, where $\oplus_p$ denotes addition modulo p (see e.g. [11]). To evaluate a single $\oplus_p$-gate we consider the following hypergraph $H = (V, E)$, where

$$V = \{a_i, b_i, c_i \mid i = 0, \ldots, p - 1\},$$
$$E = \{\{a_i, b_j, c_k\} \mid i \oplus_p j = k\}.$$

The following figure shows H for the case $p = 3$, where we use triangles to depict hyperedges.

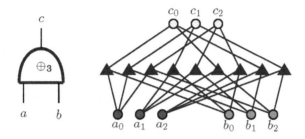

Observe that for every pair of input nodes a_i, b_j, there is exactly one hyperedge e in E which contains a_i and b_j:

$$e = \{a_i, b_j, c_k\}, \text{ where } k = i \oplus_3 j$$

Hence, it follows for any automorphism f that

$$f(a_0) = a_i \wedge f(b_0) = b_j$$
$$\Rightarrow f(\{a_0, b_0, c_0\}) = \{a_i, b_j, f(c_0)\} \in E$$
$$\Rightarrow f(c_0) = c_k, \text{ where } k = i \oplus_3 j.$$

By replacing all $\oplus_p$-gates by such a gadget we can transform C into a hypergraph H which has a nontrivial automorphism g if and only if C evaluates to 1. Note that in order to force g to fix the p input nodes corresponding to a 0 input we can color them differently. Also, in order to force g to move the p input nodes corresponding to a 1 input we additionally insert p parallel feedback edges from the p output nodes to these nodes. Finally, by inserting additional vertices and edges we can force g to cyclically shift the p output nodes.   □

Next we give a polynomial time algorithm for computing a generating set of $Aut(X)$ for a b-bounded hypergraph $X = (V, E)$. More precisely, we give an $n^{O(b)}$ algorithm for the problem, where $n = \|V\|$. We remark that if the hyperedges are all of constant size, i.e. $\|e\| \leq k$ for all $e \in E$ and constant k, then the problem is deterministic logspace reducible to BCGI which is known to be in NC [8]. However, when hyperedges are of unbounded size, it is not clear whether HGI_b is reducible to BCGI. Our polynomial time algorithm for HGI_b applies ideas from a different result of Luks [7]. We recall some definitions.

Definition 13. Let $k \geq 1$. A finite group G is said to be in the class Γ_k if all nonabelian composition factors of G are isomorphic to subgroups of S_k.

Theorem 14. [7] Let $\Delta \subseteq \Omega$ and let $G \leq \mathrm{Sym}(\Omega)$ be given by a generating set S. If $G \in \Gamma_k$ then there is an $n^{O(k)}$ algorithm for computing the set stabilizer subgroup $G_\Delta = \{g \in G \mid g(\Delta) = \Delta\}$.

For a colored hypergraph $X = (V, E, \mathcal{C})$ with $\mathcal{C} = (C', C'')$ let X' denote the hypergraph (C', E') where $E' = \{e \cap C' \mid e \in E\}$. X'' is defined accordingly. We first give an algorithm for the following problem, that we repeatedly use as a subroutine. Suppose $X = (V, E, \mathcal{C})$ is such a hypergraph where the automorphism groups $\mathrm{Aut}(X')$ and $\mathrm{Aut}(X'')$ are in Γ_k. The problem is to compute in polynomial time a generating set of $\mathrm{Aut}(X)$ from generating sets of $\mathrm{Aut}(X')$ and $\mathrm{Aut}(X'')$.

We first consider the embedding map

$$\varphi : \mathrm{Aut}(X') \to \mathrm{Sym}(C') \times \mathrm{Sym}(E') \text{ given by } \varphi(\pi) = (\pi, \tau),$$

where τ is simply defined as the action of π on E'. Similarly, we have the embedding map $\psi : \mathrm{Aut}(X'') \to \mathrm{Sym}(C'') \times \mathrm{Sym}(E'')$. Since φ and ψ are easy to compute, we can easily compute the generating set for $\varphi(\mathrm{Aut}(X'))$ as the image of the given generating set of $\mathrm{Aut}(X')$. Similarly, we can compute a generating set for $\psi(\mathrm{Aut}(X''))$. Thus, we can compute a generating set for the product group $\varphi(\mathrm{Aut}(X')) \times \psi(\mathrm{Aut}(X''))$ as a permutation group acting on $C' \cup E' \cup C'' \cup E''$.

Furthermore, since this permutation group action extends uniquely to $E' \times E''$, we can easily compute a generating set S for the product group $\varphi(\mathrm{Aut}(X')) \times \psi(\mathrm{Aut}(X''))$ as a permutation group acting on $C' \cup E' \cup C'' \cup E'' \cup (E' \times E'')$.

Notice that we can see $\mathrm{Aut}(X)$ as a subgroup of $\varphi(\mathrm{Aut}(X')) \times \psi(\mathrm{Aut}(X''))$. To compute $\mathrm{Aut}(X)$ we construct a bipartite graph Z with $W = E' \cup E''$ as the vertex set and edge set F, where for $e' \in E'$ and $e'' \in E''$ we include (e', e'') in F if and only if $e' \cup e'' \in E$.

Now, in order to invoke Theorem 14, let Ω denote the set

$$\Omega = C' \cup E' \cup C'' \cup E'' \cup (E' \times E'')$$

and let $G \leq \mathrm{Sym}(\Omega)$ be the group $\varphi(\mathrm{Aut}(X')) \times \psi(\mathrm{Aut}(X''))$. Since $\mathrm{Aut}(X')$ and $\mathrm{Aut}(X'')$ are in Γ_k and since Γ_k is closed under homomorphic images and products [7], it follows that $\varphi(\mathrm{Aut}(X')) \times \psi(\mathrm{Aut}(X''))$ is also in Γ_k. Hence, letting $\Delta = F$, Theorem 14 implies that we can compute G_Δ (which is $\mathrm{Aut}(X)$) in time $n^{O(k)}$. Notice that it suffices to retain the action of G_Δ on $V = C' \cup C''$ and we can discard the remaining part.

We now consider the problem of computing $\mathrm{Aut}(X)$ from scratch. For each $s = 1, \ldots, m$ we define hypergraphs $X_{1,s} = (C_{1,s}, E_{1,s})$, where $C_{1,s} = \bigcup_{i=1}^{s} C_i$ and $E_{1,s} = \{e \cap C_{1,s} \mid e \in E\}$. Notice that X itself is $X_{1,m}$. We also define $X_s = (C_s, E_s)$, where $E_s = \{e \cap C_s \mid e \in E\}$. Notice that X_s is a constant-size hypergraph for each s and $\mathrm{Aut}(X_s)$ can be computed easily in polynomial time. Our polynomial time algorithm starts by first computing $\mathrm{Aut}(X_{1,1}) = \mathrm{Aut}(X_1)$.

Then for increasing values of $s = 1, \ldots, m$ it progressively computes the group $\mathrm{Aut}(X_{1,s})$ from the already computed groups $\mathrm{Aut}(X_{1,s-1})$ and $\mathrm{Aut}(X_s)$ by using the algorithm explained above. Notice that the groups $\mathrm{Aut}(X_{1,s-1})$ and $\mathrm{Aut}(X_s)$ are in Γ_k because their orbits are bounded by k. This completes the proof of the following result.

Theorem 15. *Given a hypergraph* $X = (V, E)$ *with color classes of size bounded by* k, *there is an* $n^{O(k)}$ *time algorithm that computes* $\mathrm{Aut}(X)$ *as a generating set in* $\mathrm{Sym}(V)$. *In particular,* HGI *and* HGA *are in* P.

References

1. V. Arvind, P. Kurur, and T. Vijayaraghavan. Bounded color multiplicity graph isomorphism is in the #L hierarchy. In *Proc. 20th Annual IEEE Conference on Computational Complexity*, pages 13–27. IEEE Computer Society Press, 2005.
2. L. Babai and E. Luks. Canonical labeling of graphs. In *Proc. 15th ACM Symposium on Theory of Computing*, 171–183, 1983.
3. G. Buntrock, C. Damm, U. Hertrampf, and C. Meinel. Structure and importance of logspace-MOD classes. *Mathematical Systems Theory*, 25:223–237, 1992.
4. J. Cai, M. Fürer, N. Immerman. An optimal lower bound on the number of variables for graph identifications. *Combinatorica*, 12(4):389-410, 1992.
5. M. Furst, J. Hopcroft, and E. Luks. Polynomial time algorithms for permutation groups. In *Proc. 21st IEEE Symposium on the Foundations of Computer Science*, pages 36–41. IEEE Computer Society Press, 1980.
6. B. Jenner, J. Köbler, P. McKenzie, and J. Torán. Completeness results for graph isomorphism. *Journal of Computer and System Sciences*, 66:549–566, 2003.
7. E. Luks. Isomorphism of bounded valence can be tested in polynomial time. *Journal of Computer and System Sciences*, 25:42–65, 1982.
8. E. Luks. Parallel algorithms for permutation groups and graph isomorphism. In *Proc. 27th IEEE Symposium on the Foundations of Computer Science*, pages 292–302. IEEE Computer Society Press, 1986.
9. E. Luks. Hypergraph isomorphism and structural equivalence of boolean functions. In *Proc. 31st ACM Symposium on Theory of Computing*, 652–658. ACM Press, 1999.
10. O. Reingold. Undirected st-connectivity in log-space. In *Proc. 37th ACM Symposium on Theory of Computing*, pages 376–385. ACM Press, 2005.
11. J. Torán. On the hardness of graph isomorphism. *SIAM Journal on Computing*, 33(5):1093–1108, 2004.

Forbidden Substrings, Kolmogorov Complexity and Almost Periodic Sequences

A.Yu. Rumyantsev and M.A. Ushakov

Logic and algorithms theory division, Mathematics Department,
Moscow State University, Russia

Abstract. Assume that for some $\alpha < 1$ and for all nutural n a set F_n of at most $2^{\alpha n}$ "forbidden" binary strings of length n is fixed. Then there exists an infinite binary sequence ω that does not have (long) forbidden substrings.

We prove this combinatorial statement by translating it into a statement about Kolmogorov complexity and compare this proofl with a combinatorial one based on Laslo Lovasz local lemma.

Then we construct an almost periodic sequence with the same property (thus combines the results from [1] and [2]).

Both the combinatorial proof and Kolmogorov complexity argument can be generalized to the multidimensional case.

1 Forbidden Strings

Fix some positive constant $\alpha < 1$. Assume that for each natural n a set F_n of binary strings of length n is fixed. Assume that F_n consists of at most $2^{\alpha n}$ strings.

We look for an infinite binary sequence ω that does not contain forbidden substrings.

Proposition 1. *There exists an infinite binary sequence ω and a constant N such that for any $n > N$ the sequence ω does not have a substring x of length n that belongs to F_n.*

One may consider strings in F_n as "forbidden" strings of length n; proposition then says that there exists an infinite sequence without (sufficiently long) forbidden substrings.

For example, we can forbid strings having low Kolmogorov complexity. Let F_n be the set of all strings of length n whose complexity is less than αn. Then $\#F_n$ does not exceed $2^{\alpha n}$ (there are at most $2^{\alpha n}$ programs of size less than αn).

Therefore Proposition 1 implies the following statement that was used in [1]:

Proposition 2. *For any $\alpha < 1$ there exists a number N and an infinite binary sequence ω such that any its substring x of length greater than N has high complexity:*

$$\mathrm{K}(x) \geq \alpha|x|.$$

B. Durand and W. Thomas (Eds.): STACS 2006, LNCS 3884, pp. 396–407, 2006.
© Springer-Verlag Berlin Heidelberg 2006

Here $K(x)$ stands for Kolmogorov complexity of x (the length of the shortest program producing x, the definition is given in [3]); it does not matter which version of Kolmogorov compexity (prefix, plain, etc.) we consider since the logarithmic difference between them can be compensated by a small change in α. The notation $|x|$ means the length of string x.

Our observation is that the reverse implication is true, i.e., Proposition 2 implies Proposition 1. It is easy to see if we consider a stronger version of Proposition 2 when K is replaced by a relativized version K_A where A is an arbitrary oracle (an external procedure that can be called). Indeed, consider the set of all forbidden strings as an oracle. Then the relativized complexity of any string in F_n does not exceed $\alpha n + O(1)$ since its ordinal number in the length-sorted list of all forbidden strings is at most $\sum_{k \leq n} 2^{\alpha k} = O(2^{\alpha n})$. The constant $O(1)$ can be absorbed by a small change in α, and we get the statement of Proposition 1.

More interestingly, we can avoid relativization and derive Proposition 1 from (non-relativized) Propostion 2. It can be done as follows.

First note that we may assume (without loss of generality) that α is rational. Assume that for some set F of forbidden strings the statement of Proposition 1 is false. Then for each $c \in \mathbb{N}$ there exists a set F^c with the following properties:

(a) F^c consists of strings of length greater than c;
(b) F^c contains at most $2^{\alpha k}$ strings of length k for any k;
(c) any infinite binary string has at least one substring that belongs to F^c.

(Indeed, let F^c be the set of all forbidden strings that have length more than c.)

The statement (c) can be reformulated as follows: the family of open sets S_x for all $x \in F^c$ covers the set Ω of all binary sequences, where S_x is a set of all sequences that have substring x. The standard compactness argument implies that F^c can be replaced by its finite subset, so we assume without loss of generality that F^c is finite.

The properties (a), (b) and (c) are enumerable (for finite F^c): each S_x is an enumerable union of intervals, so if the sets S_x for $x \in F^c$ cover Ω, this can be discovered at a finite step. (In fact, they are decidable, but this does not matter.) So the first set F_c encountered in the enumeration (for a given c) is a computable function of c.

Now we can construct a decidable set of forbidden strings that does not satisfy the statement of Proposition 1. Indeed, construct a sequence $c_1 < c_2 < c_3 < \ldots$ where c_{i+1} is greater than the length of all strings in F^{c_i} and take the union of all F^{c_i}. We obtain the decidable set $\hat{F}$ such that $\hat{F}$ contains at most $2^{\alpha k}$ strings of length k for any k, and any infinite binary string has (for any i) at least one substring of length greater that c_i that belongs to $\hat{F}$. For this decidable set we need no special oracle, q.e.d.

The proof of Proposition 2 given in [1] uses prefix complexity. See below Section 3 where we prove the stronger version of this Proposition needed for our purposes.

2   Combinatorial Proof

The statement of Proposition 1 has nothing to do with Kolmogorov complexity.
So it would be natural to look for a combinatorial proof.

The simplest idea is to use the random bits as the elements of the sequence.
Then the probability of running into a forbidden string in a given k positions

$$\omega_n \omega_{n+1} \ldots \omega_{n+k-1}$$

is bounded by $2^{-(1-\alpha)k}$, i.e., exponentially decreases when $k \to \infty$. However,
the number of positions where a forbidden string of a given length can appear is
infinite, and the sum of probablities is infinite too. And, indeed, a truly random
sequence contains any string as its substring, so we need to use something else.

Note that two non-overlapping fragments of a random sequence are independent. So the dependence can be localized and we can apply the following
well-known statement:

Proposition 3 (Laslo Lovasz local lemma). *Let G be a graph with vertex
set $V = \{v_1, \ldots, v_n\}$ and edge set E. Let A_i be some event associated with vertex
v_i. Assume that for each i the event A_i is independent with the random variable
"outcomes of all Λ_j such that v_j is not connected to v_i by an edge". Let $p_i \in (0,1)$
be a number associated with A_i in such a way that*

$$\Pr[A_i] \leq p_i \prod_{v_j \sim v_i} (1 - p_j)$$

*where the product is taken over all neighbour vertices v_j (connected to v_i by an
edge). Then*

$$\Pr[\text{neither of } A_i \text{ happens}] \geq \prod_{i=1}^{n} (1 - p_i)$$

and, therefore, this event is non-empty.

The proof of this Lemma could be found, e.g., in [4], p. 115.

To apply this Lemma to our case consider a finite random string of some
fixed length N where all bits are independent and unbiased (both outcomes
have probability $1/2$). Consider a graph whose vertices are intervals of indices
(i.e., places where a substring is located) of length at least L (some constant to
be chosen later). Two intervals are connected by an edge if they are not disjoint
(share some bit). For each interval v consider the event A_v: "substring of the
random string located at v is forbidden". This event is independent with all
events that deal with bits outside v, so the independence condition is fulfilled.

Let $p_v = 2^{-\delta|v|}$ for all v and some δ (to be chosen later). To apply the lemma,
we need to prove that

$$\Pr[A_v] \leq p_v \prod_{\substack{v \text{ and } w \text{ are} \\ \text{not disjoint}}} (1 - p_w).$$

Let $l \geq L$ be the length of the string v and let

$$R = \prod_{\substack{v \text{ and } w \text{ are} \\ \text{not disjoint}}} (1 - p_w).$$

Then

$$R \geq \prod_{k=L}^{N} (1 - 2^{-\delta k})^{l+k}$$

(strings w have length between L and N and there are at most $l + k$ strings of length k that share bits with v), and

$$R \geq \left[\prod_{k \geq L} (1 - 2^{-\delta k}) \right]^{l} \prod_{k \geq L} (1 - 2^{-\delta k})^{k}$$

(we split the product in two parts and replace finite products by infinite ones). The product $\prod(1 - \varepsilon_i)$ converges if and only if the series $\sum \varepsilon_i$ converges. The corresponding series

$$\sum_{k \geq L} 2^{-\delta k} \text{ and } \sum_{k \geq L} k \cdot 2^{-\delta k}$$

do converge. Therefore both products converge and for a large L both products are close to 1:

$$R \geq C_1^l C_2 \geq D^l$$

where C_1, C_2 and D are some constants that could be made close to 1 by choosing a large enough L (not depending on l). Then

$$p_v R \geq 2^{-\delta l} D^l \geq 2^{-\delta l} 2^{-\gamma l} = 2^{-(\delta + \gamma)l},$$

where $\gamma = -\log D$ could be arbitrarily small for some L. We choose δ and L in such a way that $\delta < (1 - \alpha)/2$ and $\gamma < (1 - \alpha)/2$. Then

$$p_v R \geq 2^{-(1-\alpha)l} \geq \Pr[A_v]$$

(forbidden strings form a $2^{-(1-\alpha)l}$-fraction of all strings having length l) and conditions of Lovasz lemma are fulfilled.

So we see that for some large L and for all sufficiently large N there exists a string of length N that does not contain forbidden strings of length L or more. Standard compactness argument shows that there exists an infinite binary string with the same property.

This finishes the combinatorial proof of Proposition 1.

Note that this combinatorial proof hardly can be considered as a mere translation of Kolmogorov complexity argument. Another reason to consider it as a different proof is that it has a straightforward generalization for several dimensions. (The Kolmogorov complexity argument has this too, as we see in Section 5, but requires significant changes.)

A d-dimensional sequence is a function $\omega \colon \mathbb{Z}^d \to \{0,1\}$. Instead of substrings we consider d-dimensional "subcubes" in the sequence, i.e., restrictions of ω to some cube in $\mathbb{Z}^d$. For any n there are 2^{n^d} different cubes with side n. Assume that for every $n > 1$ a set F_n of not more than $2^{\alpha n^d}$ "forbidden cubes" is fixed.

Proposition 4. *There exists a number L and d-dimensional sequence that does not contain forbidden subcube with side greater than L.*

The proof repeats the combinatorial proof of Proposition 1 with the following changes. The bound for R now is

$$R \geq \prod_{k=L}^{N} (1 - 2^{-\delta k^d})^{(l+k)^d},$$

since there are at most $(l + k)^d$ cubes with side k intersecting a given cube with side l. Then we represent $(l + k)^d$ as a sum of $d + 1$ monomials and get a representation of this bound as a product of infinite products, each for one monomial. Every product has the following form (for some i in $0 \ldots d$ and for some c_i that depends on d and i, but not k and l):

$$\prod_{k \geq L} (1 - 2^{-\delta k^d})^{c_i l^i k^j} = \left[\prod_{k \geq L} (1 - 2^{-\delta k^d})^{k^j} \right]^{c_i l^i}.$$

The corresponding series obviously converge (due to the same reasons as before), and again we can make expression $[\ldots]$ as close to 1 as needed by choosing L (and again the choice of L does not depend on l). Then the estimate for R takes the form:

$$R \geq \prod_{i=0}^{d} D_i^{c_i l^i} \geq \prod_{i=1}^{d} D_i^{C l^d} \geq \left[\prod_{i=1}^{d} D_i^{C} \right]^{l^d} \geq D^{l^d},$$

where c_i, D_i, C and D are some constants, and C and D could be made as close to 1 as needed.

Then the proof goes exactly as before.

3   Construction of Almost Periodic Sequences

A sequence is called *almost periodic* if each of its substrings has infinitely many occurences at limited distances, i.e., for any substring x there exists a number k such that any substring y of ω of length k contains x.

The following result is proven in [2] (in the paper almost periodic sequences were called strongly almost periodic sequences):

Proposition 5. *Let $\alpha < 1$ be a constant. There exists an almost periodic sequence ω such that any sufficiently long prefix x of ω has large complexity: $K(x) \geq \alpha |x|$.*

Comparing this statement with Proposition 2, we see that there is an additional requirement for the sequence to be almost periodic; on the other hand high complexity is guaranteed only for prefixes (and not for all substrings).

Now we combine these two results:

Proposition 6. *Let $\alpha < 1$ be a constant. There exists an almost periodic sequence ω such that any sufficiently long substring x of ω has large complexity: $K(x) \geq \alpha|x|$.*

The paper [2] provides a universal construction for almost periodic sequences. Now we suggest another, less general construction that is more suitable for our purposes.

Namely, we define some equivalence relation on the set of indices ($\mathbb{N}$). Then we construct a sequence

$$\omega = \omega_0\omega_1\omega_2\ldots$$

with the following property: $i \equiv j \Rightarrow \omega_i = \omega_j$. In other words, all the places that belong to one equivalence class carry the same bit. This property guarantees that ω is almost periodic if the equivalence relation is chosen in a proper way.

Let $n_0, n_1, n_2, \ldots$ be an increasing sequence of natural numbers such that n_{i+1} is a multiple of n_i for each i. The prefix of length n_0, i.e., the interval $[0, n_0)$, is repeated with period n_1. This means that for any i such that $0 \leq i < n_0$ the numbers

$$i, i + n_1, i + 2n_1, i + 3n_1, \ldots$$

belong to the same equivalence class. In the similar way the interval $[0, n_1)$ is repeated with period n_2: for any i such that $0 \leq i < n_2$ the numbers

$$i, i + n_2, i + 2n_2, i + 3n_2, \ldots$$

are equivalent. (Note that n_2 is a multiple of n_1, therefore the equivalence classes constructed at the first step are not changed.) And so on: for any $i \in [0, n_s)$ and for any k the numbers i and $i + kn_{s+1}$ are equivalent.

The following statement is almost evident:

Proposition 7. *If a sequence ω respects this equivalence relation, i.e., the equivalent positions have equal bits, then the sequence in almost periodic.*

Indeed, in the definition of an almost periodic sequence we may require that each prefix of the sequence has infinitely many occurences at limited distances (since each substring is a part of some prefix). And this is guaranteed: any prefix of length $l < n_s$ appears with period n_{s+1}.

The same construction can be explained in a different way. Consider the positional system where the last digit of integer x is $x \bmod n_0$, the previous digit is $(x \text{ div } n_0) \bmod n_1$ etc. Then all numbers of the form $\ldots 0z$ (for any given $z \in [0, n_0)$) are equivalent; we say that they have rank 1. Then we make (for any y, z such that $y \neq 0$) all numbers of the form $\ldots 0yz$ equivalent and assign rank 2 to them, etc.

Fig. 1. Primary (shaded) and secondary bits in a sequence

If the sequence of periods $n_0 < n_1 < n_2 < \ldots$ is growing fast enough, then the equivalence relation does not restrict significantly the freedom of bit choice: going from left to right, we see that most of the bits are "primary" bits (are leftmost bits in their equivalence class, not copies of previous bits; these copies are called "secondary" bits, see Fig. 1).

Indeed, bits of rank 1 start with n_0 primary bits, these bits are repeated as secondary bits with period n_1, so secondary bits of rank 1 form a n_0/n_1-fraction of all bits in the sequence; secondary bits of rank 2 form a n_1/n_2-fraction etc. So the sum $\sum_i \frac{n_i}{n_{i+1}}$ is the upper bound of the density of "non-fresh" bits. More precise estimate: prefix of any length N has at least DN fresh bits where

$$D = \prod_i (1 - n_i/n_{i+1}).$$

This gives a simple proof of Proposition 5. For a given α choose a computable sequence $n_0 < n_1 < n_2 < \ldots$ that grows fast enough and has $D > \alpha$. Then take a Martin-Löf random sequence ξ and place its bits (from left to right) at all free positions (duplicating bits as required by the equivalence relation). We get an almost periodic sequence ω; at least DN bits of ξ can be algorithmically reconstructed from ω's prefix of length N. It remains to note that algorithmic transformation cannot increase complexity and that complexity of m-bit prefix of a random sequence is at least $m - o(m)$ (it would be at least m for monotone or prefix complexity, but could be $O(\log m)$ smaller for plain complexity).

4   Proof of the Main Result

Could we apply the same argument (with sequence ω from Proposition 2 instead of a random sequence) to prove Proposition 6? Not directly. To explain the difficulty and the way to overcome it, consider the simplified picture where only the equivalence of rank 1 is used. Then the sequence constructed has the form

$$\omega = A\,B_0\,A\,B_1\,A\,B_2\,A\,B_3\,A\ldots$$

where A is the group of primary bits of rank 1 (repeated with period n_1); A and B_i are taken from a sequence

$$\xi = A\,B_0\,B_1\,B_2\,B_3\ldots$$

(provided by Proposition 2). If some substring x of ω is located entirely in A or some B_i, its high complexity is guaranteed by Proposition 2. However, if x

appears on the boundary between A and B_i for some $i > 0$, then x is composed from two substrings of ξ and its complexity is not guaranteed to be high.

To overcome this difficulty, we need the following stronger version of Proposition 2.

Proposition 8. *For any $\alpha < 1$ there exists a number N and an infinite binary sequence ω such that any its substring*

$$x = \omega_n \omega_{n+1} \omega_{n+2} \ldots \omega_{n+k-1}$$

of length $k > N$ has high conditional complexity with respect to previous bits:

$$K(\omega_n \omega_{n+1} \omega_{n+2} \ldots \omega_{n+k-1} \mid \omega_0 \omega_1 \omega_2 \ldots \omega_{n-1}) \geq \alpha k.$$

The proof follows the scheme from [1]. Let $\beta < 1$ be greater than α. Let m be some integer number (we will fix it later). Let the first m bits of ω be the sequence x of length m with maximal prefix complexity (denoted by KP). Then add the next m bits to get the maximal prefix complexity of the entire sequence. This increase would be at least $m - O(\log m)$.

[Indeed, for any strings x and y we have

$$KP(x, y) = KP(x) + KP(y \mid x, KP(x)) + O(1);$$

(Kolmogorov – Levin theorem); if y has been chosen to maximize the second term in the sum, then $KP(y \mid \ldots) \geq |y|$ and $KP(x, y) \geq KP(x) + |y| - O(1)$. Therefore, for this y

$$KP(xy) \geq KP(x, y) - KP(|y|) - O(1) \geq KP(x) + |y| - O(\log |y|),$$

since (x, y) can be reconstructed from xy and $|y|$ and $KP(|y|) = O(\log |y|)$. See [1] for details.]

Then we add string z of length m that maximizes $KP(xyz)$ and so on.

In this way we construct a sequence $\omega = xyz \ldots$ such that the prefix complexity of its initial segments increases by $m - c \log m$ for every added block of m bits. We can choose m such that $m - c \log m - O(1) > \beta m$.

Then the statement of the Proposition follows from Kolmogorov – Levin theorem if the substring is "aligned" (starts and ends on the boundaries of length m blocks). Since m is fixed, the statement is true for non-aligned blocks of large enough length (boundary effects are compensated by the difference between α and β).

Proposition 8 is proven.

Let us explain why this modification helps in the model situation considered above. If a substring x of the sequence $AB_0 AB_1 AB_2 \ldots$ is on the boundary between A and some B_i, then it can be split into two parts x_A and x_B. The string x_A is a substring of A and therefore has high complexity. The string x_B is a substring of some B_i and therefore also has high complexity and even high conditional complexity with respect to some prefix containing A. If we prove that x_A is simple relatively this prifix we can use Kolmogorov – Levin theorem to prove that x has high complexity.

Similar arguments work in general case when we have to consider bits of all ranks. To finish the proof we need the following Lemma:

Lemma. *Let ω be the sequence satisfying the statement of Proposition 8. Then*

$$K(V(a_0, b_0), V(a_1, b_1), \ldots, V(a_{s-1}, b_{s-1})) \geq$$
$$\alpha L - O(s \log L) - K(a_0 \mid a_1) - K(a_1 \mid a_2) - \ldots - K(a_{s-2} \mid a_{s-1})$$

for any $a_0 < b_0 \leq a_1 < b_1 \leq \ldots \leq a_{s-1} < b_{s-1}$, where $V(a, b)$ stands for $\omega_a \omega_{a+1} \ldots \omega_{b-1}$ and $L = (b_0 - a_0) + (b_1 - a_1) + \ldots + (b_{s-1} - a_{s-1})$.

In fact, for Proposition 6 we need only the case $s = 3$ of this Lemma.

The proof of Lemma is based on Kolmogorov – Levin theorem about complexity of pairs. The statement of Proposition 8 guarantees the following inequality:

$$K(V(a_{s-1}, b_{s-1}) \mid V(0, a_{s-1})) \geq \alpha(b_{s-1} - a_{s-1}) - O(\log L). \qquad (*)$$

We will prove the following inequality of any $i = 0, 1, \ldots, s - 2$:

$$K(V(a_i, b_i), V(a_{i+1}, b_{i+1}), \ldots, V(a_{s-1}, b_{s-1}) \mid V(0, a_i)) -$$
$$K(V(a_{i+1}, b_{i+1}), \ldots, V(a_{s-1}, b_{s-1}) \mid V(0, a_{i+1})) \geq \qquad (**)$$
$$\alpha(b_i - a_i) - O(\log L) - K(a_i \mid a_{i+1}).$$

If we add up $(**)$ for all $i = 0, 1, \ldots, s - 2$ with $(*)$ we obtain the required inequality (and even stronger one with relative complexity in the left-hand side). Let us prove the inequality $(**)$ now. By W we denote the sequence $(V(a_{i+1}, b_{i+1}), \ldots, V(a_{s-1}, b_{s-1}))$. The following inequality follows from the Kolmogorov – Levin theorem and the statement of Proposition 8:

$$K(V(a_i, b_i), W \mid V(0, a_i)) - K(W \mid V(0, a_i), V(a_i, b_i)) =$$
$$K(V(a_i, b_i) \mid V(0, a_i)) - O(\log L) \geq \alpha(b_i - a_i) - O(\log L).$$

To finish the proof of Lemma, let us prove the inequality

$$K(W \mid V(0, a_{i+1})) \leq K(W \mid V(0, a_i), V(a_i, b_i)) + K(a_i \mid a_{i+1}) + O(\log L).$$

One can obtain W from $V(0, a_{i+1})$ in the following way: find a_{i+1} using the length of the string $V(0, a_{i+1})$, convert a_{i+1} into a_i by the shortest program, compute b_i by adding difference $b_i - a_i$ to a_i, cut intervals $[0, a_i)$ and $[a_i, b_i)$ from string $V(0, a_{i+1})$ and execute the shortest program that converts $(V(0, a_i), V(a_i, b_i))$ into W. This needs $K(W \mid V(0, a_i), V(a_i, b_i)) + K(a_i \mid a_{i+1}) + O(\log L)$ bits to obtain W from $V(0, a_{i+1})$. The inequality is proven, q.e.d.

The proof of Proposition 6 uses the same construction as proof of Proposition 5 but it takes a sequence satisfying the statement of Proposition 8 instead of a random sequence.

Let v be a sequence satisfying the statement of Proposition 8 with some $\alpha' > \alpha$ and ω be the resulting sequence (if we apply the construction of an almost periodic sequence to the sequence v). It has been proved before that ω is an almost periodic sequence. We need only to prove the following estimate of a complexity of any substring of ω:

$$K(\omega_m \omega_{m+1} \omega_{m+2} \ldots \omega_{m+k-1}) \geq \alpha k.$$

for any sufficiently long k and for any m.

Suppose that sequence $\{n_i\}$ grows fast enough, i.e. $n_{i+1} > \frac{4n_i}{\alpha'-\alpha}$. Suppose i is the smallest index such that $n_i \geq k$. Due to our construction of sequence ω any element of ω corresponds to some element of v. Different elements of $\omega_m \omega_{m+1} \ldots \omega_{m+k-1}$ of rank not less than i (i.e. elements repeated with period n_i or greater by our construction) correspond to different elements of v because the distance between elements of the given substring of ω is less than n_i (and less than the period). It is easy to prove that in this substring the density of elements of small rank (less than i) is not greater than $\alpha' - \alpha$.

So the substring $\omega_m \ldots \omega_{m+k-1}$ corresponds to some intervals in v. Throw away all elements of small ranks from these intervals of v and denote the remaining intervals by $[a_0, b_0), \ldots, [a_{s-1}, b_{s-1})$, where $a_0 < b_0 \leq \ldots \leq a_{s-1} < b_{s-1}$. The number of these intervals is at most 3. Indeed, we can enumerate all elements of $\omega_m \ldots \omega_{m+k-1}$ from left to right, not counting elements of small ranks, and for each element find the corresponding element of v. The index of corresponding element will increase by 1 every time except when we cross a point of type $n_i j$ or $n_i j + n_{i-1}$ (where j is integer). But there are at most 2 points of this type in the interval of length k so there are at most 3 corresponding intervals.

Substrings $V(a_0, b_0), \ldots, V(a_{s-1}, b_{s-1})$ (defined as in Lemma) can be computed by an algorithm using the given substring of ω. The algorithm needs only to know the value of $m \bmod n_{i-1}$ for finding elements with small rank (less than i) and the relative positions of elements of $\omega_m \ldots \omega_{m+k-1}$ corresponding to v_{a_j} and v_{b_j-1} where $j = 0, 1, \ldots, s - 1$. Because $s \leq 3$ only a logarithmical amount of additional bits is needed. So we can prove the following inequality to finish the proof of Proposition 6:

$$K(V(a_0, b_0), \ldots, V(a_{s-1}, b_{s-1})) \geq \alpha k - O(\log k).$$

We can use Lemma for this because $\alpha' L > \alpha k$, where $L = (b_0 - a_0) + (b_1 - a_1) + \ldots + (b_{s-1} - a_{s-1})$ (we have already proved that in this substring the density of elements of small rank is not greater than $\alpha' - \alpha$, hence $k - L \leq (\alpha' - \alpha)k$).

If we prove that $K(a_j \mid a_{j+1}) = O(\log k)$ we will finish the proof of the proposition. Suppose we know a_{j+1}. We can find a_j in the following way. Find the element of the given substring of ω corresponding to $v_{a_{j+1}}$. Add to the index of the found element the difference between the indexes of the elements of the given substring corresponding to v_{a_j} and $v_{a_{j+1}}$ (this difference is not greater than the lenght of the given substring, i.e., we use only a logarithmical amount of memory). We get an element of ω corresponding to v_{a_j}. It can be used to calculate a_j. But the first step of this algorithm uses knowing the position of the given substring which needs an unlimited amount of memory. We can avoid using this position if we notice that the rank i of elements of ω corresponding to v_{a_j} is not greater than the rank I of elements of ω corresponding to $v_{a_{j+1}}$ (because $a_j < a_{j+1}$). So n_I is a multiple of n_i. Hence at the first step we can take any element of ω corrensponding to $v_{a_{j+1}}$ (for example, the first one). We get the same result since the elements corresponding to v_{a_j} repeat with period n_i and the elements corresponding to $v_{a_{j+1}}$ repeat with period n_I.

Therefore we construct the algorithm proving that $K(a_j \mid a_{j+1}) = O(\log k)$, and so the proof of Proposition 6 is complete.

Remarks.

1. Proposition 6 implies the existence of a bi-infinite almost periodic sequence with complex substrings (using the standard compactness argument; this argument can be even simplified for the special case of almost periodic sequences).
2. The proof of Proposition 6 works for relativized version of complexity. Therefore we get (as explained above) the following (pure combinatorial) strong version of Proposition 1:

Corollary. *Assume that for each n a set F_n of forbidden substrings of length n is fixed, and the size of F_n is at most $2^{\alpha n}$. Then there exists an infinite almost periodic binary sequence ω and a constant N such that for any $n > N$ the sequence ω does not have a substring x that belongs to F_n.*

5   Multidimensional Case

Similar but more delicate arguments could be applied to multidimensional case too.

A d-dimensional sequence $\omega : \mathbb{Z}^d \to \{0,1\}$ is *almost periodic* if for any cube x that appears in ω there exists a number k such that any subcube with side k contains x inside.

Proposition 9. *Fix an integer $d \geq 1$. Let α be a positive number less than 1. There exists an almost periodic d-dimensional sequence ω such that any sufficiently large subcube x of ω has large complexity:*

$$K(x) \geq \alpha \cdot volume(x)$$

Here volume is the number of points, i.e., $side^d$.

In the multidimensional case the complexity argument needs Proposition 8 even if we do not insist that ω is almost periodic.

Informally, the idea of the proof can be explained as follows. Consider, for example, the case $d = 2$. Take a sequence ξ from Proposition 8 and write down its terms along a spiral.

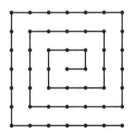

Then we need to bound the complexity of a cube (i.e., square). This square contains several substrings of the sequence ξ. (Unlike the previous case where only 3 substrings were needed, now the number of substrings is proportional to the side of the square.) Then we apply the Lemma to these substrings to get the bound for the complexity of the entire square.

This works if we do not require ω to be almost periodic (so the argument above could replace the combinatorial proof using Lovasz lemma). It needs additional modifications to get the almost periodic sequence. Similar to one-dimensional construction, the cube of size n_0 is duplicated periodically in all directions with shifts being multiples of n_1; the cube of size n_1 is duplicated with shifts being multiples of n_2, etc.

As in one-dimensional case, it is easy to see that this construction guarantees that ω is almost periodic. Then we fill the positions in a spiral order respecting the duplication structure described.

6   Remarks

Kolmogorov complexity is often used in combinatorial constructions as the replacement of counting arguments. (Instead of proving that the total number of objects is larger that the number of "bad" objects we prove that an object of maximal complexity is "good".) Sometimes people even say that the use of Kolmogorov complexity is just a simple reformulation that often hides the combinatorial essence of the argument.

In our opinion this is not always true. Even without the almost periodicity requirement the two natural proofs of Proposition 1 (using complexity argument and Lovasz lemma) are quite different. The proof of Proposition 2 uses prefix complexity and cannot be directly translated into a counting argument. On the other hand, the use of Lovasz lemma in a combinatorial proof cannot be easily reformulated in terms of Kolmogorov complexity. (Moreover, for almost periodic case we don't know how to apply Lovasz lemma argument and complexity proof remains the only one known to us.)

Acknowledgements

The authors would like to thank Alexander Shen and Yury Pritykin for help.

References

1. Bruno Durand, Leonid Levin, Alexander Shen, *Complex tilings*, STOC Proceedings, 2001, p. 732–739; enhanced version: http://arXiv.org/abs/cs.CC/0107008
2. Andrei Muchnik, Alexei Semenov and Maxim Ushakov, Almost periodic sequences, *Theoretical Computer Science*, 304 (issue 1-3, July 2003), p. 1–33.
3. Li M., Vitanyi P, *An Introduction to Kolmogorov Complexity and Its*, Applications, 2nd ed. N.Y.: Springer, 1997.
4. Rajeev Motwani, Prabhakar Raghavan, *Randomized algorithms*, Cambridge University Press, New York, NY, 1995.

Online Learning and Resource-Bounded Dimension: Winnow Yields New Lower Bounds for Hard Sets[*]

John M. Hitchcock

Department of Computer Science, University of Wyoming

Abstract. We establish a relationship between the online mistake-bound model of learning and resource-bounded dimension. This connection is combined with the Winnow algorithm to obtain new results about the density of hard sets under adaptive reductions. This improves previous work of Fu (1995) and Lutz and Zhao (2000), and solves one of Lutz and Mayordomo's "Twelve Problems in Resource-Bounded Measure" (1999).

1 Introduction

This paper has two main contributions: (i) establishing a close relationship between resource-bounded dimension and Littlestone's online mistake-bound model of learning, and (ii) using this relationship along with the Winnow algorithm to resolve an open problem in computational complexity. In this introduction we briefly describe these contributions.

1.1 Online Learning and Dimension

Lindner, Schuler, and Watanabe [14] studied connections between computational learning theory and resource-bounded measure, primarily working with the probably approximately correct (PAC) model. They also included the observation that any "admissible" subclass of P/poly that is polynomial-time learnable in Angluin's exact learning model [2] must have p-measure 0. The proof of this made use of the essential equivalence between Angluin's model and Littlestone's online mistake-bound model [15].

In the online mistake-bound model, a learner is presented a sequence of examples, and is asked to predict whether or not they belong to some unknown target concept. The concept is drawn from some concept class, which is known to the learner, and the examples may be chosen by an adversary. After making its prediction about each example, the learner is told the correct classification for the example, and learner may use this knowledge in making future predictions. The mistake bound of the learner is the maximum number of incorrect predictions the learner will make, over any choice of target concept and sequence of examples.

---

[*] This research was supported in part by National Science Foundation grant 0515313.

B. Durand and W. Thomas (Eds.): STACS 2006, LNCS 3884, pp. 408–419, 2006.

We push the observation of [14] much further, developing a powerful, general framework for showing that classes have resource-bounded *dimension* 0. Resource-bounded measure and dimension involve betting on the membership of strings in an unknown set. To prove that a class has dimension 0, we show that it suffices to give a reduction to a family of concept classes that has a good mistake-bound learning algorithm. It is possible that the reduction can take exponential-time and that the learning algorithm can also take exponential-time, as long as the mistake bound of the algorithm is subexponential. If we have a reduction from the unknown set to a concept in learnable concept class, we can view the reduction as generating a sequence of examples, apply the learning algorithm to these examples, and use the learning algorithm's predictions to design a good betting strategy. Formal details of this framework are given in Section 3.

1.2   Density of Hard Sets

The two most common notions of polynomial-time reductions are many-one ($\leq_m^P$) and Turing ($\leq_T^P$). A many-one reduction from A to B maps instances of A to instance of B, preserving membership. A Turing reduction from A to B makes many, possibly adaptive, queries to B in order to solve A. Many-one reductions are a special case of Turing reductions. In between $\leq_m^P$ and $\leq_T^P$ is a wide variety of polynomial-time reductions of different strengths.

A common use of reductions is to demonstrate hardness for a complexity class. Let $\leq_r^P$ be a polynomial-time reducibility. For any set B, let $P_r(B) = \{A \mid A \leq_r^P B\}$ be the class of all problems that $\leq_r^P$-reduce to B. We say that B is $\leq_r^P$-*hard* for a complexity class $\mathcal{C}$ if $\mathcal{C} \subseteq P_r(B)$, that is, every problem in $\mathcal{C}$ $\leq_r^P$-reduces to B. For a class $\mathcal{D}$ of sets, a useful notation is $P_r(\mathcal{D}) = \bigcup_{B \in \mathcal{D}} P_r(B)$.

A problem B is *dense* if there exists $\epsilon > 0$ such that $|B_{\leq n}| > 2^{n^\epsilon}$ for all but finitely many n. All known hard sets for the exponential-time complexity classes $E = \text{DTIME}(2^{O(n)})$ or $\text{EXP} = \text{DTIME}(2^{n^{O(1)}})$ are dense. Whether every hard set must be dense has been often studied. First, Meyer [24] showed that every $\leq_m^P$-hard set for E must be dense, and he observed that proving the same for $\leq_T^P$-reductions would imply that E has exponential circuit-size complexity. Since then, a line of research has obtained results for a variety of reductions between $\leq_m^P$ and $\leq_T^P$, specifically the conjunctive ($\leq_c^P$) and disjunctive ($\leq_d^P$) reductions, and for various functions $f(n)$, the bounded query $\leq_{f(n)-tt}^P$ and $\leq_{f(n)-T}^P$ reductions:

1. Watanabe [26,9] showed that every hard set for E under the $\leq_c^P$, $\leq_d^P$, or $\leq_{O(\log n)-tt}^P$ reductions is dense.
2. Lutz and Mayordomo [19] showed that for all $\alpha < 1$, the class $P_{n^\alpha - tt}(\text{DENSE}^c)$ has p-measure 0, where DENSE is the class of all dense sets. Since E does not have p-measure 0, their result implies that every $\leq_{n^\alpha - tt}^P$-hard set for E is dense.
3. Fu [8] showed that for all $\alpha < 1/2$, every $\leq_{n^\alpha - T}^P$-hard set for E is dense, and that for all $\alpha < 1$, every $\leq_{n^\alpha - T}^P$-hard set for EXP is dense.

4. Lutz and Zhao [21] gave a measure-theoretic strengthening of Fu's results, showing that for all $\alpha < 1/2$, $P_{n^\alpha - T}(\text{DENSE}^c)$ has p-measure 0, and that for all $\alpha < 1$, $P_{n^\alpha - T}(\text{DENSE}^c)$ has p_2-measure 0.

This contrast between E and EXP in the last two references was left as a curious open problem, and exposited by Lutz and Mayordomo [20] as one of their "Twelve Problems in Resource-Bounded Measure":

Problem 6. For $\alpha \leq \frac{1}{2} < 1$, is it the case that $P_{n^\alpha - T}(\text{DENSE}^c)$ has p-measure 0 (or at least, that $E \not\subseteq P_{n^\alpha - T}(\text{SPARSE}))$?

We resolve this problem, showing the much stronger conclusion that the classes in question have p-dimension 0. But first, in Section 4, we prove a theorem about disjunctive reductions that illustrates the basic idea of our technique. We show that the class $P_d(\text{DENSE}^c)$ has p-dimension 0. The proof uses the learning framework of Section 3 and Littlestone's Winnow algorithm [15]. Suppose that $A \leq^P_d S$, where S is a nondense set. Then there is a reduction g mapping strings to sets of strings such that $x \in A$ if and only if at least one string in $g(x)$ belongs to S. We view the reduction g as generating examples that we can use to learn a disjunction based on S. Because S is subexponentially dense, the target disjunction involves a subexponential number of variables out of exponentially many variables. This is truly a case "when irrelevant attributes abound" [15] and the Winnow algorithm perfoms exceedingly well to establish our dimension result. In the same section we also use the learning framework to show that $P_c(\text{DENSE}^c)$ has p-dimension 0. These results give new proofs of Watanabe's aforementioned theorems about $\leq^P_d$-hard and $\leq^P_c$-hard sets for E.

Our main theorem, presented in Section 5, is that for all $\alpha < 1$, $P_{n^\alpha - T}(\text{DENSE}^c)$ has p-dimension 0. This substantially improves the results of [19,8,21]. The resource-bounded measure proofs in [19,21] use the concept of *weak stochasticity*. As observed by Mayordomo [23], this stochasticity approach can be extended to show a -1st-*order scaled dimension* [11] result, but it seems a different technique is needed for an (unscaled) dimension result. Our learning framework turns out to be just what is needed. We reduce the class $P_{n^\alpha - T}(\text{DENSE}^c)$ to a family of learnable disjunctions. For this, we make use of a technique that Allender, Hemaspaandra, Ogiwara, and Watanabe [1] used to prove a surprising result converting bounded-query reductions to sparse sets into disjunctive reductions to sparse sets: $P_{\text{btt}}(\text{SPARSE}) \subseteq P_d(\text{SPARSE})$. Carefully applying the same technique on a sublinear-query Turing-reduction to a nondense set results in a disjunction with a nearly exponential blowup, but it can still be learned by Winnow in our dimension setting.

The density of complete and hard sets for NP has also been studied often, with motivation coming originally from the Berman-Hartmanis isomorphism conjecture [5]: all many-one complete sets are dense if the isomorphism conjecture holds. Since no absolute results about the density of NP-complete or NP-hard sets can be proved without separating P from NP, the approach has been to prove conditional results under a hypothesis on NP. Mahaney [22] showed that if $P \neq NP$, then no sparse set is $\leq^P_m$-hard for NP. Ogiwara and Watanabe

[25] extended Mahaney's theorem to the $\leq_{btt}^P$-hard sets. Deriving a result from $P \neq NP$ about NP-hard sets under unbounded truth-table reductions is still an open problem, but a measure-theoretic assumption yields very strong consequences. Lutz and Zhao [21] showed that under the hypothesis "NP does not have p-measure 0," every $\leq_{n^\alpha - T}^P$-hard set for NP must be dense, for all $\alpha < 1$. In Section 6 we present the same conclusion under the weaker hypothesis "NP has positive p-dimension," and some additional consequences.

2   Preliminaries

The set of all binary strings is $\{0,1\}^*$. The length of a string $x \in \{0,1\}^*$ is $|x|$. We write λ for the empty string. For $n \in \mathbb{N}$, $\{0,1\}^n$ is the set of strings of length n and $\{0,1\}^{\leq n}$ is the set of strings of length at most n.

A *language* is a subset $L \subseteq \{0,1\}^*$. We write $L_{\leq n} = L \cap \{0,1\}^{\leq n}$ and $L_{=n} = L \cap \{0,1\}^n$. We say that L is *sparse* if there is a polynomial $p(n)$ such that for all n, $|L_{=n}| \leq p(n)$. We say that L is *(exponentially) dense* if there is a constant $\epsilon > 0$ such that $|L_{\leq n}| > 2^{n^\epsilon}$ for all sufficiently large n. We write SPARSE and DENSE for the classes of all sparse languages and all dense languages. The complement $DENSE^c$ of DENSE is the class of all *nondense* languages.

2.1   Resource-Bounded Measure and Dimension

Resource-bounded measure and dimension were introduced in [16,18,4]. Here we briefly review the definitions and basic properties. We refer to the original sources and also the surveys [17,20,12] for more information.

The *Cantor space* is $C = \{0,1\}^\infty$. Each language $A \subseteq \{0,1\}^*$ is identified with its characteristic sequence $\chi_A \in C$ according to the standard (lexicographic) enumeration of $\{0,1\}^*$. We typically write A in place of χ_A. In this way a complexity class $\mathcal{C} \subseteq \mathcal{P}(\{0,1\}^*)$ is viewed as a subset $\mathcal{C} \subseteq C$. We use the notation $S \upharpoonright n$ to denote the first n bits of a sequence $S \in C$.

Let $s > 0$ be a real number. An *s-gale* is a function $d : \{0,1\}^* \to [0,\infty)$ such that for all $w \in \{0,1\}^*$, $d(w) = \frac{d(w0)+d(w1)}{2^s}$. A *martingale* is a 1-gale.

The goal of an s-gale is to obtain large values on sequences:

Definition 2.1. *Let d be an s-gale and $S \in C$.*

1. *d succeeds on S if $\limsup_{n\to\infty} d(S \upharpoonright n) = \infty$.*
2. *d succeeds strongly on S if $\liminf_{n\to\infty} d(S \upharpoonright n) = \infty$.*
3. *The success set of d is $S^\infty[d] = \{S \in C \mid d \text{ succeeds on } S\}$.*
4. *The strong success set of d is $S_{str}^\infty[d] = \{S \in C \mid d \text{ succeeds strongly on } S\}$.*

Notice that the smaller s is, the more difficult it is for an s-gale to obtain large values. Succeeding martingales ($s = 1$) imply measure 0, and the infimum s for which an s-gale can succeed (or strongly succeed) gives the dimension (or strong dimension):

Definition 2.2. *Let $X \subseteq C$.*

1. *X has p-measure 0, written $\mu_p(X) = 0$, if there is a polynomial-time computable martingale d such that $X \subseteq S^\infty[d]$.*
2. *The p-dimension of X, written $\dim_p(X)$, is the infimum of all s such that there exists a polynomial-time computable s-gale d with $X \subseteq S^\infty[d]$.*
3. *The strong p-dimension of X, written $\mathrm{Dim}_p(X)$, is the infimum of all s such that there exists a polynomial-time computable s-gale d with $X \subseteq S_{\mathrm{str}}^\infty[d]$.*

We now summarize some of the basic properties of the p-dimensions and p-measure.

Proposition 2.3. *([18,4]) Let $X, Y \subseteq C$.*

1. *$0 \leq \dim_p(X) \leq \mathrm{Dim}_p(X) \leq 1$.*
2. *If $\dim_p(X) < 1$, then $\mu_p(X) = 0$.*
3. *If $X \subseteq Y$, then $\dim_p(X) \leq \dim_p(Y)$ and $\mathrm{Dim}_p(X) \leq \mathrm{Dim}_p(Y)$.*

The following theorem indicates that the p-dimensions are useful for studies within the complexity class E.

Theorem 2.4. *([16,18,4])*

1. *$\mu_p(E) \neq 0$. In particular, $\dim_p(E) = \mathrm{Dim}_p(E) = 1$.*
2. *For all $c \in \mathbb{N}$, $\mathrm{Dim}_p(\mathrm{DTIME}(2^{cn})) = 0$.*

2.2   Online Mistake-Bound Model of Learning

A *concept* is a set $C \subseteq U$, where U is some *universe*. A concept C is often identified with its characteristic function $f_C : U \to \{0, 1\}$. A *concept class* is a set of concepts $\mathcal{C} \subseteq \mathcal{P}(U)$.

Given a concept class $\mathcal{C}$ and a universe U, a learning algorithm tries to learn an unknown target concept $C \in \mathcal{C}$. The algorithm is given a sequence of examples $x_1, x_2, \ldots$ in U. When given each example x_i, the algorithm must predict if $x_i \in C$ or $x_i \notin C$. The algorithm is then told the correct answer and given the next example. The algorithm makes a *mistake* if its prediction for membership of x_i in C is wrong. This proceeds until every member of U is given as an example.

The goal is to minimize the number of mistakes. The *mistake bound* of a learning algorithm A for a concept class $\mathcal{C}$ is the maximum over all $C \in \mathcal{C}$ of the number of mistakes A makes when learning C, over all possible sequences of examples. The *running time* of A on $\mathcal{C}$ is the maximum time A takes to make a prediction.

2.3   Disjunctions and Winnow

An interesting concept class is the class of monotone disjunctions, which can be efficiently learned by Littlestone's Winnow algorithm [15]. A monotone disjunction on $\{0, 1\}^n$ is a formula of the form $\phi_V = \bigvee_{i \in V} x_i$, where $V \subseteq \{1, \ldots, n\}$ and we write a string $x \in \{0, 1\}^n$ as $x = x_1 \cdots x_n$. The concept ϕ_V can also be viewed as the set $\{x \in \{0, 1\}^n \mid \phi_V(x) = 1\}$ or equivalently as $\{A \subseteq \{1, \ldots, n\} \mid A \cap V \neq \emptyset\}$.

The Winnow algorithm has two parameters α (a weight update multiplier) and θ (a threshold value). Initially, each variable x_i has a weight $w_i = 1$. To classify a string x, the algorithm predicts that x is in the concept if $\sum_i w_i x_i > \theta$, and not in the concept otherwise. The weights are updated as follows whenever a mistake is made.

- If a negative example x is incorrectly classified, then set $w_i := 0$ for all i such that $x_i = 1$. (Certainly these x_i's are not in the disjunction.)
- If a positive example x is incorrectly classified, then set $w_i := \alpha \cdot w_i$ for all i such that $x_i = 1$. (It is considered more likely that these x_i's are in the disjunction.)

A useful setting of the parameters is $\alpha = 2$ and $\theta = n/2$. With these parameters, Littlestone proved that Winnow will make at most $2k \log n + 2$ mistakes when the target disjunction has at most k literals. Also, the algorithm uses $O(n)$ time to classify each example and update the weights.

3   Learning and Dimension

In this section we present a framework relating online learning to resource-bounded dimension. This framework is based on reducibility to learnable concept class families.

Definition 3.1. *A sequence* $\mathcal{C} = (\mathcal{C}_n \mid n \in \mathbb{N})$ *of concept classes is called a concept class family.*

We consider two types of reductions:

Definition 3.2. *Let* $L \subseteq \{0,1\}^*$, $\mathcal{C} = (\mathcal{C}_n \mid n \in \mathbb{N})$ *be a concept class family, and* $r(n)$ *be a time bound.*

1. *We say* L *strongly reduces to* $\mathcal{C}$ *in* $r(n)$ *time, and we write* $L \leq_{str}^r \mathcal{C}$, *if there exists a sequence of target concepts* $(c_n \in \mathcal{C}_n \mid n \in \mathbb{N})$ *and a reduction* f *computable in* $O(r(n))$ *time such that for all but finitely many* n, *for all* $x \in \{0,1\}^n$, $x \in L$ *if and only if* $f(x) \in c_n$.
2. *We say* L *weakly reduces to* $\mathcal{C}$ *in* $r(n)$ *time, we write* $L \leq_{wk}^r \mathcal{C}$ *if there a reduction* f *computable in* $O(r(n))$ *time such that for infinitely many* n, *there is a concept* $c_n \in \mathcal{C}_n$ *such that for all* $x \in \{0,1\}^{\leq n}$, $x \in L$ *if and only if* $f(0^n, x) \in c_n$.

It is necessary to quantify both the time complexity and mistake bound for learning a concept class family:

Definition 3.3. *Let* $t, m : \mathbb{N} \to \mathbb{N}$. *We say that* $\mathcal{C} \in \mathcal{L}(t, m)$ *if there is an algorithm that learns* $\mathcal{C}_n$ *in* $O(t(n))$ *time with mistake bound* $m(n)$.

Combining the two previous definitions we arrive at our central technical concept:

Definition 3.4. *Let $r, t, m : \mathbb{N} \to \mathbb{N}$.*

1. *$\mathcal{RL}_{\text{str}}(r, t, m)$ is the class of all languages that $\leq^r_{str}$-reduce to some concept class family in $\mathcal{L}(t, m)$.*
2. *$\mathcal{RL}_{\text{wk}}(r, t, m)$ is the class of all languages that $\leq^r_{wk}$-reduce to some concept class family in $\mathcal{L}(t, m)$.*

A remark about the parameters in this definition is in order. If $A \in \mathcal{RL}_{\text{str}}(r, t, m)$, then $A \leq^r_{str} \mathcal{C}$ for some concept class family $\mathcal{C} = (\mathcal{C}_n \mid n \in \mathbb{N})$. Then $x \in A_{=n}$ if and only if $f(x) \in c_n$, where $c_n \in \mathcal{C}_n$ is the target concept and f is the reduction. We emphasize that the complexity of learning $\mathcal{C}_n$ is measured in terms of $n = |x|$, and not the size of c_n or $f(x)$. Instead $\mathcal{C}_n$ is learnable in time $O(t(n))$ with mistake bound $m(n)$.

The following theorem is the main technical tool in this paper. Here we consider exponential-time reductions to concept classes that can be learned in exponetial-time, but with subexponentially-many mistakes.

Theorem 3.5. *Let $c \in \mathbb{N}$.*

1. *$\mathcal{RL}_{\text{str}}(2^{cn}, 2^{cn}, o(2^n))$ has strong p-dimension 0.*
2. *$\mathcal{RL}_{\text{wk}}(2^{cn}, 2^{cn}, o(2^n))$ has p-dimension 0.*

4   Disjunctive and Conjunctive Reductions

In this section, as a warmup to our main theorem, we present two basic applications of Theorem 3.5. First, we consider disjunctive reductions.

Theorem 4.1. *$\mathrm{P}_{\mathrm{d}}(\text{DENSE}^c)$ has p-dimension 0.*

Proof. We will show that $\mathrm{P}_{\mathrm{d}}(\text{DENSE}^c) \subseteq \mathcal{RL}_{\text{wk}}(2^{2n}, 2^{2n}, o(2^n))$. For this, let $A \in \mathrm{P}_{\mathrm{d}}(\text{DENSE}^c)$ be arbitrary. Then there is a set $S \in \text{DENSE}^c$ and a reduction $f : \{0,1\}^* \to \mathcal{P}(\{0,1\}^*)$ computable in polynomial time $p(n)$ such that for all $x \in \{0,1\}^*$, $x \in A$ if and only if $f(x) \cap S \neq \emptyset$. Note that on an input of length n, all queries of f have length bounded by $p(n)$. Also, since S is nondense, for any $\epsilon > 0$ there are infinitely many n such that

$$|S_{\leq p(n)}| \leq 2^{n^\epsilon}. \tag{4.1}$$

Let $Q_n = \bigcup_{|x| \leq n} f(x)$ be the set of all queries made by f up through length n. Then $|Q_n| \leq 2^{n+1} p(n)$. Enumerate Q_n as $q_1, \ldots, q_N$. Then each subset of $R \subseteq Q_n$ can be identified with its characteristic string $\chi_R \in \{0,1\}^N$ according to this enumeration. We define $\mathcal{C}_n$ to be the concept class of all monotone disjunctions on $\{0,1\}^N$ that have at most 2^{n^ϵ} literals. Our target disjunction is

$$\phi_n = \bigvee_{i : q_i \in S} q_i,$$

which is a member of $\mathcal{C}_n$ whenever (4.1) holds. For any $x \in \{0,1\}^{\leq n}$,

$$x \in A \iff \phi_n(\chi_{f(x)}) = 1.$$

Given x, $\chi_{f(x)}$ can be computed in $O(2^{2n})$ time. Therefore $A \leq_{wk}^{O(2^{2n})} \mathcal{C} = (\mathcal{C}_n \mid n \in \mathbb{N})$. Since Winnow learns $\mathcal{C}_n$ making at most $2 \cdot 2^{n^\epsilon} \log |Q_n| + 2 = o(2^n)$ mistakes, it follows that $A \in \mathcal{RL}_{wk}(2^{2n}, 2^{2n}, o(2^n))$. □

Next, we consider conjunctive reductions.

Theorem 4.2. $P_c(\text{DENSE}^c)$ *has p-dimension 0.*

Proof. We will show that $P_c(\text{DENSE}^c) \subseteq \mathcal{RL}_{wk}(2^n, 2^{2n}, o(2^n))$. For this, let $A \leq_c^P S \in \text{DENSE}^c$. Then there is a reduction $f : \{0,1\}^* \to \mathcal{P}(\{0,1\}^*)$ computable in polynomial time $p(n)$ such that for all $x \in \{0,1\}^*$, $x \in A$ if and only if $f(x) \subseteq S$.

Fix an input length n, and let $Q_n = \bigcup_{|x| \leq n} f(x)$. Let $\epsilon > 0$ and consider the concept class $\mathcal{C}_n = \{\mathcal{P}(X) \mid X \subseteq Q_n \text{ and } |X| \leq 2^{n^\epsilon}\}$. Our target concept is $C_n = \mathcal{P}(S \cap Q_n)$. For infinitely many n, $|S \cap Q_n| \leq |S_{\leq p(n)}| \leq 2^{n^\epsilon}$, in which case $C_n \in \mathcal{C}_n$. For any $x \in \{0,1\}^{\leq n}$, we have $x \in A$ if and only if $f(x) \in C_n$. Therefore $A \leq_{wk}^{p(n)} \mathcal{C} = (\mathcal{C}_n \mid n \in \mathbb{N})$.

The class $\mathcal{C}_n$ can be learned by a simple algorithm that makes at most $|X|$ mistakes when learning $\mathcal{P}(X)$. The hypothesis for X is simply the union of all positive examples seen so far. More explicitly, the algorithm begins with the hypothesis $H = \emptyset$. In any stage, given an example Q, the algorithm predicts 'yes' if $Q \subseteq H$ and 'no' otherwise. If the prediction is 'no,' but Q is revealed to be a positive example, then the hypothesis is updated as $H := H \cup Q$. The algorithm will never make a mistake on a negative example, and can make at most $|X|$ mistakes on positive examples.

This algorithm shows that $\mathcal{C} \in \mathcal{L}(2^{2n}, o(2^n))$, so $A \in \mathcal{RL}_{wk}(p(n), 2^{2n}, o(2^n))$. It follows that $P_c(\text{DENSE}^c) \subseteq \mathcal{RL}_{wk}(2^n, 2^{2n}, 2^{n^\epsilon})$. □

Since $\dim_p(\mathrm{E}) = 1$, we have new proofs of the following results of Watanabe.

Corollary 4.3. (Watanabe [26]) $\mathrm{E} \not\subseteq P_d(\text{DENSE}^c)$ *and* $\mathrm{E} \not\subseteq P_c(\text{DENSE}^c)$. *That is, every $\leq_d^P$-hard or $\leq_c^P$-hard set for E is dense.*

5 Adaptive Reductions

In this section we prove our main theorem, which concerns adaptive reductions that make a sublinear number of queries to a nondense set. It turns out that this problem can also be reduced to learning disjunctions.

In a surprising result (refuting a conjecture of Ko [13]), Allender, Hemaspaandra, Ogiwara, and Watanabe [1] showed that $P_{btt}(\text{SPARSE}) \subseteq P_d(\text{SPARSE})$. The disjunctive reduction they obtain will not be polynomial-time computable if the original reduction has more than a constant number of queries. However, in the proof of the following theorem we are still able to exploit their technique, and obtain an exponential-time reduction to a disjunction. Then we can apply the Winnow algorithm as in the previous section.

Theorem 5.1. *For all $\alpha < 1$, $P_{n^\alpha - T}(\text{DENSE}^c)$ has p-dimension 0.*

Proof. Let $L \leq^P_{n^\alpha - T} S \in \text{DENSE}^c$ via some oracle machine M. We will show how to reduce L to a class of disjunctions.

Fix an input length n. For an input $x \in \{0,1\}^{\leq n}$, consider using each $z \in \{0,1\}^{n^\alpha}$ as the sequence of yes/no answers to M's queries. Each z causes M to produce a sequence of queries $w_0^{x,z}, \ldots, w_{k(x,z)}^{x,z}$, where $k(x,z) < n^\alpha$, and an accepting or rejecting decision. Let $Z_x \subseteq \{0,1\}^{n^\alpha}$ be the set of all query answer sequences that cause M to accept x. Then we have $x \in L$ if and only if

$$(\exists z \in Z_x)(\forall 0 \leq j \leq k(x,z))\ S[w_j^{x,z}] = z[j],$$

which is equivalent to $(\exists z \in Z_x)(\forall 0 \leq j \leq k(x,z))\ z[j] \cdot w_j^{x,z} \in S^c \oplus S$, where $S^c \oplus S$ is the disjoint union $\{0x \mid x \in S^c\} \cup \{1x \mid x \in S\}$.

A key part of the proof that $P_{btt}(\text{SPARSE}) \subseteq P_d(\text{SPARSE})$ in [1] is to show that $P_{1-tt}(\text{SPARSE})$ is contained in $P_d(\text{SPARSE})$. The same argument yields that $P_{1-tt}(\text{DENSE}^c) \subseteq P_d(\text{DENSE}^c)$. Therefore, there is a set $U \in \text{DENSE}^c$ such that $S^c \oplus S \leq^P_d U$. Letting g be this polynomial-time disjunctive reduction, we have $x \in L$ if and only if

$$(\exists z \in Z_x)(\forall 0 \leq j \leq k(x,z))\ g(z[j] \cdot w_j^{x,z}) \cap U \neq \emptyset.$$

For each $z \in Z_x$, let

$$H_{x,z} = \{\langle u_0, \ldots, u_{k(x,z)}\rangle \mid (\forall 0 \leq j \leq k(x,z))\ u_j \in g(z[j] \cdot w_j^{x,z})\}.$$

Define

$$A_n = \{\langle u_0, \ldots, u_k\rangle \mid k < n^\alpha \text{ and } (\forall 0 \leq j \leq k)\ u_j \in U\}.$$

Then we have $x \in L$ if and only if $(\exists z \in Z_x)(\exists v \in H_{x,z})\ v \in A_n$. Letting $H_x = \bigcup_{z \in Z_x} H_{x,z}$, we can rewrite this as

$$x \in L \iff H_x \cap A_n \neq \emptyset. \tag{5.1}$$

Let $r(n)$ be a polynomial bounding the number of queries g outputs on an input of form $z[j] \cdot w_j^{x,z}$, where $|x| \leq n$. Then $|H_{x,z}| \leq r(n)^{n^\alpha}$, so

$$|H_x| \leq |Z_x| \cdot r(n)^{n^\alpha} \leq 2^{n^\alpha \cdot (1 + \log r(n))}. \tag{5.2}$$

Also, $|A_n| \leq n^\alpha \cdot |U_{\leq r(n)}|^{n^\alpha}$. Let $\epsilon \in (0, 1-\alpha)$, and let $\delta \in (\alpha + \epsilon, 1)$. Then since U is nondense, for infinitely many n, we have $|U_{\leq r(n)}| \leq 2^{n^\epsilon}$. This implies

$$(\exists^\infty n)\ |A_n| \leq n^\alpha \cdot 2^{n^{\alpha+\epsilon}} \leq 2^{n^\delta}. \tag{5.3}$$

Let

$$H_n = \bigcup_{x \in \{0,1\}^{\leq n}} H_x.$$

Then from (5.2), $|H_n| \leq 2^{2n}$ if n is sufficiently large.

Enumerate H_n as $h_1, \cdots, h_N$. We identify any $R \subseteq H_n$ with its characteristic string $\chi_R^{(n)} \in \{0,1\}^N$ according to this enumeration. Let $\mathcal{C}_n$ be the concept class of all monotone disjunctions on $\{0,1\}^N$ that have at most 2^{n^δ} literals.

Define the disjunction $\phi_n = \bigvee_{i: h_i \in A_n} h_i$, which by (5.3) is in $\mathcal{C}_n$ for infinitely many n. For any $x \in \{0,1\}^{\le n}$, from (5.1) it follows that

$$x \in L \iff \phi_n(\chi_{H_x}^{(n)}) = 1.$$

Given $x \in \{0,1\}^{\le n}$, we can compute $\chi_{H_x}^{(n)}$ in $O(2^n \cdot \mathrm{poly}(n) + |H_n|)$ time. Therefore, letting $\mathcal{C} = (\mathcal{C}_n \mid n \in \mathbb{N})$, we have $L \le_{wk}^{2^{2n}} \mathcal{C}$. Since $\mathcal{C}_n$ is learnable by Winnow with at most $2 \cdot 2^{n^\delta} \cdot \log|H_n| + 2 = o(2^n)$ mistakes, it follows that $L \in \mathcal{RL}_{\mathrm{wk}}(2^{2n}, 2^{2n}, o(2^n))$. $\qquad\square$

As a corollary, we have a positive answer to the question of Lutz and Mayordomo [20] mentioned in the introduction:

Corollary 5.2. *For all $\alpha < 1$, $\mathrm{P}_{n^\alpha - \mathrm{T}}(\mathrm{DENSE}^c)$ has p-measure 0.*

Corollary 5.3. *For all $\alpha < 1$, $\mathrm{E} \not\subseteq \mathrm{P}_{n^\alpha - \mathrm{T}}(\mathrm{DENSE}^c)$. That is, every $\le_{n^\alpha - \mathrm{T}}^{\mathrm{P}}$-hard set for E is dense.*

We remark that if we scale down from nondense sets to sparse sets, the same proof technique can handle more queries.

Theorem 5.4. $\mathrm{P}_{o(n/\log n) - \mathrm{T}}(\mathrm{SPARSE})$ *has strong p-dimension 0.*

The following corollary improves the result of Fu [8] that $\mathrm{E} \not\subseteq \mathrm{P}_{o(n/\log n) - \mathrm{T}}(\mathrm{TALLY})$.

Corollary 5.5. $\mathrm{E} \not\subseteq \mathrm{P}_{o(n/\log n) - \mathrm{T}}(\mathrm{SPARSE})$.

6   Hard Sets for NP

The hypothesis "NP has positive p-dimension," written $\dim_\mathrm{p}(\mathrm{NP}) > 0$, was first used in [10] to study the inapproximability of MAX3SAT. This positive dimension hypothesis is apparently much weaker than Lutz's often-investigated $\mu_\mathrm{p}(\mathrm{NP}) \ne 0$ hypothesis, but is a stronger assumption than $\mathrm{P} \ne \mathrm{NP}$:

$$\mu_\mathrm{p}(\mathrm{NP}) \ne 0 \Rightarrow \dim_\mathrm{p}(\mathrm{NP}) = 1 \Rightarrow \dim_\mathrm{p}(\mathrm{NP}) > 0 \Rightarrow \mathrm{P} \ne \mathrm{NP}.$$

The measure hypothesis $\mu_\mathrm{p}(\mathrm{NP}) \ne 0$ has many plausible consequences that are not known to follow from $\mathrm{P} \ne \mathrm{NP}$ (see e.g. [20]). So far few consequences of $\dim_\mathrm{p}(\mathrm{NP}) > 0$ are known. The following corollary of our results begins to remedy this.

Theorem 6.1. *If $\dim_\mathrm{p}(\mathrm{NP}) > 0$, then every set that is hard for NP under $\le_\mathrm{d}^{\mathrm{P}}$-reductions, $\le_\mathrm{c}^{\mathrm{P}}$-reductions, or $\le_{n^\alpha - \mathrm{T}}^{\mathrm{P}}$-reductions ($\alpha < 1$) is dense, and every set that is hard under $\le_{o(n/\log n) - \mathrm{T}}^{\mathrm{P}}$-reductions is not sparse.*

The consequences in Theorem 6.1 are much stronger than what is known to follow from P $\neq$ NP. If P $\neq$ NP, then no $\leq_{btt}^P$-hard or $\leq_c^P$-hard set is sparse [25,3], but it is not known whether hard sets under disjunctive reductions or unbounded Turing reductions can be sparse.

Another result is that if NP $\neq$ RP, then no $\leq_d^P$-hard set for NP is sparse [7,6]. It is interesting to see that while the hypotheses $\dim_p(NP) > 0$ and NP $\neq$ RP are apparently incomparable, they both have implications for the density of the disjunctively-hard sets for NP.

7 Conclusion

Our connection between online learning and resource-bounded dimension appears to be a powerful tool for computational complexity. We have used it to give relatively simple proofs and improvements of several previous results.

An interesting observation is that for all reductions $\leq_\tau^P$ for which we know how to prove "every $\leq_\tau^P$-hard set for E is dense," by the results presented here we can actually prove "$P_\tau(\text{DENSE}^c)$ has p-dimension 0." Indeed, we have proven the strongest results for Turing reductions in this way.

References

1. E. Allender, L. A. Hemachandra, M. Ogiwara, and O. Watanabe. Relating equivalence and reducibility to sparse sets. *SIAM Journal on Computing*, 21(3):521–539, 1992.
2. D. Angluin. Queries and concept learning. *Machine Learning*, 2(4):319–342, 1988.
3. V. Arvind, Y. Han, L. Hemachandra, J. Köbler, A. Lozano, M. Mundhenk, A. Ogiwara, U. Schöning, R. Silvestri, and T. Thierauf. Reductions to sets of low information content. In K. Ambos-Spies, S. Homer, and U. Schöning, editors, *Complexity Theory: Current Research*, pages 1–45. Cambridge University Press, 1993.
4. K. B. Athreya, J. M. Hitchcock, J. H. Lutz, and E. Mayordomo. Effective strong dimension in algorithmic information and computational complexity. *SIAM Journal on Computing*. To appear.
5. L. Berman and J. Hartmanis. On isomorphism and density of NP and other complete sets. *SIAM Journal on Computing*, 6(2):305–322, 1977.
6. H. Buhrman, L. Fortnow, and L. Torenvliet. Six hypotheses in search of a theorem. In *Proceedings of the 12th Annual IEEE Conference on Computational Complexity*, pages 2–12. IEEE Computer Society, 1997.
7. J. Cai, A. V. Naik, and D. Sivakumar. On the existence of hard sparse sets under weak reductions. In *Proceedings of the 13th Annual Symposium on Theoretical Aspects of Computer Science*, pages 307–318. Springer-Verlag, 1996.
8. B. Fu. With quasilinear queries EXP is not polynomial time Turing reducible to sparse sets. *SIAM Journal on Computing*, 24(5):1082–1090, 1995.
9. L. A. Hemachandra, M. Ogiwara, and O. Watanabe. How hard are sparse sets? In *Proceedings of the Seventh Annual Structure in Complexity Theory Conference*, pages 222–238. IEEE Computer Society Press, 1992.
10. J. M. Hitchcock. MAX3SAT is exponentially hard to approximate if NP has positive dimension. *Theoretical Computer Science*, 289(1):861–869, 2002.

11. J. M. Hitchcock, J. H. Lutz, and E. Mayordomo. Scaled dimension and nonuniform complexity. *Journal of Computer and System Sciences*, 69(2):97–122, 2004.

12. J. M. Hitchcock, J. H. Lutz, and E. Mayordomo. The fractal geometry of complexity classes. *SIGACT News*, 36(3):24–38, September 2005.

13. K. Ko. Distinguishing conjunctive and disjunctive reducibilities by sparse sets. *Information and Computation*, 81(1):62–87, 1989.

14. W. Lindner, R. Schuler, and O. Watanabe. Resource-bounded measure and learnability. *Theory of Computing Systems*, 33(2):151–170, 2000.

15. N. Littlestone. Learning quickly when irrelevant attributes abound: A new linear-threshold algorithm. *Machine Learning*, 2(4):285–318, 1987.

16. J. H. Lutz. Almost everywhere high nonuniform complexity. *Journal of Computer and System Sciences*, 44(2):220–258, 1992.

17. J. H. Lutz. The quantitative structure of exponential time. In L. A. Hemaspaandra and A. L. Selman, editors, *Complexity Theory Retrospective II*, pages 225–254. Springer-Verlag, 1997.

18. J. H. Lutz. Dimension in complexity classes. *SIAM Journal on Computing*, 32(5):1236–1259, 2003.

19. J. H. Lutz and E. Mayordomo. Measure, stochasticity, and the density of hard languages. *SIAM Journal on Computing*, 23(4):762–779, 1994.

20. J. H. Lutz and E. Mayordomo. Twelve problems in resource-bounded measure. *Bulletin of the European Association for Theoretical Computer Science*, 68:64–80, 1999. Also in *Current Trends in Theoretical Computer Science: Entering the 21st Century*, pages 83–101, World Scientific Publishing, 2001.

21. J. H. Lutz and Y. Zhao. The density of weakly complete problems under adaptive reductions. *SIAM Journal on Computing*, 30(4):1197–1210, 2000.

22. S. R. Mahaney. Sparse complete sets for NP: Solution of a conjecture of Berman and Hartmanis. *Journal of Computer and System Sciences*, 25:130–143, 1982.

23. E. Mayordomo. Personal communication, 2002.

24. A. R. Meyer, 1977. Reported in [5].

25. M. Ogiwara and O. Watanabe. On polynomial bounded truth-table reducibility of NP sets to sparse sets. *SIAM Journal on Computing*, 20(3):471–483, 1991.

26. O. Watanabe. Polynomial time reducibility to a set of small density. In *Proceedings of the Second Structure in Complexity Theory Conference*, pages 138–146. IEEE Computer Society, 1987.

Regularity Problems for Visibly Pushdown Languages

Vince Bárány[1], Christof Löding[1], and Olivier Serre[2],[*]

[1] RWTH Aachen, Germany
[2] LIAFA, Université Paris VII & CNRS, France

Abstract. Visibly pushdown automata are special pushdown automata whose stack behavior is driven by the input symbols according to a partition of the alphabet. We show that it is decidable for a given visibly pushdown automaton whether it is equivalent to a visibly counter automaton, i.e. an automaton that uses its stack only as counter. In particular, this allows to decide whether a given visibly pushdown language is a regular restriction of the set of well-matched words, meaning that the language can be accepted by a finite automaton if only well-matched words are considered as input.

1 Introduction

The class of context-free languages (CFL) plays an important role in several areas of computer science. Besides its definition using context-free grammars it has various other characterizations, the most prominent being the one via nondeterministic pushdown automata. It is well known that CFL does not enjoy good closure properties, e.g. it is not closed under complement and intersection, and that several interesting problems are undecidable, e.g. checking whether a context free language is regular, or whether it contains all words. This situation only slightly improves when considering the subclass of deterministic context free languages, i.e. languages accepted by deterministic pushdown automata (see [10] for an overview).

Another subclass of CFL that has recently been defined in [2] is the class of visibly pushdown languages. These are languages that are accepted by pushdown automata whose stack behavior (i.e. whether to execute a push, a pop, or no stack operation) is completely determined by the input symbol according to a fixed partition of the input alphabet. These automata are called visibly pushdown automata (VPA). As shown in [2,3] this class of visibly pushdown languages enjoys many good properties similar to those of the class of regular languages, the main reason for this being that each nondeterministic VPA can be transformed into an equivalent deterministic one. Visibly pushdown automata have turned out to be useful in various context, e.g. as specification formalism for verification

[*] Supported by the European Community Research Training Network "Games and Automata for Synthesis and Validation" (GAMES). Most of this work was done when the third author was a postdoctoral researcher at RWTH Aachen.

and synthesis problems for pushdown systems [1, 11], and as automaton model for processing XML streams [14, 12].

As each nondeterministic VPA can be determinized, all problems that concern the accepted language and that are decidable for deterministic pushdown automata are also decidable for VPAs. For example, in [15] and later with improved complexity in [16] it is shown that for a given deterministic pushdown automaton it is decidable whether its accepted language is regular. Hence, this problem is also decidable for VPAs.

In the context of validating streaming XML documents a similar question has been addressed in [14]. Phrased in the terminology of finite automata on words and trees the problem of validating streaming documents is the following: given the coding of a tree by a word using opening and closing tags around each subtree, check whether the corresponding tree belongs to a given regular tree language. It is rather simple to see that this task can be solved by a deterministic pushdown automaton that pushes a symbol onto the stack for each opening tag and pops a symbol for each closing tag. One of the questions raised and analyzed in [14] is whether one can decide for a given tree language if the streaming validation task can be solved by a finite automaton. As such an automaton has to verify that the input codes a tree, the class of these tree languages is rather restricted. The question gets more involved under the assumption that the input indeed codes a tree.

Coming back to VPAs, this assumption on the input being the coding of a tree corresponds to the assumption that the input is well-matched in the sense that each symbol that is pushed is popped eventually (each opening tag has a matching closing tag). The question of regularity of the accepted language then becomes: given a VPA, is there an equivalent finite automaton, where equivalence is restricted to the set of well-matched words? Restricting the equivalence to well-matched words can also be seen as allowing the finite automaton to count the difference between opening and closing tags to know in the end if the input was well-matched. This model is what we refer to as a visibly counter automaton (VCA). The main result of this paper is that it is decidable for a given VPA whether it is equivalent to a VCA. This problem is mentioned in [16] for deterministic pushdown automata and deterministic one-counter automata, and is to our knowledge still open.

The remainder of this paper is organized as follows. In Section 2 we provide the basic definitions of visibly pushdown and counter automata and state the main questions that we address. In Section 3 we give some basic concepts and constructions on which the decidability proofs are based. In Section 4 we show that it is decidable for a given visibly pushdown automaton whether it is equivalent to a visibly counter automaton that is allowed to test its counter value up to a certain threshold, and in Section 5 we prove that it is decidable whether such a threshold can be reduced.

We thank Victor Vianu and Luc Segoufin for drawing our attention to this topic.

2   Definitions

For a finite set X we denote the set of finite words over X by X^*. We denote by ε the empty word. For $u \in X^*$, we write $u(n)$ for the nth letter in u and $u\lceil_n$ for the prefix of length n of u, i.e., $u\lceil_0 = \varepsilon$ and $u\lceil_n = u(0) \cdots u(n-1)$ for $n \geq 1$.

A pushdown alphabet is a tuple $\widetilde{\Sigma} = \langle \Sigma_c, \Sigma_r, \Sigma_{\text{int}} \rangle$ that comprises three disjoint finite alphabets: Σ_c is a finite set of *calls*, Σ_r is a finite set of *returns*, and Σ_{int} is a finite set of *internal actions*. For any such $\widetilde{\Sigma}$, let $\Sigma = \Sigma_c \cup \Sigma_r \cup \Sigma_{\text{int}}$.

We define *visibly pushdown automata* over $\widetilde{\Sigma}$. Intuitively, a visibly pushdown automaton is a pushdown automaton restricted such that it pushes onto the stack only when it reads a call, it pops the stack only on reading a return, and it does not use the stack when reading an internal action.

Definition 1 (Visibly pushdown automaton [2]). *A visibly pushdown automaton (VPA) over $\widetilde{\Sigma}$ is a tuple $\mathcal{A} = (Q, \Sigma, \Gamma, Q_{in}, F, \Delta)$ where Q is a finite set of states, $Q_{in} \subseteq Q$ is a set of initial states, $F \subseteq Q$ is a set of final states, Γ is a finite stack alphabet, and $\Delta \subseteq (Q \times \Sigma_c \times Q \times \Gamma) \cup (Q \times \Sigma_r \times \Gamma \times Q) \cup (Q \times \Sigma_{\text{int}} \times Q)$ is the transition relation.*

To represent stacks we use a special bottom-of-stack symbol $\perp$ that is not in Γ. A *stack* is a finite sequence from the set $\perp \cdot \Gamma^*$ starting with the special symbol $\perp$ on the left, and ending with the top symbol on the right.[1] The *empty stack* is the one that only contains the symbol $\perp$.

A transition (q, a, q', γ) with $a \in \Sigma_c$ is a push-transition where on reading a, γ is pushed onto the stack and the control changes from state q to q'. Similarly, (q, a, γ, q') with $a \in \Sigma_r$ is a pop-transition where γ is read from the top of the stack and popped (if the top of stack is $\perp$, then no pop-transition can be applied), and the control state changes from q to q'. Our model (in contrast to the original definition from [2]) is therefore inherently restricted to input words having no prefix of negative stack height (to be defined below). Note that on internal actions, there is no stack operation.

A *configuration* of a VPA $\mathcal{A}$ is a pair (σ, q), where $q \in Q$ and $\sigma \in \perp \cdot \Gamma^*$. There is an *a-transition* from a configuration (σ, q) to (σ', q'), denoted $(\sigma, q) \xrightarrow{a} (\sigma', q')$ ($\mathcal{A}$ will be clear from context), if the following are satisfied.

- If a is a call, then $\sigma' = \sigma\gamma$ for some $(q, a, q', \gamma) \in \Delta$.
- If a is a return, then $\sigma = \sigma'\gamma$ for some $(q, a, \gamma, q') \in \Delta$.
- If a is an internal action, then $\sigma = \sigma'$ and $(q, a, q') \in \Delta$.

For a finite word $u = a_0 a_1 \cdots a_n$ in Σ^*, a *run* of $\mathcal{A}$ on u is a sequence of configurations $(\sigma_0, q_0)(\sigma_1, q_1) \cdots (\sigma_{n+1}, q_{n+1})$, where $q_0 \in Q_{in}$, $\sigma_0 = \perp$ and for every $0 \leq i \leq n$, $(\sigma_i, q_i) \xrightarrow{a_i} (\sigma_{i+1}, q_{i+1})$ holds. In this case we also use the notation $(\sigma_0, q_0) \xrightarrow{u} (\sigma_{n+1}, q_{n+1})$. A word $u \in \Sigma^*$ is *accepted* by a VPA if there is a run over u which ends in a final configuration, that is a configuration with

---

[1] Note that we are using here the reverse of the more common notation of stacks, having the top symbol on the left and the bottom on the right.

empty stack and a control state, which is final. The language $L(\mathcal{A})$ of a VPA $\mathcal{A}$ is the set of words accepted by $\mathcal{A}$.

A VPA is *deterministic* if it has a unique initial state q_{in}, and for each input letter and configuration there is at most one successor configuration. For deterministic VPAs (DVPAs) we denote the transition relation by δ instead of Δ and write $\delta(q, a) = (q', \gamma)$ instead of $(q, a, q', \gamma) \in \delta$ if $a \in \Sigma_c$, $\delta(q, a, \gamma) = q'$ instead of $(q, a, \gamma, q') \in \delta$ if $a \in \Sigma_r$, and $\delta(q, a) = q'$ instead of $(q, a, q') \in \delta$ if $a \in \Sigma_{int}$.

Let us stress, that during the run of any VPA $\mathcal{A}$ on a given word $u \in \Sigma^*$ the automaton $\mathcal{A}$ controls only which symbols are pushed on the stack, but not when a symbol is pushed or popped. At each step, the height of the stack is pre-determined by the prefix of u read thus far. Let $\chi(a)$ be the *sign* of the symbol $a \in \Sigma$ defined as $\chi(a) = 1$ if $a \in \Sigma_c$, $\chi(a) = 0$ if $a \in \Sigma_{int}$, and $\chi(a) = -1$ if $a \in \Sigma_r$. We define the *stack height* $\mathrm{sh}(u)$ of a word $u \in \Sigma^*$ as the sum of the signs of its constituent symbols, with $\mathrm{sh}(\varepsilon) = 0$. Furthermore, let $\mathrm{minsh}(u) = \min\{\mathrm{sh}(u \restriction_n) \mid 0 \le n \le |u|\}$ and $\mathrm{maxsh}(u) = \max\{\mathrm{sh}(u \restriction_n) \mid 0 \le n \le |u|\}$. A word u is *well matched* if $\mathrm{sh}(u) = \mathrm{minsh}(u) = 0$.

Given a DVPA $\mathcal{A}$ with control states Q each well-matched word $u \in \Sigma^*$ induces a transformation $T_u^{\mathcal{A}} : Q \to Q$ defined as $\{(q, q') \mid (\bot, q) \xrightarrow{u} (\bot, q')\}$, which completely describes the behavior of $\mathcal{A}$ on reading u in any context. The set of all transformations induced by a well-matched word is denoted $\mathcal{T}_{\mathrm{wm}}^{\mathcal{A}}$. In the following we write just $\mathcal{T}_{\mathrm{wm}}$ and T_u when $\mathcal{A}$ is understood.

The fact that VPAs control only the content of their stack but not its height allows one to determinize every VPA as shown in [2]. In the rest of the paper we will therefore assume that all VPAs considered are deterministic.

Note that we have defined acceptance with empty stack. This implies, together with the noted implicit restriction imposed by the visibility condition, that only well-matched words can be accepted. Therefore, we are considering only languages that are subsets of the language $L_{\mathrm{wm}} = \{u \in \Sigma^* \mid \mathrm{sh}(u) = \mathrm{minsh}(u) = 0\}$ of well-matched words. Observe that L_{wm} is accepted by a trivial single state DVPA $\mathcal{A}_{\mathrm{wm}}$ having a single stack symbol, hence using its stack solely as a counter to keep track of the stack height of the word being read. The following definition generalizes this concept.

Definition 2 (Visibly counter automaton). *A visibly counter automaton with threshold m (m-VCA) over $\widetilde{\Sigma}$ is a tuple $\mathcal{A} = (Q, \Sigma, q_{in}, F, \delta_0, \ldots, \delta_m)$ where Q is a finite set of states, $q_{in} \in Q$ is the initial state, $F \subseteq Q$ is a set of final states, $m \ge 0$ is a threshold, and $\delta_i : Q \times \Sigma \to Q$ is a transition function for every $i = 0, \ldots, m$.*

A configuration of $\mathcal{A}$ is a pair (k, q) of counter value $k \in \mathbb{N}$ and state $q \in Q$. For $a \in \Sigma$, there is an a-transition from (k, q) to (k', q'), denoted $(k, q) \xrightarrow{a} (k', q')$, if $k' = k + \chi(a)$, and $q' = \delta_k(q, a)$ if $k < m$ and $q' = \delta_m(q, a)$ if $k \ge m$.

For a finite word $u = a_0 a_1 \cdots a_n$ in Σ^*, the *run* of $\mathcal{A}$ on u is the sequence $(k_0, q_0)(k_1, q_1) \cdots (k_{n+1}, q_{n+1})$ of configurations, where $q_0 = q_{in}$, $k_0 = 0$, and $(k_i, q_i) \xrightarrow{a_i} (k_{i+1}, q_{i+1})$ for every $0 \le i \le n$. A word $u \in \Sigma^*$ is *accepted* by a VCA

$\mathcal{A}$ if the run of $\mathcal{A}$ over u ends in a final configuration, that is a configuration with counter value 0 and control state from F. The language $L(\mathcal{A})$ of a VCA $\mathcal{A}$ is the set of words accepted by $\mathcal{A}$.

Observe that a 0-VCA has absolutely no access to its counter, which can be perceived as an auxiliary device ensuring that only well-matched words are accepted. Other than that, a zero threshold VCA is essentially a finite automaton. Indeed, it is easy to see that a language L is accepted by some 0-VCA if and only if $L = L' \cap L_{\mathrm{wm}}$ for some regular language L'. The next example shows that $(m+1)$-VCAs are more powerful than m-VCAs.

Example 1. Consider the languages $L_m = \{\Sigma_c^n \Sigma_r^{n-m} \Sigma_c^{l-m} \Sigma_r^l \mid m \leq l, n \in \mathbb{N}\}$ defined for each $m \in \mathbb{N}$. Each L_m consists of well-matched words and is clearly accepted by an appropriate $(m+1)$-VCA. Moreover, it is easy to show that there is no m-VCA accepting L_m.

Note that we have defined VCAs to be deterministic. In Section 5 we also use non-deterministic VCAs with the natural definition. The standard subset construction that is used to determinize finite automata can also be used to determinize VCAs.

Based on the preceding definitions we can now state the problems that we address:

(1) Given a DVPA $\mathcal{A}$ and $m \in \mathbb{N}$, is there an m-VCA that accepts $L(\mathcal{A})$?
(2) Given a DVPA $\mathcal{A}$, is there $m \in \mathbb{N}$ and an m-VCA that accepts $L(\mathcal{A})$?
(3) Given an m-VCA $\mathcal{A}$ and $m' \in \mathbb{N}$, is there an m'-VCA that accepts $L(\mathcal{A})$?

Note that decidability of the two last questions implies decidability of the first one. The following example illustrates that for (2) and (3) an exponential blow-up in the size of the automaton is unavoidable.

Example 2. Let Σ be the alphabet with $\Sigma_c = \{c_a, c_b\}$, $\Sigma_r = \{r_a, r_b\}$ and $\Sigma_{\mathrm{int}} = \emptyset$. For a given $m \in \mathbb{N}$ let $L_m = \{c_{x_1} \cdots c_{x_m} w r_{x_m} \cdots r_{x_1} \mid x_1, \ldots, x_m \in \{a, b\}$ and $w \in L_{\mathrm{wm}}\}$ be the set of well-matched words starting with m initial calls and ending with m corresponding returns. For each m, it is easily seen that L_m is accepted by a DVPA with $\mathcal{O}(m)$ states that stores the first m calls on its stack and then compares them to the m final returns. Instead of storing the initial calls on the stack it is also possible to memorize them in the control state, leading to an $(m+1)$-VCA with $\mathcal{O}(2^m)$ states. A pumping argument shows that this exponential blow-up is unavoidable.

For each m, let L'_m be the set of well-matched words that end with a sequence of m returns, where the first return in this sequence is r_a: such a language is easily accepted by an m-VCA with two states. It can also be accepted by an exponentially larger 0-VCA that remembers in its control states the last m returns. Again, a pumping argument shows that this exponential blow-up is unavoidable.

The rest of the paper is devoted to the proof of the following result (cf. Theorem 2 in Section 4 and Theorem 3 in Section 5).

Theorem 1. *Questions (1), (2) and (3) are decidable and lead to effective constructions.*

From a prior remark concerning languages accepted by 0-VcAs and from the decidability of (1) for $m = 0$ we obtain the following result.

Corollary 1. *It is decidable, whether a given VPA $\mathcal{A}$ accepts a regular restriction of the set of well-matched words, i.e. whether $L(\mathcal{A}) = L \cap L_{\text{wm}}$ for some regular language L. When so, then a finite automaton recognizing L can be effectively constructed.*

Concerning the restriction that we only consider languages that are subsets of L_{wm}, note that the case where acceptance is defined only via final states can be reduced to our setting as follows. By adding a fresh symbol to Σ_r used to close unmatched calls, one can pass to a language consisting of well-matched words only. This new language can be recognized by a VcA (accepting with final states and counter value 0) iff the original language can be recognized by a VcA accepting with final states only.

3   Basic Tools and Constructions

We shall now introduce the basic concepts and tools that we are using. Throughout the rest of the paper let $\mathcal{A} = (Q, \Sigma, \Gamma, q_{in}, F, \delta)$ be a given DVPA.

We use finite single-tape and multi-tape letter-to-letter automata to represent sets and relations of configurations respectively. Therefore we assume w.l.o.g. that Q and Γ are disjoint, and identify each configuration (σ, q) of $\mathcal{A}$ with the word σq. Letter-to-letter 2-tape finite automata accept precisely the length-preserving rational relations. Basic results on length-preserving and synchronized rational relations can be found in [9]. Letter-to-letter multi-tape finite automata can be seen as classical single-tape finite automata over the product alphabet. Hence, all classical constructions and results of automata theory apply. Below we often use this fact without explicit reference. In various estimates we use the binary function $\exp(k, n)$ denoting a tower of exponentials of height k defined inductively by letting $\exp(0, n) = n$ and $\exp(k + 1, n) = 2^{\exp(k,n)}$ for all k and n.

When considering language acceptance only those configurations of $\mathcal{A}$ are of concern that are *reachable* from the initial configuration. Accordingly, in our constructions we restrict our attention to the set $V_{\mathcal{A}}$ of configurations of $\mathcal{A}$ reachable from the initial configuration. The fact, first observed by Büchi [7], that $V_{\mathcal{A}}$ is regular is therefore essential. Moreover, an obvious adaptation of the construction of [6] (see also [8]) shows that a non-deterministic finite automaton recognizing $V_{\mathcal{A}}$ with $\mathcal{O}(|Q|)$ states can be constructed in polynomial time. From now on by configuration we always mean reachable configuration, unless explicitly stated otherwise.

First we define equivalence (denoted $\sim$) of configurations of $\mathcal{A}$ in a standard way according to the languages they accept, and observe a necessary condition (2') for a positive answer for question (2). Next we show that $\sim$, when considered as a binary relation on words describing the configurations, can be accepted by a letter-to-letter two-tape automaton. This allows us not only to decide (2') but also to prove its sufficiency.

The *configuration graph* of $\mathcal{A}$ is the edge-labelled graph $G_{\mathcal{A}} = (V_{\mathcal{A}}, E_{\mathcal{A}})$, where $V_{\mathcal{A}}$ is, as above, the set of reachable configurations of $\mathcal{A}$ and $E_{\mathcal{A}}$ is the set that contains all triples of the form $((\sigma, q), a, (\sigma', q'))$ such that $(\sigma, q), (\sigma', q') \in V_{\mathcal{A}}$, $a \in \Sigma$, and $(\sigma, q) \xrightarrow{a} (\sigma', q')$. Below we often suppress the index $\mathcal{A}$.

Definition 3 (Equivalence of configurations). *Two configurations $\sigma q, \sigma' q'$ of $\mathcal{A}$ are equivalent, in symbols $\sigma q \sim \sigma' q'$, if $|\sigma| = |\sigma'|$ and for every word $u \in \Sigma^*$ there is an accepting run of $\mathcal{A}$ labelled by u from (σ, q) to a final configuration iff there is one from (σ', q').*

Because $\mathcal{A}$ is deterministic $\sim$ is in fact a congruence with respect to the transition relations $\xrightarrow{a}$ ($a \in \Sigma$) restricted to the set of reachable configurations. This allows us to define the quotient graph $G/_{\sim}$ as follows.

Definition 4 (Quotient of the configuration graph). *We define the quotient of the configuration graph $G = (V, E)$ with respect to the congruence $\sim$ as $G/_{\sim} = (V/_{\sim}, E/_{\sim})$, where $V/_{\sim}$ consists of equivalence classes of V under $\sim$ and for all $C_1, C_2 \in V/_{\sim}$, and for any letter $a \in \Sigma$, $(C_1, a, C_2) \in E/_{\sim}$ if and only if there are some (equivalently for all) $v_1 \in C_1$ and $v_2 \in C_2$ such that $(v_1, a, v_2) \in E$.*

Note that by definition $\sigma q \sim \sigma' q'$ implies that $|\sigma| = |\sigma'|$. In other words, $\sim$ refines the equivalence defined according to stack height, i.e. $\sim$ is length-preserving. If we denote by $V|_n$ the set of reachable configurations that contain n stack symbols, i.e. $V|_n = V \cap (\perp \cdot \Gamma^n Q)$, then $\sim$ induces a certain number of equivalence classes on each set $V|_n$. In case $\mathcal{A}$ is equivalent to some m-VCA, this number of equivalence classes must be bounded by a bound independent of n, because configurations of a VCA that have the same counter value can only be distinguished by finitely many control states.

Proposition 1. *The following is necessary for (2) to have a positive answer.*

(2') $\exists K \forall n \ V|_n$ *is partitioned into at most K $\sim$-equivalence classes.*

It is, however, not immediate that the above condition is also sufficient. Both to show equivalence of (2) and (2') and to prove their decidability the following observation is crucial.

Lemma 1. *One can effectively construct a letter-to-letter 2-tape automaton $\mathcal{A}_{\sim}$ having at most $2^{\mathcal{O}(|Q|^2)}$ states and recognizing $\sim$.*

This lemma can be shown by noting that an automaton can guess a separating word for two configurations of the same length n. Such a word consists of n returns interleaved with well-matched words. As not the particular well-matched words u but only the transformations T_u (from the finite set $\mathcal{T}_{\text{wm}}$) induced by them are interesting, a finite automaton can check whether two configurations are *not* equivalent. Then one can conclude using the closure properties of finite letter-to-letter 2-tape automata.

We are interested in the number of equivalence classes of $\sim$ for each stack height and therefore want to elect representatives for these classes. For this purpose we fix some linear ordering of the symbols of Γ and Q, thus determining the lexicographic ordering $<_{\mathrm{lex}}$ of all configurations. Note that $<_{\mathrm{lex}}$ is synchronized rational, hence, its restriction to words of equal length is recognized by a letter-to-letter automaton. Using the automata recognizing $<_{\mathrm{lex}}$, $\sim$, and V we can further construct an automaton recognizing the set Rep = $\{\sigma q \in V \mid \neg\exists \sigma' q' \in V(\sigma q \sim \sigma' q' \wedge \sigma' q' <_{\mathrm{lex}} \sigma q)\}$ of lexicographically smallest representatives of each $\sim$-class as follows: One can construct a letter-to-letter automaton recognizing pairs of equivalent reachable configurations, such that the first component precedes the second one in the lexicographic ordering. After projection onto the second component, determinization, and complementation (with respect to V) one obtains a deterministic automaton $\mathcal{A}_{\mathrm{Rep}}$ recognizing Rep. The largest one of the components is the automaton $\mathcal{A}_\sim$ and the costliest operation is, of course, determinization potentially causing an exponential increase in the number of states. Thus, we obtain $\exp(2, \mathcal{O}(|Q|^2))$ as an upper bound on the size of $\mathcal{A}_{\mathrm{Rep}}$.

We now observe that (2') is equivalent to the slenderness of Rep. Following [13] and [4] we say that a language $L \subseteq \Gamma^*$ is *slender* if there is a constant K such that $|L \cap \Gamma^n| \leq K$ for all $n \in \mathbb{N}$, in which case we may also say that L is K-*thin*. Let us therefore introduce the notation $\mathrm{Rep}_n = \mathrm{Rep} \cap V|_n$. Analogously, we say that the graph $G/_\sim$ is *slender* if there is a constant K such that $|(V|_n)/_\sim| \leq K$ for all $n \geq 0$. Relying on results of [4] and [13] we immediately obtain the following.

Proposition 2. *Condition (2') is decidable, moreover, if Rep is K-thin, then $K \leq |\Gamma|^{N-2} \cdot |Q| = \exp(3, \mathcal{O}(|Q|^2))$, where N is the number of states of the minimal deterministic automaton recognizing Rep.*

Let us assume that $G/_\sim$ is slender. We identify each of its nodes C with the pair $(\mathrm{sh}(C), \mathrm{index}(C)) \in \mathbb{N} \times \{1, \cdots, K\}$, where the stack height of a class C is the stack height of any (hence all) of the configurations belonging to C and the index of C is the position of its representative $w \in C \cap \mathrm{Rep}$ with respect to $<_{\mathrm{lex}}$ among $\mathrm{Rep}_{\mathrm{sh}(C)}$. In the next section we will show that in case $G/_\sim$ is slender it is (in the above representation) actually the configuration graph of a VCA. The following lemma constitutes an important step in the proof of this result.

Lemma 2. *Assuming condition (2') holds with slenderness bound K we can effectively construct an automaton $\mathcal{C}_\sim$ reading stack contents and whose states $q_\sim$ encode mappings $\rho_{q_\sim} : Q \to \{0, \ldots, K\}$ where K is the slenderness index of $G/_\sim$. After reading a stack content σ the automaton $\mathcal{C}_\sim$ is in a state $q_\sim$ such that $\mathrm{index}((\sigma, q)) = \rho_{q_\sim}(q)$ for all $q \in Q$. Moreover $\exp(5, \mathcal{O}(|Q|^2))$ is an upper bound on the number of states of $\mathcal{C}_\sim$.*

4   From Pushdown to Counter Automata: Decidability of Question (2)

In this section we prove that slenderness is actually a sufficient condition for (2) to hold. As it is also necessary and decidable, it shows the decidability of question (2). Effectiveness follows from the proof.

Assume that $\mathcal{A}$ is a DVPA (with the usual components) such that $G_{\mathcal{A}}/\!\!\sim$ is slender, and let K be a slenderness bound, i.e. there are at most K classes on each level of $G_{\mathcal{A}}/\!\!\sim$.

The proof and the construction are split in two steps. First we show that $G_{\mathcal{A}}/\!\!\sim$ can be effectively described by an ultimately periodic word. Then, in the second step, it easily follows that $\mathcal{A}$ is equivalent to an m-VCA with m being the offset of the ultimately periodic word.

The infinite word describing $G_{\mathcal{A}}/\!\!\sim$ is such that the nth letter codes the edges of $E_{\mathcal{A}}/\!\!\sim$ that leave the vertices from the nth level, i.e., the outgoing edges from the vertex set $\{(n,i) \mid i \in \{1,\ldots,K\}\}$. These edges are fully described by a (partial) mapping assigning to each pair (i,a) of class index and input letter the index of the class reached from class i on level n when reading an a. If there are less than i classes on level n, then the value for (i,a) is undefined.

More formally, the description $\tau_n : \{1,\ldots,K\} \times \Sigma \to \{1,\ldots,K\}$ of the nth level of $G_{\mathcal{A}}/\!\!\sim$ is defined by $\tau_n(i,a) = j$ iff $((n,i), a, (n+\chi(a), j)) \in E_{\mathcal{A}}/\!\!\sim$ and $\tau_n(i,a)$ is undefined if (n,i) is not a vertex of $G_{\mathcal{A}}/\!\!\sim$.

The sequence $\alpha := \tau_0 \tau_1 \ldots$ completely describes $G_{\mathcal{A}}/\!\!\sim$. Using the automaton $\mathcal{C}_\sim$ (cf. Lemma 2) it is possible to construct a finite state machine that outputs this sequence. This implies the main technical result of this section, namely that α is ultimately periodic.

Lemma 3. *The description* $\alpha = \tau_0\tau_1\tau_2,\ldots$ *of* $G_{\mathcal{A}}/\!\!\sim$ *is an ultimately periodic sequence that can be constructed effectively.*

As α is ultimately periodic there are numbers m and k such that $\alpha = \tau_0 \cdots \tau_{m-1}(\tau_m \cdots \tau_{m+k-1})^\omega$. We call m the offset and k the period of α. It is not difficult to verify that a VCA that knows whether it is in the offset part of α (using its threshold) or in the periodic part (using a modulo k counter to keep track of the position) can simulate $\mathcal{A}$. This is established in the following proposition.

Proposition 3. *If the description* $\alpha = \tau_0\tau_1 \ldots$ *of* $G_{\mathcal{A}}/\!\!\sim$ *is ultimately periodic with offset m and period k, then one can build an m-VCA $\mathcal{B}$ such that* $L(\mathcal{A}) = L(\mathcal{B})$.

Combining Propositions 1, 2, and 3 we get the following theorem answering question (2) from Section 3.

Theorem 2. *It is decidable if for a given VPA there exists an equivalent VCA. If such a VCA exists it can be effectively constructed and has* $\mathcal{O}((|\Gamma| \cdot |Q_{\mathcal{C}_\sim}| \cdot K)^{2K} \cdot K)$ *states and its threshold is bounded by* $\mathcal{O}((|\Gamma| \cdot |Q_{\mathcal{C}_\sim}| \cdot K)^{2K})$.

5 Reducing the Threshold: Decidability of Question (3)

In all this section, we assume that $\mathcal{A} = (Q, \Sigma, q_{in}, F, \delta_0, \ldots, \delta_m)$ is an m-VCA for some threshold m. Given $m' < m$, we want to decide whether there is an m'-VCA $\mathcal{B}$ such that $L(\mathcal{A}) = L(\mathcal{B})$. If such a $\mathcal{B}$ exists, we want to provide an effective construction of it.

The decision procedure that we present consists of two steps. First, we build an m'-VCA $\mathcal{A}'$ and show that if $\mathcal{A}$ is equivalent to some m'-VCA then $L(\mathcal{A}) = L(\mathcal{A}')$. Intuitively, $\mathcal{A}'$ is a canonical candidate to be equivalent to $\mathcal{A}$. Then, we have to check whether $L(\mathcal{A}) = L(\mathcal{A}')$ holds, which is known to be decidable [2].

As the technical details of the construction of $\mathcal{A}'$ and the correctness proofs are quite involved we restrict ourselves in the following to an explanation of the underlying ideas.

The difference between $\mathcal{A}$ and an m'-VCA is that for a word w of stack height h with $m' \leq h < m$ the automaton $\mathcal{A}$ exactly knows the current stack height because it uses δ_h to compute the next configuration, whereas the m'-VCA only knows that the stack height is at least m'. Such a situation is depicted in Figure 1 (where for now we ignore all annotations except m and m').

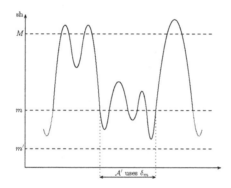

Fig. 1. A critical situation when simulating threshold m by threshold m'

The main idea is to show that, under the assumption that $\mathcal{A}$ is indeed equivalent to some m'-VCA, this additional information gained by $\mathcal{A}$ when using δ_h is not used (under certain conditions) so that instead of using δ_h to compute the next configuration one could also have used δ_m. The conditions under which it is possible to use δ_m instead of the correct transition function δ_h are also illustrated in Figure 1. If the input exceeds a certain stack height (denoted by M, a parameter depending on the size of $\mathcal{A}$), then comes back into the area between m and m', and then again goes beyond M, then one can also use δ_m when the stack height is between m and m', without changing the acceptance behavior of $\mathcal{A}$. The condition on the stack height is needed for the correctness proof to be able to apply pumping arguments without changing the transformation on the state space that is induced by the input.

This allows to construct a nondeterministic m'-VCA $\mathcal{A}'$ that maintains in its state space a counter up to M that is updated according to the stack height. As long as the stack height stays below M, $\mathcal{A}'$ can exactly simulate $\mathcal{A}$. If the stack height exceeds M, $\mathcal{A}'$ starts using δ_m for its transitions, and it guesses the points where it can switch back to exact simulation of $\mathcal{A}$. These are the points where the stack height falls below M and reaches a value less than m' before exceeding M again. These guesses can be verified as correct by $\mathcal{A}'$ at the moment where the stack height goes below m' (because then it can compare the counter value maintained in the state space with the real stack height).

As nondeterministic VCAs can be determinized as explained in Section 2, we obtain the following lemma.

Lemma 4. *From $\mathcal{A}$ one can construct an m'-VCA $\mathcal{A}'$ such that $L(\mathcal{A}) \neq L(\mathcal{A}')$ implies that there is no m'-VCA that is equivalent to $\mathcal{A}$.*

Finally, using the fact that equivalence for VPAs (hence for VCAs) is decidable, we obtain the following result answering question (3) from Section 2.

Theorem 3. *It is decidable, given an m-VCA $\mathcal{A}$ and $m' < m$, whether $\mathcal{A}$ is equivalent to some m' VCA, in which case such an m'-VCA $\mathcal{A}'$ can be constructed effectively.*

Concerning complexity, we note that the number of states of (the nondeterministic) $\mathcal{A}'$ is in $\mathcal{O}(|Q|^{2|Q|})$ (stemming from the definition of M). To check equivalence of $\mathcal{A}'$ with $\mathcal{A}$, one determinizes $\mathcal{A}'$ (exponential blow-up) and transforms it into a VPA: hence the complexity is doubly exponential in $|Q|$.

6   Conclusion

We have introduced the notion of visibly counter automaton as a direct adaption of standard one-counter automata to the framework of visibly pushdown automata. We have shown that it is decidable for a given VPA if it is equivalent to some VCA, even if we allow the counter to be tested up to a certain threshold, and provided an algorithm to construct such a counter automaton if it exists. This solves a special case of a problem that was posed in [16] for general deterministic pushdown automata.

A drawback of the presented proof is the high complexity of the resulting construction. The upper bound on the size of the VCA that we construct is 6-fold exponential in the size of the given visibly pushdown automaton, whereas the lower bound (Example 2) that we can prove is only singly exponential.

As a corollary of our main result we obtain that it is decidable for a given VPA whether it accepts a regular restriction of the set of well-matched words, i.e. whether its language is of the form $L \cap L_{\mathrm{wm}}$ for a regular language L. To answer the question from [14] one would have to solve the corresponding problem with L_{wm} replaced by another language: If we consider inputs as obtained when coding trees by words using opening and closing tags for the subtrees, then L_{wm}

describes those words for which each opening tag is closed by *some* closing tag. To be a valid coding of a tree (in the sense of [14]) each opening tag has to be closed by a unique corresponding tag, i.e. the word has to be *strongly* well matched (see also [5]). Hence, to decide whether membership for a set $L(\mathcal{A})$ of coded trees can be tested by a finite automaton under the assumption that the input is well formed in the above sense, one has to check if $L(\mathcal{A})$ is of the form $L \cap L_{\mathrm{swm}}$ for some regular language L and for L_{swm} being the set of strongly well-matched words.

Currently, we are working on the following generalization of these problems: Given two VPAs $\mathcal{A}$ and $\mathcal{B}$, is the language accepted by $\mathcal{A}$ a regular restriction of the language accepted by $\mathcal{B}$, i.e. $L(\mathcal{A}) = L \cap L(\mathcal{B})$ for some regular language L?

References

1. R. Alur, K. Etessami, and P. Madhusudan. A temporal logic of nested calls and returns. In *TACAS'04, LNCS* 2988, 467–481. Springer, 2004.
2. R. Alur and P. Madhusudan. Visibly pushdown languages. In *Proceedings of STOC'04*, pages 202–211. ACM, 2004.
3. R. Alur, P. Madhusudan, V. Kumar, and M. Viswanatha. Congruences for visibly pushdown languages. In *ICALP'05, LNCS* 3580, pages 1102–1114, 2005.
4. M. Andraşiu, G. Păun, J. Dassow, and A. Salomaa. Language-theoretic problems arising from Richelieu cryptosystems. *Theor. Comp. Sci.*, 116(2):339–357, 1993.
5. Jean Berstel and Luc Boasson. Formal properties of XML grammars and languages. *Acta Informatica*, 38(9):649–671, 2002.
6. A. Bouajjani, J. Esparza, and O. Maler. Reachability analysis of pushdown automata: Application to model-checking. In *CONCUR'97, LNCS* 1243, pages 135–150. Springer, 1997.
7. J. R. Büchi. Regular canonical systems. *Archiv für Mathematische Grundlagenforschung*, 6:91–111, 1964.
8. J. Esparza, D. Hansel, P. Rossmanith, and S Schwoon. Efficient algorithms for model checking pushdown systems. In *CAV'00, LNCS* 1855, pp. 232–247. Springer.
9. C. Frougny and J. Sakarovitch. Synchronized rational relations of finite and infinite words. *Theoretical Computer Science*, 108(1):45–82, 1993.
10. J. E. Hopcroft and J. D. Ullman. *Formal Languages and their Relation to Automata*. Addison-Wesley, 1969.
11. C. Löding, P. Madhusudan, and O. Serre. Visibly pushdown games. In *FST&TCS'04, LNCS* 3328, pages 408–420. Springer, 2004.
12. C. Pitcher. Visibly pushdown expression effects for XML stream processing. In *Programming Language Technologies for XML, PLAN-X'05*, pages 5–19, 2005.
13. G. Păun and A. Salomaa. Thin and slender languages. *Discrete Applied Mathematics*, 61(3):257–270, 1995.
14. L. Segoufin and V. Vianu. Validating streaming XML documents. In *Proceedings of PODS'02*, pages 53–64. ACM, 2002.
15. R. E. Stearns. A regularity test for pushdown machines. *Information and Control*, 11(3):323–340, 1967.
16. L. G. Valiant. Regularity and related problems for deterministic pushdown automata. *Journal of the ACM*, 22(1):1–10, 1975.

Regular Expressions and NFAs Without ε-Transitions

(Extended Abstract)

Georg Schnitger[*]

Institut für Informatik, Johann Wolfgang Goethe-Universität,
Robert Mayer Straße 11–15, 60054 Frankfurt am Main, Germany
georg@thi.informatik.uni-frankfurt.de

Abstract. We consider the problem of converting regular expressions of length n over an alphabet of size k into ε-free NFAs with as few transitions as possible. Whereas the previously best construction uses $O(n \cdot \min\{k, \log_2 n\} \cdot \log_2 n)$ transitions, we show that $O(n \cdot \log_2 2k \cdot \log_2 n)$ transitions suffice. For small alphabets we further improve the upper bound to $O(n \cdot \log_2 2k \cdot k^{L_k(n)+1})$, where $L_k(n) = O(\log_2^* n)$. In particular, $n \cdot 2^{O(\log_2^* n)}$ transitions and hence almost linear size suffice for the binary alphabet! Finally we show the lower bound $\Omega(n \cdot \log_2^2 2k)$ and as a consequence the upper bound $O(n \cdot \log_2^2 n)$ of [7] for general alphabets is best possible. Thus the conversion problem is solved for large alphabets ($k = n^{\Omega(1)}$) and almost solved for small alphabets ($k = O(1)$).

Classification. Automata and formal languages, descriptional complexity, nondeterministic automata, regular expressions.

1 Introduction

One of the central tasks on the border between formal language theory and complexity theory is to describe infinite objects such as languages by finite formalisms such as automata, grammars, expressions etc., and to investigate the descriptional complexity and capability of these formalisms. Formalisms like expressions and finite automata have proven to be very useful in building compilers, and techniques converting a regular expression into an ε-free nondeterministic finite automaton were used as basic tools in the design of computer systems such as UNIX ([5], p. 123, and [11]). The descriptional complexity of an expression R is its length, i.e., the number of symbols occurring in R, and the descriptional complexity of a nondeterministic finite automaton (NFA) is the number of its transitions, where identical edges with distinct labels are differentiated.

All classical conversions [1,3,9,11] produce ε-free NFAs with worst-case size quadratic in the length of the given regular expression and for some time this was assumed to be optimal [10]. But then Hromkovic, Seibert and Wilke [7] constructed ε-free NFAs with just $O(n \cdot (\log_2 n)^2)$ transitions for regular expressions

---

[*] Work supported by DFG grant SCHN 503/2-2.

B. Durand and W. Thomas (Eds.): STACS 2006, LNCS 3884, pp. 432–443, 2006.

of length n and this transformation can even be implemented to run in time $O(n \cdot \log_2 n + m)$, where m is the size of the output [4]. Subsequently Geffert [2] showed that even ε-free NFAs with $O(n \cdot k \cdot \log_2 n)$ transitions suffice for alphabets of size k, improving the bound of [7] for small alphabets.

We considerably improve the upper bound of [2] for alphabets of (small) size k. To describe our result we define $l_0 = 2^k$, $l_{j+1} = (2k)^{l_j / \log_2 l_j}$ and set $L_k(n)$ to be the smallest i with $l_i \geq n$. Observe that $L_k(n) = O(\log^* n)$ for all $k \geq 2$.

Theorem 1. *Every regular expression R of length n over an alphabet of size k can be recognized by an ε-free NFA with at most*

$$O(n \cdot \min\{\log_2 n, k^{L_k(n)+1}\} \cdot \log_2 2k)$$

transitions.

As a first consequence we obtain ε-free NFAs of size $O(n \cdot \log_2 n \cdot \log_2 2k)$ for regular expressions of length n over an alphabet of size k. For small alphabets, for instance if $k = O(\log_2 \log_2 n)$, the upper bound $O(n \cdot k^{L_k(n)+1} \cdot \log_2 2k)$ is better. In particular, $n \cdot 2^{O(\log_2^* n)}$ transitions and hence almost linear size suffice for the binary alphabet.

[8] shows that regular expressions of length n require ε-free NFAs with at least $\Omega(n(\log_2 n)^2 / \log_2 \log_2 n)$ transitions, improving the lower bound $\Omega(n \cdot \log_2 n)$ of [7]. We also improve the lower bound.

Theorem 2. *There are regular expressions of length n over an alphabet of size k such that any equivalent ε-free NFA has at least $\Omega(n \cdot \log_2^2 2k)$ transitions.*

Thus the construction of [7] is optimal for large alphabets, i.e., if $k = n^{\Omega(1)}$. Since Theorem 1 is almost optimal for alphabets of fixed size, improvements for alphabets of intermediate size, i.e., $\omega(1) = k = n^{o(1)}$, are still required.

In Section 2 we show how to construct small ε-free NFAs for a given regular expression R using ideas from [2,7]. We obtain the upper bound $O(n \cdot \log_2 n \cdot \log_2 2k)$ by short-cutting ε-paths within the canonical NFA. Whereas the canonical NFA is derived from the expression tree of R, the shortcuts are derived from a decomposition tree, a balanced version of the expression tree. The subsequent improvement for small alphabets is based on repeatedly applying the previous upper bound to larger and larger subexpressions. We give a brief sketch of the lower bound for E_n in Section 3. Conclusions and open problems are stated in Section 4.

2   Small ε-Free NFAs for Regular Expressions

Assume that the regular expression R is given. Then we have $R = R_1 \overset{+}{\circ} R_2$ or $R = S^*$ for subexpressions R_1, R_2, S of R. Fig. 1 shows this recursive expansion in NFA-notation. Thus, after completing this recursive expansion, we arrive at an NFA N_R with a unique initial state q_0 and a unique final state q_f. Moreover,

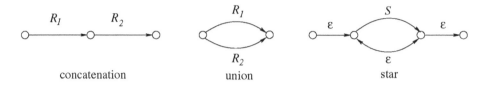

Fig. 1. The initial step in determining the NFA N_R for a regular expression R. The undirected version of N_R is a series-parallel graph.

no transition enters q_0 and no transition leaves q_f. Finally observe that N_R has at most $O(n)$ transitions for any regular expression R of length n.

Let T_R be the **expression tree** of R with root r. We say that C is a partial cut iff for no pair (x, y) of different vertices in C is x a descendant of y. We define $T_R^v(C)$, for a vertex v and a partial cut C of T_R, as the subtree of T_R with root v, where we remove all children-links for vertices in C. Hence vertices in C which are also descendants of v are (artificial) leaves in $T_R^v(C)$. Moreover we label the leaf for $x \in C$ with an artificial symbol denoting the regular subexpression determined by x in T_R. $R^v(C)$ denotes the subexpression specified by $T_R^v(C)$; in particular $R^v(C)$ contains for each artificial leaf the corresponding artificial symbol. Finally $N_R^v(C)$ is the NFA obtained by recursively expanding $R^v(C)$ except for the artificial symbols of $R^v(C)$ which correspond to artificial transitions.

We introduce the **decomposition tree** T_R^* for R as a balanced, small depth version of T_R. We begin by determining a **separating vertex** v of T_R, namely a vertex of T_R with a subtree of at least $\frac{n}{3}$, but less than $\frac{2n}{3}$ leaves. Then $T_1 = T_R^v(\emptyset)$ is the subtree of T_R with root v and T_1 determines the regular expression $S = R^v(\emptyset)$. We remove the edge connecting v to its parent, reattach v as an artificial leaf labeled with the artificial symbol S and obtain the second subtree $T_2 = T_R^r(\{v\})$. T_2 specifies the regular expression $R^r(\{v\})$ and we obtain the original expression R after replacing the artificial symbol S in $R^r(\{v\})$ by the expression $R^v(\emptyset)$. $N_R^r(\{v\})$ contains a unique transition $q_1 \xrightarrow{S} q_2$ with label S. We obtain N_R from $N_R^r(\{v\})$ after identifying the unique initial and final states of $N_R^v(\emptyset)$ with q_1 and q_2 respectively and then replacing the transition $q_1 \xrightarrow{S} q_2$ by $N_R^v(\emptyset)$.

To define the decomposition tree T_R^* we create a root s. We say that $q_1 \xrightarrow{S} q_2$ is the **artificial transition of s** and label s with the quadruple $(r, \emptyset, q_1^s, q_2^s)$ where $q_1^s = q_1$ and $q_2^s = q_2$. (In general, if we label a vertex t of T_R^* with (u, C, q_1^t, q_2^t), then t represents the expression tree $T_R^u(C)$. q_1^t, q_2^t are the endpoints of the artificial transition of t.) We introduce a left child representing $T_R^v(\emptyset)$ as well as a right child representing $T_R^r(\{v\})$. We recursively repeat this expansion process for each child of r. However from now on separating vertices have to have at least $\frac{N}{3}$, but less than $\frac{2N}{3}$ *original* (i.e., non-artificial) leaves in their subtrees, where N is the current number of original leaves. We continue to expand until all trees contain at most one original leaf. If we have reached a vertex t of T_R^* whose expression tree $T_R^u(C)$ has exactly one original leaf representing the original transition $q_1^t \xrightarrow{a} q_2^t$, then we label t by the quadruple (u, C, q_1^t, q_2^t). We summarize the important properties of T_R^*.

Proposition 1. *Let R be a regular expression of length n.*

(a) *Each ε-free transition of N_R appears exactly once as an artificial transition of a leaf of T_R^*.*
(b) *A left child in T_R^* represents a subtree of T_R.*
(c) *Let $p \xrightarrow{a} q$ and $r \xrightarrow{b} s$ be two transitions of N_R corresponding to leaves l_1 and l_2 respectively. If t is the lowest common ancestor of l_1 and l_2 and if there is an ε-path $q \xrightarrow{*} r$ in N_R, then the path traverses an endpoint of the artificial transition of t.*
(d) *The depth of T_R^* is bounded by $O(\log_2 n)$.*

Proof. (a) Any ε-free transition of N_R appears exactly once as a leaf of the expression tree T_R. The claim follows, since each expansion step for T_R^* decomposes T_R. (b) Assume that t is a left child and let v be the separating vertex of its parent. Then t represents $T_R^v(\emptyset)$, the subtree of T_R with root v. (c) The lowest common ancestor t represents the expression $R^u(C)$ for some vertex u and a cut C. If w is the separating vertex of t, then $R^u(C)$ is decomposed into the expressions $R^w(C)$, recognized by $N_R^w(C)$, and $R^u(C \cup \{w\})$, recognized by $N_R^u(C \cup \{w\})$. Both endpoints of, say, $p \xrightarrow{a} q$ belong to $N_R^w(C)$, the endpoints of $r \xrightarrow{b} s$ lie outside. But $N_R^w(C)$ can only be entered through its initial or final state which coincides with an endpoint of the artificial transition of t. (d) follows, since the number of original leaves is reduced each time by at least $\frac{2}{3}$.   □

2.1   Constructing ε-Free NFAs from the Decomposition Tree

We use ideas from [7] to convert a regular expression of length n over an alphabet of size k into an ε-free NFA with $O(n \cdot \log_2 n \cdot \log_2 2k)$ transitions. Thus we improve upon the conversion of [2], where $O(n \cdot \log_2 n \cdot k)$ transitions are shown to suffice.

Lemma 1. *Let R be a regular expression of length n over an alphabet of size k. Then there is an ε-free NFA N for R with $O(n \cdot \log_2 n \cdot \log_2 2k)$ transitions. N has a unique initial state. If $\varepsilon \notin L(R)$, then N has one final state and otherwise N has at most two final states.*

Proof. Assume that q_0 is the unique initial state of N_R and q_f its unique final state. We moreover assume that all states of N_R have an ε-loop. We choose q_0 and q_f as well as all endpoints of artificial transitions of T_R^* as states for our ε-free NFA. q_0 is still the initial state and q_f is (jointly with q_0, whenever $\varepsilon \in L(R)$) the only final state. Thus the ε-free NFA results from N_R after removing all ε-transitions (and all states which are incident to ε-transitions only) and inserting new ε-free transitions.

Let $p \xrightarrow{a} q$ be a transition of N_R and let l be the corresponding leaf of T_R^*. We define the set A to contain q_0, q_f as well as all ancestors of l including l itself. We assume that q_0 and q_f are roots of imaginary trees such that *all* leaves of

T_R^* belong to the "right subtree" of q_0 as well as to the "left subtree" of q_f. Consider any two vertices $s, t \in A$ and let q_1^s, q_2^s and q_1^t, q_2^t be the endpoints of the artificial transitions for s and t respectively. (We set $q_1^s = q_2^s = q_0$ for $s = q_0$ and $q_1^t = q_2^t = q_f$ for $t = q_f$.) We interconnect states for $i, j \in \{1, 2\}$ as follows.

Insert the transition $\mathbf{q_i^s} \xrightarrow{\mathbf{a}} \mathbf{q_j^t}$, if there are ε-paths $q_i^s \xrightarrow{*} p$ and $q \xrightarrow{*} q_j^t$ in N_R.

Let N be the NFA obtained from N_R after these insertions and after removing all ε-transitions. Obviously any accepting path from q_0 to q_f in N can be extended by ε-transitions to an accepting path in N_R and hence $L(N) \subseteq L(N_R)$. Now consider an accepting path

$$q_0 \xrightarrow{\varepsilon} \cdots \xrightarrow{\varepsilon} p_1 \xrightarrow{a_1} q_1 \xrightarrow{\varepsilon} \cdots \xrightarrow{\varepsilon} p_r \xrightarrow{a_r} q_r \xrightarrow{\varepsilon} \cdots \xrightarrow{\varepsilon} q_f$$

for the word $a_1 \circ \cdots \circ a_r$ in N_R. Since all states of N_R have ε-loops, we may assume that all ε-free transitions are separated by non-trivial ε-paths. Let l_i be the leaf of T_R^* corresponding to the transition $p_i \xrightarrow{a_i} q_i$. To obtain an accepting path in N, let $v_0 = q_0, v_1, \ldots, v_{r-1}, v_r = q_f$ be the sequence of vertices, where v_i ($1 \leq i \leq r - 1$) is the lowest common ancestor of l_i and l_{i+1} in T_R^*. By Proposition 1(c) the ε-paths from q_{j-1} to p_j and from q_j to p_{j+1} have to hit endpoints $q_{i_{j-1}}^{v_{j-1}} \in \{q_1^{v_{j-1}}, q_2^{v_{j-1}}\}$ and $q_{i_j}^{v_j} \in \{q_1^{v_j}, q_2^{v_j}\}$ of the respective artificial transition. But then N contains the transition $q_{i_{j-1}}^{v_{j-1}} \xrightarrow{a_j} q_{i_j}^{v_j}$ and

$$q_0 \xrightarrow{a_1} q_{i_1}^{v_1} \xrightarrow{a_2} \cdots \xrightarrow{a_{j-1}} q_{i_{j-1}}^{v_{j-1}} \xrightarrow{a_j} q_{i_j}^{v_j} \xrightarrow{a_{j+1}} q_{i_{j+1}}^{v_{j+1}} \xrightarrow{a_{j+2}} \cdots \xrightarrow{a_r} q_f$$

is an accepting path in N. Hence $L(N_R) \subseteq L(N)$ and N and N_R are equivalent.

We still have to count the number of transitions of N. We introduce transitions $q_i^s \xrightarrow{a} q_j^t$ (resp. $q_i^t \xrightarrow{a} q_j^s$) only for transitions $p \xrightarrow{a} q$ which are represented by leaves belonging to the subtrees of s and t. Hence s must be an ancestor or a descendant of t in T_R^* and, for given vertices s, t, we introduce at most $\min\{|s|, |t|, k\}$ transitions, where $|s|$ and $|t|$ are the number of leaves in the subtrees of s and t respectively.

We fix s. There are $O(|s|/k)$ descendants t of s with $|t| \geq k$ and $O(|s|)$ transitions correspond to those "large" vertices t. The remaining small vertices are partitioned into $O(\log_2 k)$ levels, where one level produces $O(|s|)$ transitions. Thus the number of transitions between q_i^s and q_j^t for descendants t of s is bounded by $O(|s| \cdot (1 + \log_2 k))$ and hence by $O(|s| \cdot \log_2 2k)$. Finally we partition all vertices s of T_R^* into $O(\log_2 n)$ levels, where one level requires $O(n \cdot \log_2 2k)$ transitions, and overall $O(n \cdot \log_2 n \cdot \log_2 2k)$ transitions suffice. $\square$

2.2   A Recursive Construction of Small ε-Free NFAs

How can we come up with even smaller ε-free NFAs? Assume that we have partitioned the regular expression R into (very small) subexpressions of roughly same size η. We apply the construction of Lemma 1 to all subexpressions and introduce at most $O(n \cdot \log_2 \eta \cdot \log_2 2k)$ transitions, a significant reduction if η is drastically smaller than n. However now we have to connect different subexpressions with "global" transitions and Lemma 1 inserts the vast majority of transitions, leading to a total of $O(n \cdot \log_2 n \cdot \log_2 2k)$ transitions. But we can do far better, if we are willing to increase the size of ε-free NFAs for every subexpression S.

Definition 1. *Let N be an ε-free NFA with initial state q_0 and let F be the set of final states. We say that a transition (q_0, r) is an **initial transition** and that r is a **post-initial state**. Analogously a transition (r, s) is a **final transition**, provided $s \in F$, and r is a **pre-final state**.*

Observe that it suffices to connect a global transition for a subexpression S with a post-initial or pre-final state of an ε-free NFA for S. As a consequence, the number of global transitions for S is reduced drastically, provided we have only few post-initial and pre-final states. But, given an ε-free NFA, how large are equivalent ε-free NFAs with relatively few initial or final states?

Proposition 2. *Let N be an ε-free NFA with s transitions over an alphabet Σ of size k. Then there is an equivalent ε-free NFA N' with $O(k^2 + k \cdot s)$ transitions and at most $3k + k^2$ initial or final transitions. N' has one initial state. If $\varepsilon \notin L(N)$, then N' has one final state and otherwise at most two final states.*

Proof. Assume that q_0' is the initial state of N, F is the set of final states and $\Sigma = \{1, \ldots, k\}$. Let $\rho_1, \ldots, \rho_p$ be the post-initial states of N and $\sigma_1, \ldots, \sigma_q$ be the pre-final states of N. Moreover let R_i be the set of post-initial states in N receiving an i-transitions from q_0' and let S_i be the set of pre-final states of N sending an i-transitions into a state of F.

We introduce a new initial state q_0 and a new final state q_f. (q_0 is the second accepting state, if $\varepsilon \in L(N)$.) For every $a \in \Sigma \cap L(N)$ we insert the transition from $q_0 \xrightarrow{a} q_f$. We introduce new post-initial states $r_1, \ldots, r_k$, new pre-final states $s_1, \ldots, s_k$ and insert i-transitions from q_0 to r_i as well as from s_i to q_f.

If $\rho_j \in R_i$ and if (ρ_j, s) is a transition with label b, then insert the transition (r_i, s) with label b. Analogously, if $\sigma_j \in S_i$ and if (r, σ_j) is a transition with label b, then we insert the transition (r, s_i) with label b. Thus the new states r_i and s_i inherit their outgoing respectively incoming transitions from the states they "are responsible for". Finally, to accept all words of length two in $L(N)$, we introduce at most k^2 further initial and final transitions incident with q_0, q_f and post-initial states.

Observe that the new NFA N' is equivalent with N, since, after leaving the new states r_i and before reaching the new states s_j, N' works like N. The states q_0 and q_f are incident with at most $k + 2 \cdot k + k^2$ transitions: up to k transitions link q_0 and q_f, $2 \cdot k$ transitions connect q_0 and the r_i's (or the s_i's and q_f) and at most k^2 transitions accept words of length two. Finally at most ks transitions leave states r_i, not more than ks transitions enter states s_j and hence the number of transitions increases from s to at most $s \cdot (2k + 1) + 3 \cdot k + k^2$. $\qquad\square$

We now observe that the combination of Lemma 1 and Proposition 2 provides a significant savings for alphabets of small size.

Proof of Theorem 1. As in the proof of Lemma 1 we assume that all states of N_R have an ε-loop. We begin by growing the decomposition tree T_R^*. Assume that vertex u is labeled with the quadruple (v, C, q_1^u, q_2^u) and hence u represents the subtree $T_R^v(C)$ and the expression $R^v(C)$. If $T_R^v(C)$ has no original leaves, then we stop expansion and remove u. If $T_R^v(C)$ has exactly one original leaf l, then we perform one more expansion step with l as separating vertex and remove the right child. (Any left child of T_R^* represents a subtree of the expression tree T_R, see Proposition 1b). Let $T_R^*(u)$ be the subtree of T_R^* with root u.

2.2.1   The Initial Phase 0
We first consider all vertices u of T_R^* whose subtree $T_R^*(u)$ has at least L_0, but less than $3 \cdot L_0$ original leaves. (We set $L_0 = 2^k$). Then we *process* all left children of vertices on the rightmost path in $T_R^*(u)$ starting in root u. If $l(u)$ is such a left child, then it represents a subtree $T_{l(u)}$ of T_R. We apply the procedure of Lemma 1 to $T_{l(u)}$ and obtain an ε-free NFA $N_{l(u)}$ recognizing the expression defined by $T_{l(u)}$. According to Lemma 1 we insert $O(N \cdot \log_2 N \cdot \log_2 2k)$ transitions for regular expressions of length N and hence u triggers the insertion of at most $O(x_0)$ transitions, where

$$x_0 = L_0 \cdot \log_2 L_0 \cdot \log_2 2k.$$

2.2.2   Phase j
We consider all vertices u of T_R^* with at least L_j, but less than $3 \cdot L_j$ original leaves in their subtrees $T_R^*(u)$. (We require $L_{j-1} < L_j$ and fix L_j later). Again we *process* all left children of vertices on the rightmost path in $T_R^*(u)$ starting in root u. Pick any such left child $l(u)$. We build an ε-free NFA $N_{l(u)}$ from the ε-free NFAs N_w for those descendants w of $l(u)$ in $T_R^*(u)$ which we processed in the previous phase.

We call any such descendant w a $(j-1)$-descendant of $l(u)$. N_w is an ε-free NFA with $O(x_{j-1})$ transitions which recognizes the expression represented by w. We apply Proposition 2 to N_w and obtain an ε-free NFA N_w^* with at most $O(k^2)$ initial or final transitions and size bounded by $O(k \cdot x_{j-1})$. Observe that the same asymptotic bound holds for the entire chain of left children to which w belongs, since sizes decrease almost geometrically.

We utilize the few initial or final transitions to cheaply interconnect N_w^* with (endpoints of artificial transitions assigned to) ancestors of w within $T_R^*(u)$. We proceed as in Lemma 1, but now we are working with full-fledged ε-free NFAs instead of ε-free transitions. In particular we have to differentiate three cases, namely firstly the new case $\varepsilon \in L(N_w^*)$, then the original case considered in Lemma 1, namely $a \in L(N_w^*)$ for some letter a, and finally the second new case, namely that N_w^* contributes at least two subsequent (ε-free) transitions. Let q_1, q_2 be the unique initial and final states of N_w^*.

(0) Assume $\varepsilon \in L(N_w^*)$. This case establishes ε-paths and is of interest only for the two remaining cases.

(1) Assume $a \in L(N_w^*)$ for some letter a. If vertices t_1, t_2 of $T_R^*(u)$ are ancestors of w and if there are ε-paths $q_i^{t_1} \xrightarrow{*} q_1$ and $q_2 \xrightarrow{*} q_j^{t_2}$ for $i, j \in \{1, 2\}$, then introduce the transition $q_i^{t_1} \xrightarrow{a} q_j^{t_2}$.

(2) Let $p \xrightarrow{a} q$ be an arbitrary initial transition of N_w^*. Then, for any ancestor t of w in $T_R^*(u)$, for any $i \in \{1, 2\}$ and for any ε-path $q_i^t \xrightarrow{*} p$ introduce the transition $q_i^t \xrightarrow{a} q$. Analogously, if $r \xrightarrow{a} s$ is an arbitrary final transition of N_w^* and if there is an ε-path $s \xrightarrow{*} q_i^t$, then introduce the transition $r \xrightarrow{a} q_i^t$.

Assume that $l(u)$ is labeled by the quadruple (v, C, q_1^u, q_2^u). The states of the ε-free NFA $N_{l(u)}$ are either states of some NFA N_w^*, for a $(j-1)$-descendant w of $l(u)$, or are endpoints of artificial transitions assigned to ancestors of w within $T_R^*(u)$. Transitions are defined by the transitions of $N^*(w)$ and by the links introduced above. q_1^u and q_2^u are the initial and final state respectively.

We claim that $N_{l(u)}$ is equivalent with the NFA $N_R^v(C)$ of $l(u)$. We insert a transition $p \xrightarrow{a} q$ into $N_{l(u)}$ only if there is a path $p \xrightarrow{*} r \xrightarrow{a} s \xrightarrow{*} q$ in N_R. Thus $L(N_{l(u)}) \subseteq L(N_R^v(C))$ follows and $L(N_R^v(C)) \subseteq L(N_{l(u)})$ remains to be shown.

Any accepting path $\mathcal{P}$ in $N_R^v(C)$ traverses sub-NFAs corresponding to some sequence $(w_1, \ldots, w_s)$ of $(j-1)$ descendants of $l(u)$, where we require that at least one letter is read for each w_i. We have assumed that each state of N_R is equipped with an ε-loop and hence we know that there are ε-paths from (an endpoint of the artificial transition of) w_i to (an endpoint of the artificial transition of) w_{i+1} in $N_R^v(C)$. Moreover, if s and t are the least common ancestors of w_{i-1} and w_i in $T_R^*(u)$, respectively of w_i and w_{i+1}, then there are ε-paths $s \xrightarrow{*} w_i$ and $w_i \xrightarrow{*} t$. If at least two letters are read for w_i then the corresponding accepting path $\mathcal{Q}$ in $N_{l(u)}$ runs from s to w_i and then to t; otherwise it runs from s directly to t.

2.2.3   Accounting

How many transitions did we introduce overall? To count transitions from class (1), observe that $T_R^*(l(u))$ has at most $O(\log_2 L_j)$ levels. At most $O(k \cdot L_j/L_{j-1})$ new transitions connect vertices of a fixed level of $T_R^*(l(u))$ with vertices from some descendant level above the $(j-1)$-descendants of $l(u)$. However the $O(L_j/L_{j-1})$ chains of left children contribute additionally up to $O(k \cdot \log_2 L_{j-1})$ transitions to the fixed ancestor level and hence the total number of transitions in the first class is bounded by $O(k \cdot L_j/L_{j-1} \cdot \log_2 L_{j-1} \cdot \log_2 L_j)$.

The transitions in class (2) connect one of the $O(k^2)$ post-initial and pre-final states of some N_w^* with endpoints of artificial transitions for at most $O(\log_2 L_j)$ ancestors within $T_R^*(l(u))$. Since the vertices w come from $O(L_j/L_{j-1})$ chains of left children, we have introduced at most $O(k^2 \cdot \frac{L_j}{L_{j-1}} \cdot \log_2 L_{j-1} \cdot \log_2 L_j)$ transitions from the second class. Observe that all $(j-1)$ descendants have at most $O(k \cdot \frac{L_j}{L_{j-1}} \cdot x_{j-1})$ transitions even if we include the blow-up due to Proposition 2, and hence $N_{l(u)}$ has at most $O(x_j)$ transitions, where

$$x_j = k \cdot \frac{L_j}{L_{j-1}} \cdot x_{j-1} + (k \cdot \frac{L_j}{L_{j-1}} \cdot \log_2 L_j + k^2 \cdot \frac{L_j}{L_{j-1}} \cdot \log_2 L_j) \cdot \log_2 L_{j-1}$$

$$\leq k \cdot \frac{L_j}{L_{j-1}} \cdot x_{j-1} + 2k^2 \cdot \frac{L_j}{L_{j-1}} \cdot \log_2 L_j \cdot \log_2 L_{j-1}. \tag{1}$$

We iterate recurrence (1) and get for $1 \le r \le j$

$$x_j \le k^r \cdot \frac{L_j}{L_{j-r}} \cdot x_{j-r} + \sum_{s=0}^{r-1} 2k^{s+2} \cdot \frac{L_j}{L_{j-1-s}} \cdot \log_2 L_{j-s} \cdot \log_2 L_{j-s-1}. \qquad (2)$$

Thus, if we assume that $n = L_i$ and set $j = i = r$ in (2), then we introduce at most $O(x_i)$ transitions, where

$$x_i = k^i \cdot \frac{n}{L_0} \cdot x_0 + \sum_{s=0}^{i-1} 2k^{s+2} \cdot \frac{n}{L_{i-s-1}} \cdot \log_2 L_{i-s} \cdot \log_2 L_{i-s-1}$$

The first term coincides with $O(k^i \cdot n \cdot \log_2 L_0 \cdot \log_2 2k)$, since $x_0 = L_0 \cdot \log_2 L_0 \cdot \log_2 2k$. We set $L_j = (2k)^{L_{j-1} \cdot \log_2 L_0/(k \cdot \log_2 L_{j-1})}$ and both terms in (2) are bounded by $O(k^i \cdot n \cdot \log_2 L_0 \cdot \log_2 2k) = O(k^{i+1} \cdot n \cdot \log_2 2k)$, since $L_0 = 2^k$. Thus $x_i = O(k^{i+1} \cdot n \cdot \log_2 2k)$ and $L_j = (2k)^{L_{j-1}/\log_2 L_{j-1}}$. $\qquad \square$

3   A Sketch of the Lower Bound

Our main result is the following lower bound for the regular expression

$$E_n = (1 + \varepsilon) \circ (2 + \varepsilon) \circ \cdots \circ (n + \varepsilon)$$

which improves upon the $\Omega(n \cdot \log_2^2 n/\log_2 \log_2 n)$ bound of [8].

Lemma 2. ε-free NFAs for E_n have at least $\Omega(n \cdot \log_2^2 n)$ transitions.

Before giving a proof sketch we show that Theorem 2 is a consequence of Lemma 2. We set $R_{n,k} = (E_k)^{n/k}$ and assume that $N_{n,k}$ is an ε-free NFA recognizing $R_{n,k}$. We say that a transition e of $N_{n,k}$ belongs to copy i iff e is traversed by an accepting path with label sequence $(1 \circ 2 \circ \cdots \circ k)^{i-1} \circ \sigma \circ (1 \circ 2 \circ \cdots \circ k)^{n/k-i}$ while reading $\sigma \ne \varepsilon$. Now assume that there is a transition e which belongs to two different copies i, j with $i < j$. Then we can construct an accepting path with label sequence $(1 \circ 2 \circ \cdots \circ k)^{j-1} \circ \tau \circ (1 \circ 2 \circ \cdots \circ k)^{n/k-i}$ and $N_{n,k}$ accepts a word outside of $R_{n,k}$. Thus any transition belongs to at most one copy. As a consequence of Lemma 2 $N_{n,k}$ has $\Omega(k \cdot \log_2^2 k)$ transitions for each copy and hence $N_{n,k}$ has at least $\Omega(\frac{n}{k} \cdot k \cdot \log_2^2 k)$ transitions. Observe that the unary regular expression 1^n requires NFAs of linear size and hence we actually get the lower bound $\Omega(n \cdot \log_2^2 2k)$.

A Proof Sketch of Lemma 2. Let $n = 2^k$. We define the ordered complete binary tree T_n with vertex set $\{1, \ldots, n-1\}$ and depth $k-1$. We assign names to vertices such that an inorder traversal of T_n produces the sequence $(1, \ldots, n-1)$. We label the root r of T_n with the set $L(r) = \{1, \ldots, n\}$. If vertex v is labeled with the set $L(v) = \{i + 1, \ldots, i + 2t\}$, then we label its left child v_l with $L(v_l) = \{i+1, \ldots, i+t\}$ and its right child v_r with $L(v_r) = \{i+t+1, \ldots, i+2t\}$. $L(v_l)$ coincides with the set of vertices in the left subtree of v including v and $L(v_r)$ coincides with the set of vertices in the right subtree including the lowest ancestor w with v in its left subtree. Finally define $|v| = |L(v)|$ as the size of v. Observe that $|v| = 2$ holds for every leaf v and we interpret its children v_l, resp. v_r as "virtual leaves".

Let N_n be an arbitrary ε-free NFA for E_n. Our basic approach follows the argument of [8]. In particular, we assume that N_n is in *normal form*, i.e., $\{0, 1, \ldots, n\}$ is the set of states of N_n and any transition $i \xrightarrow{l} j$ satisfies $i < l \leq j$. We say that vertex v of T_n is *crossed from the left* in N_n iff for all $i \in L(v_l)$ and all sequences σ with $\sigma \circ i \in E_n$ there is a path in N_n with label sequence $\sigma \circ i$ which ends in a state $y \in L(v_r)$; in particular the last transition of the path "crosses" v, since it ends in $L(v_r)$ and is labeled with a letter from $L(v_l)$. If all sequences $i \circ \tau \in E_n$ with arbitrary $i \in L(v_r)$ have a path which starts in some state $x \in L(v_l)$, then we say that v is crossed from the right.

Proposition 3. *[8] Let v be an arbitrary vertex of T_n. Then, for any ε-free NFA in normal form, v is crossed from the left or v is crossed from the right.*

Proof. Assume that v is not crossed from the left. Then there is a word $\sigma \circ i \in E_n$ with $i \in L(v_l)$ such that no path in N_n with label sequence $\sigma \circ i$ has a final transition crossing v. If v is also not crossed from the right, then there is a word $j \circ \tau \in E_n$ with $j \in L(v_r)$ such that no path in N_n with label sequence $j \circ \tau$ has an initial transition crossing v. But then N_n rejects $\sigma \circ i \circ j \circ \tau \in E_n$. $\square$

Let C be the set of vertices $v \in T_n$ which are crossed from the left. We assume that "more" vertices are crossed from the left and hence we concentrate on C. Assume that $w \in T_n$ belongs to C and that vertex v belongs to Left(w), the set of vertices of T_n which belong to the left subtree of w. Then any sequence $\sigma \circ j \in E_n$ with $j \in L(v_r)$, and hence $j \in L(w_l)$, has a path $p(\sigma, j)$ in N_n with label sequence $\sigma \circ j$ which ends in a state $y \in L(w_r)$. Observe that the last transition $e = (x, y)$ of $p(\sigma, j)$ identifies w as the unique tree vertex with $j \in L(w_l)$ and $y \in L(w_r)$. Moreover, if $x \in L(v_l)$, then e also identifies v as the unique tree vertex with $x \in L(v_l)$ and $j \in L(v_r)$.

For $h' < h \leq k-1$ define $N(h, h')$ as the number of pairs (i, w), where $w \in T_n$ has height h and i belongs to the right subtree of a vertex $v \in $ Left(w) with height h'. Then $N(h, h') = n/4$ and $\sum_{h' < h \leq k-1} N(h, h') = \Omega(n \cdot \log_2^2 n)$ holds. Hence it suffices to show that each pair (v, w) with $w \in C$ and $v \in $ Left(w) has $\Omega(|v_r|)$ transitions which identify v as well as w.

Labels $j \in L(v_r)$ are problematic if all j-transitions $e = (x, y)$ with $x \in L(v)$ and $y \in L(w_r)$ are *short* for (v, w), i.e., any such transition e starts in $x \in L(v_r)$. If label $j \in L(v_r)$ is short, then j-transitions "into" $L(w_r)$ depart close to j. Thus, if v_r has many short labels for (v, w), then many copies of a preceding i-transition, for $i \in L(u)$ and $u \in $ Left(v), are required to reach the many different starting points of j-transitions for short labels $j \in L(v_r)$.

To formalize this intuition we determine how far to the left short j-transitions extend, but not relative to transitions starting in $L(v)$ and ending in $L(w_r)$ for some specific w, but rather relative to a worst-case sequence $\tau = j \circ \sigma \circ k \in E_n$(with $k \in L(w_l)$ for an arbitrary $w \in C$) such that any path with sequence τ starts very close to j, if we require the path to start in $L(v)$ and to end in $L(w_r)$.

Definition 2. *(a) For vertices $v \in T_n$, $w \in C$ (with $v < w$) and labels $j \in L(v_r)$, $k \in L(w_l)$ define*

$$d_{v,w}(j,k) = \min_{\tau = j \circ \sigma \circ k \in E_n} \max_{x \in L(v), y \in L(w_r)} \{j - x \mid \exists \text{ path } x \xrightarrow{*} y \text{ with sequence } \tau\}.$$

(b) For a real number $s \geq 1/2$ let $X_s(v)$ be the set of all labels $j \in L(v_r)$ with $\min_{w \in C, k \in L(w_l)} d_{v,w}(j,k) \leq \frac{|v_r|}{s}$. Finally define $s(v)$ to be the maximal s with $|X_s(v)| \geq \frac{|v_r|}{2}$.

Thus $X_s(v)$ is the set of all letters $j \in L(v_r)$ for which distance at most $|v_r|/s$ between left endpoint and label j can be enforced. Moreover distance at most $|v_r|/s(v)$ is obtained for all labels in $X_{s(v)}(v)$ and hence for at least one half of all labels in $j \in L(v_r)$.

If v is crossed from the left and if u belongs to Left(v), then at least $\Omega(s(v))$ i-transitions with fixed label $i \in L(u)$ end in v_r, since all labels $j \in X_{s(v)}(v)$ have to be approached with distance at most $|v_r|/s(v)$. All in all $\Omega(|u| \cdot s(v))$ i-transitions for $i \in L(u)$ are required. Any such transition identifies v, however there will be some double counting, since the same i-transition may be counted for several vertices $u \in$ Left(v) with $i \in L(u)$. In particular we show

Lemma 3. N_n *has at least* $\Omega(\sum_{v \in C} \sum_{u \in \text{Left}(v)} |u| \cdot \frac{s(v)}{\log_2^2(4s(u))})$ *transitions.*

We omit the proof due to space limitations. Lemma 2 follows after some further accounting arguments.

4   Conclusions and Open Problems

We have shown that every regular expression R of length n over an alphabet of size k can be recognized by an ε-free NFA with $O(n \cdot \min\{\log_2 n, k^{L_k(n)+1}\} \cdot \log_2 2k)$ transitions. For alphabets of fixed size (i.e., $k = O(1)$) our result implies that $O(n \cdot 2^{O(\log_2^* n)})$ transitions and hence almost linear size suffice. We have also shown the lower bound $\Omega(n \cdot \log_2^2 2k)$ and hence the construction of [7] is optimal for large alphabets, i.e., if $k = n^{\Omega(1)}$.

A first important open question concerns the binary alphabet: do ε-free NFAs of linear size exist or is it possible to show a super-linear lower bound? Although we have considerably narrowed the gap between lower and upper bounds, the gap for alphabets of intermediate size, i.e., $\omega(1) = k = n^{o(1)}$ remains to be closed and this is the second important open problem. For instance, for $k = \log_2 n$ the lower bound $\Omega(n \cdot (\log_2 \log_2 n)^2)$ and the upper bound $O(n \cdot \log_2 n \cdot \log_2 \log_2 n)$ are still by a factor of $\log_2 n / \log_2 \log_2 n$ apart.

Thirdly the size blowup when converting an NFA into an equivalent ε-free NFA remains to be determined. In [6] a family N_n of NFAs is constructed which has equivalent ε-free NFAs of size $\Omega(n^2/\log_2^2 n)$ only. However the alphabet of N_n has size $n/\log_2 n$ and the upper bound is $O(n^2 \cdot |\Sigma|)$.

Acknowledgement. Thanks to Gregor Gramlich and Juraj Hromkovic for many helpful discussions and to the referees for many helpful comments.

References

1. R. Book, S. Even , S. Greibach, and G. Ott, Ambiguity in graphs and expressions, *IEEE Trans. Comput.* 20, pp. 149-153, 1971.
2. V. Geffert, Translation of binary regular expressions into nondeterministic ε-free automata with $O(n \log n)$ transitions, *JCSS* 66, pp. 451-472, 2003.
3. V.M. Glushkov, The abstract theory of automata, *Russian Math. Surveys* 16, pp. 1-53, 1961. Translation by J. M. Jackson from *Usp. Mat. Naut.* 16, pp. 3-41, 1961.
4. C. Hagenah and A. Muscholl, Computing ϵ-free NFA from regular expressions in $O(n \log^2(n))$ Time, *R.A.I.R.O. Theoret. Inform. Appl.*, 34, pp. 257-277, 2000.
5. J.E. Hopcroft, R. Motwani, J.D. Ullman, Introduction to Automata Theory, Languages and Computation, Addison-Wesley, 2001.
6. J. Hromkovič, G. Schnitger, NFAs with and without ε-transitions. ICALP 2005, pp. 385-396.
7. J. Hromkovič, S. Seibert, T. Wilke, Translating regular expression into small ε-free nondeterministic automata, *J. of Comput. and Syst. Sci.*, 62, pp. 565-588, 2001.
8. Y. Lifshits, A lower bound on the size of ε-free NFA corresponding to a regular expression, *Inf. Process. Lett.* 85(6), pp. 293-299, 2003.
9. M.O. Rabin and D.Scott, Finite automata and their decision problems, *IBM J. Res. Develop.* 3, pp. 114-125, 1959.
10. S. Sippu and E. Soisalon-Soininen, "Parsing Theory, Vol. I: Languages and Parsing", Springer-Verlag, 1988.
11. K. Thompson, Regular expression search, *Commun. ACM* 11, pp. 419-422, 1968.

Redundancy in Complete Sets*

Christian Glaßer[1], A. Pavan[2,**],
Alan L. Selman[3,***], and Liyu Zhang[3]

[1] Universität Würzburg
glasser@informatik.uni-wuerzburg.de
[2] Iowa State University
pavan@cs.iastate.edu
[3] University at Buffalo
{selman, lzhang7}@cse.buffalo.edu

Abstract. We show that a set is m-autoreducible if and only if it is m-mitotic. This solves a long standing open question in a surprising way. As a consequence of this unconditional result and recent work by Glaßer et al. [12], complete sets for all of the following complexity classes are m-mitotic: NP, coNP, $\oplus$P, PSPACE, and NEXP, as well as all levels of PH, MODPH, and the Boolean hierarchy over NP. In the cases of NP, PSPACE, NEXP, and PH, this at once answers several well-studied open questions. These results tell us that complete sets share a redundancy that was not known before. In particular, every NP-complete set A splits into two NP-complete sets A_1 and A_2.

We disprove the equivalence between autoreducibility and mitoticity for all polynomial-time-bounded reducibilities between 3-tt-reducibility and Turing-reducibility: There exists a sparse set in EXP that is polynomial-time 3-tt-autoreducible, but not weakly polynomial-time T-mitotic. In particular, polynomial-time T-autoreducibility does not imply polynomial-time weak T-mitoticity, which solves an open question by Buhrman and Torenvliet.

We generalize autoreducibility to define poly-autoreducibility and give evidence that NP-complete sets are poly-autoreducible.

1 Introduction

It is a well known observation that for many interesting complexity classes, all *known* complete sets contain "redundant" information. For example, consider SAT. Given a boolean formula ϕ one can produce two different formulas ϕ_1 and ϕ_2 such that the question of whether ϕ is satisfiable or not is equivalent to the question of whether ϕ_1 or ϕ_2 are satisfiable. Thus ϕ_1 and ϕ_2 contain information about ϕ. Another example is the Permanent. Given a matrix M, we can reduce the computation of the permanent of M to computing the permanent of $M + R$, $M + 2R, \ldots, M + nR$, where R is a randomly chosen matrix. Thus information

---

* A full version of this paper is available as ECCC Technical Report TR05-068.
** Research supported in part by NSF grants CCCF-0430807.
*** Research supported in part by NSF grant CCR-0307077.

about the permanent of M is contained in a few random looking matrices. We interpret this as "SAT and Permanent contain redundant information".

In this paper we study the question of how much redundancy is contained in complete sets of complexity classes. There are several ways to measure "redundancy". We focus on the two notions *autoreducibility* and *mitoticity*.

Trakhtenbrot [18] defined a set A to be *autoreducible* if there is an oracle Turing machine M such that $A = L(M^A)$ and M on input x never queries x. For complexity classes like NP and PSPACE refined measures are needed. In this spirit, Ambos-Spies [2] defined the notion of polynomial-time autoreducibility and the more restricted form m-autoreducibility. A set A is *polynomial-time autoreducible* if it is autoreducible via a oracle Turing machine that runs in polynomial-time. A is *m-autoreducible* if A is polynomial-time many-one reducible to A via a function f such that $f(x) \neq x$ for every x. Both notions demand information contained in $A(x)$ to be present among strings different from x. In the case of m-autoreducibility, the redundancy in A is even more apparent—if a set A is m-autoreducible, then x and $f(x)$ have the same information about A.

A stronger form of redundancy is described by the notion of *mitoticity* which was introduced by Ladner [15] for the recursive setting and by Ambos-Spies [2] for the polynomial-time setting. A set A is *m-mitotic* if there is a set $S \in$ P such that A, $A \cap S$, and $A \cap \overline{S}$ are polynomial-time many-one equivalent. Thus if a set is m-mitotic, then A can be split into two parts such that both parts have exactly the same information as the original set has.

Ambos-Spies [2] showed that if a set is m-mitotic, then it is m-autoreducible and he raised the question of whether the converse holds. In this paper we resolve this question and show that every m-autoreducible set is m-mitotic. This is our main result. Since its proof is very involved, we present our main idea with help of a simplified graph problem which will be described in Section 3. This simplification drops many of the important details from our formal proof, but still captures the spirt of the core problem. Our main result is all the more surprising, because it is known [2] that polynomial-time T-autoreducibility does not imply polynomial-time T-mitoticity. We improve this and disprove the equivalence between autoreducibility and mitoticity for all polynomial-time-bounded reducibilities between 3-tt-reducibility and Turing-reducibility: There exists a sparse set in EXP that is polynomial-time 3-tt-autoreducible, but not weakly polynomial-time T-mitotic. In particular, polynomial-time T-autoreducible does not imply polynomial-time weakly T-mitotic. This result settles another open question raised by Buhrman and Torenvliet [9].

Our main result relates local redundancy to global redundancy in the following sense. If a set A is m-autoreducible, then x and $f(x)$ contain the same information about A. This can be viewed as local redundancy. Whereas if A is m-mitotic, then A can be split into two sets B and C such that A, B, and C are polynomial-time many-one equivalent. Thus the sets B and C have exactly the same information as the original set A. This can be viewed as global redundancy in A. Our main result states that local redundancy is the same as global redundancy.

As a consequence of this result and recent work of Glaßer et al. [13, 12], we can show that all complete sets for many interesting classes such as NP, PSPACE, NEXP, and levels of PH are m-mitotic. Thus they all contain redundant information in a strong sense. This resolves several longstanding open questions raised by Ambos-Spies [2], Buhrman, Hoene, and Torenvliet [8], and Buhrman and Torenvliet [9].

Our result can also be viewed as a step towards understanding the isomorphism conjecture [5]. This conjecture states that all NP-complete sets are isomorphic to each other. In spite of several years of research, we do not have any concrete evidence either in support or against the isomorphism conjecture[1]. It is easy to see that if the isomorphism conjecture holds for classes such as NP, PSPACE, and EXP, then complete sets for these classes are m-autoreducible as well as m-mitotic. Given our current inability to make progress about the isomorphism conjecture, the next best thing we can hope for is to make progress on statements that the isomorphism conjecture implies. We note that this is not an entirely new approach. For example, if the isomorphism conjecture is true, then NP-complete sets cannot be sparse. This motivated researchers to consider the question of whether complete sets for NP can be sparse. This line of research led to the beautiful results of Mahaney [16] and Ogiwara and Watanabe [17] who showed that complete sets for NP cannot be sparse unless P = NP. Our results show that another consequence of isomorphism, namely "NP-complete sets are m-mitotic" holds. Note that this is an unconditional result.

Buhrman et al. [7] and Buhrman and Torenvliet [10, 11] argue that it is critical to study the notions of autoreducibility and mitoticity. They showed that resolving questions regarding autoreducibility of complete sets leads to unconditional separation results. For example, consider the question of whether truth-table complete sets for PSPACE are non-adaptive autoreducible. An affirmative answer separates NP from NL, while a negative answer separates the polynomial-time hierarchy from PSPACE. They argue that this approach does not have the curse of *relativization* and is worth pursuing. We refer the reader to the recent survey by Buhrman and Torenvliet [11] for more details.

In Section 4, we extend the notion of autoreducibility and define *poly-autored-ucibility*. A motivation for this is to understand the isomorphism conjecture and the notion of paddability. Recall that the isomorphism conjecture is true if and only if all NP-complete sets are paddable. Paddability implies the following: If L is paddable, then given x and a polynomial p, we can produce $p(|x|)$ distinct strings such that if x is in L, then all these strings are in L and if x is not in L, then none of these strings are in L. Autoreducibility implies that given x we can produce a *single* string y different from x such that $L(x) = L(y)$. A natural question that arises is whether we can produce more strings whose membership in L is the same as the membership of x in L. This leads us to the notion of f-autoreducibility: A set L is f-autoreducible, if there is a polynomial-time algorithm that on input x outputs $f(|x|)$ distinct strings (different from x) whose membership in L is the same as the membership of x in L. It is obvious

---

[1] It is currently believed that if one-way functions exist, then the isomorphism conjecture is false. However, we do not have a proof of this.

that paddability implies poly-autoreducibility. The question of whether "NP complete sets are poly-autoreducible" is weaker than the question of whether "NP-complete sets are paddable."

We provide evidence for poly-autoreducibility of NP-complete sets. We show that if one-way permutations exist, then NP-complete sets are log-autoreducible. Moreover, if one-way permutations and quick pseudo-random generators exist, then NP-complete sets are poly-autoreducible. We also show that if NP-complete sets are poly-autoreducible, then they have infinite subsets that can be decided in linear-exponential time. A complete version of this paper can be found at ECCC [14].

1.1 Previous Work

The question of whether complete sets for various classes are autoreducible has been studied extensively [19, 4, 7]. Beigel and Feigenbaum [4] showed that Turing complete sets for the classes that form the polynomial hierarchy, Σ_i^P, Π_i^P, and Δ_i^P, are Turing autoreducible. Thus, all Turing complete sets for NP are Turing autoreducible. Buhrman et al. [7] showed that Turing complete sets for EXP and Δ_i^{EXP} are autoreducible, whereas there exists a Turing complete set for EESPACE that is not Turing auto-reducible. Regarding NP, Buhrman et al. [7] showed that truth-table complete sets for NP are probabilistic truth-table autoreducible. Recently, Glaßer et al. [13, 12] showed that complete sets for classes such as NP, PSPACE, Σ_i^P are m-autoreducible.

Buhrman, Hoene, and Torenvliet [8] showed that EXP complete sets are weakly many-one mitotic. This result was recently improved independently by Kurtz [11] and Glaßer et al. [13, 12]. Buhrman and Torenvliet [11] observed that Kurtz' proof can be extended to show that 2-tt complete sets for EXP are 2-tt mitotic. This cannot be extended to 3-tt reductions: There exist 3-tt complete sets for EXP that are not btt-autoreducible and hence not btt-mitotic [6]. Glaßer et al. also showed that NEXP complete sets are weakly m-mitotic and PSPACE-complete sets are weak Turing-mitotic.

2 Preliminaries

We use standard notation and assume familiarity with standard resource-bounded reductions. We consider words in lexicographic order. All used reductions are polynomial-time computable.

Definition 1 ([2]). *A set A is polynomially T-autoreducible (T-autoreducible, for short) if there exists a polynomial-time-bounded oracle Turing machine M such that $A = L(M^A)$ and for all x, M on input x never queries x. A set A is polynomially m-autoreducible (m-autoreducible, for short) if $A \leq_m^p A$ via a reduction function f such that for all x, $f(x) \neq x$.*

Definition 2 ([2]). *A recursive set A is polynomial-time T-mitotic (T-mitotic, for short) if there exists a set $B \in P$ such that $A \equiv_T^p A \cap B \equiv_T^p A \cap \overline{B}$. A is polynomial-time m-mitotic (m-mitotic, for short) if there exists a set $B \in P$ such that $A \equiv_m^p A \cap B \equiv_m^p A \cap \overline{B}$.*

Definition 3 ([2]). *A recursive set A is* polynomial-time weakly T-mitotic *(weakly T-mitotic, for short) if there exist disjoint sets A_0 and A_1 such that $A_0 \cup A_1 = A$, and $A \equiv_T^p A_0 \equiv_T^p A_1$. A is* polynomial-time weakly m-mitotic *(weakly m-mitotic, for short) if there exist disjoint sets A_0 and A_1 such that $A_0 \cup A_1 = A$, and $A \equiv_m^p A_0 \equiv_m^p A_1$.*

Definition 4. *Let f be a function from $\mathbb{N}$ to $\mathbb{N}$. A set L is f-autoreducible, if there is a polynomial-time algorithm $\mathcal{A}$ that on input x outputs $y_1, y_2, \cdots, y_m$ such that $f(|x|) = m$, if $x \in L$, then $\{y_1, y_2, \cdots, y_m\} \subseteq L$, and if $x \notin L$, then $\{y_1, y_2, \cdots, y_m\} \cap L = \emptyset$. A set is* poly-autoreducible, *if it is n^k-autoreducible for every $k \geq 1$.*

A language is DTIME($T(n)$)-*complex* if L does not belong to DTIME($T(n)$) almost everywhere; that is, every Turing machine M that accepts L runs in time greater than $T(|x|)$, for all but finitely many words x. A language L is *immune* to a complexity class $\mathcal{C}$, or $\mathcal{C}$-*immune*, if L is infinite and no infinite subset of L belongs to $\mathcal{C}$. A language L is *bi-immune* to a complexity class $\mathcal{C}$, or $\mathcal{C}$-*bi-immune*, if both L and $\overline{L}$ are $\mathcal{C}$-immune. Balcázar and Schöning [3] proved that for every time-constructible function T, L is DTIME($T(n)$)-complex if and only if L is bi-immune to DTIME($T(n)$).

3   m-Autoreducibility Equals m-Mitoticity

It is easy to see that if a nontrivial language L is m-mitotic, then it is m-autoreducible. If L is m-mitotic, then there is a set $S \in P$ such that $L \cap S \leq_m^p L \cap \overline{S}$ via some f and $L \cap \overline{S} \leq_m^p L \cap S$ via some g. On input x, the m-autoreduction for L works as follows: If $x \in S$ and $f(x) \notin S$, then output $f(x)$. If $x \notin S$ and $g(x) \in S$, then output $g(x)$. Otherwise, output a fixed element from $\overline{L} - \{x\}$.

So m-mitoticity implies m-autoreducibility. The main result of this paper shows that surprisingly the converse holds true as well, i.e., m-mitoticity and m-autoreducibility are equivalent notions.

Theorem 1. *Let L be any set such that $|\overline{L}| \geq 2$. L is m-autoreducible if and only if L is m-mitotic.*

We mention the main ideas and the intuition behind the proof and describe the combinatorial core of the problem.

Assume that L is m-autoreducible via reduction function f. Given x, the repeated application of f yields a sequence of words $x, f(x), f(f(x)), \ldots$, which we call the trajectory of x. These trajectories either are infinite or end in a cycle of length at least 2. Note that as f is an autoreduction, $x \neq f(x)$.

At first glance it seems that m-mitoticity can be easily established by the following idea: In every trajectory, label the words at even positions with $+$ and all other words with $-$. Define S to be the set of strings whose label is $+$. With this 'definition' of S it seems that f reduces $L \cap S$ to $L \cap \overline{S}$ and $L \cap \overline{S}$ to $L \cap S$.

However, this labeling strategy has at least two problems. First, it is not clear that $S \in$ P; because given a string y, we have to compute the parity of the position of y in a trajectory. As trajectories can be of exponential length, this might take exponential time. The second and more fundamental problem is the following: The labeling generated above is *inconsistent* and not well defined. For example, let $f(x) = y$. To label y which trajectory should we use? The trajectory of x or the trajectory of y? If we use trajectory of x, y gets a label of $+$, whereas if we use the trajectory of y, then it gets a label of $-$. Thus S is not well defined and so this idea does not work. It fails because the labeling strategy is a global strategy. To label a string we have to consider all the trajectories in which x occurs. Every single x gives rise to a labeling of possibly infinitely many words, and these labelings may overlap in an inconsistent way.

We resolve this by using a *local labeling strategy*. More precisely, we compute a label for a given x just by looking at the neighboring values x, $f(x)$, and $f(f(x))$. It is immediately clear that such a strategy is well-defined and therefore defines a consistent labeling. We also should guarantee that this local strategy strictly alternates labels, i.e., x gets $+$ if and only if $f(x)$ gets $-$. Such an alternation of labels would help us to establish the m-mitoticity of L.

Thus our goal will be to find a local labeling strategy that has a nice alternation behavior. However, we settle for something less. Instead of requiring that the labels strictly alternate, we only require that given x, at least one of $f(x), f(f(x)), \cdots, f^m(x)$ gets a label that is different from the label of x, where m is polynomially bounded in the length of x. This suffices to show m-mitoticity.

The most difficult part in our proof is to show that there exists a local labeling strategy that has this weaker alternation property.

We now formulate the core underlying problem. To keep this proof sketch simpler, we make several assumptions and ignore several technical but important details. If we assume (for simplicity) that on strings $x \notin 1^*$ the autoreduction is length preserving such that $f(x) > x$, then we arrive at the following graph labeling problem.

Core Problem. Let G_n be a directed graph with 2^n vertices such that every string of length n is a vertex of G_n. Assume that 1^n is a sink, that nodes $u \neq 1^n$ have outdegree 1, and that $u < v$ for edges (u, v). For $u \neq 1^n$ let $s(u)$ denote u's unique successor, i.e., $s(u) = v$ if (u, v) is an edge. Find a strategy that labels each node with either $+$ or $-$ such that:

(i) Given a node u, its label can be computed in polynomial time in n.
(ii) There exists a polynomial p such that for every node u, at least one of the nodes $s(u), s(s(u)), \ldots, s^{p(n)}(u)$ gets a label that is different from the label of u.

We exhibit a labeling strategy with these properties. To define this labeling, we use the following *distance function*: $d(x, y) \stackrel{df}{=} \lfloor \log |y - x| \rfloor$ (our formal proof uses a variant of this function). The core problem is solved by the following local strategy.

```
0    // Strategy for labeling node x
1    let y = s(x) and z = s(y).
2    if d(x,y) > d(y,z) then output −
3    if d(x,y) < d(y,z) then output +
4    r := d(x,y)
5    output + iff ⌊x/2^{r+1}⌋ is even
```

Clearly, this labeling strategy satisfies condition (i). We give a sketch of the proof that it also satisfies condition (ii). Define $m = 5n$ and let $u_1, u_2, \ldots, u_m$ be a path in the graph. It suffices to show that not all the nodes $u_1, u_2, \ldots, u_m$ obtain the same label. Assume that this does not hold, say all these nodes get label $+$. So no output is made in line 2 and therefore, the distances $d(u_i, u_{i+1})$ do not decrease. Note that the distance function maps to natural numbers. If we have more than n increases, then the distance between u_{m-1} and u_m is bigger than n. Therefore, $u_m - u_{m-1} > 2^{n+1}$, which is impossible for words of length n. So along the path $u_1, u_2, \ldots, u_m$ there exist at least $m - n = 4n$ positions where the distance stays the same. By a pigeon hole argument there exist four consecutive such positions, i.e., nodes $v = u_i$, $w = u_{i+1}$, $x = u_{i+2}$, $y = u_{i+3}$, $z = u_{i+4}$ such that $d(v, w) = d(w, x) = d(x, y) = d(y, z)$. So for the inputs v, w, and x, we reach line 4 where the algorithm will assign $r = d(v, w)$. Observe that for all words w_1 and w_2, the value $d(w_1, w_2)$ allows an approximation of $w_2 - w_1$ up to a factor of 2. More precisely, $w - v$, $x - w$, and $y - x$ belong to the interval $[2^r, 2^{r+1})$. It is an easy observation that this implies that not all of the following values can have the same parity: $\lfloor v/2^{r+1} \rfloor$, $\lfloor w/2^{r+1} \rfloor$, and $\lfloor x/2^{r+1} \rfloor$. According to line 5, not all words v, w, and x obtain the same label. This is a contradiction which shows that not all the nodes $u_1, u_2, \ldots, u_m$ obtain the same label. This proves (ii) and solves the core of the labeling problem.

The labeling strategy allows the definition of a set $S \in P$ such that whenever we follow the trajectory of x for more than $5|x|$ steps, then we find at least one alternation between S and $\overline{S}$. This establishes m-mitoticity for L.

Call a set L *nontrivial* if $\|L\| \geq 2$ and $\|\overline{L}\| \geq 2$. We have the following corollaries of the main theorem.

Corollary 1. *Every nontrivial set that is many-one complete for one of the following complexity classes is m-mitotic.*

- NP, coNP, $\oplus$P, PSPACE, EXP, NEXP
- *any level of* PH, MODPH, *or the Boolean hierarchy over* NP

Proof. Glaßer et al. [12] showed that all many-one complete sets of the above classes are m-autoreducible. By Theorem 1, these sets are m-mitotic. □

Corollary 2. *A nontrivial set L is NP-complete if and only if L is the union of two disjoint P-separable NP-complete sets.*

So unions of disjoint P-separable NP-complete sets form exactly the class of NP-complete sets. What class is obtained when we drop P-separability? Does this class contain a set that is not NP-complete? In other words, is the union of disjoint NP-complete sets always NP-complete? We leave this as an open question.

Ambos-Spies [2] defined a set A to be ω-m-mitotic if for every $n \geq 2$ there exists a partition $(Q_1, \ldots, Q_n)$ of Σ^* such that each Q_i is polynomial-time decidable and the following sets are polynomial-time many-one equivalent: A, $A \cap Q_1, \ldots, A \cap Q_n$.

Corollary 3. *Every nontrivial infinite set that is many-one complete for a class mentioned in Corollary 1 is ω-m-mitotic.*

We note that the proof of the main theorem actually yields the following theorem.

Theorem 2. *Every 1-tt-autoreducible set is 1-tt-mitotic.*

The following theorem shows in a strong way that T-autoreducible does not imply weakly T-mitotic. Hence, our main theorem cannot be generalized.

Theorem 3. *There exists $L \in$ SPARSE $\cap$ EXP such that*

- *L is 3-tt-autoreducible, but*
- *L is not weakly T-mitotic.*

Thus there exist 3-tt-autoreducible sets that are not even T-mitotic, whereas every 1-tt-autoreducible set is 1-tt mitotic. We do not know what happens when we consider 2-tt reductions. Is every 2-tt-autoreducible set 2-tt-mitotic or does there exist a 2-tt-autoreducible set that is not 2-tt-mitotic? We leave this as an open question.

4 Poly-Autoreducibility

In this section we consider the question of whether NP-complete sets are f-autoreducible, for some growing function f.

Lemma 1. *Let L be an NP-complete language. For every polynomial $q(.)$ there is a polynomial-time algorithm $\mathcal{A}$ such that $\mathcal{A}$ on input x, $|x| = n$,*

- *either decides the membership of x in L*
- *or outputs strings $y_1, \cdots, y_m$ such that*
  - *$x \in L \Rightarrow \{y_1, y_2, \cdots, y_m\} \subseteq L$,*
  - *$x \notin L \Rightarrow \{y_1, y_2, \cdots, y_m\} \cap L = \emptyset$,*
  - *$m = q(n)$, and $x \neq y_1, \neq y_2 \neq \cdots \neq y_m$.*

This above lemma comes close to showing that NP-complete sets are poly-autoreducible, except for a small caveat. Let L be any NP-complete language. If the algorithm from Lemma 1 neither accepts x or rejects x, then it produces polynomially many equivalent strings. However, to show L is poly-autoreducible, we must produce polynomially-many equivalent strings even when the algorithm accepts or rejects.

This boils down to the following problem: Let L be an NP-complete language. Given 0^n as input, in polynomial time output polynomially many distinct strings such that all of them are in L. Similarly, output polynomially many distinct strings such that none of them are in L.

Below, we show that if one-way permutations exist, then we can achieve this task. We start with a result by Agrawal [1].

Definition 5. *Let f be a many-one reduction from A to B. We say f is $g(n)$-sparse, if for every n, no more than $g(n)$ strings of length n are mapped to a single string via f.*

Lemma 2. *([1]) If one-way permutations exist, then NP-complete sets are complete with respect reductions that are $2^n/2^{n^\gamma}$ sparse. Here γ is a fixed constant less than 1.*

Lemma 3. *Let L be NP-complete. If one-way permutations exist, then there exists a polynomial-time algorithm that on input 0^n outputs $\log n$ distinct strings in L and $\log n$ strings out of L.*

If we consider probabilistic algorithms, then we obtain a stronger consequence.

Lemma 4. *Let L be NP-complete. Assume one-way permutations exist. For every polynomial q, there exists a polynomial-time probabilistic algorithm $\mathcal{B}$ that on input 0^n outputs $q(n)$ distinct strings from L and $q(n)$ distinct strings from $\overline{L}$ with high probability.*

If we assume quick pseudo-random generators exist, then we can derandomize the above procedure.

Lemma 5. *Let L be any NP-complete language. If one-way permutations and quick pseudo-random generators exist, then for every polynomial $q(n)$, there is a polynomial-time algorithm that on input 0^n outputs $q(n)$ many distinct strings from L and $q(n)$ many distinct strings out of L.*

Combining Lemmas 1 and 3, we obtain the following result.

Theorem 4. *If one-way permutations exist, then every NP-complete language is $\log n$-autoreducible.*

Combining Lemmas 1 and 5, we obtain the following result.

Theorem 5. *If one-way permutations and quick pseudo-random generators exist, NP-complete sets are poly-autoreducible.*

Finally, we consider another hypothesis from which poly-autoreducibility of NP-complete sets follows.

Theorem 6. *If there exists a UP machine M that accepts 0^* such that no P-machine can compute infinitely many accepting computations of $M(0^n)$, then NP-complete sets are poly-autoreducible.*

Next we consider the possibility of an unconditional proof that NP-complete sets are poly-autoreducible. We relate this with the notion of immunity. We show that if NP-complete sets are poly-autoreducible, then they are not E-immune. It is known that NP-complete sets are not generic [12]. This proof is based on the fact that NP-complete sets are autoreducible. Genericity is stronger notion than immunity, i.e., if a language L is not immune, then it can not be generic. Our result says that improving the autoreducibility result for NP-complete sets gives a stronger consequence—namely they are not immune.

Theorem 7. *If every* NP*-complete set is poly-autoreducible, then no* NP*-complete set is* E*-immune.*

References

1. M. Agrawal. Pseudo-random generators and structure of complete degrees. In *17th Annual IEEE Conference on Computational Complexity*, pages 139–145, 2002.
2. K. Ambos-Spies. P-mitotic sets. In E. Börger, G. Hasenjäger, and D. Roding, editors, *Logic and Machines, Lecture Notes in Computer Science 177*, pages 1–23. Springer-Verlag, 1984.
3. J. Balcázar and U. Schöning. Bi-immune sets for complexity classes. *Mathematical Systems Theory*, 18(1):1–10, June 1985.
4. R. Beigel and J. Feigenbaum. On being incoherent without being very hard. *Computational Complexity*, 2:1–17, 1992.
5. L. Berman and J. Hartmanis. On isomorphism and density of NP and other complete sets. *SIAM Journal on Computing*, 6:305–322, 1977.
6. H. Buhrman, L. Fortnow, D. van Melkebeek, and L. Torenvliet. Separating complexity classes using autoreducibility. *SIAM Journal on Computing*, 29(5):1497–1520, 2000.
7. H. Buhrman, L. Fortnow, D. van Melkebeek, and L. Torenvliet. Using autoreducibility to separate complexity classes. *SIAM Journal on Computing*, 29(5):1497–1520, 2000.
8. H. Buhrman, A. Hoene, and L. Torenvliet. Splittings, robustness, and structure of complete sets. *SIAM Journal on Computing*, 27:637–653, 1998.
9. H. Buhrman and L. Torenvliet. On the structure of complete sets. In *Proceedings 9th Structure in Complexity Theory*, pages 118–133, 1994.
10. H. Buhrman and L. Torenvliet. Separating complexity classes using structural properties. In *Proceedings of the 19th IEEE Conference on Computational Complexity*, pages 130–138, 2004.
11. H. Buhrman and L. Torenvliet. A Post's program for complexity theory. *Bulleting of the EATCS*, 85:41–51, 2005.
12. C. Glaßer, M. Ogihara, A. Pavan, A. L. Selman, and L. Zhang. Autoreducibility, mitoticity, and immunity. In *Proceedings 30th International Symposium on Mathematical Foundations of Computer Science*, volume 3618 of *Lecture Notes in Computer Science*, pages 387–398. Springer-Verlag, 2005.
13. C. Glaßer, M. Ogihara, A. Pavan, A. L. Selman, and L. Zhang. Autoreducibility, mitoticity, and immunity. Technical Report TR05-11, ECCC, 2005.
14. C. Glaßer, A. Pavan, A. L. Selman, and L. Zhang. Redundancy in complete sets. Technical Report 05-068, Electronic Colloquium on Computational Complexity (ECCC), 2005.
15. R. Ladner. Mitotic recursively enumerable sets. *Journal of Symbolic Logic*, 38(2):199–211, 1973.
16. S. Mahaney. Sparse complete sets for NP: Solution of a conjecture of Berman and Hartmanis. *Journal of Computer and Systems Sciences*, 25(2):130–143, 1982.

17. M. Ogiwara and O. Watanabe. On polynomial-time bounded truth-table reducibility of NP sets to sparse sets. *SIAM Journal of Computing*, 20(3):471–483, 1991.
18. B. Trakhtenbrot. On autoreducibility. *Dokl. Akad. Nauk SSSR*, 192, 1970. Translation in Soviet Math. Dokl. 11: 814– 817, 1970.
19. A. Yao. Coherent functions and program checkers. In *Proceedings of the 22n Annual Symposium on Theory of Computing*, pages 89–94, 1990.

Sparse Selfreducible Sets and Polynomial Size Circuit Lower Bounds

Harry Buhrman[1,2], Leen Torenvliet[2], and Falk Unger[1]

[1] CWI Amsterdam
[2] Universiteit van Amsterdam

Abstract. It is well-known that the class of sets that can be computed by polynomial size circuits is equal to the class of sets that are polynomial time reducible to a sparse set. It is widely believed, but unfortunately up to now unproven, that there are sets in EXP^{NP}, or even in EXP that are not computable by polynomial size circuits and hence are not reducible to a sparse set. In this paper we study this question in a more restricted setting: what is the computational complexity of sparse sets that are *selfreducible*? It follows from earlier work of Lozano and Toran [10] that EXP^{NP} does not have sparse selfreducible hard sets. We define a natural version of selfreduction, tree-selfreducibility, and show that NEXP does not have sparse tree-selfreducible hard sets. We also show that this result is optimal with respect to relativizing proofs, by exhibiting an oracle relative to which all of EXP is reducible to a sparse tree-selfreducible set. These lower bounds are corollaries of more general results about the computational complexity of sparse sets that are selfreducible, and can be interpreted as super-polynomial circuit lower bounds for NEXP.

Keywords: Computational Complexity, Sparseness, Selfreducibility.

1 Introduction

Finding techniques to separate complexity classes is one of the, if not *the*, main open problem in complexity theory. Our understanding towards solving problems like the P versus NP problem is very limited. Not only do we not know how to separate P from NP, we don't even know how to separate EXP^{NP} from the class of sets that have polynomial size circuits. Work on derandomization assumes much stronger separations than this, like for example that EXP requires exponential size circuits.

It is long known that the class of sets that have polynomial size circuits equals the class of sets that are polynomial time Turing reducible to a sparse set [11]. In this paper we address the question of whether EXP^{NP} and smaller classes are Turing reducible to a sparse set by restricting the sparse set to be selfreducible and even *tree selfreducible*. A set S is selfreducible if there exists a polynomial time machine that can decide membership of x in S by making queries to S that are smaller than x in some well defined way (see Definition 2.2). A set is tree-selfreducible if the underlying query graph of the selfreduction is a tree.

B. Durand and W. Thomas (Eds.): STACS 2006, LNCS 3884, pp. 455–468, 2006.
© Springer-Verlag Berlin Heidelberg 2006

We do not know of any examples of selfreductions that are not essentially tree-selfreductions. These restrictions on the sparse set can also be interpreted as restricted versions of polynomial size circuits.

It follows from work of Lozano and Toran [10] that EXP^{NP} does *not* have sparse selfreducible hard sets. We extend this result by showing that NEXP does not have sparse sets hard that are tree-selfreducible. This result is optimal with respect to relativizing proof techniques, since we also obtain a relativized world where EXP has a sparse tree-selfreducible hard set. These results imply super-polynomial lower bounds for NEXP with respect to a restricted class of circuits.

These lower bounds are consequences from more general results on the complexity of sparse selfreducible sets. Lozano and Toran showed that sparse selfreducible sets are in P^{NP}, we give a different proof of this result that allows us to generalize it to sets of smaller density. We further show that tree-selfreducible sparse sets are in $\text{P}^{\text{NP}[O(\log n)]}$, the class of languages that can be decided with logarithmically many queries to an NP oracle. It follows from this result, that NEXP does not have sparse tree-selfreducible hard sets. Connecting this with recent results of Fortnow et al. [6, 18] it follows that if EXP has a sparse tree-selfreducible hard set, then it is in NP/log. We next exhibit a relativized world where there exists a sparse 2-parity selfreducible set in $\text{P}^{\text{NP}[O(\log n)]}$ that is not in any lower complexity class, i.e., requires $O(\log n)$ queries to NP. This solves an open question from [10]. We also show a relativized world where there is a sparse Turing selfreducible set that is not truth-table selfreducible, and present some absolute results about the complexity of selfreducible sets that have sub-polynomial densities. Summarizing our results:

- Every sparse set that is tree-selfreducible can be computed in $\text{P}^{\text{NP}[O(\log n)]}$. This allows us to prove that NEXP does not have sparse tree-selfreducible hard sets. On the other hand we show a relativized world where EXP *does* have tree-selfreducible sparse hard sets.
- We construct a relativized world where there exists a sparse selfreducible set in $\text{P}^{\text{NP}[\log n]}$, that can not be computed with fewer queries to NP. This partially answers an open question from [10].
- Every log-sparse selfreducible set is in $\text{P}^{\text{NP}[O(\log n)^2]}$, and every log-sparse btt-selfreducible set is in P.

2   Definitions and Notation

We assume the reader to be familiar with standard complexity theory notation, as for example in [15]. Let $\Sigma = \{0, 1\}$. We write λ for the empty word. For a set $A \subseteq \Sigma^*$, let $A^{=n}$ be the set of strings from A of length n and $A^{\leq n} = \bigcup_{i=0}^{n} A^{=i}$. Note that $\Sigma^n = (\Sigma^*)^{=n}$ by this notation. The empty string will be denoted by λ. Pairing functions will be denoted by $\langle ., . \rangle$ and concatenation of strings x and y by xy. Implicitly using a standard mapping between numbers and strings in binary, we will use numbers as arguments to functions where strings are required and vice versa.

Definition 2.1. *A partial order $\prec$ on Σ^* is called* polynomially related *if and only if there exists a k such that for all $x, y \in \Sigma^*$*

1. $y \prec x \rightarrow |y| \leq |x|^k$
2. $x \prec y$ *is decidable in time* $(|x| + |y|)^k$
3. *Every descending chain starting with x has length at most $|x|^k$.*

Let $\prec$ be polynomially related. The Directed Acyclic Graph that represents the weak initial segment dominated by x, i.e., $\{y \mid y \in \Sigma^* \wedge y \prec x\} \cup \{x\}$ is denoted by S_x. The nodes of S_x are named by the strings in this weak initial segment, and the edges run between strings that are related by the ordering. In this paper we will happily make use of type-conflicting notations, like $Y \subseteq S_x$ where Y is a set of strings and S_x is the graph just defined. Here we mean that the strings from Y are nodes in S_x. First we define selfreducibility for some ordering $\prec$.

Definition 2.2. *Let r be some reduction type. A set $S \subseteq \Sigma^*$ is called $\leq_r^P$-selfreducible with respect to the polynomially related ordering $\prec$ on Σ^* if and only if*

1. $S \leq_r^P S$ *and*
2. *For any input $x \in \Sigma^*$ the reduction queries only elements y with $y \prec x$*

An example of a selfreducible set is SAT, the set of satisfiable boolean formulas. There exists a well-known $\leq_{2d}^P$-selfreduction for SAT, which is even length decreasing, where queries are formed by assigning values to variables.

We will say that a reduction M that witnesses the selfreducibility of S *obeys* $\prec$. We will denote the set of strings that is queried by oracle machine M on input x, the *query set* of M on input x, by $Q_M(x)$. If M is a non-adaptive machine then this query set is independent of the oracle. If M is an adaptive machine then this notation is sometimes enriched with the oracle, e.g., $Q_M^A(x)$, or, if the oracle is left out, the set of all *potential* queries is meant by this notation (of exponential size for polynomial time bounded oracle machines, but sometimes even bigger). This notation can also be used to denote an even bigger set. If V is a set of strings, then $Q_M(V) = \bigcup_{v \in V} Q_M(v)$.

We now define a very natural extension of self-reductions, see further below for some comments.

Definition 2.3. *Let S be a self-reducible set, witnessed by the deterministic polynomial time oracle machine M, which obeys the ordering $\prec$.*

For a string x define the graph G as follows:

1. *The nodes of G are all strings y with $y \prec x$.*
2. *for $y, z \in G$, there is a edge from y to z if and only if $z \in Q_M^O(y)$ for some oracle O.*

Let S_x^M be the (connected) component of G that contains x. We say that M is a tree-selfreduction *if for all x, S_x^M is always a tree.*

If $L(M^S) = S$ for some tree-selfreduction M, then this S is tree-selfreducible.

Note that S_x^M contains all strings y which could be possibly queried in the self-reduction of x, no matter which (possibly wrong) answers M gets.

Definition 2.4. *Let* $\prec$ *be a polynomially related order and let* $T_x \subseteq S_x$ *be a tree that has root* x. *For* $y \in T_x$ *we define the* depth *of* y *as* $d_x(y) = \#nodes$ *on the path from* x *to* y *in* T_x. *Consider a labeling* $l : T_x \mapsto \{0,1\}$. *Let* T_D *be the set of nodes from* $l^{-1}(1)$ *that are minimal w.r.t.* $\prec$, *i.e.,* $T_D = \{y \in l^{-1}(1)|$ *there is no* $z \in l^{-1}(1) - \{y\}$ *such that* y *is on the path from* z *to the root* $x\}$. *We define the* weight *of* T_x *as* $weight(T_x) = \sum_{y \in T_D} d_x(y)$.

Definition 2.5. *Let* M *be a selfreduction obeying* $\prec$ *and* T *some set. Consider a labeling* $l : T \mapsto \{0,1\}$ *of* T. *We call* l consistent *with* M, *or* M-consistent, *if and only if* $(\forall y \in T)[l(y) = 1 \Leftrightarrow M(y)$ *accepts when queries of* $M(y)$ *in* T *are answered according to* l *and queries outside* T *are answered NO]*.

Note that for each set T there is a unique M-consistent labeling for T, which can be easily found in a bottom-up fashion.

We want to mention that all selfreducible sets we know of can be made (or even are) tree-selfreducible. Take SAT as a simple example. Consider the standard reduction which on input $\phi(x_1, \ldots, x_n)$ with $n > 0$ queries $\phi(0, x_2, \ldots, x_n)$ and $\phi(1, x_2, \ldots, x_n)$, but does not simplify the terms, and accepts iff one of the queries is true. For $n = 0$ it outputs the truth value of ϕ. Then this reduction is obviously a tree-selfreduction.

We call a set $S \subseteq \Sigma^*$ *sparse* if and only if $\|S^{\leq n}\| \in O(Pol(n))$. S is called *log-sparse* if and only if $\|S^{\leq n}\| \in O(\log(n))$.

It is well-known that the class $P^{NP[O(\log n)]}$ (P-machines that can make $O(\log n)$ adaptive queries to an NP-oracle) is equivalent to the class $P^{NP_\|}$ (P-machines that can only make non-adaptive queries to an NP-oracle), see [19]. This class is commonly referred to as Θ_2^P. The class P^{NP} is commonly referred to as Δ_2^P.

3   Sparse Selfreducible Sets

We start by citing a result from [10].

Theorem 3.1. *If* $S \subseteq \Sigma^*$ *is sparse and selfreducible then* S *is in* Δ_2^P.

We want to note that we will later state a theorem (Theorem 5.1), whose proof can be easily adapted to yield the same result, thus giving an alternative proof for Theorem 3.1.

An open question from [10] is whether Δ_2^P in Theorem 3.1 is optimal. We will first show in Theorem 3.3 that for the natural case of sparse, tree-selfreducible sets we can find a better bound than Δ_2^P, namely Θ_2^P. Later, in Corollary 3.6, we will show that this is probably tight by exhibiting a relativized world and a sparse, tree-selfreducible set S in which $S \in \Theta_2^P$ but S is not lower. This will follow from Theorem 3.4 and Lemma 3.5. We will first isolate and prove the crucial lemma.

Lemma 3.2. *Let M be a tree-selfreduction obeying $\prec$ and witnessing the self-reducibility of some set S and let x be some input. Let $T \subseteq S_x^M$ be a tree with root x which has maximal weight among all trees $T \subseteq S_x^M$, which can be labeled M-consistently. Then it holds that $S \cap S_x^M \subseteq T$.[1]*

Proof. Let l be the M-consistent labeling of T. Assume that $(S \cap S_x^M) - T$ is nonempty. Let y be the deepest node in $(S \cap S_x^M) - T$. Note, that this implies that y has no children in S. Let p be the (unique) path from y to the root x in S_x^M. Let $p = p_{out}p_{in}$ such that p_{out} is outside of T and p_{in} is inside. Note that p_{out} contains at least y. Let T' be the same as T, but with path p_{out} added and let l' be the (unique) M-consistent labeling of T'. Note that $l'(y) = 1$, because y has no children in S. For nodes in T, labelings l and l' differ at most on the path p_{in}. But the total weight that p_{in} contributes to the weight of T can be at most $|p_{in}|$. But path p_{out} contributes $|p| > |p_{in}|$ to the weight of T', so the weight of T' is larger than that of T. □

Theorem 3.3. *If S is sparse and $\leq_T^P$-tree-selfreducible then $S \in \Theta_2^P$.*

Proof. Fix x and a selfreduction machine M. We will give a Θ_2^P-algorithm to compute $x \in S$. First find the maximum weight $w_{max}(x)$ of any M-consistent labeled T. It is clear that $w_{max}(x) \in O(Pol(|x|))$. Further, there is a $k > 0$ such that for any x there is a maximally weighted tree $T \subseteq S_x^M$ with $\|T\| \leq |x|^k$. We can find $w_{max}(x)$ with logarithmically many queries to an NP oracle of the following type: "Given a weight w, guess a tree T of size at most $|x|^k$, a labeling l such that $weight(T) \geq w$. Accept if the labeling of l is M-consistent."

Lemma 3.2 guarantees that any tree T of maximum weight will contain all nodes in $S_x^M \cap S$. Recall that there is only one M-consistent labeling for any T. But the true labeling of such maximally weighted T, i.e. $l(y) = 1 \leftrightarrow y \in S$, is of course M-consistent. So the (unique) M-consistent labeling of T labels all nodes correctly.

The final query will then be "Guess a tree T of size at most $|x|^k$ and an M-consistent labeling l such that $weight(T) = w_{max}(x)$. Accept iff $l(x) = 1$." Our Θ_2^P-algorithm then just outputs the result of this query. □

Let us now prove a relativized lower bound on sparse, tree-selfreducible sets, see Corollary 3.6. The proof follows easily from Theorem 3.4 and Lemma 3.5. In Theorem 3.4 we show that if there are NE-machines with a certain structural property, then one can easily derive an S as desired. In Lemma 3.5 we will then show that there is a relativized world in which NE-machines have this property.

Theorem 3.4. *Assume there is an NE-machine M and a set B such that*

1. *M has at most 2^n accepting paths for all inputs of length n*
2. *$x \in B$ if and only if the number of accepting paths of $M(x)$ is odd*
3. *$B \notin EXP^{NP[n]}$.*

Then there is a sparse, tree-selfreducible set with $S \in P^{NP[\log n+1]} - P^{NP[\log n]}$.

---

[1] Here S_x^M resp. T denote the nodes of the graphs $S_x^{S,M}$, T.

Proof. Define
$$S' := \{\langle x, Pad'(x)\rangle \mid x \in B\},$$
where $Pad'(x)$ and the pairing function are chosen such that $|\langle x, Pad'(x)\rangle| = 2^{|x|}$. First note that S' is sparse.

Conditions 1 and 2 suggest the following $\mathrm{P^{NP[\log n+1]}}$-algorithm A for S' on input $\langle x, y\rangle$: If $y \neq Pad'(x)$ reject. Set $n = |x|$ and $m = 2^n = |\langle x, Pad'(x)\rangle|$. Then $M(x)$ runs in time m and has at most m accepting paths. Therefore the number of accepting paths of $M(x)$ can be computed with $\log m + 1$ queries to a suitable NP-oracle. Accept if this number is odd, otherwise reject.

Let us now prove $S' \notin \mathrm{P^{NP[\log n]}}$. Assume there was a $\mathrm{P^{NP[\log n]}}$-algorithm A' which decides S'. Then the following $\mathrm{EXP^{NP[n]}}$-algorithm for B shows a contradiction to condition 3. On input x compute $\langle x, Pad'(x)\rangle$, which has length $m = 2^n$. Start A' on $\langle x, Pad'(x)\rangle$ which can by assumption decide $x \in B$ with $\log m = n$ queries to an NP-oracle.

Now define we define the selfreducible set S.
$$S := \{\langle x, Pad(x), v\rangle \mid \oplus\|\{w \mid vw \text{ is an accepting path of } M(x)\}\| = 1\},$$
where this time Pad and $\langle \cdot, \cdot, \cdot \rangle$ are chosen such that $\langle x, Pad(x), \lambda\rangle = 2^{|x|}$. We have
$$\chi_S(\langle x, Pad(x), v\rangle) = \chi_S(\langle x, Pad(x), v0\rangle) \oplus \chi_S(\langle x, Pad(x), v1\rangle),$$
which means that S is 2-parity-selfreducible (χ_S is the characteristic function of S). The fact that $S \in \mathrm{P^{NP[\log n+1]}} - \mathrm{P^{NP[\log n]}}$ follows immediately from the proof for S'. $\qquad\square$

Lemma 3.5. *There is an oracle O, an NE-machine M and a set B such that*

1. *M^O has at most 2^n accepting paths for all inputs of length n*
2. *$x \in B$ if and only if the number of accepting paths of $M^O(x)$ is odd*
3. *$B \notin \mathrm{EXP}^{O,\mathrm{NP}^O[n]}$*

Proof. First we define the NE-machine M. On input 0^n it non-deterministically guesses all paths y with $|y| = 2^n$ and accepts on path y iff $y \in O$. On inputs other than 0^n it always rejects. Each oracle O defines B uniquely by condition 2.

Now we use a diagonalization argument to construct O such that conditions 1 and 3 hold. For any oracle O let K_O be the standard linear time NP^O-complete set
$$\langle x, i, t\rangle \in K_O \leftrightarrow \text{the } i\text{-th NP}^O\text{-machine accepts } x \text{ after } \leq t \text{ steps.}$$
Let N^O be an NP^O-machine accepting K_O which runs in time $O(n)$ on inputs of length n. We will prove that $B \notin \mathrm{EXP}^{O,N^O[n]}$ and by our choice of N this is equivalent to condition 3.

Let $\{M_i\}_i$ be an enumeration of all exponential time bounded oracle machines such that M_k on inputs of length n runs in time 2^{n^k} and for any oracle O makes at most 2^{n^k} queries to O and at most n queries to N^O.

We now describe the k-th stage of the diagonalization. Set the function $m(k)$ to

$$m(k) = 2^{2^{m(k-1)}} + m(k-1) \tag{1}$$

where $m(0) = 2$. For ease of notation we also write m for $m(k)$. We will ensure that

$$M_k^{O,N^O}(0^m) \text{ accepts } \leftrightarrow 0^m \notin B. \tag{2}$$

As will be clear from the construction none of the later stages will change this property. This implies condition 3.

Initially set $F := \emptyset$. In each stage of the diagonalization we will add elements to F. These elements are "frozen" and are not allowed to be put into O later.

Let u be an m-bit string. Since M_k asks at most m queries to N^O, this string induces a computation of $M_k^{O,N^O}(0^m)$, if we define that the answer of the i-th query to N^O is given by the i-th bit of u. Let Q_m be the set of all possible queries of $M_k^{O,N^O}(0^m)$ to N^O, for any such u. Note that $|Q_m| \leq 1 + 2 + 4 + \ldots 2^{m-1} = 2^m - 1$. For any such u, freeze all direct queries from M_k to O in $M_k^{O,N^O}(0^m)$ and put them into F.

We now put some elements of length 2^m into O such that we get (2).

1. $Q := Q_m$
2. WHILE there is $w \in (\Sigma^{2^m} - F) \cup \{\lambda\}$ and $q \in Q$ such that $N^{O \cup \{w\}}(q) = 1$ DO

   (a) $Q := Q - \{q\}; O := O \cup \{w\}$
   (b) Add all queries on the left-most accepting path of $N^{O \cup \{w\}}(q)$ to F

3. IF $\left(M_k^{O,N^O}(0^m)\text{accepts and}|O^{=2^m}| = \text{odd}\right)$ or $\left(M_k^{O,N^O}(0^m)\text{rejects and} |O^{=2^m}| = \text{even}\right)$
   THEN take any $w \in \Sigma^{2^m} - F$ and set $O := O \cup \{w\}$

The idea behind this algorithm is very simple: In the WHILE-loop we try to find as many potential queries $q \in Q$ to N^O, for which $N^O(q)$ already accepts ($w = \lambda$) or $N^O(q)$ becomes accepting if we add one element w of length 2^m to O. We do not want to undo the acceptance of $N^O(q)$ in later iterations, so we "freeze" all queries on the left-most accepting path and put them into F.

Note that the WHILE-loop terminates after at most $|Q_m| \leq 2^m - 1$ iterations. Observe that after the completion of the WHILE-loop adding one of these unfrozen elemens of length 2^m to O also cannot change the acceptance of $N^O(q)$ for any of the remaining $q \in Q$. Since all direct queries from M_k to O in $M_k^{O,N^O}(0^m)$ were also frozen initially, the acceptance of $M_k^{O,N^O}(0^m)$ cannot change in step 3. On the other hand, adding a 2^m-long element to O adds an accepting path to $M^O(0^m)$, which changes the predicate $0^m \in B$. Thus, line 3 ensures (2) if we show that

Claim 1. *There is at least one unfrozen string of length 2^m at the beginning of line 3.*

Proof. We first observe that by (1) none of the 2^{2^m} strings of length 2^m were frozen in one of the previous stages. We then freeze at most $2^m \times 2^{m^k}$ direct queries of M_k to O and in each of the at most $2^m - 1$ iterations of the WHILE-loop at most 2^{m^k} strings, altogether less than 2^{m^k+m+1} strings. This is smaller than 2^{2^m} by (1). ☐

Now, condition 3 follows from (2), since in the k-th stage we add only elements to O, which by (1) are too long to be queried anywhere in any $M_l^{O,N^O}(0^{m_l})$ for $l < k$. Furthermore, our procedure adds at most 2^m elements of size 2^m to O and thus by construction of M condition 1 also holds. Condition 2 holds by definition. ☐

From Theorem 3.4 and Lemma 3.5 we get

Corollary 3.6. *There is an oracle O and a sparse set S, which is 2-parity-tree-selfreducible in the relativized world O, but $S \notin P^{O,NP^O[\log n]}$.*

4   Applications: Lower Bounds for NEXP

In this section we apply the results about sparse tree selfreducible sets to obtain lower bounds for NEXP. It is well-known that $P^{SPARSE} = P/poly$. However the question whether EXP^{NP} is contained in $P/poly$ is still open. The best known lower bound along these lines shows that MA_{exp}, the class of languages that allow for exponential long Arthur-Merlin games, is not in $P/poly$ [4].

We will show that Theorem 3.3 can be used directly to show that NEXP does not reduce to a sparse set that is tree selfreducible. The class of sets that reduce to sparse tree-selfreducible sets can be interpreted as sets that are computed by some restricted form of polynomial size circuits, and hence this result yields some lower bound for NEXP with respect to this class of polynomial size circuits. We will show moreover that there exists a relativized world where EXP has a sparse tree-selfreducible hard set.

Theorem 4.1. *Let K be a Turing complete set for NEXP. There is no sparse tree selfreducible set S such that $K \leq_T^p S$.*

Proof. This follows directly from [12], where it is shown that NEXP is not contained in $P^{NP[O(\log n)]}$ and Theorem 3.3. ☐

The same idea can be used to obtain a sub-exponential lower bound for the density of tree-selfreducible hard sets for $EXP^{NP_\parallel}$. Details will follow in the full version of the paper.

We next show that with relativizing techniques this is as far as one gets:

Theorem 4.2. *There exists an oracle A and a sparse tree-selfreducible set S such that for every set $B \in \text{EXP}^A$, $B \leq_T^{p^A} S$.*

Proof. It is sufficient to show that K^A, the standard 2^n-time complete set for EXP^A, reduces to S. The proof goes along the same lines as [20], where an oracle is constructed relative to which EXP is in P/poly.

For length n we will code for all the 2^n strings x_i of length n, $K^A(x_i)$ into A. Assume that we correctly coded all strings of length $\leq n - 1$ into A. Let M be such that $L(M^X) = K^X$ for all X. Since M^A runs in time 2^n it can query at most 2^n strings to A on any input x_i of length $\leq n$. Let $Q = Q_M^A((\Sigma^*)^{\leq n})$.

Then $\|Q\| \leq 2^{3n}$ and so $(\exists z_n \in \Sigma^{4n})(\forall v)[\langle z_n, v \rangle \notin Q]$. Now we are able to code for every string x_i of length n, $K^A(x_i)$ into A as follows.

$$\langle z_n, x_i \rangle \in A \leftrightarrow K^A(x_i) = 1$$

It is clear that the above construction will yield a z_n for every length n. We now will code z_n into a sparse tree selfreducible set S as follows:

$$S = \{\langle 0^n, v \rangle \mid \exists w : vw = z_n\}$$

It is easy to see that given access to S one can recover z_n and then decide for every string x of length n whether it is in K^A by querying $\langle z_n, x \rangle$. In order to make the set S tree-selfreducible we put $\langle z_n, \lambda \rangle$ in A as well. The selfreduction for S is now as follows: on input $\langle 0^n, v \rangle$ query whether $\langle 0^n, v0 \rangle$ or $\langle 0^n, v1 \rangle$ is in S, if $|v| < 4n$, otherwise decide $\langle 0^n, v \rangle$ for $|v| = 4n$ by querying whether $\langle z_n, \lambda \rangle \in A$. □

We don't know how to prove that EXP does have a sparse tree-selfreducible hard set, but we can connect this question to a recent line of research by Fortnow, Klivans, Shaltiel, and Umans [6, 18]: If EXP has a sparse tree selfreducible hard set then EXP $\in$ NP/log. From Theorem 3.3 we would have under this assumption that EXP $\in P^{NP[O(\log n)]}$ and by the results in [6] that EXP $\in$ NP/log.

5   Log-Sparse Selfreducible Sets

We will next prove that log-sparse selfreducible sets are in $L \in P^{NP[O(\log^2 n)]}$. The proof of the theorem can easily be adapted to yield Theorem 3.1, which was first proven in [10] with a different proof. The proof idea is the following. Given a log-sparse selfreducible set S and a string x, then $S_x \cap S$ has at most $O(\log|x|)$ elements. We will show a $P^{NP[O(\log^2 n)]}$ algorithm that recovers these elements in a "depth first" fashion. The structure S_x is now no longer a tree, since different paths can lead to the same element, but with the help of an NP oracle, the longest path to such a string can be recovered. The length of such a longest path is the *depth* of this string. Note that there can be different strings with the same depth in S_x, but if there are, then their longest paths from x split. Having recovered all elements in $S_x \cap L$ in this way, there can be only one string in this set of depth 0, namely x.

Theorem 5.1. *If $L \subseteq \Sigma^*$ is log-sparse and $\leq_T^P$-selfreducible then $L \in \mathrm{P}^{\mathrm{NP}[O(\log^2 n)]}$.*

Proof. Choose a constant c' such that $\|L \cap (\Sigma^*)^{\leq n}\| \leq c' \log n$. Let $\prec$ be the underlying polynomially related ordering and let M be a polynomial-time oracle machine witnessing the selfreduction. Fix some input x. Choose s such that $S_x \subseteq (\Sigma^*)^{\leq |x|^s}$. Now $\|S_x \cap L\| \leq c's \log n$. Let $c's = c$.

For $y \prec x$ let $d_x(y) = \max\{d \mid x \succ a_1 \succ \cdots \succ a_{d-1} \succ y\}$, where $d_x(x) = 0$. We call $d_x(y)$ the depth of y.

We define oracle O as $\langle x, d_1, \ldots, d_l \rangle \in O$ if and only if there are distinct strings $a_1, \ldots, a_l$ such that $(\forall 1 \leq i < j \leq l)[d_i < d_j \wedge a_i \prec x \wedge d_x(a_i) \geq d_i \wedge M^{\{a_1, \ldots, a_{i-1}\}}(a_i) = 1]$.

Clearly, $O \in \mathrm{NP}$. We now give a $\mathrm{P}^{\mathrm{NP}[\log^2 n]}$-algorithm that decides whether $x \in L$.

1. $i := 0$
2. WHILE $\langle x, d_1, \ldots, d_i, 0 \rangle \in O$ DO
   (a) $i := i + 1$
   (b) Use binary search to find the maximum value d_i such that $\langle x, d_1, \ldots, d_{i-1}, d_i \rangle \in O$
3. ACCEPT if $i > 0$ and $d_i = 0$; otherwise REJECT

The following claim is immediate.

Claim 2. *After the i-th iteration of line 2b, the algorithm has recovered $d_1, \ldots, d_i$ such that $(\forall y \in L \cap S_x)[(\exists j \leq i)[d_j = d(y)]$ or $d(y) < d_i]$.*

Claim 3. *The algorithm stops after at most $c \log |x|$ iterations.*

Proof. After $c \log |x|$ iterations, it has built a string of $c \log |x|$ values d_i, The query in step 2b requires $L \cap S_x$ to have $c \log |x|$ distinct strings that are accepted by M using this set of strings as an oracle. By Claim 2 these are the deepest strings in $L \cap S_x$ since the second part of the disjunct can no longer be true. Hence acceptance of M means that these strings are indeed in $L \cap S_x$. So after $\|S_x\| - 1 < c \log |x|$ iterations, the next query requires recovering *all* strings in $L \cap S_x$. Furthermore, there is at most one string of depth 0. □

The proof of the theorem is now completed by observing that the depth of any string in S_x is at most polynomial in $|x|$. Hence binary search can be performed in $O(\log |x|)$ steps. □

If in Theorem 5.1 we assume $\leq_{btt}^P$-selfreducibility then we get a stronger conclusion.

Theorem 5.2. *If L is log-sparse and $\leq_{btt}^P$-selfreducible then $L \in \mathrm{P}$.*

Proof. Let c be a constant such that $\|L^{\leq n}\| \leq c \log n$ and let s be a constant such that $(\forall x)[S_x \subseteq (\Sigma^*)^{\leq |x|^s}]$. This implies that $\|S_x \cap L\| \leq cs \log |x|$. Let M be an oracle machine that witnesses the $\leq_{btt}^P$ reduction and assume that M asks no more than b queries on any input. Because of the fact that queries are asked non-adaptively, we can limit the structure S_x to queries "of interest," i.e., we can assume that the set of nodes that are direct descendants of x is $Q_M(x)$, the set of nodes that are direct descendants of these nodes is $Q_M(Q_M(x))$ etc. This is what will make the algorithm below polynomial time bounded.

Of course, a string may be queried on different paths and therefore S_x is still a DAG. For a node y in S_x this time define the depth $d_x(y)$ as the *minimal* length of a path from x to y in S_x, where $d_x(x) = 0$. Let S_x^k be the part of S_x which contains all nodes up to depth k (inclusive). Set $level_x(k) = \{y \in S_x : d_x(y) = k\}$. Note that for $i \neq j$ it holds $level_x(i) \cap level_x(j) = \emptyset$. Nodes in level k of S_x^k can only have nodes in level k or $k - 1$ as ancestors. Therefore nodes in level k are the sinks in the DAG S_x^k.

Claim 4. *Consider a partial labeling* $l : S_x^k \mapsto \{0, 1\}$. *Call l correct if* $l(x) = 1 \Leftrightarrow x \in L$. *If all nodes in level k are labeled correctly, then S_x^k can be labeled correctly using M.*

Proof. With induction on the number of unlabeled nodes remaining. It is clear that if this number is 1, i.e., only x remains unlabeled, then we know the answer to the queries $Q_M(x)$, so we can label x correctly. If this number is m, then starting from x we can walk down a path to end up in a node y that has only (correctly labeled) sinks as descendants, i.e., we know the answers to the queries $Q_M(y)$ and therefore can label y correctly. The DAG $S_x^{k'}$ that is S_x^k with y additionally labeled has one less unlabeled node. □

Surprisingly, the fact that makes the proof complete is that, for large enough k, x can also be labeled correctly if some or all nodes in level k of S_x^k are labeled *incorrectly*. Therefore, the following algorithm decides whether $x \in L$.

1. FOR $k = 2cs \log|x|$ to $3cs \log|x|$ DO
   (a) Compute the DAG S_x^k.
   (b) Label all nodes in S_x^k as follows. Label all nodes in S_x^k of depth k with 0. Compute from that the labels of all nodes in S_x^k with lower depth using the selfreduction.
   (c) IF the root, i.e., x is labeled 1 and S_x^k does not contain more than $cs \log|x|$ 1-nodes THEN accept and HALT.
2. Reject.

Note that each S_x^k contains at most $\frac{b^{3cs \log|x|+1}-1}{b-1} \in O(Pol(|x|))$ nodes. Thus, this algorithm works in polynomial time. We now show that it is also correct.

Claim 5. *If $x \in L$ then A accepts.*

Proof. It is clear that (the correct) selfreduction-DAG S_x always contains a level k with $2cs \log|x| \leq k \leq 3cs \log|x|$ such that $level_x(k) \cap L = \emptyset$, because there can be at most $cs \log|x|$ elements from L in S_x. For such k A labels the nodes in $level_x(k)$ correctly. The claim now follows from Claim 4. □

Claim 6. *If A accepts then $x \in L$.*

Proof. Assume contrarily that A accepts an $x \notin L$, during iteration k. By Claim 4 this can only happen if $level_x(k)$ is not correctly labeled, which means:

$$level_x(k) \cap L \neq \emptyset.$$

Let S_x^k be labeled as given by step 1b. Since S_x^k contains $k \geq 2cs \log |x| + 1$ levels and A accepts x, the condition in 1c implies that S_x^k contains at least $cs \log |x| + 1$ levels whose nodes are all labeled with 0. Assume we now change the labels of the nodes in $level_x(k) \cap L$ (correctly) from 0 to 1 and compute from that the labels of all other levels in S_x^k. By Claim 4 S_k^x is then correctly labeled. We now want to prove that this cannot have an effect on the label of x.

Suppose it changes the label of the root x. Changing the label of a node y only has an effect if this changes the label of at least one of the parents of y and nodes in level i can only have parents in levels $\leq i-1$. So for each $i = k-1, \ldots, 0$ there must be at least one node in $level_x(i)$ which is changed. Thus, the changed S_x^k has at least one 1-node in each of the $\geq cs \log |x| + 1$ levels which before were completely labeled with 0. But this contradicts that S_x contains at most $cs \log |x|$ 1-nodes. □

This completes the proof of Theorem 5.2. □

The same proof idea also establishes that log-sparse sets L, which are $\leq_{tt}^P$-selfreducible can be decided in time $O(n^{\log n})$.

6   Conclusions

Selfreducible sets are all in PSPACE. Sparse sets can be of arbitrary complexity. The intersection of these classes turns out to be of considerably less computational complexity. Since selfreducibility is a property that many problems share and it is a crucial property that allows for recursive programs and divide and conquer strategies, it is interesting to investigate properties of selfreducible sets in different complexity classes and of different densities. Many open questions remain here, especially with respect to different forms of selfreducibility and the corresponding upper bounds on the computational complexity of problems. This paper is just a starting point that shows some interesting and sometimes unexpected cases.

Our results are also somewhat surprising with respect to structural properties of complexity classes. Sparse sets show, concerning their structural and computational properties, great resemblance to P-selective sets [7]. Sparse sets and P-selective sets are equivalent with respect to polynomial-time Turing reductions [16, 17]. Both P-selective sets [8] and Sparse sets have polynomial size circuits [11] and their difference in lowness (if any) is limited (see [9]). Both P-selective sets [1, 2, 13] and Sparse sets [14] have the property that if NP btt reduces to such a set, then P = NP. The situation becomes drastically different when we limit these classes to their selfreducible subclasses. Where the class of selfreducible P-selective sets is just another name for P [5], the computational complexity of the sparse selfreducible sets is quite a different matter as we have shown in this paper.

Some specific open problems are the following:

1. Does there exist a sparse selfreducible set that is not in $P^{NP[\log n]}$? This would yield a sparse selfreducible set that is not tree selfreducible.
2. Does there exist a relativized world where NEXP has a sparse selfreducible hard set?
3. Prove that EXP does not have a sparse tree selfreducible hard set. This proof needs to be non-relativizing, but it may be within reach using non-relativizing techniques from for example the MIP = NEXP proof.
4. Can the super polynomial lower bounds for NEXP be used to prove some kind of derandomization result? Is the selfreducibility restriction on the sparse set a real restriction or could one show that if NEXP has a sparse hard set then there also is a sparse hard set that is (tree) selfreducible?

References

[1] Agrawal, M., Arvind, V.: Quasi-linear truth-table reductions to p-selective sets. Theoretical Computer Science **158** (1996) 361–370
[2] Beigel, R., Kummer, M., Stephan, F.: Approximable sets. Information and Computation **120** (1995) 304–314
[3] Berman, L., Hartmanis, H.: On isomorphisms and density of NP and other complete sets. SIAM J. Comput. **6** (1977) 305–322
[4] Buhrman, H., Fortnow, L., Thierauf, T.: Nonrelativizing separations. In: IEEE Conference on Computational Complexity, IEE Computer Society Press (1998) 8–12
[5] Buhrman, H., Torenvliet, L.: P-selective self-reducible sets: A new characterization of P. J. Computer and System Sciences **53** (1996) 210–217
[6] Fortnow, L., Klivans, A.: NP with small advice. In: Proceedings of the 20th IEEE Conference on Computationa Complexity, IEEE Computer Society Press (2005) to appear.
[7] Hemaspaandra, L., Torenvliet, L.: Theory of Semi-Feasible Algorithms. Monographs in Theoretical Computer Science. Springer-Verlag, Heidelberg (2002)
[8] Ko, K.I.: On self-reducibility and weak P-selectivity. J. Comput. System Sci. **26** (1983) 209–211
[9] Ko, K., Schöning, U.: On circuit-size and the low hierarchy in NP. SIAM J. Comput. **14** (1985) 41–51
[10] Lozano, A., Torán, J.: Self-reducible sets of small density. Mathematical Systems Theory (1991)
[11] Meyer, A.: oral communication. cited in [3] (1977)
[12] Mocas, S.: Separating Exponential Time Classes from Polynomial Time Classes. PhD thesis, Northeastern University (1993)
[13] Ogihara, M.: Polynomial-time membership comparable sets. SIAM Journal on Computing **24** (1995) 1168–1181
[14] Ogiwara, M., Watanabe, O.: On polynomial time bounded truth-table reducibility of NP sets to sparse sets. SIAM J. Comput. **20** (1991) 471–483

[15] Papadimitriou, C.: Computational Complexity. Addison Wesley (1994)
[16] Selman, A.: P-selective sets, tally languages, and the behavior of polynomial time reducibilities on NP. Math. Systems Theory **13** (1979) 55–65
[17] Selman, A.: Analogues of semicursive sets and effective reducibilities to the study of NP complexity. Information and Control **52** (1982) 36–51
[18] Shaltiel, R., Umans, C.: Pseudorandomness for approximate counting and sampling. Technical Report TR04-086, ECCC (2004)
[19] Wagner, K.: Bounded query computations. In: Proc. 3rd Structure in Complexity in Conference, IEEE Computer Society Press (1988) 260–278
[20] Wilson, C.: Relativized circuit complexity. J. Comput. System Sci. **31** (1985) 169–181

Linear Advice for Randomized Logarithmic Space

Lance Fortnow[1] and Adam R. Klivans[2],[*]

[1] Department of Computer Science, University of Chicago,
Chicago, IL 60637
fortnow@cs.uchicago.edu
[2] Department of Computer Science, The University of Texas at Austin,
Austin, TX 78712
klivans@cs.utexas.edu

Abstract. We show that RL ⊆ L/$O(n)$, i.e., any language computable in randomized logarithmic space can be computed in deterministic logarithmic space with a linear amount of non-uniform advice. To prove our result we use an ultra-low space walk on the Gabber-Galil expander graph due to Gutfreund and Viola.

1 Introduction

The question of whether RL, randomized logarithmic space, can be simulated in L, deterministic logarithmic space, remains a central challenge in complexity-theory. The best known deterministic simulation of randomized logarithmic space is due to Saks and Zhou [16] who, building on seminal work due to Nisan [13], proved that BPL ⊆ L$^{3/2}$. Recently in a breakthrough result, Reingold [14] proved that the *s-t* connectivity problem on undirected graphs could be solved in L; this implies SL, the symmetric analogue of NL, equals L. The possibility of extending his techniques to prove RL = L has subsequently been investigated by Reingold, Trevisan and Vadhan [15].

1.1 Randomness and Non-uniformity

The relationship between randomness and non-uniformity is a topic of fundamental importance in complexity theory. Derandomizing complexity classes frequently involves a consideration of the smallest non-uniform complexity class or circuit family which contains a particular randomized class. Adleman's well known result on BPP [1], for example, shows that BPP can be computed by polynomial-size circuits. Goldreich and Wigderson [7] have recently proved an interesting relationship between the non-uniform complexity of RL and the existence of deterministic algorithms for RL which work on almost every input. More specifically they showed that if RL is computable in log-space with $o(n)$ bits of non-uniform advice and furthermore most of the advice strings are "good," (i.e.

---

[*] Initial work done while visiting TTI-Chicago.

result in a correct simulation of the RL machine by the advice-taking deterministic logarithmic-space machine) then there exists a deterministic log-space simulation of RL which errs on a $o(1)$ fraction of inputs for every input length. In other words, finding a log-space simulation with $o(n)$ bits of non-uniform advice is a step towards showing that RL is "almost" in L.

1.2   Our Results

We prove that every language in RL can be computed in L with $O(n)$ bits of additional non-uniform advice:

Theorem 1. *Every language in* RL *can be computed by a deterministic, log-space Turing machine which receives* $O(n)$ *bits of non-uniform advice on a two-way read-only input tape.*

In Section 3, we state as a corollary of Nisan's well-known pseudo-random generator for space bounded computation the inclusion RL $\subseteq$ L$/O(n \log n)$. What is more difficult is to show that RL $\subseteq$ L$/O(n)$. To do this we use a non-standard, space-efficient walk on an expander graph when the edge labels are presented on a two-way read-only input tape. Such a walk was used by Gutfreund and Viola [8] in the context of building pseudorandom generators computable in AC⁰:

Theorem 2 (Gutfreund-Viola). *There exists an* $O(\log(n))$-*space algorithm for taking a walk of length* $O(n)$ *on a particular constant-degree expander graph with* $2^{O(n)}$ *nodes if the algorithm has access to an initial vertex and edge labels describing the walk via a two-way read-only advice tape.*

We present a proof of Theorem 2 for completeness. Note that a naive implementation of a walk on a graph of size $2^{O(n)}$ would require $O(n)$ space just to remember the current vertex label. The main tool for using even less space is a Gabber-Galil graph [6] in conjunction with log-space algorithms for converting to and from Chinese Remainder Representations of integers [4].

We can then apply the above walk on an expander graph to amplify the success probability of the RL algorithm using only $O(n)$ random bits and $O(\log n)$ space. Using Adleman's trick [1] we can conclude that there must exist a good advice string of length $O(n)$ which works for all inputs.

1.3   Related Work

Bar-Yossef, Goldreich, and Wigderson [3] initiated a study of on-line, space-efficient generation of walks on expander graphs. Their work shows how to amplify a space S algorithm using r random bits with error probability $1/3$ to an $O(kS)$-space algorithm that uses $r + O(k)$ random bits and has error probability $\epsilon^{\Omega(k)}$ for any constant $\epsilon > 0$. Note, however, that we need $\epsilon < 2^{-n}$ and hence must take $k \geq n$. The space of their resulting algorithm will then be $\Omega(nS)$, which is much too large for our application here. Our savings comes from the fact that we have an initial vertex and the edge labels of a particular walk on an expander graph on an advice tape– we thus do not need to remember an initial vertex or the edge labels.

2   Preliminaries

Karp and Lipton [10] give a general definition of complexity classes with advice.

Definition 1 (Karp-Lipton). *For a complexity class C, the class $C/f(n)$ is the set of languages A such that there exists a language B in C and a sequence of strings $u_0, u_1, \ldots$, with $|a_n| = f(n)$ and for all x in A,*

$$x \in A \Leftrightarrow (x, a_{|x|}) \in B.$$

Our main result shows that for every language A in RL there is a constant c such that A is in L/cn. By Definition 1, the logarithmic space machine accepting B will have access to the advice as part of the 2-way read-only input tape.

A randomized space algorithm on input x has read-once access to a string of random bits r. As in the work of Nisan [13], we view a randomized space algorithm as a *deterministic* branching program where the input x to the randomized space algorithm has been fixed and the only remaining input to the branching program is r (hence to derandomize a randomized space algorithm we need only estimate the acceptance probability of a fixed deterministic branching program):

Definition 2. *A space($S(n)$) algorithm is a deterministic branching program of size $2^{S(n)}$ mapping inputs of length n to $\{0, 1\}$. Often we simply write space(S) and omit the input length n.*

To achieve our result we start with pseudorandom generators for space-bounded computations:

Definition 3. *A generator $G : \{0,1\}^m \rightarrow \{0,1\}^n$ is called a* pseudorandom *generator for space(S) with parameter ϵ if for every space(S) algorithm A with input y we have*

$$|\Pr(A(y) \text{ accepts}) - \Pr(A(G(x)) \text{ accepts})| \leq \epsilon$$

where y is chosen uniformly at random in $\{0, 1\}^n$ and x uniformly in $\{0, 1\}^m$.

We will also use expander graphs.

Definition 4. *An graph $G = (V, E)$ is an ϵ-expander if there exists a constant $\epsilon > 0$ such that for all U such that $|U| \leq |V|/2$, $|U \cup N(U)| \geq (1 + \epsilon)|U|$ where $N(U)$ is the neighborhood of U. A set of graphs $\{G_1, G_2, \ldots\}$ is a family of constant-degree expander graphs if there are fixed constants d and ϵ such that for all n, G_n has n vertices, the degree of every vertex of G_n is d and G_n is an ϵ-expander.*

We will use the now well-known fact that explicitly constructible constant-degree expander graphs can be used to reduce the error of randomized decision procedures:

Theorem 3 ([5, 9]). *Let* $L \in$ BPP *be decided by a probabilistic turing machine* M *using* $r(n) = n^{O(1)}$ *random bits. Then there exists a probabilistic polynomial-time algorithm for deciding* L *using* $O(r(n)+t)$ *random bits with error probability* 2^{-t}. *The algorithm chooses a random vertex of an expander graph and walks for* t *steps substituting the labels of the vertices in place of the truly random bits* M *would have used.*

For details on the history, constructions, spectral theory, and applications of expander graphs see Motwani and Raghavan [12] or the lecture notes for a course by Nati Linial and Avi Wigderson [11].

3   Starting Point: Nisan's Generator

We will use Nisan's well-known generator for space bounded computation [13] as the starting point for the proof of our main result:

Theorem 4 (Nisan). *For any* R *and* S *there exists a pseudorandom generator*

$$G : \{0, 1\}^{O(S \log(R/S))} \rightarrow \{0, 1\}^R$$

for space(S) with error parameter 2^{-S}. *Furthermore, if the seed is written on a two-way read-only tape, the ith bit of the output of the generator can be computed in* $O(S)$ *space.*

Corollary 1. *For any language* A *in* RL *there is a probabilistic algorithm solving* L *with one-sided error* $1/n$ *using logarithmic space and* $O(\log^2 n)$ *random bits on a 2-way read-only tape.*

Fix an RL language A. To prove the existence of a good advice string of length $O(n \log n)$ for A we first apply Corollary 1 to get a randomized algorithm using $O(\log^2 n)$ random bits that for any fixed input x in A results in an answer which is correct with probability at least $1 - 1/n$ and always rejects on inputs not in A.

Running the algorithm $2n/(\log n)$ times independently results in a randomized algorithm with error probability strictly less than 2^{-n}. Thus, by a union bound over all inputs, there must exist a sequence of n seeds to G which results in a correct classification of *any* input x.

Hence the total advice string is of length $O(n \log n)$. Now assume that a log-space machine is given access to this advice string on a read-only (multiple access) input tape. Since the ith bit of Nisan's generator is computable in log-space, we can carry out Adleman's trick in logarithmic space. Thus, as a corollary to Nisan's generator we have the following:

Corollary 2. RL $\subseteq$ L/$O(n \log n)$.

We wish to reduce the size of the advice to $O(n)$ bits. Note that any attempt to construct an advice string using only Adleman's trick cannot hope to achieve an advice string of length less than n. This is because using k bits of randomness in a black-box fashion can only drive the error probability of the algorithm down to 2^{-k} (without derandomizing the algorithm altogether).

4   Space-Efficient Walks on Gabber-Galil Graphs

Consider a language $L \in \mathsf{RL}$ and its associated randomized Turing machine M. A now standard approach for derandomizing algorithms [5, 9] is to "re-cycle" random bits via a walk on a suitable expander graph. For example, if we have a pseudorandom generator G with seed length s, and a suitable constant degree expander graph E with 2^s vertices, we can use the following randomness-efficient algorithm to compute L:

- Associate each vertex of the graph with a seed of G (note each vertex has a label of length s).
- Use s random bits to choose an initial vertex of E and walk randomly (by choosing t random edge labels) for t more steps to select t more seeds for G.
- Simulate L using the output of the generator G instead of truly random bits. Do this t times independently and accept if any of the t simulations result in accept.

This algorithm uses $O(s + t)$ random bits and requires space $O(s)$ (plus the space required to compute a neighbor of the current vertex). Applying Theorem 3 we see that the error probability of this algorithm is at most 2^{-t}.

We would like to use the above algorithm where the vertices of an expander graph correspond to seeds for Nisan's pseudorandom generator. Unfortunately, to simulate languages in RL using the output of Nisan's generator the seed must be of size $\Omega(\log^2 n)$, which means that the above algorithm will use at least $\Omega(\log^2 n)$ space just to keep track of the current vertex. As such, we will have to use a very space-efficient method for traversing an expander graph due to Gutfreund and Viola [8]. The walk uses a result due to Gabber and Galil [6]:

Theorem 5 (Gabber-Galil). *Let Z_m be the integers modulo m and let E be a graph with vertex set $Z_m \times Z_m$ and edge relations $(x, y) \Rightarrow (x, y) \cup (x, x + y) \cup (x, x + y + 1) \cup (x + y, y) \cup (x + y + 1, y)$. (i.e. each vertex is connected to 5 other vertices via the above simple arithmetic operations mod m). Then E is an expander graph with m^2 vertices and degree 5.*

We now present the walk due to Gutfreund and Viola [8], and for completeness we include a proof of correctness.

Let $m = \Pi_{i=1}^{k} p_i$ where each p_i is a distinct prime. Then we can view each vertex in the above expander graph via the Chinese remainder theorem as $(a_1, \ldots, a_k) \times (b_1, \ldots, b_k)$ where each $a_i, b_i \in Z_{p_i}$. Since we are interested in representing seeds of length $O(\log^2 n)$, we can think of m as the product of $O(\log n)$ primes, each of bit-length $O(\log n)$.

The idea is that to walk on this graph, we need only keep track of an index, and two residues $a, b \in Z_p$ for p a prime of bit-length at most $O(\log n)$. That is to say, we will store a *residue* of the Chinese Remainder Representation of m, rather than the integer m itself, and we can update the current residue during each step of the walk in log-space (the following lemma is implicit in Gutfreund and Viola [8]):

Lemma 1. *Let* E *be the Gabber-Galil graph* $Z_m \times Z_m$ *as above where* $m = \Pi_{i=1}^{a \log n} p_i$, *the product of* $a \log n$ *primes (for sufficiently large* a*) each of bit length at most* $O(\log n)$*. There exists an* $O(\log n)$ *space algorithm* W *such that on input* s*, a starting vertex of* E*, a sequence of edge labels* $t = (t_1, \ldots, t_\ell)$*, and* $1 \le i \le a \log n$*, a residue index,* W *outputs the* i*th residues of the two integers representing the vertex reached by starting at* s *and traversing the edge labels indicated by* t*.*

Proof. Assume that s can be represented via the Chinese Remainder Theorem as $(a_1, \ldots, a_\ell) \times (b_1, \ldots, a_\ell)$ where each a_i, b_i has bit-length $O(\log n)$. The edge relations of the Gabber-Galil graph involve only an addition and may be carried out component-wise. For example, if we are currently storing a_i and b_i and the next edge label t_j is equal to 1, then the new components we store are a_i and $a_i + b_i \mod p_i$. Hence we need only remember the two current residues, a prime p_i, and the number of steps we have already taken. This requires $O(\log n)$ space.

Thus, although we cannot store an entire vertex of our expander graph, we can compute residues of vertices explored by a random walk on the graph. Unfortunately, Nisan's generator requires a seed described by the original representation of vertices of the Gabber-Galil graph. As such we will require a space-efficient routine for computing the ith bit of an integer m if we are only given access to its residues modulo distinct primes. Following Gutfreund and Viola, we can apply a recent space-efficient algorithm for coverting to and from the Chinese Remainder Representation due to Chiu, Davida and Litow [4]:

Theorem 6 (Chiu-Davida-Litow). *Let* $a_1, \ldots, a_\ell$ *be the Chinese Remainder Representation of an integer* m *with respect to primes* $p_1, \ldots, p_\ell$*. There exists a log-space algorithm* D *such that on input* $a_1, \ldots, a_\ell$*, primes* $p_1, \ldots, p_\ell$*, and index* i*,* D *outputs the* i*th bit of the binary representation of the integer* m*.*

For a further discussion of space-efficient, uniform algorithms for arithmetic operations such as division and converting from the Chinese Remainder Representation see Allender et al. [2].

5   Putting It All Together

We can combine these space-efficient tools to prove our main result:

Theorem 7. $\mathsf{RL} \subseteq \mathsf{L}/O(n)$

Proof. From the discussion at the beginning of Section 4, we know that for polynomial-time advice taking Turing machines simulating RL, for every input length n there must exist a good advice string $A(n)$ of length $O(n)$ consisting of an initial vertex on a suitable Gabber-Galil expander graph with $n^{O(\log n)}$ vertices and a sequence of $2n$ edge labels. Let us assume that the vertices of the graph equal $Z_m \times Z_m$ where m is a product of $O(\log n)$ primes $p_1, \ldots, p_k$ each of bit length $O(\log n)$ (such primes are guaranteed to exist by the Prime

Number Theorem). Augment our advice string $A(n)$ with primes $p_1, \ldots, p_k$. Call this advice string $A'(n)$.

Our claim is that $A'(n)$ is a good advice string for a log-space Turing machine M which computes the above simulation of RL using the following procedures:

1. Simulate the RL machine $2n$ times using the output from Nisan's generator on seeds corresponding to the vertices reached by the walk given on the advice string. If any simulation results in accept then accept.
2. When Nisan's generator requires a bit from the seed, walk on the Gabber-Galil graph to obtain the ith bit of the binary representation of that vertex.
3. When the ith bit of a vertex from the graph is required, apply the Chinese Remainder Representation algorithm given from Theorem 6 and use Lemma 1 to obtain any residue required by the CRR algorithm of Theorem 6.

Note that each procedure can be performed in log-space, and the entire algorithm is a composition of three log-space procedures. Therefore, the entire simulation carried out by M can be performed in log-space. We use the fact that the initial vertex and edge labels are written on a tape in a critical way: whenever the algorithm needs a bit of the ith vertex label from the walk on the graph, we can move our tape head back to the initial vertex and walk from the beginning for i steps.

We note that our results extend to BPL, the two-sided error version of RL.

6   Connectivity for Expander Graphs in NC^1?

We conclude with the following question: is the undirected connectivity problem for expander graphs in NC^1? We have an NC^1 algorithm for taking walks of length $O(\log n)$ on the Gabber-Galil graphs used for our main theorem. Since expander graphs have diameter $O(\log n)$, this yields an NC^1 algorithm for connectivity on Gabber-Galil graphs. Is it possible that for a general expander graph connectivity is in RNC^1? Such a result would lead to the intriguing possibility that more general connectivity problems are not only in L but in NC^1.

Acknowledgments

Thanks to Jaikumar Radhakrishnan and Rahul Santhanam for helpful discussions. Thanks to Emanuele Viola for making us aware of his work with Danny Gutfreund.

References

1. L. Adleman. Two theorems on random polynomial time. In *Proceedings of the 19th IEEE Symposium on Foundations of Computer Science*, pages 75–83. IEEE, New York, 1978.
2. E. Allender, D. Barrington, and W. Hesse. Uniform circuits for division: Consequences and problems. In *Annual IEEE Conference on Computational Complexity*, volume 16, 2001.

3. Z. Bar-Yossef, O. Goldreich, and A. Wigderson. Deterministic amplification of space-bounded probabilistic algorithms. In *Proceedings of the 14th IEEE Conference on Computational Complexity*, pages 188–199. IEEE, New York, 1999.

4. A. Chiu, G. Davida, and B. Litow. Division in logspace-uniform NC^1. *RAIRO - Theoretical Informatics and Applications*, 35:259–275, 2001.

5. A. Cohen and A. Wigderson. Dispensers, deterministic amplification, and weak random sources (extended abstract). In *Proc. 30th Ann. IEEE Symp. on Foundations of Computer Science*, pages 14–25, Research Triangle Park, NC, October 1989. IEEE Computer Society Press.

6. O. Gabber and Z. Galil. Explicit constructions of linear-sized superconcentrators. *Journal of Computer and System Sciences*, 22(3):407–420, June 1981.

7. O. Goldreich and A. Wigderson. Derandomization that is rarely wrong from short advice that is typically good. In *Proceedings of the 6th International Workshop on Randomization and Approximation Techniques*, volume 2483 of *Lecture Notes in Computer Science*, pages 209–223. Springer, Berlin, 2002.

8. D. Gutfreund and E. Viola. Fooling parity tests with parity gates. In *Proceedings of the 8th International Workshop on Randomization and Computation (RANDOM)*. 2004.

9. R. Impagliazzo and D. Zuckerman. How to recycle random bits. In *Proceedings of the 30th IEEE Symposium on Foundations of Computer Science*, pages 248–253. IEEE, New York, 1989.

10. R. Karp and R. Lipton. Some connections between nonuniform and uniform complexity classes. In *Proceedings of the 12th ACM Symposium on the Theory of Computing*, pages 302–309. ACM, New York, 1980.

11. N. Linial and A. Wigderson. Lecture notes on expander graphs and their applications. http://www.math.ias.edu/~boaz/ExpanderCourse/index.html, 2002.

12. Rajeev Motwani and Prabhakar Raghavan. *Randomized Algorithms*. Cambridge University Press, 1997.

13. N. Nisan. Pseudorandom generators for space-bounded computation. *Combinatorica*, 12(4):449–461, 1992.

14. O. Reingold. Undirected st-connectivity in log-space. In *Proceedings of the 36th ACM Symposium on the Theory of Computing*. ACM, New York, 2005. To appear.

15. O. Reingold, L. Trevisan, and S. Vadhan. Pseudorandom walks in biregular graphs and the RL vs. L problem. Technical Report TR05-022, Electronic Colloquium on Computational Complexity, 2005.

16. M. Saks and S. Zhou. $BP_H SPACE(S) \subseteq DPSPACE(S^{3/2})$. *Journal of Computer and System Sciences*, 58(2):376–403, April 1999.

Nested Pebbles and Transitive Closure

Joost Engelfriet and Hendrik Jan Hoogeboom

Leiden University, Institute of Advanced Computer Science,
P.O.Box 9512, 2300 RA Leiden, The Netherlands

Abstract. First-order logic with k-ary deterministic transitive closure
has the same power as two-way k-head deterministic automata that use
a finite set of nested pebbles. This result is valid for strings, ranked trees,
and in general for families of graphs having a fixed automaton that can
be used to traverse the nodes of each of the graphs in the family. Other
examples of such families are grids, toruses, and rectangular mazes.

1 Introduction

The complexity class DSPACE($\log n$) of string languages accepted in logarithmic
space by deterministic Turing machines, has two well-known distinct characterizations. The first one is in terms of deterministic two-way automata with several
heads working on the input tape (and no additional storage). Second, Immerman
[20] showed that these languages can be specified using first-order logic with an
additional deterministic transitive closure operator – it is one of the main results in the field of descriptive complexity [11, 21]. Similar characterizations of
NSPACE($\log n$) hold for their nondeterministic counterparts.

The two characterizations each have a natural parameter indicating the relative complexity of the mechanism used. For multi-head automata the parameter
is the number of heads used to scan the input. It is known that $k + 1$ heads
are better than k, even for a single-letter input alphabet [26]. For transitive closure logics, the parameter is the arity of the transitive closure operators used. It
seems to be open whether $(k + 1)$-ary transitive closure is more powerful than
k-ary transitive closure.

Bargury and Makowsky [2] characterize k-head automata by a "k-regular"
subset of first-order logic with k-ary transitive closure but their characterization
only works in the nondeterministic case: "the modification of the k-regular formulas needed to take out the nondeterminism will spoil their elegant form, and
we do not pursue this further".

Here we set out from the other side and present an automata-theoretic characterization of first-order logic with deterministic k-ary transitive closure. Our
deterministic two-way automaton model has k heads, as expected, but is augmented with the possibility to put an arbitrary finite number of pebbles on its
input tape, to mark positions for further use. If these pebbles can be used at
will it is folklore that we obtain again DSPACE($\log n$), a family too large for
our purpose. Instead we only allow pebbles that are used in a nested (or LIFO)
fashion: all pebbles can be 'seen' by the automaton as usual, but only the last

B. Durand and W. Thomas (Eds.): STACS 2006, LNCS 3884, pp. 477–488, 2006.
© Springer-Verlag Berlin Heidelberg 2006

one dropped can be picked up [19, 13, 25, 16]. On the other hand our pebbles are more flexible than the usual ones: they can be 'retrieved from a distance', i.e., a pebble can be picked up even when no head is scanning its position.

Our equivalence result (Theorem 5) is stated and proved for ranked trees in general, of which strings are a special case. The automaton model is the deterministic tree-walking automaton (with nested pebbles) which generalizes two-way automata on strings. One consequence of the result is that the class of tree languages accepted by these automata is closed under complement [27].

In Section 3 we translate logical formulas into automata, following [13] and additionally using the technique of Sipser [32] to deterministically search a computation space. Section 4 considers the reverse. As in [2] we adapt Kleene's construction to obtain regular formulas from automata, thus getting rid of the states of the automaton, but we need to iterate that construction: once for each nested pebble. In Section 5 we discuss the main result for single-head tree-walking automata, which are relevant as a model of XML [25, 28, 23]. Finally, in Section 6 we show how to extend our results to more general graph-like structures, such as unranked trees (important for XML), grids (as in [2]; important for picture recognition [3, 18, 24]), toruses, and, for $k \geq 2$, mazes [4, 8].

Due to space limitations many examples, technical details, explanations, footnotes, and references had to be omitted. The reader may find them in [14].

2   Preliminaries

A *ranked alphabet* is a finite set Σ together with a mapping rank : $\Sigma \to \mathbb{N}$. Terms over Σ are recursively defined: if $\sigma \in \Sigma$ is of rank n, and $t_1, \ldots, t_n$ are terms, then $\sigma(t_1, \ldots, t_n)$ is a term. As usual, terms are visualized as *trees*, which are special labelled graphs; $\sigma(t_1, \ldots, t_n)$ as a tree which has a root labelled by σ and outgoing edges labelled by $1, \ldots, n$ leading to the roots of trees for $t_1, \ldots, t_n$. The root of subtree t_i has child number i; for the root of the full tree it is 0.

For $k \geq 1$, a k-head *tree-walking automaton* or *twa* is a finite-state device that moves its k heads from node to node along the edges of the input tree. It determines its next move based on its present state, and the label and child number of the nodes visited. Accordingly, it changes state and, for each of its heads, it stays at the node, or moves either up to the parent of the node, or down to a specified child. If the automaton has no next move, we say it *halts*. The *language accepted* by the k-head twa $\mathcal{A}$ is the set of all trees on which $\mathcal{A}$ has a computation starting with all its heads at the root of the tree in the initial state and halting in an accepting state, again at the root of the tree. The family of languages accepted by k-head (deterministic) twa is denoted by $\mathsf{NW}^k\mathsf{A}$ ($\mathsf{DW}^k\mathsf{A}$).

A twa is able to make a systematic search of the tree (a preorder traversal), even using a single head, as follows. When a node is reached for the first time (entering it from above) the automaton continues in the direction of the first child; when a leaf is reached, the automaton goes up again. If a node is reached from below, from a child, it goes down again, to the next child, if that exists;

otherwise it moves to the parent of the node. The search ends when the root is entered from its last child. This traversal underlies our basic constructions.

In both [29] and [30], as an example, the authors explicitly construct a deterministic 1-head twa that evaluates boolean trees, i.e., terms with binary operators 'and' and 'or' and constants 0 and 1.

Strings form a special case. Tree-walking automata on monadic trees (each symbol has rank one except a special symbol with rank zero) are equivalent to the usual two-way automata on strings.

For an overview of the theory of first-order and monadic second-order logic on strings and trees in relation to formal language theory, see [34]. Here we consider *first-order* logic, describing properties of trees. The logic has node variables $x, y, \ldots$, which for a given tree range over its nodes. There are four types of atomic formulas over Σ: $\mathrm{lab}_\sigma(x)$, for every $\sigma \in \Sigma$, meaning that x has label σ; $\mathrm{edg}_i(x, y)$, for every i at most the rank of a symbol in Σ, meaning that the i-th child of x is y; $x \leq y$, meaning that x is an ancestor of y; and $x = y$. Formulas are built using the connectives $\neg$, $\wedge$, and $\vee$, and quantifiers $\exists$ and $\forall$ as usual.

If t is a tree with nodes $u_1, \ldots, u_n$, and ϕ is a formula such that its free variables are $x_1, \ldots, x_n$, then we write $t \models \phi(u_1, \ldots, u_n)$ if formula ϕ holds for t where the free x_i are valuated as u_i.

For fixed $k \geq 1$, by overlined symbols like $\overline{x}$ we denote k-tuples of objects of the type referred to by x, like logical variables, nodes in a tree, or pebbles.

Let $\phi(\overline{x}, \overline{y})$ be a formula where $\overline{x}, \overline{y}$ are distinct k-tuples of variables occurring free in ϕ. We use $\phi^*(\overline{x}, \overline{y})$ to denote the *k-ary transitive closure* of ϕ with respect to $\overline{x}, \overline{y}$. Informally, $\phi^*(\overline{x}, \overline{y})$ means that we can make a series of jumps from nodes $\overline{x}$ to nodes $\overline{y}$ such that each pair of consecutive k-tuples $\overline{x}', \overline{y}'$ connected by a jump satisfies $\phi(\overline{x}', \overline{y}')$. The formula ϕ may have additional free variables.

A predicate $\phi(\overline{x}, \overline{y})$ with free variables $\overline{x}, \overline{y}$ is *functional* (in $\overline{x}, \overline{y}$) if for every tree t and k-tuple of nodes $\overline{u}$ there is at most one k-tuple $\overline{v}$ such that $t \models \phi(\overline{u}, \overline{v})$. If ϕ has more free variables than $\overline{x}, \overline{y}$, this should hold for each fixed valuation of those variables. The transitive closure $\phi^*(\overline{x}, \overline{y})$ is *deterministic* if ϕ is functional (in the variables with respect to which the transitive closure is taken).

The *tree language* defined by a closed formula ϕ consists of all trees t such that $t \models \phi$. The family of all tree languages that are first-order definable is denoted by FO; if one additionally allows k-ary (deterministic) transitive closure we have the family FO+TC^k (FO+DTC^k). For strings, general (deterministic) transitive closure (i.e., over unbounded values of k) characterizes the complexity class NSPACE($\log n$) (DSPACE($\log n$)), see [11, 21].

3   From Logic to Nested Pebbles

A k-head tree-walking automaton with *nested pebbles* is a k-head twa that is additionally equipped with a finite set of pebbles. During the computation it may drop these pebbles (one by one) on nodes visited by its heads, to mark specific positions. It may test the currently visited nodes to see which pebbles are present. Moreover, it may retrieve a pebble from anywhere in the tree, provided the life times of the pebbles are nested (which means that only the last one dropped can be retrieved). This can be formalized by keeping a (bounded) stack in the configuration of the automaton, pushing and popping pebbles when they are dropped and retrieved. Pebbles can be reused any number of times (but there is only one copy of each pebble). Computations should start and end with all heads at the root without pebbles on the input tree. The family of tree languages accepted by (deterministic) k-head twa with nested pebbles is denoted by $\mathsf{NPW}^k\mathsf{A}$ ($\mathsf{DPW}^k\mathsf{A}$).

Note that pebbles (1) are nested, as in e.g., [19, 13, 25, 16]; without this restriction again the classes $\mathsf{DSPACE}(\log n)$ and $\mathsf{NSPACE}(\log n)$ are obtained, (2) behave as *pointers*: we can store the address of a node when we visit it, and we can later whipe the address from memory without returning to the node itself ("abstract markers" [3] as opposed to the usual "physical markers"), (3) always remain visible to the automaton (not only the last one dropped, as in [19]).

Example 1. As in [6], consider a ranked alphabet with one binary symbol and two nullary symbols a and b, and consider the trees for which the path to each a-labelled leaf contains an even number of nodes on the 'branching structure' of the tree, i.e., nodes for which both the left and right subtree contain an a-labelled leaf. This is a first-order definable tree language that cannot be accepted by any single-head nondeterministic twa (without pebbles) [6].

However, it can be accepted by a (single-head) deterministic twa with two nested pebbles as follows. Using a preorder traversal, the first pebble is placed consecutively on a-labelled leaves. For each such leaf we follow the path upwards to the root counting the number of nodes that belong to the branching structure. To test whether a node belongs to that structure we place the second pebble on the node and test whether its other subtree, i.e., the one that does not contain the first pebble, contains an a-labelled leaf (using again a traversal of that subtree, the root of which can be recognized through the second pebble). □

We now generalize the inclusion $\mathsf{FO} \subseteq \mathsf{DPW}^1\mathsf{A}$ from [13], introducing k-ary transitive closure, as well as allowing k heads. Note that here we use 'pointer-like' pebbles, rather than the usual pebbles.

Lemma 2. *For ranked trees,* $\mathsf{FO}+\mathsf{DTC}^k \subseteq \mathsf{DPW}^k\mathsf{A}$.

Proof. By induction on the structure of the formula ϕ we construct an automaton $\mathcal{A}$ that always halts on its input tree t. Generally speaking, each variable of ϕ acts as a pebble for $\mathcal{A}$. In case of k-ary transitive closure we need $3k$ pebbles to test the formula. Most features can be simulated using a single head, moving pebbles around, only for transitive closure we need all the k heads.

For intermediate formulas with free variables we fix the valuation of these variables by putting pebbles on the tree, one for each variable, and $\mathcal{A}$ should evaluate the formula according to this valuation; it may test these pebbles but is not allowed to retrieve them. Automaton $\mathcal{A}$ is started in the initial state with all heads at the root of the tree t, it may use additional pebbles (in a nested fashion), and it should halt again with all heads at the root.

For the atomic formulas (single-head) automata are easily constructed. As an example, for $edg_i(x, y)$ the automaton searches for pebble x, determines whether x has an i-th child (the arity of the node can be seen from its label), moves to that child, and checks there whether pebble y is present.

For the negation $\phi = \neg\phi_1$ we use the automaton for ϕ_1, changing its accepting states to the complementary set. This works thanks to the fact that the automata we build are always halting. A similar argument works for conjunction and disjunction, running the automata for the two constituents consecutively.

For quantification $\phi = (\forall x)\phi_1$ the automaton $\mathcal{A}$ makes a systematic traversal through the tree, using a single head. Reaching a node it drops a pebble x, returns to the root, and runs the automaton for ϕ_1 as a subroutine; the free variable x of ϕ_1 is marked by the pebble, as requested by the inductive hypothesis. When the test for $\phi_1(x)$ is positive, $\mathcal{A}$ returns to the node marked x (searching for it), picks up the pebble, and places it on the next node of the traversal; $\mathcal{A}$ accepts if it has succesfully run the test for ϕ_1 for each node. Existential quantification is treated similarly.

For transitive closure $\phi = \phi_1^*$ we need to walk from one k-tuple of nodes $\overline{x}$ to another k-tuple $\overline{y}$ with 'jumps' specified by the $2k$-ary formula ϕ_1. Doing this in a straightforward way, we might end 'jumping around' in a cycle. To obtain an automaton that always halts we use the technique of Sipser [32], and run this walk backwards. It is based on the observation that the computation space is actually a tree. Consider all k-tuples of nodes of the input tree t, and connect vertex[1] $\overline{u}$ to vertex $\overline{v}$ if the pair $(\overline{u}, \overline{v})$ in t satisfies $\phi_1(\overline{x}, \overline{y})$. As ϕ_1 is functional, for each vertex there is at most one outgoing arc. Choosing vertex $\overline{y}$ as root we obtain a directed tree $t_k(\overline{y})$, with arcs defined by ϕ_1 pointing towards the root $\overline{y}$; there is no bound on the number of arcs incident to each vertex. It consists of all vertices $\overline{u}$ that satisfy $t \models \phi(\overline{u}, \overline{y})$.

The automaton $\mathcal{A}$ traverses that tree $t_k(\overline{y})$ and tries to find the vertex marked by pebbles $\overline{x}$. However, $t_k(\overline{y})$ is not explicitly available and has to be reconstructed while walking on the input tree t, using the automaton $\mathcal{A}_1$ for ϕ_1 as a subroutine. Note that k-tuples of nodes of t can be enumerated (ordered) using the lexicographical ordering based on the preorder in t. To find the successor of a k-tuple $\overline{z}$ we act like adding one to a k-ary number: change the last coordinate of the tuple $\overline{z}$ into its successor (here the preorder successor in t) if that exists, otherwise reset that coordinate to the first element in the ordering (here the root of t), and consider the last-but-one coordinate, etc. In fact, this can be done by a single-head twa using the pebbles marking $\overline{z}$ in a nested fashion.

We traverse the tree $t_k(\overline{y})$, with $2k$ pebbles $\overline{x}$ and $\overline{y}$ fixed, with the help of $3k$ additional pebbles $\overline{x}'$, $\overline{y}'$, and $\overline{z}'$. Starting in $\overline{y}$ we determine whether vertex $\overline{x}$ belongs to $t_k(\overline{y})$. During this traversal, $\mathcal{A}$ keeps track of the current vertex of $t_k(\overline{y})$ with its k heads. The order of dropping the pebbles $\overline{x}'$ and $\overline{y}'$ differs in the two algorithmic steps below: in the first we have to check $\phi_1(\overline{x}', \overline{y}')$ 'backwards', finding $\overline{x}'$ given $\overline{y}'$, while in the second it is the other way around.

---

[1] For clarity we distinguish 'node' in the input tree from 'vertex' in the computation space, i.e., a k-tuple of nodes. Similarly we use 'edge' and 'arc'.

Step one: check whether the current vertex has a first child in $t_k(\overline{y})$, and go there if it exists. We drop pebbles $\overline{y}'$ to fix the current vertex, and we 'lexicographically' place pebbles $\overline{x}'$ on each candidate vertex (except $\overline{v}$). For each k-tuple $\overline{x}'$ we check $\phi_1(\overline{x}', \overline{y}')$ using automaton $\mathcal{A}_1$ as a subroutine. If the formula is true, we have found the first child in $t_k(\overline{y})$ and we move the k heads to the nodes marked by $\overline{x}'$, lift pebbles $\overline{x}'$, and retrieve pebbles $\overline{y}'$ (from a distance). Otherwise we move $\overline{x}'$ to the next candidate vertex. If none of the candidates $\overline{x}'$ satisfies $\phi_1(\overline{x}', \overline{y}')$, the vertex $\overline{y}'$ apparently has no child in $t_k(\overline{y})$.

Step two: check for a right sibling in $t_k(\overline{y})$, and go there if it exists, or go up (to the parent of the current vertex) otherwise. The problem here is to adhere to the proper nesting of the pebbles. First drop pebbles $\overline{x}'$ on the current vertex. Then determine its parent in $t_k(\overline{y})$; this is the unique vertex $\overline{y}'$ that satisfies $\phi_1(\overline{x}', \overline{y}')$, thanks to the functionality of ϕ_1. It can be found in a 'lexicographic' traversal of all k-tuples of nodes of t using pebbles $\overline{y}'$ and subroutine $\mathcal{A}_1$. Leave $\overline{y}'$ on the parent and return to $\overline{x}'$ (by searching for it in t). Using the third set of k pebbles $\overline{z}'$, traverse the k-tuples of nodes of t from $\overline{x}'$ onwards and try to find the next k-tuple that satisfies $\phi_1(\overline{z}', \overline{y}')$ when $\overline{z}'$ is dropped. If found, it is the right sibling of $\overline{x}'$; return there, lift $\overline{z}'$, and retrieve $\overline{y}'$ and $\overline{x}'$. Otherwise, the current vertex has no right sibling; go up in the tree $t_k(\overline{y})$, i.e., return to $\overline{y}'$, lift $\overline{y}'$, and retrieve $\overline{x}'$.   □

4   From Nested Pebbles to Logic

The classical result of Kleene shows how to transform a finite-state automaton into a regular expression, which basically means that we have a way to dispose of the states of the automaton. It is observed in [2] that this technique can also be used to transform multi-head automata on grids into equivalent formulas with transitive closure: transitive closure may very well specify sequences of consecutive positions on the input, but has no direct means to store states. A similar technique is used here. As our model includes pebbles, this imposes an additional problem, which we solve by iterating the construction for each pebble. Unlike [2] we have managed to find a formulation that works well for both the nondeterministic and deterministic case.

If the step relation of a deterministic finite-state device with k heads is specified by logical formulas, then its computation relation can be expressed using k-ary deterministic transitive closure. This is formalized as follows.

Let Φ be a $Q \times Q$ matrix of predicates $\phi_{p,q}(\overline{x}, \overline{y})$, $p, q \in Q$ for some finite set Q (of states), where $\overline{x}, \overline{y}$ each are k distinct variables occurring free in all $\phi_{p,q}$. We define the *computation closure* of Φ with respect to $\overline{x}, \overline{y}$ as the matrix $\Phi^{\#}$ consisting of predicates $\phi_{p,q}^{\#}(\overline{x}, \overline{y})$ where $t \models \phi_{p,q}^{\#}(\overline{u}, \overline{v})$ iff there exists a sequence of k-tuples of nodes $\overline{u}_0, \overline{u}_1, \ldots, \overline{u}_n$ and a sequence of states $p_0, p_1, \ldots, p_n$, $n \geq 1$, such that $\overline{u} = \overline{u}_0$, $\overline{v} = \overline{u}_n$, $p = p_0$, $q = p_n$, and $t \models \phi_{p_i, p_{i+1}}(\overline{u}_i, \overline{u}_{i+1})$ for $0 \leq i < n$.[2] Intuitively $t \models \phi_{p,q}^{\#}(\overline{u}, \overline{v})$ means that there is a Φ-path of consecutive steps (as specified by Φ) leading from nodes $\overline{u}$ in state p to nodes $\overline{v}$ in state q.

---

[2] For simplicity, we disregard the remaining free variables of the $\phi_{p,q}$ and $\phi_{p,q}^{\#}$.

We say that Φ is *deterministic* if its predicates are both functional and exclusive, i.e., for any $p, q, q' \in Q$ and $3k$ nodes $\overline{u}, \overline{v}, \overline{v}'$ of any tree t, if both $t \models \phi_{p,q}(\overline{u}, \overline{v})$ and $t \models \phi_{p,q'}(\overline{u}, \overline{v}')$ then $q = q'$ and $\overline{v} = \overline{v}'$.

Lemma 3. *If Φ is in* FO+DTC^k *and deterministic, then $\Phi^{\#}$ is in* FO+DTC^k.

Proof. Assume that $Q = \{1, 2, \ldots, m\}$. We construct matrices $\Phi^{(\ell)}$ of formulas $\phi_{p,q}^{(\ell)}$ in FO+TC^k which are defined as $\phi_{p,q}^{\#}$, except that the intermediate states $p_1, \ldots, p_{n-1}$ are chosen from $\{1, \ldots, \ell\}$. In particular, $\Phi^{(0)} = \Phi$, and $\Phi^{(m)} = \Phi^{\#}$. Inductively we obtain $\Phi^{(\ell+1)}$ as follows: $\phi_{p,q}^{(\ell+1)}(\overline{x}, \overline{y}) = \phi_{p,q}^{(\ell)}(\overline{x}, \overline{y}) \vee (\exists \overline{x}' \ \overline{y}')[\ \phi_{p,\ell+1}^{(\ell)}(\overline{x}, \overline{x}') \wedge (\phi_{\ell+1,\ell+1}^{(\ell)})^{*}(\overline{x}', \overline{y}') \wedge \phi_{\ell+1,q}^{(\ell)}(\overline{y}', \overline{y})\ ]$. The transitive closure is deterministic: $\phi_{\ell+1,\ell+1}^{(\ell)}$ is functional because Φ is deterministic and because Φ-paths ending in $\ell+1$ cannot be extended following the definition of $\Phi^{(\ell)}$. $\square$

Lemma 4. *For ranked trees,* DPW^{k}A $\subseteq$ FO+DTC^k.

Proof. Consider a k-head twa $\mathcal{A}$ with n pebbles $x_n, \ldots, x_1$, used in the order given, i.e., x_n is always placed on the bottom of the pebble stack. View $\mathcal{A}$ as consisting of $n+1$ 'levels' $\mathcal{A}_n, \ldots, \mathcal{A}_1, \mathcal{A}_0$ such that $\mathcal{A}_\ell$ is a k-head twa with ℓ pebbles $x_\ell, \ldots, x_1$, available for dropping and retrieving, whereas pebbles $x_n, \ldots, x_{\ell+1}$ have a fixed position on the tree and $\mathcal{A}_\ell$ may test for their presence. Basically, $\mathcal{A}_\ell$ acts as a twa that drops pebble x_ℓ, then queries $\mathcal{A}_{\ell-1}$ where to go in the tree, moves there, and retrieves pebble x_ℓ (from a distance).

The number of pebbles dropped can be kept in the finite control of $\mathcal{A}$, so we can unambiguously partition its state set as $Q = Q_n \cup \cdots \cup Q_1 \cup Q_0$, where Q_ℓ consists of states where ℓ pebbles are still available. Automaton $\mathcal{A}_\ell$ is the restriction of $\mathcal{A}$ to the states in Q_ℓ.

For $\mathcal{A}_\ell$ a matrix $\Phi^{(\ell)}$ is constructed with predicates $\phi_{p,q}^{(\ell)}$ for $p, q \in Q_\ell$. These predicates represent the single steps of $\mathcal{A}_\ell$, so $t \models \phi_{p,q}^{(\ell)\#}(\overline{u}, \overline{v})$ iff $\mathcal{A}_\ell$ has a nonempty computation from configuration $[p, \overline{u}]$ to configuration $[q, \overline{v}]$. Note that $\Phi^{(\ell)}$ has additional free variables $x_n, \ldots, x_{\ell+1}$ that will hold the positions of the pebbles already placed on the tree.

First assume that pebble x_ℓ has not been dropped. For each of its heads, $\mathcal{A}_\ell$ may test the presence of pebbles $x_n, \ldots, x_{\ell+1}$, and the node label and child number of the current node, and then it may move each of its heads. These steps, relations between the current and next configurations $[p, \overline{u}]$ and $[q, \overline{v}]$, are easily expressed in first-order logic. E.g., if the automaton can move head 5 to the first child of a node with label σ, while the node under head 6 has child number 2 and does not contain pebble $x_{\ell+1}$, then $\phi_{p,q}^{(\ell)}(\overline{u}, \overline{v})$ is the conjunction of the formulas $\text{lab}_\sigma(u[5])$, $(\exists u')\,\text{edg}_2(u', u[6])$, $u[6] \neq x_{\ell+1}$, $\text{edg}_1(u[5], v[5])$, and $u[j] = v[j]$ for $j \neq 5$ (where $u[i]$ denotes the ith component of $\overline{u}$).

Additionally when $\ell \geq 1$, $\mathcal{A}_\ell$ may drop pebble x_ℓ at the position of head i in state p, call $\mathcal{A}_{\ell-1}$, and retrieve pebble x_ℓ returning to state q. Such a 'macro step' from configuration $[p, \overline{u}]$ to $[q, \overline{v}]$ is only possible when there is a pair of pebble instructions $(p, \text{drop}_i(x_\ell), p')$ and $(q', \text{retrieve}(x_\ell), q)$, such that $\mathcal{A}_{\ell-1}$ has a (nonempty) computation from $[p', \overline{u}]$ to $[q', \overline{v}]$, i.e., $t \models \phi_{p',q'}^{(\ell-1)\#}(\overline{u}, \overline{v})$. (^{3}) Hence, $\mathcal{A}_\ell$ can take that step iff the disjunction of $\phi_{p',q'}^{(\ell-1)\#}(\overline{u}, \overline{v})$ over all such q' holds, where the free variable x_ℓ in that formula is replaced by $u[i]$, the position at which the pebble is dropped.

The resulting step matrix $\Phi^{(\ell)}$ is deterministic thanks to the determinism of $\mathcal{A}$ and $\Phi^{(\ell-1)}$. It is in FO+DTC^k by Lemma 3. The computational behaviour of $\mathcal{A}_\ell$ is expressed by $\Phi^{(\ell)\#}$, and that of $\mathcal{A}$ by the disjunction of all formulas $\phi_{p,q}^{(n)\#}(\overline{\text{root}}, \overline{\text{root}})$ with p initial and q accepting.   □

Combining Lemmas 2 and 4, we immediately get the main result of this paper. Note that it includes the case of strings.

Theorem 5. *For ranked trees,* DPW^{k}A = FO+DTC^k.

As a corollary we may transfer two obvious properties of FO+DTC^k, closure under complement and union, to deterministic twa with nested pebbles, where the result is nontrivial. In the proof of Lemma 2 we have constructed automata that are always halting. As all our constructions are effective this means that 'always-halting' is a normal form for deterministic twa with nested pebbles. In fact, the two closure properties follow rather directly from this normal form. This is further studied with regard to the number of pebbles needed in [27].

When the twa is not deterministic we no longer can assure the determinism of the formulas $\Phi^{(\ell)}$ in the proof of Lemma 4. However, they are in FO+TC^k. The proof of Lemma 4 uses negation only on atomic predicates, to model negative tests of the automaton (to check there is no specific pebble on a node). Since negation is not used in the proof of Lemma 3, we obtain *positive formulas*, allowing transitive closure only within the scope of an even number of negations (see, e.g., [11, 21]).

Conversely, for positive formulas there is also a result similar to Lemma 2. For disjunction and existential quantification the automaton now uses nondeterminism in the obvious way. For transitive closure $\phi = \phi_1^*$ the Sipser technique is not needed: $\mathcal{A}$ checks nondeterministically the existence of a path from vertex $\overline{x}$ to vertex $\overline{y}$ in the directed graph determined by ϕ_1.

Denoting the positive restriction of FO+TC^k by FO+posTC^k, we thus obtain a characterization for the nondeterministic case. We do not know whether NPW^{k}A is closed under complement (i.e., whether 'pos' can be dropped from this result).

Theorem 6. *For ranked trees,* NPW^{k}A = FO+posTC^k.

5   Single Head on Trees

Single-head tree-walking automata (with output) were introduced as a device for syntax-directed translation [1] (see [17]). Quite recently they came into fashion again as a model for translation of XML specifications [25, 28, 23, 7].

---

^{3} Here we assume that instructions dropping and retrieving pebbles have no tests.

The control of a single-head tree-walking automaton is at a single node of the input tree, i.e., sequential. Thus it differs from the classic bottom-up/top-down tree automata, which are inherently parallel in the sense that the control is fused/split for every branching of the tree.

The power of the classic model is well known: it accepts the regular tree languages. For twa however, the situation was unclear for a long time. They recognize regular tree languages only, but it was conjectured in [12] (and later in [15, 13, 7]) that they cannot recognize them all. Recently this has been proved for deterministic and nondeterministic twa in [5] and [6], respectively.

To strengthen the power of the single-head twa, keeping its sequential nature, in [13] the single-head twa was equipped with nested pebbles. We showed there that the tree languages accepted by such twa are still regular.

As observed before, $\mathsf{DSPACE}(\log n)$ is the class of languages accepted by single-head two-way automata with (nonnested) pebbles. Thus, for $k = 1$ (single-head automata vs. unary transitive closure), our main characterization for tree languages, Theorem 5, can be seen as a 'regular' restriction of the result of Immerman characterizing $\mathsf{DSPACE}(\log n)$; on the one hand only automata with *nested* pebbles are allowed, while on the other hand we consider only *unary* transitive closure, i.e., transitive closure for $\phi(x, y)$ where x, y are single variables. Note that unary transitive closure can be simulated in monadic second-order logic (MSO), which defines the family REG of regular tree languages [10, 33].

In the diagram we compare the family $\mathsf{FO}+\mathsf{DTC}^1 = \mathsf{DPW}^1\mathsf{A}$ with several next of kin. Lines without question mark denote proper inclusion. By LFO we denote the family of languages definable in *local* first-order logic, i.e., dropping the atomic formula $x \leq y$. The regular language $(aa)^*$ shows that $\mathsf{DW}^1\mathsf{A} \not\subseteq \mathsf{FO}$.

Consider the binary trees that among their leaves have (exactly) three positions marked by a special symbol a in such a way that there is an internal node, the left subtree of which contains a single a, while its right subtree contains the other two. This example from [5] shows that $\mathsf{DW}^1\mathsf{A} \subset \mathsf{NW}^1\mathsf{A}$ and moreover that $\mathsf{FO} \not\subseteq \mathsf{DW}^1\mathsf{A}$.

The example to prove $\mathsf{REG} \not\subseteq \mathsf{NW}^1\mathsf{A}$ [6] shows even that $\mathsf{FO} \not\subseteq \mathsf{NW}^1\mathsf{A}$, cf. Example 1. Logical characterizations of $\mathsf{DW}^1\mathsf{A}$ and $\mathsf{NW}^1\mathsf{A}$ are given in [29], using transitive closure in a restricted way. In [31] several logics for regular

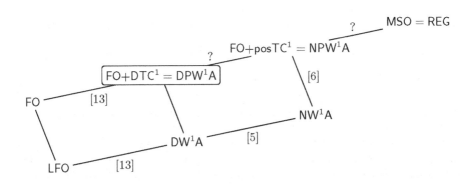

tree languages are studied; it is stated as an open problem whether all regular tree languages can be defined using monadic transitive closure, i.e., whether $\mathsf{FO+TC^1 = REG}$.

Some of the inclusions between the families of trees we have studied are not known to be strict. In particular, can single-head twa with pebbles recognize all regular tree languages? If, instead of with pebbles, they are equipped with a synchronized pushdown or, equivalently, with 'marbles', then they do recognize all regular tree languages [22, 15]. There are several other open questions. Is there a hierarchy for tree languages accepted by (deterministic) twa with respect to the number of pebbles these automata use? Are our 'pointer' pebbles more powerful than the usual 'physical' ones? For nonnested pebbles the two types have the same power, even when the number of pebbles is fixed [3].

6 Walking on Graphs

We generalize our results on trees (and strings) to more general families of graphs (with both node and edge labels). To have a meaningful notion of graph-walking automaton we only consider (connected) graphs with a natural locality condition: a node cannot have two incident edges with the same label and the same direction. Trees over a ranked alphabet fall under this definition since we label the edge from a parent to its i-th child by i. Unranked trees satisfy this condition when represented with 'first child' and 'next sibling' edges. Two-dimensional grids satisfy it by distinguishing between horizontal and vertical edges.

A k-head *graph-walking automaton with nested pebbles* is like its relative for trees, but it may additionally check whether one of its current nodes has an incident incoming/outgoing edge with a specific label (generalizing the concepts of child number and rank). Generally graphs do not have a distinguished node (like the root for trees); thus for acceptance of an input graph we require that the automaton has an accepting computation when started with all its heads on *any* node of the input graph. Not all automata satisfy this requirement.

The first-order logic for graphs over the label alphabet Σ has atomic formulas $\mathrm{lab}_\sigma(x)$, $\sigma \in \Sigma$, for a node x with label σ, $\mathrm{edg}_\sigma(x, y)$, $\sigma \in \Sigma$, for an edge from x to y with label σ, and $x = y$. We do not allow the predicate $x \leq y$, although for trees that can be defined in first-order logic with deterministic transitive closure.

For arbitrary families of graphs the computation of an automaton can be specified in logic, like in Section 4. (We keep the notation for the families.)

Lemma 7. *For every family of graphs,* $\mathsf{DPW^kA \subseteq FO+DTC^k}$.

The other direction holds for all families of graphs for which there exists a (fixed) single-head deterministic graph-walking automaton (with nested pebbles) that can traverse each graph of the family, visiting each node at least once. Such a family is called *searchable*, and the fixed automaton a *guide*.

Unranked (ordered) trees, without bound on the number of children of a node, are a searchable graph family in their representation as binary trees. The (single-head) automata in this representation may move to the first child or to the next sibling of a node (and back), exactly as customary in the literature [28, 30] (albeit without pebbles). Rectangular (directed) grids, edges pointing to the right or downwards, with edge labels distinguishing these two types of edges, form another example of a searchable family. This can be generalized to higher dimensions [2].

Cyclic grids, or toruses, where the last node of each row has an edge to the first node of that row, and similarly for columns, can be searched using two pebbles. We search the grid row-by-row: the first pebble marks the position we start with (in order to stop when all rows are visited; this pebble is not moved), the second pebble moves down in the first column to mark the position in which we started the row (in order to stop when we finish the row; we then move the pebble down to the next row until we meet the first pebble).

Theorem 8. *For every searchable family of graphs,* $\mathsf{DPW}^k\mathsf{A} = \mathsf{FO}+\mathsf{DTC}^k$.

The family of all graphs is not searchable, not even with nonnested pebbles or with several heads. This follows from results of Cook and Rackoff [9].

It is open whether we can search a maze (a connected subgraph of a grid) with a single head using nested pebbles. However with two heads we can search a maze [4]. To cover this family we need to extend the notion of searchability: a family of graphs is k-*searchable* if there is a deterministic guide as before, now having k heads. We have to extend our automaton model with a *new instruction* that moves a given head to a given pebble (like the 'jumping' instruction from [9]). This is quite natural if we see pebbles as pointers, storing the address of a node. With this assumption we get a result as above for k-searchable families.

Some unresolved questions were stated in Section 5. Another question is whether our results can be generalized to alternating automata and the alternating transitive closure operator of [20].

References

1. A.V. Aho, J.D. Ullman. Translations on a context free grammar, Inform. Control. 19, 439–475, 1971.
2. Y. Bargury, J.A. Makowsky. The expressive power of transitive closure and 2-way multihead automata, Proceedings CSL '91, LNCS 626, 1–14, 1992.
3. M. Blum, C. Hewitt. Automata on a 2-dimensional tape, Proceedings 8th IEEE SWAT, 155–160, 1967.
4. M. Blum, D. Kozen. On the power of the compass (or, why mazes are easier to search than graphs), Proceedings 19th FOCS, 132–142, 1978.
5. M. Bojańczyk, T. Colcombet. Tree-walking automata cannot be determinized, Proceedings ICALP 2004 (J. Diaz et al., eds.), LNCS 3142, 246–256, 2004.
6. M. Bojańczyk, T. Colcombet. Tree-walking automata do not recognize all regular languages, Proceedings 37th STOC (H.N. Gabow, R. Fagin, eds.), 2005.
7. A. Brüggemann-Klein, D. Wood. Caterpillars: A context specification technique, Markup Languages 2, 81–106, 2000.
8. L. Budach. Automata and labyrinths, Math. Nachr. 86, 195–282, 1978.
9. S.A. Cook, C.W. Rackoff. Space lower bounds for maze threadability on restricted machines, SIAM J. Comput. 9, 636–652, 1980.

10. J. Doner. Tree acceptors and some of their applications, J. Comp. Syst. Sci. 4, 406–451, 1970.

11. H.-D. Ebbinghaus, J. Flum. *Finite Model Theory*, second edition, Perspectives in Mathematical Logic, Springer-Verlag, Berlin, 1999.

12. J. Engelfriet. Context-free grammars with storage, Leiden University, Technical Report 86–11, 1986.

13. J. Engelfriet, H.J. Hoogeboom. Tree-walking pebble automata, *Jewels are forever* (J. Karhumäki et al., eds.), Springer-Verlag, 72–83, 1999.

14. J. Engelfriet, H.J. Hoogeboom. Automata with nested pebbles capture first-order logic with transitive closure. LIACS Tech. Rep. 2005-02, Leiden University, 2005.

15. J. Engelfriet, H.J. Hoogeboom, J.-P. van Best. Trips on trees, Acta Cyb. 14, 51–64, 1999.

16. J. Engelfriet, S. Maneth. A comparison of pebble tree transducers with macro tree transducers, Acta Inf. 39, 613–698, 2003.

17. J. Engelfriet, G. Rozenberg, G. Slutzki. Tree transducers, L systems, and two-way machines, J. Comp. Syst. Sci. 20, 150–202, 1980.

18. D. Giammarresi, A. Restivo. Two-dimensional languages, *Handbook of Formal Languages*, Vol. 3 (G. Rozenberg, A. Salomaa, eds.), Springer-Verlag, 1997.

19. N. Globerman, D. Harel. Complexity results for two-way and multi-pebble automata and their logics, TCS 169, 161–184, 1996.

20. N. Immerman. Languages that capture complexity classes, SIAM J. Comput. 16, 760–778, 1987.

21. N. Immerman. *Descriptive Complexity*, Springer-Verlag, New York, 1999.

22. T. Kamimura, G. Slutzki. Parallel and two-way automata on directed ordered acyclic graphs, Inform. Control. 49, 10–51, 1981.

23. N. Klarlund, T. Schwentick, D. Suciu. XML: Model, Schemas, Types, Logics, and Queries, *Logics for Emerging Applications of Databases* (J. Chomicki et al., eds.), Springer-Verlag, 1–41, 2004.

24. O. Matz, N. Schweikardt, W. Thomas. The monadic quantifier alternation hierarchy over grids and graphs, Inform. Comput. 179, 356–383, 2002.

25. T. Milo, D. Suciu, V. Vianu. Typechecking for XML transformers, J. Comp. Syst. Sci. 66, 66–97, 2003.

26. B. Monien. Two-way multihead automata over a one-letter alphabet, RAIRO – ITA 14, 67–82, 1980.

27. A. Muscholl, M. Samuelides, L. Segoufin. Complementing deterministic tree-walking automata, Manuscript, 2005. To appear in IPL.

28. F. Neven. Automata, logic, and XML, Proceedings CSL 2002 (J.C. Bradfield, ed.), LNCS 2471, 2–26, 2002.

29. F. Neven, T. Schwentick. On the power of tree-walking automata, Inform. Comput. 183, 86–103, 2003.

30. A. Okhotin, K. Salomaa, M. Domaratzki. One-visit caterpillar tree automata, Fund. Inf. 52, 2002, 361–375.

31. A. Potthoff. Logische Klassifizierung regulärer Baumsprachen, PhD thesis, Institut für Informatik und Praktische Mathematik, Universität Kiel, 1994.

32. M. Sipser. Halting space-bounded computations, TCS 10, 335–338, 1980.

33. J.W. Thatcher, J.B. Wright. Generalized finite automata theory with an application to a decision problem of second-order logic, MST 2, 57–81, 1968.

34. W. Thomas. Languages, automata, and logic, *Handbook of Formal Languages*, Vol. 3 (G. Rozenberg, A. Salomaa, eds.), Springer-Verlag, 1997.

Definability of Languages by Generalized First-Order Formulas over (N,+)

Amitabha Roy and Howard Straubing

Computer Science Department, Boston College, Chestnut Hill,
Massachusetts 02476, USA

Abstract. We consider an extension of first-order logic by modular quantifiers of a fixed modulus q. Drawing on collapse results from finite model theory and techniques of finite semigroup theory, we show that if the only available numerical predicate is addition, then sentences in this logic cannot define the set of bit strings in which the number of 1's is divisible by a prime p that does not divide q. More generally, we completely characterize the regular languages definable in this logic. The corresponding statement, with addition replaced by arbitrary numerical predicates, is equivalent to the conjectured separation of the circuit complexity class ACC from NC^1. Thus our theorem can be viewed as proving a highly uniform version of the conjecture.

1 Background

The circuit complexity class $ACC(q)$ is the family of languages recognized by constant-depth polynomial-size families of circuits containing unbounded fan-in AND, OR, and MOD_q gates for some fixed modulus $q > 0$. It is known that if q is a prime power, and p is a prime that does not divide q, then $ACC(q)$ does not contain the language L_p consisting of all bit strings in which the number of 1's is divisible by p (Razborov [15], Smolensky [17]). But for moduli q that have distinct prime divisors, little is known, and the task of separating ACC, the union of the $ACC(q)$, from NC^1 is an outstanding unsolved problem in circuit complexity.

$ACC(q)$ has a model-theoretic characterization as the family of languages definable in an extension of first-order logic which contains predicate symbols for arbitrary relations on the natural numbers, and in which special "modular quantifiers" of modulus q occur along with ordinary quantifiers. (Barrington, et. al. [2], Straubing [18].) Since there are languages that are complete for NC^1 under constant-depth reductions, in order to separate NC^1 from ACC, it is sufficient to show that for each $q > 1$ there is a language in NC^1 that does not belong to $ACC(q)$. This suggests that one might be able to attack the problem by model-theoretic means. However, the problem has resisted solution by this or any other method, and little progress has been made since the appearance of Smolensky's work.

Recently, K.J. Lange [9] raised the possibility of proving the separation for logics with a restricted class of numerical predicates. It is already known (Straubing, Thérien and Thomas [19]) that if the only available numerical predicate is $<$,

B. Durand and W. Thomas (Eds.): STACS 2006, LNCS 3884, pp. 489–499, 2006.
© Springer-Verlag Berlin Heidelberg 2006

then all the languages definable with ordinary and modular quantifiers of modulus q are regular, and all the groups in the syntactic monoids of these languages are solvable, of cardinality dividing a power of q. This implies, for example, that if q is odd, then one cannot define the set of bit strings with an even number of 1's in this logic. The natural next step is to allow the ternary relation $x + y = z$ on the natural numbers. One can prove the analogue of the separation between AC^0 and NC^1 in this setting by purely model-theoretic means, without recourse to results from circuit complexity. (Originally proved by Lynch [12]. The question is discussed at length in Barrington, *et. al.* [4].) In the present paper we extend this to formulas with ordinary and modular quantifiers over the numerical predicate $x + y = z$. This can be viewed as proving the separation between ACC and NC^1 in a highly uniform setting.

We note that natural uniform versions of AC^0 and ACC result when one allows both addition and multiplication as numerical predicates. (Barrington, Immerman and Straubing [3].) These formulas behave very differently, and are much harder to analyze by model-theoretic means. So separating ACC from NC^1 even in this natural uniform setting still appears to be a very difficult problem.

We obtain our result by first showing, in Section 3, that it is sufficient to consider sentences that only quantify over positions in a bit string that contain a 1. The underlying quantifier-elimination procedure, while rather complicated in the case of modular quantifiers, is based on an idea that goes back to Presburger [14]. In Section 4, we use another model-theoretic collapse, this one based on Ramsey's Theorem, to show that it is sufficient to consider sentences in which the only numerical predicate is $<$, which can be analyzed by known semigroup-theoretic methods. Semigroup theory has been used in the past to obtain rather weak lower bounds for computations by circuits and branching programs (e.g., Barrington and Straubing [5]). By coupling the algebra in this way with ideas from model theory, we are able to extend its reach.

Nurmonen [13] establishes different nonexpressibility results for sentences with modular quantifiers, using a version of Ehrenfeucht-Fraïssé games. Schweikardt [16] proves nonexpressibility results for logics with different generalized quantifiers over the base $(\mathbb{N}, +)$. Extension of the Ramsey property to generalized quantifiers is discussed in Benedikt and Libkin [8].

Of course, we are most interested in proving the separation over arbitrary numerical predicates, or, at the very least, over a class of numerical predicates that includes both addition and multiplication. In the final section we discuss both the prospects for generalizing the present work, and the obstacles to doing so.

Considerations of length oblige us to give only an outline of the main argument; the complete details will be given in the full paper.

We have relied heavily on the account of collapse results for embedded finite models contained in two works on finite model theory by L. Libkin: the survey article [10] and the book [11]. We acknowledge helpful discussions with Klaus-Jörn Lange, Denis Thérien, David Mix Barrington, and the late Clemens Lautemann.

2 Notation and Statement of Result

We consider first-order logic $FO[+]$ with a single ternary relation $x + y = z$. Formulas are interpreted in the natural numbers $\mathbf{N}$. We supplement this with a single unary relation π. The resulting formulas are interpreted in bit strings, with $\pi(x)$ taken to mean that the bit in position x is 1. In fact we can consider several such interpretations: in finite bit strings ($w \in \{0,1\}^*$), in infinite bit strings ($w \in \{0,1\}^{\mathbf{N}}$) and in infinite bit strings with a finite number of 1's ($w \in \{0,1\}^*0^\omega$, where 0^ω denotes an infinite sequence of 0's). A sentence ϕ in this logic accordingly defines three sets of strings:

$$L_\phi^{fin} = \{w \in \{0,1\}^* : w \models \phi\},$$

$$L_\phi^\infty = \{w \in \{0,1\}^{\mathbf{N}} : w \models \phi\},$$

and

$$L_\phi^{fs} = \{w \in \{0,1\}^*0^\omega : w \models \phi\}.$$

(The letters "fs" stand for "finite support.")

For example, let ϕ be the sentence

$$\exists x \exists y ((x = y + y) \wedge \pi(x)),$$

which asserts that there is a 1 in an even numbered position. Note that for this sentence L_ϕ^{fs} is a proper subset of L_ϕ^∞, and that $L_\phi^{fs} = L_\phi^{fin}0^\omega$.

We denote this logic by $FO[\pi, +]$. More generally, if $\mathcal{R}$ is any set of relations on $\mathbf{N}$, we denote the analogous logic by $FO[\pi, \mathcal{R}]$. We define the languages L_ϕ^{fin}, etc., in exactly the same way.

To this apparatus we adjoin *modular quantifiers* $\exists^{r \bmod q}$ for a fixed modulus q. The interpretation of $\exists^{r \bmod q} x \, \phi$ is, informally, 'the number of positions x for which ϕ holds is congruent to r modulo q.' More precisely, let $\phi(x, y_1, \ldots, y_k)$ be a formula with free variables $x, y_1, \ldots, y_k$. Let $w \in \{0,1\}^*$ or $w \in \{0,1\}^{\mathbf{N}}$, and let $a_1, \ldots, a_k < |w|$. (If w is infinite this last condition is automatically satisfied for any natural numbers a_i.) Then we define

$$w \models (\exists^{r \bmod q} x \, \phi)(a_1, \ldots, a_k)$$

if and only if

$$|\{b < |w| : w \models \phi(b, a_1, \ldots, a_k)\}| \equiv r \pmod{q}.$$

(In particular, for infinite strings w, this implies that the set $\{b < |w| : w \models \phi(b, a_1, \ldots, a_k)\}$ is finite.) For example, the sentence

$$\exists^{0 \bmod 2} x \, \pi(x)$$

defines, in all three interpretations, the set of strings with an even number of 1's.

We denote this logic by $(FO + MOD_q)[\pi, +]$.

Here is our main result. Let $m > 1$, and let L_m denote the set of all finite bit strings in which the number of 1's is divisible by m.

Theorem 1. *If m is a prime that does not divide q, then there is no sentence* ϕ *in* $(FO + MOD_q)[\pi, +]$ *such that* $L_\phi^{fin} = L_m$, *or* $L_\phi^\infty = L_m 0^\omega$.

Remark. If we consider instead the family $\mathcal{N}$ of all relations on $\mathbf{N}$, then the family of languages in $\{0,1\}^*$ definable by sentences in $(FO + MOD_q)(\pi, \mathcal{N})$ is precisely the nonuniform circuit complexity class $ACC(q)$ ([2, 18]). If we let $\times$ denote multiplication in $\mathbf{N}$, then $(FO + MOD_q)[\pi, +, \times]$ is the natural uniform version of $ACC(q)$ ([3]). For these logics, the analogues of Theorem 1 are equivalent to the conjectured separation of $ACC(q)$ and NC^1 in the nonuniform and uniform cases, respectively. Thus our theorem can be thought of as establishing this separation in a highly uniform setting.

In our proof of Theorem 1, we will use some notions from the algebraic theory of finite automata: To each regular language $L \subseteq \Sigma^*$ there is associated a finite monoid $M(L)$ (the *syntactic monoid* of L) and a homomorphism $\mu_L : \Sigma^* \to M(L)$ (the *syntactic morphism* of L) such that the value $\mu_L(w)$ determines whether or not $w \in L$. That is, there is a subset X of $M(L)$ such that $L = \mu_L^{-1}(X)$. ($M(L)$ is the *smallest* monoid with this property: it is the monoid of transformations on the states of the minimal automaton of L induced by elements of Σ^*. The homomorphism μ_L maps a word w to the transformation it induces, and X is the set of transformations that take an initial state to an accepting state.)

If $L \subseteq \Sigma^*$, and $\lambda \in \Sigma$, we say λ is a *neutral letter* for L if for any $u, v \in \Sigma^*$, $u\lambda v \in L$ if and only if $uv \in L$. In other words, deleting or inserting occurrences of λ does not affect a word's membership in L. In the algebraic setting, λ is a neutral letter for L if and only if $\mu_L(\lambda)$ is the identity of $M(L)$.

3   Collapse to Active Domain Formulas

While our goal is to prove a result about definability sets of finite strings, most of our argument concerns definability of sets of infinite strings. An easy reduction makes the connection between the two models.

Lemma 1. *Let* ϕ *be a sentence of* $(FO + MOD_q)[\pi, +]$ *and let* $L = L_\phi^{fin}$. *Then there is a sentence* ϕ' *of* $(FO + MOD_q)[\pi, +]$ *such that*

$$L_{\phi'}^{fs} = L_{\phi'}^\infty = L0^\omega.$$

Proof. We define a formula $\phi[\leq x]$ with a single free variable x by rewriting it from the innermost quantifier outward, replacing each instance of

$$Qz\alpha,$$

where Q is the quantifier $\exists$ or $\exists^{r \bmod q}$, by

$$Qz((z \leq x) \wedge \alpha).$$

Then $L0^\omega$ is defined by the sentence

$$\exists x(\forall y(\pi(y) \to y \leq x) \wedge \phi[\leq x]).$$

Remark. Obviously, Lemma 1 holds for any of the logics $(FO + MOD_q)[\pi, \mathcal{R}]$ in which $\leq$ is definable.

An *active domain formula* in $(FO + MOD_q)[\pi, +]$ is one in which every quantifier occurs in the form

$$Qx(\pi(x) \wedge \alpha),$$

where Q is either the ordinary existential quantifier or a modular quantifier, and α is a formula. We call these *active domain quantifiers*. In other words, we allow quantification only over positions that contain the bit 1. Libkin [10] sketches a proof that one can replace every formula in $FO[\pi, +]$ by an equivalent active-domain formula, provided one extends the signature. (The "natural-active collapse".) Here we generalize this result to formulas that contain modular quantifiers.

We consider the logic

$$(FO + MOD_q)[\pi, +, <, 0, 1, \{\equiv_s: s > 1\}],$$

in which $+$ is now treated as a binary function, 0 and 1 are constants, and $\equiv_s$ is a binary relation symbol denoting congruence modulo s. Of course, all these new constants and relations are definable in $FO[+]$, but we need to include them formally as part of the language in order to carry out the quantifier elimination.

Theorem 2. *Let ϕ be a formula of $(FO + MOD_q)[\pi, +, <, 0, 1, \{\equiv_s: s > 1\}]$, with free variables in $\{x_1, \ldots, x_r\}$. Then there is an active-domain formula ψ in the same logic such that for all $w \in \{0,1\}^*0^\omega$ and $a_1, \ldots, a_r \in \mathbf{N}$, we have $w \models \phi(a_1, \ldots, a_r)$, if and only if $w \models \psi(a_1, \ldots, a_r)$.*

Proof. The proof, which we only sketch, is by induction on the construction of ϕ. There is nothing to prove in the base case of quantifier-free formulas. For the inductive step, we assume

$$\phi = Qz\, \phi' \tag{1}$$

where Q is either an existential quantifier ($\exists$) or a modular quantifier ($\exists^{k \bmod q}$) and ϕ' is a formula such that any quantifier appearing in ϕ' is an active domain quantifier. We assume that ϕ' has free variables $x_1, x_2, \ldots x_r$ and bound variables (hence active domain variables) $y_1, y_2, \ldots, y_s$.

Notation. We shall write $\hat{\mathbf{v}}^m$ to denote the tuple $(v_1, v_2, \ldots, v_m)$. When m is obvious from the context or is irrelevant, we simply write $\hat{\mathbf{v}}$ and refer to the i-th coordinate as $\hat{\mathbf{v}}_i$.

By application of standard techniques, we may assume ϕ has the form

$$Q z \left((z \equiv_d c) \wedge \phi' \right)$$

for some $d > 0$, where Q is an ordinary existential or ordinary modular quantifier and ϕ' is an active domain formula in which every atomic formula involving z is either of the form $z < \rho$ or $z = \rho$ or $z > \rho$. Moreover ρ has the form

$$a_0 + a_1 w_1 + \cdots + a_k w_k,$$

where the a_i are integers and the w_i are variables different from z. Strictly speaking, such an expression is not a term in our logic, since we do not assume that subtraction is available. We thus must regard a formula like $z < \rho$ as a kind of shorthand for $z + \rho_1 = \rho_2$, where ρ_1 and ρ_2 are terms.

We now fix an instantiation of $\hat{\mathbf{x}}^r$, the free variables of ϕ, by a tuple $\hat{\mathbf{a}}^r \in \mathbf{N}^r$. To simplify the notation, we will not make explicit reference to $\hat{\mathbf{a}}^r$ in the remainder of the proof. Each ρ appearing on the right-hand side of one of our atomic formulas accordingly defines a partial function g from $\mathbf{N}^s$ into $\mathbf{N}$, where s is the number of active-domain variables. We let $\{g_i : i \in I\}$ denote the set of these partial functions.

Let $w \in \{0, 1\}^*0^\omega$, and let $D \subseteq \mathbb{N}$ denote the set of positions in w that contain 1's. (That is, D is the active domain of w.) Let

$$\mathcal{B} = \bigcup_{i \in I} \{g_i(\hat{\mathbf{y}}) | \hat{\mathbf{y}} \in D^s\}.$$

Write $\mathcal{B}$ as an ordered set $\{b_0, b_1, \ldots, b_{p-1}\}$ where $b_0 < b_1 < b_2 \cdots < b_{p-1}$. We denote by (a, b) the set $\{x \in \mathbf{N} : a < x < b\}$. By an *interval* in $\mathcal{B}$, we will mean either the leftmost interval $(-1, b_0)$, or intervals of the form (b_i, b_{i+1}) for $0 \le i < p - 2$ or the rightmost interval (b_{p-1}, ∞).

Following Libkin [10], we show:

Lemma 2. *If there exists an integer z_0 in an interval in $\mathcal{B}$ such that*

$$w \models \phi'(z_0).$$

then

$$w \models \phi'(z_0').$$

for every z_0' in the interval.(That is, if an interval contains a witness, then every point in the interval is a witness.)

The proof given in [10] only considers the case of ordinary quantifiers, not modular quantifiers, but in this lemma the modular quantifiers do not introduce any new complication.

Corollary 1. *Let (l, r) be an interval in $\mathcal{B}$ such that $l \equiv_{dq} \alpha$. Then*

$$w \models \{(z_0 \equiv_d c) \wedge \phi'(z_0)\}$$

for some $z_0 \in (l, r)$ iff

$$w \models \phi'(l + ((c - \alpha) \bmod d))$$

We also have a special property concerning the infinite interval (b_{p-1}, ∞):

Corollary 2. *Let $b_{p-1} \equiv_{dq} \alpha$ If*

$$w \models \exists^{k \bmod q} z \ \{(z \equiv_d c) \wedge \phi'\}$$

then

$$w \not\models \phi'(b_{p-1} + ((c - \alpha) \bmod d))$$

We note as well the following fact;

Lemma 3. *Let* $l, r \in \mathbf{Z}$ *and* $c, d, q, \alpha, \beta \in \mathbf{N}$ *be such that*

$$l \equiv \alpha \bmod dq \text{ and } r \equiv \beta \bmod dq.$$

Let $\eta(\alpha, \beta)$ *denote the number of integers* x *in* (l, r) *such that* $x \equiv_d c$. *Then,*

$$\eta(\alpha, \beta) \equiv 1 + \frac{\beta - \alpha - ((c - \alpha) \bmod d) - ((\beta - c) \bmod d)}{d} \quad (\bmod q).$$

(The point of the foregoing lemma is that given c, d, q, α and β, $\eta(\alpha, \beta)$ is determined by the constants α, β, c, d, q. The exact form of the expression is irrelevant.)

We now proceed to the quantifier elimination by building an active domain formula equivalent to $\phi = \exists^{k \bmod q} z((z \equiv_d c) \wedge \phi'(z))$. The idea is to write a formula that counts, modulo q, the number of witnesses to $(z \equiv_d c) \wedge \phi'(z)$ in each interval of $\mathcal{B}$. (We may restrict our argument to elimination of the modular quantifier, since the much simpler case of the ordinary quantifier is treated in [11].)

Let us provisionally admit into our formulas a predicate "$x \in \mathcal{B}$". We first show that we can replace ϕ by an equivalent formula that only quantifies over elements of $\mathcal{B}$; that is, in which each occurrence of a quantifier is of the form $\mathcal{Q}x((x \in \mathcal{B}) \wedge \psi)$. Indeed, with such restricted quantification we can say that there are $t \bmod q$ elements x of $\mathcal{B}$ such that $x \equiv_d c$ and $\phi'(x)$, and also that there are $t' \bmod q$ intervals (x, y) in $\mathcal{B}$ such that $x \equiv_{dq} \alpha$, $y \equiv_{dq} \beta$, and $\phi'(x+(c-\alpha) \bmod d)$. It follows from Corollaries 1 and 2, and Lemma 3, that with a boolean combination of such formulas we can express ϕ. (We need Corollary 2 to say that the infinite interval (b_{p-1}, ∞) contains no witnesses.)

Second, we can express quantification over elements of $\mathcal{B}$ by quantification over tuples of active-domain elements. Of course, we can say "there exists $x \in \mathcal{B}$ such that $\psi(x)$" by saying there exists a tuple $\hat{\mathbf{y}}$ such that $\psi(g_i(\hat{\mathbf{y}}))$ for some i. But it is not clear how to extend this to modular quantification over active domain elements, since a single element of $\mathcal{B}$ could be the image of many different tuples $\hat{\mathbf{y}}$ under more than one of the partial function g_i. The trick is to consider the tuples $\hat{\mathbf{y}}$ such that no lexicographically larger tuple has the same image under g_i, and no g_j with $j > i$ maps to $g_i(\hat{\mathbf{y}})$. It is easy to express that a tuple is maximal in this sense, and thus we can replace our formula by one in which we quantify over such maximal tuples.

Finally, we can express modular quantification over tuples $\hat{\mathbf{y}}$ in terms of modular quantification over individual elements by noting that

$$\exists^{k \bmod q}(y_1, y_2) \; \alpha$$

is equivalent to the disjunction, over all functions $f : Z_q \to Z_q$, where $\sum_{i=0}^{q-1} i f(i) = k$, of the following:

$$\bigwedge_{i=0}^{q-1} \exists^{i \bmod q} y_1 \exists^{f(i) \bmod q} y_2 \; \alpha.$$

We can extend this inductively to quantification over tuples of arbitrary size.

4   Collapse to Formulas with $<$ as the Only Numerical Predicate

4.1   Ramsey Property

Our discussion here closely parallels that of Libkin [11]. Let $\mathcal{R}$ be any set of relations on $\mathbf{N}$, and let $\phi(x_1, \ldots, x_k)$ be an active-domain formula in $(FO + MOD_q)[\pi, \mathcal{R}]$. We say that ϕ has the *Ramsey property* if for each infinite subset X of $\mathbf{N}$, there exist an infinite subset Y of X and an active-domain formula $\psi(x_1, \ldots, x_k)$ in $(FO + MOD_q)[\pi, <]$ that satisfies the following condition: If $w \in \{0, 1\}^* 0^\omega$ and all the 1's in w are in positions belonging to Y, then for all $a_1, \ldots, a_k \in Y$,

$$w \models \phi(a_1, \ldots, a_k) \text{ iff } w \models \psi(a_1, \ldots, a_k).$$

Lemma 4. *Let $\mathcal{N}$ be the set of all relations on $\mathbf{N}$. Every active-domain formula in $(FO + MOD_q)[\pi, \mathcal{N}]$ has the Ramsey property.*

The proof is essentially the same as the one for formulas with ordinary first-order quantifiers, given in [11]. The introduction of modular quantifiers does not alter the argument.

The Ramsey property allows us to capture a subset of a language expressible by a formula ϕ (which satisfies the Ramsey property) by a new formula over a very limited vocabulary (the only numerical predicate allowed is $<$). This limited vocabulary restricts the kind of language that can be expressed.

Lemma 5. *Let $L_\psi = \{w \mid w \in \{0, 1\}^*\}$ be the set of finite bit strings defined by an active-domain sentence $\psi \in (FO + MOD_q)[\pi, <]$.*

 (i) The language L_ψ is regular. Moreover, the syntactic monoid $M(L_\psi)$ contains only solvable groups whose order divides a power of q.
 (ii) L_ψ has 0 as a neutral letter.
 (iii) Let $z \in \Sigma^*$. Then $z \in L_\psi$ iff $z0^\omega \models \psi$.

Proof. Condition (i) follows from a result of Straubing, Thérien and Thomas [19]. Inserting or deleting 0's from any string satisfying ψ does not alter the truth value of any atomic formula of the form $x < y$ provided the variables represent positions containing 1, which is the case here, since ψ is active-domain. Conditions (ii) and (iii) follow from an easy induction on the quantifier depth.

4.2   Proof of Theorem 1

Let m be a prime that does not divide q, and suppose, contrary to the claim in the theorem, that L_m is defined by a sentence ϕ of $(FO + MOD_q)[\pi, +]$. By Lemma 1, Theorem 2, and Lemma 4, there exists an active-domain sentence ψ of $(FO + MOD_q)[\pi, <]$ and an infinite subset Y of $\mathbf{N}$ such that for all $w \in \{0, 1\}^* 0^\omega$ in which all 1's are in positions belonging to Y, $w \models \phi$ if and only if $w \in L_m 0^\omega$. Let L_ψ denote the set of finite bit strings that satisfy ψ. We prove:

Lemma 6. $L_m = L_\psi$.

Proof. We first show that $L_\psi \subseteq L_m$. Let $z' \in L_\psi$. We pad z' with 0's so that the 1's in the new padded string z'' appear in positions included in the set Y. Since $z'' \in L_\psi$ (by Lemma 5 (ii)), we conclude that $z''0^\omega \models \psi$ (by Lemma 5 (iii)). Since the 1's in $z''0^\omega$ appear in positions in Y, $z''0^\omega \models \phi$. Hence $z''0^\omega \in L_m 0^\omega$, so $z'' \in L_m$. Removing additional neutral letter 0's introduced while padding z', we conclude that $z' \in L_m$.

The opposite inclusion $(L_m \subseteq L_\psi)$ is proved by reversing each step above.

Since the syntactic monoid of L_m is the cyclic group Z_m and that of L_ψ has groups of order dividing a power of q (via Lemma 5), we have a contradiction since $(m, q) = 1$. Thus L_m cannot be defined by a sentence in $(FO + MOD_q)[\pi, +]$. This completes the proof.

4.3 Other Non-definability Results

Here we show how to extend Theorem 1 to prove nonexpressibility results for other languages. We begin by removing the restriction to binary alphabets.

Let Σ be a finite alphabet and let us consider languages definable in the logic $\mathcal{L}_{q,\Sigma,+} = (FO + MOD_q)[\{\pi_\sigma : \sigma \in \Sigma\}, +]$, where each π_σ is a unary predicate: $\pi_\sigma x$ is interpreted to mean that the letter in position x is σ. We designate a special letter $\lambda \in \Sigma$, and say that a formula is active-domain (with respect to λ) if every existential and modular quantifier Q occurs in the form $Qx(\pi_\sigma x \wedge \alpha)$, where $\sigma \neq \lambda$. Note that we need never use the atomic formula $\pi_\lambda x$, even in non-active-domain formulas, as it is equivalent to the conjunction of the $\neg\pi_\sigma x$ over all letters σ not equal to λ. All the preceding results hold in this broader setting, with no changes to their proofs. We thus have:

Theorem 3. *Let $L \subseteq \Sigma^*$, with $\lambda \in \Sigma$ a neutral letter for L. If L is definable in $\mathcal{L}_{q,\Sigma,+}$, then it is definable by a sentence of $(FO + MOD_q)[\{\pi_\sigma : \sigma \in \Sigma\}, <]$. In particular, L is regular, and every group in $M(L)$ is solvable, with cardinality dividing a power of q.*

The foregoing theorem allows us to give an effective characterization of all the *regular* languages in $\mathcal{L}_{q,\Sigma,+}$.

Theorem 4. *Let $L \subseteq \Sigma^*$, be regular. L is definable in $\mathcal{L}_{q,\Sigma,+}$ if and only if for all $t > 0$, every group in $\mu_L(\Sigma^t)$ is solvable, and has cardinality dividing a power of q.*

The reduction to the neutral letter case is somewhat involved, so we omit the proof. The same property is known to characterize the regular languages in $ACC(q)$, provided that the conjectured separation of $ACC(q)$ and NC^1 holds. ([2]). (The condition is effective, because there are only finitely many distinct sets $\mu_L(\Sigma^t)$, and these can be effectively enumerated.)

Here is an application of Theorem 4. Let G be a finite group and let $\Sigma \subseteq G$ be a set of generators of G. We treat G as a finite alphabet; to each word

$w \in \Sigma^*$ we assign the group element $\phi(w)$ that results by multiplying together the letters of w. The *word problem for G* (with respect to Σ) is the language $\{w \in \Sigma^* : \phi(w) = 1\}$. Barrington [1] showed that the word problem for any finite nonsolvable group is complete for NC^1 with respect to constant-depth reductions, so that the conjectured separation of ACC from NC^1 is equivalent to the assertion that no such word problem belongs to ACC. We can verify directly that no such word problem L is definable in $\mathcal{L}_{q,\Sigma,+}$: L is a regular language, and it is easy to check that $M(L) = G$ and $\mu_L = \phi$. If G is nonsolvable then its commutator subgroup G' is also non-solvable and thus every element of G' is the image of a word over Σ of length divisible by G (each commutator is an image of a word of the form $uvu^{|G|-1}v^{|G|-1}$ where $u, v \in \Sigma$). We can pad each of these words with a sufficient number of copies of $\sigma^{|G|}$ (for some fixed $\sigma \in \Sigma$) so that they all have the same length t. Thus $G' \subseteq \phi(\Sigma^t)$.

Theorem 5. *No word problem of a finite nonsolvable group is definable in any* $\mathcal{L}_{q,\Sigma,+}$.

Note that it is precisely the nonsolvability of G, rather than the relation between $|G|$ and q, that is at issue here: For instance, a word problem of the alternating group of degree 5, whose cardinality is 60, is not definable in $\mathcal{L}_{30,\Sigma,+}$. even though the cardinality and modulus are consistent. On the other hand, the word problem for any solvable group of order 60 is definable in this logic.

5   Directions for Further Research

In many steps of the algorithm for reducing a sentence defining L_m to an active-domain sentence, we introduced ordinary quantifiers even when the original formula had only modular quantifiers. If there were a way to avoid this, we could also prove, by the same techniques, that the language $0^*1\{0,1\}^*$ cannot be defined by a formula over $(\mathbf{N}, +)$ having only modular quantifiers. If addition is replaced by arbitrary numerical predicates, this statement is equivalent to the conjecture that the circuit complexity class CC^0 does not contain the language 1^*. (CC^0 is the class of languages recognized by constant-depth, polynomial-size circuit families in which every gate is a $MODq$ gate for a fixed modulus q. See Barrington, et. al. [6].)

Of course, we would really like to prove our result over a base of arbitrary numerical predicates, or at the very least, over the base $\{+, \times\}$. Note, however, that in these logics it is possible to define the set of infinite strings with an even number of 1's in first-order logic without using modular quantifiers! Let $E(x)$ be the numerical predicate "the binary expansion of x contains an even number of 1's, and $B(x, y)$ the predicate "bit y in the binary expansion of x is 1". Then the set of infinite bit strings with an even number of 1's is defined by

$$\exists x(E(x) \wedge \forall y(\pi(y) \leftrightarrow B(x, y))).$$

Both E and B are definable over $(+, \times)$. This shows that we cannot extend the natural-active collapse argument to these richer logics. It also shows (since we know, from circuit complexity, that first-order sentences cannot define PARITY for *finite* strings) that there are important differences between finite and infinite strings as regards definability.

One possible approach to more general formulas is to try to prove the collapse for sentences that define regular languages.

We have not mentioned circuits at all, even though this work was inspired by a problem in circuit complexity. It would be interesting to know if there is any natural interpretation of the classes $FO[\pi, +]$ or $(FO + MOD)[\pi, +]$ in terms of circuits.

References

1. D. Mix Barrington, "Bounded-Width Polynomial-Size Branching Programs Recognize Exactly Those Languages in NC^1", *J. Comp. Syst. Sci.* **38** (1989) 150–164.
2. D. Mix Barrington, K. Compton, H. Straubing, and D. Thérien, "Regular Languages in NC^1", *J. Comp. Syst. Sci.* **44** (1992) 478–499.
3. D. Mix Barrington, N. Immerman, and H. Straubing, "On Uniformity in NC^1", *J. Comp. Syst. Sci.* **41** (1990) 274–306.
4. D.M. Barrington, N. Immerman, C. Lautemann, N. Schweikardt, and D. Thérien The Crane Beach Conjecture, LICS '01, 187-196.
5. D. Mix Barrington and H. Straubing, "Superlinear Lower Bounds for Bounded-Width Branching Programs",*J. Comp. Syst. Sci.* **50** (1995) 374–381.
6. D. Mix Barrington, H. Straubing, and D. Thérien, "Nonuniform Automata over Groups", *Information and Computation* **89** (1990) 109–132.
7. D. Mix Barrington and D. Thérien, "Finite Monoids and the Fine Structure of NC^1", *JACM* **35** (1988) 941–952 .
8. M. Benedikt and L. Libkin, "Relational Queries over Interpreted Structures", *Journal of the ACM*, volume 47, no. 4, 644–680, 2000.
9. K.-J. Lange, "Some Results on Majority Quantifiers over Words", *19th IEEE Conference on Computational Complexity,* (2004) 123-129.
10. L. Libkin, "Embedded finite models and constraint databases", in E. Grädel, et. al., eds., *Finite Model Theory and its Applications,* Springer, New York (2005).
11. L. Libkin, *Elements of Finite Model Theory,* Springer, New York (2004).
12. J. F. Lynch, "On Sets of Relations Definable by Addition", . *J. Symbolic Logic,* **47** (1982) 659-668.
13. J. Nurmonen, "Counting Modulo Quantifiers on Finite Structures", *Information and Computation* **160** (2000), 183–207.
14. M. Presburger, "Ueber die Vollstaendigkeit eines gewissen Systems der Arithmetik ganzer Zahlen, in welchem die Addition als einzige Operation hervortritt." In Comptes Rendus du I congrès de Mathématiciens des Pays Slaves. Warsaw, Poland: pp. 92-101, 1929.
15. A. A. Razborov, "Lower Bounds for the Size of Circuits of Bounded Depth with Basis $\{\wedge, \oplus\}$", *Math. Notes of the Soviet Academy of Sciences* **41** (1987) 333–338.
16. N. Schweikardt, "Arithmetic, first-order logic, and counting quantifiers", *ACM Transactions on Computational Logic*, volume 6, number 3, 634–671, July, 2005.
17. R. Smolensky, "Algebraic Methods in the Theory of Lower Bounds for Boolean Circuit Complexity", *Proc. 19th ACM STOC* (1987) 77–82.
18. H. Straubing, *Finite Automata, Formal Logic and Circuit Complexity*, Birkhaüser, Boston (1994).
19. H. Straubing, D. Thérien, and W. Thomas, "Regular Languages Defined with Generalized Quantifiers", *Information and Computation* **118** (1995) 289-301.

Generalized Modal Satisfiability[*]

Michael Bauland[1], Edith Hemaspaandra[2],
Henning Schnoor[1], and Ilka Schnoor[1]

[1] Theoretische Informatik, Universität Hannover, Appelstr. 4,
30167 Hannover, Germany
{bauland, henning.schnoor, ilka.schnoor}@thi.uni-hannover.de
[2] Department of Computer Science, Rochester Institute of Technology,
Rochester, NY 14623, USA
eh@cs.rit.edu

Abstract. It is well-known that modal satisfiability is PSPACE-complete [Lad77]. However, the complexity may decrease if we restrict the set of propositional operators used. Note that there exist an infinite number of propositional operators, since a propositional operator is simply a Boolean function. We completely classify the complexity of modal satisfiability for every finite set of propositional operators, i.e., in contrast to previous work, we classify an infinite number of problems. We show that, depending on the set of propositional operators, modal satisfiability is PSPACE-complete, coNP-complete, or in P. We obtain this trichotomy not only for modal formulas, but also for their more succinct representation using modal circuits.

Keywords: computational complexity, modal logic.

1 Introduction

Modal logic has been a subject of research for a long time. The first steps in introducing modal logics into mathematics were taken by Lewis [Lew18, LL32]. Further important works are [vW51] and [Göd33]. In the 1960s modal logic was enriched with relational semantics, and finally completeness results such as presented in [LS77] and [Seg71] were feasible. This resulted in an increase of research in this area. Especially Kripke contributed much to the relational approach (see, e.g., [Kri63a, Kri63b]), that is why the most general logic in our context is called K. A historical overview can be found in [BdRV01] and a comprehensive view of its evolution is given by [Gol03].

Modal logic has been used as a powerful tool to reason about knowledge and belief in artificial intelligence since the 1970s [Moo79, MSHI78]. Some decades ago it was discovered that it can also be used in cryptographic and other protocols (see, e.g., [HMT88, FI86, LR86]). Nowadays modal logic is still an important issue. For example, Bennett and Galton introduced a new modal language called Versatile Event Logic in [BG04] and Liau characterizes the relationship among belief, information acquisition and trust by use of modal logic [Lia03]. With the

[*] Supported in part by grants NSF-CCR-0311021 and DFG VO 630/5-1.

recognition of the usefulness of modal logic for computer science, complexity issues became interesting and still are: For example, in the context of knowledge updates of an agent on the basis of the modal logic S5 [BZ05] or a tableau calculus for a description logic [DM00]. The first results in this area were obtained by Ladner, who showed that modal satisfiability is PSPACE-complete [Lad77]. A good guide to different aspects of the complexity of modal logic is the paper by Halpern and Moses [HM92]. They not only look at the propositional single-agent case, but also at the multi-agent case, which is still PSPACE-complete.

A natural question arising here is: Are there any restrictions of modal logic that are easier from a complexity point of view? There are several possible restrictions to look at. When considering the usual possible-worlds semantics for modal logic, an often-studied restriction is to consider special classes of frames (reflexive, transitive, etc.). These restrictions are important in many applications, since, for example, a modal operator that represents "always" would typically require the accessibility relation in the frame to be reflexive and transitive; for a modal operator that represents "a processor knows" the relation would be reflexive, transitive, and symmetric. It should be noted that such restrictions do not necessarily decrease the complexity; for many common restrictions, the complexity remains the same [Lad77, HM92] and it is even possible that the complexity increases.

Another natural restriction of modal logic is to restrict the set of formulas. Halpern was able to decrease the complexity of different modal logics to linear time by allowing only a finite set of variables and finite nesting of modal operators [Hal95]. Restricted modal languages where only a subset of the relevant modal operators are allowed have been studied in the context of linear temporal logic (see, e.g., [SC85]). Some description logics can be viewed as modal logic with a restriction on the propositional operators that are allowed. For the complexity of description logics, see, e.g., [SS91, DHL$^+$92, DLNN97]. For the complexity of modal logic with other restrictions on the set of operators, see [Hem01].

There exist an infinite number of propositional operators, since a propositional operator is simply a Boolean function. We completely classify the complexity of modal satisfiability for every finite set of propositional operators, i.e., we classify an infinite number of problems. The restriction on the propositional operators leads to a classification following the structure of Post's lattice [Pos41]. For propositional logic, Lewis showed that this problem is dichotomic: Depending on the set of operators, propositional satisfiability is either NP-complete or solvable in polynomial time [Lew79]. For modal satisfiability, we achieve a trichotomy: We show that modal satisfiability is PSPACE-complete, coNP-complete, or in P.

When considering sets of operations which do not include negation, the complexity for the cases where one modal operator is allowed sometimes differs from the case where we allow both operator $\Diamond$ and its dual operator $\Box$. With only one of these, modal satisfiability is PSPACE-complete exactly in those cases in which propositional satisfiability is NP-complete. When we allow both modal operators, the jump to PSPACE-completeness happens earlier, i.e., with a set of operations with less expressive power.

We also look at *modal circuits*, which are a succinct way of representing modal formulas. We show that this does not give us a significantly different complexity than the formula case.

2 Preliminaries

Modal logic is an extension of classical propositional logic that talks about "possible worlds." Modal formulas in our context are basically propositional formulas with an additional unary operator $\Diamond$. A model for a given formula consists of a graph with propositional assignments. To be more precise, a *frame* consists of a set W of "worlds," and a "successor" relation $R \subseteq W \times W$. For $(w, w') \in R$, we say w' is a *successor* of w. A *model* M consists of a frame (W, R), a set X of propositional variables and a function $\pi \colon X \to \mathcal{P}(W)$. For $x \in X$, $\pi(x)$ denotes the set of worlds in which the variable x is true. $\Box\phi$ is defined as $\neg\Diamond\neg\phi$. Intuitively, $\Diamond\phi$ means "there is a successor world in which ϕ holds," and $\Box\phi$ means "ϕ holds in all successor worlds." For a class $\mathcal{F}$ of frames, we say a model M is an $\mathcal{F}$-model if the underlying frame is an element of $\mathcal{F}$.

For a formula ϕ built over the variables X, propositional operators $\wedge$ and $\neg$, and the unary modal operator $\Diamond$, we define what "ϕ holds at world w" means for model M (or M, w *satisfies* ϕ) with assignment function π, written as $M, w \models \phi$.

If ϕ is a propositional variable x, then $M, w \models \phi$ if and only if $w \in \pi(x)$. As usual, $M, w \models \phi_1 \wedge \phi_2$ if and only if $(M, w \models \phi_1$ and $M, w \models \phi_2)$ and $M, w \models \neg\phi$ iff $M, w \not\models \phi$. Finally, $M, w \models \Diamond\phi$ if and only if there is a world $w' \in W$ such that $(w, w') \in R$ and $M, w' \models \phi$.

For a class $\mathcal{F}$ of frames, we say a formula ϕ is $\mathcal{F}$-satisfiable if there exists an $\mathcal{F}$-model $M = (W, R, \pi)$ and a world $w \in W$ such that $M, w \models \phi$. For ϕ and ψ modal formulas, we write $\phi \equiv_\mathcal{F} \psi$ if for every world in every $\mathcal{F}$-model, ϕ holds if and only if ψ holds. Formula ϕ is an $\mathcal{F}$-tautology if $\phi \equiv_\mathcal{F} 1$, and ϕ is $\mathcal{F}$-constant if $\phi \equiv_\mathcal{F} 0$ or $\phi \equiv_\mathcal{F} 1$.

We now define some classes of frames commonly used in applications of modal logic. To see how these frames correspond to axioms and proof systems, see, for example, [BdRV01]. K is the class of all frames, KD is the class of frames in which every world has a successor, T and K4 are the class of reflexive, resp. transitive frames. The class of of reflexive and transitive frames is called S4, and S5 is the class of reflexive, symmetric, and transitive frames. The *reflexive singleton* is the frame consisting of one world w, and the relation $\{(w, w)\}$. Note that all classes of frames $\mathcal{F}$ described above contain the reflexive singleton.

We also consider a more general notion of modal formulas, whose propositional analog has been studied extensively. For a finite set B of Boolean functions, a *modal B-circuit* is a propositional Boolean circuit with gates for functions from B and additional gates representing the modal operators $\Diamond$ or $\Box$. To be more precise, a modal B-circuit consists of a directed acyclic graph (interpreted as gates and wires in the circuit), and a marking defining the type of gates, i.e., describing which function or operator they represent. An order on the edges is defined for non-commutative operations appearing as gate functions. For a precise definition of Boolean circuits, see, e.g., [Vol99]. A modal B-formula can be seen as a modal B-circuit where each gate has out-degree ≤ 1. This corresponds to the intuitive idea of a formula: Such a circuit can be written down as a formula, without growing significantly in size. Semantically we interpret a circuit as a succinct representation of its formula expansion. For a modal B-circuit C, the *modal depth of C*, $md(C)$, is the maximal number of gates representing modal operators on a directed path in the graph. If, in addition to input gates, gates representing functions from B, and modal gates, we also allow negative literals to occur (as happens, for example, in description logics), we say C is a *modal B-circuit with atomic negation*. Since every Boolean function can be expressed using only $\neg$ and $\wedge$, the semantics for circuits allowing arbitrary connectives is immediate.

We now define the various modal satisfiability problems we are interested in.

Definition 2.1. *Let B be a finite set of Boolean functions, $\mathcal{F}$ a class of frames, and $M \subseteq \{\Diamond, \Box\}$. Then $\mathrm{FORM}_M(B)$ $(\mathrm{CIRC}_M(B))$ is the set of modal B-formulas (B-circuits) using modal operators in M. $\mathrm{FORM}_M^{\sim}(B)$ $(\mathrm{CIRC}_M^{\sim}(B))$ are the corresponding modal formulas (circuits) with atomic negation. $\mathcal{F}$-$\mathrm{MSAT}_M^F(B)$ $(\mathcal{F}$-$\mathrm{MSAT}_M^C(B))$ is the problem, given a formula (circuit) from $\mathrm{FORM}_M(B)$ $(\mathrm{CIRC}_M(B))$, to determine if it is $\mathcal{F}$-satisfiable. The problems $\mathcal{F}$-$\mathrm{MSAT}_M^{\tilde{F}}(B)$ and $\mathcal{F}$-$\mathrm{MSAT}_M^{\tilde{C}}(B)$ are defined analogously, where the circuits and formulas may also contain atomic negation.*

We now state Ladner's theorem in our notation.

Theorem 2.2 ([Lad77]). *The problems K-$\mathrm{MSAT}_\Box^F(\wedge, \neg)$, T-$\mathrm{MSAT}_\Box^F(\wedge, \neg)$, $\mathrm{S4}$-$\mathrm{MSAT}_\Box^F(\wedge, \neg)$ are PSPACE-complete and $\mathrm{S5}$-$\mathrm{MSAT}_\Box^F(\wedge, \neg)$ is NP-complete.*

In [Hem01], Hemaspaandra examined the complexity of K-$\mathrm{MSAT}_M^F(B)$ and K-$\mathrm{MSAT}_M^{\tilde{F}}(B)$ for all $M \subseteq \{\Box, \Diamond\}$ and $B \subseteq \{\wedge, \vee, \neg, 0, 1\}$. In this paper, we classify the complexity of modal satisfiability for all finite sets of Boolean functions, i.e., in contrast to previous results on the complexity of modal logics, we classify an infinite number of modal logics.

To obtain our classification, we define some notions about Boolean functions. A set B of Boolean functions is called a *clone* if it is closed under *superposition*, that is, B contains all projection functions and is closed under permutation, identification of variables, and arbitrary composition. The set of clones forms a lattice, which has been completely classified by Post [Pos41] (see [Rei01, p. 24]). For a set B of Boolean functions, let $[B]$ be the smallest clone containing B.

We define the clones that arise in our complexity classification. For more information about Post's lattice and its use in complexity classifications of propositional logic, see, for example, [BCRV03, BCRV04].

The smallest clone contains only projections and is named $\mathrm{I}_2 = [\{id\}]$. $\mathrm{I}_1 = [\{id, 1\}]$ (where id is the Boolean identity function). The largest clone $\mathrm{BF} = [\{\wedge, \neg\}]$ is the set of all Boolean functions. The set of all monotonic functions forms a clone denoted by $\mathrm{M} = [\{\vee, \wedge, 0, 1\}]$. D consists of all *self-dual* functions, i.e., $f \in \mathrm{D}$ if and only if $f(x_1, \ldots, x_n) = \neg f(\neg x_1, \ldots, \neg x_n)$. $\mathrm{L} = [\{\oplus, 1\}]$ is the clone of all linear Boolean functions (where $\oplus$ is the Boolean exclusive OR). The clone of all Boolean functions that can be written using only disjunction and constants is called $V = [\{\vee, 1, 0\}]$; $V_0 = [\{\vee, 0\}]$ and $V_2 = [\{\vee\}]$. Similarly, the clone $E = [\{\wedge, 0, 1\}]$ contains the Boolean functions that can be written as conjunctions of variables and constants; $E_0 = [\{\wedge, 0\}]$ and $E_2 = [\{\wedge\}]$. R_1 is built from all 1-*reproducing* functions, i.e., all functions f with $f(1, \ldots, 1) = 1$. The clone $N = [\{\neg, 1\}]$ consists of the projections, their negations, and all constant Boolean functions. $S_1 = [\{x \wedge \overline{y}\}]$ and $S_{11} = S_1 \cap M$.

Clones have an important property: The functions in a clone $[B]$ are exactly those that can be composed from variables and functions from B only. If we interpret Boolean formulas as Boolean functions, then $[B]$ consists of all formulas that are equivalent to a formula built with variables and connectives from B. Therefore, this framework can be used to investigate problems related to Boolean formulas depending on which connectives are allowed. Lewis presented the result that the satisfiability problem for Boolean formulas with connectives from B is NP-complete if $S_1 \subseteq [B]$ and in P otherwise [Lew79]. Another example is the classification of the equivalence problem given by Reith: The problem to decide whether formulas with connectives from B are equivalent is in LOGSPACE, if $[B] \subseteq V$ or $[B] \subseteq E$ or $[B] \subseteq L$, and coNP-complete in all other cases [Rei01]. Dichotomy results for counting the solutions of formulas [RW00], finding the minimal solutions of formulas [RV00], and learnability of Boolean formulas and circuits [Dal00] were achieved as well. Post's lattice has been a helpful tool in the constraint satisfaction context. This is surprising, because constraint satisfaction problems are not related to Post's lattice by definition, but clones appear through a Galois connection [JCG97]. Nordh used this technique in the context of nonmonotonic logics and presented a trichotomy theorem for the inference problem in propositional circumscription [Nor05].

An important point in the present paper is the result that the complexity of the various satisfiability problems for sets B only depends on the clone generated by B. This is not obvious, since expressing some function in a clone with the base functions of the respective clone can lead to a formula with exponential length. A crucial tool in restricting the length of the resulting formula is the following lemma showing that for certain sets B there are always short formulas representing the functions AND and OR. Part (2) and (3) follow directly from the proofs in [Lew79], part (1) is from [Sch05].

Lemma 2.3. *1. Let B be a finite set of Boolean functions such that $V \subseteq [B] \subseteq M$ ($E \subseteq [B] \subseteq M$, resp.). Then there exists a B-formula $f(x, y)$ such that f represents $x \vee y$ ($x \wedge y$, resp.) and each of the variables x and y occurs exactly once in $f(x, y)$.*

2. Let B be a finite set of Boolean functions such that $[B] = BF$. Then there are B-formulas $f(x, y)$ and $g(x, y)$ such that f represents $x \vee y$, g represents $x \wedge y$, and both variables occur in each of these formulas exactly once.

3. Let B be a finite set of Boolean functions such that $N \subseteq [B]$. Then there is a B-formula $f(x)$ such that f represents $\neg x$ and the variable x occurs in f only once.

3 Results

Our main result is the following trichotomy for modal satisfiability.

Theorem 3.1. *Let B be a finite set of Boolean functions.*

1. *If $[D] \supseteq S_{11}$, then K-MSAT$^F_{\Box,\Diamond}(D)$ and K-MSAT$^C_{\Box,\Diamond}(B)$ are PSPACE-complete (Corollary 3.9).*
2. *If $[B] \in \{E, E_0\}$, then K-MSAT$^F_{\Box,\Diamond}(B)$ and K-MSAT$^C_{\Box,\Diamond}(B)$ are coNP-complete (Lemma 3.10, Lemma 3.5, and Lemma 3.11).*
3. *In all other cases, K-MSAT$^F_{\Box,\Diamond}(B)$ and K-MSAT$^C_{\Box,\Diamond}(B)$ are in P (Section 3.4 and the structure of Post's lattice).*

In the cases where only one modal operator is allowed we achieve a dichotomy.

Theorem 3.2. *Let B be a finite set of Boolean functions.*

1. *If $[B] \supseteq S_1$, then K-MSAT$^F_{\Box}(B)$, K-MSAT$^F_{\Diamond}(B)$, K-MSAT$^C_{\Box}(B)$, and K-MSAT$^C_{\Diamond}(B)$ are PSPACE-complete (Corollary 3.9 and Corollary 3.6);*
2. *otherwise, they are in P (Section 3.4 and the structure of Post's lattice).*

This dichotomy is a natural analog of Lewis's result that the satisfiability problem for Boolean formulas with connectives from B is NP-complete if $S_1 \subseteq [B]$ and in P otherwise [Lew79].

From these theorems, we conclude that using the more succinct representation of modal circuits does not increase the polynomial degree of the complexity of the satisfiability problem (for two problems A and B, we write $A \equiv^p_m B$ if $A \leq^p_m B$ and $B \leq^p_m A$).

Corollary 3.3. *Let B be a finite set of Boolean functions. Then* K-MSAT$^C_{\Box,\Diamond}(B)$ $\equiv^p_m$ K-MSAT$^F_{\Box,\Diamond}(B)$, K-MSAT$^C_{\Box}(B)$ $\equiv^p_m$ K-MSAT$^F_{\Box}(B)$, *and* K-MSAT$^C_{\Diamond}(B) \equiv^p_m$ K-MSAT$^F_{\Diamond}(B)$.

The remainder of this section is devoted to proving these theorems. We will generalize many of these results to other classes of frames in Section 4. *Because of space limitations, most of the proofs are omitted; all proofs can be found in the full version of this paper [BHSS05].*

3.1 General Upper Bounds

It is well-known that the $\mathcal{F}$-satisfiability problem for modal formulas using the operators $\Box, \wedge$, and $\neg$ is solvable in PSPACE for a variety of classes $\mathcal{F}$ of frames (see [Lad77]). The following theorem shows that the circuit case can be reduced to the formula case, thus putting the circuit problems in PSPACE as well.

Theorem 3.4. *Let $\mathcal{F}$ be a class of frames. Then $\mathcal{F}$-MSAT$^C_{\Box}(\wedge, \neg)$ $\leq^{\log}_m$ $\mathcal{F}$-MSAT$^F_{\Box}(\wedge, \neg)$.*

To handle bases other than $\{\wedge, \neg\}$, we need the following simple lemma. The proof uses the standard gate replacement technique.

Lemma 3.5. *Let B_1, B_2 be finite sets of Boolean formulas, let $\mathcal{F}$ be a class of frames, and let $M \subseteq \{\Box, \Diamond\}$. If $[B_1] \subseteq [B_2]$, then $\mathcal{F}$-MSAT$_M^C(B_1) \leq_m^{\log}$ $\mathcal{F}$-MSAT$_M^C(B_2)$ and $\mathcal{F}$-MSAT$_M^{\tilde{C}}(B_1) \leq_m^{\log} \mathcal{F}$-MSAT$_M^{\tilde{C}}(B_2)$.*

Corollary 3.6. *Let B be a finite set of Boolean formulas, and let $\mathcal{F}$ be a class of frames. Then*

1. *S5-MSAT$_{\Box,\Diamond}^{\tilde{C}}(B)$ is in NP, and*
2. *K-MSAT$_{\Box,\Diamond}^{\tilde{C}}(B)$, KD-MSAT$_{\Box,\Diamond}^{\tilde{C}}(B)$, T-MSAT$_{\Box,\Diamond}^{\tilde{C}}(B)$, and S4-MSAT$_{\Box,\Diamond}^{\tilde{C}}(B)$ are in PSPACE.*

Proof. For K, T, S4, and S5, this follows from Theorem 3.4, the results in [Lad77] (see Theorem 2.2), and Lemma 3.5. The KD upper bound easily follows from [Lad77]. □

3.2 PSPACE-Completeness

We now show how to express general modal formulas and circuits using one modal operator and any set B of Boolean functions such that $[B] \supseteq S_1$. This implies that the satisfiability problems for such sets B of Boolean functions are as hard as the general case. The proof of the following theorem uses a generalization of ideas from the proof of the main result in [Lew79].

Theorem 3.7. *Let $\mathcal{F}$ be a class of frames, B a finite set of Boolean functions, $S_1 \subseteq [B]$. Let $M \subseteq \{\Box, \Diamond\}$ such that $M \neq \emptyset$. Then $\mathcal{F}$-MSAT$_{\Box,\Diamond}^F(\wedge, \neg) \leq_m^{\log}$ $\mathcal{F}$-MSAT$_M^F(B)$ and $\mathcal{F}$-MSAT$_{\Box,\Diamond}^C(\wedge, \neg) \leq_m^{\log} \mathcal{F}$-MSAT$_M^C(B)$.*

Proof. Let $\phi \in \text{FORM}_{\Box,\Diamond}(\wedge, \neg)$. Without loss of generality, assume that ϕ contains only modal operators from M. Let $B' := B \cup \{1\}$. Then $[B'] = \text{BF}$ (since I_1 is the smallest clone containing 1, and BF is the smallest clone containing I_1 and S_1). It follows from Lemma 2.3 that there is a B' formula $f_\neg(x)$ that represents $\neg x$, and x occurs in $f_\neg(x)$ only once, and there exist B' formulas $f_\wedge(x, y)$ and $f_\vee(x, y)$ such that $f_\wedge$ represents $\wedge$, $f_\vee(x, y)$ represents $\vee$, and x and y occur exactly once in $f_\wedge(x, y)$ and exactly once in $f_\vee(x, y)$. In ϕ, replace every occurrence of $\wedge$ with $f_\wedge$, every occurrence of $\vee$ with $f_\vee$, and every occurrence of $\neg$ with $f_\neg$. Call the resulting formula ϕ'. Clearly, ϕ' is a formula in $\text{FORM}_M(B')$ and ϕ' is equivalent to ϕ. By choice of $f_\vee, f_\wedge$, and $f_\neg$, ϕ' is computable in polynomial time.

Now replace every occurrence of 1 with a new variable t and force t to be 1 in every relevant world by adding $\wedge \bigwedge_{i=0}^{md(\phi)} \Box^i t$. This is a conjunction of linearly many terms (since $md(\phi) \leq |\phi|$). We insert parentheses in such a way that we get a tree of $\wedge$'s of logarithmic depth. Now express the $\wedge$'s in this tree with the equivalent B-formula (which exists, since $[B] \supseteq S_1 \supset E_2 = [\wedge]$) with the result only increasing polynomially in size.

For circuits, use the same construction; this is in fact easier, since short formulas are not required. □

The following theorem implies that for the class K, PSPACE-completeness already holds for a lower class in Post's lattice. It can be proven in a similar way as the preceding theorem.

Theorem 3.8. *Let $\mathcal{F}$ be a class of frames, let B be a finite set of Boolean functions such that $S_{11} \subseteq [B]$, and let $M \subseteq \{\Box, \Diamond\}$. Then $\mathcal{F}\text{-MSAT}^{F}_{M}(\wedge, \vee, 0) \leq^{\log}_{m}$ $\mathcal{F}\text{-MSAT}^{F}_{M}(B)$ and $\mathcal{F}\text{-MSAT}^{C}_{M}(\wedge, \vee, 0) \leq^{\log}_{m} \mathcal{F}\text{-MSAT}^{C}_{M}(B)$.*

Corollary 3.9.   *1. Let $\mathcal{F}$ be a class of frames such that $S4 \subseteq \mathcal{F} \subseteq K$, and let B be a finite set of Boolean functions such that $[B] \supseteq S_{1}$. Let $\emptyset \neq M \subseteq \{\Box, \Diamond\}$. Then $\mathcal{F}\text{-MSAT}^{F}_{M}(B)$ and $\mathcal{F}\text{-MSAT}^{C}_{M}(B)$ are $\leq^{\log}_{m}$-hard for PSPACE.*
   2. Let B be a finite set of Boolean functions such that $S_{11} \subseteq [B]$. Then $K\text{-MSAT}^{F}_{\Box,\Diamond}(B)$ and $K\text{-MSAT}^{C}_{\Box,\Diamond}(B)$ are PSPACE-complete.

Proof. Part 1 follows directly from Theorem 3.7 and [Lad77]. In [Hem01, Theorem 6.5], it is shown that $K\text{-MSAT}^{F}_{\Box,\Diamond}(\wedge, \vee, 0)$ is PSPACE-hard. Thus part (2) follows from Theorem 3.8 and Corollary 3.6.   □

3.3   coNP-Completeness

In [Hem01], the analogous result of the following lemma was shown for formulas. We prove that this coNP upper bound also holds for circuits.

Lemma 3.10. $K\text{-MSAT}^{\tilde{C}}_{\Box,\Diamond}(\wedge, 0, 1) \in \text{coNP}$.

In [Hem05], it is shown that $K\text{-MSAT}^{F}_{\Box,\Diamond}(\wedge, 0)$ is coNP-hard. Applying Lemma 2.3, we obtain the following result.

Lemma 3.11. *Let B be a finite set of Boolean functions such that $[B] \supseteq E_{0}$. Then $K\text{-MSAT}^{F}_{\Box,\Diamond}(B)$ is coNP-hard.*

3.4   Polynomial Time

Every propositional formula ϕ representing a function belonging to the clone R_{1} is satisfiable, since $\phi(1, \ldots, 1) = 1$. Similarly, formulas ϕ representing self-dual functions (i.e., the functions in the clone D) are always satisfiable, since if $\phi(0, \ldots, 0) = 0$, then $\phi(1, \ldots, 1) = 1$. These results transfer to the modal case: For the reflexive singleton, an assignment to the propositional variables satisfying the "de-modalized" version of a formula (where all modal operators are replaced with identities) satisfies the original modal formula. This leads to the following lemma.

Lemma 3.12. *Let $[B] \subseteq R_{1}$ or $[B] \subseteq D$, and let $\mathcal{F}$ be a class of frames that contains the reflexive singleton. Then every formula from $\text{FORM}_{\Box,\Diamond}(B)$ is $\mathcal{F}$-satisfiable. In particular, $\mathcal{F}\text{-MSAT}^{F}_{\Box,\Diamond}(B), \mathcal{F}\text{-MSAT}^{C}_{\Box,\Diamond}(B) \in \text{LOGSPACE}$.*

The following can easily be proven when considering that $[\{\neg, 1\}] = N$ and that a modal $\{\neg, 1\}$-circuit is basically a linear graph.

Lemma 3.13. *Let B be a finite set of Boolean functions such that $[B] \subseteq N$. Then K-MSAT$_{\square,\Diamond}^C(B) \in$ LOGSPACE.*

Another easy problem is when the set B contains only disjunctions.

Lemma 3.14. *Let B be a finite set of Boolean functions such that $V_2 \subseteq [B] \subseteq V$ and let $M \subseteq \{\Diamond, \square\}$. Then K-MSAT$_M^C(B)$ and K-MSAT$_M^{\tilde{C}}(B)$ are complete for NLOGSPACE under $\leq_m^{\log}$ reductions if $0 \in [B]$, and in LOGSPACE otherwise. The corresponding formula problems are solvable in LOGSPACE.*

Our satisfiability problems for formulas having only $\oplus$ and constants in the propositional base are easy. For the propositional case, this holds because unsatisfiable formulas of this kind are of a very easy form: Every variable and the constant 1 appear an even number of times. In the modal case, unsatisfiable formulas over these connectives are of a similarly regular form. The result also holds for modal circuits.

Theorem 3.15. *Let B be a finite set of Boolean functions such that $[B] \subseteq L$. Then K-MSAT$_{\square,\Diamond}^C(B) \in$ P.*

We now look at satisfiability problems with only one modal operator present. The following lemma can be proven in a similar way as [Hem01, Theorems 6.4.5 and 6.4.6].

Lemma 3.16. *Let B be a finite set of Boolean functions such that $[B] \subseteq M$. Then K-MSAT$_\Diamond^C(B)$ and K-MSAT$_\square^C(B)$ are solvable in polynomial time.*

4   Other Classes of Frames

While K-satisfiability for variable-free formulas using both modal operators, constants, and the Boolean connectives $\land$ and $\lor$ is complete for PSPACE, this problem (even with variables) is solvable in polynomial time if we look only at frames in which each world has a successor. The proof uses a reduction to the propositional case, and then applies the results from [Rei01, Theorem 3.8].

Theorem 4.1. *Let $\mathcal{F}$ be a non-empty class of frames such that $\mathcal{F} \subseteq$ KD, and let B be a finite set of Boolean functions such that $[B] \subseteq M$.*

1. *If $B \subseteq R_1$, then $\mathcal{F}$-MSAT$_{\square,\Diamond}^C(B)$ is in LOGSPACE.*
2. *If $[B] \in \{V_0, V, E_0, E\}$, then $\mathcal{F}$-MSAT$_{\square,\Diamond}^C(B)$ is $\leq_m^{\log}$-complete for NLOGSPACE.*
3. *If $[B] \supseteq S_{11}$, then $\mathcal{F}$-MSAT$_{\square,\Diamond}^C(B)$ is $\leq_m^{\log}$-complete for P.*
4. *$\mathcal{F}$-MSAT$_{\square,\Diamond}^F(B)$ is in LOGSPACE.*

For KD, the other polynomial time results carry over from K. The PSPACE upper bound follows from Corollary 3.6 and PSPACE-hardness follows from Corollary 3.9 (since S4 $\subseteq$ KD $\subseteq$ K). This leads to the following classification.

Theorem 4.2. *Let B be a finite set of Boolean functions.*

1. *If* $[B] \supseteq S_1$, *then* KD-MSAT$_\Box^F(B)$, KD-MSAT$_\Diamond^F(B)$, KD-MSAT$_{\Box,\Diamond}^F(B)$, KD-MSAT$_\Box^C(B)$, KD-MSAT$_\Diamond^C(B)$, *and* KD-MSAT$_{\Box,\Diamond}^C(B)$ *are* PSPACE-*complete;*
2. *otherwise, they are in* P.

From our proofs it also follows that the classes S4 and T behave like KD, except that a full classification for sets B such that $[B] \subseteq L$ is still open. We conjecture that Theorem 4.2 holds with KD replaced by T and with KD replaced by S4. The class S5 shows behavior similar to S4, except that the cases which are PSPACE-complete for S4 are NP-complete for S5.

5 Further Research

A natural next question to look at is the classification for other classes of frames. For $\mathcal{F} \in \{T, S4, S5\}$, our proofs already give a complete classification with the exception of the complexity of the problems $\mathcal{F}$-MSAT$_{\Box,\Diamond}^C(B)$ and $\mathcal{F}$-MSAT$_{\Box,\Diamond}^F(B)$ where $B \subseteq L$. We conjecture that these cases are solvable in polynomial time as well. Another interesting question is the exact complexity of our polynomial cases, most notably the case where the propositional operators represent linear functions. The complete classification for the cases with atomic negation is also still open.

There are many other interesting directions for future research. For example, one can look at other decision problems (e.g., global satisfiability and formula minimization), one can look at extensions of modal logic (e.g., multi-modal logic, description logic, temporal logic, logics of knowledge, etc.), and one can try to generalize modal logic modally as well as propositionally.

Acknowledgements

We thank the anonymous referees for their helpful comments and suggestions.

References

[BCRV03] E. Böhler, N. Creignou, S. Reith, and H. Vollmer. Playing with Boolean blocks, part I: Post's lattice with applications to complexity theory. *SIGACT News*, 34(4):38–52, 2003.

[BCRV04] E. Böhler, N. Creignou, S. Reith, and H. Vollmer. Playing with Boolean blocks, part II: Constraint satisfaction problems. *SIGACT News*, 35(1):22–35, 2004.

[BdRV01] P. Blackburn, M. de Rijke, and Y. Venema. *Modal logic*. Cambridge University Press, New York, NY, USA, 2001.

[BG04] B. Bennett and A. Galton. A unifying semantics for time and events. *Artificial Intelligence*, 153(1-2):13–48, 2004.

[BHSS05]   M. Bauland, E. Hemaspaandra, H. Schnoor, and I. Schnoor. Generalized modal satisfiability. Technical report, Theoretical Computer Science, University of Hannover, 2005.

[BZ05]   C. Baral and Y. Zhang. Knowledge updates: Semantics and complexity issues. *Artificial Intelligence*, 164(1-2):209–243, 2005.

[Dal00]   V. Dalmau. *Computational Complexity of Problems over Generalized Formulas*. PhD thesis, Department de Llenguatges i Sistemes Informàtica, Universitat Politécnica de Catalunya, 2000.

[DHL$^+$92]   F. Donini, B. Hollunder, M. Lenzerini, D. Nardi, W. Nutt, and A. Spaccamela. The complexity of existential quantification in concept languages. *Artificial Intelligence*, 53(2-3):309–327, 1992.

[DLNN97]   F. Donini, M. Lenzerini, D. Nardi, and W. Nutt. The complexity of concept languages. *Information and Computation*, 134:1–58, 1997.

[DM00]   F. Donini and F. Massacci. EXPTIME tableaux for ALC. *Artificial Intelligence*, 124(1):87–138, 2000.

[FI86]   M. Fischer and N. Immerman. Foundations of knowledge for distributed systems. In *TARK '86: Proceedings of the 1986 Conference on Theoretical Aspects of Reasoning About Knowledge*, pages 171–185, San Francisco, CA, USA, 1986. Morgan Kaufmann Publishers Inc.

[Göd33]   K. Gödel. Eine Interpretation des intuitionistischen Aussagenkalküls. *Ergebnisse eines mathematischen Kolloquiums*, 4:34–40, 1933.

[Gol03]   R. Goldblatt. Mathematical modal logic: A view of its evolution. *Journal of Applied Logic*, 1(5-6):309–392, 2003.

[Hal95]   J. Halpern. The effect of bounding the number of primitive propositional and the depth of nesting on the complexity of modal logic. *Artificial Intelligence*, 75(2):361–372, 1995.

[Hem01]   E. Hemaspaandra. The complexity of poor man's logic. *Journal of Logic and Computation*, 11(4):609–622, 2001. Corrected version: [Hem05].

[Hem05]   E. Hemaspaandra. The Complexity of Poor Man's Logic, *CoRR*, cs.LO/9911014, 1999. Revised 2005.

[HM92]   J. Halpern and Y. Moses. A guide to completeness and complexity for modal logics of knowledge and belief. *Artificial Intelligence*, 54(2):319–379, 1992.

[HMT88]   J. Halpern, Y. Moses, and M. Tuttle. A knowledge-based analysis of zero knowledge. In *STOC '88: Proceedings of the 20th Annual ACM Symposium on Theory of Computing*, pages 132–147, New York, NY, USA, 1988. ACM Press.

[JCG97]   P. Jeavons, D. Cohen, and M. Gyssens. Closure properties of constraints. *Journal of the ACM*, 44(4):527–548, 1997.

[Kri63a]   S. Kripke. A semantical analysis of modal logic I: Normal modal propositional calculi. *Zeitschrift für Mathematische Logik und Grundlagen der Mathematik*, 9:67–96, 1963.

[Kri63b]   S. Kripke. Semantical considerations on modal logic. *Acta Philosophica Fennica*, 16:83–94, 1963.

[Lad77]   R. Ladner. The computational complexity of provability in systems of modal propositional logic. *SIAM Journal on Computing*, 6(3):467–480, 1977.

[Lew18]   C. Lewis. *A Survey of Symbolic Logic*. University of California Press, Berkley, 1918.

[Lew79]   H. Lewis. Satisfiability problems for propositional calculi. *Mathematical Systems Theory*, 13:45–53, 1979.

[Lia03]   C.-J. Liau. Belief, information acquisition, and trust in multi-agent systems – a modal logic formulation. *Artificial Intelligence*, 149(1):31–60, 2003.

[LL32]   C. Lewis and C. Langford. *Symbolic Logic*. Dover, 1932.

[LR86]   R. Ladner and J. Reif. The logic of distributed protocols: Preliminary report. In *TARK '86: Proceedings of the 1986 Conference on Theoretical Aspects of Reasoning About Knowledge*, pages 207–222, San Francisco, CA, USA, 1986. Morgan Kaufmann Publishers Inc.

[LS77]   E. Lemmon and D. Scott. An introduction to modal logic - the 'Lemmon Notes', 1977.

[Moo79]   R. Moore. Reasoning about knowledge and action. Technical Report 191, AI Center, SRI International, 333 Ravenswood Ave., Menlo Park, CA 94025, 1979.

[MSHI78]   J. McCarthy, M. Sato, T. Hayashi, and S. Igarashi. On the model theory of knowledge. Technical report, Stanford, CA, USA, 1978.

[Nor05]   G. Nordh. A trichotomy in the complexity of propositional circumscription. In *Proceedings of the 11th International Conference on Logic for Programming*, volume 3452 of *Lecture Notes in Computer Science*, pages 257–269. Springer Verlag, 2005.

[Pos41]   E. Post. The two-valued iterative systems of mathematical logic. *Annals of Mathematical Studies*, 5:1–122, 1941.

[Rei01]   S. Reith. *Generalized Satisfiability Problems*. PhD thesis, Fachbereich Mathematik und Informatik, Universität Würzburg, 2001.

[RV00]   S. Reith and H. Vollmer. Optimal satisfiability for propositional calculi and constraint satisfaction problems. In *Proceedings of the 25th International Symposium on Mathematical Foundations of Computer Science*, volume 1893 of *Lecture Notes in Computer Science*, pages 640–649. Springer Verlag, 2000.

[RW00]   S. Reith and K. Wagner. The complexity of problems defined by Boolean circuits. Technical Report 255, Institut für Informatik, Universität Würzburg, 2000. To appear in *Proceedings of the International Conference on Mathematical Foundation of Informatics*, Hanoi, Oct. 25–28, 1999.

[SC85]   A. Sistla and E. Clarke. The complexity of propositional linear temporal logics. *Journal of the ACM*, 32(3):733–749, 1985.

[Sch05]   H. Schnoor. The complexity of the Boolean formula value problem. Technical report, Theoretical Computer Science, University of Hannover, 2005.

[Seg71]   K. Segerberg. *An Essay in Classical Modal Logic*. Filosofiska studier 13, University of Uppsala, 1971.

[SS91]   M. Schmidt-Schauss and G. Smolka. Attributive concept descriptions with complements. *Artificial Intelligence*, 48(1):1–26, 1991.

[Vol99]   H. Vollmer. *Introduction to Circuit Complexity – A Uniform Approach*. Texts in Theoretical Computer Science. Springer Verlag, Berlin Heidelberg, 1999.

[vW51]   G. von Wright. *An Essay in Modal Logic*. North-Holland Publishing Company, Amsterdam, 1951.

Strategy Improvement and Randomized Subexponential Algorithms for Stochastic Parity Games

Krishnendu Chatterjee[1] and Thomas A. Henzinger[1,2]

[1] EECS, UC Berkeley, USA
[2] EPFL, Switzerland
{c_krish, tah}@eecs.berkeley.edu

Abstract. A stochastic graph game is played by two players on a game graph with probabilistic transitions. We consider stochastic graph games with ω-regular winning conditions specified as parity objectives. These games lie in NP $\cap$ coNP. We present a strategy improvement algorithm for stochastic parity games; this is the first non-brute-force algorithm for solving these games. From the strategy improvement algorithm we obtain a randomized subexponential-time algorithm to solve such games.

1 Introduction

Graph games. A stochastic graph game [5] is played on a directed graph with three kinds of states: player-1, player-2, and probabilistic states. At player-1 states, player 1 chooses a successor state; at player-2 states, player 2 chooses a successor state; at probabilistic states, a successor state is chosen according to a given probability distribution. The outcome of playing the game forever is an infinite path through the graph. If there are no probabilistic states, we refer to the game as a *2-player graph game*; otherwise, as a $2\frac{1}{2}$-*player graph game*.

Parity objectives. The theory of graph games with ω-regular winning conditions is the foundation for modeling and synthesizing reactive processes with fairness constraints. In the case of $2\frac{1}{2}$-player graph games, the two players represent a reactive system and its environment, and the probabilistic states represent uncertainty. The *parity* objectives provide an adequate model, as the fairness constraints of reactive processes are ω-regular, and every ω-regular winning condition can be specified as a parity objective [11]. The solution problem for a $2\frac{1}{2}$-player game with parity objective Φ asks for each state s, for the maximal probability with which player 1 can ensure the satisfaction of Φ if the game is started from s (this probability is called the *value* of the game at s). An *optimal strategy* for player 1 is a strategy that enables player 1 to win with that maximal probability. The existence of *pure memoryless* optimal strategies for $2\frac{1}{2}$-player games with parity objectives was established recently in [4] (a pure memoryless strategy chooses for each player-1 state a unique successor state). The existence of pure memoryless optimal strategies implies that the solution problem for $2\frac{1}{2}$-player games with parity objectives lies in NP $\cap$ coNP.

B. Durand and W. Thomas (Eds.): STACS 2006, LNCS 3884, pp. 512–523, 2006.

Previous algorithms. Emerson and Jutla [7] had showed in 1988 that 2-player parity games (*without* probabilistic states) can be solved in NP ∩ coNP. However, to date no polynomial-time algorithm is known to solve these games. In 2000, Vöge and Jurdziński [12] gave a *strategy improvement* algorithm for 2-player parity games. A strategy improvement scheme iterates local optimizations of a pure memoryless strategy; this works if the iteration can be shown to converge to a global optimum [8]. Although the best known bound for the worst-case running time of Vöge and Jurdziński is exponential, it behaves very well in practice. Moreover, Björklund et al. [1] used the strategy improvement scheme to derive a randomized subexponential-time algorithm for 2-player parity games. And recently, Jurdziński et al. [9] found a deterministic subexponential-time algorithm for 2-player parity games.

For $2\frac{1}{2}$-player games (*with* probabilistic states), Condon [5] proved containment in NP ∩ coNP in 1992 for the restricted case of *reachability* objectives, and she gave a strategy improvement algorithm for this subclass of $2\frac{1}{2}$-player games. Again, no polynomial-time algorithm is known to solve these games, but using strategy improvement, Ludwig [10] derived a randomized subexponential-time algorithm for $2\frac{1}{2}$-player reachability games on *binary* game graphs (game graphs with maximum out-degree 2). The techniques of [1] also yield a randomized subexponential-time algorithm for the nonbinary class of $2\frac{1}{2}$-player reachability games. However, the techniques of [9] do not extend to give a deterministic subexponential-time algorithm for $2\frac{1}{2}$-player reachability games. For the full class of $2\frac{1}{2}$-player games with general parity objectives, no algorithm has been known which is better than a brute-force enumeration of the set of all possible pure memoryless strategies, and chosing the best one.

Our results. We present the first strategy improvement algorithm for $2\frac{1}{2}$-player parity games. Our algorithm combines both techniques for 2-player parity games and for $2\frac{1}{2}$-player reachability games, employing a novel reduction from $2\frac{1}{2}$-player parity games (with quantitative winning criteria) to 2-player parity games (with qualitative winning criteria). We then show how the techniques of [1] can be extended to our strategy improvement algorithm to obtain a randomized subexponential algorithm for $2\frac{1}{2}$-player parity games. Given a game graph with n states and a parity objective with d priorities, the expected running time of our algorithm is $2^{O\left(\sqrt{d \cdot n \cdot \log(n)}\right)}$. The algorithm is subexponential if $d = O\left(\frac{n^{1-\varepsilon}}{\log(n)}\right)$ for some $\varepsilon > 0$. Thus, for the special case of reachability objectives, the expected running time of our algorithm matches the bound of the best known algorithm.

2   Definitions

We consider turn-based probabilistic games and some of its subclasses.

Game graphs. A *turn-based probabilistic game graph* ($2\frac{1}{2}$-*player game graph*) $G = ((S, E), (S_1, S_2, S_\bigcirc), \delta)$ consists of a directed graph (S, E), a partition $(S_1, S_2, S_\bigcirc)$ of the finite set S of states, and a probabilistic transition function δ: $S_\bigcirc \to \mathcal{D}(S)$, where $\mathcal{D}(S)$ denotes the set of probability distributions over the state space S. The states in S_1 are the *player*-1 states, where player 1 decides the successor state; the states in S_2 are the *player*-2 states, where player 2 decides the successor state; and the states in $S_\bigcirc$ are the *probabilistic* states, where the successor state is chosen according to the probabilistic transition function δ. We assume that for $s \in S_\bigcirc$ and $t \in S$, we have $(s,t) \in E$ iff $\delta(s)(t) > 0$, and we often write $\delta(s,t)$ for $\delta(s)(t)$. For technical convenience we assume that every state in the graph (S, E) has at least one outgoing edge. For a state $s \in S$, we write $E(s)$ to denote the set $\{t \in S \mid (s,t) \in E\}$ of possible successors. The *turn-based deterministic game graphs* (*2-player game graphs*) are the special case of the $2\frac{1}{2}$-player game graphs with $S_\bigcirc = \emptyset$. The *Markov decision processes* ($1\frac{1}{2}$-*player game graphs*) are the special case of the $2\frac{1}{2}$-player game graphs with $S_1 = \emptyset$ or $S_2 = \emptyset$. We refer to the MDPs with $S_2 = \emptyset$ as *player*-1 MDPs, and to the MDPs with $S_1 = \emptyset$ as *player*-2 MDPs.

Plays and strategies. An infinite path, or a *play*, of the game graph G is an infinite sequence $\omega = \langle s_0, s_1, s_2, \ldots \rangle$ of states such that $(s_k, s_{k+1}) \in E$ for all $k \in \mathbb{N}$. We write Ω for the set of all plays, and for a state $s \in S$, we write $\Omega_s \subseteq \Omega$ for the set of plays that start from the state s. A *strategy* for player 1 is a function $\sigma \colon S^* \cdot S_1 \to \mathcal{D}(S)$ that assigns a probability distribution to all finite sequences $\boldsymbol{w} \in S^* \cdot S_1$ of states ending in a player-1 state (the sequence represents a prefix of a play). Player 1 follows the strategy σ if in each player-1 move, given that the current history of the game is $\boldsymbol{w} \in S^* \cdot S_1$, she chooses the next state according to the probability distribution $\sigma(\boldsymbol{w})$. A strategy must prescribe only available moves, i.e., for all $\boldsymbol{w} \in S^*$, $s \in S_1$, and $t \in S$, if $\sigma(\boldsymbol{w} \cdot s)(t) > 0$, then $(s,t) \in E$. The strategies for player 2 are defined analogously. We denote by Σ and Π the set of all strategies for player 1 and player 2, respectively.

Once a starting state $s \in S$ and strategies $\sigma \in \Sigma$ and $\pi \in \Pi$ for the two players are fixed, the outcome of the game is a random walk $\omega_s^{\sigma,\pi}$ for which the probabilities of events are uniquely defined, where an *event* $\mathcal{A} \subseteq \Omega$ is a measurable set of paths. Given strategies σ for player 1 and π for player 2, a play $\omega = \langle s_0, s_1, s_2, \ldots \rangle$ is *feasible* if for every $k \in \mathbb{N}$ the following three conditions hold: (1) if $s_k \in S_\bigcirc$, then $(s_k, s_{k+1}) \in E$; (2) if $s_k \in S_1$, then $\sigma(s_0, s_1, \ldots, s_k)(s_{k+1}) > 0$; and (3) if $s_k \in S_2$ then $\pi(s_0, s_1, \ldots, s_k)(s_{k+1}) > 0$. Given two strategies $\sigma \in \Sigma$ and $\pi \in \Pi$, and a state $s \in S$, we denote by $\mathrm{Outcome}(s, \sigma, \pi) \subseteq \Omega_s$ the set of feasible plays that start from s given the strategies σ and π. For a state $s \in S$ and an event $\mathcal{A} \subseteq \Omega$, we write $\Pr_s^{\sigma,\pi}(\mathcal{A})$ for the probability that a path belongs to $\mathcal{A}$ if the game starts from the state s and the players follow the strategies σ and π, respectively.

Strategies that do not use randomization are called pure. A player-1 strategy σ is *pure* if for all $\boldsymbol{w} \in S^*$ and $s \in S_1$, there is a state $t \in S$ such that $\sigma(\boldsymbol{w} \cdot s)(t) = 1$. A *memoryless* player-1 strategy does not depend on the history of the play but only on the current state; it can be represented as a function $\sigma \colon S_1 \to \mathcal{D}(S)$. A *pure memoryless strategy* is a strategy that is both pure and memoryless. A pure memoryless strategy for player 1 can be represented as a function $\sigma \colon S_1 \to S$. We denote by Σ^{PM} the set of pure memoryless strategies for player 1. The pure memoryless player-2 strategies Π^{PM} are defined analogously.

Given a pure memoryless strategy $\sigma \in \Sigma^{PM}$, let G_σ be the game graph obtained from G under the constraint that player 1 follows the strategy σ. The corresponding definition G_π for a player-2 strategy $\pi \in \Pi^{PM}$ is analogous, and we write $G_{\sigma,\pi}$ for the game graph obtained from G if both players follow the pure memoryless strategies σ and π, respectively. Observe that given a $2\frac{1}{2}$-player game graph G and a pure memoryless player 1 strategy σ, the result G_σ is a player-2 MDP. Similarly, for a player-1 MDP G and a pure memoryless player-1 strategy σ, the result G_σ is a Markov chain. Hence, if G is a $2\frac{1}{2}$-player game graph and the two players follow pure memoryless strategies σ and π, the result $G_{\sigma,\pi}$ is a Markov chain.

Objectives. We specify objectives for the players by providing a set of *winning plays* $\Phi \subseteq \Omega$ for each player. We say that a play ω *satisfies* the objective Φ if $\omega \in \Phi$. We study only zero-sum games, where the objectives of the two players are complementary; i.e., if player 1 has the objective Φ, then player 2 has the objective $\Omega \setminus \Phi$. We consider ω-*regular objectives* [11], specified as parity conditions. We also define the special case of reachability objectives.

- *Reachability objectives.* Given a set $T \subseteq S$ of "target" states, the reachability objective requires that some state of T be visited. The set of winning plays is $\text{Reach}(T) = \{ \omega = \langle s_0, s_1, s_2, \ldots \rangle \in \Omega \mid s_k \in T \text{ for some } k \geq 0 \}$.
- *Parity objectives.* For $c, d \in \mathbb{N}$, we write $[c..d] = \{ c, c+1, \ldots, d \}$. Let p: $S \to [0..d]$ be a function that assigns a *priority* $p(s)$ to every state $s \in S$, where $d \in \mathbb{N}$. For a play $\omega = \langle s_0, s_1, \ldots \rangle \in \Omega$, we define $\text{Inf}(\omega) = \{ s \in S \mid s_k = s \text{ for infinitely many } k \}$ to be the set of states that occur infinitely often in ω. The *even-parity objective* is defined as $\text{Parity}(p) = \{ \omega \in \Omega \mid \min\big(p(\text{Inf}(\omega))\big) \text{ is even } \}$, and the *odd-parity objective* as $\text{coParity}(p) = \{ \omega \in \Omega \mid \min\big(p(\text{Inf}(\omega))\big) \text{ is odd } \}$.

Sure winning, almost-sure winning, and optimality. Given a player-1 objective Φ, a strategy $\sigma \in \Sigma$ is *sure winning* for player 1 from a state $s \in S$ if for every strategy $\pi \in \Pi$ for player 2, we have $\text{Outcome}(s, \sigma, \pi) \subseteq \Phi$. The strategy σ is *almost-sure winning* for player 1 from the state s for the objective Φ if for every player-2 strategy π, we have $\text{Pr}_s^{\sigma,\pi}(\Phi) = 1$. The sure and almost-sure winning strategies for player 2 are defined analogously. Given an objective Φ, the *sure winning set* $\langle\!\langle 1 \rangle\!\rangle_{sure}(\Phi)$ for player 1 is the set of states from which player 1 has a sure winning strategy. The *almost-sure winning set* $\langle\!\langle 1 \rangle\!\rangle_{almost}(\Phi)$ for player 1 is the set of states from which player 1 has an almost-sure winning strategy. The sure winning set $\langle\!\langle 2 \rangle\!\rangle_{sure}(\Omega \setminus \Phi)$ and the almost-sure winning set $\langle\!\langle 2 \rangle\!\rangle_{almost}(\Omega \setminus \Phi)$ for player 2 are defined analogously. It follows from the definitions that for all $2\frac{1}{2}$-player game graphs and all objectives Φ, we have $\langle\!\langle 1 \rangle\!\rangle_{sure}(\Phi) \subseteq \langle\!\langle 1 \rangle\!\rangle_{almost}(\Phi)$. A game is sure (resp. almost-sure) winning for player i if player i wins surely (resp. almost-surely) from every state in the game.

Given objectives $\Phi \subseteq \Omega$ for player 1 and $\Omega \setminus \Phi$ for player 2, we define the *value* functions $\langle\!\langle 1 \rangle\!\rangle_{val}$ and $\langle\!\langle 2 \rangle\!\rangle_{val}$ for the players 1 and 2, respectively, as the following functions from the state space S to the interval $[0, 1]$ of reals: for all states $s \in S$, let $\langle\!\langle 1 \rangle\!\rangle_{val}(\Phi)(s) = \sup_{\sigma \in \Sigma} \inf_{\pi \in \Pi} \text{Pr}_s^{\sigma,\pi}(\Phi)$ and $\langle\!\langle 2 \rangle\!\rangle_{val}(\Omega \setminus \Phi)(s) =$

$\sup_{\pi \in \Pi} \inf_{\sigma \in \Sigma} \Pr_s^{\sigma,\pi}(\Omega \setminus \Phi)$. In other words, the value $\langle\!\langle 1 \rangle\!\rangle_{val}(\Phi)(s)$ gives the maximal probability with which player 1 can achieve her objective Φ from state s, and analogously for player 2. The strategies that achieve the value are called optimal: a strategy σ for player 1 is *optimal* from the state s for the objective Φ if $\langle\!\langle 1 \rangle\!\rangle_{val}(\Phi)(s) = \inf_{\pi \in \Pi} \Pr_s^{\sigma,\pi}(\Phi)$. The optimal strategies for player 2 are defined analogously.

Consider a family $\Sigma^C \subseteq \Sigma$ of special strategies for player 1. We say that the family Σ^C *suffices* with respect to a player-1 objective Φ on a class $\mathcal{G}$ of game graphs for *sure winning* if for every game graph $G \in \mathcal{G}$ and state $s \in \langle\!\langle 1 \rangle\!\rangle_{sure}(\Phi)$, there is a player-1 strategy $\sigma \in \Sigma^C$ such that for every player-2 strategy $\pi \in \Pi$, we have Outcome$(s, \sigma, \pi) \subseteq \Phi$. Similarly, the family Σ^C *suffices* with respect to the objective Φ on the class $\mathcal{G}$ of game graphs for *almost-sure winning* if for every game graph $G \in \mathcal{G}$ and state $s \in \langle\!\langle 1 \rangle\!\rangle_{almost}(\Phi)$, there is a player-1 strategy $\sigma \in \Sigma^C$ such that for every player-2 strategy $\pi \in \Pi$, we have $\Pr_s^{\sigma,\pi}(\Phi) = 1$; and for *optimality*, if for every game graph $G \in \mathcal{G}$ and state $s \in S$, there is a player-1 strategy $\sigma \in \Sigma^C$ such that $\langle\!\langle 1 \rangle\!\rangle_{val}(\Phi)(s) = \inf_{\pi \in \Pi} \Pr_s^{\sigma,\pi}(\Phi)$. We now state the classical determinacy results for 2-player and $2\frac{1}{2}$-player parity games.

Theorem 1 (Qualitative determinacy). [7] *For all 2-player game graphs and parity objectives* Φ, *we have* $\langle\!\langle 1 \rangle\!\rangle_{sure}(\Phi) - S \setminus \langle\!\langle 2 \rangle\!\rangle_{sure}(\Omega \setminus \Phi)$. *Moreover, on 2-player game graphs, the family of pure memoryless strategies suffices for sure winning with respect to parity objectives.*

Theorem 2 (Quantitative determinacy). [4] *For all* $2\frac{1}{2}$-*player game graphs, all parity objectives* Φ, *and all states* s, *we have* $\langle\!\langle 1 \rangle\!\rangle_{val}(\Phi)(s) + \langle\!\langle 2 \rangle\!\rangle_{val}(\Omega \setminus \Phi)(s) = 1$. *The family of pure memoryless strategies suffices for optimality with respect to parity objectives on* $2\frac{1}{2}$-*player game graphs.*

Since in $2\frac{1}{2}$-player games with parity objectives pure memoryless strategies suffice for optimality, in the sequel we consider only pure memoryless strategies.

3   Strategy Improvement Algorithm

The main result of this paper is a strategy improvement algorithm for $2\frac{1}{2}$-player games with parity objectives. Before presenting the algorithm, we recall a few key properties of $2\frac{1}{2}$-player parity games, which were proved in [2,3].

Useful properties. We first present a reduction of $2\frac{1}{2}$-player parity games to 2-player parity games, preserving the ability of player 1 to win almost-surely.

Reduction. Given a $2\frac{1}{2}$-player game graph $G = ((S, E), (S_1, S_2, S_\bigcirc), \delta)$ with a priority function $p \colon S \to [0..d]$, we construct a 2-player game graph $\overline{G} = ((\overline{S}, \overline{E}), (\overline{S}_1, \overline{S}_2), \overline{\delta})$ together with a priority function $\overline{p} : \overline{S} \to [0..d]$. The construction is specified as follows. For every nonprobabilistic state $s \in S_1 \cup S_2$, there is a corresponding state $\overline{s} \in \overline{S}$ such that (1) $\overline{s} \in \overline{S}_1$ iff $s \in S_1$, and (2) $\overline{p}(\overline{s}) = p(s)$, and (3) $(\overline{s}, \overline{t}) \in \overline{E}$ iff $(s, t) \in E$. From the state $\overline{s}$ with $\overline{p}(\overline{s}) = p(s)$, the players play the following 3-step game in $\overline{G}$. First, in state $\overline{s}$ player 2 chooses a successor

$(\widetilde{s}, 2k)$, for $k \in \{0, 1, \ldots, j\}$, where $p(s) = 2j$ or $p(s) = 2j - 1$. For every state $(\widetilde{s}, 2k)$, we have $\overline{p}(\widetilde{s}, 2k) = p(s)$. For $k > 1$, in state $(\widetilde{s}, 2k)$ player 1 chooses from two successors: state $(\widehat{s}, 2k - 1)$ with $\overline{p}(\widehat{s}, 2k - 1) = 2k - 1$, or state $(\widehat{s}, 2k)$ with $\overline{p}(\widehat{s}, 2k) = 2k$. The state $(\widetilde{s}, 0)$ has only one successor $(\widehat{s}, 0)$, with $\overline{p}(\widehat{s}, 0) = 0$. Finally, in each state $(\widehat{s}, k)$ the choice is between all states $\overline{t}$ such that $(s, t) \in E$, and it belongs to player 1 if k is odd, and to player 2 if k is even. We denote by $\mathrm{Tr}_{almost}(G)$ the 2-player game graph $\overline{G}$, as defined by this reduction. Also given a pure memoryless strategy $\overline{\sigma}$ for the 2-player game graph $\overline{G}$, a strategy $\mathrm{Tr}_{almost}(\overline{\sigma}) = \sigma$ for the $2\frac{1}{2}$-player game graph G is defined as follows: $\sigma(s) = t$ iff $\overline{\sigma}(\overline{s}) = \overline{t}$, for all $s \in S_1$. Similar definitions hold for player 2.

Lemma 1. [3] *Given a $2\frac{1}{2}$-player game graph G with the parity objective* $\mathrm{Parity}(p)$ *for player 1, let $\overline{U}_1$ and $\overline{U}_2$ be the sure winning sets for players 1 and 2, respectively, in the 2-player game graph $\overline{G} = \mathrm{Tr}_{almost}(G)$ with the modified parity objective* $\mathrm{Parity}(\overline{p})$. *Define the sets U_1 and U_2 in the original $2\frac{1}{2}$-player game graph G by $U_1 = \{ s \in S \mid \overline{s} \in \overline{U}_1 \}$ and $U_2 = \{ s \in S \mid \overline{s} \in \overline{U}_2 \}$. Then the following assertions hold: (a) $U_1 = \langle\!\langle 1 \rangle\!\rangle_{almost}(\mathrm{Parity}(p)) = S \setminus U_2$; and (b) if $\overline{\sigma}$ is a pure memoryless sure winning strategy for player 1 from $\overline{U}_1$ in $\overline{G}$, then $\mathrm{Tr}_{almost}(\overline{\sigma})$ is an almost-sure winning strategy for player 1 from U_1 in G.*

Subgames. A set $U \subseteq S$ of states is δ-*closed* if for every probabilistic state $u \in U \cap S_\bigcirc$, if $(u, t) \in E$, then $t \in U$. The set U is δ-*live* if for every nonprobabilistic state $s \in U \cap (S_1 \cup S_2)$, there is a state $t \in U$ such that $(s, t) \in E$. A δ-closed and δ-live subset U of S induces a *subgame graph* of G, denoted by $G \upharpoonright U$.

Boundary probabilistic states. Given a set U of states, let $\mathrm{BP}(U) = \{ s \in U \cap S_\bigcirc \mid \exists t \in E(s). t \notin U \}$ be the set of *boundary* probabilistic states, which have an edge out of U. Given a set U of states and a parity objective $\mathrm{Parity}(p)$ for player 1, we define a transformation $\mathrm{Tr}_{win1}(U)$ of U as follows: every state s in $\mathrm{BP}(U)$ is converted to an *absorbing* state (a state with a self-loop) and assigned the even priority $2 \cdot \lfloor \frac{d}{2} \rfloor$; thus, every state in $\mathrm{BP}(U)$ is changed to a sure winning state for player 1. Observe that if U is δ-live, then $\mathrm{Tr}_{win1}(G \upharpoonright U)$ is a game graph.

Value classes. Given a parity objective Φ, for every real $r \in \mathbb{R}$ the *value class* with value r, denoted $\mathrm{VC}(r) = \{ s \in S \mid \langle\!\langle 1 \rangle\!\rangle_{val}(\Phi)(s) = r \}$, is the set of states with value r for player 1. It follows easily that for every $r > 0$, the value class $\mathrm{VC}(r)$ is δ-live. The following lemma establishes a connection between value classes, the transformation Tr_{win1}, and the almost-sure winning states.

Lemma 2. [2] *For every real $r > 0$, for the value class $\mathrm{VC}(r)$ with parity objective* $\mathrm{Parity}(p)$ *for player 1, the game $\mathrm{Tr}_{win1}(G \upharpoonright \mathrm{VC}(r))$ is almost-sure winning for player 1.*

It follows from Lemma 1 and Lemma 2 that for every value class $\mathrm{VC}(r)$ with $r > 0$, the game $\mathrm{Tr}_{almost}(\mathrm{Tr}_{win1}(G \upharpoonright \mathrm{VC}(r)))$ is sure winning for player 1.

Strategy improvement algorithm. Given a strategy π and a set U of states, we denote by $\pi \upharpoonright U$ a strategy that for every state in U follows the strategy π.

Values and value classes given by strategies. Given a player-2 strategy π and a parity objective Φ for player 1, we denote the value of player 1 given the strategy π as follows: $\langle\langle 1 \rangle\rangle^{\pi}_{val}(\Phi)(s) = \sup_{\sigma \in \Sigma^{PM}} \Pr^{\sigma,\pi}_{s}(\Phi)$. Similarly, we define the value classes given strategy π as $\mathrm{VC}^{\pi}(r) = \{\, s \in S \mid \langle\langle 1 \rangle\rangle^{\pi}_{val}(\Phi)(s) = r\,\}$, for all $r \in \mathbb{R}$.

Witnesses for player 2. Given a $2\frac{1}{2}$-player game graph G and a parity objective Φ for player 1, a *witness* $w_2 = (\pi, \overline{\pi}_Q)$ for player 2 is specified as follows: (a) the strategy π is a player-2 strategy for the $2\frac{1}{2}$-player game graph G; and (b) for every value class $\mathrm{VC}^{\pi}(r)$, the strategy $\overline{\pi}_Q \upharpoonright \mathrm{VC}^{\pi}(r)$ is a player-2 strategy for the 2-player game graph $\overline{G}_r = \mathrm{Tr}_{almost}(\mathrm{Tr}_{win1}(G \upharpoonright \mathrm{VC}^{\pi}(r)))$. We require that $\pi = \mathrm{Tr}_{almost}(\overline{\pi}_Q)$. The witness $w_2 = (\pi, \overline{\pi}_Q)$ for player 2 is an *optimal* witness if π is an optimal strategy for player 2.

Ordering of witnesses. We define an ordering relation $\prec$ on witnesses as follows: given two witnesses $w_2 = (\pi, \overline{\pi}_Q)$ and $w_2' = (\pi', \overline{\pi}'_Q)$ for player 2, let $w_2 \prec w_2'$ iff one of the following two conditions holds:

1. for all states s, we have $\langle\langle 1 \rangle\rangle^{\pi}_{val}(\Phi)(s) \geq \langle\langle 1 \rangle\rangle^{\pi'}_{val}(\Phi)(s)$, and for some state s, we have $\langle\langle 1 \rangle\rangle^{\pi}_{val}(\Phi)(s) > \langle\langle 1 \rangle\rangle^{\pi'}_{val}(\Phi)(s)$; or
2. for all states s, we have $\langle\langle 1 \rangle\rangle^{\pi}_{val}(\Phi)(s) = \langle\langle 1 \rangle\rangle^{\pi'}_{val}(\Phi)(s)$, and in some value class $\mathrm{VC}^{\pi}(r) = \mathrm{VC}^{\pi'}(r)$, we have $(\overline{\pi}_Q \upharpoonright \mathrm{VC}^{\pi}(r)) \prec_Q (\overline{\pi}'_Q \upharpoonright \mathrm{VC}^{\pi}(r))$ in the 2-player parity game $\mathrm{Tr}_{almost}(\mathrm{Tr}_{win1}(G \upharpoonright \mathrm{VC}^{\pi}(r)))$, where $\prec_Q$ denotes the ordering of strategies for a strategy improvement algorithm for 2-player parity games (e.g., as defined in [1,12]).

Profitable switches. Given a witness $w_2 = (\pi, \overline{\pi}_Q)$ for player 2, we specify a procedure `ProfitableSwitch` to "improve" the witness according to the witness ordering $\prec$. The procedure is described in Algorithm 1. An informal description of the procedure is as follows: given a witness $w_2 = (\pi, \overline{\pi}_Q)$, the algorithm computes the values $\langle\langle 1 \rangle\rangle^{\pi}_{val}(\Phi)(s)$ for all states. If there is a state $s \in S_2$ such that the strategy can be "value improved," i.e., there is a state $t \in E(s)$ with $\langle\langle 1 \rangle\rangle^{\pi}_{val}(\Phi)(t) < \langle\langle 1 \rangle\rangle^{\pi}_{val}(\Phi)(s)$, then the witness is modified by setting $\pi(s)$ to t. This step is similar to the strategy improvement step of [6] and is achieved in Step 2.1 of `ProfitableSwitch`. Otherwise, in every value class $\mathrm{VC}^{\pi}(r)$, the strategy $\overline{\pi}_Q$ is "improved" for the game $\mathrm{Tr}_{almost}(\mathrm{Tr}_{win1}(G \upharpoonright \mathrm{VC}^{\pi}(r)))$ with respect to the ordering $\prec_Q$ of strategies for 2-player parity games. This is achieved in Step 2.2.

Proposition 1. *Given a strategy π for player 2, for all states $s \in \mathrm{VC}^{\pi}(r) \cap S_1$ and $t \in E(s)$, we have $\langle\langle 1 \rangle\rangle^{\pi}_{val}(\Phi)(t) \leq r$, that is, $E(s) \subseteq \bigcup_{0 \leq q \leq r} \mathrm{VC}^{\pi}(q)$.*

Proposition 2. *Given a strategy π for player 2, for all strategies σ for player 1, if there is a closed recurrent class C in the Markov chain $G_{\sigma,\pi}$ with $C \subseteq \mathrm{VC}^{\pi}(r)$ for $r < 1$, then $\min(p(C))$ is odd.*

Lemma 3. *Let $w_2 = (\pi, \overline{\pi}_Q)$ be an input to Algorithm 1, and let $w_2' = (\pi', \overline{\pi}'_Q)$ be the corresponding output, that is, $w_2' = $ `ProfitableSwitch`(G, w_2). If the set I in Step 2 of Algorithm 1 is nonempty, then (a) $\langle\langle 1 \rangle\rangle^{\pi}_{val}(\Phi)(s) \geq \langle\langle 1 \rangle\rangle^{\pi'}_{val}(\Phi)(s)$ for all $s \in S$, and (b) $\langle\langle 1 \rangle\rangle^{\pi}_{val}(\Phi)(s) > \langle\langle 1 \rangle\rangle^{\pi'}_{val}(\Phi)(s)$ for all $s \in I$.*

Algorithm 1. ProfitableSwitch

Input: $2\frac{1}{2}$-player game G, parity objective Φ for pl. 1, witness $w_2 = (\pi, \overline{\pi}_Q)$ for pl. 2.
Output: a witness w_2' for player 2 such that either $w_2 = w_2'$ or $w_2 \prec w_2'$.
1. (Step 1.) Compute $\langle\!\langle 1 \rangle\!\rangle_{val}^\pi(\Phi)(s)$ for all states s.
2. (Step 2.) Consider the set $I = \{\, s \in S_2 \mid \exists t \in E(s).\ \langle\!\langle 1 \rangle\!\rangle_{val}^\pi(\Phi)(s) > \langle\!\langle 1 \rangle\!\rangle_{val}^\pi(\Phi)(t) \,\}$.
   2.1 (Value improvement) If $I \neq \emptyset$ then set π' as follows.
      $\pi'(s) = \pi(s)$ for $s \in S_2 \setminus I$,
      $\pi'(s) = t$ for $s \in I$ and $t \in E(s)$ such that $\langle\!\langle 1 \rangle\!\rangle_{val}^\pi(\Phi)(s) > \langle\!\langle 1 \rangle\!\rangle_{val}^\pi(\Phi)(t)$;
      and set $\overline{\pi}_Q'$ to be an arbitrary strategy such that $\pi' = \text{Tr}_{almost}(\overline{\pi}_Q')$.
   2.2 (Qualitative improvement) **else** for every value class $\text{VC}^\pi(r)$:
      let $\overline{G}_r$ be the 2-player game graph $\text{Tr}_{almost}(\text{Tr}_{win1}(G \restriction \text{VC}^\pi(r)))$;
      set $(\overline{\pi}_Q' \restriction \text{VC}^\pi(r)) = \texttt{TwoPlSwitch}(\overline{G}_r, \overline{\pi}_Q \restriction \text{VC}^\pi(r))$ and $\pi' = \text{Tr}_{almost}(\overline{\pi}_Q')$,
      where $\texttt{TwoPlSwitch}$ is a strategy improvement step for 2-player parity games.
3. **return** $w_2' = (\pi', \overline{\pi}_Q')$.

Proof. Consider a switch of the strategy of player 2 from π to π', as constructed in Step 2.1 of Algorithm 1. Consider a strategy σ for player 1 and a closed recurrent class C in $G_{\sigma,\pi'}$ such that $C \subseteq \bigcup_{r<1} \text{VC}^\pi(r)$. Let $z = \min\{\, r < 1 \mid C \cap \text{VC}^\pi(r) \neq \emptyset \,\}$, that is, $\text{VC}^\pi(z)$ is the least value class with nonempty intersection with C. A state $s \in \text{VC}^\pi(z) \cap C$ satisfies the following conditions:

1. If $s \in S_1$, then $\sigma(s) \in \text{VC}^\pi(z)$. This follows since, by Proposition 1, we have $E(s) \subseteq \bigcup_{0 \le q \le z} \text{VC}^\pi(q)$ and $C \cap \text{VC}^\pi(q) = \emptyset$ for $q < z$.
2. If $s \in S_2$, then $\pi'(s) \in \text{VC}^\pi(z)$. This follows since, by construction, we have $\pi'(s) \in \bigcup_{0 \le q \le z} \text{VC}^\pi(q)$ and $C \cap \text{VC}^\pi(q) = \emptyset$ for $q < z$. Also, since $s \in \text{VC}^\pi(z)$ and $\pi'(s) \in \text{VC}^\pi(z)$, it follows that $\pi'(s) = \pi(s)$.
3. If $s \in S_\bigcirc$, then $E(s) \subseteq \text{VC}^\pi(z)$. This follows since for $s \in S_\bigcirc$, if $E(s) \subsetneq \text{VC}^\pi(z)$, then $E(s) \cap (\bigcup_{0 \le q < z} \text{VC}^\pi(q)) \neq \emptyset$. Since C is closed, and $C \cap \text{VC}^\pi(q) = \emptyset$ for $q < z$, the claim follows.

It follows that $C \subseteq \text{VC}^\pi(z)$ and for all states $s \in C \cap S_2$, we have $\pi'(s) = \pi(s)$. Hence by Proposition 2, $\min(p(C))$ is odd.

   It follows that if player 2 switches to the strategy π', as constructed when Step 2.1 of Algorithm 1 is executed, then for all strategies σ for player 1 the following assertion holds: if there is a closed recurrent class $C \subseteq S \setminus \text{VC}^\pi(1)$ in the Markov chain $G_{\sigma,\pi'}$, then C is winning for player 2, i.e., $\min(p(C))$ is odd. Hence, given strategy π', an optimal counter-strategy for player 1 maximizes the probability to reach $\text{VC}^\pi(1)$. The desired result follows from arguments similar to $2\frac{1}{2}$-player games with reachability objectives [6], with $\text{VC}^\pi(1)$ as the target for player 1, and the value improvement step (Step 2.1 of Algorithm 1). ∎

Lemma 4. *Let $w_2 = (\pi, \overline{\pi}_Q)$ be an input to Algorithm 1, and let $w_2' = (\pi', \overline{\pi}_Q')$ be the corresponding output, that is, $w_2' = \texttt{ProfitableSwitch}(G, w_2)$, such that $w_2 \neq w_2'$. If the set I in Step 2 of Algorithm 1 is empty, then (a) $\langle\!\langle 1 \rangle\!\rangle_{val}^{\pi'}(\Phi)(s) \ge \langle\!\langle 1 \rangle\!\rangle_{val}^\pi(\Phi)(s)$ for all $s \in S$, and (b) if $\langle\!\langle 1 \rangle\!\rangle_{val}^{\pi'}(\Phi)(s) = \langle\!\langle 1 \rangle\!\rangle_{val}^\pi(\Phi)(s)$ for all $s \in S$, then $(\overline{\pi}_Q \restriction \text{VC}^\pi(r)) \prec_Q (\overline{\pi}_Q' \restriction \text{VC}^\pi(r))$ for some value class $\text{VC}^\pi(r)$.*

Proof. (sketch) It follows from Proposition 2 that for all strategies σ for player 1, if C is a closed recurrent class in $G_{\sigma,\pi}$ and $C \subseteq \mathrm{VC}^\pi(r)$ for $r < 1$, then $\min(p(C))$ is odd. Let π' be the strategy constructed from π in Step 2.2 of Algorithm 1. Since π' is obtained as qualitative improvement of π, it can be shown that if C is a closed recurrent class in $G_{\sigma,\pi'}$ and $C \subseteq \mathrm{VC}^\pi(r)$, then $\min(p(C))$ is odd. Arguments similar to Lemma 3 show that: for all strategies σ for player 1, if there is a closed recurrent class $C \subseteq S \setminus \mathrm{VC}^\pi(1)$ in $G_{\sigma,\pi'}$, then C is winning for player 2, that is, $\min(p(C))$ is odd. Since in strategy π' player 2 chooses every edge in the same value class as π, it can be shown that for all states s, we have $\langle\!\langle 1 \rangle\!\rangle_{val}^\pi(\Phi)(s) \geq \langle\!\langle 1 \rangle\!\rangle_{val}^{\pi'}(\Phi)(s)$. If $\langle\!\langle 1 \rangle\!\rangle_{val}^\pi(\Phi)(s) = \langle\!\langle 1 \rangle\!\rangle_{val}^{\pi'}(\Phi)(s)$ for all states s, then $\mathrm{VC}^\pi(r) = \mathrm{VC}^{\pi'}(r)$ for all r, that is, the value classes given π and π' coincide. Then, by the properties of the procedure `TwoPlSwitch` and since $w_2 \neq w_2'$, condition 2 of the lemma holds.   ∎

Lemma 5. *For every $2\frac{1}{2}$-player game graph G and witness w_2 for player 2, if $w_2 \neq$ `ProfitableSwitch`(G, w_2), then $w_2 \prec$ `ProfitableSwitch`(G, w_2).*

Lemma 3 and Lemma 4 yield Lemma 5. The key argument to establish that if $w_2 =$ `ProfitableSwitch`(G, w_2), then w_2 is an optimal witness for player 2, goes as follows. Let w_2 be a player-2 witness such that $w_2 =$ `ProfitableSwitch`(G, w_2), and let $w_1 = (\sigma, \overline{\sigma}_Q)$ be the optimal counter-witness for player 1. Consider a value class $\mathrm{VC}^\pi(r)$ with $r > 0$, and the game graph $\overline{G}_r = \mathrm{Tr}_{almost}(\mathrm{Tr}_{win1}(G \upharpoonright \mathrm{VC}^\pi(r)))$. Since $\overline{\pi}_Q$ cannot be improved against $\overline{\sigma}_Q$ with respect to the ordering $\prec_Q$ in any value class, it follows that $\overline{\sigma}_Q$ is a sure winning strategy in $\overline{G}_r$. Hence it follows from Lemma 1 that σ is an almost-sure winning strategy for player 1 in $\mathrm{Tr}_{win1}(G \upharpoonright \mathrm{VC}^\pi(r))$, since $\sigma = \mathrm{Tr}_{almost}(\overline{\sigma}_Q)$. Consider any strategy π' for player 2 against σ, and consider the Markov chain $G_{\sigma,\pi'}$. Since σ is almost-sure winning in $\mathrm{Tr}_{win1}(G \upharpoonright \mathrm{VC}^\pi(r))$ for all $r > 0$, it follows that for any closed recurrent class C of $G_{\sigma,\pi'}$ such that $C \subseteq \bigcup_{r>0} \mathrm{VC}^\pi(r)$, the set C is winning for player 1 (i.e., $\min(p(C))$ is even). Moreover, since the strategy π cannot be "value improved," it follows from arguments similar to [6] for $2\frac{1}{2}$-player reachability games that for all player-2 strategies π' and all states $s \in \mathrm{VC}^\pi(r)$, we have $\mathrm{Pr}_s^{\sigma,\pi'}(\Phi) \geq r$. Hence $\langle\!\langle 1 \rangle\!\rangle_{val}(\Phi)(s) \geq r$. Since σ is an optimal strategy against π, we have $r = \langle\!\langle 1 \rangle\!\rangle_{val}^\pi(\Phi)(s) \geq \langle\!\langle 1 \rangle\!\rangle_{val}(\Phi)(s)$ for all states $s \in \mathrm{VC}^\pi(r)$. This establishes the optimality of π.

Lemma 6. *For every $2\frac{1}{2}$-player game graph G and witness w_2 for player 2, if $w_2 =$ `ProfitableSwitch`(G, w_2), then w_2 is an optimal witness.*

A strategy improvement algorithm using the `ProfitableSwitch` procedure is described in Algorithm 2. The correctness of the algorithm follows from Lemma 6. Let $I_2(k)$ and $I_R(k)$ denote bounds on the number of iterations of strategy improvement algorithms for 2-player parity games and $2\frac{1}{2}$-player reachability games, respectively, for game graphs with k states. The number of iterations of Algorithm 2 between two value improvement steps can be bounded by $n \cdot I_2(n \cdot d)$, and hence the total number of iterations of Algorithm 2 can be bounded by $n \cdot I_2(n \cdot d) \cdot I_R(n \cdot d)$. Given an optimal strategy π for player 2, the values for both the players can be computed in polytime by computing the values of the MDP G_π [4].

Algorithm 2. StrategyImprovementAlgorithm

Input: a $2\frac{1}{2}$-player game graph G with parity objective Φ for player 1.
Output: a witness w_2^* for player 2.
1. Choose an arbitrary witness w_2 for player 2.
2. while $w_2 \neq$ ProfitableSwitch(G, w_2) do $w_2 =$ ProfitableSwitch(G, w_2)
3. **return** $w_2^* = w_2$.

Theorem 3 (Correctness of Algorithm 2). *The output w_2^* of Algorithm 2 is an optimal witness for player 2.*

4 Randomized Subexponential Algorithm

We now present a randomized subexponential-time algorithm for $2\frac{1}{2}$-player parity games, by combining an algorithm of Björklund et al. [1] and the witness improvement procedure ProfitableSwitch.

Games and improving subgames. Given $l, m \in \mathbb{N}$, let $\mathcal{G}(l, m)$ be the class of $2\frac{1}{2}$-player game graphs with the set S_2 of player 2 states partitioned into two sets as follows: (a) $O_1 = \{\, s \in S_2 \mid |E(s)| = 1 \,\}$, i.e., the set of states with out-degree 1; and (b) $O_2 = S_2 \setminus O_1$, with $O_2 \leq l$ and $\sum_{s \in O_2} |E(s)| \leq m$. There is no restriction for player 1. Given a game $G \in \mathcal{G}(l, m)$, a state $s \in O_2$, and an edge $e = (s, t)$, we define the subgame $\widetilde{G}_e$ by deleting all edges from s other than the edge e. Observe that $\widetilde{G}_e \in \mathcal{G}(l - 1, m - |E(s)|)$, and hence also $\widetilde{G}_e \in \mathcal{G}(l, m)$. If $w_2 = (\pi, \overline{\pi}_Q)$ is a witness for player 2 in $G \in \mathcal{G}(l, m)$, then a subgame $\widetilde{G}$ is w_2-*improving* if some witness $w_2' = (\pi', \overline{\pi}_Q')$ in $\widetilde{G}$ satisfies $w_2 \prec w_2'$.

Informal description of Algorithm 3. The algorithm takes a $2\frac{1}{2}$-player parity game and an initial witness w_2^0, and proceeds in three steps. In Step 1, it constructs r pairs of w_2^0-improving subgames $\widetilde{G}$ and corresponding improved witnesses w_2 in $\widetilde{G}$. This is achieved by the procedure ImprovingSubgames. The parameter r will be chosen to obtain a suitable complexity analysis. In Step 2, the algorithm selects uniformly at random one of the improving subgames $\widetilde{G}$ with corresponding witness w_2, and recursively computes an optimal witness w_2^* in $\widetilde{G}$ from w_2 as the initial witness. If the witness w_2^* is optimal in the original game G, then the algorithm terminates and returns w_2^*. Otherwise it improves w_2^* by a call to ProfitableSwitch, and continues at Step 1 with the improved witness ProfitableSwitch(G, w_2^*) as the initial witness.

The procedure ImprovingSubgames constructs a sequence of game graphs $G^0, G^1, \ldots, G^{r-l}$ with $G^i \in \mathcal{G}(l, l + i)$ such that all $(l + i)$-subgames $\widetilde{G}_e^i$ of G^i are w_2^0-improving. The subgame G^{i+1} is constructed from G^i as follows: we compute an optimal witness w_2^i in G^i, and if w_2^i is optimal in G, then we have discovered an optimal witness; otherwise we construct G^{i+1} by adding any *target* edge e

---

Algorithm 3. RandomizedAlgorithm ($2\frac{1}{2}$-player parity games)

Input: a $2\frac{1}{2}$-player game graph $G \in \mathcal{G}(l, m)$, a parity objective Parity(p) for pl. 1
and an initial witness w_2^0 for pl. 2.
Output : an optimal witness $w_2^* = (\pi^*, \overline{\pi}_Q^*)$ for player 2.
1. (Step 1) Collect a set I of r pairs $(\widetilde{G}, w_2)$ of subgames $\widetilde{G}$ of G, and
    corresponding witnesses w_2 in $\widetilde{G}$ such that $w_2^0 \prec w_2$.
    (This is achieved by the procedure ImprovingSubgames below).
2. (Step 2) Select a pair $(\widetilde{G}, w_2)$ from I uniformly at random.
    2.1 Find an optimal witness in $w_2^* \in \widetilde{G}$ by applying the algorithm recursively,
        with w_2 as the initial witness.
3. (Step 3) **if** w_2^* is an optimal witness in the original game G **then return** w_2^*.
        else let $w_2 = $ ProfitableSwitch(G, w_2^*), and
                goto Step 1 with G and w_2 as the initial witness.
procedure ImprovingSubgames
1. Construct sequence $G^0, G^1, \ldots, G^{r-l}$ of subgames with $G^i \in \mathcal{G}(l, l+i)$ as follows:
    1.1 G^0 is the game where each edge is fixed according to w_2^0.
    1.2 Let w_2^i be an optimal witness in G^i;
        1.2.1 **if** w_2^i is an optimal witness in the original game G
            then return w_2^i.
        1.2.2 **else** let e be any target of ProfitableSwitch(G, w_2^i);
            the subgame G^{i+1} is G^i with the edge e added.
2. **return** r subgames (fixing one of the r edges in G^{r-l}) and associated witnesses.

---

of ProfitableSwitch(G, w_2^i) in G^i, i.e., e is an edge required in the witness
ProfitableSwitch(G, w_2^i) that is not in the witness w_2^i.

The correctness of the algorithm can be seen as follows. Observe that every time Step 1 is executed, the initial witness is improved with respect to the ordering $\prec$ on witnesses. Since the number of witnesses is bounded, the termination of the algorithm is guaranteed. Step 3 of Algorithm 3 and Step 1.2.1 of procedure ImprovingSubgames ensure that on termination of the algorithm, the returned witness is optimal. Lemma 7 bounds the expected number of iterations of Algorithm 3. The analysis is similar to the results of [1].

Lemma 7. *Algorithm 3 computes an optimal witness. The expected number of iterations $T(\cdot, \cdot)$ of Algorithm 3 for a game $G \in \mathcal{G}(l, m)$ is bounded by the following recurrence:* $T(l, m) \leq \sum_{i=l}^{r} T(l, i) + T(l-1, m-2) + \frac{1}{r} \cdot \sum_{i=1}^{r} T(l, m-i) + 1$.

For a game graph G with $|S| = n$, we obtain a bound of n^2 for m. Applying a symmetric version of Algorithm 3 for player 1 if $|S_1| \leq |S_2|$, we can bound l by $\min\{|S_1|, |S_2|\}$. Using this fact and an analysis of Kalai for linear programming, Björklund et al. [1] showed that $m^{O\left(\sqrt{l/\log(l)}\right)} = 2^{O\left(\sqrt{l \cdot \log(l)}\right)}$ is a solution to the recurrence of Lemma 7, by choosing $r = \max\{l, \frac{m}{2}\}$, where $l = \min\{|S_1|, |S_2|\}$.

Lemma 8. *Procedure* ProfitableSwitch *can be computed in polynomial time.*

A call to `ProfitableSwitch` requires solving an MDP with parity objectives quantitatively (Step 1 of `ProfitableSwitch`; for a polynomial-time procedure, see [4]) and computing a profitable switch for 2-player parity games (Step 2.2 of `ProfitableSwitch`; for a polynomial-time procedure, see [1,12]). Thus Lemma 8 follows. We obtain Theorem 4 as follows. Observe that the reduction Tr_{almost} of $2\frac{1}{2}$ player games to 2-player games causes a blow-up by a factor of d for states in $S_\bigcirc$. This fact, along with the solution for the recurrence of Lemma 7, using $l = d \cdot n_0 + \min\{n_1, n_2\}$ in the solution, yields that the expected number of iterations of Algorithm 3 is bounded by $2^{O\left(\sqrt{z \cdot \log(z)}\right)}$, where $z = d \cdot n_0 + \min\{n_1, n_2\}$. Each iteration of the algorithm requires a call to `ProfitableSwitch`. This analysis with Lemma 8 proves Theorem 4.

Theorem 4. *Given a $2\frac{1}{2}$-player game graph G with a priority function $p\colon S \to [0..d]$, the value $\langle\!\langle 1 \rangle\!\rangle_{val}(\mathrm{Parity}(p))(s)$ can be computed for all states $s \in S$ in time $2^{O\left(\sqrt{z \cdot \log(z)}\right)} \cdot O\big(poly(n)\big)$, where $n_1 = |S_1|$, $n_2 = |S_2|$, $n_0 = |S_\bigcirc|$, $z = (n_0 \cdot d + \min\{n_1, n_2\})$, and poly represents a polynomial function.*

Acknowledgments. This research was supported in part by the AFOSR MURI grant F49620-00-1-0327 and the NSF ITR grant CCR-0225610.

References

1. H. Bjorklund, S. Sandberg, and S. Vorobyov. A discrete subexponential algorithm for parity games. In *STACS*, pages 663–674. LNCS 2607, Springer, 2003.
2. K. Chatterjee, L. de Alfaro, and T.A. Henzinger. The complexity of stochastic Rabin and Streett games. In *ICALP*, pages 878–890. LNCS 3580, Springer, 2005.
3. K. Chatterjee, M. Jurdziński, and T. A. Henzinger. Simple stochastic parity games. In *CSL*, pages 100–113. LNCS 2803, Springer, 2003.
4. K. Chatterjee, M. Jurdziński, and T.A. Henzinger. Quantitative stochastic parity games. In *SODA*, pages 114–123. SIAM, 2004.
5. A. Condon. The complexity of stochastic games. *Information and Computation*, 96:203–224, 1992.
6. A. Condon. On algorithms for simple stochastic games. In *Advances in Computational Complexity Theory*, pages 51–73. American Mathematical Society, 1993.
7. E.A. Emerson and C. Jutla. The complexity of tree automata and logics of programs. In *FOCS*, pages 328–337. IEEE Computer Society Press, 1988.
8. A. Hoffman and R. Karp. On nonterminating stochastic games. *Management Science*, 12:359–370, 1966.
9. M. Jurdziński, M. Paterson, and U. Zwick. A deterministic subexponential algorithm for solving parity games. In *SODA (To appear)*, 2006.
10. W. Ludwig. A subexponential randomized algorithm for the simple stochastic game problem. *Information and Computation*, 117:151–155, 1995.
11. W. Thomas. Languages, automata, and logic. In *Handbook of Formal Languages*, volume 3, Beyond Words, chapter 7, pages 389–455. Springer, 1997.
12. J. Vöge and M. Jurdziński. A discrete strategy improvement algorithm for solving parity games. In *CAV*, pages 202–215. LNCS 1855, Springer, 2000.

DAG-Width and Parity Games

Dietmar Berwanger[1], Anuj Dawar[2], Paul Hunter[3], and Stephan Kreutzer[3]

[1] LaBRI, Université de Bordeaux 1
dwb@labri.fr
[2] University of Cambridge Computer Laboratory
anuj.dawar@cl.cam.ac.uk
[3] Logic and Discrete Systems, Institute for Computer Science,
Humboldt-University Berlin
{hunter, kreutzer}@informatik.hu-berlin.de

Abstract. Tree-width is a well-known metric on undirected graphs that measures how tree-like a graph is and gives a notion of graph decomposition that proves useful in algorithm development. Tree-width is characterised by a game known as the cops-and-robber game where a number of cops chase a robber on the graph. We consider the natural adaptation of this game to directed graphs and show that monotone strategies in the game yield a measure with an associated notion of graph decomposition that can be seen to describe how close a directed graph is to a directed acyclic graph (DAG). This promises to be useful in developing algorithms on directed graphs. In particular, we show that the problem of determining the winner of a parity game is solvable in polynomial time on graphs of bounded DAG-width. We also consider the relationship between DAG-width and other measures such as entanglement and directed tree-width. One consequence we obtain is that certain NP-complete problems such as Hamiltonicity and disjoint paths are polynomial-time computable on graphs of bounded DAG-width.

1 Introduction

The groundbreaking work of Robertson and Seymour in their graph minor project has focused much attention on tree-decompositions of graphs and associated measures of graph connectivity such as tree-width [13]. Aside from their interest in graph structure theory, these notions have also proved very useful in the development of algorithms. The tree-width of a graph is a measure of how tree-like the graph is, and it is found that small tree-width allows for graph decompositions along which recursive algorithms can work. Many problems that are intractable in general can be solved efficiently on graphs of bounded tree-width. These include such classical NP-complete problems as finding a Hamiltonian cycle in a graph or detecting if a graph is three-colourable. Indeed, a general result of Courcelle [4] shows that any property definable in monadic second-order logic is solvable in linear time on graphs of fixed tree-width.

The idea of designing algorithms that work on tree-decompositions of the input has been generalised from graphs to other kinds of structures. Usually the tree-width of a structure is defined as that of the underlying connectivity (or Gaifman) graph. For instance, the tree-width of a directed graph is simply that of the undirected graph we get by forgetting the direction of edges, a process which leads to some loss of information.

B. Durand and W. Thomas (Eds.): STACS 2006, LNCS 3884, pp. 524–536, 2006.
© Springer-Verlag Berlin Heidelberg 2006

This loss may be significant if the algorithmic problems we are interested in are inherently directed. A good example is the problem of detecting Hamiltonian cycles. While we know that this can be solved easily on graphs with small tree-width, there are also directed graphs with very simple connectivity structure which have large tree-width. A directed acyclic graph (DAG) is a particularly simple structure, but we lose sight of this when we erase the direction on the edges and find the underlying undirected graph to be dense. Several proposals have been made (see [12,8,2,14]) which extend notions of tree-decompositions and tree-width to directed graphs. In particular, Johnson et al. [8] introduce the notion of *directed tree-width* where directed acyclic graphs have width 0 and they show that Hamiltonicity can be solved for graphs of bounded directed tree-width in polynomial time. However, the definition and characterisations of this measure are somewhat unwieldy and they have not, so far, resulted in many further developments in algorithms.

We are especially interested in one particular problem on directed graphs, that of determining the winner of a *parity game*. This is an infinite two-player game played on a directed graph where the nodes are labelled by priorities. The players take turns pushing a token along edges of the graph. The winner is determined by the parity of the least priority occurring infinitely often in this infinite play. Parity games have proved useful in the development of model-checking algorithms used in the verification of concurrent systems. The modal μ-calculus, introduced in [10], is a widely used logic for the specification of such systems, encompassing a variety of modal and temporal logics. The problem of determining, given a system $\mathcal{A}$ and a formula φ of the μ-calculus, whether or not $\mathcal{A}$ satisfies φ can be turned into a parity game (see [6]). The exact complexity of solving parity games is an open problem that has received a large amount of attention. It is known [9] that the problem is in NP $\cap$ co-NP and no polynomial time algorithm is known. Obdržàlek [11] showed that for each k there is a polynomial time algorithm that solves parity games on graphs of tree-width at most k. He points out that the algorithm would not give good bounds, for instance, on directed acyclic graphs even though solving the games on such graphs is easy. He asks whether there is a structural property of directed graphs that would allow a fast algorithm on both bounded tree-width structures and on DAGs.

In this paper, we give just such a generalisation. We introduce a new measure of the connectivity of graphs that we call DAG-width[1]. It is intermediate between tree-width and directed tree-width, in that for any graph $\mathcal{G}$, the directed tree-width of $\mathcal{G}$ is no greater than its DAG-width which, in turn, is no greater than its tree-width. Thus, the class of structures of DAG-width $k + 1$ or less includes all structures of tree-width k and more (in particular, DAGs of arbitrarily high tree-width all have DAG-width 1).

The notion of DAG-width can be understood as a simple adaptation of the game of *cops and robber* (which characterises tree-width) to directed graphs. The game is played by two players, one of whom controls a set of k cops attempting to catch a robber controlled by the other player. The cop player can move any set of cops to any nodes on the graph, while the robber can move along any path in the graph as long as there is no cop currently on the path. Such games have been extensively studied

---

[1] We understand that Obdržàlek has defined a similar measure in a paper to appear at SODA'06. We have not yet had an opportunity to see that paper.

(see [15,5,7,1,2]). It is known [15] that the cop player has a winning strategy on an undirected graph $\mathcal{G}$ using $k + 1$ cops if, and only if, $\mathcal{G}$ has tree-width k. We consider the natural adaptation of this game to directed graphs, by constraining the robber to move along directed paths. We show that the class of directed graphs where there is a monotone (in a sense we make precise) strategy for k cops to win is characterised by its width in a decomposition that is a generalisation of tree-decompositions. We are then able to show that the problem of determining the winner of a parity game is solvable in polynomial time on the class of graphs of DAG-width k, for any fixed k.

In Section 2, we introduce some notation. Section 3 introduces the cops and robber game, DAG-decompositions and DAG-width and shows the equivalence between the existence of monotone winning strategies and DAG-width. Also in Section 3 we discuss some algorithmic aspects of DAG-width. Section 4 relates DAG-width to other measures of graph connectivity, and Section 5 demonstrates a polynomial time algorithm for solving parity games on graphs with bounded DAG-width. All proofs appear in the full version of the paper, available on the authors' homepages.

2   Preliminaries

We first fix some notation used throughout the paper. All graphs used are finite, directed and simple unless otherwise stated.

We write ω for the set of finite ordinals, i.e. natural numbers. For every $n \in \omega$, we write $[n]$ for the set $\{1, \ldots, n\}$. For every set V and every $k \in \omega$, we write $[V]^k$ for the set of all k-element subsets of V, that is, $[V]^k := \{\{x_1, \ldots, x_k\} \subseteq V : x_i \neq x_j$ whenever $i \neq j\}$. We write $[V]^{\leq k}$ for the set of all $X \subseteq V$ with $|X| \leq k$.

Let $\mathcal{G}$ be a directed graph. We write $V^{\mathcal{G}}$ for the set of its vertices and $E^{\mathcal{G}}$ for the set of its edges. E^{op} denotes the set of edges that results from reversing the edges in $E \subseteq E^{\mathcal{G}}$, i.e. $E^{op} = \{(w, v) : (v, w) \in E\}$. The graph $\mathcal{G}^{op}$ is defined to be $(V^{\mathcal{G}}, (E^{\mathcal{G}})^{op})$.

A tree-decomposition of a graph $\mathcal{G}$ is a labelled tree $(\mathcal{T}, (X_t)_{t \in V^{\mathcal{T}}})$ where $X_t \subseteq V^{\mathcal{G}}$ for each vertex $t \in V^{\mathcal{T}}$, for each edge $(u, v) \in E^{\mathcal{G}}$ there is a $t \in V^{\mathcal{T}}$ such that $\{u, v\} \subseteq X_t$, and for each $v \in V^{\mathcal{G}}$, the set $\{t \in V^{\mathcal{T}} : v \in X_t\}$ forms a connected subtree of $\mathcal{T}$. The *width* of a tree-decomposition is the cardinality of the largest X_t minus one. The tree-width of $\mathcal{G}$ is the smallest k such that $\mathcal{G}$ has a tree-decomposition of width k.

Let $\mathcal{D} := (D, A)$ be a directed, acyclic graph (DAG). The partial order $\preceq_{\mathcal{D}}$ (or $\preceq_A$) on D is the reflexive, transitive closure of A. A *root* of a set $X \subseteq D$ is a $\preceq_{\mathcal{D}}$-minimal element of X, that is, $r \in X$ is a root of X if there is no $y \in X$ such that $y \preceq_{\mathcal{D}} r$. Analogously, a *leaf* of $X \subseteq D$ is a $\preceq_{\mathcal{D}}$-maximal element.

3   Games, Strategies and Decompositions

This section contains the graph theoretical part of this paper. We define DAG-width and its relation to graph searching games. As mentioned in the introduction, the notion of tree-width has a natural characterisation in terms of a cops and robber game. Directed tree-width has also been characterised in terms of such games [8], but these games appear to be less intuitive. In this paper, we consider the straightforward extension of the cops and robber game to directed graphs. We show that these games give a characterisation of the graph connectivity measure that we call DAG-width and introduce in Section 3.2. We comment on algorithmic properties in Section 3.3.

3.1 Cops and Robber Games

The Game.   The Cops and Robber game on a digraph is a game where k cops try to catch a robber who may run along paths in the digraph. While the robber is confined to moving along paths in the graph, the cops may move to any vertex at any time. A formal definition follows.

Definition 3.1 (Cops and Robber Game). Given a graph $\mathcal{G} := (V, E)$, the *k-cops and robber* game on $\mathcal{G}$ is played between two players, the *cop* and the *robber* player, as follows:

- At the beginning, the cop player chooses $X_0 \in [V]^{\leq k}$, and the robber player chooses a vertex r_0 of $V \setminus X_0$, giving position (X_0, r_0).
- From position (X_i, r_i), the cop player chooses $X_{i+1} \in [V]^{\leq k}$, and the robber player chooses a vertex r_{i+1} of $V \setminus X_{i+1}$ such that there is a path from r_i to r_{i+1} which does not pass through a vertex in $X_i \cap X_{i+1}$. If no such vertex exists then the robber player loses.

A *play* in the game is a (finite or infinite) sequence $\pi := (X_0, r_0)(X_1, r_1) \ldots$ of positions such that the transition from (X_i, r_i) to (X_{i+1}, r_{r+1}) is a valid move by the rules above and such that the play is finite if, and only if, $r_n \in X_n$ for the final position (X_n, r_n). A play is winning for the robber player if it is infinite.

As always when dealing with games we are less interested in a single play in the game as in strategies that allow a player to win every play in the game. Winning strategies for the cop player play a crucial role throughout this paper. We therefore give a precise definition of this notion.

Definition 3.2. Let $\mathcal{G} := (V, E)$ be a directed graph. A *(k-cop) strategy* for the cop player is a function f from $[V]^{\leq k} \times V$ to $[V]^{\leq k}$. A play $(X_0, r_0), (X_1, r_1), \ldots$ is *consistent* with a strategy f if $X_{i+1} = f(X_i, r_i)$ for all i. The strategy f is called a *winning strategy*, if every play consistent with f is winning for the cop player.

Definition 3.3 (Game-width). The *game-width* $gw(\mathcal{G})$ of $\mathcal{G}$ is the least k such that the cop player has a strategy to win the k-cops and robber game on $\mathcal{G}$.

Variants of the game where the robber moves first or only one cop can be moved at a time or the cops are lifted and placed in separate moves are all equivalent in that the game-width of a graph does not depend on the variant.

Lemma 3.4. *For every finite, non-empty, directed graph $\mathcal{G}$ the game-width $gw(\mathcal{G})$ is at least one and $gw(\mathcal{G}) = 1$ if, and only if, $\mathcal{G}$ is acyclic.*

Games similar to the one defined above have been used to give game characterisations of concepts like undirected tree-width [15] and also the directed tree-width of [8]. Directed tree-width is invariant under reversing the edges of a graph. As we see below, this is not true of the game-width we have defined. One exception are graphs of game-width 1, i.e. acyclic graphs.

Proposition 3.5. $gw(\mathcal{G}) = 1$ if, and only if $gw(\mathcal{G}^{op}) = 1$.

Proposition 3.6. For any j, k with $2 \leq j \leq k$, there exists a graph T_k^j such that $gw(T_k^j) = j$ and $gw((T_k^j)^{op}) = k$.

In the sequel we consider a restriction of the cop player to monotone strategies.

Definition 3.7 (Monotone strategy).

(i) A strategy for the cop player is *cop-monotone* if in playing the strategy, no vertex is visited twice by cops. That is, if $(X_0, r_0), (X_1, r_1) \ldots$ is a play consistent with the strategy, then for every $0 \leq i < n$ and $v \in X_i \setminus X_{i+1}, v \notin X_j$ for all $j > i$.
(ii) A strategy for the cop player is *robber-monotone* if in playing the strategy, the set of vertices reachable by the robber is non-increasing.

Lemma 3.8. *If the cop player has a cop-monotone or robber-monotone winning strategy then it also has a winning strategy that is both, cop- and robber-monotone.*

From this lemma we can define a *monotone winning strategy* in the obvious way.

3.2   DAG-Decompositions and DAG-Width

In this section we define the notion of DAG-width which measures how close a given graph is to being acyclic. We present a decomposition of directed graphs that is somewhat similar in style to tree-decompositions of undirected graphs. We show then that a graph has DAG-width k if, and only if, the cop player has a monotone winning strategy in the k-cops and robber game played on that graph. We conclude with some properties enjoyed by DAG-width.

Definition 3.9. Let $\mathcal{G} := (V, E)$ be a graph. A set $W \subseteq V$ *guards* a set $V' \subseteq V$ if whenever there is an edge $(u, v) \in E$ such that $u \in V'$ and $v \notin V'$ then $v \in W$.

Definition 3.10 (DAG-decomposition). Let $\mathcal{G} := (V, E)$ be a directed graph. A DAG-*decomposition* is a tuple $\mathfrak{D} = (D, (X_d)_{d \in V^D})$ such that

(D1) D is a DAG.
(D2) $\bigcup_{d \in V^D} X_d = V$.
(D3) For all $d \preceq_D d' \preceq_D d''$, $X_d \cap X_{d''} \subseteq X_{d'}$.
(D4) For a root d, X_d is guarded by $\varnothing$.
(D5) For all $(d, d') \in E^D$, $X_d \cap X_{d'}$ guards $\mathcal{X}_{d'} \setminus X_d$, where $\mathcal{X}_{d'} := \bigcup_{d' \preceq_D d''} X_{d''}$.

The width of $\mathfrak{D}$ is defined as $\max\{|X_d| : d \in V^D\}$. The DAG-*width* of a graph is defined as the minimal width of any of its DAG-decompositions.

The main result of this section is an equivalence between monotone strategies for the cop player and DAG-decompositions.

Theorem 3.11. *For any graph $\mathcal{G}$ there is a DAG-decomposition of $\mathcal{G}$ of width k if, and only if, the cop player has a monotone winning strategy in the k-cops and robber game on $\mathcal{G}$.*

For algorithmic purposes, it is often useful to have a normal form for decompositions. The following is similar to one for tree-decompositions as presented in [3].

Definition 3.12. A DAG-decomposition $(\mathcal{D}, (X_d)_{d \in V^{\mathcal{D}}})$ is *nice* if

(N1) $\mathcal{D}$ has a unique root.
(N2) Every $d \in V^{\mathcal{D}}$ has at most two successors.
(N3) If d_1, d_2 are two successors of d_0, then $X_{d_0} = X_{d_1} = X_{d_2}$.
(N4) If d_1 is the unique successor of d_0, then $|X_{d_0} \Delta X_{d_1}| \leq 1$, where Δ is the symmetric set difference operator $(A \Delta B = (A \setminus B) \cup (B \setminus A))$.

Nice decompositions can be seen as corresponding to strategies where we place or remove only one cop at a time. It should therefore not be surprising that we can transform any DAG-decomposition into one which is nice.

Theorem 3.13. *If $\mathcal{G}$ has a DAG-decomposition of width k, it has a nice DAG-decomposition of width k.*

Tree-width on undirected graphs also has a useful characterisation in terms of balanced separators. We are able to obtain one direction of a similar characterisation for DAG-width by showing that graphs of small DAG-width admit small balanced *directed separators*. The definition and proofs can be found in the full version. We also show that the DAG-width of graphs is closed under directed unions, which is considered (see [8]) an important property of a reasonable decomposition of directed graphs.

3.3 Algorithmic Aspects of Bounded DAG-Width

We now consider algorithmic applications of DAG-width as well as the complexity of deciding the DAG-width of a graph and computing an optimal decomposition. The following is a direct consequence of the similar result for tree-width.

Theorem 3.14. *Given a digraph $\mathcal{G}$ and a natural number k, deciding if the DAG-width of $\mathcal{G}$ is at most k is NP-complete.*

However, for any fixed k, it is possible, in polynomial time, to decide if a graph has DAG-width at most k and to compute a DAG-decomposition of this width if it has. We give an algorithm for this that is based on computing monotone winning strategies in the k-cops and robber game.

Theorem 3.15. *Let $\mathcal{G}$ be a directed graph and let $k < \omega$. There is a polynomial time algorithm for deciding if the cop player has a monotone winning strategy in the k-cops and robber game on $\mathcal{G}$ and for computing such a strategy.*

Note also that the translation of strategies into decompositions is computationally easy, i.e. can be done in polynomial time. Since winning strategies can be computed in polynomial time in the size of the graph, we get the following.

Proposition 3.16. *Given a graph $\mathcal{G}$ of* DAG-*width k, a* DAG-*decomposition of $\mathcal{G}$ of width k can be computed in time $\mathcal{O}(|\mathcal{G}|^{\mathcal{O}(k)})$.*

Algorithms on Graphs of Bounded DAG-*Width.* As the directed tree-width of a graph is bounded above by a constant factor of its DAG-width (see Proposition 4.1), any graph property that can be decided in polynomial time on classes of graphs of bounded directed tree-width can be decided on classes of graphs of bounded DAG-width also. This implies that properties such as Hamiltonicity that are known to be polynomial time on graphs of bounded directed tree-width [8] can be solved efficiently on graphs of bounded DAG-width too. We give a nontrivial application of DAG-width in Section 5 where we show that parity games can be solved on graphs of bounded DAG-width, something which is not known for directed tree-width.

As for the relation to undirected tree-width, it is clear that not all graph properties that can be decided in polynomial time on graphs of bounded tree-width can also be decided efficiently on graphs of bounded DAG-width. For instance, the 3-colourability problem is known to be decidable in polynomial time on graphs of bounded tree-width. However, the problem does not depend on the direction of edges. So if the problem was solvable in polynomial time on graphs of bounded DAG-width then for every given graph we could simply direct the edges so that it becomes acyclic, i.e. of DAG-width 1, and solve the problem then. This shows that 3-colourability is not solvable efficiently on graphs of bounded DAG-width unless PTIME $=$ NP. It also implies that Courcelle's theorem does fail for DAG-width, as 3-colourability is easily seen to be MSO-definable.

The obvious question that arises is whether one can define a suitable notion of "directed problem" and then show that every MSO-definable "directed" graph problem can be decided efficiently on graphs of bounded DAG-width. This is part of ongoing work.

4   Relation to Other Graph Connectivity Measures

As a structural measure for undirected graphs, the concept of tree-width is of unrivalled robustness. On the realm of directed graphs, however, its heritage seems to be split among several different concepts. Comparing DAG-width with tree-width, it is easily seen that every tree-decomposition of an undirected graph $\mathcal{G}$ is a DAG-decomposition of the directed graph formed by replacing every edge by two edges, one in each direction. Conversely, the DAG-width of the graph formed in this way is exactly its tree-width. On the other hand a clique with an acyclic orientation provides an example of a digraph with small DAG-width but arbitrarily large tree-width. In the sequel we compare DAG-width with other connectivity measures for digraphs, namely directed tree-width introduced by Johnson et al. [8], and entanglement proposed by Berwanger and Grädel [2].

Directed Tree-Width. Aiming to reproduce the success of tree-decompositions in allowing divide-and-conquer algorithms, directed tree-width is associated to a tree-shaped representation of the input graph. It was proved that this representation leads to efficient algorithms for solving a particular class of NP-complete problems, including, e.g., Hamiltonicity, when directed tree-width is bounded. Unfortunately this generic method does not cover many interesting problems. In particular, the efficient solution of parity games on bounded tree-width has failed so far to generalise to directed tree-width.

In terms of games, directed tree width is characterised by a restriction of the cops-and-robber game for DAG-width, in which the robber is only permitted to move to vertices where there exists a directed cop-free path from his intended destination back to his current position. On the basis of the game characterisation, it is clear that the directed tree-width of a graph provides a lower bound for its DAG-width. Conversely, the DAG-width of a graph cannot be bounded in terms of its directed tree-width.

Proposition 4.1.

(i) *If a graph has* DAG-*width* k, *its directed tree-width is at most* $3k + 1$.
(ii) *There are graphs with arbitrarily large* DAG-*width and directed tree-width* 1.

Entanglement. The notion of entanglement measures the nesting depth of directed cycles in a graph. In terms of cops-and-robber games, it is obtained by restricting the mobility of both the robber and the cops so that in any round, the cop player may send one cop to the robber's current position (or do nothing) while the robber can only move to a successor of his current residence.

Unlike the other graph widths considered here, entanglement is not associated to an efficient tree-shaped graph representation. Nevertheless, it was shown that parity games on arenas of bounded entanglement can be solved in polynomial time.

The following proposition shows that having bounded DAG-width is more general than having bounded entanglement. On the other hand, the gap between DAG-width and entanglement can be at most logarithmic in the number of graph vertices.

Proposition 4.2.

(i) *If a graph has entanglement* k, *its* DAG-*width is at most* $k + 1$.
(ii) *There are graphs with arbitrarily large entanglement but with* DAG-*width* 2.
(iii) *If a graph* $\mathcal{G}$ *has* DAG-*width* k, *its entanglement is at most* $(k + 1) \cdot \log |V^{\mathcal{G}}|$.

We conclude that, despite their conceptual affinity, directed tree-width, entanglement, and DAG-width are rather different measures.

5   Parity Games on Graphs of Bounded DAG-Width

A parity game $\mathcal{P}$ is a tuple (V, V_0, E, Ω) where (V, E) is a directed graph, $V_0 \subseteq V$ and $\Omega : V \to \omega$ is a function assigning a priority to each node. There is no loss of generality in assuming that the range of Ω is contained in $[n]$ where $n = |V|$ and we will make this assumption from now on.

Intuitively, two players called Odd and Even play a parity game by pushing a token along the edges of the graph with Even playing when the token is on a vertex in V_0 and Odd playing otherwise. Formally, a play of the game $\mathcal{P}$ is an infinite sequence

$\pi = (v_i \mid i \in \omega)$ such that $(v_i, v_{i+1}) \in E$ for all i. We say π is winning for Even if $\liminf_{i \to \infty} \Omega(v_i)$ is even and π is winning for Odd otherwise.

A *strategy* is a map $f : V^{<\omega} \to V$ such that for any sequence $(v_0 \cdots v_i) \in V^{<\omega}$, $(v_i, f(v_0 \cdots v_i)) \in E$. A play $\pi = (v_i \mid i \in \omega)$ is consistent with Even playing f if whenever $v_i \in V_0$, $v_{i+1} = f(v_0 \cdots v_i)$. Similarly, π is consistent with Odd playing f if whenever $v_i \notin V_0$, $v_{i+1} = f(v_0 \cdots v_i)$. A strategy f is winning for Even if every play consistent with Even playing f is winning for Even. A strategy is *memoryless* if whenever $u_0 \cdots u_i$ and $v_0 \cdots v_j$ are two sequences in $V^{<\omega}$ with $u_i = v_j$, then $f(u_0 \cdots u_i) = f(v_0 \cdots v_j)$. It is known that parity games are determined, i.e. for any game and starting position, either Even or Odd has a winning strategy and indeed, a memoryless one. However, we do not assume in our construction that the strategies we consider are memoryless.

The following ordering on $[n]$ is useful in evaluating competing strategies. For priorities $i, j \in [n]$ we say $i \sqsubseteq j$ if either

(i) i is odd and j is even, or
(ii) i and j are both odd and $i \leq j$, or
(iii) i and j are both even and $j \leq i$.

Intuitively, $i \sqsubseteq j$ if the priority i is "better" for player Odd than j.

We are interested in the problem of determining, given a parity game and starting node, which player has a winning strategy. The complexity of this problem in general remains a major open question, as explained in Section 1. We demonstrate that parity games are solvable on arenas of bounded DAG-width by an algorithm similar in spirit to that of Obdržàlek [11]. That algorithm relies on the fact that in a tree-decomposition, a set of k nodes guards all entries and exits to the part of the graph below it, and thus all cycles must pass through this set. In the case of a DAG-decomposition, while the small set guards all exits from the subgraph below it, there may be an unlimited number of edges going into this subgraph. This is the main challenge that our algorithm addresses, and is specifically solved in Lemmas 5.1, 5.2 and 5.3.

For a parity game $\mathcal{P} = (V, V_0, E, \Omega)$ consider $U \subseteq V$ and a set W that guards U. Fix a pair of strategies f and g. For any $v \in U$, there is exactly one play $\pi = (v_i : i \in \omega)$ that is consistent with Even playing f and Odd playing g. Let π' be the maximal initial segment of π that is contained in U. The *outcome* of the pair of strategies (f, g) (given U and v) is defined as follows.

$$\text{out}_{f,g}(U, v) := \begin{cases} \text{winEven} & \text{if } \pi' = \pi \text{ and } \pi \text{ is winning for Even;} \\ \text{winOdd} & \text{if } \pi' = \pi \text{ and } \pi \text{ is winning for Odd;} \\ (v_{i+1}, p) & \text{if } \pi' = v_0 \cdots v_i \text{ and } p = \min\{\Omega(v_j) \mid 0 \leq j \leq i+1\}. \end{cases}$$

By construction, if $\text{out}_{f,g}(U, v) = (w, p)$ then $w \in W$. More generally, for any set $W \subseteq V$, define the set of potential outcomes in W, written pot-out(W), to be the set $\{\text{winEven}, \text{winOdd}\} \cup \{(w, p) : w \in W \text{ and } p \in [n]\}$. We define a partial order $\trianglelefteq$ on pot-out(W) which orders potential outcomes according to how good they are for player Odd. It is the least partial order satisfying the following conditions:

(i) winOdd $\trianglelefteq o$ for all outcomes o;

(ii) $o \trianglelefteq$ winEven for all outcomes o;

(iii) $(w, p) \trianglelefteq (w, p')$ if $p \sqsubseteq p'$ for all $w \in W$.

In particular, (w, p) and (w', p') are incomparable if $w \neq w'$. The idea is that if g and g' are strategies such that $\text{out}_{f,g}(U, v) \trianglelefteq \text{out}_{f,g'}(U, v)$ then player Odd is better off playing strategy g rather than g' in response to Even playing according to f.

A single outcome is the result of fixing the strategies played by both players in the sub-game induced by a set of vertices U. If we fix the strategy of player Even to be f but consider all possible strategies that Odd may play, we can order these strategies according to their outcome. If one strategy achieves outcome o and another o' with $o \trianglelefteq o'$, there is no reason for Odd to consider the latter strategy. Thus, we define $\text{result}_f(U, v)$ to be the set of outcomes that are achieved by the best strategies that Odd may follow, in response to Even playing according to f. More formally, $\text{result}_f(U, v)$ is the set of $\trianglelefteq$-minimal elements in the set $\{o : o = \text{out}_{f,g}(U, v) \text{ for some } g\}$. Thus, $\text{result}_f(U, v)$ is an anti-chain in the partial order $(\text{pot-out}(W), \trianglelefteq)$, where W is a set of guards for U. We write pot-res(W) for the set of *potential results* in W. To be precise, pot-res(W) is the set of all anti-chains in the partial order $(\text{pot-out}(W), \trianglelefteq)$. By definition of the order $\trianglelefteq$, if either of winEven or winOdd is in the set $\text{result}_f(U, v)$, then it is the sole element of the set. Also, for each $w \in W$, there is at most one p such that $(w, p) \in \text{result}_f(U, v)$ so the number of distinct values that $\text{result}_f(U, v)$ can take is at most $(|U| + 1)^{|W|} + 2$ (in fact, $(d + 1)^{|W|}$, where d is the number of different priorities in U). This is the cardinality of the set pot-res(W).

We also abuse notation and extend the order $\trianglelefteq$ to the set pot-res(W) pointwise. That is, for $r, s \in \text{pot-res}(W)$ we write $r \trianglelefteq s$ if, for each $o \in s$, there is an $o' \in r$ with $o' \trianglelefteq o$. With this definition, the order $\trianglelefteq$ on pot-res(W) admits greatest lower bounds. Indeed, the greatest lower bound $r \sqcap s$ of r and s can be obtained by taking the set of $\trianglelefteq$ minimal elements in the set of outcomes $r \cup s$. One further piece of notation we use is that we write $\text{Res}(U, v)$ for the set $\{\text{result}_f(U, v) : f \text{ is a strategy}\}$.

Suppose now that $\mathcal{P} = (V, V_0, E, \Omega)$ is a parity game and we are given a DAG decomposition $(\mathcal{D}, (X_d)_{d \in V^{\mathcal{D}}})$ of (V, E) of width k that is nice in the sense of Definition 3.12. For each $d \in V^{\mathcal{D}}$, we write V_d for the set $\mathcal{X}_d \setminus X_d$. The key to the algorithm is that we construct the set of results $\text{Res}(V_d, v)$ for each $v \in V_d$. Since V_d is guarded by X_d, $|X_d| \leq k$ and $|V_d| \leq n$, the number of distinct values of $\text{result}_f(V_d, v)$ as f ranges over all possible strategies is at most $(n + 1)^k + 2$.

We define the following, which is our key data structure: Frontier$(d) = \{(v, r) : v \in V_d \text{ and } r = \text{result}_f(V_d, v) \text{ for some strategy } f\}$. Note that in the definitions of $\text{result}_f(U, v)$ and Frontier(d), f and g range over *all* strategies and not just memoryless ones. The bound on the number of possible values of $\text{result}_f(V_d, v)$ guarantees that $|\text{Frontier}(d)| \leq n((n + 1)^k + 2)$. We aim to show how Frontier(d) can be constructed from the set of frontiers of the successors of d in polynomial time. When d is a leaf, $V_d = \varnothing$ and thus Frontier$(d) = \varnothing$. There are four inductive cases to consider.

Case 1: d has two successors e_1 and e_2. In this case, $X_d = X_{e_1} = X_{e_2}$ by (N2). We claim that Frontier$(d) = \text{Frontier}(e_1) \cup \text{Frontier}(e_2)$.

Case 2: d has one successor e and $X_d = X_e$. In this case, Frontier(d) = Frontier(e).

Case 3: d has one successor e and $X_d \setminus X_e = \{u\}$. Then, by (D3), $u \notin V_e$. Also, by definition of V_d, $u \notin V_d$. We conclude that $V_d = V_e$. Moreover, since X_e guards V_e, there is no path from any element of V_e to u except through X_e. Hence, Frontier(d) = Frontier(e).

Case 4: d has one successor e and $X_e \setminus X_d = \{u\}$. This is the critical case. Here $V_d = V_e \cup \{u\}$ and in order to construct Frontier(d) we must determine the results of all plays beginning at u.

Consider the set of vertices v in $\mathcal{X}_d$ such that $(u, v) \in E^{\mathcal{G}}$. These fall into two categories. Either $v \in X_d$ or $v \in V_e$. Let $x_1, \ldots, x_s$ enumerate the first category and let $v_1, \ldots, v_m$ enumerate the second. Let $O = \{(x_i, \min\{\Omega(x_i), \Omega(u)\}) : 1 \le i \le s\}$. This is the set of outcomes obtained if play in the parity game proceeds directly from u to an element of X_d. Note that as no two outcomes in O are comparable with respect to $\trianglelefteq$, $O \in \text{pot-res}(X_d)$. We write $\mathcal{O}$ for $\{\{o\} : o \in O\}$ That is $\mathcal{O}$ is the set of singleton results obtained from O. For each v_i we know, from Frontier(e), the set Res(V_e, v_i). For each result $r \in \text{Res}(V_e, v_i)$, we write mod$(r)$ for the set of outcomes defined by modifying r as follows. First, if r contains an outcome (u, p), we replace it by winEven if $\min\{p, \Omega(u)\}$ is even and winOdd if it is odd. Secondly, for any pair $(w, p) \in r$ where $w \ne u$, we replace it with $(w, \min\{p, \Omega(u)\})$. Finally, we take the set of $\trianglelefteq$-minimal elements from the resulting set. This is mod(r). Note that mod$(r) \in \text{pot-res}(X_d)$. The intuition is that mod$(\text{result}_f(V_e, v_i))$ defines the set of best possible outcomes for player Odd, if starting at u, the play goes to v_i and from that point on, player Even plays according to strategy f. For each $1 \le i \le m$, let $M_i = \{\text{mod}(r) : r \in \text{Res}(V_e, v_i)\}$.

We now wish to use the sets of results M_i, O and $\mathcal{O}$ to construct the Res(V_d, u). We need to distinguish between the cases when $u \in V_0$ (i.e. player Even plays from u in the parity game) and $u \in V \setminus V_0$ (i.e. player Odd plays).

The simpler case is when $u \in V_0$.

Lemma 5.1. *If* $u \in V_0$, *then* $Res(V_d, u) = \bigcup_i M_i \cup \mathcal{O}$.

The case when $u \notin V_0$ is somewhat trickier. To explain how we can obtain Res(V_d, u) in this case, we formulate the following lemma.

Lemma 5.2. *If* $u \notin V_0$, *then* $r \in Res(V_d, u)$ *if, and only if, there is a function c on the set $[m]$ with $c(i) \in M_i$ such that $r = O \sqcap \bigcap_{i \in [m]} c(i)$.*

Lemma 5.2 suggests constructing Res(V_d, u) by considering all possible choice functions c. However, as each set M_i may have as many as $(n+1)^k + 2$ elements, there are $m^{(n+1)^k + 2}$ possibilities for c and our algorithm would be exponential. We consider an alternative way of constructing Res(V_d, u). Recall that Res$(V_d, u) \subseteq \text{pot-res}(X_d)$ and the latter set has at most $(n+1)^k + 2$ elements. We check, for each $r \in \text{pot-res}(X_d)$, in polynomial time, whether there is a choice function c as in Lemma 5.2 that yields r. In particular, we take the following alternative characterisation of Res(V_d, u).

Lemma 5.3. *If* $u \notin V_0$, *then* $r \in Res(V_d, u)$ *if, and only if, there is a set $D \subseteq [m]$ with $|D| \le |r|$ and a function d on D with $d(i) \in M_i$ such that*

(i) $r = O \sqcap \bigsqcap_{i \in D} d(i)$; and

(ii) *for each $i \notin D$ there is an $r_i \in M_i$ with $r \trianglelefteq r_i$.*

Now, any $r \in$ pot-res(X_d) has at most k elements. Thus, to check whether such an r is in Res(V_d, u) we cycle through all sets $D \subseteq [m]$ with k or fewer elements (and there are $\mathcal{O}(n^k)$ such sets) and for each one consider all candidate functions d (of which there are $\mathcal{O}(n^{k^2})$). Having found a d which gives $r = O \sqcap \bigsqcap_D d(i)$, we then need to find a suitable r_i in each $i \in [m] \setminus D$. For this we must, at worst, go through all elements of all the sets M_i and compare them to r. This can be done in time $\mathcal{O}(n^{k+1})$.

We have now obtained the set Res(V_d, u). One barrier remains to completing the construction of Frontier(d). Elements (v, r) of Frontier(e) may have outcomes in r of the form (u, p). Since u is not in X_d, these must be resolved by combining them with results from Res(V_d, u). To be precise, let $r \in$ Res(V_e, v) for some $v \in V_e$ and $s \in$ Res(V_d, u). Define the combined result $c(r, s)$ as follows:

- if r does not contain an outcome of the form (u, p), then $c(r, s) = r$;
- otherwise, r contains a pair (u, p). Let s' be obtained from s by replacing every pair (w, q) by $(w, \min\{p, q\})$. $c(r, s) = r \sqcap s'$.

Intuitively, if $r = \text{result}_f(V_e, v)$ and $s = \text{result}_{f'}(V_d, u)$ then $c(r, s)$ is the set of $\trianglelefteq$-minimal outcomes that can be obtained if player Even plays according to f starting at v *until* the node u is encountered and then switches to strategy f'.

Lemma 5.4. *For $v \in V_e$, $Res(V_d, v) = \{c(r, s) : r \in Res(V_e, v) \text{ and } s \in Res(V_d, u)\}$.*

We now obtain Frontier$(d) = \{(v, r) : r \in \text{Res}(V_d, v)\}$.

Theorem 5.5. *For each k, there is a polynomial p and an algorithm running in time $\mathcal{O}(p(n))$ which determines the winner of parity games on all graphs with* DAG-*width at most k.*

References

1. J. BARÁT, *Directed path-width and monotonicity in digraph searching.* To appear in *Graphs and Combinatorics.*
2. D. BERWANGER AND E. GRÄDEL, *Entanglement – a measure for the complexity of directed graphs with applications to logic and games*, in LPAR, 2004, pp. 209–223.
3. H. L. BODLAENDER, *Treewidth: Algorithmic techniques and results*, in MFCS, 1997, pp. 19–36.
4. B. COURCELLE, *Graph rewriting: An algebraic and logic approach*, in Handbook of Theoretical Computer Science, Volume B: Formal Models and Sematics (B), J. van Leeuwan, ed., 1990, pp. 193–242.
5. N. D. DENDRIS, L. M. KIROUSIS, AND D. M. THILIKOS, *Fugitive-search games on graphs and related parameters*, TCS, 172 (1997), pp. 233–254.
6. E. EMERSON, C. JUTLA, AND A. SISTLA, *On model checking for the μ-calculus and its fragments*, TCS, 258 (2001), pp. 491–522.
7. G. GOTTLOB, N. LEONE, AND F. SCARCELLO, *Robbers, marshals, and guards: Game theoretic and logical characterizations of hypertree width*, in PODS, 2001, pp. 195–201.

8. T. JOHNSON, N. ROBERTSON, P. D. SEYMOUR, AND R. THOMAS, *Directed tree-width*, Journal of Combinatorial Theory, Series B, 82 (2001), pp. 138–154.

9. M. JURDZIŃSKI, *Deciding the winner in parity games is in UP ∩ co-UP*, Information Processing Letters, 68 (1998), pp. 119–124.

10. D. KOZEN, *Results on the propositional mu-calculus*, TCS, 27 (1983), pp. 333–354.

11. J. OBDRŽÁLEK, *Fast mu-calculus model checking when tree-width is bounded*, in Proceedings of 15th International Conference on Computer Aided Verification, vol. 2725 of LNCS, Springer, 2003, pp. 80–92.

12. B. A. REED, *Introducing directed tree width*, in 6th Twente Workshop on Graphs and Combinatorial Optimization, vol. 3 of Electron. Notes Discrete Math, Elsevier, 1999.

13. N. ROBERTSON AND P. SEYMOUR, *Graph Minors. III. Planar tree-width*, Journal of Combinatorial Theory, Series B, 36 (1984), pp. 49–63.

14. M. SAFARI, *D-width: A more natural measure for directed tree width*, in MFCS 2005, vol. 3618 of LNCS, Springer, 2005, pp. 745–756.

15. P. SEYMOUR AND R. THOMAS, *Graph searching, and a min-max theorem for tree-width*, Journal of Combinatorial Theory, Series B, 58 (1993), pp. 22–33.

Reliable Computations Based
on Locally Decodable Codes

Andrei Romashchenko*

LIP de Lyon (CNRS) and IITP of Moscow
anromash@mccme.ru

Abstract. We investigate the coded model of fault-tolerant computations introduced by D. Spielman. In this model the input and the output of a computational circuit is treated as words in some error-correcting code. A circuit is said to
compute some function correctly if for an input which is a encoded argument of
the function, the output, been decoded, is the value of the function on the given
argument.

We consider two models of faults. In the first one we suppose that an elementary processor at each step can be corrupted with some small probability,
and faults of different processors are independent. For this model, we prove that
a parallel computation running on n elementary non-faulty processors in time
$t = \mathrm{poly}(n)$ can be simulated on $\mathcal{O}(n \log n / \log \log n)$ faulty processors in
time $\mathcal{O}(t \log \log n)$. Note that we get a sub-logarithmic blow up of the memory,
which cannot be achieved in the classic model of faulty boolean circuit, where
the input is not encoded.

In the second model, we assume that at each step some fixed fraction of elementary processors can be corrupted by an adversary, who is free to chose these
processors arbitrarily. We show that in this model any computation can be made
reliable with an exponential blow up of the memory.

Our method employs a sort of *mixing mappings*, which enjoy some properties of expanders. Based on mixing mappings, we implement an effective self-
correcting procedure for an array of faulty processors.

1 Introduction

The problem of reliable computations with faulty elements was investigated for several
types of computational models. In the most popular models, the computation is implemented by a circuit of boolean gates; each gate can fail with a small probability. A
circuit is said fault-tolerant if it returns a correct result with high probability. For the first
time such a model of computation was proposed by J. von Neumann [1]. Later ideas by
von Neumann was developed by Dobrushin and Ortyukov [3]. Futher N. Pippenger in
[5] presented an *effective* transformation of every boolean circuit with non-faulty gates
into a fault-tolerant circuit.

The construction by Pippenger requires only logarithmic increasing of the number
of gates. In general, this result cannot be improved, and logarithmic redundancy is in-
evitable [4, 6, 7, 8]. But this lower bound is caused by the need to encode the input with

---

* Supported in part by RSSF and by RFBR grant # 03-01-00475.

B. Durand and W. Thomas (Eds.): STACS 2006, LNCS 3884, pp. 537–548, 2006.

some error-correcting code and then to decode the answer. This obstacle can be eliminated if we allow to get the input and to return the result as an encoded word. Such a model was used in the work of D. Spielman [10]. Let us define this model (with minor modifications) in detail.

The Computational Model. The computational array consists of N elementary processors $s_1, \ldots, s_N$. At each moment, every processor contains one bit of information (a processor is said to have an internal state 0 or 1). We fix two functions,

$$E : \{0, 1\}^n \to \{0, 1\}^N \quad \text{and} \quad D : \{0, 1\}^N \to \{0, 1\}^n,$$

which are normally the encoding and decoding functions of some error-correcting code. We say that a circuit gets an input $x \in \{0, 1\}^n$ if the initial state of the memory $(s_1, \ldots, s_N)$ is equal to $E(x)$.

Denote by $s_1^{(t)}, \ldots, s_N^{(t)}$ the internal states of the processors at moment t. We call by *a circuit of depth T* a list of instructions $F_i^{(t)}$, $t = 1, \ldots, T$ which define how each of the processors should update its internal state at each moment. More precisely, the state of a processor s_i at moments t is defined by the rule

$$s_i^{(t)} = F_i^{(t)}(s_{j_1}^{(t-1)}, s_{j_2}^{(t-1)}, \ldots, s_{j_r}^{(t-1)}),$$

where $F_i^{(t)}$ is a boolean function (the indexes $j_1, \ldots, j_r$ depend on i and t). The arity r is supposed to be fixed in advance. We shall always suppose that r is a constant independent of n.

We say that a circuit of depth T computes a function $f : \{0, 1\}^n \to \{0, 1\}^n$ if for all x the equality $D(s_1^{(T)}, \ldots, s_N^{(T)}) = f(x)$ holds provided that $(s_1^{(0)}, \ldots, s_N^{(0)}) = E(x)$. In other words, we use the encoding E to provide the circuit with an input, and then use the decoding D to retrieve the result. We shall say also that such a circuit computes f *in T steps.*

In the model of random faults we suppose that at each step every processor can be randomly corrupted, i.e., with some small probability ϵ it can change its internal state contrary to the rule above. Faults at different positions i and moments t (i.e., the events that a processor s_i is corrupted at moment t) are supposed to by independent. For this model, we say that some circuit correctly computes a function f with probability $(1 - \epsilon')$, if

$$\text{Prob}[D(\bar{s}^{(T)}) = f(x)|\bar{s}^{(0)} = E(x)] = 1 - \epsilon',$$

where $\bar{s}^{(0)} = (s_1^{(0)}, \ldots, s_N^{(0)})$ is an initial state of the computational array and $\bar{s}^{(T)} = (s_1^{(T)}, \ldots, s_N^{(T)})$ is the corresponding final state of the processors.

In another model, faults are caused by an malevolent adversary. This means that at each moment t any ϵN processors can be corrupted. We say that a circuit computes some function f if for every choice of the positions where processors are corrupted, the final result is correct, i.e., for all x the equality $D(s_1^{(T)}, \ldots, s_N^{(T)}) = f(x)$ holds provided that $(s_1^{(0)}, \ldots, s_N^{(0)}) = E(x)$.

Note that the defined model is trivial if the functions E and D may depend on f. Our goal is to implement reliable computations of *all* functions given *one* pair of natural

(E, D). In the sequel for each n we fix some E_n and D_n (one pair for the model with random faults or another for the model with an adversary) and show that every function f that can be computed on space n and in time $poly(n)$ in the model without faults, can be also computed with these encoding and decoding functions by a reliable circuit in a faulty model.

The rest of the paper is organized as follows. In Section 2 we introduce the mixing mappings, the main combinatorial tool of our proofs. In Section 3 we consider the model of random faults and prove that every circuit of depth T with n processors can be converted in a reliable circuit with $\mathcal{O}(n \log(nT)/\log\log(nT))$ processors, which computes the same function in time $\mathcal{O}(T \log\log(nT))$. Comparative to [10], our bounds for time and space are better (the construction by D. Spielman requires a poly-logarithmic blow up and poly-logarithmic slow down); however, we get only a constant probability of an error, whereas in [10] the probability of an error is exponentially small. Let us note that for $T = poly(n)$, the blow up in our construction is equal to $\mathcal{O}(\log n/\log\log n)$, i.e., it is below the $\log n$ barrier, which is strict for the usual model of faulty boolean circuits (where the input of a circuit is not encoded). To the best of our knowledge, it is the first construction of reliable circuits with sub-logarithmic blow up of the memory. Our construction is effective, i.e., the fault-tolerant circuit can be constructed from the original one by a polynomial algorithm. In Section 4 we deal with the model where faults are chosen by an adversary. We prove that every circuit of depth T with n processors can be converted in a reliable circuit with $2^{\mathcal{O}(n)}$ processors and depth $\mathcal{O}(nT)$.

2   Mixing Functions

In this section we define mixing mappings and prove some of their properties.

Definition 1. *We call a mapping* $F : \{0,1\}^m \times \{0,1\}^\tau \to \{0,1\}^m$ *a* (m, τ, α, β)-*mixer if for all* $A, B \subset \{0,1\}^m$ *such that* $|A| \geq \alpha 2^m$ *the condition*

$$\left\| |\{(x,u) \ : \ x \in A, F(x,u) \in B\}| - \frac{2^\tau |A| \cdot |B|}{2^m} \right\| \leq \beta 2^\tau |A|$$

holds.

This definition was inspired by the well-known Expander Mixing Lemma, see e.g, [11]. The Expander Mixing Lemma implies that an expander is a mixer with appropriate parameters.

We need mixers with some additional structure. First of all, we consider $L = \{0,1\}^m$ as an m-dimensional linear space over $\mathbb{Z}/2\mathbb{Z}$ (for $u, v \in L$ the vector $u + v$ is just the bitwise sum of u and v modulo 2).

Definition 2. *We call a mapping* $F : \{0,1\}^m \times \{0,1\}^\tau \to \{0,1\}^m$ *a linear* (m, τ, α, β)-*mixer if (1)* F *is a* (m, τ, α, β)-*mixer, and (2) the mapping* $G : \{0,1\}^m \times \{0,1\}^\tau \to \{0,1\}^m$ *defined as* $G(x, u) = F(x, u) + x$, *is also a* (m, τ, α, β)-*mixer.*

Standard probabilistic arguments imply that a linear mixer exists:

Lemma 1. *For all* $\alpha, \beta \in (0, 1)$ *there exists a* τ *such that for all* m *there exists a linear* (m, τ, α, β)-*mixer.*

The following properties of a mixer easily follows from the definition:

Claim 1. *If $0 < \delta' < \delta < 1/8$ then for every $(m, \tau, \delta', \delta)$-mixer F, for every $B \subset \{0,1\}^m$ of size at most $2^m/8$ there are less than $\delta 2^m$ elements $x \in \{0,1\}^m$ such that for at least 25% of $u \in \{0,1\}^\tau$ we have $F(x, u) \in B$.*

Claim 2. *Let $0 < \delta' < \delta < 1/128$ and F be an $(m, \tau, \delta', \delta)$-mixer. Then for every $B \subset \{0,1\}^m$ of size at most $2^m/128$ there are less than $\delta 2^m$ elements $x \in \{0,1\}^m$ such that for at least $1/64$ of all $u \in \{0,1\}^\tau$ we have $F(x, u) \in B$.*

We omit the proofs; both claims can be easily deduced from the definition.

As we mentioned above, any expander is a mixer with appropriate parameters; but an expander may be not a *linear* mixer. Thus, if we need an effective construction of a linear mixer, we should develop new technique. Below we explain an explicit construction of linear mixers required for our proofs.

Lemma 2. *Let F_1 be an $(m_1, \tau_1, \alpha_1, \beta_1)$-mixer and F_2 be an $(m_2, \tau_2, \alpha_2, \beta_2)$-mixer. Then the tensor product $F = F_1 \otimes F_2$*

$$F : \{0,1\}^{m_1+m_2} \times \{0,1\}^{\tau_1+\tau_2} \rightarrow \{0,1\}^{m_1+m_2}$$

is an $(m_1 + m_2, \tau_1 + \tau_2, \alpha, \beta)$-mixer for $\alpha = \sqrt{\alpha_2}$ and $\beta = \mathcal{O}(\frac{\alpha_1+\beta_1+\beta_2}{\alpha_2\sqrt{\alpha_2}} + \sqrt{\alpha_2})$

This lemma is interesting for $\alpha_1 \ll \beta_1 \ll \beta_2 \ll \alpha_2$. The bound in this lemma is quite rough, but it is enough for our applications below. We omit the proof of the Lemma due to the lack of space.

Remark that if F_1 and F_2 are *linear* mixers then $F_1 \otimes F_2$ is also a *linear* mixer.

Lemma 3. *For every α, β and a large enough τ there exists an algorithm which for an input m constructs a linear (m, τ, α, β)-mixer in time $\mathrm{poly}(2^{2^m})$ (the time bound is not very nice, but it is better than exhaustive search!).*

Proof. Denote $N = 2^{2^m}$. From Lemma 1 it follows that for all α', β' and a large enough τ', for all n a linear $(n, \tau', \alpha', \beta')$-mixer exists. If $2^{n2^n} = \mathrm{poly}(N)$, we can construct a linear $(n, \tau', \alpha', \beta')$-mixer in time $\mathrm{poly}(N)$ using a brute force search. In particular, for any $\alpha', \beta', \alpha'', \beta''$ we can get in time $\mathrm{poly}(N)$ some linear mixers with parameters $(m/2, \tau', \alpha', \beta')$ and $(m/2, \tau'', \alpha'', \beta'')$. Further, construct a tensor product of these two mixers. From Lemma 2 it follows that we obtain a linear $(m, \tau' + \tau'', \sqrt{\alpha''}, \mathcal{O}(\frac{\alpha'+\beta'+\beta'}{\alpha''\sqrt{\alpha''}} + \sqrt{\alpha''}))$-mixer. It remains to choose appropriate α', α'' and β', β''. □

3 Computations Resistant to Random Faults

In this section we show how to implement reliable computations with a circuit based on faulty processors. We assume that each processor at each step of computation is corrupted with small enough probability $\epsilon > 0$, and that faults at different processors and at different moments of time are independent.

We use encoding based on the Hadamard code. Remind that the Hadamard code is a mapping

$$Had \; : \; \{0,1\}^n \to \{0,1\}^{2^n}$$

where $Had(a_1, \ldots, a_n)$ is the table of all values of the linear function of n variables

$$f(x_1, \ldots, x_n) - \sum_{1 \leq i \leq n} a_i x_i$$

(the coefficients a_i and the variables x_i ranges over the field $\mathbb{Z}/2\mathbb{Z}$). The code is linear, so for all $x, y \in \{0,1\}^n$ we have $Had(x \oplus y) = Had(x) \oplus Had(y)$. Here and in the sequel we denote by $x \oplus y$ the bitwise sum modulo 2.

Theorem 1. *For every circuit S of depth t with n processors, for every $\epsilon_{res} > 0$ there exists a fault tolerant circuit $\hat{S}$ of depth $\mathcal{O}(t \log \log(tn))$, with $\mathcal{O}(n \log(nt)/\log\log(nt))$ processors that computes the same function as S, for the input encoded with some function E and the output decode with some D. The circuit $\hat{S}$ is reliable in the following sense: if every processor at each step is corrupted with a small enough probability ϵ (faults at different processors and different steps are independent) then the result is correct with probability at least $(1 - \epsilon_{res})$.*

The functions E, D depend on n and t but not on the function computed by the circuit.

The transformation of a circuit to the reliable form is effective, i.e., there exists a polynomial algorithm that constructs such a circuit $\hat{S}$ given S.

Remark 1. Remind that the computations in our model are defined by instructions of the form $s_i^{(t)} = F_i^{(t)}(s_{j_1}^{(t-1)}, s_{j_2}^{(t-1)}, \ldots, s_{j_r}^{(t-1)})$, where the arity r is bounded. It is clear that any computation can be implemented with instructions of arity 2. It is not hard to convert a circuit S based on any r-ary elements into an equivalent circuit S' based on 2-ary gates; we need only a constant blow-up of the memory and a constant slow down. In the model of faulty computations, this transformation also changes the probability to get an error at the output. If the original circuit S returns the correct result with probability $(1 - \epsilon_{res})$ provided that each processor at each step fails with probability ϵ_0, then the new circuit S' returns the correct value with the same probability $(1 - \epsilon_{res})$ provided that each processor at each step fails with some smaller probability ϵ_1 (the value ϵ_1 depends on ϵ_0 and r).

Thus, the arities of basic gates of S and $\hat{S}$ are not essential for our proof. For simplicity, we shall assume that arity of gates for the original circuit S is at most 2; we shall construct $\hat{S}$ an equivalent fault-tolerant circuit $\hat{S}$ with a large enough (but bounded) arity $r = \mathcal{O}(1)$.

Proof. Denote by $y^{(j)} = (y_1^{(j)}, y_2^{(j)}, \ldots, y_n^{(j)})$ the state of the memory of S at the j-th step of the computation, $j = 0, \ldots, t$. We shall define an encoding $E : \{0,1\}^n \to \{0,1\}^N$ and the corresponding decoding $D : \{0,1\}^N \to \{0,1\}^n$, and a circuit $\hat{S}$ with N processors so that for every j the internal state $z^{(j)} = (z_1^{(i)}, \ldots, z_N^{(i)})$ of $\hat{S}$ with high probability is close to $E(y^{(i)})$. More precisely, with high probability for each j the equality $D(z^{(j)}) = y^{(j)}$ holds. In particular, for the final result $z^{(t)}$ we get $D(z^{(t)}) = y^{(t)}$.

Let us implement the plan presented above. We split the set of variables $(x_1, \ldots, x_n)$ into blocks of size $k = \log(\log n + \log t) + C$:

$$b_1 = x_1, \ldots, x_k,$$
$$b_2 = x_{k+1}, \ldots, x_{2k},$$
$$\ldots \quad \ldots$$

(the constant C will be chosen below). The total number of blocks b_i is $r = \lceil n/k \rceil$.

A Restriction on Parallelism. Let us restrict the power of the parallelism in S. We convert the circuit S in a circuit S' such that at each step in every block b_i only one processor changes its internal state[1]. It is easy to construct from S a new circuit S' that satisfy the conditions above; S' needs $\mathcal{O}(n)$ processors and runs in time $T = \mathcal{O}(t \log \log(nt))$.

To prove the theorem, we show that S' can be simulated by a fault-tolerant circuit $\hat{S}$ with a blow up of the memory $\mathcal{O}(2^k/k)$ in real time, i.e., without any slow down.

First of all, we specify encoding and decoding. Define encoding $E : \{0,1\}^n \rightarrow \{0,1\}^N$ as follows:

$$E(x_1, \ldots, x_n) = Had(b_1) \ldots Had(b_r).$$

Denote $\hat{b}_i = Had(b_i)$. Note that the length of each $\hat{b}_i$ is $2^k = 2^C \log(tn)$ and the length of the codewords is $N = r2^k = \mathcal{O}(n \log(tn)/\log\log(tn))$.

Define manipulations with encoded data. Denote by $z^{(j)} = (z_1^{(j)}, z_2^{(j)}, \ldots, z_N^{(j)})$ the state of memory of the circuit $\hat{S}$ at j-th stage of computation. We split $z^{(j)}$ into blocks of length 2^k and denote them $\hat{b}_1^{(j)}, \ldots, \hat{b}_r^{(j)}$.

Further we define the transition rule: how $z^{(j+1)}$ is computed from $z^{(j)}$. We define it so that with high probability for all j each block $\hat{b}_i^{(j)}$ differs from $Had(b_i^{(j)})$ in a fraction at most $1/8$ of all bits.

In our model, at each step $j = 1, \ldots, T$ each value $z_1, \ldots, z_N$ is computed as a function of the internal states of $\mathcal{O}(1)$ processors at the previous step. Remind that some of cells can be corrupted by random faults. Faults at different cells are independent and each one occurs with probability ϵ. We say that a random perturbation is ϵ_0-*normal* if for every block $\hat{b}_i$ at each stage of computation, there are at most $\epsilon_0 \cdot 2^k$ faults.

Lemma 4. *For all $\epsilon_0 \in (\epsilon, 1/8)$ and large enough constant C (which defines the length of the blocks b_i) a perturbation is ϵ_0-normal with probability greater than $(1 - \epsilon_{res})$.*

Proof of the Lemma. For each block $\hat{b}_i$ at each step the average number of faults is equal to $\epsilon 2^k$. From the Chernoff bound it follows that for some $c > 0$

$$\text{Prob[number of faults} > \epsilon_0 2^k] < e^{-c(\epsilon-\epsilon_0)^2 2^k}.$$

Sum up this probability for all $i = 1, \ldots, r$ and all steps $j = 1, \ldots, T$. The sum is less than ϵ_{res} if the constant C is large enough.   □

---

[1] Remind that we assume also that every time when a processor of S (and S' as well) changes its state, its new state is a boolean function of internal states of at most *two* processors from the previous step.

Let us fix some $\epsilon_0 \in (\epsilon, 1/8)$; in the sequel we construct $\hat{S}$ that always returns a correct result for an ϵ_0-normal perturbation. From Lemma 4 it follows that such a circuit returns a correct result with probability at least $(1 - \epsilon_{res})$.

Fix some $\delta_0 > 0$ such that $8\delta_0 + \epsilon_0 < 1/8$. Let Mix be a linear $(k, \tau, \delta_0/2, \delta_0)$-mixer. As we showed in Lemma 3, such a mixer can be found in time poly(n, t).

Now we are ready to define the computation rules for $\hat{S}$. In the sequel we prove by induction the following property: if at step j each $\hat{b}_i^{(j)}$ differs from the corresponding $Had(b_i^{(j)})$ in at most $1/8$ of bits, and the perturbation is ϵ_0-normal, then each $\hat{b}_i^{(j+1)}$ also differs from the corresponding $Had(b_i^{(j+1)})$ in at most $1/8$ of bits.

We define the computation step in two stages. First, we define $\hat{b}_i'$, $i = 1, \ldots, r$; each bit of $\hat{b}_i'$ is a function of $\mathcal{O}(1)$ bits from $\hat{b}_i^{(j)}$. At the second stage we define $\hat{b}_i''$, $i = 1, \ldots, r$, each bit of a $\hat{b}_i''$ is a function of $\mathcal{O}(1)$ bits from $(\hat{b}_1', \ldots, \hat{b}_r')$. Then we set $\hat{b}_i^{(j+1)} = \hat{b}_i''$, $i = 1, \ldots, r$. We stress that this separation into two stages is used only to make the explanation more clear and pictorial. In the circuit we do not need an intermediate stage: the states of the processors at the $(j + 1)$-th step $\hat{b}_i^{(j+1)}$ are directly defined as functions of $\hat{b}_i^{(j)}$.

The First Stage of the Construction. Fix i and j, and let $\hat{b}_i^{(j)} = (u_1, \ldots, u_{2^k})$. We identify an integer $q \in \{1, \ldots, 2^k\}$ and its k-digit binary representation (with leading zeros). Thus, we can use q as the first argument of the mixer Mix. Further, for an integer $q \leq 2^k$ and $\eta \in \{0, 1\}^\tau$ we can consider $u_{Mix(q,\eta) \oplus q}$, where $Mix(q, \eta) \oplus q$ is the bitwise sum of two k-bit words: $Mix(q, \eta)$ and the binary representation of q.

Assume that $\hat{b}_i^{(j)}$ differs from $E(b_i^{(j)})$ in at most $2^k/8$ bits. For every q the bit u_q' is computed as

$$u_q' = \underset{\eta \in \{0,1\}^\tau}{\text{majority}}\{(u_{Mix(q,\eta) \oplus q} - u_{Mix(q,\eta)})\}$$

Let us bound the number of bits in $\hat{b}' = (u_1', \ldots, u_{2^k}')$ that differ from the q-th bits of $Had(b_i^{(j)})$. There may be two reasons why a u_q' differs from the corresponding bit of $Had(b_i^{(j)})$:

1. in the sequence $u_{Mix(q,0) \oplus q}, \ldots, u_{Mix(q,\tau) \oplus q}$ at least 25% of values differ from the corresponding values of $Had(b_i^{(j)})$;
2. in the sequence $u_{Mix(q,0)}, \ldots, u_{Mix(q,\tau)}$ at least 25% of values differ from the corresponding values of $Had(b_i^{(j)})$;

From Claim 1 it follows that there are at most $2\delta_0 2^k$ positions q where at least one of these two conditions hold. Thus, we proved the following bound:

Claim 3. *There are at most $2\delta_0 2^k$ positions $q = 1, \ldots, 2^k$, where u_q' differs from the corresponding bit of $Had(b_i^{(j)})$.*

The Second Stage of the Construction. Remind that we assume that at each step of the computation in the original circuit S' exactly one bit of every block is updated. Let on the j-th step in a block $\hat{b}_{i_0}$ the internal state of a processor c_0 was updated: $x_{c_0}^{(j+1)} = F_j(x_{c_1}^{(j)}, x_{c_2}^{(j)})$, where F_j is some boolean function. Let the bits x_{c_1} and x_{c_2}

be from the blocks $\hat{b}_{i_1}$ and $\hat{b}_{i_2}$ respectively. The difference between $Had(b_{i_0}^{(j)})$ and $Had(b_{i_0}^{(j+1)})$ is a function of $x_{c_0}, x_{c_1}, x_{c_2}$. As the Hadamard code is linear, if

$$\xi = F_j(x_{c_1}^{(j)}, x_{c_2}^{(j)}) - x_{c_0}^{(j)},$$

then the difference

$$Had(b_{i_0}^{(j)}) - Had(b_{i_0}^{(j+1)})$$

is equal to $Had(0, \ldots, 0, \xi, 0, \ldots, 0)$.

Now we define u_q''. For $\hat{b}_{i_0}''$ make the computations as follows. Let $\hat{b}_{j_0}' = (u_1', \ldots, u_{2^k}')$, $\hat{b}_{j_1}' = (v_1', \ldots, v_{2^k}')$ and $\hat{b}_{j_2}' = (w_1', \ldots, w_{2^k}')$. For each $q = 1, \ldots, 2^k$ we estimate the value x_{c_0} as

$$\tilde{x}_{c_0} = \underset{\eta \in \{0,1\}^\tau}{\text{majority}}\{(u_{Mix(q,\eta) \oplus c_0}' - u_{Mix(q,\eta)}')\},$$

and the values x_{c_1} and x_{c_2} as

$$\tilde{x}_{c_1} = \underset{\eta \in \{0,1\}^\tau}{\text{majority}}\{(v_{Mix(q,\eta) \oplus c_1}' - v_{Mix(q,\eta)}')\},$$
$$\tilde{x}_{c_2} = \underset{\eta \in \{0,1\}^\tau}{\text{majority}}\{(w_{Mix(q,\eta) \oplus c_2}' - w_{Mix(q,\eta)}')\},$$

respectively. Set $\tilde{\xi} = F_j(\tilde{x}_{c_1}, \tilde{x}_{c_2}) - \tilde{x}_{c_0}$ and add the q-th digit of the vector $Had(0, \ldots, 0, \tilde{\xi}, 0, \ldots, 0)$ to the bit u_q' (as usual, all operations are in the field $\mathbb{Z}/2\mathbb{Z}$). Note that $\tilde{x}_{c_0}$ is estimated correctly unless at least 25% of the values $v_{Mix(q,0)+c_0}, \ldots, v_{Mix(q,\tau)+c_0}$ are corrupted or at least 25% of the values $v_{Mix(q,0)}, \ldots, v_{Mix(q,\tau)}$ are corrupted. From Claim 1, there are at most $2\delta_0 2^k$ positions q where one of these two conditions holds. The same is true for $\tilde{x}_{c_1}$ and $\tilde{x}_{c_2}$.

Remark 2. Here we have used the fact that the Hadamard code is locally decodable: the value x_{c_0} can be calculated from two digits of the codeword ($u_{Mix(q,\eta) \oplus c_0}'$ and $u_{Mix(q,\eta)}'$), if only these digits are not corrupted.

Let us bound the number of positions q where u_q'' differs from the q-th bit of $Had(b_1^{(j+1)})$. All but $2\delta_0 2^k$ positions u_q' are equal to the corresponding bits of the (2^k)-bit string $Had(b_1^{(j)})$. All three values x_{c_0}, x_{c_1}, and x_{c_2} are estimated correctly for all but $6\delta_0 2^k$ positions. Further, at most $\epsilon_0 2^k$ bits of one block can be corrupted due to random faults. In total, the fraction of position where u_q'' is not equal to the corresponding bit of $Had(b_0^{(i+1)})$ is at most $8\delta_0 + \epsilon_0 < 1/8$. □

We proved that any computation can be made reliable so that the result is correct with some fixed probability $(1 - \epsilon_{res})$. We might want to get a circuit which fails with exponentially small probability. To decrease the probability of a failure, we can increase the size k of blocks b_i used in the proof of Theorem 1. If we let $k = C' \log \log(tn)$ for some $C' > 1$, our construction results in a circuit which fails with probability $e^{-\Omega(\log^{C'}(tn))}$; the blow up of the memory in this circuit is $\mathcal{O}(2^k/k) = \mathcal{O}(\log^{\mathcal{O}(1)}(tn))$. To implement this construction, we need a linear mixer with parameters $(\log(\log^{\mathcal{O}(1)}(tn)), \tau, \alpha, \beta)$.

Such a mixer can be constructed in time $\text{poly}(t, n)$; really, it is enough to get the tensor product of $\mathcal{O}(1)$ linear mixers with parameters $(\frac{1}{2} \log \log(nt), \mathcal{O}(1), \mathcal{O}(1), \mathcal{O}(1))$.

To get a circuit that fails with probability $e^{-\Omega(n)}$, we need a linear $(\log(tn), \tau, \alpha, \beta)$-mixer. Such a mixer exists, though we have no *effective* algorithm to construct it. But if we omit the condition that $\hat{S}$ can be received from S effectively then we can apply the same arguments as in Theorem 1 and get the following result.

Theorem 2. *For every circuits S of depth t with n processors there exists a circuit $\hat{S}$ of depth $\text{poly}(t, n)$ with $\text{poly}(t, n)$ processors that computes the same function for the encoded input so that if every processor at each step is corrupted with a small enough probability ϵ then the result is correct with probability at least $(1 - e^{-\Omega(n)})$.*

In contrast to Theorem 1, our Theorem 2 is not a really new result. The same statement can be proved using technique from [5]. Actually, a stronger statement was proved by D. Spielman in [10]: he showed how to construct a circuit that fails with probability $e^{-\Omega(n)}$ and has only poly-logarithmic excess of memory, while our method requires polynomial blow-up of the circuit. We presented here Theorem 2 to compare our method with previous works and to show the limits of our technique.

4 Computations Resistant to an Adversary

Now we consider the model where at each step an adversary chooses arbitrarily the fraction ϵ of all processors and corrupts them. We show that any computation circuit can be made resistant to such an adversary with an exponential blow up of the memory.

Theorem 3. *For a small enough $\epsilon > 0$, for every circuit S of depth t with n processors there exists a circuit $\hat{S}$ of depth $\mathcal{O}(tn)$ with $N = 2^{\mathcal{O}(n)}$ processors such that $\hat{S}$ computes correctly the same function (for the encoded input) if at each step the fraction at most ϵ of all processors are corrupted.*

Encoding of the input and decoding of the output depend only on n, not on a particular function.

Proof. First of all, we convert the given circuit S into another circuit S' (which computes the same function) such that the following conditions hold:

- at each step of the computation only one processor of S' can update its internal state (the other keep the state of the previous step),
- when the internal state of a processor is updated, its new state is calculated as a boolean function of internal states of *one or two* processors at the previous step.

The price for this transformation is an n time slow down and a constant blow up of the memory.

Thus, we have a circuit S', which has $\mathcal{O}(n)$ processors and runs in time $T = \mathcal{O}(tn)$. Further we construct a fault-tolerant (in the adversary model) circuit $\hat{S}$ that simulates S' in real time. To simplify the notations, we shall assume that S' has *exactly* n processors (the constant blow up of the memory is not essential for our arguments in the sequel).

Denote by $y^{(j)} = (y_1^{(j)}, \ldots, y_n^{(j)})$ the state of the memory of S' at the j-th step of computation, $j = 0, \ldots, T$. Define an encoding function

$$E \; : \; \{0,1\}^n \rightarrow \{0,1\}^{2^{2n}}$$

as follows:

$$E(x) = Had(x), \ldots, Had(x),$$

i.e., E is just the Hadamard code repeated 2^n times. The decoding function $D \; : \; \{0,1\}^{2^{2n}} \rightarrow \{0,1\}^n$ is defined as follows: to get $D(x)$, we split x into 2^n blocks of length 2^{2n}; decode each block using the Hadamard decoding; then for each position $i = 1, \ldots, n$ take the majority of the i-th bits in all 2^n results.

We shall define the computation process so that at each step j the memory of $\hat{S}$ contains a value $z^{(j)} = (z_1^{(j)}, \ldots, z_N^{(j)})$ which differers from $E(y^{(j)})$ in at most δN positions for $\delta = (1/2)^{14}$. Note that this condition implies $D(z^{(t)}) = y^{(t)}$.

Let us split the processors $(z_1^{(j)}, \ldots, z_N^{(j)})$ into 2^n blocks of size 2^n and denote these blocks $b_1^{(j)}, \ldots, b_{2^n}^{(j)}$.

We shall employ an $(n, \tau, \delta_0/2, \delta_0)$-mixer Mix, where $\delta_0 = (\delta - \epsilon)/7$. Such a mixer exists for large enough $\tau = \tau(\delta_0)$. We don't need it to be a *linear* mixer, so we can employ an expander with small enough second eigenvalue. There are known constructions of *effective* expanders with required parameters, e.g. [12].

We describe the transformation from $z^{(j)}$ to $z^{(j+1)}$ in four stages.

Stage 1. Fix $i \in \{1, \ldots, 2^n\}$. For each $\eta \in \{0,1\}^\tau$ take the block $b_{Mix(i,\eta)}^{(j)} = (u_1(\eta), \ldots, u_{2^n}(\eta))$ and compute the vector

$$w_i(\eta) = (u_{i \oplus 1}(\eta) - u_1(\eta), \ldots, u_{i \oplus 2^n}(\eta) - u_{2^n}(\eta))$$

(here for $i, s \in \{1, \ldots, 2^n\}$ we denote by $i \oplus s$ the bitwise sum of n-bits binary representations of i and s with leading zeros). For each position $r = 1, \ldots, 2^n$ get the majority of the r-th bits in all $w_i(\eta)$, $\eta \in \{0,1\}^\tau$. Denote by $c_i^{(j)}$ the obtained result (which is a vector in $\{0,1\}^{2^n}$).

Let us call a block $b_i^{(j)}$ *harmed* if it differs from $Had(y^{(j)})$ in more than $\sqrt{\delta}2^n$ positions. We assumed that $z^{(j)}$ differs from $E(y^{(j)})$ in at most δN position. Hence, there are at most $\sqrt{\delta}2^n$ harmed locks $b_i^{(j)}$.

From Claim 2 it follows that for the fraction at least $(1 - \delta_0)$ of all indexes i there are at least $(1 - 1/64) \cdot 2^\tau$ non-harmed blocks $b_{Mix(i,\eta)}^{(j)}$ ($\eta \in \{0,1\}^\tau$). Note that if for some i at least the fraction $(1 - 1/64)$ of blocks $b_{Mix(i,\eta)}^{(j)}$ are not harmed, then all but $4 \cdot (1/64 + \sqrt{\delta}) \cdot 2^n < 2^n/8$ bits in the resulted block $c_i^{(j)}$ are equal to $y_i^{(j)}$.

Stage 2. Fix $i \in \{1, \ldots, 2^n\}$. Let the block $c_i^{(j)}$ consists of bits $(v_1, \ldots, v_{2^n})$. For each $r = 1, \ldots, 2^n$ calculate the majority

$$v_r' = \underset{\eta \in \{0,1\}^\tau}{\mathrm{majority}} \{v_{Mix(r,\eta)}\},$$

Set $d_i^{(j)} = (v_1', \ldots, v_{2^n}')$.

From Claim 1 it follows that if a block $c_i^{(j)}$ contains at least $7/8 \cdot 2^n$ bits equal to $y_i^{(j)}$, then in the corresponding block $d_i^{(j)}$ at least $(1 - \delta_0)2^n$ bits are equal to $y_i^{(j)}$. Of course, we guarantee nothing for a block $d_i^{(j)}$ if in the corresponding $c_i^{(j)}$ more than $1/8$ of all bits differ from $y_i^{(j)}$.

Stage 3. This stage is trivial: we just make a permutation of bits in $(d_1^{(j)}, \ldots, d_{2^n}^{(j)})$. For each l, m we get the l-th bit from $c_m^{(j)}$ and put it to the m-th position in the l-th block. Denote the result $(f_1^{(j)}, \ldots, f_{2^n}^{(j)})$.

Stage 4. Assume that at the j-th step of the computation in the original circuit S' the bit y_{i_0} is modified: $y_{i_0}^{(j+1)} = F(y_{i_1}^{(j+1)}, y_{i_2}^{(j+1)})$, where F is some boolean function (we assumed that that the boolean function F has at most two arguments). Fix $i \in \{1, \ldots, 2^n\}$ and denote $f_i^{(j)} = (u_1, \ldots, u_{2^n})$. For each $q = 1, \ldots, 2^n$ calculate

$$\tilde{y}_{i_0} = u_{q \oplus i_0} - u_q$$
$$\tilde{y}_{i_1} = u_{q \oplus i_1} - u_q$$
$$\tilde{y}_{i_2} = u_{q \oplus i_2} - u_q$$

Then set $\xi_q = F(\tilde{y}_{i_1}, \tilde{y}_{i_2}) - \tilde{y}_{i_0}$, and calculate $\triangle_q = Had(0, \ldots, 0, \xi_q, 0, \ldots, 0)$ (the value ξ_q is placed at the i_0-th position). Further, get the q-th position of $\triangle_q$ and add it to the value u_q. Denote the resulted block $b_i^{(j+1)}$, and set $z^{(j+1)} = (b_1^{(j+1)}, \ldots, b_{2^n}^{(j+1)})$.

If $f_i^{(j)}$ differs from $Had(y_i^{(j)})$ in $\gamma 2^n$ positions (for some fraction $\gamma \in (0, 1)$) then $b_i^{(j+1)}$ differs from $Had(y^{(j+1)})$ in at most $6\gamma 2^n$ positions. Hence, the whole vector $z^{(j+1)}$ differs from $E(y^{(j+1)})$ in at most $(\delta_0 + 6\delta_0 + \epsilon)2^n$ positions. Note that $7\delta_0 + \epsilon = \delta$, and we are done.

In our construction each bit $z^{(j+1)}$ depends on $\mathcal{O}(1)$ bit from $z^{(j)}$. Thus, we have well defined the transition rule $z^{(j)} \mapsto z^{(j+1)}$. $\qquad \square$

5 Conclusion

We proved that any parallel computation fulfilled on memory n in time t can be simulated by a reliable circuit with memory $\mathcal{O}(n \log(nt)/ \log \log(nt))$ in time $\mathcal{O}(t \log \log(tn))$. Such a reliable circuit returns a correct result with high probability, even if the elementary processors are faulty elements (i.e., each logical gate at each step faults with some small probability ϵ, and faults of different elements are independent). Our construction employs encoding based on the Hadamard code. Actually similar arguments can be applied for a code base on any other *linear locally decodable code*. For example, we can use the Reed-Muller code instead of the Hadamard code; then essentially the same construction provides a bit stronger bound: any computation which was done on a non-faulty circuit with memory n in time t, can be simulated on faulty elements with memory $\mathcal{O}(n \log(nt)/ \log^C \log(nt))$, where the constant C can be made arbitrarily large. By this method, we cannot obtain much better bounds, because for any linear locally decodable error correcting code the codeword length must be exponential in the block length [13]. Thus, the main question, which remains open, is if reliable polynomial computations can be fulfilled on memory $\mathcal{O}(n)$.

Our second result, which concerns computations resistant to an adversary who can corrupt at each step some fraction of memory cells, seems quite weak. We presented a construction with exponential blow-up of the memory. Again, the proved bound cannot be essentially improved with our method, because it is based on a linear locally decodable error correcting code. Remind that if we want just to *store* some information (without computations), this can be done with a constant blow-up of the memory, even if at each step an adversary corrupts some fraction of memory cells [2]. To our knowledge, there are no results achieving a polynomial blow up of the memory for circuits computing an arbitrary function and tolerating a constant fraction of processors being corrupted at every step. In [9] this problem was solved only for a special class of boolean functions. Thus, the second important open question is if *any computation* can be made resistant to an adversary, with linear or at least polynomial increasing of the memory. Another interesting question is if a linear (m, τ, α, β)-mixer can be effectively constructed in time $\text{poly}(2^m)$. If such a construction exists, the proof of Theorem 2 can be made effective.

References

1. von Neuman, J.: Probabilistic logics and the synthesis of reliable organisms from unreliable components. In C. Shannon and J. McCarthy, editors, Automata Studies. Princeton University Press, 1956.
2. Kuznetsov, A.V.: Information storage in a memory assembled from unreliable components. Problems of Information Transmission. **9**:3 (1973) 254–264
3. Dobrushin, R.L. and Ortyukov S.L.: Upper bound for the redundancy of self-correcting arrangement of unreliable functional elements. Problems for Information Transmission, **13**:1 (1977) 203–218
4. Dobrushin, R.L. and Ortyukov S.L.: Lower bound on the redundancy of self-correcting arrangement of unreliable functional elements. Problems for Information Transmission, **13**:1 (1977) 201–208
5. Pippenger, N.: On Networks of Noisy gates. In Proc. of the 26-th IEEE FOCS Symposium (1985) 30–38
6. Pippenger, N., Stamoulis, G.D., and Tsitsikilis J.N.: On a lower bound on for the redundancy of reliable networks with noisy gates. IEEE Trans. Inform. Theory, **37**:3 (1991) 639–643
7. Reischuk, R., and Schmeltz, B.: Reliable computation with noisy circuits and decision trees – a general $n \log n$ lower bound. In Proc. of the 32-th IEEE FOCS Symposium (1991) 602–611
8. Gács, P. and Gál, A.: Lower Bounds for the Complexity of Reliable Boolean Circuits with Noisy Gates. IEEE Transactions Information Theory. **40** (1994) 579–583
9. Gál, A., and Szegedy, M.: Fault Tolerant Circuits and Probabilistically Checkable proofs. In Proc. of 10th Annual Structure in Complexity Theory Conference (1995) 65–73
10. Spielman, D.A.: Highly fault-Tolerant parallel Computation. Proc. of the 37-th IEEE FOCS Symposium (1996) 154–163
11. Goldreich, A. and Wigderson, A.: Tiny Families of Functions with Random Properties: a Quality-Size Trade-off for Hashing. Random Struct. Algorithms **11**:4 (1997) 315–343
12. Reingold, O., Vadhan, S., and Wigderson, A.: Entropy waves, the zig-zag product, and new constant degree expanders. Annals of Mathematics, **155**:1 (2002) 157–187
13. Goldreich, O., Karloff, H.J., Schulman, L.J., and Trevisan, L.: Lower Bounds for Linear Locally Decodable Codes and Private Information Retrieval. IEEE Conference on Computational Complexity (2002) 175–183

Convergence of Autonomous Mobile Robots with Inaccurate Sensors and Movements

(Extended Abstract)

Reuven Cohen* and David Peleg**

Department of Computer Science and Applied Mathematics,
The Weizmann Institute of Science, Rehovot 76100, Israel
{r.cohen, david.peleg}@weizmann.ac.il

Abstract. The common theoretical model adopted in recent studies on algorithms for systems of autonomous mobile robots assumes that the positional input of the robots is obtained by perfectly accurate visual sensors, that robot movements are accurate, and that internal calculations performed by the robots on (real) coordinates are perfectly accurate as well. The current paper concentrates on the effect of weakening this rather strong set of assumptions, and replacing it with the more realistic assumption that the robot sensors, movement and internal calculations may have slight inaccuracies. Specifically, the paper concentrates on the ability of robot systems with inaccurate sensors, movements and calculations to carry out the task of convergence. The paper presents several impossibility results, limiting the inaccuracy allowing convergence. The main positive result is an algorithm for convergence under bounded measurement, movement and calculation errors.

1 Introduction

Background. Distributed systems consisting of autonomous mobile robots (a.k.a. *robot swarms*) are motivated by the idea that instead of using a single, highly sophisticated and expensive robot, it may be advantageous in certain situations to employ a group of small, simple and relatively cheap robots. This approach is of interest for a number of reasons. Multiple robot systems may be used to accomplish tasks that *cannot* be achieved by a single robot. Such systems usually have decreased cost due to the simpler individual robot structure. These systems can be used in a variety of environments where the acting (human or artificial) agents may be at risk, such as military operations, exploratory space missions, cleanups of toxic spills, fire fighting, search and rescue missions, and other hazardous tasks. In such situations, a multiple robot system has a better chance of successfully carrying out its mission (while possibly accepting the loss or destruction of some of its robots) than a single irreplaceable robot. Such systems may also be useful for carrying out simple repetitive tasks that humans may find extremely boring or tiresome.

---

* Supported by the Pacific Theatres Foundation.
** Supported in part by a grant from the Israel Science Foundation.

B. Durand and W. Thomas (Eds.): STACS 2006, LNCS 3884, pp. 549–560, 2006.

Subsequently, studies of autonomous mobile robot systems can be found in different disciplines, from engineering to artificial intelligence. (A survey on the area is presented in [4].)

A number of recent studies on autonomous mobile robot systems focus on algorithms for distributed control and coordination from a distributed computing point of view (cf. [10, 13, 12, 2]). The approach is to propose suitable computational models and analyze the minimal capabilities the robots must possess in order to achieve their common goals. The basic model studied in the these papers can be summarized as follows. The robots execute a given algorithm in order to achieve a prespecified task. Each robot in the system is assumed to operate individually in simple cycles consisting of three steps:

(1) "Look": determine the current configuration by identifying the locations of all visible robots and marking them on your private coordinate system,

(2) "Compute": execute the given algorithm, resulting in a goal point p_G, and

(3) "Move": travel towards the point p_G. The robot might stop before reaching its goal point p_G, but is guaranteed to traverse at least some minimal distance unit (unless reaching the goal first).

Weak and Strong Model Assumptions. Due to the focus on cheap robot design and the minimal capabilities allowing the robots to perform some tasks, most papers in this area (cf. [10, 13, 9, 5]) assume the robots to be rather limited. Specifically, the robots are assumed to be indistinguishable, so when looking at the current configuration, a robot cannot tell the identity of the robots at each of the points (apart from itself). Furthermore, the robots are assumed to have no means of direct communication. This gives rise to challenging "distributed coordination" problems since the only permissible communication is based on "positional" or "geometric" information exchange, yielding an interesting variant of the classical (direct-communication based) distributed model.

Moreover, the robots are also assumed to be *oblivious* (or memoryless), namely, they cannot remember their previous states, their previous actions or the previous positions of the other robots. Hence the algorithm employed by the robots for the "compute" step cannot rely on information from previous cycles, and its only input is the current configuration. While this is admittedly an over-restrictive and unrealistic assumption, developing algorithms for the oblivious model still makes sense in various settings, for two reasons. First, solutions that rely on non-obliviousness do not necessarily work in a dynamic environment where the robots are activated in different cycles, or robots might be added to or removed from the system dynamically. Secondly, any algorithm that works correctly for oblivious robots is inherently self-stabilizing, i.e., it withstands transient errors that alter the robots' local states.

On the other hand, the robot model studied in the literature includes the following *overly strong* assumptions:

- when a robot observes its surroundings, it obtains a perfect map of the locations of the other robots relative to itself,
- when a robot performs internal calculations on (real) coordinates, the outcome is exact (infinite precision) and suffers no numerical errors, and

– when a robot decides to move to a point p, it progresses on the straight line connecting its current location to p, stopping either precisely at p or at some earlier point on the straight line segment leading to it.

All of these assumptions are unrealistic. In practice, the robot measurements suffer from nonnegligible inaccuracies in both distance and angle estimations. (The most common range sensors in mobile robots are sonar sensors. The accuracy in range estimation of the common models is about $\pm 1\%$ and the angular separation is about $3°$; see, e.g., [11]. Other possible range detectors are based on laser range detection, which is usually more accurate than the sonar, and on stereoscopic vision, which is usually less accurate.) The same applies to the precision of robot movements. Due to various mechanical factors such as unstable power supply, friction and force control, the exact distance a robot traverses in a single cycle is hard to control, or even predict with high accuracy. This makes most previous algorithms proposed in the literature inapplicable in most practical settings. Finally, the robots' internal calculations cannot be assumed precise, for a variety of well-understood reasons such as convergence rates of numerical procedures, truncated numeric representations, rounding errors and more.

In this paper we address the issue of imperfections in robot measurements, calculations and movements. Specifically, we replace the unrealistic assumptions described above with more appropriate ones, allowing for measurement, calculation and movement inaccuracies, and show that efficient algorithmic solutions can still be obtained in the resulting model.

We focus on the *gathering* and *convergence* problems, which have been extensively studied in the common (fully accurate) model (cf. [13, 10, 5]). The *gathering* problem is defined as follows. Starting from any initial configuration, the robots should occupy a single point within a finite number of steps. The closely related *convergence* problem requires the robots to converge to a single point, rather than reach it (namely, for every $\epsilon > 0$ there must be a time t_ϵ by which all robots are within distance of at most ϵ of each other).

It is important to note that analyzing the effect of errors is not merely of theoretical value. In Section 3 we show that gathering cannot be guaranteed in environments with errors, and illustrate how certain existing geometric algorithms, including ones designed for fault tolerance, fail to guarantee even convergence in the presence of small errors. We also show (in Theorem 9) that the standard center of gravity algorithm may also fail to converge when errors occur.

Related Work. A number of problems concerning coordination in autonomous mobile robot systems have been considered so far in the literature. The gathering problem was first discussed in [13] in the semi-synchronous model. It was proven that it is impossible to gather *two* oblivious autonomous mobile robots that have no common sense of orientation under the semi-synchronous model. Also, an algorithm was presented in [13] for gathering $N \geq 3$ robots in the semi-synchronous model. In the asynchronous model, a gathering algorithm has recently been described in [5]. Fault tolerant gathering algorithms (in the crash and Byzantine fault models) were studied in [1]. The gathering problem was also studied in a system where the robots have limited visibility. The visibility conditions are modeled by means of a *visibility graph*, representing the (symmetric) visibility relation of the robots with respect to one another, i.e., an edge exists between robots i and j if and only if i and j are visible to each other. (Note that in this model visibility is a boolean predicate and does not involve imprecisions, namely, if robot j is visible to robot i then its precise coordinates are measured accurately.) It was shown that the problem is unsolvable in case the visibility graph is not connected [9]. In [2] a convergence algorithm was provided for any N, in limited visibility systems. The natural gravitational algorithm based on going to the center of gravity, and its convergence properties, were studied in [6].

Other problems studied, e.g., in [12, 13, 7, 8, 10, 3], concern formation of various geometric patterns, flocking (or "following the leader"), distributed search after (static or moving) targets, achieving even distribution, partitioning and wake-up via the freeze-tag paradigm.

Our Results. In this paper we study the convergence problem in the common semi-synchronous model where the robots' only inputs are obtained by inaccurate visual sensors, and their movements and internal calculations may be inaccurate as well. In Section 3 we present several impossibility theorems, limiting the inaccuracy allowing convergence, and prohibiting a general algorithm for gathering in a finite number of steps. In Section 4 we present an algorithm for convergence under bounded error, and prove its correctness, first in the fully synchronous model, and then in the semi-synchronous model. Finally, we compare the proposed algorithm with the ordinary center of gravity algorithm.

2   The Model

Each of the N robots i in the system is assumed to operate individually in simple cycles. Every cycle consists of three steps, "look", "compute" and "move". The result of the "look" step taken by i is a multiset of points $P = \{p_1, \ldots, p_N\}$ (with $p_i = 0$ in i's local coordinate system) defining the current *configuration* and used by the robot in calculating its next goal point p_G. Note that the "look" and "move" steps are carried out identically in every cycle, independently of the algorithm used. The differences between different algorithms occur in the "compute" step. Moreover, the procedure carried out in the "compute" step is identical for all robots. If the robots are oblivious, then the algorithm cannot rely on information from previous cycles, thus the procedure can be fully specified by describing a single "compute" step, and its only input is the current configuration P, giving the locations of the robots.

As mentioned earlier, our computational model for studying and analyzing problems of coordinating and controlling a set of autonomous mobile robots follows the well studied *semi-synchronous* ($\mathcal{SSYNC}$) model. This model is partially synchronous, in the sense that all robots operate according to the same clock cycles, but not all robots are necessarily active in all cycles. Those robots which are awake at a given cycle make take a measurement of the positions of all other robots. Then they may make a computation and move instantaneously accordingly. The activation of the different robots can be thought of as managed by a hypothetical scheduler, whose only fairness obligation is that each robot must be activated and given a chance to operate infinitely often in any infinite execution. On the way to establishing the result on the $\mathcal{SSYNC}$ model, we prove it first in the fully synchronous ($\mathcal{FSYNC}$) model. Finally, we also discuss its performance in the fully asynchronous ($\mathcal{ASYNC}$) model.

Our model assumes that the robot's location estimation is imprecise, with imprecision bounded by some accuracy parameter ϵ known at the robot's design. In general, this imprecision can affect both distance and angle estimations. In particular, distance imprecision means that if the true location of an observed point in i's coordinate system is $\overline{V}$ and the measurement taken by i is $\bar{v}$, then this measurement will satisfy $(1-\epsilon)V < v < (1+\epsilon)V$. (Throughout, for a vector $\bar{v}$, we denote by v its scalar length, $v = |\bar{v}|$. Also, capital letters are used for exact quantities, whereas lowercase ones denote the robots' views).

The accuracy in angle measurements is θ_0 (where it can always be assumed that $\theta_0 \leq \pi$). I.e., the angle θ between the actual distance vector $\overline{V}$ and the measured distance vector $\bar{v}$ satisfies $\theta \leq \theta_0$, or alternatively, $\cos\theta = \frac{\overline{V}\bar{v}}{Vv} \geq \cos\theta_0$. In what follows, we consider the model $\mathcal{ERR}$ in which both types of imprecision are possible, and the model $\mathcal{ERR}^-$ where only distance estimates are inaccurate. This gives rise to six composite timing/error models, denoted $\langle \mathcal{T}, \mathcal{E} \rangle$, where $\mathcal{T}$ is the timing model under consideration ($\mathcal{FSYNC}$, $\mathcal{SSYNC}$ or $\mathcal{ASYNC}$) and $\mathcal{E}$ is the error model ($\mathcal{ERR}$ or $\mathcal{ERR}^-$).

While in reality each robot uses its own private coordinate system, for simplicity of presentation it is convenient to assume the existence of a global coordinate system (which is unknown to the robots) and use it for our notation. Throughout, we denote by $\bar{R}_j$ the location of robot j in the global coordinate system. In addition, for every two robots i and j, denote by $\overline{V}_j^i = \bar{R}_j - \bar{R}_i$ the true location of robot j from the position of robot i (i.e., the true vector from i to j), and by $\bar{v}_j^i$ the location of robot j as measured by i, *translated* to the global coordinate system. Likewise, our algorithm and its analysis will be described in the global coordinate system, although each of the robots will apply it in its own local coordinate system. As the functions computed by the algorithm are all invariant under translations, this representation does not violate the correctness of our analysis.

If the robots may have inaccuracies in distance estimation but not in directions, then i will measure $\overline{V}_j^i$ as $\bar{v}_j^i = (1 + \epsilon_j^i)\overline{V}_j^i$, where $-\epsilon < \epsilon_j^i < \epsilon$ is the local error factor in distance estimation at robot i. For robots with inaccuracy in angle measurement as well, if the true distance is V_j^i, then i will measure it as $v_j^i = (1 + \epsilon_j^i)V_j^i$, where $-\epsilon < \epsilon_j^i < \epsilon$ and the angle θ between $\overline{V}_j^i$ and $\bar{v}_j^i$ will satisfy $|\theta| \leq \theta_0$. Values computed at time-slot t are denoted by a parameter $[t]$. Also, the actual error factor is time dependent and its value at time t is denoted by $\epsilon_j^i[t]$. The parameter t is omitted whenever clear from the context.

Inaccuracies in movement and calculations should also be taken into account. For movement, we may assume that if a robot wants to move from its current location $\bar{R}_i$ to some goal point p_G, then it will move on a vector at an angle of at most ϕ_0 from the vector $\overrightarrow{r_i p_G}$ and to any distance d in the interval $d \in [1 - \epsilon, 1 + \epsilon] \cdot |\overrightarrow{r_i p_G}|$. Also, when it calculates a goal point $p_G = (x, y)$, it will have a multiplicative error of up to ϵ. In the center of gravity algorithms presented below, the calculation error is bounded linearly in the calculated terms. Hence it can be seen that relative movement and calculation errors can be replaced with errors in measurement causing the same effect, so these errors can be treated using the same algorithm by recalibrating ϵ. (Note that *absolute* errors in movement or calculation can not be treated, since even when the robots have already almost converged, such errors may cause them to spread again.) Therefore, throughout most of the ensuing technical development, we will assume only measurement inaccuracies.

We use the following technical lemma. (Most proofs are deferred to full paper.)

Lemma 1. *For two vectors $\bar{a}$ and $\bar{b}$ with $a \leq 1 \leq b$, let $x = |\bar{a} - \bar{b}|$ and $y = |\bar{a} - \bar{b}/b|$. Then (1) $x^2 - y^2 \geq (b-1)^2 + 2(1-a)(b-1) \geq (b-1)^2$, and (2) $y \leq x$.*

3   The Effect of Measurement Errors

To appreciate the importance of error analysis one must realize two facts. First, computers are limited in their computational power, and therefore cannot perform perfect precision calculations. This may seem insignificant, since floating point arithmetic can be made to very high accuracy with modern computers. However, this may prove to be a practical problem. For instance, the point that minimizes the sum of distances to the robots' locations (also known as the Weber point) may be used to achieve gathering. However, this point is not computable, due to its infinite sensitivity to location errors. Second, the correctness of algorithms that use geometric properties of the plane is usually proven using theorems from Euclidean geometry. However, these theorems are, in many cases, inappropriate when measurement or calculation errors occur.

Impossibility Results. We start with some impossibility results. The proofs of these results are based on the ability of the adversary to partition the space of possible initial configurations into countably many regions, each of uncountably many configurations (say, on the basis of the initial distance between the robots), such that within each region, the outcome of the algorithm (i.e., the movement instructions to the robots) is the same. The following theorem holds even in a rather strong setting where the timing model is fully synchronous, and the robots have unlimited memory and are allowed to use randomness.

Theorem 1. *Even in the strong setting outlined above, gathering is impossible (1) for two robots on the line with inexact distance measurements, (2) for any number of robots assuming inaccuracies in both the distance and angle measurements.*

It seems reasonable to conjecture that even convergence is impossible for robots with large measurement errors. The exact limits are not completely clear. The following theorem gives some rather weak limits on the possibility of convergence.

In the theorem we assume that the robot has no sense of direction in a strong way, *i.e.*, at every cycle the adversary can choose each robot's axes independent of previous cycles.

Theorem 2. *For a configuration of $N = 3$ robots having an error parameter $\theta_0 \geq \pi/3$ in angle measurement, there is no deterministic algorithm for convergence even assuming exact distance estimation, fully synchronous model and unlimited memory.*

Problems with Existing Algorithms. To illustrate the second point raised in the beginning of this section, consider the algorithm 3 − Gather presented in [1]. This algorithm achieves gathering of three robots using several simple rules. One of these rules states that if the robots form an obtuse triangle, then they move towards the vertex with the obtuse angle. As shown above, no algorithm can guarantee gathering when measurement errors occur. Furthermore, although this algorithm is designed to robustness and achieves gathering even if one of the robots fails, one can verify that it might fail to achieve even convergence in the presence of angle measurement errors of at least $15°$.

Likewise, for a group of $N > 3$ robots the algorithm N − Gather is presented in [1]. In this algorithm the smallest enclosing circle of the robot group is calculated, and in case there is a single robot inside this circle, it does not move. In the presence of measurement inaccuracies, this rule can potentially cause deadlock, implying that the algorithm might fail to achieve even convergence in the presence of angle and distance measurement errors of $\epsilon > 0$.

4   The Convergence Algorithm

Algorithm Go_to_COG. A natural algorithm for autonomous robot convergence is the gravitational algorithm, where each robot computes the average position (center of gravity) of the group, $\bar{v}^i_{cog} = \frac{1}{N} \sum_j \bar{v}^i_j$, and moves towards it.

The properties of Algorithm Go_to_COG in a model with fully accurate measurements have been studied in [6]. In particular, it is proven that a group of N robots executing Algorithm Go_to_COG will converge in the $\mathcal{ASYNC}$ model with no measurement errors. If measurements are not guaranteed to be accurate, Algorithm Go_to_COG may not guarantee convergence. Nevertheless, convergence *is* guaranteed in the fully synchronous model, i.e., we have the following.

Lemma 2. *In the $\langle \mathcal{FSYNC}, \mathcal{ERR}^- \rangle$ model with $\epsilon < \frac{1}{2}$, a group of N robots performing Algorithm Go_to_COG converges.*

The convergence of Algorithm Go_to_COG in the $\mathcal{SSYNC}$ model is not clear at the moment. However, as shown below, in the $\mathcal{ASYNC}$ model there are scenarios where robots executing Algorithm Go_to_COG fail to converge. This leads us to propose the following slightly more involved algorithm.

Algorithm RCG. Our algorithm, named RCG, is based on calculating the center of gravity (CoG) of the group of robots, and also estimating the maximum possible error in the CoG calculation. The robot makes no movement if it is within the maximum possible error from the CoG. If it is outside the circle of error, it moves towards the CoG, but only up to the bounds of the circle of error. We fix a conservative error estimate parameter, $\epsilon_0 > \epsilon$.

Following is a more detailed explanation of the algorithm. In step 1, the measured center of gravity is estimated using the conducted measurements. In step 2 the distance to the furthest robot is found. Notice that this distance may not be accurate, and that this needs not be even the real furthest robot. The result of step 2 is used in step 3 to give an estimate of the possible error in the CoG calculation. In step 4 the robots decide to hold if it is within the circle of error, or calculates its destination point, which is on the boundary of the error circle centered at the calculated CoG. A formal description of the algorithm is given next. Note that Algorithm Go_to_COG is identical to Algorithm RCG with parameter $\epsilon_0 = 0$.

Code for robot i

1. Estimate the measured center of gravity, $\bar{v}^i_{cog} = \frac{1}{N}\sum_j \bar{v}^i_j$
2. Let $d^i_{max} = \max_j\{v^i_j\}$ /* max distance measured to another robot
3. Let $\rho^i = \dfrac{\epsilon_0}{1-\epsilon_0} \cdot d^i_{max}$ /* estimate for max error in calculated CoG
4. If $v^i_{cog} > \rho^i$ then move to the point $\bar{c}_i = (1 - \rho^i/v^i_{cog}) \cdot \bar{v}^i_{cog}$.
   Otherwise do not move.

Analysis of RCG in the Semi-synchronous Model. We first prove the convergence of Algorithm RCG in the $\langle \mathcal{FSYNC}, \mathcal{ERR}^- \rangle$ model. Denote the true center of gravity of the robots in the global coordinate system by $\bar{R}_{cog} = \frac{1}{N}\sum_j \bar{R}_j$, and the vector from robot i to the center of gravity by $\overline{V}^i_{cog} = \bar{R}_{cog} - \bar{R}_i = \frac{1}{N}\sum_j \overline{V}^i_j$, where $\overline{V}^i_j = \bar{R}_j - \bar{R}_i$. Denote the distance from the true center of gravity of the robots to the robot farthest from it by $D_{cog} = \max_i\{V^i_{cog}\}$. Also, denote the true distance from i to the robot farthest from it by $D^i_{max} = \max_j\{V^i_j\}$. We use the following two properties.

Fact 3. *For every i:* (a) $D^i_{max} \le 2D_{cog}$,
(b) $(1 - \epsilon_0)D^i_{max} < (1 - \epsilon)D^i_{max} \le d^i_{max} \le (1 + \epsilon)D^i_{max} < (1 + \epsilon_0)D^i_{max}$.

For the synchronous model, we define the tth round to begin at time t and end at time $t + 1$. The robots all perform their Look phase simultaneously. The robots' *moment of inertia* at time t is defined as

$$I[t] = \frac{1}{N}\sum_j \left(\overline{V}^j_{cog}[t]\right)^2 = \frac{1}{N}\sum_j \left(\bar{R}_j[t] - \bar{R}_{cog}[t]\right)^2 .$$

Defining $I_{\bar{x}}[t] \equiv \frac{1}{N}\sum_j (\bar{R}_j[t] - \bar{x})^2$, we use the following fact.

Fact 4. $I_{\bar{x}}[t]$ *attains its minimum on* $\bar{x} = \bar{R}_{cog}[t]$.

For ease of presentation, we assume a slightly simpler model where the move step of a robot is ensured to bring it to its goal point p_G. A slightly more involved analysis, deferred to the full paper, applies to the usual setting where it is assumed that the robot might stop before reaching p_G, but is guaranteed to traverse at least some minimal distance unit (unless reaching the goal first).

Our main lemma is the following.

Lemma 3. *For fixed $\epsilon_0 < 0.2$, in the $\langle \mathcal{FSYNC}, \mathcal{ERR}^- \rangle$ model, Algorithm RCG guarantees that at every round t:*

1. *at least one robot can move,*
2. *every robot i decreases its distance from the true center of gravity at time t, i.e., $|\bar{R}_i[t+1] - \bar{R}_{cog}[t]| < |\bar{R}_i[t] - \bar{R}_{cog}[t]|$,*
3. *the robots' moment of inertia decreases, i.e., $I[t+1] < I[t]$.*

Proof. Consider some time t. Denote by $\overline{err}^i = \frac{1}{N} \sum_j \epsilon_j^i \overline{V}_j^i$ the error component in the center of gravity calculation by robot i. Then the center of gravity computed by robot i can be expressed as

$$\bar{v}_{cog}^i \;=\; \frac{1}{N} \sum_j \bar{r}_j^i \;=\; \frac{1}{N} \sum_j (\bar{R}_j + \epsilon_j^i \overline{V}_j^i) \;=\; \overline{V}_{cog}^i + \overline{err}^i \;.$$

By the bounded error assumption and Fact 3(a),

$$err^i \;=\; \frac{1}{N} \sum_j \epsilon_j^i \cdot V_j^i \;\leq\; \epsilon D_{max}^i \;\leq\; 2\epsilon D_{cog} \;<\; 2\epsilon_0 D_{cog}. \tag{1}$$

By the two parts of Fact 3, the calculated value ρ^i is bounded by

$$\rho^i \;\leq\; \frac{\epsilon_0(1+\epsilon_0)}{1-\epsilon_0} \cdot D_{max}^i \;\leq\; \frac{\epsilon_0(1+\epsilon_0)}{1-\epsilon_0} \cdot 2D_{cog}. \tag{2}$$

Combining (1) and (2), we have for each i,

$$err^i + \rho^i \;\leq\; f(\epsilon_0) \cdot D_{cog} \;<\; D_{cog} \;, \tag{3}$$

where $f(\epsilon_0) = 4\epsilon_0/(1 - \epsilon_0)$, and the last inequality follows from the assumption that $\epsilon_0 < 0.2$.

For $k = \arg\max_j\{V_{cog}^j\}$, the robot farthest from the center of gravity, we have $V_{cog}^k = D_{cog}$ and $\bar{v}_{cog}^k = \overline{V}_{cog}^k + \overline{err}^k$, hence by (3) and the triangle inequality, $\rho^k < V_{cog}^k - err^k \leq v_{cog}^k$. This implies that at round t, robot k is allowed to move in Step 4 of the algorithm, proving Part 1 of the Lemma.

To prove Part 2, consider a round t and a robot i which moved in round t. Fix $\bar{x} = \bar{R}_{cog}[t]$ and take $\bar{a} = \overline{err}^i[t]/\rho^i[t]$ and $\bar{b} = \bar{v}_{cog}^i[t]/\rho^i[t]$. Note that by (1) and Fact 3(b), $err^i[t] \leq \frac{\epsilon}{1-\epsilon} \cdot d_{max}^i[t] < \frac{\epsilon_0}{1-\epsilon_0} \cdot d_{max}^i[t] = \rho^i[t]$, hence $a \leq 1$. Also, at round $t+1$, robot i moves if and only if $v_{cog}^i[t] > \rho^i[t]$, hence if i moved then $b = v_{cog}^i[t]/\rho^i[t] > 1$. Hence Lemma 1(1) can be applied. Noting that $\bar{b}/b = (\bar{R}_i[t+1] - \bar{R}_{cog}[t])/\rho^i[t]$, we get $|\bar{R}_i[t+1] - \bar{x}| < |\bar{R}_i[t] - \bar{x}|$, yielding Part 2 of the Lemma. It remains to prove Part 3. Note that for a robot that did not move, $|\bar{R}_i[t+1] - \bar{x}| = |\bar{R}_i[t] - \bar{x}|$. Using this fact that and Part 2, we have that $I_{\bar{x}}[t+1] < I_{\bar{x}}[t] = I[t]$. Finally, by Fact 4, $I[t+1] \leq I_{\bar{x}}[t+1]$, yielding Part 3 of the Lemma. ∎

Theorem 5. *In every execution of Algorithm* RCG *in the* $\langle \mathcal{FSYNC}, \mathcal{ERR}^- \rangle$ *model, the robots converge.*

Proof. By Part 1 of Lemma 3, the robot k most distant from the center of gravity can always move if Algorithm Go_to_COG is applied. By Part 2 of Lemma 3, in round t every robot decreases its distance from the old center of gravity, $\bar{x} = \bar{R}_{cog}[t]$. Therefore, to bound from below the decrease in I, we are only required to examine the behavior of the most distant robot. By Lemma 1(1) with $a = \bar{R}_k[t+1] - \bar{R}_{cog}[t]$ and $b = \bar{R}_k[t] - \bar{R}_{cog}[t]$, for $\epsilon_0 < 0.2$ we have $v_{cog}^k > \rho^k$ and $(\bar{R}_k[t] - \bar{R}_{cog}[t])^2 - (\bar{R}_k[t+1] - \bar{R}_{cog}[t])^2 \geq (v_{cog}^k - \rho^k)^2$. Since $\bar{v}_{cog}^k = \overline{V}_{cog}^k + \overline{err}^k$, and using the triangle inequality,

$$(\bar{R}_k[t] - \bar{R}_{cog}[t])^2 - (\bar{R}_k[t+1] - \bar{R}_{cog}[t])^2 \geq \left(V_{cog}^k - (\rho^k + err^k)\right)^2 . \quad (4)$$

Recall that since k is the most distant robot, $V_{cog}^k = D_{cog}$. Denoting $\gamma = 1 - f(\epsilon_0)$, we have by (3) that

$$V_{cog}^k - (\rho^k + err^k) \geq \gamma \cdot D_{cog} . \quad (5)$$

As mentioned above, if $\epsilon_0 < 0.2$, then $\gamma > 0$. We also use the fact that $I[t] = I_{\bar{x}}[t] \leq D_{cog}^2$. Together with Fact 4 and inequalities (4) and (5), we have that

$$I[t+1] \leq I_{\bar{x}}[t+1] = \frac{1}{N}\left((\bar{R}_k[t+1] - \bar{R}_{cog}[t])^2 + \sum_{j \neq k}(\bar{R}_j[t+1] - \bar{R}_{cog}[t])^2\right)$$

$$\leq \frac{1}{N}(\bar{R}_k[t+1] - \bar{R}_{cog}[t])^2 + \frac{1}{N}\sum_{j \neq k}(\bar{R}_j[t] - \bar{R}_{cog}[t])^2$$

$$\leq \frac{1}{N}(\bar{R}_k[t+1] - \bar{R}_{cog}[t])^2 - \frac{1}{N}(\bar{R}_k[t] - \bar{R}_{cog}[t])^2 + I[t]$$

$$\leq I[t] - \frac{1}{N}\left(V_{cog}^k - (\rho^k + err^k)\right)^2 \leq I[t] - \frac{\gamma^2}{N} \cdot D_{cog}^2 \leq I[t]\left(1 - \frac{\gamma^2}{N}\right)$$

and therefore the system converges, proving the theorem. ∎

We now turn to the $\mathcal{ERR}$ model, allowing also inaccuracies in angle measurements, and observe that Theorem 5 can be extended to hold true in this model as well, with a suitable choice of ϵ_0.

Theorem 6. *Taking* $\epsilon_0 > \sqrt{2(1-\epsilon)(1-\cos\theta_0) + \epsilon^2}$, *in every execution of Algorithm* RCG *in the* $\langle \mathcal{FSYNC}, \mathcal{ERR} \rangle$ *model, the robots converge.*

Turning to the semi-synchronous model, we observe that the results of Theorem 6 hold true also for the $\langle \mathcal{SSYNC}, \mathcal{ERR} \rangle$ model, yielding the following.

Theorem 7. *In every execution of Algorithm* RCG *(with ϵ_0 as in Theorem 6) in the $\langle \mathcal{SSYNC}, \mathcal{ERR} \rangle$ model, the robots converge.*

Finally, let us turn to robots with movement and calculation inaccuracies. In case of robots with inaccurate movements, we assume it is always possible to tune the algorithm such that the distance traveled is always less than or equal to the distance aimed i.e., instead of moving by a vector $\bar{v}$, move by a vector $\nu\bar{v}$. Suppose now that D, the distance traveled by the robot is bounded by $(1 - \alpha)R \leq D \leq (1 + \alpha)R$, where R is the norm of the output of the algorithm, and α is a constant denoting the accuracy of the robot's movement. ν can be chosen such that $(1 + \alpha)\nu \leq 1$. The result can be obtained following the same line of proof. The details are deferred to the full paper. As for calculation inaccuracies, since we assume a multiplicative inaccuracy, and by the linearity of the calculation, it can be treated as a measurement inaccuracy with the proper addition to ϵ.

Analysis of RCG **in the Fully Asynchronous Model.** So far, we have not been able to establish the convergence of Algorithm RCG in the fully asynchronous model. In this section we prove its convergence in the restricted one-dimensional case and with no angle inaccuracies, i.e., in the $\langle \mathcal{ASYNC}, \mathcal{ERR}^- \rangle$ model.

Denote by $\bar{c}_i[t]$ the calculated destination of robot i at time t. If robot i has not gone through a look yet, or has reached its previous destination then, by definition, $\bar{c}_i[t] = \bar{R}_i[t]$. Notice that we set $\bar{c}_i[t]$ to be the destination of the robot's motion after the look phase even if the robot has not yet completed its computation, and is still unaware of this destination.

Theorem 8. *In the $\langle \mathcal{ASYNC}, \mathcal{ERR}^- \rangle$ model, N robots performing Algorithm* RCG *converge on the line.*

Conjecture 1. Algorithm RCG converges in the $\langle \mathcal{ASYNC}, \mathcal{ERR} \rangle$ model for sufficiently small error in the angle and distance measurements.

Separating Go_to_COG **from** RCG **in the** $\mathcal{ASYNC}$ **Model.** This section establishes the advantage of Algorithm RCG over the basic Algorithm Go_to_COG. In the fully synchronous case there is no justification for using the more involved Algorithm RCG, since the simpler Algorithm Go_to_COG also guarantees convergence as shown above in Lemma 2.

However, a gap between the two algorithms can be established in the fully asynchronous model. Specifically, we now show that the ordinary center of gravity algorithm Go_to_COG does not converge in the $\langle \mathcal{ASYNC}, \mathcal{ERR}^- \rangle$ model, even when the robots are positioned on a straight line. Contrasting this result with Theorem 8 yields the claimed separation between the two algorithms.

Theorem 9. *In the $\langle \mathcal{ASYNC}, \mathcal{ERR}^- \rangle$ model, for every ϵ and $N > 1/\epsilon$ there exists an activation schedule for which Algorithm* Go_to_COG *does not converge, even when the robots are restricted to a line.*

References

1. N. Agmon and D. Peleg. Fault-tolerant gathering algorithms for autonomous mobile robots. In *Proc. 15th SODA*, 1063–1071, 2004.
2. H. Ando, Y. Oasa, I. Suzuki, and M. Yamashita. A distributed memoryless point convergence algorithm for mobile robots with limited visibility. *IEEE Trans. Robotics and Automation*, 15:818–828, 1999.
3. E. Arkin, M. Bender, S. Fekete, J. Mitchell, and M. Skutella. The freeze-tag problem: How to wake up a swarm of robots. In *Proc. 13th SODA*, 2002.
4. Y.U. Cao, A.S. Fukunaga, and A.B. Kahng. Cooperative mobile robotics: Antecedents and directions. *Autonomous Robots*, 4(1):7–23, March 1997.
5. M. Cieliebak, P. Flocchini, G. Prencipe, and N. Santoro. Solving the robots gathering problem. In *Proc. 30th ICALP*, 1181–1196, 2003.
6. R. Cohen and D. Peleg. Convergence properties of the gravitational algorithm in asynchronous robot systems. *SIAM J. on Computing*, 34:1516–1528, 2005.
7. X. Defago and A. Konagaya. Circle formation for oblivious anonymous mobile robots with no common sense of orientation. In *Proc. 2nd ACM Workshop on Principles of Mobile Computing*, 97–104. ACM Press, 2002.
8. P. Flocchini, G. Prencipe, N. Santoro, and P. Widmayer. Hard tasks for weak robots: The role of common knowledge in pattern formation by autonomous mobile robots. In *Proc. 10th Int. Symp. on Algo. and Computation*, 93–102, 1999.
9. P. Flocchini, G. Prencipe, N. Santoro, and P. Widmayer. Gathering of autonomous mobile robots with limited visibility. In *Proc. 18th STACS*, 247–258, 2001.
10. B. V. Gervasi and G. Prencipe. Coordination without communication: The case of the flocking problem. *Discrete Applied Mathematics*, 143:203–223, 2004.
11. SensComp Inc. *Spec. of 6500 series ranging modules.* http://www.senscomp.com.
12. K. Sugihara and I. Suzuki. Distributed algorithms for formation of geometric patterns with many mobile robots. *J. of Robotic Systems*, 13(3):127–139, 1996.
13. I. Suzuki and M. Yamashita. Distributed anonymous mobile robots: Formation of geometric patterns. *SIAM J. on Computing*, 28:1347–1363, 1999.

A Faster Algorithm for the
Steiner Tree Problem*

Daniel Mölle, Stefan Richter, and Peter Rossmanith

Department of Computer Science,
RWTH Aachen University, Fed. Rep. of Germany
{moelle, richter, rossmani}@cs.rwth-aachen.de

Abstract. For decades, the algorithm providing the smallest proven worst-case running time (with respect to the number of terminals) for the Steiner tree problem has been the one by Dreyfus and Wagner. In this paper, a new algorithm is developed, which improves the running time from $O(3^k n + 2^k n^2 + n^3)$ to $(2+\delta)^k \cdot poly(n)$ for arbitrary but fixed $\delta > 0$. Like its predecessor, this algorithm follows the dynamic programming paradigm. Whereas in effect the Dreyfus–Wagner recursion splits the optimal Steiner tree in two parts of arbitrary sizes, our approach looks for a set of nodes that separate the tree into parts containing only few terminals. It is then possible to solve an instance of the Steiner tree problem more efficiently by combining partial solutions.

1 Introduction

As of today, NP-hard problems cannot be solved in polynomial time. Nevertheless, we have to deal with many of them in everyday applications. Earlier work has resulted in numerous ways to address this dilemma, among them approximation, randomized algorithms, parameterized complexity, heuristics and many more. Recently, there has been renewed vigor in the field of exact algorithms, and the exponential runtime bounds for many problems have been improved. Some examples for such improvement have been achieved with new algorithms for 3-SATISFIABILITY [5], INDEPENDENT SET [1, 8], DOMINATING SET [4], and MAX-CUT [6, 9, 10].

The Steiner tree problem on networks is to find a subgraph of minimum total edge weight that connects all nodes in a given node subset. Since we assume positive weights for all edges, this subgraph must be a tree. The respective decision problem, in which we ask for the existence of such a subgraph whose weight does not exceed a given limit, is known to be NP-complete. The optimization problem is APX-complete, even if the edge weights are restricted to $\{1, 2\}$ [2]. On the positive side, it is polynomial-time approximable within $1 + (\ln 3)/2 \approx 1.55$, or within 1.28 in the aforementioned restricted variation [7].

The best exact algorithm for the Steiner tree problem known today is due to Dreyfus and Wagner [3]. Given a graph with n nodes and m edges, a set of

---

* Supported by the DFG under grant RO 927/6-1 (TAPI).

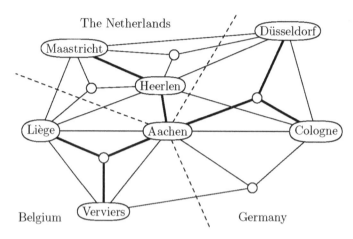

Fig. 1. A geographic example for Steiner trees

k terminals, and the lengths of the edges, it computes a minimum Steiner tree in time $O(3^k n + 2^k n^2 + n^2 \log n + nm)$ using dynamic programming. In this paper, we present an algorithm achieving the same in time $(2 + \delta)^k \cdot poly(n)$ for arbitrarily small but fixed $\delta > 0$.

For an intuitive introduction to some important concepts we use, look at the cities in Figure 1. The need for a high-speed road network linking these cities arises, and some of the existing roads must be enlarged. When we assign a prospective renovation cost to each existing road, the cheapest solutions to the problem are exactly the optimal Steiner trees in this network for the terminal set of cities. One approach consists of solving the problem independently in Belgium, the Netherlands, and Germany. This method works in our example, because there is an optimal Steiner tree for the entire network—drawn fat in Figure 1— in which you cross borders only in Aachen. That is, if every globally optimal solution contained the highway from Lige to Maastricht, regionally optimal trees could not be combined to form such a solution.

Abstracting from geography, in Section 2 we will formally define *regions* to be maximal subtrees of a Steiner tree where every leaf is a terminal and vice versa. In our example, the Belgium towns (including Aachen) would form one region. The nationally constrained parts of the tree in Figure 1 are connected unions of regions, and as such will formally be called *confederations*. At the end of Section 2, we will prove that the heart of our algorithm can construct optimal Steiner trees for confederations by combining regional solutions. Note that this entails optimal Steiner trees for the entire network.

As our method relies on exhaustive search, it is only fast when all regions contain few terminals. Fortunately, there is a way to accomplish this. If, for an illustrative example, we find the Belgium region too large, promoting the central node in it to the rank of terminal will cut the region in three smaller pieces. Section 3 will formalize the process of incorporating additional terminals, which allows us to derive the aforementioned runtime bounds.

2   Networks, Regions, and Confederations

We adhere to the notation commonly used for the Steiner tree problem. For a graph G, let $V[G]$ and $E[G]$ denote the set of nodes and the set of edges in G, respectively. The edge weights of an input graph G are assumed to be given by a function $\ell \colon E[G] \to \mathbf{N}$, where $\mathbf{N} = \{1, 2, 3, \dots\}$. Formally, an instance of the Steiner tree problem consists of a network (G, ℓ) and a terminal set $Y \subseteq V[G]$.

We define the union operation on graphs G_1, G_2 as follows: $G_1 \cup G_2 := (V[G_1] \cup V[G_2], E[G_1] \cup E[G_2])$. For any subgraph G' of G from a network (G, ℓ), let $\ell(G') := \sum_{e \in E[G']} \ell(e)$.

In order to explain our algorithm, let us imagine that we know an optimal Steiner tree T. We will examine some properties of T that enable us to find T— or an equivalent tree—with relative efficiency. As outlined before, T consists of *regions* (also called *components* or *full Steiner trees* in the literature), which we will define now as maximal subtrees that do not contain terminals as inner nodes.

Definition 1. Let (G, ℓ) be a network, $Z \subseteq V[G]$ a set of terminals, and T be an optimal Steiner tree for Z. A (T, Z)-*region* is an inclusion-maximal subtree R of T in which every terminal is a leaf. The set of all (T, Z)-regions in T is denoted by $\mathcal{R}(T, Z)$.

See Figure 2 for an illustration of this concept. Regions are the smallest building blocks in our approach. From the tree structure of T, it is easy to see that neighboring regions overlap in exactly one node, which has to be a terminal. They can then be united to form *confederations*.

Definition 2. A connected union T' of (T, Z)-regions is called a (T, Z)-*confederation*. Accordingly, the corresponding terminal set $Z' = Z \cap V[T']$ is called (T, Z)-*confederate*.

As already indicated in the introduction, (T, Z)-regions are optimal for their terminal sets, provided T is optimal for Z. The following lemma formalizes a generalization of this fact for confederations.

Lemma 1. Let (G, ℓ) be a network and T be an optimal Steiner tree for $Z \subseteq V[G]$. Any (T, Z)-confederation T' constitutes an optimal Steiner tree for its terminal set.

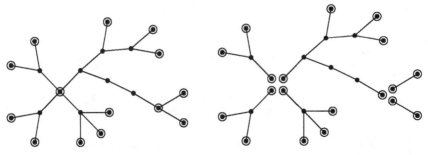

Fig. 2. A Steiner tree T with emphasized terminals Z and its (T, Z)-regions

Proof. Assume that $T' = R_1 \cup \cdots \cup R_s$, where $\mathcal{R}(T, Z) = \{R_1, \ldots, R_t\}$, and that there is a Steiner tree S' for $Z \cap V[T']$ that is cheaper than T'. The graph S obtained from T by replacing T' with S' is connected: All the nodes shared by T' and $R_{s+1} \cup \cdots \cup R_t$ are contained in $Z \cap V[T']$. Moreover, S' connects all the nodes from $Z \cap V[T']$.

Since S is a connected graph that contains all the nodes from Z, it is a Steiner graph for Z which is cheaper than T, a contradiction. $\qquad\square$

The vital property stated in the above lemma lies at the heart of our method. It ensures that we can build up an optimal Steiner tree T for Z—which is a (T, Z)-confederation itself—region by region. That is, when we add a neighboring region to a confederation, a larger confederation arises, and thus a larger optimal Steiner tree.

While optimal Steiner trees for (T, Z)-confederate terminal sets can be constructed recursively, we do not have a means to find or detect (T, Z)-confederate sets efficiently. This problem will be addressed by going through all subsets Z' of Z. We can then be sure to hit the confederate ones in particular. Notice, however, that we still do not know with regard to which optimal Steiner tree T the respective Z' are (T, Z)-confederate. Therefore we must be certain that replacing a (T, Z)-confederation by some other optimal Steiner tree for its terminal set does not destroy the optimality of the whole Steiner tree.

In general, it is not true that combining two optimal Steiner trees for terminal sets Z_1 and Z_2 with $|Z_1 \cap Z_2| = 1$ yields an optimal Steiner tree for $Z_1 \cup Z_2$, as exemplified by the network in Figure 3. The following lemma shows, however, that if this *is* the case for two edge-disjoint optimal Steiner trees for Z_1 and Z_2, then it is the case for *any* two such trees.

Lemma 2. *Let (G, ℓ) be a network and $Z_1, Z_2 \subseteq V[G]$ with $|Z_1 \cap Z_2| = 1$. If there are edge-disjoint optimal Steiner trees T_1 for Z_1 and T_2 for Z_2 such that $T_1 \cup T_2$ is an optimal Steiner tree for $Z_1 \cup Z_2$, then the union $T_1' \cup T_2'$ of any two optimal Steiner trees T_1' for Z_1 and T_2' for Z_2 is also an optimal Steiner tree for $Z_1 \cup Z_2$.*

Proof. If there are edge-disjoint optimal Steiner trees T_1 for Z_1 and T_2 for Z_2 such that $T_1 \cup T_2$ is an optimal Steiner tree for $Z_1 \cup Z_2$, then the respective

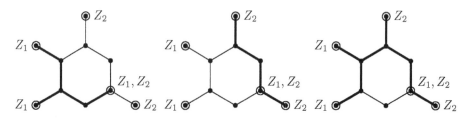

Fig. 3. Two optimal Steiner trees for Z_1, Z_2 that do not combine to form an optimal Steiner tree for $Z_1 \cup Z_2$ (assuming unit-weight edges), compared to an optimal Steiner tree for $Z_1 \cup Z_2$

optimal weights are $\ell(T_1)$, $\ell(T_2)$, and $\ell(T_1 \cup T_2) = \ell(T_1) + \ell(T_2)$. Let T_1' and T_2' be arbitrary optimal Steiner trees for Z_1 and Z_2, respectively. Observe that because Z_1 and Z_2 share a node, $T_1' \cup T_2'$ is a Steiner graph for $Z_1 \cup Z_2$. Of course, $\ell(T_1') = \ell(T_1)$ and $\ell(T_2') = \ell(T_2)$. This implies

$$\ell(T_1' \cup T_2') \leq \ell(T_1') + \ell(T_2') - \ell(T_1) + \ell(T_2) - \ell(T_1 \cup T_2)$$

By optimality of the Steiner tree $T_1 \cup T_2$, we know that $T_1' \cup T_2'$ is also optimal.
□

We are now able to formulate a very important theorem regarding the correctness of our algorithm for the Steiner tree problem.

As mentioned earlier, (T, Z)-confederations are very helpful for the construction of optimal Steiner trees, but probably hard to find and detect in general. Our algorithm in Figure 4 will thus use the concept of (T, Z)-confederations only implicitly inside a dynamic programming approach. It builds a table of Steiner graphs for each $Z' \subseteq Z$ in order of cardinality. For subsets of size up to q, the Dreyfus–Wagner algorithm is employed. From then on, the cheapest bipartition of Z' is chosen. However, we only look at partitions where the smaller part is of size at most q. Lemma 1 shows that we get an optimal Steiner tree this way when we set q to be the maximum number of terminals in a single (T, Z)-region for some optimal Steiner tree T. Let us first formalize this concept.

Definition 3. Let (G, ℓ) be a network and $Z \subseteq V[G]$. We say that Z is q-granular iff there exists an optimal Steiner tree T for Z in G such that $|V[R] \cap Z| \leq q$ for every $R \in \mathcal{R}(T, Z)$.

Theorem 1. *Let (G, ℓ) be a network and $Z \subseteq V[G]$ a set of terminals. If Z is q-granular, then the algorithm from Figure 4 computes an optimal Steiner tree for Z.*

Proof. Let us prove the following statement: If Z' is (T, Z)-confederate for an optimal Steiner tree T, then $S(Z')$ is an optimal Steiner tree for Z'. This will entail the claim, since Z is (T, Z)-confederate for any Steiner tree.

```
1    for all
2      Z' ⊆ Z such that |Z'| ≤ q do
3        S(Z') ← Dreyfus–Wagner((G,ℓ), Z')
4      od; for all Z' ⊆ Z (ascending size) such that
5        |Z'| > q do
6          S(Z') ← G;
7          for all Z'' ⊆ Z' and v ∈ Z'' such that 2 ≤ |Z''| ≤ q do
8            if ℓ(S(Z'')) + ℓ(S(Z' − Z'' ∪ {v})) < ℓ(S(Z'))
9            then S(Z') ← S(Z'') ∪ S(Z' − Z'' ∪ {v}) fi
10         od
11     od; return S(Z)
```

Fig. 4. Computing a Steiner tree for a q-granular set Z of terminals

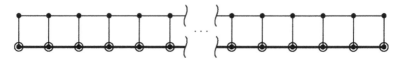

Fig. 5. The $n \times 2$-grid as a network with emphasized terminals Z, edges of weight 1 (thin) and edges of weight $2 + 1/(n-1) + 1/n^2$ (thick). All the thin edges form the unique optimal Steiner tree T for Z, implying that Z is q-granular only for $q \geq n$. Observe that optimal Steiner trees for any Z_1, Z_2 with $Z_1 \cup Z_2 = Z$, $|Z_1 \cap Z_2| = 1$, and $|Z_1|, |Z_2| \geq 2$ do *not* combine to form an optimal Steiner tree for Z.

Let T' be the (T, Z)-confederation that corresponds to Z'. We use induction on the number of regions in T'. Since Z is q-granular, the claim holds if T' is a single region, by correctness of the Dreyfus–Wagner algorithm. For a (T, Z)-confederation $T' = R_1 \cup \cdots \cup R_t$ with $t \geq 2$, we can find a region—say R_1—that shares only a single node v with the other regions, such that $R_2 \cup \cdots \cup R_t$ is still a (T, Z)-confederation. By Lemma 1, R_1 as well as $R_2 \cup \cdots \cup R_t$ constitute optimal Steiner trees for their respective terminals. Because these trees are edge-disjoint, Lemma 2 applies. Using the induction hypothesis, we thus get that $S' := S(Z \cap V[R_1]) \cup S(Z \cap V[R_2 \cup \cdots \cup R_t])$ is an optimal Steiner tree for Z'. Because Z is q-granular, at some time in line seven $Z'' = Z \cap V[R_1]$ and $Z' - Z'' \cup \{v\} = Z \cap V[R_2 \cup \cdots \cup R_t]$. Therefore, the algorithm will eventually consider S'. Any graph $S(Z')$ computed by the algorithm is a Steiner graph for Z'. Thus, if $S(Z') = S'$ at any time during the computation, S' cannot be replaced because it is optimal. With the same argument, $S(Z')$ will be set to S' at some point, unless $S(Z')$ already contains another optimal Steiner tree for Z'. $\square$

The network shown in Figure 5 illustrates the importance of the concept of q-granularity for this theorem: If no two optimal Steiner trees for Z_1, Z_2 with $Z_1 \cup Z_2 = Z$, $|Z_1 \cap Z_2| = 1$, and $|Z_1|, |Z_2| \geq 2$ combine to form an optimal Steiner tree for Z, the algorithm from Figure 4 is bound to fail for $q < n$.

3   Extending a Terminal Set

In the previous section, we established an algorithm that finds an optimal Steiner tree for any q-granular terminal set Z in a network (G, ℓ). In general, however, a terminal set Y can be $|Y|$-granular in the worst-case. Neither is this case absurd—it just needs all terminals to be leafs in every optimal Steiner tree—nor is it obvious how to predict the granularity of an instance. That seems to render the algorithm completely useless (in these instances, the Dreyfus–Wagner implementation would do all the work).

Fortunately, it is possible to reduce granularity by extending terminal sets. The following lemma shows how we can achieve nearly arbitrary granularities by incorporating only a limited number of additional terminals.

Lemma 3. *Let T be an optimal Steiner tree for some $Y \subseteq V[T]$, and q an integer with $2 \leq q \leq |Y|$. There is a set $Z \subseteq V[T]$, $Y \subseteq Z$, such that $|Z| \leq |Y| + \frac{|Y|}{q-1}$ and $|V[R] \cap Z| \leq q$ for every (T, Z)-region, implying that Z is q-granular.*

Proof. Consider the following algorithm:

1. Let $Z := Y$.
2. While T contains a (T, Z)-region R with $2 \le |V[R] \cap Z| \le q$ that only shares a single node v with the other (T, Z)-regions, remove R save v from T.
3. If T is only a single (T, Z)-region with $|V[T] \cap Z| \le q$, then output Z and stop.
4. Choose a root r in T and let $v := r$.
5. While a subtree rooted in a child w of v contains more than $q - 1$ nodes from Z, let $v := w$.
6. Add v to Z and go to step (2).

See Figure 6 for an illustration of how the above algorithm constructs such a set Z.

To prove the correctness, let us first show that each time a node v is added to Z in step (6), at least q nodes from Z are removed in step (2) immediately. It is easy to see that subtrees rooted at some node v can be turned into (T, Z)-regions by adding v to Z, and that these regions will then be removed by step (2) (save v). Moreover, as soon as $|V[T] \cap Z| \le q$, step (3) will terminate the algorithm.

Whenever the **while**-loop in step (5) terminates, each subtree rooted at a child w of v contains at most $q - 1$ nodes from Z. Hence, by adding v to Z, each of these subtrees becomes a (T, Z)-region containing at most q nodes from Z. On the other hand, the union of all the subtrees rooted at v contains more than $q - 1$ nodes from Z, because v would not have been chosen by the **while**-loop in step (5) otherwise. Furthermore, v cannot be in Z yet—otherwise, step (2) would have removed the regions originating from the subtrees rooted at v. Thus, $|Z| \le |Y| + \frac{|Y|}{q-1}$ follows.

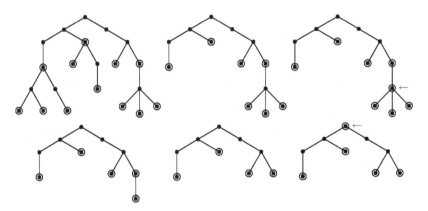

Fig. 6. We apply the algorithm from Lemma 3 to a tree T with emphasized terminals Z, using $q = 3$ and always choosing the upper node as a root. Executing step (2) leads to the removal of four subtrees. Then, step (5) chooses another node for addition to Z. In the next iterations of step (2), three more subtrees are deleted. Another iteration removes yet another subtree. After that, step (5) chooses the root for addition to Z. Finally, after applying steps (2) and (3), the algorithm terminates.

It remains to show that $|V[R] \cap Z| \leq q$ for every $R \in \mathcal{R}(T, Z)$. This is obvious, because the algorithm only removes sufficiently small regions (save some articulation nodes) from T repeatedly, until the tree is a single small region or a single terminal.

Since q or more nodes from Z are removed whenever a node v is added to Z, the size of Z decreases by at least $q-1$ in every such case. However, the algorithm does not add any more nodes to Z as soon as $|V[T] \cap Z| \leq q$. The size of Z is thus bounded from above by $|Y| + \lceil (|Y| - q)/(q - 1) \rceil \leq |Y| + |Y|/(q - 1)$.   □

Observe that every optimal Steiner tree for Z is also an optimal Steiner tree for Y: Because Z is a superset of Y and $Z \subseteq V[T]$, T is also an optimal Steiner tree for Z. Any other optimal Steiner tree T' for Z has the same weight and connects all the nodes from Y, too.

Of course, Lemma 3 only serves to show the existence of a small set of extension terminals, even though the proof is algorithmical in nature. Without an optimal Steiner tree, we do not seem to have other means than exhaustive search to find the right extension nodes. Let us now accumulate the threads and analyze the running time for the new method:

Theorem 2. *For every $\delta \in (0, 1]$ there is an algorithm that computes an optimal Steiner tree for an arbitrary network (G, ℓ) and terminal set $Y \subseteq V[G]$ in at most $(2 + \delta)^{|Y|} poly(|V[G]|)$ steps.*

Proof. Choose $\varepsilon > 0$ such that $12\varepsilon \ln(1/3/\varepsilon) = \delta$ and let $q = \lceil \varepsilon|Y| \rceil + 1$. If $|Y| < 1/\varepsilon$, then the number of terminals is bounded by a constant, allowing us to find an optimal Steiner tree in polynomial time employing the Dreyfus–Wagner algorithm.

Otherwise, there is a q-granular $Z \subseteq V[G]$, $Y \subseteq Z$, of size $\lfloor |Y| + |Y|/(q - 1) \rfloor$ such that any optimal Steiner tree for Z also constitutes an optimal Steiner tree for Y, due to Lemma 3. Moreover, by Lemma 1, we may employ the algorithm from Figure 4 to obtain an optimal Steiner tree for Z, solving the entire problem.

Unfortunately, we do not know Z. We can, however, just go through all $X \subseteq V[G]$ with $|X| = \lfloor |Y|/(q - 1) \rfloor$ and run the algorithm for $Z = Y \cup X$. Since it always outputs Steiner graphs for Z and it finds an optimal Steiner tree for Z at least once, the cheapest Steiner graph returned is indeed an optimal Steiner tree for Z.

The runtime can be estimated as follows. It is easy to see that $q \leq |Z|/2$, since $\varepsilon < 1/20$ even for $\delta = 1$. For $n = |V[G]|$, the running time of lines 1–3 of the algorithm is

$$\sum_{i=1}^{q} \binom{|Z|}{i} 3^i poly(n) \leq \binom{|Z|}{q} 3^q poly(n),$$

because the running time of the Dreyfus–Wagner algorithm is $3^i poly(n)$ if there are i terminals. The running time of lines 4–10 can be estimated by

$$\sum_{i=q}^{|Z|} \binom{|Z|}{i} \sum_{j=2}^{q} \binom{i}{j} poly(n) \le 2^{|Z|} \binom{|Z|}{q} poly(n).$$

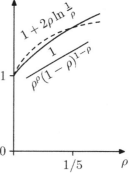

Now let $\rho = q/|Z|$ and assume $\rho < 3/20$, which will be proven correct later. By a well-known estimation and some technical formula manipulations we get

$$\binom{|Z|}{\rho|Z|} \le \left(\frac{1}{\rho^{\rho}(1-\rho)^{1-\rho}} \right)^{|Z|} \le \left(1 + 2\rho \ln(1/\rho)\right)^{|Z|}.$$

We also have that $|Z| = |X| + |Y| \le |Y| + 1/\varepsilon$, because $|X| \le |Y|/(q-1) \le |Y|/\varepsilon|Y|$. Therefore, there are no more than $n^{1/\varepsilon} = poly(n)$ possibilities to choose X. Taking this into consideration, the overall time to compute an optimal Steiner tree becomes

$$2^{|Y|+1/\varepsilon}\left(1 + 2\rho\ln(1/\rho)\right)^{|Y|+1/\varepsilon} poly(n) \le \left(2 + 12\varepsilon\ln(1/3\varepsilon)\right)^{|Y|} poly(n),$$

where the inequality holds because by $|Y| \ge 1/\varepsilon$,

$$\rho = \frac{q}{|Z|} = \frac{\lceil \varepsilon|Y| \rceil + 1}{|Y| + \lfloor |Y|/(q-1) \rfloor}$$

$$= \frac{\lceil \varepsilon|Y| \rceil + 1}{|Y| + \lfloor |Y|/\lceil \varepsilon|Y| \rceil \rfloor} \le \frac{\varepsilon|Y| + 2}{|Y|} \le \varepsilon + \frac{2}{1/\varepsilon} = 3\varepsilon < \frac{3}{20}.$$

In total, we get an algorithm with running time $(2 + \delta)^{|Y|} poly(n)$.   □

4   Remarks on Running Time

Of course, the constants in the above worst-case analysis become very large even for moderate δ. On the other hand, we used many extremely rough estimations. This is because we put the emphasis on the asymptotical behavior, as we aimed to improve the running time from $O(3^k \cdot n^3)$ to $c^k \cdot poly(n)$ for c as close to 2 as we like, concentrating on obtaining a simple analysis rather than the best possible bounds.

The polynomial degree in the running time contains $1/\varepsilon$ as an additive term. This can be improved to $O(\varepsilon^{-2/3})$ by dividing the process into two stages using different granularities. However, the analysis becomes quite involved at this point. It remains an open question whether the problem can be solved in $2^k \cdot poly(n)$.

The fact that we need to go through many extension sets X seems to play a dominant role in the running time of our algorithm for practical values of k. For many instances, however, a much smaller set X might suffice. To make use of this, the following property would be very useful: If the algorithm outputs a non-optimal Steiner graph for all sets X of size q, then it finds a cheaper Steiner graph for some set X of size $q + 1$. Unfortunately, the grid in Figure 5 proves to be a counter-example to this conjecture. Still, finding relaxed versions of this property might be an interesting question for future research.

On a final note, the idea of joining more than two partial solutions to speed up dynamic programming algorithms seems to lend itself to generalization. The application of this principle might result in better runtime bounds for many problems.

Acknowledgement. We thank the anonymous referees for many helpful comments.

References

1. R. Beigel. Finding maximum independent sets in sparse and general graphs. In *Proc. of 10th SODA*, pages 856–857, 1999.
2. M. Bern and P. Plassmann. The Steiner problem with edge lengths 1 and 2. *Information Processing Letters*, 32:171–176, 1989.
3. S. E. Dreyfus and R. A. Wagner. The Steiner problem in graphs. *Networks*, 1:195–207, 1972.
4. F. V. Fomin, F. Grandoni, and D. Kratsch. Measure and conquer: Domination – A case study. In *Proc. of 32d ICALP*, LNCS. Springer, 2005.
5. K. Iwama and S. Tamaki. Improved upper bounds for 3-SAT. In *Proc. of 15th SODA*, pages 328–328, 2004.
6. J. Kneis, D. Mölle, S. Richter, and P. Rossmanith. Algorithm based on the treewidth of sparse graphs. In *Proc. of 31st WG*, LNCS. Springer, 2005. To appear.
7. G. Robins and A. Zelikovsky. Improved Steiner tree approximation in graphs. In *Proc. of 11th SODA*, pages 770–779, 2000.
8. J. M. Robson. Algorithms for maximum independent sets. *Journal of Algorithms*, 7:425–440, 1986.
9. A. Scott and G. B. Sorkin. Faster algorithms for Max-CUT and Max-CSP, with polynomial expected time for sparse instances. In *Proc. of 7th RANDOM*, number 2764 in LNCS, pages 382–395. Springer, 2003.
10. R. Williams. A new algorithm for optimal constraint satisfaction and its implications. In *Proc. of 31st ICALP*, number 3142 in LNCS, pages 1227–1237. Springer, 2004.

Generating Randomized Roundings with Cardinality Constraints and Derandomizations

Benjamin Doerr

Max–Planck–Institut für Informatik, Saarbrücken, Germany

Abstract. We provide a general method to generate randomized roundings that satisfy cardinality constraints. Our approach is different from the one taken by Srinivasan (FOCS 2001) and Gandhi et al. (FOCS 2002) for one global constraint and the bipartite edge weight rounding problem.

Also for these special cases, our approach is the first that can be derandomized. For the bipartite edge weight rounding problem, in addition, we gain an $\tilde{O}(|V|)$ factor run-time improvement for generating the randomized solution.

We also improve the current best result on the general problem of derandomizing randomized roundings. Here we obtain a simple $O(mn \log n)$ time algorithm that works in the RAM model for arbitrary matrices with entries in $\mathbb{Q}_{\geq 0}$. This improves over the $O(mn^2 \log(mn))$ time solution of Srivastav and Stangier.

1 Introduction and Results

Many combinatorial optimization problems can easily be formulated as integer linear programs (ILPs). Unfortunately, solving ILPs is NP–hard, whereas solving linear programs (without integrality constraints) is easy, both in theory and practice. Therefore, a natural and widely used technique is to solve the linear relaxation of the ILP and then transform its solution into an integer one.

Typically, this requires rounding a vector x to an integer one y in such a way that the rounding errors $|(Ax)_i - (Ay)_i|$, $i \in [m] := \{1, \ldots, m\}$, are small for some given $m \times n$ matrix A.

1.1 Randomized Rounding

A very successful approach to such rounding problems is the one of *randomized rounding* introduced by Raghavan and Thompson [RT87, Rag88]. Here the integer vector y is obtained from x by rounding each component j independently with probabilities derived from the fractional part of x_j. In particular, if $x \in [0, 1]^n$, we have $\Pr(y_j = 1) = x_j$ and $\Pr(y_j = 0) = 1 - x_j$ independently for all $j \in [n]$.

Since the components are rounded independently, the rounding error $|(Ax)_i - (Ay)_i|$ in constraint i is a sum of independent random variables. Thus it is highly concentrated around its mean, which by choice of the probabilities is zero. Large deviation bounds like the Chernoff inequality allow to quantify such violations and thus yield performance guarantees. The derandomization problem is to transform this randomized approach into deterministic rounding algorithms that keep the rounding errors $|(Ax)_i - (Ay)_i|$ below some threshold.

B. Durand and W. Thomas (Eds.): STACS 2006, LNCS 3884, pp. 571–583, 2006.

1.2 Hard Constraints

Whereas the independence in rounding the variables ensures that the rounding errors $|(Ax)_i - (Ay)_i|$ are small, it is very weak in guaranteeing that a constraint is satisfied without error. We call a constraint *hard constraint* if we require our solution to satisfy it without violation. In this paper, we are mainly concerned with *cardinality constraints*. These are constraints on unweighted sums of variables. Let us give an example where such hard constraints naturally occur.

The *integer splittable flow problem* is the following routing problem. Given an undirected graph and several source–sink pairs (s_i, t_i) together with integral demands d_i, we are looking for integer flows f_i from s_i to t_i having flow value d_i such that the maximum edge congestion is minimized. Solving the non-integral relaxation and applying path stripping (cf. [GKR+99]), we end up with this rounding problem: Round a solution $(x_P)_P$ of the linear system

$$\text{Minimize } W \text{ s. t. } \sum_{P \ni e} x_P \leq W, \; \forall e$$

$$\sum_{P \in \mathcal{P}_i} x_P = d_i, \; \forall i$$

$$x_P \geq 0, \; \forall P$$

to an integer one such that the first set of constraints is violated not too much and the second one is satisfied without any violation (hard constraints).

Further examples of rounding problems with hard constraints include other routing applications ([RT91, Sri01]), many flow problems ([RT87, RT91, GKR+99]), partial and capacitated covering problems ([GKPS02, GHK+03]) and the assignment problem with extra constraints ([AFK02]).

For the special case of the above problem where all d_i are one, Raghavan and Thompson [RT87] presented an easy solution: For each i, they pick one $P \in \mathcal{P}_i$ with probability x_P and then set $y_P = 1$ and $y_{P'} = 0$ for all $P' \in \mathcal{P}_i \setminus \{P\}$. For the general case, however, this idea and all promising looking extensions fail. Guruswami et al. [GKR+99] state on the integral splittable flow problem (ISF) in comparison to the unsplittable flow problem that "standard roundings techniques are not as easily applied to ISF".

At FOCS 2001, Srinivasan [Sri01] presented a way to compute randomized roundings that respect the constraint that the sum of all variables remains unchanged (one *global cardinality constraint*) and fulfill some negative correlation properties.

Gandhi, Khuller, Parthasarathy and Srinivasan [GKPS02] combined the deterministic "pipage rounding" algorithm of Ageev and Sviridenko [AS04] with Srinivasan's approach to obtain randomized roundings of edge weights in bipartite graphs that are degree preserving. By this we mean that the sum of weights of all edges incident with some vertex is not changed by the rounding. The roundings of Gandhi et al. also fulfill negative correlation properties, but only on sets of edges incident with a common vertex.

Both Srinivasan [Sri01] and Gandhi et al. [GKPS02] do not consider the derandomization problem. A first derandomization of Srinivasan's [Sri01] roundings was given in [Doe05]. For the bipartite edge weight rounding problem, Ageev and

Sviridenko [AS04] state that any randomized rounding algorithm "will be too sophisticated to admit derandomization".

1.3 Our Contribution

In this paper, we extend the work of [Doe05] in several directions.

Randomized Roundings with Constraints. We show that for all sets of cardinality constraints, the general problem of generating randomized roundings can be reduced to the one for $\{0, \frac{1}{2}\}$ vectors. This immediately yields a simpler way to generate the randomized roundings used in Srinivasan [Sri01], Gandhi et al. [GKPS02], Sadakane, Takki-Chebihi and Tokuyama [STT01] and [Doe04]. For the rounding problem of [GKPS02], we even gain an $\tilde{O}(|V|)$ factor in the run-time.

Derandomizations with Constraints. Since our approach is structurally simpler than the earlier ones, we do obtain the corresponding derandomizations. In fact, we may even use classical derandomizations like Raghavan's. In consequence, derandomizing randomized rounding approaches for the bipartite edge weight rounding problem is much easier than what is conjectured in Ageev and Sviridenko [AS04]. Note that this derandomization is more than a re-invention of the original algorithm of Ageev and Sviridenko. It also keeps those rounding errors small for which the randomized approach allowed the use of Chernoff type large deviation bounds.

Counter-Examples. We also show that a number of natural properties of independent randomized roundings may not hold in the presence of constraints. For example, let $f : [0, 1]^n \to \mathbb{R}$ be non-decreasing, $x, x' \in [0, 1]^n$ and y, y' be independent randomized roundings of x, x' respectively. Then $x \le x'$ (component-wise) implies $E(f(y)) \le E(f(y'))$. We show that already a single cardinality constraint may inflict that no randomized roundings respecting this constraint have the above property.

General Randomized Rounding Derandomization. Our final result is not to be overlooked due to its simple less than a page proof. Here we give an easy $O(mn \log n)$ time derandomization for arbitrary constraint matrices $A \in ([0, 1] \cap \mathbb{Q})^{m \times n}$ that works in the RAM model. This improves over the $O(mn^2 \log(mn))$ time (and 30 pages) landmark solution of Srivastav and Stangier [SS96]. Note that Raghavan's derandomization [Rag88] needs to compute the exponential function and in consequence in the RAM model only works for binary matrices (as pointed out in Section 2.2 of his paper).

2 Randomized Rounding, Constraints and Correlation

For a number r write $\lceil r \rceil = \{n \in \mathbb{N} \mid n \le r\}$, $\lfloor r \rfloor = \max\{z \in \mathbb{Z} \mid z \le r\}$, $\lceil r \rceil = \min\{z \in \mathbb{Z} \mid z \ge r\}$ and $\{r\} = r - \lfloor r \rfloor$. We write $z \approx r$ if $z \in \{\lfloor r \rfloor, \lceil r \rceil\}$. We use these notations for vectors as well (component-wise).

Let $x \in \mathbb{R}$. A real-valued random variable y is called *randomized rounding of x* if $\Pr(y = \lfloor x \rfloor + 1) = \{x\}$ and $\Pr(y = \lfloor x \rfloor) = 1 - \{x\}$. Since only the fractional parts of x and y are relevant, we usually have $x \in [0, 1]$. In this case, we have

$$\Pr(y = 1) = x,$$
$$\Pr(y = 0) = 1 - x.$$

For $x \in \mathbb{R}^n$, we call $y = (y_1, \ldots, y_n)$ randomized rounding of x if y_j is a randomized rounding of x_j for all $j \in [n]$.

The algorithmic concept of randomized rounding can be formulated as follows: Fix a number $n \in \mathbb{N}$, the number of variables to be rounded. Let $X \subseteq [0,1]^n$. This is the set of vectors for which we allow randomized rounding. Typically, this will be $[0,1]^n$ or a suitably rich subset thereof. A family $(\Pr_x)_{x \in X}$ of probability distributions on $\{0,1\}^n$ is called *randomized rounding*, if for all $x \in X$, a sample y from $\Pr_x$ is a randomized rounding of x.

As described in the introduction, we are interested in roundings that satisfy some hard constraints. Though usually we will only regard cardinality constraints (requiring the sum of some variable to be unchanged), it will be convenient to encode hard constraints in a matrix B. Our aim then is that a rounding y of x satisfies $By = Bx$. Of course, if Bx is not integral, this can never be satisfied. We therefore relax the condition to $By \approx Bx$. In a randomized setting, we often obtain the slightly stronger condition that By is a randomized rounding of Bx.

Besides satisfying hard constraints we still want to keep other rounding errors small (as does independent randomized rounding). A useful concept here is the one of negative correlation, which implies Chernoff type large deviation inequalities.

We call a set $\{X_j \mid j \in S\}$ of binary random variables *negatively correlated* if for all $S_0 \subseteq S, b \in \{0,1\}$, we have $\Pr(\forall j \in S_0 : X_j = b) \leq \prod_{j \in S_0} \Pr(y_j = b)$. As shown in [PS97], this implies the usual Chernoff-Hoeffding bounds on large deviations. The following version is not strongest possible, but sufficient for most purposes.

Lemma 1. *Let $\{X_j \mid j \in S\}$ be a set of negatively correlated binary random variables and $a_j \in [0,1], j \in S$. Put $X = \sum_{j \in S} a_j X_j$ and $\mu = E(X)$. Then for all $\delta \in [0,1]$,*

$$\Pr(X \geq (1+\delta)\mu) \leq \exp(-\tfrac{1}{3}\mu\delta^2),$$
$$\Pr(X \leq (1-\delta)\mu) \leq \exp(-\tfrac{1}{2}\mu\delta^2).$$

It turns out that hard constraints and negative correlation cannot always be achieved simultaneously. We therefore restrict ourselves to negative correlation on certain sets of variables. Let $\mathcal{S} \subseteq 2^{[n]}$ be closed under taking subsets, that is, $S_0 \subseteq S \in \mathcal{S}$ implies $S_0 \in \mathcal{S}$.

Definition 1. *We call $(\Pr_x)$ randomized rounding with respect to B and $\mathcal{S}$, if for all x a sample y from $\Pr_x$ satisfies the following.*

(A1) *y is a randomized rounding of x.*
(A2) *By is a randomized rounding of Bx.*
(A3) *For all $S \in \mathcal{S}, \forall b \in \{0,1\} : \Pr(\forall j \in S : y_j = b) \leq \prod_{j \in S} \Pr(y_j = b)$.*

In this language, we know the following. Clearly, independent randomized rounding is a randomized rounding with respect to the empty matrix B and $\mathcal{S} = 2^{[n]}$. Srinivasan [Sri01] showed that for the $1 \times n$ matrix $B = (1 \ldots 1)$, randomized roundings with respect to B and $\mathcal{S} = 2^{[n]}$ exist and can be generated in time $O(n)$. Let $G = (V, E)$ be a bipartite graph and $B = (b_{ij})_{\substack{i \in V \\ j \in E}}$ its vertex-edge-incidence matrix. For $v \in V$ let $E_v = \{e \in E \mid v \in e\}$. Gandhi et al. [GKPS02] showed that there are randomized roundings with respect to B and $\mathcal{S} = \{E_0 \mid \exists v \in V : E_0 \subseteq E_v\}$. They can be generated in time $O(mn)$. From [Doe03, Doe04], we have that if B is totally unimodular, then randomized roundings with respect to B and $\mathcal{S} = \emptyset$ exist. Recall that a matrix is totally unimodular if each square submatrix has determinant -1, 0 or 1. If B is not totally unimodular, then not even for $X = \{0, \frac{1}{2}\}^n$ a randomized rounding $(\mathrm{Pr}_x)_{x \in X}$ with respect to B and $\mathcal{S} = \emptyset$ exists.

Throughout the paper let $A \in [0, 1]^{m_A \times n}$ and $x \in [0, 1]^n$. Let B be a totally unimodular $m_B \times n$ matrix.

3 Binary Reductions

A central step of our method is a reduction to the problem of rounding $\{0, \frac{1}{2}\}$ vectors, similar as in Beck and Spencer [BS84]. This reduced rounding problem turns out to be structurally and computationally much simpler than the general one. We start by describing the connection between the reduced and the general problem.

3.1 Randomized Roundings

Let $\mathcal{S} \subseteq 2^{[n]}$ be closed under taking subsets. Let $(\mathrm{Pr}_x)_{x \in \{0, \frac{1}{2}\}^n}$ be a family of probability distributions on $\{0, 1\}^n$. We call the family (Pr_x) *basic randomized rounding with respect to B and $\mathcal{S}$*, if for all $x \in \{0, \frac{1}{2}\}^n$ a sample y from Pr_x satisfies (A1) to (A3) and

(A4) $\mathrm{Pr}_x(y) = \mathrm{Pr}_x(2x - y)$.

The key result of this subsection is that any basic randomized rounding can be extended to a randomized rounding $(\overline{\mathrm{Pr}}_x)$, where x ranges over all vectors in $[0, 1]^n$ having finite binary length. The simple idea is to iterate basic randomized rounding digit by digit:

Digit by digit rounding: Let $x \in [0, 1]^n$ having binary length ℓ (that is, all x_i can be written as $x_i = \sum_{j=0}^{\ell} d_j 2^{-j}$ with $d_j \in \{0, 1\}$). There is nothing to show for $\ell = 0$, so assume $\ell \geq 1$. Write $x = x' + 2^{-\ell+1} x''$ with $x'' \in \{0, \frac{1}{2}\}^n$ and $x' \in [0, 1]^n$ having binary length at most $\ell - 1$. Let y'' be a sample from the basic randomized rounding $\mathrm{Pr}_{x''}$. Set $\tilde{x} := x' + 2^{-\ell+1} y''$. Note that $\tilde{x}$ has binary length at most $\ell - 1$. Repeat this procedure until a binary vector is obtained. For each x having finite binary expansion, this defines a probability distribution $\overline{\mathrm{Pr}}_x$ on $\{0, 1\}^n$.

Theorem 1. *Let (Pr_x) be a basic randomized rounding with respect to B and $\mathcal{S}$. Then $(\overline{\mathrm{Pr}}_x)$ with x ranging over all $[0, 1]$ vectors having finite binary length is a randomized rounding with respect to B and $\mathcal{S}$.*

Proof. We proceed by induction. If $x \in \{0, \frac{1}{2}, 1\}^n$, we simply have $\mathrm{Pr}_x = \overline{\mathrm{Pr}}_x$. Let x therefore have binary length $\ell > 1$. Let $x = x' + 2^{-\ell+1} x''$ with $x'' \in \{0, \frac{1}{2}\}^n$ and $x' \in [0, 1]^n$ having binary length at most $\ell - 1$. Let y'' be a sample from $\mathrm{Pr}_{x''}$. Set $\tilde{x} := x' + 2^{-\ell+1} y''$. Let y be a sample from $\overline{\mathrm{Pr}}_{\tilde{x}}$. By construction, y has distribution $\overline{\mathrm{Pr}}_x$.

(A1): Let $j \in [n]$. By induction,

$$\Pr(y_j = 1) = \sum_{\varepsilon \in \{0,1\}} \Pr(y_j'' = \varepsilon) \Pr(y_j = 1 \,|\, y_j'' = \varepsilon)$$

$$= \sum_{\varepsilon \in \{0,1\}} \Pr(y_j'' = \varepsilon)(x_j' + 2^{-\ell+1}\varepsilon) = x_j.$$

(A2): Let $i \in [m_B]$. If $Bx'' \in \mathbb{Z}$, then $Bx = B\tilde{x}$ with probability one and By is a randomized rounding of Bx for both y being a sample from $\overline{\Pr}_x$ and $\overline{\Pr}_{\tilde{x}}$. If $Bx'' \notin \mathbb{Z}$, then $\Pr((Bx'')_i = (Bx'')_i + \frac{1}{2}) = \Pr((Bx'')_i = (Bx'')_i - \frac{1}{2}) = \frac{1}{2}$ by (A2). By induction, we have

$$\Pr((By)_i = \lfloor (Bx)_i \rfloor + 1)$$
$$= \Pr\left((B\tilde{x})_i = (Bx)_i + 2^{-\ell}\right) \Pr\left((By)_i = \lfloor (Bx)_i \rfloor + 1 \,\big|\, (B\tilde{x})_i = (Bx)_i + 2^{-\ell}\right) +$$
$$\Pr\left((B\tilde{x})_i = (Bx)_i - 2^{-\ell}\right) \Pr\left((By)_i = \lfloor (Bx)_i \rfloor + 1 \,\big|\, (B\tilde{x})_i = (Bx)_i - 2^{-\ell}\right)$$
$$= \tfrac{1}{2}(\{(Bx)_i\} + 2^{-\ell}) + \tfrac{1}{2}(\{(Bx)_i\} - 2^{-\ell}) = \{(Bx)_i\}.$$

(A3): Let $S \in \mathcal{S}$. Note that $\prod_{j \in S}(x_j' + 2^{-\ell+1}\varepsilon_j) + \prod_{j \in S}(x_j' + 2^{-\ell+1}(2x_j'' - \varepsilon_j)) \leq 2 \prod_{j \in S} x_j$ holds for all roundings ε of x''. Hence by induction and (A4),

$$\Pr(\forall j \in S : y_j = 1)$$
$$= \sum_{\varepsilon \in \{0,1\}^n} \Pr(y'' = \varepsilon) \Pr((\forall j \in S : y_j = 1) \,|\, y'' = \varepsilon)$$
$$\leq \sum_{\varepsilon \in \{0,1\}^n} \Pr(y'' = \varepsilon) \prod_{j \in S}(x_j' + 2^{-\ell+1}\varepsilon_j)$$
$$= \tfrac{1}{2} \sum_{\varepsilon \in \{0,1\}^n} \Pr(y'' = \varepsilon) \left(\prod_{j \in S}(x_j' + 2^{-\ell+1}\varepsilon_j) + \prod_{j \in S}(x_j' + 2^{-\ell+1}(2x_j'' - \varepsilon_j)) \right)$$
$$\leq \tfrac{1}{2} \sum_{\varepsilon \in \{0,1\}^n} \Pr(y'' = \varepsilon) \cdot 2 \prod_{j \in S} x_j = \prod_{j \in S} x_j.$$

A similar argument shows the claim for $b = 0$. □

3.2 Derandomizations

In this subsection, we extend the binary expansion method to the derandomization problem. A *randomized rounding derandomization* (with constant c) is an algorithm that computes for given $A \in [0,1]^{m_A \times n}$ and $x \in [0,1]^n$ a $y \in \{0,1\}^n$ such that for all $i \in [m_A]$,

$$|(Ax)_i - (Ay)_i| \leq c\sqrt{\max\{(Ax)_i, \ln(2m_A)\} \ln(2m_A)}.$$

It thus achieves (with minor loss) the existential bounds given by randomized rounding.

The following derandomizations are known.

(i) If $A \in \{0, 1\}^{m_A \times n}$ and $x \in \{0, \frac{1}{2}, 1\}^n$, then Spencer's [Spe94] method of conditional probabilities yields a straight-forward $O(m_A n)$–time derandomization with constant $c = \sqrt{\frac{1}{2}}$. Note that the conditional probabilities in this special case are easy to compute via binomial coefficients.

(ii) Raghavan's derandomization [Rag88] via so-called pessimistic estimators is more complicated, but allows a wider range of vectors. Still in time $O(m_A n)$, it achieves the constant $c = e - 1$. In the RAM model, it works for all $A \in \{0, 1\}^{m_A \times n}$ and $x \in ([0, 1] \cap \mathbb{Q})^n$. If one allows precise computations with real numbers in constant time (in particular exponential functions), then this extends to arbitrary $A \in [0, 1]^{m_A \times n}$.

(iii) Srivastav and Stangier [SS96] give a derandomization for all $A \in ([0, 1] \cap \mathbb{Q})^{m_A \times n}$ in the RAM model, though at the price of an increased run-time of $O(m_A n^2 \log(m_A n))$. Also, it is quite complicated from the view-point of implementation. The constant c is not explicitly stated in the paper, but by plugging in the inequality of Angluin-Valiant given there, one achieves $c = \sqrt{3}$.

(iv) In the last section of this paper, we show how to use the binary expansion ideas to obtain a relatively simple derandomization that works for all $A \in ([0, 1] \cap \mathbb{Q})^{m_A \times n}$ and $x \in ([0, 1] \cap \mathbb{Q})^n$ in time $O(m_A n \log n)$ in the RAM model. The constant in this case is $4(e - 1)(1 + o(1))$.

For $\ell \in \mathbb{N}$ and $c \in \mathbb{R}_{\geq 0}$ let $f(\ell, c) = c \sum_{i=1}^{\ell} 2^{-(i-1)/2} \prod_{j=i+1}^{\ell} (1 + 2^{-(j-1)/2} c)^{1/2}$.

Theorem 2 (Digit by digit derandomization). *Let $\mathcal{A}$ be an algorithm which for some matrix A and any $x \in \{0, \frac{1}{2}, 1\}^n$ computes a rounding y of x such that $By \approx Bx$ and*

$$\forall i \in [m_A] : |(Ax)_i - (Ay)_i| \leq c\sqrt{\max\{(Ax)_i, \ln(2m_A)\} \ln(2m_A)}.$$

Then for each $x \in [0, 1]^n$ having binary length ℓ, a rounding y such that $By \approx Bx$ and

$$\forall i \in [m_A] : |(Ax)_i - (Ay)_i| \leq f(\ell, c)\sqrt{\max\{(Ax)_i, \ln(2m_A)\} \ln(2m_A)}$$

can be computed by ℓ times invoking $\mathcal{A}$.

We omit the proof for reasons of space. Similar as in the proof of Theorem 1, we use induction over the length of the binary expansion. Some care has to be taken to control the size of $(A\tilde{x})_i$ for the intermediate roundings $\tilde{x}$.

We end this section with some rough estimates of the constants $f(c, \ell)$.

Lemma 2. $f(c) := \lim_{\ell \to \infty} f(\ell, c)$ *exists for all c and satisfies $f(c) = c^{O(\log c)}$. We have $f(\sqrt{\frac{1}{2}}) \leq 4$, $f(e - 1) \leq 18$, and $f(\sqrt{3}) \leq 19$.*

Let us remark that the increase in the constant in most cases in not as bad as $f(c) = c^{\Theta(\log c)}$ suggests. If $\log m_A = o((Ax)_i)$, then a closer look at the proof of Theorem 2 yields $|(Ax)_i - (Ay)_i| \leq 2(\sqrt{2} + 1)(1 + o(1))c\sqrt{(Ax)_i \ln(2m_A)}$. Hence asymptotically we only lose a factor of less than 5 in the large deviation bound. In fact, already if $(Ax)_i \geq c^2 \ln(2m_A)$, we obtain a bound of $|(Ax)_i - (Ay)_i| \leq 7c\sqrt{(Ax)_i \ln(2m_A)}$.

4   Randomized Roundings with Disjoint Constraints

We now use the binary expansion method developed in the previous section to generate randomized roundings that satisfy disjoint cardinality constraints. Hence throughout this section ;et $B \in \{0,1\}^{m_B \times n}$ and $\|B\|_1 := \max_j \sum_i |b_{ij}| = 1$. For the generation of the roundings, this is a microscopic extension of Srinivasan's [Sri01] result. The reader's focus should therefore be on the simplicity of our approach.

As should be clear by now, all we have to do is analyze the $\{0, \frac{1}{2}\}$ case. Let us assume that B is stored in some $O(n)$ space datastructure allowing amortized linear time enumerations of the sets $\{j \in [n] \mid b_{ij} = 1\}$ for all $i \in [m_B]$.

Lemma 3. *There are basic randomized roundings* $(\mathrm{Pr}_{x,B})$ *with respect to B and $2^{[n]}$. A sample from* $(\mathrm{Pr}_{x,B})$ *can be generated in time $O(n)$.*

Proof. Let $x \in \{0, \frac{1}{2}\}^n$. For $i \in [m_B]$ let $E_i := \{j \in [n] \mid x_j = \frac{1}{2}, b_{ij} = 1\}$. Choose a set $\mathcal{M}$ of disjoint 2–subsets of $[n]$ such that $|E_i \setminus \bigcup \mathcal{M}| \leq 1$ and $|M \cap E_i| \neq 1$ hold for all $i \in [m_B]$ and $M \in \mathcal{M}$. In other words, $\mathcal{M}$ is a maximal collection of disjoint 2–sets of $[n]$ that all intersect all E_i in a trivial way[1].

For each $\{j_1, j_2\}$ independently we flip a coin to decide whether $(y_{j_1}, y_{j_2}) = (1, 0)$ or $(y_{j_1}, y_{j_2}) = (0, 1)$. For all $j \in [n] \setminus \bigcup \mathcal{M}$ let y_j be a randomized rounding of x_j independent from all other random choices. The above defines a basic randomized rounding $(\mathrm{Pr}_{x,B})$ with respect to B and $2^{[n]}$. $\qquad\square$

From Theorem 1 and 3, the following is immediate.

Theorem 3. *There are randomized roundings* $(\overline{\mathrm{Pr}}_{x,B})$ *with respect to B and $2^{[n]}$. A sample from* $(\overline{\mathrm{Pr}}_{x,B})$ *can be generated in time $O(n\ell)$, where ℓ is the binary length of x.*

We now derandomize the construction above. Here the simpler, compared to previous work more sequential construction proves to be advantageous. As before, we only have to analyze the $0, \frac{1}{2}$ case.

Lemma 4. *Let A be an $m_A \times n$ matrix. Let $x \in \{0, \frac{1}{2}\}^n$. Then a binary vector y such that $By \approx Bx$ and*

$$\forall i \in [m_A] : |(Ax)_i - (Ay)_i| \leq 2c\sqrt{\max\{(Ax)_i, \ln(4m_A)\} \ln(4m_A)}$$

can be computed by applying a derandomization to a $2m_A \times n$ matrix with entries from $\{a_{ij} \mid i \in [m_A], j \in [m]\}$.

The proof is again omitted for reasons of space. The main idea is to note that the rounding errors inflicted by the matching rounding of Lemma 3 can be written as a weighted sum of binary random variables representing the coin flips.

Combining Theorem 2 and Lemma 4 with the derandomizations cited in Section 3.2, we obtain the following derandomized version of Srinivasan's results.

---

[1] As we will see, the particular choice of $\mathcal{M}$ is completely irrelevant. Assume therefore that we have fixed some deterministic way to choose it (e.g., greedily in the natural order of $[n]$).

Theorem 4. *Let* $A \in [0,1]^{m_A \times n}$. *Let* $x \in [0,1]^n$. *Then for all* $\ell \in \mathbb{N}$, *a binary vector* y *can be computed such that* $By \approx Bx$ *and*

$$\forall i \in [m_A] : |(Ax)_i - (Ay)_i| \leq f(\ell, 2c)\sqrt{\max\{(Ax)_i, \ln(4m_A)\}\ln(4m_A)} + n2^{-\ell}$$

This has a time complexity of ℓ *times the one of applying a derandomization to a* $2m_A \times n$ *matrix with entries in* $\{a_{ij} \mid i \in [m], j \in [n]\}$.

Some bounds on constants that are relevant in connection with the derandomizations mentioned in Subsection 3.2 are $f(2\sqrt{\frac{1}{2}}) \leq 13$, $f(2e-2) \leq 90$, and $f(2\sqrt{3}) \leq 92$. However, the remark following Lemma 2 also applies to the theorem above, i.e., for $(Ax)_i$ large compared to $\ln(4m_A)$, the increase in the constants become less significant.

5   Bipartite Edge Weight Rounding

In this section, we consider sets of cardinality constraints where each variable may be contained in up to two constraints. Throughout this section, let $B = \binom{B_1}{B_2}$, where the B_i are 0, 1 matrices such that $\|B_i\|_1 = 1$. We assume that B is represented by a datastructure allowing constant time queries of type "given i, find j such that $b_{ij} = 1$" and "given j, find i such that $b_{ij} = 1$".

For such constraints, negative correlation on $\mathcal{S} = 2^{[n]}$ is too much to ask for. We restrict ourselves to $\mathcal{S}_B = \{S \subseteq [n] \mid \exists i \in [m_B] \, \forall j \in S : b_{ij} = 1\}$.

Problems of this type have been regarded in Gandhi et al. [GKPS02]. They used a formulation as rounding problem for edge weights in bipartite graphs. We briefly fix the connection.

Bipartite edge weight rounding problem: Given a bipartite graph $G = (U \dot\cup V, E)$ and edge weights $x \in [0,1]^E$, find $y \in \{0,1\}^E$ such that (B1) y_e is a randomized rounding of x_e for all $e \in E$, (B2) $\sum_{e \ni v} y_e \approx \sum_{e \ni v} x_e$ for all $v \in U \cup V$ and (B3) for all $v \in U \cup V$, $S \subseteq \{e \in E \mid v \in E\}$ and $b \in \{0,1\}$, we have $\Pr(\forall i \in S : y_e = b) \leq \prod_{e \in S} \Pr(y_e = b)$.

The bipartite edge weight rounding problem is easily seem to be captured by our setting: Define $B_1 = (b_{ue}) \in \{0,1\}^{U \times E}$ through $b_{ue} = 1$ if and only if $u \in e$ as well as $B_2 = (b_{ve}) \in \{0,1\}^{V \times E}$ through $b_{ve} = 1$ if and only if $v \in e$. Then $By \approx Bx$ for some $x \in [0,1]^E$, $y \in \{0,1\}^E$ is just the degree preservation condition (B2). Also, negative correlation on $\mathcal{S}_B$ is equivalent to (B3).

The bipartite edge weight rounding problem for edge weights 0 ('no edge', if you like) and $\frac{1}{2}$ is easily solved. Here the pipage rounding idea of [AS04] fixes each variable to an integer value in amortized constant time. This saves an $O(|V|)$ run-time factor compared to the general case.

Lemma 5. *There are basic randomized roundings with respect to* B *and* $\mathcal{S}$. *They can be sampled in time* $O(n)$.

The lemma above together with the general reduction of Theorem 1 yields the following version of the bipartite edge weight rounding result of Gandhi et al. Note that the time complexity here is superior to the $O(|E||V|)$ bound of Gandhi et al. [GKPS02] (unless we are working with an overly high precision ℓ).

Theorem 5. *There are randomized roundings* $(\overline{\mathrm{Pr}}_{x,B})$ *with respect to B and* $2^{[n]}$. *A sample from* $(\overline{\mathrm{Pr}}_{x,B})$ *can be generated in time $O(n\ell)$, where ℓ is the binary length of x.*

Again, the randomized algorithm above can be derandomized.

Lemma 6. *Let $A \in [0,1]^{m_A \times n}$. Assume that for each $i_A \in [m_A]$ there is an $i_B \in [m_B]$ such that for all $j \in [n]$, $b_{i_B j} = 1$ whenever $a_{i_A j} \neq 0$. Let $x \in \{0, \frac{1}{2}\}^n$. Then a binary vector y such that $By \approx Bx$ and*

$$|(Ax)_i - (Ay)_i| \leq 2c\sqrt{\max\{(Ax)_i, \ln(4m_A)\}\ln(4m_A)}$$

for all $i \in [m_A]$ can be computed by applying a derandomization to a matrix of dimension at most $2m_A \times n$ with entries from $\{a_{ij} \mid i \in [m_A], j \in [n]\}$.

Combining the Lemma 6 with Theorem 2, we obtain the following derandomization of the result of Gandhi et al.

Theorem 6. *Let $A \in [0,1]^{m_A \times n}$ such that for each $i_A \in [m_A]$ there is an $i_B \in [m_B]$ such that for all $j \in [n]$, $b_{i_B j} = 1$ whenever $a_{i_A j} \neq 0$. Then for all $\ell \in \mathbb{N}$, a binary vector y can be computed such that $By \approx Bx$ and*

$$\forall i \in [m_A] : |(Ax)_i - (Ay)_i| \leq f(\ell, 2c)\sqrt{\max\{(Ax)_i, \ln(4m_A)\}\ln(4m_A)} + n2^{-\ell}.$$

The time complexity is ℓ times the one of a derandomization for $2m \times n$ matrices with entries from $\{a_{ij} \mid i \in [m_A], j \in [m]\}$.

6   Other Constraints

It is relatively easy to see that Theorems 3, 4, 5 and 6 can be extended to include hard constraint matrices $B \in \{-1, 0, 1\}$ as long as B is totally unimodular. An extension to further values, however, is not possible. Also, Theorems 3 and 4 can be extended to other sparsely intersecting constraints than the ones of Section 5. We now give two examples involving substantially different hard constraints.

Sequence Rounding. In connection with an image processing application, Sadakane, Takki-Chebihi and Tokuyama [STT01] compute roundings of sequences such that the rounding errors in all intervals are less than one. This is in fact a classical problem, but the new aspect in their work is that they need a randomized solution as this is less likely to produce unwanted structures in the images. The approach taken in [STT01] is via efficiently computing several roundings and then taking a random one.

A simpler way using the framework of this paper is to compute a randomized rounding y of $x \in [0,1]^n$ with the additional constraints that for each interval $I \subseteq [n]$, $\sum_{i \in I} y_i$ is a randomized rounding of $\sum_{i \in I} x_i$. To do so, we have to understand this problem for $0, \frac{1}{2}$ sequences, which is trivial.

Matrix Rounding. Asano et al. [AKOT03] model the digital halftoning problem as matrix rounding problem. For $X \in [0,1]^{m \times n}$ and $Y \in \{0,1\}^{m \times n}$, they set

$$d(X,Y) := \sum_{\substack{i \in [m-1] \\ j \in [n-1]}} \left| \sum_{k,\ell \in \{0,1\}} (x_{i+k,j+\ell} - y_{i+k,j+\ell}) \right|.$$

They claim that the image represented by Y is a good halftoning of the image represented by X, if $d(X, Y)$ is small. The current best solution for computing good roundings with respect to this error measure uses dependent randomized roundings [Doe04]. Let Y be a randomized rounding of X with respect to the constraints[2]

$$y_{i,j} + y_{i,j+1} \approx x_{i,j} + x_{i,j+1}, i \in [m], j \in [n-1],$$
$$y_{i,j} + y_{i+1,j} \approx x_{i,j} + x_{i+1,j}, i \in [m-1] \text{ odd}, j \in [n],$$
$$y_{i,j} + y_{i,j+1} + y_{i+1,j} + y_{i+1,j+1} \approx x_{i,j} + x_{i,j+1} + x_{i+1,j} + x_{i+1,j+1},$$
$$i \in [m-1] \text{ odd}, j \in [n-1].$$

Then $E(d(X, Y)) \leq 0.55$ holds for all X. The existence of such dependent roundings easily follows from the totally unimodularity condition. However, actually computing them in linear time involves tedious case distinctions.

With the reduction of Theorem 1, life is much easier since we only have to regard $X \in \{0, \frac{1}{2}\}^{m \times n}$. In this case, a constraint of the third type is either implied by constraints of the first two kinds, or it contains exactly two non-integral variables. All constraints thus yield a bipartite graph $G = ([m] \times [n], E)$ with $\{(i_1, j_1), (i_2, j_2)\} \in E$ telling us that exactly one of y_{i_1,j_1} and y_{i_2,j_2} has to become one, and these are all constraints. This makes it easy to compute such a rounding: For each connected component of G, flip a fair coin to decide which of the two classes of the bipartition shall be rounded to one, and set the other variables to zero.

7   A Word of Warning

We have to note that dependencies like cardinality constraints inflict that some natural properties are unexpectedly not satisfied. Call a function $f : \{0,1\}^n \to \mathbb{R}$ non-decreasing if $y \leq z$ (component-wise) implies $f(y) \leq f(z)$.

(i) There are $S \subseteq [n]$ such that the roundings of Section 4 and 5 make $x \mapsto \Pr_x(\forall i \in S : y_i = 1)$ *not* non-decreasing.

(ii) The compared to (A3) stronger property that for all disjoint $S, T \subseteq [n]$, one has $\Pr_x(y_{|S} \equiv 1 \,|\, y_{|T} \equiv 1) \leq \Pr_x(y_{|S} \equiv 1)$ also does not hold for the roundings of Section 4 and 5. This is the reason why in [GKPS02] this property could only be proven for a single cardinality constraint and only by prescribing a particular order for the individual roundings.

(iii) There are non-decreasing functions f such that *any* randomized rounding with respect to a single cardinality constraint makes $x \mapsto E_x(f)$ not non-decreasing.

All these phenomena, of course, are not possible for independent randomized roundings.

Let us also mention the following. A distribution on $\{0,1\}^n$ is negatively associated (NA), if for all non-decreasing $f, g : \{0,1\}^n \to \mathbb{R}_{\geq 0}$ we have $E(fg) \leq E(f)E(g)$. The distributions of Section 4 and 5 are not (NA).

---

[2] Here we use the notation $y \approx x$ to denote that y is a randomized rounding of x.

8   General Derandomization

In this section, we improve and simplify the derandomization result of Stangier and Srivastav [SS96]. Recall from Section 3.2 that Raghavan's derandomization in the RAM model only works for binary matrices. This problem was solved in [SS96], though at the price of a significantly higher time complexity of $O(mn^2 \log(mn))$. Also, this approach is hard to implement due to its technical demands. We overcome these difficulties by reducing the general problem to Raghavan's setting and obtain the following result.

Theorem 7. *Let* $A \in ([0,1] \cap \mathbb{Q})^{m \times n}$, $x \in ([0,1] \cap \mathbb{Q})^n$ *and* $\ell \in \mathbb{N}$. *Then a* $y \in \{0,1\}^n$ *such that*

$$|(Ax)_i - (Ay)_i| \leq 2(e-1)\sqrt{\max\left\{(Ax)_i, \ln(2\ell m)\right\} \ln(2\ell m)} + 2^{-\ell} n$$

holds for all $i \in [m]$ *can be computed in time* $O(mn\ell)$ *in the RAM model.*

Proof (Sketch). Use the binary expansion $A = \sum_{k=1}^{\ell} 2^{-k} A^{(k)}$ of A and apply Raghavan's derandomization to the $\ell m \times n$ matrix obtained from stacking the $A^{(i)}$.   □

Note that if we choose $\ell = \lceil \log_2 n \rceil$, the additive extra term is at most one. Note also, that then the $\ln(2\ell m)$ term is just a factor away from the usual $\ln(2m)$: We may assume $m \geq \log n$. Otherwise using linear algebra we may transform x into a vector x' such that $Ax = Ax'$ and at most m components of x' are not 0 or 1. But if $\ell = \lceil \log_2 n \rceil \leq 2m$, then $\ln(2\ell m) \leq 2\ln(2m)$.

Finally, note the following. By combining Lemma 4 with the elementary derandomization for $\{0, \frac{1}{2}, 1\}$ vectors in Section 3.2, we obtain a very elementary and simple to implement algorithm for arbitrary vectors and binary matrices.

References

[AFK02]   S. Arora, A. Frieze, and H. Kaplan. A new rounding procedure for the assignment problem with applications to dense graph arrangement problems. *Math. Program.*, 92:1–36, 2002.

[AKOT03]   T. Asano, N. Katoh, K. Obokata, and T. Tokuyama. Matrix rounding under the L_p-discrepancy measure and its application to digital halftoning. *SIAM J. Comput.*, 32:1423–1435, 2003.

[AS04]   A. A. Ageev and M. I. Sviridenko. Pipage rounding: a new method of constructing algorithms with proven performance guarantee. *J. Comb. Optim.*, 8:307–328, 2004.

[BS84]   J. Beck and J. Spencer. Integral approximation sequences. *Math. Programming*, 30:88–98, 1984.

[Doe03]   B. Doerr.   Non-independent randomized rounding.   In *Proceedings of the 14th Annual ACM-SIAM Symposium on Discrete Algorithms (SODA)*, pages 506–507, 2003.

[Doe04]   B. Doerr.   Nonindependent randomized rounding and an application to digital halftoning. *SIAM Journal on Computing*, 34:299–317, 2004.

[Doe05]   B. Doerr. Roundings respecting hard constraints. In *Proceedings of the 22nd Annual Symposium on Theoretical Aspects of Computer Science (STACS'05)*, volume 3404 of *Lecture Notes in Computer Science*, pages 617–628, Berlin–Heidelberg, 2005. Springer-Verlag.

[GHK+03]   R. Gandhi, E. Halperin, S. Khuller, G. Kortsarz, and A. Srinivasan. An improved approximation algorithm for vertex cover with hard capacities. In J. Baeten, J. Lenstra, J. Parrow, and G. Woeginger, editors, *Automata, Languages and Programming. 30th International Colloquium, ICALP 2003*, volume 2719 of *Lecture Notes in Computer Science*, pages 164–175, Berlin–Heidelberg, 2003. Springer Verlag.

[GKPS02]   R. Gandhi, S. Khuller, S. Parthasarathy, and A. Srinivasan. Dependent rounding in bipartite graphs. In *Proc. IEEE Symposium on Foundations of Computer Science (FOCS)*, pages 323–332, 2002.

[GKR+99]   V. Guruswami, S. Khanna, R. Rajaraman, B. Shepherd, and M. Yannakakis. Near-optimal hardness results and approximation algorithms for edge-disjoint paths and related problems. In *Annual ACM Symposium on Theory of Computing (STOC)*, pages 19–28, New York, 1999. ACM.

[PS97]   A. Panconesi and A. Srinivasan. Randomized distributed edge coloring via an extension of the Chernoff-Hoeffding bounds. *SIAM J. Comput.*, 26:350–368, 1997.

[Rag88]   P. Raghavan. Probabilistic construction of deterministic algorithms: Approximating packing integer programs. *J. Comput. Syst. Sci.*, 37:130–143, 1988.

[RT87]   P. Raghavan and C. D. Thompson. Randomized rounding: A technique for provably good algorithms and algorithmic proofs. *Combinatorica*, 7:365–374, 1987.

[RT91]   P. Raghavan and C. D. Thompson. Multiterminal global routing: a deterministic approximation scheme. *Algorithmica*, 6:73–82, 1991.

[Spe94]   J. Spencer. *Ten lectures on the probabilistic method*, volume 64 of *CBMS-NSF Regional Conference Series in Applied Mathematics*. Society for Industrial and Applied Mathematics (SIAM), Philadelphia, PA, 1994.

[Sri01]   A. Srinivasan. Distributions on level-sets with applications to approximations algorithms. In *Proc. 41th Ann. IEEE Symp. on Foundations of Computer Science (FOCS)*, pages 588–597, 2001.

[SS96]   A. Srivastav and P. Stangier. Algorithmic Chernoff-Hoeffding inequalities in integer programming. *Random Structures & Algorithms*, 8:27–58, 1996.

[STT01]   K. Sadakane, N. Takki-Chebihi, and T. Tokuyama. Combinatorics and algorithms on low-discrepancy roundings of a real sequence. In *ICALP 2001*, volume 2076 of *Lecture Notes in Computer Science*, pages 166–177, Berlin Heidelberg, 2001. Springer-Verlag.

Online Sorting Buffers on Line

Rohit Khandekar[1] and Vinayaka Pandit[2]

[1] University of Waterloo, ON, Canada
rkhandekar@gmail.com
[2] IBM India Research Laboratory, New Delhi, India
pvinayak@in.ibm.com

Abstract. We consider the online scheduling problem for sorting buffers on a line metric, motivated by an application to disc scheduling. Input is an online sequence of requests. Each request is a block of data to be written on a specified track of the disc. To write a block on a particular track, the scheduler has to bring the disc head to that track. The cost of moving the disc head from a track to another is the distance between those tracks. A sorting buffer that can store at most k requests at a time is available to the scheduler. This buffer can be used to rearrange the input sequence. The objective is to minimize the total cost of head movement while serving the requests. On a disc with n uniformly-spaced tracks, we give a randomized online algorithm with a competitive ratio of $O(\log^2 n)$ in expectation against an oblivious adversary. We show that any deterministic strategy which makes scheduling decisions based only on the contents of the buffer has a competitive ratio of $\Omega(k)$ or $\Omega(\log n / \log \log n)$.

1 Introduction

Disc scheduling is a fundamental problem in the design of storage systems. Standard text books on operating systems discuss various heuristics for scheduling the movement of the disc head. It is to be noted that the difference in the performance of different scheduling strategies can be seen only when a buffer which can hold more than one request in order to rearrange the sequence is available (See section 13.2 in [10]). In this paper, we consider the problem of efficiently scheduling the head movement of a disc when a buffer of limited size is available to rearrange an online sequence of requests. The access time in disc scheduling has two components, namely *seek time* and *rotational latency*. Seek time is the time required to move between tracks. The seek time can be reliably estimated using a straight line metric. The rotational latency is the time required for the required sector to move underneath the disc head. It is difficult to estimate rotational latency reliably. So, we consider disc scheduling with only seek time and model the disc as a number of tracks arranged on a straight line. The input is a sequence of requests. Each request is a block of data and specifies the track on which the data needs to be written. Available to the scheduler is a buffer which can store at most k requests at a time. The buffer can be used to rearrange the input sequence. To write a block of data onto a track, the disc head has to be

B. Durand and W. Thomas (Eds.): STACS 2006, LNCS 3884, pp. 584–595, 2006.

moved to that track. The cost of moving the head from track i to track j is assumed to be $|i - j|$. The goal of the scheduler is to serve all the requests while minimizing the overall cost of head movement. We call this problem the Sorting Buffers problem (SBP) on a line metric to be consistent with the previous work. Note that, when unlimited buffer size is available, the optimal schedule can be found offline by simply sorting the requests.

The Disc Scheduling problem is well studied in the design of storage systems. Shortest Seek First (SSF), Shortest Time First(STF), and CSCAN are some of the popular heuristics. SSF strategy, at each step, schedules the request with the shortest seek time among all the pending requests. STF strategy, at each step, schedules the request with the minimum of seek time plus rotational latency among all the pending requests. In the CSCAN schedule, the head starts from one end of the disc and travels to the other end, servicing all the requests on a track while passing it. After reaching the other end, it moves back to its starting point and repeats. Note that the CSCAN may violate buffer constraint. However, this problem has not been studied from the point of view of approximation guarantees. Andrews et al. [1] studied a related disc scheduling problem in the offline setting. They consider a model in which a convex reachability function specifies the time taken for the head to move between two tracks. Given a set of requests, they consider the problem of minimizing the time required to serve all the requests. Unlike the SBP, their problem does not have buffer constraint. They gave a $\frac{3}{2}$ approximation for the disc scheduling problem with convex reachability function in which there is no buffer constraint. They show that the problem can be solved optimally in polynomial time if the reachability function is linear. They leave the online problem with buffer constraint as an open problem.

1.1 Previous Work

The SBP can be defined on any metric space. The input is a sequence of requests each of which corresponds to a point in the metric space. To serve a request after its arrival, the server has to visit the corresponding point. The cost of moving the server from a point to another is the distance between those points. The sorting buffer of size k can be used to rearrange the sequence so as to minimize the total movement of the server to serve all the requests. Let N denote the total number of requests. It is easy to see that if $k = N$ and there is one request to each point, then this problem is essentially the Hamiltonian path problem on the given metric. Thus the offline version of SBP on general metrics is NP-hard. On a line metric, however, it is not known if the offline version is NP-hard for a general k. To the best of our knowledge, no non-trivial lower or upper bounds on the approximation (resp. competitive) ratio of either deterministic or randomized offline (resp. online) algorithms are known for a line metric.

The SBP on a uniform metric (in which all pairwise distances are 1) has been studied before. This problem is interesting only when multiple requests are allowed for the points in the metric space. Räcke et al. [9] presented a deterministic online algorithm, called *Bounded Waste* that has $O(\log^2 k)$ competitive ratio. They also showed that some natural strategies like FIFO, LRU etc. have $\Omega(\sqrt{k})$

competitive ratio. Englert and Westermann [6] considered a generalization of the uniform metric in which moving to a point p from any other point in the space has a cost c_p. They proposed an algorithm called Maximum Adjusted Penalty (MAP) and showed that it gives an $O(\log k)$ approximation, thus improving the competitive ratio of the SBP on uniform metric. Kohrt and Pruhs [8] also considered the uniform metric but with different optimization measure. Their objective was to maximize the reduction in the cost from that of the schedule without a buffer. They presented a 20-approximation algorithm for this variant and this ratio was improved to 9 by Bar-Yehuda and Laserson [2]. It is not known if the offline version of SBP on the uniform metric is NP-hard.

The offline version of the sorting buffers problem on any metric can be solved optimally using dynamic programming in $O(N^{k+1})$ time where N is the number of requests in the sequence. This follows from the observation that the algorithm can pick k requests to hold in the buffer from first i requests in $\binom{i}{k}$ ways when the $(i+1)$th request arrives. Suppose there is a constraint that a request has to be served within D time steps of it being released, then, the dynamic program can be modified to compute the optimal schedule in $O(D^{k+1})$ time.

The dial-a-ride problem with finite capacity considered by Charikar and Raghavachari [5] is related to the SBP. In this problem, the input is a sequence of requests each of which is a source-destination pair in a metric space on n points. Each request needs an object to be transferred from its source to its destination. The goal is to serve all the requests using a vehicle of capacity k so that total length of the tour is minimized. The non-preemptive version requires that once an object is picked, it can be dropped only at its destination while in the preemptive version the objects can be dropped at intermediate locations and picked up later. Charikar and Raghavachari [5] give approximation algorithms for both preemptive and non-preemptive versions using Bartal's metric embedding result. Note, however, that the SBP enforces a sequencing constraint that is not imposed by the dial-a-ride problem. In the SBP, if we are serving the ith request, then it is necessary that at most k requests from first to $(i-1)$th request be outstanding. Whereas, in the dial-a-ride problem, the requests can be served in any order as long as the schedule meets the capacity constraint. Therefore their techniques are not applicable to the SBP. They give $O(\log n)$ approximation for the preemptive case and $O(\sqrt{k}\log n)$ approximation for the non-preemptive case. The capacitated vehicle routing problem considered by Charikar et al. [4] is a variant in which all objects are identical and hence an object picked from a source can be dropped at any of the destinations. They give the best known approximation ratio of 5 for this problem and survey the previous results.

1.2   Our Results

In Section 3, we show that natural strategies like FIFO and Nearest-First have a competitive ratio of $\Omega(k)$. We also show that any deterministic algorithm that takes decisions just based on the unordered set of requests which are currently in the buffer has a competitive ratio of $\Omega(k)$. The same proof implies a lower bound of $\Omega(\log n / \log \log n)$ for deterministic algorithms that take decisions just based on the unordered set of requests currently in the buffer.

Next in Section 4, we provide the first non-trivial competitive ratio for the online SBP on a line metric. For a line metric $\{1, \ldots, n\}$ with distance between i and j being $|i - j|$, we present a randomized online algorithm with a competitive ratio of $O(\log^2 n)$ in omputation against an oblivious adversary. This algorithm also yields a competitive ratio of $O(\alpha^{-1} \log^2 n)$ if we are allowed to use a buffer of size αk for any $1 \leq \alpha \leq \log n$.

Our algorithm is based on the probabilistic embedding of the line metric into the so-called hierarchical well-separated trees (HSTs) first introduced by Bartal [3]. He showed that any metric on n points can be probabilistically approximated within a factor of $O(\log n \log \log n)$ by metrics on HSTs. This factor was later improved to $O(\log n)$ by Fakcharoenphol et al. [7]. It is easy to see that the line metric $\{1, \ldots, n\}$ can be probabilistically approximated within a factor of $O(\log n)$ by the metrics induced by binary trees of depth $1 + \log n$ such that the edges in level i have length $n/2^i$. We provide a simple lower bound on the cost of the optimum on a tree metric by counting how many times it must cross a particular edge in the tree. Using this lower bound, we prove that the expected cost of our algorithm is within the factor of $O(\log^2 n)$ of the optimum.

Our algorithm generalizes to "line-like" metrics. More precisely, consider a metric such that for every subset of points in the metric space, cost of the minimum spanning tree on that subset is within α times the diameter of that subset. For such a metric on n points, our algorithm is $O(\alpha \log n \log D)$ competitive in expectation, where D is the aspect ratio, i.e., the ratio of the maximum pairwise distance to the minimum pairwise distance in the metric.

2   The Sorting Buffers Problem

Let (V, d) be a metric on n points. The input to the Sorting Buffers Problem (SBP) consists of a sequence of N requests. The ith request is labeled with a point $p_i \in V$. There is a server, initially located at a point $p_0 \in V$. To serve ith request, the server has to visit p_i after its arrival. There is a sorting buffer which can hold up to k requests at a time. The first k requests arrive initially. The $(i + 1)$th request arrives after we have served at least $i + 1 - k$ requests among the first i requests for $i \geq k$. Thus we can keep at most k requests pending at any time. The output is such a legal schedule of serving the requests. More formally, the output is given by a permutation π of $1, \ldots, N$ where the ith request to be served is denoted by $\pi(i)$. Since we can keep at most k requests pending at a time, a legal schedule must satisfy that the ith request to be served must be among first $i + k - 1$ requests arrived, i.e., $\pi(i) \leq i + k - 1$. The cost of the schedule is the total distance that the server has to travel, i.e., $C_\pi = \sum_{i=1}^{N} d(p_{\pi(i)}, p_{\pi(i-1)})$ where $\pi(0) = 0$ corresponds to the starting point. The goal in SBP is to find a legal schedule π that minimizes C_π where the $(i + 1)$th request is revealed only after serving at least $i - k + 1$ requests for $i \geq k$.

2.1   The Disc Model

Overlooking rotational latency, a disc is modeled as an arrangement of tracks numbered $1, \ldots, n$. The time taken to move the head from track i to track j is assumed to be $|i - j|$. The Disc Scheduling problem is the SBP on the line metric space $(\{1, \ldots, n\}, d)$ where $d(i, j) = |i - j|$. It is not known if the offline problem is NP-hard. We argue in Section 4.1 that the algorithm that serves the requests in the order they arrive is an $O(k)$-approximation. To the best of our knowledge, no algorithm with a better guarantee was known before.

3   Why Some Natural Strategies Fail

Many natural deterministic strategies suffer from one of the following drawbacks.

1. Some strategies block large part of the sorting buffer with requests that are kept pending for a long time. Doing this, the effective buffer size drops well below k, thereby yielding a bad competitive ratio. For example, consider the Nearest-First (also called Shortest-Trip-First (STF)) strategy that always serves the request nearest to the current head location. Suppose that the initial head location is 0. Let the input sequence be $3, \ldots, 3, 1, 0, 1, 0, \ldots$ where there are $k-1$ requests to 3. The STF strategy keeps the $k-1$ requests to 3 pending and serves the 1s and 0s alternatively using an effective buffer of size 1. The optimum schedule, on the other hand, gets rid of the requests to 3 by making a single trip to 3 and uses the full buffer to serve the remaining sequence. It is easy to see that if the sequence of 1s and 0s is large enough, the cost of STF is $\Omega(k)$ times that of the optimum.

2. Some other strategies, in an attempt to free the buffer slots, travel too far too often. The optimum, however, saves the distant requests and serves about k of them at once by making a single trip. Consider, for example, the First-In-First-Out (FIFO) strategy that serves the requests in the order they arrive. Suppose again that the initial head location is 0 and the input sequence is a repetition of the following block of requests: $n, 1, \ldots, 1, 0, \ldots, 0$ where there are k requests to 1 and k requests to 0 in each block. The FIFO strategy makes a trip to n for each block while the optimum serves 1s and 0s for $k-1$ blocks and then serves the accumulated $k-1$ requests to n by making a single trip to n. Note that, in doing this, the effective buffer size for the optimum reduces from k to 1. However, for the sequence of k 1s followed by k 0s, having a buffer of size k is no better than having buffer of size 1. It is now easy to see that if $n = k$, then FIFO is $\Omega(k)$ worse than the optimum.

Thus a good strategy must necessarily strike a balance between clearing the requests soon to free the buffer and not traveling too far too often. Obvious combinations of the two objectives, like making decisions based on the ratios of the distance traveled to the number of requests served also fail on similar input instances. We refer the reader to Räcke et al. [9] for more discussion.

3.1 Memoryless Deterministic Algorithms

We call a deterministic algorithm *memoryless* if it makes its scheduling decisions based purely on the set of pending requests in the buffer. Such an algorithm can be completely specified by a function ρ such that for every possible (multi-)set S of k requests in the buffer, the algorithm picks a request $\rho(S) \in S$ to serve next. The Nearest-First strategy is an example of a memoryless algorithm.

Theorem 1. *Any memoryless algorithm is $\Omega(k)$-competitive.*

Proof. We consider $k+1$ points $S = \{1, k, k^2, \ldots, k^k\}$ on a line. Consider any deterministic memoryless algorithm A. Suppose the head is initially at 1. We start with a request at each of the points $k, \ldots, k^k$. Whenever A moves the head from location p_i to location p_j, we release a new request at p_i. Thus, at all times, the pending requests and the head location together span all the $k+1$ points in S. Construct a directed graph G on S as vertices as follows. Add an edge from p_i to p_j if from the configuration with the head at p_i (with pending requests at all other points), A moves the head to p_j. Observe that the out-degree of every vertex is one and that G must have a cycle reachable from 1.

Suppose there is a cycle of length two between points $p_i, p_j \in S$ that is reachable from 1. In this case, we give an input sequence as follows. We first follow the path from 1 to p_i so that the head now resides at p_i. We then give a long sequence of the requests of the form $p_j, p_i, p_j, p_i, \ldots$. A serves p_js and p_is alternatively, keeping the other $k-1$ requests pending in the buffer. The optimum algorithm will instead clear the other $k-1$ requests first and use the full buffer to save an $\Omega(k)$ factor in the cost. Note that this situation is "blocking-the-buffer" (item 1) in the discussion on why some strategies fail.

Suppose, on the other hand, all the cycles reachable from 1 are of length greater than two. Consider such a cycle C on $p_1, p_2, \ldots p_c \in S$ where $c > 2$ and $p_1 > p_2, \ldots, p_c$. Note that the edges (p_1, p_2) and (p_c, p_1) have lengths that are $\Omega(k)$ times the total lengths of all the other edges in C. We now give an input sequence as follows. We first make A bring the head at p_1 and then repeat the following block of requests several times: $p_2, p_3, \ldots, p_c$. For each such block, A makes a trip of C while the optimum serves $p_2, \ldots, p_c$ repeatedly till it accumulates $k-1$ requests to p_1 and then clears all of them in one trip to p_1. Thus overall it saves an $\Omega(k)$ factor in the cost over A. Note that this situation is "too-far-too-often" (item 2) in the discussion on why some strategies fail.

Thus, any deterministic memoryless strategy has a competitive ratio of $\Omega(k)$.

Note that $n = k^k$ in the above example. Thus the lower bound we proved in terms of n is $\Omega(\frac{\log n}{\log \log n})$.

4 Algorithm for Sorting Buffers on a Line

4.1 A Lower Bound on **OPT** for a Tree Metric

Consider first an instance of SBP on a two-point metric $\{0, 1\}$ with $d(0, 1) = 1$. Let us assume that $p_0 = 0$. There is a simple algorithm that behaves optimally on this metric space. It starts by serving all requests at 0 till it accumulates k requests to 1. It then makes a transition to 1 and keeps serving requests to 1 till k requests to 0 are accumulated. It then makes a transition to 0 and repeats. It is easy to see that this algorithm is optimal. Consider the First-In-First-Out algorithm that uses a buffer of size 1 and serves any request as soon as it arrives. It is easy to see that this algorithm is $O(k)$-competitive since it makes $O(k)$ trips for every trip of the optimum algorithm.

We use $\mathsf{OPT}(k)$ to denote both the optimum algorithm with a buffer of size k and its cost. The following lemma states the relationship between $\mathsf{OPT}(\cdot)$ over different buffer sizes for a two-point metric.

Lemma 1. *For a two-point metric, for any $1 \le l \le k$, we have $\mathsf{OPT}(k) \ge \mathsf{OPT}(\lceil k/l \rceil)/2l$.*

Proof. Let $p_0 = 0$. By the time $\mathsf{OPT}(\lceil k/l \rceil)$ makes l trips to 1, we know that $\mathsf{OPT}(k)$ must have accumulated at least k requests to 1. Therefore $\mathsf{OPT}(k)$ must travel to 1 at least once. Note that the l (round) trips to 1 cost at most $2l$ for $\mathsf{OPT}(\lceil k/l \rceil)$. We can repeat this argument for every trip of $\mathsf{OPT}(k)$ to conclude the lemma.

Using the above observations, we now present a lower bound on the optimum cost for the sorting buffers problem on a tree metric.[1] Consider a tree T with lengths $d_e \ge 0$ assigned to the edges $e \in T$. Let $p_1, \ldots, p_N$ denote the input sequence of points in the tree. Refer to Figure 1. Fix an edge $e \in T$. Let L_e and

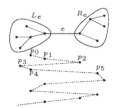

Fig. 1. Lower bound contributed by edge e in the tree

R_e be the two subtrees formed by removing e from T. We can shrink L_e to form a super-node 0 and shrink R_e to form a super-node 1 to obtain an instance of SBP on a two-point metric $\{0, 1\}$ with $d(0, 1) = d_e$. Let LB_e denote the cost of the optimum on this instance. It is clear that any algorithm must spend at least LB_e for traveling on edge e. Thus

$$\mathsf{LB} = \sum_{e \in T} \mathsf{LB}_e \tag{1}$$

is a lower bound on the cost OPT of the optimum on the original tree instance. Again, the algorithm that serves the requests in the order they arrive in, is $O(k)$-competitive.

---

[1] Recall that a tree metric is a metric on the set of vertices in a tree where the distances are defined by the path lengths between the pairs of points.

Let $\mathsf{LB}(k)$ denote the above lower bound on $\mathsf{OPT}(k)$, the optimum with a buffer of size k. The following lemma follows from Lemma 1.

Lemma 2. *For any $1 \leq l \leq k$, we have $\mathsf{LB}(k) \geq \mathsf{LB}(\lceil k/l \rceil)/2l$.*

4.2   An Algorithm on a Binary Tree

Consider a rooted binary tree T on $n = 2^h$ leaves. The height of this tree is $h = \log n$. The edges are partitioned into levels 1 to h according to their distance from the root; the edges incident to the root are in level 1 while the edges incident to the leaves are in level h. Figure 2 shows such a tree with $n = 8$ leaves. Let each edge in level i have cost $n/2^i$. Consider a metric on the leaves of this tree

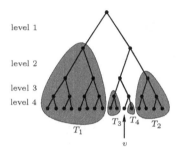

Fig. 2. Partition of the leaves in a phase of the algorithm

defined by the path lengths. In this section, we present a deterministic online algorithm for SBP on this metric that has a competitive ratio of $O(\log^2 n)$. Since the First-In-First-Out algorithm has a competitive ratio of $O(k)$, we assume that $k > h = \log n$. We also assume for simplicity that h divides k.

Algorithm. The algorithm goes in phases. Suppose that in the beginning of a phase, the server is present at a leaf v as shown in Figure 2. We partition the leaves other than v into h subsets as shown in the figure. Consider the path P_v from v to the root. Let T_i be the tree hanging to the path P_v at level i for $1 \leq i \leq h$. Let V_i be the set of leaves in T_i. We think of the sorting buffer of size k as being divided into h sub-buffers of size k/h each. We associate the ith sub-buffer with V_i, i.e., we accumulate all the pending requests in V_i in the ith sub-buffer. The algorithm maintains the following invariant.

> Invariant. Each of the h sub-buffers has at most k/h requests, i.e., there are at most k/h pending requests in any V_i.

We input new requests till one of the sub-buffers overflows. Suppose that the jth the sub-buffer overflows. The algorithm then clears all the pending requests in the subtrees $T_j, T_{j+1}, \ldots, T_h$ by performing an Eulerian tour of the trees $T_j, \ldots, T_h$. At the end of the tour, the head is at an arbitrary leaf of T_j. The algorithm, then, enters the next phase.

To prove the correctness of the algorithm we have to argue that at most k requests are pending at any point in the algorithm.

Lemma 3. *The invariant is satisfied in the beginning of any phase.*

Proof. Initially, the invariant is trivially satisfied. Suppose that it is satisfied in the beginning a phase and that the jth sub-buffer overflows in that phase. The division into trees and sub-buffers changes after the move. However since the server resides in T_j, all the trees $T_1, \ldots, T_{j-1}$ and their corresponding sub-buffers remain unchanged. Also since the algorithm clears all the pending requests in the trees $T_j, \ldots, T_h$, the jth to hth sub-buffers after the move are all empty. Thus the invariant is also satisfied in the beginning of the next phase.

Next we argue that the algorithm is $O(\log^2 n)$ competitive.

Theorem 2. *The total distance traveled by the server in the algorithm is* $O(\mathsf{OPT} \cdot \log^2 n)$.

Proof. For an edge $e \in T$, let $\mathsf{LB}_e(k)$ be the lower bound on $\mathsf{OPT}(k)$ contributed by e as defined in Section 4.1. Let $\mathsf{LB}(k) = \sum_e \mathsf{LB}_e(k)$. Let $\mathsf{LB}_e(k/h)$ and $\mathsf{LB}(k/h)$ be the corresponding quantities assuming a buffer size of k/h. We know from Lemma 2 that

$$\mathsf{OPT}(k) \geq \mathsf{LB}(k) \geq \mathsf{LB}(k/h)/2h = \mathsf{LB}(k/h)/2\log n.$$

To prove the lemma, next we argue that the total cost of the algorithm is $O(\mathsf{LB}(k/h) \cdot \log n)$.

To this end, consider a phase t. Suppose jth sub-buffer overflows in this phase. Let v be the leaf corresponding to the current position of the server and let u be the vertex on the path from v to the root between the levels j and $j - 1$. In this phase, the algorithm spends at most twice the cost of subtree below u. Let e be the parent edge of tree T_j, i.e., the edge that connects T_j to u. Note that the cost of e is $n/2^j$ while the total cost of the subtree below u is $2(\log n - j) \cdot n/2^j = O(\log n \cdot n/2^j)$. With a loss of factor $O(\log n)$, we charge the cost of clearing the requests in $T_j \cup \cdots \cup T_h$ to the cost paid in traversing e in this phase. We say that the phase t transfers a charge of $n/2^j$ to e.

Now fix an edge $e \in T$. Let C_e denote the total charge transfered to e from all the phases. We show that $C_e \leq \mathsf{LB}_e(k/h)$. Refer to Figure 1. Let L_e and R_e be the two subtrees formed by removing e from T. Now C_e is the total cost paid by our algorithm for traversing e in the phases which transfer a charge to e. Note that in these phases, we traverse e to go from L_e to R_e or vice-versa only when there are at least k/h pending requests on the other side. Thus C_e is a at most $\mathsf{LB}_e(k/h)$, the lower bound contributed by e assuming a buffer size of k/h. Thus, $\sum_e C_e \leq \sum_e \mathsf{LB}_e(k/h) = \mathsf{LB}(k/h)$ and the proof is complete.

Lemma 4. *For any $1 \leq \alpha \leq \log n$, there is an $O(\alpha^{-1} \log^2 n)$ competitive algorithm for SBP on the binary tree metric defined above if the algorithm is allowed to use a buffer of size αk.*

The algorithm is similar to the above one except that it assigns a sub-buffer of size $\alpha k/h$ to each of the h subtrees. The proof that it has the claimed competitive ratio is similar to that of Theorem 2 and is omitted.

4.3 An Algorithm on a Line Metric

Our algorithm for a line metric is based on the probabilistic approximation of the line metric by a binary tree metric considered in the previous section.

Definition 1 (Bartal [3]). *A set of metric spaces S over a set of points V, α-probabilistically approximates a metric space M over V, if*

- *every metric space in S dominates M, i.e., for each $N \in S$ and $u, v \in V$ we have $N(u, v) \geq M(u, v)$, and*
- *there exists a distribution over the metric spaces $N \in S$ such that for every pair $u, v \in V$, we have $\mathbb{E}[N(u, v)] \leq \alpha M(u, v)$.*

Definition 2. *A r-hierarchically well-separated tree (r-HST) is a edge-weighted rooted tree with the following properties.*

- *The weights of edges between a node to any of its children are same.*
- *The edge weights along any path from root to a leaf decrease by at least a factor of r.*

Bartal [3] showed that any connected edge-weighted graph G can be α-probabilistically approximated by a family r-HSTs where $\alpha = O(r \log n \log \log n)$. Fakcharoenphol, Rao and Talwar [7] later improved this factor to $\alpha = O(r \log n)$. We prove the following lemma for completeness.

Lemma 5. *A line metric on uniformly-spaced n points can be $O(\log n)$ probabilistically approximated by a family of binary 2-HSTs.*

Proof. Assume for simplicity that $n = 2^h$ for some integer h. Let M be a metric on $\{1, \ldots, n\}$ with $M(i, j) = |i - j|$. Consider a binary tree T on $2n$ leaves. Label the leaves from left to right as $l_1, \ldots, l_{2n}$. Partition the edges into levels as shown in Figure 2, i.e., the edges incident to the root are in level 1 and those incident to the leaves are in level $1 + \log n$. Assign a weight of $n/2^i$ to each edge in level i. Now pick r uniformly at random from the set $\{0, 1, \ldots, n - 1\}$. Let N be the metric induced on the leaves $l_{r+1}, l_{r+2}, \ldots, l_{r+n}$ and consider a bijection from $\{1, \ldots, n\}$ to $\{l_{r+1}, \ldots, l_{r+n}\}$ that maps i to l_{r+i}.

It is easy to see that under this mapping, the metric space N dominates M. Now consider any pair i and $i + 1$. It is easy to see that $\mathbb{E}[N(l_{r+i}, l_{r+i+1})] = O(\log n)$. By linearity of expectation, we have that for any pair $1 \leq i, j \leq n$, we have $\mathbb{E}[N(l_{r+i}, l_{r+j})] = O(\log n \cdot |i - j|)$. Therefore this distribution over the binary 2-HSTs forms $O(\log n)$-approximation to the line metric as desired.

It is now easy to extend our algorithm on binary 2-HSTs to the line metric. We first pick a binary 2-HST from the distribution that gives $O(\log n)$ probabilistic approximation to the line metric. Then we run our deterministic online $O(\log^2 n)$-competitive algorithm on this binary tree. It is easy to see that the resulting algorithm is a randomized online algorithm that achieves a competitive ratio of $O(\log^3 n)$ in expectation against an oblivious adversary. It is necessary that the adversary be oblivious to our random choice of the 2-HST metric. Again for $1 \leq \alpha \leq \log n$, we can improve the competitive ratio to $O(\alpha^{-1} \log^3 n)$ by using a buffer of size αk.

4.4   Improved Analysis

Previous analysis can be improved to show that the same algorithm is in fact $O(\log^2 n)$ competitive on the line metric. The main idea is to consider the actual distance traveled on the line at the end of each phase instead of using the tree approximation as a black box.

   Note that in our algorithm, the disc head moves only at the end of a phase when the buffer for one of the levels overflows. Suppose the buffer for level j overflows at the end of a phase. So, the contribution by the parent edge of the tree T_j to the lower bound of the tree instance is $n/2^j$. In this move, we clear all the requests belonging to the trees $T_j, T_{j+1}, \ldots, T_{1+\log n}$. Our embedding guarantees that the total distance traveled *on the line metric*, in order to clear all these requests, is $O(n/2^j)$. By adding over all the phases, it is clear that the total distance traveled by the algorithm is $O(LB)$ where LB is the lower bound on the tree instance with buffer size $k/\log n$. This, in turn, implies a randomized $O(\log^2 n)$ competitive ratio for our algorithm against an oblivious adversary. Again, for $1 \le \alpha \le \log n$, we can improve the competitive ratio to $O(\alpha^{-1} \log^2 n)$ by using a buffer of size αk.

4.5   Extension to "Line-Like" Metrics

Our algorithm generalizes to other "line-like" metrics. Consider a metric space such that for every subset of points, the cost of the minimum spanning tree on that subset is at most α times the maximum pairwise distance in that subset. For such a metric space on n points, we next argue that our algorithm yields $O(\alpha \log n \log D)$ approximation, where D is the aspect ratio of the metric.

   In the algorithm for such metric spaces, we first approximate the given metric by 2-HSTs of height $h = O(\log D)$. Note that these HSTs need *not* be binary trees. Our algorithm on such an HST assigns a buffer of size k/h to each subtree as before. We incur a factor of $\log D$ from here. Recall that at the end of each phase, we clear requests in subtrees $T_j, \ldots, T_h$ for some j and charge it to the lower bound contributed by the parent edge of T_j. The distance traveled *in the original metric* is at most a constant times the cost of the minimum spanning tree on the subset of points, say S, that are mapped into these subtrees. This cost, by our assumption, is at most $\alpha \cdot diam(S)$, where $diam(S)$. Now, we observe that the cost of the parent edge of T_j is in fact $\Omega(diam(S))$. This follows from the construction of the HSTs in Fakcharoenphol et al. [7]. Thus our cost in this move is at most α times the contribution to the lower bound. As before, it can be shown that we get an overall approximation guarantee of $O(\alpha \log n \log D)$.

5   Conclusions

For both uniform and line metrics, no hardness results are known for the offline versions and no lower bounds known on competitive ratio of online algorithms. Any results in this direction will be interesting. Unlike the uniform metric, it looks unlikely that a deterministic strategy will give a non-trivial approximation ratio for a line metric. An $\Omega(k)$ lower bound on the competitive ratio of *any* deterministic online algorithm for a line metric would be interesting. On the other hand, any polylog k approximation on a line metric will also be interesting. Can we prove some non-trivial results for general HSTs, and hence for a general metric?

Acknowledgments

We thank Peter Sanders for suggesting the problem. Thanks to Varsha Dani and Tom Hayes for useful discussions and for discovering a simple proof of the lower bound for memoryless algorithms. We thank Kunal Talwar for suggesting extensions in the direction of "line-like" metrics.

References

1. M. Andrews, M. Bender, and L. Zhang. New algorithms for disc scheduling. *Algorithmica*, 32(2):277–301, 2002.
2. R. Bar-Yehuda and J. Laserson. 9-approximation algorithm for the sorting buffers problem. In *3rd Workshop on Approximation and Online Algorithms*, 2005.
3. Y. Bartal. Probabilistic approximations of metric spaces and its algorithmic applications. In *IEEE Symposium on Foundations of Computer Science*, pages 184–193, 1996.
4. M. Charikar, S. Khuller, and B. Raghavachari. Algorithms for capacitated vehicle routing. *SIAM Journal of Computing*, 31:665–682, 2001.
5. M. Charikar and B. Raghavachari. The finite capacity dial-a-ride problem. In *IEEE Symposium on Foundations of Computer Science*, pages 458–467, 1998.
6. M. Englert and M. Westermann. Reordering buffer management for non-uniform cost models. In *Proceedings of the 32nd International Colloquium on Algorithms, Langauages, and Programming*, pages 627–638, 2005.
7. J. Fakcharoenphol, S. Rao, and K. Talwar. A tight bound on approximating arbitrary metrics by tree metrics. In *35th Annual ACM Symposium on Theory of Computing*, pages 448–455, 2003.
8. J. Kohrt and K. Pruhs. A constant approximation algorithm for sorting buffers. In *LATIN 04*, pages 193–202, 2004.
9. H. Räcke, C. Sohler, and M. Westermann. Online scheduling for sorting buffers. In *Proceedings of the European Symposium on Algorithms*, pages 820–832, 2002.
10. A. Silberschatz, P. Galvin, and G. Gagne. *Applied Operating System Concepts*, chapter 13. pages 435–468. John Wiley and Sons, 1st ed., 2000.

Optimal Node Routing

Yossi Azar[1,*] and Yoel Chaiutin

School of Computer Science, Tel Aviv University, Tel Aviv 69978, Israel
`azar@tau.ac.il, yoel1717@yahoo.com`

Abstract. We study route selection for packet switching in the competitive throughput model. In contrast to previous papers which considered competitive algorithms for packet scheduling, we consider the packet routing problem (output port selection in a node). We model the node routing problem as follows: a node has an arbitrary number of input ports and an arbitrary number of output queues. At each time unit, an arbitrary number of new packets may arrive, each packet is associated with a subset of the output ports (which correspond to the next edges on the allowed paths for the packet). Each output queue transmits packets in some arbitrary manner. Arrival and transmission are arbitrary and controlled by an adversary. The node routing algorithm has to route each packet to one of the allowed output ports, without exceeding the size of the queues. The goal is to maximize the number of the transmitted packets. In this paper, we show that all non-refusal algorithms are 2-competitive. Our main result is an almost optimal $\frac{e}{e-1} \approx 1.58$-competitive algorithm, for a large enough queue size. For packets with arbitrary values (allowing preemption) we present a 2-competitive algorithm for any queue size.

1 Introduction

Overview. A general network consists of nodes (routers) and communication links through which packets of information flow. Generally, a node consists of input ports, a switching module and an output buffer connected to each output port. A packet is received at an input port and then forwarded through the switching module to an appropriate output buffer. If the output buffer is full, the switching module must drop some packets. Most of previous works for competitive packet routing/switching considered the packet scheduling problem in the buffers assuming the path of the packet through the network is fixed and known. Moreover, to the best of our knowledge there are no results for the routing problem (i.e. path or output selection) in the competitive throughput model. In this paper we consider the simplest packet routing problem which is choosing a route in a node.

Traditionally, similar problems were analyzed while assuming either a specific distribution of the arrival rates (see [15,21]), or some predefined structure of the sequence of arriving packets such as in the Adversarial Queueing Theory

---

[*] Research supported in part by the German-Israeli Foundation and by the Israel Science Foundation.

(AQT), in which the adversary injects packets that obey some capacity constraints, so that packet dropping is not necessary. It is interesting to note that the first papers on AQT assumed fixed paths [5,16,17,22] and only later papers began considering the path selection problem [1,6,8,9,17]. Recently, throughput problems for various types of switches, in special graphs and arbitrary graphs were studied while avoiding any a-priori assumption on the input. As already mentioned, all these papers considered the packet scheduling but not the packet routing or output port selection. Here we consider the simplest routing (path selection or output port selection) which is node routing. Packets arrive at the input ports, while each packet is associated with a subset of the output ports. The packet has to be routed into one of these output ports.

We model the problem of node routing as follows: a node has an arbitrary number of input ports and an arbitrary number of output queues (denoted by m). All the output queues are of size B. At each time unit, an arbitrary number of new packets may arrive, each packet is associated with a subset of the output ports (which corresponds to the next edges on the allowed paths for the packet). Each output queue transmits packets in some arbitrary manner. Arrival and transmission are arbitrary and controlled by an adversary. In contrast to the models where the path is fixed, and hence each packet needs to be routed into a unique output queue, in our model the output queue needs to be selected. In particular the main decision problem is into which output queue to send the packet (among the allowed destination output queues). If the buffers are full, the packet must be rejected. The goal of the routing algorithm is to maximize the number of the transmitted packets. We also consider the model of packets with arbitrary values. In this model we allow preemption.

Our Results. We show that all non-refusal algorithms are 2-competitive for any queue size, and are optimal for queues of size 1. Our main result is an optimal deterministic $\frac{e}{e-1} \approx 1.58$-competitive algorithm for node routing, for a large enough size of the queues. This is done by designing an optimal $\frac{e}{e-1}$-competitive *fractional* algorithm and transforming it into a discrete algorithm with a competitive ratio of $\frac{e}{e-1}(1 + \frac{2m+1}{B})$. We show that our algorithm is almost optimal by providing a lower bound of $\frac{e}{e-1} - \Theta(\frac{1}{m})$. Actually the discretization is more general. In fact, we present a generic technique for transforming any *fractional* c-competitive algorithm for the node routing problem to a discrete $c(1 + \frac{2m+1}{B})$-competitive algorithm. We also show an optimal $\frac{e}{e-1} - o(1)$-competitive *randomized* algorithm for any B. For the preemptive model with arbitrary values, we present a 2-competitive algorithm for any m and B. This is done using the zero-one principle presented in [13]. The algorithm is optimal for $B = 1$. We also show a 3-competitive algorithm for another model called "the multi queue switch routing model" which is omitted due to space limitation.

Our Techniques. We start by constructing an online reduction from the fractional node routing problem to the problem of finding a maximum fractional matching in a bipartite graph (defined in [10]). Thus, using algorithm WL from [10] which is $\frac{e}{e-1}$-competitive, we obtain an $\frac{e}{e-1}$-competitive fractional algorithm. Then we present a generic technique for transforming any *fractional* algorithm for the node routing problem to a discrete algorithm. Specifically, we present an algorithm with larger queues and no packet loss that maintains a running simulation of the fractional algorithm. This is done using the vector rounding algorithm of [3]. We then transform this algorithm to an algorithm with queues of size B. For the model of packets with arbitrary values, we obtain the upper bound by using the zero-one principle presented in [13]. Due to space limitation most proofs are omitted.

Related Results for Packet Switching. There are many results for packet switching that appear in [4,20,14,10,13,12]. For general graphs the main results appear in [2,13,19,7].

2   Problem Definition and Notations

We model the problem as follows: a node has an arbitrary number of input ports and m output queues $\{Q_1, ..., Q_m\}$ each of size B. At each time unit, an arbitrary number of new packets may arrive. Each packet p is associated with a subset of the output queues $Q^p \subseteq \{Q_1, ..., Q_m\}$ (which corresponds to the next edges on the allowed paths for the packet). All packets are of equal size and value. W.l.o.g. we assume that all m queues are empty at the beginning. Each time unit is divided into two phases: in the *arrival* phase a set of packets arrive, each packet p is associated with a specific subset of the output queues Q^p. For each packet p, the main decision problem is into which output queue to send it (among the allowed destination output queues Q^p). Each packet might be rejected at its arrival. If the queues Q^p are full, the packet must be rejected. Clearly, in this model there is no need for preemption since the preference of one packet over the other is pointless. At the *transmission* phase, a subset of the queues (could be also none or all the queues) is selected by the system, and a packet is transmitted from each of the heads of these queues (only from the non empty queues). Both arrival and transmission are controlled by an adversary, but we assume no starvation for each of the output queues, i.e., eventually all queues will transmit and become empty. The goal of the algorithm is to maximize the number of the transmitted packets.

We use the term *non-refusal algorithm* for algorithms that accept every packet which the queues have space for. Obviously, every algorithm can be transformed into a non-refusal algorithm without worsening its competitiveness.

The model above is the *synchronous model* in which all the packets of a single arrival phase arrive together (and all the transmissions in the transmission phase are made together), so the algorithm has full knowledge about the arrival phase before it has to make its routing decisions. We also consider the *event driven* model where arrivals and transmissions occur in arbitrary times. We denote by σ the sequence events. Each event in σ can be either an arrival event or a transmission event. The algorithm must respond after every arrival event and route the packet (or discard it). Note that the event driven model is stronger than the synchronous one. Nevertheless our upper bounds hold even for the event driven model, and the lower bounds hold even for the synchronous model (except for the arbitrary B lower bound).

We also consider the *arbitrary-value* model where each packet is associated with a non-negative value. The goal is to maximize the sum of values of the transmitted packets. In this model we allow preemption. A preempted packet is a packet which is dropped after residing in a queue. Our results hold for both FIFO and non FIFO queues (e.g. priority queues). Note that the throughput of the optimal algorithm is the same for FIFO and non FIFO models.

Notations. We use the terms insert, accept and assign for packet insertion into one of the queues. For packet rejection we use the terms drop and discard. We use the event driven model throughout the paper, except for the lower bound sections. We denote by *event t* the t-th event, which can be either an arrival or a transmission event. We use the term *arrival event j* to denote the j-th arrival event, i.e., the arrival of the j-th packet. Similarly we denote by *transmission event k* the k-th transmission event. For event t, we use the term *load* of queue Q_i to refer to the number of packets residing in that queue, at that time. Given an online routing algorithm A we denote by $A(\sigma)$ the value of A given the sequence σ, i.e., the total amount of transmitted packets after all the packets of the sequence have arrived and all the queues have been emptied. We denote the optimal (offline) algorithm by *Opt*. A deterministic online algorithm A is *c-competitive* for all maximization problems described in this paper iff for every packet sequence σ we have: $Opt(\sigma) \leq c \cdot A(\sigma)$.

3   Optimal Algorithm for Node Routing

First we discuss non-refusal algorithms. We note that every non-refusal algorithm is 2-competitive, for every $B \geq 1$ (this is optimal for $B = 1$, as shown in section 4). The general non-refusal algorithm is defined as follows:

Non-refusal Algorithm. For each incoming packet p: insert p into any queue $Q_i \in Q^p$ which is not full. If such a queue does not exist, discard p.

It can be proved directly that every non-refusal algorithm is 2-competitive, but it also follows from Remark 1 or Theorem 10.

Next we describe our optimal algorithm which is $\frac{e}{e-1}(1 + \frac{2m+1}{B})$-competitive. In subsection 3.1 we present algorithm FR for the fractional model, which is $\frac{e}{e-1}$-competitive. In subsection 3.2, we present a generic technique to transform any c-competitive online fractional algorithm for our problem into a discrete algorithm with a competitive ratio of $c(1 + \frac{2m+1}{B})$. We begin with the fractional routing algorithm in the following subsection.

3.1   Algorithm for the Fractional Version of Node Routing

In this subsection we consider the fractional model, which is a relaxation of the discrete model that was presented in section 2. In the fractional model we allow the online algorithm to accept fractional packets into the queues as well as to transmit fractional packets. More formally, at arrival event t of packet p, the algorithm may split the packet into fractions and insert them into the queues Q^p (only into queues with sufficient free space), and may also discard any of the fractions. The value of each fraction is equal to its size. We note by k_i^t the total amount of fractions of p inserted into queue Q_i, at arrival event t. The total amount of fractions inserted into all the queues, must not exceed the unit value, i.e., $\sum_{i=1}^m k_i^t \leq 1$. In addition, $Q_i \notin Q^p$ implies that $k_i^t = 0$. At transmission event from queue Q_i, Q_i transmits a volume of a unit from its head (if there are enough fractions in queue Q_i). If the total fractions in the queue are less then a unit, they will all be transmitted. We assume that sequence σ consists of *integral* packets. In this subsection we show a fractional algorithm for the problem with a competitive ratio of $\frac{e}{e-1} \approx 1.58$, even against an optimal algorithm which is allowed to split incoming packets. We begin by introducing the problem of an online unweighted fractional matching in a bipartite graph. Then we introduce a translation of our problem (the fractional model) into the problem of the online unweighted fractional matching in a bipartite graph.

The online unweighted matching in a bipartite graph problem is defined as follows (as appears in [10]): first, consider an online version of the maximum bipartite matching on a graph $G = (S, R, E)$, where S and R are the disjoint vertex sets and E is the set of edges. We refer to sets S and R as the servers and the requests, respectively. The objective is to match a request to a server. At step t, a vertex $r_t \in R$ along with all of its incident edges, arrives online. Algorithm A can either reject r_t, or irreversibly match it to an unmatched vertex $s_i \in S$ adjacent to r_t. The goal of the online algorithm is to maximize the size of the matching.

The fractional version of the online unweighted matching is as follows: each request r_t has a size x_t which is the amount of work needed to service it. Algorithm A can match a fraction of size $k_i^t \in [0, x_t]$ to each vertex $s_i \in S$ adjacent to r_t. If request t is matched partially to some server i with weight k_i^t then $\sum_{i=1}^{|S|} k_i^t \leq x_t$. But the load of each server i, which is $\sum_{t=1}^n k_i^t$ where n is the length of σ, must be at most 1. We use the terms *match* and *assign* interchangeably. The goal of the online algorithm is to maximize the sum of matched fractions, i.e., to maximize $\sum_{t,i} k_i^t$.

Our translation of the fractional routing problem into the problem of online unweighted fractional matching in a bipartite graph, opposes, in some sense, the translations in [10,11]. In our translation, the incoming packets are the requests and they are matched to the queues which are the servers, while in [10,11] the requests are the transmission events and they are matched to the packets, which are the servers.

Given a sequence σ, we translate it into the bipartite graph $G^\sigma = (R, S, E)$, which is defined as follows:

- Let T denote the total number of packets. We define the set of requests as $R = \{r_1, ..., r_T\}$ all with unit sizes, i.e., $x_i = 1$ for each $1 \leq i \leq T$. Each request corresponds to a packet.

- For each queue Q_i we define a set of servers S_i, which represents the queue over time. Specifically, each S_i ($1 \leq i \leq m$) contains $T + B$ servers: $S_i = \{s_i^1, ..., s_i^{T+B}\}$. The S_i's are disjoint and the full set of servers is defined as $S = \bigcup_{i=1}^m S_i$.
- Let y_i^t denote the number of times queue Q_i was selected for transmission until *arrival event* t. We denote by S_i^t the B servers in S_i that represent queue Q_i at arrival event t. We define $S_i^t = \{s_i^{y_i^t+1}, ..., s_i^{y_i^t+B}\} \subseteq S_i$. Consider packet p arriving at arrival event t, associated with the subset Q^p. In the bipartite graph problem, at step t a request r_t arrives along with its incident edges, which are the edges connecting it to all servers in S_i^t for each i such that $Q_i \in Q^p$. More formally, r_t arrives with the following incident edges: $\{(s, r_t) | s \in S_i^t, Q_i \in Q^p\}$.

Definition 1. *A route RT for a sequence of arriving packets σ, for the fractional node routing problem, is a set of triplets of the form (t, Q_i, k_i^t) for each $1 \leq i \leq m$ and $1 \leq t \leq T$. Each triplet expresses that at arrival event t, queue Q_i gets the fraction $k_i^t \geq 0$ of the packet p. The size of the route, denoted by $|RT|$, is the total amount of fractions transmitted. Since all accepted fractions of packets are transmitted, clearly $|RT| = \sum_{t,i} k_i^t$.*

Definition 2. *A route RT for a sequence σ is called legal if for every triplet (t, Q_i, k_i^t), queue Q_i has free space of at least k_i^t at arrival event t (the empty space is a function of accepted fractions and transmissions which are in σ) and $\sum_{i=1}^m k_i^t \leq 1$, and $Q_i \notin Q^p$ implies that $k_i^t = 0$.*

The following lemmas connect bipartite fractional matching to our problem.

Lemma 1. *Every legal fractional route RT for the sequence σ can be mapped, in an online fashion, to a fractional matching M in G^σ such that $|RT| = |M|$.*

Lemma 2. *Every fractional matching M in G^σ can be translated, in an online fashion in polynomial time, to a legal fractional route RT for σ such that $|RT| = |M|$.*

The following corollary result directly from Lemmas 1 and 2.

Corollary 1. *For any sequence σ, the size of the optimal fractional route for σ is equal to the size of a maximum fractional matching in G^σ.*

Remark 1. Actually, Lemmas 1 and 2 hold also for integral node routing and integral online matching. This implies that every non-refusal algorithm for node routing is 2-competitive.

Remark 2. We note that by using the above reduction from the integral node routing to the integral online matching, and applying the randomized algorithm $RANKING$ of Karp *et al.* [18], we obtain a randomized $\frac{e}{e-1} - o(1)$-competitive algorithm. Our main focus is on a deterministic algorithm.

We use algorithm WL for the unweighted fractional matching problem, presented in [10], and its results.

Algorithm WL. For each request r_t adjacent to servers $\{s_1, ..., s_n\}$, match a fraction of size k_j^t for each adjacent s_j, where $k_j^t = (h - l_j)_+$ and $h \leq 1$ is the maximum number such that $\sum_{j=1}^{n} k_j^t \leq 1$. By $(f)_+$ we mean $\max\{f, 0\}$.

Theorem 1. [10] *Algorithm WL is $\frac{e}{e-1}$-competitive and optimal for the bipartite unweighted fractional matching problem.*

Now we present algorithm FR for the fractional node routing problem:

Algorithm Fractional Routing (FR)

- Maintain a running simulation of WL in the online constructed graph G^σ.
- **Routing:** For arrival event t, translate the matching of r_t to servers made by WL, using Lemma 2, to legal routing triplets (t, Q_i, k_i^t). Recall that each triplet corresponds to assigning a k_i^t fraction of p to queue Q_i.

Note that FR is not a non-refusal algorithm (because in G^σ, the total load of the servers in S_i^t might be bigger than the load of queue Q_i at arrival event t).

Theorem 2. *For every sequence σ, $Opt(\sigma) \leq \frac{e}{e-1} FR(\sigma)$ for the fractional node routing.*

3.2   The Discretization of the Fractional Algorithm

In this subsection we introduce a generic technique for the discretization of any fractional algorithm for the node routing problem. In particular, we will use this technique for the discretization of algorithm FR, which was presented in subsection 3.1. We present in subsection 3.2.1 a general technique for the discretization of the routing decisions of any fractional algorithm A, for the *unbounded queues* version of the problem. We will use this technique in subsection 3.2.2 for the discretization of the routing of any fractional algorithm A, for our version of the node routing problem (bounded queues).

3.2.1 The Unbounded Queues Node Routing Problem

Consider the node routing problem, but assume that the queues are unbounded. Thus all packets are accepted into the queues. The algorithm *must* accept each packet p into one of the queues Q^p. In this case, we may consider the minimization of the maximum queue size (notice that it is a *cost* problem). We define the cost of the online algorithm A (denoted by $cost(A)$), given a finite sequence σ, as the maximum length of a queue taken over all queues and times. The goal is to minimize the maximum cost, i.e., minimize the maximum queue size over time.

We now turn to consider a relaxation of the model and allow an online algorithm to split packets and assign fractions of packets to the queues, provided that for every packet p, the total size of the accepted fractions of p into queues Q^p is exactly 1, i.e., the algorithm is not allowed to discard any fraction of the packet. We denote by A the online fractional algorithm. We now introduce a general technique for the discretization of the routing decisions of algorithm A, given a finite sequence of *integral* packets σ while adding an additive cost of at most $2m$ to the cost of A.

Now, we define a discrete algorithm M which gets the fractional algorithm A as a parameter and denote it by M^A. At each packet arrival, in order to decide which queue to serve, M^A computes the load seen by A and uses this information for making its decisions. Given arrival event t, let l_i^A and l_i be the simulated load of A and the actual load of M^A on queue Q_i at that event, respectively.

For the decisions of M^A we use a theorem for the **vector rounding problem** which applies to our problem. We address the vector rounding problem as presented in [3][1]. The m-dimensional vector rounding problem is this: the input is a list of vectors $(V_1, V_2, ...)$ arriving online, where each $V_t = (v_t^1, ..., v_t^m)$ is a vector of length m over the reals which suffices $\sum_{i=1}^{m} v_t^i \in Z^+$. The output is a list of integer vectors $(Z_1, Z_2, ...)$ where $Z_t = (z_t^1, ..., z_t^m)$ is a rounding of V_t that preserves the sum, i.e., for all $1 \le i \le m$ we have that $z_t^i \in \{\lfloor v_t^i \rfloor, \lceil v_t^i \rceil\}$ and that $\sum_{i=1}^{m} z_t^i = \sum_{i=1}^{m} v_t^i$. The goal is to make the accumulated difference in each entry as small as possible, i.e., for every t we want $max_{1 \le i \le m} |\sum_{j=1}^{t} z_j^i - \sum_{j=1}^{t} v_j^i|$ to be as small as possible. The cost of the algorithm is the unfairness, which is the maximum accumulated difference in each entry over all time units. The goal is to minimize the unfairness.

The major special case we will be interested in is when vectors V_t satisfy $0 \le v_t^i \le 1$ and $\sum_{i=1}^{m} v_t^i = 1$. Since the output vector Z_t satisfies $z_t^i \in \{\lfloor v_t^i \rfloor, \lceil v_t^i \rceil\}$ and $\sum_{i=1}^{m} z_t^i = \sum_{i=1}^{m} v_t^i = 1$, we conclude that Z_t will be all zeros except for one entry $z_t^{i'} = 1$ where i' satisfies $v_t^{i'} > 0$.

In [3], the authors show a way to build an algorithm (we call it VR) for the online vector rounding problem.

Theorem 3. [3] *For the vector rounding problem, algorithm VR's unfairness is at most m, i.e., for every t, $max_{1 \le i \le m} |\sum_{j=1}^{t} z_j^i - \sum_{j=1}^{t} v_j^i| \le m$.*

Now we present the discrete algorithm M^A.

Algorithm M^A. Maintain a running online simulation of algorithm A. For each arrival event t of a packet p let k_i^t denote the fractions inserted into the queue Q_i by A, i.e., $\sum_{Q_i \in Q^p} k_i^t = 1$, and $Q_i \notin Q^p$ implies that $k_i^t = 0$. For each arrival event t of a packet p do:

- Build an input vector V_t for the vector rounding problem $V_t = (v_t^1, ..., v_t^m)$ where $v_t^i = k_i^t$.
- Get the output vector $Z_t = (z_t^1, ..., z_t^m)$ from the simulated algorithm VR given the input vectors $(V_1, ..., V_t)$.
- Insert p into queue Q_i which satisfies $z_t^i = 1$.

Note that our algorithm is not affected by the transmissions. We also note that the insertion of p is legal, because all the input vectors of VR are of the form $V_t = (v_t^1, ..., v_t^m) = (k_1^t, ..., k_m^t)$ and satisfy $0 \le k_i^t \le 1$ and $\sum_{i=1}^{m} k_i^t = 1$,

---

[1] The vector rounding problem is a generalization of the car pool problem.

and as mentioned, this implies that Z_t consists of zeros except for one entry $z_t^{i'} = 1$, where i' satisfies $k_{i'}^t > 0$, this implies that $Q_{i'} \in Q^p$.

Now we introduce our main theorem for this subsection.

Theorem 4. *For every sequence σ, $cost(M^A) \leq cost(A) + 2m$. Alternatively, for sequences in which A needs queues of size B, M^A needs queues of size $B+2m$.*

3.2.2 The Discretization of the Fractional Algorithm

In this subsection we return to the standard node routing model (with bounded queues). We show a general technique to transform any c-competitive *fractional* algorithm for the node routing problem into a discrete algorithm with a competitive ratio of $c(1 + \frac{2m+1}{B})$. Specifically, we will apply this technique for the discretization of algorithm FR, which was presented in subsection 3.1.

For our discretization process we rely on the results of the unbounded problem, which were presented in subsection 3.2.1. We want to address the packets which were accepted by FR as the input sequence σ for the cost problem studied in subsection 3.2.1, in a way which we will present later. Recall that algorithm FR might accept fractional packets due to insufficient queue space. Since in the model of the unbounded problem we study the case where σ consists of integral packets, we want FR to only accept packets integrally, i.e., for every packet p, to accept k_i^t such that $\sum_{i=1}^m k_i^t = 1$. Hence, we continue by considering the following problem: assume we are given an online c-competitive algorithm A. We want to produce a competitive algorithm $\hat{A}$ which assigns only integral packets. We provide algorithm $\hat{A}$ which has queues of size $B + 1$ (algorithm A maintains queues of size B); we will get rid of this assumption later on. Intuitively, $\hat{A}$ emulates the routing of A and accepts only integral packets.

Definition 3. *We denote fractional algorithms that accept integral packets whenever there is sufficient space for a whole packet, and otherwise discard the whole packet, as* discrete non-refusal *algorithms.*

We now define the transformation of a given algorithm A with queues of size B into algorithm $\hat{A}$ with queues of size $B + 1$, which is a discrete non-refusal algorithm. Let l_i^t and $\hat{l}_i^t$ be the loads of queue Q_i at event t in A and $\hat{A}$, respectively.

Algorithm $\hat{A}$

- Maintain a running simulation of A. Let k_i^t be the fraction size which was inserted by algorithm A at arrival event t into queue Q_i.
- If the queues in Q^p do not have sufficient space (the total free space is less than a unit), discard p.
- Otherwise, assign to queue Q_i amount of $min(B + 1 - \hat{l}_i, k_i^t)$, i.e., assign k_i^t if there is enough space or just fill the queue. After assigning to all the queues in Q^p, if the total inserted volume is less than a unit, insert the rest of p arbitrarily into any subset of the queues in Q^p which has sufficient empty space.

Obviously, $\hat{A}$ is a discrete non-refusal algorithm. Now we prove that $A(\sigma) \leq \hat{A}(\sigma)$ for every σ.

Theorem 5. *For a given algorithm A with queues of size B, algorithm $\hat{A}$ with queues of size $B + 1$ has at least the same throughput as A, given the same sequence σ.*

Now, we consider algorithm M presented in subsection 3.2.1. Recall that M simulates some fractional algorithm which fully accepts every incoming packet (since the queues are unbounded). We want to use algorithm M on algorithm $\hat{A}$ (denoted by $M^{\hat{A}}$) which is a discrete non-refusal algorithm. Since $\hat{A}$ has queues of size $B + 1$, algorithm $M^{\hat{A}}$ will use queues of size $B + 1 + 2m$. Still, we cannot use M on $\hat{A}$ since $\hat{A}$ rejects packets when there is no sufficient free space in queues Q^p. Therefore we extend algorithm M to work on algorithms that reject whole packets. This is done by skipping the events in which packets are rejected. We now present the following lemma:

Lemma 3. *Algorithm $M^{\hat{A}}$ with queues of size $B+1+2m$ has the same throughput as algorithm $\hat{A}$ with queues of size $B + 1$, given the same sequence σ.*

We now return to our original model where queues are of size B. By Theorem 5 and Lemma 3, if $M^{\hat{A}}$ had queues of size $B + 1 + 2m$, then algorithm $M^{\hat{A}}$ would have had at least the same throughput as A. Unfortunately, this is not the case, so we continue by emulating an algorithm with large queues with an algorithm with small queues.

We use the emulation presented in [10]: assume we are given an online competitive discrete algorithm A with queues of size y and we want to produce a competitive algorithm E^A with queues of size $y' < y$. We present algorithm E^A, and the corresponding theorem from [10].

Algorithm E^A. Maintain a running simulation of algorithm A. Accept a packet into queue Q_i if A accepts it to queue Q_i and the queue is not full.

Theorem 6. *[10] If A and E^A have queues of size y and y' ($y > y'$) respectively, then $A(\sigma) \leq \frac{y}{y'} E^A(\sigma)$.*

We prove the main result of this section with the next theorem.

Theorem 7. *Given any c-competitive fractional algorithm A for the node routing problem, algorithm $E^{M^{\hat{A}}}$ is a $c(1 + \frac{2m+1}{B})$-competitive discrete algorithm for the node routing problem.*

Corollary 2. *Applying Theorem 7 on algorithm FR, produces algorithm $E^{M^{\hat{FR}}}$ which is $\frac{e}{e-1}(1 + \frac{2m+1}{B})$-competitive. For $B \gg m$ the competitive ratio of $E^{M^{\hat{FR}}}$ approaches $\frac{e}{e-1} \approx 1.58$.*

4   Lower Bounds

In this section we show some lower bounds for the node routing problem. We first show that an $\frac{e}{e-1}$-competitive algorithm for node routing is optimal. This lower bound holds even for randomized algorithms. Then we show a lower bound of 2 for deterministic algorithms when $B = 1$, even for the synchronous model.

Theorem 8. *The competitive ratio of every algorithm for the node routing problem is at least $\frac{e}{e-1} - \Theta(\frac{1}{m})$, for every B.*

Theorem 9. *The competitive ratio of every algorithm for the node routing problem is at least 2, for $B = 1$ and any $m \geq 2$.*

5   Algorithm for Arbitrary-Value Node Routing

In this section we consider the *arbitrary-value* node routing problem. We introduce algorithm *greedy routing* (*GR*) which is 2-competitive, for the problem.

Algorithm Greedy Routing (*GR*)
When packet p associated with the subset Q^p arrives:

- If there exists queue Q_i ($Q_i \in Q^p$) which is not full, insert p into Q_i.
- Otherwise, we note by p' a packet with the smallest value in Q^p and by queue $Q_{i'}$ ($Q_{i'} \in Q^p$) we note its queue.
  - If $value(p') < value(p)$, preempt p' and insert p into queue $Q_{i'}$.
  - Otherwise, drop p.

We use the zero-one principle of [13] to show that GR is 2-competitive.

Theorem 10. *Algorithm GR is 2-competitive for the arbitrary-value node routing problem.*

Corollary 3. *For the unit-value model, algorithm GR includes all non-refusal algorithms, hence every non-refusal algorithm is 2-competitive.*

References

1. W. Aiello, E. Kushilevitz, and R. Ostrovsky. Adaptive packet routing for bursty adversarial traffic. In *Proc. of the 30th ACM Symp. on Theory of Computing (STOC)*, pages 359–368, 1998.
2. W. Aiello, R. Ostrovsky, E. Kushilevitz, and A. Rosén. Dynamic routing on networks with fixed-size buffers. In *Proc. 14th ACM-SIAM Symp. on Discrete Algorithms*, pages 771–780, 2003.
3. Miklos Ajtai, James Aspnes, Moni Naor, Yuval Rabani, Leonard J. Schulman, and Orli Waarts. Fairness in scheduling. *Journal of Algorithms*, 29(2):306–357, November 1998.
4. S. Albers and M. Schmidt. On the performance of greedy algorithms in packet buffering. In *Proc. 36th ACM Symp. on Theory of Computing*, pages 35–44, 2004.
5. M. Andrews, B. Awerbuch, A. Fernández, J. Kleinberg, T. Leighton, and Z. Liu. Universal stability results for greedy contention-resolution protocols. In *Proc. 37th IEEE Symp. on Found. of Comp. Science*, pages 380–389, 1996.
6. B. Awerbuch, P. Berenbrink, A. Brinkmann, and C. Scheideler. Simple online strategies for adversarial systems. In *Proc. of the 42nd IEEE Symp. on Foundation of Comuputer Science (FOCS)*, 2001.

7. B. Awerbuch, A. Brinkmann, and C. Scheideler. Anycasting and multicasting in adversarial systems: Routing and admission control. In *Proc. 30th ICALP*, pages 1153–1168, 2003.

8. B. Awerbuch and F. Leighton. Improved approximation algorithms for the multicommodity flow problem and local competitive routing in dynamic networks. In *Proc. of the 26th ACM Symp. on Theory of Computing (STOC)*, pages 487–496, 1994.

9. B. Awerbuch, Y. Mansour, and N. Shavit. End-to-end communication with polynomial overhead. In *Proc. of the 30th IEEE Symp. on Foundation of Comupter Science (FOCS)*, pages 358–363, 1989.

10. Y. Azar and M. Litichevskey. Maximizing throughput in multi-queue switches. In *Proc. 12th Annual European Symposium on Algorithms*, pages 53–64, 2004.

11. Y. Azar and Y. Richter. Management of multi-queue switches in QoS networks. In *Proc. 35th ACM Symp. on Theory of Computing*, pages 82–89, 2003.

12. Y. Azar and Y. Richter. An improved algorithm for CIOQ switches. In *Proc. 12th Annual European Symposium on Algorithms*, pages 65–76, 2004.

13. Y. Azar and Y. Richter. The zero-one principle for switching networks. In *Proc. 36th ACM Symp. on Theory of Computing*, 2004. 64–71.

14. N. Bansal, L. Fleischer, T. Kimbrel, M. Mahdian, B. Schieber, and M. Sviridenko. Further improvements in competitive guarantees for QoS buffering. pages 196–207, 2004.

15. A. Birman, H. R. Gail, S. L. Hantler, Z. Rosberg, and M. Sidi. An optimal service policy for buffer systems. *Journal of the Association Computing Machinery (JACM)*, 42(3):641–657, 1995.

16. A. Borodin, J.Kleinberg, P. Raghavan, M. Sudan, and D. Williamson. Adversarial queuing theory. In *Proc. 28th ACM Symp. on Theory of Computing*, pages 376–385, 1996.

17. D. Gamarnik. Stability of adaptive and non-adaptive packet routing policies in adversarial queueing networks. In *Proc. of the 31st ACM Symp. on Theory of Computing (STOC)*, pages 206–214, 1999.

18. R. Karp, U. Vazirani, and V. Vazirani. An optimal algorithm for on-line bipartite matching. In *Proceedings of 22nd Annual ACM Symposium on Theory of Computing*, pages 352–358, may 1990.

19. A. Kesselman, Z. Lotker, Y. Mansour, and B. Patt-Shamir. Buffer overflows of merging streams. In *Proc. 11th Annual European Symposium on Algorithms*, pages 349–360, 2003.

20. A. Kesselman, Z. Lotker, Y. Mansour, B. Patt-Shamir, B. Schieber, and M. Sviridenko. Buffer overflow management in QoS switches. In *Proc. 33rd ACM Symp. on Theory of Computing*, pages 520–529, 2001.

21. M. May, J. C. Bolot, A. Jean-Marie, and C. Diot. Simple performance models of differentiated services for the internet. In *Proceedings of the IEEE INFOCOM '1999*, pages 1385–1394.

22. C. Scheideler and B. Vocking. From static to dynamic routing: efficient transformations of store-and-forward protocols. In *Proc. of the 31st ACM Symp. on Theory of Computing (STOC)*, pages 215–224, 1999.

Memoryless Facility Location in One Pass

Dimitris Fotakis

Department of Information and Communication Systems Engineering,
University of the Aegean, 83200 Karlovasi, Samos, Greece
fotakis@aegean.gr

Abstract. We present the first one-pass memoryless algorithm for metric Facility Location which maintains a set of facilities approximating the optimal facility configuration within a constant factor. The algorithm considers the demand points one-by-one in arbitrary order, is randomized and very simple to state and implement. It runs in linear time and keeps in memory only the facility locations currently open. We prove that its competitive ratio is less than 14 in the special case of uniform facility costs and less than 49 in the general case of non-uniform facility costs.

1 Introduction

In many applications dealing with data streams, we have to maintain a set of representatives for a sequence of points in a metric space. This is performed by a *streaming* algorithm which considers the points one-by-one in arbitrary order and maintains a set of representatives of reasonable accuracy using a limited amount of computational resources (see e.g. [13] for a survey on streaming computation).

This scenario can be naturally relaxed to a Facility Location problem. Given a (multi)set of n demand points in a metric space, the Facility Location problem seeks for a set of locations to open facilities minimizing the total cost of opening facilities and assigning every demand point to its nearest facility. To model the streaming scenario above, we let facilities play the role of representatives and require that the algorithm be *memoryless* and operate in *one pass*. In addition, we consider the special case of *uniform facility costs*, where the cost of opening a facility is the same for all points.

The algorithm is charged for the inaccuracy of its facilities (assignment cost) and the cost of maintaining them in main memory (facility cost). The assignment cost is the sum of distances of the points considered so far to their closest facility. As for the facility cost, we assume that the algorithm *buys* its memory space. Thus keeping a point in memory as a representative is equivalent to making the point a new facility. We assume that buying memory is an *irrevocable decision* and charge the algorithm for its *worst-case memory* consumption.

A *one-pass* algorithm considers the points one-by-one in arbitrary order. When a point is considered, a *memoryless* algorithm can either discard its location or make it a new facility. In the latter case, either the new facility replaces some existing one (i.e. the existing facility is discarded and the new facility takes its place in main memory) or the algorithm buys additional memory space and incurs the irrevocable cost of increasing its facilities by one. Discarding a demand point is also an irrevocable decision

B. Durand and W. Thomas (Eds.): STACS 2006, LNCS 3884, pp. 608–620, 2006.

because the algorithm has no access to previously considered points unless it keeps them in memory (as open facilities). For each discarded point, the algorithm incurs an assignment cost equal to the distance to its closest facility. The assignment cost is not irrevocable since it may increase or decrease every time the algorithm's facility configuration changes. Thus a one-pass memoryless algorithm can be regarded as a streaming algorithm using only $O(|\Gamma_{\max}|)$ space, where $|\Gamma_{\max}|$ denotes the maximum number of facilities maintained by the algorithm.

Related Work and Motivation. There has been a significant amount of work on streaming algorithms for k-Median[1], a problem closely related to Facility Location. Guha et al. [10] initiated the study of streaming algorithms for k-Median. Their algorithms run in $\tilde{O}(nk)$ time[2] and $O(n^\epsilon)$ space and achieve an approximation ratio of $2^{O(1/\epsilon)}$ for k much smaller than n^ϵ. The best known streaming algorithm for k-Median is the randomized algorithm of Charikar et al. [5], which runs in $O(nk \log^2 n)$ time and $O(k \log^2 n)$ space and achieves a constant approximation ratio.

Indyk [11] considered k-Median, Facility Location, and other geometric problems in the discrete d-dimensional space and the turnstile model, where the data stream consists of both insertions and deletions of points. For k-Median, Indyk gave a randomized $O(1)$-approximation algorithm using $\tilde{O}(k)$ space and having $\tilde{O}(k)$ update time per insertion/deletion and $\tilde{O}(nk)$ query time. Frahling and Sohler [8] recently improved the approximation ratio to $(1 + \epsilon)$ by a randomized algorithm using $\tilde{O}(k^2/\epsilon^{O(d)})$ space and having polylogarithmic update time and $O(k^5 \log^9 n + k^2 \log^5 n \exp(\epsilon^{-d}))$ query time.

In Facility Location, we have a facility cost instead of the constraint on the number of facilities (medians). Hence it may be advantageous to open a large number of facilities. In fact there are instances where every (near-)optimal solution needs to open $\Omega(n)$ facilities and every $O(1)$-approximation algorithm needs to run for $\Omega(n^2)$ time (see e.g. the lower bound in [14, Section 4]). Thus we cannot hope for an algorithm, even a randomized one, that maintains an $O(1)$-approximation to the optimal facility configuration in $o(n^2)$ time or $o(n)$ space.

To overcome this limitation, we may let the algorithm approximate only the *optimal cost* (i.e. maintain only the cost of an approximate facility configuration and not the configuration itself). In the discrete d-dimensional space and the turnstile model, Indyk [11] gave a randomized algorithm that approximates the optimal cost within a factor of $O(d \log^2 n)$ in $\tilde{O}(n)$ time and $O(d^2 \log^2 n)$ space. In the *offline* setting, where there is random access to the input, Bădoiu et al. [3] proved that the optimal cost of Facility Location with uniform facility costs can be approximated within a constant factor in $O(n \log^2 n)$ time. For non-uniform facility costs, they showed that there is no $o(n^2)$-time algorithm approximating the optimal cost within any factor.

Nevertheless in practical applications dealing with data streams, we usually need an algorithm that maintains an approximate facility configuration and not just an estimation of the optimal cost (see e.g. [9] for applications on clustering large data sets). Then one naturally turns to streaming algorithms using a *minimal* amount of resources.

---

[1] Given a (multi)set of n demand points in a metric space, the k-Median problem seeks for k points (medians) that minimize the sum of distances from each demand to its nearest median.

[2] The $\tilde{O}$-notation is the same as the O-notation except it hides polylogarithmic terms instead of constants, e.g. $\tilde{O}(nk)$ is $O(nk \operatorname{poly}(\log n))$.

Arguably a one-pass memoryless algorithm is of this sort. It maintains an approximate facility configuration using just the memory required to store its facilities.

The *online* algorithm of Meyerson [12] was the first one-pass memoryless algorithm for Facility Location. In the online setting, the decisions (and the costs) of opening a facility at a particular location and assigning a demand to a particular facility are irrevocable. Meyerson's algorithm is randomized and achieves a competitive ratio of $O(\log n)$. If the demand points arrive in *random order*, the algorithm achieves an expected constant competitive ratio. However, the latter version is *not a one-pass memoryless algorithm* because computing a random permutation of the demand sequence requires random access to it. Despite its logarithmic competitiveness, Meyerson's algorithm found significant applications because it is simple, memoryless, and one-pass. Also motivated by practical applications, Anagnostopoulos *et al.* [1] presented a $\Theta(2^d \log n)$-competitive deterministic online algorithm for the d-dimensional space, which can be regarded as a memoryless algorithm.

On the other hand, no online algorithm for Facility Location can achieve a competitive ratio better than $\Omega(\frac{\log n}{\log \log n})$ even on the line [6]. To overcome the logarithmic lower bound, we considered the incremental setting introduced in [4], where the algorithm can also merge existing facilities and only the decision of clustering some demands together is irrevocable. In [7], we presented an $O(1)$-competitive deterministic incremental algorithm for Facility Location with uniform facility costs. That algorithm runs in $O(n^2 |F_{\max}|)$ time and makes non-trivial use of previous demands' locations for opening and merging facilities. Thus it is *not a memoryless* algorithm.

Contribution. In previous work, Anagnostopoulos *et al.* [1] and Meyerson [12] presented one-pass memoryless algorithms of logarithmic competitive ratio and we presented a one-pass (but not memoryless) algorithm of constant competitive ratio [7]. In this work, we present the first one-pass memoryless algorithm that maintains a set of facilities approximating the optimal facility configuration within a constant factor.

The algorithm is randomized and extremely simple to state and implement. It is based on the notion of *final distance*, an upper bound on the distance of a demand to the nearest facility at any future point in time. We use a modification of the randomized rule of [12] for opening new facilities. More specifically, each new demand becomes a new facility with probability proportional to the ratio of its *final distance* to the facility cost. To avoid facility proliferation leading to logarithmic competitiveness, we develop a memoryless version of the merge rule of [7]. In particular, a *replacement ball* around each new facility is determined when the facility opens. Then the facility is replaced by the first new facility in its replacement ball.

The algorithm keeps in memory only the locations and the replacement radii of the facilities currently open. It runs in $O(n|F_{\max}|)$ time and $O(|F_{\max}|)$ space. The worst-case running time is linear in the size of the input because representing the metric space needs $\Theta(n^2)$ bits. We prove that the algorithm's competitive ratio is less than 14.

We should note that the algorithm maintains an $O(1)$-competitive facility configuration but it does not maintain *the assignment cost* of the demands considered so far with respect to it. A provably good facility configuration suffices in most practical applications (e.g. location problems, clustering problems). If an estimation of the assignment cost is required, we can use the techniques of [5] and maintain a small set of weighted points whose assignment cost is within a constant factor of the algorithm's cost. This increases the algorithm's time and space complexity by a factor of $O(\log^2 n)$.

We do not know how to exclude the possibility of a one-pass *memoryless* algorithm which maintains both a competitive facility configuration and its assignment cost. However, the lower bound of [6] implies that every one-pass algorithm which maintains $o(k \log n)$ facilities incurs an *initial* assignment cost of $\omega(1)$ times the optimal cost, where k denotes the number of optimal facility locations and the initial assignment cost of a demand is the distance to its nearest facility just after the demand is considered. In other words, the initial assignment cost is not an $O(1)$-approximation to the optimal assignment cost unless the algorithm maintains $\Omega(k \log n)$ facilities. Therefore, to establish a constant competitive ratio, we consider the updated assignment cost every time a new facility significantly closer to an optimal facility opens. Even though this cost can be bounded in terms of the optimal cost, its value cannot be approximated unless we maintain sufficient information about the demand locations. Hence a logarithmic competitive ratio seems unavoidable for every one-pass memoryless algorithm that maintains both a competitive facility configuration and its assignment cost.

We also present a generalized version of the algorithm dealing with non-uniform facility costs and prove that its competitive ratio is less than 49. The algorithm runs in $O(n|M|)$ (i.e. linear) time and $O(|M|)$ space, where $|M|$ is the number of potential facility locations. In case of non-uniform facility costs, every algorithm must be aware of the potential facility locations and their opening costs. Nevertheless, we believe that this result is interesting because it shows that even for the general case of non-uniform facility costs, a near optimal facility configuration can be maintained by a one-pass algorithm that does not keep any information about the demand sequence other than the locations and the replacement radii of its facilities.

Problem Definition and Notation. We evaluate the performance of the algorithms presented in this paper using *competitive analysis* (see e.g. [2]). An algorithm is c-competitive if for all sequences of demand points, the cost incurred by the algorithm is at most c times the cost incurred by an optimal offline algorithm on the same instance.

We consider a metric space $\mathcal{M} = (M, d)$, where M is the set of points and $d : M \times M \mapsto \mathbb{R}_+$ is the distance function, which is non-negative, symmetric and satisfies the triangle inequality. For a point $u \in M$ and a subset of points $M' \subseteq M$, let $d(M', u) \equiv \min_{v \in M'}\{d(v, u)\}$ denote the distance between u and the closest point in M'. Let $d(\emptyset, u) \equiv \infty$. Let $B_u(r) \equiv \{v : d(u, v) \leq r\}$ denote the ball of radius r around u. We slightly abuse the notation by letting the same symbol denote both a demand (facility) and the corresponding point of the metric space.

For each point $w \in M$, we are also given the cost f_w of opening a facility at w. We distinguish between the special case of *uniform* facility costs, where the cost of opening a facility, denoted f, is the same for all points, and the general case of *non-uniform* facility costs, where the cost of opening a facility at $w \in M$, denoted f_w, depends on w and there are no restrictions on the facility costs.

The demand sequence consists of points, which are not necessarily distinct. We only consider unit demands and allow multiple demands to be located at the same point. We use n to denote the total number of demands. Given a (multi)set D of demand points, we seek for a facility configuration $F \subseteq M$ that minimizes $\sum_{w \in F} f_w + \sum_{u \in D} d(F, u)$.

2   The Algorithm for Uniform Facility Costs

The algorithm FFL (Fig. 1) maintains its *facility configuration*, denoted F, and its *facility replacement configuration* consisting of the *replacement radius* $m(w)$ of each facility w. The algorithm is based on the notion of *final distance*. For every point p and facility w, the *final distance* of p to w, denoted $g(w, p)$, is equal to $d(w, p) + 2m(w)$. The *final distance* of p to the facility configuration F, denoted $g(F, p)$, is

$$g(F, p) = \min_{w \in F}\{g(w, p)\} = \min_{w \in F}\{d(w, p) + 2m(w)\}. \tag{1}$$

If F is the current facility configuration, we refer to $g(F, p)$ as the final distance of p. For each demand u, we usually refer to $g(F, u)$ as the *final assignment cost* of u. Intuitively, the current algorithm's configuration gives p the guarantee that there will exist a facility within p's final distance at any future point in time.

$F \leftarrow \emptyset;$
for each new demand u:
$\quad g(F, u) \leftarrow \min_{z \in F}\{d(z, u) + 2m(z)\};$
$\quad$ with probability $\min\{g(F, u)/(\alpha f), 1\}$ do
$\quad\quad$ let w be the location of u;
$\quad\quad$ updateConfiguration(F, w);

updateConfiguration(F, w)
$\quad m(w) \leftarrow \min\{g(F, u), \alpha f\}/6;$
$\quad$ for each $z \in F$ do
$\quad\quad$ if $d(z, w) \leq m(z)$ then
$\quad\quad\quad F \leftarrow F \setminus \{z\};$
$\quad F \leftarrow F \cup \{w\};$

Fig. 1. The algorithm *Fast Facility Location* (FFL) parameterized by α

When a new demand u arrives, it computes its final assignment cost $g(F, u)$ with respect to the current algorithm's configuration. Then u opens a new facility w located at the same point as u with probability $\min\{g(F, u)/(\alpha f), 1\}$, where α is a parameter trading off the facility against the assignment cost. The replacement radius $m(w)$ of the new facility is set to $\min\{g(F, u), \alpha f\}/6$. For every existing facility z, if w is included in z's replacement ball $B_z(m(z))$, z is *replaced by* w (i.e. z is removed from F when w opens). Intuitively, w replaces every existing facility whose removal from $F \cup \{w\}$ does not increase the final distance of any points.

Let $|F_{\max}|$ be the maximum number of facilities maintained by FFL. It is straightforward to implement FFL in $O(n |F_{\max}|)$ time and $O(|F_{\max}|)$ space. Then we prove the following theorem.

Theorem 1. *For $\alpha = \frac{19}{8}$, the competitive ratio of FFL is less than 14.*

Preliminaries. For an arbitrary fixed sequence of demands, we compare the algorithm's cost with the cost of a fixed optimal facility configuration, denoted F^*. To avoid confusing the algorithm's facilities with the facilities in F^*, we use the term *optimal center*, or simply *center*, to refer to an optimal facility in F^* and the term *facility* to refer to an algorithm's facility in F.

In the optimal solution, each demand is assigned to the nearest center in F^*. For each demand u, we let c_u denote the optimal center to which u is assigned, let

$d_u^* \equiv d(c_u, u) = d(F^*, u)$ denote the optimal assignment cost of u, let $\mathrm{Asg}^* = \sum_u d_u^*$ denote the optimal assignment cost, and let $\mathrm{Fac}^* = |F^*| f$ denote the optimal facility cost. For the analysis, we use the clustering induced by F^* to map the demands and the algorithm's facilities to the optimal centers. In particular, a demand u (resp. facility w) is mapped to c_u (resp. c_w), i.e. the optimal center nearest to u (resp. w). If a new demand u opens a new facility w, c_w is the same as c_u.

Every time we want to explicitly refer to the algorithm's configuration at the moment when a demand arrives (resp. facility opens), we use the demand's (resp. facility's) identifier as a subscript. E.g. for a demand u (resp. facility w), F_u (resp. F_w) denotes the facility configuration just before u arrives (resp. w opens). Similarly, $g(F_u, u)$ (resp. $g(F_u, c_u)$) denotes the final distance of u (resp. c_u) when u arrives. If a new demand u opens a new facility w, F_w is the same as F_u.

Let $u_1, \ldots, u_n$ be the demand sequence considered by FFL. For every $j \geq 1$, let $F_{u_{j+1}}$ denote FFL's facility configuration after the demand u_j is considered. At that point, the facility cost of FFL is $\max_{1 \leq i \leq j}\{|F_{u_{i+1}}| f\}$ and the assignment cost is $\sum_{i=1}^{j} d(F_{u_{j+1}}, u_i)$. In the following, we prove that after the demand u_j is considered, FFL's expected cost is within a constant factor of the optimal facility cost and the optimal assignment cost for the demands $u_1, \ldots, u_j$.

2.1 A Sketch of the Analysis

In this section, we outline the proof of Theorem 1. For simplicity, we assume that the optimal solution consists of a single optimal center, denoted c, throughout the sketch of the proof. After explaining the role of the final distance in the analysis, we present the main ideas required for bounding the expected facility and assignment costs of FFL.

Final Distance. We prove that the final distance of every point p is non-increasing with time and that $B_p(g(F, p))$ will always contain a facility (cf. Proposition 1). Consequently, the (actual) assignment cost of a demand never exceeds its final assignment cost (with respect to the current or any of the past configurations). Thus we ignore facility replacements and use the demands' final assignment cost to bound their assignment cost throughout the execution of the algorithm.

Facility Cost. The facilities are divided into *supported* (by the optimal solution) and *unsupported* ones. A facility opened by a demand u is unsupported if $d_u^* < \min\{\frac{1}{\lambda} g(F_u, c), \frac{\alpha}{\lambda+1} f\}$ and supported otherwise, where λ is a positive constant chosen sufficiently large.

By definition, the optimal assignment cost of a demand u opening a supported facility is large enough to compensate for u's expected contribution to the facility cost. More specifically, the expected contribution of every demand u to the cost of supported facilities is at most $\frac{\lambda+1}{\alpha} d_u^*$ and the expected cost of supported facilities is at most $\frac{\lambda+1}{\alpha} \mathrm{Asg}^*$ (cf. Lemma 1).

The parameter λ is chosen so that every demand u opening an unsupported facility w is sufficiently close to the optimal center c compared to the final distance of c. Therefore, w's replacement ball is sufficiently large to include every unsupported

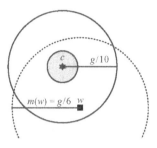

Fig. 2. Bounding the number of unsupported facilities. Let $\lambda = 10$, and let $g \equiv g(F_w, c)$ denote the final distance of c just before an unsupported facility w opens. Since w is unsupported, $d(w, c)$ is small compared to g (i.e. $d(w, c) < g/10$). As a result, $m(w) \approx g/6$ and $g(w, c) = d(w, c) + 2m(w) < g/2$. Hence, every unsupported facility opening after w lies in a ball of radius less than $g/20$ around c (the small grey ball) which is included in the replacement ball of w.

facility opening after w (cf. Fig. 2). Hence w is replaced no later than the moment when the first new unsupported facility opens (cf. Lemma 2). In general, the number of unsupported facilities is bounded by the number of optimal centers and their cost never exceeds Fac^*.

Putting the cost of supported and unsupported facilities together, we obtain that the expected facility cost is at most $\text{Fac}^* + \frac{\lambda+1}{\alpha} \text{Asg}^*$. We should highlight that the analysis treats every increase in the facility cost as an irrevocable decision.

Assignment Cost. We break down the analysis of the assignment cost into *disjoint phases* according to the final distance of c (this is possible because c's final distance is non-increasing). For every integer j, let $r(j) = f/2^j$. The j-th phase lasts as long as the final distance of c is in $[r(j), 2r(j))$. After the beginning of phase j, there will always exist a facility within a distance of $2r(j)$ from c. We distinguish between *inner* and *outer* demands. A demand u arriving in phase j is inner if $d_u^* < r(j)/\mu$ and outer otherwise (cf. Fig. 3), where μ is a positive constant.

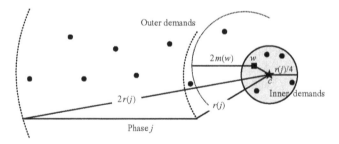

Fig. 3. Inner and outer demands in phase j (for $\mu = 4$). The optimal center c is in phase j as long as c's final distance is in $[r(j), 2r(j))$. A demand arriving in phase j is inner if it is included in the grey ball around c and outer otherwise. Let w be a new facility opened by an inner demand. Just before w opens, the final distance of w is less than $9r(j)/4$. Thus, $m(w) < 9r(j)/24$ and $g(w, c) < r(j)/4 + 9r(j)/12 = r(j)$. As a result, phase j ends when w opens.

The parameter μ is chosen sufficiently large so that inner demands arriving in the current phase can be regarded as essentially located at the same point as c (cf. Fig. 3). Using this intuition, we prove that (i) the current phase ends when an inner demand opens a new facility (cf. Proposition 3), and (ii) every inner demand arriving in phase j opens a new facility with probability at least $\frac{\mu-1}{\alpha\mu} 2^{-j}$ (cf. Proposition 4). Therefore, the expected number of inner demands arriving in phase j is at most $\frac{1}{\mu-1} 2^j$ and the expected number of inner demands arriving up to the end of phase j is at most $\frac{\alpha\mu}{\mu-1} 2^{j+1}$ (cf. the proof of Lemma 3).

Inner demands arriving up to the end of phase j are charged with their final assignment cost with respect to any algorithm's configuration in phase j. As long as c is in phase j, the final assignment cost of every demand is at most $f/2^{j-1}$ plus its optimal assignment cost. Thus the expected final assignment cost (with respect to any configuration in phase j) of the inner demands arriving up to the end of phase j is at most $\frac{4\alpha\mu}{\mu-1} f$ plus their optimal assignment cost (cf. Lemma 3).

On the other hand, when an outer demand u arrives, its final assignment cost $g(F_u, u)$ is at most $(2\mu + 1)$ times its optimal assignment cost (cf. Lemma 4). By the properties of the final distance, the (actual) assignment cost of u never exceeds $(2\mu+1)d_u^*$. Putting the assignment cost of inner and outer demands together, we obtain that the algorithm's expected assignment cost is at most $\frac{4\alpha\mu}{\mu-1} \mathrm{Fac}^* + (2\mu + 1)\mathrm{Asg}^*$.

2.2   Final Distance

Proposition 1. *Let F be the current facility configuration and let F' be any future facility configuration. For every point p, $g(F', p) \leq g(F, p)$ and $d(F', p) \leq g(F, p)$.*

Proof. Let w be the facility in F of minimum final distance to p, i.e. $g(w, p) = g(F, p)$. The proposition is true as long as w is open. We prove that if w is replaced by a new facility w', then $g(w', p) \leq g(w, p)$. We first show that $2m(w') \leq m(w)$:

$$2m(w') \leq \tfrac{1}{3} g(w, w') = \tfrac{1}{3}[d(w, w') + 2m(w)] \leq m(w) \qquad (2)$$

The first inequality holds because $m(w') \leq g(F_{w'}, w')/6$, since w' and the demand opening it are located at the same point, and $w \in F_{w'}$, since w is replaced by w'. The last inequality follows from $d(w, w') \leq m(w)$, because w is replaced by w'. Using (2) and $d(w, w') \leq m(w)$, we obtain that

$$g(w', p) \leq d(w, p) + d(w, w') + m(w) \leq d(w, p) + 2m(w) = g(F, p)$$

Consequently, the final distance of p does not increase when w is replaced by w'. In addition, $d(w', p) \leq g(w', p) \leq g(F, p)$ and $B_p(g(F, p))$ still contains a facility.   □

It is easy to establish the equivalent of the triangle inequality for the final distance.

Proposition 2. *For every facility configuration F and points p_1, p_2,*

$$g(F, p_1) \leq g(F, p_2) + d(p_1, p_2).$$

2.3 Facility Cost

Let u be a demand opening a new facility w. We recall that the facility w is *unsupported* if $d_u^* < \min\{\frac{1}{\lambda} g(F_u, c_u), \frac{\alpha}{\lambda+1} f\}$ and *supported* otherwise. Since u and w are located at the same point and are mapped to the same optimal center, $d(w, c_w) \equiv d_u^*$. Hence, it is equivalent to say that the facility w is unsupported if $d(w, c_w) < \min\{\frac{1}{\lambda} g(F_w, c_w), \frac{\alpha}{\lambda+1} f\}$ and supported otherwise.

Lemma 1. *The expected cost of supported facilities is at most $\frac{\lambda+1}{\alpha} \text{Asg}^*$.*

Proof. A new demand u opens a new facility with probability $\min\{g(F_u, u)/(\alpha f), 1\}$. The new facility is supported only if $d_u^* \geq \min\{\frac{1}{\lambda} g(F_u, c_u), \frac{\alpha}{\lambda+1} f\}$. Consequently, the expected contribution of u to the cost of supported facilities is at most $\frac{\lambda+1}{\alpha} d_u^*$. ☐

Lemma 2. *Let w be an unsupported facility, and let w' be a new unsupported facility mapped to c_w (i.e. to the same center as w). Then $d(w, w') \leq m(w)$ and w is replaced no later than w''s opening.*

Proof. We first show that for every unsupported facility w, $d(w, c_w) < \frac{6}{\lambda-1} m(w)$. By the definition of unsupported facilities, $d(w, c_w) < \min\{\frac{1}{\lambda} g(F_w, c_w), \frac{\alpha}{\lambda+1} f\}$. If $m(w) = \alpha f/6$, then $d(w, c_w) < \frac{\alpha}{\lambda+1} f = \frac{6}{\lambda+1} m(w) < \frac{6}{\lambda-1} m(w)$.

If $m(w) = g(F_u, u)/6 \equiv g(F_w, w)/6$, then

$$m(w) \geq \tfrac{1}{6} [g(F_w, c_w) - d(w, c_w)] > \tfrac{1}{6} [\lambda\, d(w, c_w) - d(w, c_w)] = \tfrac{\lambda-1}{6} d(w, c_w)$$

Next we prove that $d(w, c_w) < \frac{6}{\lambda-1} m(w)$ implies the lemma:

$$
\begin{aligned}
d(w, w') &\leq d(w, c_w) + d(w', c_w) \\
&< d(w, c_w) + \tfrac{1}{\lambda} [d(w, c_w) + 2m(w)] && \text{as } d(w', c_w) < \tfrac{1}{\lambda} g(w, c_w) \\
&< [\tfrac{\lambda+1}{\lambda} \tfrac{6}{\lambda-1} + \tfrac{2}{\lambda}] m(w) \leq m(w) && \text{as } d(w, c_w) < \tfrac{6}{\lambda-1} m(w)
\end{aligned}
$$

The second inequality holds because w' is an unsupported facility mapped to c_w and the final distance of c_w is non-increasing. The last inequality holds for every $\lambda \geq 9.5$. ☐

Hence, there always exists at most one unsupported facility mapped to each optimal center. Therefore, the total expected facility cost never exceeds $\text{Fac}^* + \frac{\lambda+1}{\alpha} \text{Asg}^*$.

2.4 Assignment Cost

We break down the analysis of the assignment cost of the demands mapped to every optimal center c into disjoint phases according to the final distance of c. For every integer $j \geq -2$, let $r(j) = f/2^j$, and let $r(-3) = \infty$. The j-th phase, starts just after the final distance of c becomes less than $r(j-1)$ and ends when the final distance of c becomes less than $r(j)$. We use the convention that phase -2 starts when the first demand arrives. The optimal center c is in phase j while $g(F, c) \in [r(j), r(j-1))$. For every $j \geq -1$, $g(F, c) < 2r(j)$ after c enters phase j.

A demand u arrives in phase j if $g(F_u, c_u) \in [r(j), r(j-1))$, i.e. when u arrives, the optimal center to which u is mapped is in phase j. A demand u arriving in phase j is called *inner* if $d_u^* < r(j)/\mu$ and *outer* otherwise. We use the convention that every demand arriving in phase -2 is inner.

Inner Demands. Inner demands are charged with their final assignment cost with respect to any algorithm's configuration in the current phase.

Proposition 3. *Let u be an inner demand which arrives in phase j and opens a new facility w. Then, $g(w, c_u) < r(j)$ and the j-th phase ends as soon as w opens.*

Proof. If u arrives in phase $j \geq 1$, $g(F_u, c_u) < 2r(j)$. Since u and w are located at the same point, $d(w, c_u) \equiv d_u^* < r(j)/\mu$. Therefore, for every $\mu \geq 4$,

$$
\begin{aligned}
g(w, c_u) &= d(w, c_u) + 2m(w) \\
&\leq d(w, c_u) + \tfrac{1}{3}[g(F_u, c_u) + d_u^*] && \text{as } m(w) \leq g(F_u, u)/6 \\
&< [\tfrac{4}{3}\tfrac{1}{\mu} + \tfrac{2}{3}]r(j) \leq r(j) && \text{as } d_u^* < r(j)/\mu \text{ and } g(F_u, c_u) < 2r(j)
\end{aligned}
$$

If u arrives in phase -2, $g(w, c_u) \leq (1 + \tfrac{\alpha}{3})f$ because $d(w, c_u) \equiv d_u^* \leq f$ and $m(w) \leq \alpha f/6$. Hence, $g(w, c_u) < 4f$ for every $\alpha < 9$. In both cases, the final distance of c_u becomes less than $r(j)$ and the j-th phase ends when w opens. $\square$

Proposition 4. *Every inner demand arriving in phase j opens a new facility with probability at least $\min\{\tfrac{\mu-1}{\alpha\mu} 2^{-j}, 1\}$.*

Proof. If u is an inner demand arriving in phase j, $g(F_u, c_u) \geq r(j) = f/2^j$ and $d_u^* < r(j)/\mu$. Then,

$$
g(F_u, u)/(\alpha f) \geq (g(F_u, c_u) - d_u^*)/(\alpha f) > \tfrac{\mu-1}{\mu} r(j)/(\alpha f) = \tfrac{\mu-1}{\alpha\mu} 2^{-j}   \qquad \square
$$

Lemma 3. *For every optimal center c, the expected final assignment cost of the inner demands mapped to c and considered up to the end of the current phase is at most $\tfrac{4\alpha\mu}{\mu-1} f$ plus their optimal assignment cost.*

Proof. By Propositions 3 and 4, every inner demand arriving in phase j opens a new facility and concludes the current phase with probability at least $\min\{\tfrac{\mu-1}{\alpha\mu} 2^{-j}, 1\}$. Let us assume that $\mu \geq 4$ (as required by Proposition 3) and $\alpha \in [2, 3]$ (for simplicity).

If a demand mapped to c arrives in phase -2, it concludes phase -2 with certainty. Hence, there is at most one (inner) demand arriving in phase -2. In addition, phase -2 has ended when the assignment cost of the first demand mapped to c is considered.

For every phase $j \geq -1$, the probability that an inner demand opens a new facility and concludes phase j is at least $\tfrac{\mu-1}{\alpha\mu} 2^{-j}$. Therefore, the expected number of inner demands arriving in phase j is at most $\tfrac{\alpha\mu}{\mu-1} 2^j$.

Let $j \geq -1$ be the current phase of c. By linearity of expectation, the expected number of inner demands arriving up to the end of phase j is at most $\tfrac{\alpha\mu}{\mu-1} 2^{j+1}$. After c enters phase j, the final assignment cost of a demand u mapped to c (with respect to the current configuration F) is

$$
g(F, u) \leq g(F, c) + d_u^* < 2r(j) + d_u^* = f/2^{j-1} + d_u^*
$$

Since the expected number of inner demands mapped to c and considered up to the end of phase j is at most $\tfrac{\alpha\mu}{\mu-1} 2^{j+1}$, their expected final assignment cost never exceeds $\tfrac{4\alpha\mu}{\mu-1} f$ plus their optimal assignment cost. $\square$

Outer Demands. Outer demands are charged with their final assignment cost with respect to the algorithm's configuration at their arrival time. The following lemma is an immediate consequence of the definition of outer demands.

Lemma 4. *The final assignment cost of every outer demand u is at most $(2\mu + 1)d_u^*$.*

Combining Lemmas 3 and 4 and Proposition 1, we obtain that the algorithm's expected assignment cost is at most $\frac{4\alpha\mu}{\mu-1}\operatorname{Fac}^* + (2\mu + 1)\operatorname{Asg}^*$.

Concluding the Proof of Theorem 1. For every $\lambda \geq 9.5$, $\mu \geq 4$, and $\alpha \in [2, 3]$, the expected cost of FFL is at most

$$\left(\tfrac{4\alpha\mu}{\mu-1}+1\right)\operatorname{Fac}^* + \left(\tfrac{\lambda+1}{\alpha}+2\mu+1\right)\operatorname{Asg}^* \leq \max\{\tfrac{4\alpha\mu}{\mu-1}+1, \tfrac{\lambda+1}{\alpha}+2\mu+1\}\left(\operatorname{Fac}^* + \operatorname{Asg}^*\right)$$

Using $\alpha = \frac{19}{8}$, $\lambda = 10$, and $\mu = 4$, we obtain a competitive ratio of $13\frac{2}{3}$. The competitive ratio can be improved to 12 by a careful analysis.

3   The Algorithm for Non-uniform Facility Costs

Then, we present a generalized version of FFL dealing with non-uniform facility costs. As before, the algorithm FNFL (Fig. 4) maintains its facility configuration F and the replacement radius $m(w)$ of each facility w.

Every new demand u computes its final assignment cost $g(F, u)$ with respect to the current algorithm's configuration. In case of non-uniform facility costs, opening a facility at the same point as u may be too expensive or even infeasible. Hence u finds the facility w in $B_u(\frac{1}{\lambda-1}g(F, u))$ that minimizes $f_w + d(w, u)$, where the parameter λ is chosen so that w is located sufficiently close to u. Then u opens a new facility at w with probability $\min\{g(F, u)/(\alpha f_w), 1\}$. The replacement radius $m(w)$ of the new facility is set to $\min\{g(F, w), 3f_w + \frac{\lambda-3}{2}d(w, u)\}/6$. Every existing facility z that includes w in its replacement ball $B_z(m(z))$ is replaced by w.

Let $|M|$ be the number of potential facility locations. FNFL can be implemented in $O(n|M|)$ time and $O(|M|)$ space. Then we sketch the proof of the following theorem.

Theorem 2. *For $\lambda = 20$ and $\alpha = \frac{38}{7}$, the competitive ratio of FNFL is less than 49.*

$F \leftarrow \emptyset$;	updateConfiguration(F, w)
for each new demand u:	$g(F, w) \leftarrow \min_{z \in F}\{d(z, w) + 2m(z)\}$;
$\quad g(F, u) \leftarrow \min_{z \in F}\{d(z, u) + 2m(z)\}$;	$m(w) \leftarrow \frac{1}{6}\min\{g(F, w), 3f_w + \frac{\lambda-3}{2}d(w, u)\}$;
$\quad$ let w be the facility in $B_u(\frac{1}{\lambda-1}g(F, u))$	for each $z \in F$ do
$\quad\quad$ that minimizes $f_w + d(w, u)$;	$\quad$ if $d(z, w) \leq m(z)$ then
$\quad$ with probability $\min\{g(F, u)/(\alpha f_w), 1\}$	$\quad\quad F \leftarrow F \setminus \{z\}$;
$\quad\quad$ do updateConfiguration(F, w);	$F \leftarrow F \cup \{w\}$;

Fig. 4. The algorithm Fast Non-Uniform Facility Location (FNFL) parameterized by α and λ

Preliminaries. We use the same notation as in the analysis of FFL. As for the facility costs, f_p denotes the cost of opening a new facility at point p. The optimal facility cost is $\text{Fac}^* = \sum_{c \in F^*} f_c$. For every demand u, f_{c_u} denotes the facility cost of the optimal center to which u is mapped.

The outline of the proof is similar to that in Section 2.1. We first observe that Propositions 1 and 2 also hold for FNFL. A basic property of FNFL is that for every demand u with $d_u^* < g(F_u, c_u)/\lambda$ (i.e. u is sufficiently close to c_u), $c_u \in B_u(\frac{1}{\lambda-1} g(F_u, u))$. Therefore, for every facility w opened by a demand u with $d_u^* < g(F_u, c_u)/\lambda$, the opening cost of w is no greater than $f_{c_u} + d_u^*$.

Facility Cost. Supported and unsupported facilities are now defined in a different way. Let u be a demand opening a new facility w. The facility w is *unsupported* if $d_u^* < g(F_u, c_u)/\lambda$ and $m(w) \geq \frac{\lambda-3}{12} d(w, c_w)$, and *supported* otherwise. If w is an unsupported facility, $f_w \leq f_{c_u} + d_u^*$ because $c_u \in B_u(\frac{1}{\lambda-1} g(F_u, u))$. As in Section 2.3, we prove that the expected facility cost of FNFL is at most $\text{Fac}^* + \frac{\lambda+1}{\alpha} \text{Asg}^*$.

Assignment Cost. For every optimal center c, we break down the analysis of the assignment cost of the demands mapped c into disjoint phases according to the final distance of c. The definition of phases takes the facility cost of c into account. For every integer $j \geq -3$, let $r_c(j) = f_c/2^j$, and let $r_c(-4) = \infty$. For every $j \geq -3$, the optimal center c is in phase j as long as $g(F, c) \in [r_c(j), r_c(j-1))$. A demand u arriving in phase j is *inner* if $d_u^* < r_{c_u}(j)/\lambda$ and *outer* otherwise.

The assignment cost of every outer demand u never exceeds $(2\lambda + 1)d_u^*$. Inner demands are charged with their final assignment cost with respect to the current configuration. Similarly to the proof of Lemma 3, we show that the expected final assignment cost of inner demands mapped to c and considered up to the end of phase j is no greater than $\frac{4\alpha(\lambda+8)}{\lambda-1} f_c$ plus their optimal assignment cost. Consequently, the expected assignment cost of FNFL is at most $\frac{4\alpha(\lambda+8)}{\lambda-1} \text{Fac}^* + (2\lambda + 1)\text{Asg}^*$.

Concluding the Proof of Theorem 2. The expected cost of FNFL is at most

$$\left(\frac{4\alpha(\lambda+8)}{\lambda-1} + 1\right) \text{Fac}^* + \left(1 + \frac{1}{\alpha}\right)(2\lambda + 1) \text{Asg}^*$$

for every $\lambda \in [20, 34]$ and $\alpha \in [\frac{19}{7}, \frac{38}{7}]$. For $\lambda = 20$ and $\alpha = \frac{38}{7}$, the competitive ratio of FNFL is less than 48.6.

Acknowledgements. The author thanks an anonymous referee whose detailed comments greatly assisted in improving the final presentation.

References

1. A. Anagnostopoulos, R. Bent, E. Upfal, and P. Van Hentenryck. A Simple and Deterministic Competitive Algorithm for Online Facility Location. *Information and Computation*, 194:175–202, 2004.
2. A. Borodin and R. El-Yaniv. *Online Computation and Competitive Analysis*. Cambridge University Press, 1998.

3. M. Bădoiu, A. Czumaj, P. Indyk, and C. Sohler. Facility Location in Sublinear Time. *ICALP '05*, LNCS 3580, pp. 866–877, 2005.
4. M. Charikar, C. Chekuri, T. Feder, and R. Motwani. Incremental Clustering and Dynamic Information Retrieval. *STOC '97*, pp. 626–635, 1997.
5. M. Charikar, L. O'Callaghan, and R. Panigrahy. Better Streaming Algorithms for Clustering Problems. *STOC '03*, pp. 30–39, 2003.
6. D. Fotakis. On the Competitive Ratio for Online Facility Location. *ICALP '03*, LNCS 2719, pp. 637–652, 2003.
7. D. Fotakis. Incremental Algorithms for Facility Location and k-Median. *ESA '04*, LNCS 3221, pp. 347–321, 2004.
8. G. Frahling and C. Sohler. Coresets in Dynamic Geometric Data Streams. *STOC '05*, pp. 209–217, 2005.
9. S. Guha, A. Meyerson, N. Mishra, R. Motwani, and L. O'Callaghan. Clustering Data Streams: Theory and Practice. *IEEE Transactions on Knowledge and Data Engineering*, 15(3):515–528, 2003.
10. S. Guha, N. Mishra, R. Motwani, and L. O'Callaghan. Clustering Data Streams. *FOCS '00*, pp. 359–366, 2000.
11. P. Indyk. Algorithms for Dynamic Geometric Problems over Data Streams. *STOC '04*, pp. 373–380, 2004.
12. A. Meyerson. Online Facility Location. *FOCS '01*, pp. 426–431, 2001.
13. S. Muthukrishnan. Data Streams: Algorithms and Applications. *SODA '03*, invited talk. Available at http://athos.rutgers.edy/~muthu/stream-1-1.ps, 2003.
14. M. Thorup. Quick k-Median, k-Center, and Facility Location for Sparse Graphs. *SIAM J. on Computing*, 34(2):405–432, 2005.

Energy-Efficient Algorithms for Flow Time Minimization

Susanne Albers[1] and Hiroshi Fujiwara[2,*]

[1] Institut für Informatik, Albert-Ludwigs-Universität Freiburg,
Georges-Köhler-Allee 79, 79110 Freiburg, Germany
salbers@informatik.uni-freiburg.de
[2] Department of Communications and Computer Engineering,
Graduate School of Informatics, Kyoto University, Japan
fujiwara@lab2.kuis.kyoto-u.ac.jp

Abstract. We study scheduling problems in battery-operated computing devices, aiming at schedules with low total energy consumption.
While most of the previous work has focused on finding feasible schedules
in deadline-based settings, in this paper we are interested in schedules
that guarantee good response times. More specifically, our goal is to
schedule a sequence of jobs on a variable speed processor so as to minimize the total cost consisting of the power consumption and the total
flow time of all the jobs. We first show that when the amount of work,
for any job, may take an arbitrary value, then no online algorithm can
achieve a constant competitive ratio. Therefore, most of the paper is
concerned with unit-size jobs. We devise a deterministic constant competitive online algorithm and show that the offline problem can be solved
in polynomial time.

1 Introduction

Embedded systems and portable devices play an ever-increasing role in every
day life. Prominent examples are mobile phones, palmtops and laptop computers
that are used by a significant fraction of the population today. Many of these
devices are battery-operated so that effective power management strategies are
essential to guarantee a good performance and availability of the systems. The
microprocessors built into these devices can typically perform tasks at different
speeds – the higher the speed, the higher the power consumption is. As a result,
there has recently been considerable research interest in dynamic speed scaling
strategies; we refer the reader to [1, 2, 3, 7, 10, 13] for a selection of the papers
that have been published in algorithms conferences.

Most of the previous work considers a scenario where a sequence of jobs,
each specified by a release time, a deadline and an amount of work that must
be performed to complete the task, has to be scheduled on a single processor.
The processor may run at variable speed. At speed s, the power consumption is

---

* This work was done while visiting the University of Freiburg.

B. Durand and W. Thomas (Eds.): STACS 2006, LNCS 3884, pp. 621–633, 2006.
© Springer-Verlag Berlin Heidelberg 2006

$P(s) = s^\alpha$ per time unit, where $\alpha > 1$ is a constant. The goal is to find a feasible schedule such that the total power consumption over the entire time horizon is as small as possible. While this basic framework gives insight into effective power conservation, it ignores the important aspect that users typically expect good response times for their jobs. Furthermore, in many computational systems, jobs are not labeled with deadlines. For example, operating systems such as Window and Unix installed on laptops do not employ deadline-based scheduling.

Therefore, in this paper, we study algorithms that minimize energy usage and at the same time guarantee good response times. In the scientific literature, response time is modeled as *flow time*. The flow time of a job is the length of the time interval between the release time and the completion time of the job. Unfortunately, energy minimization and flow time minimization are orthogonal objectives. To save energy, the processor should run at low speed, which yields high flow times. On the other hand, to ensure small flow times, the processor should run at high speed, which results in a high energy consumption. In order to overcome this conflict, Pruhs et al. [10] recently studied the problem of minimizing the average flow time of a sequence of jobs when a *fixed amount of energy* is available. They presented a polynomial time offline algorithm for unit-size jobs. However, it is not clear how to handle the online scenario where jobs arrival times are unknown.

Instead, in this paper, we propose a different approach to integrate energy and flow time minimization: We seek schedules that minimize the total cost consisting of the power consumption and the flow times of jobs. More specifically, a sequence of jobs, each specified by an amount of work, arrives over time and must be scheduled on one processor. Preemption of jobs is not allowed. The goal is to dynamically set the speed of the processor so as to minimize the sum of (a) the total power consumption and (b) the total flow times of all the jobs. Such combined objective functions have been studied for many other bicriteria optimization problems with orthogonal objectives. The papers [5, 8], e.g., consider a TCP acknowledgement problem, minimizing the sum of acknowledgement costs and acknowledgement delays incurred for data packets. In [6] the authors study network design and minimize the total hardware and QoS costs. More generally, in the classical facility location problem, one minimizes the sum of the facility installation and total client service costs, see [4, 9] for surveys.

For our energy/flow-time minimization problem, we are interested in both online and offline algorithms. Following [11], an online algorithm A is said to be c-competitive if there exists a constant a such that, for all job sequences σ, the total cost $A(\sigma)$ satisfies $A(\sigma) \leq c \cdot \mathrm{OPT}(\sigma) + a$, where $\mathrm{OPT}(\sigma)$ is the cost of an optimal offline algorithm.

Previous work: In their seminal paper, Yao et al. [13] introduced the basic problem of scheduling a sequence of jobs, each having a release time, a deadline and a certain workload, so as to minimize the energy usage. Here, preemption of jobs is allowed. Yao et al. showed that the offline problem can be solved optimally in polynomial time and presented two online algorithms called *Average Rate* and *Optimal Available*. They analyzed *Average Rate*, for $\alpha \geq 2$, and proved an upper bound of $2^\alpha \alpha^\alpha$ and a lower bound of α^α on the competitiveness. Bansal et al. [2] studied *Optimal Available* and showed that its competitive ratio is exactly α^α. Furthermore, they developed a new algorithm that achieves a competitiveness of $2(\alpha/(\alpha - 1))^\alpha e^\alpha$ and proved that any randomized online algorithm has a performance ratio of at least $\Omega((4/3)^\alpha)$.

Irani et al. [7] studied an extended scenario where the processor can be put into a low-power sleep state when idle. They gave an offline algorithm that achieves a 3-approximation and developed a general strategy that transforms an online algorithm for the setting without sleep state into an online algorithm for the setting with sleep state. They obtain constant competitive online algorithms, but the constants are large. For the famous cube root rule $P(s) = s^3$, the competitive ratio is 540. The factor can be reduced to 84, see [2]. Settings with several sleep states were considered in [1]. Speed scaling to minimize the maximum temperature of a processor was addressed in [2, 3].

As mentioned above, Pruhs et al. [10] study the problem of minimizing the average flow time of jobs given a fixed amount of energy. For unit-size jobs, they devise a polynomial time algorithm that simultaneously computes, for each possible energy level, the schedule with smallest average flow time.

Our contribution: We investigate the problem of scheduling a sequence of n jobs on a variable speed processor so as to minimize the total cost consisting of the power consumption and the flow times of jobs. We first show that when the amount of work, for any job, may take an arbitrary value, then any deterministic online algorithm has a competitive ratio of at least $\Omega(n^{1-1/\alpha})$. This result implies that speed scaling does not help to overcome bad scheduling decisions: It is well-known that in standard scheduling, no online algorithm for flow time minimization can be better than $\Omega(n)$-competitive. Our lower bound, allowing speed scaling, is almost as high.

Because of the $\Omega(n^{1-1/\alpha})$ lower bound, most of our paper is concerned with unit-size jobs. We develop a deterministic phase-based online algorithm that achieves a constant competitive ratio. The algorithm is simple and requires scheduling decisions to be made only every once in a while, which is advantageous in low-power devices. Initially, the algorithm computes a schedule for the first batch of jobs released at time 0. While these jobs are being processed, the algorithm collects the new jobs that arrive in the meantime. Once the first batch of jobs is finished, the algorithm computes a schedule for the second batch. This process repeats until no more jobs arrive. Within each batch the processing speeds are easy to determine. When there are i unfinished jobs in the batch, the speed is set to $\sqrt[\alpha]{i/c}$, where c is a constant that depends on the value of α. We prove that the competitive ratio of our algorithm is upper bounded by $8.3e(1 + \Phi)^\alpha$, where $\Phi = (1 + \sqrt{5})/2 \approx 1.618$ is the Golden Ratio. We remark that a phase-based scheduling algorithm was also used in makespan minimization on parallel machines [12]. However, for our problem, the scheduling strategy within the phases and the analysis techniques employed are completely different.

Furthermore, in this paper we develop a polynomial time algorithm for computing an optimal offline schedule. We would like to point out that we could use the algorithm by Pruhs et al. [10], but this would yield a rather complicated algorithm for our problem. Instead, we design a simple, direct algorithm based on dynamic programming. Our approach can also be used to address the problem of Pruhs et al., i.e. we are able to determine a schedule with minimum flow time given a fixed amount of enery. This can be seen as an additional advantage of our new objective function.

2   Preliminaries

Consider a sequence of jobs $\sigma = \sigma_1, \ldots, \sigma_n$ which are to be scheduled on one processor. Job σ_i is released at time r_i and requires p_i CPU cycles. We assume $r_1 = 0$ and $r_i \leq r_{i+1}$, for $i = 1, \ldots, n - 1$. A schedule $\mathcal{S}$ specifies, for each job σ_i, a time interval I_i and a speed s_i such that σ_i is processed at speed s_i continuously, without interruption, throughout I_i. Let $P(s) = s^\alpha$ be the power consumption per time unit of the CPU depending on s. The constant $\alpha > 1$ is a real number. As $P(s)$ is convex, we may assume w.l.o.g. that each σ_i is processed at a constant speed s_i. A schedule $\mathcal{S}$ is feasible if, for any i, interval I_i starts no earlier than r_i, and the processing requirements are met, i.e. $p_i = s_i |I_i|$. Here $|I_i|$ denotes the length of I_i. Furthermore, in a feasible schedule $\mathcal{S}$ the intervals I_i must be non-overlapping. The energy consumption of $\mathcal{S}$ is $E(\mathcal{S}) = \sum_{i=1}^{n} P(s_i)|I_i|$. For any i, let c_i be the completion time of job i, i.e. c_i is equal to the end of I_i. The flow time of job i is $f_i = c_i - r_i$ and the flow time of $\mathcal{S}$ is given by $F(\mathcal{S}) = \sum_{i=1}^{n} f_i$. We seek schedules $\mathcal{S}$ that minimize the sum $g(\mathcal{S}) = E(\mathcal{S}) + F(\mathcal{S})$.

3   Arbitrary Size Jobs

We show that if the jobs' processing requirements may take arbitrary values, then no online algorithm can achieve a bounded competitive ratio. The proof of the following theorem is omitted due to space constraints.

Theorem 1. *The competitive ratio of any deterministic online algorithm is* $\Omega(n^{1-1/\alpha})$ *if the processing requirements* $p_1, \ldots, p_n$ *may take arbitrary values.*

4   An Online Algorithm for Unit-Size Jobs

In this section we study the case that the processing requirements of all jobs are the same, i.e. $p_i = 1$, for all jobs. We develop a deterministic online algorithm that achieves a constant competitive ratio, for all α. The algorithm is called *Phasebal* and aims at balancing the incurred power consumption with the generated flow time. If α is small, then the ratio is roughly $1 : \alpha - 1$. If α is large, then the ratio is $1 : 1$. As the name suggests, the algorithm operates in phases.

Let n_1 be the number of jobs that are released initially at time $t = 0$. In the first phase *Phasebal* processes these jobs in an optimal or nearly optimal way, ignoring jobs that may arrive in the meantime. More precisely, the speed sequence for the n_1 jobs is $\sqrt[\alpha]{n_1/c}, \sqrt[\alpha]{(n_1 - 1)/c}, \ldots, \sqrt[\alpha]{1/c}$, i.e. the j-th of these n_1 jobs is executed at speed $\sqrt[\alpha]{(n_1 - j + 1)/c}$ for $j = 1, \ldots, n_1$. Here c is a constant that depends on α. Let n_2 be the number of jobs that arrive in phase 1. *Phasebal* processes these jobs in a second phase. In general, in phase i *Phasebal* schedules the n_i jobs that arrived in phase $i - 1$ using the speed sequence $\sqrt[\alpha]{(n_i - j + 1)/c}$, for $j = 1, \ldots, n_i$. Again, jobs that arrive during the phase are ignored until the end of the phase. A formal description of the algorithm is as follows.

Algorithm Phasebal: If $\alpha < (19 + \sqrt{161})/10$, then set $c := \alpha - 1$; otherwise set $c := 1$. Let n_1 be the number of jobs arriving at time $t = 0$ and set $i = 1$. While $n_i > 0$, execute the following two steps: (1) For $j = 1, \ldots, n_i$, process the j-th job using a speed of $\sqrt[\alpha]{(n_i - j + 1)/c}$. We refer to this entire time interval as phase i. (2) Let n_{i+1} be the number of jobs that arrive in phase i and set $i := i + 1$.

Theorem 2. *Phasebal has a competitiveness of at most* $(1+\Phi)(1+\Phi^{\frac{\alpha}{(2\alpha-1)}})^{(\alpha-1)}$ $\frac{\alpha^\alpha}{(\alpha-1)^{\alpha-1}} \min\{\frac{5\alpha-2}{2\alpha-1}, \frac{4}{2\alpha-1} + \frac{4}{\alpha-1}\}$, *where* $\Phi = (1 + \sqrt{5})/2 \approx 1.618$.

Before proving Theorem 2, we briefly discuss the competitiveness. We first observe that $\frac{\alpha^\alpha}{(\alpha-1)^{\alpha-1}} \leq e\alpha$. Moreover, $\frac{\alpha(5\alpha-2)}{2\alpha-1}$ is increasing in α, while $\frac{4\alpha}{2\alpha-1} + \frac{4\alpha}{\alpha-1}$ is decreasing in α. Standard algebraic manipulations show that the latter two expressions are equal for $\alpha_0 = (19 + \sqrt{161})/10$. Thus, the competitive ratio is upper bounded by $(1 + \Phi)^\alpha e^{\frac{\alpha_0(5\alpha_0 - 2)}{2\alpha_0 - 1}} < (1 + \Phi)^\alpha e \cdot 8.22$.

In the remainder of this section we will analyze *Phasebal*. The global analysis consists of two cases. We will first address $c = 1$ and then $c = \alpha - 1$. In each case we first upper bound the total cost incurred by *Phasebal* and then lower bound the cost of an optimal schedule. In the case $c = 1$ we will consider a pseudo-optimal algorithm that operates with similar speeds as *Phasebal*. We will prove that the cost of such a pseudo-optimal algorithm is at most a factor of 2 away from the true optimum. In any case we will show that an optimal or pseudo-optimal algorithm finishes jobs no later than *Phasebal*. This property will be crucial to determine the time intervals in which optimal schedules process jobs and to lower bound the corresponding speeds. These speed bounds will then allow us to estimate the optimal cost and to finally compare it to the online cost.

Let $t_0 = 0$ and t_i be the time when phase i ends, i.e. the n_i jobs released during phase $i - 1$ (released initially, if $i = 1$) are processed in the time interval $[t_{i-1}, t_i)$, which constitutes phase i. Given a job sequence σ, let $\mathcal{S}_{PB}$ be the schedule of *Phasebal* and let $\mathcal{S}_{OPT}$ be an optimal schedule.

Case 1: $c = 1$ We start by analyzing the cost and time horizon of $\mathcal{S}_{PB}$. Suppose that there are k phases, i.e. no new jobs arrive in phase k. In phase i the algorithm needs $1/\sqrt[\alpha]{n_i - j + 1}$ time units to complete the j-th job. Thus the power consumption in the phase is

$$\sum_{j=1}^{n_i}(\sqrt[\alpha]{n_i-j+1})^{\alpha}/\sqrt[\alpha]{n_i-j+1} = \sum_{j=1}^{n_i}(n_i-j+1)^{1-1/\alpha}$$

$$\leq \tfrac{\alpha}{2\alpha-1}(n_i^{2-1/\alpha}-1)+n_i^{1-1/\alpha}.$$

The length of phase i is

$$T(n_i) = \sum_{j=1}^{n_i}1/\sqrt[\alpha]{n_i-j+1} \leq \tfrac{\alpha}{\alpha-1}n_i^{1-1/\alpha}. \tag{1}$$

As for the flow time, the n_i jobs scheduled in the phase incur a flow time of

$$\sum_{j=1}^{n_i}(n_i-j+1)/\sqrt[\alpha]{n_i-j+1} \leq \tfrac{\alpha}{2\alpha-1}(n_i^{2-1/\alpha}-1)+n_i^{1-1/\alpha},$$

while the n_{i+1} jobs released during the phase incur a flow time of at most n_{i+1} times the length of the phase. We obtain

$$g(\mathcal{S}_{PB}) \leq \sum_{i=1}^{k}(\tfrac{2\alpha}{2\alpha-1}(n_i^{2-1/\alpha}-1)+2n_i^{1-1/\alpha}) + \sum_{i=1}^{k-1}n_{i+1}\tfrac{\alpha}{\alpha-1}n_i^{1-1/\alpha}.$$

The second sum is bounded by $\sum_{i=1}^{k-1}\tfrac{\alpha}{\alpha-1}\max\{n_i,n_{i+1}\}^{2-1/\alpha} \leq \sum_{i=1}^{k}\tfrac{2\alpha}{\alpha-1}n_i^{2-1/\alpha}$ and we conclude

$$g(\mathcal{S}_{PB}) \leq 2\sum_{i=1}^{k}(\tfrac{\alpha}{2\alpha-1}(n_i^{2-1/\alpha}-1)+n_i^{1-1/\alpha}+\tfrac{\alpha}{\alpha-1}n_i^{2-1/\alpha}). \tag{2}$$

We next lower bound the cost of an optimal schedule. As mentioned before, it will be convenient to consider a pseudo-optimal schedule $\mathcal{S}_{POPT}$. This is the best schedule that satisfies the constraint that, at any time, if there are ℓ *active* jobs, then the processor speed is at least $\sqrt[\alpha]{\ell}$. We call a job active if it has arrived but is not yet finished. In the next lemma we show that the objective function value $g(\mathcal{S}_{POPT})$ is not far from the true optimum $g(\mathcal{S}_{OPT})$.

Lemma 1. *For any job sequence, $g(\mathcal{S}_{POPT}) \leq 2g(\mathcal{S}_{OPT})$.*

Proof. Consider the optimal schedule $g(\mathcal{S}_{OPT})$. We may assume w.l.o.g. that in this schedule the speed only changes when a jobs gets finished of new jobs arrive. We partition the time horizon of $\mathcal{S}_{OPT}$ into a sequence of intervals $I_1,\ldots,I_m$ such that, for any such interval, the number of active jobs does not change. Let $E(I_i)$ and $F(I_i)$ be the energy consumption and flow time, respectively, generated in I_i, $i=1,\ldots,m$. We have $E(I_i)=s_i^{\alpha}\delta_i$ and $F(I_i)=\ell_i\delta_i$, where s_i is the speed, ℓ_i is the number of active jobs in I_i and δ_i is the length of I_i. Clearly $g(\mathcal{S}_{OPT})=\sum_{i=1}^{m}(E(I_i)+F(I_i))$.

Now we change $\mathcal{S}_{OPT}$ as follows. In any interval I_i with $s_i < \sqrt[\alpha]{\ell_i}$ we increase the speed to $\sqrt[\alpha]{\ell_i}$, incurring an energy consumption of $\ell_i\delta_i$, which is equal to

$F(I_i)$ in original schedule $\mathcal{S}_{OPT}$. In this modification step, the flow time of jobs can only decrease. Because of the increased speed, the processor may run out of jobs in some intervals. Then the processor is simply idle. We obtain a schedule whose cost is bounded by $\sum_{i=1}^{m}(E(I_i) + 2F(I_i)) \leq 2g(\mathcal{S}_{OPT})$ and that satisfies the constraint that the processor speed it at least $\sqrt[\alpha]{\ell}$ in intervals with ℓ active job. Hence $q(\mathcal{S}_{POPT}) < 2g(\mathcal{S}_{OPT})$. □

The next lemma shows that in $\mathcal{S}_{POPT}$ jobs finish no later than in $\mathcal{S}_{PB}$.

Lemma 2. *For $c = 1$, in $\mathcal{S}_{POPT}$ the n_1 jobs released at time t_0 are finished by time t_1 and the n_i jobs released during phase $i - 1$ are finished by time t_i, for $i = 2, \ldots, k$.*

Proof. We show the lemma inductively. As for the n_1 jobs released at time t_0, the schedule $\mathcal{S}_{POPT}$ processes the j-th of these jobs at a speed of at least $\sqrt[\alpha]{n_1 - j + 1}$ because there are at least $n - j + 1$ active jobs. Thus the n_1 jobs are completed no later than $\sum_{j=1}^{n_1} 1/\sqrt[\alpha]{n_1 - j + 1}$, which is equal to the length of the first phase, see (1).

Now suppose that jobs released by time t_{i-1} are finished by time t_i and consider the n_{i+1} jobs released in phase i. At time t_i there are at most these n_{i+1} jobs unfinished. Let $\overline{n}_{i+1}$ be the actual number of active jobs at that time. Again, the j-th of these jobs is processed at a speed of at least $(\overline{n}_{i+1} - j + 1)^{1/\alpha}$ so that the execution of these $\overline{n}_{i+1}$ jobs ends no later than $\sum_{j=1}^{\overline{n}_{i+1}}(\overline{n}_{i+1} - j + 1)^{-1/\alpha}$ and this sum is not larger than the length of phase $i + 1$, see (1). □

Lemma 3. *If a schedule has to process ℓ jobs during a time period of length $T \leq \ell\sqrt[\alpha]{\alpha - 1}$, then its total cost is at least $FLAT(\ell, T) \geq (\ell/T)^{\alpha}T + T$.*

The proof is omitted.

Lemma 4. *For $\alpha \geq 2$, there holds $g(\mathcal{S}_{POPT}) \geq C^{1-\alpha}(1+\Phi)^{-1}(1+\Phi^{\alpha/(2\alpha-1)})^{1-\alpha}$ $\sum_{i=1}^{k} n_i^{2-1/\alpha} + \sum_{i=1}^{k} T(n_i)$, where $C = \alpha/(\alpha - 1)$ and $\Phi = (1 + \sqrt{5})/2$.*

Proof. By Lemma 2, for $i \geq 2$, the n_i jobs arriving in phase $i - 1$ are finished by time t_i in $\mathcal{S}_{POPT}$. Thus $\mathcal{S}_{POPT}$ processes these jobs in a window of length at most $T(n_{i-1}) + T(n_i)$. Let $T'(n_i) = \min\{T(n_{i-1}) + T(n_i), n_i\sqrt[\alpha]{\alpha - 1}\}$. Applying Lemma 3, we obtain that the n_i jobs incur a cost of at least

$$\frac{n_i^{\alpha}}{(T'(n_i))^{\alpha-1}} + T'(n_i) \geq \frac{n_i^{\alpha}}{(T(n_{i-1}) + T(n_i))^{\alpha-1}} + T'(n_i)$$

$$\geq \frac{n_i^{\alpha}}{(T(n_{i-1}) + T(n_i))^{\alpha-1}} + T(n_i).$$

The last inequality holds because $T(n_i) \leq n_i \leq n_i\sqrt[\alpha]{\alpha - 1}$, for $\alpha \geq 2$ and hence $T'(n_i) \geq T(n_i)$. Similarly, for the n_1 jobs released at time $t = 0$, the cost it at least $n_1^{\alpha}/(T(n_1))^{\alpha-1} + T(n_1)$. Summing up, the total cost of $\mathcal{S}_{POPT}$ is at least

$$\frac{n_1^{\alpha}}{(T(n_1))^{\alpha-1}} + \sum_{i=2}^{k} \frac{n_i^{\alpha}}{(T(n_{i-1}) + T(n_i))^{\alpha-1}} + \sum_{i=1}^{k} T(n_i).$$

In the following we show that the first two terms in the above expression are at least $C^{1-\alpha}(1 + \Phi)^{-1}(1 + \Phi^{\alpha/(2\alpha-1)})^{1-\alpha} \sum_{i=1}^{k} n_i^{2-1/\alpha}$, which establishes the lemma to be proven. Since $T(n_i) \leq Cn_i^{1-1/\alpha}$, it suffices to show

$$(1 + \Phi)(1 + \Phi^{\alpha/(2\alpha-1)})^{\alpha-1} \left(\frac{n_1^\alpha}{\left(n_1^{1-1/\alpha}\right)^{\alpha-1}} + \sum_{i=2}^{k} \frac{n_i^\alpha}{\left(n_{i-1}^{1-1/\alpha} + n_i^{1-1/\alpha}\right)^{\alpha-1}} \right)$$

$$\geq \sum_{i=1}^{k} n_i^{2-1/\alpha}. \tag{3}$$

To this end we partition the sequence of job numbers $n_1, \ldots, n_k$ into subsequences such that, within each subsequence, $n_i \geq \Phi^{\alpha/(2\alpha-1)} n_{i+1}$. More formally, the first subsequence starts with index $b_1 = 1$ and ends with the smallest index e_1 satisfying $n_{e_1} < \Phi^{\alpha/(2\alpha-1)} n_{e_1+1}$. Suppose that $l - 1$ subsequences have been constructed. Then the l-st sequence starts at index $b_l = e_{l-1} + 1$ and ends with the smallest index $e_l \geq b_l$ such that $n_{e_l} < \Phi^{\alpha/(2\alpha-1)} n_{e_l+1}$. The last subsequence ends with index k.

We will prove (3) by considering the individual subsequences. Since within a subsequence $n_{i+1} \leq n_i \Phi^{-\alpha/(2\alpha-1)}$, we have $n_{i+1}^{2-1/\alpha} \leq n_i^{2-1/\alpha}/\Phi$. Therefore, for any subsequence l, using the limit of the geometric series

$$\sum_{i=b_l}^{e_l} n_i^{2-1/\alpha} \leq n_{b_l}^{2-1/\alpha}/(1 - 1/\Phi) = (1 + \Phi)n_{b_l}^{2-1/\alpha}, \tag{4}$$

which upper bounds terms on the right hand side of (3). As for the left hand side of (3), we have for the first subsequence,

$$(1 + \Phi)(1 + \Phi^{\alpha/(2\alpha-1)})^{\alpha-1} \left(\frac{n_1^\alpha}{\left(n_1^{1-1/\alpha}\right)^{\alpha-1}} + \sum_{i=2}^{e_1} \frac{n_i^\alpha}{\left(n_{i-1}^{1-1/\alpha} + n_i^{1-1/\alpha}\right)^{\alpha-1}} \right)$$

$$\geq (1 + \Phi)n_1^{2-1/\alpha}.$$

For any other subsequence l, we have

$$(1 + \Phi)(1 + \Phi^{\alpha/(2\alpha-1)})^{\alpha-1} \sum_{i=b_l}^{e_l} \frac{n_i^\alpha}{\left(n_{i-1}^{1-1/\alpha} + n_i^{1-1/\alpha}\right)^{\alpha-1}}$$

$$\geq (1 + \Phi)(1 + \Phi^{\alpha/(2\alpha-1)})^{\alpha-1} \frac{n_{b_l}^\alpha}{\left(n_{b_l-1}^{1-1/\alpha} + n_{b_l}^{1-1/\alpha}\right)^{\alpha-1}}$$

$$\geq (1 + \Phi)(1 + \Phi^{\alpha/(2\alpha-1)})^{\alpha-1} \frac{n_{b_l}^\alpha}{\left((\Phi^{(\alpha-1)/(2\alpha-1)} + 1)n_{b_l}^{1-1/\alpha}\right)^{\alpha-1}}$$

$$\geq (1 + \Phi)n_{b_l}^{2-1/\alpha}.$$

The second to last inequality holds because n_{b_l-1} and n_{b_l} belong to different subsequences and hence $n_{b_l-1} < \Phi^{\alpha/(2\alpha-1)} n_{b_l}$. The above inequalities together with (4) imply (3). □

Lemma 5. *For $\alpha \geq 2$ and $c = 1$, the competitive ratio of Phasebal is at most*
$(1 + \Phi)(1 + \Phi^{\alpha/(2\alpha-1)})(\alpha-1)\frac{\alpha^\alpha}{(\alpha-1)^{\alpha-1}}(\frac{4}{2\alpha-1} + \frac{4}{\alpha-1}).$

Proof. Using (2) as well as Lemmas 1 and 4 we obtain that the competitive ratio of *Phasebal* is bounded by

$$(1 + \Phi)(1 + \Phi^{\alpha/(2\alpha-1)})(\alpha-1)\frac{4\sum_{i=1}^{k}((\frac{\alpha}{2\alpha-1} + \frac{\alpha}{\alpha-1})n_i^{2-1/\alpha} + n_i^{1-1/\alpha})}{\sum_{i=1}^{k}((\frac{\alpha}{\alpha-1})^{1-\alpha}n_i^{2-1/\alpha} + T(n_i))}.$$

Considering the terms of order $n^{2-1/\alpha}$, we obtain the performance ratio we are aiming at. It remains to show that $n_i^{1-1/\alpha}/T(n_i)$ does not violate this ratio. Note that $T(n_i) \geq 1$. Thus, if $n_i^{1-1/\alpha} \leq 2$ we have

$$n_i^{1-1/\alpha}/T(n_i) \leq 2 \leq 4(\frac{\alpha}{\alpha-1})^{\alpha-1}(\frac{\alpha}{2\alpha-1} + \frac{\alpha}{\alpha-1}). \tag{5}$$

If $n_i^{1-1/\alpha} > 2$, then we use the fact that $T(n_i) = \sum_{j=1}^{n_i} 1/\sqrt[\alpha]{n_i - j + 1} \geq \frac{\alpha}{\alpha-1}((n_i + 1)^{1-1/\alpha} - 1) \geq \frac{1}{2}\frac{\alpha}{\alpha-1}n_i^{1-1/\alpha}$ and we can argue as in (5), since $(\alpha - 1)/\alpha < 1$. □

Case 2: $c = \alpha - 1$ The global structure of the analysis is the same as in the case $c = 1$ but some of the calculations become more involved. Moreover, with respect to the optimum cost, we will consider the true optimum rather than the cost of a pseudo-optimal algorithm.

We start again by analyzing the cost and time of *Phasebal*. As before we assume that there are k phases. In phase i, *Phasebal* uses $1/\sqrt[\alpha]{(n_i - j + 1)/(\alpha - 1)}$ time units to process the j-th job. This yields a power consumption of

$$\sum_{j=1}^{n_i}\left(\frac{n_i - j + 1}{\alpha - 1}\right)^{1-1/\alpha} \leq C_E(n_i^{2-1/\alpha} - 1) + (\alpha - 1)^{1/\alpha-1}n_i^{1-1/\alpha}$$

with $C_E = (\alpha-1)^{\frac{1}{\alpha}-1}\frac{\alpha}{2\alpha-1}$. The phase length is $T(n_i) = \sum_{j=1}^{n_i} 1/\left(\frac{n_i-j+1}{\alpha-1}\right)^{1/\alpha}$. Here we have

$$C_T((n_i + 1)^{1-1/\alpha} - 1) < T(n_i) < C_T(n_i^{1-1/\alpha} - 1/\alpha) \tag{6}$$

with $C_T = \alpha(\alpha-1)^{\frac{1}{\alpha}-1}$. In phase i the n_i jobs processed during the phase incur a flow time of

$$\sum_{j=1}^{n_i}(n_i - j + 1)/\left(\frac{n_i - j + 1}{\alpha - 1}\right)^{1/\alpha} = (\alpha - 1)^{1/\alpha}\sum_{j=1}^{n_i}(n_i - j + 1)^{1-1/\alpha}$$

$$\leq C_F(n_i^{2-1/\alpha} - 1) + (\alpha - 1)^{1/\alpha}n_i^{1-1/\alpha}$$

with $C_F = (\alpha - 1)^{\frac{1}{\alpha}} \frac{\alpha}{2\alpha - 1}$, while the n_{i+1} jobs arriving in the phase incur a cost of at most $n_{i+1} T(n_i)$. We obtain

$$g(\mathcal{S}_{PB}) \leq (C_E + C_F) \sum_{i=1}^{k} (n_i^{2-1/\alpha} - 1) + 2C_T \sum_{i=1}^{k} n_i^{2-1/\alpha} + \alpha(\alpha - 1)^{1/\alpha - 1} \sum_{i=1}^{k} n_i^{1-1/\alpha}. \tag{7}$$

We next lower bound the cost of an optimal schedule. Again we call a job *active* if it has arrived but is still unfinished. The proofs of the next three lemmas are omitted.

Lemma 6. *There exists an optimal schedule $\mathcal{S}_{OPT}$ having the property that, at any time, if there are ℓ active jobs, then the processor speed is at least $\sqrt[\alpha]{\ell/(\alpha - 1)}$.*

Lemma 7. *For $c = \alpha - 1$, in $\mathcal{S}_{OPT}$ the n_1 jobs released at time t_0 are finished by time t_1 and the n_i jobs released during phase $i - 1$ are finished by time t_i, for $i = 2, \ldots, k$.*

Lemma 8. *There holds $g(\mathcal{S}_{OPT}) \geq C_T^{1-\alpha}(1 + \Phi)^{-1}(1 + \Phi^{\alpha/(2\alpha-1)})^{(1-\alpha)}$ $\sum_{i=1}^{k} n_i^{2-1/\alpha} + \sum_{i=1}^{k} T(n_i)$, where $\Phi = (1 + \sqrt{5})/2.$.*

Lemma 9. *For $c = \alpha - 1$, the competitive ratio of Phasebal is at most $(1 + \Phi)(1 + \Phi^{\alpha/(2\alpha-1)})^{(\alpha-1)} \frac{\alpha^\alpha}{(\alpha-1)^{\alpha-1}} \frac{5\alpha-2}{2\alpha-1}.$*

Proof. Using (6), (7) and Lemma 8, we can determine the ratio of the online cost to the optimal offline cost as in Lemma 5. The desired competitive ratio can then be derived using algebraic manipulations. The calculations are more involved than in the proof of Lemma 5. Details are given in the full version of the paper. □

Theorem 2 now follows from Lemmas 5 and 9, observing that $\alpha_0 = (19 + \sqrt{161})/10 \geq 2$ and that, for $\alpha > \alpha_0$, we have $\frac{4}{2\alpha-1} + \frac{4}{\alpha-1} < \frac{5\alpha-2}{2\alpha-1}$.

5   An Optimal Offline Algorithm for Unit-Size Jobs

We present a polynomial time algorithm for computing an optimal schedule, given a sequence of unit-size jobs that is known offline. Our algorithm is based on dynamic programming and constructs an optimal schedule for a given job sequence σ by computing optimal schedules for subsequences of σ. A schedule for σ can be viewed as a sequence of subschedules $S_1, S_2, \ldots, S_m$, where any S_j processes a subsequence of jobs $j_1, \ldots, j_k$ starting at time r_{j_1} such that $c_i > r_{i+1}$ for $i = j_1, \ldots, j_k - 1$ and $c_{j_k} \leq r_{j_k+1}$. In words, jobs j_1 to j_k are scheduled continuously without interruption such that the completion time of any job i is after the release time of job $i + 1$ and the last job j_k is finished no later than the release time of job $j_k + 1$. As we will prove in the next two lemmas, the optimal speeds in such subschedules S_j can be determined easily. For convenience, the lemmas are stated for a general number n of jobs that have to be scheduled in an interval $[t, t')$. The proofs are omitted.

Lemma 10. *Consider n jobs that have to be scheduled in time interval $[t,t')$ such that $r_1 = t$ and $r_n < t'$. Suppose that in an optimal schedule $c_i > r_{i+1}$, for $i = 1,\ldots,n-1$. If $t' - t \geq \sum_{i=1}^{n} \sqrt[\alpha]{(\alpha-1)/(n-i+1)}$, then the i-th job in the sequence is executed at speed $s_i = \sqrt[\alpha]{(n-i+1)/(\alpha-1)}$.*

Lemma 11. *Consider n jobs that have to be scheduled in time interval $[t,t')$ such that $r_1 = t$ and $r_n < t'$. Suppose that in an optimal schedule $c_i > r_{i+1}$, for $i = 1,\ldots,n-1$. If $t' - t < \sum_{i=1}^{n} \sqrt[\alpha]{(\alpha-1)/(n-i+1)}$, then the i-th job in the sequence is executed at speed $s_i = \sqrt[\alpha]{(n-i+1+c)/(\alpha-1)}$, where c is the unique value such that $\sum_{i=1}^{n} \sqrt[\alpha]{(\alpha-1)/(n-i+1+c)} = t' - t$.*

Of course, an optimal schedule for a given σ need not satisfy the condition that $c_i > r_{i+1}$, for $i = 1,\ldots,n-1$. In fact, this is the case if the speeds specified in Lemmas 10 and 11 do not give a feasible schedule, i.e. there exists an i such that $c_i = \sum_{j=1}^{i} t_j \leq r_{i+1}$, with $t_i = 1/s_i$ and s_i as specified in the lemmas. Obviously, this infeasibility is easy to check in linear time.

We are now ready to describe our optimal offline algorithm, a pseudo-code of which is presented in Figure 1. Given a jobs sequence consisting of n jobs, the algorithm constructs optimal schedules for subproblems of increasing size. Let $P[i, i+l]$ be the subproblem consisting of jobs i to $i+l$ assuming that the processing may start at time r_i and must be finished by time r_{i+l+1}, where $1 \leq i \leq n$ and $0 \leq l \leq n - i$. We define $r_{n+1} = \infty$. Let $C[i, i+l]$ be the cost of an optimal schedule for $P[i, i+l]$. We are eventually interested in $C[1,n]$. In an initialization phase, the algorithm starts by computing optimal schedules for $P[i, i]$ of length $l = 0$, see lines 1 to 3 of the pseudo-code. If $r_{i+1} - r_i \geq \sqrt[\alpha]{\alpha - 1}$, then Lemma 10 implies that the optimal speed for job i is equal to $\sqrt[\alpha]{1/(\alpha - 1)}$. If $r_{i+1} - r_i < \sqrt[\alpha]{\alpha - 1}$, then by Lemma 11 the optimal speed is $1/(r_{i+1} - r_i)$. Note that this value can be infinity if $r_{i+1} = r_i$. The calculation of $C[i,i]$ in line 3 will ensure that in this case an optimal schedule will not complete job i by r_{i+1}.

Algorithm Dynamic Programming
1. **for** $i := 1$ **to** n **do**
2.   **if** $r_{i+1} - r_i \geq \sqrt[\alpha]{\alpha - 1}$ **then** $S[i] := \sqrt[\alpha]{1/(\alpha - 1)}$ **else** $S[i] := 1/(r_{i+1} - r_i)$;
3. $C[i,i] := (S[i])^{\alpha-1} + 1/S[i]$;
4. **for** $l := 1$ **to** $n - 1$ **do**
5.   **for** $i := 1$ **to** $n - l$ **do**
6. $C[i, i+l] := \min_{i \leq j < i+l}\{C[i,j] + C[j+1, i+l]\}$;
7.   Compute an optimal schedule for $P[i, i+l]$ according to Lemmas 10 and 11 assuming $c_j > r_{j+1}$ for $j = i,\ldots,i+l-1$ and let $s_i,\ldots,s_{i+l}$ be the computed speeds;
8.   **if** schedule is feasible **then** $C := \sum_{j=i}^{i+l} s_j^{\alpha-1} + \sum_{j=i}^{i+l}(i+l-j+1)/s_j$ **else** $C := \infty$;
9.   **if** $C < C[i, i+l]$ **then** $C[i, i+l] := C$ and $S[j] := s_j$ for $j = i,\ldots,i+l$;

Fig. 1. The dynamic programming algorithm

After the initialization phase the algorithm considers subproblems $P[i, i+l]$ for increasing l. An optimal solution to $P[i, i+l]$ has the property that either (a) there exists an index j with $j < i+l$ such that $c_j \le r_{j+1}$ or (b) $c_j > r_{j+1}$ for $j = i, \ldots, i+l-1$. In case (a) an optimal schedule for $P[i, i+l]$ is composed of optimal schedules for $P[i, j]$ and $P[j+1, i+l]$, which is reflected in line 6 of the pseudo-code. In case (b) we can compute optimal processing speeds according to Lemmas 10 and 11, checking if the speeds give indeed a feasible schedule. This is done in lines 7 and 8 of the algorithm. In a final step the algorithm checks if case (a) or (b) holds. The algorithm has a running time of $O(n^3 \log \rho)$, where ρ is the inverse of the desired precision. Note that in Lemma 11, c can be computed only approximately using binary search.

We briefly mention that we can use our dynamic programming approach to compute a schedule that minimizes the total flow time of jobs, given a fixed amount A of energy. Here we simply consider the minimization of a weighted objective function $g_\beta(\mathcal{S}) = \beta E(\mathcal{S}) + (1 - \beta) F(\mathcal{S})$, where $0 < \beta < 1$. By suitably choosing β, we obtain an optimal schedule $\mathcal{S}_{OPT}$ for g_β with $E(\mathcal{S}_{OPT}) = A$. This schedule minimizes the flow time. Details can be found in the full version of the paper.

References

1. J. Augustine, S. Irani and C. Swamy. Optimal power-down strategies. *Proc. 45th Annual IEEE Symposium on Foundations of Computer Science*, 530-539, 2004.
2. N. Bansal, T. Kimbrel and K. Pruhs. Dynamic speed scaling to manage energy and temperature. *Proc. 45th Annual IEEE Symposium on Foundations of Computer Science*, 520–529, 2004.
3. N. Bansal and K. Pruhs. Speed scaling to manage temperature. *Proc. 22nd Annual Symposium on Theoretical Aspects of Computer Science (STACS)*, Springer LNCS 3404, 460–471, 2005.
4. G. Cornuéjols, G.L. Nemhauser and L.A. Wolsey. The uncapacitated facility location problem. In P. Mirchandani and R. Francis (eds.), *Discrete Location Theory*, 119–171, John Wiley & Sons, 1990.
5. D.R. Dooly, S.A. Goldman, and S.D. Scott. On-line analyis of the TCP acknowledgment delay problem. *Journal of the ACM*, 48:243–273, 2001.
6. A. Fabrikant, A. Luthra, E. Maneva, C.H. Papadimitriou and S. Shenker. On a network creation game. *Proc. 22nd Annual ACM Symposium on Principles of Distributed Computing*, 347–351, 2003.
7. S. Irani, S. Shukla and R. Gupta. Algorithms for power savings. *Proc. 14th Annual ACM-SIAM Symposium on Discrete Algorithms*, 37–46, 2003.
8. A.R. Karlin, C. Kenyon and D. Randall. Dynamic TCP acknowledgement and other stories about $e/(e-1)$. *Proc. 33rd ACM Symposium on Theory of Computing*, 502–509, 2001.
9. P. Mirchandani and R. Francis (eds.). *Discrete Location Theory*. John Wiley & Sons, 1990.
10. K. Pruhs, P. Uthaisombut and G. Woeginger. Getting the best response for your erg. *Proc. 9th Scandinavian Workshop on Algorithm Theory (SWAT)*, Springer LNCS 3111, 15–25, 2004.

11. D.D. Sleator und R.E. Tarjan. Amortized efficiency of list update and paging rules. *Communications of the ACM*, 28:202-208, 1985.
12. D. Shmoys, J. Wein and D.P. Williamson. Scheduling parallel machines on-line. *SIAM Journal on Computing*, 24:1313-1331, 1995.
13. F. Yao, A. Demers and S. Shenker. A scheduling model for reduced CPU energy. *Proc. 36th Annual Symposium on Foundations of Computer Science*, 374–382, 1995.

Efficient Qualitative Analysis of Classes of Recursive Markov Decision Processes and Simple Stochastic Games

Kousha Etessami[1] and Mihalis Yannakakis[2]

[1] LFCS, School of Informatics, University of Edinburgh
[2] Department of Computer Science, Columbia University

Abstract. Recursive Markov Decision Processes (RMDPs) and Recursive Simple Stochastic Games (RSSGs) are natural models for recursive systems involving both probabilistic and non-probabilistic actions. As shown recently [10], fundamental problems about such models, e.g., termination, are undecidable in general, but decidable for the important class of 1-exit RMDPs and RSSGs. These capture controlled and game versions of multi-type Branching Processes, an important and well-studied class of stochastic processes. In this paper we provide efficient algorithms for the *qualitative termination* problem for these models: does the process terminate almost surely when the players use their optimal strategies? Polynomial time algorithms are given for both maximizing and minimizing 1-exit RMDPs (the two cases are not symmetric). For 1-exit RSSGs the problem is in NP∩coNP, and furthermore, it is at least as hard as other well-known NP∩coNP problems on games, e.g., Condon's *quantitative* termination problem for finite SSGs ([3]). For the class of linearly-recursive 1-exit RSSGs, we show that the problem can be solved in polynomial time.

1 Introduction

In recent work [10], we introduced and studied Recursive Markov Decision Processes (RMDPs) and Recursive Simple Stochastic Games (RSSGs), which provide natural models for recursive systems (e.g., programs with procedures) involving both probabilistic and non-probabilistic actions. They define infinite-state MDPs and SSGs that extend Recursive Markov Chains (RMCs) ([8, 9]) with non-probabilistic actions that are controlled by a controller and/or the environment (the "players"). Informally, a recursive model (RMC, RMDP, RSSG) consists of a (finite) collection of finite state component models (resp. MC, MDP, SSG) that can call each other in a potentially recursive manner.

In this paper we focus on the important class of *1-exit RMDPs* and *1-exit RSSGs*, which we will denote by *1-RMDP* and *1-RSSG*. These are RMDPs and RSSGs where every component contains exactly 1 exit node. Without players, 1-RMCs correspond tightly to both Stochastic Context-Free Grammars (SCFGs) and Multi-Type Branching Processes (MT-BPs). Branching processes are an important class of stochastic processes, dating back to the early work of Galton and

B. Durand and W. Thomas (Eds.): STACS 2006, LNCS 3884, pp. 634–645, 2006.

Watson in the 19th century, and continuing in the 20th century in the work of Kolmogorov, Sevastianov, Harris and others for MT-BPs and beyond (see, e.g., [14]). MT-BPs model the growth of a population of objects of distinct types. In each generation each object of a given type gives rise, according to a probability distribution, to a multi-set of objects of distinct types. These stochastic processes have been used in a variety of applications, including in population genetics ([16]), nuclear chain reactions, ([7]), and RNA modeling in computational biology (based on SCFGs) ([22]). SCFGs are also fundamental models in statistical natural language processing (see, e.g., [19]). 1-RMDPs correspond to a controlled version of MT-BPs (and SCFGs): the reproduction of some types can be controlled, while the dynamics of other types is probabilistic as in ordinary MT-BPs; or the controller may be able to influence the reproduction of some types by choosing among a set of probability distributions (e.g., the branching Markov decision chains of [21]). The goal of the controller is either to maximize the probability of extinction or to minimize it (maximize survival probability). This model would also be suitable for analysis of population dynamics under worst-case (or best-case) assumptions for some types and probabilistic assumptions for others. Such controlled MT-BPs can be readily translated to 1-RMDPs, where the types of the MT-BP correspond to the components of the RMDP, extinction in the MT-BP corresponds to termination in the RMDP, and our results can be used for the design of strategies to achieve or prevent extinction.

Among our results in [10], we showed that for maximizing (minimizing) 1-RMDPs, the *Qualitative Termination Problem* (**Qual-TP**), is in NP (coNP, respectively), and that the same problem for 1-RSSGs is in $\Sigma_2^P \cap \Pi_2^P$. **Qual-TP** is the problem of deciding whether player 1 (the maximizer) has a strategy to force termination with probability 1, regardless of the strategy employed by player 2 (the minimizer). (In a maximizing 1-RMDP, the only player present is the maximizer, and in a minimizing 1-RMDP the only player present is the minimizer.)

In this paper we improve significantly on the above results. We show that **Qual-TP**, both for maximizing 1-RMDPs and for minimizing 1-RMDPs, can in fact be decided in polynomial time. It follows easily from this and strong determinacy results from [10], that for 1-RSSGs **Qual-TP** is in NP∩coNP. We show that one can not easily improve on this upper bound, by providing a polynomial time reduction from the *Quantitative Termination Problem* (**Quan-TP**) for finite SSGs ([3]) to the **Qual-TP** problem for 1-RSSGs. Condon [3] showed that for finite SSGs the **Quan-TP** problem, specifically the problem of deciding whether player 1 has a strategy to force termination with probability $\geq 1/2$, is in NP∩coNP. Whether the problem can be solved in P is a well-known open problem, that includes as special cases several other longstanding problems (e.g., "parity games" and "mean-payoff" games). We note (as is already known) that for finite SSGs, **Qual-TP** itself is in polynomial time. We in fact show a more general result, namely, that **Qual-TP** is in polynomial time for the class of 1-RSSGs that are *linearly-recursive*.

Thus, we provide a new class of infinite-state SSGs whose qualitative decision problem is at least as hard as the quantitative decision problem for finite SSGs, and quite possibly harder, but which we can still decide in NP∩coNP. We already showed in [10, 8] that the even harder **Quan-TP** problem for 1-RSSGs can be decided in PSPACE, and that improving that upper bound even to NP, even for 1-RMCs, would resolve a long standing open problem in the complexity of numerical computation, namely the square-root sum problem ([12]).

Most proofs are omitted due to space.

Related Work. Both MDPs and Stochastic Games have a vast literature (see [20, 11]). As mentioned, we introduced RMDPs and RSSGs and studied both quantitative and qualitative termination problems in [10]. We showed that for multi-exit models these problems are undecidable, and that (qualitative) model checking is undecidable even in the 1-exit case. Our earlier work [8, 9] developed the theory and algorithms for RMCs and [6, 2] studied the related model of probabilistic Pushdown Systems (pPDSs).

Our algorithms here were partly inspired by recent unpublished work by Denardo and Rothblum [4, 5] on *Multi-Matrix Multiplicative Systems*. They study families of square nonnegative matrices, which arise from choosing each matrix row independently from a choice of rows, and they give LP characterizations of when the spectral radius of all matrices in the family will be ≥ 1 or > 1. None of our results follow from theirs, but we use techniques similar to theirs, along with other techniques, to obtain our upper bounds.

2   Definitions and Background

A *Recursive Simple Stochastic Game (RSSG)*, A, is a tuple $A = (A_1, \ldots, A_k)$, where each *component* $A_i = (N_i, B_i, Y_i, En_i, Ex_i, \mathsf{pl}_i, \delta_i)$ consists of:

- A set N_i of *nodes*, with a distinguished subset En_i of *entry* nodes and a (disjoint) subset Ex_i of *exit* nodes.
- A set B_i of *boxes*, and a mapping $Y_i : B_i \mapsto \{1, \ldots, k\}$ that assigns to every box (the index of) a component. To each box $b \in B_i$, we associate a set of *call ports*, $Call_b = \{(b, en) \mid en \in En_{Y(b)}\}$, and a set of *return ports*, $Return_b = \{(b, ex) \mid ex \in Ex_{Y(b)}\}$. Let $Call^i = \cup_{b \in B_i} Call_b$, $Return^i = \cup_{b \in B_i} Return_b$, and let $Q_i = N_i \cup Call^i \cup Return^i$ be the set of all nodes, call ports and return ports; we refer to these as the *vertices* of component A_i.
- A mapping $\mathsf{pl}_i : Q_i \mapsto \{0, 1, 2\}$ that assigns to every vertex a player (Player 0 represents "chance" or "nature"). We assume $\mathsf{pl}_i(ex) = 0$ for all $ex \in Ex_i$.
- A transition relation $\delta_i \subseteq (Q_i \times (\mathbb{R} \cup \{\perp\}) \times Q_i)$, where for each tuple $(u, x, v) \in \delta_i$, the source $u \in (N_i \setminus Ex_i) \cup Return^i$, the destination $v \in (N_i \setminus En_i) \cup Call^i$, and x is either (i) a real number $p_{u,v} \in (0, 1]$ (the transition probability) if $\mathsf{pl}_i(u) = 0$, or (ii) $x = \perp$ if $\mathsf{pl}_i(u) = 1$ or 2. For computational purposes we assume that the given probabilities $p_{u,v}$ are rational. Furthermore they must satisfy the consistency property: for every $u \in \mathsf{pl}_i^{-1}(0)$, $\sum_{\{v' \mid (u, p_{u,v'}, v') \in \delta_i\}} p_{u,v'} = 1$, unless u is a call port or exit node, neither of which have outgoing transitions, in which case by default $\sum_{v'} p_{u,v'} = 0$.

We use the symbols $(N, B, Q, \delta, \text{etc.})$ without a subscript, to denote the union over all components. Thus, e.g., $N = \cup_{i=1}^{k} N_i$ is the set of all nodes of A, $\delta = \cup_{i=1}^{k} \delta_i$ the set of all transitions, etc.

An RSSG A defines a global denumerable Simple Stochastic Game (SSG) $M_A = (V = V_0 \cup V_1 \cup V_2, \Delta, \texttt{pl})$ as follows. The global *states* $V \subseteq B^* \times Q$ of M_A are pairs of the form $\langle \beta, u \rangle$, where $\beta \in B^*$ is a (possibly empty) sequence of boxes and $u \in Q$ is a *vertex* of A. More precisely, the states $V \subseteq B^* \times Q$ and transitions Δ are defined inductively as follows:

1. $\langle \epsilon, u \rangle \in V$, for $u \in Q$. (ϵ denotes the empty string.)
2. if $\langle \beta, u \rangle \in V$ & $(u, x, v) \in \delta$, then $\langle \beta, v \rangle \in V$ and $(\langle \beta, u \rangle, x, \langle \beta, v \rangle) \in \Delta$.
3. if $\langle \beta, (b, en) \rangle \in V$ & $(b, en) \in Call_b$, then $\langle \beta b, en \rangle \in V$ & $(\langle \beta, (b, en) \rangle, 1, \langle \beta b, en \rangle) \in \Delta$.
4. if $\langle \beta b, ex \rangle \in V$ & $(b, ex) \in Return_b$, then $\langle \beta, (b, ex) \rangle \in V$ & $(\langle \beta b, ex \rangle, 1, \langle \beta, (b, ex) \rangle) \in \Delta$.

The mapping $\texttt{pl} : V \mapsto \{0, 1, 2\}$ is given as follows: $\texttt{pl}(\langle \beta, u \rangle) = \texttt{pl}(u)$ if u is in $Q \setminus (Call \cup Ex)$, and $\texttt{pl}(\langle \beta, u \rangle) = 0$ if $u \in Call \cup Ex$. The set of vertices V is partitioned into V_0, V_1, and V_2, where $V_i = \texttt{pl}^{-1}(i)$. We consider M_A with various *initial states* of the form $\langle \epsilon, u \rangle$, denoting this by M_A^u. Some states of M_A are *terminating states* and have no outgoing transitions. These are states $\langle \epsilon, ex \rangle$, where ex is an exit node.

An RSSG where $V_2 = \emptyset$ ($V_1 = \emptyset$) is called a maximizing (minimizing, respectively) *Recursive Markov Decision Process* (RMDP); an RSSG where $V_1 \cup V_2 = \emptyset$ is called a *Recursive Markov Chain* (RMC) ([8, 9]); an RSSG where $V_0 \cup V_2 = \emptyset$ is called a Recursive Graph or *Recursive State Machine*(RSM) ([1]). Define *1-RSSGs* to be those RSSGs where every component has 1 exit, and likewise define *1-RMDPs* and *1-RMCs*. W.l.o.g., we can assume every component has 1 entry, because multi-entry RSSGs can be transformed to equivalent 1-entry RSSGs with polynomial blowup (similar to RSM transformations [1]). This is not so for exits, e.g., qualitative termination is undecidable for multi-exit RMDPs, whereas it is decidable for 1-RSSGs (see [10]). *This entire paper is focused on 1-RSSGs and 1-RMDPs.* Accordingly, some of our notation is simpler than that used for general RSSGs in [10]. We shall call a 1-RSSG (1-RMDP, etc.) *linear* if there in no path of transitions in any component from any return port to a call port.

Our basic goal is to answer qualitative termination questions for 1-RSSGs: "Does player 1 have a strategy to force the game to terminate at exit ex (i.e., reach $\langle \epsilon, ex \rangle$), starting at $\langle \epsilon, u \rangle$, with probability 1, regardless of how player 2 plays?". A *strategy* σ for player i, $i \in \{1, 2\}$, is a function $\sigma : V^* V_i \mapsto V$, where, given the history $ws \in V^* V_i$ of play so far, with $s \in V_i$ (i.e., it is player i's turn to play a move), $\sigma(ws) = s'$ determines the next move of player i, where $(s, \bot, s') \in \Delta$. (We could also allow randomized strategies.) Let Ψ_i denote the set of all strategies for player i. A pair of strategies $\sigma \in \Psi_1$ and $\tau \in \Psi_2$ induce in a straightforward way a Markov chain $M_A^{\sigma, \tau} = (V^*, \Delta')$, whose set of states is the set V^* of histories. Given an initial vertex u, suppose ex is the *unique* exit node of u's component. Let $q_u^{*, \sigma, \tau}$ be the probability that, in $M_A^{\sigma, \tau}$, starting at initial state $\langle \epsilon, u \rangle$ we will eventually terminate, by reaching some $w \langle \epsilon, ex \rangle$, for $w \in V^*$. From general determinacy results (e.g., [18]) it follows that $\sup_{\sigma \in \Psi_1} \inf_{\tau \in \Psi_2} q_u^{*, \sigma, \tau} = \inf_{\tau \in \Psi_2} \sup_{\sigma \in \Psi_1} q_u^{*, \sigma, \tau}$. This is the *value* of the game

starting at u, which we denote by q_u^*. We are interested in the following problem:
Qual-TP: Given A, a 1-RSSG (or 1-RMDP), and given a vertex u in A, is $q_u^* = 1$?

For a strategy $\sigma \in \Psi_1$, let $q_u^{*,\sigma} = \inf_{\tau \in \Psi_2} q_u^{*,\sigma,\tau}$, and for $\tau \in \Psi_2$, let $q_u^{*,\cdot,\tau} = \sup_{\sigma \in \Psi_1} q_u^{*,\sigma,\tau}$. We showed in [10] that 1-RSSGs satisfy a strong form of memoryless determinacy, namely, call a strategy *Stackless & Memoryless (S&M)* if it depends neither on the history of the game nor on the current call stack, i.e., only depends on the current vertex. We call a game *S&M-determined* if both players have S&M optimal strategies.

Theorem 1. *([10]) Every 1-RSSG termination game is S&M-determined. (Moreover, there is an S&M strategy $\sigma^* \in \Psi_1$ that maximizes the value of $q_u^{*,\sigma}$ for all u, and likewise a $\tau^* \in \Psi_2$ that minimizes the value of $q_u^{*,\cdot,\tau}$ for all u.)*

For multi-exit RMDPs and RSSGs things are very different. We showed that even memoryless determinacy fails badly (there may not exist any optimal strategy at all, only ϵ-optimal ones)), and furthermore **Qual-TP** is undecidable (see [10]).

Note that there are finitely many S&M strategies for player i: each picks one edge out of each vertex belonging to player i. For 1-RMCs, where there are only probabilistic vertices, we showed in [8] that **Qual-TP** can be decided in polynomial time, using a spectral radius characterization for certain moment matrices associated with 1-RMCs. It followed, by guessing strategies, that **Qual-TP** for both maximizing and minimizing 1-RMDPs is in NP, and that **Qual-TP** for 1-RSSGs is in $\Sigma_2^P \cap \Pi_2^P$. We obtain far stronger upper bounds in this paper. We will also use the following fact from [10].

Proposition 1. *([10]) We can decide in P-time if the value of a 1-RSSG termination game (and optimal termination probability in a maximizing or minimizing 1-RMDP) is exactly 0.*

In ([10]) we defined a monotone system S_A of nonlinear min-max equations for 1-RSSGs A, and showed that its *Least Fixed Point* solution yields the desired probabilities q_u^*. These systems generalize both the linear Bellman's equations for MDPs, as well as the nonlinear system of polynomial equation for RMCs studied in [8]. Here we recall these systems of equations (with a slightly simpler notation). Let us use a variable x_u for each unknown q_u^*, and let $\mathbf{x}$ be the vector of all $x_u, u \in Q$. The system S_A has one equation of the form $x_u = P(\mathbf{x})$ for each vertex u. Suppose that u is in component A_i with (unique) exit ex. There are 5 cases based on the "*Type*" of u.

1. $u \in Type_1$: $u = ex$. In this case: $x_u = 1$.
2. $u \in Type_{rand}$: $\mathtt{pl}(u){=}0$ & $u \in (N_i \backslash \{ex\}) \cup Return^i$: $x_u{=}\sum_{\{v | (u, p_{u,v}, v) \in \delta\}} p_{u,v} x_v$.
   (If u has no outgoing transitions, this equation is by definition $x_u = 0$.)
3. $u \in Type_{call}$: $u = (b, en)$ is a call port: $x_{(b,en)} = x_{en} \cdot x_{(b,ex')}$, where $ex' \in Ex_{Y(b)}$ is the unique exit of $A_{Y(b)}$.
4. $u \in Type_{max}$: $\mathtt{pl}(u) = 1$ & $u \in (N_i \backslash \{ex\}) \cup Return^i$: $x_u = \max_{\{v | (u, \perp, v) \in \delta\}} x_v$.
   (If u has no outgoing transitions, we define $\max(\emptyset) = 0$.)
5. $u \in Type_{min}$: $\mathtt{pl}(u) = 2$ and $u \in (N_i \backslash \{ex\}) \cup Return^i$: $x_u = \min_{\{v | (u, \perp, v) \in \delta\}} x_v$.
   (If u has no outgoing transitions, we define $\min(\emptyset) = 0$.)

In vector notation, we denote the system S_A by $\mathbf{x} = P(\mathbf{x})$.

Given 1-RSSG A, we can easily construct S_A in linear time. For vectors $\mathbf{x}, \mathbf{y} \in \mathbb{R}^n$, define $\mathbf{x} \leq \mathbf{y}$ to mean $x_j \leq y_j$ for every coordinate j. Let $\mathbf{q}^* \in \mathbb{R}^n$ denote the n-vector of q_u^*'s.

Theorem 2. *([10]) Let $\mathbf{x} = P(\mathbf{x})$ be the system S_A associated with 1-RSSG A. Then $\mathbf{q}^* = P(\mathbf{q}^*)$, and for all $\mathbf{q}' \in \mathbb{R}_{\geq 0}^n$, if $\mathbf{q}' = P(\mathbf{q}')$, then $\mathbf{q}^* \leq \mathbf{q}'$ (in other words, $\mathbf{q}^*$ is the* Least Fixed Point, *of $P : \mathbb{R}_{\geq 0}^n \mapsto \mathbb{R}_{\geq 0}^n$).*

3 Qualitative Termination for 1-RMDPs in P-Time

We show that, for both maximizing 1-RMDPs and minimizing 1-RMDPs, qualitative termination can be decided in polynomial time. Please note: the two cases are not symmetric. We provide distinct algorithms for each of them. An important result for us is this:

Theorem 3. *([8])* `Qual-TP` *for 1-RMCs is decidable in polynomial time.*

We briefly indicate the key elements of that upper bound (please see [8] for more details). Our algorithm employed a spectral radius characterization of *moment matrices* associated with 1-RMCs. Given the system of polynomial equations $x = P(x)$ for a 1-RMC (no min and max types), its moment matrix B is the square Jacobian matrix of $P(x)$, whose (i,j)'th entry is the partial derivative $\partial P_i(x)/\partial x_j$, evaluated at the all 1 vector (i.e., $x_u \leftarrow 1$ for $u \in Q$). We showed in [8] that if the system $x = P(x)$ is decomposed into strongly connected components (SCCs) in a natural way, and we associate a moment matrix B_C to each SCC, C, then $q_u^* = 1$ for every u where x_u is in C, iff either u is of $Type_1$, or [C has successor SCCs and $q_v^* = 1$ for all nodes v in any successor SCC of C, and $\rho(B_C) \leq 1$, where $\rho(M)$ is the spectral radius of a square matrix M].

Theorem 4. *Given a maximizing 1-RMDP, A, and a vertex u of A, we can decide in polynomial time whether $q_u^* = 1$. In other words, for maximizing 1-RMDPs,* `Qual-TP` *is in* **P**.

Proof. Given a maximizing 1-RMDP, A, we shall determine for all vertices u, whether $q_u^* = 1$, $q_u^* = 0$, or $0 < q_u^* < 1$. The system of equations $x = P(x)$ for A defines a labeled *dependency graph*, $G_A = (Q, \rightarrow)$, as follows: the nodes Q of G_A are the vertices of A, and there is an edge $u \rightarrow v$ iff x_v appears on the right hand side of the equation $x_u = P_u(x)$. Each node u is labeled by its *Type*. If $u \in Type_{rand}$, i.e., u is a probabilistic vertex, and x_v appears in the weighted sum $P_u(x)$ as a term $p_{u,v} x_v$, then the edge from u to v is labeled by the probability $p_{u,v}$. Otherwise, the edge is unlabeled.

We wish to partition the nodes of the dependency graph into three classes: $Z_0 = \{u \mid q_u^* = 0\}$, $Z_1 = \{u \mid q_u^* = 1\}$, and $Z_\$ = \{u \mid 0 < q_u^* < 1\}$. In our algorithm we will use a fourth partition, $Z_?$, to denote those nodes for which we have not yet determined to which partition they belong. We first compute Z_0. By proposition 1, this can be done easily in P-time even for 1-RSSGs. Once we have computed Z_0, the remaining nodes belong either to Z_1 or $Z_\$$. Clearly, $Type_1$ nodes belong to Z_1.

Initialize: $Z_1 \leftarrow Type_1$; $Z_\$ \leftarrow \emptyset$; and $Z_? \leftarrow Q \setminus (Z_1 \cup Z_0)$;

Next, we do one "preprocessing" step to categorize some remaining "easy" nodes into Z_1 and $Z_\$$, as follows:

> **repeat**
>> **if** $u \in Z_? \cap (Type_{rand} \cup Type_{call})$ has all of its successors in Z_1
>>> **then** $Z_? \leftarrow Z_? \setminus \{u\}$; $Z_1 \leftarrow Z_1 \cup \{u\}$;
>>
>> **if** $u \in Z_? \cap Type_{max}$ has some successor in Z_1
>>> **then** $Z_? \leftarrow Z_? \setminus \{u\}$; $Z_1 \leftarrow Z_1 \cup \{u\}$;
>>
>> **if** $u \in Z_? \cap (Type_{rand} \cup Type_{call})$ has some successor in $Z_0 \cup Z_\$$
>>> **then** $Z_? \leftarrow Z_? \setminus \{u\}$; $Z_\$ \leftarrow Z_\$ \cup \{u\}$;
>>
>> **if** $u \in \dot{Z}_? \cap Type_{max}$ has all successors in $(Z_0 \cup Z_\$)$
>>> **then** $Z_? \leftarrow Z_? \setminus \{u\}$; $Z_\$ \leftarrow Z_\$ \cup \{u\}$;
>
> **until** (there is no change to $Z_?$)

The preprocessing step will not, in general, empty $Z_?$, and we need to categorize the remaining nodes in $Z_?$. We will construct a set of linear inequalities (an LP without an objective function) which has a solution iff there are any remaining node in $Z_?$ which belongs in Z_1, and if so, the solution we obtain to the LP will let us find and remove from $Z_?$ some more nodes that belong in Z_1.

Note that, if we can do this, then we can solve our problem, because all we need to do is iterate: we repeatedly do a preprocessing step, followed by the LP step to remove nodes from $Z_?$, until no more nodes can be removed, at which point we are done: the remaining nodes in $Z_?$ all belong to $Z_\$$.

For the LP step, restrict attention to the vertices remaining in $Z_?$. These vertices induce a subgraph of G_A, call it G'_A. Call a remaining probabilistic node u in $Z_?$ *leaky* if it does not have full probability on its outgoing transitions inside G'_A. Note that this happens if and only if some of u's outedges in G_A lead to nodes in Z_1 (otherwise, if u had an outedge to a node in Z_0 or $Z_\$$, it would already have been removed from $Z_?$ during preprocessing). Let $\mathcal{L}$ denote the set of remaining leaky nodes in $Z_?$. We add an extra terminal node t to G'_A, and for every $u \in \mathcal{L}$ we add a probabilistic edge $u \xrightarrow{p_{u,t}} t$, where $p_{u,t} = 1 - \sum_{v \in Z_?} p_{u,v}$.

W.l.o.g., assume that both entries of components and return nodes are probabilistic nodes (this can easily be assured by minor adjustments to the input 1-RSSG). The LP has a variable y_i for every node $i \in Z_?$ that is not $Type_{max}$, and has a variables $y_{i,j}$ for every $Type_{max}$ node $i \in Z_?$ and successor $j \in Z_?$ of i. In addition there are flow variables $f_{i,j,k}$ for each node $i \in Z_?$, and every edge $j \to k$ in G'_A. The constraints are as follows.

1. For every $j \in Type_{rand} \cup Type_{call}$ that is not a component entry or a return:

$$y_j \geq \sum_{i \to j \,\wedge\, i \in Type_{rand}} p_{i,j} y_i + \sum_{i \to j \,\wedge\, i \in Type_{max}} y_{i,j}$$

2. For every $j \in Type_{max}$:

$$\sum_k y_{j,k} \geq \sum_{i \to j \,\wedge\, i \in Type_{rand}} p_{i,j} y_i + \sum_{i \to j \,\wedge\, i \in Type_{max}} y_{i,j}$$

3. For every node i that is the entry of a component, say A_r:

$$y_i \geq \sum_{j=(b,en) \in Type_{call} \,\wedge\, Y(b)=r} y_j$$

4. For every node i that is a return node, say of box b: $y_i \geq y_j$, where j is the call node of b.
5. $\sum_i y_i + \sum_{i,j} y_{i,j} = 1$.
6. $y \geq 0$.

Regard the dependency graph as a network flow graph with capacity on each edge $i \to j$ coming out of a max node equal to $y_{i,j}$ and capacity of edges $i \to j$ coming out of the other vertices equal to y_i. We set up one flow problem for each $i \in Z_?$, with source i, sink t and flow variables f_{ijk}.

7. For every vertex i, we have flow conservation constraints on the variables $f_{i,j,k}$, i.e., $\sum_k f_{i,j,k} = \sum_k f_{i,k,j}$, for all nodes $j \in Z_?$, $j \neq i, t$.
8. Nonnegativity constraints: $f_{i,j,k} \geq 0$ for all i, j, k.
9. Capacity constraints: $f_{i,j,k} \leq y_{j,k}$ for every $j \in Type_{max}$ with successor k, and for every node i; and $f_{i,j,k} \leq y_j$ for every $j \in Type_{rand} \cup Type_{call}$ and successor k in G'_A and every node i.
10. Source constraints: $\sum_k f_{i,i,k} = y_i/2^{2m}$, for every $i \in Type_{rand} \cup Type_{call}$, and $\sum_k f_{i,i,k} = \sum_j y_{i,j}/2^{2m}$, for $i \in Type_{max}$, where m is defined as follows. Suppose our LP in constraints (1.-6.) has r variables and constraints, and that its rational entries have numerator and denominator with at most l bits. If there is a solution to (1.-6.), then (see, e.g., [13]), there is a rational solution whose numerators and denominators require at most $m = poly(r,l)$ bits to encode, where $poly(r,l)$ is a polynomial in r and l. Note $r \in O(|G'_A|)$, l is bounded by the number of bits required for the transition probabilities $p_{u,v}$ in A, hence m is polynomial in the input size.

The purpose of constraints (7-10) is to ensure that every vertex with a nonzero y variable can reach a leaky vertex in the subgraph of G'_A induced by the support of the y solution vector.

Lemma 1. *There exists a vertex $u \in Z_?$ such that $q_u^* = 1$ if and only if the LP constraints in (1.–10.) are feasible. Moreover, from a solution to the LP we can find a (partial) strategy for the maximizing player that forces termination from some such u with probability $= 1$.*

So to summarize, we set up and solve the LP. If there is no solution, then for all remaining vertices $u \in Z_?$, $q_u^* < 1$, and thus $u \in Z_\$$. If there is a solution, use the above partial (randomized) strategy for some of the max nodes, leaving the strategy for other nodes unspecified. This allows us to set to 1 some vertices (vertices in the bottom SCC's of the resulting 1-RMC), and thus to move them to Z_1. We can then iterate the preprocessing step and then the LP step until we reach a fixed point, at which point we have categorized all vertices u into one of Z_0, Z_1 or $Z_\$$.   □

Theorem 5. *Given a minimizing 1-RMDP, A, and a vertex u of A, we can decide in polynomial time whether $q_u^* = 1$. In other words, for minimizing 1-RMDPs, Qual-TP is in* **P**.

Proof. As in the previous theorem, we want to classify the vertices into $Z_0, Z_\$, Z_1$, this time under optimal play of the minimizing player. We again consider the dependency graph G_A of A. We will again use $Z_?$ to denote those vertices that have not yet been classified.

Initialize: $Z_1 \leftarrow Type_1$; $Z_\$ \leftarrow \emptyset$; and $Z_? \leftarrow Q\backslash (Z_1 \cup Z_0)$;

Next, we again do a "preprocessing" step, which is "dual" to that of the preprocessing we did for maximizing 1-RMDPs, and categorizes some remaining "easy" nodes into Z_1 and $Z_\$$:

repeat
  if $u \in Z_? \cap (Type_{rand} \cup Type_{call})$ has all of its successors in Z_1
    then $Z_? \leftarrow Z_? \setminus \{u\}$; $Z_1 \leftarrow Z_1 \cup \{u\}$;
  if $u \in Z_? \cap Type_{min}$ has some successor in $Z_\$$
    then $Z_? \leftarrow Z_? \setminus \{u\}$; $Z_\$ \leftarrow Z_\$ \cup \{u\}$;
  if $u \in Z_? \cap (Type_{rand} \cup Type_{call})$ has some successor in $Z_0 \cup Z_\$$
    then $Z_? \leftarrow Z_? \setminus \{u\}$; $Z_\$ \leftarrow Z_\$ \cup \{u\}$;
  if $u \in Z_? \cap Type_{min}$ has all successors in (Z_1)
    then $Z_? \leftarrow Z_? \setminus \{u\}$; $Z_1 \leftarrow Z_1 \cup \{u\}$;
  until (there is no change to $Z_?$)

Note that, after the preprocessing step, for every edge $u \to v$ in G_A from $u \in Z_?$ to $v \notin Z_?$, it must be the case that $v \in Z_1$ (otherwise, u would have already been moved to $Z_\$$ or Z_0). After preprocessing, we formulate a (different) LP, which has a solution iff there are more nodes currently in $Z_?$ which belong in $Z_\$$. Restrict attention to nodes in $Z_?$, and consider the subgraph G_A' of G_A induced by the nodes in $Z_?$. The LP has a variable y_i for every remaining vertex $i \in Z_?$ such that $i \notin Type_{min}$, and has a variable y_{ij} for every (remaining) node $i \in Type_{min}$, and successor j of i in G_A'. We shall need the following lemma:

Lemma 2. *Consider a square nonnegative matrices B with at most n rows and having rational entries with at most l bits each. If $\rho(B) > 1$ then $\rho(B) \geq 1 + 1/2^m$ where $m = poly(n, l)$ and $poly(n, l)$ is some polynomial in n and l.*

Let $d = (1 + 1/2^m)$. The constraints of our LP are as follows. For the LP we restrict attention to only those nodes j, i in $Z_?$.

1. For every $j \in Type_{rand}$ that is not a component entry or a return, as well as for every $j \in Type_{call}$:

$$dy_j \leq \sum_{i \in Type_{rand} \wedge i \to j} p_{i,j} y_i + \sum_{i \in Type_{min} \wedge i \to j} y_{i,j}$$

2. For every $j \in Type_{min}$:

$$d \sum_k y_{j,k} \leq \sum_{i \in Type_{rand} \wedge i \to j} p_{i,j} y_i + \sum_{i \in Type_{min} \wedge i \to j} y_{i,j}$$

3. For every node i that is the entry of a component, say A_r:

$$dy_i \leq \sum_{j=(b,en) \in Type_{call} \wedge Y(b)=r} y_j$$

4. For every node i that is a return node, say of box b: $dy_i \leq y_j$, where j is the entry node of b.
5. $\sum_i y_i + \sum_{i,j} y_{i,j} = 1$.
6. $y \geq 0$.

Lemma 3. *There exists a vertex $u \in Z_?$ such that $q_u^* < 1$ if and only if the LP in (1. – 6.) is feasible. Moreover, from a solution to the LP we can find a (partial) strategy that forces termination from some such u with probability < 1.*

To summarize, we find Z_0, then do preprocessing to determines the "easy" Z_1 and $Z_\$$ nodes. Then, we set up and solve the LP, finding some more $Z_\$$ vertices, removing them, and iterating again with a preprocessing and LP step, until we exhaust $Z_?$ or there is no solution to the LP; in the latter case the remaining vertices all belong to Z_1. As for a strategy that achieves these assignments, in each iteration when we solve the LP we fix the strategy for certain of the min nodes in a way that ensures that some new vertices will be added to $Z_\$$ and leave the other min nodes undetermined. Moreover, in preprocessing, if $Type_{min}$ nodes get assigned $Z_\$$ based on an outedge, we fix the strategy at that node accordingly.   □

4   Qualitative Termination for 1-RSSGs in NP∩coNP

The following is a simple corollary of Theorems 1, 4, and 5.

Corollary 1. *Given a 1-RSSG, A, and given a vertex u of A, we can decide in both NP and coNP whether $q_u^* = 1$. In other words, the qualitative termination problem for 1-RSSGs is in NP∩coNP.*

As the following theorem shows, it will not be easy to improve this upper bound. Note that finite SSGs, defined by Condon [3], are a special case of 1-RSSGs (we can simply identify the terminal node "1" of the SSG with the unique exit of a single component with no boxes). Define the quantitative termination problem for finite SSGs to be the problem of deciding, given a finite SSG G, and a vertex u of G, whether $q_u^* \geq 1/2$. Condon [3] showed that this problem is in NP ∩ coNP, and it has been a major open problem whether this upper bound can be improved to P-time.

Theorem 6. *There is a P-time reduction from the quantitative termination problem for finite SSGs to the qualitative termination problem for 1-RSSGs.*

It is not at all clear whether there is a reduction from qualitative termination for 1-RSSGs to quantitative termination for finite SSGs. Thus, `Qual-TP` for 1-RSSGs appears to constitute a new harder game problem in NP∩coNP.

5   Qualitative Termination for Linear 1-RSSGs in P-Time

We now show that for linear 1-RSSGs, there is a P-time algorithm for deciding `Qual-TP`. This generalizes of course the case of flat games.

Theorem 7. *Given a linear 1-RSSG, and a vertex u, there is a polynomial time algorithm to decide whether $q_u^* = 1$.*

Proof. Given a linear 1-RSSG, A, consider its dependency graph G_A. The nodes of partitioned partitioned into 5 types: $Type_{max}$, $Type_{min}$, $Type_{rand}$, $Type_{call}$, and $Type_1$. Let Q be the set of all vertices of G_A. Our algorithm is depicted in Figure 1. We claim that a call to Prune(Q) returns precisely those vertices in $Z_1 = \{u \mid q_u^* = 1\}$. The proof is omitted due to space.   □

Prune(Q)
   $W \leftarrow Q$;
   repeat
      $W \leftarrow W \backslash \text{PruneMin}(W)$;
      $W \leftarrow \text{PruneMax}(W)$;
   until (there is no change in W);
   return W;

PruneMin(W)
   $S \leftarrow W \setminus Type_1$;
   repeat
      if there is a node u in $S \cap (Type_{rand} \cup Type_{max})$ that has a
         successor in $W \setminus S$, **then** $S \leftarrow S \setminus \{u\}$;
      if there is a node u in $S \cap (Type_{min} \cup Type_{call})$ that has no
         successor in S, **then** $S \leftarrow S \setminus \{u\}$;
   until (there is no change in S);
   return S;

PruneMax(W)
   $S \leftarrow W$;
   repeat
      if there is a node u in $S \cap (Type_{rand} \cup Type_{min} \cup Type_{call})$ that has a
         successor in $Q \setminus S$, **then** $S \leftarrow S \setminus \{u\}$;
      if there is a node u in $S \cap Type_{max}$ that has no
         successor in S, **then** $S \leftarrow S \setminus \{u\}$;
   until (there is no change in S);
   return S;

Fig. 1. P-time qualitative termination algorithm for linear 1-RSSGs

The algorithm applies more generally to *piecewise linear* 1-RSSGs, where every vertex $v \in Type_{call}$ has at most one successor in the dependency graph G_A that is in the same SCC as v.

Acknowledgment. Thanks to Uri Rothblum for providing us a copy of his unpublished work [4, 5]. Research partially supported by NSF Grant CCF-04-30946.

References

1. R. Alur, M. Benedikt, K. Etessami, P. Godefroid, T. W. Reps, and M. Yannakakis. Analysis of recursive state machines. In *ACM Trans. Progr. Lang. Sys.*, 27:786-818, 2005.
2. T. Brázdil, A. Kučera, and O. Stražovský. Decidability of temporal properties of probabilistic pushdown automata. In *Proc. of STACS'05*, 2005.
3. A. Condon. The complexity of stochastic games. *Inf. & Comp.*, 96(2):203–224, 1992.
4. E. V. Denardo and U. G. Rothblum. Totally expanding multiplicative systems. unpublished manuscript, submitted, 2005.
5. E. V. Denardo and U. G. Rothblum. A turnpike theorem for a risk-sensitive Markov decision process with stopping. unpublished manuscript, submitted, 2005.
6. J. Esparza, A. Kučera, and R. Mayr. Model checking probabilistic pushdown automata. In *Proc. of 19th IEEE LICS'04*, 2004.
7. C. J. Everett and S. Ulam. Multiplicative systems, part i., ii, and iii. Technical Report 683,690,707, Los Alamos Scientific Laboratory, 1948.
8. K. Etessami and M. Yannakakis. Recursive Markov chains, stochastic grammars, and monotone systems of nonlinear equations. In *Proc. of 22nd STACS'05*. Springer, vol. 3404 of LNCS , 2005.
9. K. Etessami and M. Yannakakis. Algorithmic verification of recursive probabilistic state machines. In *Proc. 11th TACAS'05*, Springer, vol. 3440 of LNCS, 2005.
10. K. Etessami and M. Yannakakis. Recursive Markov decision processes and recursive stochastic games. In *Proc. of 32nd Int. Coll. on Automata, Languages, and Programming (ICALP'05)*, Springer, vol. 3580 of LNCS, 2005.
11. J. Filar and K. Vrieze. *Competitive Markov Decision Processes*. Springer, 1997.
12. M. R. Garey, R. L. Graham, and D. S. Johnson. Some NP-complete geometric problems. In *8th ACM STOC*, pages 10–22, 1976.
13. M. Grötschel, L. Lovász, and A. Schrijver. *Geometric Algorithms and Combinatorial Optimization*. Springer-Verlag, 2nd edition, 1993.
14. T. E. Harris. *The Theory of Branching Processes*. Springer-Verlag, 1963.
15. R. J. Horn and C. R. Johnson. *Matrix Analysis*. Cambridge U. Press, 1985.
16. P. Jagers. *Branching Processes with Biological Applications*. Wiley, 1975.
17. K. Mahler. An inequality for the discriminant of a polynomial. *Michigan Math J.*, 11:257–262, 1964.
18. D. A. Martin. Determinacy of Blackwell games. *J. Symb. Logic*, 63(4):1565–1581, 1998.
19. C. Manning and H. Schütze. *Foundations of Statistical Natural Language Processing*. MIT Press, 1999.
20. M. L. Puterman. *Markov Decision Processes*. Wiley, 1994.
21. U. G. Rothblum, P. Whittle. Growth optimality for branching Markov decision chains. Math. of Operations Research, 7: 582-601, 1982.
22. Y. Sakakibara, M. Brown, R Hughey, I.S. Mian, K. Sjolander, R. Underwood, and D. Haussler. Stochastic context-free grammars for tRNA modeling. *Nucleic Acids Research*, 22(23):5112–5120, 1994.

Datalog and Constraint Satisfaction
with Infinite Templates

Manuel Bodirsky[1] and Víctor Dalmau[2]

[1] Humboldt-Universität zu Berlin
bodirsky@informatik.hu-berlin.de
[2] Universitat Pompeu Fabra
victor.dalmau@upf.edu

Abstract. We relate the expressive power of Datalog and constraint satisfaction with infinite templates. The relationship is twofold: On the one hand, we prove that every non-empty problem that is closed under disjoint unions and has Datalog width one can be formulated as a constraint satisfaction problem (CSP) with a countable template that is ω-*categorical*. Structures with this property are of central interest in classical model theory. On the other hand, we identify classes of CSPs that can be solved in polynomial time with a Datalog program. For that, we generalise the notion of the *canonical Datalog program* of a CSP, which was previously defined only for CSPs with finite templates by Feder and Vardi. We show that if the template Γ is ω-categorical, then CSP(Γ) can be solved by an (l, k)-Datalog program if and only if the problem is solved by the canonical (l, k)-Datalog program for Γ. Finally, we prove algebraic characterisations for those ω-categorical templates whose CSP has Datalog width $(1, k)$, and for those whose CSP has strict Datalog width l.

Topic: Logic in Computer Science, Computational Complexity.

1 Introduction

In a constraint satisfaction problem we are given a set of variables and a set of constraints on these variables, and want to find an assignment of values to the variables such that all the constraints are satisfied. The computational complexity of the constraint satisfaction problem depends on the constraint language that is used in the instances of the problem. Constraint satisfaction problems can be modeled as homomorphism as shown below; here we refer to [18]. For detailed formal definitions of relational structures and homomorphisms, see Section 2. Let Γ be a (finite or infinite) structure with a relational signature τ. Then the *constraint satisfaction problem (CSP)* for Γ is the following computational problem.

CSP(Γ)
INSTANCE: A finite τ-structure S.
QUESTION: Is there a homomorphism from S to Γ?

The structure Γ is called the *template* of the constraint satisfaction problem CSP(Γ). For example, if the template is the dense linear order of the rational

B. Durand and W. Thomas (Eds.): STACS 2006, LNCS 3884, pp. 646–659, 2006.
© Springer-Verlag Berlin Heidelberg 2006

numbers $(\mathbb{Q}, <)$, then it is easy to see that $\text{CSP}(\Gamma)$ is the well-known problem of digraph-acyclicity.

A class $\mathcal{C}$ is said to be *closed under inverse homomorphisms* (sometimes also *anti-monotone*) if $B \in \mathcal{C}$ implies that $A \in \mathcal{C}$, whenever there is a homomorphism from A to B [19]. Clearly, the set of all positive instances of a constraint satisfaction problem is closed under inverse homomorphisms. Conversely, if the set of positive instances of a computational problem is closed under inverse homomorphisms and disjoint unions, then it can be formulated as a constraint satisfaction problem with a countably infinite template. This is an easy observation due to Feder and Vardi, which shows in particular that every constraint satisfaction problem can be formulated with an at most countable structure. For *finite* templates Γ, the complexity of $\text{CSP}(\Gamma)$ attracted a lot of attention in recent years [14, 3, 20, 18, 30, 24, 23, 13, 15, 7, 25]; this list covers only a very small fraction of relevant publications, and we refer to a recent survey paper for a more complete account [26]. Feder and Vardi [18] conjectured that every constraint satisfaction problem is either in P or NP-complete. This so-called *dichotomy conjecture* is still open.

Datalog. Datalog is the language of logic programs without function symbols, see e.g. [25, 17]. Datalog is an important algorithmic tool to study the complexity of constraint satisfaction problems; for constraint satisfaction with finite templates, this was for instance investigated by Feder and Vardi [18]. Let τ be a relational signature; the relation symbols in τ will also be called the *input relation symbols*. A Datalog program consists of a finite set of Horn clauses $C_1, \dots, C_k$ (they are also called the *rules* of the Datalog program) containing atomic formulas with relation symbols from the signature τ, together with atomic formulas with some new relation symbols. These new relation symbols are called *IDBs* (short for *intensional database*). Each clause is a set of literals where at most one of these literals is positive. The positive literals do not contain an input relation. The semantics of a Datalog program can be specified using *fixed point operators*, as e.g. in [17, 25]. We show an example.

$$
\begin{aligned}
\mathsf{tc}(x, y) &\leftarrow \mathsf{edge}(x, y) \\
\mathsf{tc}(x, y) &\leftarrow \mathsf{tc}(x, u), \mathsf{tc}(u, y) \\
\mathsf{false} &\leftarrow \mathsf{tc}(x, x)
\end{aligned}
$$

Here, the binary relation edge is the only input relation, tc is a binary relation computed by the program, and false is a 0-ary relation computed by the program. The Datalog program computes with the help of the relation tc the transitive closure of the edges in the input relation, and derives false if and only if the input (which can be seen as a digraph defined on the variables) contains a directed cycle. In general, we say that a problem is *solved* by a Datalog program, if the distinguished 0-ary predicate false is derived on an instance of the problem if and only if the instance has no solution.

We say that a Datalog program Φ has *width* (l, k), $0 \le l \le k$, if it has at most l variables in rule heads and at most k variables per rule (we also say that Φ is an (l, k)-*Datalog program*). A problem is *of width* (l, k), if it can be solved by an (l, k)-Datalog program. The problem of acyclicity has for instance width $(2, 3)$, as demonstrated above. A problem has *width* l if it is of width (l, k) for some $k \geq l$, and it is of *bounded width*, if it has width l for some $l \geq 0$. It is easy to see that all bounded width problems are tractable, since the rules can derive only a polynomial number of facts. It is an open question whether there is an algorithm that decides for a given finite template T whether $\mathrm{CSP}(T)$ can be solved by a Datalog program of width (l, k). Similarly, we do not know how to decide width l, or bounded width, with the notable exception of width one [18] (see also [15]).

Results. We prove that every non-empty problem that is closed under disjoint unions and has Datalog width one can be formulated as a constraint satisfaction problem with an ω-categorical template $\Gamma(\Phi)$. A structure is called ω-*categorical*, if its first-order theory has only one countable model, up to isomorphism. For ω-categorical templates we can apply the so-called *algebraic approach to constraint satisfaction* [24, 7, 26]. This approach was originally developed for constraint satisfaction with finite templates, but several fundamental theorems also hold for ω-categorical templates [6, 4].

Next, we investigate which constraint satisfaction problems can be solved with a Datalog program in polynomial time. An important tool to characterize the expressive power of Datalog for constraint satisfaction is the notion of *canonical Datalog programs*. This concept was introduced by Feder and Vardi for finite templates; we present a generalization to ω-categorical templates. We prove that a constraint satisfaction problem with an ω-categorical template can be solved by an (l, k)-Datalog program if and only if the canonical (l, k)-Datalog program for Γ solves the problem.

Next, we prove a characterization of constraint satisfaction problems with ω-categorical templates Γ having width $(1, k)$, generalizing a result from [15]. A special case of width 1 problems are problems that can be decided by establishing *arc-consistency* (sometimes also called *hyperarc-consistency*), which is a well-known and intensively studied technique in artificial intelligence. We show that if a constraint satisfaction problem with an ω-categorical template can be decided by establishing arc-consistency, then it can also be formulated as a constraint satisfaction problem with a finite template.

Finally, we characterize *strict width* l, a notion that was again introduced for finite templates and for $l \geq 2$ in [18]; for a formal definition see Section 6. Jeavons et al. [23] say that in this case *establishing strong k-consistency ensures global consistency*. For finite templates, strict width l can be characterized by an algebraic closure condition [18, 23]. In Section 6 we generalize this result to ω-categorical templates.

Applications and Related Work. The results presented here in particular apply to computational problems that arise in the literature on *binary relation algebras* [28, 16]. Well-known examples of such binary relation algebras are the *point algebra*, the *containment algebra*, *Allen's interval algebra*, and the *left linear point algebra*. It is well-known [2, 16] that these relation algebras have concrete representations that are ω-categorical structures with a binary relational signature. The constraint satisfaction problems for these structures received a lot of attention [12, 21, 28], and one of the most studied algorithms in this area is the path-consistency algorithm. Translated into this setting, our results imply an algebraic characterization of those relation algebras with an ω-categorical representation where the path-consistency algorithm establishes global consistency (this corresponds to strict width two).

2   Definitions

A *relational signature* τ is a (here always at most countable) set of *relation symbols* R_i, each associated with an *arity* k_i. A *(relational) structure* Γ over *relational signature* τ (also called τ-*structure*) is a set D_Γ (the *domain*) together with a relation $R_i \subseteq D_\Gamma^{k_i}$ for each relation symbol of arity k_i. If necessary, we write R^Γ to indicate that we are talking about the relation R belonging to the structure Γ. For simplicity, we denote both a relation symbol and its corresponding relation with the same symbol. For a τ-structure Γ and $R \in \tau$ it will also be convenient to say that $R(u_1, \dots, u_k)$ *holds in* Γ iff $(u_1, \dots, u_k) \in R$. We sometimes use the shortened notation $\overline{x}$ for a vector $x_1, \dots, x_n$ of any length.

Let Γ and Γ' be τ-structures. A *homomorphism* from Γ to Γ' is a function f from D_Γ to $D_{\Gamma'}$ such that for each n-ary relation symbol in τ and each n-tuple $\overline{a}$, if $\overline{a} \in R^\Gamma$, then $(f(a_1), \dots, f(a_n)) \in R^{\Gamma'}$. In this case we say that the map f *preserves* the relation R. Two structures Γ_1 and Γ_2 are called *homomorphically equivalent*, if there is a homomorphism from Γ_1 to Γ_2 and a homomorphism from Γ_2 to Γ_1. A *strong homomorphism* f satisfies the stronger condition that for each n-ary relation symbol in τ and each n-tuple $\overline{a}$, $\overline{a} \in R^\Gamma$ if and only if $(f(a_1), \dots, f(a_n)) \in R^{\Gamma'}$. An *embedding* of a Γ in Γ' is an injective strong homomorphism, an *isomorphism* is a surjective embedding. Isomorphisms from Γ to Γ are called *automorphisms*.

The *disjoint union* of two τ-structures Γ and Γ' is a τ-structure that is defined on the disjoint union of the domains of Γ and Γ'. A relation holds on vertices of the disjoint union if and only if it either holds in Γ or in Γ'. A τ-structure is called *connected* iff it is not the disjoint union of two τ-structures with a non-empty domain.

We can use first-order formulas over the signature τ to define relations over a given τ-structure Γ: for a formula φ with k free variables the corresponding relation R is the set of all k-tuples satisfying the formula φ in Γ. If we add relations to a given τ-structure Γ, then the resulting structure Γ' with a larger signature $\tau' \supset \tau$ is called a τ'-*expansion* of Γ, and Γ is called a τ-*reduct* of Γ'. This should not be confused with the notions of *extension* and *restriction*. Recall from [22]: If Γ and Γ' are structures of the same signature, with $D_\Gamma \subseteq D_{\Gamma'}$, and the inclusion map is an embedding, then we say that Γ' is an *extension* of Γ, and that Γ a *restriction* of Γ'.

A first-order formula φ is said to be *primitive positive* (we say φ is a *p.p.-formula*, for short) iff it is of the form $\exists \overline{x}(\varphi_1(\overline{x}) \wedge \cdots \wedge \varphi_k(\overline{x}))$ where $\varphi_1, \dots, \varphi_k$ are atomic formulas (which might be equality relations of the form $x = y$).

3   Countably Categorical Templates

The concept of ω-categoricity is of central interest in model theory [22, 8]. A countable structure Γ is called ω-categorical, if all countable models of the first-order theory of Γ are isomorphic to Γ. The following is a well-known and deep connection that shows that ω-categoricity of Γ is a property of the automorphism group of Γ, without reference to concepts from logic (see [22]). An *orbit of n-subsets in* Γ is a largest set O of subsets of Γ of cardinality n such that for all $S_1, S_2 \in O$ there is an automorphism a of Γ such that $a(S_1) = S_2$.

Theorem 1 (Engeler, Ryll-Nardzewski, Svenonius; see e.g. [22]). *The following properties of a countable structure Γ are equivalent:*

1. *the structure Γ is ω-categorical;*
2. *for each $n \geq 1$, there are finitely many orbits of n-subsets in Γ;*
3. *for each $n \geq 1$, there are finitely many inequivalent first-order formulas with n free variables over Γ.*

Examples. An example of an ω-categorical directed graph is the set of rational numbers with the dense linear order $(\mathbb{Q}, <)$. The (tractable) constraint satisfaction problem for this structure is digraph acyclicity. Clearly, there is only one orbit of subsets of cardinality n, and by Theorem 1 the structure is ω-categorical.

Another important example it the *universal triangle free graph* $\mathcal{A}$. This structure is the up to isomorphism unique countable K_3-free graph with the following *extension property*: whenever S is a subset and T is a disjoint independent subset of the vertices in $\mathcal{A}$, then $\mathcal{A}$ contains a vertex $v \notin S \cup T$ that is linked to no vertex in S and to all vertices in T. Since the extension property can be formulated by an (infinite) set of first-order sentences, it follows that $\mathcal{A}$ is ω-categorical [22]. The structure $\mathcal{A}$ is called the *universal* triangle free graph, because every other countable triangle free graph embeds into $\mathcal{A}$. Hence, CSP($\mathcal{A}$) is clearly tractable. However, this simple problem can not be formulated as a constraint satisfaction problem with a finite template [18, 30].

4   Constraint Satisfaction for Datalog

In this section we show that every class of structures with Datalog width one can be formulated as a constraint satisfaction problem with an ω-categorical template. A Datalog program of width one accepts a class of structures that can be described by a sentence of a fragment of existential second order logic called *monotone monadic SNP without inequalities (MMSNP)*. We show that every problem in MMSNP can be formulated as the constraint satisfaction problem for an ω-categorical template.

An *SNP sentence* is an existential second-order sentence with a universal first-order part. The first order part might contain the existentially quantified relation symbols and additional relation symbols from a given signature τ (the *input* relations). We shall assume that SNP formulas are written in *negation normal form*, i.e., the first-order part is written in conjunctive normal form, and each disjunction is written as a negated conjunction of positive and negative literals. The class SNP consists of all problems on τ-structures that can be described by an SNP sentence.

The class *MMSNP*, defined by Feder and Vardi, is the class of problems that can be described by an SNP sentence with the additional requirements that the existentially quantified relations are monadic, that every input relation symbol occurs negatively in the SNP sentence, and that it does not contain inequalities. Every problem in MMSNP is under randomized Turing reductions equivalent to a constraint satisfaction problem with a finite template [18]; a deterministic reduction was recently announced by Kun [27]. It is easy to see that MMSNP contains all constraint satisfaction problems with finite templates. Thus, MM-SNP has a dichotomy if and only if CSP has a dichotomy.

It is easy to see that $(1, k)$-Datalog is contained in MMSNP: We introduce an existentially quantified unary predicate for each of the unary IDBs in the Datalog program. It is then straightforward to translate the rules of the Datalog program into first-order formulas with at most k first-order variables. We now want to prove that every problem in MMSNP can be formulated as a constraint satisfaction problem with a countably categorical template. In full generality, this cannot be true for two reasons. Firstly, there are MMSNP sentences that are false in all structures, whereas the instance without any constraints is contained in every constraint satisfaction problem with a non-empty template. Secondly, whereas constraint satisfaction problems are always closed under disjoint union, this is not necessarily the case for problems in MMSNP (a simple example of a MMSNP problem not closed under disjoint union is the one defined by the formula $\forall x, y \, \neg(P(x) \wedge Q(x))$). Hence we shall assume that we are dealing with a *non-empty* problem in MMSNP that is *closed under disjoint union*.

To prove the claim under this assumptions, we need a recent model-theoretic result of Cherlin, Shelah and Shi [10]. Let $\mathcal{N}$ be a finite set of finite structures with a relational signature τ. In this paper, a τ-structure Δ is called $\mathcal{N}$-*free* if there is no homomorphism from any structure in $\mathcal{N}$ to Δ. A structure Γ in a class of countable structures $\mathcal{C}$ is called *universal* for $\mathcal{C}$, if it contains all structures in $\mathcal{C}$ as an induced substructure.

Theorem 2 (of [10]). *Let $\mathcal{N}$ be a finite set of finite connected τ-structures. Then there is an ω-categorical universal structure Δ that is universal for the class of all countable $\mathcal{N}$-free structures.*

Cherlin, Shelah and Shi proved this statement for (undirected) graphs, but the proof does not rely on this assumption on the signature, and works for arbitrary relational signatures. The statement in its general form also follows from a result in [11]. We use this ω-categorical structure to prove the following.

Theorem 3. *Every non-empty problem in MMSNP that is closed under disjoint unions can be formulated as $CSP(\Gamma)$ with an ω-categorical template Γ.*

Proof. Let Φ be a MMSNP sentence with input signature τ whose set $\mathcal{M}$ of finite models is closed under disjoint unions. We have to find an ω-categorical

τ-structure Γ, such that $\mathcal{M}$ equals $\mathrm{CSP}(\Gamma)$. Recall the assumption that Φ is written in negation normal form. Let $P_1, \ldots, P_k$ be the existential monadic predicates in Φ. By monotonicity, all such literals with input relations are positive. For each existential monadic relation P_i we introduce a relation symbol P_i', and replace negative literals of the form $\neg P_i(x)$ in Φ by $P_i'(x)$. We shall denote the formula obtained after this transformation by Φ'. Let τ' be the signature containing the input relations from τ, the existential monadic relations P_i, and the symbols P_i' for the negative occurrences of the existential relations. We define $\mathcal{N}$ to be the set of τ'-structures containing for each clause $\neg(L_1 \wedge \cdots \wedge L_m)$ in Φ' the canonical database [9] of $(L_1 \wedge \cdots \wedge L_m)$. We shall use the fact that a τ'-structure S satisfies a clause $\neg(L_1 \wedge \cdots \wedge L_m)$ if and only if the the canonical database of $(L_1 \wedge \cdots \wedge L_m)$ is not homomorphic to S.

We can assume without loss of generality that Φ is minimal in the sense that if we remove a literal from some of the clauses the formula obtained is inequivalent. We shall show that then all structures in $\mathcal{N}$ are connected. Let us suppose that this is not the case. Then there is a clause C in Φ that corresponds to a non connected structure in $\mathcal{N}$. The clause C can be written as $\neg(E \wedge F)$ where the set X of variables in E and the set Y of variables in F do not intersect. Consider the formulas Φ_E and Φ_F obtained from Φ by replacing C by $\neg E$ and C by $\neg F$, respectively. By minimality of Φ there is a structure M_E that satisfies Φ but not Φ_E, and similarly there exists a structure M_F that satisfies Φ but not Φ_F. By assumption, the disjoint union M of M_E and M_F satisfies Φ. Then there exists a τ''-expansion M'' of M where $\tau'' = \tau \cup \{P_1, \ldots, P_k\}$ that satisfies the first-order part of Φ. Consider the substructures M_E'' and M_F'' of M'' induced by the vertices of M_E and M_F. We have that M_E'' does not satisfy the first-order part of Φ_E (otherwise M_E would satisfy Φ_E). Consequently, there is an assignment s_E of the universal variables that falsifies some clause. This clause must necessarily be $\neg E$ (since otherwise M'' would not satisfy the first-order part of Φ). By similar reasoning we can infer that there is an assignment s_F of the universal variables of Φ to elements of M_F that falsifies $\neg F$. Finally, fix any assignment s that coincides with s_E over X and with s_F over Y (such an assignment exists because X and Y are disjoint). Clearly, s falsifies C and M does not satisfy Φ, a contradiction. Hence, we shall assume that every structure in $\mathcal{N}$ is connected.

Then Theorem 2 asserts the existence of a $\mathcal{N}$-free ω-categorical τ'-structure Δ that is universal for all $\mathcal{N}$-free structures. We use Δ to define the template Γ for the constraint satisfaction problem. To do this, restrict the domain of Δ to those points that have the property that either P_i or P_i' holds (but not both P_i and P_i') for all existential monadic predicates P_i. The resulting structure Δ' is non-empty, since the problem defined by Φ is non-empty. Then we take the reduct of Δ' that only contains the input relations from τ. It is well-known [22] that reducts and first-order restrictions of ω-categorical structures are again ω-categorical. Hence the resulting τ-structure Γ is ω-categorical.

We claim that an τ-structure S satisfies Φ if and only if $S \in \mathrm{CSP}(\Gamma)$. Let S be an structure that has a homomorphism h to Γ. Let S' be the τ'-expansion of S such that for each $i = 1, \ldots, k$ the relation $P_i(x)$ holds in S' if and only if $P_i(h(x))$ holds in Δ', and $P_i'(x)$ holds in S' if and only if $P_i'(h(x))$ holds in Δ'. Clearly, h defines a homomorphism from S' to Δ' and also from S to Δ. In consequence, none of the structures from $\mathcal{N}$ maps to S' (otherwise it would also map to Δ). Hence, the τ''-reduction of S' satisfies all the clauses of the first-order part of Φ and hence S satisfies Φ.

Conversely, let S be a structure satisfying Φ. Consequently, there exists a τ'-expansion S' of S that satisfies the first-order part of Φ' and where for every element x exactly one of $P_i(x)$ or $P_i'(x)$ holds. Clearly, no structure in $\mathcal{N}$ is homomorphic to the expanded structure, and by universality of Γ the τ'-structure S' is an induced substructure of Δ. Since for every point of S' exactly one of P_i an P_i' holds, S' is also an induced substructure of Δ'. Consequently, S is homomorphic to its τ-reduct Γ. This completes the proof.   □

In particular, we proved the following.

Theorem 4. *Every non-empty problem in $(1, k)$-Datalog that is closed under disjoint unions can be formulated as a constraint satisfaction problem with an ω-categorical template.*

For a typical example of a constraint satisfaction problem in MMSNP, which cannot be described with a finite template [30], and which is not in (l, k)-Datalog for all $1 \leq l \leq k$, consider the following computational problem. Given is a finite graph S, and we want to test whether we can partition the vertices of S in two parts such that each part is triangle-free. The ω-categorical template that is used in the proof of Theorem 3 consists of two copies C_1 and C_2 of $\nleq$, where we add an undirected edge from all vertices in C_1 to all vertices in C_2. The corresponding constraint satisfaction problem is NP-hard [4, 1].

5   Canonical Datalog Programs

In this section we define the canonical Datalog program of a constraint satisfaction problem with an ω-categorical template, and prove that such a problem can be solved by an (l, k)-Datalog program if and only if the canonical (l, k)-Datalog program solves the problem.

For *finite* templates T with a relational signature τ the canonical Datalog program of for CSP(T) was defined in [18]. This motivates the following definition of canonical Datalog programs for constraint satisfaction problems with ω-categorical templates Γ. The *canonical (l, k)-Datalog program* contains an IDB for every at most l-ary primitive positive definable relation. The empty 0-ary relation serves as false. We introduce a rule $R \leftarrow R_1, \ldots, R_j$ into the canonical Datalog program if R is an IDB, the corresponding implication is valid in Γ, and contains at most k variables. If Γ is ω-categorical, Theorem 1 asserts that there are finitely many inequivalent such implications, and hence the canonical (l, k)-Datalog program is finite. Theorem 1 also implies that on a given instance the Datalog program can only derive a finite number of facts. This number is polynomial in the size of the instance, and thus the Datalog program can be evaluated in polynomial time. Observe that all stages during the evaluation of a canonical Datalog program on a given instance give rise to another instance S' of $CSP(\Gamma')$, where Γ' is the expansion of Γ by all at most l-ary primitive positive definable relations, and where S' contains all the tuples from these relations that are inferred so far [17].

Due to space limitations, the proof of the following theorem is omitted in this extended abstract.

Theorem 5. *A constraint satisfaction problem $CSP(\Gamma)$ with an ω-categorical template Γ can be solved with an (l, k)-Datalog program if and only if the canonical (l, k)-Datalog program solves $CSP(\Gamma)$.*

6   Datalog for Constraint Satisfaction

We characterize the ω-categorical templates whose constraint satisfaction problems have bounded width. These characterizations generalism algebraic characterizations of Datalog width that are known for constraint satisfaction with finite templates. However, not all results remain valid for infinite templates: It is well-known [18] that the constraint satisfaction of a finite template has Datalog width one if and only if the so-called *arc-consistency procedure* solves the problem. This is no longer true for infinite templates. We characterize both width one and the expressive power of the arc-consistency procedure for infinite ω-categorical templates, and present an example that shows that the two concepts are different. We also present an algebraic characterization of strict width l. Note that width one and strict width k are the only concepts of bounded Datalog width that are known to be decidable for finite templates.

First, we prove a crucial property of ω-categorical structures needed several times later. The proof contains a typical proof technique for ω-categorical structures.

Lemma 1. *Let Γ be a finite or infinite ω-categorical structure with relational signature τ, and let Δ be a countable relational structure with the same signature τ. If there is no homomorphism from Δ to Γ, then there is a finite substructure of Δ that does not homomorphically map to Γ.*

Proof. Suppose every finite substructure of Δ homomorphically maps to Γ. We show the contraposition of the lemma, and prove the existence of a homomorphism from Δ to Γ. Let $a_1, a_2, \ldots$ be an enumeration of Δ. We construct a directed acyclic graph with finite out-degree, where each node lies on some level $n \geq 0$. The nodes on level n are equivalence classes of homomorphisms from the substructure of Δ induced by $a_1, \ldots, a_n$ to Γ. Two such homomorphisms f and g are equivalent, if there is an automorphism α of Γ such that $f\alpha = g$. Two equivalence classes of homomorphisms on level n and $n+1$ are adjacent, if there are representatives of the classes such that one is a restriction of the other. Theorem 1 asserts that there are only finitely many orbits of subsets of cardinality

k in the automorphism group of the ω-categorical structure Γ, for all $k \geq 0$ (clearly, this also holds if Γ is finite). Hence, the constructed directed graph has a finite out-degree. By assumption, there is a homomorphism from the structure induced by $a_1, a_2, \ldots, a_n$ to Γ for all $n \geq 0$, and hence the directed graph has vertices on all levels. König's Lemma asserts the existence of an infinite path in the tree, which defines a homomorphism from Δ to Γ.   □

6.1 Width Zero

An example of a template whose constraint satisfaction problem has width 0 is the universal triangle-free graph $\not\triangle$. Since there is a primitive positive sentence that states the existence of a triangle in a graph, and since every graph without a triangle is homomorphic to $\not\triangle$, there is a (finite) Datalog program of width 0 that solves CSP($\not\triangle$). In general, it is easy to see that a constraint satisfaction problem has width 0 if and only if there is a finite set of *obstructions* for CSP(Γ), i.e., a finite set $\mathcal{N}$ of finite τ-structures such that every finite τ-structure A is homomorphic to Γ if and only if no substructure in $\mathcal{N}$ is homomorphic to A. For finite templates, this is closely related to results in [14, 31, 3].

6.2 Width One

Let Γ be an ω-categorical structure with relational signature τ, and Φ be the canonical $(1, k)$-Datalog program for Γ. By Theorem 4, the class of τ-structures accepted by Φ is itself a constraint satisfaction problem with an ω-categorical template. We denote this template by $\Gamma(1, k)$. The proof follows easily from Theorem 5, Theorem 4, and Lemma 1, and can be found in the long version of this paper.

Theorem 6. *Let Γ be ω-categorical. A constraint satisfaction problem CSP(Γ) can be solved by an $(1, k)$-Datalog program if and only if there is a homomorphism from $\Gamma(1, k)$ to Γ.*

The *arc-consistency procedure (AC)* is an algorithm for constraint satisfaction problems that is intensively studied in Artificial Intelligence. It can be described as the subset of the canonical Datalog program of width one that consists of all rules with bodies containing at most one non-IDB. An instance that is stable under inferences of this Datalog program is called *arc-consistent*. For finite templates T it is known that the arc-consistency procedure solves CSP(T) if and only if CSP(T) has width one [18]. For infinite structures, this is no longer true: consider for instance CSP($\not\triangle$), which has width 0, but cannot be solved by the arc-consistency procedure. The reason is that $\not\triangle$ has only one orbit, and we thus have to consider at least three relations in the input to infer that the input contains triangle.

Theorem 7. *Let Γ be an ω-categorical relational structure. If CSP(Γ) is solved by the arc-consistency algorithm, then Γ is homomorphically equivalent to a finite structure.*

Proof. Since Γ is ω-categorical, the automorphism group of Γ has a finite number of orbits (i.e., orbits of 1-subsets) $O_1, \ldots, O_n$. We define the *orbit structure* of Γ, which is a finite relational τ-structure whose vertices $S_1, \ldots, S_{2^n-1}$ are the nonempty subsets of $\{O_1, \ldots, O_n\}$, and where a k-ary relation R from τ holds on $S_{i_1}, \ldots, S_{i_k}$ if for every vertex v_j in an orbit from S_{i_j} there are vertices $v_1, \ldots, v_{j-1}, v_{j+1}, \ldots, v_k$ from $S_{i_1}, \ldots, S_{i_{j-1}}, S_{i_{j+1}}, \ldots, S_{i_k}$, respectively, such that R holds on $v_1, \ldots, v_k$ in Γ. Every unary relation that can be inferred by the arc-consistency procedure corresponds to a list of orbits of Γ, because Γ is ω-categorical. Since every rule application of the procedure involves a single input relation, the definition of the orbit structure implies that the Datalog program cannot infer any new relations on the orbit structure, and therefore the orbit structure is arc-consistent. Since the arc-consistency procedure solves the constraint satisfaction problem for Γ, the orbit structure is homomorphic to Γ.

Next we show that there is also a homomorphism from Γ to the orbit structure. Every finite substructure S of Γ is a satisfiable instance of CSP(Γ), and hence the arc-consistency procedure does not derive **false** on it. Consider the arc-consistent instance computed by the arc-consistency procedure on S. For each variable v we have computed a set of unary predicates, that corresponds to a non-empty subset of orbits, and that tells us to which element of the orbit structure we can map v. By definition of the orbit structure, this mapping is a homomorphism, because if there is a constraint that is not supported in the orbit structure, then the arc-consistency algorithm would have removed at least one of the orbits in the orbit list for the involved variable. We have thus shown that all finite substructures S of Γ homomorphically map to the orbit structure. Since Γ is countable and the orbit structure is finite, we conclude with Lemma 1 that there is a homomorphism from Γ to the orbit structure.   □

6.3   Strict Width l

In this section we present an algebraic characterization of strict bounded width for constraint satisfaction problems with an ω-categorical template Γ. An instance S of a constraint satisfaction problem that is computed by the canonical (l, k)-Datalog program is called *globally consistent*, if every homomorphism from an induced substructure of S to Γ can be extended to homomorphism from S to Γ. If for some $k \geq l$ all instances of CSP(Γ) that are computed by the canonical (l, k)-program are globally consistent, we say that Γ has *strict width l*.

The algebraic approach to constraint satisfaction rests on the notion of *polymorphisms*. Let Γ be a relational structure with signature τ. A *polymorphism* is a homomorphism from Γ^l to Γ, for some l, where Γ^l is a relational τ-structure defined as follows. The vertices of Γ^l are l-tuples over elements from V_Γ, and k such l-tuples $(v_1^i, \ldots, v_l^i)$, $1 \leq i \leq l$, are joined by a k-ary relation R from τ iff $(v_j^1, \ldots, v_j^k)$ is in R_Γ, for all $1 \leq j \leq l$.

We say that an operation f is a *weak near-unanimity operation* (short, *wnu-operation*), if it satisfies the identities $f(x, \ldots, x, y, x, \ldots, x) = f(x, \ldots, x)$, i.e., in the case that the arguments have the same value x except at one argument position, the function has the value $f(x, \ldots, x)$. In other words, the value y of the exceptional argument does not influence the value of the function f. For every subset A of Γ, we say that an operation is *idempotent on* A if $f(a, \ldots, a) = a$ for all $a \in A$. If a wnu-operation f is idempotent on the entire domain, then f is called a *near-unanimity operation* (short, *nu-operation*).

Feder and Vardi [18] proved that a finite template Γ has a $l + 1$-ary near-unanimity operation (they call it the $l + 1$ mapping property) if and only if $CSP(\Gamma)$ has strict width l. Another proof of this theorem was given in [23]. It is stated there that the proof extends to arbitrary infinite templates, if we want to characterize bounded strict width on instances of the constraint satisfaction problem that might be infinite. However, we would like to describe the complexity of constraint satisfaction problems with finite instances.

In fact, there are structures that do not have a nu-operation, but have A-idempotent wnu-operations for all finite subsets A of Γ. One example for such a structure is the universal triangle-free graph $\not\triangle$. A theorem by Larose and Tardif shows that every finite or infinite graph with a nu-operation is bipartite [29]. Since the universal triangle-free graph contains all cycles of length larger than three, it therefore cannot have a nu-operation. However the universal triangle-free graph has strict width 2. Indeed, for any instance S accepted by the canonical $(2, 3)$-Datalog program, every partial mapping from S to $\not\triangle$ satisfying all the facts derived by the program (and in particular not containing any triangle) can be extended to a complete homomorphisms – this follows from the extension properties of the template.

The following theorem characterizes strict width l, $l \geq 2$, for constraint satisfaction with ω-categorical templates. The proof is based on the same ideas as the proofs in [18] and [23]. However, we need some model theoretic adjustments to make these ideas work on ω-categorical templates. Due to space limitations, we omit the proof in this extended abstract. The equivalence of (1) and (2) is also shown in [5].

Theorem 8. *Let Γ be an ω-categorical structure with relational signature τ of bounded maximal arity. Then the following are equivalent, for $l \geq 2$:*

1. *For every finite subset A of Γ there is a $l + 1$-ary wnu-operation that is idempotent on A.*
2. *Every primitive positive formula is in Γ equivalent to a conjunction of at most l-ary primitive positive formulas.*
3. *$CSP(\Gamma)$ has strict width l.*

Acknowledgements. We are grateful to anonymous referees.

References

1. D. Achlioptas. The complexity of G-free colourability. *Discrete Mathematics*, 165:21–30, 1997.
2. H. Andréka and R. D. Maddux. Representations for small relation algebras. *Notre Dame Journal of Formal Logic*, 35(4):550–562, 1994.

3. A. Atserias. On digraph coloring problems and treewidth duality. In *20th IEEE Symposium on Logic in Computer Science (LICS)*, pages 106–115, 2005.
4. M. Bodirsky. The core of a countably categorical structure. In *Proceedings of the 22nd Annual Symposium on Theoretical Aspects of Computer Science (STACS'05)*, Stuttgart, LNCS 3404, pages 100–110, Springer, 2005.
5. M. Bodirsky and H. Chen. Oligomorphic clones. Preprint, 2005.
6. M. Bodirsky and J. Nešetřil. Constraint satisfaction with countable homogeneous templates. In *Proceedings of CSL'03*, pages 44–57, Vienna, 2003.
7. A. Bulatov, A. Krokhin, and P. G. Jeavons. Classifying the complexity of constraints using finite algebras. *SIAM Journal on Computing*, 34:720–742, 2005.
8. P. J. Cameron. *Oligomorphic Permutation Groups*. Cambridge University Press, 1990.
9. A. K. Chandra and P. M. Merlin. Optimal implementation of conjunctive queries in relational data bases. In *Proceddings of STOC'77*, pages 77–90, 1977.
10. G. Cherlin, S. Shelah, and N. Shi. Universal graphs with forbidden subgraphs and algebraic closure. *Advances in Applied Mathematics*, 22:454–491, 1999.
11. J. Covington. Homogenizable relational structures. *Illinois Journal of Mathematics*, 34(4):731–743, 1990.
12. M. Cristiani and R. Hirsch. The complexity of the constraint satisfaction problem for small relation algebras. *Artificial Intelligence Journal*, 156:177–196, 2004.
13. V. Dalmau. A new tractable class of constraint satisfaction problems. *Ann. Math. Artif. Intell.*, 44(1-2):61–85, 2005.
14. V. Dalmau, A. A. Krokhin, and B. Larose. First-order definable retraction problems for posets and reflexive graph. In *LICS'04*, pages 232–241, 2004.
15. V. Dalmau and J. Pearson. Closure functions and width 1 problems. *CP'99*, pages 159–173, 1999.
16. I. Düntsch. Relation algebras and their application in temporal and spatial reasoning. *Artificial Intelligence Review*, 23:315–357, 2005.
17. H.-D. Ebbinghaus and J. Flum. *Finite Model Theory*. Springer, 1999. 2nd edition.
18. T. Feder and M. Vardi. The computational structure of monotone monadic SNP and constraint satisfaction: A study through Datalog and group theory. *SIAM Journal on Computing*, 28:57–104, 1999.
19. T. Feder and M. Vardi. Homomorphism closed vs. existential positive. In *Symposium on Logic in Computer Science (LICS'03)*, pages 311–320, 2003.
20. P. Hell and J. Nešetřil. On the complexity of H-coloring. *Journal of Combinatorial Theory, Series B*, 48:92–110, 1990.
21. R. Hirsch. Expressive power and complexity in algebraic logic. *Journal of Logic and Computation*, 7(3):309 – 351, 1997.
22. W. Hodges. *A shorter model theory*. Cambridge University Press, 1997.
23. P. Jeavons, D. Cohen, and M. Cooper. Constraints, consistency and closure. *AI*, 101(1-2):251–265, 1998.
24. P. Jeavons, D. Cohen, and M. Gyssens. Closure properties of constraints. *Journal of the ACM*, 44(4):527–548, 1997.
25. P. G. Kolaitis and M. Y. Vardi. Conjunctive-query containment and constraint satisfaction. In *Proceedings of PODS'98*, pages 205–213, 1998.
26. A. Krokhin, A. Bulatov, and P. Jeavons. The complexity of constraint satisfaction: An algebraic approach (survey paper). *Structural Theory of Automata, Semigroups and Universal Algebra, NATO Science Series II: Mathematics, Physics, and Chemistry*, 207:181–213, 2005.
27. G. Kun. Every problem in MMSNP is polynomial time equivalent to a CSP. Personal communication, 2005.

28. P. B. Ladkin and R. D. Maddux. On binary constraint problems. *Journal of the Association for Computing Machinery*, 41(3):435–469, 1994.
29. B. Larose and C. Tardif. Strongly rigid graphs and projectivity. *Multiple-Valued Logic 7*, pages 339–361, 2001.
30. F. Madelaine and I. A. Stewart. Some problems not definable using structure homomorphisms. *MCS technical report. University of Leicester*, 99(18), 1999.
31. J. Nešetřil and C. Tardif. Duality theorems for finite structures (characterising gaps and good characterisations). *Journal of Combininatorial Theory Series B*, 80:80–97, 2000.

Evaluating Monotone Circuits on Cylinders, Planes and Tori

Nutan Limaye, Meena Mahajan, and Jayalal Sarma M.N.

The Institute of Mathematical Sciences, Chennai 600 113, India
{nutan, meena, jayalal}@imsc.res.in

Abstract. We revisit monotone planar circuits MPCVP, with special attention to circuits with cylindrical embeddings. MPCVP is known to be in NC^3 in general, and in LogDCFL for the special case of upward stratified circuits. We characterize cylindricality, which is stronger than planarity but strictly generalizes upward planarity, and make the characterization partially constructive. We use this construction, and four key reduction lemmas, to obtain several improvements. We show that monotone circuits with embeddings that are stratified cylindrical, cylindrical, planar one-input-face and focused can be evaluated in LogDCFL, $AC^1(LogDCFL)$, LogCFL and $AC^1(LogDCFL)$ respectively. We note that the NC^3 algorithm for general MPCVP is in $AC^1(LogCFL) = SAC^2$. Finally, we show that monotone circuits with toroidal embeddings can, given such an embedding, be evaluated in NC.

1 Introduction

The Circuit Value Problem CVP is a well-studied problem in complexity theory. When each gate is labelled AND, OR or NOT, CVP is complete for the complexity class P. It remains complete if the circuits are monotone (no NOT gates); it also remains complete if the underlying graph has a planar embedding. But if the circuit is monotone and planar (MPCVP), then evaluating it is in NC.

In [8,9], planar CVP and monotone CVP are shown to be P-complete, and a special case of MPCVP, upward stratified (see Section 2 for definitions) is shown to be in NC^2. Subsequently, the upper bound for this special case was improved to LogCFL[7], and the result was extended in [13] by showing that a less restrictive special case, namely layered upward planar monotone circuits, is also in NC, in fact in NC^3. Independently and in parallel, the work of [6] an Yang [23] showed that MPCVP in its full generality is in NC^4 and in NC^3 respectively. More recently, [4] showed that for monotone upward stratified circuits — the special case from [9,7] — there is in fact an upper bound of LogDCFL. Here, LogDCFL and LogCFL are the classes of languages logspace-many-one-reducible to deterministic and arbitrary context-free languages respectively. Recall that $L \subseteq NL \subseteq LogCFL$, $L \subseteq LogDCFL \subseteq LogCFL$, and $LogCFL = SAC^1 \subseteq AC^1 \subseteq NC^2$. See any standard text on circuit complexity (e.g. [21]) for more details.

Using the insights developed in recent works [4,11,10,2] to exploit restricted topology in circuits, we review the previous work on MPCVP, and improve the

B. Durand and W. Thomas (Eds.): STACS 2006, LNCS 3884, pp. 660–671, 2006.

known upper bounds for different cases of the problem. Our main improvements are in the case of circuits with cylindrical embeddings, which have been studied in the context of small-width circuits in [11,10]. We also extend the NC upper bound on MPCVP to toroidal (genus one) monotone circuits.

Our upper bounds use the class PDLP: problems logspace many-one reducible to finding the length of a longest path in a planar DAG. Since finding longest paths in DAGs reduces to finding shortest paths[1], L ⊆ PDLP ⊆ L(PDLP) ⊆ NL.

Our contributions are as follows: (1) We characterize cylindrical graphs as spanning subgraphs of single-source single-sink planar DAGs (Theorem 1). This generalizes Theorem 2 of [10], which is the analogous result for layered cylindrical graphs. Layering, in general, could be harder than logspace; nonetheless we obtain a partial logspace-constructive version, even when the given DAG is not layered. (2) We present four reductions among some of the topological restrictions, as shown in the diagram below. The normal arrows go from stronger to weaker constraints, the dotted arrows indicate logspace reductions, and the dashed arrows indicate the reductions in L(PDLP) (3) Using the reduction lemmas, we improve some upper bounds; see table below.

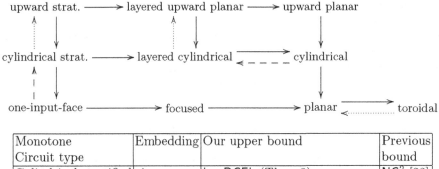

Monotone Circuit type	Embedding	Our upper bound	Previous bound
Cylindrical stratified	given	LogDCFL (Thm. 2)	NC^2 [23]
One input face	not needed	L(PDLP ⊕ LogDCFL) (Thm. 3)	NC^2 [23]
Cylindrical	given	AC^1(LogDCFL) (Thm, 4)	–
Planar	not needed	AC^1(LogCFL) = SAC^2(Thm. 5)	NC^3 [23]
Toroidal	given	AC^1(LogCFL) = SAC^2(Thm. 6)	P

2 Basic Definitions

2.1 Circuits

We consider circuits (directed acyclic graphs or DAGs) with gates labeled AND, OR, NO-OP, 0, 1 having fan-in 2,2,1,0,0, respectively. (Fan-in 0 nodes are called source nodes.) We assume w.l.o.g. that source nodes have fan-out 1 and that no gate has fan-out greater than 2. We do not assume that there is a single sink.

---

[1] It is conceivable that the class PDLP is strictly contained in NL. Hence whenever we need longest paths in planar DAGs, we state our upper bounds in terms of PDLP.

A circuit with variables is a circuit in which some fan-in zero gates are labeled by variables. By generalized circuits we mean circuits which also have constant gates with non-zero fan-in and possible fan-out more than 1; the output of such a gate is independent of its inputs, but the input wires could play a role in determining the planar embeddings. Generalized circuits, with or without variables, arise in the recursive steps of the algorithms from [6,23].

A circuit is said to be *layered* if there is a partition $V = V_0 \cup V_1 \cup \ldots \cup V_h$ such that all edges go from some V_i to V_{i+1}. The layered circuit is said to be *stratified* if all source nodes are in layer V_0. (Thus, stratified implies layered.)

2.2 Topological Embeddings and Drawings

A graph is said to be *planar* if it can be drawn on a plane in such a way that the representations of no two edges cross, except at shared endpoints. By the results of [16,3,17], deciding if a given graph is planar is in L.

A planar embedding is *bimodal* if at every vertex v, all outgoing (incoming) edges appear consecutively around v. It is easy to see ([18], [10] Lemma 5, [23] Lemmas 3.1 and 3.2) that in a planar DAG with a single source and a single sink, every embedding is bimodal.

A planar embedding of a DAG is said to be a *one-input-face embedding* if all source nodes lie on the same face. Testing if a planar DAG is one-input-face, and if so, uncovering such an embedding, is easy: add a new source node with edges to all the old sources, and test for planarity.

A drawing of a digraph on the plane is *upward* if the drawing of every edge is monotonically increasing in the vertical direction. Every DAG has an upward embedding, which can be recovered by a topological sort. (Also, only DAGs have upward embeddings, since a cycle cannot be embedded in an upward way.) A digraph is *upward planar* if it has an embedding that is upward and planar. Though all DAGs are upward, not all planar DAGs are upward planar.

A digraph is *cylindrical* if it can be embedded on a cylinder surface, in a way such that all edges are monotonically increasing in the direction of the axis of the cylinder. (Clearly such a digraph must also be acyclic, a DAG.) Note that the surface of the cylinder can be embedded on a plane in a straightforward way. A cylindrical embedding will give rise to a planar embedding where all edges flow in an inward direction towards a central face. Hence every cylindrical embedding is also bimodal, even if it is not single-source single-sink. Cylindricality strictly generalizes upward planarity, but cannot capture all planar DAGs. (See the full version for illustrative figures.) A *layered cylindrical embedding* is a cylindrical embedding where layers correspond to disjoint circles of the cylinder.

A DAG is said to be *upward stratified* (*cylindrical stratified*) if it is layered, stratified, and has an upward planar (cylindrical respectively) embedding. It follows that an upward/cylindrical stratified circuit has a one-input-face embedding. (A layered upward planar DAG need not be stratified.) In [6], the term *restricted stratified* is used to denote circuits which are *cylindrical stratified* as defined above (without the *restricted*, the authors of [6] mean *generalized* circuits). On the other hand, in [4], *stratified* refers to *upward stratified* as described here.

A planar embedding of a DAG G is *focused* if there is a subset S of source nodes, all of which are embedded on a single face, and every node of G not reachable from S is itself a source node. This is a topological analogue of a skewness condition on circuits.

2.3 Representing Embeddings

A planar embedding of a planar graph can be constructed in L [16,3,17]. The algorithm constructs a planar combinatorial embedding, specifying the cyclic (clockwise, say) ordering of edges around each vertex in some plane embedding. Checking whether a given combinatorial embedding corresponds to an embedding on the plane can be done in logspace.

We briefly discuss how faces are specified in any planar embedding. Recall that embeddings ignore directions on edges. In fact, for each (undirected) edge (u, v), the embedding will specify where arc (u, v) figures in the circular list around u, and where arc (v, u) figures in the circular list around v. The arcs (u, v) and (v, u) are expected to be superimposed in the corresponding geometric embedding. We use the term edges to refer to directed edges of the original graph, while we use the term arcs to refer to the directed arcs in the combinatorial embedding. For every arc $e = (u, v)$, there are faces $L(e)$ and $R(e)$ to the left and right, respectively, of the edge. If G is a connected graph when directions on edges are ignored, then for every face f, the set of edges with $f \in \{L(e), R(e)\}$ form a connected graph. This set can be traversed systematically as follows. Start with an arc $e = (u, v)$ such that, say, $f = R(e)$. Let $e' = (v, w)$ be the arc preceding (v, u) in the cyclic ordering around v. Then $f = R(e')$. Keep advancing in this way until the starting arc is encountered again; in the process, the entire boundary of f will be traversed. We assume that f is "named" by the lexicographically smallest arc $a = (u, v)$ such that $f = R(a)$. See [15,22] for more about representing embeddings.

For layered cylindrical or layered upward planar embeddings, we assume that the embedding is given in the following form: (a) the cyclic ordering of edges around each vertex (the planar combinatorial embedding) corresponding to the geometric embedding, and (b) the circular or left-to-right ordering of vertices at each layer. It is straightforward to see that given such information, we can verify in logspace that it indeed corresponds to some layered cylindrical or layered upward planar geometric embedding.

For cylindrical embeddings of non-layered graphs, we need to specify some more information. Imagine circles drawn on the surface of the cylinder, through each vertex. The ordering of the circles along the axis of the cylinder imposes a partial order on the vertices (total, if no two vertices lie on the same circle); consider any total order extending this. This ordering corresponds to non-decreasing distance of vertices from the left end of the cylinder. For each vertex u, we can talk of its left face and its right face: the left face is the face between u's leftmost incoming edge (last incoming edge in clockwise ordering) and leftmost outgoing edge (first outgoing edge in clockwise ordering), while the right face is the face between its rightmost incoming and outgoing edges. If u is a source, then the left and the right face are the same, and it is the face containing the (initial segment of) the ray drawn out of u against the cylinder axis. Similarly, if u is a sink, it is the face containing the (initial segment of) the ray drawn out of u along the cylinder axis. Given the clockwise ordering of edges around each vertex, the left and right faces can be determined for each u that is not a source or sink. For a source / sink u, if we explicitly specify the leftmost outgoing / incoming edge, then this face can be determined. We call this edge $L(u)$.

With this background, we now assume that the following information about the cylindrical embedding is available: (a) the cyclic ordering of edges around each vertex (the planar combinatorial embedding), (b) a total order $v_1, v_2, \ldots, v_n$ of the vertices, extending the partial order induced by the cylindrical embedding, and (c) for each source / sink u, the edge $L(u)$. In particular, the edges $L(v_1)$ and $L(v_n)$ specify the faces f_l and f_r corresponding to the left and right ends of the cylinder.

It is not clear that cylindricality can be verified, given (a), (b), (c) above. However, this information is sufficient for the results of this paper.

3   Graphs on Cylinders

Upward planar graphs have been characterized independently in [12] and [5]: A DAG is upward planar if and only if it is a subgraph of a planar st-digraph, that is, a planar DAG with a single source s, a single sink t, and an edge from s to t. Extending this proof, [10] characterizes layered cylindricality: a layered digraph is layered cylindrical if and only if it is a subgraph of a layered planar DAG with a unique source and a unique sink. While the result is implicit in the work of [19], the major contribution in the proof of [10] is to make the transformation uniform. In a similar vein, we characterize cylindricality (without the layered property); while the topological ideas are already there in the proofs of [19,10], we prove it in a different way to obtain suitable uniformity bounds. We then use these to evaluate cylindrical circuits.

One direction of our characterization crucially uses a layered embedding algorithm independently due to [23] and [6]. The algorithm of [23] is stated for single-sink digraphs where there is a one-input-face planar embedding (an embedding in which all sources appear on the same face), while that of [6] is stated for focused circuits. We will use the version from [23] and we observe that this includes, as a special case, single-source single-sink planar DAGs (SSPDs). ([23] uses the notation layered one-input-face for cylindrical stratified (all source nodes at the first layer)). An important property of such embeddings is that all vertices are bimodal; thus left and right faces of a vertex are defined. The algorithm appears in [23] Section 3 and in [6] Section 4, and is described below.

Input: A one-input-face single-sink planar directed acyclic graph H.

Output: A layered cylindrical embedding of a graph H', obtained from H by subdividing edges into directed paths.

Let t be the sink of H.

1. $\forall v \in H$, find $d(v)$, the longest distance to t. Set $d = \max d(v)$. The input nodes are in V_0. A non-input node u is in layer $l(u) = d - d(u)$.

2. For $e = (u, v)$, let $k_e = l(v) - l(u) - 1$. If $k_e > 0$, then create dummy nodes $n_1, n_2 \ldots n_k$ and add edges $(u, n_1), (n_1, n_2) \ldots (n_k, v)$. (i.e. subdivide e into a directed path of length $l(v) - l(u)$.) The node n_i will be in layer $l(u) + i$.
3. For each node u (including dummy nodes), the left (or right) neighbour of u is the node on the boundary of the left (or right, respectively) face of u with the same layer number as u.

Steps 1-2 of the algorithm provide the layering, step 3 provides the cylindrical embedding. For correctness, see Section 3 of [23] or [6]. We observe the following:

Proposition 1. *The layered embedding algorithm above runs in* L(PDLP).

Now we establish our characterization by the following two lemmas. The first follows directly from the proposition above; the second lemma is proved here.

Lemma 1. *If G is a spanning subgraph of an SSPD H, then G has a cylindrical embedding which, given G and H, can be constructed in* L(PDLP).

Lemma 2. *If a planar DAG G has a cylindrical embedding, then it is a spanning subgraph of a cylindrical SSPD H.*

Proof. Consider the layout of the graph on the cylinder surface, with vertices in order $v_1, v_2, \ldots, v_n$. If any vertex v_i other than v_n is a sink, we need to add an edge from it to some v_j with $j \geq i$ without destroying cylindricality. Such a v_j can always be found as follows: imagine a particle moving out of v_i along the direction of the cylinder axis. It aims to avoid intersecting any edge. So if it meets an edge, it moves parallel to and infinitesimally close to the edge. Since all edges are cylindrical, its movement is still monotonic with respect to the axis. As soon as it reaches (infinitesimally close to) a vertex, we declare that vertex to be v_j. If it never meets an edge or a vertex, then it will exit at the right end of the cylinder. In this case we declare v_n to be v_j. The movement ensures that the edge (v_i, v_j) can be added preserving cylindricality. A similar procedure applied after this will make all sources other than v_1 have incoming edges. □

Theorem 1. *Let G be any planar DAG. The following are equivalent.*
(1) G has a cylindrical embedding.
(2) G is a spanning subgraph of a cylindrical SSPD H.
(3) G is a spanning subgraph of a SSPD H.

It follows that testing for cylindricality is in NP. However, it may not be NP-hard, though it generalizes upward planarity, testing for which is NP-complete.

One direction of the theorem above is already constructive using Lemma 1. We make the proof of Lemma 2 constructive via a more complicated construction. This construction works only for one stage (multiple sinks to single sink or multiple sources to single source), and yields only a planar (not cylindrical) embedding of H. The advantage is that it is implementable in logspace.

Lemma 3. *Let G be a connected (in the undirected sense) cylindrical DAG with S sources and T sinks. Given a cylindrical embedding G, we can construct, in L, a planar single-source DAG H_s with T sinks and a planar single-sink DAG H_t with S sources such that G is a spanning subgraph of both.*

Proof. We describe how to construct H_s; the construction of H_t is symmetric. Since G is connected, for every face f, the edges incident on f form a connected graph. For each face f, let i be the smallest index such that v_i is on the boundary of the face. Then there is some edge $e = (v_i, v_j)$ such that $f = R(e)$. Start traversing the boundary of f, starting with such an edge $e = (v_i, v_j)$. For each v_k encountered on the boundary with in-degree 0, add edge (v_i, v_k).

Clearly this preserves acyclicity, since all new edges are from a lower indexed to larger indexed vertex. This also preserves planarity. The new edges are inserted, in the order encountered, into the cyclic ordering around v_i immediately after the arc (v_i, v_j). A new edge (v_i, v_k) is inserted into the cyclic ordering around v_k immediately after the arc (v_l, v_k) which led to the discovery of v_k on this face boundary. Thus we obtain the new planar combinatorial embedding.

Since G is connected, every source has a path to v_1. Hence every source lies on the boundary of at least one face with a lower indexed vertex, and hence acquires an incoming edge. Thus at the end, only v_1 is a source.   □

4   Circuits on Cylinders

We now show that for circuit evaluation, any technique applicable to layered upward planar circuits also applies to cylindrical circuits, with a uniformity requirement in $\mathsf{L(PDLP)} \subseteq \mathsf{NL}$. The result is obtained in two stages: first we show how to deal with layered cylindrical circuits, and then we show how to layer arbitrary cylindrical circuits. We also show that one-input-face circuits reduce to upward stratified circuits, with a similar uniformity requirement.

Lemma 4. *Given a monotone circuit C with a layered cylindrical embedding $\mathcal{E}$, we can in logspace obtain an equivalent monotone circuit C' with a layered upward planar embedding $\mathcal{E}'$. Further, if $\mathcal{E}$ is stratified, so is $\mathcal{E}'$.*

Proof Sketch. Intuitively, what we want to do is as follows. Consider a geometric embedding of C on the plane, with layers corresponding to concentric circles and edges travelling inwards. By rotating a ray shooting out of the root, we can find an angular position where it does not contain the embedding of any node. By deforming edge representations if necessary, we can ensure that each edge intersects the ray (at this angular position) in at most one point. Now simply "cut" the circuit C along the ray. This gives rise to dangling in-edges and out-edges and a circuit D which is layered upward planar. Patch multiple copies of D side-by-side, feeding zeroes to the dangling edges of the extremal copies, and let the root of the middle copy be the new root. (Or, evaluate the OR of the roots of all copies.) The full proof describes how to perform all this in logspace.   □

In the above proof, the layering being given appears crucial. We observe below that without layering, the same conversion can be performed in $\mathsf{L(PDLP)}$.

Lemma 5. *Evaluating a monotone circuit C with a cylindrical embedding $\mathcal{E}$ reduces in $\mathsf{L(PDLP)}$ to evaluating a monotone layered cylindrical circuit C' with embedding $\mathcal{E}'$. Further, if $\mathcal{E}$ is one-input-face, then $\mathcal{E}'$ is stratified.*

Proof. We proceed in four steps.

1. We remove from C all nodes with no directed path to the output gate of C. This gives an equivalent circuit G with a single sink, and with an inherited cylindrical embedding.
2. From the given cylindrical embedding of G, we construct the SSPD H with the same vertices as G and containing all the edges of G.
3. Using the algorithm from Section 3, we obtain a layered cylindrical embedding of an SSPD H', obtained by subdividing edges of H into directed paths.
4. We get a layered cylindrical embedding of a digraph G' from that of H' by throwing away all directed paths corresponding to edges in $H \setminus G$. Labelling all the new subdivision vertices with type NO-OP makes G' a circuit.

Since C is a planar DAG, Step (1) can be performed in $\mathsf{L(PDLP)}$. Step (2) uses Lemma 3, and can be performed in logspace. Step (3) uses Lemma 1, and runs in $\mathsf{L(PDLP)}$. It is easy to see that Step (4) can be performed in logspace. □

Note that the layered embedding algorithm needs a single-sink one-input-face embedding. In the above proof, the one-input-face condition is achieved in step 2 by exploiting cylindricality. However, if the given circuit already has a one-input-face embedding, then cylindricality is not needed. Thus we have:

Lemma 6. *Evaluating a monotone circuit C with a one-input-face embedding $\mathcal{E}$ reduces in $\mathsf{L(PDLP)}$ to evaluating a monotone stratified cylindrical circuit C' with embedding $\mathcal{E}'$.*

5   Improved Upper Bounds for MPCVP

In this section we revisit some of the MPCVP algorithms in the literature. We observe that some of these algorithms have tighter bounds than claimed. Wherever possible, we apply reduction lemmas of Section 4 to expand the class of circuits for which the algorithm applies, and also try to weaken the input requirements.

Goldschlager [9] considered upward stratified circuits. He showed that in this special case, if the corresponding embedding is given with the input, then MPCVP is in NC^2. This upper bound was improved to LogCFL by Dymond and Cook [7] by giving a polynomial time AuxPDA algorithm.

Using a bottom up approach, Barrington et.al.[4] showed that monotone upward stratified circuits, presented along with such an embedding, can be evaluated in $\mathsf{LogDCFL}$. The DAuxPDA algorithm repeatedly transforms the input by (a) detecting when a 0- or 1- interval at the input layer fails to propagate high enough, and (b) replacing the interval by all 1s or all 0s. The transformation thus preserves the value of the output gate. The stack is used to keep track of the frontier upto which simplifying transformations have been made.

Polynomial running time is ensured, amongst other things, by the placement of a virtual blocking interval of 0s on either extreme at each level. The algorithm requires the upward stratified embedding to be supplied as input. Though not stated explicitly, it also works for circuits with multiple sinks[2].

Since virtual blocking intervals cannot be placed at extremes of each layer for a cylindrical embedding, we do not see how to extend this algorithm to work for stratified cylindrical circuits. However, we can still obtain this upper bound by using Lemma 4 in conjunction with this algorithm:

Theorem 2. *Given a monotone planar circuit C with a stratified cylindrical embedding, determining whether C evaluates to 1 is in* LogDCFL.

What if the embedding needed for Theorem 2 is not explicitly given? Note that stratified cylindrical embeddings are one-input-face, though the converse may not hold. But one-input-face embeddings need not be given; they can be constructed in logspace. With such an embedding, we can apply Lemma 6 and Theorem 2 to get a slightly weaker upper bound for a more general class:

Theorem 3. *A monotone planar circuit C which has a one-input-face embedding can be evaluated in* $L(PDLP \oplus LogDCFL) \subseteq L(NL \oplus LogDCFL) \subseteq LogCFL$.

Layered one-input-face circuits were considered by Yang [23] as a step towards placing general MPCVP in NC. Note that these are precisely cylindrical stratified circuits. In Section 2 of [23], an upper bound of NC^2 is obtained for evaluating such circuits. Rather than use a tool like Lemma 4 followed by the bound of [9], Yang devised a somewhat different algorithm, since a modification of it was used in a later section. The essence of his algorithm was the same as in [7]: evaluating the given circuit C is equivalent to evaluating a circuit C' which tries to determine, for each interval or segment of gates at each level, whether this interval evaluates to all 1s. It can be seen that C' has polynomial algebraic degree, and hence is in NC^2 by [14]. However, it is now known that circuits of degree polynomial in circuit size can be evaluated in LogCFL [20]. Thus we have

Proposition 2. *The algorithm of Section 2, [23], for evaluating instances of MPCVP presented with stratified cylindrical embeddings, is in* LogCFL.

This bound was independently obtained by Delcher and Kosaraju [6], who observed that the algorithm of [7] works also for the cylindrical stratified case. This is because even for such embeddings, the proving sub-circuit for validity of intervals has a tree structure which is polynomial-sized.

In [13], the requirement that the circuit be stratified was dropped for the first time. The input is required to be a monotone layered upward planar circuit, with the witnessing embedding supplied. Now intervals of contiguous 1s at an intermediate layer can be split by an input node; hence the previous algorithms are inapplicable. The idea in [13] is to repeatedly split the circuit horizontally at a layer such that both pieces are between 1/4 and 3/4 of the entire circuit

---

[2] Intervals of 1s may merge though separated not just by a 0 interval but by 0- & 1-intervals; the discussion leading to Proposition 8 of [4] still holds.

in size. Evaluate each piece recursively, replacing cut off wires by variables. The details appear in [13] and in [6].

Due to monotonicity, if a gate evaluates to 1 (0) even when all variables are set to 0 (1, respectively), then the gate evaluates to 1 (0, resp.) for all settings of the variables. So by evaluating such a circuit on two settings — all variables 1, and all variables 0 — the gates can be partitioned into three sets: evaluating to 1, or 0, or depending on the input variables. Once the recursive evaluation is done, the bottom piece is entirely evaluated and the top piece has some variable gates. But now the values of all its variable inputs are known from the bottom piece, so this piece can be fully evaluated.

Clearly, the recursion depth is logarithmic, and the base case of recursion is a monotone upward stratified circuit with variables. In [13], an upper bound of NC^3 is obtained by using the fact that the NC^2 bound of [9] applies to the base case, despite the presence of variables, to obtain the three-part partition.

Note that at internal stages of the recursion, the circuits could contain constant gates with non-zero fan-in (e.g. an OR gate could get as inputs one 1 and one variable from the preceding level of recursion). So, to apply Goldschlager's algorithm to the base case, the constant gates with non-zero fan-in are explicitly removed. That is, to patch up the two pieces, only the sub-circuit induced by gates which depend on variables is considered. Also, since the strategy evaluates every gate in the circuit, it is is insensitive to multiple sinks.

Kosaraju's upper bound can be tightened by noting that a log-recursion-depth algorithm, using the algorithm of [4] rather than [9] for the base case, yields an implementation in $AC^1(LogDCFL)$. Further, the class of circuits for which this bound applies can be expanded:

Theorem 4. *An instance of MPCVP, presented with a cylindrical embedding, can be solved in* $AC^1(LogDCFL)$.

Proof. Let C be the given circuit with a cylindrical embedding. Using Lemma 5, we obtain in $L(PDLP) \subseteq AC^1$ an equivalent circuit C' with a layered cylindrical embedding $\mathcal{E}$. Applying Lemma 4 gives, in L, an equivalent layered upward planar circuit C'', which, by the preceding discussion, can be evaluated in $AC^1(LogDCFL)$. Note that for subcircuits evaluated at recursive steps, embeddings are inherited from $\mathcal{E}$.   □

Circuits with focused embeddings are considered in [6], since they arise in recursive stages of their final algorithm for general MPCVP. Such a circuit C can be converted to an equivalent upward stratified one C' (constructing such an embedding) by simplifying the neighbours of the inputs not on the special face and then using Lemma 6 followed by Lemma 4. One consequence is that some internal nodes may be constant nodes; e.g. an OR gate with a skew 1 input from outside f simplifies to a constant gate, but still has another input wire feeding into it. We could cut off such wires, but only *after* obtaining the stratified cylindrical embedding using Lemma 6. Now we can use Theorem 4. Since C' can be obtained from C in $L(PDLP) \subseteq AC^1$, and since C'' can be obtained from C in logspace, C can be evaluated $AC^1(LogDCFL)$.

The final algorithms of both [23] and [6] make no assumptions about the embedding; given an instance of MPCVP with *any* planar embedding, they show that evaluation is in NC. They repeatedly evaluate carefully chosen smaller circuits with special embeddings (cylindrical stratified or focused), which for the smaller circuits can always be obtained in NC, from the given planar embedding.

Yang's analysis proceeds by showing that $O(\log n)$ iterations of the following suffice: For each face f containing some inputs, consider the subcircuit C_f reachable (in a directed sense) from f. C_f can be converted to get a circuit with variables and a focused embedding. Evaluate this circuit as far as possible using a generalization of the scheme leading to Theorem 3. Then perform some obvious Proposition 2. Then perform some obvious simplifications, and reiterate. The generalization does not permit the use of [4] or Theorem 2. However, the strategy is the same as originally used by Yang for one-input-face embeddings; namely, there is an equivalent polynomial degree circuit doing this partial evaluation. Hence, by [20], it can be performed in LogCFL. Hence, a careful analysis of Yang's algorithm allows us to conclude that MPCVP is in $AC^1(LogCFL)$. However, it can be seen that this class is the same as SAC^2. Thus we have:

Theorem 5. *Evaluating a given monotone planar circuit C is in SAC^2.*

6   Monotone Toroidal (Genus-One) Circuits

A digraph is *toroidal* if it can be embedded on a torus. We look at circuits whose underlying DAG is toroidal. We assume that the toroidal embedding is given as a combinatorial embedding; that it is toroidal can be verified in logspace.

Any closed curve separates the plane into disconnected regions, but a closed curve can disconnect the surface of a torus or leave it connected. In the latter case, it is called a surface non-separating curve. Using the following result from [1], we establish a reduction lemma which, along with Theorem 5, immediately gives the main result of this section.

Lemma 7 ([1]). *Given a non-planar graph G with an embedding on the torus, a surface non-separating cycle in G can be found in log space.*

Lemma 8. *A monotone circuit C with a toroidal embedding can be converted in log space to an equivalent monotone circuit C' with a planar embedding.*

Proof Sketch. The lemma is proved by essentially using the idea from [1]. Intuitively what we do is as follows. Consider a given toroidal embedding. Using Lemma 7, find a cycle (in the undirected sense) such that "cutting" the circuit along the cycle makes the remaining graph planar. Now paste together several copies as in the cylindrical case such that at least one copy evaluates to the same function as the original circuit. The pasting is done preserving planarity.   □

Theorem 6. *A monotone circuit, given with an embedding on a torus, can be evaluated in SAC^2.*

Acknowledgment. The authors thank the referees for helpful comments.

References

1. E. Allender, S. Datta, and S. Roy. The directed planar reachability problem. In *Proc. 0Fth FCTTCC, LNCC vol. 0001, pages 200 249, 2005.*
2. E. Allender, S. Datta, and S. Roy. Topology inside NC^1. In *Proc. 20th IEEE Conference on Computational Complexity*, pages 298–307, 2005.
3. E. Allender and M. Mahajan. The complexity of planarity testing. *Information and Computation*, 189(1):117–134, 2004.
4. D. A. Mix Barrington, C.-J. Lu, P. Bro Miltersen, and S. Skyum. On monotone planar circuits. In *IEEE Conf. Computational Complexity*, pages 24–31, 1999.
5. G. Di Battista and R. Tamassia. Algorithms for plane representations of acyclic digraphs. *Theoretical Computer Science*, 61:175–198, 1988.
6. A. L. Delcher and S. R. Kosaraju. An NC algorithm for evaluating monotone planar circuits. *SIAM Journal of Computing*, 24(2):369–375, 1995.
7. P. W. Dymond and S. A. Cook. Complexity Theory of Parallel Time and Hardware. *Information and Computation*, 80(3):205–226, 1989.
8. L. M. Goldschlager. The monotone and planar circuit value problems are logspace complete for P. *SIGACT News*, 9(2):25–29, 1977.
9. L. M. Goldschlager. A space efficient algorithm for the monotone planar circuit value problem. *Information Processing Letters*, 10(1):25–27, 1980.
10. K. Hansen. Constant width planar computation characterizes ACC^0. In *Proc. 21st STACS, LNCS vol. 2996*, pages 44–55, 2004.
11. K. Hansen, P. Bro Miltersen, and V Vinay. Circuits on cylinders. In *Proc. 14th FCT, LNCS vol. 2751*, pages 171–182, 2003.
12. D. Kelly. Fundamentals of planar ordered sets. *Discrete Mathematics*, 63(2,3):197–216, 1987.
13. S. R. Kosaraju. On the parallel evaluation of classes of circuits. In *Proc. 10th FSTTCS Conference, LNCS vol. 472*, pages 232–237, 1990.
14. G.L. Miller, V. Ramachandran, and E. Kaltofen. Efficient parallel evaluation of straight-line code and arithmetic circuits. *SIAM Jl. Computing*, 17:687–695, 1988.
15. B. Mohar and C. Thomassen. *Graphs on Surfaces*. John Hopkins Univ. Press, 2001.
16. V. Ramachandran and J. Reif. Planarity testing in parallel. *Journal of Computer and System Sciences*, 49:517–561, 1994.
17. O. Reingold. Undirected st-conenctivity in logspace. In *Proc. 37th STOC*, pages 376–385, 2005.
18. R. Tamassia and I. G. Tollis. A unified approach to visibility representations of planar graphs. *Discrete and Computational Geometry*, 1(1):312–341, 1986.
19. R. Tamassia and I. G. Tollis. Tessellation representations of planar graphs. In *Proc. 27th Allerton Conf. Commun., Control & Computing, UIUC*, pages 48–57, 1989.
20. H. Venkateswaran. Properties that characterize LogCFL. *Journal of Computer and System Sciences*, 42:380–404, 1991.
21. H. Vollmer. *Introduction to Circuit Complexity: A Uniform Approach*. Springer, 1999.
22. A. T. White. *Graphs, Groups and Surfaces*. North-Holland, Amsterdam, 1973.
23. H. Yang. An NC algorithm for the general planar monotone circuit value problem. In *Proc. 3rd IEEE Symp. Parallel & Distributed Processing*, pages 196–203, 1991.

Constant-Depth Circuits for Arithmetic in Finite Fields of Characteristic Two

Alexander Healy* and Emanuele Viola**

Division of Engineering and Applied Sciences,
Harvard University, Cambridge, MA 02138
ahealy@fas.harvard.edu, viola@eecs.harvard.edu

Abstract. We study the complexity of arithmetic in finite fields of characteristic two, $\mathbb{F}_{2^n}$. We concentrate on the following two problems:

– Iterated Multiplication: Given $\alpha_1, \ldots, \alpha_t \in \mathbb{F}_{2^n}$, compute $\alpha_1 \cdots \alpha_t$.
– Exponentiation: Given $\alpha \in \mathbb{F}_{2^n}$ and a t-bit integer k, compute α^k.

We first consider the explicit realization $\mathbb{F}_{2^n} = \mathbb{F}_2[x]/(x^{2 \cdot 3^l} + x^{3^l} + 1)$, where $n = 2 \cdot 3^l$. We exhibit *Dlogtime*-uniform poly(n, t)-size TC^0 circuits computing exponentiation. To the best of our knowledge, prior to this work it was not even known how to compute exponentiation in logarithmic space, i.e. space $O(\log(n + t))$, over any finite field of size $2^{\Omega(n)}$. We also exhibit, for every $\epsilon > 0$, *Dlogtime*-uniform poly$(n, 2^{t^\epsilon})$-size $AC^0[\oplus]$ circuits computing iterated multiplication and exponentiation, which we prove is optimal.

Second, we consider arbitrary realizations of $\mathbb{F}_{2^n}$ as $\mathbb{F}_2[x]/(f(x))$, for an irreducible $f(x) \in \mathbb{F}_2[x]$ that is given as part of the input. We exhibit, for every $\epsilon > 0$, *Dlogtime*-uniform poly$(n, 2^{t^\epsilon})$-size $AC^0[\oplus]$ circuits computing iterated multiplication, which is again tight. We also exhibit *Dlogtime*-uniform poly$(n, 2^t)$-size $AC^0[\oplus]$ circuits for exponentiation.

Our results over $\mathbb{F}_2[x]/(x^{2 \cdot 3^l} + x^{3^l} + 1)$ have several consequences:

We prove that *Dlogtime*-uniform TC^0 equals the class AE of functions computable by certain arithmetic expressions. This answers a question raised by Frandsen, Valence and Barrington (Mathematical Systems Theory '94). We also show how certain optimal constructions of k-wise independent and ϵ-biased generators are explicitly computable in *Dlogtime*-uniform $AC^0[\oplus]$ and TC^0. This addresses a question raised by Gutfreund and Viola (RANDOM '04).

1 Introduction

Finite fields have a wide variety of applications in computer science, ranging from Coding Theory to Cryptography to Complexity Theory. In this work we study the complexity of arithmetic operations in finite fields.

---

* Research supported by NSF grant CCR-0205423 and a Sandia Fellowship.
** Research supported by NSF grant CCR-0133096, US-Israel BSF grant 2002246, ONR grant N-00014-04-1-0478.

When considering the complexity of finite field arithmetic, there are two distinct problems one must consider. The first is the problem of actually *constructing* the desired finite field, $\mathbb{F}$; for example, one must find a prime p in order to realize the field $\mathbb{F}_p$ as $\mathbb{Z}/p\mathbb{Z}$. The second is the problem of performing arithmetic operations, such as addition, multiplication and exponentiation in the field $\mathbb{F}$. In this work, we focus on this second problem, and restrict our attention to fields $\mathbb{F}$ where a realization of the field can be found very easily, or where a realization of $\mathbb{F}$ is given as part of the input.

Specifically, we will focus on finite fields of characteristic two; that is, finite fields $\mathbb{F}_{2^n}$ having 2^n elements. In particular, the question we address is: *To what extent can basic field operations (e.g., multiplication, exponentiation) in $\mathbb{F}_{2^n}$ be computed by constant-depth circuits?* In our work, we consider three natural classes of unbounded fan-in constant-depth circuits: circuits over the bases $\{\wedge, \vee\}$ (i.e., AC^0), $\{\wedge, \vee, Parity\}$ (i.e., $AC^0[\oplus]$), and $\{\wedge, \vee, Majority\}$ (i.e., TC^0). Moreover, we will focus on *uniform* constant-depth circuits, although we defer the discussion of uniformity until the paragraph "Uniformity" later in this section. Recall that, for polynomial-size circuits, $AC^0 \subsetneq AC^0[\oplus] \subsetneq TC^0 \subseteq$ log-depth circuits $\subseteq$ logarithmic space, where the last two inclusion holds under logarithmic-space uniformity and the separations follow from works by Furst et al. [FSS] and Razborov [Raz], respectively (and hold even for non-uniform circuits). See, e.g., [Hås, Vol] for background on constant-depth circuits.

Field Operations. Recall that the finite field $\mathbb{F}_{2^n}$ of characteristic two is generally realized as $\mathbb{F}_2[x]/(f(x))$ where $f(x) \in \mathbb{F}_2[x]$ is an irreducible polynomial of degree n. Thus, field elements are polynomials of degree at most $n-1$ over $\mathbb{F}_2$, addition of two field elements is addition in $\mathbb{F}_2[x]$ and multiplication of field elements is carried out modulo the irreducible polynomial $f(x)$. Throughout, we identify a field element $\alpha = a_{n-1}x^{n-1} + \cdots + a_1 x + a_0 \in \mathbb{F}_{2^n}$ with the n-dimensional bit-vector $(a_0, a_1, \ldots, a_{n-1}) \in \{0,1\}^n$, and we assume that all field elements that are given as inputs or returned as outputs of some computation are of this form.

In such a realization of $\mathbb{F}_{2^n}$, addition of two field elements is just component-wise XOR and therefore trivial, even for AC^0 circuits. It is also easy to establish the complexity of Iterated Addition, i.e. given $\alpha_1, \alpha_2, \ldots, \alpha_t \in \mathbb{F}_{2^n}$, computing $\alpha_1 + \alpha_2 + \cdots + \alpha_t \in \mathbb{F}_{2^n}$. This is easily seen to be computable by $AC^0[\oplus]$ circuits of size $\text{poly}(n, t)$. On the other hand, since parity is a special case of Iterated Addition, the latter requires AC^0 circuits of size $\text{poly}(n, 2^{t^\varepsilon})$ (see, e.g., [Hås]). Thus, we concentrate on the following *multiplicative* field operations:

- Iterated Multiplication: Given $\alpha_1, \alpha_2, \ldots, \alpha_t \in \mathbb{F}_{2^n}$, compute $\alpha_1 \cdot \alpha_2 \cdots \alpha_t$.
- Exponentiation: Given $\alpha \in \mathbb{F}_{2^n}$, and a t-bit integer k, compute α^k.

The goal is to compute these functions as efficiently as possible for given parameters n, t. We note that these functions can be computed in time $\text{poly}(n, t)$ (using the repeated squaring algorithm for exponentiation). In this work we ask what the smallest constant-depth circuits are for computing these functions. Note that computing Iterated Multiplication immediately implies being able to compute the product of two given field elements. While solving this latter problem is already non-trivial (for *Dlogtime-*, or even logspace-uniform constant-depth circuits), we will not address it separately.

Our Results. We present two different types of results. The first concerns field operations in a *specific* realization of $\mathbb{F}_{2^n}$, which we denote $\tilde{\mathbb{F}}_{2^n}$. The second type concerns field operations in an *arbitrary* realization of $\mathbb{F}_{2^n}$ as $\mathbb{F}_2[x]/(f(x))$, where we assume that the irreducible polynomial $f(x)$ is given as part of the input. We describe both of these kinds of results in more detail below. Then we discuss some applications of our results.

Results in the specific representation $\tilde{\mathbb{F}}_{2^n}$: In this setting, we assume that n is of the form $n = 2 \cdot 3^l$, for some integer $l \geq 0$, and we employ the explicit realization of $\mathbb{F}_{2^n}$ given by $\mathbb{F}_2[x]/(f(x))$ where $f(x)$ is the irreducible polynomial $x^{2 \cdot 3^l} + x^{3^l} + 1 \in \mathbb{F}_2[x]$. Our results are summarized in the top half of Table 1.

We show that exponentiation can be computed by uniform TC^0 circuits of size $\text{poly}(n, t)$ (i.e. what is achievable by standard unbounded-depth circuits). To the best of our knowledge, prior to this work it was not even known how to compute exponentiation in logarithmic space, i.e. space $O(\log(n + t))$, over any finite field of size $2^{\Omega(n)}$. As a corollary, we improve upon a theorem of Agrawal et al. [AAI$^+$] concerning exponentiation in uniform AC^0. In the case of iterated multiplication of t field elements, results of Hesse et al. [HAB] imply that this problem can be solved by uniform TC^0 circuits of size $\text{poly}(n, t)$.

We also show that, for every $\epsilon > 0$, iterated multiplication and exponentiation can be computed by uniform $AC^0[\oplus]$ circuits of size $\text{poly}(n, 2^{t^\epsilon})$. Moreover, we show that this is tight: neither iterated multiplication nor exponentiation can be computed by (nonuniform) $AC^0[\oplus]$ circuits of size $\text{poly}(n, 2^{t^{o(1)}})$.

Results in arbitrary representation $\mathbb{F}_2[x]/(f(x))$: In this setting we assume that the irreducible polynomial $f(x)$ is arbitrary, but is given to the circuit as part of the input. Our results are summarized in the bottom half of Table 1.

We show (with a more complicated proof than in the specific representation case) that iterated multiplication can be computed by uniform $AC^0[\oplus]$ circuits of size $\text{poly}(n, 2^{t^\epsilon})$, and this is again tight. We show that exponentiation can be computed by uniform $AC^0[\oplus]$ circuits of size $\text{poly}(n, 2^t)$, but we do not know how to match the size $\text{poly}(n, 2^{t^\epsilon})$ achieved in the specific representation case. More dramatically, we do not know if there exist $\text{poly}(n, 2^{o(t)})$-size TC^0 circuits for exponentiation. On the other hand, we also observe that there are $AC^0[\oplus]$ circuits of size n^{t^ϵ} for exponentiation; this bound is, in general, incomparable to our previous bound of $\text{poly}(n, 2^t)$.

While we cannot establish a lower bound for exponentiation, we show that testing whether a given $\mathbb{F}_2[x]$ polynomial of degree n is irreducible can be $AC^0[\oplus]$-reduced to computing exponentiation in a given representation of $\mathbb{F}_{2^n}$, for exponents with $t = n$ bits. Specifically, a modification of Rabin's irreducibility test [Rab, MS] gives a TC^0 reduction; we show a finer analysis that gives a $AC^0[\oplus]$ reduction. Thus, the task of improving exponentiation modulo a given (irreducible) polynomial is closely related to obtaining upper bounds on the complexity of testing irreducibility of a given $\mathbb{F}_2[x]$ polynomial. Some lower bounds for the latter problem are given in the recent work of Allender et al. [ABD$^+$].

Table 1. Complexity of Operations in $\mathbb{F}_{2^n}$

Problem	AC^0	$AC^0[\oplus]$	TC^0
Operations in $\tilde{\mathbb{F}}_{2^n} \equiv \mathbb{F}_2[x]/(x^{2 \cdot 3^l} + x^{3^l} + 1)$			
Iterated Multiplication:	$\text{poly}(2^{n^\epsilon}, 2^{t^\epsilon})$	$\text{poly}(n, 2^{t^\epsilon})$	$\text{poly}(n, t)$
$\alpha_1, \ldots, \alpha_t \in \tilde{\mathbb{F}}_{2^n} \rightarrow \prod_i \alpha_i$	Cor. to [HAB]	[Thm. 3]	[HAB]
Exponentiation:	$\text{poly}(2^{n^\epsilon}, 2^{t^\epsilon})$	$\text{poly}(n, 2^{t^\epsilon})$	$\text{poly}(n, t)$
$\alpha \in \tilde{\mathbb{F}}_{2^n}$, t-bit $k \in \mathbb{Z} \rightarrow \alpha^k$	[Cor. 1]	[Thm. 3]	[Thm. 2]
Operations in $\mathbb{F}_{2^n} \equiv \mathbb{F}_2[x]/(f(x))$ for given $f(x)$ of degree n			
Iterated Multiplication:	$\text{poly}(2^{n^\epsilon}, 2^{t^\epsilon})$	$\text{poly}(n, 2^{t^\epsilon})$	$\text{poly}(n, t)$
$\alpha_1, \ldots, \alpha_t \in \mathbb{F}_{2^n} \rightarrow \prod_i \alpha_i$	Cor. to [HAB]	[Thm. 5]	[HAB]
Exponentiation:	$\text{poly}(2^{n^\epsilon}, 2^{2^{\epsilon t}})$	$\text{poly}(n, 2^t)$	$\text{poly}(n, 2^t)$
$\alpha \in \mathbb{F}_{2^n}$, t-bit $k \in \mathbb{Z} \rightarrow \alpha^k$	Cor. to [HAB]	[Thm. 5]	[HAB]
In the above, $\epsilon > 0$ is arbitrary, but the circuits have depth $O(1/\epsilon)$.			

However, it is still open whether irreducibility of a given degree-n polynomial in $\mathbb{F}_2[x]$ can be decided by $AC^0[\oplus]$ circuits of size $\text{poly}(n)$.

Applications. Our results in $\tilde{\mathbb{F}}_{2^n}$ have several applications, discussed below.

$AE = Dlogtime\text{-}uniform\ TC^0$: Frandsen, Valence and Barrington [FVB] study the relationship between uniform TC^0 and the class AE of functions computable by certain arithmetic expressions (defined in Section 2.3). Remarkably, they show that *Dlogtime*-uniform TC^0 is contained in AE. Conversely, they show that AE is contained in P-uniform TC^0, but they leave open whether the inclusion holds under *Dlogtime* uniformity. We show that AE is in fact contained in *Dlogtime*-uniform TC^0, thus proving that $AE = Dlogtime$-uniform TC^0. (See paragraph "Uniformity" for a discussion of *Dlogtime*-uniformity.)

Pseudorandom Generators: We implement certain "pseudorandom" generators in *Dlogtime*-uniform constant-depth circuits. Specifically, we show how an optimal construction of k-wise independent generators can be implemented in uniform $AC^0[\oplus]$, and how an optimal construction of ϵ-biased generators can be implemented in uniform TC^0. These constructions are explicit, i.e. the circuits are given the seed and an index i, and compute the i-th output bit of the generator. These results address a problem posed by Gutfreund and Viola [GV].

Overview of Techniques. Our results for the specific representation of $\mathbb{F}_{2^n}$ as $\tilde{\mathbb{F}}_{2^n} := \mathbb{F}_2[x]/(x^{2 \cdot 3^l} + x^{3^l} + 1)$ exploit the special structure of the irreducible polynomial $x^{2 \cdot 3^l} + x^{3^l} + 1 \in \mathbb{F}_2[x]$. The crucial observation (Fact 9) is that the order of x modulo $x^{2 \cdot 3^l} + x^{3^l} + 1$ is small and is easily computed, namely it is 3^{l+1}. Thus, we are able to compute large powers of the element $x \in \tilde{\mathbb{F}}_{2^n}$ by considering the exponent k modulo the order of x. To better illustrate this idea we now sketch a proof of the fact that exponentiation over $\tilde{\mathbb{F}}_{2^n}$ can be computed by uniform TC^0 circuits of size $\text{poly}(n, t)$. Let $\alpha \in \tilde{\mathbb{F}}_{2^n}$ and an exponent $0 \leq k < 2^t$

be given. We think of α as a polynomial $\alpha(x) \in \mathbb{F}_2[x]$. Writing k in binary as $k = k_{t-1}k_{t-2} \cdots k_0 = \sum_{i<t} k_i 2^i$ where $k_i \in \{0,1\}$, we have:

$$\alpha(x)^k = \alpha(x)^{\sum_{i<t} k_i 2^i} = \prod_{i<t} \alpha(x)^{k_i 2^i} = \prod_{i<t} \alpha\left(x^{2^i}\right)^{k_i}$$

where the last equality follows from the fact that we are working in characteristic 2. Using the fact that the iterated product of t field elements is computable by uniform TC^0 circuits of size $\mathrm{poly}(n,t)$ (which follows from results in [HAB]), all that is left to do is to show how to compute $\alpha(x^{2^i})^{k_i}$. Since $k_i \in \{0,1\}$, the only hard step of this is computing x^{2^i} which can be done using the fact, discussed above, that the order of x is 3^{l+1}. Specifically, first we reduce $2^i \mod 3^{l+1}$ using results about the complexity of integer arithmetic by Hesse et. al. [HAB]. After the exponent is reduced, computing the corresponding power of x is easy.

To prove that $AE = Dlogtime$-uniform TC^0 we also show that $\widetilde{\mathbb{F}}_{2^n}$ has an easily computable *dual basis* (as a vector-space over $\mathbb{F}_2$).

The other techniques we use are based on existing algorithms in the literature, e.g. [Kun, Sie, Rei, Ebe]. Our main contribution is noticing that for some settings of parameters they can be implemented in $AC^0[\oplus]$ and moreover that they give tight results for $AC^0[\oplus]$. We now describe these techniques in more detail.

In the case of arbitrary realizations of $\mathbb{F}_{2^n}$ as $\mathbb{F}_2[x]/(f(x))$, the main technical challenge is reducing polynomials modulo $f(x)$. Previous work has addressed this problem and shown how (over arbitrary fields) this can be solved by uniform log-depth circuits (of fan-in 2) [Rei, Ebe], and even by uniform TC^0 circuits [HAB]. The usual approach is to give a parallel implementation of the Kung-Sievking algorithm [Kun, Sie] to reduce polynomial division to the problem of computing small powers of polynomials. However, this reduction requires summations of $\mathrm{poly}(n)$ polynomials, which is why previous results only give implementations in log-depth or by TC^0 circuits. We take the same approach; however, we observe that in our setting we may compute large summations using parity gates. This allows us to implement polynomial division over $\mathbb{F}_2[x]$ in $AC^0[\oplus]$.

Both in our results for $\widetilde{\mathbb{F}}_{2^n}$ and for arbitrary realizations of $\mathbb{F}_{2^n}$, we make use of the Discrete Fourier Transform (DFT). This allows us to reduce the problem of multiplication or exponentiation of polynomials to the problem of multiplying or exponentiating field elements in fields of size $\mathrm{poly}(n)$ (and these problems are feasible for AC^0 circuits). Eberly [Ebe] and Reif [Rei] have also employed the DFT in their works on performing polynomial arithmetic in log-depth circuits. However, as with polynomial division in $\mathbb{F}_2[x]$, the fact that we are working with polynomials over $\mathbb{F}_2$ allows us to compute the DFT and inverse DFT in uniform $AC^0[\oplus]$ (and not just in log-depth or TC^0).

Other Related Work. Works by Reif [Rei] and Eberly [Ebe] show how basic field arithmetic can be computed by log-depth circuits. Fich and Tompa [FT] show that modular exponentiation of polynomials over a finite field of polynomial size can be computed in NC^2, i.e. by polynomial-size circuits of depth $O(\log^2 n)$. The results of Hesse, Allender and Barrington [HAB] imply that some field arithmetic can be accomplished by uniform TC^0. Indeed, the main result of [HAB] states that integer division can be computed by (uniform) TC^0 circuits, and hence addition and multiplication in the field $\mathbb{F}_p \simeq \mathbb{Z}/p\mathbb{Z}$ can be accomplished (in TC^0) by adding or multiplying elements as integers and then reducing modulo p. Other results from [HAB] imply that uniform TC^0 circuits can compute iterated multiplication in (arbitrary realizations of) $\mathbb{F}_{2^n}$. Some results on the complexity of arithmetic in finite fields of *unbounded* characteristic are given in [SF].

Uniformity. In the previous discussion we refer to uniform circuits for the various problems we consider. When working with restricted circuit classes, such as AC^0, $AC^0[\oplus]$ and TC^0, one must be careful not to allow the machine constructing the circuits to be more powerful than the circuits themselves. Indeed, one of the significant technical contributions of [HAB] is showing that integer division is in uniform TC^0 under a very strong notion of uniformity, namely *Dlogtime*-uniformity [BIS]. A circuit is *Dlogtime*-uniform if, given indices of two gates in the circuit, one can determine the types of the gates and whether they are connected in linear time in the length of the indices (which is logarithmic in the size of the circuit). *Dlogtime*-uniformity has become the generally-accepted convention for uniformity in constant-depth circuits. One reason for this is that *Dlogtime*-uniform constant-depth circuits have several elegant characterizations (see, e.g., [BIS]); in fact, our results will prove yet another such characterization, namely *Dlogtime*-uniform $TC^0 = AE$. *Unless otherwise specified, in this work "uniform" always means "Dlogtime-uniform".*

If one is willing to relax the uniformity condition to polynomial-time, then some of our results over $\tilde{\mathbb{F}}_{2^n}$ can be proved more easily. For instance, the exponentiation result requires computing $x^{2^i} \in \tilde{\mathbb{F}}_{2^n}$ for a given i. Instead of actually computing x^{2^i} in the circuit, these values could be computed in polynomial time and then hardwired into the circuit. In contrast, in the case of our results in arbitrary realizations of $\mathbb{F}_{2^n}$, we do not know how to improve any of our results, even if we allow non-uniformity. On the other hand, if one allows non-uniformity that *depends on the irreducible polynomial $f(x)$*, then one can simplify some proofs, and improve the exponentiation result to match the parameters that we achieve in $\tilde{\mathbb{F}}_{2^n}$ (by hardwiring the values x^{2^i} into the circuit, as above).

Open Problems. We now mention two open problems, both of which are also open for nonuniform circuits: Given an irreducible polynomial $f(x)$ of degree n and $\alpha \in \mathbb{F}_2[x]/(f(x))$, is it possible to compute α^{2^i} for any $i = \omega(\log n)$ in TC^0? It turns out that this is what limits our results about exponentiation in $\mathbb{F}_2[x]/(f(x))$. Given $\alpha \in \tilde{\mathbb{F}}_{2^n}$, can α^{-1} be computed in $AC^0[\oplus]$?

Organization. This paper is organized as follows. In Section 2 we formally state our results. In Section 3 we sketch the proofs of our results for performing field operations in $\tilde{\mathbb{F}}_{2^n}$ in $AC^0[\oplus]$. We omit the rest of the proofs due to space limitations. The full version of this paper is available on the Electronic Colloquium on Computational Complexity (TR05-087, http://eccc.uni-trier.de/eccc/).

2   Our Results

In this section we formally state our results. In Section 2.1 we discuss our results in the specific case where n is of the form $n = 2 \cdot 3^l$, and $\mathbb{F}_{2^n}$ is realized as $\mathbb{F}_2[x]/(x^{2 \cdot 3^l} + x^{3^l} + 1)$, i.e. using the explicit irreducible polynomial $x^{2 \cdot 3^l} + x^{3^l} + 1 \in \mathbb{F}_2[x]$. In Section 2.2 we discuss our results in realizations of $\mathbb{F}_{2^n}$ as $\mathbb{F}_2[x]/(f(x))$ for an *arbitrary* irreducible polynomial $f(x) \in \mathbb{F}_2[x]$ that is given as part of the input. Then we discuss applications of our results. In Section 2.3 we prove that uniform $TC^0 = AE$. In Section 2.4 we exhibit k-wise independent and ϵ-biased generators explicitly computable in uniform $AC^0[\oplus]$ and TC^0.

2.1   Field Arithmetic in $\tilde{\mathbb{F}}_{2^n}$

Below we summarize our main results over the field $\tilde{\mathbb{F}}_{2^n}$, defined below.

Fact 1 ([vL], Theorem 1.1.28). *The polynomial $x^{2 \cdot 3^l} + x^{3^l} + 1 \in \mathbb{F}_2[x]$ is irreducible for all integers $l \geq 0$.*

Definition 1. *For n of the form $n = 2 \cdot 3^l$, we define $\tilde{\mathbb{F}}_{2^n}$ to be the specific realization of $\mathbb{F}_{2^n}$ given by $\tilde{\mathbb{F}}_{2^n} := \mathbb{F}_2[x]/(x^{2 \cdot 3^l} + x^{3^l} + 1)$.*

The next theorem states our results about field arithmetic over $\tilde{\mathbb{F}}_{2^n}$ in uniform TC^0. The first item follows from results of Hesse, Allender and Barrington [HAB]; nonetheless, we state it for the sake of comparison with our other results.

Theorem 2. *Let $n = 2 \cdot 3^l$. There exist uniform TC^0 circuits of size $\mathrm{poly}(n, t)$ that perform the following:*

1. *[HAB] Given $\alpha_1, \alpha_2, \ldots, \alpha_t \in \tilde{\mathbb{F}}_{2^n}$, compute $\alpha_1 \cdot \alpha_2 \cdots \alpha_t \in \tilde{\mathbb{F}}_{2^n}$.*
2. *Given $\alpha \in \tilde{\mathbb{F}}_{2^n}$ and a t-bit integer k, compute $\alpha^k \in \tilde{\mathbb{F}}_{2^n}$.*

In particular, uniform TC^0 circuits of polynomial size are capable of performing iterated multiplication and exponentiation in $\tilde{\mathbb{F}}_{2^n}$ that match the parameters that can be achieved by unbounded-depth circuits.

We now state our results about field arithmetic over $\tilde{\mathbb{F}}_{2^n}$ in uniform $AC^0[\oplus]$.

Theorem 3. *Let $n = 2 \cdot 3^l$. Then, for every constant $\epsilon > 0$, there exist uniform $AC^0[\oplus]$ circuits of size $\mathrm{poly}(n, 2^{t^\epsilon})$ that perform the following:*

1. *Given $\alpha_1, \alpha_2, \ldots, \alpha_t \in \tilde{\mathbb{F}}_{2^n}$, compute $\alpha_1 \cdot \alpha_2 \cdots \alpha_t \in \tilde{\mathbb{F}}_{2^n}$.*
2. *Given $\alpha \in \tilde{\mathbb{F}}_{2^n}$ and a t-bit integer k, compute $\alpha^k \in \tilde{\mathbb{F}}_{2^n}$.*

While these parameters are worse than for TC^0 circuits, they are tight:

Theorem 4. *For every constant d there is an $\epsilon > 0$ such that, for sufficiently large t and $n = 2 \cdot 3^l$, the following cannot be computed by (nonuniform) $AC^0[\oplus]$ circuits of depth d and size $2^{2^{\epsilon n}} \cdot 2^{t^\epsilon}$:*

1. *Given $\alpha_1, \alpha_2, \ldots, \alpha_t \in \tilde{\mathbb{F}}_{2^n}$, compute $\alpha_1 \cdot \alpha_2 \cdots \alpha_t \in \tilde{\mathbb{F}}_{2^n}$.*
2. *Given $\alpha \in \tilde{\mathbb{F}}_{2^n}$ and a t-bit integer k, compute $\alpha^k \in \tilde{\mathbb{F}}_{2^n}$.*

In fact, Item (1) in the above negative result holds not only for $\tilde{\mathbb{F}}_{2^n}$, but for any sufficiently large field, and Item (2) holds for fields of a variety of different sizes.

It is known that every function in NL can be computed by uniform AC^0 circuits of size 2^{n^ϵ} and depth $O(1/\epsilon)$ (see, e.g., Lemma 21 in [AHM$^+$]). Since NL contains uniform TC^0, we obtain the following corollary to Theorem 2.

Corollary 1. *Let $n = 2 \cdot 3^l$. Then, for every constant $\epsilon > 0$, there exist uniform AC^0 circuits of size $\mathrm{poly}(2^{n^\epsilon}, 2^{t^\epsilon})$ that perform the following:*

1. *Given $\alpha_1, \alpha_2, \ldots, \alpha_t \in \tilde{\mathbb{F}}_{2^n}$, compute $\alpha_1 \cdot \alpha_2 \cdots \alpha_t \in \tilde{\mathbb{F}}_{2^n}$.*
2. *Given $\alpha \in \tilde{\mathbb{F}}_{2^n}$ and a t-bit integer k, compute $\alpha^k \in \tilde{\mathbb{F}}_{2^n}$.*

This improves on a theorem of Agrawal et al. [AAI$^+$] who show that field exponentiation is computable by uniform AC^0 circuits of size $\mathrm{poly}(2^n, 2^t)$. Corollary 1 is also tight for many settings of parameters (see Theorem 4).

2.2 Field Arithmetic in Arbitrary Realizations of $\mathbb{F}_{2^n}$

As noted above, one of the advantages of working with the field $\tilde{\mathbb{F}}_{2^n}$ is that we achieve tight results for TC^0, $AC^0[\oplus]$ and AC^0. However, the use of $\tilde{\mathbb{F}}_{2^n}$ requires that $n = 2 \cdot 3^l$, and thus does not allow for the construction of $\mathbb{F}_{2^n}$ for all n; moreover some applications may require field computations in a specific field $\mathbb{F}_2[x]/(f(x))$ for some given irreducible polynomial $f(x)$ other than $x^{2 \cdot 3^l} + x^{3^l} + 1$. Thus we are led to study the complexity of arithmetic in the ring $\mathbb{F}_2[x]/(f(x))$ where the polynomial $f(x) \in \mathbb{F}_2[x]$ is *given as part of the input*. If, in addition, we have the promise that $f(x)$ is irreducible, then this corresponds to arithmetic in the field $\mathbb{F}_{2^n} \simeq \mathbb{F}_2[x]/(f(x))$.

Theorem 5.
1. *For every constant $\epsilon > 0$, there exist uniform $AC^0[\oplus]$ circuits of size $\mathrm{poly}(n, 2^{t^\epsilon})$ that perform the following: Given $f(x) \in \mathbb{F}_2[x]$ of degree n and $\alpha_1, \alpha_2, \ldots, \alpha_t \in \mathbb{F}_2[x]/(f(x))$, compute $\alpha_1 \cdot \alpha_2 \cdots \alpha_t \in \mathbb{F}_2[x]/(f(x))$.*
2. *There exist uniform $AC^0[\oplus]$ circuits of size $\mathrm{poly}(n, 2^t)$ that perform the following: Given $f(x) \in \mathbb{F}_2[x]$ of degree n, $\alpha \in \mathbb{F}_2[x]/(f(x))$ and a t-bit integer k, compute $\alpha^k \in \mathbb{F}_2[x]/(f(x))$.*

Since Item 1 of Theorem 4 actually holds for any realization of $\mathbb{F}_{2^n}$, and not just for $\tilde{\mathbb{F}}_{2^n}$, Item 1 of Theorem 5 is tight.

Unlike Item 2 in Theorem 3, Exponentiation now requires size $\mathrm{poly}(n, 2^t)$, instead of $\mathrm{poly}(n, 2^{t^\epsilon})$. We do not know how to improve this to size $\mathrm{poly}(n, 2^{o(t)})$, even for TC^0 circuits. However, we observe that Exponentiation is also computable by (uniform) $AC^0[\oplus]$ circuits of size n^{t^ϵ}. (This can be shown using techniques similar to those mentioned at the end of this section.)

On the other hand, we show that testing irreducibility of a $\mathbb{F}_2[x]$ polynomial is $AC^0[\oplus]$ reducible to exponentiating modulo a given irreducible polynomial.

Theorem 6. *The problem of determining whether a given polynomial $f(x) \in \mathbb{F}_2[x]$ of degree n is irreducible, is $\mathrm{poly}(n)$-size $AC^0[\oplus]$-reducible to the following problem: Given an irreducible polynomial $f(x) \in \mathbb{F}_2[x]$ of degree n, compute the conjugates $x, x^2, x^{2^2}, \ldots, x^{2^{n-1}} \pmod{f(x)}$.*

Alternatively, Theorem 6 can be interpreted as a positive result. Indeed, since computing x^{2^i} (mod $f(x)$) can be shown to be in $\oplus$L, it implies that irreducibility testing is also in $\oplus$L. In turn, this implies that irreducibility testing has $AC^0[\oplus]$ circuits of size 2^{n^ϵ}, simply because all of $\oplus$L does. This latter claim follows from the same techniques that give that every function in NL can be computed by uniform AC^0 circuits of size 2^{n^ϵ} and depth $O(1/\epsilon)$ (see, e.g., Lemma 21 in [AHM+]).

2.3 $AE = Dlogtime$ Uniform TC^0

Frandsen, Valence and Barrington [FVB] study the relationship between uniform TC^0 and the class AE of functions computable by certain arithmetic expressions (defined below). Remarkably, they show that $Dlogtime$-uniform TC^0 is contained in AE. Conversely, they show that AE is contained in P-uniform TC^0, but they leave open whether the inclusion holds for $Dlogtime$ uniformity. We show that AE is in fact contained in $Dlogtime$-uniform TC^0, thus proving that $AE = Dlogtime$-uniform TC^0. (All these inclusions between classes hold in a certain technical sense that is made clear below.)

We now briefly review the definition of AE and then state our results.

Definition 2 ([FVB]). *Let I be an infinite set of formal indices. The set of formal arithmetic expressions is defined as follows. The basic expressions are x (we think of this as the field element x), and* Input *(we think of this as the input field element). If e, e' are expressions (possibly containing the index $i \in I$ as a free variable), then we may form new composite expressions $\sum_{i=1}^{u} e, \prod_{i=1}^{u} e, e + e', e \cdot e', e^{2^i}$, where $i \in I$ and u is either an index, i.e. $u \in I$, or is any polynomial in n (we think of n as the input length).*

An arithmetic expression is well-formed *if all indices are bound and they are bound in a semantically sound way (we omit details). We associate to every well-formed arithmetic expression e a family of functions $f_n^e : \tilde{\mathbb{F}}_{2^n} \to \tilde{\mathbb{F}}_{2^n}$, for every n of the form $n = 2 \cdot 3^l$ (note that all computations are performed over the field $\tilde{\mathbb{F}}_{2^n}$). The class AE consists of those families of functions $f_n : \tilde{\mathbb{F}}_{2^n} \to \tilde{\mathbb{F}}_{2^n}$ that are described by arithmetic expressions (for every n of the form $n = 2 \cdot 3^l$).*

For example, the *trace* function, $\mathrm{tr}(\mathrm{Input}) := \sum_{i=0}^{n-1} \mathrm{Input}^{2^i}$, is in AE.

Theorem 7. *$AE = Dlogtime-uniform\ TC^0$ in the following sense:*

Let $f : \{0, 1\}^n \to \{0, 1\}^n$ be in $Dlogtime$-uniform TC^0. Then there is $f' : \tilde{\mathbb{F}}_{2^n} \to \tilde{\mathbb{F}}_{2^n}$ in AE such that for every n of the form $2 \cdot 3^l$, and for every x of length n, $f(x) = f'(x)$.

Conversely, let $f : \tilde{\mathbb{F}}_{2^n} \to \tilde{\mathbb{F}}_{2^n}$ be in AE. Then there is $f' : \{0, 1\}^n \to \{0, 1\}^n$ in $Dlogtime$-uniform TC^0 such that for every n of the form $2 \cdot 3^l$, and for every x of length n, $f(x) = f'(x)$.

Our definition of arithmetic expressions is slightly different from the definition in [FVB]. In the full version of this paper we compare the two definitions and argue that our definition only makes our results stronger.

2.4   k-Wise and ε-Biased Generators

We use our results on computing field operations to give constant-depth implementations of certain "pseudorandom" generators, namely k-wise independent and $ε$-biased generators. Informally, a generator $G : \{0,1\}^s \to \{0,1\}^m$ is k-wise independent if every fixed k output bits are distributed uniformly and independently (over random choice of the input $r \in \{0,1\}^s$), while it is $ε$-biased if every fixed subset of the output bits has the property that the parity of the bits in the subset is 1 with probability $p \in [1/2 - ε, 1/2 + ε]$ (over random choice of the the input). We refer the reader to the book by Goldreich [Gol], or to the full version of this paper, for background and discussion of these generators.

We say that a generator $G : \{0,1\}^s \to \{0,1\}^m$ is *explicitly computable* in uniform TC^0 (resp., $AC^0[\oplus]$) if there is a uniform TC^0 (resp., $AC^0[\oplus]$) circuit of size $\text{poly}(s, \log m)$ that, given $x \in \{0,1\}^s$ and $i \le m$, computes the i-th output bit of $G(x)$. Using our previous results we obtain the following theorem. In both cases, the seed length s is optimal up to constant factors (see, e.g., [Gol]).

Theorem 8.
1. *For every k and m there is a k-wise independent generator $G : \{0,1\}^s \to \{0,1\}^m$, with $s = O(k \log m)$ that is explicitly computable by uniform $AC^0[\oplus]$ circuits of size $\text{poly}(s, \log m) = \text{poly}(s)$.*
2. *For every $ε$ and m, there is an $ε$-biased generator $G : \{0,1\}^s \to \{0,1\}^m$ with $s = O(\log m + \log(1/ε))$ that is explicitly computable by uniform TC^0 circuits of size $\text{poly}(s, \log m) = \text{poly}(s)$.*

Remark 1. A previous and different construction of k-wise independent generators in [GV] matches (up to constant factors) Item 1 in Theorem 8 for the special case $k = O(1)$. The construction in Item 1 in Theorem 8 improves on the construction in [GV] for $k = ω(1)$. Also, in [GV] they exhibit a construction of $ε$-biased generators computable by uniform $AC^0[\oplus]$ circuits (while the construction in Item 2 in Theorem 8 uses TC^0 circuits). However, the construction in [GV] has worse dependence on $ε$.

3   Arithmetic in $\tilde{\mathbb{F}}_{2^n}$

In this section we sketch the proof of our positive results on arithmetic in $\tilde{\mathbb{F}}_{2^n} = \mathbb{F}_2[x]/(x^{2 \cdot 3^l} + x^{3^l} + 1)$ in $AC^0[\oplus]$, i.e. Theorem 3. One useful property of $\tilde{\mathbb{F}}_{2^n}$ is that the order of $x \in \tilde{\mathbb{F}}_{2^n}$ is small, specifically it is $3^{l+1} = O(n)$. (A priori, it could have been as large as $2^n - 1$.)

Fact 9. *The order of $x \in \tilde{\mathbb{F}}_{2^n}$ is 3^{l+1}.*

Proof. Observe that $x^{3^{l+1}} \equiv 1 \pmod{x^{2 \cdot 3^l} + x^{3^l} + 1}$. Thus the order of x must divide 3^{l+1}. Noting that $x^{3^l} \not\equiv 1 \pmod{x^{2 \cdot 3^l} + x^{3^l} + 1}$, the result follows.   □

A crucial way in which Fact 9 is useful is that it allows us to compute high powers, $α^k$, of field elements $α \in \tilde{\mathbb{F}}_{2^n}$, in the special case when k is a power of 2.

Lemma 1. *Let n be of the form $n = 2 \cdot 3^l$. For every constant $\epsilon > 0$, there exist uniform $AC^0[\oplus]$ circuits of size $\mathrm{poly}(n, 2^{t^\epsilon})$ that, on input $\alpha \in \tilde{\mathbb{F}}_{2^n}$ and $i \leq t$ (in unary), computes $\alpha^{2^i} \in \tilde{\mathbb{F}}_{2^n}$.*

Proof sketch. We use a result by Hesse et al. that, for every constant $\epsilon > 0$, integer multiplication and division of m-bit numbers can be computed by uniform AC^0 circuits of size 2^{m^ϵ} ([HAB], Theorem 5.1).

Since $\alpha^{2^n} = \alpha$ for all $\alpha \in \tilde{\mathbb{F}}_{2^n}$, we first reduce i modulo n. From this point on we assume that $i \leq n$. Let $\alpha(x) \in \mathbb{F}_2[x]$ be the polynomial representing α. Thus, it suffices to compute $\alpha(x)^{2^i} \equiv \alpha(x^{2^i})$ modulo $x^{2 \cdot 3^l} + x^{3^l} + 1$. In particular, it suffices to compute $x^{h \cdot 2^i}$ in $\tilde{\mathbb{F}}_{2^n}$ for every $h, i \leq n$, since then we can then compute $\alpha(x^{2^i})$ by simply summing the appropriate terms.

We compute $x^{h \cdot 2^i} \in \tilde{\mathbb{F}}_{2^n}$ as follows. First, recall that the order of x modulo $x^{2 \cdot 3^l} + x^{3^l} + 1$ is 3^{l+1} by Fact 9. Thus we compute $k \equiv h \cdot 2^i \bmod 3^{l+1}$, and then we are left with the task of computing x^k modulo $x^{2 \cdot 3^l} + x^{3^l} + 1$. Clearly, if $k < 2 \cdot 3^l$, then the result is simply x^k; on the other hand, if $2 \cdot 3^l \leq k < 3 \cdot 3^l$, then $x^k \equiv x^{k-3^l} + x^{k-2 \cdot 3^l}$. □

In the full version of this paper we prove the above lemma for circuits of size $\mathrm{poly}(n, t)$, as opposed to $\mathrm{poly}(n, 2^{t^\epsilon})$. The weaker version stated here suffices for the results presented below. We now outline how the above lemma is used to prove our positive results on arithmetic in $\tilde{\mathbb{F}}_{2^n}$ in $AC^0[\oplus]$.

Proof idea for Theorem 3. (1) The idea is to reduce the problem to computing iterated multiplication over a smaller field $\mathbb{F}'$ of size $poly(n, t)$ via the Discrete Fourier Transform (DFT). We first show how to compute the DFT in $AC^0[\oplus]$. Then we show how to compute iterated multiplication over $\mathbb{F}'$ in AC^0 by exploiting the fact that the field $\mathbb{F}'$ is small. Finally, we recover the product by applying the inverse DFT, which can also be computed in $AC^0[\oplus]$.

(2) As outlined in the "Overview of Techniques" paragraph of the introduction, we can reduce this problem to computing the product of t field elements. Specifically, this reduction needs to compute α^{2^i} for $i \leq t$, which can be computed in uniform $AC^0[\oplus]$ by Lemma 1. For the iterated product we use the previous item. □

Acknowledgements. We would like to thank Eric Allender for various encouraging discussions, and in particular for pointing out that field multiplication in $\tilde{\mathbb{F}}_{2^n}$ is in uniform $AC^0[\oplus]$. We also thank Gudmund Frandsen for helpful discussions on [FVB]. We thank the anonymous CCC '05 referees, Kristoffer Hansen, Dan Gutfreund and Salil Vadhan for helpful comments. Samir Datta independently proved some of our negative results from Theorem 4.

References

[AAI+]   M. Agrawal, E. Allender, R. Impagliazzo, T. Pitassi, and S. Rudich. Reducing the complexity of reductions. *Comput. Complexity*, 10(2):117–138, 2001.

[ABD+]   E. Allender, A. Bernasconi, C. Damm, J. von zur Gathen, M. Saks, and I. Sh-
parlinski. Complexity of some arithmetic problems for binary polynomials.
Comput. Complexity, 12(1-2):23–47, 2003.

[AHM+]   E. Allender, L. Hellerstein, P. McCabe, T. Pitassi, and M. Saks. Minimizing DNF Formulas and AC0 Circuits Given a Truth Table. *Electronic Colloquium on Computational Complexity*, TR05-126, 2005. http://
www.eccc.uni trier.de/eccc.

[BIS]   D. A. M. Barrington, N. Immerman, and H. Straubing. On uniformity within
NC^1. *J. Comput. System Sci.*, 41(3):274–306, 1990.

[BFS]   J. Boyar, G. Frandsen, and C. Sturtivant. An arithmetic model of computation
equivalent to threshold circuits. *Theoret. Comput. Sci.*, 93(2):303–319, 1992.

[Ebe]   W. Eberly. Very fast parallel polynomial arithmetic. *SIAM Journal on
Computing*, 18(5):955–976, 1989.

[FT]   F. E. Fich and M. Tompa. The parallel complexity of exponentiating polynomials over finite fields. *J. Assoc. Comput. Mach.*, 35(3):651–667, 1988.

[FVB]   G. S. Frandsen, M. Valence, and D. A. M. Barrington. Some results on uniform
arithmetic circuit complexity. *Math. Systems Theory*, 27(2):105–124, 1994.

[FSS]   M. L. Furst, J. B. Saxe, and M. Sipser. Parity, Circuits, and the Polynomial-
Time Hierarchy. *Mathematical Systems Theory*, 17(1):13–27, April 1984.

[Gol]   O. Goldreich. *Modern cryptography, probabilistic proofs and pseudorandomness*, vol. 17 of *Algorithms and Combinatorics*. Springer-Verlag, Berlin, 1999.

[GV]   D. Gutfreund and E. Viola. Fooling Parity Tests with Parity Gates. In *8th
International Workshop on Randomization and Computation*, Lecture Notes
in Computer Science, Vol. 3122, 381–392. Springer-Verlag, 2004.

[Hås]   J. Håstad. *Computational limitations of small-depth circuits*. MIT Press,
1987.

[HAB]   W. Hesse, E. Allender, and D. A. M. Barrington. Uniform constant-depth
threshold circuits for division and iterated multiplication. *J. Comput. System
Sci.*, 65(4):695–716, 2002.

[Kun]   H. T. Kung. On computing reciprocals of power series. *Numerical Math*,
22:341–348, 1974.

[MS]   M. Morgensteren and E. Shamir. Parallel Algorithms for Arithmetics, Irreducibility and Factoring of GFq-Polynomials. Stanford University Technical
Report STAN-CS-83-991, December 1983.

[Rab]   M. O. Rabin. Probabilistic Algorithms in Finite Fields. *SIAM Journal on
Computing*, 9(2):273–280, 1980.

[Raz]   A. A. Razborov. Lower bounds on the dimension of schemes of bounded
depth in a complete basis containing the logical addition function. *Mat.
Zametki*, 41(4):598–607, 623, 1987.

[Rei]   J. Reif. Logarithmic depth circuits for algebraic functions. *SIAM Journal
on Computing*, 15(1):231–242, 1986.

[Sie]   M. Sieveking. An algorithm for division of power series. *Computing*, 10:
153–156, 1972.

[SF]   C. Sturtivant and G. S. Frandsen. The computational efficacy of finite-field
arithmetic. *Theoret. Comput. Sci.*, 112(2):291–309, 1993.

[vL]   J. H. van Lint. *Introduction to coding theory*. Springer-Verlag, 3rd edition,
1999.

[Vol]   H. Vollmer. *Introduction to circuit complexity*. Springer-Verlag, Berlin, 1999.

Weighted Asynchronous Cellular Automata

Dietrich Kuske*

Institut für Informatik, Universität Leipzig

Abstract. We study weighted distributed systems whose behavior is described as a formal power series over a free partially commutative or trace monoid. We characterize the behavior of such systems both, in the deterministic and in the non-deterministic case. As a consequence, we obtain a particularly simple class of sequential weighted automata that have already the full expressive power.

1 Introduction

Mazurkiewicz [7] used free partially commutative or trace monoids to relate the interleaving and the partial-order semantics of a distributed system (see [4] for surveys on the many results on trace monoids). Two of the fundamental results in this field have been found by Ochmański [8] and Zielonka [12]. Ochmański's theorem states the coincidence of recognizable and c-rational sets in trace monoids; hence it is a generalization of Kleene's theorem [6].

Another generalization of Kleene's theorem is due to Schützenberger [10] who considers weighted automata where transitions carry weights in some semiring like $(\mathbb{N}, +, \cdot)$ or $(\mathbb{N}, \max, +)$. The behavior of a weighted automaton is a function from the free monoid Σ^* into the semiring, i.e., a formal power series. Schützenberger's theorem states that a formal power series is the behavior of some weighted automaton iff it is rational.

Droste & Gastin [5] found a common formulation of the two distinct generalizations of Kleene's theorem by Schützenberger and Ochmański. Technically, they consider formal power series over the trace monoid $\mathbb{M}$, i.e., functions from $\mathbb{M}$ into the semiring. If the semiring is commutative, then, indeed, a formal power series over $\mathbb{M}$ is recognizable iff it is mc-rational. From this, strengthenings of Ochmański's, of Kleene's as well as of Schützenberger's theorem for commutative semirings follow.

Ochmański's automata as well as Droste & Gastin's weighted automata can be thought of as sequential automata that do not distinguish between interleavings of the same partial-order execution (i.e., are trace closed). If, e.g., the actions a and b use disjoint resources, then the total weight of executing the words ab and ba should be the same. This is in particular the case in the automaton below over the semiring $(\mathbb{N}, +, \cdot)$ (both words get the value 12).

But the two a-transitions in this automaton have different weights. Hence it seems impossible to consider it as the global state space of a distributed system where a and b are executed by independent subprocesses. This phenomenon seems to be unavoidable in the proofs from [5].

---

* This work was begun when the author visited Centre de Mathématiques et Informatique, Marseille. He thanks Rémi Morin and all his hosts for their hospitality and the fruitful discussions during his stay.

B. Durand and W. Thomas (Eds.): STACS 2006, LNCS 3884, pp. 684–695, 2006.

While Ochmański's and Droste & Gastin's automata represent the interleaving behavior of a system, asynchronous cellular automata are a distributed model whose semantics is more naturally described in a partial-order setting (see e.g. [11] where this view is exhibited explicitly). The interleaving behavior of such an asynchronous cellular automaton is trace closed since it is defined to be the set of interleavings of its partial-order behavior. Zielonka proved that recognizability and trace-closure are not only necessary, but also sufficient for a language to be the interleaving behavior of some (deterministic) distributed finite-state system.

It is the aim of this paper to extend Zielonka's theorem to weighted distributed systems. Theorem 4.1 states that a formal power series over the trace monoid $\mathbb{M}$ is recognizable iff it is the behavior of some weighted asynchronous cellular automaton. As a consequence, we show that any recognizable formal power series can be realized by an automaton as in [5] not exhibiting the above contra-intuitive phenomenon. The expressive power of deterministic and nondeterministic weighted sequential automata are distinct. We can therefore not expect that every weighted asynchronous cellular automata can be transformed into an equivalent deterministic one. Theorem 5.3 describes the formal power series that can be realized by deterministic weighted asynchronous cellular automata.

2 Distributed Alphabets and Asynchronous Cellular Automata

Let $\mathbb{T}$ be a nonempty and finite set of *action types* and $D \subseteq \mathbb{T} \times \mathbb{T}$ a symmetric and reflexive *dependence relation*; its complement in $\mathbb{T}^2$ is denoted I. For $\ell \in \mathbb{T}$, let $D(\ell) = \{m \in \mathbb{T} \mid (m, \ell) \in D\}$. *We fix the pair $(\mathbb{T}, D)$ throughout this paper.* A *dependence alphabet* is a tuple $\Sigma = (\Sigma_\ell)_{\ell \in \mathbb{T}}$ of nonempty and mutually disjoint alphabets. Abusing notation, we denote the set $\bigcup_{\ell \in \mathbb{T}} \Sigma_\ell$ by Σ as well. For $a \in \Sigma$, let $\mathrm{tp}(a) \in \mathbb{T}$ denote the unique type $\ell \in \mathbb{T}$ with $a \in \Sigma_\ell$. Furthermore, for $a, b \in \Sigma$, we write $(a, b) \in D$ and $(a, b) \in I$ as shorthand for $(\mathrm{tp}(a), \mathrm{tp}(b)) \in D$ and $(\mathrm{tp}(a), \mathrm{tp}(b)) \in I$, resp.

Let $\Sigma = (\Sigma_\ell)_{\ell \in \mathbb{T}}$ be a dependence alphabet. Then $\sim$ denotes the least congruence relation on the free monoid Σ^* with $ab \sim ba$ for all $a, b \in \Sigma$ with $(a, b) \notin D$. The quotient $\Sigma^*/\sim$ is the *trace monoid generated by Σ*. Its elements are equivalence classes $[u]$ of words. These equivalence classes can be represented naturally as follows: A *trace over Σ* is a finite labeled poset $t = (V, \leq, \lambda)$ with $\lambda : V \to \Sigma$ such that the following hold for all $x, y \in V$:

- if $(\lambda(x), \lambda(y)) \in D$, then $x \leq y$ or $y \leq x$
- if $x < y$ and there is no node in between, then $(\lambda(x), \lambda(y)) \in D$.

The set of traces over Σ is denoted by $\mathbb{M}(\Sigma)$. For a trace $t = (V, \leq, \lambda)$ and a node $x \in V$, let $\mathrm{tp}(x) = \mathrm{tp}(\lambda(x))$.

Let $t = (V, \leq, \lambda)$ be a trace. A linear extension of t is a structure $(V, \sqsubseteq, \lambda)$ such that $\sqsubseteq$ is a linear order on V extending $\leq$. Such a linear extension can naturally be considered as a word over Σ, hence we define $\mathrm{Lin}(t) \subseteq \Sigma^*$ as the set of linear extensions of the trace t. Now a basic result in trace theory asserts that Lin maps $\mathbb{M}(\Sigma)$ bijectively onto the trace monoid generated by Σ. In the following, we will identify an equivalence class $[w]$ with the trace $\mathrm{Lin}^{-1}([w])$, i.e., consider the set $\mathbb{M}(\Sigma)$ as underlying set of the trace monoid. This allows to define a set of traces $L \subseteq \mathbb{M}(\Sigma)$ to be *recognizable* if there exists a homomorphism $f : \mathbb{M}(\Sigma) \rightarrow S$ into some finite monoid S such that $L = f^{-1}f(L)$.

Now fix some linear order $\lesssim$ on the set Σ. It induces, in the natural way, the lexicographic order on Σ^* that we denote by $\lesssim$ as well. Since any equivalence class $[w]$ is finite, it contains a $\lesssim$-minimal word $\mathrm{lnf}([w])$, called the lexicographic normal form of $[w]$. Let $\mathrm{LNF} = \{\mathrm{lnf}(t) \mid t \in \mathbb{M}(\Sigma)\}$ denote the set of lexicographic normal forms. The language $\mathrm{LNF} \subseteq \Sigma^*$ is known to be recognizable.

Let $t = (V, \leq, \lambda)$ be a trace and $y \in V$. Then $\Downarrow y = \{x \in V \mid x < y\}$ is a subset of V. The restriction of t to $\Downarrow y$ is a trace $(\Downarrow y, \leq \cap (\Downarrow y)^2, \lambda{\restriction}\Downarrow y)$. Now let $\ell \in \mathbb{T}$. Then the set of nodes $x \in V$ with $\mathrm{tp}(x) = \ell$ is linearly ordered by $\leq$, hence (if not empty) this set contains a largest element that we denote by $\delta_\ell(t)$. Occasionally, we will identify the node $\delta_\ell(t)$ with the set $\Downarrow\delta_\ell(t)$ or the trace $(\Downarrow\delta_\ell(t), \leq, \lambda)$. This allows to define, for $A \subseteq \mathbb{T}$, the set $\delta_A(t) = \bigcup_{\ell \in A} \delta_\ell(t)$ which again gives rise to a trace, namely the restriction of t to this set.

Definition 2.1. *Let Σ be a dependence alphabet. An* asynchronous cellular automaton *or ACA is a tuple*

$$\mathcal{A} = ((Q_m)_{m \in \mathbb{T}}, (T_m)_{m \in \mathbb{T}}, I, F)$$

where

- Q_m *is a finite set of local states for any $m \in \mathbb{T}$,*
- $T_\ell \subseteq \prod_{m \in D(\ell)} Q_m \times \Sigma_\ell \times Q_\ell$ *is a local transition relation for any $\ell \in \mathbb{T}$,*
- $I, F \subseteq \prod_{\ell \in \mathbb{T}} Q_\ell$ *are sets of global initial and final states, resp.*

The ACA $\mathcal{A}$ is deterministic *if I is a singleton and $((p_m)_{m \in D(\ell)}, a, q_\ell^i) \in T_\ell$ for $i = 1, 2$ implies $q_\ell^1 = q_\ell^2$.*

In the terminology of [13], these ACAs are special "finite asynchronous automata". Zielonka reserves the name "finite asynchronous cellular automaton" to those ACAs that satisfy $|\Sigma_\ell| = 1$ for all $\ell \in \mathbb{T}$.

Let $t = (V, \leq, \lambda)$ be a trace over Σ. A function $r : V \rightarrow \bigcup_{\ell \in \mathbb{T}} Q_\ell$ is a *run* provided $r(x) \in Q_{\mathrm{tp}(x)}$ for $x \in V$. Let $\iota \in \prod_{\ell \in \mathbb{T}} Q_\ell$ be a global state. For $x \in V$ and $m \in \mathbb{T}$ let $r_m^-(\iota, x) = r(\partial_m(\Downarrow x))$ if $\partial_m(\Downarrow x)$ is defined, and $r_m^-(\iota, x) = \iota_m$ otherwise. Similarly, define $\mathrm{final}_m(\iota, r, t) = r(\partial_m(t))$ if $\partial_m(t)$ is defined, and $\mathrm{final}_m(\iota, r, t) = \iota_m$ otherwise. The pair (ι, r) is an *accepting run* provided $\iota \in I$, $((r_m^-(\iota, x))_{m \in D(\ell)}, \lambda(x), r(x)) \in T_\ell$ for any $x \in V$ with $\mathrm{tp}(x) = \ell$, and $\mathrm{final}(\iota, r, t) \in F$. Let the language $L(\mathcal{A}) \subseteq \mathbb{M}(\Sigma)$ of the ACA $\mathcal{A}$ comprise all traces t that allow a successful run (ι, r) of $\mathcal{A}$ on t.

Zielonka [12] showed that a set of traces is recognizable iff it is the language of some (deterministic) ACA where $\mathrm{tp}(a) = a$ for $a \in \Sigma$. His results from 1987 have been extended in [13], Prop. 7.6.2 and Thm. 7.6.11 from that paper yield the following:

Theorem 2.2 ([13]). *Let $L \subseteq \mathbb{M}(\Sigma)$ be a trace language. Then L is recognizable iff there exists an ACA $\mathcal{A}$ with $L(\mathcal{A}) = L$ iff there exists a deterministic ACA $\mathcal{A}$ with $L(\mathcal{A}) = L$.*

3   Weighted Automata over Traces

In this section, we define weighted sequential and distributed automata that associate costs with any trace. Their behavior is described by functions mapping traces to elements of a semiring.

3.1   Semirings

A *semiring* is an algebraic structure $(K, +, \cdot, 0, 1)$ with two binary operations such that $(K, +, 0)$ is a commutative monoid, $(K, \cdot, 1)$ a monoid, multiplication distributes over addition, and $x \cdot 0 = 0 \cdot x = 0$ for any $x \in K$. It is *commutative* if $(K, \cdot)$ is a commutative monoid. Examples of commutative semirings are rings, Boolean algebras like the two-elements Boolean algebra $\{t\!t, f\!f\}$ with conjunction and disjunction, but also structures like $(\mathbb{N} \cup \{\infty\}, \min, +, \infty, 0)$ or $([0, 1], \max, \min, 0, 1)$. For a semiring K and $n \in \mathbb{N}$, let $K^{n \times n}$ denote the set of $n \times n$-matrices over K. For these matrices, we can define addition $+$ and multiplication $\cdot$ as usual using the semiring operations $+$ and $\cdot$. The resulting structure $(K^{n \times n}, +, \cdot, \underline{0}, E)$ (where $\underline{0}$ is the 0-matrix and E the unit matrix) is again a semiring (that need not be commutative even if K is commutative). *Throughout this paper, we fix a commutative semiring K.*

3.2   Presentations

We start with a sequential model of weighted automata, so called presentations. Let $[n] = \{1, 2, \ldots, n\}$.

Definition 3.1. *Let M be some monoid. A triple* $(\text{in}, \mu, \text{out})$ *consisting of two functions* $\text{in}, \text{out} : [n] \to K$ *and a homomorphism* $\mu : M \to (K^{n \times n}, \cdot, E)$ *is an* n-dimensional presentation. *It represents the* formal power series *or* fps $S : M \to K$ *defined by*

$$(S, t) = \sum_{\iota, \text{fin} \in [n]} \text{in}(\iota) \cdot \mu(t)_{\iota, \text{fin}} \cdot \text{out}(\text{fin}) = \text{in} \cdot \mu(t) \cdot \text{out}$$

for $t \in M$ (where we consider in *as a row and* out *as a column vector). A function* $S : M \to K$ *is* recognizable *if there exists a presentation representing S.*

Example 3.2. Let $\Sigma_1 = \{a\}$ and $\Sigma_2 = \{b\}$ with $\mathbb{T} = \{1, 2\}$ and $(1, 2) \notin D$. Then

$$\mu(a) = \begin{pmatrix} 0 & 1 & 0 & 0 \\ 0 & 0 & 0 & 0 \\ 0 & 0 & 0 & 1 \\ 0 & 0 & 0 & 0 \end{pmatrix} \quad \text{and } \mu(b) = \begin{pmatrix} 0 & 0 & 2 & 0 \\ 0 & 0 & 0 & 2 \\ 0 & 0 & 0 & 0 \\ 0 & 0 & 0 & 0 \end{pmatrix}$$

define a monoid homomorphism $\mathbb{M}(\Sigma) \to (\mathbb{N}, +, \cdot)^{4 \times 4}$ since $(\mu(a) \cdot \mu(b))_{14} = 2 = (\mu(b) \cdot \mu(a))_{14}$ and $(\mu(a) \cdot \mu(b))_{ij} = 0 = (\mu(b) \cdot \mu(a))_{ij}$ for all other pairs (i, j). Furthermore, define

$$\text{in}(i) = \begin{cases} 3 & \text{if } i = 1 \\ 0 & \text{otherwise} \end{cases} \quad \text{and out}(i) = \begin{cases} 2 & \text{if } i = 4 \\ 0 & \text{otherwise} \end{cases}$$

Then $(\text{in}, \mu, \text{out})$ is a presentation of the function $S : \mathbb{M}(\Sigma) \to \mathbb{N}$ with $(S, [ab]) = 12$ and $(S, t) = 0$ for $t \neq [ab]$.

Presentations can be interpreted as automata whose transitions are provided with costs (hence the notion *"recognizable function"*); see the automaton on the right. Note that in that automaton, both a-transitions carry the same weight (and the same holds for the b-transitions); thus the two a-transitions

can be thought of as being "the same". This mirrors the idea that the two paths (labeled ab and ba) represent a concurrent execution of the trace $[ab] = [ba]$.

We give another presentation of the same function (it is depicted in the introduction):

$$\mu'(a) = \begin{pmatrix} 0\ 3\ 0\ 0 \\ 0\ 0\ 0\ 0 \\ 0\ 0\ 0\ 2 \\ 0\ 0\ 0\ 0 \end{pmatrix} \text{ and } \mu'(b) = \begin{pmatrix} 0\ 0\ 6\ 0 \\ 0\ 0\ 0\ 4 \\ 0\ 0\ 0\ 0 \\ 0\ 0\ 0\ 0 \end{pmatrix}$$

$$\text{in}'(i) = \begin{cases} 1 & \text{if } i = 1 \\ 0 & \text{otherwise} \end{cases} \text{ and } \text{out}(i) = \begin{cases} 1 & \text{if } i = 4 \\ 0 & \text{otherwise} \end{cases}$$

Note that the initial weights in this second presentation are from $\{0, 1\}$ and that there are unique states ι and fin with $\text{in}'(\iota) = 1$ and $\text{out}'(\text{fin}) = 1$. Droste & Gastin [5–Prop. 9] show that any presentation of a mono-alphabetic function $\mathbb{M}(\Sigma) \to K$ can be transformed into an equivalent one satisfying these properties (and we followed their proof when constructing the second presentation from the first one).

Note finally that the two a-transitions in the second presentation have different weights which violates the intuition discussed above that they should be "the same" in a distributed system.

Definition 3.3. *A homomorphism $\mu : \mathbb{M}(\Sigma) \to (K^{n \times n}, \cdot, E)$ is I-consistent if, for any $a, b \in \Sigma$ with $(a, b) \in I$ and any $p, q, r \in [n]$, we have*

1. *If $\mu(a)_{p,q} \neq 0 \neq \mu(b)_{q,r}$, then there exists $q' \in [n]$ with $\mu(a)_{p,q} = \mu(a)_{q',r}$ and $\mu(b)_{q,r} = \mu(b)_{p,q'}$.*
2. *If $\mu(a)_{p,q} \neq 0 \neq \mu(b)_{p,r}$, then there exists $s \in [n]$ such that $\mu(b)_{q,s} \neq 0$.*

A presentation $(\text{in}, \mu, \text{out})$ is I-consistent if μ is I-consistent.

The first of these requirements avoids the contra-intuitive situation discussed in the example above. It is a weighted version of the I-diamond property known from trace theory. Similarly, the second one is a weighted version of the F-diamond property: it requires that execution of an action a cannot toggle the status of an independent action b from "enabled" into "disabled" (the other direction is already taken care of by the first requirement).

3.3   Weighted Distributed Automata

Next we define a distributed model of weighted automata on traces, so called weighted asynchronous cellular automata.

Definition 3.4. *Let Σ be a dependence alphabet. A* weighted asynchronous cellular automaton *or* wACA *for short is a tuple $\mathcal{A} = ((Q_m)_{m \in T}, (c_\ell)_{\ell \in T}, \text{in}, \text{out})$ where*

- Q_m is a finite set of local states for any $m \in \mathbb{T}$,
- $c_\ell : \prod_{m \in D(\ell)} Q_m \times \Sigma_\ell \times Q_\ell \to K$ is a local weight function for any $\ell \in \mathbb{T}$,
- in, out : $\prod_{m \in \mathbb{T}} Q_m \to K$ are functions describing the cost for entering and leaving the system.

The weighted ACA $\mathcal{A}$ is deterministic if

- there is at most one state $\iota \in \prod_{m \in \mathbb{T}} Q_m$ with $\mathrm{in}(\iota) \neq 0$, and
- for any $\ell \in \mathbb{T}$, $a \in \Sigma_\ell$, $p_m \in Q_m$ for $m \in D(\ell)$, and $q_\ell, r_\ell \in Q_\ell$ with $c_\ell((p_m)_{m \in D(\ell)}, a, q_\ell) \neq 0 \neq c_\ell((p_m)_{m \in D(\ell)}, a, r_\ell)$, we have $q_\ell = r_\ell$.

If the dependence alphabet $(\Sigma_\ell)_{\ell \in \mathbb{T}}$ consists of singletons, only, then these wACAs can be seen as a weighted version of Zielonka's "finite asynchronous cellular automata". For technical convenience, we chose this more liberal notion, but since our results hold for all dependence alphabets, they hold in particular for "finite asynchrounous cellular automata with weights".

Let $t = (V, \leq, \lambda)$ be a trace over Σ. Runs $r : V \to \bigcup_{\ell \in \mathbb{T}} Q_\ell$ and states $r_m^-(\iota, x)$ and $\mathrm{final}_m(\iota, r, t)$ are defined as before for ACAs. Then the *running cost* of the run r *starting in* ι is given by

$$\mathrm{rcost}(\iota, r, t) = \prod_{x \in V} c_{\mathrm{tp}(x)}((r_m^-(\iota, x))_{m \in D(\mathrm{tp}(x))}, \lambda(x), r(x))$$

and the *cost* of the run r starting in ι is

$$\mathrm{cost}(\iota, r, t) = \mathrm{in}(\iota) \cdot \mathrm{rcost}(\iota, r, t) \cdot \mathrm{out}(\mathrm{final}(\iota, r, t))$$

The function $\|\mathcal{A}\| : \mathbb{M}(\Sigma) \to K$ is the *behavior of* $\mathcal{A}$; for any trace $t = (V, \leq, \lambda)$, it is given by

$$(\|\mathcal{A}\|, t) = \sum \{\mathrm{cost}(\iota, r, t) \mid \iota \in \prod_{m \in \mathbb{T}} Q_m, r : V \to \bigcup_{m \in \mathbb{T}} Q_m \text{ run}\}$$

Let $K = (\{tt, ff\}, \vee, \wedge)$ be the Boolean semiring and $((Q_\ell)_{\ell \in \mathbb{T}}, (T_\ell)_{\ell \in \mathbb{T}}, I, F)$ be an ACA $\mathcal{A}$. Define c_ℓ, in, and out to be the characteristic functions of T_ℓ, I, and F, resp. This defines a wACA $\mathcal{A}' = ((Q_\ell)_{\ell \in \mathbb{T}}, (c_\ell)_{\ell \in \mathbb{T}}, \mathrm{in}, \mathrm{out})$. It is routine to show for any trace $t \in \mathbb{M}(\Sigma)$, that $t \in L(\mathcal{A})$ iff $(\|\mathcal{A}'\|, t) = tt$. Since the inverse conversion of wACAs over the Boolean semiring into ACAs is equally well possible, the concept of a weighted ACA generalizes that of an ACA.

4   Presentations and wACAs

This section is devoted to the fact that a function $\mathbb{M}(\Sigma) \to K$ is recognizable iff it is the behavior of some wACA. We start showing that wACAs can be transformed into equivalent I-consistent presentations (Proposition 4.3 and Lemma 4.2). Later, we will transform arbitrary presentations into equivalent wACAs (Prop. 4.8). In summary, we will obtain the following theorem whose proof can be found at the end of this section.

Theorem 4.1. *Let $S : \mathbb{M}(\Sigma) \to K$ be some function. Then the following are equivalent*

(1) S is the behavior of some wACA
(2) S is recognizable
(3) S has an I-consistent presentation

Thus, as a byproduct, we obtain that any presentation can be transformed into an equivalent I-consistent one.

4.1   wACA-Recognizable Series Are Recognizable

In this section, we show how to transform wACAs into I-consistent presentations.

Let $\mathcal{A} = ((Q_\ell)_{\ell \in \mathbb{T}}, (c_\ell)_{\ell \in \mathbb{T}}, \text{in}, \text{out})$ be a wACA and let $Q = \prod_{m \in \mathbb{T}} Q_m$ be the set of global states of $\mathcal{A}$. We define the mapping $\mu : \Sigma \to K^{Q \times Q}$ by

$$\mu(a)_{p,q} = \begin{cases} c_{\mathrm{tp}(a)}((p_\ell)_{\ell \in D(\mathrm{tp}(a))}, a, q_{\mathrm{tp}(a)}) & \text{if } p_m = q_m \text{ for all } m \neq \mathrm{tp}(a) \\ 0 & \text{otherwise} \end{cases}$$

where $p = (p_m)_{m \in \mathbb{T}}$ and $q = (q_m)_{m \in \mathbb{T}}$ are global states from Q.

We consider the elements of $K^{Q \times Q}$ (i.e., the functions from Q^2 into K) as $Q \times Q$-matrices. For these matrices, multiplication is defined in the standard way:

$$(M \cdot N)_{p,r} = \sum_{q \in Q} M_{p,q} \cdot N_{q,r}$$

Lemma 4.2. *The mapping μ extends uniquely to a monoid homomorphism μ from $\mathbb{M}(\Sigma)$ into $(K^{Q \times Q}, \cdot, E)$. This homomorphism is I-consistent.*

Let $n = |Q|$. Then in and out can be considered as functions from $[n]$ to K. Furthermore, μ can be thought of as a homomorphism from $\mathbb{M}(\Sigma)$ into $(K^{n \times n}, \cdot, E)$. Hence the triple $(\text{in}, \mu, \text{out})$ is a presentation over $\mathbb{M}(\Sigma)$ called the *canonical presentation* *associated with* $\mathcal{A}$. We will show that it presents the behavior of $\mathcal{A}$.

Proposition 4.3. *Let $(\text{in}, \mu, \text{out})$ be the canonical presentation associated with the weighted asynchronous-cellular automaton $\mathcal{A}$. Then the function $\|\mathcal{A}\|$ is presented by $(\text{in}, \mu, \text{out})$.*

Proof. The crucial point is that, for any trace t and any $p, q \in Q$, we have $\mu(t)_{p,q} = \sum \mathrm{rcost}(p, r, t)$ where the sum is taken over all runs r on t with final$(p, r) = q$.   □

4.2   Recognizable Series Are wACA-Recognizable

The proof of the implication $(2) \Rightarrow (1)$ from Theorem 4.1, uses some preliminary results. To formulate them, let $\Gamma = (\Gamma_m)_{m \in \mathbb{T}}$ be some dependence alphabet.

Lemma 4.4. *For any homomorphism $c : \mathbb{M}(\Gamma) \to (K, \cdot, E)$, there exists a deterministic wACA $\mathcal{A}_c$ with $(\|\mathcal{A}_c\|, t) = c(t)$ for $t \in \mathbb{M}(\Gamma)$.*

Proof. Let $Q_m = \{1\}$ for $m \in \mathbb{T}$ and define $\text{in}(\iota) = \text{out}(\iota) = 1$ for the only global state ι. Furthermore, $c_\ell((p_m)_{m \in D(\ell)}, a, q_m) = c(a)$ for any $a \in \Sigma_\ell$, $p_m \in Q_m$, and $q_\ell \in Q_\ell$. Checking $\text{cost}(\iota, r, t) = c(t)$ for the only global state ι and the only run r on the trace t is routine. $\qquad\square$

Corollary 4.6 below claims that the restriction of the behavior of some wACA to a recognizable trace language can again be described by a wACA. For handling these restrictions, we use the more general Hadamard-product: For $S, T : \mathbb{M}(\Gamma) \to K$, the *Hadamard-product* $S \odot T : \mathbb{M}(\Gamma) \to K$ is defined by $(S \odot T, t) = (S, t) \cdot (T, t)$ for $t \in \mathbb{M}(\Gamma)$.

Lemma 4.5. *Let $\mathcal{A}^1$ and $\mathcal{A}^2$ be wACAs. Then the Hadamard product $\|\mathcal{A}^1\| \odot \|\mathcal{A}^2\|$ is the behavior of some wACA $\mathcal{A}$. If $\mathcal{A}^1$ and $\mathcal{A}^2$ are deterministic, then $\mathcal{A}$ can be chosen deterministic.*

Proof. The wACA $\mathcal{A}$ is the "direct product" of the wACAs $\mathcal{A}^1$ and $\mathcal{A}^2$. The point is that any local state from $\mathcal{A}$ consists of a pair of local states from $\mathcal{A}^1$ and $\mathcal{A}^2$, any run r of $\mathcal{A}$ is a pair (r_1, r_2) of runs of $\mathcal{A}^1$ and $\mathcal{A}^2$. $\qquad\square$

For a function $S : \mathbb{M}(\Sigma) \to K$ and a language $L \subseteq \mathbb{M}(\Sigma)$, let $T = S\!\restriction_L$ denote the function defined by $(T, t) = (S, t)$ for $t \in L$ and $(T, t) = 0$ otherwise.

Corollary 4.6. *Let $L \subseteq \mathbb{M}(\Gamma)$ be a recognizable trace language and let $\mathcal{A}$ be some wACA. Then there exists a wACA $\mathcal{A}'$ with $\|\mathcal{A}'\| = \|\mathcal{A}\|\!\restriction_L$. If $\mathcal{A}$ is deterministic, then $\mathcal{A}'$ can be chosen deterministic.*

Proof. By Theorem 2.2, there exists a deterministic ACA $\mathcal{A}_L$ with $L(\mathcal{A}) = L$. Then the characteristic function of L is the behavior of some deterministic wACA $\mathcal{A}_1$. Since $\|\mathcal{A}\|\!\restriction_L = \|\mathcal{A}\| \odot \|\mathcal{A}_1\|$, the result follows from Lemma 4.5. $\qquad\square$

Informally, Proposition 4.7 states that the inverse image of the projection of the behaviour of some wACA can again be realized by some wACA.

A mapping $\pi : \Gamma \to \Sigma$ is *type-preserving* if $\text{tp}(a) = \text{tp}(\pi(a))$ for any $a \in \Sigma$. Note that any type-preserving mapping π extends uniquely to a homomorphism $\pi : \mathbb{M}(\Gamma) \to \mathbb{M}(\Sigma)$.

Proposition 4.7. *Let $\mathcal{A} = ((Q_m)_{m \in \mathbb{T}}, (c_m)_{m \in \mathbb{T}}, \text{in}, \text{out})$ be a wACA over Γ with behavior $\|\mathcal{A}\| = S : \mathbb{M}(\Gamma) \to K$ and $\pi : \Gamma \to \Sigma$ be some type-preserving mapping. Then there exists a wACA $\mathcal{A}'$ over Σ with behavior $(T, t) = \sum_{u \in \pi^{-1}(t)} (S, u)$.*

Proof. We construct the wACA $\mathcal{A}'$ as follows

- $Q'_m = Q_m \times \Gamma_m$ for $m \in \mathbb{T}$
- Let $\ell \in \mathbb{T}$, $a \in \Sigma_\ell$, $(q_\ell, b) \in Q'_\ell$ and, for $m \in D(\ell)$, let $(p_m, b_m) \in Q'_m$. Then set

$$c'_\ell((p_m, b_m)_{m \in D(\ell)}, a, (q_\ell, b)) = \begin{cases} c_\ell((p_m)_{m \in D(\ell)}, b, q_\ell) & \text{if } \pi(b) = a \\ 0 & \text{otherwise} \end{cases}$$

– for $m \in \mathbb{T}$ let $b_m^0 \in \Gamma_m$ be fixed. Then

$$\mathrm{in}'((q_m, b_m)_{m\in\mathbb{T}}) = \begin{cases} \mathrm{in}((q_m)_{m\in\mathbb{T}}) & \text{if } b_m = b_m^0 \text{ for all } m \in \mathbb{T} \\ 0 & \text{otherwise} \end{cases}$$

– $\mathrm{out}'((q_m, b_m)_{m\in\mathbb{T}}) = \mathrm{out}((q_m)_{m\in\mathbb{T}})$

Now let $t = (V, \leq, \lambda) \in \mathbb{M}(\Sigma)$ be a trace over Σ. Consider the following two sets

– $R' = \{(q', r') \mid q' \in Q', r' \text{ run of } \mathcal{A}' \text{ on } t \text{ with } \mathrm{cost}(q', r', t) \neq 0\}$
– $R = \{(q, r, u) \mid q' \in Q', u \in \pi^{-1}(t), r' \text{ run of } \mathcal{A} \text{ on } u \text{ with } \mathrm{cost}(q, r, u) \neq 0\}$

Then $(\|\mathcal{A}'\|, t) = \sum_{(q', r') \in R'} \mathrm{cost}(q', r', t)$ and $(T, t) = \sum_{(q, r, u) \in R} \mathrm{cost}(q, r, u)$. Let $u = (V, \leq, \gamma)$ with $\pi(u) = t$, $q = (q_m)_{m\in\mathbb{T}}$, and r a run of $\mathcal{A}$ on u. For $m \in \mathbb{T}$, let $q_m' = (q_m, b_m^0)$. Further, for $x \in V$, define $r'(x) = (r(x), \gamma(x))$. It can be shown that $(q', r') \in R'$ and $\mathrm{cost}_{\mathcal{A}'}(q', r', t) = \mathrm{cost}_{\mathcal{A}}(q, r, u)$. Furthermore, this construction yields a bijection from R onto R'. □

Now we come to the most crucial statement in this section, namely that any recognizable fps is the behavior of some wACA.

Proposition 4.8. *Let* $(\mathrm{in}, \mu, \mathrm{out})$ *be a presentation of the fps* $S : \mathbb{M}(\Sigma) \to K$ *with* $(S, [\varepsilon]) = 0$. *Then there exists a wACA* $\mathcal{A}$ *with* $S = \|\mathcal{A}\|$.

Proof. Let $T : \Sigma^* \to K$ be defined by $(T, w) = (S, [w])$ if $w \in \mathrm{LNF}$ and $(T, w) = 0$ otherwise. Note that $(T, \varepsilon) = 0$. Hence there exist $n' \in \mathbb{N}$, $\iota, \mathrm{fin} \in [n']$, $\mathrm{in}', \mathrm{out}' : [n] \to K$ and a homomorphism $\mu' : \Sigma^* \to (K^{n' \times n'}, \cdot, E)$ such that, for any $w \in \Sigma^*$, we have $(T, w) = \mu'(w)_{\iota, \mathrm{fin}}$ by [9–p. 32] and [5–Prop. 29].

For $\ell \in \mathbb{T}$ set $\Gamma_\ell = [n'] \times \Sigma_\ell \times [n']$ and let the language $L \subseteq \Gamma$ over $\Gamma = (\Gamma_\ell)_{\ell\in\mathbb{T}}$ comprise of those words $(i_k, a_k, j_k)_{0 \leq k \leq N}$ over Γ satisfying

1. $i_0 = \iota$, $j_k = i_{k+1}$ for $1 \leq k < N$ and $j_N = \mathrm{fin}$
2. $a_0 a_1 \ldots a_N \in \mathrm{LNF}$.

Since $\mathrm{LNF} \subseteq \Sigma^*$ is recognizable, the languages L and therefore $[L] = \{[w] \mid w \in L\} \subseteq \mathbb{M}(\Gamma)$ are recognizable as well [4].

Now consider the homomorphisms $c : \mathbb{M}(\Gamma) \to (K, \cdot, E)$ and $\pi : \mathbb{M}(\Gamma) \to \mathbb{M}(\Sigma)$ given by $c(i, a, j) = \mu(a)_{ij}$ and $\pi(i, a, j) = a$ for $(i, a, j) \in \Gamma$ (these homomorphisms exist since $(K, \cdot, E)$ is commutative and since π is type-preserving, resp.).

By Lemma 4.4, the homomorphism c is the behavior of some wACA. Hence, by Corollary 4.6, the formal power series $c{\restriction}[L]$ is also the behavior of some wACA. Now Proposition 4.7 implies the same for the fps $\pi^{-1}(c{\restriction}[L])$. Showing $S = \pi^{-1}(c{\restriction}[L])$ completes the proof. □

Proof of Theorem 4.1. The implication $(1){\Rightarrow}(3)$ is Prop. 4.3, $(3){\Rightarrow}(2)$ is trivial. So let $(\mathrm{in}, \mu, \mathrm{out})$ be some presentation representing the fps $S : \mathbb{M}(\Sigma) \to K$. Then $K = \{[\varepsilon]\}$ and $L = \mathbb{M}(\Sigma) \setminus \{[\varepsilon]\}$ are recognizable. Hence the functions $S \restriction K$ and $S \restriction L$ are recognizable and therefore the behavior of some wACAs $\mathcal{A}_1$ and $\mathcal{A}_2$ by Prop. 4.8. Considering the "disjoint union" of these two wACAs, we obtain a wACA $\mathcal{A}$ whose behaviour equals S. □

5 Deterministic wACAs and Presentations

Recall that any recognizable trace language can be accepted by a deterministic ACA. Furthermore, any presentation can be thought of as an automaton with weighted transitions. There are presentations of functions $\Sigma^* \to K$ that do not admit a presentation whose underlying automaton is deterministic. We can therefore not expect that every presentable function $\mathbb{M}(\Sigma) \to K$ is the behavior of some *deterministic* wACA. It is the aim of this section to identify a class of presentations that correspond to deterministic wACAs.

Definition 5.1. *Let* (in, μ, out) *be an n-dimensional I-consistent presentation over* $\mathbb{M}(\Sigma)$. *We call this presentation* deterministic *if it meets the following two requirements.*

(1) There is a unique state $\iota \in [n]$ with in$(\iota) \neq 0$.
(2) Any row of $\mu(a)$ contains at most one non-zero entry (i.e., $\mu(a)_{i,j} \neq 0 \neq \mu(a)_{i,k}$ implies $j = k$ for any $a \in \Sigma$ and $i, j, k \in [n]$).

A first attempt to transform a deterministic presentation into an equivalent determinstic wACA would be to change the proof of Prop. 4.8 accordingly. But this turns out to be problematic since Prop. 4.7 does not necessarily preserve determinism (all the other steps in the proof go through verbatim). Our alternative proof strategy follows the ideas from [1] where a deterministic asynchronous cellular automaton is constructed from an asynchronous mapping.

From now on, let (in, μ, out) *be a deterministic n-dimensional presentation over* $\mathbb{M}(\Sigma)$. *Furthermore, let $\iota \in [n]$ be the unique state with* in$(\iota) \neq 0$.

Then (in, μ, out) defines a partial mapping $. : [n] \times \Sigma^* \to [n]$ by $p.\varepsilon = p$ for $p \in [n]$, $p.a = q$ iff $\mu(a)_{p,q} \neq 0$ for $a \in \Sigma$ and $p, q \in [n]$, and $p.aw = (p.a).w$ for $[n] \in Q$, $a \in \Sigma$ and $w \in \Sigma^*$ provided the right hand side is defined. Since μ is I-consistent, equivalent words lead to the same state (if at all). Thus, we can define the partial mapping $. : [n] \times \mathbb{M}(\Sigma, I) \to [n]$ by $q.[u] = q.u$ where u is some word over Σ.

Let Part$([n], [n])$ denote the monoid of partial functions from $[n]$ into itself with composition as operation. Then $\eta : \mathbb{M}(\Sigma) \to$ Part$([n], [n])$ defined by $\eta(t)(i) = i.t$ is a monoid homomorphism. By [3–Cor. 8.3.18] there exists a mapping $\varphi : \mathbb{M}(\Sigma) \to S$ into some finite set S such that

- $\varphi(s) = \varphi(t)$ implies $\eta(s) = \eta(t)$ for any two traces s and t,
- for any $s \in \mathbb{M}(\Sigma)$ and $a \in \Sigma$, the value $\varphi(\partial_{D(\mathrm{tp}(a))}(s)a)$ is determined by $\varphi_{D(\mathrm{tp}(a))}(s)$ and by $a \in \Sigma$ and
- for any $s \in \mathbb{M}(\Sigma)$ and any $A, B \subseteq \mathbb{T}$, the value $\varphi(\partial_{A \cup B}(s))$ is determined by $\varphi(\partial_A(s))$, $\varphi(\partial_B(s))$, A, and B.

A mapping satisfying the second and third stipulation above is called *asynchronous*. From this asynchronous mapping, we will now define a deterministic weighted asynchronous cellular automaton $\mathcal{A}$ following the ideas from [1] (see [2] for an exposition).

For $m \in \mathbb{T}$, let $Q_m = S$. For $q = (q_m)_{m \in \mathbb{T}} \in \prod_{m \in \mathbb{T}} Q_m$ set in$_{\mathcal{A}}(q) = $ in(ι) if $q_m = \varphi([\varepsilon])$ for all $m \in \mathbb{T}$. Otherwise, set in$_{\mathcal{A}}(q) = 0$. To define mapping out$_{\mathcal{A}}$, let

$q = (q_m)_{m \in \mathbb{T}}$. If there is a trace t with $q_m = \varphi \partial_m(t)$ for all $m \in \mathbb{T}$ such that $\iota.t$ is defined, then $\mathrm{out}_{\mathcal{A}}(q) = \mathrm{out}(\iota.t)$. Otherwise, $\mathrm{out}_{\mathcal{A}}(q) = 0$. Since φ is asynchronous, the mapping $\mathrm{out}_{\mathcal{A}}$ is welldefined.

Now let $a \in \Sigma_\ell$, $p_m \in Q_m$ for $m \in D(\ell)$, and $q_\ell \in Q_\ell$. Suppose there exists a trace $t \in \mathbb{M}(\Sigma)$ be a trace satisfying

- $p_m = \varphi \partial_m(t)$ for all $m \in D(\ell)$,
- $q_\ell = \varphi \partial_\ell(ta)$, and
- such that $\iota.ta$ is defined.

Then set $c_\ell(((p_m)_{m \in D(\ell)}, a, q_\ell)) = \mu(a)_{\iota.t, \iota.ta}$. If no such trace t exists, then the value $c_\ell(((p_m)_{m \in D(\ell)}, a, q_\ell))$ is set to 0. Again, this is welldefined since φ is asynchronous. Hence we defined a deterministic wACA $\mathcal{A} = ((Q_\ell)_{\ell \in \mathbb{T}}, (c_\ell)_{\ell \in \mathbb{T}}, \mathrm{in}_{\mathcal{A}}, \mathrm{out}_{\mathcal{A}})$.

We now want to show that the behavior of the deterministic wACA $\mathcal{A}$ equals the function represented by the deterministic presentation $(\mathrm{in}, \mu, \mathrm{out})$. Let $(\mathrm{in}_{\mathcal{A}}, \mu_{\mathcal{A}}, \mathrm{out}_{\mathcal{A}})$ be the canonical presentation associated with the wACA $\mathcal{A}$. In the following, let $p^0 = (p_\ell^0)_{\ell \in \mathbb{T}}$ be a global state of the wACA $\mathcal{A}$ with $p_\ell^0 = \varphi([\varepsilon])$ for all $\ell \in \mathbb{T}$. Let furthermore $a_1, a_2, \ldots, a_n \in \Sigma$ be letters and write t^i for the word $a_1 a_2 \ldots a_i$. By induction on i, one first shows the following:

Claim. If $p^1, p^2, \ldots, p^n \in \prod_{\ell \in \mathbb{T}} Q_\ell$ such that $\prod_{1 \leq i \leq n} \mu_{\mathcal{A}}(a_i)_{p^{i-1}, p^i} \neq 0$, then

(1) $p_\ell^i = \varphi \partial_\ell(t^i)$ for $\ell \in \mathbb{T}$,
(2) $\iota.t^i$ is defined, and
(3) $\mu_{\mathcal{A}}(a_i)_{p^{i-1}, p^i} = \mu(a_i)_{\iota.t^{i-1}, \iota.t^i}$ for $1 \leq i \leq n$.

Proposition 5.2. Let $(\mathrm{in}, \mu, \mathrm{out})$ be a deterministic k-dimensional presentation and let $\mathcal{A}$ be the deterministic wACA constructed above. Then $(\|\mathcal{A}\|, t) = \sum_{p, q \in [k]} \mathrm{in}(p) \cdot \mu(t)_{p,q} \cdot \mathrm{out}(q)$ for any trace $t \in \mathbb{M}(\Sigma)$.

Proof. Let ι be the only element of $Q = [k]$ with $\mathrm{in}(\iota) \neq 0$ and let $t = [a_1 a_2 \ldots a_n] \in \mathbb{M}(\Sigma)$ be some trace. Furthermore, let $(\mathrm{in}_{\mathcal{A}}, \mu_{\mathcal{A}}, \mathrm{out}_{\mathcal{A}})$ be the presentation associated with the deterministic wACA $\mathcal{A}$.

First assume $(\|\mathcal{A}\|, t) \neq 0$, i.e., $\sum_{p, q \in \prod Q_m} (\mathrm{in}_{\mathcal{A}}(p) \cdot \mu_{\mathcal{A}}(t)_{p,q} \cdot \mathrm{out}_{\mathcal{A}}(q)) \neq 0$. Then there are global states p^i of $\mathcal{A}$ with $\mathrm{in}_{\mathcal{A}}(p^0) \cdot \prod_{1 \leq i \leq n} \mu_{\mathcal{A}}(a_i)_{p^{i-1}, p^i} \cdot \mathrm{out}_{\mathcal{A}}(p^n) \neq 0$. Hence, using the claim, one can verify

$$(\|\mathcal{A}\|, t) = \mathrm{in}_{\mathcal{A}}(p^0) \cdot \prod_{1 \leq i \leq n} \mu_{\mathcal{A}}(a_i)_{p^{i-1}, p^i} \cdot \mathrm{out}_{\mathcal{A}}(p^n)$$

$$= \mathrm{in}(\iota) \cdot \prod_{1 \leq i \leq n} \mu(a_i)_{\iota.t^{i-1}, \iota.t^i} \cdot \mathrm{out}(\iota.t^n).$$

Since $(\mathrm{in}, \mu, \mathrm{out})$ is deterministic, this equals $\mathrm{in} \cdot \mu(t) \cdot \mathrm{out}$. Thus, we showed the equality from the proposition for all traces t with $(\|\mathcal{A}\|, t) \neq 0$.

If $\mathrm{in} \cdot \mu(t) \cdot \mathrm{out} \neq 0$, we can argue similarly. □

The following theorem summarizes the results of this section

Theorem 5.3. *Let* $S : \mathbb{M}(\Sigma) \to K$ *be a function. Then the following are equivalent*

- *S is the behavior of some deterministic wACA*
- *S has a deterministic presentation*

Recall that any deterministic presentation is I-consistent by the very definition. Hence this theorem is an analogue of the equivalence of (1) and (3) in Theorem 4.1. Recall that the second presentation in Example 3.2 satisfies (1) and (2) from Definition 5.1 but is not I-consistent. As any presentation, there exists an equivalent I-consistent one. But this transformation via nondeterministic wACAs can in general destroy property (2). Thus, it is an open question as to whether I-consistency is necessary for Prop. 5.2 to hold.

References

1. R. Cori, Y. Métivier, and W. Zielonka. Asynchronous mappings and asynchronous cellular automata. *Information and Computation*, 106:159–202, 1993.
2. V. Diekert. *Combinatorics on Traces*. Lecture Notes in Comp. Science vol. 454. Springer, 1990.
3. V. Diekert and A. Muscholl. Construction of asynchronous automata. In *[4]*, pages 249–267. 1995.
4. V. Diekert and G. Rozenberg. *The Book of Traces*. World Scientific Publ. Co., 1995.
5. M. Droste and P. Gastin. The Kleene-Schützenberger theorem for formal power series in partially commuting variables. *Information and Computation*, 153:47–80, 1999.
6. S.C. Kleene. Representation of events in nerve nets and finite automata. In C.E. Shannon and J. McCarthy, editors, *Automata Studies*, Annals of Mathematics Studies vol. 34, pages 3–40. Princeton University Press, 1956.
7. A. Mazurkiewicz. Concurrent program schemes and their interpretation. Technical report, DAIMI Report PB-78, Aarhus University, 1977.
8. E. Ochmański. Regular behaviour of concurrent systems. *Bull. Europ. Assoc. for Theor. Comp. Science*, 27:56–67, 1985.
9. A. Salomaa and M. Siottola. *Automata-Theoretic Aspects of Formal Power Series*. Texts and Monographs in Computer Science. Springer, 1978.
10. M.P. Schützenberger. On the definition of a family of automata. *Inf. Control*, 4:245–270, 1961.
11. W. Thomas. On logical definability of trace languages. In V. Diekert, editor, *Proceedings of a workshop of the ESPRIT BRA No 3166: Algebraic and Syntactic Methods in Computer Science (ASMICS) 1989*, Report TUM-I9002, Technical University of Munich, pages 172–182, 1990.
12. W. Zielonka. Notes on finite asynchronous automata. *R.A.I.R.O. - Informatique Théorique et Applications*, 21:99–135, 1987.
13. W. Zielonka. Asynchronous automata. In *[4]*, pages 205–248. 1995.

On the Complexity of the "Most General" Firing Squad Synchronization Problem

Darin Goldstein and Kojiro Kobayashi

[1] Department of Computer Engineering and Computer Science,
California State University, Long Beach
daring@cecs.csulb.edu
[2] Department of Information Systems Science,
Faculty of Engineering, Soka University
kobayasi@t.soka.ac.jp

Abstract. We show that if a minimal-time solution exists for a fundamental distributed computation primitive, synchronizing a general directed network of finite-state processors, then there must exist an extraordinarily fast $O(ED \log_2 D (\log_2 n)^2)$ algorithm in the RAM model of computation for exactly determining the diameter of a general directed graph. The proof is constructive.

This result interconnects two very distinct areas of computer science: distributed protocols on networks of intercommunicating finite-state machines and standard algorithms on the usual RAM model of computation.

1 Introduction

The Firing Squad Synchronization Problem (or FSSP, for short) is a famous problem originally posed almost half a century ago. A prisoner is about to be executed by firing squad. The firing squad is made up of soldiers who have formed up in a straight line with muskets aimed at the prisoner. The general stands on the left side of the line, ready to give the order, but he knows that he can only communicate with the soldier to his right. In fact, each soldier can only communicate with the soldier to his immediate left and/or right, but nobody else. Soldiers have limited memory and can only pass along simple instructions. Is it possible to come up with a protocol, independent of the size of the line, for getting all of the soldiers to fire at the prisoner simultaneously if their only means of communication are small, whispered instructions only to adjacent soldiers? (The possibility of counting the number of soldiers in the line can be discounted because no soldier can remember such a potentially large amount of information; each soldier can only remember messages that are independent of the size of the line.)

The problem itself is interesting as a mathematical puzzle. More importantly, there are also applications to the synchronization of small, fast processors in large networks. In the literature on the subject (e.g. [26, 18]), the problem has been referred to as "macrosynchronization given microsynchronization" and "realizing global synchronization using only local information exchange." The synchronization of multiple small but fast processors in general networks is a

B. Durand and W. Thomas (Eds.): STACS 2006, LNCS 3884, pp. 696–711, 2006.
© Springer-Verlag Berlin Heidelberg 2006

fundamental problem of parallel processing and a computing primitive of distributed computation.

The FSSP has a rich history that spans decades. J. Mazoyer provides an overview of the problem (up to 1986, at least) in addition to some of its history in [20]. The problem was originally introduced by J. Myhill in 1957, though the first published reference is [??] from 1962. J. McCarthy and M. Minsky [00] first solved the problem. Since then, many variations of FSSP have been proposed and studied (e.g. [21, 28, 19, 24, 12, 3, 2] to name just a few). Despite this long history of study and extensive compilation of results, one fundamental problem for FSSP has remained open. It concerns existence of optimal-time solutions of variations of FSSP.

For one variation of FSSP that has a solution and a problem instance (a network) X of the variation, the *minimum firing time* of X is the minimum of the firing times of solutions A for the instance X, where A ranges over all solutions of the variation. A *minimal-time solution* of the variation is a solution whose firing time for X is the minimum firing time of X for any problem instance X of the variation. Variations of FSSP are classified to the following three types:

Type A: The variation has a minimal-time solution.
Type B: The variation has a solution but has no minimal-time solution.
Type C: The variation has no solution.

We know many examples of variations of Type A. The original FSSP is one of them. The minimum firing time of a linear array of n soldiers is $2n - 2$ and minimal-time solutions were found by Goto [11] and Waksman [29]. Other examples are linear arrays with arbitrary position of the general, bilateral rings, one-way rings, rectangles, and squares (the position of the general may be either one of the corners or may be at any position). We also know examples of variations of Type C. One is the variation for directed networks such that an out-port may have arbitrary fan-out ([17]) and another is the variation for undirected networks with unbounded number of generals ([14, 13]). However, at present we know no example of variations of Type B. Neither do we know that variations of Type B do not exist. Therefore, the problem to decide whether variations of Type B exist or not is a very important open problem. Here we restrict ourselves to variations of FSSP that are obtained from the original FSSP by generalization only with respect to network topologies. Without this restriction, we know of some variations of Type B: [21] (generalization with respect to the communication delay time) and [27] (generalization with respect to the number and the role of generals).

In the followings, we list eight variations for which solutions are known but minimal-time solutions are not known.

2PATH: The variations for paths in the two-dimensional grid space. A path may bend but may not touch itself. The general is at one of the end points.
g-2PATH: The same as 2PATH but the general may be at any position (g is for "*g*eneralized").
2REG: The variation for finite connected regions in the two-dimensional grid space. The general may be at any position. Regions may have holes.

3PATH, g-3PATH, 3REG: The variations analogous to 2PATH, g-2PATH, 2REG respectively for the three-dimensional grid space.

UN: The variation for general undirected networks. Networks must be connected.

DN: The variation for general directed networks. Networks must be strongly connected.

We know a solution with firing time $3r - 1$ for UN ([25]) (r denotes the radius of the network) and solutions for DN with an exponential firing time ([16]), with a firing time $O(n^2)$ ([7]), and with a firing time $O(Dn)$ ([26], D denotes the diameter of the network). However, these solutions are not minimal-time. The variation DN is especially important because it is the "most general" FSSP.

We have made the conjecture that all of these eight variations have no minimal-time solutions, and hence are of Type B. The attempt to prove this conjecture was initiated by one of the authors. In [18], the second author introduced a problem which he called 2PEP (the two-dimensional path extension problem). The problem is to decide, for each given path in the two-dimensional grid space, whether we can extend the path from the specified end point so that the length of the path is doubled. At present we know no polynomial time algorithm for 2PEP, but at the same time we are unable to prove that it is NP-complete. It was proved in [18] that if 2PATH has a minimal-time solution then 2PEP has an $O(n^2)$ time algorithm. This result readily applies also to g-2PATH and 2REG. Hence, finding a minimal-time solution of 2PATH, g-2PATH, 2REG is at least as hard as finding an $O(n^2)$ time algorithm for 2PEP. Next, in [10] and [9], we could prove a stronger result for the other three variations: if P $\neq$ NP then 3PATH, g-3PATH, 3REG have no minimal-time solutions. Hence, finding a minimal-time solution of 3PATH, g-3PATH, 3REG is at least as hard as finding a polynomial time algorithm for SAT. UN and DN remain to be studied.

In the present paper, we show that finding a minimal-time solution of DN is at least as hard as finding an "incredibly" fast algorithm in the RAM model of computation for the problem to determine the diameter of a general directed graph. We formalize this with the following theorem.

Theorem 1. *If there exists a minimal-time solution for the general directed network topology, then there exists a deterministic algorithm in the RAM model of computation that can exactly determine the diameter of a general unweighted directed graph in time* $O(ED \log_2 D (\log_2 n)^2)$ *where* E *is the number of edges,* D *is the diameter of the graph, and* $n = |V|$ *is the number of vertices.*

To give some idea on how the algorithm mentioned in Theorem 1 compares with other algorithms, in Table 1 we show the running time $O(En + n^2 \log n)$ of Dijkstra's algorithm and the running time $O(ED \log D(\log n)^2)$ of our algorithm for some special cases of E, D. Our algorithm is slower only for the case $D = \Theta(n)$ (narrow and long graphs).

The literature for algorithms that determine the diameter of a graph is vast and varied. There are *numerous* references for finding the diameter of directed graphs, most of which reduce to the problems of matrix multiplication (that

Table 1. Comparison of the running times of Dijkstra's algorithm and the algorithm of Theorem 1

	$O(En + n^2 \log n)$	$O(ED \log D (\log n)^2)$
$E = \Theta(n^2), D = \Theta(n)$	$O(n^3)$	$O(n^3 (\log n)^3)$
$E = \Theta(n^2), D = \Theta(\sqrt{n})$	$O(n^3)$	$O(n^2\sqrt{n}(\log n)^3)$
$E = \Theta(n^2), D = \Theta(1)$	$O(n^3)$	$O(n^2(\log n)^2)$
$E = \Theta(n\sqrt{n}), D = \Theta(n)$	$O(n^2\sqrt{n})$	$O(n^2\sqrt{n}(\log n)^3)$
$E = \Theta(n\sqrt{n}), D = \Theta(\sqrt{n})$	$O(n^2\sqrt{n})$	$O(n^2(\log n)^3)$
$E = \Theta(n\sqrt{n}), D = \Theta(1)$	$O(n^2\sqrt{n})$	$O(n\sqrt{n}(\log n)^2)$
$E = \Theta(n), D = \Theta(n)$	$O(n^2 \log n)$	$O(n^2(\log n)^3)$
$E = \Theta(n), D = \Theta(\sqrt{n})$	$O(n^2 \log n)$	$O(n\sqrt{n}(\log n)^3)$
$E = \Theta(n), D = \Theta(1)$	$O(n^2 \log n)$	$O(n(\log n)^2)$

ignores addition as an elementary operation when determining the complexity, which our model does not) or solving the All Pairs Shortest Path (APSP) problem; the problem of finding approximate diameters (e.g. [1]) has also been examined. To date, the best methods for determining the diameter of an arbitrary directed network that do not involve "fast matrix multiplication" [4] rely on solving the All Pairs Shortest Path problem [5]. Applying Dijkstra's algorithm for each vertex in the network leads to a running time of $O(En + n^2 \log n)$. This algorithm was improved by Karger et al. [15] in 1993 to $O(nE^* + n^2 \log n)$ where E^* represents the total number of edges that participate in the shortest length paths. Fredman [8] in 1976 introduced a method in the algebraic computation tree model by which the min-plus matrix multiplication necessary to compute the APSP could be determined using only $O(n^{2.5})$ additions and comparisons, but the best known implementation by Zwick [31] in the RAM model runs in time slightly $o(n^3)$. We have presented only a very superficial sampling of the results in this area[1], but Zwick [30] presents a fairly complete survey on the APSP problem and most of its reasonable variations.

To the authors' knowledge, this is the first result that directly relates solutions for distributed networks of finite-state automata to algorithms that run in the "usual" RAM model of computation. To this point, there is no established connection between the diameter of a graph and the minimum firing time of a network of finite-state machines. The time bound of $O(ED \log_2 D (\log_2 n)^2)$ mentioned above in Theorem 1 beats every known result in the relevant sections of Zwick's survey by large factors, especially for somewhat sparse graphs with small diameter. One can therefore look at our results in one of two ways. One can view Theorem 1 as evidence that a minimal-time solution for the FSSP on general directed networks does not exist, given the size of the asymptotic gap between the conclusions of the theorem and the current state of research of the diameter problem after decades of searching for better algorithms. One can also view Theorem 1 as motivation for searching for a minimal-time solution, for if

---

[1] Indeed, there were at least six results leading up to Zwick's recent $o(n^3)$ algorithm just for implementing Fredman's original 1976 idea.

such a solution were found, it would be a major leap forward in the search for better "shortest-path" algorithms.

The rest of the paper will be organized as follows. In Section 2, we introduce the automata model we will use for the remainder of the paper. Section 3 contains the proof of Theorem 1 including proof sketches of the relevant lemmas. Finally, in Section 4, we present conclusions and open problems.

2   The Model

As mentioned previously, we wish to model the operation of a large network of processors whose computations are all governed synchronously by the same global clock. The model is intended to mathematically abstract a physical switching network or a very large-scale parallel processing machine. The processors are designed to be small, fast, and unable to access large memory caches. Each processor is identical and assumed to have a fixed constant number of ports which can both send and receive a constant amount of data per clock cycle. ("Constant" quantities must be independent of the size and structure of the network.)

More formally, the problem is to construct a deterministic finite-state automaton with a transition function that satisfies certain conditions. We assume that each processor in the network is identical and initially in a special "quiescent" state, in which, at each time-step, the processor sends a "blank" character through all of its ports. A processor remains in the quiescent state until a non-blank character is received by one of its in-ports. We consider connected networks of such identical synchronous finite-state automata with vertex degree uniformly bounded by a constant. These automata are meant to model very small, fast processors. The network itself may have a specific topology (see below) but potentially unbounded size. The network is formed by connecting ports of a given processor to those of other processors with wires. Not all ports of a given processor need necessarily be connected to other processors. The network has a global clock, the pulses between which each processor performs its computations. Processors synchronously, within a single global clock pulse, perform the following actions in order: read in the inputs from each of their ports, process their individual state changes, and prepare and broadcast their outputs. As mentioned, our network structure is specifically designed to model the practical situation of many small and fast processors performing a synchronous distributed computation. The goal of the protocol is to cause every process in the network to enter the same special "firing" state for the first time *simultaneously.*

A *solution A* for a given network topology (e.g. the bidirectional line, as above) is defined to be the instantiation of an automaton with a transition function that satisfies the firing conditions outlined above for any network size. (So, by this definition, a *solution* for the bidirectional line must function for a bidirectional line of **any** size.) Assuming a solution A is specified, the *firing time* of A on a given network will refer to the number of clock cycles it takes for this network of processors programmed with the solution A to complete the protocol and simultaneously fire. The *minimum firing time* of a given network (of a specified topology) will refer to the minimum over all solutions A of the firing time of the network of processors programmed with solution A. A *minimal-time solution* A_{min} for a given network topology will be a solution such that the firing time for a network of any given size (with the given topology) programmed with the algorithm A_{min} will equal the minimum firing time of the network. Note that even though the network can be of arbitrary size, the size of the algorithm A_{min} must be fixed. The topology we consider in this paper is the most general possible, an arbitrary directed network of bounded degree[2].

3   The Results

To prove Theorem 1, we will first prove two lemmas.

Definition 1. *Given a graph G, define $I(v)$ to be the set of edges with terminal vertex v. For $e \in I(v')$, define $d_e(v, v')$ to be the length of the shortest path from v to v' such that the final edge in the path is e. Then for two vertices $v, v' \in G$, define $d^*(v, v')$ to be $\max_{e' \in I(v')} d_{e'}(v, v')$.*

Lemma 1. *Let $G = (V, E)$ be a general directed network with degree at most δ. Then the minimum firing time for the general directed network topology with degree bound δ running on G is $\max_{v \in V}\{d^*(root, v) + \max_{v' \in V} d(v, v')\}$.*

Proof. We need to prove two things. First, we show that any solution must take at least this long on any given network. To get a contradiction, fix a solution A that runs in time strictly less than this on some network N, and let v and v' be two vertices that achieve the above maximum. So the running time of A on N is $d^*(root, v) + d(v, v') - 1$. Let $e = (v_{prev}, v) \in I(v)$ be the edge that satisfies the maximum in the definition of d^*. Replace the edge e with a string of edges and vertices as follows: $v_{prev} \rightarrow v$ should transform into $v_{prev} \rightarrow v_1 \rightarrow v_2 \rightarrow \ldots \rightarrow v_n \rightarrow v$ for some large number n. We claim that the computational transcripts of all nodes common to both networks are exactly the same through time $d^*(root, v) - 1$. Note that up until this time in both networks, the edge e has been transmitting only the quiescent value to v. Thus up until time $d^*(root, v) - 1$, all common nodes in both networks will have exactly the same computational transcripts. So because there are no other modifications, for each i, at time $d^*(root, v) + i$, the nodes at a distance i from v in network N' might have different states from their counterparts in network N. However, this implies that v' must fire at the same time in both networks. If we choose n large enough, because the argument is independent of the value of n, this is clearly impossible.

Next, we must show that, given a fixed network N, there exists a solution that fires in time at most $\max_{v, v' \in V}\{d^*(root, v) + d(v, v')\}$. Consider a solution A for

---

[2] The topology must be of bounded degree so that the automata can distinguish between their various in and outports.

FSSP that runs on general arbitrary strongly-connected digraphs. We modify this solution to get another solution A' that runs in minimal-time by running a second protocol in parallel to the original A. Release tokens that check (in a breath-first manner) whether the network follows the structure of N. Note that if we allow the tokens to keep track of the network structure that they have "seen" thus far, it is possible for each non-quiescent vertex to have pre-knowledge of when the next token should come in and what it should have thus far seen. If a token reaches or does not reach (in the case that a token in expected to arrive and does not) a vertex that is not consistent with the structure of N, then the vertex immediately sends out "USE A" breadth-first tokens. We claim that such tokens, if released from a vertex, have the time to reach every vertex in N before or at time $\max_{v \in V}\{d^*(root, v) + \max_{v' \in V} d(v, v')\}$. Assume that there exists a vertex or edge that is inconsistent with the structure of N. It will take time at most $d^*(root, v)$ for the breadth first tokens to first be released by the offending vertex v. Once released it takes time $\max_{v' \in V}\{d(v, v')\}$ for the tokens to reach every vertex in the network. Thus, either all network nodes receive "USE A" tokens or they do not before or at time $\max_{v \in V}\{d^*(root, v) + \max_{v' \in V} d(v, v')\}$. If the vertices receive a "USE A" token, they use the protocol from the original algorithm A and if not, they simply fire at time $\max_{v \in V}\{d^*(root, v) + \max_{v' \in V} d(v, v')\}$.   □

Lemma 2. *The two statements (a) "There exists a constant δ such that there exists a minimal-time solution for general directed networks with degree bounded by δ." and (b) "For any constant δ, there exists a minimal-time solution for general directed networks with degree bounded by δ." are equivalent.*

Proof Sketch. (b) implies (a) is obvious. We therefore concentrate on showing (a) implies (b). Fix the constant δ in statement (a). We claim that for networks with degree smaller than δ, the same solution will work because the minimum firing time of G is independent of δ by Lemma 1. For networks with degree δ' strictly greater than δ, we can form a new solution from the old solution as follows: For every node N, split N into as many nodes of degree δ as necessary. In other words, form an inward binary search tree towards the node and an outgoing binary search tree away from the node with both binary search trees degree-bounded by δ. We can now connect the wires corresponding to the links in the original network. See Figure 1. In the figure, we assume that we know the transition function of the A automata. The transition function of the automaton A', the conglomeration of all of the A's, can be determined as follows.

The state of A' is defined uniquely by the states of the A's and the inputs and outputs of each A within the conglomeration. Given a vector of input values to the automata A', the output vector after a single time step will consist of the computational transcripts of every A automata within A' if we allowed $2 \log_\delta \delta'$ "internal" time steps to elapse. Note that we are allowed to perform this simulation because both δ and δ' are constants of the network. The A' automata fires if and only if its internal A automata fire during some "internal" time step.

It is then possible to show that the firing time arising from this construction is in fact minimal via Lemma 1. The basic idea behind this calculation is to

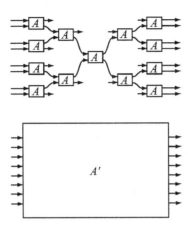

Fig. 1. This figure illustrates the transformation from multiple copies of a known automata solution A to the conglomeration A'

note that each edge is elongated by a factor of $2\log_\delta \delta'$ ($\log_\delta \delta'$ inwards and $\log_\delta \delta'$ outwards) which causes the minimum firing time to gain a similar factor. However, the internal simulation of $2\log_\delta \delta'$ time steps down to a single time step by A' compresses the running time by exactly the same factor. The actual calculations are somewhat tedious and are therefore omitted.   □

The remainder of this section will be devoted to proving Theorem 1.

> *If there exists a minimal-time solution for the general directed network topology, then there exists a deterministic algorithm in the RAM model of computation that can exactly determine the diameter of a general unweighted directed graph in time $O(ED\log_2 D\,(\log_2 n)^2)$ where E is the number of edges, D is the diameter of the graph, and $n = |V|$ is the number of vertices.*

Proof (rest of Section 3). We will assume that we are given an arbitrary directed graph $G = (V, E)$ and are operating in the RAM model of computation. Tarjan's strongly connected components algorithm can be used to determine in time $O(V + E)$ time whether the graph has infinite diameter or not. We will assume that the diameter is found to be finite. (In other words, Tarjan's algorithm returns a single strongly connected component.) We now proceed with a series of transformations of the graph G. Note that via Lemma 2, as long as we can show the resulting transformation is degree-bounded by some constant, it doesn't matter what the constant is.

3.1   Transforming G to G'

The first transformation takes the inputs of v, for each $v \in V$, and extends them to be of length $\frac{3}{2}\log_2 n$ that cascade inwards so as to form a incoming binary tree towards v. Note that $\log_2 n$ is sufficient for this operation; the reason for extending the natural length will become clear below. We perform a similar operation on the outgoing edges. We then connect the wires corresponding to the edges in G. The result is similar to the illustration in Figure 1 except for the fact that the final outgoing and incoming edges are extended by an additional $\frac{1}{2}\log_2 n$. Note that we have reduced the degree of the graph to at most 2. Call this transformed graph $G' = (V', E')$. We will later refer to the vertices of G inside of G'. This will refer to the set of vertices at the roots of the incoming (and hence outgoing as well) binary trees.

3.2   Transforming G' to $G''(k)$

We now create a graph $G''(k)$ from G'. First, create n bidirectional paths of length k, for some number $k \geq 3\log_2 n$ to be determined later. For ease of communication, we will henceforward refer to these bidirectional paths as *connectors*. Connect one end of each of these connectors to a vertex in G. (Note that each vertex in G is both the end of a connector and a vertex in G' as well.) Create a root node and have the root cascade outputs outwards, each branch of length $\log_2 n$. Have these link up with the other end of the connectors. In a similar way, have edges cascade inwards towards the root, each of length $\log_2 n$, from each path linking up with the same vertices. Note that the two binary trees going into and out from the root do not touch except at their leaf vertices (at the ends of the connectors) and at the root. Finally, delete all nodes of incoming and outgoing binary trees that are not used for going from a node in G or the root to another node in G or the root. Then we obtain a strongly-connected directed graph, and this is $G''(k)$. Note also that $G''(k)$ has bounded degree. See Figure 2 for an illustration of this process.

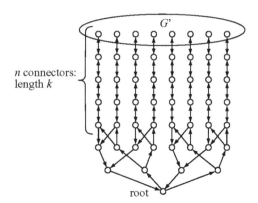

Fig. 2. This figure illustrates the transformation from G' to $G''(k)$

3.3   Calculations

Consider the quantity $f(v) = d^*(root, v) + \max_{v' \in V''(k)} d(v, v')$ for each vertex in $v \in V''(k)$. The firing time of the network is $\max_{v \in V''(k)} f(v)$ by Lemma 1. Our overall goal will be to determine this value.

First, we split the vertices in $V''(k)$ into four major subdivisions and consider $f(v)$ for v in each subdivision. (A brief spoiler: It will turn out that the maximum value of $f(v)$ will occur in the fourth subdivision we consider. We need to calculate the function values for each of the other subdivisions in order to be sure.)

1. All vertices in the lower two binary trees, the final set of edges linking up with the root. It is clear that $d^*(root, v) = d(root, v)$ for vertices in this subdivision (with the exception of the root itself). Within this subdivision, consider vertices v that are oriented by their edges away from the root such that $d(root, v) = i \le \log_2 n$. Note that the vertex $v' \in V''(k)$ that maximizes $d(v, v')$ will be somewhere in G'. The "cheapest" way to get there will be to head back to the root because the expense of jumping from one vertex to another in the graph G' is $3 \log_2 n$ and this is more than the jump back to the root and to another path. Thus, $d(v, v') = (\log_2 n - i) + 2 \log_2 n + k + (3 \log_2 n - 1) = 6 \log_2 n + k - i - 1$ and $f(v) = i + d(v, v') = 6 \log_2 n + k - 1$.

Now consider vertices v in the same subdivision that are oriented towards the root such that $d(root, v) = \log_2 n + i$. By the same arguments, the same bound applies and thus $f(v) = 6 \log_2 n + k - 1$ as well. Finally, note that $f(root) = 6 \log_2 n + k - 1$ as well. (These values will turn out to be smaller than other potential values of $f(v)$ for other subdivisions; hence we will be able to assume that v is not located in either binary tree and these calculations will not contribute to the final determination of the firing time.)

2. The set of vertices in the connectors. (Note that this set has nonempty intersection with V.) Via the following lemma, we will show that only vertices in V need to be examined to find the maximal f value. These vertices will therefore be separated into their own subdivision below.

Lemma 3. *Within the connector subdivision, the function f is nondecreasing with increasing distance from the root.*

Proof. Consider any vertex v within a connector that is not the furthest possible from the root. Let v_{next} be the vertex one unit further from the root. Consider $f(v)$. Let v' be a vertex satisfying the maximum in the definition of f. Then $d(v_{next}, v') \ge d(v, v') - 1$. Because $d^*(root, v_{next}) \ge d^*(root, v) + 1$, we have

$$f(v_{next}) = d^*(root, v_{next}) + \max_{v'' \in V''(k)} d(v_{next}, v'')$$

$$\ge d^*(root, v) + 1 + d(v_{next}, v') \ge d^*(root, v) + 1 + d(v, v') - 1 = f(v)$$

$\square$

By the above lemma, we can assume without loss of generality that when locating the maximum within the connector subdivision, v is the vertex on the connector furthest from the root. However, this reduces us to only searching vertices V in the original graph $G \subset G'$. This will be the next subdivision we consider.

3. The vertices in the original graph G, namely V. We claim that $\forall v \in V, \exists v' \in V' - V$ such that $f(v') \geq f(v)$. In fact, note that the vertex v' on the terminal of any input from v will satisfy this claim. We omit the details due to space constraints.

4. $V' - V$. Let v be a vertex that maximizes $f(v)$ within the set $V' - V$ (i.e. $f(v) = \max_{v' \in V' - V} f(v')$). We will assume throughout this section without loss of generality that $d(G, v) = i$ for some $i \geq 1$. Then $d^*(root, v) = \log_2 n + k + i$. Let v' satisfy the maximum in the definition of f. Our goal is to determine the various values that $f(v)$ might take depending on where v' winds up being. First, however, we need a preliminary lemma. For the remainder of the paper, let the quantity $d_r(v, v')$ (resp. $d_n(v, v')$) be the length of the shortest path from v to v' that does (resp. does not) pass through the root.

Lemma 4. *Assume $v \in V' - V$. Then the vertex v' that maximizes $d(v, v')$ cannot be in either of the bottom binary trees assuming $k \geq 3 \log_2 n$ and $D \geq 2$.*

Proof. If $k \geq 3 \log_2 n$, then there exists a vertex v' in the connector located $2 \log_2 n$ above the bottom of the connector. Note that

$$d_r(v, v') = 3 \log_2 n - i + k + 2 \log_2 n + 2 \log_2 n = 7 \log_2 n + k - i$$

and

$$d_n(v, v') = 3 \log_2 n - i + 3D \log_2 n + (k - 2 \log_2 n) \geq 7 \log_2 n + k - i$$

Thus $d(v, v') = 7 \log_2 n + k - i$ whereas the furthest vertex from v located within either binary tree is a distance $3 \log_2 n - i + k + \log_2 n + \log_2 n - 1 = 5 \log_2 n + k - i - 1$. $\square$

If $v \in V' - V$, there are several cases to consider for the location of v' and the behavior of the shortest path from v to v'. By the above lemma (Lemma 4), we only need to consider v' that lie within G' or the connectors.

- Assume that $v \in V' - V$, $v' \in V'$, and the shortest path from v to v' passes through the root. Then $f(v) = \log_2 n + k + i + (3 \log_2 n - i) + k + \log_2 n + \log_2 n + k + j$ where $d(G, v') = j < 3 \log_2 n$. Maximizing over the possible values of v' yields a total value of $9 \log_2 n + 3k - 1$.
- Assume that $v \in V' - V$, $v' \in V'$, and the shortest path from v to v' does not pass through the root. Then the path from v to v' must remain entirely inside G'. $f(v) = \log_2 n + k + i + (3 \log_2 n - i) + 3D \log_2 n + j$ where $d(G, v') = j$. Once again maximizing over the values of v', we get $f(v) = (7 + 3D) \log_2 n + k - 1$ in this case.

Remark: It is worth noting at this point that by examination of the calculations above, $\exists v \in V' - V, \exists v' \in V'$ such that

$$f(v) = \min\{d^*(root, v) + d_n(v, v'), d^*(root, v) + d_r(v, v')\} =$$

$$\min\{(7 + 3D)\log_2 n + k - 1, 9\log_2 n + 3k - 1\}$$

In fact, we can pinpoint two vertices that fit the bill: Assume that two vertices a and b in G are the initial and terminal vertices of a path that equals the diameter of C. In $C''(h)$, consider the equivalent vertices. If we let v be any where on an incoming binary tree to a (*not* a itself) and we let v' be a furthest vertex from b on its outgoing binary tree, then the pair v and v' will satisfy the equalities above.

- Assume that $v \in V' - V$, $v' \notin V'$, and the shortest path from v to v' passes through the root. Note that v' is therefore in a connector. In this case, $f(v) = \log_2 n + k + 3\log_2 n + k + \log_2 n + d(root, v')$. We also have that $d(root, v') = \log_2 n + k - d(v', G)$. So finally we get $f(v) = 6\log_2 n + 3k - d(v', G)$ in this case.
- Assume that $v \in V' - V$, $v' \notin V'$, and the shortest path from v to v' does not pass through the root. Again, v' is in a connector. In this case, $f(v) = \log_2 n + k + 3\log_2 n + 3D\log_2 n + d(G, v') = (4 + 3D)\log_2 n + k + d(G, v')$.

Note that each of these values is larger than the values calculated in any of the other subdivisions. Thus, we can conclude that *the function $f(v)$ is maximized just when $v \in V' - V$*. For the remainder of the paper, we will concentrating exclusively on this case.

Lemma 5. *Assume that $v \in V' - V$, and $v' \notin V'$. A vertex v' that maximizes $d(v, v')$ will be such that*

$$|d_n(v, v') - d_r(v, v')| \le 1$$

Proof. By Lemma 4, v' is within a connector. Consider any v' within a connector such that $|d_n(v, v') - d_r(v, v')| \ge 2$. Assume without loss of generality that $d_n(v, v') > d_r(v, v')$. (The other case is symmetric.) Then $d_n(v, v') \ge d_r(v, v')+2$. Consider the vertex v'' that is one "higher" than v' (i.e. the vertex that is one unit closer to G'). Then we have that $d_r(v, v'') = d_r(v, v') + 1$ and

$$d_n(v, v'') = d_n(v, v') - 1 \ge d_r(v, v') + 1 \Rightarrow d(v, v'') = d_r(v, v') + 1$$

Because $d_n(v, v') > d_r(v, v')$, we have that $d(v, v') = d_r(v, v') < d(v, v'')$. □

Lemma 6. *For some $v \in V' - V$ and some value of k. Assume that $v' \in V'$ is the vertex that maximizes the value of $d(v, v')$. If the shortest path from v to v' does not pass through the root, then if the value of k is increased, for a fixed v, as long as $v' \in V'$, the shortest path from v to v' will not pass through the root.*

Proof. Figure 2. Assume $d_n(v, v') < d_r(v, v')$. Increasing k can only increase the value of $d_r(v, v')$ but does not affect the value of $d_n(v, v')$. □

Lemma 7. *Assume that $v \in V' - V$ and $v' \notin V'$ for some value of k and that v' is the vertex that maximizes the value of $d(v, v')$ (and is therefore in a connector by Lemma 4). Then with increasing values of k, v' will remain in the connector subdivision.*

Proof. Note that if $v' \notin V'$ then the distance from v to v' is greater than from v to any vertex in V'. Increasing the value of k only increases the distances of vertices outside of V'. $\qquad\square$

Lemma 8. *Assume that v is the vertex that maximizes the function $f(v)$. (Then, by the above discussion, $v \in V' - V$.) Let v' be the vertex that maximizes the value of $d(v, v')$. If $k > 4D \log_2 n$ and $D \geq 2$, then $v' \notin V'$.*

Proof. Figure 2. If $v \in V' - V$, then an obvious upper bound on the distance from v to any other vertex in V' is $(3D + 2) \log_2 n$. $\qquad\square$

3.4   The Algorithm

We now outline the algorithm we use to determine D. First, we determine whether $D = 1$ by checking whether every vertex is connected to every other vertex. This takes time $O(V + E)$. Now, starting from $k = 3 \log_2 n$, we simulate the minimal-time solution on the network $G'''(k)$ and note the firing time. Note that once we examine the firing times assuming the four possible locations of the vertex v, we conclude that the maximum value of $f(v)$ is achieved when v is in $V' - V$. Thus, for the remainder of this section, we will implicitly assume this. The firing time must satisfy one of the four values calculated above for this situation.

If the firing time is $9 \log_2 n + 3k - 1$, then double the value of k and note the firing time again. (In the unlikely case that this value happens to be equal to one of the other possible three values, another doubling of k will suffice to distinguish.) By Lemmas 6 and 7, eventually we will either have $v' \in V'$ and a running time of $(7 + 3D) \log_2 n + k - 1$ or $v' \notin V'$. Note that if $v' \in V'$ at this point, then we must have (by the Remark in Section 3.3)

$$(7 + 3D) \log_2 n + k - 1 \leq 9 \log_2 n + 3k - 1 \Rightarrow k \geq \frac{3D - 2}{2} \log_2 n$$

If we quadruple the value of k, we are therefore guaranteed that $k > 4D \log_2 n$. Thus, quadruple the value of k and again note the firing time. At this point, we must have $v' \notin V'$ by Lemma 8.

By Lemma 5, we know that v' satisfies the inequality $|d_n(v, v') - d_r(v, v')| \leq 1$. Therefore either prediction (a) $6 \log_2 n + 3k - d(v', G)$ or (b) $(4 + 3D) \log_2 n + k + d(G, v')$ for the running time is at most off by 1. Note that because $v' \notin V'$, v' must be in a connector and therefore $d(G, v') = d(v', G)$. Because we know the value of the running time, we can estimate the value of $d(G, v')$ to within 1 using (a). Using (b) and this value, we can then estimate the value of $(4 + 3D) \log_2 n$ to within 2. The range for $(4 + 3D) \log_2 n$ is therefore 5 at most. If $\log_2 n > 1$, then at most a single value of D will satisfy the inequalities.

3.5   Time Analysis

To analyze the running time of the algorithm, we note that for a given value of k, to simulate the minimal-solution on the network $V''(k)$ in the RAM model of computation requires time proportional to the product of the number edges and the firing time. The firing time for any given value of k is clearly $O(D \log_2 n)$ because $k = O(D \log_2 n)$. The number of edges in the graph has been increased from E to $O(E \log_2 n)$. Thus the total simulation time is $O(ED \log_2 D (\log_2 n)^2)$.

$\square$

4 Conclusions and Open Problems

In this paper, we have given a constructive proof that if a minimal-time solution exists for a fundamental distributed computation primitive, synchronizing a general directed network of finite-state processors, then there must exist an extraordinarily fast $O(ED \log_2 D (\log_2 n)^2)$ algorithm in the RAM model of computation for exactly determining the diameter of a general directed graph. This result opens up a number of very promising areas of research, the most obvious of which seem to be the following.

- Is this the best possible correspondence? If a minimal-time solution for the FSSP on the general directed network topology does happen to exist, is $O(ED \log_2 D (\log_2 n)^2)$ the best we can possibly do? Note that practically no effort was made to improve the simulation of the FSSP solution.
- Are results possible in the other direction? To date, there have been no results (to the author's knowledge, at least) that relate fast algorithms in standard models of computation such as the RAM model or the Turing Machine model to protocols for intercommunicating finite-state automata. Such a result would be extremely interesting.
- The approximation angle looks like a very promising area of research. If there exists an "almost" minimal-time solution, does it say anything about the algorithm for the diameter problem?
- What other problems can be related to the FSSP on various other topologies? It seems as if other natural topologies might (probably, should) have similar "natural" correspondences like this one.

References

[1] Aingworth, Chekuri, and Motwani. Fast estimation of diameter and shortest paths (without matrix multiplication). In *SODA: ACM-SIAM Symposium on Discrete Algorithms (A Conference on Theoretical and Experimental Analysis of Discrete Algorithms)*, 1996.

[2] R. Balzer. An 8-state minimal time solution to the firing squad synchronization problem. *Information and Control*, 10:22–42, 1967.

[3] A. Berthiaume, T. Bittner, L. Perković, A. Settle, and J. Simon. Bounding the firing synchronization problem on a ring. *Theoretical Computer Science*, 320:213–228, 2004.

[4] D. Coppersmith and S. Winograd. Matrix multiplication via arithmetic progressions. *Journal of Symbolic Computation*, 9:251–280, 1990.

[5] Thomas H. Cormen, Charles E. Leiserson, and Ronald L. Rivest. *Introduction to Algorithms*. MIT Press/McGraw-Hill, 1990.

[6] K. Culik. Variations of the firing squad problem and applications. *Information Processing Letters*, 30:153–157, 1989.

[7] S. Even, A. Litman, and P. Winkler. Computing with snakes in directed networks of automata. *J. Algorithms*, 24(1):158–170, 1997.

[8] M.L. Fredman. New bounds on the complexity of the shortest path problem. *SIAM J. of Computing*, 5:83–89, 1976.

[9] D. Goldstein and K. Kobayashi. On the complexity of network synchronization, to appear in *SIAM J. of Computing*.

[10] D. Goldstein and K. Kobayashi. On the complexity of network synchronization. *Algorithms and Computation, Proc. 15th International Symposium, ISAAC 2004, HongKong, China*, 5(3):289–308, 2004.

[11] E. Goto. A minimal time solution of the firing squad problem. *Course Notes for Applied Mathematics 298, Harvard University*, pages 52–59, 1962.

[12] A. Grasselli. Synchronization of cellular arrays: The firing squad problem in two dimensions. *Information and Control*, 28:113–124, 1975.

[13] T. Jiang. The synchronization of nonuniform networks of finite automata. In *Proc. of the 30th Annual ACM Symp. on Foundations of Computer Science*, pages 376–381, 1989.

[14] T. Jiang. The synchronization of nonuniform networks of finite automata. *Information and Computation*, 97:234–261, 1992.

[15] David R. Karger, Daphne Koller, and Steven J. Phillips. Finding the hidden path: Time bounds for all-pairs shortest paths. In *IEEE Symposium on Foundations of Computer Science*, pages 560–568, 1991.

[16] K. Kobayashi. The firing squad synchronization problem for a class of polyautomata networks. *J. Comput. System Sci.*, 17(3):300–318, 1978.

[17] K. Kobayashi. On the minimal firing time of the firing squad synchronization problem for polyautomata networks. *Theoretical Computer Science*, 7:149–167, 1978.

[18] K. Kobayashi. On time optimal solutions of the firing squad synchronization problem for two-dimensional paths. *Theoretical Computer Science*, 259:129–143, 28 May 2001.

[19] S. LaTorre, M. Napoli, and M. Parente. Synchronization of one-way connected processors. *Complex Systems*, 10:239–255, 1996.

[20] J. Mazoyer. An overview of the firing synchronization problem. In *Automata Networks, LITP Spring School on Theoretical Computer Science, Angelès-Village, France, May 12-16, 1986, Proceedings*, volume 316 of *Lecture Notes in Computer Science*. Springer, 1988.

[21] J. Mazoyer. Synchronization of two interacting finite automata. *International Journal of Algebra and Computation*, 5(3):289–308, 1995.

[22] M. Minsky. *Computation: Finite and Infinite Machines*. Prentice-Hall, Englewood Cliffs, NJ, 1967.

[23] E. F. Moore. *Sequential Machines, Selected Papers*. Addison Wesley, Reading, MA, 1962.

[24] F. R. Moore and G. G. Langdon. A generalized firing squad problem. *Information and Control*, 12:212–220, 1968.

[25] Y. Nishitani and N. Honda. The firing squad synchronization problem for graphs. *Theoretical Computer Science*, 14(1):39–61, 1981.

[26] R. Ostrovsky and D. Wilkerson. Faster computation on directed networks of automata. In *Proceedings of the Fourteenth Annual ACM Symposium on Principles of Distributed Computing*, pages 38–46, Ottawa, Ontario, Canada, 2–23 August 1995.

[27] H. Schmid and T. Worsch. The firing squad synchronization problem with many generals for one-dimensional CA. *3rd IFIP International Conference on Theoretical Computer Science*, pages 111–124, 2004.

[28] I. Shinahr. Two- and three-dimensional firing-squad synchronization problem. *Information and Control*, 24:163–180, 1974.

[29] A. Waksman. An optimum solution to the firing squad synchronization problem. *Information and Control*, 9:66–78, 1966.

[30] Uri Zwick. Exact and approximate distances in graphs.

[31] Uri Zwick. A slightly improved sub-cubic algorithm for the all pairs shortest paths problem with real edge lengths. In Rudolf Fleischer and Gerhard Trippen, editors, *ISAAC*, volume 3341 of *Lecture Notes in Computer Science*, pages 921–932. Springer, 2004.

Author Index

Lecture Notes in Computer Science

For information about Vols. 1–3778

please contact your bookseller or Springer